BOTANY FOR DEGREE STUDENTS

B.Sc. Third Year

As per UGC Model Curriculum

- **Plant Physiology, Biochemistry and Biotechnology**
- **Ecology and Utilization of Plants**

B P PANDEY
MSc, PhD, FPSI
Former Head
Department of Botany
J.V. College, Baraut, Uttar Pradesh

S Chand And Company Limited
(ISO 9001 Certified Company)

S Chand And Company Limited
(ISO 9001 Certified Company)
Head Office: Block B-1, House No. D-1, Ground Floor, Mohan Co-operative Industrial Estate, New Delhi – 110 044 | Phone: 011-66672000
Registered Office: A-27, 2nd Floor, Mohan Co-operative Industrial Estate, New Delhi – 110 044
Phone: 011-49731800
www.**schandpublishing.com**; e-mail: **info@schandpublishing.com**

Branches

Chennai : Ph: 23632120; chennai@schandpublishing.com
Guwahati : Ph: 2738811, 2735640; guwahati@schandpublishing.com
Hyderabad : Ph: 40186018; hyderabad@schandpublishing.com
Jalandhar : Ph: 4645630; jalandhar@schandpublishing.com
Kolkata : Ph: 23357458, 23353914; kolkata@schandpublishing.com
Lucknow : Ph: 4003633; lucknow@schandpublishing.com
Mumbai : Ph: 25000297; mumbai@schandpublishing.com
Patna : Ph: 2260011; patna@schandpublishing.com

First Edition 2007
Reprints 2009, 2010, 2011, 2013, 2014, 2015 (Twice), 2016, 2019, 2020, 2021

Reprint 2022 (Twice)

ISBN: 978-81-219-2845-8 **Product Code:** H6BTY68BOTN10ENAA07O

PRINTED IN INDIA

By Vikas Publishing House Private Limited, Plot 20/4, Site-IV, Industrial Area Sahibabad, Ghaziabad – 201 010 and Published by S Chand And Company Limited, A-27, 2nd Floor, Mohan Co-operative Industrial Estate, New Delhi – 110 044.

PREFACE

The present book 'Botany for Degree Students' B.Sc., Third Year, has been written strictly according to **UGC Model Syllabus** for all Indian Universities. The syllabus of B.Sc. III has been divided into **Course-I** and **Course II.** Course I contains **'Plant Physiology, Biochemistry and Biotechnology'** where subject matter has been divided into *four* units, and expanded into *nine* chapters; while course II contains **'Ecology and Utilization of Plants' (Economic Botany),** having *two* units and *sixteen* chapters.

Thus, whole subject matter of B.Sc. III (Botany) has been expanded into *six* units and *twenty five* chapters. The subject matter has been written in simple understandable language that can easily be grasped by students. To make subject easy and understandable, self-explanatory labelled diagrams have been added. Each unit or chapter as it may be is followed by various types of questions, such as very short answer questions, short answer questions, long answer questions, diagrammatic questions and multiple choice questions, asked repeatedly in various Indian Universities and competitive examinations. Every effort has been put to provide latest and up-to-date information of the subject. I think, this book would be helpful for those to whom, this has been written.

I am thankful to Shri Ravindra Kumar Gupta, CMD, Shri Navin Joshi, General Manager (S & M) and Shri R. S. Saxena, Advisor, S. Chand & Company Ltd., who have assigned this work to me to complete.

I am also thankful to those to whom I know or do not know but who have helped me a lot directly or indirectly. I thank all of them.

Suggestions from all corners would be welcomed and incorporated in subsequent editions.

B. P. Pandey

'Neelkanth Kuteer'
Subhasnagar, BARAUT,
250 611

UGC MODEL SYLLABUS

B.Sc. Third Year

BOTANY

COURSE-I : PLANT PHYSIOLOGY, BIOCHEMISTRY AND BIOTECHNOLOGY

1. Basics of enzymology : Discovery and nomenclature; characteristics of enzymes; concept of holoenzyme, apoenzyme, coenzyme and cofactors; regulation of enzyme activity; mechanism of action.
2. Plant-water relations : Importance of water to plant life; physical properties of water; diffusion and osmosis; absorption, transport of water and transpiration; physiology of stomata.
3. Mineral nutrition : Essential macro- and micro-elements and their role; mineral uptake deficiency and toxicity symptoms.
4. Mechanism of phloem transport; source-and relationship factors affecting translocation.
5. Photosynthesis : Transport of organic substances way Significance; historical aspects; photosynthetic pigments; action spectra and enhancement effects; concepts of two photosystems; Z-scheme: photophosphorylation: Calvin cycle; C4 pathway; CAM plants; photorespiration.
6. Respiration: ATP – the biological energy currency; aerobic and anaerobic respiration; Kreb cycle; electron transport mechanism (chemi-osmotic theory); redox potential; oxidative phosphorylation; pentose phosphate pathway.
7. Nitrogen and lipid metabolism : Biology of nitrogen fixation: importance of nitrate reductase and its regulation; ammonium assimilation; structure and function of lipids: fatty acid biosynthesis; β-oxidation; saturated and unsaturated fatty acids; storage and mobilization of fatty acids.
8. Growth and development: Definitions; phases of growth and development; kinetics of growth; seed dormancy. Seed germination and factors of their regulation: plant movements; the concept of photoperiodism: physiology of flowering: florigen concept : biological clocks; physiology of senescence. Fruit ripening: plant hormones – auxins, gibberellins, cytokinins, abscisic acid and ethylene, history of their discovery, biosynthesis and mechanism of action; photomorphogenesis; phytochromes and cryptochromes, their discovery, physiological role and mechanism of action
9. Genetic engineering: Tools and techniques of recombinant DNA technology; cloning vectors; genomic and cDNA library: transposable elements; techniques of gene mapping and chromosome walking.
10. Biotechnology: Functional definition; basic aspects of plant tissue culture; cellular totipotency, differentiation and morphogenesis; biology of *Agrobacterium;* vectors for gene delivery and marker genes; salient achievements in crop biotechnology.

COURSE-II : ECOLOGY AND UTILIZATION OF PLANTS

ECOLOGY

1. Plants and environment : Atmosphere (gaseous composition), water (properties of water cycle), light (global radiation, photosynthetically active radiation), temperature, soil (development, soil profiles, physico-chemical properties), and biota.
2. Morphological, anatomical and physiological responses of plants to water (hydrophytes and xerophytes), temperature (thermoperiodicity and vernalization), light (photoperiodism, heliophytes and sciophytes) and salinity.
3. Population ecology : Growth curves: ecotypes: ecads.
4. Community ecology : community characteristics, frequency, density, cover, life forms, biological spectrum; ecological succession.
5. Ecosystems : Structure: abiotic and biotic components: food chain, food web, ecological pyramids : energy flow; biogeochemical cycles of carbon nitrogen and phosphorus.
6. Biogeographical regions of India.
7. Vegetation types of India : Forests and grasslands.

UTILIZATION OF PLANTS

8. Food plants: Rice, wheat, maize potato, sugarcane.
9. Fibres: cotton and jute.
10. Vegetable oils : Groundnut, mustard and coconut.
11. General account of sources of firewood, timber and bamboos.
12. Spices : General account
13. Medicinal plants : General account
14. Beverages: Tea and coffee
15. Rubber.

CONTENTS

COURSE : I

PLANT PHYSIOLOGY, BIOCHEMISTRY AND BIOTECHNOLOGY

Unit-1 : Biochemistry

Chapter-1 : Basics of Enzymology - Enzymes 3 – 22

Unit-2 : Plant Physiology

Chapter-2 : Plant Water Relations 23 – 66

Chapter-3 : Mineral Nutrition and Nitrogen Metabolism 67 – 96

Chapter-4 : Respiration 97 – 133

Chapter-5 : Photosynthesis and Transport of Organic Substances 134 – 192

Chapter-6 : Plant Growth and Development 193 – 235

Chapter-7 : Lipid (Fat) Metabolism 236 – 251

Unit-3 : Genetic Engineering

Chapter-8 : Genetic Engineering, Cloning and Genomics 252 – 282

Unit-4 : Biotechnology

Chapter-9 : Plant Tissue Culture and Biotechnology 283 – 317

COURSE : II

ECOLOGY AND UTILIZATION OF PLANTS

Unit-5 : Ecology

Chapter-10 : Organisms and the Environment 321 – 355

Chapter-11 : Population, Biotic Community and Succession 356 – 385

Chapter-12 : Ecosystem : Structure and Function 386 – 438

Chapter-13 : Natural Resources and Vegetation Types of India 439 – 472

Chapter-14 : Biodiversity and Biogeographical Regions of India 473 – 512

Chapter-15 : Environmental Changes 513 – 557

Unit-6 : Utilization of Plants (Economic Botany)

Chapter-16 : Introduction 558 – 562

Chapter-17 : Food Plants 563 – 581

Chapter-18 : Fibres and Fibre Yielding Plants 582 – 592

Chapter-19 : Vegetable and Fatty Oils 593 – 597

Chapter-20 : Wood and Cork 598 – 618

Chapter-21 : Bamboos in India 619 – 626

Chapter-22 : Spices and Condiments 627 – 666

Chapter-23 : The Medicinal Plants : Pharmacognosy 667 – 688

Chapter-24 : Non-Alcoholic Beverages and Beverage Plants 689 – 702

Chapter-25 : Rubber and its Products 703 – 706

Question Bank 707 – 714

COURSE – I

PLANT PHYSIOLOGY, BIOCHEMISTRY AND BIOTECHNOLOGY

UNIT - 1
Biochemistry : Enzymology

1
CHAPTER

Basics of Enzymology – Enzymes

An **enzyme** is a specialized protein produced within an organism or cell which is capable of catalysing a specific chemical reaction. Since the enzyme acts as a catalyst, it is sometimes referred to as **biocatalyst**. A catalyst influences the rate of a chemical reaction, usually without undergoing any change itself. Thus, an enzyme differs from a normal catalyst. The enzyme may participate in a reaction by combining with the substrate. Ultimately, however, it is set free. The DNA in each cell has the necessary message (blue print) for the production of all the enzymes required by it. The cell uses this information as and when necessary to produce the enzymes required to catalyse specific reactions at any point of time.

Enzymes are synthesised by living cells. But they retain their catalytic action even when extracted from cells. For example, Rennet tablets containing the enzyme rennin from the calf's stomach have long been used for coagulating milk protein to obtain casein (cheese from milk), for various uses, including the preparation of eatables.

Discovery of enzymes. The term enzyme was coined by **Kuhne** (1878) for catalytically active substances previously called **ferments**. Enzymes were discovered by accident. In 1897 a German chemist Edward Buchner was preparing yeast extract for medicinal purposes. The extracts he prepared invariably turned bad. He added some sugar to prevent spoilage. To his surprise the sugar fermented into alcohol. Pasteur had already demonstrated that living yeast cells promoted fermentation. Buchner discovered that living cells were not necessary but that yeast extract could bring about fermentation. The word **enzyme** was coined. It literally means **'in yeast'** but is now used as a general name for biocatalysts.

Definition. Mayrback (1952) defined 'enzymes are simple or compound proteins acting a specific catalysts'. They may also be defined as 'organic substances capable of catalysing chemical reactions in the living systems.'

However, enzymes speed up the rate of reaction without affecting the equilibrium. They exhibit a striking specificity to manifest themselves in their catalytic reactions, enhancing and controlling the course of these processes.

In 1878, **Friedrich W. Kuhne** coined the word **enzyme** (Gk. *en* + *zyme* = in yeast), and replaced the older word 'ferment' by it.

NOMENCLATURE AND CLASSIFICATION OF ENZYMES

Enzymes were variously named in the past. Enzymes, such as ptyalin (salivary amylase). pepsin and trypsin give no indication of their action. While other enzymes such as amylase, sucrase, protease and lipase were named after the substrates on which they acted — amylose (starch), sucrose, protein and lipids respectively. Still others were named according to the source from which they were obtained — papain from papaya, bromelain from pineapple (of family Bromeliaceae). Some like DNA polymerase indicate its specific action, polymerisation. To facilitate scientific communication and to maintain uniformity, enzymes are now categorised and named by an **international code of enzyme nomenclature**.

The name of each enzyme ends with an **ase** and consists of two parts. The first part indicates its **substrates** and the second the reaction catalysed. For example, glutamate pyruvate transaminase transfers an amino group from the substrate glutamate to another substrate pyruvate.

Enzymes are grouped into *six* major classes :

Class 1. Oxireductases. These catalyse oxidation or reduction of their substrates and act by removing or adding electrons from or to substrates, *e.g.*, cytochrome oxidase oxidises cytochrome.

Class 2. Transferases. These transfer specific groups from one substrate to another. The chemical group transferred in the process is not in a free state, *e.g.*, glutamate pyruvate transaminase.

Class 3. Hydrolases. These break down large molecules into smaller ones by the introduction of water (hydrolysis) and breaking of specific covalent bonds. Most digestive enzymes belong to this catagory, *e.g.*, amylase which hydrolyses starch.

Class 4. Lyases. These catalyse the cleavage of specific covalent bonds and removal of groups without hydrolysis, *e.g.*, histidine decarboxylase cleaves C–C bond in histidine to form carbon dioxide and histamine.

Class 5. Isomerases. These catalyse the rearrangement of molecular structure to form isomeres, *e.g.*, phosphohexose isomerase changes gulcose-6-phosphate to fructose-6-phosphate (both are hexose phosphates).

Class 6. Ligases. Thse catalyse covalent bonding of two substrates to form a large molecule. The energy for reaction is derived from the hydrolysis of ATP. Pyruvate carboxylase combines pyruvate and carbon dioxide to form oxaloacetate at the expense of ATP.

ENDOENZYMES AND EXOENZYMES

The enzymes wnich are synthesized by the living cells usually operate inside them within the boundary of their membrane. Such enzymes are called **endoenzymes** or **intracellular enzymes.** Most of the enzymes belong to this category. But in heterotrophic microorganisms such as bacteria and fungi, some of the enzymes, *e.g.*, **protease, cellulase, amylase**, etc., produced by the cells are secreted out and are active in the outer environment of cells where they convert complex insoluble chemical compounds into simple soluble compounds, the latter are then readily absorbed by them. Such enzymes which operate externally to the cells in which they originate are called **exoenzymes** or **extracellular enzymes** exoenzymes are digestive or hydrolytic in nature and obviously their function is to make insoluble substances available as substrates to the cells for their growth. The insectivorous plants which also exhibit heterotrophic nutrition have special secretory cells to produce such enzymes especially proteolytic ones to digest their prey. During the germination of cereal grains, hydrolytic enzymes are secreted into the endosperm by **scutella** and **aleurone cells** to convert insoluble reserve food material into soluble one so that it may be made available to the growing seedling. It is believed that generally the plant cells have some limited capacity to secrete enzymes outer to them which might be located on the cell walls or they may be released into the surrounding environment. Secretion of hydrolytic enzymes into the other medium by cells of higher plants in tissue culture experiments is well known that supports this belief.

STRUCTURE OF ENZYMES

All enzymes are proteins but all proteins are not enzymes. Many enzymes contain a non-protein part called the **prosthetic group.** These are tightly bound to the enzyme. Some prosthetic groups are metal compounds. For example, iron-porphyrin complexes form the prosthetic groups of cytochromes. In addition, certain organic compounds and inorganic ions are required for enzyme activity. They are loosely bound to the enzyme and are called **co-factors.**

Nucleotides of the vitamin nicotinamide (NAD) and riboflavin (FAD) are organic **co-factors** or **coenzymes** of many oxidising enzymes of the mitochondria. In addition certain metals, especially those occurring in trace amounts, facilitate enzyme reaction. Iron (Fe^{++}) is a co-factor responsible for the catalytic action of catalase.

All enzymes have a specific three dimonsional structrue, a part of which is known as the **active site**. An enzyme may have more than one active site.

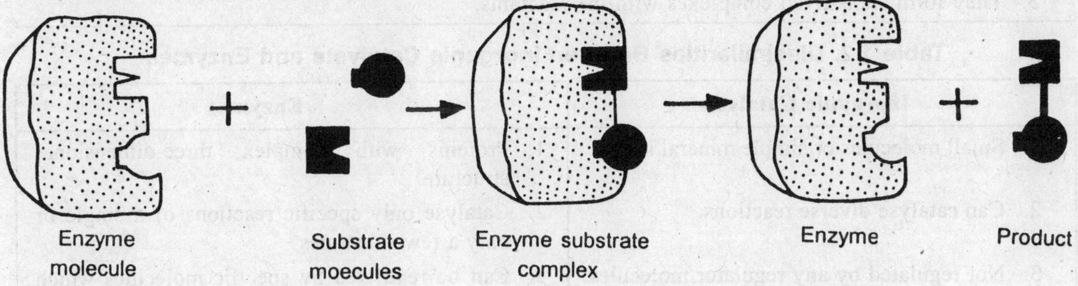

Fig. 1.1. Lock and key hypothesis of enzyme action.

The active site serves as a **'lock'** into which the reactant, commonly referred to as the substrate fits in like a **'key'** (Fig. 1.1). The points where the substrate is bound on the active site is known as the substrate-binding site. Substrate binding causes a lowering of the activation energy and allows the reaction to proceed (Fig. 1.2). Once the reaction is completed the enzyme releases the **products** and is ready to catalyse again.

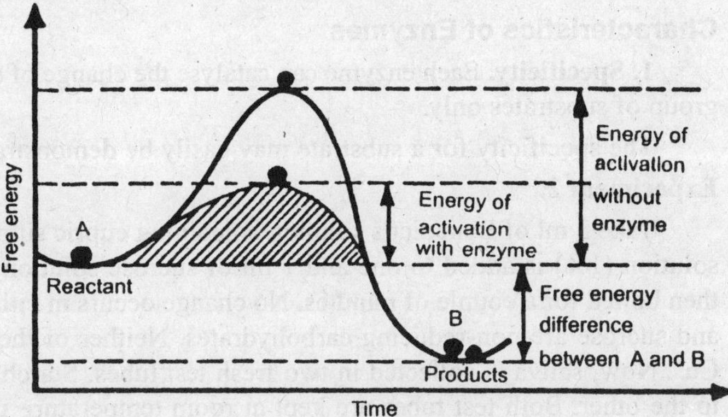

Fig. 1.2. Activation energy requirement of non-catalysed and enzyme-catalysed reactions.

A single molecule of the enzyme carbonic anhydrase, the fastest enzyme known, hyrates 36 million (36 × 10^6) molecules of carbon dioxide per minute.

$$CO_2 + H_2O \xrightarrow{\text{carbonic anhydrase}} H_2CO_3$$

The catalysed reaction is 10 million (10^7) times faster than the non-catalysed reaction.

Experiment 1.

Take 2 ml of hydrogen peroxide (H_2O_2) solution in two test tubes. Add a pinch of manganese dioxide (MnO_2) powder to one and a small piece of fresh liver or potato to the other. Keep the test tubes at room temerature in summer or in a beaker of warm water (at 38°C) in winter. Bubbles of oxygen evolve in the two test tubes. Thus both manganese dioxide and a tissue enzyme (peroxidase or catalase) can catalyse the evolution of oxygen from hydrogen-peroxide. Repeat the same experiment, but use boiled and cooled manganese dioxide (MnO_2) suspension and a boiled or cooled liver/potato slice. Gas evolution still occurs in the hydrogen peroxide solution treated with boiled and cooled manganese dioxide. But the tube containing the boiled tissue slices does not evolve oxygen.

This experiment demonstrates that heat inactivates the enzyme but not the inorganic catalyst.

Table 1.1. Similarities Between Inorganic Catalysts and Enzymes

1. Inorganic catalysts and enzymes remain unchanged after reaction, chemically and quantitatively and can be used over and over again.
2. Both are required in minute quantities compared to the substrate.
3. They do not alter the equilibrium of a reversible reaction.
4. They do not initiate a reaction, but only enhance its rate by lowering the activation energy.
5. They form short-lived complexes with the reactants.

Table 1.2. Dissimilarities Between Inorganic Catalysts and Enzymes

Inorganic Catalysts	Enzymes
1. Small molecules or simple mineral ions.	1. Proteins with complex three-dimensional structure.
2. Can catalyse diverse reactions.	2. Catalyse only specific reactions of a single or only a few substrates
3. Not regulated by any regulator molecules.	3. Can be regulated by specific molecules which can change conformation, and therefore, activity.
4. They are less sensitive to temperature and pH.	4. They are more sensitive to change in temperature and pH.

Characteristics of Enzymes

1. Specificity. Each enzyme can catalyse the change of either a specific substrate or a specific group of substrates only.

The specificity for a substrate may easily be demonstrated by an experiment as follows :

Experiment 2.

Take 3 ml of Benedict's solution containing cupric sulphate, in two test tubes. 1 ml of starch solution (1%) is added to one and 1 ml of sucrose solution (1%) to the other. The solutions are then boiled for a couple of minutes. No change occurs in either of the test tubes. Thus, both starch and sucrose are non-reducing carbohydrates. Neither of them can reduce Cu^{++} of the reagent to Cu^{+}. Now, saliva is collected in two fresh test tubes. Starch is added to one of them and sucrose to the other. Both test tubes are kept at room temperature in summer for about one hour and in warm water (about 38^{0}C) in winter. Benedicti's solution is then added to both test tubes and their contents are boiled for a couple of minutes. The blue solution changes to yellow or red precipitate in the test tube containing starch.

This indicates that the enzyme salivary amylase hydrolyses starch to reducing sugars. The latter reduces Cu^{++} to Cu^{+} which forms the coloured precipitate of cuprous oxide (Cu_2O). But the test tube containing sucrose shows no change. Thus, salivary amylase cannot hydrolyse sucrose. No reducing sugar is produced and the colour of benedict's soloution remains unchanged.

2. Optimum temperature. Enzymes generally work over a narrow range of temperatures. Usually it corresponds to the body temperature of the organism. For instance, human enzymes work at the normal body temperature. Each enzyme shows its highest activity at a particular temperature. This is called the **optimum temperature**. Activity declines both above and below the optimum temperatrure. Cold preserves the enzyme in a temporarily inactive state. Food may be preserved for long in a frozen state because neither microbial enzyme nor enzymes in the food can act at low temperatures to cause its spoilage. As demonstrated earlier (experiment 1), very high temperatures destroy enzyme activity. This is because proteins are denatured by heat. For this reason, few cells can tolerate temperatures above 45°C. Some heat resistant species like microoganisms living in hot springs at temperatures close to 100°C, possess heat-resistant enzymes.

3. Optimum pH. Each enzyme shows its highest activity at a specific pH. This is called the **optimum pH**. Activity declines both above and below the optimum pH. Most intracellular enzymes function best around neutral pH. Some digestive enzymes have their optimum in the acidic or alkaline range. For example, the protein digesting enzyme pepsin found in the stomach, has an optimum pH of 2.0. Another protein-digesting enzyme, trypsin, found in the duodenum, functions best in an alkaline pH of 8.5.

Experiment 3.

Take 2 ml starch solution (1%) in two test tubes. Some saliva is collected mixed with the contents of each tube. A few drops of dilute hydrochloric acid are added to one of the tubes. Both test tubes are then kept for one hour at room temperature in summer or in beaker containing warm water (about 38^{0}C) in winter. Subsequently a drop of dilute iodine solution is added to each tube. In the test tube whose contents were acidified, a blue colour develops due to the reactions of unchanged starch with iodine. But the content of the other test tube is not coloured blue with iodine, indicating that starch has been hydrolysed by salivary amylase.

This shows that amylase hydrolyses starch at salivery pH, but fails to act at a more acid pH.

In the stomach, salivary amylase swallowed with food is inactivated by the hydrochloric acid present in the gastric juice.

4. Enzyme-substrate complex. Each enzyme (E) possesses a substrate-binding site on the active site in its molecule. A highly reactive enzyme-substrate complex is consequently produced. The latter almost immediately dissociates into the product or products (P) and the unchanged enzyme.

$$E + S \rightleftharpoons \underset{\text{short-lived}}{ES} \longrightarrow EP \longrightarrow E + P$$

Formation of the enzyme-substrate complex is essential for catalysis. The higher the affinity of the enzyme for its substrate, the greater is its catalytic activity.

5. Effect of substrate concentration. With the increase in substrate concentration (S), the velocity (V) of the enzymatic reaction rises at first. But the rise in V decreases progressively with the rise in (S). The reaction ultimately reaches a maximum velocity, which is not exceeded by any further rise in substrate concentration (Fig. 1.3). This happens because enzyme molecules are fewer than the substrate molecules. Any more increase in substrate concentration will saturate all the enzyme molecules. No enzyme is left free to bind with additional molecules of the substrate.

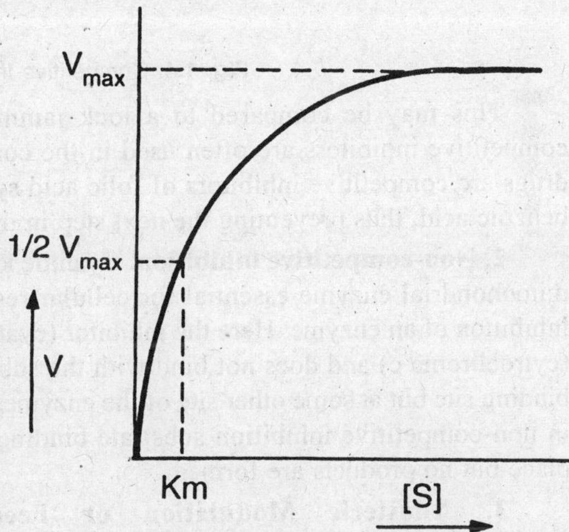

Fig. 1.3. Effect of substrate concentration on the velocity (*V*) of enzymatic reactiony V_{max} = maximum velocity.

Michaelis constant (K_m). Michaelis constant (K_m) of an enzyme is the substrate concentration at which the reaction attains half its maximum velocity (Fig. 1.3). It is a measure of the affinity of the enzyme for its substrate. The lower the K_m, the higher is the substrate affinity of the enzyme. An enzyme possesses different affinities for different substrates. Thus, its K_m value differs from substrate to substrate. Proteases can act on a variety of proteins. The K_m value of the protease will vary with the type of protein.

Inhibition or Regulation of Enzyme Activity

Enzyme action can be inhibited in **four** different ways. Inhibition of enzyme action by denaturation of proteins has already been discussed. The other **three** ways in which enzymes can be inhibited are as follows :

1. Competitive Inhibition. The action of an enzyme may be reduced or inhibited in the presence of a substance that closely resembles the substrate in molecular structure. Such an inhibition is called a **competitive inhibition** of that enzyme. Due to its closed structural similarity with the substrate, the inhibitor competes with the latter for the substrate-binding site of the enzyme (Fig. 1.4). Consequently, the enzyme cannot participate in catalysing the change of the substrate. As a result, the enzyme action declines, *e.g.*, the inhibition of succinate dehydrogenase by malonate, which closely resembles succinate in structure.

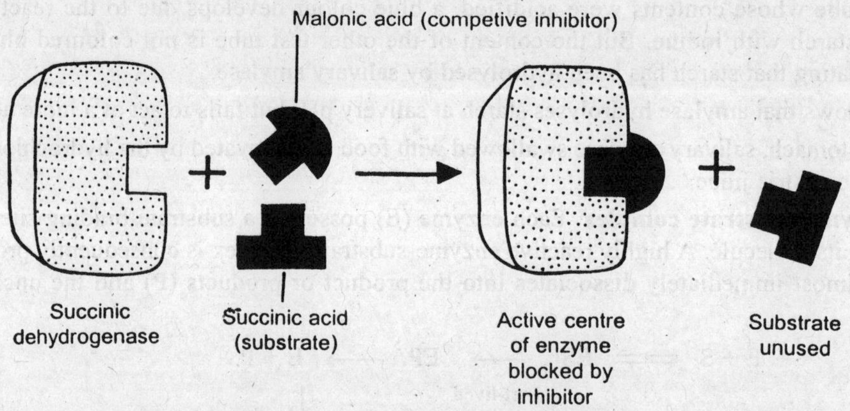

Fig. 1.4. Competitive inhibition of enzyme action

This may be compared to a lock jammed by a key similar to the original key. Such competitive inhibitors are often used in the control of bacterial pathogens. For instance, sulpha drugs are competitive inhibitors of folic acid synthesis in bacteria as they substitute for p-amino benzoic acid, thus preventing the next step in the synthesis.

2. Non-competitive inhibition. Cyanide kills an animal by inhibiting cytochrome oxidase, a mitochondrial enzyme essential for cellular respiration. This is an example of non-competitive inhibition of an enzyme. Here the inhibitor (cyanide) has no structural similarity wtih the substrate (cytrochrome c) and does not bind with the substrate-binding site but at some other site of the enzyme. Thus, in non-competitive inhibition substrate binding takes place but no products are formed.

3. Allosteric Modulation or Feedback Inhibition. The activites of some enzymes, particularly those which form a part of a chain of reactions (metabolic pathway), are regulated internally. Some specific low molecular weight substance, such as the product/products of another enzyme further on in the chain, acts as the inhibitor. Such a modulator substance binds with a specific site of the enzyme different from its substrate-binding site. This binding increases or decreases the enzyme action. Such enzymes are called **allosteric enzymes** (Fig. 1.5).

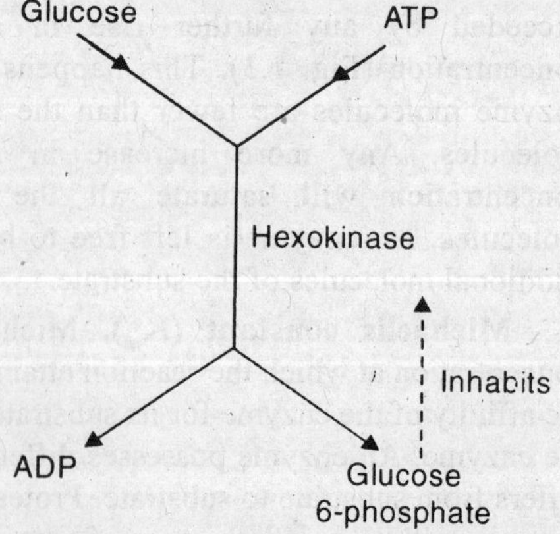

Fig. 1.5. Feedback allosteric inhibition

Coenzymes or Cofactors

For some chemical reactions a substance other than an enzyme and substrates are necessary. These substances are called **coenzymes** or **cofactors.** For example, the formation of **acetyl-CoA** can occur only in the presence of **coenzyme-A.** The **coenzymes** are therefore, necessary if a particular enzyme is to be active. Like enzymes, coenzymes also remain unchanged at the end of a reaction.

Coenzymes, however, differ from enzymes in that they are not proteins. They are of low molecular weight (500 or so) and are heat stable. Enzymes, which are proteins have a comparatively higher molecular weight and undergo denaturation at temperatures, 50° to 60°C. However, an enzyme may function independently of a coenzyme but a coenzyme cannot function without an enzyme.

COFACTORS (COENZYMES)

In many cases some non-protein chemical substances are required in the reaction mixture to catalyse the enzyme reaction. These substances which are **not tightly bound** to the enzyme are called **cofactors**. Majority of the cofactrors may be divided into 2 groups :

(A) Specific Coenzymes

These are usually organic compounds of complicated structrue often acting as carriers of some chemical group. Many coenzymes are either *vitamins* or derivatives of *vitamins.*

Depending upon the nature of chemical group which they carry, the coenzymes are classified as follows :

(1) Hydrogen Carriers (Oxidation-Reduction Reactions)

(*i*) **Coenzyme I** (Co I) or **NAD⁺** (Nicotinamide Adenine Dinucleotide). Its reduced form is **NADH + H⁺**.

Previously known as DPN (Diphospho Pyridine Dinucleotide) contains vitamin *Nicotinic acid* (*Niacin*) in amide form.

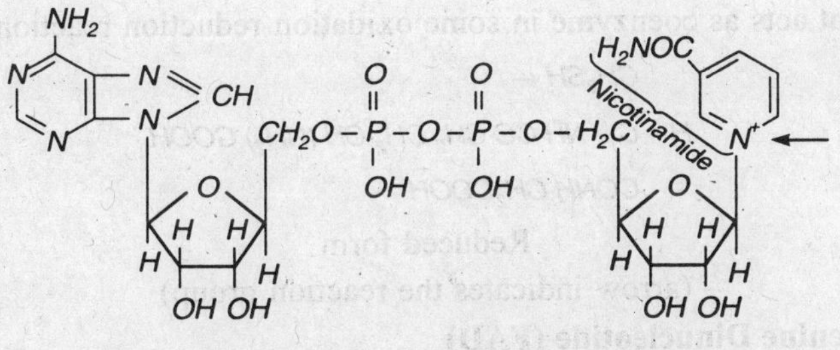

NAD^+ (DPN^+) – oxidised form.*
(arrow indicates the position where reduction occurs)

(*ii*) **Coenzyme II** (Co II) or **$NADP^+$** (Nicotinamide Adenine Dinucleotide Phosphate). Its reduced form is **$NADPH+H^+$.**

Previously known as **TPN** (Triphospho Pyridine Dinucleotide).

NADP$^+$ (TPN$^+$)-oxidised form. *

(arrow indicates the position where reduction occurs)

***(iii)* Lipoic Acid** (also helps in oxidative decarboxylation of α-keto acids)

$CH_2CH_2CH(CH_2)_4COOH$ with S—S (Oxidised)

$CH_2CH_2CH(CH_2)_4COOH$ with HS SH (Reduced)

***(iv)* Ascorbic Acid** (*Vitamin C*)

Reduced | Oxidised

***(v)* Glutathione** (GSH reduced form)/GSSG (oxidised form). It is a tripeptide (γ-glutamyl cyseinly glycine) that acts as coenzyme in some oxidation-reduction reactions in cells.

$$\begin{array}{l} \quad CH_2SH \leftarrow \\ H-C-NH\cdot CO\cdot CH_2\cdot CH_2\cdot CH(NH_2)\,COOH \\ \quad CONH\cdot CH_2COOH \end{array}$$

Reduced form

(arrow indicates the reaction group)

***(vi)* Flavin Adenine Dinucleotide (FAD)**

It is derivative of *Vitamin B_2 i.e., Riboflavin*. Often it acts as prosthetic group.

FAD (oxidised form)

(arrows indicate the positions where reduction takes place)

(*vii*) Flavin Mono Nucleotide (FMN)

It is also derivative of *Vitamin* B_2 and often acts as prosthetic group.

$CH_2 \cdot CH(OH) \cdot CH(OH) \cdot CH(OH) \cdot CH_2OPO_3H_2$

N N

H_3C =N

H_3C NH

N OH

FMN (Oxidised form)

(arrows indicate the positions where reduciton takes place)

(*viii*) Coenzyme Q (CO-Q) or Ubiquinone

O CH_3

H_3C-O $(CH_2CH=C-CH_2)H$ 6–10

H_3C-O CH_2

O

CO-Q (Oxidised form)

(arrows indicate the position where reduction occurs)

(2) Acyl Group Carrier

(*i*) Coenzyme-A (CoA or CoA-SH)

It contains phosphopantetheine which is a derivative of the *vitamin pantothenic acid*. It was discovered by **Fritz Albert Lipman.** Nobel Laureate of 1953 in Physiology or Medicine category.

$HO-P(=O)-OCH_2C(CH_3)_2-CH(OH)-C(=O)-NHCH_2CH_2C(=O)NHCH_2CH_2SH$

O

NH_2

N C

C N

HC

C CH

5′

$HO-P-OCH_2$ N N

O

O

4′ H C 1′

H 2′ H

O OH

$HO-P-OH$

O

ADENOSINE-3′, 5′-BISPHOSPHATE

Coenzyme A

(arrow indicates the reactive group)

(3) Amino Group Carrier

(*i*) Pyridoxal Phosphate

It is derivative of *Vitamin* B_6 *i.e., pyridoxin.*

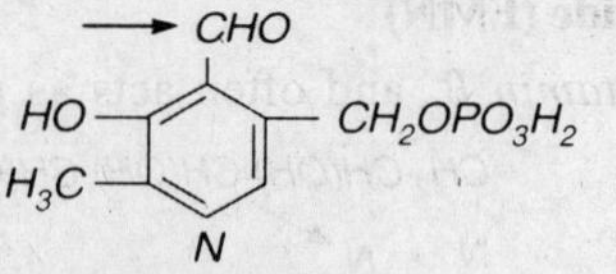

Pyridoxal phosphate
(arrow indicates the reactive group)

(4) CO_2 Carrier

(*i*) Biotin (Coenzyme R)

This is a *vitamin* and often acts as prosthetic group.

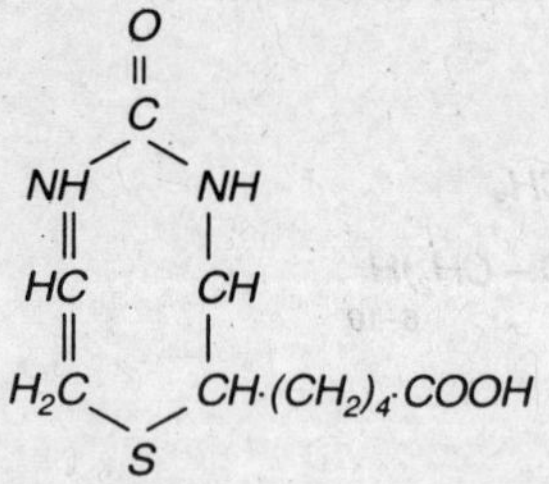

Biotin

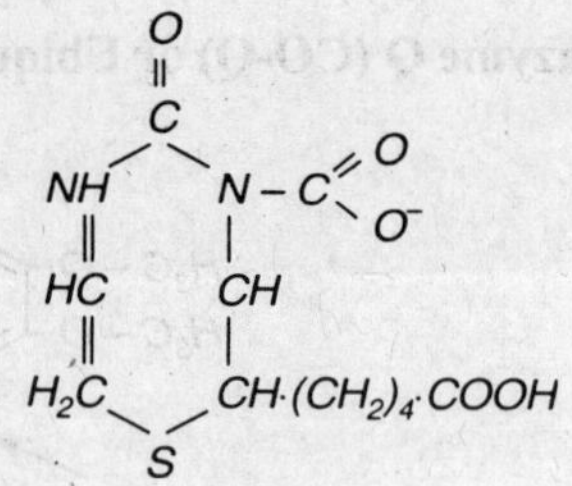

Biotin CO_2-Compound

(5) Thiamine Pyrophosphate (TPP) or Cocarboxylase

This is a derivative of *Vitamin B_1 i.e., thiamine* and plays an important role as coenzyme in the oxidative decarboxylation of α-keto acids (*e.g.*, pyruvic acid α-ketoglutaric acid).

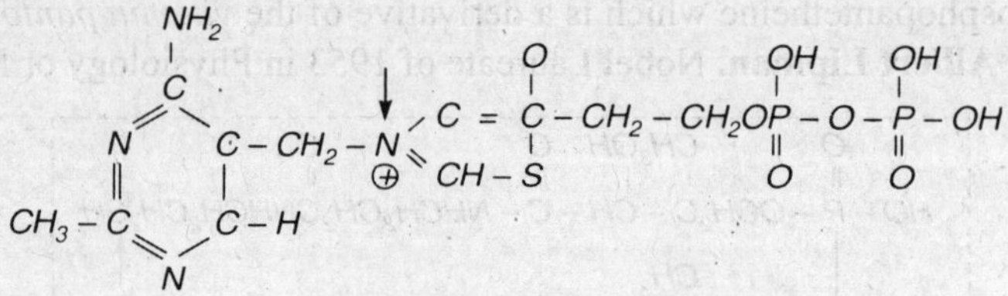

TPP (Thiamine Pyro phosphate)
(arrow indicates the reactive part)

(6) Carrier of One-Carbon Compounds (*e.g.*, formic acid, formaldehyde).

(*i*) Coenzyme F (Tetrahydrofolic acid)

It is derivative of *vitamin folic acid.*

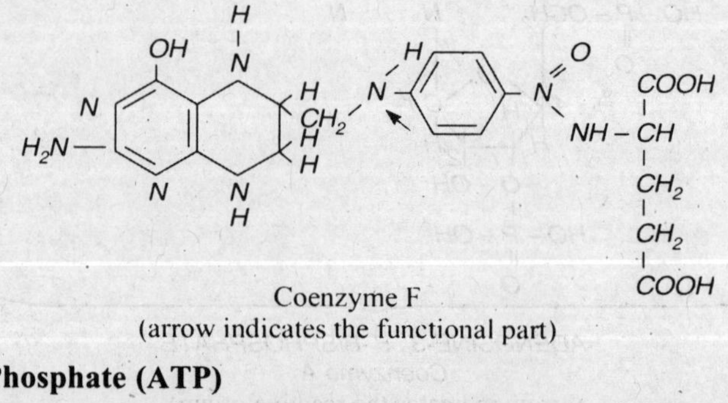

Coenzyme F
(arrow indicates the functional part)

(7) Adenosine Tri Phosphate (ATP)

It acts as coenzyme and helps in catalysing many enzyme reactions by activating either the substrate or the enzyme and providing energy. It also acts as carrier of phosphate group.

energy-rich bonds

$$OH-\overset{\overset{O}{\|}}{\underset{\underset{OH}{|}}{P}}\sim O-\overset{\overset{O}{\|}}{\underset{\underset{OH}{|}}{P}}\sim O-\overset{\overset{O}{\|}}{\underset{\underset{OH}{|}}{P}}-O-CH_2 \text{ (adenosine; } NH_2\text{, ribose with OH OH)}$$

(8) UDP (Uridine Diphosphate)

This is carrier of sugar molecules such as glucose, galactose or their derivatives. For instance, when UDP picks up a glucose molecule, it becomes converted into **UDPG** (Uridine Diphosphate Glucose). See Chapter 13 for Structure of UDPG.

(9) PAPS (3' phosphoadenosine –5'–phosphosulphate)

This is carrier of sulphate group.

(B) General Activators

These are usually some metallic ions, the presence of which in some way activates the enzyme reaction.

HOLOENZYMES AND APOENZYMES

An enzyme and a coenzyme together form an enzymatically active conjugated protein called the **holoenzyme.** The enzyme (protein) component of the holoenzyme is called an **apoenzyme.** The non-protein coenzyme is called the **prosthetic group.** Thus :

Holoenzyme = Apoenzyme + Prosthetic group.

The major coenzymes of dehydrogenases are the two dinucleotides, NAD and NADP.

The coenzyme (prosthetic group) may be tightly bound to the apoenzyme or may easily dissociate from it.

The **prosthetic group** remains attached to the **apoenzyme** while undergoing oxidation and reduction. For example, the porphyrin groups of cytochromes are more or less permanently attached to their apoenzymes. The **coenzyme** on the other hand may undergo reduction while attached to one apoenzyme, and then migrate to another apoenzyme where it can be oxidised. Thus NAD, NADP and CoA are considered to be coenzymes while hemes, flavins and biotin are considered to be prosthetic groups.

The coenzyme usually acts as an accepter or donor of an atom or a group, *e.g.*, NAD, NADP, FMN and CoQ are hydrogen carrires, ATP transfers the phosphate group and CoA the acyl group.

Many coenzymes are closely related to **vitamins** and are the derivation of vitamins. The B group vitamins (except biotin and lipoic acid) function as parts of some coenzyme, *e.g.*, CoA is a derivative of **pantothenic acid.** When there is vitamin deficiency the coenzyme concentration decreases. Consequently enzyme function is depressed. Nucleotides may also function as coenzymes in certain metabolic reactions.

Activators

In same cases enzymes may be activated by simple substance like metal ions which are then called **activators.** The substrate forms a complex with the metal ions and then reacts with the enzyme. Examples of metal ions which function as **cofactors** are :

Na^+, K^+, Ca^{++}, Co^{++}, Mg^{++}, Mn^{++}, Cd^{++}, Fe^{++}, Cr^{+++} and Al^{+++}

MECHANISM OF THE ENZYME ACTION

Swante August Arrhenius, the Nobel Prize winner of 1903 in Chemistry, first pointed out that all the molecules in a given population do not have the same kinetic energy. Some molecules due to collisions have more energy and are **energy-rich molecules** while the others are **energy-poor molecules**.

In an ordinary chemical reaction only energy rich molecules can take part at normal temperature due to an **energy barrier** to reaction, hence the rate of reaction is lower. The higher is the energy barrier for a molecule, the greater is its stability or inactiveness to take part in reaction. The energy required to hurdle molecules over this energy barrier is called the **energy of activation.**

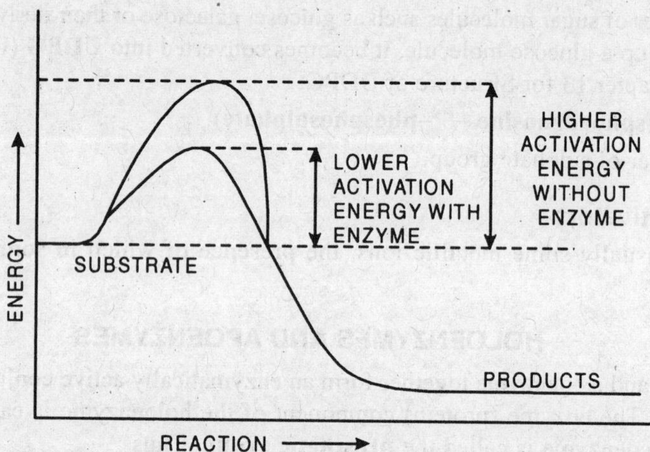

Fig. 1.6. Activation energy of an enzyme catalysed reaction is lower than that of an uncatalysed reaction.

At higher temperature the rate of chemical reaction becomes faster because increased temprature brings about an increase in the number of **activated molecules** by increasing their movement and number of collisions due to thermal agitation.

However, in case of **enzyme catalysed reactions** the rate of reaction is **optimum** at normal body temeratures. It is because all the molecules (energy-rich and energy-poor) can combine with the active sites of enzyme to form **enzyme substrate complex** which later on breaks into enzyme and the product. Thus, the enzymes act by lowering the energy of activation of the reactions (Fig. 1.6).

STRUCTURE OF ENZYMES

1. Protein part the enzyme (apoenzyme)

Major part of the enzyme consists of protein whose molecular weight may vary from few thousands to over a million. For example :

Enzyme	Molecular weight
Ribonuclease	approx 12700
Papain	" 20700
Glutamic dehydrogenase	" over a million

Usually the enzymic proteins consist of only one polypeptide chain, but sometimes there may be more than one polypeptide chains.

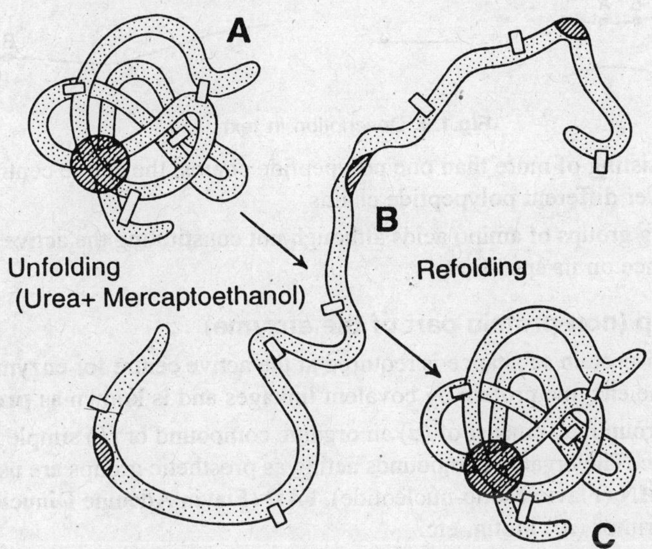

Fig. 1.7. A, Native ribonuclease showing four disulphide bonds, crossed hatched area is the active site; B, unfolded chain with broken disulphide bonds and active site; C, refolding of chain and restoration of disulphide bonds.

Enzymic proteins consist of the same 20 different types of the amino acids which constitute other proteins.

Sequence of the amino acids is **specific** in specific enzyme proteins. Their tertiary structure is also very specific and important for their biological activity. Loss of the tertiary structure renders the enzyme inactive. For example, in enzyme **ribonuclease** different folds of the polypeptide chain are held together by **four disulphide linkages** (S-S) to give it a compact shape (Fig. 1.7 A). If these linkages are reduced with mercaptoethanol, the disulphide groups are converted to **sulphydryl groups** (– SH) resulting in unfolding of the polypetide chain and loss of enzymic activity (Fig. 1.7 B). Reoxidation of the protein in air results in information of the disulphide bonds and return of its enzymic activity (1.7 C).

2. Active Centre or Active Site

Active centre or the active site is that part of the enzyme which takes part in the reaction. Specific substrate combines with it and this combination brings about the biochemical reaction.

Active centre is very **specific.** There may be one or more active centre on a single enzyme molecule.

It is believed that the active centre consists of amino acids residues which are very close to each other in native structure of the enzyme because of the folding of the polypeptide chain, but in fact may be far apart in the primary sequence (Fig. 1.8). This also explains why unfolding of the polypeptide chain by denaturing agents results in loss of its activity.

Fig. 1.8. Description in the text.

If the active centre is present only on one fold of the polypeptide chain and does not extend over the different folds as shown in fig. 1.9, the unfolding of the polypeptide chain may not necessarily result in loss of enzme activity.

Fig.1.9. Description in text.

In enzymes consisting of more than one polypeptide chains, the active centre may extend an individual chain or over different polypeptide chains.

The neighbouring groups of amino acids although not constituting the active centre, may also have profound influence on its specificity.

3. Prosthetic Group (non-protein part of the enzyme)

Sometimes a non-protein substance is required at the active centre for enzyme activity which is **bound tightly** to the enzyme protein by covalent linkages and is known as **prosthetic group.**

The prosthetic group may consist of (*a*) an organic compound or (*b*) simple metal ions, such as Cu, Zu, Mn, Mo, etc. The organic compounds acting as prosthetic groups are usually (*i*) a flavin compound, such as **FMN** (Flavin Mono-nucleotide), **FAD** (Flavin Adenine Dinucleotide) (*ii*) heme iron, *i.e.,* iron-porphyrin or (*iii*) biotin, etc.

PROPERTIES OF ENZYMES

(1) Catalytic Property

Like inorganic catalysts the enzymes are active in very small or catalytic amounts and remain unchanged after the reaction. Only a small amount of enzyme is enough to convert large quantity of the substrate into products. Their **turnover number (*k* cat)** *i.e.*, the number of substrate molecules converted to product by one molecule of enzyme per second per active site may range from 0.5 to 4×10^7. The turnover number for ***catalase*** has been calculated as 4×10^7 and is among the highest known. ***Lysozyme*** on the other extreme, has lowest value of turnover number *i.e.,* 0.5.

(2) Specificity

Enzymes are very specific in their action. A particular enzyme usually acts on a particular substrate to catalyse a particular type of reaction. In some cases the specificity of the enzymes may not be so strong but rather general and a particular enzyme may act on a group of substrates.

(3) Reversibility

In most of the cases the reactions catalysed by the enzymes are reversible depending upon the requirements of the cell. But, in some cases there are separate enzymes for forward and backward reactions. Or, a reaction catalysed by a particular enzyme may not be reversible at all.

(4) Sensitiveness to Heat and Inhibitors

(*i*) The enzymes are very sensitive to heat, *i.e.*, they are **themolabile.** They are inactivated at very low temperatures. At very high temperatures 60°-70°C usually they are destroyed (denatured). Low molecular weight enzymes are comparatively more heat stable.

(*ii*) Enzymes are also sensitive to inhibitors. While some inhibitors may partially inhibit their activity, other inhibitors like poisons destroy them permanently and inhibit their activity.

(5) Colloidal Nature

The enzymes are of colloidal dimensions and present large surfaces for reactants in water to facilitate the enzyme reaction.

FACTORS AFFECTING ENZYME ACTIVITY

(1) Temperature

Usually the activity of the enzymes is **optimum** at normal body temperature.

At very low temperature (*i.e.*, about 0°C) the activity of the enzymes is **minimum.**

An increase in temperature up to a certain limit increases the enzyme activity, **maximum** being at about 45°C after whcih the enzyme activity is retarded (Fig.1.10) Beyond 60°–70°C usually their activity is permanently stopped due to the **denaturation** of enzymes.

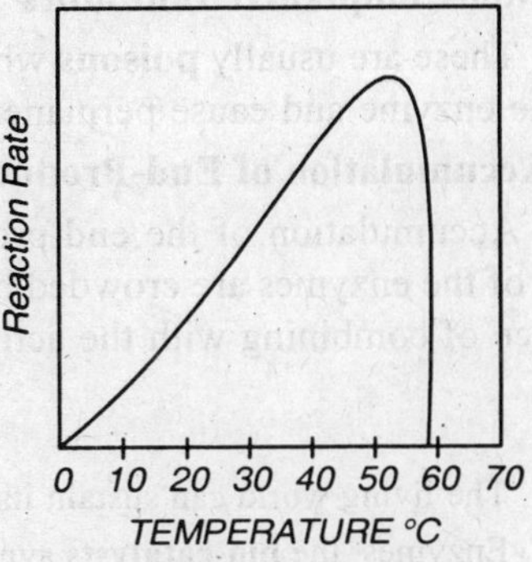

***Fig. 1.10.** Effect of temperature on enzymic reaction.*

(2) Hydrogen Ion Concentration (pH)

Enzymes are active only over a limited range of pH. Some enzymes, *e.g.*, *trypsin* are active in alkaline medium (high pH.), *diastase* in neutral medium, while *pepsin* shows optimum activity in acidic medium (low pH).

(3) Water

In absence of water the enzyme activity is suppressed so much so that in dry seeds the enzymes are almost inactive.

Proper hydration of the cells is neccessary for enzyme activity because (*i*) water provides medium for enzyme reaction to take place, and (*ii*) in many cases it is one of the reactants.

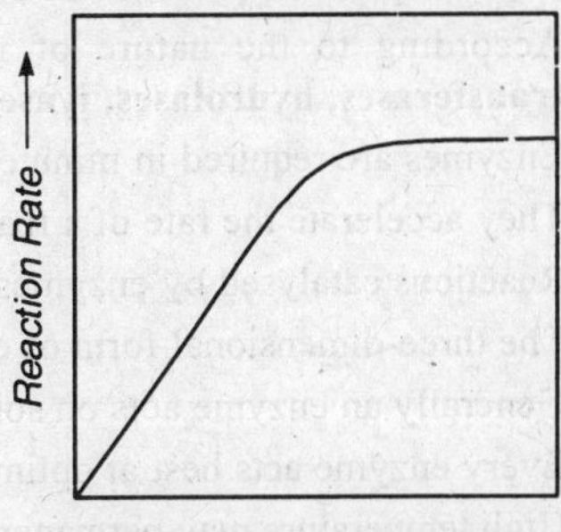

***Fig. 1.11.** Effect of substrate concentration on enzymic reaction.*

(4) Concentration of the Substrate

Increase in the concentration. of the substrate brings about an increase in the activity of the enzyme till all the **active sites** of the enzyme molecules are saturated with substrate. After this the rate of enzyme reaction becomes steady and addition of the substrate will not have positive effect (Fig. 1.11).

(5) Enzyme Concentration

Usually a very small amount of the enzyme can consume large amount of the substrate. Increase in the concentration of the enzyme will increase the rate of reaction catalysed by it provided there is enough concentration of substrate (Fig. 1.12). It is because (*i*) increased number of enzyme molecules will have more active sites, and (*ii*) at higher concentration of the enzyme the inhibitors will fall short.

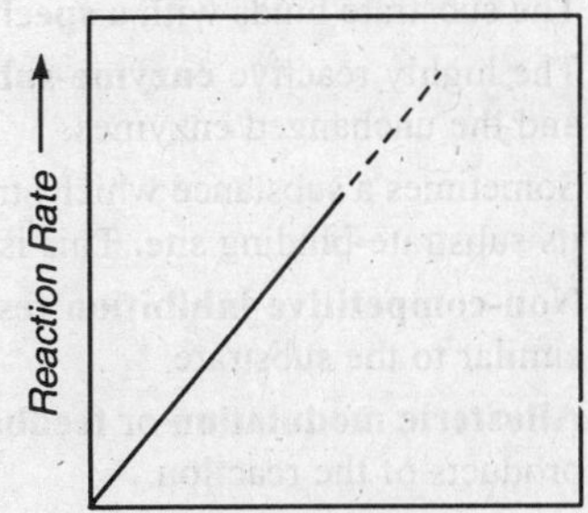

***Fig. 1.12.** Effect of enzyme concentration on the rate of reaction.*

(6) Inhibitors

Presence of inhibitors in the reaction mixture inhibits the activity of the enzymes partially or completely depending upon the nature of the inhibitors. Inhibitors are less effective when the concentration of the enzyme and substrate is higher. Inhibitors are of two types.

(*i*) Competitive Inhibitors

Such inhibitors have structural similarity with the substrate both of which **compete** for the same active site of enzyme. If competitive inhibitor pre-occupies the active site, the substrate molecule will be unable to combine with the enzyme and hence, the enzyme activity will be inhibited. But, this inhibition is of **reversible type** because removal of the competitive inhibitor restores the activity of the enzyme.

(*ii*) Non-Competitive Inhibitors

These are usually **poisons** which do not compete for the active sites but destroy the structure of the enzyme and cause permanent or **irreversible inhibition** of the acitvity of the enzyme.

(7) Accumulation of End-Products

Accumulation of the end-products retards the enzymic activity mainly because the active sites of the enzymes are crowded by them and substrate molecules will have comparatively lesser chance of combining with the active sites.

SUMMARY

1. The living world can sustain its busy pace entirely because of the presence of **enzymes.**
2. Enzymes, the **bio-catalysts** synthesised in living cells, are **proteins.**
3. Some have a non-protein part called the **prosthetic group** or **coenzyme.**
4. Many require the presence of **co-factors** like mineral ions to function.
5. According to the nature of reactions catalysed, enzymes are classified into **oxireductases, transferases, hydrolases, lyases, isomerases** and **ligases.**
6. Enzymes are required in minute amounts and are not used up or changed at the end of a reaction.
7. They accelerate the rate of a reaction by reducing the energy required for it.
8. Reactions catalysed by enzymes are often a million times faster than the non-catalysed ones.
9. The three-dimensional form of enzymes makes the enzyme more versatile than inorganic catalysts.
10. Generally an enzyme acts on some specific **substrates.**
11. Every enzyme acts best at optimum pH and at optimim temperature.
12. High temperature may permanently destroy enzyme action.
13. Cold preserves the enzyme in a temporarily inactive state.
14. Enzyme activity is also controlled by substrate concentration.
15. The substrate binds with a **specific site** of the enzyme.
16. The highly reactive **enzyme-substrate complex** almost immediately dissociates to give the product and the unchanged enzymes.
17. Sometimes a substance which structurally resembles the substrate, inhibits the enzymes by occupying its substrate-binding site. This is called **competitive inhibition.**
18. **Non-competitive inhibition** results from the interaction with a substance which is not structurally similar to the substrate.
19. **Allosteric modulation** or **feedback inhibition** is due to inhibition of enzyme action by the product/ products of the reaction.

TERMS TO REMEMBER

Activation energy. An initial input of energy to start the reaction.

Active site. All enzymes have specific three-dimensional structrue, a part of which is known as **active site** which serves as a **'lock'** into which the reactant (substrate) fits in like a **'key'**.

Allosteric enzymes. Where a modulator substance binds with a specific site of the enzyme different from its substrate binding site. This binding increases or decreases the enzyme action. Such enzymes are called allosteric enzymes.

Catalyst. A small amount of a substance that influences significantly the rate of a chemical reaction. The substance itself is not used up or changed at the end of reaction.

Catalysis. The phenomenon is called catalysis.

Coenzymes or Co-factors. Certain organic compounds and inorganic ions are required for enzyme activity ; they are loosely bound to the enzyme, such as NAD and FAD (*i.e.*, nucleotides of the vitamin nicotinamide and riboflavin).

Enzymes. Molecules synthesized by the cells; they are biological catalysts.

Feedback inhibition. Decline in enzyme activity by the allosteric effect of the product, *e.g.*, allosteric inhibition of hexokinase by glucose-6-phosphate.

Michaelis constant (K_m). Substrate concentration at which the reaction attains half its maximum velocity.

QUESTION BANK

N.B. Undermentioned questions are repeatedly asked in various Indian Universities and competitive examinations.

Short and Long Answer Questions

1. What are enzymes ? Give a brief account of the enzyme you have studied.
2. What are enzymes ? State their characteristics and discuss the role in digestion.
3. What are enzymes ? Give an account of the general properties and nomenclature of enzymes.
4. Wrtie short note on prosthetic group.
5. Write note on coenzymes.
6. Write an esssay on enzymes.
7. What are enzymes ? Discuss their mechanism of action.
8. What are enzymes ? Discuss their properties and mode of action.
9. Discuss the chemical natrue, mechanism of action and properties of enzymes.
10. Discuss the structure and the mechanism of the enzyme action.
11. Enzymes are 'biological middle men'. Justify the statement.
12. Write short note on enzymes.
13. Boiled yeast cannot ferment sugar. Comment.
14. Give an example and importance of coenzyme.
15. Where are the following enzymes located in the plant cell? In what process do they participate ?
    (*a*) hydrolases (*b*) dehydrogenases
    (*c*) lyases (*d*) transferases.
16. What kind of reaction does each of the following group of enzymes catalyse ?
17. List threee important properties of enzymes.
18. How does competitive inhibition differ from allosteric inhibition ?
19. How does competitive inhibition of an enzyme occur ? How is such inhibiton useful ?
20. What role does temperature play in food preservation ? Explain.
21. How are enzymes named ? How have enzymes been classified ?
22. What does an enzyme do in terms of energy requirement of a reaction ? What would happen if the enzyme did not play this role ?
23. What is the name given to that part of the enzyme where catalytic work is carried out ?
24. Explain differences in the following :
    (*a*) coenzyme (*b*) prosthetic group (*c*) co-factor
25. Explain the following :
    Frozen and boiled leaf extract hardly exhibit any enzyme activity but when brought to room temperature, the former shows enzyme activity and the later fails to do so.
26. What are enzymes ? Write about their :
    (*a*) Mode of action (*b*) Role in plant metabolism (*c*) Classification

27. Give a short account of the mechanism of action of enzymes.
28. Differentiate between :
    (*a*) Competitive and non-competitive inhibitors
    (*b*) Coenzymes and co-factors
    (*c*) NAD^+ and $NADP^+$
29. Describe chemical nature and characteristics of enzymes.
30. What is Michaelis-Meten Constant ? What is its significance.
31. What are coenzymes ? Illustrate your answer with two examples.
32. What are enzymes? Describe the various factors affecting the enzymes activity.
33. Give an account of classification and mechanism of action of enzymes.
34. What kind of reaction does each of the following group of enzymes catalyse
    (*a*) hydrolases (*b*) dehydrogenases
    (*c*) lyases (*d*) transferases
35. What is the name given to that part of the enzyme where catalytic work is carried out ?
36. How does competitive inhibition differ from allosteric inhibition ?
37. How does competitive inhibition of an enzyme occur ? How is such inhibition useful ?
38. List three important properties of enzymes.
39. How are enzymes named ? How have enzymes been classified recently ?
40. What role does temperature play in food preservation ? Explain.

Multiple Choice Questions

1. Which of the following statements is not correct ?
   (*a*) Enzymes are simple or combined proteins acting as specific catalysts.
   (*b*) Enzymes are organic catalysts.
   (*c*) Enzymes are biological catalysts.
   (*d*) None of the above.
2. Protein part of the enzyme is called
   (*a*) holoenzyme (*b*) prosthetic group (*c*) apoenzyme (*d*) none of the above
3. The specific region of the enzymic protien which is involved in biochemical reaction is called :
   (*a*) active region (*b*) active site (*c*) both (*a*) and (*b*) (*d*) none of the above
4. The enzymes differ from inorganic catalysts in being :
   (*a*) products of living cells
   (*b*) proteinaceous in nature
   (*c*) remarkably specific toward their substrates
   (*d*) all of above
5. The enzymes which operate externally to the cells in which they originate are called :
   (*a*) exoenzymes (*b*) extracellular enzymes
   (*c*) exocellular enzymes (*d*) all of above
6. One of the first enzymes discovered in isozyme form is :
   (*a*) aspartate kinase (*b*) lactate dehydrogenase
   (*c*) malate dehydrogenase (*d*) none of the above
7. If an enzyme substrate system has lower value of Km, then binding of substrate to its enzyme will be :
   (*a*) more strong (*b*) less strong (*c*) inhibited (*d*) none of the above
8. Concentration of substrate to half saturate the enzyme or to cause half the maximal recaction rate is called :
   (*a*) Michaelis constant (*b*) Michaelis-Menten constant
   (*c*) Km (*d*) all of above

9. Allosteric enzymes have
(*a*) Active sites (*b*) allosteric sites (*c*) both (*a*) and (*b*) (*d*) none of the above
10. Competitive inhibitor of the activity of the enzyme succinic dehydrogenase is :
(*a*) malonic acid (*b*) succinic acid (*c*) fumaric acid (*d*) none of the above
11. Which of the following statements is not correct ?
(*a*) Enzymes act in very small amounts
(*b*) Enzymes are sensitive to heat, pH and inhibitors
(*c*) Enzymes are specific in their action
(*d*) Enzymes catalyse only reversible reactions
12. The enzymes act by :
(*a*) loweirng the energy of activation of reactions
(*b*) increasing the energy of activation of reactions
(*c*) conserving the energy of activation of reactions
(*d*) none of the above
13. Which of the following coenzymes is a carrier of acyl group ?
(*a*) Coenzyme A (*b*) Coenzyme Q (*c*) Coenzyme II (*d*) Coenzyme F
14. Which of following organic compounds act as prosthetic groups in some enzymic reactions and as cofactor in some others ?
(*a*) FMN (*b*) FAD (*c*) Biotin (*d*) All of above
15. Non-protien chemical substances which are required for enzymic reatcion but are not tightly bound to the enzyme (by covalent linkages) are called :
(*a*) prosthetic groups (*b*) cofactors
(*c*) general activators (*d*) coenzymes
16. Which of the following hydrolysing enzymes do not hydrolyse ester bonds
(*a*) Phosphatases (*b*) Phosphatidases (*c*) Lipases (*d*) Peptidases
17. According to new system of classification the enzymes are grouped into :
(*a*) four major groups (*b*) five major groups
(*c*) six major groups (*d*) seven major groups
18. First enzyme to be isolated in pure crystalline form was :
(*a*) zymase (*b*) urease (*c*) invertase (*d*) diastase
19. The word enzyme was coined by :
(*a*) F.W. Kuhne (*b*) Louis Pastur (*c*) Alexopoulos (*d*) C.E. Bessey
20. Except for small group of catalytic RNA molecules, all enzymes are :
(*a*) proteins (*b*) lipids (*c*) protiens and lipids (*d*) none of the above
21. Different enzymes show optimum activity in :
(*a*) acidic medium (*b*) alkaline medium (*c*) neutral medium (*d*) either of the above
22. Coenzyme is
(*a*) always a protein (*b*) often a metal
(*c*) always an inorganic compound (*d*) often a vitamin
23. Enzymes generally have :
(*a*) same pH and temperature optima
(*b*) same pH but different temperature optima
(*c*) different pH but same temperature optima
(*d*) different pH and different temperature optima
24. The term feed-back refers to :
(*a*) the effect of substrate on the rate of enzymatic reaction

(*b*) the effect of end product on the rate of enzymatic reaction
(*c*) the effect of enzyme concentration on its rate of reaction
(*d*) the effect of an external compound on the rate of enzymatic reaction.

25. Which of the following is the best evidence for the lock and key theory of enzyme action ?
(*a*) All isolated enzymes have been identified as proteins
(*b*) Compounds similar in structure to the substrate inhibit the reaction.
(*c*) Enzymes are found in living organisms and speed up certain reactions.
(*d*) Enzymes determine the direction of a reaction.

ANSWERS

1. (*d*)	2. (*c*)	3. (*b*)	4. (*d*)	5. (*d*)	6. (*b*)	7. (*a*)	8. (*d*)	9. (*c*)	10. (*a*)
11. (*d*)	12. (*a*)	13. (*a*)	14. (*d*)	15. (*b*)	16. (*d*)	17. (*c*)	18. (*b*)	19. (*a*)	20. (*a*)
21. (*d*)	22. (*b*)	23. (*d*)	24. (*b*)	25. (*c*)					

UNIT - 2
Plant Physiology

2
CHAPTER

Plant Water Relations

The plants absorb water and soluble mineral salts from the soil by their root system. This function of absorption is facilitated by the unicellular root hairs present on the roots. These unicellular root hairs enter in the interspaces of the soil particles, irregularly. These root hairs absorb the water found in the form of thin films around these soil particles.

Cell as a physiological unit. Cells are the building blocks from which living things are made. A single cell may make up an entire organism, or groups of cells may be loosely organised and live together.

Scientists have known about cells for a long time. About two hundred years back biologists realised that 'the cell is the basic unit of all living organisms, hence **cell theory** was proposed by two German scientists, 'Schleiden and Schwann'.

The major ideas of cell theory are:

1. Living things are made up of cells and cell products.
2. Cells are very much alike in their structure and composition.
3. Cells have a set of functions that they carry out in order to stay alive.
4. New cells arise from old cells by cell reproduction.

However, plant cells are composed of cell wall and protoplast.

The term **protoplast** is generally used to refer collectively to the **plasma membrane** and **protoplasm**.

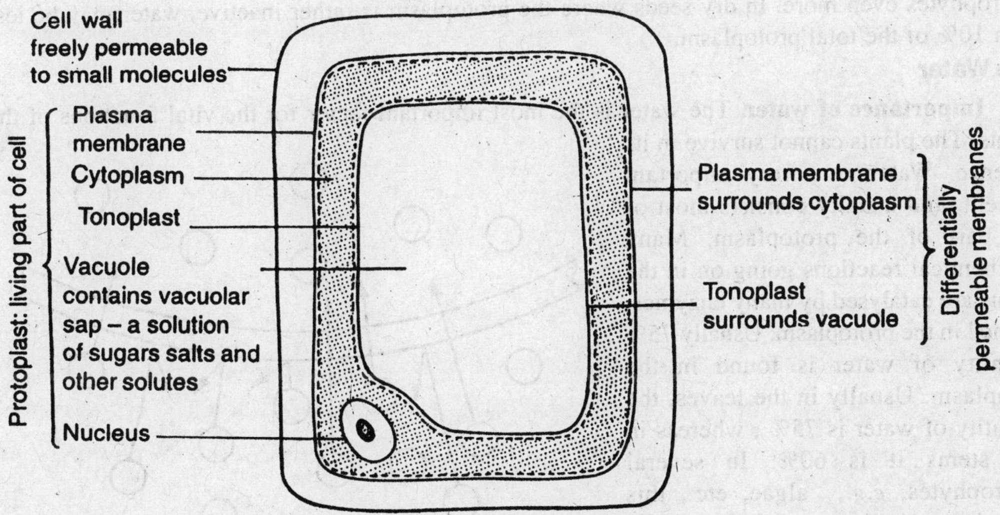

Fig. 2.1. Structure of a plant cell.

The protoplasm refers to the living contents of the cells and consists of **cytoplasm** and **nucleus**.

Generally, the plant cell possesses three compartments:

(*i*) Vacuole, (*ii*) protoplasm and (*iii*) cell wall. These compartments are separated from each other by **plasma membranes**, *i.e.*, the tonoplast is found in between the vacuole and the protoplasm, while plasmalemma between the protoplasm and the cell wall. Plasmodesmata, connect the protoplasm of one cell to that of other cell.

The plasma membrane is **selectively permeable**. This means, it allows some materials to pass through and not others.

The plasma membrane is made up almost of protein and lipid molecules.

Protoplasm. According to modern view, the protoplasm is a colourless, semi-transparent and viscous substance. It is considered as complex colloidal system of many phases.

Table 2.1. Differences Between Plant Cells and Animal Cells

Plant Cells	Animal Cells
1. Plant cells have a rigid cell wall made of **cellulose**.	1. Cellulose cell wall absent.
2. **Plastids** present. Photosynthesis takes place in **chloroplasts**.	2. Plastids absent. No photosynthesis, no chloroplasts.
3. Each plant cell has a large vacuole containing water and salt.	3. Vacuoles absent.
4. Centrioles absent.	4. They have **centrioles**, which are active during reproduction of cell. Centrioles contain microtubules which help to separate and to sort chromosome material during cell division.

However, it is well established that water is chief component of all physiologically active protoplasm and in such cases may constitute upto 90% of the protoplasm and in cases of hydrophytes even more. In dry seeds where the protoplasm is rather inactive, water may be less than 10% of the total protoplasm.

The Water

Importance of water. The water is the most important factor for the vital functions of the plants. The plants cannot survive in its absence. Water is a very important solvent, and usually consists most of the part of the protoplasm. Many biochemical reactions going on in the plants are catalysed by many enzymes, formed in the protoplasm. Usually 75% quantity of water is found in the cytoplasm. Usually in the leaves, the quantity of water is 75% , whereas in the stems it is 60%. In several hydrophytes, *e.g.*, algae, etc., this quantity exceeds up to 98%. In the same way, the quantity of water in the xerophytes is usually 60% or lesser than this. In the dormant seeds, this

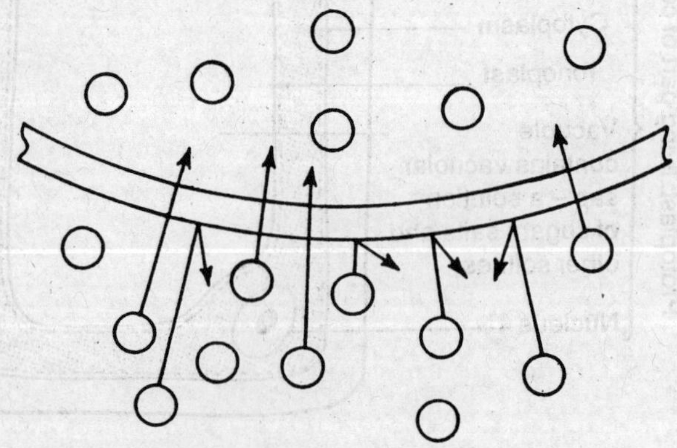

Fig. 2.2. Passage of molecules of different substances through a selectively permeable membrane.

quantity of water is only about 10%. This is, of course very much less for any vital function, and therefore, the germination of these seeds is only then possible, when they get sufficient moisture. The cytoplasm can remain viable only in the presence of sufficient amount of water. In the scarcity of water, it dies. The water also plays an important role in the function of photosynthesis. The water is a good solvent, the minerals cannot be absorbed unless they are not soluble in water. The cell of the plants remain turgid by water; and give temporary mechanical support to the young plants.

The water found in the plants, is a very small part of the water absorbed by the plants, rest of the larger part of the water goes out from the surface of the plants, by a vital process, called **transpiration**.

Availability of the water in the soil. About three forms of water are found in the soil. Just after the rains, because of gravitation some amount of water along with some mineral salts goes in the lower strata of the earth. This is called **gravitational** or **free water** and cannot be used by the plants. Besides this, every particle of soil holds some imbibed water in it. This water is hold up in the soil particle with such a great imbibition force, that it cannot be separated from it for the use of the plant. This is called **hygroscopic water**, which also cannot be absorbed from the soil by the plants. Besides these each soil particle is surrounded by a loose film of water; this film is attracted by the capillary force to the soil particle, and such water film is called **capillary water**. Now it is to be noticed, that only this capillary water may be absorbed by the root hairs of the plants. The mineral salts are also found in this water in soluble state, and are absorbed by the plants along with water. If soil lacks this capillary water, the plants very soon begin to wither and ultimately die.

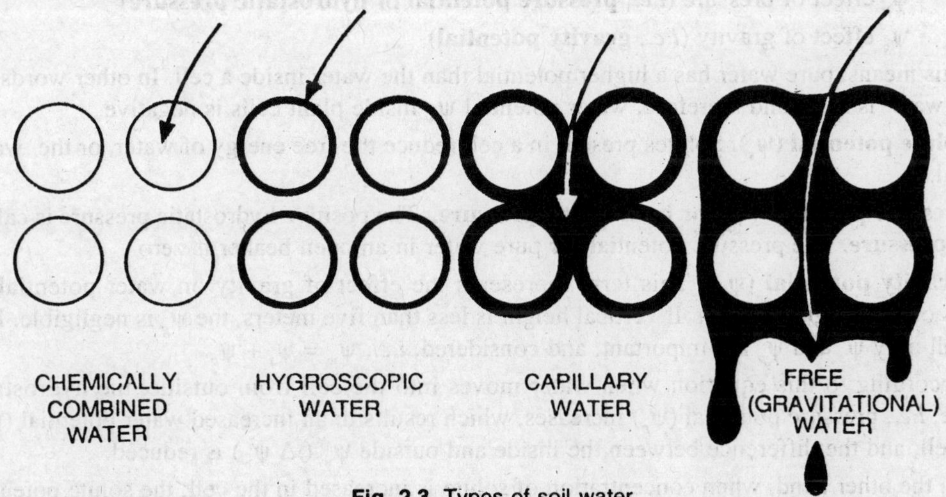

Fig. 2.3. Types of soil water.

WATER POTENTIAL ψ (Psi)

All plants and other living organisms require free energy to grow and reproduce.

In thermodynamics, free energy represents potential to do work. According to thermodynamic laws every component of a system possesses **free energy** capable of doing work under constant temperature conditions.

The potential energy of water is called **water potential**. Water potential is regarded as the tendency of water to leave a system. It is often used while explaining the direction in which water will flow from one cell to another, or from one part of the plant to another such as from soil to root, from root to leaves, from leaves to air, or from soil to air.

Water always moves from a region of higher water potential to that of lower water potential. It can also be said that the difference in water potential between two points is a measure of the amount of work, *i.e.*, energy, needed to move water from one point to the other.

Osmotic movement of water involves certain work done and in fact the main driving force behind this movement is the difference between free energies of water on two sides of the selectively permeable membrane. For water free energy molecule is known as water potential (ψ_w).

Water potential is measured in terms of pressure. Common measurement unit of water potential is **Pascal**, *Pa* ; *1* Megapascal represents 10 bars, *i.e.*, 1 **Megapascal** = 10 **bars**. One bar is close to one atmosphere of pressure, *i.e.*, 1 bar = 0.987 atmosphere of pressure.

At atmospheric pressure water potential of pure water is zero, and therefore, all solutions at atmospheric pressure have lower water potentials than water, *i.e.*, they have a negative value.

Water potential ψ_w is measured in relative quantity, and expressed as the difference between the potential of a solution in a given state and the potential of the same solution in a standard state.

Water potential is lowered by the addition of solutes and as water potential value is zero for pure water, all other water potential values will be negative. Thus, the movement of water will take place in osmotic or other systems from a region of **higher water potential** (*i.e.*, less negative) to a region of **lower water potential** (*i.e.*, more negative).

Water potential of any solution (ψ_w) is influenced by three factors: (*i*) concentration, (*ii*) pressure and (*iii*) gravity. This can be represented by following equation:

$$\psi_w = \psi_s + \psi_p + \psi_g$$

i.e., ψ_s effect of solutes (*i.e.*, **solute potential** or **osmotic potential**)

ψ_p effect of pressure (*i.e.*, **pressure potential** or **hydrostatic pressure**)

ψ_g effect of gravity (*i.e.*, **gravity potential**)

This means, pure water has a higher potential than the water inside a cell. In other words ψ_w of pure water is zero, and therefore, water potential ψ_w inside plant cells is negative.

Solute potential (ψ_s). Solutes present in a cell reduce the free energy of water, or the water potential.

Pressure potential (ψ_p) or hydrostatic pressure. The positive hydrostatic pressure is called **turgor pressure**. The pressure potential for pure water in an open beaker is zero.

Gravity potential (ψ_g). This term represents the effect of gravity on water potential. It depends on the height of water. If vertical height is less than five meters, the ψ_g is negligible. In a plant cell only ψ_s and ψ_p are important, and considered, *i.e.*, $\psi_w = \psi_s + \psi_p$.

According to this equation when water moves into the cell from outside, the hydrostatic pressure, *i.e.*, pressure potential (ψ_p) increases, which results in an increased water potential (ψ_w) of the cell, and the difference between the inside and outside ψ_w ($\Delta\ \psi_w$) is reduced.

On the other hand, when concentration of solute is increased in the cell, the solute potential (ψ_s) is lowered, and therefore, water potential (ψ_w) is decreased.

Table 2.2. Differences Between DPD and Water Potential

Diffusion Pressure Deficit (DPD)	Water Potential ψ (Psi)
1. It is measure of actual pressure.	1. It is measure of free energy.
2. It is measure of decrease in diffusion pressure of water (pure solvent) due to addition of solute molecules.	2. It is measure of difference of energy of pure water and an aqueous solution.
3. It is equal in magnitude and opposite in direction to water potential (ψ Psi).	3. It is equal in magnitude and opposite in direction to diffusion.

Thus, water moves into the cell from outside due to a water potential gradient. If a pressure is created on the cell, water moves out of it. Here, external pressure raises the water potential (ψ_w) of the cell, and thus the difference in water potential (ψ_w) inside and outside ($\Delta \psi_w$) will be such that water will flow out of cell.

There are two basic factors that affect the water potential : (*i*) amount of solute and (*ii*) external pressure.

(*i*) **Amount of solute.** Pure water at atmospheric pressure has a zero water potential. An addition of solutes lowers the water potential (ψ_w), *i.e.*, it becomes negative in value.

(*ii*) **External pressure.** Effect of pressure on water potential is just opposite to the effect of solutes, *i.e.*, the increase in pressure increases the water potential (ψ_w).

ABSORPTION AND MOVEMENT OF WATER

The absorption of water and solutes from the soil seems to have been the first physiological processes to receive the attention of early writers on botanical subjects. According to Aristotle plant foods are complex substances which have been so prepared in the soil that they can be absorbed and used by the plant without any changes, hence without the necessity of excreting any

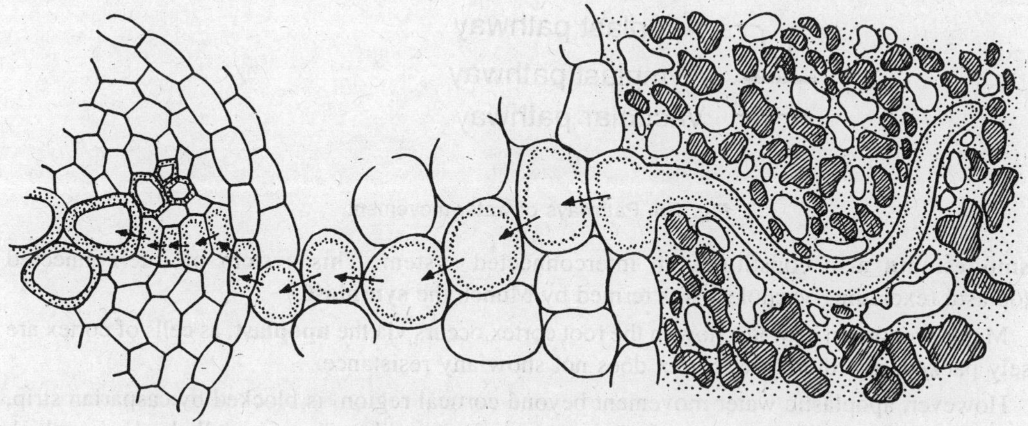

Fig. 2.4. Path of water in the root.

water or by products, as do animals. The earliest known attempt to explain absorption process was made by Andrea Cesalpino (1603). He, of course, accepted the Aristotle's theory that plants absorb their food in solution from the soil.

The pathway of water through the root. Water enters the roots principally through the walls of the root and epidermal cell of the root tips. Absorption of water by individual root hair has been demonstrated experimentally (Rosene, 1949). From the epidermal cells the water passes through successive rows of thin walled cortical cells and then through the cells of endodermis.

After passing through the endodermis water moves into xylem tissue ducts. Once in the xylem ducts its general direction of movement is upward. The xylem tissue is continuous from just back of the tips of roots, through the roots, into and through the stems, the petioles of leaves and ultimately terminates in the mesophyll of the leaf. The xylem tissue through which the water moves thus a continuous system with the body of the plant.

Water in the root moves through two pathways:

(*i*) **Apoplast pathway** and

(*ii*) **Symplast pathway.**

E. Munch of Germany introduced **apoplast-symplast** concept in 1932. He suggested that the interconnecting cell walls and intercellular spaces including the water-filled (or air-filled) xylem elements should be considered as a single system and called the **apoplast**.

The remaining part of the plant consists of the protoplasts of the cells. Evidence based upon electron micrograph now indicates that the cytoplasm connected from cell to cell via the

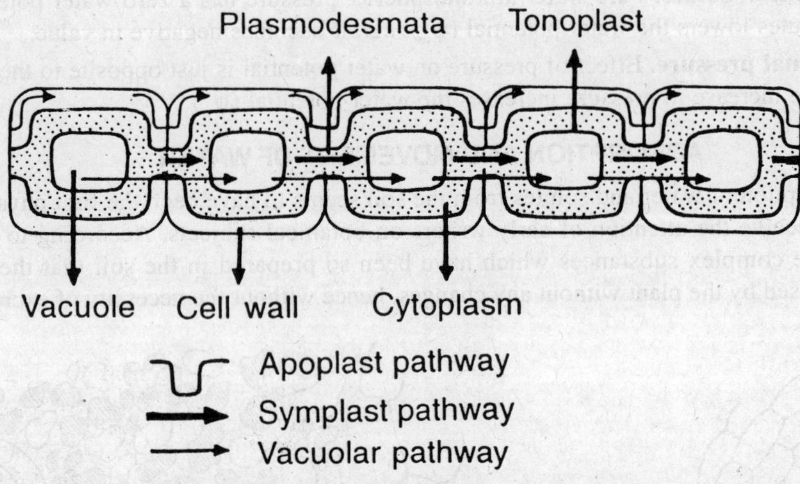

Fig. 2.5. Pathways of water movement.

plasmodesmata also constitutes an interconnected system. This system of interconnected protoplasm (excluding vacuoles) was termed by Munch the **symplast**.

Major proportion of water flow in the root cortex occurs via the **apoplast**, as cells of cortex are loosely packed, and therefore, cortex does not show any resistance.

However, apoplastic water movement beyond cortical region is blocked by casparian strip, present in endodermis. The casparian strip consists of a wax-like substance called suberin, which blocks water and solute movement through the cell wall of endodermis.

Thus, beyond endodermis, water is forced to move through cell membranes. Such movement of water through cell membrane is called **transmembrane pathway**. In this pathway water may also cross through the **tonoplast** surrounding the vacuole.

Once the water reaches root xylem, transpiration pull drives the water to move the leaves through the stem.

Diffusion

If we keep hydrogen sulphide gas (H_2S) exposed in any corner of the room, after a very short time the whole room is filled up with bad odour of the gas. In fact, this happens because the molecules of hydrogen sulphide gas gradually spread in the whole room. In other words, it can be said that the molecules of the gas gradually mix with the molecules of the air present in the room, uniformly. This process is called **diffusion**. During respiration the oxygen is used in the leaf, and the carbon dioxide is accumulated in the intercellular spaces of the leaf. Now this carbon dioxide comes outside in the atmosphere, by diffusion and the oxygen is taken in the cells by diffusion. This process of diffusion may take place in all kinds of matters such as solid, liquid and gas. In other words, it can be said that the molecules of these matters may spread uniformly in the solvent.

If the volatile substances, such as petroleum, alcohol, ether and perfumes, etc., are kept exposed in a room, then in a meanwhile, the whole room fills up with the odour of the volatile substance. It means, the molecules of these substances spread in the whole room, and gradually mix with the molecules of the air present in the room. Besides these, if the solid substances such as naphthalene balls and camphor are kept exposed in the room, very soon their molecules spread, and mix with air molecules, and the whole room fills up with their odour. The solid substances are also diffused in the liquid substance, but with a slow speed. For example, a crystal of potassium permanganate ($KMnO_4$) is dropped in a beaker filled up with water. In a mean while the water of the beaker around this crystal becomes deep red, and after some time the whole water of the beaker becomes of uniform red colour, and the crystal is fully dissolved. In the beginning, the molecules of potassium permanganate, around the crystal are in greater concentration, but very soon, the crystal dissolves, and the molecules spread uniformly in whole of the water of the beaker and the water becomes of uniform red colour, because of uniform distribution of crystal molecules, in the solvent, by diffusion. This experiment may also be done by the soluble crystal of copper sulphate. If sugar crystals are added to water and kept for some time, then after some time the water becomes sweet — this is also a good example of diffusion. In this way, the solid, liquid and gases diffuse in different media. The above mentioned examples prove that, the gases may be diffused in other gases; the liquids may diffuse in the gases or other liquids; and the solids may also diffuse either in gas or in liquid.

The above mentioned examples of **diffusion** may be summarised as follows:

That diffusion is a random movement of individual molecules from a region of higher concentration to a region of lower concentration.

Larger the difference in concentration more rapid is the flow of molecules.

Diffusion is more rapid in gases than in liquids.

When there is no further movement of molecules, a state of equilibrium is acquired, which means that the diffusing substance is evenly distributed.

In diffusion, the movement of molecules is random and independent of each other.

Suitable examples of diffusion in plants are: supply of carbon dioxide from atmosphere to the leaves for photosynthesis, and loss of water vapour from leaves to the atmosphere.

Permeability

The concept of permeability is inseparable from the idea of a membrane. Permeability is a property of the membrane, not of substance which diffuses through it.

Some membranes are impermeable to all substances, others allow all or most substances to diffuse through them with little resistance. Many biologically important membranes are of the type known as **differentially permeable**. Such membranes allow some substances to pass through them much more readily than others. The extent to which a membrane permits or restricts the movement of a substance is called **membrane permeability**. It depends on the membrane composition, as well as chemical nature of solute. Permeability can be measured readily by determining the rate at which solute passes through a membrane under a specific set of conditions.

Osmosis

It is usually seen, that when two liquids of different densities are separated from each other by placing a membrane between them, by which the water of the liquid of lower density flows towards the liquid of higher density, then such separating membrane is called the **semipermeable membrane**. In other words, only the solvent or water may pass through such membrane, the solute or the dissolved substance in the water cannot pass through it. The parchment paper, egg membrane, goat bladder, etc., are the good examples of semipermeable membranes. In the plant cell, the cell wall is permeable whereas the protoplasmic membrane is semipermeable. When concentrated and dilute solutions or the solutions of lower density and higher density are separated from each other by such membrane, then the solvent moves from a concentrated solution to dilute solution. This process is called the **'osmosis'**.

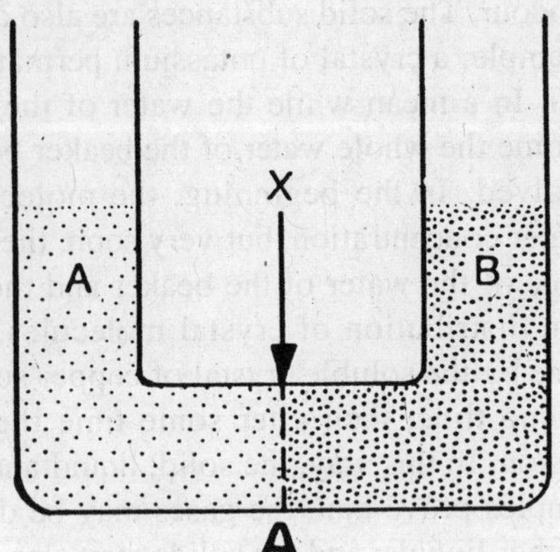

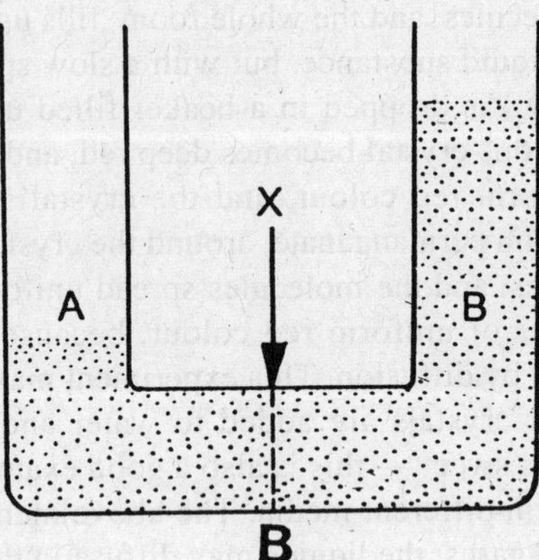

Fig. 2.6. Diagrammatic representation of osmosis, *x* indicates the semipermeable membrane.

Table 2.3. Differences Between Diffusion and Osmosis

Diffusion	Osmosis
1. Diffusion can occur in any medium.	1. It occurs only in liquid medium.
2. The diffusing molecules may be solids, liquids or gases.	2. It involves movement of solvent molecules only.
3. Semipermeable membrane is not required.	3. Semipermeable membrane is required.
4. It is dependent upon the free energy of the molecules of diffusing substance only; presence of other substances in the system is of no importance.	4. Though it is the diffusion of solvent molecules only, yet influenced by the presence of other substances (solutes) in the system.
5. It is not influenced by the pressures other than diffusion pressure.	5. It is influenced by turgor or hydrosttic pressure of the system.
6. An equilibrium in the free energy of diffusion molecules is achieved in the system.	6. Equilibrium in the free energy of solvent molecules is never achieved.
7. Factors like water potential, solute potential and pressure potential do not affect diffusion.	7. Water potential, solute potential and pressure potential do not affect osmosis in a living system.

In other words, diffusion of water through a semipermeable membrane is called **osmosis**.

The diffusion of water molecules continues to occur across the membrane till an equilibrium is acquired. At equilibrium **water potential** (ψ_w) on both sides of membrane is equal.

To demonstrate osmosis experimentally following experiments may be done:

1. Egg membrane experiment. To demonstrate osmosis by this experiment an egg is taken. On the pointed side of the egg a small hole is formed and all the white and yellow matter of the egg is taken out through this hole. Now the empty egg is placed in the dilute solution of hydrochloric acid (HCl), so that the hard shell of the egg constituted of calcium carbonate is dissolved in it. The membrane of the egg, found just beneath the cell remains intact. Now this egg membrane is thoroughly washed with tap water, and a glass tube is inserted in it through its hole and tied tightly with a piece of thread, so that the end of the glass tube remains inside the egg membrane. Now, the concentrated solution of sugar is poured in it through glass tube with the help of a thistle funnel. The sugar solution is filled up in such a quantity, so that after completely filling the egg membrane sac, it also reaches to a small height in the glass tube. Now this sugar solution filled egg membrane sac, is hanged in a beaker filled up with water, so that the level of the inner and outer liquid is same. This apparatus is kept for some time in the same position. After some time it is to be noted that the level of the liquid inside the glass tube has sufficiently risen up from the level of the water of the beaker. Water will continue to move across the membrane, resulting in the rise of solution in the funnel until an equilibrium is reached.

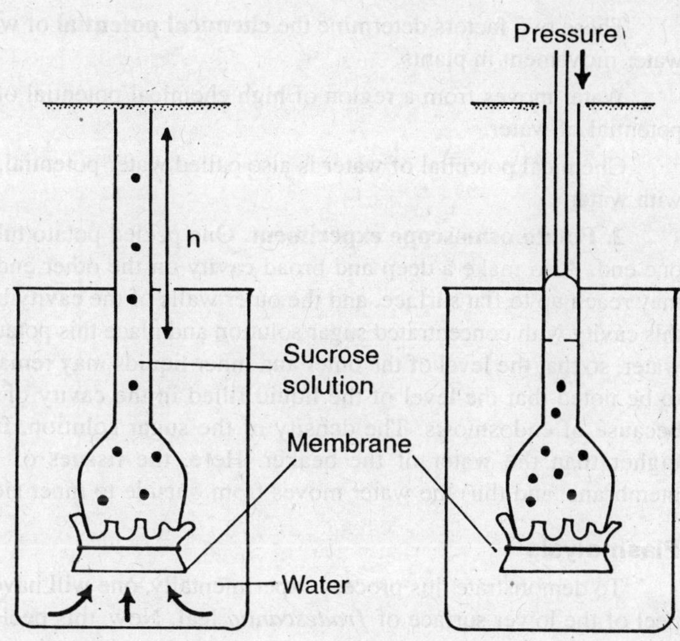

Fig. 2.7. Demonstration of osmosis by egg membrane experiment.

Pressure can be applied to the solution from upper part of funnel to prevent movement of water into it through egg membrane.

The pressure required to stop the movement of water completely is called **osmotic pressure**. This may be also referred as **osmotic potential** or **solute potential** (ψ_s).

Osmotic pressure and osmotic potential are numerically equal, but osmotic potential has a negative (–) sign.

An isolated solution without solute, and not bounded by any membrane, has no osmotic pressure. It has only the potential to result into a pressure when placed in an osmometer.

Thus, a solution has an osmotic potential (ψ_s), which is negative (–) of osmotic pressure (π, pi). The equation is

$$\psi_s = -\pi$$

If one reverses the experiment by filling up the water in the egg membrane sac and the sugar solution in the beaker then the water moves from inner side to outer side. This process is called **reverse osmosis**. The process of reverse osmosis is used for removing salts from saline water.

Process of osmosis is controlled by two important factors :

(*i*) Concentration of dissolved solutes in a solution, and

(*ii*) Pressure difference.

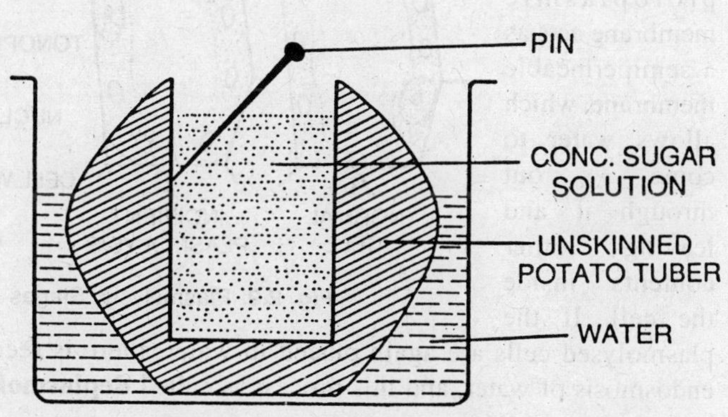

Fig. 2.8. Osmosis. Demonstration with potato osmoscope.

These two factors determine the **chemical potential** of water, which acts as driving force for water movement in plants.

Water moves from a region of high chemical potential of water to a region of low chemical potential of water.

Chemical potential of water is also called water potential, and denotes free energy associated with water.

2. Potato osmoscope experiment. One peeled potato tuber is taken. Now make it flat at its one end. Also make a deep and broad cavity on the other end of it, so that the bed of the cavity may reach up to flat surface, and the outer walls of the cavity become somewhat thin. Now fill up this cavity with concentrated sugar solution and place this potato osmoscope in a beaker containing water, so that the level of the outer and inner liquids may remain the same. After a short time, it is to be noted that the level of the liquid filled in the cavity of the potato osmoscope goes higher, because of endosmosis. The density of the sugar solution, filled up inside the potato cavity is higher than the water of the beaker. Here, the tissues of the tuber act as a semipermeable membrane, and thus the water moves from outside to inner side.

Plasmolysis

To demonstrate this process experimentally, one will have to take the red coloured epidermal peel of the lower surface of *Tradescantia* leaf. Now, this peel is mounted in a drop of water on a microscopic slide and then examined under the microscope. At this moment the cells are fully turgid, and are filled up with red coloured cell sap. Now, this epidermal peel is kept in a drop of salt or sugar solution and again examined under the microscope. The exosmosis occurs and the cell shrinks in its size. The cell soon reaches to its minimum volume. If the exosmosis continues further, there is no further shrinkage in the cell wall, but the protoplasmic membrane begins to recede or contract from the corners first. This stage is called the **incipient plasmolysis**. If exosmosis continues further, there is further contraction of the protoplasmic membrane, thus reaching the **complete plasmolysis** stage. In the case of complete plasmolysis the cell sap, protoplasm and nucleus, etc., are completely contracted in the centre of the cell leaving away the cell wall. The in-between part of the cell is filled up with the sugar or salt solution. Here, around the vacuole, the protoplasmic membrane acts as a semipermeable membrane, which allows water to come out through it and leaving other contents inside the cell. If the plasmolysed cells are again placed in water, there is recovery to the original condition due to endosmosis of water, and this process is called **deplasmolysis**.

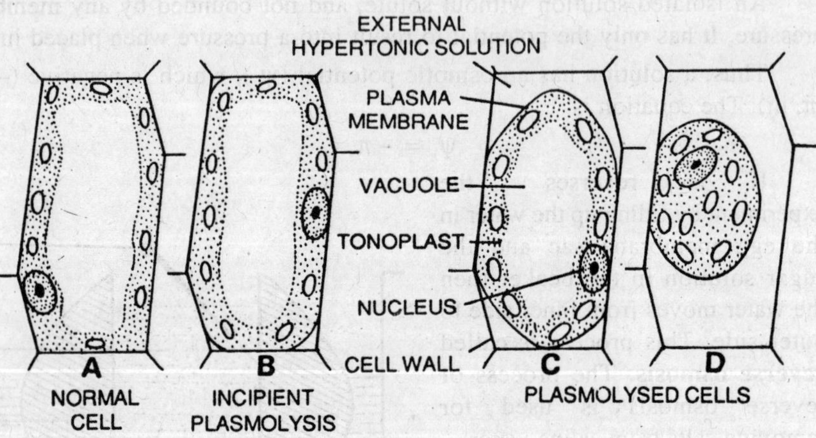

Fig. 2.9. Plasmolysis. Stages in plasmolysis of a plant cell.

Continuous loss of water from plant cells results in wilting and drooping of leaves and stems.

Imbibition

This process is a type of diffusion by which movement of water takes place along a diffusion gradient. Due to this process certain dried and half-dried matters absorb water. Such substances (*e.g.*, fibres, wood pieces, proteins and sponges) are called **adsorbants**. An adsorbant is required for the **imbibition** to occur. The cell wall and the protoplasm also absorb water by the process of imbibition. The absorption by imbibition is possible only in those conditions, where there is a great affinity between adsorbant and water, otherwise this is not possible. For example, cotton fibres absorb water by this process, whereas rubber sheet does not, as there is no affinity between rubber and water. Dry plant material or seeds act as adsorbant to imbibe water and swell. The swollen seeds produce a large pressure developed by imbibition. This pressure forces the seedlings to emerge above the ground through the soil surface. Here, with the result of imbibition pressure, the seed coat bursts and seedling comes out of ground. This pressure is called **imbibition pressure** or **matric potential** (*i.e.*, water potential of the matrix) in the context of plant water relations.

For imbibition, two conditions are important. They are:

(*i*) Water potential gradient between surface of adsorbant and the liquid imbibed.

(*ii*) Affinity between the adsorbant and the imbibed liquid.

Imbibition plays an important role in life of plants.

(*i*) The first step in the absorption of water by the roots of higher plants is the imbibition of water by the cell walls of root hairs.

(*ii*) Imbibition of water is essential for dry seeds prior to germination.

Now, imbibition pressure is called **matric potential** (ψ_m). The matric potential in an imbibant results primarily from adsorptive forces which bind water molecules to micelles or molecules of the adsorbant (imbibant) and is analogous to the osmotic potential of a solution. In a solution, the greater the amount of water present in proportion to a given amount of solute, the less negative (or higher) is the osmotic potential. Similarly, in an imbibant the greater the amount of water present in proportion to a given amount of imbibant, the less negative is the matric potential. With reference to pure water, the values of matric potentials are always negative.

The water potential of an imbibant is equal to its matric potential (always negative) plus any turgor or other pressure (pressure potential) which may be imposed upon the imbibant.

$$\psi_w = \psi_m + \psi_s$$

If the imbibant is unconfined no turgor or such pressure is involved and hence the above equation is simplified as:

$$\psi_w = \psi_m$$

THEORIES OF WATER TRANSLOCATION

A suction pump can lift water only to the barometric height, (*i.e.*, about 10.3 m or 34 feet at 1 atm pressure). Yet some trees are 130 meters (*i.e.*, 400 feet) tall from their water collecting roots in the soil to the upper tips of their branches. If the water is lifted to that height by pushing from below, some 13 atm or bars of pressure would be required in addition to that required to overcome the resistance and to maintain the flow. Different theories have been put forth to explain this phenomenon. The three important theories are:

(*i*) Capillarity

(*ii*) Root pressure

(*iii*) Cohesion theory.

Capillarity

Question arises, can the water be raised to the top of such trees by **capillarity**?

Such a rise of water in small tubes comes about because of the **adhesion** between water molecules and the wall of the tube in which the process is taking place. This adhesion forms a **meniscus** in which the water tends to climb up the side of the container. The smaller the diameter of the tube, the greater is the lifting force of the meniscus compared to the amount of water to be lifted. Hence the smaller the tube, the higher the column of capillary water will rise. Force of gravity also affects water uptake by capillarity.

According to this theory, water is first driven in due to force of adhesion between water and the walls of thin xylem vessels, *i.e.*, tracheids. As the water flows upward along the wall, strong cohesive forces between water molecules begin to function and pull the water upward. This upward pull of water continues until the forces of adhesion and cohesion are balanced by downward force of gravity.

However, capillarity alone in the usual sense does not operate in plants. This seems to be an insufficient force to lift water in tall trees. Water is lifted less than one meter in glass or cellulose tubes the size of xylem elements.

Root Pressure

Such pressures from below, produced by the roots have been observed in several species. Grapevines, for example, are capable of showing fairly high root pressure (about 5 or 6 bars), yet no one considers root pressure as a mechanism for the translocation of water in tall trees.

If a stem of plant is cut transversely above the soil surface, xylem sap will exude in the form of a drop from the cut surface. This exhibits the presence of a positive pressure in the xylem. This pressure is called **root pressure**.

Although, root pressure which is developed in the xylem of the roots can raise water to a certain height but it does not seem to be an effective force in ascent of sap due to following reasons:

(*i*) Magnitude of root pressure is quite low (about 2 bars).

(*ii*) Even in the absence of root pressure, ascent of sap continues.

(*iii*) Root pressures are not observed in conifers.

(*iv*) Mineral ions from the soil are taken up by roots and get deposited in the xylem vessels. Accumulation of ions in vessels leads to the lowering of osmotic potential.

(*v*) Root pressures are usually negligible, in temperate regions, where during summer transpiration is rapid.

Cohesion Theory

This theory was given by Henry Dixon in 1914. This theory is quite convincing and has been widely accepted. As the name suggests, the cohesion theory is based on the force of cohesion between water molecules. This forms a continuous water column throughout the xylem vessels of the plant, from the top to the root.

Water molecules remain joined to each other due to presence of **H-bonds** between them. Although H-bond is very weak, but when they are present in enormous numbers as in case of water, a very strong mutual force of attraction or cohesive force develops between water molecules and hence they remain in the form of a **continuous water column** in the xylem. The magnitude of this force is very high (about 350 atm or bars), therefore, continuous water column in the xylem cannot be broken easily due to force of gravity or any other means of resistance.

According to this theory, evaporation of water from the leaf to the atmosphere results in a decrease in the water potential of the epidermal cells, which are in direct contact with the atmosphere.

The water lost from epidermal cells is replaced by the water moving from the adjacent cells along a water potential gradient. The water lost from leaf cells is replaced by the water moving from the xylem elements in leaf veins (*i.e.*, xylem elements). Due to this pattern of water movement, a **tention** is created in water in the xylem elements of the leaves. This tention is transmitted downward to water in xylem elements of the roots through the xylem of petiole and stem and the water is pulled upward in the form of continuous unbroken **water column** to reach the transpiring surface upto the top of plants.

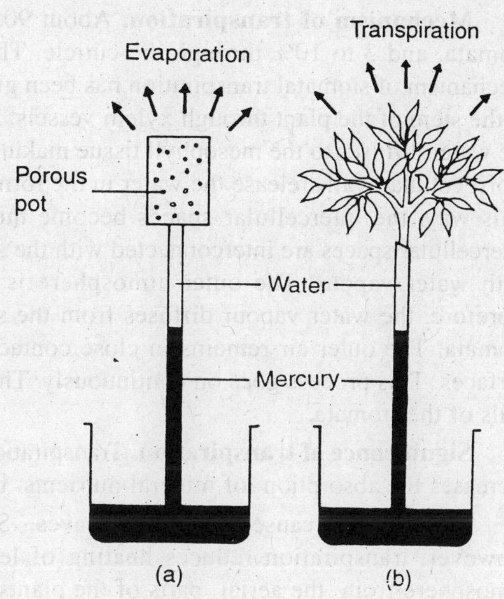

Fig. 2.10. Demonstration of cohesion theory: (*a*) evaporation and (*b*) transpiration.

Also the walls of xylem vessels are constituted of ligno-cellulose have strong affinity for water molecules. The water potential of root cells is lower than the water potential of soil, and therefore, movement of water occurs from soil to the roots. Uplift of water, is also referred to as **transpiration pull**.

TRANSPIRATION

The plants absorb sufficient quantity of water from the soil by means of their root hairs. Some part of this water is used up in the metabolic activities going on within the plant body and the rest of it evaporates from the stem and leaves. The evaporation of the water from the aerial parts of the plant is known as **transpiration**. Many authors described this phenomenon of transpiration from time to time in the following words:

'Transpiration is the loss of water in the form of water vapour from stem and leaves'.

(Miller, 1930)

'The loss of water vapour from aerial parts of living plants is known as transpiration'.

(Meyer, 1953)

'Transpiration is giving off of water vapour from the surface of a plant.'

(Thomas, 1956)

The transpiration differs from ordinary evaporation this way, that the former is controlled by the protoplasm whereas latter does not. The transpiration is a metabolic activity, which is controlled by protoplasm and may be reduced or increased where needed by the nature.

Kinds of transpiration. In general, there are three kinds of transpiration *viz.* (1) cuticular, (2) lenticular and (3) stomatal. The cuticular transpiration takes place through the cuticle found on the surface of the stem and leaves. The lenticular transpiration takes place through the lenticels found upon the stem. During night the stomata remain closed and the plants transpire through lenticels and cuticle.The stomatal transpiration is the most important. This takes place through the stomata situated on the leaves and sometimes on the green stems. The number of stomata is usually always greater on the lower surface of the leaf. About 90 to 97% of the transpiration takes place through the stomata.

Mechanism of transpiration. About 90 to 97% of the transpiration takes place through stomata, and 3 to 10% through the cuticle. The lenticular transpiration is negligible. Here the mechanism of stomatal transpiration has been given in detail. As we know well the water rises up in the stem of the plant through xylem vessels. From the xylem vessels of the veins of the leaves, the water diffuses to the mesophyll tissue making gradually the assimilatory cells, fully turgid. The moist cellular walls release the water in the form of vapour which reaches the intercellular spaces. This way, the intercellular spaces become more saturated than the outer atmosphere. As the intercellular spaces are interconnected with the sub-stomatal chambers, they also become saturated with water vapour. The outer atmosphere is comparatively drier than the inner tissues and therefore, the water vapour diffuses from the sub-stomatal chambers to the outside through the stomata. The outer air remains in close contact of the leaves, evaporating the water from their surfaces. This process goes on continuously. The stomatal transpiration is controlled by the guard cells of the stomata.

Significance of transpiration. Transpiration helps in the movement of xylem sap, which also increases the absorption of mineral nutrients by the roots from the soil.

Transpiration causes cooling of leaves. Some solar radiations cause heating of the leaves. However, transpiration reduces heating of leaves. The additional water is transpired to the atmosphere from the aerial parts of the plants.

Factors Affecting Transpiration

The transpiration is affected by several external and internal factors. They are as follows:

External Factors

1. Light. The light affects indirectly the transpiration. The rate of transpiration is always higher in light than in darkness. The main reason is that the stomata remain open in the presence of light or in the day time, and the transpiration takes place through them. In the absence of light or in the night, the stomata remain closed, and the transpiration is checked. During night only lenticular and cuticular transpiration takes place. The leaves receive heat from the sun, which accelerates the rate to some extent.

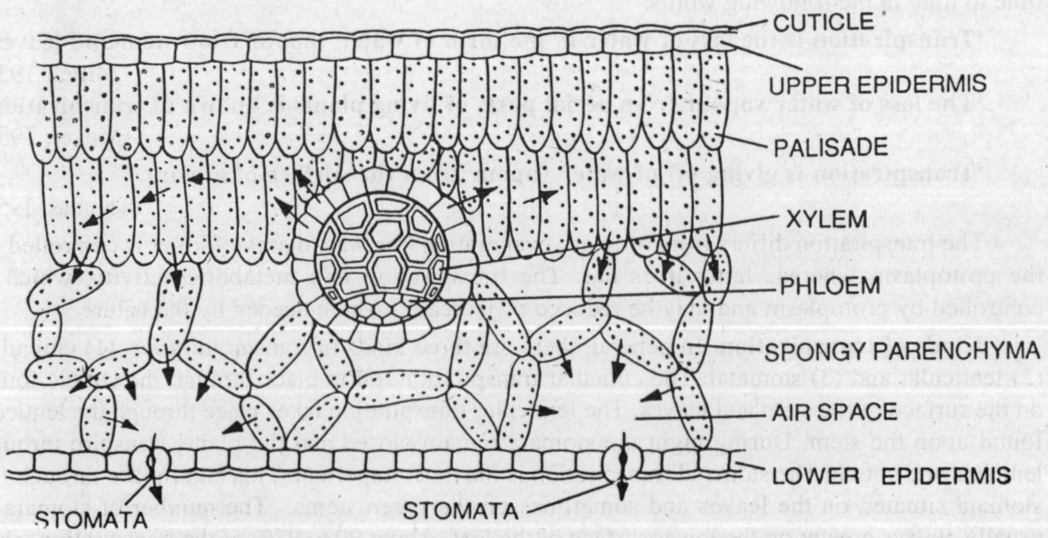

Fig. 2.11. Mechanism of stomatal transpiration. V.T.S. of a typical dicot leaf. The arrows show the movement of air.

2. Atmospheric moisture. Both the dryness and the humidity of the atmosphere affect the transpiration. When the atmosphere is very dry, then the water evaporates very rapidly from the surface of the leaves. On the other hand, in saturated atmosphere the rate of transpiration is quite slow.

3. Wind velocity. When wind blows very fast, the rate of transpiration also becomes very high, and on becoming the atmosphere calm and quiet, the transpiration lessens sufficiently due to moist air in close contact of the transpiring plants.

4. Air temperature. As the atmospheric temperature rises, the rate of transpiration also rises. In high temperatures the water evaporates from the leaf surface quite rapidly. In dry and hot air the rate of transpiration rises further.

5. Solar radiation. This factor is deeply related to the transpiration. The temperature of the leaves rises due to solar radiation, thus increasing the rate of transpiration. This way, solar radiation affects the transpiration indirectly.

6. Soil environment. This factor also affects the transpiration indirectly. The soil environment is much related to the water absorption. We know well, that the transpiring mesophyll cells of the leaves receive water from the soil. This way, the factor affecting the absorption of water from the soil, such as (1) soil water content, (2) concentration of soil solution, (3) soil temperature and (4) oxygen of soil air, etc., affect the transpiration indirectly. If there is enough of absorption of water, the transpiration takes place to a greater extent, on the other hand if the absorption of water is less the transpiration is also less.

Internal Factors

1. The root system. If the roots reach deep in the soil up to the moist layers, the root hairs absorb sufficient quantity of water and at the same time the transpiration increases from the aerial parts of the plants. On the other hand, if the root system of the plant remains confined in the upper layers of the soil, there is less absorption of water and less transpiration.

2. The stem. As we know that the water rises from the roots to the stem and leaves through the xylem vessels, and therefore, the rate of transpiration much depends upon the narrowness and broadness of the xylem vessels present in the stem. If the xylem vessels are sufficiently broad, the rate of transpiration increases, and if they are narrow there is comparatively less transpiration.

Table 2.4. Differences Between Transpiration and Evaporation

Transpiration	Evaporation
1. Transpiration is loss of water from surface of living cells.	1. It is loss from free surface of water.
2. It is a physiological process.	2. It is purely a physical process.
3. It is controlled by the processes, such as osmotic potential and water potential.	3. It is not controlled by any such process.
4. It involves living tissues.	4. It involves non-living matter.
5. Light, carbon dioxide concentration, pH and hormones regulate transpiration.	5. They do not influence evaporation.
6. It is a slow process.	6. Comparatively a fast process.
7. It is influenced by the anatomy of the transpiring organs.	7. There is no such influence.
8. It is regulated by the amount of soil water available to the plant.	8. This is continuous process, till water is present.
9. It makes the surface of leaves and young stems wet and protects them from sun burn.	9. It provides dryness to the free surface.

3. Leaf structure. The leaves of many xerophytes are too much divided and this way the total leaf area becomes very much decreased. Sometimes, some plants completely lack the leaves, and the stems of such plants become flat, leaf like, angular or rounded and green. They are called phylloclades, *e.g.*, phylloclades of *Ruscus*, *Opuntia*, etc. Less amount of water is transpired from the surface of the phylloclades. The leaves of *Asparagus* are scale-like and the cladodes are developed for photosynthesis. Very less amount of water transpires from the surface of these cladodes.

Several leaves bear the special adaptations to check the excessive transpiration. These leaves also differ anatomically from the ordinary typical leaves. Sometimes in certain leaves a waxy substance accumulates on the epidermis and the cuticle becomes quite thick, which checks the excessive transpiration. The leaves of other plants bear special cavities or depressions. The stomata are situated in these depressions which are usually surrounded by hairs. These special adaptations check transpiration to a large extent, *e.g.*, *Nerium*, *Banksia*, etc. In addition to these, the epidermis becomes multilayered, *e.g.*, *Nerium*, *Ficus*, etc., which lessens the transpiration. In several xerophytic plants, the cells of epidermis are radially elongated, which retard excessive transpiration. The sunken stomata are found in xerophytic leaves which help in retarding the transpiration. The leaves of certain xerophytes become infolded, and the stomata are protected from the direct gusts of wind, thus decreasing the rate of transpiration, *e.g.*, *Ammo-phila arenaria*. This way, the leaves bear several special adaptations which check the excessive transpirations.

STOMATAL MECHANISM

The stomata are very minute apertures, usually found on the epidermis of the leaves, Each stoma is surrounded by two kidney-shaped special epidermal cells, known as **guard cells**. The stomata may be found in all the aerial parts of the plant. They are never found on its roots. The epidermal cells surrounding the guard cells of the stoma are known as **accessory** or **subsidiary cells**. Usually the term stoma stands for the stomatal opening and the guard cells. The guard cells are always living and contain chloroplasts. These cells, however, contain much amount of protoplasm than the other ordinary cells. Usually the stomata are found scattered on the dicotyledonous leaves whereas they are arranged in parallel rows in the case of monocotyledonous leaves. The number of stomata may range from thousands to lacs per square centimeter on the surface of the leaf. The stomata may be found on both the surfaces of the leaf, but their number is always greater on the lower surface. However, the upper surface of the leaves of banyan and rubber trees lack stomata. The upper surface of the leaves of several xerophytes also lack the stomata. The free floating leaves of the water plants bear stomata only on their upper surface. In normal condition the stomata remain closed in the absence of light. They are always open in the day time or in the presence of light.

Table 2.5. Number of Stomata on Upper and Lower Surfaces of Leaves

Plants	Number of Stomata/mm^2	
	Upper surface	Lower surface
Monocot		
Wheat	50	40
Barley	70	85
Onion	175	175
Dicot		
Sunflower	120	175
Alfalfa	169	188
Geranium	29	179

Stomatal Opening and Closing

Each stoma is surrounded by two guard cells. The kidney-shaped guard cells contain chloroplasts. A respiratory cavity or chamber is found under each stoma. The mechanism of the closing and opening of the stomata depends upon the presence of sugar and starch in the guard cells. During day time or in the presence of light, the guard cells of the stomata contain sugar synthesized by their chloroplasts. The sugar is soluble and increases the concentration of the sap of guard cells. Due to higher concentration of the cytoplasm of guard cells, the water comes to them from the neighbouring cells by osmosis and they become turgid. With the result the stomata remain open. In the night or in the absence of light the sugar present in guard cells converts into the starch. The starch is insoluble, and this way the cell sap of the guard cells remains of much lower concentration than those of neighbouring cells, and the neighbouring cells take out the water from the guard cells by osmosis making them flaccid and the stomata closed. The conversion of sugar into starch during night and vice-versa in day time depends upon the acidity (pH) and alkalinity of the cell sap of guard cells. During night there is no photosynthesis and the carbon dioxide accumulates in the guard cells., converting the cell sap into weak acidic starch. During day time the carbon dioxide is used in the process of photosynthesis, the cell sap becomes alkaline and the starch converts into sugar.

Concentration of CO_2 hypothesis by Bonner and Galston. Bonner and Galston have proposed the following mechanism of opening and closing of stomata. This depends upon the concentration of the carbon dioxide (CO_2) found in the stomatal chamber and not upon the presence or absence of light.

Normally .03% of carbon dioxide is found in the atmosphere, and when the density of the CO_2 in the sub-stomatal chamber also remains .03%, then the guard cells become flaccid and the stomata closed. As the density of CO_2 retards gradually, the stoma begins to open and it opens gradually lengthwise until the density of CO_2 becomes .01%. Now the stomata are perfectly open and they are not open further beyond this density. The photosynthesis takes place in day time and much of the carbon dioxide is being used in the process, the density becomes lesser than .03% and the stomata remain open during day time. During night or in the darkness, there is no

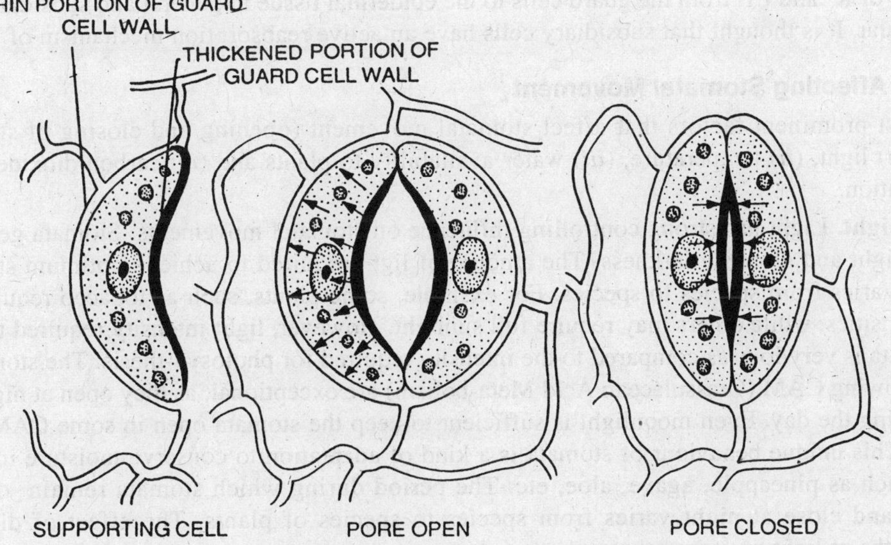

Fig. 2.12. Opening and closing of stomata.

photosynthesis, the density of carbon dioxide remains .03%, the guard cells remain flaccid and the stomata closed.

Active potassium (K^+) theory (Levitt, 1974). Role of potassium K^+ in stomatal opening is now universally accepted. This was observed for the first time by Fujino (1967) that opening of stomata occurs due to the influx of K^+ ions into the guard cells. The source of K^+ ions are nearby subsidiary and epidermal cells, thereby increasing the concentration from 50 mM to 300 mM in guard cells. The increase in K^+ ions concentration increases the osmotic concentration of guard cells thus leading to stomatal opening. The uptake of potassium K^+ controls the gradient in the water potential. This in turn triggers osmotic flow of water into the guard cells raising the turgor pressure. ATP helps in entry of K^+ ions into the guard cells. Levitt (1974) observed that proton (H^+) uptake by guard cell's chloroplasts takes place with the help of ATP. This leads to increase in value of pH in guard cells. Rise in pH converts starch into organic acid, such as malic acid.

$$\text{Starch} \rightarrow \text{phosphoglyceric acid} \xrightarrow{\text{carboxylase}} \text{Phosphophenol pyruvic acid} + CO_2$$

$$\downarrow$$

$$\text{Oxaloacetic acid}$$

$$\downarrow$$

$$\text{Malic acid}$$

Malic acid further dissociates to form H^+ and malate anion. The uptake of potassium K^+ ions is balanced by one of the following:

(*i*) uptake of Cl^-

(*ii*) transport of H^+ ions from organic acids, such as malic acid

(*iii*) by negative charges of organic acids when they lose H^+ ions.

The accumulation of large amounts of K^+ ions in guard cells is electrically balanced by the uptake of negatively charged ions, *i.e.*, chloride and malate. The high amount of malate in guard cells of open stomata accumulates by hydrolysis of starch.

The stomatal closure is considered to be brought about by a passive or highly catalysed excretion of K^+ and Cl^- from the guard cells to the epidermal tissue in general and subsidiary cells in particular. It is thought that subsidiary cells have an active reabsorption mechanism of K^+.

Factors Affecting Stomatal Movement

Most prominent factors that affect stomatal movement (opening and closing of stomata) include (*i*) light, (*ii*) temperature, (*iii*) water availability to plants and (*iv*) carbon dioxide (CO_2) concentration.

1. Light. Light has strong controlling influence on stomatal movements. Stomata generally open in light and close in darkness. The amount of light required to achieve optimum stomatal opening varies from species to species. For example, some plants, such as tobacco require low light intensities, while others may require full sunlight. However, light intensity required to open the stomata is very low, as compared to the intensity required for photosyn-thesis. The stomata of plants showing CAM (Crassulacean Acid Meta-bolism) are exceptional, as they open at night and close during the day. Even moonlight is sufficient to keep the stomata open in some CAM plant species. This unique behaviour of stomata is a kind of adaptation to conserve moisture in CAM plants, such as pineapple, agave, aloe, etc. The period during which stomata remain open in daylight and close at night varies from species to species of plants. The effect of different wavelengths of light on stomatal move-ment also varies.

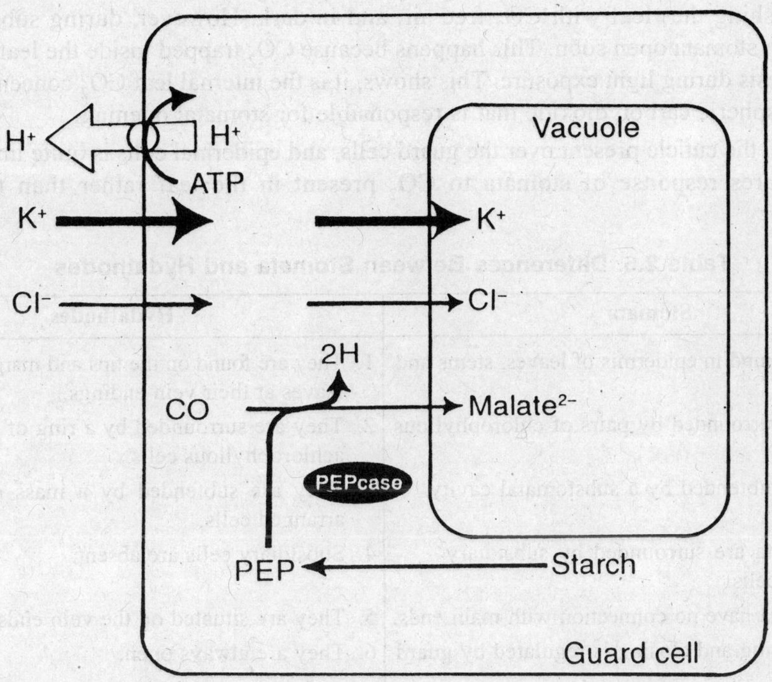

Fig. 2.13. Role of potassium K^+ ions in stomatal opening.

2. Temperature. Usually an increase in temperature leads to increase stomatal opening provided water does not become a limiting factor. In some plant species, stomata remain closed even under continuous light at 0°C. For example, in *Camellia* (tea plant), stomata do not open at very low temperature (below 0°C) even in strong light. However, if the temperature is increased, stomatal opening in such species increases. At temperatures higher than 30°C, there is decline in stomatal opening in some species.

3. Water availability. When availability of water is less, and rate of transpiration is high, plants undergo **water stress.** Water stress is also called **water deficit** or **moisture deficit**. Such plants begin to show signs of wilting and are known as **water-stressed plants**. Most of the mesophytes under such conditions close their stomata quite tightly and completely in order to protect them from the damage which may result due to extreme water shortage. The stomata reopen only when water potential of these plants is restored. This type of control of stomatal movement by water is called **hydropassive control**.

Accumulation of phytohormone **abscisic acid (ABA)** in the guard cells of several water stressed plants is now well established. The ABA causes stomata of such plants to close. When water potential of water-stressed plant is restored, the stomata reopen and ABA gradually disappears from the guard cells. The type of control of stomata by water, mediated through ABA, is called **hydroactive control**. Externally applied ABA to leaves of normal plants also induces closure of stomata.

4. Carbon dioxide (CO_2) concentration. CO_2 concentration has pronounced effect on stomatal movement. Reduced CO_2 concentration favours opening of stomata while an increase in CO_2 concentration causes stomatal closing. This happens even under the light. In certain species of plants, stomata also close merely by breathing near leaves.

The stomata which are forced to close by increased CO_2 concentration, do not reopen rapidly simply by flushing the leaf with CO_2 free air, and in dark. However, during subsequent light exposure, such stomata open soon. This happens because CO_2 trapped inside the leaf is consumed in photosynthesis during light exposure. This shows, it is the internal leaf CO_2 concentration rather than the atmospheric carbon dioxide that is responsible for stomatal opening.

However, the cuticle present over the guard cells, and epidermal cells is quite impermeable to CO_2 and ensures response of stomata to CO_2 present in the leaf rather than that of outer atmosphere.

Table 2.6. Differences Between Stomata and Hydathodes

Stomata	Hydathodes
1. They are found in epidermis of leaves, stems and fruits.	1. They are found on the tips and margins of young leaves at their vein endings.
2. They are surrounded by pairs of chlorophyllous guard cells.	2. They are surrounded by a ring of cuticularised achlorophyllous cells.
3. They are subtended by a substomatal cavity.	3. They are subtended by a mass of loosely arranged cells.
4. The stomata are surrounded by subsidiary epidermal cells.	4. Subsidiary cells are absent.
5. The stomata have no connection with main ends.	5. They are situated on the vein ends.
6. Their opening and closing is regulated by guard cells.	6. They are always open.
7. They are concerned with the process of 'transpiration'.	7. They are concerned with the process of 'guttation'.

Some endogenous factors, such as K^+, Cl^+ and H^- ions and organic acids, also influence stomatal movement.

EXPERIMENTAL WORK : Practicals

Plasmolysis

Experiment 1.

Object : To demonstrate plasmolysis.

Requirements : Red coloured *Tradescantia* leaves. Water, sugar solution, slide and microscope.

Experiment : To demonstrate plasmolysis, red coloured *Tradescantia* leaves are taken. With a sudden jerk a small segment can be peeled off from the lower surface of leaf. Cut this piece of peel into two parts. First part is put in a drop of water on a slide, second part is put in a drop of sugar solution on another slide. Now both slides are observed under the microscope respectively.

Observation : In first slide the cell structure is seen clearly. Almost most of the cells are turgid. Another slide shows that the cell contents have been separated from their cell walls and so a space is being formed nearer the cell wall.

If the sugar solutions are used with higher concentrations, then the cell contents of the cells will shrink continuously.

Explanation : First slide represents the normal condition. Only a slightly inflated condition may be observed after a considerable time.

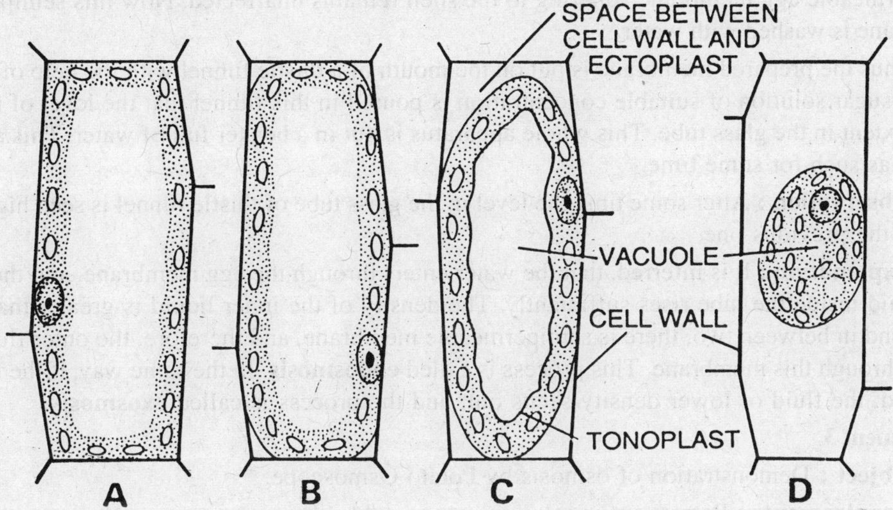

Fig. 2.14. Plasmolysis. A, turgid cell; B-D, successive stages in the shrinkage of protoplasm from the cell wall.

The second slide shows the separation of cell contents from the cell wall. It is due to loss of water from the cell. Here so-called space shows the beginning of the plasmolysis and so it is called as **incipient type of plasmolysis**.

When the concentration of outer medium is increased, the space becomes more widened due to complete loss of water from the cell. This condition is of **complete plasmolysis**.

Note : Hére in the fully plasmolysed cell the hypertonic solution; *i.e.*, the sugar solution fills the space formed between cell wall and cell contents.

If the first slide in which incipient plasmolysis has been completed, is provided by pure water or hypotonic solution, the cell will attain its normal position of turgidity. This process is known as **deplasmolysis**.

Osmosis

Experiment 2.

Object : Demonstration of osmosis by Egg Membrane Osmoscope.

Requirements : An egg, thistle funnel, dilute HCl, water beaker, sugar (sucrose) solution, stand, etc.

Experiment : To set this experiment, an egg is taken. One small hole is made at one end of the egg. The yolk and albumén are taken out of the egg shell through this hole. Now the egg shell is kept in dilute HCl for some time, with the result the hard egg shell of calcium carbonate dissolves in the dilute HCl and the

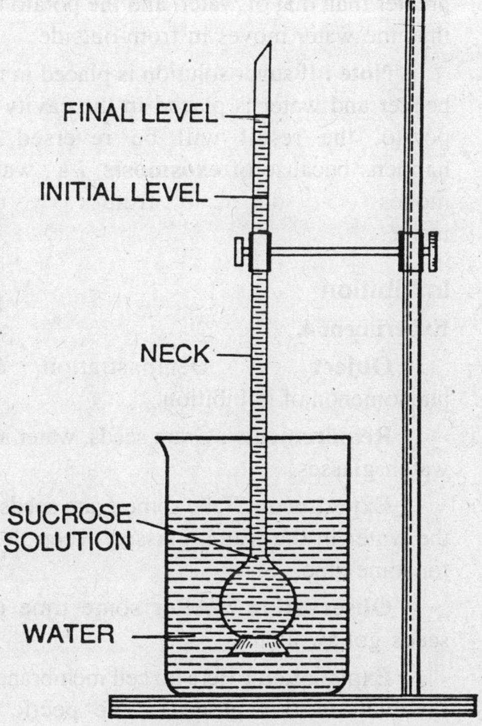

Fig. 2.15. Osmosis. Demonstration by egg membrane osmoscope.

semipermeable egg membrane adhering to the shell remains unaffected. Now this semipermeable membrane is washed with water.

Thus the prepared membrane is put on the mouth of a thistle funnel with the help of a thread. Now a sugar solution of suitable concentration is poured in this funnel that the level of it rises to some extent in the glass tube. This whole apparatus is put in a beaker full of water. This apparatus is kept as such for some time.

Observation : After some time the level in the glass tube of thistle funnel is seen higher from that of the previous one.

Explanation : It is inferred, that the water enters through the egg membrane, and the level of the liquid inside the tube rises sufficiently. The density of the inner liquid is greater than that of outer, and in between two, there is semipermeable membrane, and therefore, the outer fluid enters inside through this membrane. This process is called **endosmosis**. In the same way, if the fluids are reversed, the fluid of lower density seeps out, and the process is called **exosmosis**.

Experiment 3.

Object : Demonstration of osmosis by Potato Osmoscope.

Requirements : Potato, sugar solution, water and beaker.

Experiment : A peeled potato is made flat on its one end, and towards other end a cavity is made. This cavity must be sufficiently deep and broad, so that this may reach to the bottom, and the walls may not remain thick. Now, fill up this cavity with concentrated sugar solution and place this potato in a beaker filled up with water. The level of the outer and inner solutions must be same. After some time it is observed, that the level of the fluid inside the potato cavity rises sufficiently.

Explanation : This happens because of **endosmosis**. The density of the sugar solution is greater than that of water, and the potato tissues collectively act as semipermeable membrane, and thus the water moves in from outside.

Note : If sugar solution is placed in the beaker and water is placed in the cavity of potato, the result will be reversed. It happens because of **exosmosis**, *i.e.*, water moves out from the inside.

Imbibition

Experiment 4.

Object : Demonstration of phenomenon of imbibition.

Requirements : Gram seeds, water and watch glasses.

Experiment : Take some gram seeds in the water in the watch glasses. Keep them for some time.

Observation : After some time the seeds got swollen.

Explanation : Here no cell membrane is involved in this process. The pectic or proteinaceous surface attracts the water molecules by intermolecular forces. These pectic substances are hydrophilic in nature.

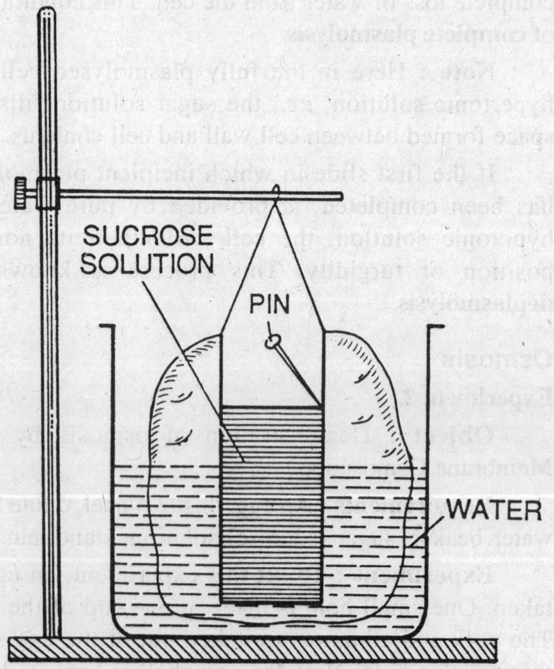

Fig. 2.16. Osmosis. Demonstration by potato osmoscope.

As the diffusion pressure of dry seeds is zero, therefore the water is absorbed in the seeds. This process remains continued till the diffusion pressure of the seeds and water becomes equal. Thus, the seeds get swollen here.

Experiment 5.

Object : Demonstration of developing pressure during imbibition.

Requirements : A corked jar, a disc fitting in it, a round scale and a pointer attached with the disc, stand, experimental gram seeds and water.

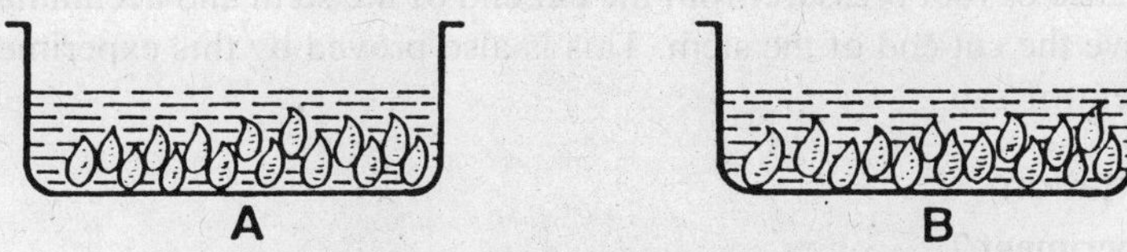

Fig. 2.17. Imbibition. Demonstration of the phenomenon.

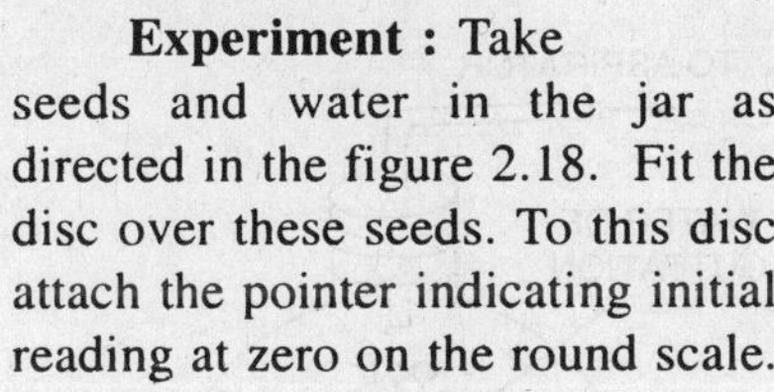

Experiment : Take seeds and water in the jar as directed in the figure 2.18. Fit the disc over these seeds. To this disc attach the pointer indicating initial reading at zero on the round scale.

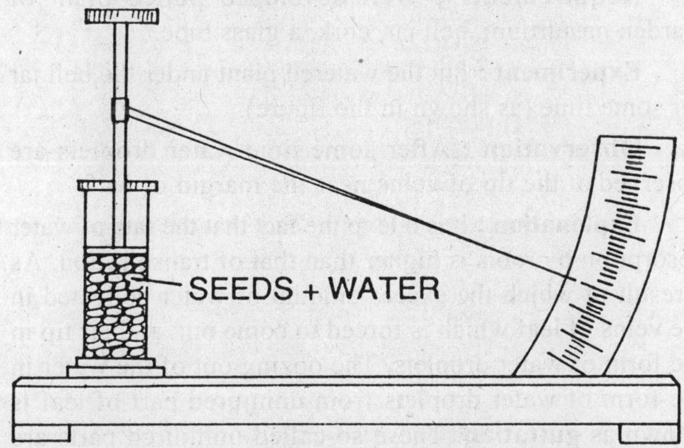

Fig 2.18. Demonstration of developing pressure during imbibition.

The whole apparatus should be airtight. Let this adjustment put as such for few hours.

Observation and expla-nation : After few hours the pointer moves down showing an upward movement of disc in the jar. Here the gram seeds swell considerably due to imbibition process. These seeds cause a pressure known as imbibition pressure.

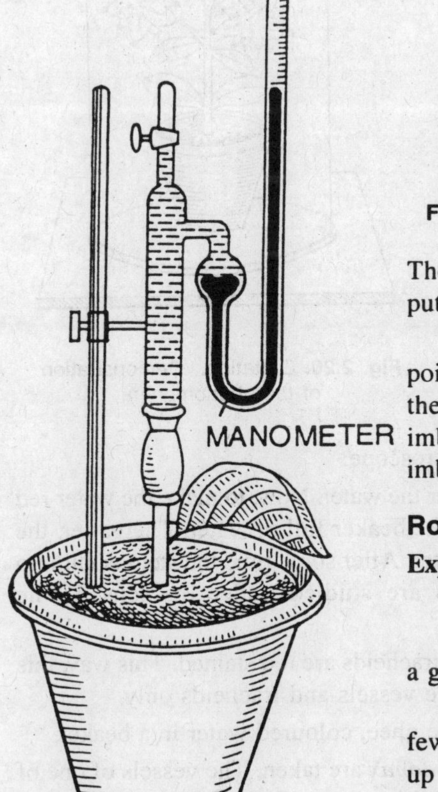

Fig. 2.19. Demonstration of root pressure.

Root Pressure

Experiment 6.

Object : Demonstration of root pressure.

Requirements : A potted balsam plant, a rubber tubing, a glass tube and a U-shaped manometer.

Experiment : The stem of a healthy balsam plant is cut a few inches above the ground level in the morning time and set up the apparatus as shown in the figure. With the help of rubber tubing a glass tube is fitted just above the stem is not dried up. A U-shaped manometer, partially filled up with mercury is connected to the glass tube. The whole apparatus must be air-tight.

Observation : When again observed the apparatus after few hours, the mercury is found on the higher level in the open tube of the manometer, and simultaneously the quantity of water is also increased in the glass tube.

Explanation : Due to accumulation of water in the glass tube, the pressure is exerted on the mercury, and it is pushed upwards in the open glass tube of manometer. This water exudates, because of root pressure from the cut end of the stem and accumulates in the glass tube fitted just above the cut end of the stem. This is also proved by this experiment that root pressure helps the ascent of sap.

Guttation

Experiment 7.

Object : Demonstration of phenomenon of Guttation.

Requirements : Well developed potted plant of Garden nasturtium, bell jar, cork, a glass tube.

Experiment : Put the watered plant under the bell jar for some time (as shown in the figure).

Observation : After some time water droplets are observed at the tip of veins near the margin of leaf.

Explanation : It is due to the fact that the rate of water absorption by roots is higher than that of transpiration. As a result of which the excess amount of water collected in the veins of leaf which is forced to come out at their tip in the form of water droplets. The oozing out of the water in the form of water droplets from uninjured part of leaf is known as **guttation**. These so-called uninjured parts are called **hydathodes** or water stomata.

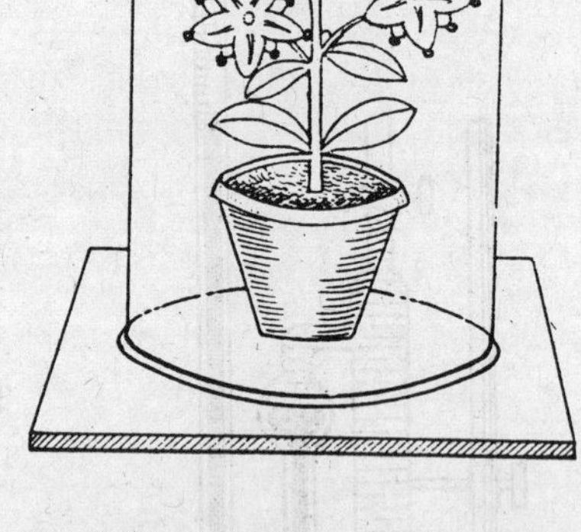

Fig. 2.20. Guttation. Demonstration of the phenomenon.

Ascent of Sap

Experiment 8.

Object : Demonstration of movement of water through xylem.

This may be shown in various ways. Some important ones are given here.

(*i*) **Requirements :** Balsam plant, water, safranin, microscope.

Experiment : The stem of the balsam plant is cut under the water. Now, to make the water red coloured, a few drops of safranin or euosine are added in the beaker full of water. Thereafter, the balsam stem, cut under the water is kept in this coloured water. After some time the stem is cut into transverse sections at various levels, and the sections are studied under the lens of the microscope.

Observation : Here, we find that only the vessels and tracheids are red stained. This way, this experiment proves, that water rises in the stem through the vessels and tracheids only.

(*ii*) **Requirements :** Plant of *Arundo donax*, vegetable ghee, coloured water in a beaker.

Experiment : Two white and linear leaves of *Arundo donax* are taken. The vessels of one of the leaves have been blocked by applying the melted vegetable ghee at its base; the other leaf is kept without blocking its vessels. Now, both the leaves are kept in the beaker full of coloured water.

Observation : After a few hours, it is observed that the leaf with blocked base is as before and the veins of the leaf with unblocked base have become coloured. This indicates the rising of water through the veins (xylem vessels).

(*iii*) **Requirements :** Two twigs of a plant, vegetable ghee, two beakers of water, etc.

Experiments : We take the two twigs of plant cut under water. The base of one of the twigs is blocked with melted vegetable ghee whereas of the other twig is not blocked. These twigs are kept in two separate beakers full of water.

Observation : After a sufficiently long time, it is observed that the leaves of the twig with blocked end have been wilted and drooped whereas the leaves of the other twig are turgid and fresh. In the second twig, the water rises through the vessels, as they are not blocked by melted vegetable ghee.

(*iv*) **The Ringing Experiment :**

Requirements : A twig, a sharp scalpel, beaker full of water, etc.

Experiment : A twig with few leaves cut under the water is taken. In the middle part of the twig, with the help of sharp scalpel, the cortex and phloem are removed, and a ring is formed. The central xylem strand is kept intact. This twig is kept in a beaker full of water.

Observation : It is observed, even after a long time that the leaves of the twig are turgid and fresh. The water supply of the leaves is not cut as the xylem strand is intact. This experiment proves that the water rises in the plant through xylem vessels.

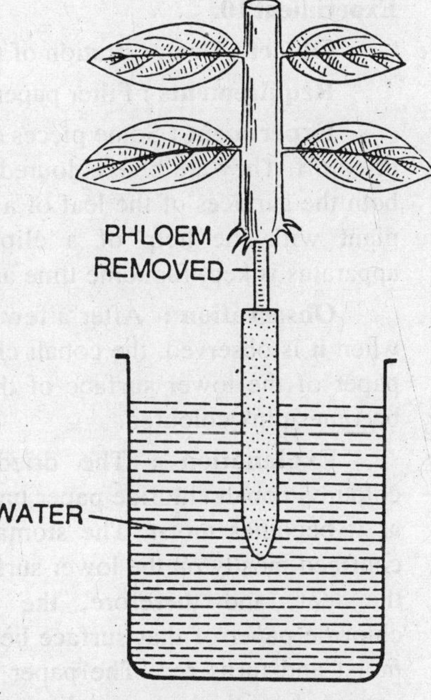

Fig. 2.21. Movement of water through xylem.

The above given experiments indicate clearly that the pathway of the ascent of sap in the plants is through the xylem vessels and the tracheids.

Transpiration

Experiment 9.

Object : Demonstration of transpiration by keeping a plant under a bell jar.

Requirements : A potted plant, rubber sheet, water, bell jar, glass plate, vaseline.

Experiment : A healthy potted plant is taken for this purpose. Now, the soil surface of the pot is being covered by the rubber sheet so that there is no evaporation of water from the moist surface of the soil. Thereafter the potted plant is kept on a glass plate is being covered over by a bell-jar. The margins of the bell-jar are vaselined, so that the atmospheric air may not enter the bell-jar. Now the covered potted plant is kept at room temperature.

Observation : After some time, the droplets of water appear on the inner surface of bell-jar.

Explanation : As all the precautions have been taken for checking the entrance of any moisture from outside,

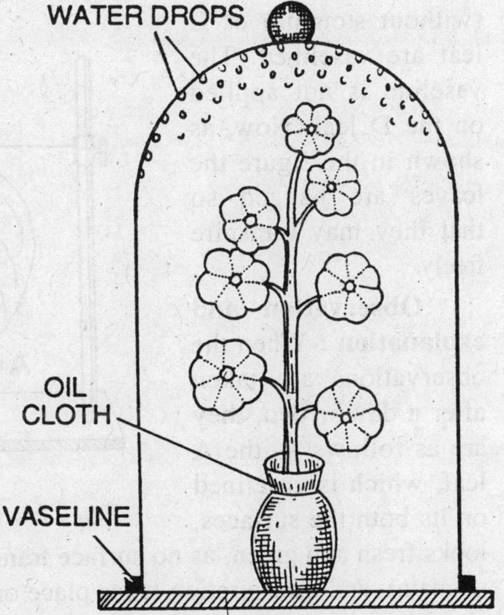

Fig. 2.22. Transpiration. Demonstration of the phenomenon.

and the release of moisture from the soil surface, these droplets of water adhered to the inner surface of the bell jar are definitely given off from the aerial parts of the plant by means of transpiration.

Experiment 10.

Object : Demonstration of transpiration by cobalt-chloride screen.

Requirements : Filter paper, cobalt-chloride solution, a potted plant, clip.

Experiment : Some pieces of the filter paper are dipped in cobalt chloride solution and then dried off. They are blue coloured. Now, two such pieces of filter paper are taken and pressed on both the surfaces of the leaf of a potted plant with the help of a clip. This apparatus is kept for some time as such.

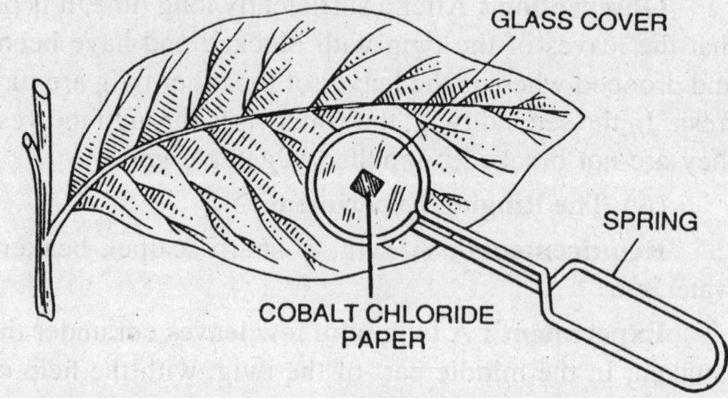

Fig. 2.23. Transpiration. Demonstration of phenomenon. (Cobaltchloride method).

Observation : After a few hours, when it is observed, the cobalt chloride paper of the lower surface of the leaf becomes pink coloured.

Explanation : The dried blue coloured cobalt chloride paper turns red as it becomes moist. The stomata are confined mostly on the lower surface of the leaf, and therefore, the cobalt chloride paper of that surface becomes moist and turns red. The paper of the upper side of the leaf may also become pink to some extent, as few stomata are found on this side.

Experiment 11.

Object : Demonstration of transpiration by four–leaf experiment.

Requirements : Four leaves, vaseline and a string.

Experiment : To demonstrate the transpiration from the leaf surface, four banyan leaves are taken. Both the surfaces of the A leaf, lower surface (with stomata) of B leaf, upper surface (without stomata) of C leaf are vaselined. The vaseline is not applied on the D leaf. Now, as shown in the figure the leaves are hanged so that they may transpire freely.

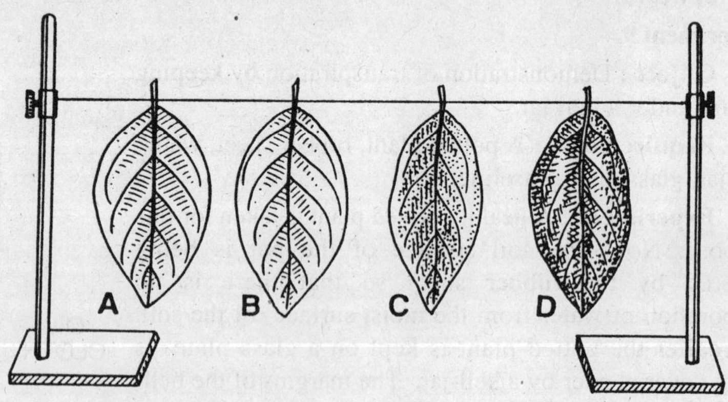

Fig. 2.24. Comparison of transpiration. Demonstration by four leaves.

Observation and explanation : When the observations are taken after a day or two, they are as follows — the A leaf, which is vaselined on its both the surfaces, looks fresh and green, as no surface transpires. The B leaf is vaselined on its lower surface (with stomata), and transpiration takes place only from the upper surface which is negligible. This leaf also remains turgid and green like the A leaf. If few stomata are present on the upper surface of the leaf, then it shrivels to some extent. The C leaf is vaselined on its upper surface, which contains less number of stomata or no stomata. The transpiration takes place from the lower stomatal surface, and the leaf shrivels to a large extent. The D leaf is not vaselined and both the surfaces transpire freely releasing much water. The leaf wilts completely in this case. This experiment proves that the rate of stomatal transpiration is fairly higher than the cuticular transpiration.

Rate of transpiration : The rate of transpiration is measured by potometer. There are many types of potometers which are designed by various scientists from time to time. These potometers are after the names of Farmer, Ganong, Garreau and Bose.

Experiment 12.

Object : Demonstration of rate of transpiration by potometers.

Requirements : Farmer or Ganong or Bose Potomer, water, plants, vaseline, stop watch, etc.

Experiment : (*i*) **Farmer's Potometer —** This type of potometer consists of a wide mouthed bottle, fitted with a three-holed rubber cork. The glass tube of water reservoir passes through one of these holes situated on one side. As shown in the figure another bent glass tube finds its way through another hole situated on the other side of the rubber cork. The central hole of the rubber cork bears a twig with few leaves cut under the water. The apparatus is completely filled up with water, and as shown in the figure the other end of the horizontal glass tube is dipped in the beaker full of water. The complete apparatus is air tight. The apparatus is kept in this position for some time. The water transpires from the surface of leaves, and to fulfil the vacuum the water rises in the horizontal glass tube. To demonstrate this process clearly, an air bubble is entered with the help of a dropper from the other end of the horizontal tube.

(*ii*) **Ganong's Potometer —** The mode of action of this potometer resembles to that of Farmer's potometer. The apparatus is completely filled up with water. A twig with few leaves cut under the water is entered in the wide mouthed bent tube on one side through the hole of the rubber cork, like that of Farmer's potometer. There is a water reservoir, from which the water may be taken inside the apparatus when needed. The horizontal tube of the potometer is well marked. The bent end of the horizontal tube remains dipped in the beaker containing water.

(*iii*) **Bose's Potometer —** We set the apparatus as shown in the figure. The water filled wide mouthed bottle fitted with three holed cork is taken. A twig with few leaves or a long petioled leaf cut under the water is inserted in the bottle through the central hole. In one hole of the cork the glass tube of water reservoir is inserted and in the other hole the U-tube with two bulbs is inserted as shown in the figure. A drop of oil is put in the U-tube. Now, the apparatus is kept in such condition so that the aerial parts of the twig transpire freely.

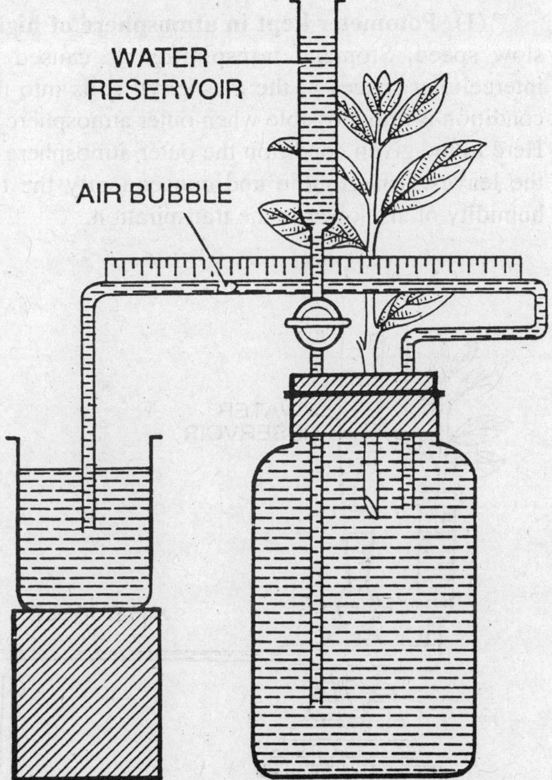

Fig. 2.25. Rate of transpiration, Demonstration by Farmer's Potometer.

Observation : (*i*) As water transpires, the air-bubble moves towards the bottle along with water. This air-bubble may be brought back by adding the water in the bottle from the reservoir which is controlled by a stop cock. There is a scale behind the horizontal tube. With the help of a stop watch, the time may be noted, which is consumed in the travelling of the air-bubble from one particular mark to another particular mark on the scale. Several readings may be taken in different conditions. The rate of transpiration may easily be calculated, if one knows the actual volume of water in between two particular points cn the horizontal tube.

(*ii*) As the leaves transpire, the water flows towards the twig in the horizontal tube. At the end of the bent horizontal tube there is a subterminal hole. With the help of a dropper or otherwise an air bubble is entered through this hole in the tube. As the leaves transpire the water from the beaker rushes towards the twig in the horizontal tube along with air bubble. The time spent in the travel of the air bubble from one particular mark to another on the horizontal tube may be noted. The air bubble may be brought back on the first mark in the tube by releasing water from the reservoir. Several observations may be taken in different transpiring conditions and the rate of the transpiration may be calculated. The apparatus must be completely air-tight.

(*iii*) The vacuum develops within the bottle because of the transpiration, and the oil drop gradually rises up in the U-tube towards the twig. As the oil drop reaches the inner bulb, it bursts and goes back again in its original position. The time taken in rising and bursting of the oil drop is being noted. Serveral such readings may be taken in different conditions. The apparatus must be completely air-tight.

Various modifications for transpiring plants : (1) Potometer kept in atmosphere of high humidity. (2) Potometer kept under the fan. (3) Potometer kept in higher temperature. (4) Potometer kept in dark.

(1) Potometer kept in atmosphere of high humidity : Here the air bubble moves with a slow speed. Stomatal transpiration is caused by the diffusion of water vapours from the intercellular spaces of the mesophyll cells into the outer atmosphere through the stomata. This condition is only possible when outer atmosphere possesses less humidity or less vapour contents. Here in the given condition the outer atmosphere is of higher humidity, so the loss of water from the leaves is negligible and consequently the transpiration is lowered. Therefore higher the humidity of air, lower is the transpiration.

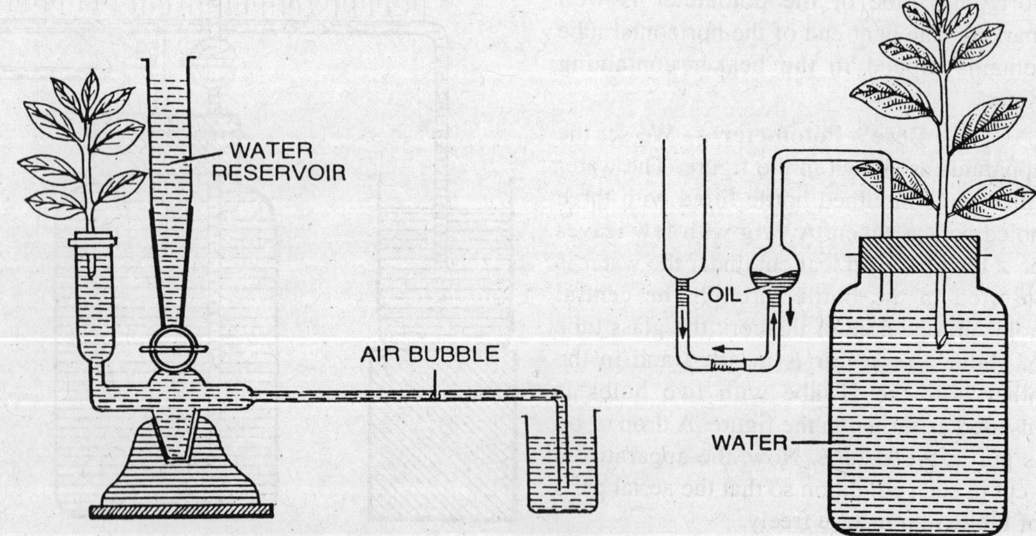

Fig. 2.26. Rate of Transpiration. Ganong's Potometer and Bose's Potometer.

(2) Potometer kept under the fan : Here the air bubble will move with a greater speed. The running fan causes to increase the rate of transpiration by sweeping away the water vapours that are present in the vicinity of transpiring plant. Therefore in this area the humidity of air becomes lower and the rate of transpiration is promoted.

(3) Potometer kept in higher temperature : Here the air bubble moves with a faster speed.

Temperature has an indirect influence on the rate of transpiration. Increase in temperature causes the following effects.

(*A*) Stomata get widened with the increase in temperature. So the loss of water is more, increasing the rate of transpiration.

(*B*) A rise in temperature decreases the relative humidity of surrounding air. The drier the air, the more rapid is transpiration.

(4) Potometer kept in dark : The air bubble does not move at all. It is due to fact that the stomata get closed when it is dark. Lowering of pH value (pH - 5) favours in closing of stomata. In dark the reducing osmotically active sugars change into the osmotically inactive starch. So the pH value lowers to 5 and the stomata are closed. On closing the stomata, no transpiration takes place, hence, the air bubble remains stable.

Experiment 13.

Object : Comparison of rate of transpiration from both the surfaces of leaf by Garreau's Potometer.

Apparatus : Garreau's potometer, stand, leaf, etc.

Experiment : *Garreau's Potometer* — With the help of this apparatus, the comparative study of the transpiration from both the surfaces of the leaf is being done. This apparatus consists of two small bell-jars, which are kept together in close contact as shown in the figure. A leaf of a potted plant is kept in between these two bell-jars. In each of these bell-jars a small test tube is kept. Each small test tube contains equal amount of anhydrous calcium chloride ($CaCl_2$). The apparatus is made air-tight, applying vaseline where they press the leaf in between. There are two manometers at the two ends of the bell-jars. They are partially filled with oil. These manometers, however, maintain the vapour of the bell-jars. If the surface of the oil, within the manometers changes, then it indicates that the apparatus is not airtight or the vapour released from the leaf surface is not completely absorbed by calcium chloride. The complete apparatus is fitted upon a vertical stand. After few hours, the calcium chloride tubes are taken out and weighed again. This way, the water transpired from both the surfaces of the leaf may be known.

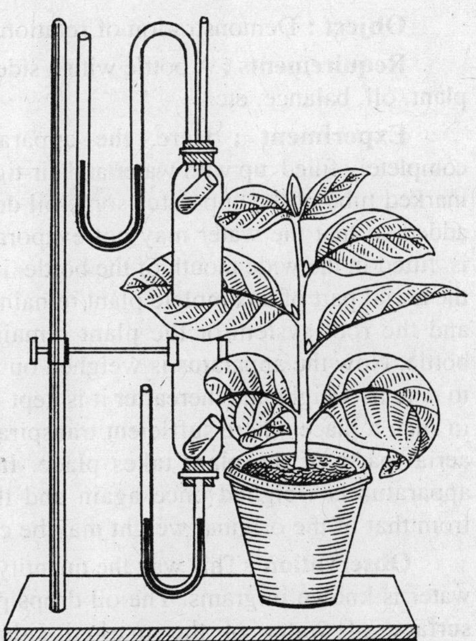

Fig. 2.27. Garreau's Potometer. Measurement of relative transpiration from two surfaces of a leaf.

Observation and explanation : The amount of water transpired from the lower surface (stomatal surface) of the leaf is always greater, as it bears large number of stomata. The cuticular transpiration takes place from the upper surface of the leaf, which is very much less.

Stomata

Experiment 14.

Object : Demonstration of the stomatal opening by Drawin's porometer.

Requirements : Darwin's porometer, water or mercury, a potted plant, stop watch, etc.

Experiment : This apparatus functions on the principle that the stomatal apertures govern the rate of flow of water. Cement the glass cup to the surface of the leaf by an adhesive (durofix, etc.) and immerse the vertical arm of tube in water or mercury. Open the clip and suck the water to a predetermined height in the vertical arm, close the clip.

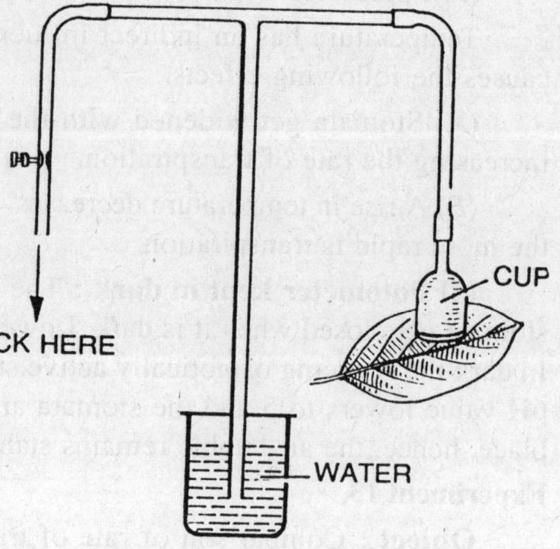

Fig. 2.28. Darwin's porometer.

Observation : Two possibilities occur here;

(*A*) The water level in arm remains unchanged.

(*B*) The water column begins to fall.

Explanation : On the closing of the clip, the air inside the porometer undergoes a reduced pressure.

(*A*) If the stomata under the cup are closed, the water level in the arm will not change.

(*B*) If the stomata under the cup are open, the air will find its way through them and water column in the arm will begin to fall.

Absorption of Water

Experiment 15.

Object : Demonstration of relation between Transpiration and Absorption.

Requirements : A bottle with a side, tube, cork, a plant, oil, balance, etc.

Experiment : Here, the apparatus must be completely filled up with water and air-tight. In the side marked tube of the apparatus some oil drops are being added so that the water may not evaporate. The cork is fitted in the wide mouth of the bottle in the way that the aerial part of a complete plant remains outside of it and the root system or the plant remains within the bottle. Now the apparatus is weighed on a pan balance to note its weight and thereafter it is kept for few hours in such a place where sufficient transpiration from the aerial parts of the plant takes place. In the end the apparatus is weighed once again and the difference from that of the original weight may be calculated.

Observation : This way the quantity of transpired water is known in grams. The oil drops present on the surface of water in the marked side tube travel somewhat downward and we know the actual quantity of absorbed water in ml by the roots. By comparison, it is noted that the quantity of absorbed water is always greater than the transpired water because some of the water is used in other metabolic activities.

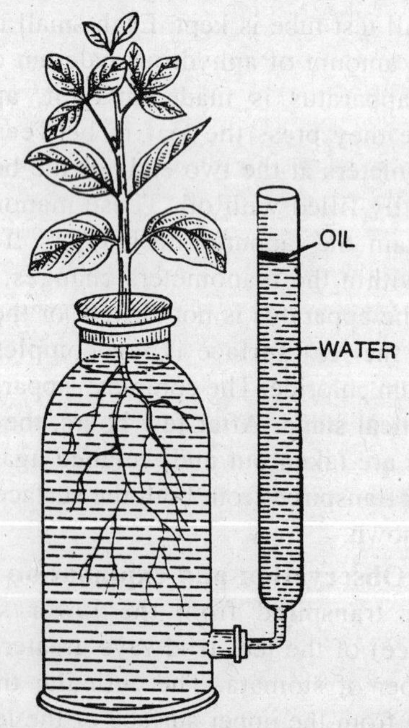

Fig. 2.29. Relation between transpiration and absorption. Demonstration of the phenomenon.

SUMMARY

1. Water is most essential for plant growth and development.
2. Water is absorbed by the root hairs of plants from soil, and translocated to the leaves.
3. Several mechanisms operate for the water to move from one cell to another, and it passes from root to stem, to leaf and ultimately to atmosphere.
4. The movement of water in the roots takes place through apoplast pathway and symplast pathway.
5. However, major portion of water in the root flows through apoplast pathway.
6. Several theories have been put forth to explain water translocation within a plant.
7. The three most important known theories are (*i*) root pressure, (*ii*) capillarity and (*iii*) Cohesion theory.
8. Transpiration is a process by which most of the water taken up by plants is lost to the number of stomata are present on the leaves, mostly on the lower leaf surface.
9. The stomatal movements, *i.e.*, opening and closing of stomata, control the flow of water from the plants to the atmosphere via transpiration.
10. Only a fraction of water absorbed is utilised for the purpose of growth, maintenance, and reproduction of the plants. The remaining quantity is lost in atmosphere by transpiration.
11. Transpiration helps in the movement of xylem sap, also called ascent of sap, and simultaneously increases absorption of mineral nutrients by the roots from the soil.
12. Transpiration also assists in cooling by evaporation of water, thus reducing heat load on the leaves.
13. There are several environmental factors which contribute to the occurrence of transpiration and control of stomatal movements.
14. The main controlling factors include — light, temperature, humidity, CO_2 concentration and availability of water.
15. Besides these factors many endogenous factors also influence both transpiration and stomatal movements.

TERMS TO REMEMBER

Absorption of water. Roots absorb water in plants. From root hair, water moves through the root cortex and to the xylem elements.

Apoplast pathway. The movement of water occurs exclusively through the cell wall without crossing any membranes.

Capillarity. Rise in water in tubes of small diameter (capillaries) kept in a vessel containing water.

Cohesion theory. Uptake of water takes place by transpirational pull. This is most accepted theory of water movement through plants.

Diffusion. Random movement of individual molecules from a region of higher concentration to a region of lower concentration.

Gravity potential ψ_g. Effect of gravity on water potential.

Imbibition. A type of diffusion where movement of water takes place along a diffusion gradient.

Imbibition pressure (matric potential). When seeds imbibe water and swell, they produce a large pressure developed by imbibition, which makes the seedlings to emerge above the ground through the soil surface.

Osmosis. Diffusion of water through a semipermeable membrane.

Osmotic potential (osmotic pressure or solute potential) ψ_s. The pressure required to stop the movement of water completely.

Pascal. The common measurement unit of water potential.

Permeability of membrane. The extent to which a membrane permits or restricts the movement of a substance.

Plasmolysis. Cell when placed in a concentrated sucrose solution loses water, and plasma membrane starts contracting.

Pressure potential (hydrostatic pressure) ψ_p. Positive hydrostatic pressure (also known as turgor pressure) is zero for pure water in an open beaker.

Root pressure. Water from soil is absorbed by root hairs and conducted through xylem vessels, which indicates the presence of a positive pressure in the xylem.

Stomata. Tiny pores, located on the surface of the epidermal layer of leaves. Each stoma has two guard cells, which contain chloroplasts.

Suberin. A wax-like substance.

Symplast pathway. Movement of water occurs from cell to cell through the plasmodesmata.

Transmembrane pathway. Movement of water through the cell membrane.

Transpiration. Loss of water in the form of water vapour from the plant to the atmosphere.

Water potential ψ_w. The potential energy of water.

QUESTION BANK

Very Short Answer Questions

1. What is protoplast ?
2. What is water potential ?
3. Give common measurement unit of water potential.
4. Where does apoplastic movement of water occur ?
5. Where does symplastic movement of water occur ?
6. Define the term 'diffusion'.
7. How would you define 'membrane permeability' ?
8. Define the term 'osmosis'.
9. What is imbibition ?
10. What is capillarity ?
11. Which theory of water translocation was given by Henry Dixon in 1914 ?
12. What is transpiration ?
13. What the term, ψ_s, is called ?
14. What the term, ψ_p is called ?
15. What the term, ψ_g is called ?
16. What is the other name of 'imbibition pressure' ?
17. Which fraction of soil water is available to plants for absorption by roots ?
18. What is 'guttation' ?
19. What will happen if a plant cell is kept in hypotonic solution for some time ?
20. When a plant cell kept in a solution is plasmolysed; what is nature of solution ?
21. Name the pore through which guttation takes place.
22. What is an isotonic solution ?
23. Name the plant process carried by hydathodes.
24. Name the pores through which guttation occurs.
25. Define plasmolysis.
26. What are antitranspirants ?
27. Who proposed K^+ exchange hypothesis for the opening and closing of stomata ?
28. Some plants close their stomata to conserve water under severe drought conditions. Name the phytohormone that helps them to do so.

ANSWERS

1. Plasma membrane + protoplasm ; 2. The potential energy of water is referred to as water potential ; 3. This is pascal, Pa (1 megapascal, Mpa = 10 bars) and is represented by Greek letter, Psi (ψ) ; 4. The apoplastic movement of water occurs exclusively through the cell wall without crossing any membranes ; 5. The symplastic movement of water occurs from cell to cell through the plasmodesmata ; 6. Diffusion is a random movement of individual molecules from a region of higher concentration to a region of lower concentration ; 7. The extent to which a membrane permits or restricts the movement of a substance is called membrane permeability; 8. Diffusion of water through a semipermeable membrane is known as osmosis ; 9. A type of diffusion by which movement of water takes place along a diffusion gradient ; 10. Capillarity refers to a rise in water in tubes of small diameter kept in a vessel containing water ; 11. Cohesion theory ; 12. Loss of water in the form of water vapour from the plant to the atmosphere is called transpiration ; 13. Solute potential or osmotic potential ; 14. Pressure potential or hydrostatic pressure ; 15. Gravity potential ; 16. Matric potential (*i.e.*, water potential of the matrix) ; 17. Capillary water; 18. The excretion of water in the form of liquid droplets from the leaves and other parts of an uninjured or intact plant is called guttation ; 19. Endosmosis ; 20. Hypertonic solution ; 21. Hydathode ; 22. When a solution has a concentration, where no gain and loss of water by osmosis separated by a semipermeable membrave from a specified solution ; 23. Guttation ; 24. Hydathodes ; 25. Shrinkage of protoplasm of a cell when placed in a hypertonic solution as a result of osmosis, is called plasmolysis ; 26. A substance applied to the plant for reducing transpiration is called anti-transpirant ; 27. Levitt (1974) ; 28. Abscisic acid.

Short Answer Questions

1. A well watered potted herbaceous plant shows wilting in the afternoon of a dry sunny day. Why ?
2. Write short notes on : (*a*) Plasmolysis (*b*) Imbibition (*c*) Osmosis (*d*) Root pressure.
3. Write brief explanatory notes on (*a*) Root pressure (*b*) Water potential (*c*) Guttation (*d*) Significance of transpiration.
4. Distinguish between the following :
   (*a*) Osmotic pressure and osmotic potential (*b*) Osmotic potential and matric potential
   (*c*) Diffusion and osmosis (*d*) Osmosis and imbibition
5. How is transpiration different from guttation ?
6. Mention any two uses of transpiration.
7. Describe in brief opening and closing of stomata.
8. List any four mechanisms that contribute to the ascent of sap in tall trees.
9. Does transpiration serve any useful function in the plant ?
10. Define guttation. How would you differentiate from transpiration ?
11. Who proposed K^+ hypothesis for opening and closing of stomata ?
12. How do potassium ions regulate the opening and closing of stomata ?
13. A farmer observed drops of water along the margins of leaves of tomato plants growing in his well irrigated field on a winter morning. Give reasons of this occurrence. Why had the water drops appeared only along the leaf margins ?
14. Write the role of osmosis in the life of plants.
15. Why do some herbaceous plants have hydathodes ? Under what conditions do they help the plants ?
16. Describe the role of osmotic potential in regulating water potential of plant cells.
17. Describe the theories related to translocation of water.
18. Define osmosis and explain how it influences other components of cell-water relations in plants.
19. Of what importance is potassium in the opening and closing of stomata ?
20. Discuss Dixon's theory of ascent of sap.
21. How do guard cells control opening and closing of stomata.
22. How do potassium ions regulate the opening and closing of stomata ?

23. How does cohesion tension theory explain the ascent of sap ?
24. Define hygroscopic and capillary waters. Which is not available to plants and why ?
25. Give the different factors that influence stomata.
26. How does transpiration differ from guttation ? What is significance of transpiration ?
27. Is there a general mechanism to explain the opening and closing of stomata ? Justify your answer.
28. Why is transpiration in higher plants called a 'necessary evil' ?
29. What is root pressure ? How can you demonstrate experimentally ?
30. How do rise in temperature and increased wind velocity affect transpiration ? Write any two adaptations shown by certain plants to reduce transpiration.
31. List the three prominent theories to explain water translocation in trees. Describe the most accepted theory out of these.

Long Answer Questions

1. Mention some of the factors that influence stomatal opening and closing. How are these factors involved in regulating stomatal behaviour ?
2. Describe the theories related to translocation of water.
3. Explain why transpiration in higher plants is considered a necessary evil ?
4. Of what importance is potassium in the opening and closing of stomata ? .
5. How do potassium ions regulate the opening and closing of stomata ?
6. What is ascent of sap ? Give Dixon's cohesive theory for it.
7. What is transpiration ? Explain various factors affecting it.

MULTIPLE CHOICE QUESTIONS (Objective Type Questions)

1. When a cell is fully turgid, which one of the following is zero ?
   (*a*) Turgor pressure (*b*) Wall pressure
   (*c*) Diffusion pressure deficit (*d*) Osmotic pressure
2. If a piece of potato tuber is placed is concentrated sugar solution
   (*a*) It would become limp due to the loss of water from its cells
   (*b*) It would become turgid by abrsorbing water from sugar solution
   (*c*) The cells will show endosmosis
   (*d*) It swells up due to imbibition
3. Bacteria cannot survive in a highly salted pickle because
   (*a*) Salt inhibits reproduction
   (*b*) Bacteria do not get enough light for photosynthesis
   (*c*) They become plasmolysed and consequently killer
   (*d*) The pickle do not contain nutrients necessary for bacteria to live
4. When beet root cylinders are washed and then placed in cold water none of the anthocyanins come out. This indicates most likely that plasma membrane is
   (*a*) Permeable to anthocyanins (*b*) Impermeable to anthocyanins
   (*c*) Differentially permeable to anthocyanins (*d*) Dead
5. Selective permeability identifies the process of
   (*a*) diffusion (*b*) osmosis (*c*) imbibition (*d*) plasmolysis
6. Seeds swell when placed in water because of
   (*a*) Osmosis (*b*) Imbibition (*c*) Hydrolysis (*d*) None
7. A cell when placed in solution gets plasmolysed. What is largely present in between the cell wall and the plasmolysed contents ?
   (*a*) Water (*b*) Cell sap
   (*c*) Hypotonic solution (*d*) Hypertonic solution

8. In the process of osmosis
   (*a*) both protoplasm and cell wall act as a single layer
   (*b*) only protoplast acts as a single layer
   (*c*) only cell membrane acts as a single layer
   (*d*) none of the above
9. The movement of water from one cell of the cortex to the adjacent one in roots is due to
   (*a*) accumulation of inorganic salts in the cells
   (*b*) accumulation of organic compounds in the cells
   (*c*) chemical potential gradient
   (*d*) water potential gradient
10. The path of water and solutes from the soil to the conducting tissue of the root is
   (*a*) soil ⟶ root hair ⟶ cortex ⟶ endodermis ⟶ pericycle ⟶ protoxylem ⟶ phloem
   (*b*) soil ⟶ root hair ⟶ cortex ⟶ pericycle ⟶ endodermis ⟶ protoxylem ⟶ metaxylem
   (*c*) soil ⟶ epidermal cell of the root ⟶ cortex ⟶ endodermis ⟶ pericycle ⟶ protoxylem ⟶ metaxylem ⟶ secondary xylem
   (*d*) soil ⟶ root hair ⟶ cortex ⟶ endodermis pericycle ⟶ protoxylem ⟶ metaxylem
11. Root cap has no role in absorption because it has
   (*a*) no direct connection with vascular system
   (*b*) loosely arranged cells
   (*c*) no cells containing chloroplasts
   (*d*) no root hairs
12. Water available to the plant is
   (*a*) run off water (*b*) gravitational water
   (*c*) hygroscopic water (*d*) capillary water
13. According to one vital force theory, ascent of sap is due to active pulsations of innermost layer of cortex just outside the endodermis. This theory was given by
   (*a*) J.C. Bose (*b*) Dixon (*c*) Strasburger (*d*) Sachs
14. Xylem transport of water in tall trees appears mainly due to
   (*a*) metabolic activity in xylem cells
   (*b*) root pressure
   (*c*) capillarity rise in xylem open pipes
   (*d*) transpiration pull and cohesion of water molecules.
15. Which one of the follcwing is the most widely accepted explanation for ascent of sap ?
   (*a*) Capillarity (*b*) Atmospheric pressure
   (*c*) Pulsation activity (*d*) Transpiration pull
16. The metal ion involved in stomatal regulation is
   (*a*) iron (*b*) magnesium (*c*) zinc (*d*) potassium
17. Increase in CO_2 concentration around leaf causes
   (*a*) rapid opening of stomata
   (*b*) partial closure of stomata
   (*c*) complete closure of stomata
   (*d*) there will be no effect on stomatal opening

18. Which one by following is used for measuring the rate of traspiration.
(*a*) potometer (*b*) porometer (*c*) osmometer (*d*) Moll's experiment
19. Which one of the following factors is the most important in the regulation of transpiration ?
(*a*) Humidity (*b*) Light (*c*) Temperature (*d*) Wind
20. Process of water exudation through hydathodes is called
(*a*) guttation (*b*) transpiration (*c*) excretion (*d*) hydrolysis
21. Increase in the CO_2 concentration of microenvironment of leaf would cause
(*a*) closure of stomata (*b*) opening of stomata
(*c*) no effect of stomata (*d*) wider opening of stomata
22. Conversion of starch to organic acid is essential for stomata
(*a*) closure (*b*) opening (*c*) initiation (*d*) growth
23. Generally the number of stomata per square cm leaf area is
(*a*) in millions (*b*) less than one hundred
(*c*) in thousands (*d*) none of the above
24. In many thin leaved mesophytes, stomata open during day and close at night, this comes under
(*a*) Barley type (*b*) Potato type (*c*) Alfalfa type (*d*) Bean type
25. Of the processes which occur in leaves, the one which may lower their temperature is
(*a*) respiration (*b*) photosynthesis (*c*) transpiration (*d*) hydrolysis
26. The following percentage of water absorbed by herbaceous plants is lost during transpiration
(*a*) 80 (*b*) 60 (*c*) 90 (*d*) 40
27. Stomata open during the day time because the guard cells
(*a*) photosynthesise and produce osmotically active sugars
(*b*) are thin walled
(*c*) are bean shaped
(*d*) have to help in gaseous exchange
28. Active K^+ exchange mechanism for opening and closing of stomata was given by
(*a*) Darwin (*b*) Levitt (*c*) Scarth (*d*) Khorana
29. In which of the following plants would metabolism be hindered if the leaves are coated with wax on their upper surface ?
(*a*) Hydrilla (*b*) Lotus (*c*) Pistia (*d*) Vallisneria
30. Root pressure is maximum when
(*a*) transpiration is high and absorption is very low
(*b*) transpiration is very low and absorption is high
(*c*) transpiration is very high and absorption is also high
(*d*) transpiration and absorption both are low
31. Water potential is equal to
(*a*) ψ (*b*) p + WP (*c*) p + TP (*d*) $y_s + y_p$
32. In rainy season, wooden door swells up due to
(*a*) imbibition (*b*) capillarity (*c*) endosmosis (*d*) exosmosis
33. Cohesion theory of ascent of sap was given by
(*a*) Dixon and Jolly (*b*) Bose (*c*) Munch (*d*) Stephen Hales
34. Instrument used in determination of transpiration is
(*a*) Tensiometer (*b*) Auxanometer (*c*) Porometer (*d*) Potometer
35. Exudation of xylem sap on cutting the shoot is due to
(*a*) transpiration (*b*) guttation (*c*) root pressure (*d*) none of the above

36. Plasmolysis occurs due to
(*a*) Absorption (*b*) Osmosis (*c*) Endosmosis (*d*) Exosmosis

37. Water exudation through hydathodes is
(*a*) Guttation (*b*) Hydrolysis (*c*) Transpiration (*d*) Excretion·

38. High carbon dioxide concentration in leaf interior will result in
(*a*) Stomatal opening (*b*) Stomatal closure
(*c*) No effect on stoma (*d*) Destruction of stomata

39. Plant cooling is due to
(*a*) Assimilation (*b*) Guttation (*c*) Photorespiration (*d*) Transpiration

40. Movement of water in the cortex of root from outside to inside is due to
(*a*) Gradient of water potential (*b*) Gradient of chemical potential
(*c*) Accumulation of organic solutes (*d*) Accumulation of inorganic salts

41. Osmotic pressure is highest in
(*a*) Hydrophytes (*b*) Mesophytes (*c*) Xerophytes (*d*) Halophytes

42. A cell placed in solution gets plasmolysed. This solution is
(*a*) Hypotonic (*b*) Hypertonic (*c*) Isotonic (*d*) Ditonic

43. Hydathodes are found in
(*a*) Stem (*b*) Leaves (*c*) Roots (*d*) All the above

44. The water potential and osmotic potential of pure water are
(*a*) 100 and zero (*b*) zero and zero (*c*) 100 and 100 (*d*) zero and 100

45. A cell will become fully turgid if it is placed in
(*a*) Hypotonic solution (*b*) Isotonic solution
(*c*) Hypertonic solution (*d*) All the above

46. Cohesion-tension theory is related to
(*a*) Respiratiom (*b*) Ascent of sap (*c*) Transpiration (*d*) Photosynthesis

47. Dry wooden stakes driven in cracks of a rock and soaked will develop pressure that will split the rock. The phenomenon is
(*a*) Osmotic pressure (*b*) Imbibition
(*c*) Turgor pressure (*d*) Deplasmolysis

48. A cell placed in hypotonic solution will
(*a*) Shrink (*b*) Show exosmosis
(*c*) Show endosmosis (*d*) Not change in shape or size

49. Which one will reduce rate of transpiration ?
(*a*) Rise in temperature (*b*) Increase in water uptake
(*c*) Increase in wind velocity (*d*) Decrease in light intensity

50. Root hair absorbs water from soil through
(*a*) Turgor pressure (*b*) Ion exchange (*c*) Osmosis (*d*) DPD

51. Wooden doors swell up and get stuck up during rainy season due to
(*a*) Endosmosis (*b*) Imbibition (*c*) Capillarity (*d*) Exosmosis

52. Rate of transpiration is measured by
(*a*) Ganong's potometer (*b*) Porometer
(*c*) Auxanometer (*d*) Respirometer

53. When water enters the plant cell one of the pressures is exerted on the cell wall :
(*a*) Turgor pressure (*b*) Suction pressure
(*c*) Root pressure (*d*) Osmotic pressure

54. Presence of glycolate in the cells will help to open stomata when there is :
(*a*) High concentration of CO_2 (*b*) Low concentration of O_2
(*c*) High concentration of CO_2 and low O_2 (*d*) High concentration of O_2 and low CO_2

55. In soil water available for plants is :
(*a*) Gravitational water (*b*) Hygroscopic water
(*c*) Capillary water (*d*) Chemically bound water

56. The water potential and osmotic potential of pure water are :
(*a*) Zero and 100 (*b*) Zero and zero (*c*) 100 and zero (*d*) 100 and 100

57. If a cell A with, DPD, 4 bars is connected to cells B, C, D whose OP and TP are respectively 4 and 4, 10 and 5 and 7 and 3 bar, the flow of water will be :
(*a*) C to A, B and D (*b*) A and D to B and C
(*c*) A to B, C and D (*d*) B to A, C and D

58. Water potential in the leaf tissue is positive (+ ve) during :
(*a*) Excessive transpiration (*b*) Low absorption
(*c*) Low transpiration (*d*) Guttation

59. With increase in turgity of a cell surrounded by water the wall pressure will :
(*a*) Increase (*b*) Decrease
(*c*) Fluctuate (*d*) Remain unchanged

60. Endosmosis of water occurs when in comparison with the outer solution, the water potential of cell sap is :
(*a*) Higher (*b*) Lower (*c*) Equal (*d*) None of these

61. The rate of absorption of water is slow at temperature near freezing point :
(*a*) It is mainly a metabolic process (*b*) Cell membrane becomes more viscous
(*c*) Growth of cells stop (*d*) Transpiration is retarded

62. What will be the nature of sugar solution, if the cell of an epidermal peel of ***Rhoeodiscolor*** shows plasmolysis ?
(*a*) Hydrophoric (*b*) Hypertonic (*c*) Isotonic (*d*) Hypotonic

63. It is desirable to add a pinch of salt to water to prolong the life of cut flower, which of the following is correct explanation ?
(*a*) Salt causes exosmosis
(*b*) Salt checks transpiration of flowering twig
(*c*) Salt inhibits the growth of microorganisms
(*d*) Salt gives the uptake of water by flower

64. If alcohol treated cell is kept in hypertonic solution it :
(*a*) Bursts (*b*) Plasmolysed (*c*) Remains same (*d*) None

65. Stomata usually open during the day time because the guard cells :
(*a*) are thin walled
(*b*) help in gas exchange
(*c*) are bean shaped
(*d*) photosynthesise and produce osmotically active sugars.

66. Plants will undergo wilting under one of these conditions :
(*a*) High temperature (*b*) High humidity
(*c*) Heavy rainfall (*d*) Absorption of water

67. If CO_2 concentration suddenly increases around the leaf, one of the following events occurs :
(*a*) Stomata open gradually
(*b*) Stomata open suddenly

(*c*) Transpiration will not be affected

(*d*) Decrease in transpiration due to sudden closure of stomata

68. Wilting in plants occurs due to increase in :

(*a*) Photosynthesis (*b*) Photoperiodism (*c*) Transpiration (*d*) Osmosis

69. Rate of transpiration can be measured by :

(*a*) Ganong's potometer (*b*) Porometer (*c*) Auxanometer (*d*) Respirometer

70. The osmotic potential of a solution is denoted by the symbol :

(*a*) y_x (*b*) DY (*c*) y_P (*d*) y_s

71. If cohesion-tention transpiration theory is correct then a break in water column in xylem vessels :

(*a*) should cause the mesophyll cells to become flaccid and result in wilting of leaves

(*b*) should increase the water content of leaves

(*c*) should have no effect at all

(*d*) should increase the rate of photosynthesis

72. Dumb-bell shaped guard cells are found in :

(*a*) Wheat (*b*) Bean (*c*) Groundnut (*c*) Sunflower

73. Who had said that 'transpiration is a necessary evil' ?

(*a*) Curtis (*b*) Steward (*c*) Anderson (*d*) J.C. Bose

74. Osmosis is defined as the process in which :

(*a*) Water diffuses from lower concentration to higher concentration

(*b*) Solutes diffuse from lower concentration to higher concentration

(*c*) Active transport of ions take place

(*d*) Passive transport of ions take place

75. Plant cells submerged in distilled water will become :

(*a*) Turgid (*b*) Flaccid (*c*) Plasmolysed (*d*) Impermeable

76. To initiate cell plasmolysis, the salt solution should be :

(*a*) Isotonic (*b*) Hypertonic (*c*) Hypotonic (*d*) None of these

77. Plant cooling occurs due to :

(*a*) Assimilation (*b*) Guttation (*c*) Photorespiration (*d*) Transpiration

78. Rate of transpiration increases :

(*a*) When there is no light (*b*) In water

(*c*) When temperature is high (*d*) In rainy season

79. Which of the following factors is most important in regulation of temperature ?

(*a*) Light (*b*) Wind (*c*) Temperature (*d*) Humidity

80. The phytohormone which is supposed to be one of the factor responsible for regulating the opening and closeing of stomata :

(*a*) ABA (*b*) GA (*c*) IBA (*d*) Kinetin

81. Transpiration pull will be maximum under which of the following conditions ?

(*a*) Open stomata, high humid atmosphere and well irrigated soil

(*b*) Open stomata, dry atmosphere and moist soil

(*c*) Open stomata, high humid atmosphere and dry soil

(*d*) Closed stomata, low light intensity and humid atmosphere

82. Osmosis is defined as :

(*a*) The flow of solvent molecules from a region of higher concentration to a region of lower concentration through a semipermeable membrane.

(*b*) The flow of low concentration liquid

(*c*) The flow of solvent molecules from a higher concentration to a lower concentration region

(*d*) The flow of solvent molecules from a lower concentration to a higher concentration region.

83. A plasmolysed cell can be deplasmolysed by placing it in :
(*a*) Pure water or hypotonic solution (*b*) Hypertonic solution
(*c*) Isotonic solution (*d*) Saturated solution

84. A cell becomes turgid when placed in :
(*a*) Isotonic solution (*b*) Hypertonic solution
(*c*) Hypotonic solution (*d*) All of these

85. Process of water exudation through hydathodes is called :
(*a*) Guttation (*b*) Transpiration (*c*) Excretion (*d*) Hydrolysis

86. Maximum transpiration occurs from :
(*a*) Protoderm and procombium (*b*) Dermal tissue
(*c*) Periderm and lenticels (*d*) Ground tissue and scale leaves

87. The degree of opening of stomata is determined by :
(*a*) Porometer (*b*) Calorimeter (*c*) Potometer (*d*) All of these

88. During absorption of water by roots the water potential of cell sap is lower than that of :
(*a*) Pure water and soil solution
(*b*) Neither pure water nor soil solution
(*c*) Pure water but higher than that of soil solution
(*d*) Soil solution but higher than that of pure water

89. A twig kept in water remain fresh for a longer time if some salt is added to it is due to :
(*a*) Absorption of more water (*b*) Exosmosis
(*c*) Decrease in transpiration rate (*d*) Decrease in bacterial degradation activity

90. The bulliform cells of leaves lose their turgidity during excessive :
(*a*) Assimilation (*b*) Transpiration (*c*) Photosynthesis (*d*) Respiration

91. Root hairs are found :
(*a*) In the zone of elongation (*b*) Adventitious roots
(*c*) On the root cap (*d*) In the zone of maturation

92. Osmotic pressure of a solution is :
(*a*) Greater than the pure solvent (*b*) Lesser than the pure solvent
(*c*) Equal to the pure solvent (*d*) More or less the pure solvent

93. Plasmodesmata connections help in :
(*a*) Cytoplasmic streaming (*b*) Synchronous mitotic divisions
(*c*) Locomotion of unicellular organisms (*d*) Movement of substances between cells

94. The cohesive force existing between molecules of water is contributing to :
(*a*) Plasmolysis (*b*) Translocation (*c*) Ascent of sap (*d*) Osmosis

95. Opening and closing of stomata is due to the :
(*a*) Hormonal change in guard cells (*b*) Change in turgor pressure of guard cells
(*c*) Gaseous exchange (*d*) Respiration

96. Stomata of a plant open due to :
(*a*) Influx of hydrogen ions (*b*) Efflux of potassium ions
(*c*) Influx of calcium ions (*d*) Influx of potassium ions

97. Water is released as droplets — this is known as :
(*a*) Root pressure (*b*) Transpiration (*c*) Guttation (*d*) None

98. If the concentration of external solution is more than the cytoplasm, the solution is known as :
(*a*) Hypertonic (*b*) Isotonic (*c*) Hypotonic (*d*) None

99. The cause of guttation is mainly :
(*a*) Root pressure (*b*) Osmosis (*c*) Transpiration pull (*d*) Cohesive force

100. The water potential and osmotic potential of pure water are :
(*a*) 100 and zero (*b*) Zero and zero (*c*) 100 and 200 (*d*) Zero and hundred

101. Which one of the following is a rapid type of absorption?
(*a*) passive absorption (*b*) active absorption
(*c*) root absorption (*d*) salt absorption

102. Active absorption of water from the soil by the root is mainly effected by
(*a*) respiratory activity of root (*b*) tension in cell sap to transpiration
(*c*) tissue organisation (*d*) none of the above

103. In soil, water available for roots is
(*a*) capillary water (*b*) gravitational water
(*c*) hygroscopic water (*d*) none of the above

104. Water will be absorbed by root hairs when the external medium is
(*a*) isotonic (*b*) hypertonic (*c*) viscous (*d*) hypotonic

105. Water absorption through roots can be increased by
(*a*) decreased transpiration (*b*) increased transpiration
(*c*) increased rate of photosynthesis (*d*) decreased absorption of ions

106. The ultimate cause for the movement of water against gravity in a tree is
(*a*) osmosis (*b*) transpiration (*c*) imbibition (*d*) photosynthesis

107. The path of water from soil upto secondary xylem is
(*a*) soil → root hair cell wall → cortex → endodermis → pericycle → protoxylem → metaxylem
(*b*) cortex → root hair → endodermis → pericycle → protoxylem → metaxylem
(*c*) pericycle → soil → root hair → cortex → endodermis → protoxylem → metaxylem
(*d*) soil → root hair cell wall → cortex → pericycle → endodermis → protoxylem → metaxylem

108. The water held tightly by the soil particles around them is known as
(*a*) capillary water (*b*) hygroscopic water (*c*) runaway water (*d*) gravitational water

109. The initial stage of water absorption by root cells is by
(*a*) absorption (*b*) imbibition (*c*) osmosis (*d*) respiration

110. Which one of the following is connected with transport of water in plants?
(*a*) phloem (*b*) xylem (*c*) epidermis (*d*) cambium

111. Osmosis is the movement of
(*a*) solute particles from higher concentration to lower concentration
(*b*) solvent particles from higher water potential to lower water potential
(*c*) solute particles from higher concentration to lower concentration through a semipermeable membrane
(*d*) solvent particle from lower water potential to higher water potential

112. Freshly cut potato chip is put into a strong solution of sugar, later it is found to be
(*a*) flaccid (*b*) turgid
(*c*) filled up with more starch (*d*) filled up with more sugar

113. The movement of water from one cell of the cortex to the adjacent one in roots is due to
(*a*) accumulation of inorganic salts in the cells
(*b*) accumulation of organic salts in the cells
(*c*) chemical potential gradient
(*d*) water potential gradient

114. A cell is plasmolysed after being kept in a hypertonic solution. What will be present between cell wall and plasmalemma ?
(*a*) Hypertonic solution (*b*) Air
(*c*) Hypotonic solution (*d*) Isotonic solution

115. Dry seeds when placed in water swell due to
(*a*) diffusion (*b*) absorption (*c*) imbibition (*d*) inhibition

116. An innovative professor, who wanted to give a live demonstration of a physiological process, filled a glass bottle with previously moistened mustard seeds and water. He screw capped the bottle and kept it away in a corner and resumed his lecture. Towards the end of his lecture, there was a sudden explosion with glass pieces of the bottle thrown around. Which of the following phenomenon did the professor want to demonstrate ?
(*a*) Imbibition (*b*) Osmosis
(*c*) Anaerobic respiration (*d*) Diffusion

117. The form and structure of growing cells are maintained because of the
(*a*) osmosis (*b*) plasmolysis (*c*) turgidity (*d*) wall pressure

118. Turgidity of cells is maintained by
(*a*) wall pressure (*b*) turgor pressure
(*c*) diffusion pressure (*d*) osmotic pressure

119. Of the process which occurs in leaves, the one which may lower their temperature is
(*a*) photosynthesis (*b*) transpiration (*c*) hydrolysis (*d*) respiration

120. Which of the following has the lowest water potential ?
(*a*) soil (*b*) root xylem (*c*) leaf cell walls (*d*) leaf air spaces

121. In hypertonic solution a cell's water potential
(*a*) increases (*b*) decreases
(*c*) first increases, then decreases (*d*) no change

122. The stomata in angiosporms open and close due to
(*a*) pressure of gases inside the leaves (*b*) their genetic constitution
(*c*) turgor pressure of the guard cells (*d*) turgor pressure of the subsidiary cells

123. When the contents of a cell are in the shrinkage state, the process is called
(*a*) hypotonic (*b*) osmosis (*c*) endosmosis (*d*) plasmolysis

124. The stage when plasmolysis just begins and the plasmalemma starts receding the cell walls, it is called
(*a*) deplasmolysis (*b*) plasmolysis
(*c*) incipient plasmolysis (*d*) none of the above

125. Due to low atmospheric pressure the rate of transpiration will
(*a*) increase (*b*) decrease slowly (*c*) decrease rapidly (*d*) remain unaffected

126. The vacuolar membrane is called
(*a*) plasmalemma (*b*) middle lamella (*c*) tonoplast (*d*) chromoplast

127. In a terrestrial habitat which of the following factors affect temperature and rainfall conditions ?
(*a*) translocation (*b*) transformation (*c*) transpiration (*d*) transduction

128. The transpiration in the plants will be lowest
(*a*) when there is high humidity in the atmosphere
(*b*) high wind velocity
(*c*) environmental conditions are very dry
(*d*) there is excess of water in the cell

129. Which of the following is used for determining the rate of transpiration in plants ?
(*a*) Auxanometer (*b*) Potentiometer (*c*) Porometer (*d*) Potometer

130. Which of the following is the most common type of transpiration ?
(*a*) Cuticular (*b*) Stomatal (*c*) Foliar (*d*) Lenticular

131. To compare the rate of transpiration from two surfaces of a leaf, a piece of dry paper, previously soaked in the solution of a chemical which changes colour when becomes moist, is taken one such chemical is
(*a*) Cobalt chloride (*b*) Calcium chloride (*c*) Sodium chloride (*d*) Cobalt nitrate

132. Root pressure is maximum when
(*a*) transpiration is very low and absorption is high
(*b*) transpiration is very high and absorption is also high
(*c*) transpiration and absorption both are low
(*d*) transpiration is high and absorption is low

133. Wilting in plant occurs when
(*a*) phloem is blocked (*b*) xylem is blocked
(*c*) epidermis is peeled off (*d*) pith is blocked

134. Loss of turgor from the cells causes drooping of leaves during day, but they recover from it at night. This is called
(*a*) temporary wilting (*b*) guttation (*c*) permanent wilting (*d*) exudation

135. In both evaporation and transpiration water is lost in the form of vapour, yet they differ because
(*a*) frequency of water loss is different in both of them
(*b*) transpiration is a physical process and evaporation is physiological process
(*c*) transpiration is a physiological process and evaporation is a physical process
(*d*) both evaporation and transpiration are similar but the rate of water loss differs

136. Increase in CO_2 concentration around leaf results in
(*a*) partial closure of stomata (*b*) complete closure of stomata
(*c*) opening of stomata (*d*) no effect on stomatal opening and closing

137. Which one of the following is produced during water stress and causes closure of stomata ?
(*a*) Abscisic acid (*b*) Auxin (*c*) Gibberellin (*d*) Ethylene

138. Which of the following plant keeps its stomata open during the night and closed during the day ?
(*a*) Water lily (*b*) Maize plant (*c*) Cactus (*d*) Cucurbit

139. Conversion of starch to organic acid is essential for
(*a*) stomatal closure (*b*) stomatal growth (*c*) stomatal opening (*d*) stomatal initiation

140. Guard cells differ from epidermal cells in having
(*a*) chloroplasts (*b*) mitochondria (*c*) cell wall (*d*) vacuole

141. Active K^+ exchange mechanism for opening and closing of stomata was given by
(*a*) Khorana (*b*) Sachs (*c*) Levitt (*d*) Strasburger

142. In succulent plants the stomata open in night and close in day. Which among the following would be best hypothesis to explain the mechanism of stomatal opening in night only ?
(*a*) CO_2 accumulates, reduces pH, stimulates enzymes, resulting in accumulation of sugars.
(*b*) decrease in CO_2 concentration in dark, starch is broken down into organic acids particularly malic acid resulting in K^+ transport and osmotic concentration of guard cells increases.
(*c*) CO_2 used up, increase pH results in accumulation of sugars.
(*d*) low CO_2 concentration accumulates organic acid resulting in the increased concentration of cell sap.

143. Stomata will open, if there is accumulation of the following element in the guard cells
(*a*) Magnesium (*b*) Potassium (*c*) Iron (*d*) Zinc

144. In many thin leaved mesophytes the stomata open during the day time and close at night. This comes under
(*a*) Alfalfa type (*b*) Potato type (*c*) Barley type (*d*) Bean type

145. The process of the escape of liquid from the tip of uninjured leaf or through hydathodes is called
(*a*) guttation (*b*) transpiration (*c*) evaporation (*d*) none of above

146. Which of the following is responsible for guttation ?
(*a*) Transpiration (*b*) Root pressure (*c*) Photosynthesis (*d*) Osmosis

147. Transpiration - cohesion - tention theory operates in
(*a*) Active absorption (*b*) Passive absorption
(*c*) Both (*a*) and (*b*) (*d*) None of the above

148. Many transplanted seedlings do not survive because
(*a*) they do not like the new soil
(*b*) they do not get required minerals
(*c*) most of root hairs are lost during transplantation
(*d*) root apex is damaged during this process

149. The most widely accepted theory for ascent of sap in trees is
(*a*) Role of atmospheric pressure
(*b*) Pulsating action of living cells
(*c*) Transpiration pull and cohesion theory of Dixon and Jolly
(*d*) Cytoplasmic theory of Hugo de Vries.

150. Which of the following contributes most to transport of water from the ground to the leaves of a tall tree ?
(*a*) Capillary rise of water through xylem vessels
(*b*) Root pressure
(*c*) Cohesion of water and transpiration pull
(*d*) Break down of ATP

ANSWERS

1. (*c*)	2. (*a*)	3. (*c*)	4. (*b*)	5. (*b*)	6. (*b*)	7. (*d*)	8. (*b*)	9. (*d*)	10. (*d*)
11. (*d*)	12. (*d*)	13. (*a*)	14. (*d*)	15. (*d*)	16. (*d*)	17. (*c*)	18. (*a*)	19. (*a*)	20. (*a*)
21. (*a*)	22. (*b*)	23. (*c*)	24. (*c*)	25. (*c*)	26. (*c*)	27. (*a*)	28. (*b*)	29. (*b*)	30. (*b*)
31. (*d*)	32. (*a*)	33. (*a*)	34. (*c*)	35. (*d*)	36. (*d*)	37. (*b*)	38. (*c*)	39. (*d*)	40. (*a*)
41. (*d*)	42. (*b*)	43. (*b*)	44. (*b*)	45. (*a*)	46. (*b*)	47. (*b*)	48. (*c*)	49. (*d*)	50. (*c*)
51. (*b*)	52. (*a*)	53. (*a*)	54. (*d*)	55. (*c*)	56. (*b*)	57. (*d*)	58. (*a*)	59. (*a*)	60. (*b*)
61. (*d*)	62. (*b*)	63. (*b*)	64. (*c*)	65. (*d*)	66. (*a*)	67. (*d*)	68. (*c*)	69. (*a*)	70. (*d*)
71. (*a*)	72. (*a*)	73. (*a*)	74. (*a*)	75. (*a*)	76. (*b*)	77. (*d*)	78. (*c*)	79. (*a*)	80. (*a*)
81. (*b*)	82. (*a*)	83. (*a*)	84. (*a*)	85. (*a*)	86. (*b*)	87. (*a*)	88. (*b*)	89. (*c*)	90. (*b*)
91. (*a*)	92. (*a*)	93. (*d*)	94. (*c*)	95. (*b*)	96. (*d*)	97. (*c*)	98. (*a*)	99. (*a*)	100. (*b*)
101. (*a*)	102. (*a*)	103. (*a*)	104. (*d*)	105. (*b*)	106. (*b*)	107. (*a*)	108. (*b*)	109. (*b*)	110. (*b*)
111. (*b*)	112. (*a*)	113. (*d*)	114. (*a*)	115. (*c*)	116. (*a*)	117. (*c*)	118. (*b*)	119. (*b*)	120. (*d*)
121. (*b*)	122. (*c*)	123. (*d*)	124. (*c*)	125. (*a*)	126. (*c*)	127. (*c*)	128. (*a*)	129. (*d*)	130. (*b*)
131. (*a*)	132. (*a*)	133. (*b*)	134. (*a*)	135. (*c*)	136. (*a*)	137. (*a*)	138. (*c*)	139. (*c*)	140. (*a*)
141. (*c*)	142. (*b*)	143. (*b*)	144. (*a*)	145. (*a*)	146. (*b*)	147. (*b*)	148. (*c*)	149. (*c*)	150. (*c*)

3

CHAPTER

Mineral Nutrition and Nitrogen Metabolism

The basic needs of all living organisms are similar and they require macromolecules, such as carbohydrates, proteins and fats for their growth and development. They also require minerals for their development.

However, green plants prepare their food from simple substances, such as water and carbon dioxide, through process of photosynthesis. This type of green organisms, which prepare their food through photosynthesis are called **autotrophs**.

While other organisms, and non-green plants cannot prepare their own food, as autotrophs do. They obtain their nutrition from autotrophs, and are called **heterotrophs**.

In the formation of carbohydrates, proteins and fats, carbon (C), hydrogen (H) and oxygen (O) are required. In addition to these elements (*i.e.*, C, H and O), plants need several other important elements, called **mineral elements**.

The mineral elements that are essential in the life of a plant, with the exception of carbon and oxygen, are obtained through the root system from the soil. Carbon and oxygen are obtained from the atmosphere. The utilization for various absorbed ions by a plant for growth and development is called the **mineral nutrition** of the plant. It has been demonstrated by water and sand culture experiments that carbon, hydrogen, oxygen, nitrogen, sulphur, phosphorus, calcium, magnesium and iron are essential for the healthy growth and development of plants. These elements are referred to as the **macronutrients** or **major elements** because of the relatively large quantities in which they are needed for plant life. Plants also require some elements in minute or trace quantities which are called the **micronutrients** or **trace** or **minor elements**. Some important micronutrients are manganese, zinc, boron, copper and molybdenum.

ESSENTIAL MINERAL ELEMENTS

An **essential element** is known, without which the plant cannot complete its life cycle. It has clear physiological role to play. The important criteria for an essential element are as follows:

(*i*) This element is absolutely necessary for supporting normal growth and reproduction of plant.

(*ii*) This element is always specific, and cannot be replaced by any other element.

(*iii*) This element is directly involved in the metabolism of the plant.

However, among the mineral elements absorbed by the plants, not all are essential.

Out of 105 or more elements discovered so far, only 20 elements have been found to be essential for growth and development of the plant.

These essential elements are classified into two broad categories called (*i*) **macronutrients** and (*ii*) **micronutrients**.

The **macronutrients** are: carbon, hydrogen, oxygen, nitrogen, phosphorus, sulphur, potassium, calcium, magnesium and silicon.

The macronutrients are generally present in plant tissues in concentrations of 1 to 10 mg per gram of dry matter.

The **micronutrients** or **trace elements** are: iron, manganese, copper, molybdenum, zinc, boron and chlorine. Recently some other such elements have also been discovered, *e.g.*, cobalt, vanadium and nickel.

The microelements are required in very low amounts, *i.e.*, about 0.1 mg per gram of dry matter.

SOURCES OF ESSENTIAL ELEMENTS FOR PLANTS

All-elements incorporated into plants are ultimately derived from the atmosphere, water and soil.

Carbon (C). It is derived from atmosphere as CO_2.

Hydrogen (H). It is mainly obtained from water.

Oxygen (O). It is either derived from the air or from the water in the form of inorganic ions.

Nitrogen (N). The plants are unable to use atmospheric nitrogen. It is inert and plants cannot make use of it directly. However, by means of atmospheric activities nitrogen combines with oxygen and is brought down by rain to the soil. Besides, there are certain highly specialised organisms called **nitrogen fixers**, such as bacteria, and cyanobacteria. They fix atmospheric nitrogen into the soil in the form of nitrites (NO_2), and nitrates (NO_3). These heterotrophic organisms, occurring in the soil convert nitrogen gas (N_2) to anionic forms such as nitrates (NO_3) or nitrite (NO_2^-), or a reduced cationic form such as NH_4^+ (ammonium). These nitrogenous compounds enter plants as nutrients through the roots and are assimilated as organic nitrogen. The plants, in turn, provide organic nitrogen to heterotrophic organisms.

The other elements are absorbed from the soil, such as phosphorus (P) as phosphate (PO_4), and sulphur (S) mainly as sulphate (SO_4).

ROLE OF ESSENTIAL ELEMENTS

The essential elements perform several important roles. They may be as follows:

The most important role of the elements is to participate in various metabolic activities, such as regulation of permeability of cell membranes; some elements are required for maintenance of osmotic pressure of cell sap, while others take part in an electron transport system. Some elements are responsible for buffer action, and some for electrical neutrality.

Macronutrients

Mineral elements include N, P, K, Ca, Mg, S and Fe. They are absorbed from soil.

Nitrogen (N). About 78% of nitrogen is found in atmospheric air, but this is of no use to plants, in its free state. This enters in the plants through stomata along with other gases, and comes out in the same state, unused.

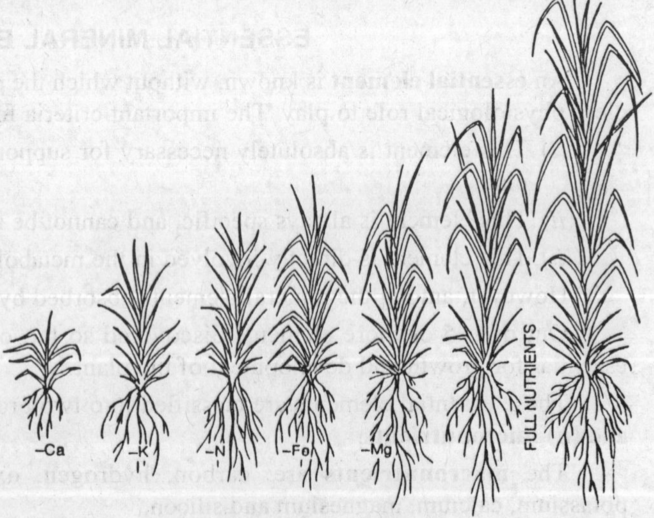

Fig. 3.1. Barley plants grown in water culture showing deficiency of certain nutrients.

The plants can take nitrogen, found in the soil in the form of NO_2^-, NO_3^- or NH_4^+. This is required by all parts of a plant, particularly the meristematic tissues. Nitrogen is found in the constituents of proteins, chlorophyll, protoplasm, nucleic acids, vitamins and hormones. For proper and normal growth of leaves of plants, it is one of the most essential elements.

Phosphorus (P). It is usually found in nucleoproteins and protoplasm. It also helps in the division of nucleus and the cell. It also helps in the decomposition of carbohydrates during respiration. Phosphorus is a constituent of cell membranes, nucleic acids and nucleotides. This is needed for all phosphorylation reactions. It plays a very important role in the ripening of the grains and fruits. This also helps a lot in the development of root system. For the development of underground parts of radish, beet root, and potato, etc., this element is required. Phosphorus is absorbed from the soil in the form of phosphate $H_2PO_4^-$ ions.

Potassium (K). This element is usually found in the growing regions of the plant. It is one of the constituents of the protoplasm. It is related to metabolic activities. This helps in the synthesis of carbohydrates and proteins. The plant grows normally in its presence, and the fruits and seeds remain quite healthy. On the scarcity of potassium, the normal growth of the leaves become yellowish, and the stem becomes somewhat thin. The plants take potassium from the soil in the form of potassium nitrate and potassium chloride, etc.

Potassium helps to determine anion-cation balance in cells, and is involved in protein synthesis, opening and closing of stomata, activation of enzymes, and maintenance of turgidity of cells. It is absorbed as K^+ ion by plants from the soil.

Calcium (Ca). This element is always found in green plants. The middle lamella of the cell wall consists of calcuim pectate. Only because of this element, the permeability of the protoplasm is maintained. Usually for the growth and development of the fruits, the calcuim is very essential. In its deficiency the cells of the plant become disorganized, and the growth of the plant is also checked. But, usually the leaves become chlorotic, when its overdose is present in the soil. Calcium is required by meristematic and differentiating tissues. It is also used in the mitotic spindle in cell division. It activates certain enzymes and plays an important role in regulating metabolic activities. Calcium accumulates in older leaves. Calcium is absorbed by the plants from the soil in the form of Ca^+ ions.

Magnesium (Mg). It is found in the chlorophyll. The chlorophyll does not form in its absence, the leaves become chlorotic, and the growth of the plant is retarded. In the seeds of pulses and cereals, this element is found in sufficient quantity. Magnesium activates enzymes in respiration and photosynthesis. It helps in synthesis of DNA and RNA. This also maintains ribosome structure. Magnesium is absorbed by the plants from the soil in the form of divalent Mg^{2+}.

Sulphur (S). It is usually found in the complex proteins of the plants. It is found in sufficient quantity in the mustard oil. It is also found in the protoplasm. In its deficiency, the leaves become chlorotic, and the stem becomes thin. The plants take this element from the soil in the form of calcium and potassium sulphate (SO_4^{2-}).

Sulphur is present in two amino acids, *i.e.*, cysteine and methionine. It is main constituent of several coenzymes, vitamins (*i.e.*, thiamine, biotin, CoA) and ferredoxin.

Iron (Fe). This also plays an important role in the formation of chlorophyll, though it is not found in the constitution of the chlorophyll. This element is always found to be present in the chromatin material of nucleus and protoplasm. It is an important constituent of proteins like ferredoxin and cytochromes which are involved in transfer of electrons. It is reversibly oxidised from Fe^{2+} to Fe^{3+} during electron transfer. It activates catalase. Iron is absorbed by the plants from the soil in the form of ferric (Fe^{3+}) ions.

Non-mineral elements include C, H and O

Carbon (C). Carbon is the main constituent of the plant material. The main part of the plant consists of it. It is usually half of plant's dry weight. Carbon is found in all organic compounds present in the plant. The plants get, carbon from the atmosphere in the form of carbon dioxide. Usually, .03% carbon dioxide is found in the atmosphere, but this percentage goes down even to .01% in day time, because of its use in the photosynthesis.

Hydrogen (H). Free hydrogen does not play any role in the plant life. When it combines with oxygen, water is formed, which is absorbed by the plants. The hydrogen is found in so many organic and inorganic compounds of the plant.

Oxygen (O). Free oxygen is taken from the atmosphere by the plants, which helps in the respiration of the living cells. Besides this, the oxygen is found in many organic compounds of the plant.

Micronutrients

Boron (B). This is also an essential element for the growth of nearly all the plants. The plants of tomato, tobacco, lemon, root, mustard, cotton and others are sufficiently benefitted by these elements. Boron also helps in the formation of the nodules of the leguminous plants. It also adds to the quantity of the sugar of beet root. The beet root becomes susceptible to various diseases, because of its deficiency. In its deficiency, the normal growth of the plant is obstructed and various leaf spots are developed. In its absence, the apices of roots and stems become brittle, and gradually they die.

Boron is also required for uptake and utilisation of Ca^{2+}, membrane function, pollen germination, cell elongation, cell differentiation and carbohydrate translocation.

Manganese (Mn). Manganese activates many enzymes which are involved in photosynthesis, respiration and nitrogen metabolism. Manganese helps in splitting of water to liberate oxygen during photosynthesis. Various vitamins are formed in the fruits, only because of its presence. The cabbage plants, pine trees and the leguminous plants require its optimum doses. Manganese is absorbed by plants in the form of manganous cation (Mn^{2+}).

Zinc (Zn). Zinc helps in the formation of chloroplasts. It activates various enzymes, especially carboxylases. Zinc is required in synthesis of auxin. It is absorbed by plants from soil in the form of (Zn^{2+}) ions.

Copper (Cu). Copper helps in formation of starch. It is required for the overall metabolism in plants. It is associated with certain enzymes involved in redox reactions and is reversibly oxidised from Cu^{+} to Cu^{2+}. Copper is absorbed by plants as cupric ion (Cu^{2+}).

Molybdenum (Mo). This helps in formation of proteins. It also helps in fixation of nitrogen in soil by *Azotobacter* and *Rhizobium*. This is a constituent of several enzymes, including nitrogenase and nitrate reductase which take part in nitrogen metabolism. This is absorbed by plants from soil in the form of molybdate ion (MoO_2^{2+}).

Chlorine (Cl). With Na^{+} and K^{+}, chlorine helps in determining solute concentration and anion-cation balance in cells. It is required for cell division in roots and leaves. It is needed to perform water-splitting reaction in photosynthesis, which evolves oxygen.

DEFICIENCY SYMPTOMS

Deficiency Symptoms (Macronutrients)

Nitrogen (N). Nitrogen deficiency causes yellowing of older leaves (chlorosis). The plant growth is stunted as protein content, cell division and cell enlargement are decreased. It also causes dormancy of lateral buds, late flowering, purple colouration and shoot axis surface and wrinkling of cereal grains.

Phosphorus (P). Phosphorus deficiency causes premature leaf fall. Dead necrotic areas develop on leaves or fruits, and leaves turn dark to blue-green in colour. It also causes delay in seed germination.

Potassium (K). Mottled chlorosis of leaves occurs. Necrotic areas are developed at the tips and margins of leaves, and they curve downward. Internodes become short, and plants adopt bushy habit. It also causes loss of cambial activity, disintegration of plastids and increase in rate of respiration.

Calcium (Ca). Calcium deficiency causes disintegration of growing meristematic regions of the root, stem and leaves. Chlorosis occurs along the margins of the younger leaves, and malformation occurs. This also leads to stunted growth of the plant.

Magnesium (Mg). Magnesium deficiency causes interveinal chlorosis of the leaves. The older leaves are affected first, and dead necrotic patches appear on the leaves. This also causes premature leaf abscission.

Sulphur (S). Sulphur deficiency causes yellowing (*i.e.*, chlorosis) of leaves, younger leaves are affected first, tips and margins of leaves roll inward, and stem becomes hard due to development of sclerenchyma. These symptoms are similar to those of nitrogen deficiency symptoms, as sulphur and nitrogen are constituents of proteins.

Iron (Fe). Iron deficiency causes rapid chlorosis of the leaves which is generally interveinal.

Deficiency Symptoms (Micronutrients)

Manganese (Mn). Manganese deficiency causes chlorotic and necrotic spots in the interveinal areas of leaves.

Zinc (Zn). Zinc deficiency causes chlorosis of older leaves which starts from tips and margins. Malformation of leaves takes place, and stunted growth of plant occurs. Its deficiency causes **mottle leaf** disease in apple, citrus, walnut, etc., and **khaira** disease in rice.

Copper (Cu). Copper deficiency causes necrosis of the tips of young leaves. It causes **die-back** of citrus and other fruit trees where leaves wither and fall, bark becomes rough and splits exuding gummy substances. **Reclamation disease** of cereals and leguminous plants is also caused due to its deficiency.

Boron (B). Boron deficiency causes death of the shoot tip. Flower formation is suppressed, root growth is stunted and shoot apices die. Fruits become of small size and root nodules in leguminous plants are not formed, and leaves become coppery in texture.

Molybdenum (Mo). Molybdenum deficiency causes chlorotic interveinal mottling of the older leaves. This may cause nitrogen deficiency, as it is component of enzymes involved in nitrogen metabolism. Flower formation is inhibited, and also causes **whip tail** disease in cauliflower plants.

Chlorine (Cl). The deficiency of chlorine in plants causes wilting of leaves, stunted root growth and reduced fruiting.

MECHANISM OF ABSORPTION OF ELEMENTS

Here the term mineral salt is used for those substances which contain nitrogen, sulphur and phosphorus in inorganic and organic combinations. The upward translocation of nitrogen, sulphur and phosphorus occurs in both organic and inorganic combinations.

Earlier this was universally agreed that upward translocation of mineral salts takes place through the xylem. But most studies of the sap present in the xylem indicated that it contains more inorganic solutes than organic ones. During the seasons, appreciable concentrations of mineral salts are commonly present in the sap vessels when upward flow of water occurs at its most rapid rate.

Table 3.1. Roles of Mineral Elements in Plants.

Element	Obtained as	Regions of plant in which required	Functions	Deficiency symptoms
Nitrogen N	NO_2^-, NO_3^- or NH_4^+	In all tissues, particularly in meristematic tissues.	Constitutent of proteins, nucleic acids, vitamins, hormones, coenzymes, ATP, chlorophyll.	Stunted growth, chlorosis.
Phosphorus P	$H_2PO_4^-$	Young tissues, withdrawn from the older metabolically less active cells.	Constituent of cell membrane. Required for the synthesis of nucleic acids, nucleotides, ATP, NAD and NADP.	Poor growth of plant, leaves dull green.
Potassium K	K^+	Meristematic tissues, buds, leaves, root tips.	Involved in protein synthesis; involved in formation of cell membrane and in opening and closing of stomata; increases hardness; helps in determination of anion cation balance in cells; activates enzymes and helps in maintenance of turgidity of cells.	Stunted growth; yellow edges to leaves; premature death and mottled appearance of leaves.
Calcium Ca	Ca^{++}	Meristematic and differentiating tissues, accumulates in older leaves.	Involved in selective permeability of cell membranes; activates certain enzymes; required for development of stem and root apices, and as calcium pectate in the middle lamella of the cell wall.	Stunted growth; chlorosis of young leaves; die back of shoots.
Magnesium Mg	Mg^{++}	All the leaves of plants; withdrawn from ageing leaves and directed to developing seeds.	Activates enzymes in phosphate metabolism; constituent of chlorophyll; maintains ribosome structure.	Chlorosis
Sulphur S	SO_4^{++}	Young leaves; stem and root tips; remobilized during senescence.	Constituent of amino acids, cysteine, methionine and proteins, coenzyme A, and ferredoxin.	Chlorosis

Element	Obtained as	Regions of plant in which required	Functions	Deficiency symptoms
Iron Fe	Fe^{+++}	Everywhere in leaves, along leaf veins.	Constituent of ferredoxin and cytochromes; required for synthesis of chlorophyll.	Chlorosis
Manganese (trace) Mn	Mn^{++}	Leaves and seeds	Activates certain enzymes (carboxylases).	Chlorosis; grey spots on leaves
Molybdenum (trace) Mo	Mo^{+++}	All parts; particularly in roots.	Activates certain enzymes in nitrogen metabolism.	Retardation of growth to some extent.
Boron (trace) B	B^{---}	Leaves and seeds	Increases the uptake of water and Ca; essential for meristematic activity and growth of pollen tube; involved in translocation of carbohydrates.	Death of stem and root apices; leaves become curled and brittle; reduces flowering; causes brown heart disease.
Copper (trace) Cu	Cu^{++}	All parts of plant	Activates certain enzymes	
Zinc (trace) Zn	Zn^{++}	All parts of plant	Activates various enzymes especially carboxylases, carbonic anhydrate and various dehydrogenases; required for auxin synthesis.	Malformation of leaves Dieback of shoots.
Chlorine Cl	Cl^-	All parts of plant	With Na^+ and K^+ helps in determining solute concentration and anion-cation balance in cells; essential for oxygen evolution in photosynthesis.	

The experiments of Clements and Engard (1938), Phillis and Mason (1940) and others have shown that ringing stems of various species do not prevent upward transport of mineral salts through the plant. This shows that upward translocation of mineral salts can take place in the xylem. Experiments on ringing often show that mineral salts can be conducted in the xylem and experiments on intercepting wood often show that they can be conducted in the phloem.

Stout and Hoagland (1939), employed the radioactive tracer technique in experiments designed to ascertain the path of upward movement of mineral salts in plants. The results of experiments conducted by these workers showed that upward movement of mineral salts occurs in the xylem, but the possibility of some such movement in the ploem also is not excluded even though it is clear that lateral movement from xylem to phloem takes place readily. The similar experiments of Gustafson (1939) also indicate that a limited amount of upward translocation of radioctive phosphate may occur through the phloem.

(*i*) The **outer space** includes the intercellular space and the cell wall. The **inner space** refers to most of cytoplasm and vacuole.

(*ii*) Entry of ions into outer space is considered as **passive**.

(*iii*) Entry into, or exit from the inner space is **active**.

(*iv*) passive entry does not require expenditure of metabolic energy.

(*v*) Active process requires metabolic energy.

(*vi*) Movement of ions is termed **flux**. The movement into the cell is **influx**, whereas outward movement is **efflux**.

Conclusively it can be said that upward translocation of mineral salts occurs in the xylem, and it appears quite certain that this is the main pathway along which their general upward translocation of mineral salts from roots to leaves occurs. Some upward translocation of mineral salts also occurs in the phloem under certain circumstances.

Absorption of Solute (Mineral Salts)

Besides water, the plant absorbs from the environment considerable quantities of mineral salts, gases and various other salts. All these are absorbed in the form of aqueous solutions. The mineral salts are absorbed from the external solutions by the roots. It is believed that they are not absorbed as whole molecules but as ions. This is evident from the unequal accumulation of the anions and cations of the same salt in the cell sap.

For many years it was supposed that various solutes entered the roots along with soil water in the process of osmosis; in other words the salts swept in with the water current. This view is erroneous because the diffusion of solute is quite different from the diffusion of water. Bonner (1953) has shown that salts can be absorbed by a cell even if there is higher concentration of a particular salt. Hoagland (1944) also found that solutes can be absorbed from a culture solution either more less rapidly than water is absorbed depending on the kind of solutes.

Passive Absorption

Passive absorption. In most cases, the movement of mineral ions into the root occurs by diffusion. Molecules or ions diffused from a region of their higher concentration to a region of their lower concentration. As these substances diffuse they exert a pressure called **diffusion pressure**. The movement of mineral ions into root cells as a result of diffusion is called **passive absorption**.

Ion exchange theory. According to the theory of **ion exchange** ions from the external solution (in which tissue is immersed) are exchanged with the ions of similar charge absorbed on the surface of the cell wall or membrane of the tissue. The colloidal fraction of the soil has an important role in ion exchange. These colloidal particles are known as **micelles**, possess negative charge and attract cations such as calcium, magnesium, potassium, ammonium, sodium, as well as hydrogen ions arising from biological activity. In the similar way, the root surface, which has a negative charge, carries many cations on its surface. These cations can be exchanged with other cations present in the soil. Thus a new cation can be absorbed on the root surface. Further transport of cations inside the root cell takes place by diffusion. Like cations, anions can also be exchanged with free 'OH^-' ions.

Donnan equilibrium. This theory explains the passive accumulation of ions, that are non-diffusible, which may be present on one side of the membrane (Donnan, 1927). Unlike diffusible ions, the membrane is not permeable to non-diffusible ions. Such ions are called **fixed ions**. They may be anions or cations. In a system, in which there are no fixed ions, there are equal number of anions and cations on both sides of the membrane at equilibrium. But in Donnan equilibrium, in order to balance, the charge of the fixed ions (anions), more ions of the other charge (cations) would be required.

For example, there is a membrane that separates a cell from the external medium and allows exchange of some ions and not others. To the inner side of this membrane there are anions, which are fixed and non-diffusible, and therefore, the membrane becomes inpermeable to these anions. In such a situation, for equilibrium to be reached, additional cations are needed to balance the negative charges of the anions that are structurally formed to the inner side of the above membrane.

According to the theory, Donnan equilibrium is attained if the product of anions$^-$ and cations$^+$ in the internal solution becomes equal to the product of anions$^-$ and cations$^+$ in the external solution, depicted by the equation as follows:

$$\frac{\text{Cations inside, Ci}^+}{\text{Cations outside Co}^+} = \frac{\text{Anions outside, Ao}^-}{\text{Anions inside, Ai}^-}$$

For example, a membrane which is permeable to Mg^+ and Cl^- ions and to X^- ions (*i.e.*, fixed ions) present inside the cell. Here, membrane has $6X^-$ fixed ions on the inner side. Cl^- ions make across the membrane by diffusion along the concentration gradient. The concentration of anions$^-$ on the inner side is now more than that of cations$^+$. In order to balance electrochemical equilibrium within the cell sap, Mg^{++} ions move across the membrane against the concentration gradient. Similarly, if there are fixed cations the anions shall move against the concentration gradient to bring about equilibrium.

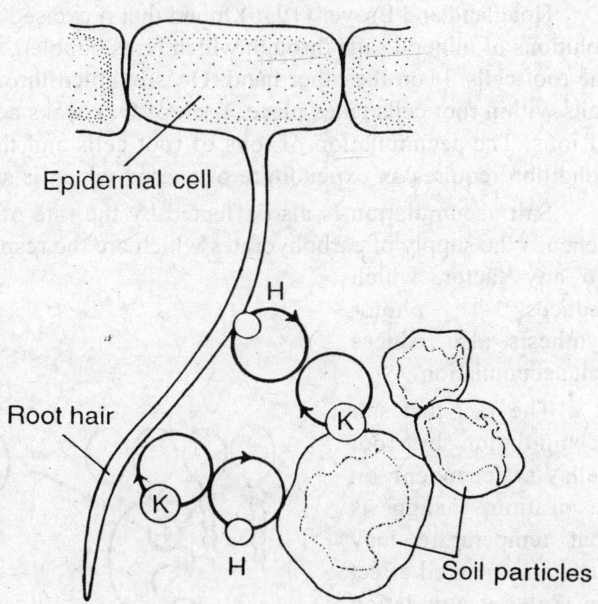

Fig. 3.2. Contact exchange theory.

Mass flow hypothesis. According to this hypothesis, roots also absorb a large quantity of ions along with absorption of water, due to transpiration. The ion absorption increases with increase in transpiration. The ions have been considered to move in a mass flow with water from the soil solution through the root and eventually to the shoot. This occurs due to passive absorption of ions by free diffusion into the free space of a tissue. Thus, mass flow of ions through root tissue occurs as a result of transpirational pull in the absence of metabolic energy. This theory was supported by Kramer (1956), Russel and Barber (1960).

Active Absorption

Generally, the lipid-protein membrane of a cell is largely permeable to free ions. The energy is considered to be involved in the transport of such free ions across the membrane. Thus *the absorption of ions, involving use of metabolic energy is called active absorption.*

The absorption of solutes takes place only in some cases by simple diffusion. Only limited quantities of minerals can pass into the cells by this method. Hoagland and Davies (1923) and many scientists have found that there is also a movement of salts against the concentration gradient, *i.e.*, from lower concentration to higher concentration for each ion. This cannot be explained on the basis of osmosis and most scientists now accept the theory of **"active absorption of solutes"** put forward by Hoagland (1923).

Active absorption of solutes or theory of salt accumulation. According to this theory Hoagland (1923) suggested that absorption of solutes takes place against higher concentration of salts. The cells near the tips of roots also have the capacity of accumulating ions (Hoagland and Broyer, 1936). If the initial salt content of the root cells is low and if other conditions are favourable, the concentration of ions in the absorbing cells may greatly be increased than that present in soil solution.This involves an expenditure of energy. This energy is supplied by the respiratory activity of the absorbing cells. The rate of accumulation is often influenced, therefore, by the previous metabolic status of the absorbing cells.

The phenomenon of salt accumulation seems confined largely to cells which have the capacity for cell division and growth. Meristematic cells and cells in the early stages of enlargement are particularly active in absorbing ions. As cells lose their capacity for growth, they also lose their capacity of mineral salt accumulation. As already referred earlier, that accumulation of salt requires expenditure of energy which is supplied by respiratory activity of cells.

Hoagland and Broyer (1936) found that if excised roots (young roots) are immersed in dilute solutions of mineral salts through which N_2 is bubbled, little or no accumulation of salts occurs in the root cells. If on the other hand, O_2 is bubbled through the solution, a rapid accumulation of salts within root cells takes place. Lack of O_2 checks aerobic respiration and prevents absorption of ions. The accumulation of ions of root cells and their retention within these cells in a free condition requires an expenditure of energy which is supplied by the process of respiration.

Salt accumulation is also affected by the rate of photosynthesis because on this process depends the supply of carbohydrates which are the respiratory substrates for efficient respiration. So any factor, which reduces photosynthesis, also reduces salt accumulation.

The fact that salt accumulation by root cells is dependent on respiration, suggests that temperature may have also marked effect on salt accumulation process.

Carrier concept. Ions, which are accumulated in cells, may move into the inner space against concentration and for this movement additional energy is required. This additional

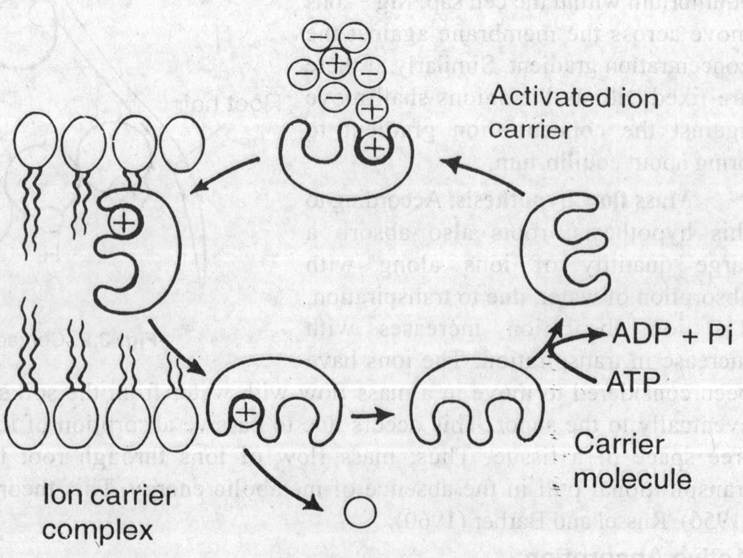

Fig. 3.3. Carrier concept.

energy is derived directly or indirectly through metabolism. This theory of active absorption has been supported by various evidences which show that active ion uptake is carried out by **carrier mechanism** for both influx and efflux of ions.

Unlike ion channels, the carrier proteins do not have pores. The membrane does not allow the ions to pass through as it is. The activated ions combine with carrier proteins and form **ion-carrier complex**, which is capable of moving across the membrane. This complex moves across the membrane and reaches the inner surface. Here, the complex breaks and releases ions into the cytoplasm of the cell. Carriers are specific, and combine with a particular type of ion.

Table 3.2. Differences Between Passive Absorption and Active Absorption

Passive Absorption	Active Absorption
1. This processes physical driving force, which is non-metallic.	1. Here, the driving force is energy derived from metabollic processes.
2. This type of absorption of ions and molecules is spontaneous and proceeds towards equilibrium.	2. This type of absorption is not spontaneous and does not proceed towards equilibrium.
3. Such absorption of a substance occurs across a protoplasmic membrane from its higher to lower chemical potential	3. The active absorption of a substance occurs across a protoplasmic membrane from its lower to higher chemical potential, *i.e.*, against concentration gradient.
4. In passive absorption, the energy yielding metabolic processes are not involved.	4. When energy yielding metabolic processes are weakened, active transport system is also checked.
5. The passive transport takes place through the protoplasmic layer in between the cell wall and the vacuole.	5. The active transport takes place across the protoplasmic membrane (*i.e.*, plasme membrane, tonoplast, etc.)

Ion Movement into the Root

Mineral nutrients absorbed by the root are carried to the xylem. This absorption takes place by two pathways, they are:

(*i*) **Apoplast pathway** and (*ii*) **Symplast pathway**

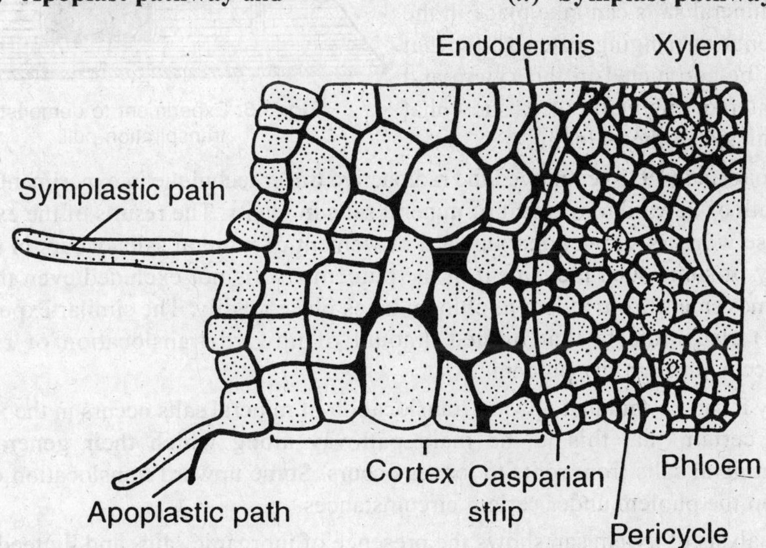

Fig. 3.4. Apoplastic and symplastic pathways of ion absorption.

(*i*) **Apoplastic pathway.** This pathway, essentially involves diffusion, and mass flow of water from cell to cell through spaces between cell wall polysaccharides.

The ions that enter the cell wall of epidermis move across cell wall of cortex, cytoplasm of endodermis, cell wall of pericycle, and finally accumulate in xylem vessels.

(*ii*) **Symplastic pathway.** In this pathway, ions that enter the cytoplasm of epidermis move across the cytoplasm, cortex, endodermis and pericycle through plasmodesmata, and finally reach to xylem vessels.

TRANSLOCATION OF SOLUTES

Upward Translocation of Mineral Salts

Here the term mineral salt is used for those substances which contain nitrogen, sulphur and phosphorus in inorganic and organic combinations. The upward translocation of nitrogen, sulphur and phosphorus occurs in both organic and inorganic combination.

Earlier this was universally agreed that upward translocation of mineral salts takes place through the xylem. But most studies of the sap present in the xylem indicated that it contains more inorganic solutes than organic ones. During the seasons appreciable concentrations of mineral salts are commonly present in the sap of vessels when upward flow of water occurs at its most rapid rate. The experiments of Clements and Engard (1938), Philiis and Mason (1940) and others have shown that ringing stems of various species do not prevent upward transport of mineral salts through the plant. This shows that upward translocation of mineral salts can take place in the xylem. Experiments on ringing often show that mineral salts can be conducted in the xylem and experiments on intercepting wood often show that they can be conducted in the phloem.

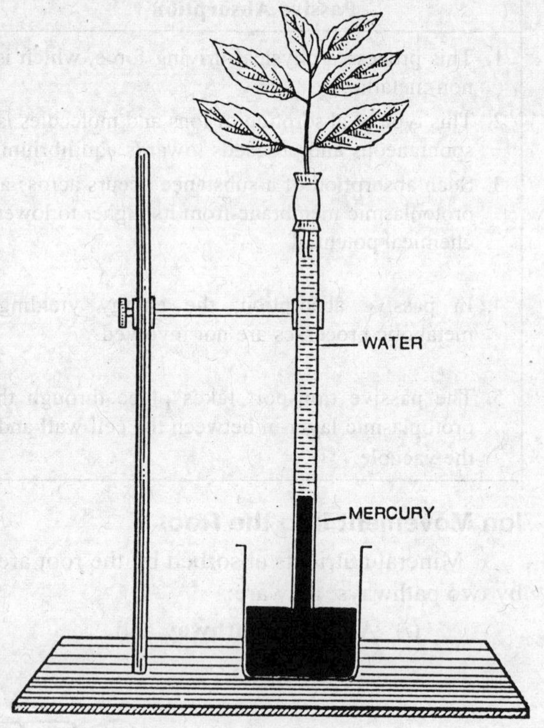

Fig. 3.5. Experiment to demonstrate transpiration pull.

Stout and Hoagland (1939) employed the radioactive tracer technique in experiments designed to ascertain the path of upward movement of mineral salts in plants. The results of the experiments conducted by these workers showed that upward movement of mineral salts occurs in the xylem, but the possibility of some such movement in the phloem also is not excluded even though it is clear that lateral movement from xylem to phloem takes place readily. The similar experiments of Gustafson (1939) also indicate that limited amount of upward translocation of radioactive phosphate may occur through the phloem.

Conclusively it can be said that upward translocation of mineral salts occurs in the xylem, and it appears quite certain that this is the main pathway along which their general upward translocation of mineral salts from roots to leaves occurs. Some upward translocation of mineral salts also occurs in the pholem under certain circumstances.

However, analysis of xylem sap shows the presence of inorganic salts, and by feeding plants with isotopes, it has exhibited that inorganic substances move up the plant through xylem.

General Aspects of the Translocation of Solutes in Plants

From the time when a young plant starts to grow until its death, there is continuous movement of solutes through the conducting elements of every organ of the plants. In the early stages of the development of the seedling, the food is generally translocated upward in the growing stems and downward in the developing roots from the storage tissues of that seed. Once the leaves are formed and plant begins to photosynthesize, the food starts flowing in downward direction, from the leaves towards the stem and the root. Likewise, as soon as the developing roots make effective contact with the soil particles, absorption of mineral salts begins, and the mineral solutes absorbed by the roots move in an upward direction, towards the shoot. Some of the solutes absorbed from the soil, especially those containing nitrogen, phosphorus or sulphur might react with the organic molecules in the root cells which have descended into the roots from the leaves. The resulting complex compounds, such as amino acids and amides may then be translocated in the reverse direction from the roots into the aerial parts of the plant.

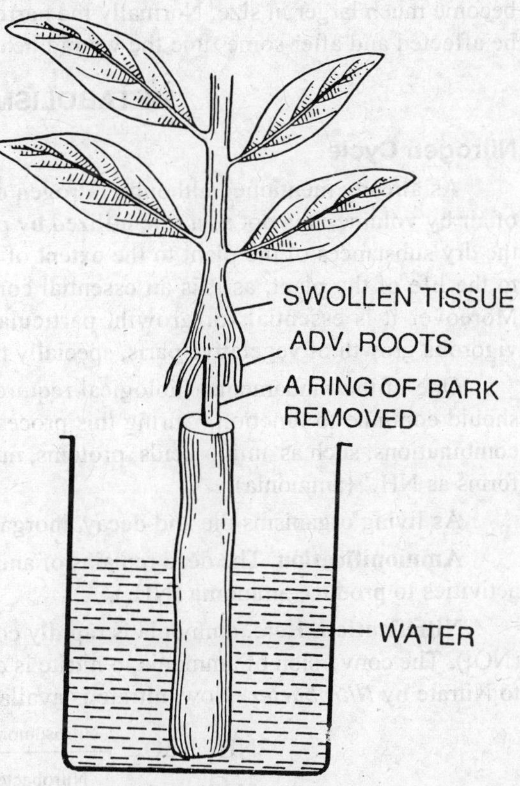

Fig. 3.6. Ringing experiment.

The mineral salts absorbed by the roots are mostly translocated to young leaves and other do not remain in the organ into which they were first translocated. Most of the mineral salts that reach the young leaves or flowers from the root may be translocated back into the stem and become redistributed to other younger parts of the plant, the translocation patterns within a plant have significant effects upon its behaviour. For example, when the young fruits and seeds develop, there is a general migration of organic materials from all parts of the plant towards the developing fruits and seeds with the result the growth of vegetative organs is checked. In cotton plants, for example, the vegetative growth decreases sufficiently, when the plant is fruiting heavily. Here the large part of synthesized carbohydrates and of the nitrogenous compounds in the above ground organs of the plant moves into the enlarging fruits. The carbohydrate supply reaching the root system becomes insufficient to maintain rapid respiration and root elongation. With the result the rate of absorption of mineral salts decreases, which in turn checks or even stops the vegetative growth of the plant (Eaton and Johan. 1944).

Evidence from ringing experiments. Much of the evidence indicating that organic solutes move towards the basal portions of the plant in the phloem has been obtained by ringing experiments. 'Ringing' means the removal of a narrow continuous band of tissues external to the xylem. Since ringing entirely encircles the stem, all tissues external to the xylem are completely intercepted. This operation is also called 'girdling'. This operation which is performed beneath the foliage region of the stem, produces two effects. They are as follows:

1. The organic food accumulates in the region above the ring which swells up as a result of this accumulation whereas the food content of the tissue below the ring is greatly diminished on

account of their being used up locally. 2. Secondly vigorous growth takes place in the region above the ring and sometimes the adventitious roots are developed. This happens, because of the accumulation of food materials in this region. The tissues below the ring are soon dried up because of starvation. Here the growth of tissues is inhibited, root development is checked and the tissues ultimately die. If the ringing is done in such a way that the phloem is not interrupted, translocation of food is not affected and the normal development of the roots beneath the ring continues.

The ringing is of much practical value in the case of fruit trees. At the time of the development of fruits the rings are being formed at different places on the trunks of the fruit tree. The ringing of the stem in such a manner prevents the downward flow of the food across the ring. With the result the food synthesized in the leaves above the ring is diverted to the fruits which become much larger in size. Normally the narrow rings are being made, so that the cambium may be affected and after some time the wound heals up.

METABOLISM OF NITROGEN

Nitrogen Cycle

As already mentioned although nitrogen occurs to the extent of about seventy eight percent of air by volume, it is not as a rule utilized by plants in its free state. However, nitrogen occurs in the dry substances of the plant to the extent of 1-3 percent only. Nevertheless, it is indispensable to the life of the plant, as it is an essential constituent of proteins, chlorophyll and protoplasm. Moreover it is essential for growth, particularly of the leaves. An excess of nitrogen causes vigorous growth of vegetative parts, specially the leaves, but delays reproduction activity.

One of the fundamental biological requirements for life to persist is that the nitrogen cycle should continue to function. During this process, the atomospheric nitrogen is fixed into organic combinations, such as amino-acids, proteins, nucleic acids, etc., in living organisms via inorganic forms as NH_4^+ (ammonia).

As living organisms die and decay, inorganic nitrogen is liberated.

Ammonification. The dead remains of animals and plants are decomposed through microbial activities to produce ammonia (NH_3).

Nitrification. Here, ammonia is rapidly converted first to nitrites (NO_2), and then to nitrates (NO_3). The conversion of ammonia to nitrite is carried out by bacteria *Nitrosomonas*, and of nitrite to Nitrate by *Nitrobacter*. Now, nitrate is available to the plant.

$$2NH_3 + 3O_2 \xrightarrow{\text{Nitrosomonas}} 2NO_2^- + 2H^+ + 2H_2O$$

$$2NO_2^- + O_2 \xrightarrow{\text{Nitrobacter}} 2NO_3^-$$

Denitrification. Where nitrate (NO_3) may convert into N_2 gas by bacteria *Pseudomonas*. This nitrogen gas may be again fixed in the form of NH_4^+ through the process of biological **nitrogen fixation.**

Nitrogen of the soil. The amount of nitrogen in the soil varies from 0.096 to 0.21 per cent. However, the soil makes the main source of nitrogen for the plant. In the plant the nitrogen exists as inorganic and organic compounds. The chief forms of inorganic compounds are the nitrates and nitrites of potassium and calcium, and also ammonia and its compounds; while the organic compounds are chiefly the proteins. Normally the ammonium compounds found in the soil are made available for the use of the green plants after conversion into nitrate by the action of certain nitrifying bacteria living in the soil. This process is known as **nitrification**. During this process the ammonium compounds are oxidized into nitrate in two stages—(*i*) these are acted upon by the nitrite-bacteria (*Nitrosomonas*) and oxidized into ($-NO_2$) and (*ii*) the nitrite thus formed is again acted upon by the nitrate-bacteria (*Nitrobacter*) and further oxidized into nitrate ($-NO_3$). The nitrate, thus produced is readily absorbed by the green plants.

Biological Nitrogen Fixation

Biological nitrogen fixation is carried out by both free-living and symbiotic bacteria.

The free-living nitrogen fixing bacteria are : Cyanobacteria, *Azotobacter* and *Clostridium.*

Symbiotic N_2 fixation. Sometimes the ammonium compounds (NH_4) are made available to the plants by the nitrogen fixers. The best known nitrogen-fixing symbiotic bacterium is *Rhizobium* (*R. leguminosarum*). This bacterium lives in soil to form root nodules in plants of the family Leguminosae such as gram, pea, groundnut, beans, etc. Root nodules are little outgrowths on roots. When a section of the fresh root nodule is examined, it looks pinkish in colour due to the presence of a pigment called **leghaemoglobin**. This pigment is closely related to **haemoglobin**, the red pigment of human blood. Like haemoglobin, legha emoglobin is an oxygen scavenger. The enzyme-nitrogenase which catalyses the fixation of nitrogen function under anaerobic conditions. Leghaemoglobin combines with oxygen and protects nitrogenase.

Nodule acts as a site for N_2 fixation. It contains all the necessary biochemical compounds, such as **nitrogenase** and **leghaemoglobin**. The enzyme nitrogenase is a Mo-Fe protein and catalyses the conversion of atmospheric N_2 to NH_3. This enzyme is extremely sensitive to oxygen to protect it from oxygen, nodules contain an **oxygen scavenger**, called **leghaemoglobin.**

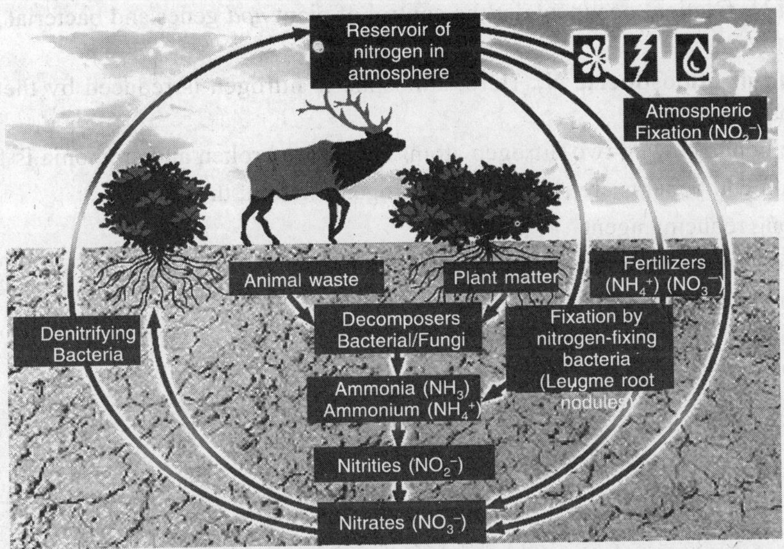

Fig. 3.7. Nitrogen cycle.

Formation of root nodules in leguminous plant. When a root hair of a leguminous plant comes in contact with *Rhizobium* (a bacterium) it is curled or deformed. Certain specific chemical substances secreted by bacteria (*Rhizobium*) are responsible for curling.

At the site of curling of root hair, rhizobia (bacteria) invade the root tissue and proliferate within the root hair.

Some of the bacteria enlarge to become membrane bound structures called **bacteroids**. The bacteroids cannot divide, and therefore, some bacteria remain untransformed, and they allow infection to spread.

An infection thread made up of plasma membrane is formed, which grows inward from the infected cell of the plant, that separates the infected tissue from rest of the plant.

Cell division is stimulated in the infected tissue and more bacteria invade the newly formed tissues.

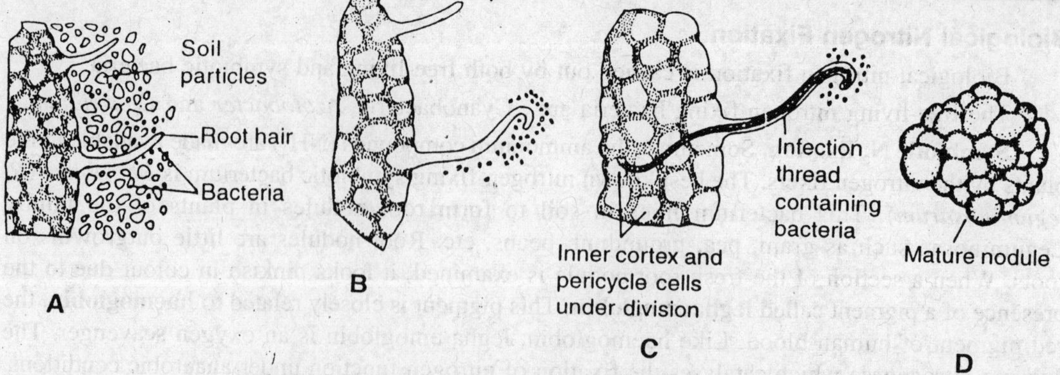

Fig. 3.8. Formation of root nodule in leguminous plant.

It is believed that a combination of **cytokinin** produced by the invading bacteria and **auxin** produced by plant cells, promotes cell division and extention, which leads to molecule formation.

The nodule thus formed, is responsible for direct vascular connection with the host for exchange of nutrients.

However, N_2 fixation, occurs under the control of plant *nod* genes and bacterial *nod*, *nif* and *fix* gene cluster.

Reduction of atmospheric N_2. Here, atmospheric nitrogen is reduced by the addition of hydrogen atoms.

The three bonds betwen two nitrogen atoms (N°N) are broken and ammonia is formed.

Three main components are required for nitrogen fixation, they are:

(*i*) A strong reducing agent.

(*ii*) ATP to transfer hydrogen atoms to dinitrogen, and

(*iii*) The enzyme systems.

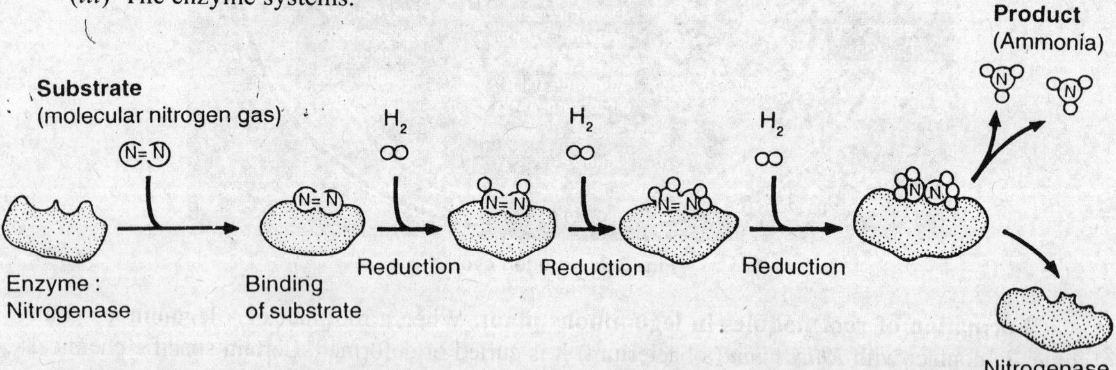

Fig. 3.9. Reduction of one molecule of N_2 in a series of reactions catalysed by nitrogenase enzyme.

The reducing agent FAD and ATP are provided by photosynthesis and respiration respectively.

Thus formed ammonia is utilised for the synthesis of amino-acids.

These amino-acids are translocated to other parts of the plant, which act as building blocks for the synthesis of various proteins.

Synthesis of Amino Acids

Amino-acids are supposed to be initial products of nitrogen assimilation.

Each amino-acid consists of at least one carboxyl (–COOH) group, and one or several amino ($-NH_2$) groups.

Synthesis of amino-acids takes place by two main methods. They are as follows:

(*i*) Reductive amination, and (*ii*) Transamination.

1. Reductive amination. Here, ammonia reacts with α-ketoglutaric acid, which results in the formation of glutamic acid. The reaction is as follows:

$$\alpha\text{-ketoglutaric acid} + NH_4^+ + \text{NAD (P) H} \xrightarrow[\text{dehydrogenase (enzyme)}]{\text{Glutamate}} \underset{\text{(glutamic acid)}}{\text{Glutamate}} + H_2O + \text{NAD (P)}$$

Enzyme responsible for this reaction is glutamate dehydrogenase.

2. Transamination. This process involves transfer of amino ($-NH_2$) group from one amino acid to the keto group of keto acid.

Glutamic acid is main from which other seventeen amino acids are formed through transamination.

The enzyme responsible for this reaction is called **transaminase**.

Amides. There are two most important amides found in plants. They are (*i*) asparagine, and (*ii*) glutamine.

They are formed from two amino acids, called

(*i*) glutamic acid, and (*ii*) aspartic acid.

During this process hydroxyl (–OH) part of the acid is replaced by another ($-NH_2$) radicle.

The enzymes responsible for such reaction may be glutamine synthetase or asparagine synthetase.

Amides have more nitrogen than amino acids, and make structural part of most proteins.

Protein Synthesis

Proteins consist of one or more polypeptide chains. Each such chain consists of hundreds of amino acids. The number of amino acids varies greatly among proteins, thus, molecular weight of proteins also varies.

The linkage of amino acids and amides in the polypeptide chain occurs through **peptide bond**, which involves the carboxyl (–COOH) group of one amino acid and the amino ($-NH_2$) group of the next.

The proteins are highly specific because of the sequence in which the amino acids are present in the protein.

Proteins have significant role in the **structural** and **functional** organisations of the cell.

Structural proteins constitute various cellular components and some extracellular parts, such as cuticle and fibres.

Functional (enzymatic and hormonal) proteins control almost all metabolic, biosynthetic, bioenergetic, growth regulating, sensory and reproductive activities of the cell.

All the proteins that are required by the cell for its different purposes are synthesized by the cell itself intracellularly.

Water Culture or Soil Less Culture (Hydroponics)

In general, soil supplies the mineral nutrients for plant growth. Since the minerals required by plants for their growth are easily absorbed in solution and therefore, it is possible to grow plants in water containing the required amount of mineral salts taking care that aerial parts of the plants are exposed to air and light. Thus, cultivation of plants by placing the roots in the nutrient solution is called **hydroponics**. It is necessary to aerate the solution to provide roots with adequate oxygen supply.

The earliest experiment using a culture solution was done by Sachs (1860) and he showed the essentiality of nitrogen for plant growth. Later Knop (1865) gave the formula of nutrient solution. His prescription for preparing a nutrient solution was used for a long time. About seventy years later Arnon and Hoagland developed a formulation to study the micronutrients as well. Iron was supplied as ferrous sulphate and often it precipitated out. This problem has now been solved by dissolving the ferrous sulphate along with a chelating agent Na-EDTA (*i.e.*, disodium salt of ethylene - diaminetetra acetic acid.)

In water culture experiments, the seedlings are grown in water containing the known nutrients in a particular proportion. The culture solutions may contain all essential elements except that whose importance has to be studied. By excluding a particular element in a culture solution, characteristic deficiency symptoms can be observed. Deficiency symptoms may vary from species to species. The result obtained from soil-less culture may then be used to determine deficiencies under field conditions.

Table 3.3. Formula of Two Nutrient Media.

Knop (1865)	*g/l*	Arnon and Hoagland (1940)	*g/l*
KNO_3	0.2	KNO_3	1.02
$Ca(NO_3)_2$	0.8	$Ca(NO_3)_2$	0.492
KH_2PO_4	0.2	$NH_4H_2PO_4$	0.23
$MgSO_4$, $7H_2O$	0.1	$MgSO_4$, $7H_2O$	0.49 mg/1
$FePO_4$	0.1	H_3BO_3	2.86
		$MnCl_2$, $4H_2O$	1.81
		$ZnSO_4$, $5H_2O$	0.08
		$CuSO_4$, $7H_2O$	0.22
		H_2MoO_4, H_2O	0.09
		$FeSO_4$, $7H_2O$ 0.5%]	0.6 ml/1
		Tartaric Acid 0.4%]	3x weekly

Application of Fertilizers

Soil fertility is defined as the ability of the soil to provide all essential plant nutrients in appropriate form and in a suitable balance. A soil can be highly fertile, *i.e.*, it has a ready supply of nutrients in suitable form. A fertile soil may be highly saline or alkaline which may not be good for agriculture. The soil poor in fertility may be made fertile by adding fertilizers in suitable dosages. Soils generally contain sufficient quantities of essential minerals. However, three important elements (*i.e.*, NPK–nitrogen, phosphorus and potassium) need to be replenished in crop fields as they are depleted by repeated cultivation. The common sources of N, P and K in this country are - sodium nitrate, ammonium sulphate, ammonium nitrate, ammonium chloride, urea, calcium ammonium nitrate, superphosphate, bonemeal, rock phosphate and calcium magnesium phosphate. The bags of NPK fertilizers are labelled 15-15-15 or 17-18-9 which refer to the percentage by weight of nitrogen, phosphorus and water soluble potassium. These fertilizers are used in different dosages according to the climatic conditions, soil and crop.

SUMMARY

1. There are three main sources which supply inorganic nutrients to plants. They are (*i*) soil, (*ii*) air and (*iii*) water.

2. Plants absorb a wide range of mineral elements.
3. So far, more than 105 elements have been discovered.
4. However, only about 20 elements are essential for the normal growth and metabolism of plants.
5. Some elements are required in large quantities by the plants, and are known as macro-elements.
6. Some elements are required in less amount, and are called micro-nutrients.
7. The elements required by the plants become the essential constituents of protein, carbohydrate, fat, nucleic acid, etc., and take part in various metabolic processes.
8. Some of these elements serve as activators or co-factors for several enzymes.
9. Deficiency of any element in plants may lead to symptoms of deficiency, such as chlorosis, necrosis, stunted growth, etc.
10. Plants absorb mineral elements through root by passive or active absorption.
11. The absorbed elements are transported from the root to the xylem by apoplastic and symplastic pathways.
12. The most essential element for sustenance of life, is nitrogen.
13. However, plants cannot use atmospheric nitrogen directly.
14. For this purpose, prior to absorption, atmospheric nitrogen is fixed into organic combination.
15. Nitrogen fixation requires — (*i*) a strong reducing agent, (*ii*) energy in the form of ATP and (*iii*) the enzyme systems.
16. This task is completed by certain nitrogen fixing bacteria and cyanobacteria. Rhizobium is associated symbiotically with the roots of leguminous plants and form nodules in the roots.
17. The atmospheric nitrogen is fixed in the form NH_4^+ (ammonia) through the process of nitrogen fixation. Ammonia converts to nitrates.
18. Plants absorb nitrates, which again convert to ammonia by enzymatic reactions.
19. Ammonium ions are then used for the synthesis of amino acids, proteins and other nitrogenous compounds in the plant.

TERMS TO REMEMBER

Active absorption. Ions accumulated in cells may move against concentration or ecp gradients.

Ammonification. The dead remains of animals and plants are decomposed through microbial activities to produce ammonia.

Autotrophs. The organisms, which can prepare their own food through photosynthesis.

Bacteroids. Some of the bacteria that enlarge to become membrane bound structures.

Denitrification. Where nitrates convert to nitrogen gas.

Donnan equilibrium. Theory, explains the passive accumulation of ions that are fixed or non-diffusible, against an ecp gradient.

Essential element. An element without which the plant cannot complete its life cycle.

Flux. Movement of ions. The movement into the cell is **influx** and the outward movement is **efflux**.

Heterotrophs. Organisms, including non-green plants, which cannot make their own food and obtain their nutrition from autotrophs.

Hydroponics. Soil-less culture of plants, where roots are immersed in nutrient solution without soil.

Ion exchange. The process of exchange between adsorbed ions and ions in solution.

Leghaemoglobin. Pink coloured substance, similar to haemoglobin of vertebrates.

Macronutrients. Generally present in plant tissues in concentration of 1 to 10 mg per gm of dry matter.

Mass flow hypothesis. Mass flow of ions through root tissue occurs as a result of transpirational pull in the absence of metabolic energy.

Micronutrients. Also known as trace elements, generally present in plant tissues in very small amounts, *i.e.*, equal to or less than 0.1 mg per gm of dry matter.

Mineral nutrition. The study of how plants obtain mineral elements, either through water, air or soil and utilise them for their growth and development.

Nitrification. Ammonia, first converts to nitrites and then to nitrates.

Passive absorption. Absorption of minerals by physical processes not involving direct expenditure of metabolic energy.

Reductive amination. In this process ammonia reacts with α-ketoglutaric acid and forms glutamic acid.

Transamination. Transfer of amino group from one amino acid to the keto group of keto acid. The enzyme responsible for such reaction is transaminase.

QUESTION BANK

Very Short Answer Questions

1. What are autotrophs ?
2. What are heterotrophs ?
3. Define mineral nutrition.
4. What is an essential element ?
5. Name the macronutrients.
6. Give names of micronutrients.
7. Define the term 'flux'.
8. What is ammonification ?
9. What is nitrification ?
10. What is denitrification ?
11. What is leghaemoglobin ?
12. What are symbionts ?
13. Name the best known symbiotic nitrogen fixing bacterium.
14. Name the enzyme involved in biological nitrogen fixation.
15. What are bacteroids ?
16. What is transamination ?
17. Name the enzyme involved in transamination.
18. Who discovered 'translocation of solutes' ?
19. Give symptom of iron deficiency in plants.
20. Give symptoms of zinc deficiency in plants.
21. What type of condition is created by leghaemoglobin in root nodules of a legume ?
22. Name the red pigment present in the root nodules of leguminous plants.
23. Name the best known symbiotic nitrogen fixing bacterium.
24. Pick out from the following list the two minerals which are not needed by the majority of plants but very much needed by almost all animals:

    Calcium, Sodium, Potassium, Iron, Iodine
25. Name one most mobile element and one most immobile element.
26. What protects nitrogenase ?
27. What is common to *Nepenthes, Utricularia* and *Drosera* as regard to a special mode of nutrition?

ANSWERS

1. Organisms, which can prepare their own food through photosynthesis; 2. Organisms, including non-green plants, which cannot prepare their own food and obtain their nutrition from autotrophs; 3. The study of how plants obtain mineral elements, either through water, air or soil, and utilise them for their

growth and development; 4. An element without which the plant cannot complete its life cycle; 5. They include carbon, hydrogen, oxygen, nitrogen, phosphorus, sulphur, potassium, calcium, magnesium and silicon; 6. They include iron, manganese, copper, molybdenum, zinc, boron and chlorine; 7. The movement of ions is termed flux; the movement into the cell is influx, while the outward movement is efflux; 8. When the dead remains of animals and plants are decomposed through microbial activities to produce ammonia; 9. When ammonia is rapidly converted first to nitrites, and then to nitrates; 10. When nitrate converts to nitrogen gas by other microorganisms, *e.g.*, *Pseudomonas*; 11. Root nodules of leguminous plant contain an oxygen scavanger of pink colour and similar to haemoglobin of vertebrates; 12. Two organisms that live in close physical association, and are of mutual benefit to each other, *e.g.*, lichen, mycorrhiza, etc; 13. *Rhizobium*; 14. Nitrogenase; 15. Some of the bacteria enlarge to become membrane-bound structures, called bacteroids; 16. This process involves the transfer of amino group from one amino acid to the keto group of keto acid; 17. Transaminase; 18. P.R. Stout and D.R. Hoagland (1939); 19. Chlorosis of leaves ; 20. Malformed leaves, inter-veinal chlorosis in leaves, and stunted growth; 21. Anaerobic; 22. Leghaemoglobin; 23. *Rhizobium*; 24. Sodium and iodine; 25. Potassium and iron; 26. The pigment leghaemoglobin; 27. All are insectivorous plants.

Short Answer Questions

1. List the macronutrients and mention their major functions.
2. Why do farmers use leguminous crops to provide nitrogen to the soil ? Explain.
3. Name the organism that fixes nitrogen in symbiotic association with a water fern. Where does it live in such plants ?
4. Prior to sowing rice, a legume crop was cultivated and ploughed back in the field, why ? Explain.
5. How are ions absorbed by the roots ? Explain.
6. What do you understand by Donnan's equilibrium ? Explain.
7. Write down deficiency symptoms produced by following elements: (*i*) Nitrogen (*ii*) Phosphorus and (*iii*) Potassium.
8. Define the following : (*i*) Micronutrients (*ii*) Macronutrients.
9. Write short notes on : (*i*) Symplastic pathway (*ii*) Apoplastic pathway.
10. Give three criteria of essentiality of an element.
11. Why do plants of the legume family usually contain more protein than other plants ?
12. Name any two autotrophic plants which also show heterotrophic mode of nutrition.
13. Prior to sowing rice, a legume crop was cultivated and ploughed back in the field, why ? Explain.
14. Explain the terms plant macroelements and plant microelements. Give one example of each.
15. Name the nitrifying bacteria of the soil. Why are they called chemoautotrophs ?
16. Mention the symptoms of any four mineral deficiencies in plants.
17. What is nitrification ?
18. A farmer adds Azotobacter culture to the soil before sowing maize. How does it increase the yield of maize.
19. Name the respective mineral nutrient element that :
    (*i*) forms the constituent of the ring structure of chlorophyll
    (*ii*) activates carboxylases
    (*iii*) forms the component of nitrogenase
    (*iv*) synthesises middle lamella in plants cells.
20. Write an explanatory note on biological nitrogen fixation.
21. Define the following : (*i*) Nutrients (*ii*) Active absorption (*iii*) Symplastic pathway.
22. Differentiate between macronutrients and micronutrients.
23. What are various steps involved in the biological nitrogen fixation in plants ?
24. Name the nitrifying bacteria of the soil. Why are they called chemoautotrophs ?
25. Why do farmers use leguminous crops to provide nitrogen to the soil ? Explain.

26. Differentiate between essential and non essential mineral elements.
27. Write short notes on:
    (*i*) Reductive amination (*ii*) Transamination.
28. Make a list of macronutrients and mention their major functions.
29. Write short notes on:
    (*i*) Donnan equilibrium (*ii*) Mass flow hypothesis.
30. State deficiency symptoms of Manganese, Boron, Iron and Zinc.
31. Why do plants need potassium and magnesium ?
32. Why do farmers use leguminous crops to provide nitrogen to the soil ? Explain.
33. Differentiate between the two manners of absorption of minerals in plants from the soil through roots.
34. What are the various steps involved in the biological nitrogen fixation in plants ?

Long Answer Questions

1. Differentiate between two types of absorption of minerals from soil through roots.
2. Describe two important functions each of K, Fe and Zn in green plants and also write deficiency symptoms of any two of them.
3. Discuss the process of biological nitrogen fixation in plants.
4. Write an essay on mineral nutrition in plants.
5. List the macronutrients and mention their major functions.
6. Differentiate between micronutrients and macronutrients. How they are helpful in plant growth ?
7. Evaluate the importance of potassium, sulphur, iron, zinc, copper and molybdenum in plants.
8. What are trace elements ? What symptoms are caused by their deficiency ?
9. Describe the role and deficiency symptoms of the following elements in plants : (*a*) Nitrogen, (*b*) Molybdenum, (*c*) Zinc, (*d*) Phosphorus, and (*e*) Copper.
10. What are macronutrients ? Discuss the role of potassium, phosphorus, sulphur and iron in plant metabolism.
11. Describe the process of symbiotic biological nitrogen fixation.
12. Describe two important functions each of the element P, B and S in green plants and also write the deficiency symptoms of any two of them.
13. Make a table showing macroelements, regions of plants in which required, functions and their deficiency symptoms.
14. Make a list of macronutrients and mention their major functions.

MULTIPLE CHOICE QUESTIONS (Objective Type Questions)

1. Nitrogen is an important constituent of
   (*a*) proteins (*b*) lipids (*c*) carbohydrates (*d*) phopholipids
2. Zinc is essential for
   (*a*) biosynthesis of chlorophyll (*b*) biosynthesis of 3- IAA
   (*c*) stomatal opening (*d*) oxidation of carbohydrates
3. The micronutrient among these is
   (*a*) Zn (*b*) N (*c*) P (*d*) Ca
4. Presence of phosphorus
   (*a*) brings about healthy root growth (*b*) promotes fruit ripening
   (*c*) retards protein formation (*d*) none of the above
5. In plants a common symptom caused by deficiencies of P, K, Ca and Mg is
   (*a*) bending of leaf tip (*b*) formation of anthocyanin
   (*c*) poor development of vasculature (*d*) appearance of dead necrotic areas

6. Chlorosis occurs when plants are grown in
(*a*) dark (*b*) shade (*c*) strong light (*d*) Fe-free medium
7. Which of the following elements is a component of Ferredoxin ?
(*a*) Copper (*b*) Manganese (*c*) Zinc (*d*) Iron
8. A trace element is an element which
(*a*) is a radioactive element and can be traced by Geiger counter
(*b*) is required in very minute amounts
(*c*) draws other elements out of protoplasm
(*d*) was one of the first to be discovered
9. Mineral salts absorbed by the roots from the soil are in the form of
(*a*) very dilute solution (*b*) dilute solution
(*c*) concentrated solution (*d*) very concentrated solution
10. The presence of an element in plant ash
(*a*) does not guarantee that it is essential
(*b*) proves that it is essential
(*c*) proves that it takes part in the physiology of plants
(*d*) indicates that it occurs in abundance in soil
11. 'Khaira' disease of rice is developed due to the deficiency of
(*a*) copper (*b*) sodium (*c*) zinc (*d*) molybdenum
12. Ash is of what significance in the study of mineral nutrition of plants?
(*a*) It tells what minerals are present in the soil
(*b*) It informs which element is essential and in which amount it is necessary for a particular plant
(*c*) It is of no practical significance
(*d*) It indicates how much irrigation is needed by the plant.
13. Active transport occurs
(*a*) Against concentration gradient and requires ATP
(*b*) Against concentration gradient but does not requires ATP
(*c*) Along concentration gradient but requires ATP
(*d*) Along concentration gradient but does not require ATP
14. Magnesium is an important component of
(*a*) Haemoglobin (*b*) Florigen (*c*) Enzymes (*d*) Chlorophyll
15. Plants use zinc as
(*a*) Zn (*b*) Zn^{2+} (*c*) $ZnSO_4$ (*d*) $Zn(NO_3)_2$
16. Loading of phloem means
(*a*) elongation of phloem cells (*b*) separation of phloem parenchyma
(*c*) strengthening of phloem fibres (*d*) pouring of sugar into phloem
17. Radioactive tracer studies with ^{14}C have shown that
(*a*) phloem transports organic nutrients (*b*) phloem transports inorganic nutrients
(*c*) xylem transports inorganic nutrients (*d*) xylem transports organic nutrients
18. Direction of conduction of food through phloem is
(*a*) from below upward (*b*) from tip to bottom
(*c*) from leaves to roots (*d*) phloem never conducts food
19. Most widely accepted theory of carbohydrates translocation is
(*a*) Mass flow theory (*b*) Root pressure theory
(*c*) Imbibition theory (*d*) Transpiration theory

20. Brown heart disease is due to deficiency
(*a*) Molybdenum (*b*) Potassium (*c*) Iron (*d*) Boron
21. Which is required for nitrogen fixation ?
(*a*) Mn (*b*) *n* (*c*) Cu (*d*) Mo
22. Premature leaf fall is due to deficiency of
(*a*) Sodium (*b*) Potassium (*c*) Zinc (*d*) Phosphorus
23. Nitrogen is an important constituent of
(*a*) Polyphosphates (*b*) Carbohydrates (*c*) Proteins (*d*) Lipids
24. Element present in middle lamella
(*a*) Zn (*b*) Ca (*c*) Cu (*d*) K
25. Plants require sulphur for
(*a*) ATP synthesis (*b*) Protein synthesis (*c*) Glucose synthesis (*d*) DNA replication
26. Nitrogen absorbed by plants is
(*a*) Converted to nitrate (*b*) Reduced to ammonia
(*c*) Changed to nitrite (*d*) Combined with oxygen
27. Which important greenhouse gas other than methane is being produced from the agricultural field ?
(*a*) Arsine (*b*) Sulphur dioxide (*c*) Ammonia (*d*) Nitrous oxide
28. Supply end in solute transport is
(*a*) root (*b*) stem
(*c*) green leaves and storage organs (*d*) xylem and phloem
29. Which is not a micronutrient
(*a*) P / Mg (*b*) Cu / B (*c*) Mo / Zn (*d*) Mn / Cu
30. Hydroponics is
(*a*) growing of aquatic plants (*b*) growing of submerged aquatic plants
(*c*) growing of floating aquatic plants (*d*) soil-less cultivation of plants
31. Increase in fertility of soil without adding any nutrients is due to :
(*a*) Floods (*b*) Crop residue
(*c*) Favourable temperature (*d*) Microbial activity
32. The presence of large amount of nitrogen in the atmosphere is due to :
(*a*) Decomposers (*b*) Nitrites (*c*) Nitrogen cycle (*d*) Ammonia
33. 80 per cent of cells has :
(*a*) Minerals (*b*) Water (*c*) Proteins (*d*) Carbohydrates
34. Plants require one of the following for formation of ATP :
(*a*) N, Ca (*b*) N, P (*c*) N, Cu (*d*) K
35. The core-element of chlorophyll is :
(*a*) Mn (*b*) Mg (*c*) Zn (*d*) Fe
36. One of the following elements of not of much importance to plants :
(*a*) Calcium (*b*) Zinc (*c*) Copper (*d*) Sodium
37. The largest and sessile pitchers are produced by :
(*a*) Nepenthes (*b*) Dischidia (*c*) Darlington (*d*) Sarracenia
38. The theory which suggests that CO_2 produced in respiration plays an important role in mineral absorption is :
(*a*) Contact exchange theory (*b*) Carbonic acid exchange theory
(*c*) Active absorption theory (*d*) None of these.
39. Cyanobacteria help farmers by :
(*a*) Reducing the alkalinity of soil (*b*) Increasing soil fertility of soil
(*c*) Neutralising alkalinity of soil (*d*) Water logging

40. Plants can be cultivated in soil less medium where all nutrients are supplied from outside in water solution and this method comes under :
(*a*) Water culture (*b*) Hydroponics
(*c*) Hybrid culture (*d*) Critical culture media

41. In fruit trees a disease exanthema is caused by the deficiency of :
(*a*) Na (*b*) Ca (*c*) Cu (*d*) P

42. The essential nutrient element required by plants in the least quantity is :
(*a*) Chlorine (*b*) Zinc (*c*) Molybdenum (*d*) Manganese

43. Enzyme catalyzed reactions can be inhibited by :
(*a*) Mg^{2++} (*b*) Zn^{2++} (*c*) Cu^{2++} (*d*) Hg^{2++}

44. Permeability of protoplasm is accelerated by :
(*a*) Na (*b*) K (*c*) Ca (*d*) P

45. The brown colour of leaves of cabbage is due to the deficiency of :
(*a*) Boron (*b*) Nitrogen (*c*) Sulphur (*d*) Calcium

46. Soil salinity is measured by :
(*a*) Porometer (*b*) Potometer (*c*) Conductivity meter (*d*) Calorimeter

47. Biofertilizer present in root nodules of non-leguminous plants is :
(*a*) *Azotobacter* (*b*) *Clostridium* (*c*) *Frankia* (*d*) *Aerohizobium*

48. The major portion of the dry weight of plants comprises of :
(*a*) Calcium, magnesium and sulphur (*b*) Carbon, nitrogen and hydrogen
(*c*) Carbon, hydrogen and oxygen (*d*) Nitrogen, phosphorus and potassium

49. Boron in green plants assists in :
(*a*) Acting as enzyme cofactor (*b*) Photosynthesis
(*c*) Sugar transport (*d*) Activation of enzymes

50. Which organism is used in rice fields for nitrogen fixation :
(*a*) Nostoc (*b*) Azolla (*c*) Trifolium (*d*) All of these

51. Zn is used up by plants in the form of :
(*a*) Zn (*b*) Zn^{++} (*c*) $ZnSO_4$ (*d*) $Zn(NO_3)_2$

52. Which of the following is not major element ?
(*a*) Zn (*b*) Ca (*c*) Mg (*d*) P

53. Function of Mg and Fe is :
(*a*) Synthesis of chlorophyll (*b*) Synthesis of proteins
(*c*) Synthesis of fat (*d*) Synthesis of organic acids

54. Which of the following is macronutrient ?
(*a*) Ca (*b*) Mn (*c*) Zn (*d*) Cu

55. Which of the following is macronutrient ?
(*a*) Zn (*b*) Ca (*c*) P (*d*) Mg

56. A facultative parasite is one which :
(*a*) Always requires a living host
(*b*) is normally a saprophytic but can also become a parasite
(*c*) Never requires a living host
(*d*) None of these

57. The movement of mineral ions into plant root cells as a result of diffusion is called :
(*a*) Endocytosis (*b*) Osmosis
(*c*) Passive absorption (*d*) Active absorption

58. The role of inorganic nutrients in plant growth was at first indicated by :
(*a*) Wood Ward (*b*) Knop (*c*) Stewart (*d*) De Saussure
59. Zinc is essential for :
(*a*) Biosynthesis of chlorophylls (*b*) Biosynthesis of 3-IAA
(*c*) Stomatal closing (*d*) Oxidation of carbohydrates
60. Cultivation by sand culture is also called :
(*a*) Soilless cultivation (*b*) Green house effect
(*c*) Photorespiration (*d*) None of these
61. Inorganic nutrients are present in the soil in the form of :
(*a*) Molecules (*b*) Atoms
(*c*) Electrically charged ions (*d*) Parasites
62. Function of Zinc is :
(*a*) Metabolism of carbohydrates (*b*) Production of IAA
(*c*) Both above (*d*) None of these
63. Which of the following is the importance of molybdenum in plant metabolism ?
(*a*) Carbon assimilation (*b*) Nitrogen reduction
(*c*) Plant breeding (*d*) Chromosome contraction
64. Phosphorus is a structural element in :
(*a*) Fat (*b*) Starch (*c*) Nucleotide (*d*) Carbohydrate
65. The leaves become yellow due to deficiency of :
(*a*) Zinc and Boron (*b*) Potassium and Calcium
(*c*) Iron and Magnesium (*d*) None of these
66. Premature leaf fall occurs due to deficiency of :
(*a*) Calcium (*b*) Iron (*c*) Phosphorus (*d*) Sulphur
67. Carnivorous plants are found on :
(*a*) nitrogen deficient soil (*b*) O_2 deficient soil
(*c*) CO_2 deficient soil (*d*) none of these
68. Hydroponics is :
(*a*) water (*b*) solution containing all the nutrients
(*c*) green house (*d*) liquid
69. The major portion of the dry weight of plants comprises of :
(*a*) Nitrogen, phosphorus and potassium (*b*) Calcium, magnesium and sulphur
(*c*) Carbon, nitrogen and hydrogen (*d*) Carbon, hydrogen and oxygen
70. Gray spots of oat are caused by deficiency of :
(*a*) Cu (*b*) Zn (*c*) Mn (*d*) Fe
71. Active transport of ions by the cell requires :
(*a*) High temperature (*b*) ATP (*c*) Alkaline pH (*d*) Salts
72. Tracer elements are :
(*a*) Microelements (*b*) Macroelements (*c*) Radio isotopes (*d*) Vitamins
73. Trace element in plants is :
(*a*) K (*b*) P (*c*) Zn (*d*) None of these
74. Plants deficients of element zinc, show its effect on the biosynthesis of plant growth hormonce :
(*a*) Auxin (*b*) Cytokinin (*c*) Ethylene (*d*) Abscisic acid
75. Which one of the following mineral elements plays an important role in biological nitrogen fixation ?
(*a*) Copper (*b*) Manganase (*c*) Zinc (*d*) Molybdenum

76. An essential element is one which
(a) is present in plant ash
(b) is available in soil
(c) is indispensable for growth
(d) improves plant growth

77. Mineral salts, which are absorbed by the roots from the soil, are present in
(a) very dilute solution
(b) concentrated solution
(c) very concentrated solution
(d) dilute solution

78. Conduction of inorganic materials in plants occurs mainly through
(a) xylem (b) phloem (c) sieve tube (d) cambium

79. In plant nutrition, elements are classified as major or minor depending on
(a) their relative production in the ash obtained after burning the plant
(b) the relative amounts required by the plants
(c) their relative importance in plant growth
(d) their availability in the soil

80. The framework elements are
(a) magnesium, copper and iron
(b) copper, carbon and oxygen
(c) manganese, calcium and nitrogen
(d) carbon, hydrogen and oxygen

81. Which one of the following plays an independent role in energy metabolism ?
(a) calcium (b) sodium (c) sulphur (d) phosphorus

82. Stem and root tips die due to deficiency of
(a) nitrogen (b) phosphorus (c) calcium (d) sulphur

83. The plant ash is an indication of
(a) mineral salts absorbed by the plant
(b) organic matter of the plant
(c) both the mineral salts and organic matter
(d) none of the above

84. Which one of the following essential elements can land plants normally take directly from the air ?
(a) Carbon (b) Nitrogen (c) Phosphorus (d) Hydrogen

85. Micronutrients
(a) are called 'micro' becuase they play only a minor role in plant nutrition
(b) may be omitted from culture media without any detrimental effect on the plants
(c) are as important in nutrition as the macronutrients
(d) are more important in nutrition than macronutrients

86. A trace element
(a) is a radioactive and can be traced by Geiger - count
(b) is required in very minute amounts
(c) draws other elements out of protoplasm
(d) was one of the first to be discovered in protoplasm

87. Which of the following is a micronutrient
(a) Mg (b) Zn (c) P (d) K

88. Which of the following is a macronutrient ?
(a) Ca (b) Mo (c) Zn (d) Mn

89. Which of the following elements has not been proved to be essential for plants ?
(a) Carbon (b) Sodium (c) Zinc (d) Potassium

90. Active mineral absorption depends on
(a) expenditure of energy
(b) ions move freely
(c) ions move passively
(d) ions are active

91. Carbon becomes available to crop plants in the form of
(*a*) carbonates (*b*) carbon dioxide (*c*) element carbon (*d*) amino acids
92. Most of the plants obtain nitrogen from the soil in the form of
(*a*) nitric acid (*b*) nitrites (*c*) nitrates (*d*) free nitrogen gas
93. The bulk of nitrogen in nature is fixed by
(*a*) chemical industries (*b*) symbiotic bacteria
(*c*) lightning (*d*) denitrifying bacteria
94. All the following contributes to the availability of nitrogen compounds in the soil except
(*a*) *Azotobacter* (*b*) thunders from clouds
(*c*) bacterial decay of animal refuge (*d*) addition of lime
95. Which of the following does NPK denote ?
(*a*) Nitrogen, Protein, Kinetin (*b*) Nitrogen, Protein, Potassium
(*c*) Nitrogen, Phosphorus, Potassium (*d*) Nitrogen, Phosphorus, Kinetin
96. In plants, a common symptom caused by deficiency of P, K, Ca and Mg is
(*a*) formation of anthocyanin (*b*) shredding of leaf
(*c*) bending of leaf tip (*d*) development of necrotic areas on leaves
97. Which of the following is a component of middle lamella in plant cells ?
(*a*) Mg (*b*) K (*c*) Ca (*d*) Fe
98. Death of stem and root tips occurs due to deficiency of
(*a*) Calcium (*b*) Nitrogen (*c*) Carbon (*d*) Phosphorus
99. Absence of magnesium from plant tissue results in
(*a*) etiolation (*b*) hydrolysis (*c*) chlorosis (*d*) necrosis
100. Which element forms part of structure of chlorophyll molecule ?
(*a*) Mg (*b*) Mn (*c*) K (*d*) Fe
101. Deficiency of iron causes
(*a*) decrease in protein synthesis (*b*) interveinal chlorosis first on young leaves
(*c*) bending of stem tip (*d*) reduced leaves and stunted growth
102. Which of the following element is a component of ferredoxin ?
(*a*) Manganese (*b*) Zinc (*c*) Iron (*d*) Copper
103. Iron is mainly absorbed in
(*a*) ferrous form (*b*) ferric form (*c*) both (*a*) and (*b*) (*d*) None of above
104. Plants requiring the two metallic compounds for chlorophyll synthesis, they are
(*a*) Fe and Mg (*b*) Cu and Ca (*c*) Ca and K (*d*) Mg and Ca
105. Phosphorus is a structural element in
(*a*) proteins (*b*) carbohydrates (*c*) DNA (*d*) cytochrome
106. Which of the following essential element is not a constituent of any enzyme but stimulate the activity of many enzymes ?
(*a*) Zn (*b*) K (*c*) Mg (*d*) Mn
107. The most important use of potassium is that
(*a*) regulates many plant processes
(*b*) it provides red colour to the fruit
(*c*) it helps in photosynthesis
(*d*) it influences enzyme activity, which regulates many plant processes

108. Deficiency of molybdenum causes
(*a*) bending of leaf tip (*b*) chlorosis of leaves
(*c*) mottling and necrosis of leaves (*d*) poor development of vasculature
109. Which of the following element is necessary for translocation of sugars in plants ?
(*a*) Boron (*b*) Molybdenum (*c*) Manganese (*d*) Iron
110. Which one of the following micronutrient can be absorbed by foliage ?
(*a*) Zinc (*b*) Iron (*c*) Phosphorus (*d*) None of the above
111. Which of the following is not an essential micronutrient ?
(*a*) Nickel (*b*) Manganese (*c*) Cobalt (*d*) Boron
112. Which of these is not a nutrient for plants
(*a*) water (*b*) mineral ions (*c*) carbon dioxide (*d*) nitrogen gas
113. Symbiotic nitrogen fixation by *Rhizobium* was discovered by
(*a*) Louis Pasteur (*b*) Lister (*c*) Beijerinck (*d*) Pfeffer
114. Which pigment is essential for nitrogen fixation by leguminous plants ?
(*a*) phycocyanin (*b*) phycoerythrin (*c*) leghemoglobin (*d*) myoglobin
115. Which root tissue primarily contributes to the formation of nodules particularly in legume family ?
(*a*) epidermis (*b*) xylem (*c*) cortex (*d*) endodermis
116. The symbiotic fixation of atmospheric nitrogen in root nodules of leguminous crops is brought about by
(*a*) *Rhizobium* (*b*) *Rhizopus* (*c*) *Rhizoclonium* (*d*) *Rhizoctonia*
117. In nitrogen cycle nitrite is converted to nitrate by
(*a*) *Azotobacter* (*b*) *Rhizobium* (*c*) *Nitrobacter* (*d*) *Nitrosomonas*
118. The enzyme responsible for the reduction of molecular nitrogen to the level of ammonia in leguminous root nodules is
(*a*) nitrogenase (*b*) nitrate reductase (*c*) nitrite reductase (*d*) hexokinase
119. A bulk of nitrogen in nature is fixed by
(*a*) Lightning (*b*) Denitrifying bacteria
(*c*) Symbiotic bacteria (*d*) None of above
120. A crop plant that can grow well in nitrogen deficient soils without addition of manure is
(*a*) *Allium sativum* (*b*) *Triticum aestivum*
(*c*) *Solanum melongena* (*d*) *Cajanus cajan*
121. Conversion of organic nitrogenous compound into ammonium compound is called
(*a*) nitrification (*b*) denitrification (*c*) amononification (*d*) none of above
122. Root hairs do not play a role in
(*a*) mineral uptake (*b*) oxygen uptake (*c*) water uptake (*d*) CO_2 uptake
123. Organisms that fix nitrogen in aquatic habitats are
(*a*) diatoms (*b*) protozoans (*c*) desmids (*d*) cyanobacteria
124 Which one of the following statements is correct ?
(*a*) Legumes fix nitrogen only through specialised bacteria that live in their leaves
(*b*) Legumes fix nitrogen independently of the specialised bacteria that live in their roots
(*c*) Legumes fix nitrogen only through the specialised bacteria that live in their roots
(*d*) Legumes are incapable of fixing nitrogen
125. Exanthema, a disease of citrus plant is caused by the deficiency of
(*a*) zinc (*b*) magnesium (*c*) manganese (*d*) copper

ANSWERS

1. (*a*)	2. (*b*)	3. (*a*)	4. (*c*)	5. (*d*)	6. (*d*)	7. (*d*)	8. (*b*)	9. (*a*)	10. (*c*)
11. (*c*)	12. (*b*)	13. (*a*)	14. (*d*)	15. (*b*)	16. (*d*)	17. (*a*)	18. (*c*)	19. (*a*)	20. (*d*)
21. (*d*)	22. (*d*)	23. (*c*)	24. (*c*)	25. (*b*)	26. (*b*)	27. (*d*)	28. (*c*)	29. (*a*)	30. (*d*)
31. (*b*)	32. (*c*)	33. (*b*)	34. (*b*)	35. (*b*)	36. (*d*)	37. (*d*)	38. (*b*)	39. (*b*)	40. (*b*)
41. (*c*)	42. (*c*)	43. (*d*)	44. (*b*)	45. (*a*)	46. (*c*)	47. (*c*)	48. (*c*)	49. (*c*)	50. (*d*)
51. (*b*)	52. (*a*)	53. (*a*)	54. (*a*)	55. (*a*)	56. (*b*)	57. (*c*)	58. (*b*)	59. (*b*)	60. (*d*)
61. (*c*)	62. (*d*)	63. (*b*)	64. (*c*)	65. (*c*)	66. (*c*)	67. (*a*)	68. (*b*)	69. (*d*)	70. (*c*)
71. (*b*)	72. (*c*)	73. (*c*)	74. (*a*)	75. (*d*)	76. (*c*)	77. (*a*)	78. (*a*)	79. (*b*)	80. (*d*)
81. (*d*)	82. (*c*)	83. (*a*)	84. (*a*)	85. (*c*)	86. (*b*)	87. (*b*)	88. (*a*)	89. (*b*)	90. (*a*)
91. (*b*)	92. (*c*)	93. (*b*)	94. (*d*)	95. (*c*)	96. (*d*)	97. (*c*)	98. (*a*)	99. (*c*)	100. (*a*)
101. (*b*)	102. (*c*)	103. (*b*)	104. (*a*)	105. (*c*)	106. (*b*)	107. (*d*)	108. (*c*)	109. (*a*)	110. (*a*)
111. (*a*)	112. (*d*)	113. (*c*)	114. (*c*)	115. (*c*)	116. (*a*)	117. (*c*)	118. (*a*)	119. (*c*)	120. (*d*)
121. (*c*)	122. (*d*)	123. (*d*)	124. (*c*)	125. (*d*)					

4

CHAPTER

Respiration

We know that during photosynthesis, light energy is converted into chemical energy, and is stored in carbohydrate molecules, such as glucose and starch. Organisms make use of such energy for their activities by oxidising these high energy food molecules into simple low energy molecules, *i.e.*, carbon dioxide and water. The reactions involved in process of oxidation are known as **respiration**. The compounds that are oxidised during process of respiration are called **respiratory substrates**.

Technically, respiration is defined as follows :

This is a process by which living cells break down complex high energy food molecules into simple low energy molecules, *i.e.*, CO_2 and H_2O, releasing the energy trapped within the chemical bonds.

The energy released during oxidation of energy rich compounds is made available for activities of cells through an intermediate compound called **adenosine triphosphate** (ATP).

During process of respiration, the whole of energy contained in respiratory substrates is not released all at a time. It is released slowly in several steps of reactions controlled by different enzymes.

Respiration takes place in all types of living cells, and generally called **cellular respiration**. During the process of respiration oxygen is utilised, and CO_2 water and energy are released as products. The released energy is utilised in various energy-requiring activities of the organisms, and the carbon dioxide released during respiration is used for biosynthesis of other molecules in the cell.

As we know, important life processes, such as synthesis of proteins, fats and carbohydrates, require a certain expenditure of energy. Where does this energy come from, how is it stored, and how is it made available to the living cell, are some of the questions, which are to be answered by process of respiration.

The reaction that occurs in common respiration of glucose may be summed up as follows:

$$C_6H_{12}O_6 + 6O_2 \xrightarrow{\text{enzymes}} 6CO_2 + 6H_2O + \text{Energy}$$

(2870 kJ or 686 kcal)

Here, 686 kcal or 2870 kJ of energy is liberated per molecule of glucose. Formerly, this calculated value was 673 kcal. One kcal is equal to 1000 calories. This means that one molecule of glucose on complete oxidation yields 686 kcal (kilocalories) of energy, (*i.e.*, 686, 000 calories).

The main facts associated with respiration are :

1. Consumption of atmospheric oxygen.
2. Oxidation **and** decomposition of a portion of the stored food resulting in a loss of dry weight as **seen** in the seeds germinating in dark.

3. Liberation of carbon dioxide and a small quantity of water (the volume of CO_2 liberated is equal to volume of O_2 consumed).
4. Release of energy by breakdown of organic food, (such as carbohydrates).

Adenosine Triphosphate (ATP)

This is an **energy intermediate**. As we know, both energy-yielding and energy-consuming reactions occur within the living cell.

The potential or stored energy of one compound, such as glucose, is released and utilised, in a most efficient manner, to drive the synthesis of other compounds, such as proteins.

This energy, now stored in the newly synthesised compound (*e.g.*, protein), can in turn made be available for other synthetic reactions.

However, energy-yielding reactions of the cell occur in many instances in the absence of energy-consuming reactions. The energy released in such situation is in the form of heat and lost to the organism.

However, nature has provided the living cell with a means of temporary energy storage in the form of adenosine triphosphate (ATP).

Thus, energy released in oxidation of compounds, such as carbohydrates, lipids, proteins, etc., is immediately utilised in the synthesis of ATP from adenosine diphosphate (ADP) and inorganic phosphate (i.P.).

The chemical energy transferred to ATP can be used to drive synthetic reactions, while ADP and iP are released during the process. ATP is called **energy currency** of the cell.

ATP is capable of receiving energy from one reaction and transferring this to drive another reaction. This is of obvious advantage to the living system, as ATP can be formed in the oxidation of a variety of compounds and can be used for synthesis of a variety of compounds.

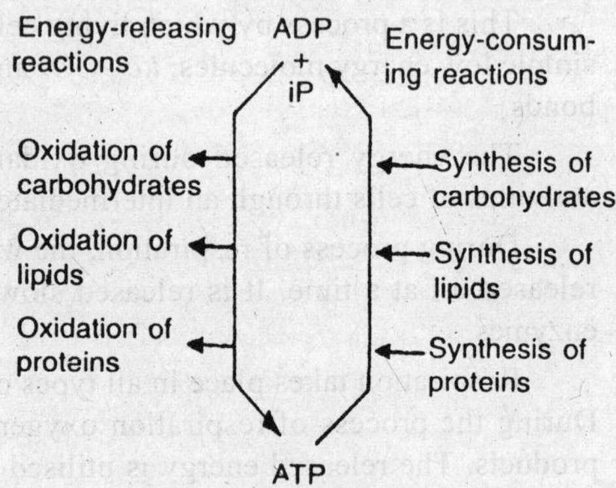

Fig. 4.1. Role of ATP as an intermediate energy transfer compounds.

Table 4.1. Differences Between Respiration and Combustion

Respiration	Combustion
1. This is a vital process; occurs inside living cells.	1. This is a chemical process; does not require living cells.
2. This is biochemical process.	2. This is physico-chemical process.
3. Energy is released step by step.	3. All the energy is released suddenly.
4. Less than 50 per cent energy is liberated as heat.	4. Most of energy is liberated as heat.
5. There is very small rise in temperature.	5. There is large increase in temperature.
6. Most of the energy is stored in ATP molecules.	6. ATP is not formed.
7. Respiratory reactions are regulated by enzymes.	7. This is a non-enzymatic process.
8. Several intermediates are formed.	8. In the process of combustion, no such intermediates are produced.
9. Oxidation takes place at the end of reaction between reduced coenzymes and oxygen.	9. Substrate directly oxidises.

In other words, oxidation of a compound, such as glucose, can provide the energy, through ATP, for synthesis of several cellular materials.

On the other hand, fuel burned in engines, where a large amount of released energy is lost in the form of heat, the oxidation of substances in the cell occurs with relatively little loss of energy. This is due to reason of the cell's very efficient energy transfer system mediated by ATP.

Here it is notable that the energy locked in a biological compound may be transferred repeatedly. Thus, in the living cell, the stored energy of glucose may be found at one time in ATP and at another time locked in the bonds of a protein molecule.

Respiratory Substrates

Respiratory substrates are those organic substances which are oxidised during respiration. They are high energy compounds and are called **respiratory substrates**. They may be carbohydrates, fats and proteins. Carbohydrates, such as glucose, fructose (hexoses), sucrose (disaccharide) or starch, inulin, hemicellulose (polysaccharide), etc., are main respiratory substrates.

Besides, fats are used as respiratory substrates by a variety of organisms as they contain more energy than carbohydrates.

In rare circumstances, when carbohydrate reserves are exhausted, fats and proteins also serve as respiratory substrates. Blackman termed the respiratory oxidation of protoplasmic protein as **protoplasmic respiration**, while oxidation of carbohydrates as **floating respiration**.

Types of Respiration

There are two main types of respiration, (*i*) **aerobic** and (*ii*) **anaerobic**.

(*i*) **Aerobic respiration.** This type of respiration leads to a complete oxidation of stored food (organic substances) in the presence of oxygen, and releases carbon dioxide, water and a large amount of energy present in respiratory substrate. Such type of respiration is generally found in higher organisms.

The overall equation is :

$$C_6H_{12}O_6 + 6O_2 \xrightarrow[\text{oxidation}]{\text{enzymes}} 6CO_2 + 6H_2O + \text{energy}$$

(2870 kJ) or (686 k cal)

(*ii*) **Anaerobic respiration.** This type of respiration occurs in complete absence of oxygen. In the absence of free oxygen, many tissues of higher plants, seeds in storage, fleshy fruits, and succulent plants, such as cacti temporarily take to a kind of respiration, called **anaerobic respiration**. Such respiration generally occurs in lower organisms like bacteria and fungi. This results in incomplete oxidation of stored food and formation of carbon dioxide and ethyl alcohol, and sometimes also various organic acids, such as malic, citric, oxalic, tartaric, etc. Very little energy is released by this process to maintain activity of protoplasm.

The equation is as follows :

$$C_6H_{12}O_6 \longrightarrow 2C_2H_5OH + 2CO_2 + \text{Energy}$$

(247 kJ) or (28 k cal)

This process of oxidation in microbes is known as **fermentation**. This is quite similar to that of anaerobic respiration in case of higher plants.

Table 4.2. Differences between Aerobic and Anaerobic Respiration

Aerobic Respiration	Anaerobic Respiration
1. Occurs in all living cells.	1. Occurs in microbes (*e.g.*, bacteria and fungi), and in germinating seeds, fleshy fruits and succulent plants.
2. This requires oxygen.	2. This does not require oxygen.
3. The end products are carbon dioxide and water.	3. The end products are carbon dioxide and alcohol.
4. Complete oxidation of one molecule of glucose produces a net of 38 ATP molecules.	4. Only two ATP molecules are produced.
5. Reactions of glycolysis occur in cytoplasm, whereas of Krebs cycle in mitochondria.	5. All reactions occur in cytoplasm. There is no need of mitochondria for this process.

Respiratory Quotient (RQ) or Respiratory Ratio

As we know, during aerobic respiration, oxygen is consumed and carbon dioxide is released. The ratio of the volume of carbon dioxide given out to the volume of oxygen absorbed in respiration, *i.e.*, $\frac{CO_2}{O_2}$ is called **respiratory quotient (RQ)** or **respiratory ratio.**

$$RQ = \frac{\text{volume of } CO_2 \text{ evolved}}{\text{volume of } O_2 \text{ consumed}}$$

The value of respiratory quotient depends upon the nature of respiratory substance used during respiration. This varies for different types of substrates.

Carbohydrates (RQ equal to unity). The value of respiratory quotient is unity (one) when hexose sugars are consumed in respiration. Here equal volumes of CO_2 and O_2 are evolved and consumed, respectively. This is obvious from overall equation of respiration.

$$C_6H_{12}O_6 + 6O_2 \longrightarrow 6CO_2 + 6H_2O$$

$$RQ = \frac{6CO_2}{6O_2} = 1.0$$

Fats (RQ less than unity). When substances poorer in oxygen than carbohydrates, *e.g.*, fats are oxidised in respiration, the value of respiratory quotient is less than one. It is explained in tripalmitin and triolein as follows :

$$\underset{\text{Tripalmitin}}{2C_{51}H_{98}O_6} + 145O_2 \longrightarrow 102CO_2 + 98H_2O$$

$$RQ = \frac{CO_2}{O_2} = \frac{102}{145} = 0.7 \text{ (less than unity)}$$

$$\underset{\text{Triolein}}{C_{57}H_{104}O_6} + 80O_2 \longrightarrow 57CO_2 + 52H_2O$$

$$RQ = \frac{CO_2}{O_2} = \frac{57}{80} = 0.7 \text{ (less than unity)}$$

RQ in above-mentioned examples is less than unity, *i.e.*, 0.7, because fats contain less oxygen than the carbohydrates, and therefore, they require relatively greater amount of oxygen for oxidation.

In case of proteins RQ is 0.5 (approximately).

Succulents (RQ is zero). Where incomplete oxidation of carbohydrates takes place RQ is zero. In such cases oxygen absorbed does not oxidise the carbohydrates completely to CO_2 and water but produces intermediate products of partial oxidation. This occurs especially in respiration of succulent plants, such as cacti (*Opuntia*). In such plants respiration of carbohydrates occurs at night when stomata are open which results in production of organic acids that utilise O_2 but do not evolve CO_2. The equation is as follows :

$$2C_6H_{12}O_6 + 3O_2 \longrightarrow \underset{\text{Malic acid}}{3C_4H_6O_5} + 3H_2O$$

$$RQ = \frac{CO_2}{O_2} = \frac{0}{3} = 0$$

Organic acids (RQ more than unity). When substances rich in oxygen are respired, the value of RQ is more than unity. This occurs in respiration of organic acids. Organic acids contain more oxygen than carbohydrates, and therefore, relatively less amount of oxygen is required for their oxidation. The examples are shown by following equations :

(*i*) $$\underset{\text{Oxalic acid}}{2(COOH)_2} + O_2 \longrightarrow 4CO_2 + 2H_2O$$

$$RQ = \frac{CO_2}{O_2} = \frac{4}{1} = 4$$

Here RQ is 4 in case of oxalic acid.

(*ii*) $$\underset{\text{Tartaric acid}}{2C_4H_6O_6} + 5O_2 \longrightarrow 8CO_2 + 6H_2O$$

$$RQ = \frac{CO_2}{O_2} = \frac{8}{5} = 1.6$$

This means, RQ is 1.6 in case of tartaric acid.

(*iii*) $$\underset{\text{Malic acid}}{C_4H_6O_5} + 3O_2 \longrightarrow 4CO_2 + 3H_2O$$

$$RQ = \frac{CO_2}{O_2} = \frac{4}{3} = 1.3$$

In case of malic acid, RQ is 1.3.

Anaerobic respiration (RQ infinity). As we know, in anaerobic respiration CO_2 is evolved but O_2 is not consumed, and therefore RQ is infinite (∞). Equation is as follows :

$$C_6H_{12}O_6 \xrightarrow[\text{enzyme}]{\text{zymase}} 2C_2H_5OH + 2CO_2 + \text{energy}$$

$$RQ = \frac{CO_2}{O_2} = \frac{2}{0} = \text{Infinity } (\infty)$$

Significance of RQ. The RQ value of a respiring tissue may provide valuable information to an investigation. From an RQ value, one can obtain a rough indication of the nature of substrate being oxidised.

It also helps in knowing the type of respiration being performed. It may also provide some information about major transformation of food materials.

If RQ is less than one, then substrate is of low oxygen content, *e.g.*, fats.

In starvation the cell material itself becomes substrate. According to Blackman, in starving conditions, if cellular proteins are the respiratory substrates then it is called **protoplasmic respiration**.

MECHANISM OF RESPIRATION

There are two major phases of respiration :

(*i*) **Glycolysis** and

(*ii*) **Krebs cycle**

During process of respiration, carbohydrates are converted into pyruvic acid through a series of enzymatic reactions. This series of reactions is known as **glycolysis** which takes place in cytosol.

Now, pyruvic acid enters mitochondria, where several enzymes catalyse the reactions, and pyruvic acid finally converts into CO_2 and water. This series of enzymatic reactions is known as **Krebs cycle** (after name of its discoverer Sir Hans Adolf Krebs (1900-1981), awarded Nobel Prize in 1953), or **tricarboxylic acid** (TCA) or **citric acid cycle**.

Glycolysis

Glycolysis is a term used to describe the sequential series of reactions present in a wide variety of tissues that starts with a hexose sugar (usually glucose) and ends with pyruvic acid. This term has originated from Greek words, *glycos* = sugar and *lysis* = splitting.

The scheme of glycolysis was discovered by three German Scientists, Gustav Embden, Otto Meyerhof and J. Parnas, and therefore, referred as **EMP pathway**, after the abbreviation of their last names.

Glycolysis is the first stage in the breakdown of glucose and is common to all organisms. This means, glycolysis is common to both aerobic and anaerobic modes of respiration. In anaerobic organisms, this is only process in respiration. Glycolysis occurs in cytoplasm of cells. During this process, glucose undergoes partial oxidation to form two molecules of pyruvic acid.

In plants, glucose is derived from sucrose, which is the end product of photosynthetic carbon reactions (also known as dark reactions) or from storage carbohydrates.

Sucrose is converted into glucose and fructose by the enzyme **invertase**. Now, these two monosaccharides (*i.e.*, glucose and fructose) enter glycolysis or EMP pathway. The main steps of glycolytic pathway are as follows:

Glycolysis is carried out in following different steps.

1. Phosphorylation of sugar (*i.e.*, first phosphorylation). Glucose and fructose are phosphorylated to give rise to glucose-6-phosphate and fructose-6-phosphate, respectively, by the activity of enzyme **hexokinase**, in presence of ATP. The phosphorylated form of glucose then isomerises to produce fructose-6-phosphate. Isomerisation takes place with the help of enzyme **phosphohexose isomerase**.

Further steps of metabolism of glucose and fructose are quite similar.

Equations are as follows :

$$\text{Glucose (6C)} + \text{ATP} \xrightarrow[\text{Mg}^{2+}]{\text{hexokinase}} \text{Glucose-6-phosphate} + \text{ADP}$$

Now isomerisation occurs :

$$\text{Glucose-6-phosphate} \xrightleftharpoons[\text{isomerase Mg}^{2+}]{\text{phosphohexose}} \text{Fructose-6-phosphate}$$

2. Phosphorylation of fructose-6-phosphate (i.e., second phosphorylation). Now, fructose-6-phosphate is phosphorylated and fructose-1, 6-bisphosphate produced by the action of enzyme **phosphofructokinase** in presence of ATP.

$$\text{Fructose-6-phosphate} + \text{ATP} \xrightarrow[\text{Mg}^{2+}]{\text{phosphofructokinase}} \text{Fructose-1, 6-bisphosphate} + \text{ADP}$$

3. Splitting. Now, fructose-1, 6-bisphosphate splits into two molecules of triose phosphate, *i.e.*, 3-phosphoglyceraldehyde (PGAL) and dihydroxyacetone phosphate (Di HAP), which are interconvertible.

$$\text{Fructose-1, 6-bisphosphate} \xrightleftharpoons{\text{aldolase}} \text{3-phosphoglyceraldehyde (PGAL)} + \text{dihydroxyacetone phosphate (Di HAP)}.$$

4. Oxidative dehydrogenation. After formation of 3-phosphoglyceraldehyde (PGAL), the glycolytic pathway enters the energy conserving phase. Here, it is oxidised to a carboxylic acid, *i.e.*, 1, 3-bisphosphoglycerate, and NAD is reduced to NADH.

$$\text{3-phosphoglyceraldehyde} + \text{NAD} \xrightleftharpoons[\text{dehydrogenase}]{\text{glyceraldehyde phosphate}} \text{1, 3-bisphosphoglycerate} + \text{NADH} + \text{H}_+$$

5. Formation of ATP. In next step of glycolysis, 3-phosphoglycerate is formed from 1, 3-bisphosphoglycerate by enzymatic activity of **phosphoglycerate kinase**, and ATP is generated during this process. Direct synthesis of ATP from intermediate metabolites is called **substrate level phosphorylation**.

$$\text{1, 3-bisphosphoglycerate} + \text{ADP} \xrightleftharpoons[\text{kinase} + \text{Mg}^{2+}]{\text{phosphoglycerate}} \text{3-phosphoglycerate} + \text{ATP}$$

This type of formation of ATP, where a phosphate group is directly transferred from a substrate to ADP to form ATP, is different from the ATP produced by ATP synthase during oxidative phosphorylation in mitochondria or in chloroplasts (during photophosphorylation in photosynthesis).

6. Isomerisation. In next step 3-phosphoglycerate converts into its isomer 2-phosphoglycerate by catalytic activity of enzyme phosphoglyceromutase.

$$\text{3-phosphoglycerate} \xrightleftharpoons[\text{glyceromutase} + \text{Mg}^{2+}]{\text{phospho-}} \text{2-phosphoglycerate}$$

7. Dehydration. In subsequent step 2-phosphoglycerate converts into phosphoenol pyruvate (PEP) in the presence of enzyme pyruvate kinase and liberates ATP.

$$\text{2-phosphoglycerate} \xrightleftharpoons{\text{pyruvate kinase}} \text{phosphoenol pyruvate} + \text{ATP}$$

$$\text{2-phosphoenol pyruvate} + 2\ \text{ADP} \rightleftharpoons \text{pyruvic acid} + 2\ \text{ATP}$$

8. Generation and utilisation of ATP during glycolysis. During glycolytic pathway, the molecules of ATP are produced as follows :

(*i*) Direct transfer of phosphate to ATP.

(*ii*) Oxidation of NADH produced during glycolytic pathway to NAD^+.

9. In the end of glycolysis net gain of ATP.

(*i*) During glycolysis two triose phosphate molecules are formed from one glucose molecule, and 4 ATP molecules are produced.

(*ii*) Out of 4 ATP molecules, 2 ATP molecules are utilised in first few steps in converting glucose to fructose-1, .6-bisphosphate.

(*iii*) Moreover, three ATP molecules are produced from oxidation of each of two molecules of NADH produced during catabolism of glucose.

(*iv*) In all, a net gain of 8 molecules occurs during process of glycolysis.

(*v*) However, in anaerobic respiration, NADH + H^+ is not converted to ATP, and therefore, only 2 ATP molecules are produced.

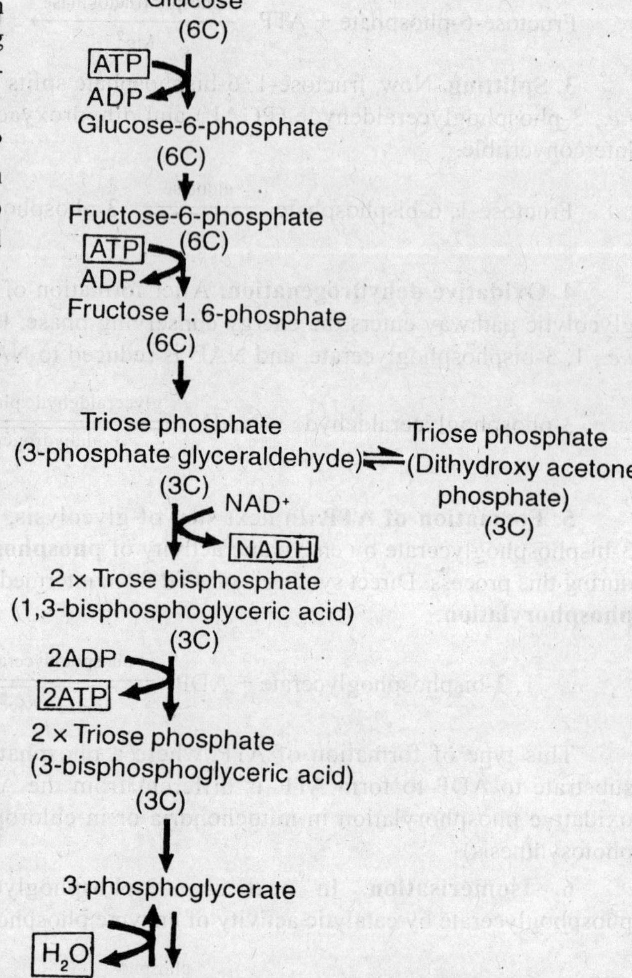

Fig. 4.2. Main steps of glycolysis or EMP pathway.

Oxidative Decarboxylation of Pyruvic Acid

(Aerobic Oxidation of Pyruvic Acid)

Now, pyruvic acid generated in cytoplasm through glycolysis is transferred to mitochondria. This is initiation of second phase of respiration. As soon as, pyruvic acid enters the mitochondria, one of the three carbon atoms of pyruvic acid is oxidised to carbon dioxide in a reaction called **oxidative decarboxylation**. Here, pyruvate is first decarboxylated, and thereafter oxidised by enzyme **pyruvate dehydrogenase**. This enzyme is made up of a decarboxylase, lipoic acid, TPP, transacetylase and Mg^{2+}.

Acetyl Co-A acts as substrate entrant for Krebs cycle. The equation is as follows :

$$\text{Pyruvate} + \text{NAD+} + \text{Co-A} \xrightarrow[\text{dehydrogenase}]{\text{pyruvate}} \text{Acetyl Co-A} + \text{NADH} + H^+ + CO_2^-$$

Acetyl Co-A can enter into mitochondria while pyruvic acid cannot.

Krebs Cycle

Sir Hans Adolf Krebs, discovered role of pyruvate in conversion of glucose hydrogens into fumarate. He discovered, in 1937, **tricarboxylic acid cycle** (*i.e.*, TCA cycle), also known as **Citric acid cycle** or **Krebs cycle**. Citric acid cycle occurs in matrix of mitochondria. This cycle involves two decarboxylations and four dehydrogenations. Various steps of these reactions are as follows :

The starting point of Krebs cycle is entrance of acetyl Co-A into a reaction to form citric acid. Krebs elucidated this cycle, and explained how pyruvate is broken down to CO_2 and H_2O. For this pioneer work Krebs was awarded Nobel Prize in 1953.

In the first reaction of Krebs cycle, one molecule of acetyl Co-A combines with 4-carbon oxaloacetic acid (OAA); with the result 6-carbon citric acid is produced, and Co-A is released. This reaction is catalysed by enzyme citrate synthase.

$$\underset{(4C)}{\text{Acetyl Co-A + oxaloacetic acid}} \xrightarrow[\text{enzyme}]{\text{citrate synthase}} \underset{(6C)}{\text{Citric acid + Co-A}}$$

Now, citrate (citric acid) is isomerised to isocitrate (isocitric acid).

$$\text{Citric acid} \xrightarrow[\text{enzyme}]{\text{aconitase}} \text{Cis-aconitic acid} + H_2O$$

$$\text{Cis-aconitic acid} + H_2O \xrightarrow[\text{enzyme}]{\text{aconitase}} \underset{\text{(Isocitrate)}}{\text{Isocitric acid}}$$

Cis-aconitic acid is converted into isocitric acid with the addition of water in the presence of iron containing enzyme aconitase.

$$\underset{\text{(Isocitrate)}}{\text{Isocitric acid}} + NAD^+ \xrightarrow[\text{dehydrogenase}]{\text{isocitrate}} \underset{\text{(oxalosuccinate)}}{\text{Oxalosuccinic acid}} + NADH + H^+$$

$$\underset{\text{(oxalosuccinate)}}{\text{Oxalosuccinic acid}} \xrightarrow[\text{decarboxylase}]{\text{oxalosuccinic}} \alpha\text{-Ketoglutaric acid} + CO_2$$

$$\alpha\text{-ketoglutaric acid + Co-A} + NAD^+ \xrightarrow[\text{dehydrogenase}]{\alpha\text{- ketoglutaric}} \text{Succinyl Co-A} + NADH + H^+ + CO_2$$

$$\text{Succinyl Co-A} + H_2O + \text{GDP + iP} \xrightarrow[\text{thiokinase}]{\text{succinic}} \text{Succinic acid + Co-A + GTP}$$

$$\text{GTP + ADP} \longrightarrow \text{GDP + ATP (substrate phosphorylation)}$$

$$\text{Succinic acid + FAD} \xrightarrow[\text{dehydrogenase}]{\text{Succinic}} \text{Fumaric acid} + FADH_2$$

$$\text{Fumaric acid} + H_2O \xrightarrow[\text{enzyme}]{\text{Fumarase}} \text{Malic acid}$$

$$\text{Malic acid} + NAD^+ \xrightarrow[\text{dehydrogenase}]{\text{Malic}} \text{Oxaloacetic acid (OAA)} + NADH + H^+$$

During citric acid cycle (Krebs cycle) 3 molecules of NAD^+ and one molecule of FAD (Flavin Adenine Dinucleotide) are reduced to produce NADH and $FADH_2$, respectively.

During citric acid cycle NADH and $FADH_2$ are produced. Now, they are linked with electron transport system (ETS) and produce ATP by oxidative phosphorylation. This may be summarised in following equation :

$$\text{Pyruvic acid} + 4NAD^+ + FAD + 2H_2O + ADP + iP$$
$$\downarrow \text{ mitochondrial matrix}$$
$$3CO_2 + 4NADH + 4H^+ + FADH_2 + ATP$$

In the end of Krebs cycle, glucose molecule is completely oxidised. From one glucose molecule, two pyruvic acid molecules are formed. After oxidation of one pyruvic acid molecule, three CO_2 molecules are released. Thus, in all 6 molecules of CO_2 are released.

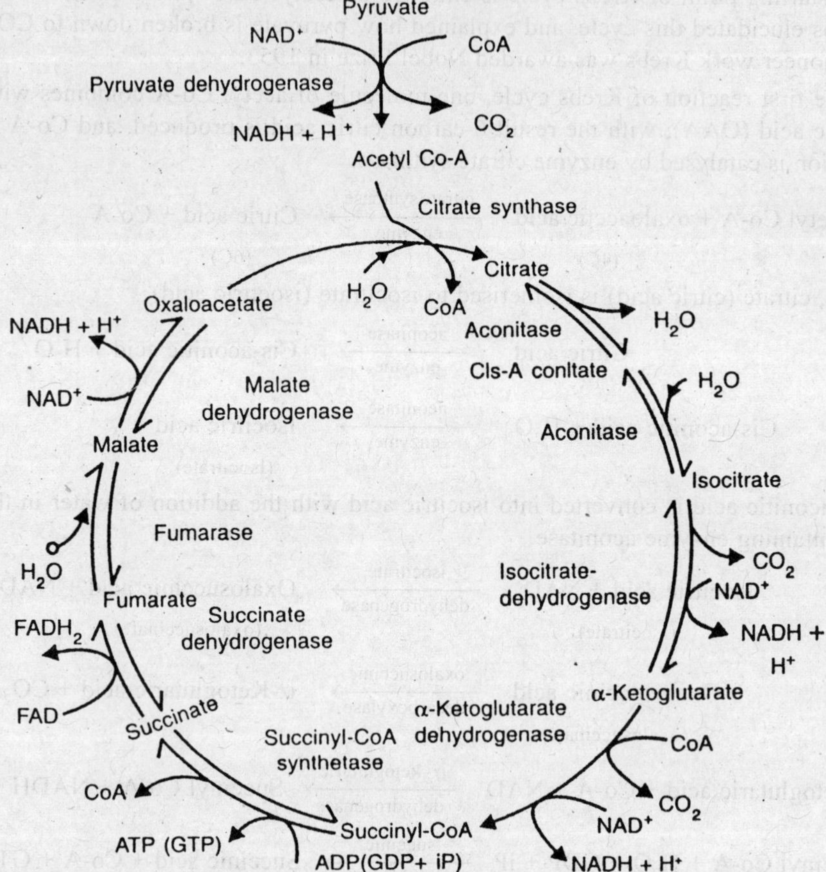

Fig. 4.3. Krebs cycle or citric acid cycle.

Electron Transport System (ETS)

By the end of Krebs cycle, glucose molecule oxidises completely, but the energy does not release till NADH and $FADH_2$ oxidise through electron transport system (ETS). The metabolic pathway through which electron passes from one carrier to another, is called **electron transport system** (ETS). The electron transport system is also known as **electron transport chain** or **mitochondrial respiratory chain**. The electron transport system consists of a series of coenzymes and cytochromes that take part in passage of electrons from a chemical to its ultimate acceptor. The passage of electrons from one-enzyme or cytochrome to the next takes place with a loss of energy at each step. Electron transport system is operative in the inner mitochondrial membrane.

The electron carriers include flavins, iron sulphur complexes, quinones and cytochromes. Most of them are prosthetic groups of proteins.

Electron transport system in mitochondria consists of **four complexes** which are found in bases of stalked particles in the **inner mitochondrial membrane**, and also **ubiquinone** (UQ) or coenzyme Q and **cytochrome** *c* which are not bound to stalked particles but act as mobile electron carriers between the complexes.

Complex-I consists of NADH-dehydrogenase or NADH-Q reductase which contains a flavoprotein FMN (flavin mononucleotide) and is associated with iron-sulphur (Fe-S) proteins. This complex is responsible for passing electrons (also protons) from mitochondrial NADH to ubiquinone (UQ), located within inner mitochondrial membrane.

$$NADH + H^+ + FMN \longrightarrow FMNH_2 + NAD^+$$

$$FMNH_2 + 2Fe^{3+} S \longrightarrow FMN + 2Fe^{2+} S + 2H^+$$

$$2Fe^{2+} S + Q + 2H^+ \longrightarrow 2Fe^{3+} S + QH_2$$

Complex-II consists of succinate dehydrogenase which contains a flavoprotein FAD (flavin adenine dinucleotide) in its prosthetic group and is associated with non heme iron-sulphur (Fe S) proteins. This complex receives electrons (also protons) from succinic acid (which is oxidised in Krebs cycle to form fumaric acid) and passes them to **ubiquinone** (UQ). Ubiquinone also receives reducing equivalents via $FADH_2$ that is generated during oxidation of succinate, through the activity of energy succinate dehydrogenase, in Krebs cycle.

$$FADH_2 + 2Fe^{3+} S \longrightarrow 2Fe^2 + S + 2H^+ + FAD$$

$$2Fe^{2+} S + Q + 2H^+ \longrightarrow 2Fe^{3+} S + QH_2$$

Complex-III consists of ubiquinol, cytochrome c and cytochrome bc_1. The reduced ubiquinone is called ubiquinol. **Here ubiquinol** is oxidised with the **transfer** of electrons to cytochrome *c* via cytochrome bc_1. Cytochrome *c* is a small protein attached to outer surface of the inner mitochondrial membrane and acts as a mobile carrier for transfer of electrons between complex III and complex IV.

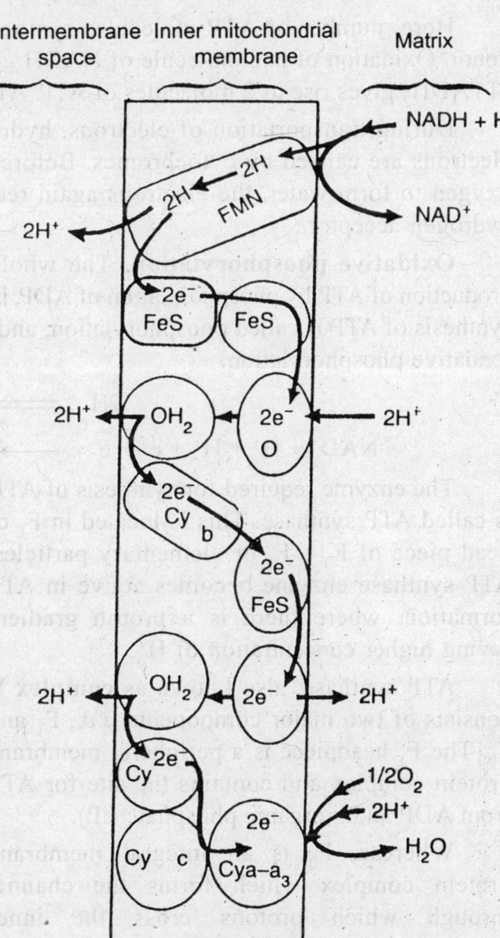

Fig. 4.4. Electron transport system (ETS)

This complex is called QH_2-cytochrome *c* reductase complex. This bears three components, *i.e.*, cytochrome *b*, non-heme iron sulphur (Fe - S), and cytochrome c_1. Coenzyme Q is also involved between Fe-S and cytochrome c_1. The equations are as follows :

$$QH_2 + 2Fe^{3+} \text{ cy } b \longrightarrow Q + 2H^+ + 2Fe^{2+} \text{ cy } b$$

$$2Fe^{2+} \text{ cy } b + 2Fe^{3+} S \longrightarrow 2Fe^{3+} \text{ cy } b + 2Fe^{2+} S$$

$$2Fe^{2+} S + Q + 2H^+ \longrightarrow 2Fe^{3+} S + QH_2$$

$$QH_2 + 2Fe^{3+} \text{ cy } c_1 \longrightarrow Q + 2H^+ + 2Fe^{2+} \text{ cy } c_1$$

Now, cytochrome c_1 transfers electrons to cy *c*. Like coenzyme Q, cy *c* is also mobile carrier of electrons.

Complex-IV is known as cytochrome *c* oxidase complex. This contains cytochromes *a* and a_3, along with two copper centres. This complex receives electrons from cytochrome *c* and passes them to 1/2 O_2. Two protons are needed and H_2O molecule is formed (terminal oxidation). Here, O_2 is ultimate acceptor of electrons. It combines with protons to form metabolic water or respiratory water.

$$\underset{(Fe^{++})}{\text{2 Ferrocytochrome}} + 2H^+ + 1/2\ O_2 \rightleftharpoons \underset{(Fe^{+++})}{\text{2 ferricytochrome } c} + H_2O$$

Complex-V. When electrons are transferred from one carrier to next carrier via complexes I to IV in electron transport system (ETS), they are coupled to ATP synthase enzyme complex for production of ATP from ADP and inorganic phosphate (iP).

Here, number of ATP molecules synthesised during ETS, depends on nature of electron donor. Oxidation of one molecule of NADH gives rise to 3 molecules of ATP, and one molecule of $FADH_2$ gives rise to 2 molecules of ATP. ATP synthase complex is called **complex V**.

During transportation of electrons, hydrogen atoms split into protons and electrons. The electrons are carried by cytochromes. Before last stage, where hydrogen atom is accepted by oxygen to form water, the electrons again recombine with their protons. Oxygen acts as final hydrogen acceptor.

Oxidative phosphorylation. The whole process, where oxygen effectively allows the production of ATP by phosphorylation of ADP, is called **oxidative phosphorylation**. In other words, synthesis of ATP is called phosphorylation, and as it takes place in presence of oxygen, it is called oxidative phosphorylation.

$$2H \rightleftharpoons 2H^+ + 2e^-$$

$$NAD^+ + H^+ + H^+ + e^- + e^- \longrightarrow NADH + H^+$$

The enzyme required for synthesis of ATP, is called ATP synthase. This is located in F_1 or head piece of $F_0 - F_1$ or elementary particles. ATP synthase enzyme becomes active in ATP formation, where there is a proton gradient saving higher concentration of H_2.

ATP synthase, also known as **complex V** consists of two major components, *i.e.*, F_1 and F_0. The F_1 headpiece is a peripheral membrane protein complex and contains the site for ATP from ADP and inorganic phosphate (iP).

Whereas, F_0 is an integral membrane protein complex which forms the channel through which protons cross the inner membrane. The passage of protons through the channel is coupled to the catalytic site of the F_1 component for the production of ATP.

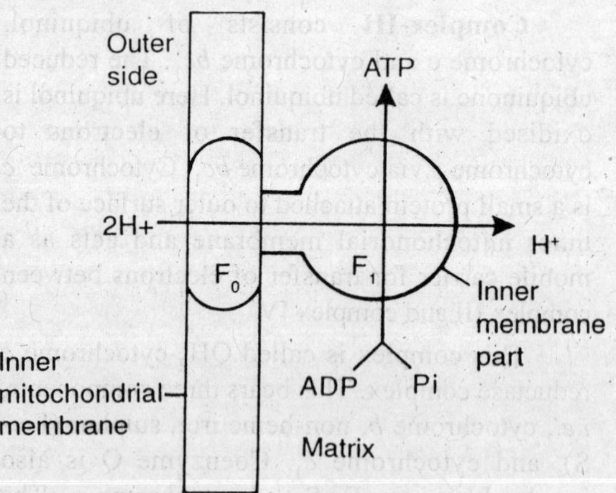

Fig. 4.5. Synthesis of ATP by inner membrane particles of mitochondrion.

Oxidation of one molecule of $NADH_2$ produces 3 ATP molecules whereas a similar oxidation of $FADH_2$ produces 2 ATP molecules.

Net gain of ATP. Complete oxidation of glucose to CO_2 and water shows that there is a net gain of 38 ATP. Each NADH + H^+ produces 3 ATP molecules, while $FADH_2$ forms only 2 ATP molecules at the end of reaction.

$$NADH + H^+ + 1/2\ O_2 + 3\ ADP + 3\ iP \longrightarrow NAD + 3\ ATP + H_2O$$

$$FADH_2 + 1/2\ O_2 + 2\ ADP + 2\ iP \longrightarrow FAD + 2\ ATP + H_2O$$

Thus, total gain of ATP in aerobic respiration is as follows :

Glycolysis ⟶ 8 ATP

Pyruvic acid ⟶ Acetyl Co-A ⟶ 6 ATP

Krebs cycle ⟶ 24 ATP

Total ⟶ 38 ATP

$$C_6H_{12}O_6 + 6O_2 + 8H_2O + 38\ (ADP + H_3PO_4) \longrightarrow 6CO_2 + 14H_2O + 38\ (ATP + H_2O).$$

However, in most eukaryotic cells, 2 molecules of ATP are required for transport of NADH produced in glycolysis into mitochondrion for further oxidation, and therefore, net gain of ATP is 36 molecules.

Significance of Krebs Cycle

1. During Krebs cycle, carbon skeletons are obtained for use in growth and maintenance of the cell.
2. Many intermediate compounds are formed which are used in synthesis of other biomolecules, such as amino acids, nucleotides, chlorophyll, cytochromes and fats.
3. During this pathway amino acids are synthesised from α-ketoglutaric acid, pyruvic acid and oxaloacetic acid.
4. Here succinyl Co-A acts as starting molecule for synthesis of chlorophyll.
5. Krebs cycle is major pathway for generation of ATP molecules, which make **energy currency** of the cell.
6. Energy is released from glucose, and is used in various biochemical reactions.
7. Phenol, anthocyanin, etc., are produced from acetyl Co-A, whereas fatty acids are formed from glycerol.
8. Glutamic acid is formed from α-ketoglutaric acid; aspartic acid from oxaloacetic acid, and alanine from aspartic acid.
9. Amino acids are used in synthesis of proteins, nucleic acids, purines and pyrimidines.
10. Succinyl Co-A carries synthesis of pyrrole compounds of chlorophyll, cytochrome and phytochrome.
11. Krebs cycle is directly related to nitrogen metabolism. α-ketoglutaric acid, an intermediate of Krebs cycle is first acceptor molecule of NH_3 forming an amino acid, the glutamic acid. From glutamic acid various transamination reactions begin to form different amino acids which ultimately condense to form proteins.
12. Krebs cycle is also intimately related with fat metabolism. Dihydroxyacetone phosphate produced in glycolysis may be converted into glycerol via glycerol-3-phosphate and *vice versa*. After β-oxidation, fatty acids give rise to active 2-C units, the acetyl Co-A which enter the Krebs cycle.

PENTOSE PHOSPHATE PATHWAY (PPP)

(Hexose Monophosphate Shunt)

As we know, major pathway for aerobic respiration of glucose is through **glycolysis** and **Krebs cycle**, however, an alternate pathway exists in many organisms. This pathway, which requires the presence of oxygen, is called **pentose phosphate pathway (PPP)** or **hexose monophosphate shunt (HMS).** As shown in figure that reduced NADP is formed in the reactions forming 6-phosphogluconic acid and ribulose-5-P. If the equivalent of a molecule of glucose is oxidised to CO_2 and H_2O via this cyclic pathway (six turns of cycle), then 12 molecules of reduced NADP would be formed. In the presence of enzyme transhydrogenase the hydrogens of NADPH can be transferred to NAD to form NADH. Hence the formation of 12 molecules of reduced NADP via hexose monophosphate shunt ultimately can lead to synthesis of 36 molecules of ATP. Thus, capture of energy released in oxidation of glucose via this pathway (hexose monophosphate shunt) is as effective as that of glycolytic-Krebs cycle pathway.

This pathway (PPP) is also known as **Warburg-Limpam-Dickens cycle** and **phosphogluconate shunt**. This was first studied by Warburg (1935) and Dickens (1938). This pathway occurs within the cytosol, where all enzymes of pentose phosphate pathway (PPP) are present.

Starting from 6 molecules of glucose 6-phosphate, the various reactions of PPP are as follows :

1. 6 molecules of glucose-6-phosphate in the presence of coenzyme NADP are oxidised into 6 molecules of 6-phosphogluconolactone by the enzyme **glucose-6-phosphate dehydrogenase**. 6 molecules of NADP are reduced in the reaction which is reversible.

$$\text{Glucose-6-phosphate + NADP} \underset{\text{dehydrogenase}}{\overset{\text{glucose-6-phosphate}}{\rightleftharpoons}} \text{6-Phosphogluconolactone + NADPH}_2$$

2. 6-phosphogluconolactone is hydrolysed by the enzyme **lactonase** to produce 6 molecules of 6-phosphogluconic acid.

$$\text{6-phosphogluconolactone} + H_2O \xrightarrow[\text{Mg}^{2+}]{\text{lactonase}} \text{6-phosphogluconic acid}$$

3. Phosphogluconic acid is oxidatively decarboxylated by the enzyme **6-phosphogluconic acid dehydrogenase.** 6 molecules of NADP are reduced, 6 molecules of CO_2 are released, and 6 molecules of ribulose-5-phosphate are produced.

$$\text{6-phosphogluconic acid + NADP} \xrightarrow[\text{dehydrogenase}]{\text{phosphogluconic acid}} \text{Ribulose-5-phosphate + NADPH}_2 + CO_2$$

4. 6 molecules of ribulose-5-phosphate isomerise into 4 molecules of xylulose-5-phosphate and 2 molecules of ribose-5-phosphate in the presence of enzymes **ribulose phosphate-3-epimerase** and **pentose phosphate isomerase**, respectively.

$$\text{Ribulose-5-phosphate} \overset{\text{epimerase}}{\rightleftharpoons} \text{xylulose-5-phosphate}$$

$$\text{Ribulose-5-phosphate} \overset{\text{isomerase}}{\rightleftharpoons} \text{ribose-5-phosphate}$$

That is,

$$\text{Ribulose-5-phosphate} \underset{\text{isomerase}}{\overset{\text{epimerase and}}{\rightleftharpoons}} \text{xylulose-5-phosphate + ribose-5-phosphate}$$

5. 2 molecules of xylulose-5-phosphate and 2 molecules of ribose-5-phosphate combine in the presence of the enzyme **transketolase** to form 2 molecules of sedoheptulose-7-phosphate and 2 molecules of phosphoglyceraldehyde.

Ribose-5-phosphate + xylulose-5-phosphate

⇅ transketolase, TPP Mg^{2+}

sedoheptulose-7-phosphate + glyceraldehyde-3-phosphate

6. 2 molecules of sedoheptulose-7-phosphate and 2 molecules of 3-phosphoglyceraldehyde combine in the presence of enzyme **transaldolase** to form 2 molecules of fructose-6-phosphate and 2 molecules of erythrose-4-phosphate.

Sedoheptulose-7-phosphate + glyceraldehyde-3-phosphate

⇅ transaldolase

Fructose-6-phosphate + erythrose-4-phosphate

7. 2 molecules of erythrose-4-phosphate react with remaining 2 molecules of xylulose-5-phosphate (reactions 4 and 5) in the presence of **transketolase** to form 2 molecules of fructose-6-phosphate and 2 molecules of 3-phosphoglyceraldehyde.

i.e., xylulose-5-phosphate + erythrose-4-phosphate

⇅ transketolase

Fructose-6-phosphate + glyceraldehyde-3-phosphate

8. One molecule of phosphoglyceraldehyde isomerises into dihydroxyacetone phosphate, in the presence of enzyme **phosphotriose isomerase.**

$$\text{Glyceraldehyde-3-phosphate} \xrightleftharpoons[\text{isomerase}]{\text{phosphotriose}} \text{dihydroxyacetone-3-phosphate}$$

9. Remaining one molecule of 3-phosphoglyceraldehyde unites with dihydroxyacetone phosphate in presence of enzyme **aldolase** to form one molecule of fructose 1, 6-diphosphate, which in the presence of **phosphatase** forms one molecule of fructose-6-phosphate.

3-phosphoglyceraldehyde + dihydroxyacetone phosphate

⇅ aldolase

Fructose-1, 6-diphosphate

10. 5 molecules of fructose-6-phosphate produced in reactions 6, 7 and 9, isomerise into 5 molecules of glucose-6-phosphate in presence of enzyme **phosphohexose isomerase**.

$$\text{Fructose-6-phosphate} \xrightleftharpoons[\text{isomerase}]{\text{phosphohexose}} \text{glucose-6-phosphate}$$

To summarise, 6 molecules of glucose-6-phosphate which enter into this pathway, produce 6 molecules of CO_2, after oxidation, and 12 molecules of reduced coenzyme $NADPH_2$, while 5 molecules of glucose-6-phosphate are regenerated.

$$\text{6 glucose-6-phosphate} + 12\ NADP^+ \longrightarrow \text{5-glucose-6-phosphate} + 12\ NADPH_2 + 6\ CO_2$$

Complete oxidation of a molecule of glucose produces 12 molecules of $NADPH_2$, which is equal to 36 ATP molecules. This capture of energy released in oxidation of glucose via this pathway (PPP) is as effective as that of glycolytic-Krebs cycle pathway, where 38 ATP molecules per glucose molecule are produced.

Significance of Pentose Phosphate Pathway (PPP)

1. This pathway provides alternative route for carbohydrate degradation.
2. This pathway (PPP) generates NADPH molecules which are used as reductants in biosynthetic processes under conditions when NADPH molecules are not generated by photosynthesis. It is, therefore, important in non-photosynthetic tissues, such as differentiating tissues, germinating seeds and during period of darkness. Production of NADPH is not linked to ATP generation in pentose phosphate pathway (PPP).

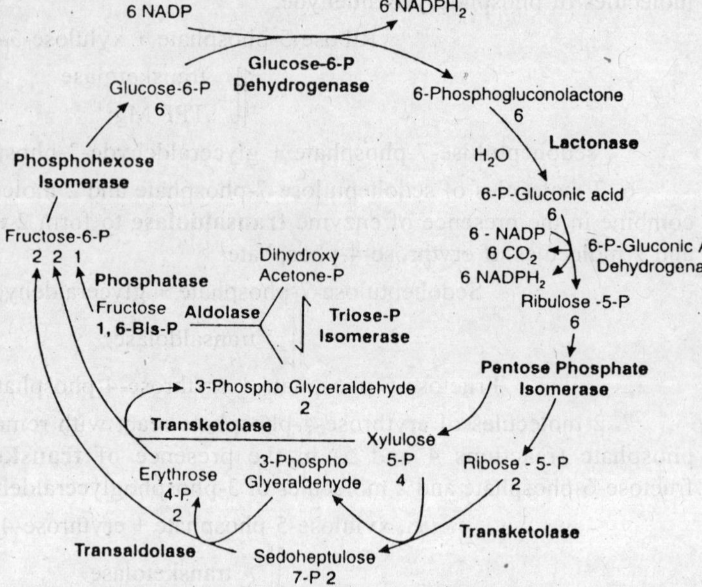

Fig. 4.6. Pentose phosphate pathway (PPP).

3. It produces ribose sugars for synthesis of nucleic acids.
4. It plays an important role in fixation of CO_2 in photosynthesis through ribulose-5-phosphate; ribulose 1, 5-phosphate derived from ribulose-5-phosphate is the primary acceptor of CO_2 in photosynthesis.
5. It provides erythrose-4-phosphate, which is required for synthesis of shikimic acid. The latter is precursor of aromatic ring compounds.
6. It produces a number of tetroses and pentoses for synthesis of nucleosides, nucleotides, nucleic acids, anthocyanins and other compounds.

ANAEROBIC RESPIRATION

When oxygen is not available for oxidation of carbohydrates then CO_2 and ethanol (C_2H_5OH) or lactic acid ($CH_3.CHOH.COOH$) are formed. This is called **anaerobic respiration**. This results in incomplete oxidation of substrate, hence energy released is much less in comparison of aerobic respiration. It generally occurs in lower organisms, such as bacteria and fungi. It also occurs in higher plants and animals under specific conditions, particularly when oxygen is limiting. Anaerobic respiration can be represented by following equation.

$$C_6H_{12}O_6 \longrightarrow 2\ C_2H_5OH + 2\ CO_2 + \text{Energy } 247 \text{ kJ or } 28 \text{ k cal}$$

Fermentation

In microorganisms, such as bacteria and yeasts, the term fermentation is more commonly used where anaerobic respiration takes place, after the name of product like alcoholic fermentation, lactic acid fermentation, etc. The enzyme complex, present in the process was named as **zymase**. However, fermentation is defined as anaerobic breakdown of carbohydrates and other organic compounds into alcohol, organic acids, gases, etc., with the help of microorganisms or their enzymes.

Table 4.3. Differences Between Aerobic and Anaerobic Respiration

Aerobic Respiration	Anaerobic Respiration
1. It requires oxygen for breaking respiratory substrate.	1. Oxygen is not required for breakdown of respiratory substrate.
2. Exchange of gases takes place between organism and environment.	2. Exchange of gases is absent.
3. Respiratory material is completely oxidised.	3. Incomplete breakdown of respiratory material.
4. The end products are CO_2 and H_2O	4. The end products are CO_2 and alcohol.
5. It involves electron transport.	5. Electron transport is absent.
6. This consists of three steps, *i.e.*, glycolysis, Krebs cycle and terminal oxidation.	6. This consists of two steps, *i.e.*, glycolysis and incomplete breakdown of pyruvic acid.
7. Reactions of glycolysis occur in cytosol and those of Krebs cycle in mitochondria.	7. All reactions occur in cytosol. Mitochondria are not required.
8. Complete oxidation of one molecule of glucose produces 38 ATP molecules.	8. Only 2 ATP molecules are produced.
9. It occurs in all living cells.	9. It occurs in certain bacteria, fungi, germinating seeds, fleshy fruits, etc.
10. Water is formed.	10. Generally, water is not formed.

The mechanism of anaerobic respiration or fermentation is similar to common pathway of aerobic respiration upto glycolysis. However, oxidation of pyruvic acid and NADH by mitochondria requires oxygen. When oxygen is limiting, NADH and pyruvic acid begin to accumulate. Under such condition, plants carry out **fermentation** (anaerobic respiration) which leads to the formation of CO_2 and either ethanol (ethyl alcohol) or lactic acid. In most of cases ethanol is produced. Two types of fermentation are common :

1. Alcoholic fermentation. In this type, the pyruvic acid is first decarboxylated to acetaldehyde in the presence of enzyme **carboxylase**. Thiamine pyrophosphate (TPP) is required as co-factor.

$$\underset{\text{Pyruvic acid}}{2\,CH_3COCOOH} \xrightarrow[\text{TPP}]{\text{carboxylase}} \underset{\text{Acetaldehyde}}{2\,CH_3CHO} + 2\,CO_2$$

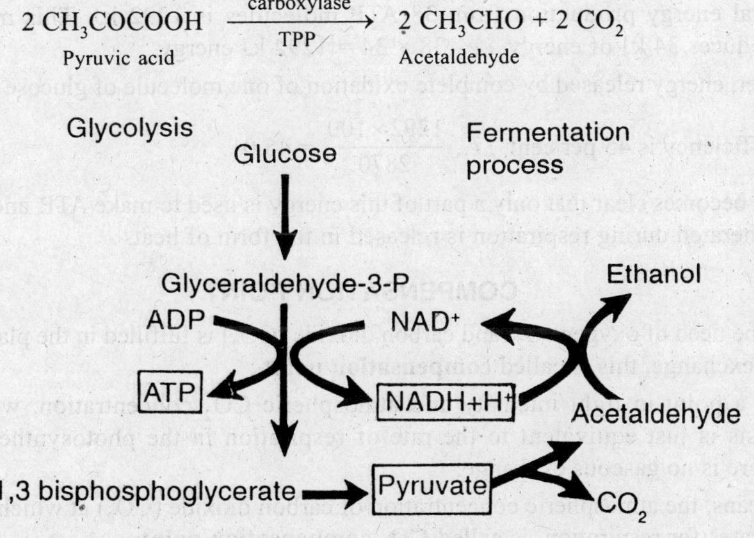

Fig. 4.7. Pathway of fermentation (anaerobic respiration) in yeast.

Acetaldehyde is then reduced to ethanol (ethyl alcohol) by the enzyme **alcohol dehydrogenase**. Coenzyme $NADH_2$ (produced in glycolysis) is oxidised.

$$\underset{\text{Acetaldehyde}}{2\,CH_3CHO} + \underset{\text{Reduced}}{NADH_2} \xrightarrow[\text{dehydrogenase}]{\text{alcoholic}} \underset{\text{Ethanol}}{CH_3CH_2OH} + \underset{\text{Oxidised}}{NAD}$$

The overall equation for anaerobic respiration involving alcoholic fermentation is as follows :

$$C_6H_{12}O_6 \longrightarrow 2\,CH_3CH_2OH + 2\,CO_2$$

2. Lactic acid fermentation. This type of fermentation occurs in lactic acid bacteria, certain fungi and muscles. Here, pyruvic acid is converted into lactic acid by the enzyme **lactic dehydrogenase**. Coenzyme $NADH_2$, produced in glycolysis, is oxidised.

$$\underset{\text{pyruvic acid}}{CH_3CO.COOH} + NADH_2 \xrightarrow[\text{dehydrogenase}]{\text{lactic}} \underset{\text{lactic acid}}{CH_3CHOH.COOH} + NAD$$

The overall equation for anaerobic respiration (lactic acid fermentation) is as follows :

$$\underset{\text{glucose}}{C_6H_{12}O_6} \longrightarrow \underset{\text{lactic acid}}{2\,CH_3CHOH.COOH}$$

There is net gain of only 2 ATP molecules (in glycolysis stage) during anaerobic respiration (or fermentation) and most of energy contained in glucose molecule is released as heat.

Importance. Many of the fermentive processes are of great importance in the household and industry. An example of the household is souring of milk or formation of curd. In the latter category are the alcoholic and acid fermentations. The **ripening** of cheese, the **retting** of flax, sannhemp, ramie, and jute, the **curing** of tobacco and tea leaves, and the **tanning** of leather are some very important examples of commercial processes dependent upon fermentations brought about by different species of bacteria. Alcoholic fermentation is used in brewing industry, where various types of beers, whisky and wines are produced, whereas CO_2 of alcoholic fermentation is used in baking industry for making bread and biscuits. Vinegar is produced by fermentation activity of acetic acid bacteria. Cleaning of raw hides is done by fermentive activity of bacteria.

Efficiency of the System

The total energy production from 38 ATP molecules is 1292 kJ. This means, one ATP molecule produces 34 kJ of energy, *i.e.*, $38 \times 34 = 1292$ kJ energy.

However, energy released by complete oxidation of one molecule of glucose is 2870 kJ. This shows, the efficiency is 45 per cent, *i.e.*, $\frac{1292 \times 100}{2870} = 45.01$.

Thus, it becomes clear that only a part of this energy is used to make ATP, and more than half of energy generated during respiration is released in the form of heat.

COMPENSATION POINT

When the need of oxygen (O_2) and carbon dioxide (CO_2) is fulfilled in the plant itself without any gaseous exchange, this is called **compensation point**.

This is a point in light intensity and atmospheric CO_2 concentration, when the rate of photosynthesis is just equivalent to the rate of respiration in the photosynthetic organs, and therefore, there is no gaseous exchange.

This means, the atmospheric concentration of carbon dioxide (CO_2) at which photosynthesis just compensates for respiration, is called **CO_2 compensation point**.

This shows, at given low concentration of CO_2 and non-limiting light intensity, the photosynthetic rate of a given plant will be equal to the total amount of respiration, *i.e.*, true respiration plus photorespiration.

The CO_2 compensation point is achieved when the amount of CO_2 uptake is equal to that generated through respiration at a non-limiting light intensity, where net photosynthesis is zero.

The value of CO_2 compensation point in C_3 plants is usually much higher (50-100 ppm) than in C_4 plants (5-10 ppm).

EXPERIMENTAL WORK : PRACTICALS

Respiration

Experiment 1

Object : Measurement of R.Q. by Ganong's Respirometer.

Requirements : Ganong's Respirometer, respiratory substrate, etc.

Experiment : The respiratory quotient may easily be measured by Ganong's respirometer. In Ganong's respirometer the amount of oxygen absorbed and carbon dioxide released out during respiration are determined simultaneously. The respirometer consists of a bulb with a graduated side tube. The volume of the apparatus is about 102 c.c. The stopper and neck of the bulb are each provided with hole. The stopper may be turned to bring the two holes opposite each other so that the air inside the bulb may communicate with the outer air and is at the atmospheric pressure. The side graduated tube is connected with a levelling tube by a rubber tubing.

For the determination of respiratory quotient the manometer is filled with mercury or a saturated solution of common salt. The water cannot be used since carbon dioxide dissolves in it. Carbon dioxide is slightly soluble in salt solution and therefore, best results can be obtained when mercury is used.

Two c.c. of plant material, the R.Q. of which is to be measured is being placed inside the bulb. In the beginning the air of the graduated tube is kept in direct communication with the atmospheric air at the atmospheric pressure. The mercury in both the tubes is brought to the same level by raising or lowering the levelling tube. Now the stopper in the neck of the bulb is turned so that the two holes get apart and the air in the apparatus is cut off from the outside air and respiration proceeds in closed chamber.

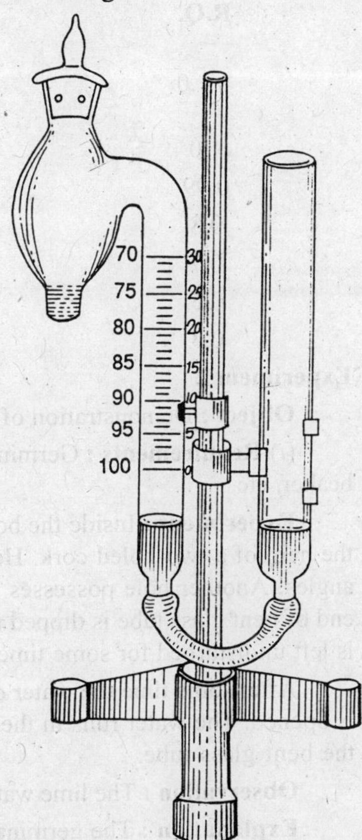

Fig. 4.8. Ganong's Respirometer.

Observation and explanation : The initial level of mercury in the graduated tube is noted. The level is again noted after about an hour. If the respiratory substrate kept in the bulb is carbohydrate, the amount of carbon dioxide released during respiration will be equal to the amount of oxygen absorbed and therefore, there will be neither a rise and nor a fall in the mercury level. If caustic potash (KOH) is now added to the apparatus, the accumulated carbon dioxide, will be absorbed, and hence a rise in the mercury level, and thus the amount of carbon dioxide can be measured. The volume of oxygen absorbed will be equal to the volume of carbon dioxide measured. Hence the R.Q. is unity.

When the respiratory substrate is a fat, the respiratory quotient is less than one. Here the amount of carbon dioxide released is less than the amount of oxygen absorbed and therefore, a vacuum is created in the closed chamber. This results in the rise of mercury level. This rise of mercury level is denoted by V_1 c.c. that is equivalent to the excess oxygen. Now the caustic potash (KOH) is added and there is a further rise of mercury level. The second rise in the mercury level is denoted by V_2 c.c. This represents the volume of carbon dioxide liberated. This way, the total amount of oxygen absorbed is equal to $V_1 + V_2$ c.c. the respiratory quotient will therefore, be less than one.

$$R.Q. = \frac{CO_2}{O_2} = \frac{V_2}{V_1 + V_2}$$

When the respiratory substrate is such that the carbon dioxide released is more than the volume of oxygen absorbed during respiration, the mercury level falls. The fall in the mercury level is denoted by V_1 c.c. that is equivalent to the excess carbon dioxide. Now the caustic potash is added to the tube. It absorbs the total amount of carbon dioxide and therefore, the mercury level rises. The rise is denoted by V_2 c.c. that is equivalent to the total volume of oxygen absorbed is then $V_2 - V_1$ c.c. The respiratory quotient will, therefore, be more than one.

$$R.Q. = \frac{CO_2}{O_2} = \frac{V_2}{V_1 - V_2}$$

R.Q. values for different respiratory substrates

R.Q.	**Respiratory substrate**
> 1.0	Carbohydrate with some anaerobic respiration. Carbohydrate synthesized from Organic acids.
1.0	Carbohydrates
0.99	Proteins with NH_3 formation
0.8	Proteins with amide formation
0.7	Fats, *e.g.*, tripalmitin
0.5	Fats with associated carbohydrate synthesis
0.3	Carbohydrates with associated organic acid synthesis

Experiment 2

Object : Demonstration of release of CO_2 during respiration.

(*i*) **Requirements :** Germinating seeds, a bottle, a bent tube, two holed cork, water neservoir, beaker, etc.

Experiment : Inside the bottle some germinating seeds are placed. The bottle is closed with the help of a two-holed cork. Here one of the holes is provided by glass tube bent twice at right angles. Another hole possesses a water reservoir which remains closed by a stopcock. The free end of bent glass tube is dipped in a beaker of water which also acts as a seal. Now this apparatus is left undisturbed for some time.

After some time the water of beaker is replaced by lime water and the stop cock of reservior is opened. The water runs in the bottle and so it drives out the air of bottle in the beaker through the bent glass tube.

Observation : The lime water turns milky.

Explanation : The germinating seeds undergo the respiration releasing CO_2 in this process. The CO_2 passes through the bent glass tube making the lime water milky.

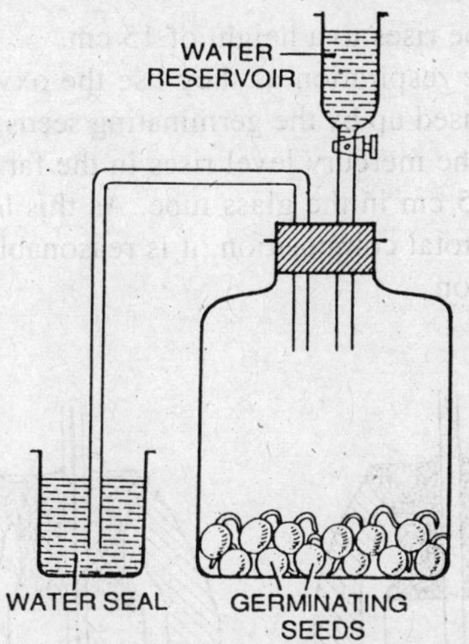

Fig. 4.9. Release of CO_2 during respiration. Demonstration of the phenomenon.

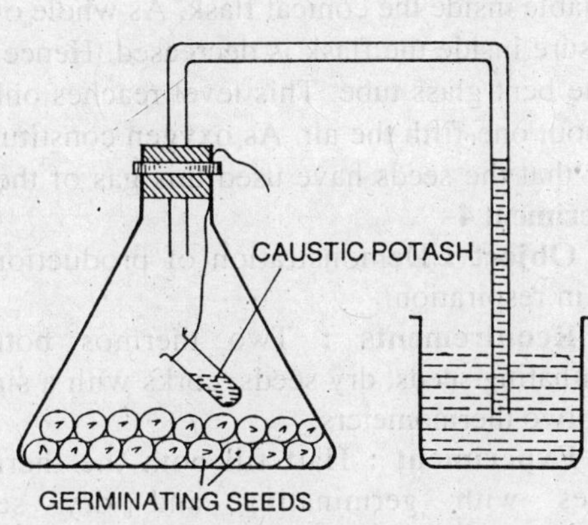

Fig. 4.10. Release of CO_2 during respiration. Demonstration for the phenomenon

(*ii*) **Requirements :** Germinating seeds, a conical flask, a single holed cork, caustic potash in a test tube, a bent glass tube, water beaker, etc.

Experiment : Inside the conical flask some germinating seeds are placed. The bottle is closed with a single-holed cork through which a glass tube bent twice at right angles is inserted. Another free end of this tube is put in a beaker of water. A test tube of caustic potash is hung inside the conical flask. Now this apparatus is put as such for some time.

Observation : Water rises in the far end of glass tube from the beaker.

Explanation : Germinating seeds undergo the anaerobic respiration, releasing CO_2 gas which is absorbed by the caustic potash. Thus a vacuum is created inside the conical flask and consequently water rises in the tube to fulfil it.

Note : The experiment should be performed in dark in case the green leaves are used as the respiratory substrate.

Experiment 3

Object : Demonstration of utility of oxygen in Respiration.

Requirements : A conical flask, a bent tube, germinating seeds, caustic potash in a small container, a mercury dish.

Experiment : Germinating seeds are taken in a conical flask in which a container of caustic potash is also put. The mouth of this conical flask is closed by a single-holed cork through which a glass tube bent twice at right angles in inserted. The far end of this tube is put in the mercury dish. Now this apparatus is left undisturbed for some time.

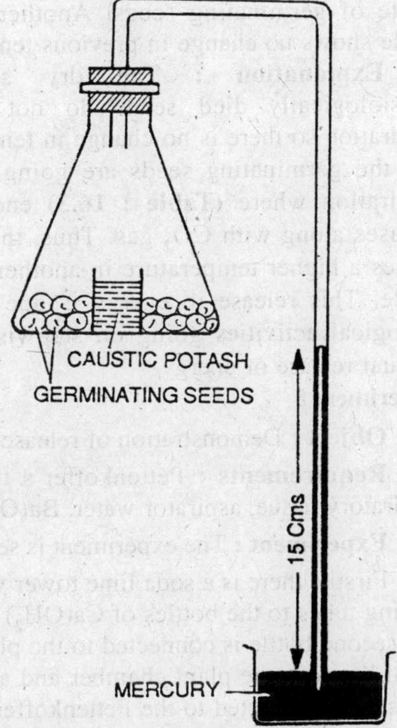

Fig. 4.11. Utility of O_2 in Respiration. Demonstration of the phenomenon.

Observation : The mercury in far end of bent glass tube rises to a height of 15 cm.

Explanation : The germinating seeds undergo aerobic respiration as they use the oxygen available inside the conical flask. As whole of the oxygen is used up by the germinating seeds, the pressure inside the flask is decreased. Hence, consequently the mercury level rises in the far end of the bent glass tube. This level reaches only a height of 15 cm in the glass tube. As this level is about one-fifth the air. As oxygen constitutes one fifth of total composition, it is reasonable to infer that the seeds have used this gas of the air in respiration.

Experiment 4

Object : Demonstration of production of heat in respiration.

Requirements : Two thermos bottles, germinating seeds, dry seeds, corks with a single hole, two thermometers.

Experiment : Half fill both the thermos bottles with germinating and dry seeds respectively. Now close their mouths with the help of single-holed corks through which a thermometer passes respectively. Put this experiment for about 24 hours.

Observation : There is a rise in temperature in the thermometer which is placed in the thermos bottle of germinating seeds. Another thermos bottle shows no change in previous temperature.

Explanation : The dry seeds or physiologically died seeds do not undergo respiration so there is no change in temperature. But the germinating seeds are going to make respiration where **(Table : 16.3)** energy also releases along with CO_2 gas. Thus, this energy causes a higher temperature in another thermos bottle. This release in energy is due to many biological activities going on stepwise with a gradual release of energy.

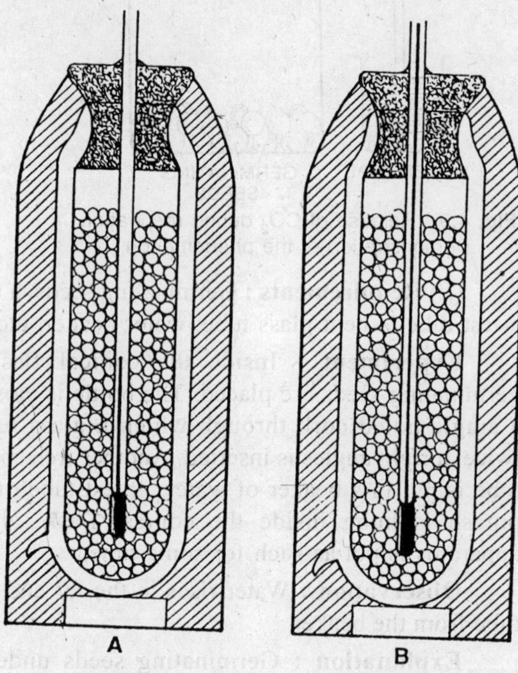

Fig. 4.12. Production of heat in respiration. Demonstration of the phenomenon.

Experiment 5

Object : Demonstration of release of CO_2 by Pettenkoffer's gas stream method.

Requirements : Pettenkoffer's tube, soda lime, $Ca(OH)_2$, passing tubes, rubber tubes, respiratory tissue, aspirator water, $Ba(OH)_2$, plant chambers, etc.

Experiment : The experiment is set as follows :

Firstly, there is a soda lime tower which is open to air by a tube. This tower is connected by passing tubes to the bottles of $Ca(OH_2)$ solution. Usually two bottles are used in the experiment. The second bottle is connected to the plant chamber by another passing tube. Pettenkoffer's tube is put between the plant chamber and aspirator by some rubber tubing connections. A two-way tap is also connected to the Pettenkoffer's tube. This tube is filled with $Ba(OH)_2$ solution.

With the help of two-way tap and aspirator the air is sucked out. Due to a lower pressure, a current of air gets in the soda lime tower. The soda lime and $Ca(OH)_2$ solution remove the whole CO_2 of air. Now the CO_2 free air reaches to plant-chamber from which it gets in the Pettenkoffer's tube. Here the $Ba(OH)_2$ solution is turned turbid. The aspirator is set in the way to regulate the slow movement of air in the form of bubbles.

After some time the turbid solution of Pettenkoffer's tube is titrated by using N/10 HCl and phenolphthalein-indicator.

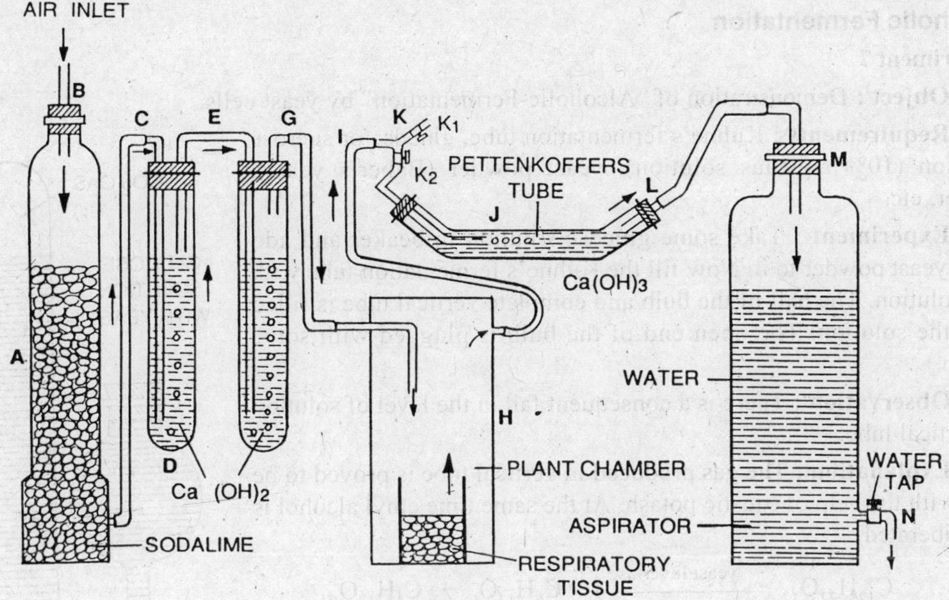

Fig. 4.13. Pettenkoffer's Tube.

Observation and explanation. The turbidity of the solution of Pettenkoffer's Tube is due to formation of $BaCO_3$. The plant tissue respires releasing CO_2 gas. The CO_2 gas reacts with $Ba(OH)_2$ solution forming $BaCO_3$.

Anaerobic Respiration

Experiment 6

Object : Demonstration of anaerobic respiration.

Requirements : Germinating seeds, a test tube, mercury, KOH crystals, petri dish, etc.

Experiment : The mercury is filled in the test tube up to the rim. Take a petri dish full of mercury and put the mercury filled test tube in the inverted condition with the help of thumb. The thumb is removed inside the petri dish. Now introduce some seeds (germinating or soaked) in the test tube with the help of a forceps. Let this apparatus be put as such for some time.

Observation : After some time the mercury level in test tube goes down. Now introduce some KOH crystals in the above test tube. After a little while the mercury level will go up to the former position.

Explanation : The lowering of mercury level in test tube indicates that some gas has been released by germinating seeds. Only CO_2 gas may be absorbed by KOH crystals. We see that with the introduction of KOH crystals the level of mercury goes high, hence the seeds have released CO_2 under the anaerobic conditions.

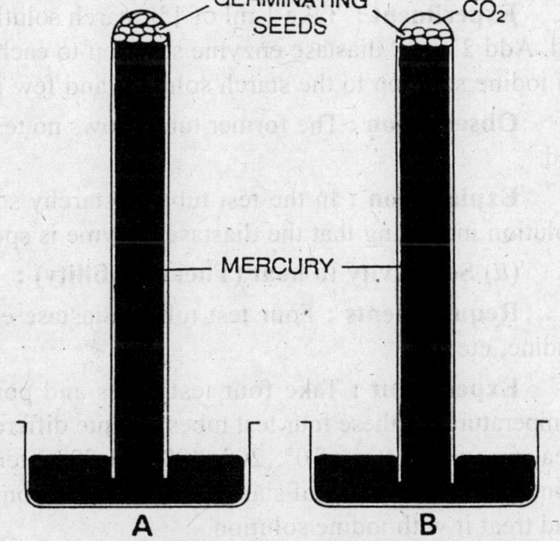

Fig. 4.14. Anaerobic Respiration. Demonstration of the phenomenon.

Alcoholic Fermentation

Experiment 7

Object : Demonstration of "Alcoholic-Fermentation" by yeast cells.

Requirements : Kuhne's fermentation tube, glucose or sucrose solution (10% aqueous solution), yeast powder (Baker's yeast), beaker, etc.

Experiment : Take some glucose solution in beaker and add 1 gm yeast powder to it. Now fill the Kuhne's fermentation tube with this solution. The half of the bulb and complete vertical tube is filled with the solution. The open end of the bulb is plugged with some cotton.

Observation : There is a consequent fall in the level of solution in vertical tube.

Explanation : The gas produced in vertical tube is proved to be CO_2 with the help of caustic potash. At the same time ethyl alcohol is also liberated :

$$C_{12}H_{22}O_{12} \xrightarrow{\text{yeast Invertase}} \underset{\text{Glucose}}{C_6H_{12}O_6} + \underset{\text{Fructose}}{C_6H_{12}O_6}$$

$$C_6H_{12}O_6 \xrightarrow{\text{Fermentation}} 2C_2H_5OH + 2CO_2$$

Alcohol may also be tested 'by lodoform test'.

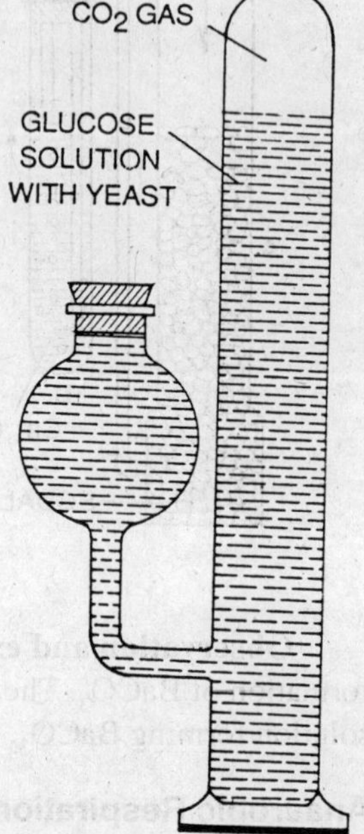

Fig. 4.15. Kuhne's Fermentation tube.

Enzymes

Experiment 8

Object : Demonstration of properties of enzyme : (*i*) specificity; (*ii*) thermolability and (*iii*) colloidal nature.

(*i*) **Specific Nature (Specificity) :**

Requirements : Test tubes, 1% starch solution, castor oil, diastase enzyme solution, Iodine, Sudan III, etc.

Experiment : Take 1 ml of 1% starch solution in a test tube and in other tube 1 ml of castor oil. Add 2 ml of diastase enzyme solution to each test tube. After about 5 minutes add few drops of iodine solution to the starch solution and few drops of Sudan III to castor oil.

Observation : The former tube shows no test of starch while in the latter castor oil is turned red.

Explanation : In the test tube of starchy solution the diastase enzyme has hydrolysed this solution indicating that the diastase enzyme is specific for starch in action.

(*ii*) **Sensitivity to heat (Thermolability) :**

Requirements : Four test tubes, diastase enzyme solution, beaker, water, starch solution, Iodine, etc.

Experiment : Take four test tubes and pour diastase enzyme solution in each of it. The temperature for these four test tubes is quite different, *i.e.*, 0°, 20°, 40° and 50°C respectively. Four beakers with water of 0°, 20°, 40° and 50°C temperature are used here to obtain the required condition. Add 10 c.c. of starch solution after some time. Now take few drops from each test tube and treat it with iodine solution.

Observation : Solution gives no colour at about 40°C.

Explanation : It shows that the rate of enzymatic reaction is optimum at this temperature. In other cases the positive iodine test is obtained.

(*iii*) **Colloidal Nature :**

Requirements : NaCl solution, enzyme solution, thistle funnel, beaker, $AgNO_3$, parchment membrane.

Experiment : A mixture of NaCl solution and enzyme solution is taken in a thistle funnel covered by the parchment membrane. Immerse it in a solution of starch. Test the filtrate with $AgNO_3$.

Observation : A white ppt. of AgCl is obtained here.

Explanation : Due to colloidal nature the enzyme molecules are large. Here only NaCl molecules are diffused through the parchment membrane, while the colloidal particles are held back on the parchment membrane.

Experiment 9

Object : Assay of invertase.

Requirements : Commercial yeast, distilled water, 3% sugar solution, Fehling's solution.

Experiment :

(*i*) **Extraction of Invertase :** Take 5 gm. of commercial yeast in a dish. Powder it and add about 100 ml water to it. After sometime filter the yeast in a beaker. The filtrate contains invertase in solution form.

(ii) **Test :** Take 3% sugar solution. Add 10 c.c. of each of the solutions in a test tube. Pour some Fehling's solution in this test tube.

Observation : No colour develops here. Put the enzyme sugar (sucrose) solution for an hour and test it again with Fehling's solution. Here red colour appears.

Explanation : Sucrose (sugar solution) is not a reducing agent so no change appears in the beginning. After some time (about an hour) it has broken down into reducing hexoses which give the test.

$$\underset{\text{Sugar solution}}{C_{12}H_{22}O_{12}} \xrightarrow[\text{Enzyme}]{H_2O} \underbrace{C_6H_{12}O_6 + C_6H_{12}O_6}_{\text{Hexoses}}$$

Note : Assay is a test of biologically active substance which is done by a living material.

Experiment 10

Object : To prepare the enzymes from wheat or rice seeds.

Requirements : Rice or wheat seeds, water, mortar and pestle, centrifuge, phosphate buffer at pH 6.9, ice bucket, etc.

Method : Take rice or wheat seeds soaked and germinated for 0, 24 and 48 hours. Place 50 seeds in a mortar and pestle with a little bit of 10 ml phosphate buffer at pH 6.9. Now grind seeds adding more fluid to extract thoroughly until 35 ml has been added. Now this process may be repeated with each set of seeds (for 0, 24 and 48 hours). Allow the homogenate to stand approximately for 10 minutes in the ice bucket. After doing so, centrifuge the debris, starch grains and other major organelles at 15,000 rpm for 10 minutes. Use the supernatant as enzyme sources.

SUMMARY

1. Respiration is a metabolic process in which carbon compounds synthesised by photosynthesis are oxidised, in the presence of various enzymes, into carbon dioxide and water with the release of energy.

2. The ratio of carbon dioxide evolved to the volume of O_2 consumed during oxidation of carbon compounds is called respiratory quotient (RQ).
3. Respiratory quotient depends upon the respiratory substrate being used in respiration.
4. Respiration is basically of two types : (*i*) aerobic respiration which occurs in the presence of oxygen, and (*ii*) anaerobic respiration that does not require oxygen.
5. Anaerobic respiration occurs in the absence of oxygen. It takes place in the cytoplasm of the cell. The end products of anaerobic respiration are ethanol or lactic acid, and very little energy is released.
6. The initial stages in respiration, which constitute the process of glycolysis, are common to both anaerobic and aerobic respiration.
7. During glycolysis, glucose converts into pyruvic acid. There is a net gain of 8 ATP molecules during glycolysis.
8. Aerobic respiration occurs in mitochondria.
9. In the mitochondria, the pyruvic acid is oxidised to acetyl-CoA, which then enters the TCA (tricarboxylic acid cycle or Krebs cycle) or citric acid cycle where it oxidises further.
10. During this process, CO_2 and NADH and $FADH_2$ are produced and take part in electron transport system (ETS) which carries the electrons in a chain to molecular oxygen, ultimately producing water.
11. The energy released during electron transport system (ETS) is utilised to synthesise ATP. This process is called oxidative phosphorylation.
12. The energy trapped in ATP is utilised in various metabolic processes by the living organisms.
13. Most organisms have an alternate route also for glucose metabolism, called oxidative pentose phosphate pathway. This pathway also operates in cytosol of the cell and produces pentose sugars and NADPH.
14. The atmospheric concentration of CO_2 at which photosynthesis just compensates for respiration, is called CO_2 compensation point.

TERMS TO REMEMBER

Aerobic respiration. This process of respiration leads to a complete oxidation of organic substances in the presence of O_2 and release CO_2, water and a large amount of energy present in the substrate.

Anaerobic respiration. This type of respiration takes place in the complete absence of oxygen.

Compensation point. The atmospheric concentration of CO_2 at which photosynthesis just compensates for respiration.

Citric acid cycle (Krebs cycle or tricarboxylic acid cycle TCA). Begins when acetyl-Co-A enters into a reaction to form citric acid.

Electron transport system (ETS). The metabolic pathway through which the electron passes from one carrier to another.

Fermentation. Process of oxidation in microbes.

Glycolysis (EMP pathway). In the process of respiration, the carbohydrates are converted into pyruvic acid through a series of enzymatic reactions, which take place in the cytosol.

Hans Adolf Krebs (1900-1981). He discovered citric acid cycle (Krebs cycle or tricarboxylic acid cycle) for production of energy in the cell in 1937. He was awarded Nobel Prize in 1953.

Hexokinase. An enzyme, which catalyses reaction where glucose and fructose are phosphorylated to give rise to glucose-6-phosphate and fructose-6-phosphate, respectively.

Invertase. An enzyme, by which sucrose is converted into glucose and fructose.

Oxidative decarboxylation. Here, pyruvate is first decarboxylated and then oxidised by the enzyme pyruvate dehydrogenase.

Oxidative phosphorylation. Production of ATP by phosphorylation of ADP.

Pentose phosphate pathway (PPP). Here, glucose-6-phosphate produced during early stages of glycolysis. This reaction takes place in the presence of the enzyme glucose-6-phosphate dehydrogenase.

Phosphofructokinase. An enzyme, which catalyses the reaction where fructose-6-phosphate is phosphorylated and fructose 1, 6-bisphosphate is produced.

Respiration. Phenomenon of release of energy by oxidation of various organic molecules, for cellular use.

Respiratory quotient (RQ) or respiratory ratio. The ratio of the volume of CO_2 evolved to the volume of O_2 consumed in respiration, *i.e.*, $RQ = \frac{CO_2}{O_2}$

Respiratory substrate. The compounds that are oxidised during respiration.

QUESTION BANK

Very short answer questions

1. Give equation of cellular respiration.
2. What are respiratory substrates?
3. What is respiratory quotient?
4. Give equation of an aerobic respiration.
5. What is glycolysis?
6. Where does glycolysis take place?
7. Who discovered citric acid cycle?
8. Expand TCA.
9. Who discovered glycolysis?
10. Which enzyme is responsible for conversion of sucrose into glucose and fructose?
11. Expand FAD.
12. What is net ATP production in glycolysis?
13. Expand PPP.
14. Define compensation point.
15. What is net gain of ATP during aerobic respiration of one molecule of glucose?

ANSWERS

1. $C_6H_{12}O_6 + 6O_2 \longrightarrow 6CO_2 + 6H_2O$ + Energy (2870 kJ) ; 2. The compounds that are oxidised during respiration ; 3. The ratio of volume of CO_2 evolved to the volume of O_2 consumed during respiration, *i.e.*, CO_2/O_2 ; 4. $C_6H_{12}O_6 \xrightarrow[\text{enzyme}]{\text{zymase}} 2C_2H_5OH + 2CO_2$ + Energy (247 KJ) ; 5. During respiration, the carbohydrates convert into pyruvic acid through a series of enzymatic reactions, they are together known as glycolysis ; 6. In cytosol ; 7. Hans Adolf Krebs ; 8. Tricarboxylic acid ; 9. Embden, Meyerhof and Parnas ; 10. Invertase ; 11. Flavin adenine dinucleotide ; 12. 2 ATP ; 13. Pentose phosphate pathway ; 14. The atmospheric concentration of CO_2 at which photosynthesis just compensates for respiration, is called CO_2 compensation point ; 15. 38 ATP molecules.

Short Answer Questions

1. Define respiration. What are respiratory substrates?
2. Why respiration is called cellular respiration?
3. Give equation of cellular respiration.
4. Define the following :
   (*a*) Respiratory quotient (*b*) Anaerobic respiration
5. How would you differentiate between aerobic respiration and anaerobic respiration ?
6. What is fermentation? Give its economic importance.

7. Distinguish between respiration and combustion.
8. Give types of respiration and their overall equations.
9. Give R.Q. value of carbohydrates and fats.
10. What is significance of R.Q ?
11. What is meant by respiration quotient (RQ)? When will the value of RQ be 1 and when will if be less than 1?
12. Distinguish between glycolysis and citric acid cycle.
13. Differentiate between glycolysis and fermentation.
14. What is oxidative phosphorylation?
15. Write short notes on :
    (*a*) Glycolysis (*b*) Electron transport chain.
16. What is respiratory substrate? Enumerate the steps involved in the breakdown of carbohydrates upto pyruvic acid.
17. Differentiate between glycolysis and Krebs cycle.
18. Distinguish between respiration and photorespiration.
19. Why does anaerobic respiration produce less energy than aerobic respiration?
20. Write the significance of citric acid cycle.

Long Answer Questions

1. Illustrate the mechanism of electron transport system.
2. Describe the pentose phosphate pathway.
3. Describe the process and role of citric acid cycle in living organism.
4. Discuss various steps of Krebs cycle. Also give the estimates of water, CO_2 and formation of ATP.
5. Define respiration. Explain the complete process of respiration in plants.
6. Where does glycolysis occur in a cell? Explain its different steps.
7. Explain non-cyclic photophosphorylation in plants. Why is this process called so?

MULTIPLE CHOICE QUESTIONS (Objective Type Questions)

1. During Krebs cycle, dehydrogenase enzyme removes hydrogen from the substance. The hydrogen atom becomes ionised into proton and electron. The electrons are passed to oxygen through a series of substances, the penultimate electron acceptor that supplies the electrons directly to oxygen is
   (*a*) DNA (*b*) ubiquinone (*c*) cytochrome a_3 (*d*) cytochrome *a*
2. The importance of Krebs cycle is
   (*a*) production of amino acids
   (*b*) production of vitamins
   (*c*) production of ATP molecules through oxidative phosphorylation.
   (*d*) to encourage gluconeogenesis
3. Which of the following is required, both in the processes of respiration as well as photosynthesis?
   (*a*) organic fuel (*b*) green cells
   (*c*) cytochrome (*d*) energy from carbon bonds
4. Glycolysis takes place in
   (*a*) cytoplasm (*b*) chloroplast (*c*) ribosome (*d*) mitochondria
5. Pentose phosphate pathway an alternative pathway of respiration, was elucidated by
   (*a*) Warburg (*b*) Dickens (*c*) Forecker (*d*) Blackman
6. Cytochromes in plant cells function mainly as
   (*a*) oxygen acceptor (*b*) CO_2 acceptor (*c*) electron acceptor (*d*) H_2O acceptor

7. The respiration of young maturing seeds is quite high but as water content decreases during further maturation respiration
(*a*) remains high (*b*) stops completely (*c*) increases steadily (*d*) decreases steadily
8. Total ATP production during EMP pathway is
(*a*) 24 ATP molecules (*b*) 8 ATP molecules (*c*) 38 ATP molecules (*d*) 6 ATP molecules
9. Respiration in cells takes place in
(*a*) ribosomes (*b*) nucleus (*c*) golgi body (*d*) mitochondria
10. Under glycolysis, the pyruvic acid is reduced to lactic acid anaerobically in
(*a*) liver (*b*) muscles (*c*) skin (*d*) brain
11. Conversion of sugar into alcohol during fermentation is due to the direct action of
(*a*) temperature (*b*) microorganisms
(*c*) concentration of sugar solution (*d*) zymase
12. The end product of glycolysis is
(*a*) pyruvic acid (*b*) ethyl alcohol (*c*) glucose (*d*) carbon dioxide
13. One of the products of anaerobic respiration is
(*a*) malic acid (*b*) lactic acid (*c*) pyruvic acid (*d*) ethyl alcohol
14. Biological oxidation in Krebs cycle involves
(*a*) N_2 (*b*) CO_2 (*c*) O_2 (*d*) SO_2
15. Adenosine diphosphate contains
(*a*) one high energy bond (*b*) two high energy bonds
(*c*) no high energy bonds (*d*) three high energy bonds
16. The formation of Acetyl coenzyme A from pyruvic acid is the result of its
(*a*) reduction (*b*) dehydration
(*c*) dephosphorylation (*d*) oxidative decarboxylation
17. The end products of aerobic respiration in plants are
(*a*) CO_2, H_2O and energy (*b*) H_2O and energy
(*c*) CO_2 and energy (*d*) CO_2 and H_2O
18. Respiratory quotient refers to
(*a*) The ratio between the amount of CO_2 produced to the amount of O_2 absorbed
(*b*) Amount of ATP produced to the amount lost during respiration
(*c*) Amount of CO_2 produced
(*d*) The number of glucose molecules breaking up to produce energy in one hour
19. Glycolysis operates in
(*a*) endoplasmic reticulum (*b*) cytoplasm
(*c*) golgi apparatus (*d*) perinuclear space
20. Common source of energy for cellular activity is
(*a*) ATP (*b*) NAD (*c*) Fat (*d*) Protein
21. Oxidation of pyruvate occurs in
(*a*) Krebs cycle (*b*) Tricarboxylic acid (*c*) Citric acid cycle (*d*) All the above
22. End product of Krebs cycle is
(*a*) Citric acid (*b*) Pyruvic acid (*c*) $CO_2 + H_2O$ (*d*) Lactic acid
23. End product of glycolysis is
(*a*) Glucose-1-phosphate (*b*) Acetyl Co-A
(*c*) Fructose-1-phosphate (*d*) Pyruvic acid
24. In aerobic respiration one gram mole of glucose yields
(*a*) 4 ATP (*b*) 24 ATP (*c*) 38 ATP (*d*) 40 ATP

25. Which is product of aerobic respiration
(*a*) pyruvic acid (*b*) malic acid (*c*) lactic acid (*d*) ethyl alcohol
26. The end product of oxidative phosphorylation is
(*a*) NADH (*b*) Oxygen (*c*) ATP (*d*) ATP + H_2O
27. Enzyme which changes glucose to glucose 6-phosphatase is
(*a*) Glucose synthetase (*b*) Hexokinase
(*c*) Glucose 6-phosphatase (*d*) Phosphorylase
28. Protein is substance in
(*a*) Anaerobic respiration (*b*) Protoplasmic respiration
(*c*) Seedling stage (*d*) Floating respiration
29. Krebs cycle occurs in
(*a*) E.R. (*b*) Chloroplasts (*c*) Mitochondria (*d*) Dictyosomes
30. Fructose 6-phosphate is changed to fructose 1, 6 diphosphate by
(*a*) Phosphoglycerate (*b*) Phosphofructokinase
(*c*) Phosphatase (*d*) Enolase
31. Metabolism of one palmitic acid molecule yields
(*a*) 36 ATP (*b*) 56 ATP (*c*) 48 ATP (*d*) 136 ATP
32. Anaerobic respiration of yeast produces
(*a*) N_2 (*b*) O_2 (*c*) CO_2 (*d*) H_2O
33. Site for glycolysis or EMP is
(*a*) cytoplasm (*b*) mitochondrion (*c*) chloroplast (*d*) nucleus
34. Anaerobic respiration is also called
(*a*) Glycolysis (*b*) Fermentation (*c*) Krebs cycle (*d*) Fragmentation
35. Production of alcohol by yeast in
(*a*) Aerobic (*b*) Light dependent (*c*) Anaerobic (*d*) both (*b*) and (*c*)
36. R.Q. is more than one in case the substrate is
(*a*) Fat (*b*) Glucose (*c*) Protein (*d*) Organic acid
37. Currency of cell is
(*a*) Granum (*b*) Mitochondria (*c*) ATP (*d*) Fat
38. ADP contains
(*a*) 1 high energy bond (*b*) 2 high energy bonds
(*c*) 3 high energy bonds (*d*) 4 high energy bonds
39. What is released at the time of seed germination ?
(*a*) Starch (*b*) Fat (*c*) Heat (*d*) Light
40. Respiratory substrate (s) in respiration is
(*a*) Glucose and sucrose (*b*) Glucose and carbon
(*c*) Glucose and oxygen (*d*) Glucose and carbon dioxide
41. Respiration occurs in
(*a*) In all living cells in light (*b*) In all living cells in dark
(*c*) Both (*a*) and (*b*) (*d*) In non green cells in dark only
42. Respiration may occur in absence of O_2 in
(*a*) Potato (*b*) *Spirogyra* (*c*) Yeast (*d*) Humans
43. Krebs cycle is also called as
(*a*) Glycolate cycle (*b*) EMP pathway (*c*) Glyoxylate cycle (*d*) Citric acid cycle
44. The net gain of energy from one gram mole of glucose during aerobic respiration is
(*a*) 38 ATP (*b*) 40 ATP (*c*) 36 ATP (*d*) 2 ATP

45. Both respiration and photosynthesis require
(*a*) Sunlight (*b*) Green pigment (*c*) Cytochromes (*d*) Phytochrome
46. Where do we observe both photosynthesis and respiration
(*a*) virus (*b*) fungi (*c*) bacteria (*d*) green plants
47. Respiratory enzymes are located in
(*a*) Matrix of mitochondria (*b*) Perimitochondrial space
(*c*) Cristae (*d*) Outer membrane
48. ATP synthesis occurs in which part of mitochondria
(*a*) Mitochondria as a whole (*b*) Inner membrane of mitochondria
(*c*) Outer membrane of mitochondria (*d*) Mitochondrial matrix
49. The amount of energy given by one molecule of ATP is
(*a*) 7.3 kcal (*b*) 7.61 kcal (*c*) 6.86 kcal (*d*) 7.06 kcal
50. EMP can produce a total of :
(*a*) 2 ATP (*b*) 6 ATP (*c*) 8 ATP (*d*) 10 ATP
51. Hexose monophosphate shunt is called :
(*a*) Pentose phosphate pathway (*b*) Total of metabolic reactions
(*c*) Conversion of glucose into pyruvate (*d*) EMP pathway
52. Pentose phosphate pathway was discovered by
(*a*) Horecker (*b*) Warburg and Dickens
(*c*) Lipmann *et al* (*d*) Kostytchen
53. Glycolysis is part of
(*a*) Aerobic respiration (*b*) Anaerobic respiration
(*c*) Both (*a*) and (*b*) (*d*) Krebs cycle
54. Reaction of glycolysis occurs in
(*a*) cytoplasm (*b*) mitochondria
(*c*) chloroplast (*d*) perimitochondrial space
55. Oxidation of $NADH_2$ yields
(*a*) 1 ATP (*b*) 2 ATP (*c*) 3 ATP (*d*) 4 ATP
56. The first substance that enters the mitochondria after cytoplasmic part of respiratory reactions is
(*a*) Oxaloacetic acid (*b*) Pyruvic acid (*c*) Citric acid (*d*) Acetyl Co-A
57. Respiratory quotient is always less than one in succulent plants because of
(*a*) Complete oxidation (*b*) Complete reduction
(*c*) Incomplete oxidation (*d*) Incomplete reduction
58. Which of the following does not involve EMP cycle
(*a*) Conversion of glucose in ethyl alcohol
(*b*) Conversion of glucose in PGAL
(*c*) Reduction of glucose into pyruvic acid
(*d*) Conversion of pyruvic acid into CO_2 and H_2O
59. How many $NADH_2$ are formed during glycolysis
(*a*) 1 (*b*) 2 (*c*) 3 (*d*) 4
60. Cheese and yogurt are products of
(*a*) Pasteurisation (*b*) Distillation (*c*) Fermentation (*d*) Dehydrogenation
61. Complete oxidation of one molecule of acetyl Co-A results in the formation of
(*a*) 12 ATP (*b*) 15 ATP (*c*) 19 ATP (*d*) 38 ATP
62. In which of the following process 2 electrons and 2 protons are used
(*a*) Dehydrogenation (*b*) Deamination (*c*) Decarboxylation (*d*) Hydrogenation

63. The process of Acetyl Co-A synthesis from pyruvic acid is called
(*a*) Reduction (*b*) Dehydrogenation
(*c*) Dephosphorylation (*d*) Oxidative decarboxylation
64. End products of aerobic respiration are
(*a*) $CO_2 + H_2O$ (*b*) $O_2 + H_2O$ + Energy
(*c*) Energy (*d*) $CO_2 + H_2O$ + Energy
65. The number of carbons in pyruvic acid is
(*a*) 3 (*b*) 4 (*c*) 6 (*d*) 2
66. Terminal cytochrome of respiratory chain which donates electrons to oxygen is
(*a*) Cyt . *b* (*b*) Cyt . *c* (*c*) Cyt . a_3 (*d*) Cyt . a_1
67. Metabolic water is
(*a*) Produced during fermentation
(*b*) Produced during transpiration
(*c*) Produced during aerobic utilisation of glucose
(*d*) Produced during condensation of glucose molecules in photosynthesis
68. Which one is a product of glycolysis ?
(*a*) FAD (*b*) NAD (*c*) NADP (*d*) NADH
69. Hydrogen atoms released at succinate level in Krebs cycle are accepted by
(*a*) FAD (*b*) NAD (*c*) NADP (*d*) AMP
70. Anaerobic respiration of organisms produces
(*a*) $CO_2 + H_2O$ (*b*) Alcohol + CO_2
(*c*) Lactic acid + H_2O (*d*) $CO_2 + H_2O$ + Energy
71. Correct sequence of electron acceptor in ATP synthesis is
(*a*) Cyt. a_1, a_3, *b*, *c* (*b*) Cyt. *b*, *c*, *a*, a_3 (*c*) Cyt. *c*, *b*, *a*, a_3 (*d*) Cyt. *b*, *c*, a_3, *a*
72. Which one transfers electrons to ETS ?
(*a*) Cytochrome (*b*) FeS
(*c*) Phytochromes (*d*) FeS and Cytochrome
73. End product of citric acid cycle or Krebs cycle is
(*a*) Citric acid (*b*) Lactic acid (*c*) $CO_2 + H_2O$ (*d*) H_2O
74. R.Q. is
(*a*) CO_2 / O_2 (*b*) O_2 / CO_2 (*c*) $1 + \frac{O_2}{CO_2}$ (*d*) $1 + \frac{CO_2}{O_2}$
75. The reactions of Krebs cycle take place
(*a*) In cytoplasm (*b*) In mitochondria
(*c*) On surface of mitochondria (*d*) In E.R. and cytoplasm
76. Glycolysis occurs in
(*a*) Muscle cells (*b*) Nerve cells (*c*) Virtually all cells (*d*) Eukaryotes
77. Respiratory Quotient is
(*a*) $\frac{CO_2}{O_2}$ (*b*) $\frac{O_2}{CO_2}$ (*c*) $\frac{N_2}{CO_2}$ (*d*) $\frac{CO_2}{N_2}$
78. Centre of phosphorylation is
(*a*) Oxysome (*b*) Peroxisome (*c*) Ribosomes (*d*) Mitochondria
79. Which of the following are essential for ATP synthesis?
(*a*) Fe and Mo (*b*) N and P (*c*) Fe and P (*d*) Mg and Mn
80. Organism used in the industrial preparation of ethanol is
(*a*) *Lactobacillus* (*b*) *Azotobacter* (*c*) *Penicillium* (*d*) *Saccharomyces*

81. Metabolism of one lipid molecule produces
(*a*) 36 ATP (*b*) 38 ATP (*c*) 147 ATP (*d*) 56 ATP
82. Anaerobic respiration of yeast produces
(*a*) N_2 (*b*) O_2 (*c*) CO_2 (*d*) H_2O
83. Who receives electron at the end of electron transport chain ?
(*a*) H_2O (*b*) H_2 (*c*) CO_2 (*d*) cytochrome a_3
84. What is the total production of ATP by breakdown of one glucose molecule during aerobic respiration ?
(*a*) 34 (*b*) 38 (*c*) 40 (*d*) 32
85. RQ of sprouting potato tubers will be
(*a*) > 1 (*b*) 0 (*c*) 1 (*d*) < 1
86. Anaerobic respiration after glycolysis is called
(*a*) Krebs cycle (*b*) Fermentation
(*c*) Zymoses (*d*) Oxidative phosphorylation
87. Production of ATP by oxidative phosphorylation is driven by energy from
(*a*) Coenzyme A
(*b*) $NADH_2$
(*c*) Pyruvic acid
(*d*) Movement of protons from inner mitochondrial membrane to cytoplasm
88. Krebs cycle is a part of .... respiration.
(*a*) Aerobic (*b*) Anaerobic (*c*) Fermentation (*d*) Facultative
89. How many ATP molecules are formed in ETS from the reduced nicotinamide adenine dinucleotide generated in one turn cycle of Krebs cycle ?
(*a*) 3 (*b*) 6 (*c*) 9 (*d*) 12
90. RQ fat is
(*a*) More than one (*b*) Zero (*c*) One (*d*) Less than one
91. Respiration is a process in which
(*a*) energy is used up
(*b*) energy is released
(*c*) energy is released and stored in the form of ATP
(*d*) energy is stored in the form of ADP
92. Chief site of respiration in plants is
(*a*) lysosomes (*b*) mitochondria (*c*) ribosomes (*d*) chloroplasts
93. In the process of respiration the potential energy stored in the organic compounds is released in the form of
(*a*) physical energy (*b*) kinetic energy (*c*) chemical energy (*d*) radiant energy
94. Why do fishes in an aquarium thrive better if green plants are growing there ? Because they
(*a*) inhale oxygen released by green plants (*b*) inhale CO_2 released by green plants
(*c*) can feed on green plants (*d*) none of the above
95. Which of the following statements regarding respiration and breathing is correct
(*a*) plants respire, while animals breathe
(*b*) higher animals breathe, as well as respire
(*c*) higher plants respire during day, and breathe at night
(*d*) none of the above
96. Complete oxidation of 1 –gm molecule yields
(*a*) 6,860,000 calories (*b*) 686,000 calories (*c*) 68600 calories (*d*) 6860 calories

97. The energy yield as a result of total oxidation of 1 glucose molecule during cellular respiration is to convert
(*a*) 32 molecules of ADP into 32 molecules of ADP
(*b*) 38 molecules of ADP into 38 molecules of ATP
(*c*) 30 molecules of ADP into 30 molecules of ATP
(*d*) 34 molecules of ADP into 34 molecules of ATP

98. The end products of respiration in plants are
(*a*) starch and O_2 (*b*) H_2O and energy
(*c*) CO_2, H_2O and energy (*d*) sugar and O_2

99. Anaerobic respiration is also known as
(*a*) intramolecular respiration (*b*) intermolecular respiration
(*c*) extramolecular respiration (*d*) molecular respiration

100. In which of the following, anaerobic respiration taken place ?
(*a*) yeast (*b*) *Spirogyra* (*c*) potato (*d*) *Spirulina*

101. How many molecules of ATP are gained during anaerobic respiration of one molecule of glucose ?
(*a*) 36 (*b*) 38 (*c*) 2 (*d*) 4

102. When a molecule of pyruvic acid is subjected to anaerobic oxidation and forms lactic acid, there is
(*a*) gain of 2 ATP molecules (*b*) loss of 3 ATP molecules
(*c*) loss of 6 ATP molecules (*d*) gain of 3 ATP molecules

103. Which one of the following process releases a CO_2 molecule ?
(*a*) Alcoholic fermentation (*b*) Glycolysis
(*c*) Lactic acid fermentation (*d*) Hydrolysis of glycogen

104. Which of the following molecules is a product of fermentation of glucose by yeast ?
(*a*) $C_6H_{12}O_6$ (*b*) $(C_6H_{10}O_5)_n$ (*c*) C_2H_5OH (*d*) CH_3OH

105. When yeast ferments glucose, the products are
(*a*) ethanol and water (*b*) water and CO_2 (*c*) methanol and CO_2 (*d*) ethanol and CO_2

106. Glycolysis is found in
(*a*) most body cells (*b*) anaerobic bacteria (*c*) eukaryotic cells (*d*) virtually all cells

107. The common phase in aerobic and anaerobic respiration is
(*a*) glycolysis (*b*) tricarboxylic acid cycle
(*c*) oxidative phosphorylation (*d*) Krebs cycle

108. The other term for glycolysis is
(*a*) TCA pathway (*b*) Carbon pathway (*c*) EMP pathway (*d*) HMS pathway

109. Enzymes taking part in glycolysis are present in
(*a*) mitochondria (*b*) cytosol (*c*) cell sap (*d*) none of above

110. What is end product of glycolysis
(*a*) C_2H_5OH (*b*) pyruvic acid (*c*) CO_2 (*d*) Glucose

111. The enzyme, which converts glucose to glucose-6-phosphate is
(*a*) glucose synthase (*b*) hexokinase (*c*) phosphorylase (*d*) arginase

112. In glycolysis net gain is two ATP molecules and two molecules of
(*a*) $NADH_2$ (*b*) $FADH_2$ (*c*) $FMNH_2$ (*d*) $NADPH_2$

113. Krebs cycle takes place in
(*a*) dictyosomes (*b*) mitochondrial matrix
(*c*) lysosomes (*d*) endoplasmic reticulum

114. Which is the connecting link between glycolysis and Krebs cycle ?
(*a*) Pyruvic acid (*b*) Oxaloacetic acid (*c*) Acetyl-CoA (*d*) Cytochrome

115. Krebs cycle starts with the formation of a 6-carbon compound by chemical reaction between
(*a*) succinic acid and acetyl-CoA (*b*) fumaric acid and pyruvic acid
(*c*) oxaloacetic acid and acetyl-CoA (*d*) malic acid and acetyl-CoA

116. What is the other name for Krebs cycle?
(*a*) Citric acid cycle (*b*) Glycolate cycle (*c*) EMP pathway (*d*) Glyoxylate cycle

117. The Krebs cycle begins with
(*a*) pyruvic acid (*b*) fumaric acid
(*c*) phosphoglyceraldehyde (*d*) lysine

118. Pyruvate dehydrogenase enzyme is used in conversion of
(*a*) glucose to pyruvate (*b*) pyruvic acid to lactic acid
(*c*) pyruvate to acetyl-CoA (*d*) pyruvate to glucose

119. Fumarase enzyme converts
(*a*) fumaric acid to malic acid (*b*) fumaric acid to isocitric acid
(*c*) malic acid to succinic acid (*d*) succinic acid to fumaric acid

120. Organelles which are regarded as 'power house' of the cell and in which the oxidative reactions of the respiratory process take place are
(*a*) lysosomes (*b*) dictyosomes (*c*) chloroplasts (*d*) mitochondria

121. Synthesis of ATP in mitochondria takes place
(*a*) in the matrix (*b*) in cristae (*c*) in outer membrane (*d*) in intracristae space

122. ATP is
(*a*) an enzyme, which brings about oxidation
(*b*) a molecule, which contains high-energy phosphate bonds
(*c*) a hormone
(*d*) a protein

123. What is 'energy currency' of living organisms ?
(*a*) chlorophyll (*b*) ATP (*c*) NADP (*d*) FAD

124. During ATP synthesis electrons pass through which one of the following ?
(*a*) CO_2 (*b*) H_2O (*c*) O_2 (*d*) cytochromes

125. An indispensable role in energy metabolism is played by
(*a*) sodium (*b*) calcium (*c*) phosphorus (*d*) boron

126. Which is also formed along with ATP in glycolysis ?
(*a*) $NADPH_2$ (*b*) $NADH_2$ (*c*) $FADH_2$ (*d*) FAD

127. Synthesis of ATP in mitochondria requires
(*a*) NADP (*b*) FMN (*c*) Oxygen (*d*) pyruvic acid

128. The atom within each cytochrome molecule that actually accepts and releases electrons is
(*a*) iron (*b*) oxygen (*c*) zinc (*d*) carbon

129. Within the mitochondrion, the proton gradient develops across the
(*a*) outer membrane (*b*) inner membrane
(*c*) intermembrane space (*d*) matrix

130. R.Q. of respiratory substrate rich in oxygen is
(*a*) equal to one (*b*) more than one (*c*) less than one (*d*) none of above

131. R.Q. of fatty substances is generally
(*a*) zero (*b*) unity (*c*) more than one (*d*) less than one

132. R.Q. is maximum, when respiratory substrate is
(*a*) fat (*b*) glucose (*c*) malic acid (*d*) protein

133. In succulent plants respiratory quotient is always less than one because of
(*a*) complete oxidation (*b*) complete reduction
(*c*) incomplete oxidation (*d*) incomplete reduction
134. If the volume of CO_2 liberated during respiration is more than the volume of O_2 used, the respiratory substrate will be
(*a*) fats (*b*) organic acids (*c*) proteins (*d*) carbohydrates
135. Hexose monophosphate shunt is
(*a*) pentose phosphate pathway
(*b*) conversion of glucose into pyruvic acid
(*c*) a process by which new products are synthesized
(*d*) glycolytic pathway
136. Photorespiration takes place in
(*a*) mitochondria (*b*) glyoxysomes (*c*) peroxisomes (*d*) Golgi bodies
137. In hospitals plants are removed from patient's room at night because
(*a*) they compete with patient for available oxygen in the room
(*b*) they produce CO_2 at night
(*c*) they produce CO_2 all the 24 hours, but give out oxygen during day
(*d*) this practice has no practical significance
138. When the bread is baked, it becomes light and spongy, becuase
(*a*) the evaporation of alcohol from fermented flour
(*b*) the evolution of CO_2 from the kneaded flour
(*c*) the presence of yeast
(*d*) none of above
139. Food products are kept under refrigeration or in cold storage because
(*a*) they remain very cool
(*b*) respiration is negligible under very low temperature and the food products maintain their flavour and freshness better and longer
(*c*) foodstuff becomes tasty
(*d*) they can be used in off-season
140. Respiration and photosynthesis, both require
(*a*) organic fuel (*b*) cytochrome
(*c*) sunlight (*d*) energy from carbon bonds
141. Which of the following statements about NAD^+ is false ?
(*a*) NAD^+ is reduced to NADH during both glycolysis and the Krebs cycle
(*b*) NAD^+ is reduced by the action of dehydrogenases
(*c*) NAD^+ can receive electrons for use in oxidative phosphorylation
(*d*) NAD^+ has more chemical energy than NADH
142. The colour of cytochrome is
(*a*) Blue (*b*) Black (*c*) Pink (*d*) Colourless
143. One stearic acid molecule releases
(*a*) 140 ATP (*b*) 147 ATP (*c*) 145 ATP (*d*) 180 ATP
144. Krebs cycle is a
(*a*) catabolic cycle (*b*) anabolic cycle (*c*) amphoteric cycle (*d*) amphibolic cycle
145. Electron acceptor in cyt a_3
(*a*) Fe (*b*) Cu (*c*) As (*d*) Both (*a*) and (*b*)

ANSWERS

1. (*c*) 2. (*c*) 3. (*c*) 4. (*a*) 5. (*c*) 6. (*c*) 7. (*c*) 8. (*b*) 9. (*d*) 10. (*b*)
11. (*b*) 12. (*a*) 13. (*d*) 14. (*c*) 15. (*b*) 16. (*d*) 17. (*a*) 18. (*a*) 19. (*b*) 20. (*a*)
21. (*d*) 22. (*c*) 23. (*d*) 24. (*c*) 25. (*b*) 26. (*d*) 27. (*b*) 28. (*b*) 29. (*c*) 30. (*b*)
31. (*d*) 32. (*c*) 33. (*a*) 34. (*b*) 35. (*c*) 36. (*d*) 37. (*c*) 38. (*a*) 39. (*c*) 40. (*c*)
41. (*c*) 42. (*c*) 43. (*d*) 44. (*a*) 45. (*c*) 46. (*d*) 47. (*a*) 48. (*b*) 49. (*a*) 50. (*c*)
51. (*a*) 52. (*b*) 53. (*c*) 54. (*a*) 55. (*c*) 56. (*d*) 57. (*c*) 58. (*d*) 59. (*b*) 60. (*c*)
61. (*a*) 62. (*a*) 63. (*d*) 64. (*d*) 65. (*a*) 66. (*c*) 67. (*c*) 68. (*d*) 69. (*a*) 70. (*c*)
71. (*b*) 72. (*d*) 73. (*c*) 74. (*a*) 75. (*b*) 76. (*c*) 77. (*a*) 78. (*a*) 79. (*c*) 80. (*d*)
81. (*c*) 82. (*c*) 83. (*d*) 84. (*b*) 85. (*c*) 86. (*b*) 87. (*b*) 88. (*a*) 89. (*c*) 90. (*d*)
91. (*c*) 92. (*b*) 93. (*c*) 94. (*a*) 95. (*b*) 96. (*b*) 97. (*b*) 98. (*c*) 99. (*a*) 100. (*a*)
101. (*c*) 102. (*c*) 103. (*a*) 104. (*c*) 105. (*d*) 106. (*d*) 107. (*a*) 108. (*c*) 109. (*b*) 110. (*b*)
111. (*b*) 112. (*a*) 113. (*b*) 114. (*c*) 115. (*c*) 116. (*a*) 117. (*a*) 118. (*c*) 119. (*a*) 120. (*d*)
121. (*b*) 122. (*b*) 123. (*b*) 124. (*d*) 125. (*c*) 126. (*b*) 127. (*c*) 128. (*a*) 129. (*b*) 130. (*a*)
131. (*d*) 132. (*c*) 133. (*c*) 134. (*b*) 135. (*a*) 136. (*c*) 137. (*c*) 138. (*b*) 139. (*b*) 140. (*b*)
141. (*d*) 142. (*c*) 143. (*b*) 144. (*d*) 145. (*a*)

5

CHAPTER

Photosynthesis and Transport of Organic Substances

The process in which certain carbohydrates are synthesized from carbon dioxide and water by chlorophyllous cells in the presence of light, oxygen being a by-product, is generally called **photosynthesis**. The summary equation of photosynthesis is as follows :

$$6CO_2 + 6H_2O \xrightarrow[\text{chlorophyllous cells}]{\text{radiant 673 kg . cal of radiant energy}} C_6H_{12}O_6 + 6O_2$$

This equation, although properly balanced, gives an erroneous impression of the mechanism by which the reaction is accomplished. By the use of H_2O and CO_2 labelled with isotopic oxygen (O^{18}), biochemists have been able to demonstrate that the oxygen released in photosynthesis comes not from CO_2 but from water.

Definition. "Photosynthesis (*Photos* = light; *synthesis* = putting together) is an anabolic process in which green plants or green parts of the plants synthesize or manufacture complex carbonaceous organic food substances (carbohydrates) with carbon dioxide and water (respectively taken from air and water) in the presence of sunlight and evolve oxygen as a by-product. Thus, in this process radiant energy is converted into chemical energy."

However, green plants apart from taking nutrients from soil, make their own food by utilising carbon dioxide, water and sunlight. This is basis of photosynthesis. During this process, oxygen is evolved and released to the atmosphere.

A simple equation given by C. B. van Niel of Stanford University, U.S.A. is as follows :

$$CO_2 + 2H_2O \xrightarrow[\text{green plants}]{\text{light energy}} (CH_2O)_n + H_2O + O_2$$

During photosynthesis, carbon dioxide CO_2 is chemically reduced to carbohydrate $(CH_2O)_n$. Here, n is an integer.

In glucose, n is 6, *i.e.*, formula of glucose is $C_6H_{12}O_6$.

Water molecule splits in presence of light. This process is called *photolysis* and O_2 is released.

Thus, released O_2 comes from water and not from CO_2.

Importance of photosynthesis. Food is the source of energy needed by animal and plant life. And this food is manufactured by green plants from inorganic substances with the aid of sunlight energy during photosynthesis. Food represents the stored energy of sunrays. Fuel in the form of wood is also the indirect result of photosynthesis activity of plants and it is also thought that coal, petroleum, etc., are also the remote consequences of photosynthesis. A number of other substances are also produced as the by-products, for example, the oxygen needed by animals and human being continues to be supplied by photosynthesis.

Historical Perspective

Study on photosynthesis originated only about 300 years ago.

For the first time Belgian physician, Jan Baptista van Helomont, on the basis of his simple experiment concluded that all the substance of the plant was produced from the water and none from the soil.

By the end of the eighteenth century, Joseph Priestley (1733-1804), in 1771-72, performed first experiment on photosynthesis, and showed that plants have ability to take up CO_2 from the atmosphere and release O_2.

Later, Dutch physician Jan Ingenhousz (1730-1799), in 1780 confirmed Prietley's work, and made following important conclusions:

(*i*) The evolution of O_2 took place only within a few hours and not weeks.

(*ii*) The evolution of O_2 took place only during day time, *i.e.*, in presence of sunlight.

(*iii*) Only green parts of plants could restore used air.

Ingenhousz was first to give an equation for photosynthesis :

$$CO_2 + H_2O \xrightarrow[\text{green plants}]{\text{light}} O_2 + \text{Organic matter}$$

In 1804, a Swiss scholar Nicholas Theodore de Saussure observed that water is essential for photosynthesis. He also correlated the importance of light during intake of CO_2 and evolution of O_2.

In 1837, Dutrochet experimentally demonstrated that chlorophyll is necessary for photosynthesis.

In 1845, Liebig told that CO_2 is the main source of all organic compounds synthesised by green plants.

In 1864, Sachs told for first time that organic matters produced by plants were carbohydrates.

SITE FOR PHOTOSYNTHESIS

Photosynthesis takes place only in the green parts of the plants, such as leaves, stems, etc.

Within a leaf photosynthesis occurs particularly in **mesophyll cells** which have **chloroplasts**.

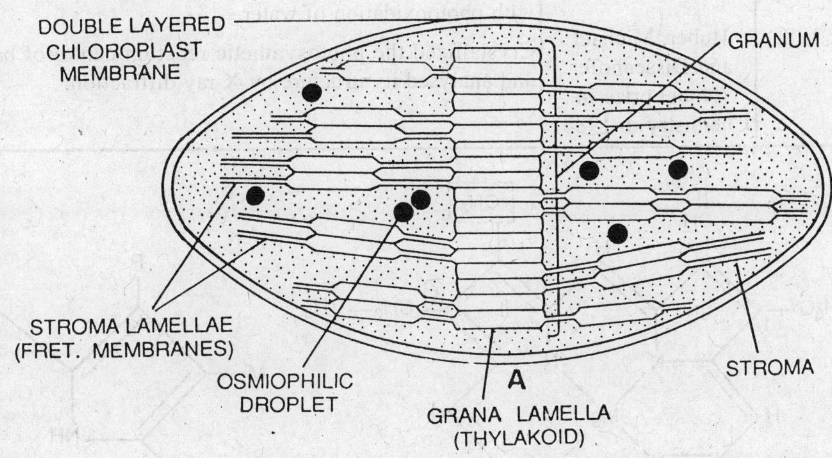

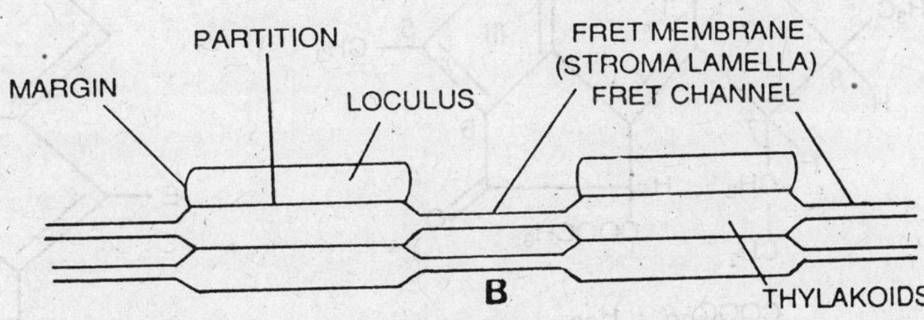

Fig. 5.1. Chloroplast. A, internal structure; B, few enlarged thylakoids from two grana.

Table 5.1. Advancement of Photosynthesis in Twentieth Century.

Year	Scientists	Work
1905	Blackman	Explained and formulated the law of limiting factors; discovered the existence of dark reaction in photosynthesis.
1920	Warburg	Introduced the unicellular green algae *Chlorella* as a suitable material to study photosynthesis.
1931	Van Neil	Reported that O_2 is released by the photolysis of water and not from CO_2.
1932	Emerson and Arnold	Reformed the 'flashing light experiment' to show the existence of light and dark reactions.
1937	Hill	Demonstrated 'photolysis of water by isolated chloroplasts' in the presence of suitable electron acceptor.
1941	Ruben and Kamen	Demonstrated that the oxygen given out during photosynthesis is from water by using O^{18} isotope.
1954	Arnon, Allen and Whatley	Used radioactive CO_2 (*i.e.*, $^{14}CO_2$) to show fixation of C by isolated chloroplasts.
1954	Calvin (Nobel Prize, 1960)	Traced the path of carbon in photosynthesis and established C_3 cycle (Calvin cycle).
1957	Emerson *et. al.*	Reported two photosystems PS I and PS II.
1963	Park and Sane	Reported that light reactions take place on the thylakoid membranes in grana and dark reactions occur in the stroma of the chloroplasts.
1965	Hatch and Slack	Reported C_4 pathway for CO_2 fixation in certain tropical grasses.
1970	Kok *et. al*	Reported that an unknown water oxidising enzyme complex is associated with photooxidation of water.
1985	Huber, Michael and Dissenhofer (Nobel prize in Chemistry in 1988)	Crystallized the photosynthetic reaction centre of bacterium *Rhodobacter* and analysed its structure by X-ray diffraction.

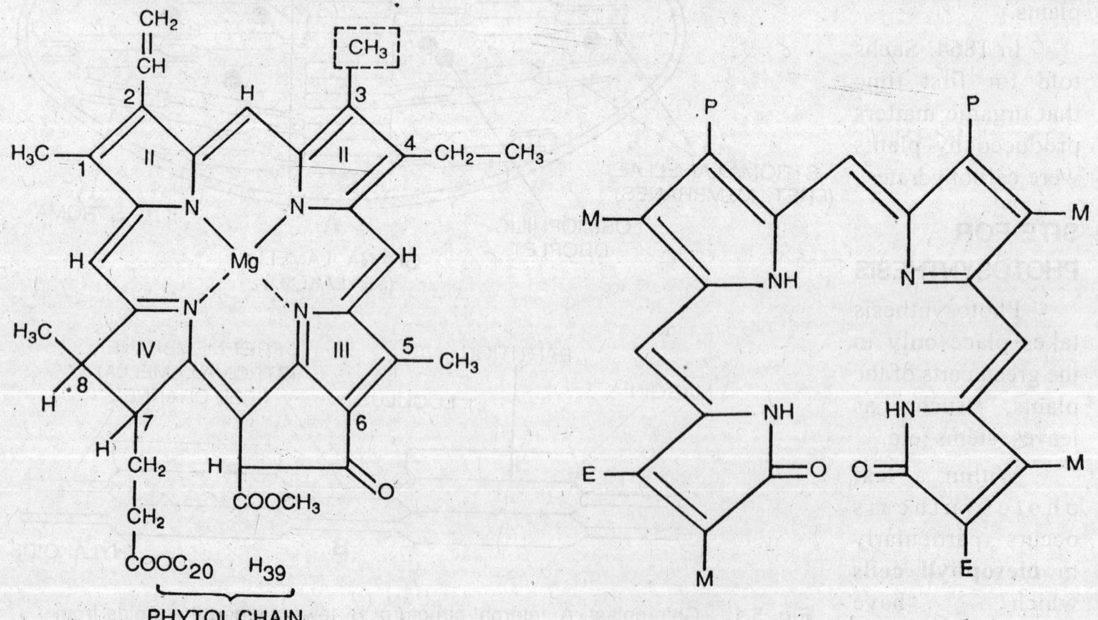

Fig. 5.2. Structural formula of Chlorophyll a.

Fig. 5.3. Structure of Phycoerythrobilin.

Chloroplasts make actual sites for photosynthesis in green plants. Chloroplasts are located at the outer margins with their broad surfaces parallel to the cell wall of the mesophyll cells.

This type of arrangement helps in easy diffusion of CO_2 required for photosynthesis from atmosphere to the inside of chloroplasts.

Chloroplast

The complex process of photosynthesis takes place from beginning to end in the chloroplast. The absorption of light energy and carbon dioxide and the conversion of the carbon dioxide to starch and evolution of oxygen all take place within the illuminated chloroplast.

Structure of the chloroplast. The contents of the fully developed chloroplast are enclosed in an envelope consisting of two membranes with an enclosed space. Each membrane being 40 to 60 Å thick and the space between them vary from 25 to 75 Å. Inside is filled with proteinaceous matrix called stroma contains starch grains and osmophilic droplets. Eye spots and pyrenoid bodies, often found in algal cells, are also found in the matrix (stroma).

Several membranes stacked on top of each other are exhibited in the cross section of the chloroplast. These membranes are paired, forming stacks of discs. On or within the lamellae the chlorophyll and other pigments are found. In the lower forms of plant life, the pigments are evenly distributed over the entire surface of the lamellae, while in the higher forms of plant life, they are restricted to certain areas of lamellae. If these concentrated areas are layered one on the top of the other, the complete stack is known as a **granum**. According to Wolken (1961) and Calvin (1959) the chloroplast is a lamellar structure composed of lipid and aqueous protein layers.

The thylakoids in the chloroplasts contain most of the machinery for photochemical reactions of photosynthesis they contain pigments required for capturing solar energy. The most important pigments are chlorophylls.

Chlorophyll Pigments

The chlorophylls, the green pigments of plants, are the most important pigments active in the process of photosynthesis. There are at least eight types of chlorophyll pigments. They are chlorophyll a, b, c, d, and e; bacteriochlorophyll-a; bacteriochlorophyll-b and chlorobium chlorophyll (bacterioviridin); of these chlorophyll a is most nearly of universal occurrence being present, in all photosynthetic organisms except the green and purple bacteria. Chlorophyll b is found in all higher plants and in the green algae, but is not found in algae of most other classes (*e.g.*, blue-green, brown and red algae). The other chlorophylls (*i.e.*, c, d and e) are found only in algae and in combination with chlorophyll a, bacteriochlorophyll (a and b) are present in the purple bacteria, whereas the green bacteria contain the pigment called bacteriovir-idin (chlorobium chlorophyll). All of the chlorophylls are very similar in chemical composition and all of them are compounds which contain magnesium.

Chlorophyll a and b are the characteristic chlorophylls of the higher plants. Neither of the two is soluble in water, but both are soluble in a number of organic reagents. Chlorophyll-a is readily soluble in absolute ethyl alcohol, ethyl ether, acetone, chloroform and carbon bisulfide. Chlorophyll-b is also soluble in above-mentioned reagents. Chlorophyll a is usually blue-green, while chlorophyll-b is yellow-green.

The chlorophyll a molecule consists of a cyclic structure composed principally of four pyrrolenuclei containing magnesium atom at its centre. Extending from one of the pyrrolerings is a long chain alcohol, the phytol part of the chlorophyll molecule. The molecular formulae for chlorophyll a and b are—$C_{55}H_{72}O_5N_4Mg$ and $C_{55}A_{70}O_6N_4Mg$

Chlorophyll b constitutes about one-fourth of the total chlorophyll content and absorbs light of different wavelength than the chlorophyll a. On absorbing light, the chlorophyll b molecule is excited and transfers its energy to the chlorophyll a molecule. Finally, the chlorophyll a molecule converts the light energy into electrical energy by bringing about charge separation. Here, chlorophyll a molecules act as **reaction centres**.

The Carotenoids (Carotenoid Pigments)

The carotenoids are lipid compounds. They are found to be distributed widely in both animals and plants. They are red, orange, yellow and brownish pigments. They are present in variable concentrations in nearly all higher plants and many microorganisms such as — red and green algae, photosynthetic bacteria and fungi.

β-carotene, the major carotenoid is orange-yellow pigment and found in plant tissues. This is generally accompanied by a-carotene.

The carotenoids which consist of carbon and hydrogen are known as **carotenes**, while the carotenoids containing oxygen are called **xanthophylls** which are found more frequently in nature than the carotenes. The major xanthophylls are generally found in green leaves.

The carotenoids are located in the chloroplasts and chromatophores. They occur there as water-insoluble protein complexes.

Like chlorophyll carotenoids are also embedded in the thylakoid membranes of chloroplasts. These accessory pigments act as antenna complexes and procure light from different regions of the spectrum than the chlorophyll. The light captured by these pigments is transferred to the reaction centres of chlorophyll-a for conversion into electrical energy. The accessory pigments (carotenoids) and the reaction centre (chlorophyll-a), together form **photosystem**.

Phycobilins

The phycobilins are found only in algae. The red and blue phycobilins are called **phycoerythrins** and **phycocyanins**, respectively. They are active in the transfer of light energy to chlorophyll for utilization in the process of photosynthesis. The role of phycobilins and carotenoids is indirect in the sense that the energy they absorb is transferred to chlorophyll before it becomes active in photosynthesis.

PHOTOCHEMICAL AND BIOSYNTHETIC PHASES

Radiant Energy

The pigment system of the chloroplast first absorbs light energy and then passes it on via intermediates to the products of photosynthesis.

When a chlorophyll molecule absorbs a light quantum (*i.e.*, photon) the molecule is excited; it means that it has been brought from its normal ground state to an excited state (*i.e.*, higher energy level). All quanta are not able to lift chlorophyll to a higher state of energy. Firstly the light is being absorbed and, therefore, the quantum absorbed possesses a sufficient amount of energy to facilitate this job. The quantum energy may be determined from the wavelength of radiation; the shorter the wavelength greater the energy.

White light, as it comes from the sun is composed of different wavelengths, ranging from the relatively long waves of red light, through successively shorter waves to violet light. **Wave length** (l) denotes the linear length by which a wave travels forward in one complete vibration. This is generally shown by units such as millimicron (mμ) or Angstrom (Å)—(1 cm = 10^{-8} Å). When white light is passed through a glass prism, it is resolved into these colours. The band of colours, is the visible spectrum. The complete visible spectrum is composed of red, orange, yellow, green, blue, indigo and violet colours. Our eyes can perceive only the **visible rays** whose wave lengths are between 390 nm to 780 nm. The wavelengths exist that we are unable to perceive with our eyes. Beyond the red are still longer, invisible rays, the **infrared**; and beyond the violet are

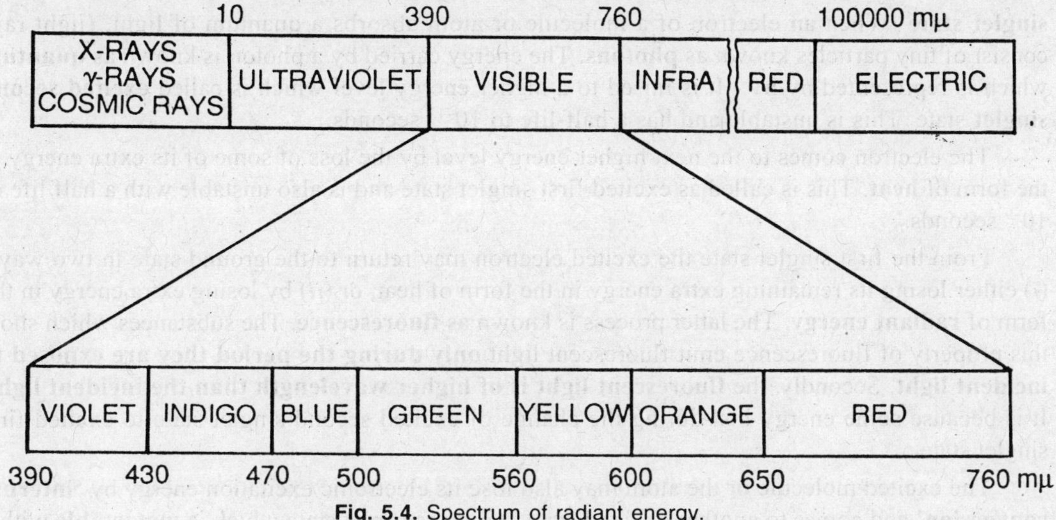

Fig. 5.4. Spectrum of radiant energy.

shorter, invisible rays, the **ultraviolet**. This way, the visible rays represent only a part of the radiant energy that comes to the earth from the sun. Here only a part of the visible spectrum is effective in photosynthesis.

Chlorophyll, however, is a pigment that gives green colour to the leaves. It absorbs light in the violet and blue wavelengths, and also in the red region of the visible spectrum of the light. This portion of the spectrum between 400 nm and 700 nm is called **photosynthetically active radiation (PAR).** As chlorophylls reflect the green light, thus they impart green colour to the leaves.

Transfer of Energy

Not all of the pigment molecules absorb light or are activated at once. Light energy absorbed by one pigment molecule is thought to be transferred through many other pigment molecules before reaching its site of action. All the pigments except chlorophyll a are known as accessory pigments. Here the transfer of light energy may be from one chlorophyll a molecule to another, from chlorophyll ƀ to chlorophyll a, from carotenoids to chlorophyll a, or from phycobilins to chlorophyll a.

The energy can be transferred from molecule to molecule as follows—the normal state of molecules is known as **ground state** or

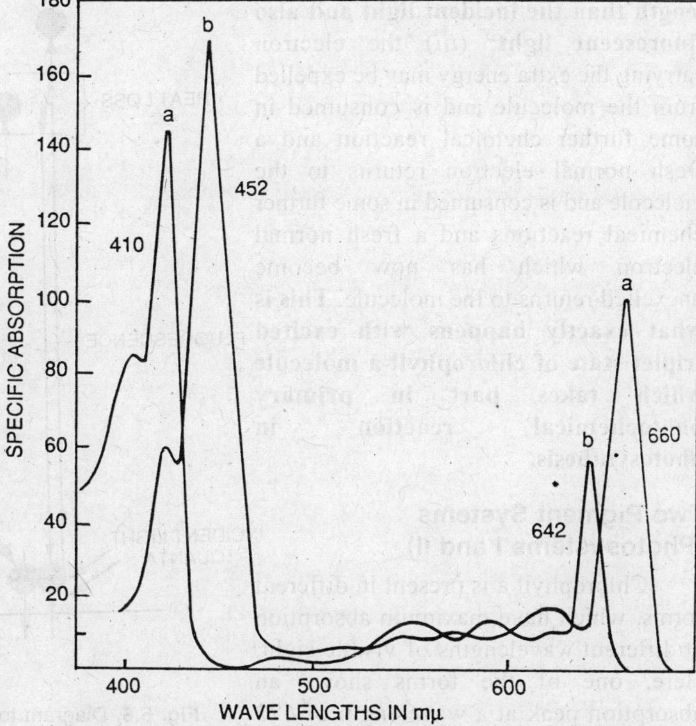

Fig. 5.5. Absorption spectra of chlorophyll a and b in either solution.

singlet state. When an electron of a molecule or atom absorbs a quantum of light, (light rays consist of tiny particles known as **photons**. The energy carried by a photon is known as **quantum** which is represented by **hv**.) It is raised to a higher energy level which is called **excited second singlet** state. This is unstable and has a half-life to 10^{-12} seconds.

The electron comes to the next higher energy level by the loss of some of its extra energy in the form of **heat**. This is called as excited first singlet state and is also unstable with a half life of 10^{-9} seconds.

From the first singlet state the excited electron may return to the ground state in two ways: (*i*) either losing its remaining extra energy in the form of heat, or (*ii*) by losing extra energy in the form of **radiant energy**. The latter process is known as **fluorescence**. The substances which show this property of fluorescence emit fluorescent light **only during the period they are exposed to incident light**. Secondly, the **fluorescent light is of higher wavelength than the incident light**. It is because some energy lost during the change of excited second singlet state to excited first singlet state.

The excited molecule or the atom may also lose its electronic excitation energy by **'internal conversion'** and comes to another exicited state known as triplet state which is **metastable** with a half life of 10^{-3} seconds. In this excited state the electron carrying extra energy can take part in further reaction.

From the triplet state the excited molecule or the atom may return to the ground state in three ways: (*i*) by losing its remaining extra energy in the form of heat: (*ii*) by losing the extra energy in the form of radiant energy. The latter process is termed **phosphorescence**. The substances which show this property of phosphorescence emit phosphorescent light **even after the incident radiant light is cut off**. Secondly, the **phosphorescent light is of higher wave length than the incident light and also fluorescent light**; (*iii*) the electron carrying the extra energy may be expelled from the molecule and is consumed in some further chemical reaction and a fresh normal electron returns to the molecule and is consumed in some further chemical reactions and a fresh normal electron which has now become unexcited returns to the molecule. **This is what exactly happens with excited triplet state of chlorophyll-a molecule which takes part in primary photochemical reaction in photosynthesis.**

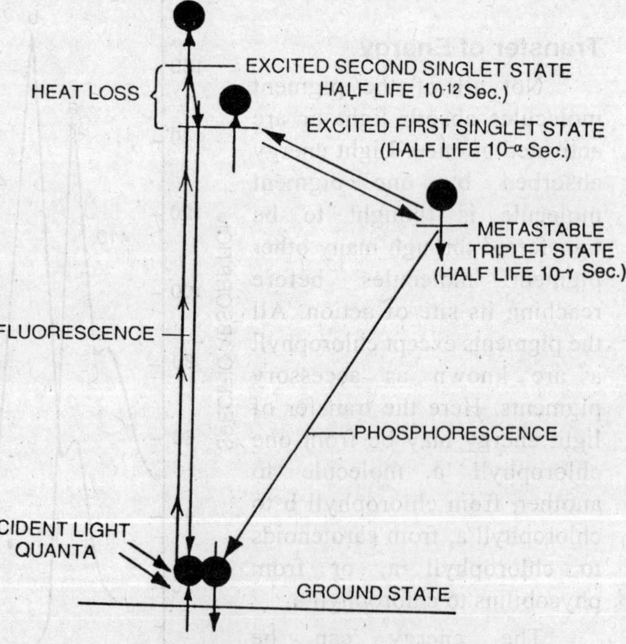

Fig. 5.6. Diagram to show excitation of molecules to fluorescence or phosphorescence.

Two Pigment Systems (Photosystems I and II)

Chlorophyll a is present in different forms, which have maximum absorption at different wavelengths of visible sight. Here, one of the forms shows an absorption peak at a wavelength of 673 nanometers (nm) and is called **chl a 673**.

Table 5.2. Differences between Photosystem I and Photosystem II

Photosystem I (PS I)	Photosystem II (PS II)
1. It is involved both in the cyclic and non-cyclic photophosphorylation.	1. It is involved only in non-cyclic photophosphorylation.
2. It receives the electrons from the photosystem II. This system produces a strong reductant which reduces $NADP^+$ to $NADPH_2$	2. PS II donates electron to PS I where $NADP^+$ is reduced.
3. Molecular oxygen is not evolved in this system.	3. This system is responsible for the photolysis of water and involves the evolution of molecular oxygen.
4. PS I is located on the outer surface of the thylakoid, *i.e.*, appressed regions of grana.	4. It is located on the inner surface of the thylakoid, *i.e.*, in the stroma and non-appressed region of grana.
5. The reaction centre of PS I is P_{700}.	5. The reaction centre of PS II is P_{680}.
6. Chlorophyll, carotenoid pigments are in the ratio 20-30 : 1.	6. Chlorophyll, carotenoid pigments are in the ratio of 3-7 : 1.
7. Chlorophyll *a* with maximum absorption at 700 nm (P_{700}).	7. Chlorophyll *a* with peak absorption at 680 nm (P_{680}).

In the similar way two other forms of chlorophyll a are **chl a 683 (P_{680})** and **chl a 703 (P_{700})**, with peak absorption at 683 and 703 nm, respectively. These pigment are anchored in thylakoids in separate units of organisation called **photosystems**. A single photosystem consists of about 250 to 400 pigment molecules.

According to Butler (1966), chlorophyll-a exists in two forms, one form with an absorption maximum at 673 nm (chl a 673) and the other with an absorption maximum at 683 nm (chl a 683). In addition to these two pigment systems or photosystems a long wave absorbing chlorophyll is also present in much smaller amounts. This is known as P 700 with an absorption maximum at 700 nm. According to Clayton (1966), this is another form of chlorophyll. Here, P stands for pigments.

The photosynthesis takes place by two photochemical processes, each process being associated with a specific group of pigments. These light absorbing group of pigments are called **pigment system I** or **photosystem I** and **pigment system II** or **photosystem II**. Light energy for pigment system (photosystem I) is collected by chl *a* 683, P 700 and the carotenoids, whereas the light energy for pigment system II (photosystem II) is collected by chl *a* 673, chlorophyll b, phycobilins (Jones and Myers, 1964). The primary function of the two photosystems (PS I and II), which interact with each other is to trap light energy and convert it to the chemical energy (ATP). This chemical energy (ATP) is utilised by living cells.

Photosynthetic Unit

Now this has been proved that the chloroplasts are composed of numerous minute photosynthetic units. A photosynthetic unit is the smallest group of collaborating pigment molecules necessary to effect a photochemical act. Park and Biggins (1946) isolated the photosynthetic unit from chloroplast lamellae. They called the photosynthetic units as **quantasomes**.

Hydrogen acceptor. There are two principal forms of hydrogen acceptor which can receive hydrogen from a substrate under the action of the appropriate dehydrogenase.

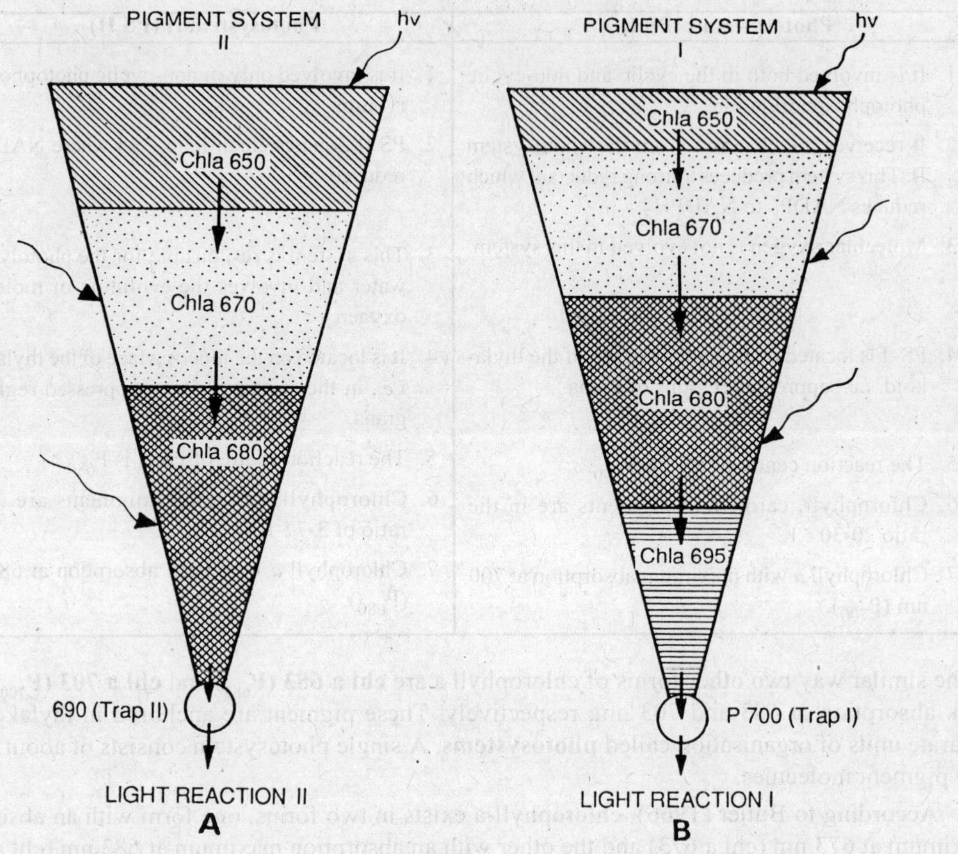

Fig. 5.7. A hypothetical scheme for the distribution of the chlorophyll in two pigment systems in green algae and higher plants.

According to the latest nomenclature, recommended by the international commission they are:

Coenzyme I–nicotinamide adenine dinucleotide (NAD)

Coenzyme II–nicotinamide adenine dinucleotide phosphate (NADP).

Adenosine triphosphate (ATP). This compound occurs in all living cells and consists of the nucleotide adenosine (A) and three phosphate radicals (—P).

A—P—P—P

For the most part it is only ATP which is able to carry out the actual transference of the high energy phosphate. (~ P) so that the ~ P formed from low energy sources is used to regenerate ATP from adenosine diphosphate (ADP) rather than be used directly.

"ATP is the main fuel of life produced in photosynthesis and oxidative phosphorylation. In both cases it is produced by an electric current, that is, energy released by a 'dropping' electron".

MECHANISM OF PHOTOSYNTHESIS

Carbon dioxide assimilation. Arnon is 1954 found that isolated chloroplasts under suitable experimental conditions, could assimilate CO_2. The enzymes involved in the reduction (assimilation) of carbon dioxide, therefore, must be present and perhaps produced within the chloroplast. The reduction of carbon dioxide is accompanied by the evolution of O_2. With the use of radioactive carbon dioxide and chromatographic techniques, D. I. Arnon and his associates could identify several soluble and insoluble products, such as—phosphate esters of glucose; fructose; ribulose; sedoheptulose; dihydroxyacetone and glyceric acid; glycolic, malic and aspartic acids; alanine, glycine; free dihydroxy-acetone; and glucose.

Stages in photosynthesis. It is now generally agreed that photosynthesis consists of two stages— (*i*) light stage and (*ii*) dark stage. The light is necessary for the light stage, whereas the dark stage is independent of the presence of light.

Evidences for the Existence of Light and Dark Stage

Three important evidences have been considered here.

Flashing light experiments. Warburg (1919) compared the rate of photosynthesis of a plant kept in continuous light for a certain time with the rate for the plant when supplied with alternate light and dark periods but receiving the same total amount of light. The rate of photosynthesis of the latter was found to be sufficiently greater than that of the plant kept in continuous light.

The explanation is as follows :

$$A \xrightarrow{\text{Photochemical}} B \xrightarrow{\text{chemical}} C \text{ (photosynthesis product).}$$

In continuous light of high intensity the reaction A—B proceeds at a faster rate than B—C so that there is a tendency for some B to accumulate. The effect of intermittent light therefore is to enable B—C to proceed in the dark at a time when there is no production of B, as well as for B—C continue in the light. This way, for given total quantity of light, there is more C produced.

Temperature experiments. Over a temperature range of about 10°–25° C, if light intensity and CO_2 concentration are relatively high, the Q_{10} (temperature coefficient) of photosynthesis is approximately two. Strictly chemical reactions have a Q_{10} of from two to three. This fact indicates that at least one of the reactions involved in photosynthesis is of purely chemical type. Since this fact was first pointed out by Blackman, this reaction is known as **Blackman reaction**. This is also called the **dark reaction**, since it does not require light and therefore, may take place in either the light or the dark.

A chemical reaction which proceeds only at the expense of absorbed light is called a photochemical reaction. The Q_{10} of a photochemical reaction is approximately one.

Temperature coefficient Q_{10}. The temperature coefficient of a physiological, chemical or physical process is the ratio of the process at any stated temperature of the rate at a lower temperature usually 10° C lower, and designed as the Q_{10} of the process. Thus, if a process is 2.3 times as fast at 25° as at 15° C, the temperature coefficient, Q_{10} is 2.3.

Temperature does not accelerate photochemical reactions. Under conditions of high light intensity, increased temperature accelerates the reaction B—C at the expense of the excess B (Q_{10} = 2).

When there is a low light intensity, A—B is slower than B—C will not affect the overall production of C. The limiting reaction A—B is photochemical (Q_{10} = 1):

Dark pick-up of CO_2. Under conditions of intense illumination but lacking CO_2, 'B' would accumulate and this would react with CO_2 in the dark.

ELECTRON TRANSPORT CHAIN

Photosystem II (PS II)

The light-driven reaction of photosynthesis also called **light reaction (Hill reaction)**, referred to as **electron transport chain**, were first propounded by Robert Hill in 1939.

The electron transport chain of photosynthesis is initiated by absorption of light by **photosystem II** (P_{680}).

When P680 absorbs light, it is excited and its electrons are transferred to an **electron acceptor molecule**.

Therefore, P_{680} becomes a strong oxidising agent, and splits a molecule of water to release oxygen. This light-dependent splitting of water molecule is called **photolysis**.

However, manganese, calcium and chloride ions play important roles in photolysis of water.

After photolysis of water, electrons are generated, which are then passed to the oxidised P_{680}.

Now, the electron deficient P_{680} (as it had already transferred its electrons to an acceptor molecule) is able to restore its electrons from the water molecule.

After accepting electron from the excited P_{680}, the primary electron acceptor is reduced. The primary electron acceptor in plants is **pheophytin**.

The reduced acceptor which is a strong reducing agent, now donates its electrons to the downstream components of the electron transport chain.

Photosystem I (PS I)

Similar to photosystem II (P_{680}), photosystem I (P_{700}) is excited on absorption of light and gets oxidised, and transfers its electrons to the primary electron acceptor (pheophytin), which, in turn gets reduced.

While the oxidised P_{700} draws electrons from photosystem II, the reduced electron acceptor of photosystem I, transfers electrons to ferredoxin and ferredoxin–NADP reductase to reduce NADP to $NADPH_2$.

$NADPH_2$ is a powerful reducing agent, and is utilised in the reduction of CO_2 to carbohydrates in the carbon reaction of photosynthesis.

The reduction of CO_2 to carbohydrates requires energy in the form of ATP, produced through electron transport chain.

Process of ATP formation from ADP in the presence of light in chloroplasts is called **photophosphorylation**.

The Light Reaction (Hill Reaction)

The light reaction is thought to be responsible for the production of a 'reducing power' and oxygen from water as a result of light energy. This is as follows :

The light energy, after absorption by chlorophyll, splits H_2O.

$$H_2O + \text{Light} \longrightarrow (H) + (OH)$$

(*i*) The (H) combines with an unidentified compound (probably ferredoxin) and is passed from this to NADP.

(*ii*) The $NADPH_2$ can cause the reduction of phosphoglyceric acid .... Phosphoglyceraldehyde, together with some ATP production.

(*iii*) The (OH) forms H_2O and oxygen :

$$4\,(OH) \longrightarrow 2H_2O + O_2$$

The light reaction gives rise to two very important production .... (*i*) a reducing agent $NADPH_2$ and (*ii*) an energy rich compound ATP. These two products of the light reaction are utilized in the dark phase of photosynthesis.

The energy transformations in photosynthesis are as follows :

(1) The radiant energy of an absorbed quantum is transformed into the energy of an activated pigment molecule

$$P + \text{Light energy} \longrightarrow P$$

(pigment molecule or activated pigment)

(2) Now the activated pigment removes an electron from the hydroxyl ion derived from the water molecule. The (OH) represents the 'free radical'. These are uncharged, but highly reactive forms.

$$H_2O \rightleftharpoons H_+ + OH^- + P$$

$$(OH)\ Pe^-$$

(3) The free radicals react in many ways; the release of oxygen and formation of free radicals of hydrogen takes place.

$$2\ (OH) \longrightarrow O_2 + 2\ (H)$$

(4) The H^+ ions from water, together with the electron attached to the pigment are transferred to certain molecules, which then carry the reducing power to other reactions.

$$NADP + H^+ + e^- \longrightarrow NADP.H$$

(reducing power)

(5) Another reaction is the recombination of the split products of water into the water molecules itself.

$$(H) + (OH) \longrightarrow H_2O$$

This reaction is strongly energy-releasing. The chloroplast puts this reaction to work by causing it to synthesize energy-rich ATP from a precursor molecule ADP and inorganic phosphate (P_i):

$$(H) + (OH) \downarrow H_2O \qquad ADP + P_i \downarrow ATP$$

(6) The energy of the ATP can now be used, in the reduction of CO_2 to sugar by the reducing power (NADP.H) generated in the light reaction.

$$CO_2 + NADP.H \longrightarrow (CH_2O) + H_2O + NADP \qquad (ATP \longrightarrow ADP + P_i)$$

This way, the radiant energy has been converted to the chemical energy of the sugar molecule by passing through a photoactivated pigment, photolyzed water fragments, and ATP. The main function of light energy in photosynthesis is to produce ATP through a complex of reactions called **photophosphorylation**.

The subsequent reactions leading to the formation of sugar from CO_2 can proceed entirely in darkness.

Photophosphorylation

Photosynthetic phosphorylation. With the discovery that CO_2 can be assimilated in isolated chloroplasts, this came into existence that the chloroplast must contain the enzymes necessary for this assimilation and must be able to produce the ATP (adenosine tri-phosphate) essential for the formation of the main photosynthesis products. Arnon and his co-workers (1954) demonstrated that the isolated chloroplasts can produce ATP in the presence of light. They gave the name to this process **photosynthetic phosphorylation**. This was revealed for the first time that mitochondria are not the only cytoplasmic particles that produce ATP. ATP formation in chloroplasts differs from that in mitochondria in that it is free from respiratory oxidations. During this process the light energy is being converted to ATP. In other words, there is a conversion to light energy of chemical energy.

ATP is only one of the necessary requirements for the reduction of carbon dioxide to the carbohydrate level. A reductant must be formed in photosynthesis that will provide the hydrogens or electrons for this reduction. Arnon (1951) demonstrated that isolated chloroplasts are capable of reducing pyridine nucleotides in light. The photochemical reaction and an enzyme system are capable of utilizing the reduced pyridine nucleotide as soon as this was formed, Arnon *et al.*, (1957) found that NADP. H_2 is the reduced pyridine nucleotide in photosynthesis. In the presence of H_2O, ADP (adenosine di-phosphate) and orthophosphate (P), substrate amounts of NADP (nicotinamide adenine dinucleotide phosphate) were reduced, accompanied by the evolution of oxygen. The equation is as follows :

$$2ADP + 2P + 2NADP + 4H_2O \longrightarrow 2ATP + O_2 + 2NADPH_2 + 2H_2O$$

As shown by the equation the evolution of one molecule of oxygen is accompanied by the reduction of two molecules of NADP and esterification of two molecules of orthophosphate. Together, ATP and $NADPH_2$ provide the energy requirements for CO_2 assimilation. Arnon gave name to this power **assimilatory power** (*i.e.*, ATP + $NADPH_2$).

According to Arnon (1967), in bacterial photosynthesis $NADH_2$ is utilized of $NADPH_2$.

Ferredoxin. In the late 1950's the reduction of $NADP^+$ was thought to be associated with a soluble protein factor found in chloroplasts. Arnon *et al.* (1957) observed that this protein reduced $NADP^+$ accompanied by the evolution of oxygen. They termed it the 'NADP reducing factor.' Thereafter the NADP reducing factor was purified and called photosynthetic pyridine nucleotide reductase (PPNR), since its catalytic activity was only apparent when chloroplasts were kept in light (San Pietro and Lang, 1958). Tagawa and Arnon (1962) recognized that PPNR is one of a family of nonhemenonflavin, iron-containing proteins that is universally present in chloroplasts. These proteins were given a generic name **ferredoxin**.

When ferredoxin was not discovered, NADP was thought to be the terminal electron acceptor of the photosynthetic light reaction. Arnon (1967) revealed that illuminated chlorophyll reacts directly with ferredoxin and not with $NADP^+$. The exposition of chlorophyll to light causes a flow of electrons to ferredoxin. Now the reduced ferredoxin causes the reduction of $NADP^+$ in an enzyme catalyzed reaction that is independent of light (Arnon, 1967). In other words, ferredoxin is termed as terminal **electron acceptor** of the photosynthetic light reaction.

The reduction of NADP takes place by ferredoxin. Under normal condition, in photosynthesis ferredoxin reduced by the acceptance of an electron is immediately reoxidized by $NADP^+$. The reduction of NADP by ferredoxin is catalyzed by ferredoxin-NADP reductase. This shows that the mechanism of $NADP^+$ reduction in photosynthesis completes in three steps (Shin and Arnon, 1965). These steps are–(*i*) photochemical reduction of ferredoxin; (*ii*) reoxidation of ferredoxin by ferredoxin $NADP^+$ reductase and (*iii*) reoxidation of ferredoxin-$NADP^+$ reductase by $NADP^+$. According to Arnon there are two types of photophosphorylation: (*i*) non-cyclic photophosphorylation and (*ii*) cyclic photophosphorylation.

Non-cyclic photophosphorylation. This is a result of an interaction of photosystem I (PSI) and photosystem II (PSII). In non-cyclic photophosphorylation, the electron is not returned to the chlorophyll molecule, but is taken up by NADP ± which thereafter reduces to NADPH. Here the electron that returns to the chlorophyll molecule is derived from an outside source which is water.

In this process oxygen is released and both $NADPH_2^-$ and ATP are formed. In green plants and many photosynthetic bacteria, however, illumination is known to produce also $NADPH_2^-$ which provides hydrogen for the reduction of carbon dioxide in the day. The electron lost by the excited chlorophyll is accepted by NADP along with a proton resulting in the formation of $NADPH_2$. The light energy is now stored in the $NADPH_2$ molecule. The proton required for the reduction of NADP is released from the dissociation of water molecule by photolysis into hydrogen $H^{\pm}$ and hydroxyl ions OH.

$$2H_2O \longrightarrow 2H^+ + 2\,(OH)$$

The hydroxyl ions react to produce water and molecular oxygen. The reaction is as follows:

$$4\,(OH) \longrightarrow 2H_2O + O_2 + 4e^-$$

Here the hydroxyl ion also releases an electron that is accepted by the cytochromes of the chloroplast. In turn, the cytochrome donates this electron to the chlorophyll molecule, which already lost an electron earlier. The energy released during this transfer of electron from the cytochrome is utilized in the formation of ATP by the photophosphorylation of ADP.

In water molecule hydrogen is strongly bound to oxygen and this can be cleaved only by the use of energy. This energy is supplied by light. This way, in non-cyclic photophosphorylation light energy takes part in two processes, *i.e.*, the activation of chlorophyll molecule and photolysis (cleavage) of water.

In non-cyclic photophosphorylation one molecule of $NADPH_2$ and one molecule of ATP are produced by the activation of chlorophyll molecule by a photon, while in cyclic photophosphorylation two molecules of ATP are produced for each photon absorbed by chlorophyll. The overall reaction of photophosphorylation is as follows:

$$2ADP + 2iP + 4H_2O + 2NADP \xrightarrow{\text{Light}} O_2 + 2ATP + 2NADPH_2 + 2H_2O$$

Cyclic photophosphorylation. When non-cyclic photophosphorylation is stopped under certain conditions, cyclic photophosphorylation takes place. The non-cyclic photophosphorylation can be stopped by illuminating isolated chloroplasts with light of wavelength greater than 680 nm.

By this way, only photosystem I (PS I) is activated, as it has a maximum absorption at 700 nm, and photosystem II (PS II), which absorbs at 680 nm, remains inactivated.

Due to inactivation of PS II, the electron flow from water to NADP is stopped, and also CO_2 fixation is retarded.

When CO_2 fixation stops, electrons are not removed from reduced NADPH. Thus, NADPH will not be oxidised and NADP will not be available as an electron acceptor.

Under above-mentioned conditions, cyclic-photophosphorylation occurs.

During cyclic-photophosphorylation, electrons from photosystem I (PS I) are not passed to NADP from the electron acceptor, as NADP is not available in oxidised state to receive electrons.

Hence, the electrons are transferred back to P_{700}.

This type of movement of electrons from an electron acceptor to P_{700} result in the formation of ATP from ADP, and the process is called cyclic photophosphorylation.

During cyclic photophosphorylation oxygen is not released, as there is no photolysis of water and $NADPH_2$ is also not produced.

In cyclic photophosphorylation the excited electron lost by the chlorophyll is returned to it through vitamin K or FMN (flavin mononucleotide) and cytochromes. The chlorophyll molecule on losing an electron assumes a positive charge and subsequently the electron is transferred to a second acceptor. This second acceptor is a group of substances collectively known as **cytochrome**

system. All the members of cytochrome system are variants of cytochrome. Ultimately these cytochromes transfer the electron to the chlorophyll molecule from where it was lost initially.

The electromagnetic energy of the light is utilized in the formation of ATP. This means that light energy is being converted into chemical energy. Here the electron after leaving a chlorophyll travels in a cyclic way and ultimately returns to the same molecule from which it initiated, and therefore, this process has been termed by Arnon as cyclic photophosphorylation. The final electron acceptor and the initial electron donor is the same substance—the chlorophyll. No outside material takes part in the process. During cyclic photophosphorylation, one electron and two ATP molecules are formed. One ATP molecule is being formed when the electron travels from the cofactor (*i.e.*, vitamin K or FMN) to the cytochromes while the other when it travels from the cytochromes back to the chlorophyll molecule.

Here the light energy is being converted into chemical energy.

In nature both processes of photophosphorylation proceed simultaneously. In green plants the non-cyclic electron transfer is essential for the production of $NADPH_2$ and ATP. The oxygen is evolved during the process. The cyclic electron transfer fulfills the requirement of the low yield of ATP during non-cyclic process. This way, the complete light phase of photophosphorylation produces ATP and $NADPH_2$ and oxygen is evolved.

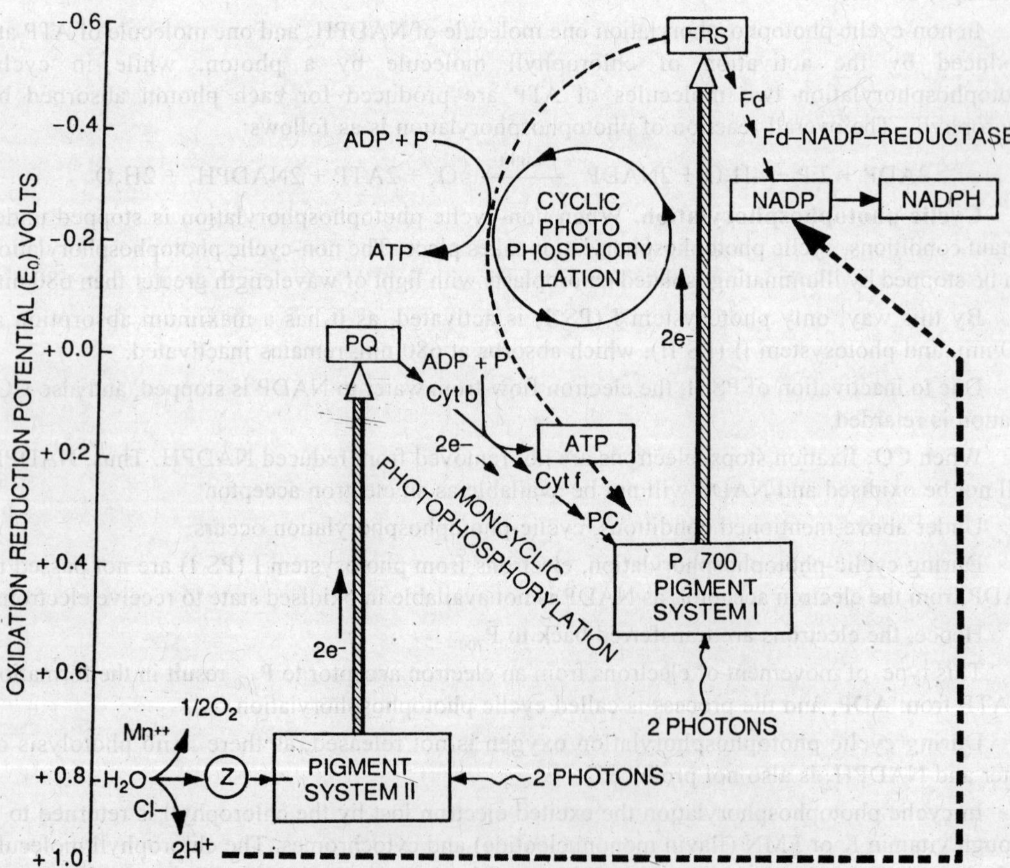

Fig. 5.8. Schematic representation of the electron transport and photophosphorylation in photosynthesis.

$NADPH_2$ is a biological reductant that brings about the reduction of carbon dioxide to carbohydrates in the dark phase of photosynthesis. Here both $NADPH_2$ and ATP provide energy for reduction. The assimilatory power of the cell is constituted by these two components. The energy of these components is derived from visible part of sunlight. In the dark phase of photosynthesis the energy, that is stored in $NADPH_2$ and ATP, is being transferred to the molecules of organic substances and stored there in the form of chemical energy.

During photosynthesis the electromagnetic energy of visible light is being converted into chemical energy. Now this energy is utilized by living cells as the driving force for various vital activities. This act of the conversion of energy is brought about by the photosynthetic cells of green plants or photosynthetic bacteria. Here the solar energy is trapped by the chlorophyll apparatus. As soon as the light energy is being transformed into chemical energy, it may be used in the formation of carbohydrates, protein synthesis and other important vital activities. The living are so designed that they can use only chemical energy for various metabolic activities. The light energy cannot be directly used for these vital activities. The light reaction of the higher plants takes place in the grana of the chloroplasts.

Table 5.3. Differences Between Cyclic and non-cyclic Photophosphorylation.

Cyclic photophosphorylation	Non-cyclic photophosphorylation
1. It is found dominantly in bacteria.	1. This is dominant in green plants.
2. It involves photosystem I only.	2. It involves both photosystems I and II.
3. Electrons move in a closed circle. Electron expelled from chlorophyll (P_{700}) returns to it after passing through a series of electron acceptors in an ETS.	3. Electrons do not move in a closed circle. Here electron expelled by reaction centre (P_{680}) are not returned to them. Here water is the ultimate source of electrons and $NADP^+$ is the final acceptor.
4. It is not involved in the photolysis of water and therefore, no oxygen is involved.	4. It is involved in the photolysis of water, and therefore oxygen is released as a by-product.
5. It sythesises only ATP, reduced NADP, *i.e.*, $NADPH_2$ is not synthesised.	5. It is involved in the synthesis of both ATP and $NADPH_2$.
6. It operates under low light intensity, anaerobic conditions and the poor availability of CO_2.	6. It takes place under optimum light, aerobic conditions and in the presence of sufficient CO_2.

Biosynthetic Phase

As already has been discussed, how ATP and $NADPH_2$ are formed during photochemical reactions in chloroplasts.

Both ATP and $NADPH_2$ are essentially needed for assimilation of CO_2 to carbohydrates. The reactions that catalyse assimilation of CO_2 to carbohydrates take place in the stroma of chloroplast where all necessary enzymes are present.

These reactions are called **carbon reactions (dark reactions)** that lead to the photosynthetic reduction of carbon to carbohydrates.

Carbon reactions (dark reactions). This part of photosynthesis is called synthesis stage and is the anabolic counterpart of respiration, *i.e.*, the synthesis of carbohydrate from carbon dioxide. This is called **photosynthetic assimilation** as it occurs in green plants. Whereas carbon can be assimilated by non-green plants and even by animal non-photosynthetically, there it is called **dark carbon** assimilation. Both these processes (dark C-assimilation and photosynthesis C-assimilation) differ in nature of end product. Carbohydrates are formed as a result of photosynthesis C-assimilation but organic acids are formed as end products in dark C-assimilation. During this process carbon dioxide and water build up carbohydrate, oxygen being released. Energy is absorbed in form of light energy and converted into chemical energy of carbohydrate molecule.

$$CO_2 + H_2O \xrightarrow[\text{Chlorophyll}]{\text{Light}} (CH_2O) + O_2$$

In first phase of carbon reactions (dark reactions), CO_2 is accepted by a 5-carbon molecule, ribulose-1, 5-bisphosphate (RuBP) and two molecules of 3-carbon compound, *i.e.*, 3-phosphoglycerate (PGA) are formed.

The 3-carbon molecule, *i.e.*, 3-phosphoglycerate, is first stable product, and therefore, this is called **C_3 pathway**.

Formation of 3-phosphoglycerate (PGA) is known as **carboxylation**. This reaction is catalysed by an enzyme called **ribulose bisphosphate carboxylase** (*i.e.*, **Rubisco**).

In addition to carboxylase activity this enzyme (Rubisco) also possesses oxygenase activity, and therefore abbreviated as **Rubisco** (*i.e.*, **ribulose bisphosphate carboxylase oxygenase**).

The oxygenase activity of the enzyme allows O_2 to complete with CO_2 for combining with RuBP (*i.e.*, **ribulose bisphosphate**) in photorespiration.

After carboxylation reaction, reduction of PGA takes place, where ATP and $NADPH_2$ formed during photochemical reactions are utilised.

With the reduction of PGA (*i.e.*, 3-phosphoglycerate), a carbohydrate, glyceraldehyde-3-phosphate, is formed.

These 3-carbon molecules (carbohydrates) are also called **triose phosphates**. They are diverted from Calvin cycle and act as precursors for synthesis of sucrose and starch.

For the completion of Calvin cycle, and for its process of continuation, on its own, **regeneration** of the initial 5-carbon acceptor molecule, *i.e.*, ribulose-1, 5-bisphosphate (RuBP), takes place.

The regeneration of RuBP from glyceraldehyde-3-phosphate requires another ATP molecule formed with the result of photophosphorylation during light reactions.

Path of carbon in photosynthesis. Our knowledge of the path of carbon in photosynthesis comes mainly from a series of brilliant researches initiated by Melvin Calvin (1946) and his co-workers. They made use of the radioactive isotope of carbon, C^{14}, from which they prepared radioactive carbon dioxide ($C^{14}\ O_2$) and then fed this to unicellular algae—*Chlorella* and *Scenedesmus*—for periods of time ranging from five seconds to several minutes. After these short periods of photosynthesis in $C^{14}O_2$ the cells were plunged into boiling ethyl alcohol in order to kill them. The cells were instantaneously killed and their enzymes denatured. The compounds into which C^{14} had been incorporated by photosynthesis were then separated from the alcoholic extract. Among the many compounds found are:

Phosphoglyceric acid (a 3-C compound) after only five seconds exposure.

Phosphoglyceraldehyde
Dihydroxyacetone phosphate] after 30 seconds exposure.

Tetroses, pentoses, hexoses and heptoses after 90 seconds exposure.

After an exposure of only 5 seconds to $C^{14}O_2$ most of the radioactive carbon is found in 3-phosphoglyceric acid (3-PGA), a three carbon compound normally considered a constituent of glycolysis. Increase in the time of exposure to $C^{14}O_2$ to 30-90 seconds found most of the isotopic carbon in phosphoglyceraldehyde dihydroxyacetone phosphate and hexose phosphates as well as 3-PGA.

The C_3 type of carbon reactions occur in the stroma of chloroplast. This **C_3 pathway** is also called **Calvin cycle** after its discoverer, Melvin Calvin, who was awarded Nobel Prize for the discovery of this pathway.

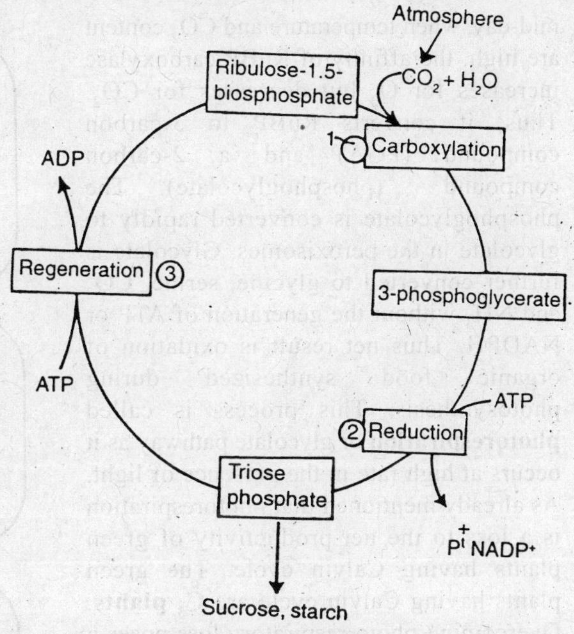

Fig. 5.9. Calvin Cycle.

Initial acceptor of carbon dioxide. The question arises as to what compound or compounds give rise to 3-PGA. In other words, what compound is the initial acceptor of the carbon dioxide molecule? It might be anticipated that the fed $C^{14}O_2$ would couple with some 2-carbon compound to produce the 3-carbon compound (3-PGA), but this seems not to be true. Calvin obtained evidence that a 5-carbon compound ribulose-1, 5, bisphosphate (RuBP), is formed from ribulose-5- phosphate. The $C^{14}O_2$ appears to attach to a 5-carbon compound ribulose-1, 5 bisphosphate (RuBP) to form an unstable 6-carbon compound, which then spontaneously decomposes to two PGA molecules. RuBP is now accepted to be the compound with which carbon dioxide first combines in photosynthesis under the influence of the enzyme, carboxydismutase. This has been confirmed that the 6-carbon compound formed by combination of CO_2 with RuBP is quite unstable and immediately splits to give rise to two molecules of PGA.

The study of the radioactive carbon labelling of the 7-carbon (C_7) and 5-carbon (C_5) sugar phosphates has shown that PGA was starting material for their synthesis and therefore, the precursor, not only of hexoses, but of RuBP.

$$\begin{array}{l} CH_2OPO_3H_2 \\ | \\ C=O \\ | \\ HCOH \\ | \\ HCOH \\ | \\ CH_2OPO_3H_2 \end{array} + C^{14}O_2 + H_2O \longrightarrow \begin{array}{l} CH_2OPO_3H \\ | \\ HCOH \\ | \\ C^{14}OOH \end{array} + \begin{array}{l} COOH \\ | \\ HCOH \\ | \\ CH_2OPO_3H \end{array}$$

Ribulose 1, 5 bisphosphate + CO_2 + H_2O $\longrightarrow$ 2, 3-phosphoglyceric acid.

The carboxylation enzyme occurs in chloroplasts together with phosphoribulokinase which catalyzes the following reaction.

Ribulose 1, 5 -phosphate + ATP $\xrightarrow{\text{Phosphoribulokinase}}$ ribulose 1, 5-diphosphate + ADP.

PHOTORESPIRATION

Respiration that is initiated in chloroplasts and occurs in light only, is called **photorespiration**.

Photorespiration or glycolate pathway. It is interesting to know that in the plants possessing Calvin cycle, the enzyme RuBP carboxylase can initiate the reversal of photosynthetic reactions. This process occurs when there is low CO_2 concentration but high O_2 concentration. At mid-day, when temperature and CO_2 content are high, the affinity of RuBP carboxylase increases for O_2 but decreases for CO_2. Thus, it converts RuBP to 3-carbon compound (PGA) and a 2-carbon compound (phosphoglycolate). The phosphoglycolate is converted rapidly to glycolate in the peroxisomes. Glycolate is further converted to glycine, serine, CO_2 and NH_3 without the generation of ATP or NADPH. Thus net result is oxidation of organic food synthesized during photosynthesis. This process is called **photorespiration** or glycolate pathway as it occurs at high rate in the presence of light. As already mentioned that photorespiration is a loss to the net productivity of green plants having Calvin cycle. The green plants having Calvin cycle are **C_3 plants**. Overcoming photo-respiratory loss poses a challenge to plants growing in the tropics.

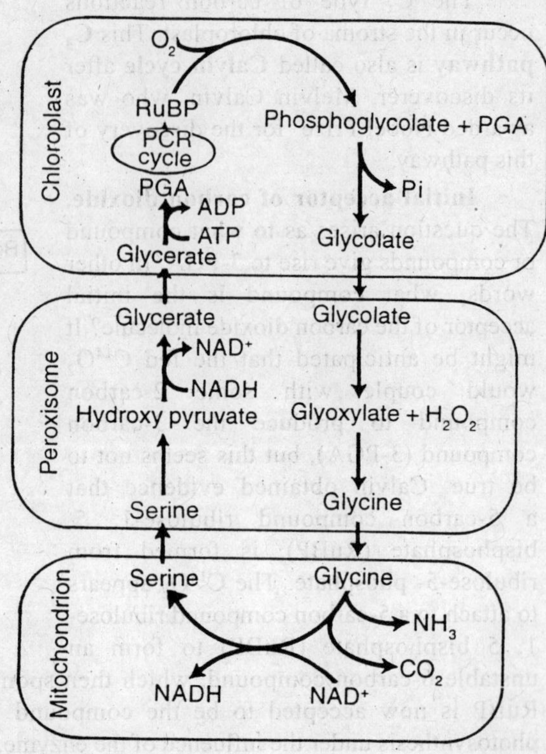

Fig. 5.10. Photorespiration.

Photorespiration occurs due to fact that the active site of enzyme Rubisco (ribulose bisphosphate carboxylase oxygenase) is same for both carboxylation and oxygenation.

The oxygenation of RuBP (ribulose bisphosphate) in the presence of O_2 is first reaction of photorespiration that leads to the formation of one molecule of phosphoglycolate, a two-carbon compound and one molecule of PGA.

Where PGA is used in Calvin cycle, and phosphoglycolate is dephosphorylated to form glycolate in the chloroplast.

From chloroplast, glycolate is diffused to **peroxisome** where it is oxidised to in glyoxylate. Here, glyoxylate is used to form amino acid, glycine. Now, glycine enters mitochondria where two glycine molecules (4 carbons) give rise to one molecule of serine (3 carbons) and one molecule of CO_2 (one carbon).

Now, serene is taken up by peroxisome, and through a series of reactions is being converted into glycerate.

This glycerate leaves the peroxisome and enters the chloroplast, where it is phosphorylated to form PGA.

Now PGA molecule enters the Calvin cycle to make carbohydrates, but one CO_2 molecule released in mitochondria during photorespiration has to be re-fixed. This means, 75 per cent of the carbon lost by the oxygenation of RuBP is recovered and 25 per cent is lost as release of one molecule of CO_2.

Photorespiration is also known as **photosynthetic carbon oxidation cycle**.

As already discussed in foregoing paragraph, this process involves an interaction of three organelles, *i.e.*, (*i*) chloroplast (*ii*) peroxisome and (*iii*) mitochondria.

Table 5.4. Differences between Photorespiration and Aerobic Respiration.

Photorespiration	Aerobic respiration
1. It occurs only in the green cells.	1. It occurs both in green and non-green cells.
2. It occurs only in the presence of light.	2. It occurs all the time irrespective of light.
3. It occurs predominantly in C_3 plants.	3. It takes place in all the plants, except anaerobic microbes.
4. It does not involve glycolysis, Krebs cycle, etc.	4. It involves glycolysis, Krebs cycle or terminal oxidation.
5. It does not produce any usable energy in the form of ATP. There is no phosphorylation.	5. It produces usable energy in the form of ATP. Substrate and oxidative phosphorylations take place.
6. The end products are CO_2 and PGA.	6. The end products are CO_2 and H_2O.
7. Its operation involves chloroplasts, peroxisomes and mitochondria.	7. Its operation involves cytoplasm, matrix and mitochondria only.
8. It maintains the carbon economy of plants.	8. There is no such role.
9. It is not essential for the normal growth of plants.	9. It is essential for the normal growth of plants.
10. In this process saturation point is attained at high oxygen concentration.	10. Here saturation point is attained at comparatively lower oxygen concentration.

Under conditions of high light and limited CO_2 supply, photorespiration plays an important role for protection of plants from photooxidative damage. This shows that if enough CO_2 is not available to utilise light energy for carboxylation, and excess energy inflicts damage to plants.

However, photorespiration being oxygenation of RuBP, utilises part of light energy and saves the plant from photooxidative damage.

The relative levels of O_2 and CO_2 are responsible for determination of the occurrence of photorespiration as both of these gases (O_2 and CO_2) compete for the same active site of enzyme Rubisco.

Increased O_2 level increases photorespiration; while increased CO_2 level decreases photorespiration, and increases C_3 photosynthesis.

Photorespiration and aerobic respiration. Photorespiration is not related to aerobic respiration. Aerobic respiration occurs throughout day and night in all types of cells, whereas photorespiration occurs in illuminated cells only. The main differences between photorespiration and aerobic respiration are as follows :

C_4 PATHWAY (HATCH-SLACK PATHWAY)

The discovery of C_4 cycle in monocots such as sugarcane, maize and sorghum has indicated that these plants have solved the problem of photorespiration. The carbon dioxide is fixed in the mesophyll cells. The initial product being a-4 carbon compound, the process is called C_4 pathway of carbon dioxide fixation.

Hatch-Slack pathway. Two Australian botanists Hatch and Slack (1966) discovered that there are two types of chloroplasts in sugarcane. One type restricted to bundle sheath cells have the normal grana. These chloroplasts carry on Hatch-Slack or C_4 cycle. Hence, Hatch-Slack cycle or C_4 cycle has been found in most monocots and some dicots. The plants having C_4 cycle are known as C_4 plants, and the plants C_3 (Calvin cycle) are C_3 plants.

Photorespiration occurs in C_3 plants (Calvin cycle), which leads to a 25 percent loss of the fixed CO_2. Photorespiration occurs in C_3 plants only, as the enzyme Rubisco catalysis both carboxylation and oxygenation reactions of the initial acceptor molecule that is RuBP.

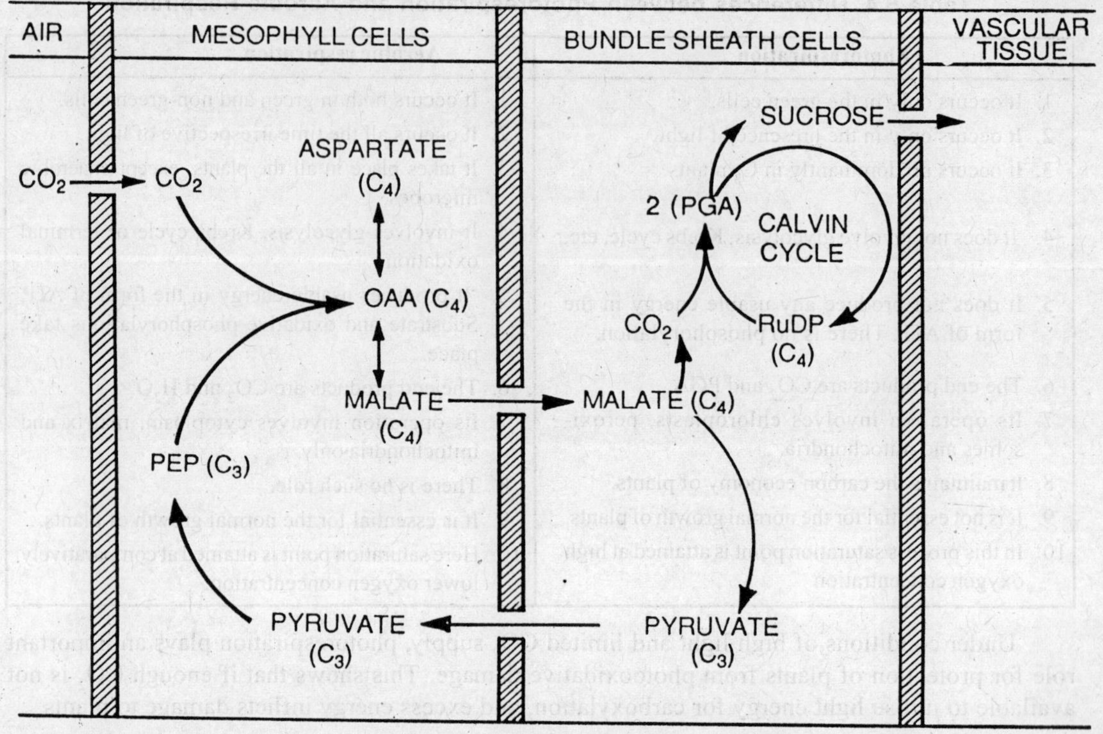

Fig. 5.11. Hatch Slack Pathway of CO_2 fixation.

In C_3 plants, photosynthesis occurs only in mesophyll cells. Photosynthesis has two types of reactions, *i.e.*, light reactions and carbon or dark reactions.

In light reactions, ATP and $NADPH_2$ are produced, and as a result of photolysis of water O_2 is released.

During carbon or dark reactions, CO_2 is assimilated and carbohydrates are produced.

As both light reactions and carbon (dark) reactions occur in mesophyll cells in C_3 plants, it becomes essential for enzyme Rubisco to catalyse both oxygenation and carboxylation reactions of RuBP, simultaneously.

However, in category of C_4 plants, nature has evolved a mechanism to avoid occurrence of photorespiration, which is thought to be a harmful process.

C_4 pathway requires the presence of two types of photosynthetic cells, *i.e.*, mesophyll cells and bundle sheath cells. The bundle sheath cells are arranged in a wreath like manner. This kind of arrangement of cells is called **Kranz anatomy** (Kranz : wreath). In Kranz anatomy, the mesophyll and bundle sheath cells are connected by plasmodesmata or cytoplasmic bridges.

The C_4 plants contain dimorphic chloroplasts. The chloroplasts in mesophyll cells are granal, whereas in bundle sheath cells they are agranal.

The granal chloroplasts contain thylakoids which are stacked to form grana, as formed in C_3 **plants**. However, in agranal chloroplasts of bundle sheath cells, grana are absent and thylakoids are present only as stroma lamellae.

The presence of two types of cells (granal and agranal) allows occurrence of **light** and **carbon (dark)** reactions separately in each type.

Here, release of O_2 takes place in one type, while fixation of CO_2 catalysed by Rubisco enzyme occurs in another type of cells.

In C_4 plants (maize, sugarcane, etc.), light reactions occur in mesophyll cells, whereas CO_2 assimilation takes place in bundle sheath cells. Such arrangement of cells does not allow O_2 released in mesophyll cells to enter in bundle-sheath cells.

Hence, Rubisco enzyme, which is present only in bundle-sheath cells, does not come into contact with O_2, and thus, oxygenation of RuBP is completely avoided.

In C_4 plants, a CO_2 concentrating mechanism is present which helps in reducing the occurrence of photorespiration (*i.e.*, oxygenation of initial acceptor RuBP). This type of CO_2 concentrating mechanism is called **C_4 pathway**.

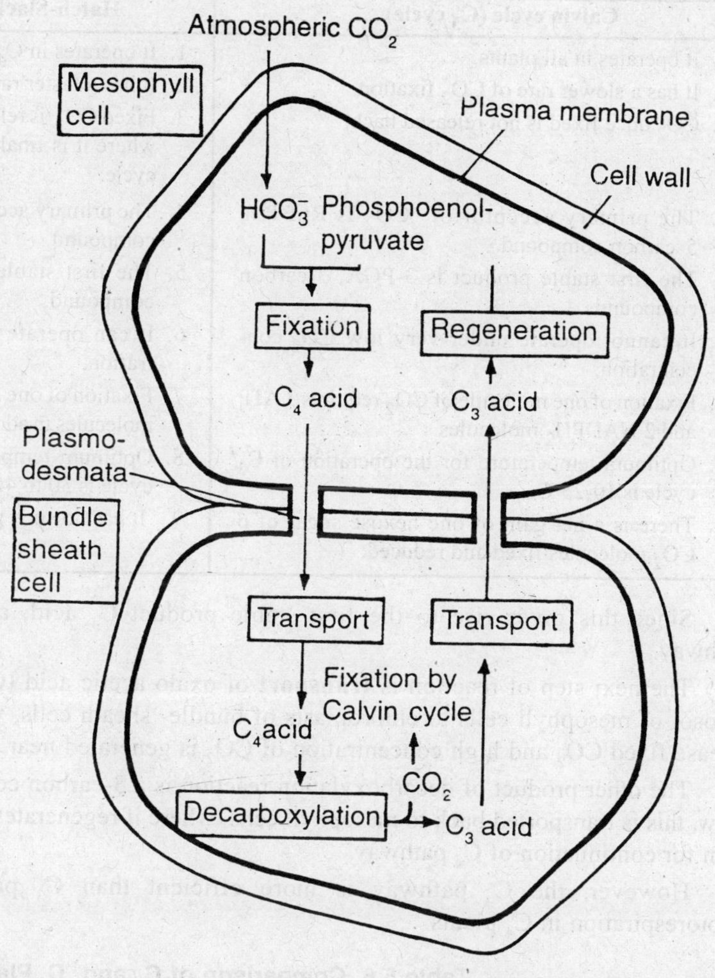

Fig. 5.12. C_4 photosynthetic carbon cycle.

For operation of C_4 pathway, both mesophyll and bundle-sheath cells are required. The main objective of C_4 pathway is to build up high concentration of CO_2 near Rubisco enzyme in bundle-sheath cells. High concentration of CO_2 near Rubisco enhances carboxylation and reduces photorespiration.

C_4 photosynthetic carbon cycle. In C_4 pathway, CO_2 from the atmosphere enters through stomata into the mesophyll cells and combines with phosphoenol pyruvate (3-carbon compound). This reaction is catalysed by an enzyme known as phosphoenol pyruvate carboxylase, *i.e.*, **PEPCase**. With the result, a C_4 acid, oxaloacetic acid (OAA) is formed.

$$CO_2 + PEP + H_2O \xrightarrow[\text{enzyme}]{\text{PEPCase}} \text{Oxaloacetic acid (OAA)} + H_3PO_4$$

The above-mentioned reaction occurs in cytosol of the mesophyll cells and is called **fixation of CO_2** or **carboxylation**.

Table 5.5. Differences between Calvin Cycle and Hatch-Slack C_4 Cycle.

Calvin cycle (C_3 cycle)	Hatch-Slack cycle (C_4 cycle)
1. It operates in all plants.	1. It operates in C_4 plants only.
2. It has a slower rate of CO_2 fixation.	2. It has a faster rate of CO_2 fixation.
3. CO_2 once fixed is not released back.	3. Fixed CO_2 is released back in bundle sheath cells where it is finally fixed and reduced by Calvin cycle.
4. The primary acceptor of CO_2 is RuBP, a 5-carbon compound.	4. The primary acceptor of CO_2 is PEP, a 3-carbon compound.
5. The first stable product is 3-PGA, 3-carbon compound.	5. The first stable product is OAA, a 4-carbon compound.
6. It cannot operate under very low CO_2 concentration.	6. It can operate under very low CO_2 concentration.
7. Fixation of one molecule of CO_2 requires 3 ATP and 2 $NADPH_2$ molecules.	7. Fixation of one molecule of CO_2 requires 2 ATP molecules in addition to that required in C_3 cycle.
8. Optimum temperature for the operation of C_3 cycle is 10-25°C.	8. Optimum temperature for the operation of C_4 cycle is 30 to 45°C.
9. There is a net gain of one hexose sugar of 6 CO_2 molecules fixed and reduced.	9. It has no such gain.

Since this gives rise to the first stable product C_4 acid, and therefore, known as C_4 pathway.

The next step of reaction is **transport** of oxalo acetic acid (OAA - 4 C compound) from cytosol of mesophyll cells to chloroplasts of bundle- sheath cells, where it is decarboxylated to release fixed CO_2 and high concentration of CO_2 is generated near Rubisco.

The other product of decarboxylation reaction is a 3-carbon compound called pyruvic acid. Now, this is transported back to mesophyll cells, where if regenerates phosphoenol pyruvate to its own for continuation of C_4 pathway.

However, the C_4 pathway is more efficient than C_3 pathway due to absence of photorespiration in C_4 plants.

Table 5.6. Comparison of C_3 and C_4 Plants.

C_3 plants	C_4 plants
1. CO_2 acceptor is RuBP (Ribulose bisphosphate).	1. CO_2—acceptor is PEP (Phosphoenol pyruvate).
2. Only one type of chloroplasts occur in the cells. Grana present in chloroplasts.	2. Two types of chloroplasts, bundle sheath chloroplasts possess C_3 cycle (Calvin cycle) and do not have grana; and mesophyll chloroplasts possess C_4 cycle (Hatch Slack cycle), and have grana.
3. First stable compound after CO_2 fixation is PGA.	3. First stable compound after CO_2 fixation is oxaloacetate.
4. Plants photosynthesize well, if CO_2 concentration is high.	4. Plants synthesize well, even if CO_2 concentration is **low**.
5. They possess photorespiration that results in a low overall efficiency in synthesis of carbohydrates.	5. They lack photorespiration and conserve more organic food and thus an overall higher photosynthetic efficiency.
6. Excess of atmospheric oxygen inhibits photosynthetic activity.	6. Excess of atmospheric oxygen has no effect on photosynthesis.

FACTORS AFFECTING PHOTOSYNTHESIS

Limiting Factors

Sachs in 1860 for the first time propounded the concept of the three **cardinal points**. According to this concept, there is a **minimum**, **optimum** and **maximum** for each factor in relation to photosynthesis. This way, with any given species there may be a minimum temperature below which no photosynthesis takes place, an optimum temperature at which the highest rate takes place and a maximum temperature beyond which no photosynthesis will take place.

In the twentieth century Blackman (1905) proposed his **principle of limiting factors**.* According to this principle the rate of photosynthesis controlled by several factors is only as rapid as the slowest factor permits. He claimed that, if all other factors are kept constant the factor under consideration will affect the rate of photosynthesis, starting at a minimum below which no photosynthesis takes place and ending with an optimum at which a horizontal would be established, that is, the rate would remain constant despite further increases in that factor. At this point some other factor becomes limiting.

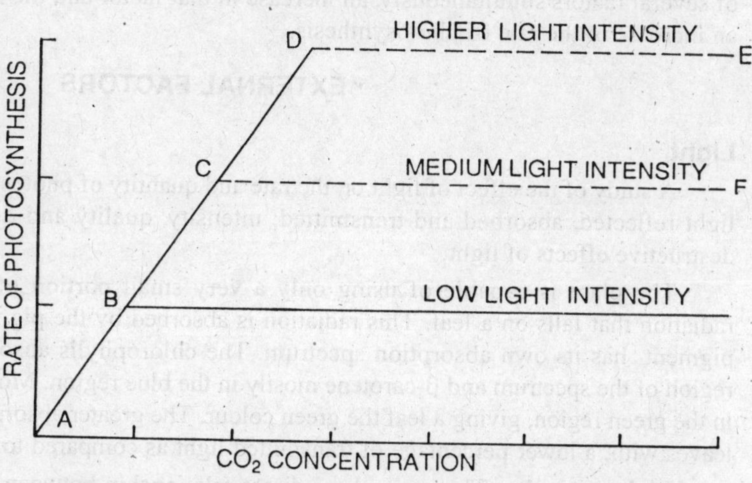

Fig. 5.13. Diagram to illustrate Blackman's law of limiting factors.

The explanation of Blackman's principle can best be presented in terms of the illustration given by Blackman himself. It is to be assumed that light intensity supplied to a leaf is just sufficient to utilize 5 mg of carbon dioxide per hour in photosynthesis. As the carbon dioxide supply is increased the rate of photosynthesis is also increased until 5 mg of carbon dioxide enters the leaf per hour. Any further increase in the supply of carbon dioxide will have no influence upon the rate of photosynthesis. Light has now become the limiting factor and further increase in the rate of photosynthesis can be brought about only by an increase in the intensity of light.

These results are indicated graphically. Here the effects of three different light intensities have been shown on the rate of photosynthesis under increasing concentrations of carbon dioxide. Under low light intensity, the rate of photosynthesis increasing concentrations of carbon dioxide. Concentration is raised until B is reached where further increase in carbon dioxide concentration is not accompanied by any increase in the rate of photosynthesis. The rate of photosynthesis becomes constant along the line BC. Any further increase in the supply of carbon dioxide will have no effect single upon the rate of photosynthesis, because light intensity has now become the limiting factor. If light intensity is now further increased, the rate of photosynthesis also increases until light again becomes a limiting factor. Here light becomes the limiting factor at the point C and there is another sharp break in the rate of photosynthesis along the line CF. Further increase in light intensity causes an increase in the rate of photosynthesis along the line CD with a

* When a process is conditioned as to its rapidity by a number of separate factors, the rate of the process limited by the pace of the slowest factors. Blackman, 1905

proportional increase in carbon dioxide concentration. The rate of photosynthesis becomes constant along the line DE where light again becomes limiting factor. At the points B, C and D the increase in the rate of photosynthesis stops abruptly because one or the other factors becomes limiting.

Light and carbon dioxide are not the only factors which can be limiting in the process of photosynthesis; other factors of photosynthesis can also become limiting under certain conditions.

It becomes clear from the above discussion that when photosynthesis is under the influence of several factors simultaneously, an increase in that factor and the limiting factor will bring about an increase in the rate of photosynthesis.

EXTERNAL FACTORS

Light

A study of the effect of light on the rate and quantity of photosynthesis would have to include light reflected, absorbed and transmitted; intensity, quality and duration of available light, and destructive effects of light.

The plant is capable of using only a very small portion of the incident electromagnetic radiation that falls on a leaf. This radiation is absorbed by the pigment complex of the leaf. Each pigment has its own absorption spectrum. The chlorophylls absorb heavily in the blue and red region of the spectrum and β-carotene mostly in the blue region. Most of the light is being reflected in the green region, giving a leaf the green colour. The greater absorption of light is found in thicker leaves with, a lower percentage of transmitted light as compared to thinner leaves.

Light intensity. There may be a direct relationship between the rate of photosynthesis and the intensity of light, provided that no other factor is limiting. For example, on bright sunny days the carbon dioxide concentration of the atmosphere is usually the limiting factor, not light intensity. However, on cloudy days light may be the limiting factor. Heinicke and Childers (1937) have found that the rate of photosynthesis steadily increased with light intensity up to about full sunlight even through saturation intensity for a single exposed leaf would be sufficiently lower. The need for higher light intensities for maximum photosynthesis of an entire tree is due to partial illumination of the inner leaves.

Both quality and intensity of light influence photosynthesis. Light between wavelength of 400 nm and 700 nm is most effective, for photosynthesis. This light is called **photosynthetically active radiation (PAR).**

Photo-oxidation. When the light intensity is increased beyond a certain point, the cells of the leaf or other photosynthesizing organ become vulnerable to chlorophyll-catalyzed photooxidations. With the result, many more chlorophyll molecules become excited than can be utilized causing the bleaching of chlorophyll and inactivation of important enzymes. The presence or absence of carotenoids and the concentration of carbon dioxide also influence the amount of photooxidation. The carotenoids have a protective role in photo-oxidation. With high concentrations of carbon dioxide, the consumption of oxygen of photo-oxidation takes place at much higher light intensities.

Light quality. White light, that comes from the sun, is composed of different wave lengths, ranging from the relatively long wave or red light (700 nm) through successively shorter waves to violet light (440 nm). When passed through a glass prism, white light is resolved into seven colours. The band of colours is the **visible spectrum**. The complete visible spectrum is composed of red, orange, yellow, green, blue, indigo and violet colours. The wave length of light that we are unable to perceive with our eyes, beyond the red and still longer, invisible rays, the **infrared**; and beyond the violet are shorter invisible rays, the **ultraviolet**. As we know, the absorption spectrum of chlorophyll shows a sharp peak at the red end and a wider but less definite peak at the blue end.

Green light is mostly transmitted or reflected and therefore, is not utilized in photosynthesis. Red and blue lights are effective in photosynthesis. Hoover (1937) while working on wheat plants found that the wavelength (655 μm) in the red is most effective and a wave length (440 μm) in the blue to be the second most effective in photosynthesis.

Duration of light period. It is assumed that a greater quantity of photosynthesis will take place in a plant exposed to longer periods of light. In general plant carries on much photosynthesis if it is provided with light for 10 to 12 hours per day. Bohning (1949) found that the leaves on young apple trees exposed to a continuous illumination and the usual atmospheric concentration of CO_2 photosynthesize at an undiminished rate of periods of at least 18 days without any noticeable damaging effect on the plant.

Carbon Dioxide

The carbon dioxide, which constitutes on the average only about 0.03 per cent by volume of the atmosphere, plays a role of the greatest significance in the biological world. As a result of the photosynthetic activity of green plants, the carbon dioxide from the atmosphere becomes chemically bound for indefinite period of the organic molecules which are the basis of all life.

Carbon dioxide supply. The green plants utilize the atmospheric carbon dioxide, while the other sources supply the gas to the atmosphere. Carbon dioxide is continually being returned to the atmosphere as a product of the respiration of plants and animals. The organic residues of plants are decomposed as a result of activities of bacteria and fungi. During decay the carbon of these residues is released in the carbon dioxide as a result of the metabolic activities of these organisms, and escapes into the atmosphere. The respiration of the soil bacteria alone results in a greater return of carbon dioxide to the atmosphere than the respiration of all animals. Carbon dioxide is

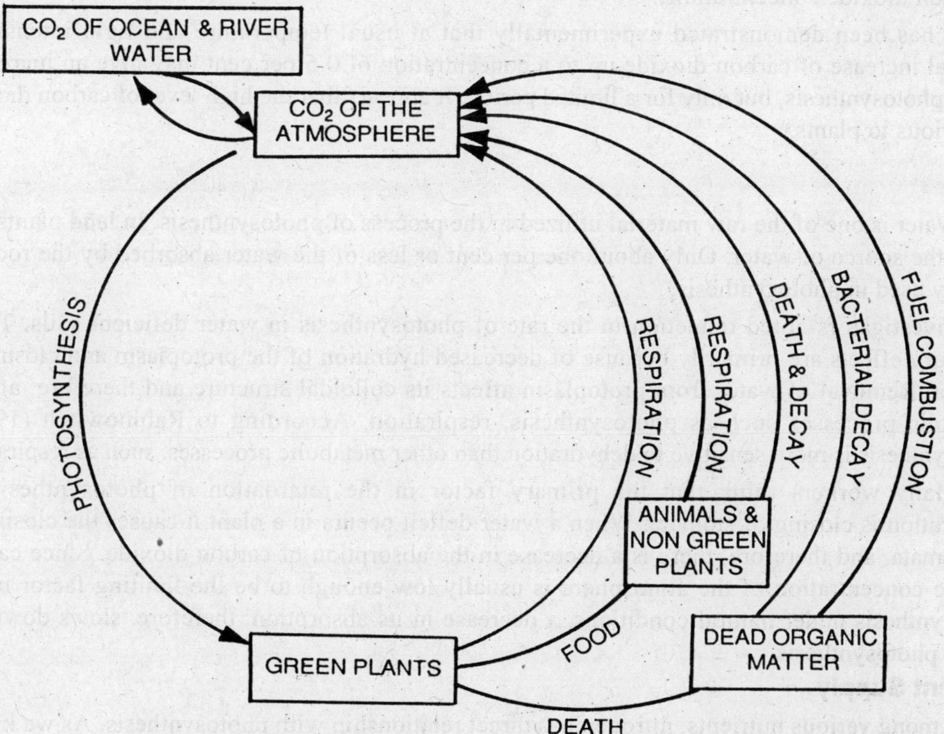

Fig. 5.14. Carbon cycle in nature.

also released into the atmosphere from volcanoes, mineral springs, and in the combustion of coal, oil, gasoline, wood and other fuel materials.

Oceans are much more important reservoirs of carbon than the atmosphere. The marine plants consume carbon dioxide in photosynthesis and release carbon dioxide in the process of respiration.

There is a constant exchange of carbon dioxide between the ocean and the atmosphere. On theoretical grounds it is assumed that the carbon dioxide concentration of the atmosphere is more or less maintained in dynamic equilibrium with that of the oceans. Carbon dioxide that escapes from the oceans whenever its atmospheric concentration falls below the usual value, and dissolves in the oceans whenever a contrary shift in atmospheric carbon dioxide concentration occurs. The maintenance of such a dynamic equilibrium between the oceans and the atmosphere is the principal factor for the constant value of carbon dioxide concentration of the atmosphere.

Absorption of carbon dioxide. Carbon dioxide in the air surrounding the leaves ultimately reaches the chloroplasts. Its inward diffusion path is through the stomata to the intercellular spaces, through walls of palisade and spongy parenchyma cells to the cytoplasm and thereafter into the chloroplasts. The walls of palisade and spongy parenchyma possess water. Carbon dioxide dissolves in water and ultimately passes through the walls in aqueous solution.

Carbon dioxide concentration. If light intensity and temperature are favourable, the carbon dioxide of the atmosphere limits the rate of photosynthesis.

The workers like Kreuslar (1887), Brown and Escombe (1902) and Pantanelli (1903) investigated that there is an increase in the rate of photosynthesis with an increase in the concentration of carbon dioxide at the lower levels of carbon dioxide concentrations. A decline in rate was observed in higher concentrations. Blackman and his co-workers claimed that after an optimum concentration is reached, the rate of photosynthesis remains constant over a wide range of carbon dioxide concentrations.

It has been demonstrated experimentally that at usual temperature and light intensities, artificial increase of carbon dioxide up to a concentration of 0.5 per cent may give an increased rate of photosynthesis, but only for a limited period. It appears that the high level of carbon dioxide is injurious to plants.

Water

Water is one of the raw material utilized in the process of photosynthesis. In land plants, the soil is the source of water. Only about one per cent or less of the water absorbed by the roots is actually used in photosynthesis.

Investigators noted reduction in the rate of photosynthesis in water deficient soils. These inhibitory effects are primarily because of decreased hydration of the protoplasm and closing of stomata. Removal of water from protoplasm affects its colloidal structure and therefore, affects metabolic processes such as photosynthesis, respiration. According to Rabinowitch (1945), photosynthesis is more sensitive to dehydration than other metabolic processes, such as respiration.

Many workers think that the primary factor in the retardation of photosynthesis by dehydration is closing of stomata. When a water deficit occurs in a plant it causes the closing of the stomata, and therefore, there is a decrease in the absorption of carbon dioxide. Since carbon dioxide concentration of the atmosphere is usually low enough to be the limiting factor in the photosynthesis under natural conditions, a decrease in its absorption, therefore, slows down the rate of photosynthesis.

Nutrient Supply

Among various nutrients, nitrogen has direct relationship with photosynthesis. As we know, nitrogen is a basic constituent of chlorophyll and all enzymes involved in carbon reactions, any reduction in nitrogen supply has an adverse effect on photosynthesis.

The major enzyme of carbon metabolism in plants is Rubisco, which alone accounts for more than half of the total leaf nitrogen.

In general, all essential elements affect the rate of photosynthesis.

Temperature

Photosynthesis can take place over a wide range of temperature. Plants of cold climates photosynthesise at much lower temperatures than do those of warm climates. The photosynthesis takes place in certain evergreen species of cold regions, even at temperatures below than 0ºC. It has been reported to occur in some species of conifers at temperatures as low as –35º, and in some kinds of lichens at –20ºC. Tropical plants normally do not photosynthesize below about 5ºC. On the other hand algae in the water of hot springs may carry on photosynthesis at a temperature as high as 75ºC. Many semi-desert and tropical species can withstand air temperatures of 55ºC. Most ordinary temperate climate plants, however, photosynthesize best between temperature of 10º and 35ºC.

If there is adequate light intensity and a normal supply of carbon dioxide (*i.e.*, no factor is limiting), the rate of photosynthesis of most ordinary land plants increases with rise in temperature up to a point (usually 25ºC) which varies somewhat from one kind of a plant to another; above this range there is rapid decline in the rate of photosynthesis primarily form injurious effects of higher temperatures on the protoplasm. At these higher temperatures the time of exposure is of importance. The higher the temperature the sooner the decline in photosynthetic rate. The decline in the rate of photosynthesis with time particularly marked at higher temperatures, is evidence of the increasingly limiting effect of some internal factor generally called the "time-factor". As regards the nature of this time factor, it is possible that it may represent the composite influence of several internal conditions. They are as follows:

(*i*) An inactivation of enzymes at higher temperatures.

(*ii*) The accumulation of the end products of the reaction which may exert a retarding effect on the rate of photosynthesis.

(*iii*) Failure of the diffusion of carbon dioxide toward the chloroplasts to keep pace with its use in photosynthesis.

(*iv*) Similar time factor effects are present in the temperature relations of other plant processes such as respiration and growth.

Oxygen Concentration

In most plants, increase in the oxygen concentration results in a decrease in the rate of photosynthesis. The normal atmospheric concentration of oxygen is high enough to induce a lower rate of photosynthesis than obtains at lower concentration of oxygen. This is demonstrated by the work of McAlister and Myers (1940) showing the effect of high and low concentrations of oxygen on the rate of photosynthesis of wheat plants. The explanation of this seems to be that oxygen actually exerts as direct inhibitory effect on photosynthesis, and the greater the concentration of oxygen the greater this effect.

Contrary to above statement the energy product in oxygen respiration is necessary for photosynthesis. Even small amount of this energy, may supplement light energy.

Chemical Compounds

Many different kinds of chemical compounds have direct or indirect effect upon the rate of photosynthesis. Especially the effects of certain substances are noticeable which markedly influence the rates of photosynthesis when present only in minute quantities. Among these chemical compounds are hydrocyanic acid, hydroxylamine, hydrogen sulphide and certain compounds containing the iodoacetyl radical.

Narcotics such as chloroform, ether and urethans, exert retarding or inhibitory effects on photosynthesis when present in very low concentrations.

INTERNAL FACTORS

Chlorophyll Content

Chlorophyll is indispensable for photosynthesis. Scientists, devised the **photosynthetic number (assimilation number)**. This assimilation number is the number of grams of carbon dioxide per hour per gram of chlorophyll. Other investigations show that there is not proportional relationship between chlorophyll content and photosynthesis in the leaves of vascular plants.

Leaf Anatomy

The rate of photosynthesis is also influenced by the anatomy of the leaf. Here the structural features of the leaf influence the amount of carbon dioxide that reaches the chloroplasts. The size, position and structure of the stomata, the size and distribution of the intercellular spaces, the relative proportions and distribution of palisade and spongy layers, thickness of the cuticle and epidermis, the amount and position of sclerenchyma, the presence of epidermal hairs, the arrangement of mesophyll cells, distribution and efficiency of vascular system, position of chloroplasts in the cells, all influence the rate of photosynthesis. The effect of the leaf anatomy upon the rate of photosynthesis is because of influences upon the rate of carbon dioxide entry, upon the light intensity penetrating to chlorenchyma cells, upon the turgidity of leaf cells and upon the rate of translocation of solutes from green cells.

Protoplasmic Factors

The rate of photosynthesis is also affected by protoplasmic factors. The most important protoplasmic factor is the enzyme complement of the protoplasmic system. A number of enzymes play an important role in the process of photosynthesis. The hydration of protoplasm is essential. If the cells lack water and protoplasm is dehydrated, the rate of photosynthesis slows down.

TRANSPORT OF ORGANIC SUBSTANCES

Food materials are manufactured mainly in the leaves and are translocated to the other regions of the plant through the phloem. Some of this is utilized for the growth of the plant, while the excess is stored. Whenever needed, the stored food material is drawn to the regions of necessity again through the phloem. The organs of the plants that store the food materials are called the storage organs. The storage of food materials occurs in fruits and seeds. In some plants root and stems are modified to perform this function, for example, rhizomes, bulbs, corms and tubers are the modifications of the stem which store the food in the form of carbohydrates, fats and proteins. The conversion of sucrose into starch at the storage region maintains the sucrose concentration low at the storage region and, thus, the concentration gradient is maintained until a sufficient quantity is accumulated. In the storage cells, starch is usually stored in the form of grains. Starch grains are formed in plastid, known as amyloplasts (leucoplasts). Starch is the common storage material in seeds and cereal grains. The growth of a starch grain takes place by apposition of successive shells around a growth centre of hilum. Some of the other polysaccharides of storage are hemicellulose and inulin.

Photosynthates are energy rich carbon compounds formed during process of photosynthesis, are transported from the leaf to non-photosynthetic organs and tissues such as roots, stem tissues and developing seeds. This type of long distance transport of photosynthates occurring through phloem is called **translocation**.

Sucrose is the principal form of carbohydrates that is translocated from leaf to the non-photosynthetic organs. Sucrose is a non-reducing sugar, and therefore, it is chemically stable. Due to this property, this does not react with other substances during translocation.

Photosynthates provide energy to non-photosynthetic tissues through respiration.

Mechanism of translocation of solutes in phloem. Several theories have been advanced to explain the mechanism of the translcoation of solutes in phloem tissues, none theory has ever received general acceptance. In order to consider the principal theories that have been proposed, a general review of various facts regarding the problem of translocation of solutes in the phloem will be useful.

1. The translocation in the conducting elements of the phloem ceases when the cells are killed. This shows that living cells are absolutely essential for the purpose.

2. The movement of solutes may be bidirectional. There is no doubt that movement of solutes in the phloem can sometimes occur in one direction and sometimes in the other. Experiments conducted by various workers have shown that both carbohydrates and phosphates move both up and down into the phloem of a stem from the point at which they enter it from the petiole of a leaf. But this phenomenon does not involve a bidirectional movement through the same phloem at the same time.

3. Relatively the large amounts of solutes are transported. The amount of carbohydrate translocated through the phloem is sufficiently large in the case of fleshy roots, tubers or fruits. Crafts and Lorenz (1944) calculated that in 33 day growing period the pumpkin fruit gained the weight to be about 5500 gms. A large amount of the increase is water but they showed that every hour the fruit gained 0.61 gram of dry weight. All this food material entered the developing fruit through one slender stalk that connects the fruit to the vine.

4. The velocity of translocation of organic solutes through the sieve tubes is quite rapid.

5. The movements of organic and inorganic solutes exhibit a periodicity. The movements of these solutes do not occur at a constant rate both day and night. There is more rapid translocation of radioactive phosphate from leaves of bean plants during the day than during the night. Mason and Mashell (1928) have reported that most of the carbohydrates that enter the developing bolls of cotton appears to move into add them during day time rather than at night. These diurnal variations occur in plants but do not follow the same pattern in all species.

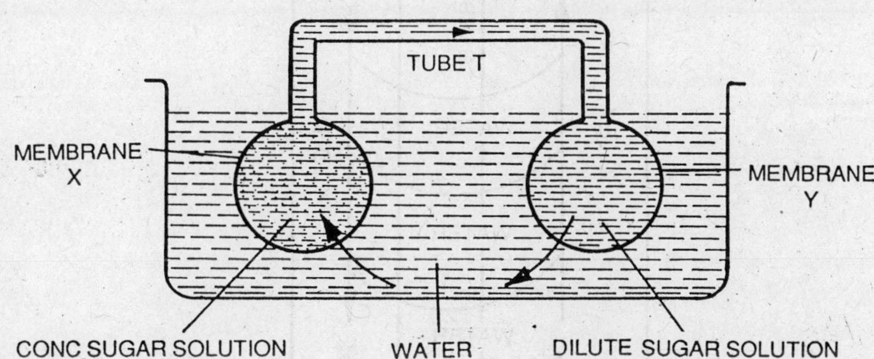

Fig. 5.15. Diagram to illustrate the principle of Munch mass flow hypothesis.

The mass flow hypothesis. This theory of translocation of solutes through phloem was advanced by Munch (1927, 1930). This theory has also put forth in modified forms by Crafts (1933, 1938) and others. The principles involved in this hypothesis can be clarified by a simple physical system. This is as follows:

As shown in the figure, two bulbs with membranes permeable only to water, are dipped in a trough containing water and are connected by a tube to form a closed system. Membrane X encloses a stronger solution of sucrose than membrane Y. Firstly, the water will enter both membranes, but greater turgor pressure develops in X which is soon to be transmitted throughout the system. With the result greater pressure in the water in Y membrane than in the pure water of the trough in which the membranes are immersed. As a result the water passes out of the membrane Y, and there is a flow of solution along the horizontal tube from X to Y. This mass flow of solution from X to Y continues unless and until the concentrations of the sugar solutions in both membranes become equal. At this point the flow of solution in the tube stops and a dynamic equilibrium establishes between the solution in the closed system and the surrounding water. If such an apparatus could be set up that we are able to arrange removal of sugar from the weaker solution (Y) as soon as it reaches it and replace the sugar lost by the stronger solution (X), flow of solution from X to Y would continue indefinitely.

According to Munch hypothesis an analogous system exists in plants. This hypothesis assumes that the translocation of solutes in plants through phloem is similar to the situations found in the experiment described in the preceding paragraph. The adjacent figure shows how it is supposed to operate as applied to the downward translocation of solutes. The photosynthetic cells of the leaf correspond to membrane X; the root cells correspond to Y and the continuous system of phloem connecting leaf and cells correspond the horizontal connecting tube. Besides xylem and cambium are also there. The osmotic pressure of the leaf cells is high as a result of photosynthesis. In the root cells the osmotic pressure is usually lower because most of the sugars translocated into them are used in metabolic activities or are converted into insoluble storage forms. Water supply to the leaf cells remains continuous through the xylem.

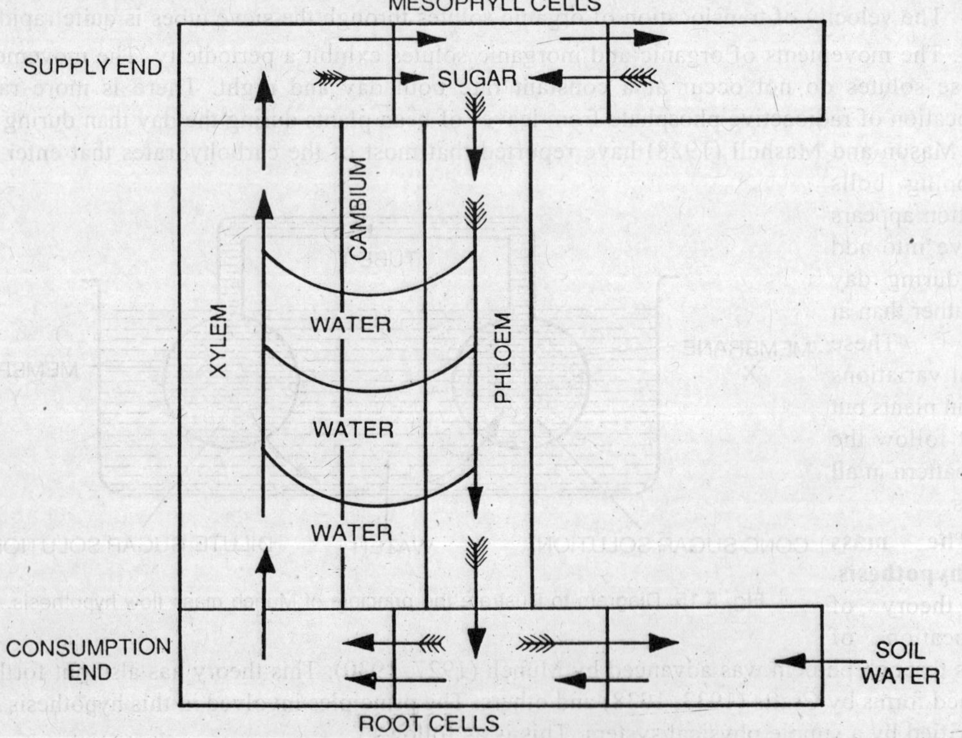

Fig. 5.16. Diagram to illustrate the mechanism of solute translocation according to the Munch hypothesis.

CRASSULACEAN ACID METABOLISM (CAM)

This type of metabolism, refers to a mechanism of photosynthesis, that is, different from C_3 and C_4 pathways. Crassulacean acid metabolism (CAM) is found only in succulents and other xerophytes or plants that grow in dry conditions.

In this type of metabolism, CO_2 is taken up by the leaves on green stems through stomata which remain open during night. However, during day time, stomata in such plants remain closed to conserve moisture.

The CO_2 taken up by succulent plants in night is fixed in the similar way as it takes place in C_4 plants to form malic acid, which is being stored in vacuole.

Hence, malic acid formed during night is used during day time as a source of CO_2 for photosynthesis to proceed through C_3 pathway.

Crassulacean metabolism is a kind of adaptation found in certain succulent plants such as pineapple to proceed photosynthesis without much loss of water, which generally occurs in plants with C_3 and C_4 pathways.

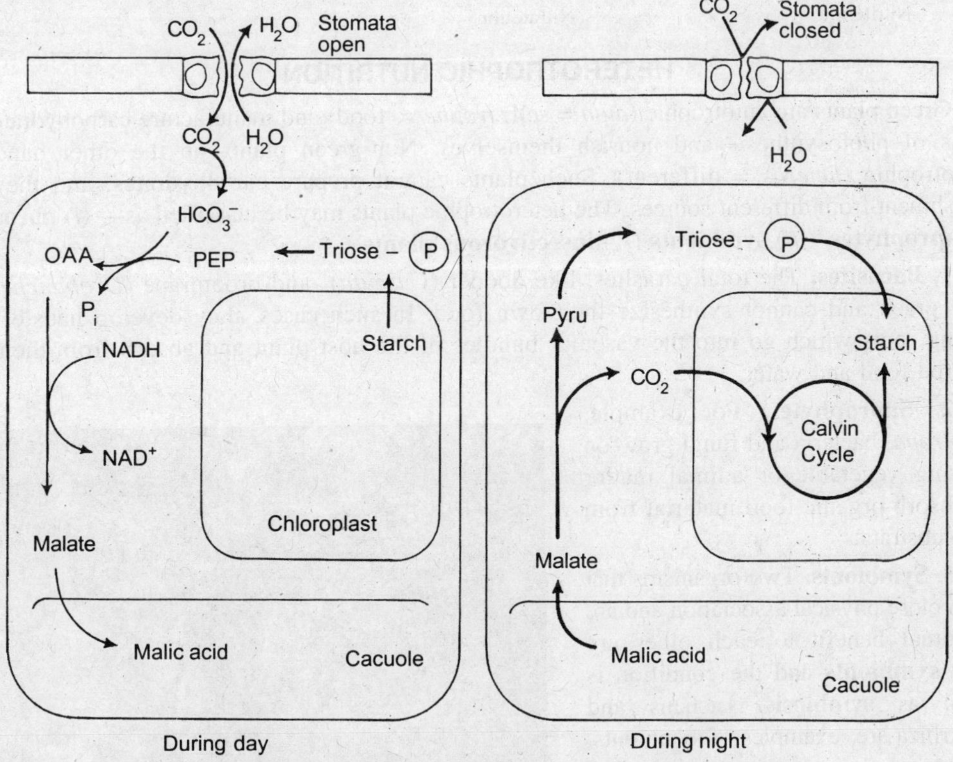

Fig. 5.17. Crassulacean acid metabolism (CAM).

CHEMOSYNTHESIS

The process of carbohydrate synthesis, in which organisms use chemical reactions to obtain energy from inorganic compounds, is called **chemosynthesis**.

The chemosynthetic bacteria are without photosynthetic pigments. For the synthesis of their own organic food (*i.e.*, carbohydrates), they obtain carbon from CO_2 of atmosphere, necessary energy from oxidation of inorganic or organic compounds such as hydrogen sulphide (H_2S), ferrous compounds (Fe^{2+}), molecular hydrogen (H_2), ammonia (NH_3), and nitrites ($-NO_2^-$).

For example bacteria of genus *Nitrosomonas* oxidise ammonia to nitrite. The energy released during oxidation is used by bacteria in the similar way as plants use energy from sunlight during photosynthesis, for converting CO_2 to carbohydrates. Such bacteria are known as **chemosynthetic autotrophs**.

The nitrifying bacteria (*Nitrosomonas*) are found in soil. Through their metabolic activates, they convert (NH_3) or ammonium ion (NH_4^+), which is formed by breakdown of animal and plant proteins during decay, to nitrates (NO_3^-). The formation of from NH_3 or NH_4^+ is termed **nitrification**. This process occurs in two steps:

(*i*) *Nitrosomonas* is most common organism found in soil that oxidises ammonia into nitrite ion ($-NO_2^-$) and (*ii*) *Nitrobacter* oxidises nitrite ion into nitrate ($-NO_3^-$) ion. These microbes obtain energy by oxidising ammonia or nitrite

(*i*) $\underset{\text{Ammonia}}{2NH_3} + 3O_2 \xrightarrow{\textit{Nitrosomonas}} \underset{\text{Nitrite ion}}{2NO_2^-} + 2H^+ + 2H_2O + \text{energy}$

(*ii*) $\underset{\text{Nitrite ion}}{2NO_2^-} + O_2 \xrightarrow{\textit{Nitrobacter}} \underset{\text{Nitrate ion}}{2NO_3^-} + \text{energy}$

HETEROTROPHIC NUTRITION

Green plants are autotrophic (*auto* = self; *trophe* = food) and manufacture carbohydrates by means of photosynthesis and nourish themselves. Non-green plants on the other hand are heterotrophic (*heteros* = different). Such plants cannot prepare carbohydrates, and they get nourishment from different sources. The heterotrophic plants may be classified as— (*i*) **parasites**; (*ii*) **saprophytes**, (*iii*) **symbionts** (*iv*) **insectivorous plants.**

1. Parasites. The total parasites, like dodder (*Cuscuta*) and broomrape (*Orobanche*) are never green and cannot synthesize their own food. In such cases, they develop haustoria or sucking roots which go into the vascular bundles of the host plant and absorb from them the prepared food and water.

2. Saprophytes. For example, *Monotropa*, bacteria and fungi grow on decaying vegetable or animal matter and absorb organic food material from their substrata.

3. Symbionts. Two organisms that live in close physical association and are of mutual benefit to each other are called **symbionts** and the condition is known as **symbiosis**. Lichens and mycorrhiza are examples of symbionts.

Lichens. A lichen is a plant consisting of two separate plants a fungus and an alga, so closely associated with each other as to appear a single plant.

Mycorrhiza. Some fungi live in close association of other higher plants where they are mutually beneficial to each other. Here fungi develop in the roots of higher plants.

Fig. 5.18. Parasite. *Cuscuta (dodder).*

4. Insectivorous plants. On the other hand the insectivorous plants are capable of manufacturing carbohydrates due to the presence of chlorophyll, but since they grow in swampy soil which are deficient in nitrogen they cannot synthesize enough proteins. They overcome this deficiency by catching small insects and digesting them. For this purpose, their leaves are specially modified in various ways. The best known examples of insectivorous plants are–the pitcher plants (*Nepenthes*), the sundew (*Drosera*), venus fly trap (*Dionaea*), butterwort (*Pinguicula*) and the bladderwort (*Utricularia*).

Fig. 5.19. Saprophyte. *Agaricus.*

(*a*)

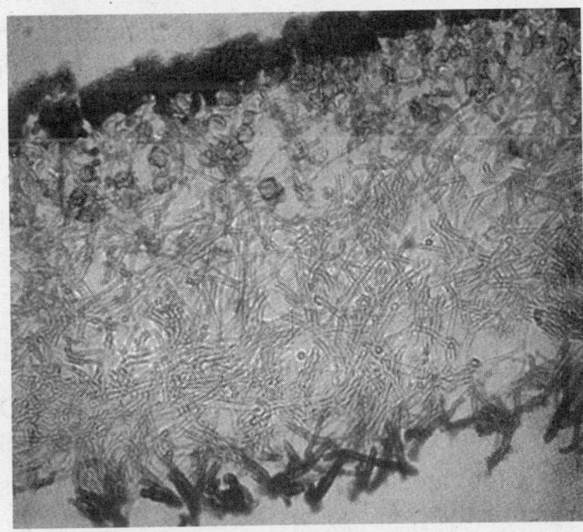

(*b*)

Fig. 5.20. Symbionts. (a) Lichens (*Cladonia*) (b) Thallus of lichen.

Pitcher plant. In India, only one species, *i.e.*, *Nepenthes khasiana* has been found. The pitcher plants are herbs or climbing under shrubs which often climb by means of tendrils. The pitcher itself is the modification of the leaf-blade. The insects are attracted to these pitchers either by their bright colours or by their nectar. Below the mouth the inside of the pitcher is covered with numerous smooth and sharp hairs, all pointing downwards. The insects that slip into water are prevented from coming out by the hairs present in the pitcher. The trapped insects eventually die and are decomposed by microorganisms. The proteins are converted into amines and are absorbed by the inner surface of the pitcher.

Fig. 5.21. Insectivorous plant : *Drosera.*

The sundew plant. Three species of *Drosera* are found in India. The sundew plant is a small herb. Each leaf of *Drosera* remains covered on the upper surface with numerous glandular hairs known as **tentacles**. Each gland secretes a viscous fluid which glitters in the sun like dew-drops and hence the name sundew. An insect landing on the leaf gets stuck and the glandular hairs bend down and prevent the insect from escaping. The enzymes secreted by the hairs digest the insect and the products are absorbed by the leaf surface. The enzyme called pepsin hydrochloric acid, acts on the insect and changes the proteins of its body into soluble and simple forms.

SIGNIFICANCE OF PHOTOSYNTHESIS

Photosynthesis is a vital process for life on this planet. This process links the physical and biological world. This process helps conversion of solar energy into organic matter that makes bulk of dry matter of any organism.

The plants biomass (dry matter), derived through photosynthesis supports humans and all other heterotrophic organisms living in biosphere.

Survival of all organisms is dependent upon food (carbohydrates, etc.), a photosynthetic product. All organisms need energy for different life activities which they derive from food (carbohydrates, fats and proteins).

Green plants are unique in the character that they are able to synthesise food for whole biokingdom.

During photosynthesis, O_2 is released, which is helpful to living organisms mainly in two ways. Firstly, in efficient utilisation of the energy rich molecules, *i.e.*, carbohydrates, in respiration.

Secondly, in making ozone (O_3) in outer layer of atmosphere, which is required to stop harmful ultraviolet (UV) rays from reaching the earth.

During photosynthesis green plants take CO_2 from atmosphere and release O_2, thus purifying air. Without O_2, life of all aerobic organisms including humans is not possible.

Crops are totally dependent on photosynthesis. This process feeds whole biokingdom directly or indirectly.

EXPERIMENTAL WORK : PRACTICALS

Photosynthesis

Experiment 1.

Object : Demonstration of starch test.

Requirements : Green leaves of a plant, burner, water, 70% alcohol, dilute iodine solution.

Experiment and observation : The green leaves of any healthy plant may be boiled in the day time thereafter, by keeping the leaves in 70% alcohol, the chlorophyll is extracted from them. Now these chlorophyll-less leaves are kept for sometime in dilute iodine solution. The leaves become deep blue or blue black in colour. This is known as 'starch test'. If the plant is kept for a long time, viz., 24 or 48 hours, in darkness, and thereafter the leaves are tested for starch test, it is always negative. The leaves do not become blue black in colour.

Explanation : As the plant was kept in darkness continuously for a long period, there was no photosynthesis, and the starch already prepared was shifted to the lower part of the plant during this period.

Experiment 2.

Object : Demonstration of release of oxygen during photosynthesis.

Requirements : Few branches of an aquatic plant, i.e., Hydrilla, etc., beaker, glass funnel, test tube, sodium bicarbonate, etc.

Experiment : The release of oxygen during photosynthetic process may be proved experimentally. A few branches of an aquatic plant, Hydrilla are kept in a big beaker full of the same pond water. Thereafter, the branches are covered with a glass funnel, and a test tube full of water is inverted at the end of the funnel as shown in the figure. If required, a small quantity of sodium bicarbonate may be added in the water, so that the supply of carbon dioxide may become adequate for the photosynthesis. Now, the apparatus is kept in the sunlight.

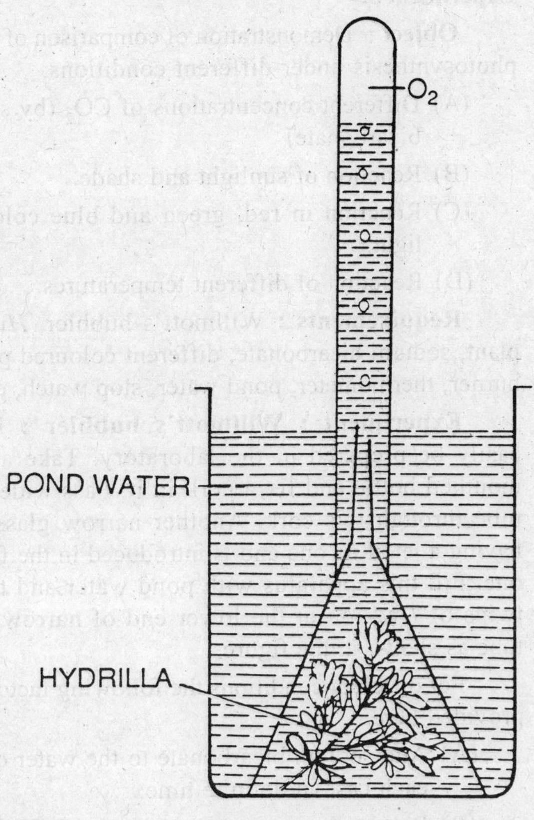

Fig. 5.22. Liberation of oxygen in photosynthesis. Demonstration of the phenomenon.

Observation : The gas bubbles may be observed from the ends of the *Hydrilla* branches kept under the glass funnel in the beaker. These gas bubbles are accumulated in the end of the test tube inverted over the end of the funnel, and the water within the tube goes downward. On test the gas is to be proved oxygen.

Note : To test the gas, the pyrogallol solution is taken in a beaker, and with the help of the thumb the tube partially filled with gas is kept inverted in the pyragallol solution. The solution enters the test tube and the tube again fills up completely because the pyragallol solution is soluble in oxygen.

Various modifications for this experiment :

(1) When the pond water of beaker is replaced by boiled or distilled water.

(2) When the above experiment is covered by a black cloth.

(3) When the *Hydrilla* twigs are replaced by terrestrial plants.

(1) When pond water of beaker is replaced by boiled or distilled water : If the pond water in the beaker is replaced by boiled or distilled water, the gas bubbles are not released from the ends of the *Hydrilla* branches kept under the glass funnel in the beaker. Why? The reason is quite clear, that during distillation or boiling of the water, the dissolved carbon dioxide goes out, which is necessary factor for photosynthesis. The photosynthesis does not take place. With the help of this modification of the experiment, the necessity of the carbon dioxide for the photosynthesis of water plants may be proved.

(2) When the above experiment is covered by a black cloth : If this apparatus is covered with black cloth, or kept in darkness, the gas bubbles are not released showing that the light is one of the essential factors for the photosynthesis in the case of water plants.

(3) When the *Hydrilla* twigs are replaced by terrestrial plants : Here the photosynthesis is completely checked. Only hydrophytes may absorb CO_2 from the water, the terrestrial plants, being of different habitat, fail to absorb CO_2 from water, and therefore, the photosynthesis is stopped here.

Experiment 3.

Object : Demonstration of comparison of rate of photosynthesis under different conditions.

(A) Different concentrations of CO_2 (by sodium bicarbonate)

(B) Reaction of sunlight and shade.

(C) Reaction in red, green and blue coloured-light.

(D) Reaction of different temperatures.

Requirements : Willmott's bubbler, *Hydrilla* plant, sodium bicarbonate, different coloured papers, burner, thermometer, pond water, stop watch, etc.

Experiment : Willmott's bubbler : It can easily be prepared in the laboratory. Take a wide mouthed bottle and fix a cork in it. Pass wide glass tube through this cork. Another narrow glass tube having a jet at its one end is introduced in the former one. Fill this apparatus with pond water and tie the twigs of *Hydrilla* at the lower end of narrow glass tube as shown in the figure.

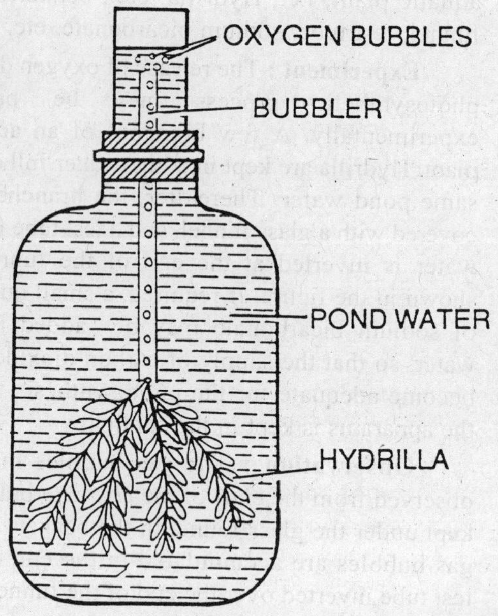

Fig. 5.23. Willmott's bubbler.

For different conditions the following factors are provided here.

(A) Add sodium bicarbonate to the water of the bottle and count the bubbles coming out in each case in definite time.

(B) Put the apparatus as such in sun and shade respectively for definite intervals and count the bubbles coming out in each case.

(C) Put the apparatus in double walled bell jar providing different coloured papers. Count the bubbles coming out in each case for definite time intervals.

(D) Take another beaker of hot water and put the apparatus in it at definite temperatures. Count the bubbles coming out in each case for definite time intervals.

Explanation :

(A) With the increasing concentration of $NaHCO_3$, the rate of photosynthesis increases. This rate of photosynthesis goes on increasing till light or some other factor acts as a limiting factor

(B) The readings show that the rate of photosynthesis is more in sun.

(C) The readings show that the rate of photosynthesis is the highest in red light and lowest in green.

(D) This experiment shows that photosynthesis occurs at a rapid rate from 10 to 35°C, provided other factors are not limiting.

Experiment 4.

Object : Demonstration of measurement of photosynthesis by Ganong's photosynthetometer.

Requirements : Ganong's photosynthetometer, green leaf, water, KOH, Kipp's apparatus, etc.

Experiment and observation : With the help of this apparatus, the quantity of released oxygen, and the quantity of utilized carbon dioxide during photosynthesis in a green leaf may easily be detected. This way, the photosynthetic quotient O_2/CO_2 may be known.

This apparatus consists of three parts A, B and C as shown in the figure. It consists of a bulb C, a measuring graduated tube A and a terminal stop cock B. The photosynthetic material to be utilized in the experiment, viz., about 2 c.c.of green leaves of garden nasturtium, etc., are kept in the bulb. The graduated tube is inverted, the stop-cock is closed and filled up with water up to that mark as much as the carbon dioxide is required. The graduated tube is closed by the hollow stopper. The hollow portion of the stopper is also filled up with water. Now, this end of the tube is to be closed with the help of the hand, and inverted in the trough full of water. Thereafter it is clamped this way that the level of water remains at par to the level of the hole of the stop-cock. Now, the stop-cock of the lower end is opened and the upper end of the graduated tube is connected with the Kipp's apparatus to receive the carbon-dioxide. The upper stop-cock is opened carefully, the carbon dioxide enters the tube, this is closed again, when the water of the tube is replaced by carbon dioxide, and its level becomes at par of the level of the outer water. Now, both the stop-cocks are being closed, and the complete tube is attached to the bulb with photosynthetic material. Now, the lower stop-cock is opened, and the carbon dioxide diffuses in the bulb containing photosynthetic material. This apparatus is kept for 3 to 4 hours in sunlight and after noting the time the lower stop-cock is closed and the tube is pulled out from the bulb. Now, this is placed in the trough filled with water, and by keeping it in the water, the hollow stopper is removed. Now, the zero mark of this measuring graduated tube is kept at par of the

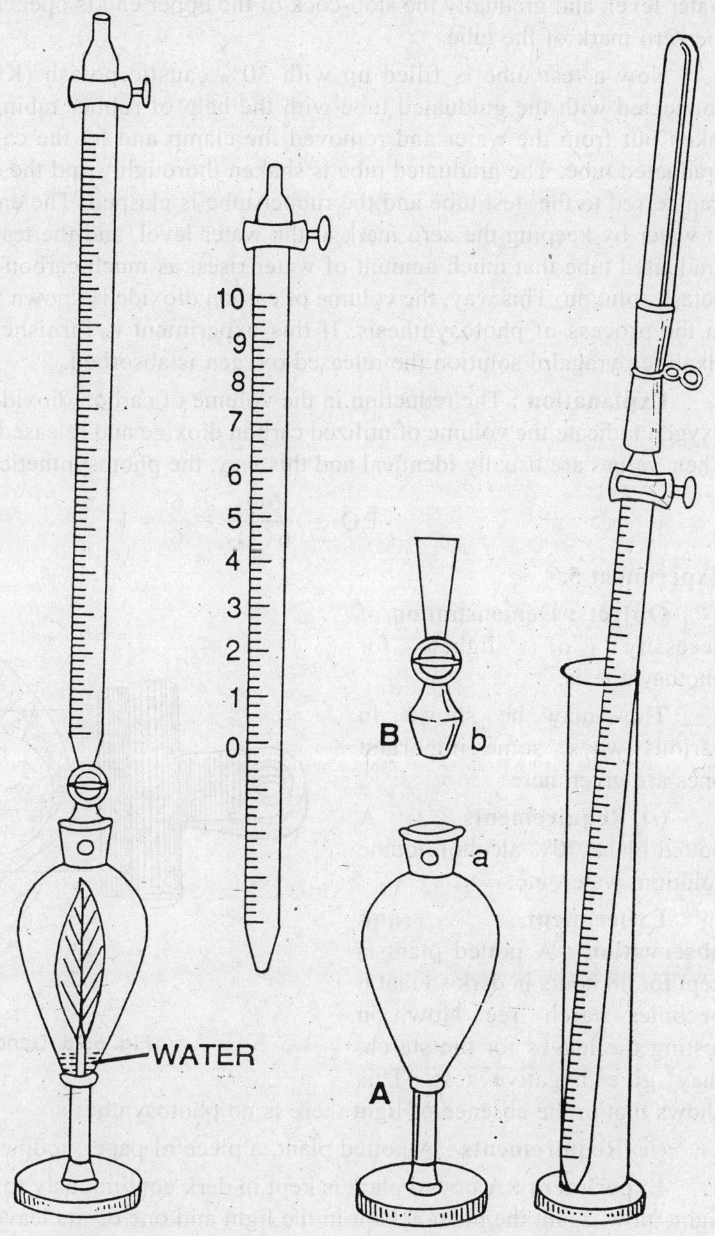

Fig. 5.24. Ganong's photosynthetometer.

water level, and gradually the stop-cock of the upper end is opened and make the water rise up to the zero mark of the tube.

Now a test tube is filled up with 30% caustic potash (KOH) solution and this tube is connected with the graduated tube with the help of rubber tubing. Thereafter, this apparatus is taken out from the water and removed the clamp and let the caustic potash solution enter the graduated tube. The graduated tube is shaken thoroughly and the caustic potash solution is again transferred to the test tube and the rubber tube is clasped. The end of the graduated tube is kept in water by keeping the zero mark at the water level, and the test tube is removed. Now, in the graduated tube that much amount of water rises, as much carbon-dioxide is absorbed by caustic potash solution. This way, the volume of carbon dioxide is known which has been used by the leaf in the process of photosynthesis. If this experiment is furnished by filling the test tube with alkaline pyragalol solution the released oxygen is absorbed.

Explanation : The reduction in the volume of carbon dioxide and addition in the volume of oxygen indicate the volume of utilized carbon dioxide and released oxygen during photosynthesis. Their values are usually identical and this way, the photosynthetic quotient is usually one.

$$\text{P.Q.} = \frac{6O_2}{6CO_2} = \frac{6}{6} = 1$$

Experiment 5.

Object : Demonstration of necessity of light for photosynthesis.

This may be shown in various ways, some important ones are given here.

(*i*) **Requirements :** A potted plant, 70% alcohol, iodine solution, water, etc.

Experiment and observation : A potted plant is kept for 48 hours in dark so that it becomes starch free. Now, on testing the leaves for the starch, they give negative test. This shows that in the absence of light there is no photosynthesis.

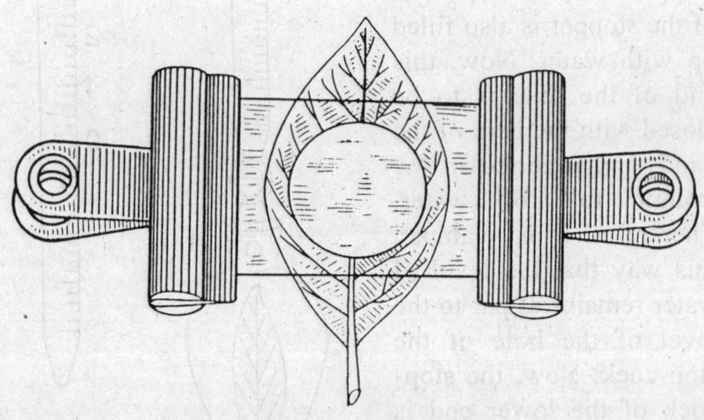

Fig. 5.25. Ganong's light screen.

(*ii*) **Requirements :** A potted plant, a piece of paper, iodine, 70% alcohol, water, etc.

Experiment : A potted plant is kept in dark continuously for 48 hours, to make it free from light. Now, again the plant is kept in the light and one of its leaves covered as in the figure. The photosynthesis begins after keeping the plant in light. After some time the partially covered leaf is detached from the plant and tested for starch.

Observation : The exposed portions of the leaf give positive test and the covered portion of the leaf gives negative test. This experiment shows that the photosynthesis takes place only in those portions of the leaf which were exposed to the light and not in covered portions.

(*ii*) **Ganong's Light Screen Test.**

Requirements : A potted plant, a Ganong's Screen, 70% alcohol, burner, iodine, water, etc.

Experiment : A potted plant is kept in darkness for about 48 hours, so that its leaves become starch free. A small Ganong's light screen is attached to a leaf of the plant as shown in the figure.

The Ganong's light screen partially covers the leaf. There is proper arrangement in the screen for the aeration of the leaf. Now, the plant along with light screen is kept in the light for photosynthesis. After 3 or 4 hours, the leaf is detached from the plant and tested for starch.

Observation : The portion of the leaf exposed to light gives positive starch test, *i.e.*, it becomes deep blue in iodine solution, whereas, the covered portion of the leaf gives negative starch test and does not become blue-black in iodine solution. This experiment proves the necessity of light for photosynthesis.

Experiment 6.

Object : Demonstration of necessity of CO_2 for photosynthesis:

Requirements : Two small sized potted plants, two bell jars, KOH solution in a petri dish, water, 70% alcohol, iodine, water, etc.

Experiment : Two small sized potted plants are taken. They are kept in the darkness at least for 48 hours, so that their leaves become starch free. Now, these potted plants are kept under two separate bell-jars. A petri dish partially filled up with KOH solution is kept under bell-jar 'A' and another petri dish parially filled with water is kept under bell-jar B. Now the apparatus is kept in sunlight for photosynthesis. After some time (3 to 4 hours), the leaves from both the potted plants are tested for starch by extracting their chlorophyll and keeping them in iodine solution.

Observation : The leaf detached from the plant kept under bell-jar, does not give positive test for starch when kept in iodine solution, whereas the leaf detached from the plant kept under bell-jar B gives positive starch test and becomes blue-black in colour, in iodine solutions.

Explanation : The KOH solution kept under bell-jar 'A' absorbs all the carbon dioxide, ceases the process of photosynthesis and starch formation. This experiment proves the necessity of the carbon dioxide for photosynthesis.

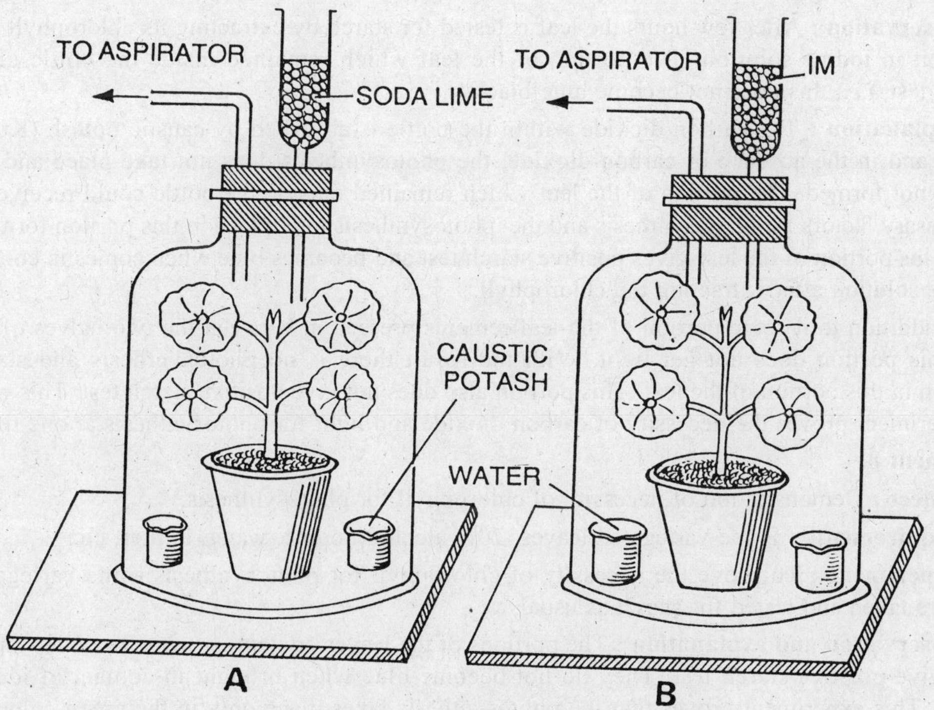

Fig. 5.26. Necessity of CO_2 for photosynthesis. Demonstration of phenomenon.

Experiment 7.

Object : Demonstration of Moll's experiment :

Requirements : A wide mouthed bottle, a split cork, conc. KOH solution, a leaf, water, beaker, wax, etc.

Experiment : A wide mouthed bottle with a split cork in two equal halves is taken. The bottle is partially filled up with concentrated caustic potash (KOH) solution. A leaf detached from the plant previously kept in darkness at least for 48 hours is pressed in between the two halves of the cork of the bottle so that half the leaf remains within the bottle and the other half outside the bottle. The petiole of the leaf remains outside which is kept in water filled beaker, so that the leaf may not become dry soon. The apparatus is made air-tight by applying melted wax so that the atmospheric air may not enter the bottle. Thereafter the apparatus is kept in the sunlight for photosynthesis.

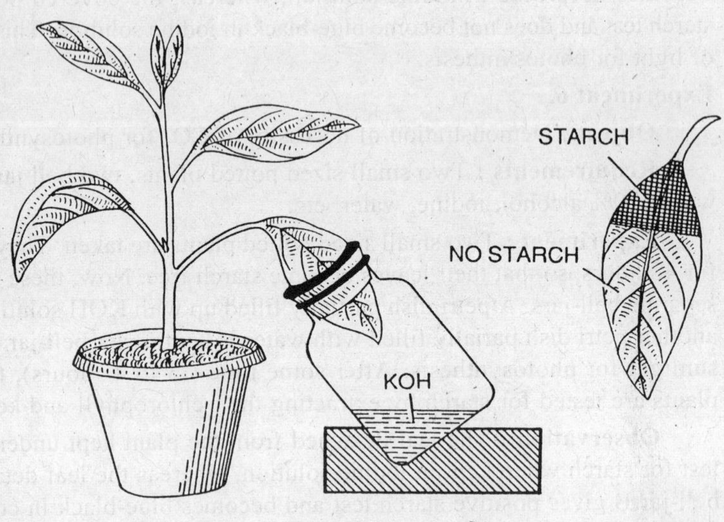

Fig 5.27. Moll's experiment.

Observation : After few hours the leaf is tested for starch by extracting its chlorophyll and keeping it in iodine solution. The portion of the leaf which remained inside the bottle gives negative test, *i.e.*, this does not become blue-black.

Explanation : The carbon dioxide within the bottle is absorbed by caustic potash (KOH) solution and in the absence of carbon dioxide, the photosynthesis does not take place and the starch is not formed. The portion of the leaf which remained outside the bottle could receive all the necessary factors for photosynthesis and the photosynthesis took place in this portion forming starch. This portion of the leaf gives positive starch test and becomes blue when comes in contact of iodine solution after extracting the chlorophyll.

In addition to it some portion of the leaf remains pressed in between the two halves of the cork. This portion does not get light. With the result there is no photosynthesis and starch formation in this portion of the leaf. This portion also does not give positive starch test. This way, this experiment proves the necessity of carbon dioxide and light for photosynthesis at one time.

Experiment 8.

Object : Demonstration of necessity of chlorophyll for photosynthesis.

Requirements : Some variegated leaves, 70% alcohol, iodine, water, burner, etc.

Experiment : To prove the necessity of chlorophyll for photosynthesis some variegated leaves are taken and tested for starch as usual.

Observation and explanation : The portions of the leaves containing white or yellow spots do not give positive starch test. They do not become blue when brought in contact of iodine solution. This experiment proves that the photosynthesis takes place only in the green coloured portion of the leaves.

Experiment 9.

Object : Demonstration of importance of light's colour for photosynthesis in land plants by Ganong's big light screen.

Requirements : Ganong's big light screen, a potted plant, 70% alcohol, iodine solution, water, burner, etc.

Experiment : The Ganong's big light screen is taken as shown in the figure. This consists of red, green and blue coloured glass screens. A long leaf is detached from a potted plant, previously kept in darkness for 48 hours and the leaf is kept under the 3 coloured screen for three or four hours in sunlight so that the photosynthesis may take place. Thereafter the leaf is tested for starch.

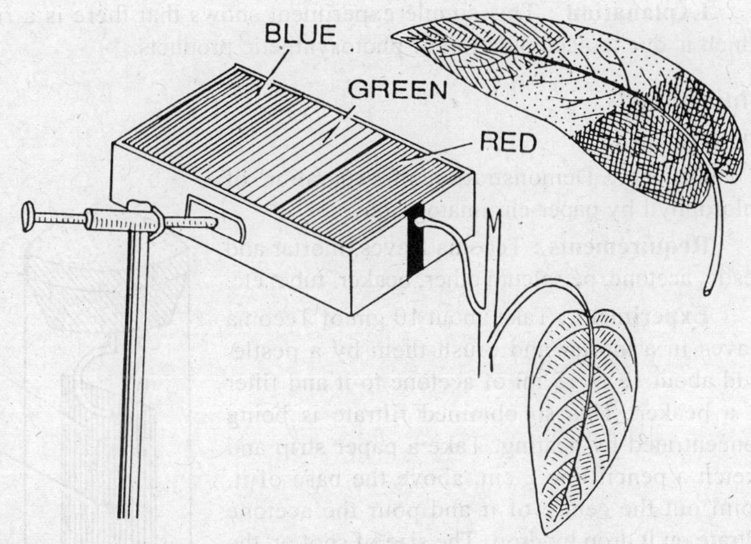

Fig. 5.28. Importance of light's colour for photosynthesis (Ganong's big light screen).

Observation : The portion of the leaf remained under red-coloured screen gives the positive starch test with deep blue colour. The portion of the leaf remained under the blue coloured screen also give positive starch test, but with fade blue colour. The part of the leaf remained under green-coloured screen either gives negative starch test or very less formation of the starch.

Explanation : This way, this experiment proves that the red and blue coloured rays are absorbed to the utmost by the chlorophyll. The absorption of the green rays by the chlorophyll is the lowest. The rate of photosynthesis is the highest under red screen and lowest under green screen.

Experiment 10.

Object : Demonstration of an increase in the mass of leaves due to formation of photosynthetic products by Ganong's leaf-area cutter.

Requirements : Ganong's leaf-area cutter, green leaves, an oven, balance, etc.

Experiment : Take two green plants. Put one in the light and another in dark. Now pick some leaves from each of the plants after putting them in their

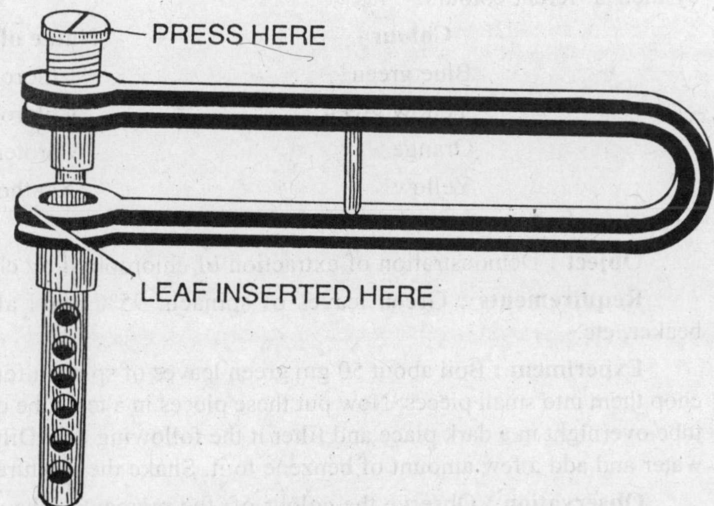

Fig. 5.29. Ganong's leaf punch.

respective conditions for few hours. Using the Ganong's leaf-area cutter, cut few circular discs from respective sets of leaves. Now put these sets in an oven and heat. After heating they are weighed properly by a balance.

Observation : The weight of the set (leaves form the plant kept in light) is found much enough than that of another set (leaves from the plants kept in dark).

Explanation : This simple experiment shows that there is a rise in dry weight of leaves which is due to accumulation of photosynthetic products.

Chlorophyll

Experiment 11.

Object : Demonstration of separation of chlorophyll by paper chromatography.

Requirements : Tecoma leaves, mortar and pestle, acetone, petroleum ether, beaker, tube, etc.

Experiment : Take about 10 gm of Tecoma leaves in a mortar and crush them by a pestle. Add about 12 to 15 ml of acetone to it and filter in a beaker. This so obtained filtrate is being concentrated by heating. Take a paper strip and sketch a pencil line 2 cm. above the base of it. Point out the centre of it and pour the acetone filtrate on it drop by drop. The size of spot on the paper strip should be small. Now, add a few drops of petroleum ether in a separate tube and place the above paper strip in a vertical position in this tube. Close the tube tightly.

Observation : Observe the paper strip after some time. The level of solvent, *i.e.*, petroleum ether and different colours should be pointed out by a pencil. Here the pigment may be identified by their different colours.

FILTER PAPER
ACETONE EXTRACT OF PIGMENTS
PETROLEUM ETHER + BENZENE (9·1 V/V)

Fig. 5.30. Paper chromatography, strip filter paper method.

Colour	Name of pigment
Blue green	Chlorophyll A
Yellow green	Chlorophyll B
Orange	Carotenes
Yellow	Xanthophyll

Experiment 12.

Object : Demonstration of extraction of chlorophyll by chemical method.

Requirements : Green leaves of spinach, 95% ethyl alcohol, distilled water, benzene, beaker, etc.

Experiment : Boil about 50 gm green leaves of spinach for some time. Dry these leaves and chop them into small pieces. Now put these pieces in a test tube containing 95% alcohol. Place this tube overnight in a dark place and filter it the following day. Dilute the filtrate with some distilled water and add a few amount of benzene to it. Shake the mixture and stand it for some time.

Observation : Observe the colour of the pigments. The upper layer is of green pigments, these are two, chlorophyll A and chlorophyll B. The lower layer is of yellow pigments, these are also two, xanthophyll and carotene.

SUMMARY

1. Photosynthesis is one of the most important phenomena of bio-kingdom that occurs in plants, algae and certain pigmented bacteria.
2. During process of photosynthesis, carbon dioxide from the atmosphere is taken in by leaves through stomata, and is used for making carbohydrates, primarily sucrose and starch.
3. This process occurs inside chloroplasts, where green pigments called chlorophyll, and other accessory pigments called carotenoids, are located on thylakoids often stacked to form grana.
4. Chlorophyll, mainly chlorophyll *a* which acts as reaction centre, takes part in the photochemical reactions.
5. Initially, light energy is absorbed by various chlorophyll molecules and accessory pigments; and thereafter supplied to the reaction centres.
6. Together, these pigments form two types of photosystems, *i.e.*, photosystem I and photosystem II.
7. In photosystem I, chlorophyll *a* molecule which absorbs light at a 700 nm acts as the reaction centre and is referred to as P_{700}.
8. In photosytem II, chlorophyll *a* molecule, which has maximum absorption at 680 nm, is the reaction centre and is referred to as P_{680}.
9. The events during light reactions involve both photosystem I and photosystem II.
10. During these photochemical reactions, splitting of water molecule initially occurs in the presence of light and the oxygen (O_2) is released at the photosystem II site.
11. Now, electrons released due to splitting of water initiate the electron transport chain, which leads to the formation of ATP and $NADPH_2$, energy required for conversion of CO_2 to carbohydrates.
12. The chemical energy, also called assimilatory power stored in the form of ATP and $NADPH_2$ is utilised for CO_2 assimilation to carbohydrates.
13. The carbon reactions for the synthesis of carbohydrates from CO_2 take place in the stroma of chloroplasts.
14. In stroma of chloroplasts, all necessary enzymes required for reactions, are present. The most notable enzyme is ribulose bisphosphate carboxylase oxygenase (Rubisco).
15. The enzyme Rubisco catalyses the initial carbon reaction, where CO_2 combines with ribulose bisphosphate (RuBP), a 5-carbon compound. The product of this reaction is a 3-carbon compound, 3-phosphoglycerate, thus, this pathway of carbon reactions is called C_3 pathway or the Calvin cycle.
16. In similar way, there is another pathway of CO_2 fixation called as C_4 pathway, which is quite different from the C_3 pathway.
17. The first product of the C_4 pathway is a 4-carbon compound, *i.e.*, oxaloacetic acid.
18. Occurrence of C_4 pathway in certain plants (called C_4 plants) in nature is an outcome of a necessity. This necessity was to get rid off an apparently wasteful process of photorespiration.
19. During photorespiration, there is a loss of fixed carbon (*i.e.*, 25 percent) and no energy rich compound is produced.
20. Certain environmental factors affect the rate of photosynthesis in plants. Such factors include (*i*) quality of light (*ii*) intensity of light, (*iii*) carbon dioxide, (*iv*) water and (*v*) temperature.
21. Certain other factors, such as (*i*) age of leaf, (*ii*) chlorophyll content, (*iii*) nutritional states of leaves also affect the rate of photosynthesis.
22. The end product of photosynthesis is sucrose, which is synthesised in the cytosol of a mesophyll cell. However, starch is synthesised inside chloroplasts only under certain conditions.
23. The sucrose moves from leaves to other organs of the plants (translocation), where it is used either for growth, or for storage in storage organs.

TERMS TO REMEMBER

Calvin cycle. Also called C_3 pathway. C_3 type of carbon reactions occur in the stroma of chloroplast.

C_3 Pathway. Where first stable product, 3-carbon molecule, 3-phosphoglycerate (PGA) is formed; reaction is catalysed by an enzyme Rubisco.

C_4 Pathway. C_4 plants possess a CO_2 concentrating mechanism.

Carbon reactions (dark reactions). Take place in the stroma of chloroplast, leading to the photosynthetic reaction of carbon to carbohydrates.

Carboxylation. Fixation of carbon dioxide. For example, formation of 3-Carbon compound, 3-phosphoglycerate (PGA).

Carotenoids. Red, orange and yellow coloured pigments.

Chemosynthesis. The process of carbohydrate synthesis, where organisms use chemical reactions to obtain energy from inorganic compounds.

Chemosynthetic autotrophs. When *Nitrosomonas* (bacteria) oxidise ammonia to nitrite, the released energy is used by the bacteria for converting CO_2 to carbohydrate. Such bacteria are chemosynthetic autotrophs.

Crassulacean acid metabolism (CAM). A different mechanism of photosynthesis that occurs in succulent plants.

Electron transport chain. The light driven reactions of photosynthesis.

Jan Ingenhousz (1730-1799). A physician, discovered that release of oxygen by plants was possible only in sunlight and only by the green parts of the plants.

Joseph Priestley (1733-1804). Discovered that plants have the ability to take up CO_2 from the atmosphere and release O_2.

Kranz anatomy. The C_4 plants contain dimorphic chloroplasts, *i.e.,* granal and agranal; granal in mesophyll cells and agranal in bundle-sheath cells.

Photolysis. Light dependent splitting of water molecule.

PEPC. Phosphoenol pyruvate carboxylase, an enzyme, which catalyses formation of a C_4 acid, oxaloacetic acid (OAA).

Photophosphorylation. The process of ATP formation from ADP in the presence of light in chloroplasts.

Photorespiration. Respiration that is initiated in chloroplasts and occurs in light only, also called **photosynthetic carbon oxidation cycle.**

Photosystem. The accessory pigments and the reaction centre together *i.e.,* PS I and PS II. Here, pigments are anchored in thylakoids in discrete units of organisation.

Photosynthesis. A process by which plants synthesise their own food in the presence of light. It takes place only in the green parts of the plant.

Photosynthetically active radiation (PAR). Portion of the spectrum between 400 nm and 700 nm.

Phytol chain. Side chain of chlorophyll molecule which extends from one of the pyrrole rings.

Pyrrole rings. Chlorophyll molecule composed of four 5-membered rings.

Reaction centre. Chlorophyll *a* molecules which convert light energy into electrical energy by bringing about electrical charge separation.

Rubisco. Ribulose bisphosphate carboxylase oxygenase, an enzyme which catalyses **carboxylation** (*i.e.,* formation of PGA).

Theodore de Saussure. He found that water is an essential requirement for photosynthesis to occur.

Translocation. Long distance transport of photosynthates which occurs through phloem.

QUESTION BANK

Very Short Answer Questions

1. Where does photosynthesis take place ?
2. What are actual sites for photosynthesis ?
3. Expand PAR.
4. Which molecules act as reaction centres in photosynthesis ?
5. What are accessory pigments ?
6. Define photophosphorylation.

7. Who formulated electron transport chain ?
8. What is photolysis ?
9. Where does photophosphorylation occur ?
10. What are essential requirements for the assimilation of CO_2 to carbohydrates ?
11. Expand Rubisco.
12. Expand RuBP.
13. Where C_3 type of carbon reactions occur ?
14. Give another appropriate name for C_3 pathway.
15. What is photorespiration ?
16. What is main function of enzyme Rubisco?
17. What is Kranz anatomy ?
18. Expand PEPC.
19. Who postulated law of limiting factors ?
20. Define chemosynthesis.
21. What are chemosynthetic autotrophs ?
22. What is end product of photosynthesis, and where it is synthesised ?
23. Which chlorophyll is known as universal photosynthetic pigment ?
24. Give name of central element found in chlorophyll molecule.
25. Mention the site of formation of glycoxylate from glycolate.
26. Name two photosynthetic pigments belonging to caretenoids.
27. Name any two C_4 plants.
28. Give single chemical equation of photosynthesis.
29. Expand the term RuBP.
30. Write one anatomical feature of C_4 plants.
31. Which of the following is not a useful function of the light reaction in photosynthesis ?
    (*a*) splitting water, (*b*) synthesis of NADPH
    (*c*) converting light energy into chemical energy (*d*) releasing oxygen for photorespiration.
32. What is the first stable product of C_4 cycle ?

ANSWERS

1. Photosynthesis takes place only in the green parts of the plant; **2.** The chloroplasts are actual sites for photosynthesis in green plants; **3.** Photosynthetically active radiation; **4.** Chlorophyll *a* molecules; **5.** These pigments, called carotenoids, are of red, orange and yellow colours; **6.** Process of ATP formation from ADP in the presence of light in chloroplasts is called photophosphorylation; **7.** Robert Hill (1939); **8.** Light-dependent splitting of water molecule is called photolysis; **9.** Photophosphorylation occurs in chloroplasts; **10.** ATP and $NADPH_2$; **11.** Ribulose bisphosphate carboxylase oxygenase; **12.** Ribulose bisphosphate; **13.** They occur in stroma of chloroplast; **14.** C_3 pathway is also called Calvin cycle after its discoverer Nobel Laureate Melvin Calvin; **15.** Respiration that is initiated in chloroplasts and occurs in light only; **16.** Photorespiration involves oxygenation of ribulose bisphosphate (RuBP) catalysed by the enzyme Rubisco; **17.** In C_4 plants where chloroplasts in mesophyll cells are granal, while in bundle-sheath cells are agranal; **18.** Phosphoenol pyruvate carboxylase; **19.** F.F. Blackman (1905); **20.** Process of carbohydrate synthesis, in which the organisms use chemical reactions to obtain energy from inorganic compound; **21.** They are certain bacteria (*e.g.*, *Nitrosomonas*), where energy released during oxidation is used for converting carbon dioxide to carbohydrates; **22.** The end product of photosynthesis is sucrose which is synthesised in the cytosol of a mesophyll cell; **23.** Chlorophyll *a*; **24.** Magnesium; **25.** Peroxisomes of mesophyll cell; **26.** Carotene, xanthophyll; **27.** Sugarcane and maize; **28.** $6CO_2 + 12H_2O \xrightarrow[\text{chlorophyll}]{\text{Light}} C_6H_{12}O_6 + 6O_2 + 6H_2O$; **29.** Ribulose bisphosphate; **30.** Leaves show 'Kranz' anatomy; **31.** (*d*) **32.** Oxaloacetic acid.

Short Answer Questions

1. What is significance of photosystem ?
2. What is Hill reaction ? Explain.
3. Describe photosynthetic unit.
4. Give a graphic outline of Calvin cycle.
5. Give a graphic outline of Hatch and Slack cycle.
6. Differentiate between C_3 and C_4 plants.
7. What are the steps that are common to C_3 and C_4 photosynthesis?
8. Differentiate between PS I and PS II.
9. Differentiate between photorespiration and aerobic respiration.
10. What is mass flow hypothesis? Explain.
11. Describe very briefly the contribution of the following scientists :
    (*i*) Jan Ingenhousz (*ii*) C. B. van Niel (*iii*) Joseph Priestley
12. Give comparison between the following:
    (*i*) C_3 and C_4 pathways
    (*ii*) Cyclic and non-cyclic photophosphorylation
    (*iii*) Carboxylation and oxygenation
    (*iv*) Anatomy of leaf in C_3 and C_4 plants.
13. Specify how a pentose phosphate is a CO_2 acceptor in the dark reaction of photosynthesis.
14. What is Rubisco? Explain its role in C_3 and C_4 photosynthesis.
15. Give the importance of C_4 plants.
16. Explain why photosynthesis is considered the most important process in the biosphere.
17. All life on earth is bottled 'sun's energy'. Justify.
18. Name the hormones which increase and decrease the rate of photosynthesis.
19. What is Blackman's law of limiting factors ?
20. Describe the importance of C_4 plants.
21. Give the path and magnitude of translocation of organic materials.
22. Name the three phases of Calvin cycle. Explain the significance of each of them.
23. Describe Blackman's law of limiting factors.
24. How would you know that the source of oxygen in photosynthesis is water and not carbon dioxide?
25. Draw a labelled diagram of the cross-section of chloroplast as seen under microscope.
26. Give the precise location where the dark and light reactions of photosynthesis take place. What are the final products of each of these reactions ?
27. Give the salient features of Kranz anatomy.
28. Distinguish between cyclic and non-cyclic photophosphorylation.
29. How would you prove experimentally that light is essential for photosynthesis ?
30. Why photosynthesis is considered the most important process in the biosphere ? Explain.
31. Write short note on Crassulacean acid metabolism.
32. What is the advantage of having more than one pigment molecule in a photocentre ?
33. What is kranz anatomy? Write a note on photosynthesis of those plants in which this anatomy is found.
34. Write a brief account on the chloroplast pigment.
35. Describe the mechanism of Hatch and Slack pathway in C_4 plants.
36. Write short notes on *Nepenthes* and *Drosera*.
37. Write a short note on chloroplast pigments.

38. List the three phases of Calvin cycle of photosynthesis. Describe in short the chemical steps in these phases.
39. List the various external and internal factors affecting photosynthesis.
40. Name any two C_4 plants. Describe how CO_2 is fixed in such plants.
41. Expand PEP. Where is it produced in C_4 plants? What is its role in the biosynthetic process ?
42. Expand RuBP. Where is it found in a C_3 plant? What is its role in the biosynthetic process ?
43. Name any two C_4 plants. Give anatomical characters of these plants.
44. Where does carboxylation take place in C_3 plants? Explain the process.
45. How did Hill prove experimentally that oxygen evolved during photosynthesis comes from water ?
46. Why does the rate of photosynthesis decrease at higher light intensities? What plays a protective role in such situations ?

Long Answer Questions

1. Describe cyclic photophosphorylation. What is the purpose of proton gradient created during the process in the thylakoids ?
2. What is Hill reaction? Write a brief note.
3. Describe photosystem I and photosystem II in green plants. How these photosystems cooperate in producing NADPH and ATP in light reactions of photosynthesis ?
4. Define carbon fixation in plants. What was the contribution of M. Calvin in regard to photosynthesis? Describe the three phases of Calvin cycle.
5. What is photorespiration? Explain how plants like sugarcane and sorghum have adapted themselves to overcome photorespiration?
6. What is the term used for long distance movement of carbohydrates in plants? How and in what form are the carbohydrates synthesised in leaves, reach long distance parts in plants? Explain the most accepted view.
7. Explain the photochemical phase of photosynthesis. Where does it take place?
8. How are ATP, NADPH and oxygen produced in non-cyclic photophosphorylation?
9. What is photophosphorylation? Distinguish between cyclic and non-cyclic photophosphorylation. Give schematic sketches of both the pathways.
10. Describe the light reaction of photosynthesis with special reference to non-cyclic and cyclic photophosphorylation.
11. Compare C_4 and C_3 pathways of CO_2 fixation during photosynthesis. Give well labelled and schematic sketch of C_4 pathways.
12. Describe the light dependent steps of photosynthesis. How are they linked to the dark reaction ?
13. Calvin cycle consists three phases. What are they ? Explain the significance of each of them.
14. Name any two C_4 plants. Describe how carbon dioxide is fixed in such plants.
15. Explain photochemical phase of photosynthesis. Where does it take place ?
16. Give an account of external factors affecting photosynthesis.
17. Explain the process of oxidation reduction occurring in photosynthesis to generate ATP and NADPH.
18. Describe cyclic photophosphorylation. What is the purpose of proton gradient created during the process in the thylakoids ?
19. Where does Calvin cycle take place ? Explain the process in detail.
20. Describe the light dependent steps of photosynthesis. How are they linked to the dark reactions ?
21. Describe the C_4 pathway in a sugarcane plant. How is this pathway an adaptive advantage to the plant ?
22. Name the pentose phosphate that fixes CO_2 in a C_3 plant. Explain Calvin cycle in such plants.
23. Mention the steps where the products of light reaction are utilised in the dark reaction.
24. Define photophosphorylation. Describe cyclic and non-cyclic photophosphorylation.
25. What is Blackman's law of limiting factors ? Explain it with reference to photosynthesis.

MULTIPLE CHOICE QUESTIONS (Objective Type Questions)

1. Who among the following proved for the first time that O_2 in photosynthesis comes out by photolysis of water ?
   (*a*) Arnon (*b*) Hill
   (*c*) Ruben, Hassid and Kamen (*d*) Blackman
2. The energy wastage occurs during
   (*a*) Dark reaction (*b*) Photorespiration (*c*) Photosynthesis (*d*) None of the above
3. Which of the following equations most adequately sums up the photosynthetic reaction ?
   (*a*) $CO_2 + H_2O = CH_2O + O_2$ (*b*) $C_6H_{12}O_6 + 6O_2 = 6CO_2 + 6H_2O$
   (*c*) $6CO_2 + 6H_2O = C_6H_{12}O_6 + 6O_2$ (*d*) $6CO_2 + 6H_2O = C_6H_{12}O_6 + 6O_2 + 6H_2O$
4. Which one of the following would not be limiting factor for photosynthesis?
   (*a*) Oxygen (*b*) Carbon dioxide (*c*) Chlorophyll (*d*) Light
5. During photosynthesis oxygen in glucose comes from
   (*a*) Water (*b*) CO_2
   (*c*) Both water and CO_2 (*d*) From air
6. Dark reaction in photosynthesis is called so because
   (*a*) It can also occur in dark (*b*) It does not require light energy
   (*c*) Of neither of these (*d*) Of both of these
7. During photosynthesis when PGA is changed into phosphoglyceraldehyde which of the following reaction occur
   (*a*) Oxidation (*b*) Reduction (*c*) Electrolysis (*d*) Hydrolysis
8. Bacteriochlorophyll differs from chlorophyll in having
   (*a*) One hydrogenated pyrrole ring (*b*) Two hydrogenated pyrrole rings
   (*c*) Three hydrogenated pyrrole rings (*d*) Four hydrogenated pyrrole rings
9. Blackman demonstrated that increasing illumination increased the photosynthetic rate upto a point when CO_2 became limiting. If light was not limiting, temperature became limiting. Emerson found that maximum CO_2 fixation would be achieved with brief flashes of light. Mark the correct statement in the following.
   (*a*) Only one factor can be limiting in photosynthesis
   (*b*) Photosynthesis consists of a light and a dark reaction
   (*c*) The trapping of light by chloroplasts is temperature dependent
   (*d*) The trapping of light by chloroplasts can occur only if CO_2 is present
10. Which of the following radioactive isotope was employed by Calvin for tracing the path of carbon in photosynthesis?
    (*a*) C^{11} (*b*) C^{14} (*c*) O^{18} (*d*) P^{35}
11. The first step in photosynthesis is
    (*a*) Joining of 3 carbon atoms to form glucose
    (*b*) Formation of ATP
    (*c*) Ionization of water
    (*d*) Excitement of an electron of chlorophyll *a* by a photon of light
12. The C_4 plants differ from C_3 plants with reference to the
    (*a*) Substance that accepts CO_2 in carbon assimilation
    (*b*) Type of end products of photosynthesis
    (*c*) Number of ATPs that are consumed in synthesising sugar
    (*d*) Type of pigments involved in photosynthesis

13. Which one of the following is a C_4 plant ?
(*a*) Papaya (*b*) Potato (*c*) Maize (*d*) Pea

14. Discovery of Emerson effect has already shown the
(*a*) Two distinct photochemical reactions (*b*) Light and dark reactions of photosynthesis
(*c*) Photophosphorylation (*d*) Photorespiration

15. The percentage of light energy utilised by higher plants for photosynthesis is
(*a*) 10 per cent (*b*) 50 per cent (*c*) 20 per cent (*d*) 1-2 per cent

16. Which of the following have high CO_2 compensation point?
(*a*) C_2 plants (*b*) C_3 plants (*c*) C_4 plants (*d*) Alpine herbs

17. Which of the following statements regarding 'compensation point' is incorrect?
(*a*) The rate of photosynthesis exactly balances that of respiration
(*b*) Neither oxygen nor carbon dioxide is given off from the leaves
(*c*) No growth occurs
(*d*) It occurs when the light intensity falls to about 20 per cent of that found on bright sunny day.

18. The percentage of light energy fixed in photosynthesis is generally around
(*a*) 35 per cent (*b*) 10 per cent (*c*) 50 per cent (*d*) 1 per cent

19. The inner envelope of chloroplast is permeable to
(*a*) Triose phosphate (*b*) Orthophosphate (*c*) Sucrose (*d*) Glutamine

20. First stable product of dark reaction in C-3 plants is
(*a*) PGA (*b*) RuBP (*c*) PEP (*d*) OAA

21. Calvin cycle is
(*a*) Dependent on light (*b*) Not dependent on light
(*c*) Inhibited by light (*d*) Partly dependent on light

22. Photosynthesis is a process in which
(*a*) CO_2 is reduced to carbohydrate (*b*) NADH is reduced to NAD
(*c*) ATP is generated (*d*) Oxidative phosphorylation occurs

23. The presumed size of chlorophyll molecule is
(*a*) Head 15 × 15Å Tail 20 Å (*b*) Head 10 × 12 Å Tail 15 Å
(*c*) Head 20 × 20 Å Tail 25 Å (*d*) Head 15 × 15Å Tail 25 Å

24. Photolysis of water is associated with
(*a*) PS I (*b*) PS II (*c*) Cyt. b (*d*) quinone

25. Photorespiration occurs in
(*a*) C_3 plants (*b*) C_4 plants (*c*) All plants (*d*) Algae

26. CO_2 fixation during C_4 pathway occurs in the chloroplast of
(*a*) guard cells (*b*) mesophyll cells (*c*) spongy parenchyma (*d*) bundle sheath cells

27. 'Kranz' type of anatomy is found in
(*a*) C_4 plants (*b*) C_3 plants (*c*) Succulents (*d*) None of the above

28. Structural component of chlorophyll among the following is
(*a*) Mg (*b*) Mn (*c*) Zn (*d*) Fe

29. In experiments on photosynthesis the isotope used by Calvin was
(*a*) O^{18} (*b*) C^{14} (*c*) C^{11} (*d*) N^{14}

30. Quantasomes occur on the surface of
(*a*) Plasmalemma (*b*) Cristae (*c*) Thylakoids (*d*) Nuclear envelope

31. At carbon-3, Chlorophyll *a* has
(*a*) – CHO group (*b*) – CH_3 group (*c*) – COOH group (*d*) No functional group

32. Photorespiration is favoured by
(*a*) High temperature and low oxygen
(*b*) High humidity and temperature
(*c*) High carbon dioxide and low oxygen
(*d*) High oxygen and low carbon dioxide

33. C_4 plants differ form C_3 plants with respect to
(*a*) Number of ATP molecules consumed
(*b*) First product
(*c*) Substrate which accepts carbon dioxide
(*d*) All the above

34. Which one of the following statements about Cytochrome P_{450} is wrong?
(*a*) It has an important role in metabolism
(*b*) It contains iron
(*c*) It is an enzyme involved in oxidation
(*d*) It is coloured cell

35. Pigments of PS I occur in
(*a*) Appressed part of granal thylakoids
(*b*) Stromal thylakoids and non-appressed parts of granal thylakoids
(*c*) Both appressed and non-appressed parts of granal thylakoids
(*d*) Stroma

36. Enzyme required in early CO_2 fixation in C_4 cycle is
(*a*) RuBP carboxylase
(*b*) RuBP oxygenase
(*c*) PEP Carboxylase
(*d*) PGA dehydrogenase

37. For fixing one molecule of CO_2 in Calvin cycle, are required
(*a*) 3 ATP + 1 $NADPH_2$
(*b*) 3 ATP + 2 $NADPH_2$
(*c*) 2ATP + 3 $NADPH_2$
(*d*) 3 ATP + $NADPH_2$

38. Carbon dioxide is fixed in
(*a*) Light reaction
(*b*) Dark reaction
(*c*) Aerobic respiration
(*d*) Anaerobic respiration

39. The pigment present universally in all green oxygenic plants is
(*a*) Chlorophyll *a* (*b*) Chlorophyll *b* (*c*) Chlorophyll *c* (*d*) Chlorophyll *m*

40. In photosynthesis, photolysis of water is used in
(*a*) Reduction of NADP
(*b*) Oxidation of NADP
(*c*) Oxidation of FAD
(*d*) None of the above

41. Scientist awarded Nobel Prize in 1960 for tracing the path of carbon is
(*a*) Hatch (*b*) Huber (*c*) Calvin (*d*) Ruben

42. Pigment system connected with oxidation of water is
(*a*) PS I (*b*) PS II (*c*) Phycobilisome (*d*) Carotenoids

43. Carotenes protect plants from
(*a*) Photo-oxidation (*b*) Desiccation (*c*) Photorespiration (*d*) Photosynthesis

44. Chemical which absorbs light energy and changes it to chemical energy is
(*a*) Xanthophyll (*b*) Chlorophyll *a* (*c*) Chlorophyll *b* (*d*) Chlorophyll *c*

45. Which of the following occurs in dark reaction of photosynthesis?
(*a*) Formation of ATP (*b*) Release of O_2 (*c*) Release of H_2 (*d*) Synthesis of PGAL

46. *Orobanche* plant is
(*a*) Partial parasite (*b*) Total root parasite (*c*) Symbiont (*d*) Total stem parasite

47. Insectivorous plants kill insects because
(*a*) insects eat their leaves
(*b*) insects eat their fruits
(*c*) they are heterotrophs
(*d*) they obtain nitrogen from killed insects

48. Insectivorous plants grow in soil which is deficient in
(*a*) calcium (*b*) magnesium (*c*) nitrogen (*d*) water

49. A pair of insectivorous plants is
(*a*) *Drosera* and *Rafflesia* (*b*) *Nepenthes* and Bladderwort
(*c*) *Dionaca* and *Viscum* (*d*) Venus fly trap and *Rafflesia*

50. Insectivorous plant is
(*a*) *Opuntia* (*b*) *Crotolaria* (*c*) *Eichhornia* (*d*) *Utricularia*

51. The mass flow hypothesis was first described by
(*a*) Swanson (*b*) Buchanan (*c*) Kurosawa (*d*) Munch

52. Which would do maximum harm to a tree ?
(*a*) The loss of half of its leaves (*b*) The loss of all its leaves
(*c*) The loss of half its branches (*d*) The loss of its bark

53. Organic substances such as sugars are translocated in the phloem can be demonstrated by
(*a*) Grafting (*b*) Root pressure (*c*) Defoliation (*d*) Ringing the stem

54. Photosynthates are transported to other part of higher plants through
(*a*) Cambial cells (*b*) Pith cells (*c*) Xylem cells (*d*) Phloem cells

55. Translocation of sugar in flowering plants occurs in the form of
(*a*) glucose (*b*) sucrose (*c*) starch (*d*) maltose

56. Rubisco (RuBP-carboxylase oxygenase), a protein which comprises
(*a*) 5% of total chloroplast portion (*b*) 11% of total chloroplast portion
(*c*) 16% of total chloroplast portion (*d*) 20% of total chloroplast portion

57. Photo-oxidation refers to
(*a*) Photolysis of water
(*b*) Photorespiration
(*c*) Light induced breakdown of substance in the presence of oxygen
(*d*) All of the above

58. Action spectrum of photosynthesis in 1883 was described by
(*a*) Calvin (*b*) Hill (*c*) Arnon (*d*) Englemann

59. The electron energy transfer between accessory pigments and chlorophyll molecules takes place by
(*a*) Transduction (*b*) Radiation (*c*) Resonance (*d*) Flourescence

60. The two enzymes responsible for primary carboxylation in C_3 and C_4 plants, respectively are
(*a*) RuBP carboxylase and pyruvate carboxylase
(*b*) PEP carboxylase and RuBP carboxylase
(*c*) RuBP carboxylase and PEP carboxylase
(*d*) PEP carboxylase and pyruvate carboxylase

61. Photorespiration is affected by
(*a*) Light intensity (*b*) Temperature (*c*) CO_2 and O_2 (*d*) All of these

62. Chlorophyll *a* is present in
(*a*) All photosynthetic eukaryotes as well as prokaryotes
(*b*) All plants except fungi
(*c*) All higher plants carrying photosynthesis
(*d*) All O_2 evolving photosynthetic organisms

63. Calvin cycle expands the following for fixation of 3-molecules of CO_2
(*a*) 9 ATP and 6 $NADPH_2$ (*b*) 8 ATP and 6 $NADPH_2$
(*c*) 9 ATP and 3 $NADPH_2$ (*d*) 6 ATP and 9 $NADPH_2$

64. Which of the following is the first product of photorespiration
(*a*) Glycolic acid (*b*) Serine (*c*) Phosphoglycolate (*d*) Glycine

65. Which element is essential for photolysis of water
(*a*) Nitrogen (*b*) Chlorine (*c*) Ingenhousz (*d*) Oxygen

66. In C_4 plants initial CO_2 fixation takes place in the chloroplasts of
(*a*) Guard cells (*b*) Spongy mesophyll (*c*) Palisade tissue (*d*) Bundle sheath

67. Photorespiration takes place only in
(*a*) Green parts of the plant (*b*) All the living cells of the plant
(*c*) Mitochondria (*d*) Root

68. The organelle associated with photorespiration is called as
(*a*) Lysosome (*b*) Peroxisome (*c*) Glyoxisome (*d*) Mesosome

69. The wavelength of light most absorbed during photosynthesis is
(*a*) 400 nm (*b*) 500 nm (*c*) 550 nm (*d*) 660 nm

70. Non-cyclic photophosphorylation involves
(*a*) Pigment system II only (*b*) Pigment system I only
(*c*) Both pigment systems I and II (*d*) None of these

71. Photophosphorylation means
(*a*) Formation of ATP from ADP (*b*) Formation of NADP
(*c*) Formation of ADP from ATP (*d*) Formation of PGA

72. Which one of the following is needed in both photosynthesis and respiration ?
(*a*) Sunlight (*b*) Chlorophyll (*c*) Glucose (*d*) Cytochrome

73. The first CO_2 acceptor in C_4 plants is
(*a*) PEP (*b*) RuBP (*c*) PGA (*d*) OAA

74. NADPH is generated through
(*a*) Photosystem I (*b*) Photosystem II
(*c*) Anaerobic respiration (*d*) Glycolysis

75. The core metal of chlorophyll is
(*a*) Fe (*b*) Mg (*c*) Ni (*d*) Cu

76. Most plants are green in colour because
(*a*) The atmosphere filters out all the colours of the visible light spectrum except green
(*b*) Green light allows maximum photosynthesis
(*c*) Green light is the most effective wavelength region of the visible spectrum in sunlight for photosynthesis
(*d*) Chlorophyll is least effective in absorbing green light

77. Protochlorophyll differs from chlorophyll in lacking
(*a*) 4 hydrogen atoms in one of its pyrrole rings
(*b*) 4 hydrogen atoms in two of its pyrrole rings
(*c*) 2 hydrogen atoms in one of its pyrrole rings
(*d*) 2 hydrogen atoms in two of its pyrrole rings

78. Which one of the following statements about cytochrome P 450 is wrong ?
(*a*) It has an important role in metabolism (*b*) It contains iron
(*c*) It is an enzyme involved in oxidation (*d*) It is a coloured cell

79. Warburg effect is decreased rate of photosynthesis at
(*a*) Low concentration of CO_2 (*b*) High concentration of CO_2
(*c*) Higher concentration of O_2 (*d*) None of these

80. Which one of the following prokaryotic algae possesses chlorophyll *b*
(*a*) *Anacystis* (*b*) *Oscillatoria* (*c*) *Prochloron* (*d*) *Segtonema*

81. Which of the following statements is correct about C_4 type of photosynthesis ?
(*a*) C_4 cycle is independent cycle
(*b*) C_4 cycle is adjunct of Calvin cycle
(*c*) RuBP carboxylase has higher affinity for CO_2
(*d*) Carboxylase is present in the bundle sheath cells

82. Assimilatory powers during photosynthesis are
(*a*) H_2O and O_2 (*b*) RuDP and RuMP
(*c*) $C_6H_{12}O_6$ and PGAL (*d*) ATP and $NADPH_2$

83. Who for the first time experimentally demonstrated that oxygen is evolved from water during photosynthesis?
(*a*) Calvin (*b*) Robert Hill (*c*) Govindji (*d*) Ruben and Kamen

84. The source of oxygen liberated in photosynthesis is
(*a*) Carbohydrate already present in the leaf (*b*) Water
(*c*) A photosynthetic enzym (*d*) Carbon dioxide

85. The isotopes that have proved useful in researches on photosynthesis are
(*a*) O^{18} and O^{14} (*b*) C^{11} and P^{32} (*c*) P^{32} and O^{18} (*d*) C^{14} and O^{18}

86. Acid concentration in CAM plants is more at :
(*a*) Right (*b*) Daytime (*c*) Down (*d*) Dusk

87. In PS-II, first known electron acceptor is :
(*a*) Cytochrome (*b*) PQ (*c*) FRS (*d*) Ferredoxin

88. In C_4 plants :
(*a*) Grana present in mesophyll cells and absent in bundle sheath cells
(*b*) Grana absent in mesophyll cells and present in bundle sheath cells
(*c*) Grana present in both mesophyll and bundle sheath cells
(*d*) Grana absent in both mesophyll and bundle and bundle sheath cells

89. Hill reaction occurs in :
(*a*) High altitude plants (*b*) Total darkness
(*c*) Absence of water (*d*) Presence of ferredoxin

90. Photorespiration in C_3 Plants starts from :
(*a*) Phosphoglycerate (*b*) Phosphoglycolate (*c*) Glycerate (*d*) Glycine

91. Which of the following helps in translocation of food in plants :
(*a*) Xylem (*b*) Phloem (*c*) Sclerenchyma (*d*) Collenchyma

92. Number of carboxylation in C_4 cycle are :
(*a*) Four (*b*) Three (*c*) Two (*d*) One

93. O_2 evolution is directly associated with :
(*a*) PS I (*b*) PS II (*c*) Phytochrome (*d*) Phycocyanin

94. Chlorophyll molecules are green in colour because they :
(*a*) Transmit green light (*b*) Transform green light
(*c*) Reflect green light (*d*) Absorb green light

95. This is not an electron transferring molecule:
(*a*) Coenzyme Q (*b*) Fe ~ S protin (*c*) ATP (*d*) NAD

96. Which one of the following is the aim of Moll's half leaf experiment? To show that :
(*a*) CO_2 is produced during anaerobic respiration (*b*) CO_2 is produced during aerobic respiration
(*c*) CO_2 is essential for photosynthesis (*d*) Chlorophyll is essential for photosynthesis

97. Orange and red colour of flower and fruit is due to :
(*a*) Leucoplast (*b*) Chloroplast (*c*) Protoplast (*d*) Chromoplast

98. Maximum photosynthesis occurs in :
(*a*) Continuous light (*b*) Green light (*c*) Red light (*d*) Blue light

99. In photosynthesis, energy from light reaction to dark reaction is transferred in the form of :
(*a*) ADP (*b*) ATP (*c*) RuBP (*d*) Chlorophyll

100. Stomata of CAM plants :
(*a*) are always open
(*b*) open during the day and close at night
(*c*) open during the night and close during the day
(*d*) never open

101. Which fraction of the visible spectrum of solar radiation are primarily absorbsed by carotenoids of the higher plants?
(*a*) Blue and green (*b*) Green and red (*c*) Red and violet (*d*) Violet and blue

102. Which one of the following in wrong in relation to photorespiration?
(*a*) It occurs in chloroplast (*b*) It occurs in daytime only
(*c*) It is characteristic of C_4 plants (*d*) It is characteristic of C_3 plants

103. Which one of the following concerns photophosphorylation?
(*a*) $ADP + AMP \xrightarrow{\text{Light energy}} ATP$ (*b*) $ADP + \text{Inorganic } PO_4 \xrightarrow{\text{Light energy}} ATP$
(*c*) $ADP + \text{Inorganic } PO_4 \longrightarrow ATP$ (*d*) $AMP + \text{Inorganic } PO_4 \xrightarrow{\text{Light energy}} ATP$

104. In sugarcane plant 14 CO_2 is fixed in malic acid in which the enzyme that fixes CO_2 is:
(*a*) Photophosnal pyruvic carboxylase (*b*) Ribulose bisphosphate carboxylase
(*c*) Ribulose phosphate Kinase (*d*) Fructose phosphatase

105. Enerson effects explain the phenomenon of :
(*a*) Transpiration (*b*) Absorption of water by roots
(*c*) Photosynthesis (*d*) Respiration

106. Photosynthesis is
(*a*) catabolic process (*b*) anabolic process (*c*) exothermic process (*d*) exergonic process

107. For photosynthesis, the plant requires
(*a*) chlorophyll (*b*) CO_2 and H_2O (*c*) light (*d*) all the above

108. CO_2 and O_2 balance in atmosphere is maintained due to
(*a*) respiration (*b*) photorespiration (*c*) photosynthesis (*d*) nitrogen metabolism

109. Which of the following equations is most adequate in summing up the photosynthesis ?
(*a*) $6CO_2 + 12H_2O + \text{light} \longrightarrow C_6H_{12}O_6 + 6O_2 + 6H_2O$
(*b*) $6CO_2 + 6H_2O \longrightarrow C_6H_{12}O_6 + 6O_2$
(*c*) $C_6H_{12}O_6 + 6O_2 \longrightarrow 6CO_2 + 6H_2O$
(*d*) $CO_2 + H_2O + \text{light} \longrightarrow CH_2O + O_2$

110. Which of the following crop plants is very efficient converter of solar energy and whose net productive value ranges from 2 kg to 4 kg / m² or even higher ?
(*a*) rice (*b*) wheat (*c*) bajra (*d*) sugarcane

111. The bulk fixation of carbon through photosynthesis takes place in
(*a*) terrestrial vegetation (*b*) tropical rain forests (*c*) crop plants (*d*) oceans

112. Nine-tenth of all photosynthesis of the world is carried by
(*a*) herbs (*b*) shrubs (*c*) algae of oceans (*d*) forests

113. The maximum evolution of oxygen is by—
(*a*) crops (*b*) forests

(*c*) siberian vegetation (*d*) phytoplanktons of sea

114. During photosynthesis the oxygen in glucose comes from
(*a*) carbon dioxide (*b*) water and carbondioxide
(*c*) oxygen in air (*d*) water

115. The main constituents of Hill reaction are
(*a*) ATP and $NADPH_2$ in chloroplast (*b*) ATP and $NADPH_2$ in mitochondria
(*c*) A reduced substance and O_2 in chloroplast (*d*) only oxygen

116. Where do the energy-capturing reactions of photosynthesis occur ?
(*a*) thylakoids (*b*) stroma (*c*) plasma membrane (*d*) cytoplasm

117. The process in which water is split, and essential for photosynthesis
(*a*) hydrolysis (*b*) photolysis (*c*) plasmolysis (*d*) hemolysis

118. The source of O_2 liberated in photosynthesis is
(*a*) photosynthetic enzyme (*b*) carbohydrates
(*c*) water (*d*) carbon dioxide

119. Which enzymes most abundantly found on earth?
(*a*) nitrogenase (*b*) invertase (*c*) rubisco (*d*) catalase

120. Chlorophyll is present
(*a*) in grana of chloroplasts (*b*) on surface of chloroplasts
(*c*) in stroma of chloroplasts (*d*) dispersed throughout chloroplasts

121. Grana refers to
(*a*) glycolysis of glucose (*b*) byproduct of photosynthesis
(*c*) stacks of quantasome (*d*) stacks of thylakoids

122. Quantasomes are found in
(*a*) chloroplast (*b*) nucleus (*c*) lysosome (*d*) mitochondria

123. Where does the primary photochemical reaction occur in the chloroplast ?
(*a*) stroma (*b*) quantasome
(*c*) inner membrane of chloroplast (*d*) outer membrane of chloroplast

124. Chemical formula for chlorophyll *a* is
(*a*) $C_{55}H_{72}O_5N_4Mg$ (*b*) $C_{55}H_{70}O_5N_4Mg$ (*c*) $C_{55}H_{77}O_6N_4Mg$ (*d*) None of above

125. The basic structure of all chlorophyll comprises of
(*a*) porphyrin system (*b*) cytochrome system (*c*) plastocyanin system (*d*) flavoprotein only

126. Which of the following wavelength of light is absorbed maximum for photosynthesis ?
(*a*) green light (*b*) red light (*c*) blue light (*d*) yellow light

127. Which wavelength of light carries out photosynthesis in bacteria ?
(*a*) blue (*b*) red (*c*) far red (*d*) ultraviolet

128. The assimilatory power in photosynthesis is
(*a*) ATP (*b*) ATP and $NADPH_2$
(*c*) NADPH (*d*) ATP, NADPH and CO_2

129. The oxygen produced during photosynthesis comes from photolysis of water proved by
(*a*) M.Calvin (*b*) Robert Major (*c*) Blackman (*d*) Ruben and Kamin

130. The process of photophosphorylation was discovered by
(*a*) Calvin (*b*) Arnon (*c*) Warburg (*d*) Weismann

131. Photosynthesis can be increased if light, instead of being administered continuously is given as brief flashes. This phenomenon was first observed by
(*a*) Emerson and Arnold (*b*) Benson and Calvin
(*c*) Hatch and Slack (*d*) Calvin and Hill

132. The two-pigment theory of photosynthesis was proposed by
(*a*) Blackman (*b*) Emerson (*c*) Arnon (*d*) Hill

133. The source of hydrogen atoms for the synthesis of glucose is
(*a*) $NADPH_2$ (*b*) $FADH_2$ (*c*) H_2O (*d*) ATP

134. The primary electron acceptor in cyclic photophosphorylation is
(*a*) carbon dioxide (*b*) a protein that contains iron and sulphur
(*c*) NADP (*d*) FAD

135. Through which of the following substances the photosystem I passes an electron to *NADP* during light reactions ?
(*a*) Plastocyanin (*b*) Cytochrome (*c*) Ferredoxin (*d*) Plastoquinone

136. ATP formation during photosynthesis is called
(*a*) substrate level phosphorylation (*b*) oxidative phosphorylation
(*c*) photophosphorylation (*d*) phosphorylation

137. Photosystem II occurs in
(*a*) stroma (*b*) grana (*c*) mitochondria (*d*) ribosomes

138. In the electron transport system in terminal oxidation the cytochrome, which donates electron to O_2 is
(*a*) cytochrome *b* (*b*) cytochrome *a* (*c*) cytochrome a_3 (*d*) cytochrome *c*

139. NADP is converted into $NADPH_2$ in
(*a*) photosystem I (*b*) photosystem II
(*c*) non-cyclic photophosphorylation (*d*) cyclic photophosphorylation

140. During cyclic photophosphorylation energy molecules are obtained in the form of
(*a*) ADP (*b*) ATP (*c*) GTP (*d*) ATP and $NADPH_2$

141. Dark reaction in photosynthesis is called, because
(*a*) it does not require light energy (*b*) cannot occur during day time
(*c*) occurs more rapidly at night (*d*) it can also occur in darkness

142. Calvin cycle occurs in
(*a*) chloroplast (*b*) mitochondria (*c*) glyoxysomes (*d*) cytoplasm

143. C_4 cycle was discovered by
(*a*) Hatch and Slack (*b*) Hill (*c*) Arnon (*d*) Calvin

144. During photochemical reactions of photosynthesis
(*a*) liberation of O_2 and formation of ATP and $NADPH_2$ takes place
(*b*) assimilation of CO_2 takes place
(*c*) liberation of O_2 and formation of $NADPH_2$
(*d*) formation of ATP and $NADPH_2$ takes place

145. CO_2 joins the photosynthetic pathway during
(*a*) dark reaction (*b*) photosystem I (*c*) photosystem II (*d*) photolysis

146. The primary acceptor during CO_2 fixation in C_3 plants is
(*a*) Ribulose bisphosphate (*b*) Glycolate
(*c*) Phosphoenolpyruvate (*d*) Triosephosphate

147. First stable compound in C_3 cycle is
(*a*) phosphoglyceraldehyde (*b*) phosphoglyceric acid
(*b*) fructose – 1, 6 diphosphate (*d*) glucose – 6 phosphate

148. Who enunciated law of limiting factor for photosynthesis ?
(*a*) Hill (*b*) Calvin (*c*) Arnold (*d*) Blackman

149. Which of the following have high CO_2 compensation point ?
(*a*) C_2 plants (*b*) C_3 plants (*c*) C_4 plants (*d*) Aquatic plants

150. Compensation point is
(*a*) where there is neither photosynthesis nor respiration
(*b*) where rate of photosynthesis is equal to rate of respiration
(*c*) when there is enough water to meet the requirements of plant
(*d*) when the entire food synthesised in photosynthesis remains unutilised

151. Photosynthesis is maximum during
(*a*) continuous weak light (*b*) continuous strong light
(*c*) intermittent light (*d*) darkness

152. What will happen to the rate of photosynthesis if sodium bicarbonate is added in the water having *Hydrilla* plant in a beaker ?
(*a*) it will be retarded (*b*) it will be remain normal
(*c*) it will be accelerated (*d*) none of above

153. Rate of photosynthesis is independent of
(*a*) intensity of light (*b*) duration of light (*c*) temperature (*d*) quality of light

154. Kranz type of anatomy is found in
(*a*) C_2 plants (*b*) C_3 plants (*c*) C_4 plants (*d*) CAM plants

155. In C_4 plants Calvin cycle
(*a*) occurs in stroma of bundle sheath chloroplasts
(*b*) occurs in chloroplasts of mesophyll
(*c*) occurs in grana of bundle sheath chloroplasts
(*d*) does not occur

156. In C_4 cycle, CO_2 is fixed by
(*a*) phosphoglyceric acid (*b*) phosphophenol pyruvate
(*c*) oxaloacetic acid (*d*) RuBP

157. The enzyme, which catalyses the photosynthetic C_4 cycle is
(*a*) RuBP carboxylase (*b*) PEP carboxylase (*c*) carbonic anhydrase (*d*) hexokinase

158. C_4 photosynthetic pathway is found in
(*a*) corn (*b*) mustard (*c*) potato (*d*) brinjal

159. Tropical plant like sugarcane shows high efficiency of CO_2 fixation, becuase of
(*a*) Hatch-Slack cycle (*b*) EMP pathway (*c*) TCA cycle (*d*) Calvin cycle

160. Which of the following plants fixes carbon dioxide by way of crassulacean acid metabolism ?
(*a*) cactus (*b*) grass (*c*) Red alga (*d*) *Hibiscus*

161. Energy waste occurs during
(*a*) photosynthesis (*b*) glycolysis
(*c*) photorespiration (*d*) pentose - phosphate pathway

162. What is substrate for photorespiration ?
(*a*) Glycolic acid (*b*) Serine (*c*) Malic acid (*d*) Indole acetic acid

163. All life on earth derives energy, directly or indirectly from sun, except
(*a*) chemosynthetic bacteria (*b*) moulds
(*c*) pathogenic bacteria (*d*) symbiotic bacteria

164. Which of the following photosynthetic bacteria have both PS I and PS II ?
(*a*) green-sulphur bacteria (*b*) purple non-sulphur bacteria
(*c*) purple-sulphur bacteria (*d*) cyanobacteria

165. The bacterial photosynthesis is different from that of other green plants as
(*a*) light is not required (*b*) oxygen is not released
(*b*) a host organism is required (*d*) energy is not fixed

166. Translocation of solutes primarily taken place through
(*a*) phloem (*b*) xylem (*c*) cortex (*d*) pith

167. The carbohydrates synthesised in the leaves are transported through sieve tubes most commonly in the form of
(*a*) glucose (*b*) sucrose (c) triose sugar (*d*) soluble starch

168. If the rate of translocation of food is slow, the the rate of photosynthesis will
(*a*) remain the same (*b*) decrease (*c*) become double (*d*) increase

169. Munch mass flow hypothesis explains
(*a*) Transcription (*b*) Translocation (*c*) Transduction (*d*) Transpiration

170. The *Orobanche* plant is
(*a*) partial stem parasite (*b*) partial root parasite (*c*) total stem parasite (*d*) total root parasite

171. *Cuscuta* is a
(*a*) partial stem parasite (*b*) partial root parasite (*c*) total stem parasite (*d*) total root parasite

172. Which of the following is an example of insectivorous plants
(*a*) *Nepenthes* (*b*) *Cuscuta* (*c*) *Rafflesia* (*d*) *Viscum*

173. Which of the following is an insectivorous plant
(*a*) *Cuscuta* (*b*) *Drosera* (*c*) *Monotropa* (*d*) *Orobanche*

174. Common name of *Utricularia* is
(*a*) Venus fly trap (*b*) bladderwort (*c*) butterwort (*d*) sundew plant

175. The example of saprophyte is
(*a*) *Nepenthes* (*b*) *Drosera* (*c*) *Cuscuta* (*d*) *Monotropa*

ANSWERS

1. (*c*) **2.** (*b*) **3.** (*d*) **4.** (*a*) **5.** (*b*) **6.** (*b*) **7.** (*b*) **8.** (*b*) **9.** (*b*) **10.** (*b*)
11. (*d*) **12.** (*a*) **13.** (*c*) **14.** (*a*) **15.** (*d*) **16.** (*b*) **17.** (*c*) **18.** (*d*) **19.** (*b*) **20.** (*a*)
21. (*b*) **22.** (*a*) **23.** (*d*) **24.** (*b*) **25.** (*b*) **26.** (*d*) **27.** (*a*) **28.** (*a*) **29.** (*b*) **30.** (*c*)
31. (*b*) **32.** (*d*) **33.** (*d*) **34.** (*d*) **35.** (*b*) **36.** (*c*) **37.** (*b*) **38.** (*b*) **39.** (*a*) **40.** (*a*)
41. (*c*) **42.** (*b*) **43.** (*a*) **44.** (*b*) **45.** (*d*) **46.** (*b*) **47.** (*d*) **48.** (*c*) **49.** (*b*) **50.** (*d*)
51. (*d*) **52.** (*d*) **53.** (*d*) **54.** (*d*) **55.** (*b*) **56.** (*c*) **57.** (*d*) **58.** (*d*) **59.** (*a*) **60.** (*c*)
61. (*d*) **62.** (*d*) **63.** (*d*) **64.** (*c*) **65.** (*b*) **66.** (*a*) **67.** (*a*) **68.** (*b*) **69.** (*d*) **70.** (*c*)
71. (*a*) **72.** (*d*) **73.** (*a*) **74.** (*b*) **75.** (*b*) **76.** (*d*) **77.** (*c*) **78.** (*d*) **79.** (*c*) **80.** (*c*)
81. (*b*) **82.** (*d*) **83.** (*b*) **84.** (*b*) **85.** (*d*) **86.** (*a*) **87.** (*b*) **88.** (*a*) **89.** (*d*) **90.** (*b*)
91. (*b*) **92.** (*c*) **93.** (*b*) **94.** (*a*) **95.** (*c*) **96.** (*c*) **97.** (*d*) **98.** (*c*) **99.** (*b*) **100.** (*c*)
101. (*d*) **102.** (*c*) **103.** (*b*) **104.** (*a*) **105.** (*c*) **106.** (*b*) **107.** (*d*) **108.** (*c*) **109.** (*a*) **110.** (*d*)
111. (*d*) **112.** (*c*) **113.** (*d*) **114.** (*a*) **115.** (*c*) **116.** (*a*) **117.** (*b*) **118.** (*c*) **119.** (*c*) **120.** (*a*)
121. (*d*) **122.** (*a*) **123.** (*b*) **124.** (*a*) **125.** (*a*) **126.** (*c*) **127.** (*c*) **128.** (*b*) **129.** (*d*) **130.** (*b*)
131. (*a*) **132.** (*b*) **133.** (*a*) **134.** (*b*) **135.** (*c*) **136.** (*c*) **137.** (*b*) **138.** (*c*) **139.** (*c*) **140.** (*b*)
141. (*a*) **142.** (*a*) **143.** (*a*) **144.** (*a*) **145.** (*a*) **146.** (*a*) **147.** (*b*) **148.** (*d*) **149.** (*b*) **150.** (*b*)
151. (*c*) **152.** (*c*) **153.** (*b*) **154.** (*c*) **155.** (*a*) **156.** (*b*) **157.** (*b*) **158.** (*a*) **159.** (*a*) **160.** (*a*)
161. (*c*) **162.** (*a*) **163.** (*a*) **164.** (*d*) **165.** (*b*) **166.** (*a*) **167.** (*b*) **168.** (*b*) **169.** (*b*) **170.** (*d*)
171. (*c*) **172.** (*a*) **173.** (*b*) **174.** (*b*) **175.** (*d*)

6

CHAPTER

Plant Growth and Development

When a seed is sown in soil, it germinates and forms a small seedling which grows over a period of time, and converts into a large and mature plant. This process is termed as **growth**.

Growth is regarded as one of the most fundamental and conspicuous characteristic of a living being. Growth can be defined as an irreversible permanent increase in size of an organ or its parts or even of an individual cell. Generally growth is accompanied by metabolic process (both anabolic and catabolic), that occur at the expense of energy. Therefore, expansion of a leaf is growth but the swelling of wood submerged in water is not.

The most complex of all the physiological processes is **growth**. The process of growth results from an increase in protoplasm, leading to an irreversible increase in size and weight. The growth involves three stages: **stage of cell division**, **stage of cell elongation** and **stage of cell differentiation**. **Growth** is associated with differentiation of cell which is a change in form and physiological activity. **Development** refers to the sequence of processes in the overall life history of a cell or an organism including growth, differentiation, maturation and senescence.

In unicellular organisms, growth involves increase in their volume and numbers. Cell division serves as a means of reproduction in these organisms, and leads to increase in population. In the multicellular organisms, however, growth is much more complex.

The life of flowering plant begins with the fertilized egg. Through a series of divisions, an embryo is formed which is enclosed in a seed. Germination then takes place leading to the formation of a seedling. After passing through the juvenile phases, a mature plant is finally formed. After a period of vegetative growth, the plant enters the reproductive stage. With the initiation and development of flowers, fruits and seeds are formed. A new generation of plants starts again. Annual plants like wheat undergo senescence and die after bearing grain. Development associated with each of three steps is controlled by several factors, both internal and external.

In plants, growth is associated with both anabolic and catabolic activities. This is a quantitative phenomenon and can be measured in relation to time. In living organisms, growth is intrinsic, while in non-living objects, it is extrinsic.

In plants, growth is confined only to **meristems**. The extreme apices of root and shoot are occupied by primary meristems, while in their older parts, secondary meristems, *i.e.*, cambia, give rise to additional vascular tissues and to protective layers of cork cells.

The activity of each meristem influences the other meristems which are close to it, giving rise to **growth correlations**. For example, while the main apical shoot meristem is active it retards the activity of more recently initiated lateral bud meristems, a phenomenon called **apical dominance**.

Besides growth, plants also show movements, both at organ-level, as well as whole-plant level. In this chapter, various aspects of growth, such as dormancy and seed germination, characteristics and conditions for growth, its measurement, growth regulators and types of movements have been discussed in detail.

Plant Growth Generally is Indeterminate

Plants are unique in that they retain the capacity for unlimited growth throughout their life. This ability is due to the presence of meristems at certain locations in their body. The cell(s) of such meristems have the capacity to divide and self-perpetuate. The form of growth wherein new cells are always being added to the plant body by the activity of the meristem is called open form of growth.

The root apical meristem (RAM) and the shoot apical meristem (SAM) are responsible for the primary growth of the plant and principally contribute to the elongation of the plant parallel to its axis. In dicotyledonous plants and gymnosperms, the lateral meristems, vascular cambium and cork cambium appear later in life. These meristems that cause increase in girth of the organ in which they are active, contributing to what is generally known as **secondary growth** of the plant.

The intercalary meristem, located just above the nodes of grasses and related plants contribute to the elongation growth. Thus, root apical meristem (RAM), shoot apical meristem (SAM), vascular cambium, cork-cambium and intercalary meristem are all called **indeterminate meristem**. However the meristem such as those involved in leaf and flower formation are active only for a short period and are primarily responsible for the formation of an organ. These are called **determinate** meristems. (Also see the chapter : Anatomy of Flowering Plants).

Growth is Measurable

Growth is measured by a variety of parameters. Some of such parameters are : (*i*) increase in fresh weight; (*ii*) increase in dry weight; (*iii*) increase in length; (*iv*) increase in area; (*v*) increase in volume; and (*vi*) increase in cell number, etc. For example, one single maize root apical meristem can give rise to more than 17,500 new cells per hour, whereas cells in a watermelon fruit may increase in size by upto 3,50,000 times. In the former, growth is expressed as increase in cell number; the latter expresses growth as increase in size of the cell. Though every cell, tissue organ and organism exhibit growth, yet all the above parameters are not applicable to all forms of growth at all times. For example, an increase in length is a good parameter for a pollen tube growth, while the growth in a dorsiventral leaf can be more easily be measured as an increase in area.

Dormancy and seed Germination

Seed. *A true seed is defined as a fertilized mature ovule that possesses embryonic plant, stored material, and a protective coat or coats.* Seed is the reproductive structure characteristic of all phanerogams. The structure of seeds may be studied in such common types of pea, gram, bean, almond or sunflower. They are all built on the same plan although there may be differences in the shape or size of the seed, the relative proportion of various parts.

There are hundreds of variations in the seed size, shape, colour and surface. The seeds range in size from tiny dust particles, as found in some orchids, to large double-coconuts. The seed surface may be smooth, wrinkled, striate, ribbed, furrowed, reticulate, tuberculate, alveolate, hairy, pulpy or having patterns like finger prints.

In the seed, life activities are temporarily suspended in order to enable the plant to successfully pass through unfavourable and injurious climatic conditions. On the approach of favourable conditions, the seed resumes active life and grows into full plant. In the form of seeds, a plant can be carried to long distances without special precautions.

The developed seeds in many plants undergo a period of dormancy before germination and further growth.

Quiescence. Generally, all plants experience a period of suspended growth. This suspension may be due to exogenous control, such as a change in environmental conditions, and is referred to as **quiescence**.

Seed dormancy. Many kinds of seeds, apparently ripe, fail to germinate even if placed under such conditions that all environmental factors are favourable. In such seeds, resumption of growth by the embryo is arrested by the conditions within the seeds themselves. The state of inhibited growth of seeds or other plant organs as a result of internal causes is usually called **dormancy** but is sometimes referred to as the rest period. Seed dormancy is of considerable advantage to the plant. It enables the embryo to safely pass the unfavourable part of the year and germinate when the conditions are suitable for the establishment of the seedlings. In nature, dormancy period coincides with the unfavourable period for the seedling of the species. Dormancy of seeds results from one or a combination of several undermentioned factors.

Impermeability of seed-coats to water. The seed coats of many species are completely impermeable to water at the time of their maturity. This condition is very common in the seeds of many legumes, (*e.g.*, clovers, alfalfa, etc.) of the water lotus, and of the morning glory. Germination fails to occur until water penetrates through the seed coats.

Mechanically resistant seed coats. In some seeds as those of mustard (*Brassica*), pigweed (*Amaranthus*), shepherd's purse (*Capsella*), etc., the seed-coats are so strong that they do not yield to the pressure of the expanding embryo. The embryos of these seeds have no dormant period and will grow readily if the seed coats are removed.

Seed-coats impermeable to oxygen. The two seeds in a fruit of cocklebur (*Xanthium*) are not equally dormant. Under natural conditions, the lower seed usually germinates in the spring following maturity while the upper seed remains dormant until the next year. The dormancy of these seeds has been demonstrated to result from the impermeability of the seed coats to oxygen.

Rudimentary embryos. In plants like ginkgo (*Ginkgo biloba*), European ash (*Fraxinus*), holly (*Ilex*) and many orchids, the embryo is unorganized when the seed is shed and attains full development before it germinates.

Dormant embryos. In many species, although the embryos are completely developed when the seed is ripe, the seed fails to germinate even when environmental conditions are favourable. Dormancy of such seed is a result of the physiological condition of the embryo. The embryos of such seeds will not grow when the seeds first ripen even if the seed coats are removed. During the period of dormancy, some physiological changes called after-ripening occur in the embryo before the seed is capable of germination. The seeds of apple, peach, iris and pine belong to this group. In nature, after-ripening occurs in winter and the seeds formed in autumn germinate the coming spring.

Germination inhibitors. Germination of seeds is sometimes checked or prevented by the presence of compounds known as **inhibitors**. These are often substances produced in one or more plant organs. For example, the juice of tomato fruits inhibits the germination of tomato seeds and many other seeds as well. Inhibitors may be present in the embryo (*e.g.*, in *Xanthium*), endosperm, (*e.g.*, in *Iris*) or in the seed coat (*e.g.*, in *Cucurbita*). Abscissic acid (ABA) is one of the most commonly detected inhibitors of germination.

Methods of Breaking the Dormancy of Seeds

The dormancy of seeds presents a practical problem of considerable economic importance. Plant growers are often interested in securing seed that will germinate soon after it is harvested. Several methods have been devised for breaking the dormancy of seeds and for shortening the period of dormancy so that they may germinate promptly.

Whenever dormancy results of any of the causes inherent in the seed coats it can be interrupted by **scarification**. For example, machine thrashed legumes seeds usually show a higher percentage of germination than those that have been harvested by hand. Strong mineral acids have been used successfully to interrupt seed dormancy caused by resistant or impermeable seed coats.

Soaking the seeds in certain chemicals like potassium nitrate, ethylene, chlorohydrine, thiourea or in certain plant hormones is known to break dormancy. After ripening of many seeds occurs more rapidly when they are kept at low temperatures than when stored at higher temperatures. Temperature from 5° to 10°C for two or three months are effective with seeds of conifers. Under natural conditions, seed dormancy is gradually overcome by processes such as weakening of the seed coat by the digestive juices in the alimentary canal of fruit-eating birds and other animals, or in still due to the action of microbes or due to mechanical abrasions. Dormancy of seeds is also broken by subjecting the seeds alternately to relatively low and high temperatures. Light is also considered as means of breaking dormancy of seeds. Seeds of sweet clover (*Melilotus alba*) and alfalfa (*Medicago sativa*) showed greatly improved germination after being subjected to hydraulic pressures of 2000 atm. at 18°C.

Germination of Seed

This is the first step towards plant growth. In favourable conditions of environment, seeds germinate. After the dormancy period is over or dormancy is broken, and other necessary conditions for growth are present, the dormant embryo resumes metabolic activities and growth.

The embryo lies dormant in the seed, but when the latter is supplied with moisture, the embryo becomes active and tends to grow and develop into a small seedling. *The resumption of metabolic activity and growth by the seed tissue, involving rehydration, utilization of reserves and gradual development of synthetic systems which enable the plant to assume an autophytic or autotrophic existence, is called germination.* In other words, *the process by which the dormant embryo wakes up and begins to grow is known as germination.* The embryo grows by absorbing the food material stored up in the cotyledons, or the endosperm when it is present. At first the seed absorbs moisture and swells up considerably, then the radicle elongates and comes out often through the micropyle and gives rise to the root. In whatever position the seed be placed in the soil, the radicle always grows downwards and gives rise to the root system. Its rate of growth is much faster than that of the plumule. As a result of swelling, the seed-coat bursts and the cotyledons partially or completely separate from each other to give way to the plumule which lies within. Prior to germination, several intricate physiological processes occur within the seed. A rapid increase in respiration occurs. This is accompanied by the appearance of some hydrolytic enzymes that break down food materials stored in the seed to simpler molecules that can be utilized by the seedling until the latter can become photosynthetically efficient.

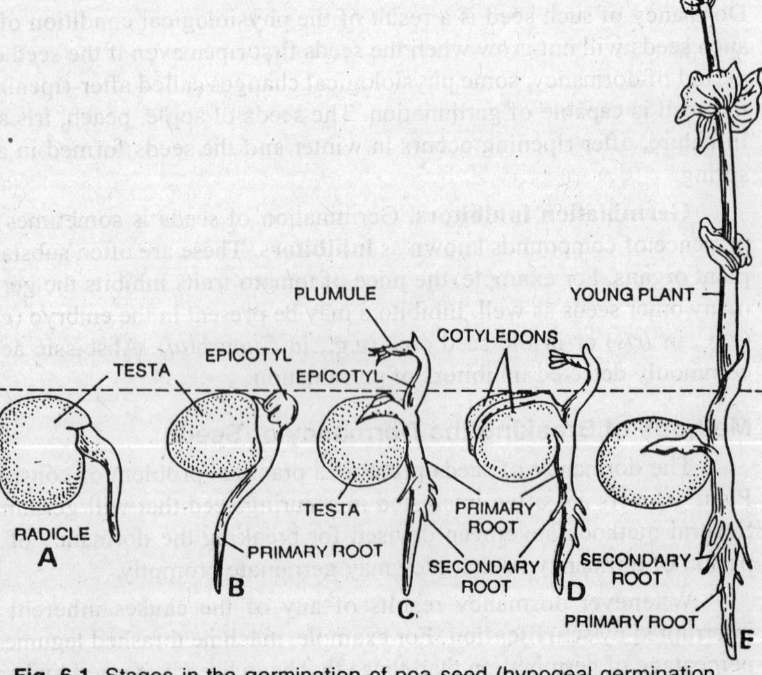

Fig. 6.1. Stages in the germination of pea seed (hypogeal germination in dicot).

Based on the behaviour of the cotyledons, germination is of two types: **epigeal** and **hypogeal**. In some seeds such as tamarind, cucumber, cotton, gourd, castor, etc., the cotyledons are seen to be pushed upwards above the soil, the germination is said to be **epigeal**. In other seeds such as gram, pea, broad bean, groundnut, mango, etc., the cotyledons remain below the soil and the germination is known as **hypogeal**. In the seeds of all monocotyledonous plants, the germination is of hypogeal type.

Vivipary

This is a special type of germination of seed. This is the phenomenon of germination of seeds within the fruit, while the fruit is still attached to the mother plant. This type of seed germination occurs in several species of the plants to mangrove vegetation, such as *Rhizophora*, *Sonneratia*, *Ceriops*, *Kandelia*, etc. The seeds in such plants do not germinate in the soil due to high salt contents and deficiency of oxygen in the marshy habitats. In such plants, the seed dormancy is absent. Instead, embryo grows continuously without any hinderance inside the fruit while it is attached to the mother plant. (Fig. 6.3)

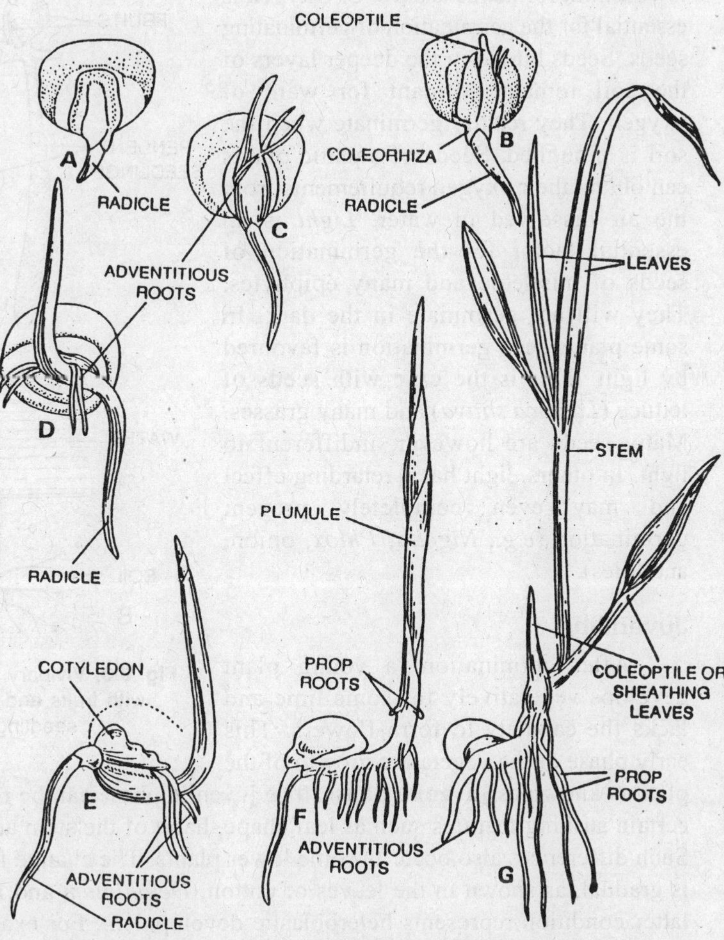

Fig. 6.2. Stages in the germination of maize grain (A-G).

During this type of germination, the hypocotyl elongates and pushes the radicle out of the seed and fruit. Thereafter, the hypocotyl grows more vigorously and becomes club-like, and sometimes attains a length of 50 cm or so, as in *Rhizophora*. The young green seedling comes out of the fruit showing the long hypocotyl and radicle. The cotyledons act as haustoria and suck food from the parent plant. When the seedling gains weight, it gets detached from the fruit, and falls down vertically like a dart. Now the radicle penetrates into the mud. The lateral roots immediately develop on this seedling, and it soon establishes as a plant in the muddy soil. The plumule remains above the water level and grows safely.

Conditions Necessary for Germination

Water, temperature, oxygen and light are the factors which are responsible for seed germination. **Water** is needed to bring about the vital activity of the dormant embryo to dissolve various salts and organic substances stored in the cotyledons or in the endosperm; to facilitate necessary chemical changes; and to help the embryo to come out easily by softening the seed-coat. **Temperature** requirement of seeds varies depending on the species. Seeds of many species germinate only if the temperature is within a certain range. Very low and very high temperatures retard or prevent germination. Wheat grains may germinate at a temperature a little above 0° C while the lowest limit for maize is 5-10°C. **Oxygen** is essential for the germination of germinating seeds. Seeds buried in the deeper layers of the soil remain dormant for want of oxygen. They readily germinate when the soil is ploughed. Seeds of aquatic plants can obtain their oxygen requirements from the air dissolved in water. *Light* is an essential factor for the germination of seeds of mistletoe and many epiphytes. They will not germinate in the dark. In some plants seed germination is favoured by light. This is the case with seeds of lettuce (*Lactuca sativa*) and many grasses. Many seeds are however, indifferent to light. In others, light has a retarding effect and may even completely prevent germination (*e.g.*, *Nigella*, *Phlox*, onion, and lilies).

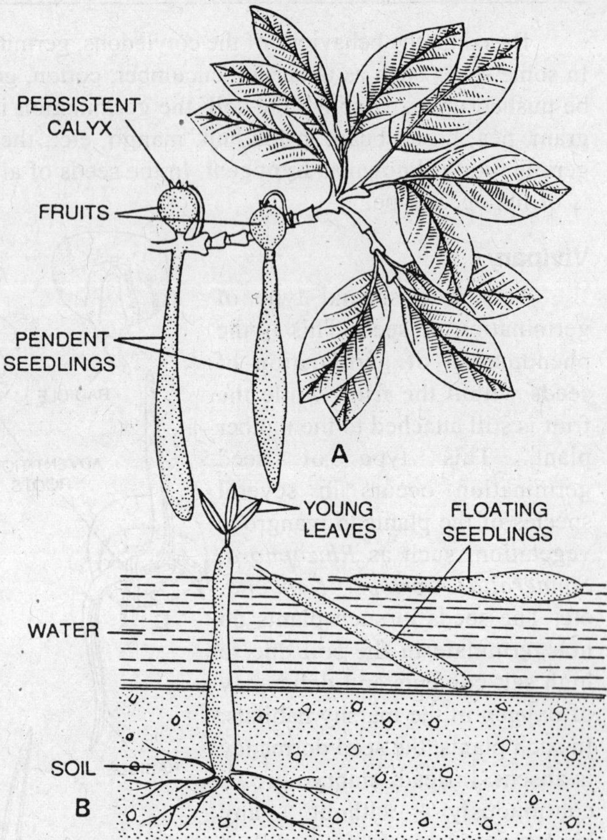

Fig. 6.3. Vivipary. A, a branch of *Rhizophora conjugata* with fruits and pendant seedlings; B, two floating seedlings and one rooted seedling.

Juvenility

After germination, a young plant develops vegetatively for some time and lacks the capacity to form flowers. This early phase of the vegetative growth of the plant is known as **juvenile phase**. The juvenile phase can be recognized from the adult phase by certain striking features such as leaf shape, habit of the stem and the size of the thorns if present. Such differences also occur in some lower plants. The change from the juvenile to the adult phase is gradual, as shown in the leaves of cotton, *Delphinium* and *Ipomoea*, or it may be abrupt. The latter condition represents heteroblastic development. For example, in some plants the first few leaves are normal, whereas the subsequent leaves may be phyllodes, (*e.g.*, *Acacia* spp).

It has been advocated that the juvenile and adult parts of the plants are also physiologically different. In the juvenile phase, the plant does not respond to stimuli that can induce flower formation. In woody plants, branches cut from the juvenile stage root more readily than those derived from the adult stage. This is a point of significance in vegetative propagation.

CHARACTERISTICS OF GROWTH

Plant growth is a complex phenomenon associated with numerous physiological processes, both of constructive and destructive types.

Constructive process leads to the formation of various nutritive substances and the protoplasm, while destructive process is responsible for their breakdown.

The protoplasm assimilates the protein food and increases in bulk, while the carbohydrates are utilised in respiration and in the formation of cell-wall substance, cellulose.

Plant growth occurs by cell division and cell enlargement followed by cell differentiation. The cell division generally occurs in apical regions of shoot and root, and therefore, meristematic cells present at root and shoot apices are responsible for growth in plants.

The cambia are present in vascular bundles of root and shoot of dicot plants. They help in increasing thickness of stem and root due to secondary growth.

In some plants, the cambium is also found just beneath the epidermis and above cortex, which is referred as cork cambium.

Thus, overall definition of plant growth is : a permanent and irreversible increase in size and form attended by an increase in weight.

Growth Curve

Whether the growth rate is studied of a cell, a plant organ, a whole plant or the whole life-cycle of plant and whether the growth rate is measured in terms of length, size, area, volume or weight. It has been found that though in initial stages, a little decrease in weight may occur due to utilization of reserved food, the growth rate increases slowly during the phase of cell elongation or enlargement and again slows down during the phase of cell maturation till it stops. The period during which the growth rate in any term is plotted against time always a S-shaped curve is obtained. This is known as the **sigmoid curve**. An analysis of the curve shows a slow start, gradually attaining a rapid growth rate followed by a period of slow growth and ultimately a decline.

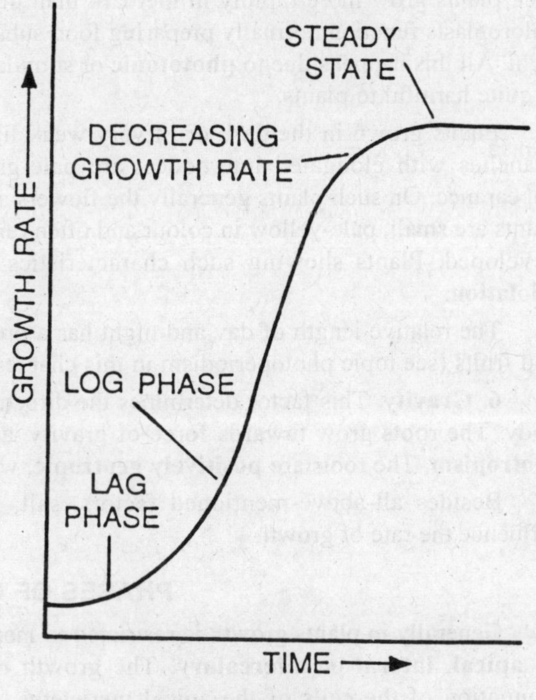

Fig. 6.4. Sigmoid curve.

The rate of plant growth is slow in the initial stages, known as **lag phase**, and increases rapidly later on, known as **log phase** (exponential phase). The growth again slows down due to the limitation of nutrients, and called **steady phase**.

The rate of growth can be measured by an increase in size or area of an organ of the plant, such as leaf, flower or fruit in a unit time. The rate of growth is called **efficiency index**.

CONDITIONS NECESSARY FOR GROWTH

As growth is brought about by the protoplasm, the conditions for growth are the same as those that maintain the activity of the protoplasm. They are as follows :

1. Supply of nutritive materials. Growth can only take place when the protoplasm of the growing region is supplied with nutritive materials. The protoplasm assimilates these materials and builds up the body of the plant. Carbohydrates make a source of energy.

2. Supply of water. An adequate supply of water is absolutely necessary to maintain the turgidity of the growing cells. Turgidity is the first step towards growth. The protoplasm can only function when it is saturated with water. However, abundant supply of water is lost due to transpiration. and only small amount is required for actual growth.

3. Supply of oxygen. Supply of free oxygen is indispensable for the respiration of all living cells. This is an oxidation process by which the **potential** energy stored in the food is released in the form of **kinetic** energy, and used by the protoplasm for hundreds of activities.

4. Suitable temperature. The protoplasm requires a suitable temperature for its activities. At a low temperature, it ceases to perform its functions or does so very slowly, whereas a temperature of 45° to 50°C coagulates and kills the protoplasm. However, the protoplasm maintains its activities within a certain range of temperature. This is called the **thermotonic** effect of temperature. The optimum temperature is generally from 28° to 30°C, while the minimum temperature is about 4°C.

5. Light. However, this condition for growth is not absolutely necessary in initial stages. In fact, plants grow more rapidly in the dark than in the light. However, stomata remain open and chloroplasts function normally preparing food substances (carbohydrates) only in the presence of light. All this happens due to **phototonic** or stimulating effect of light. Continued absence of light is quite harmful to plants.

Plants grown in the dark or in very weak light have delicate, soft and slender stems and branches with elongated internodes, are pale-green or pale-yellow in colour and sickly in appearance. On such plants generally the flowers and fruits are not produced. The leaves of such plants are small, pale-yellow in colour and often remain under-developed, and the roots are poorly developed. Plants showing such characteristics are called **etiolated**, and the phenomenon, **etiolation**.

The relative length of day and night has a profound influence on the production of flowers and fruits (see topic photoperiodism in this chapter).

6. Gravity. This factor determines the direction of growth of particular organs of the plant body. The roots grow towards force of gravity, and the stem against it. This process is called **geotropism**. The roots are **positively geotropic**, while the stems are **negatively geotropic**.

Besides all-above-mentioned factors, salt, mineral deficiencies and stress factors also influence the rate of growth.

PHASES OF GROWTH

Generally in plants growth is restricted to meristematic regions only. These **meristems** may be **apical**, **lateral** or **intercalary**. The growth in length is due to gradual enlargement and elongation of the cells of the apical meristems, *i.e.*, root-apex and shoot-apex. However, in dicotyledonous angiosperms and gymnosperms the growth in thickness is due to the activity of the lateral meristems, *i.e.*, interfascicular cambium, intrafascicular cambium and cork cambium. Growth is not a very simple process. Before completion of this process a meristematic cell has to pass to three important phases : 1. The formative or meristematic phase; 2. The phase of elongation and 3. The phase of maturation; 4. Grand period of growth.

1. The formative or meristematic phase. This is restricted to the apical meristem of root and stem. The cells of this region constantly divide and multiply in number. They have abundant protoplasm, a large nucleus and a thin cellulose wall.

The meristematic phase has constantly dividing cells both at the root apex and the shoot apex. The cells in this region are rich in protoplasm, possess large conspicuous nuclei and their cell walls are primary, thin and cellulosic with abundant plasmodesmata.

2. The phase of elongation. This lies immediately behind the formative phase. The cells do not divide in this phase, but they increase in size. The cells enlarge in size and elongate until they reach their maximum dimension. In the root, this phase occupies a length of a few millimeters, and in the stem a few centimeters.

The cells just next to meristematic zone represent the phase of elongation. Increased vacuolation cell enlargement and new cell wall deposition are the characteristics of cells in this phase.

3. The phase of maturation. This is further behind, and here, the cells start maturing to obtain a permanent size. The thickening of the cell-wall takes place in this phase.

Further, away from the apex, *i.e.*, more proximal to the phase of elongation lies the portion of the axis which is undergoing the phase of maturation. The cells occupying this zone, attain their final size, wall thickening and protoplasmic modifications. Most of the tissue/cell types represent this phase. Hence, at this phase, all the diverse tissue types observed in epidermis, cortex, vascular tissue, etc., both in root and apical end of a stem.

4. Grand period of growth. The time interval from the formative phase to maturation phase is called the grand period of growth. Every cell of the organ of plant body, shows a variation in the rate of its growth. The growth is at first slow, then it accelerates till a maximum is attained, then it falls off quickly, and gradually slows down till it comes to a standstill. This growth of organ or cell of a plant as a whole extending over the whole period is actual **grand period of growth**.

Within the grand period, variations in growth occur due to external and other causes. There is thus the **diurnal variation of growth**. Light inhibits growth, and too intense light even checks growth altogether. Thus plants grow quicker in night than in comparison of day.

There is also **seasonal variation of growth**. During winter, the growth of many plants is checked or becomes very slow, while during spring, growth proceeds rapidly.

MEASUREMENT OF GROWTH

As already mentioned, growth is a complex natural phenomenon, and generally takes place at the apical regions of the plant. Thus, the growth in length can easily be measured with the help of **auxanometer**. This phenomenon, you can understand by following simple experiments :

An auxanometer is used to measure the rate of growth in terms of shoot length.

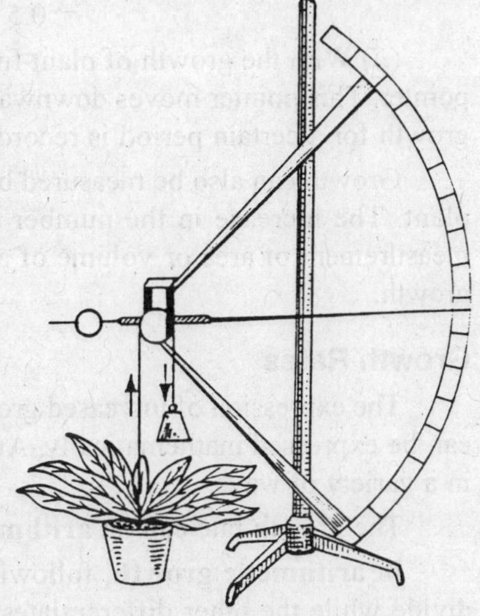

Fig. 6.5. Arc Auxanometer.

PRACTICALS

Experiment

Object : Measurement of growth in plants.

Requirements : Auxanometers : (*i*) Arc auxanometer (*ii*) Pfeffer's automatic auxanometer.

(*i*) **Arc auxanometer :** It consists of a vertical stand with a pulley to which is attached a long thread and a needle or pointer movable over a graduated arc. One end of the thread is tied with the growing point of the plant and another is stretched with a suitable weight.

(*ii*) **Pfeffer's automatic auxanometer :** It consists of a vertical stand with two pulleys. One of these two pulleys possesses a thread carrying two weights at either ends to keep it stretched. This thread on one side is also attached with a fine scratching pointer which touches a smoked drum. Another pulley is also provided with a thread of which one end is tied with the growing point of plant and another is stretched with a suitable weight.

Observation and Explanation : (*i*) With the growth of plant, the weight moves the thread down and the movement of needle can be read easily on the scale.

There exists a notable ratio between the size of pulley and the needle. If pulley is 4 inches with the needle 20 inches from the centre of the disc of pulley, growth will be magnified ten times on the scale. It can easily be illustrated by an example :

If indicator moves 4 cm in 8 hours and the magnification is 10 times, then the actual growth is:

$\because$ Growth in 8 hours = 4/10 cm

= 0.4 cm or 4 mm

$\therefore$ Growth in one hour = 4/8

= 0.5 mm/hour

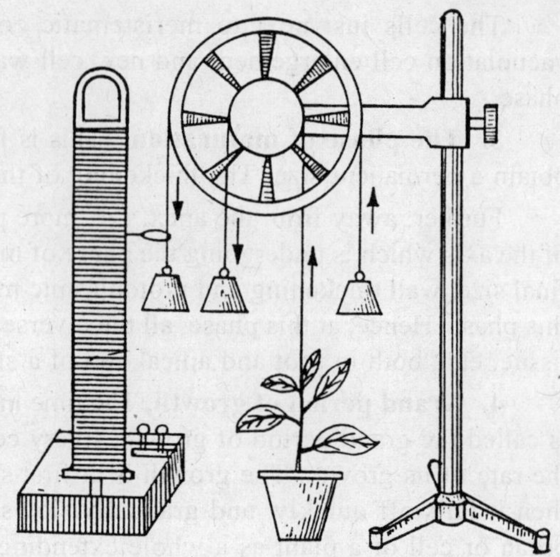

Fig. 6.6. Pfeffer's Auxanometer.

(*ii*) With the growth of plant first pulley makes moving the another which in turn moves the pointer. This pointer moves downwards and traces a spiral on the rotating smoked drum. Thus the growth for a certain period is recorded.

Growth can also be measured by an increase in weight both fresh and dry, and volume of the plant. The increase in the number of cells, especially in algae, cyanobacteria, yeast, etc. The measurement of area or volume of an organ of the plant also gives information about the rate of growth.

Growth Rates

The expression of increased growth per unit time is termed **growth rate**. Thus, rate of growth can be expressed mathematically. An organism, or a part of the organism can produce more cells in a variety of ways.

The growth rate can be **arithmetic** or **geometrical** increase :

In **arithmetic growth**, following mitotic cell division only one daughter cell continues to divide while the other differentiates and matures. The simplest expression is arithmetic growth, *e.g.*, a root elongating at a constant rate. Mathematically, it is expressed as :

$$L_t = L_o + r_t$$

L_t = length at time 't'

L_o = length at time 'zero'

r = growth rate/elongation per unit time.

Geometrical growth. However, in most systems, the initial growth is slow (lag phase), and it increases rapidly thereafter at an on exponential rate (log or exponential phase). Here, both the progeny cells following a mitotic cell division retain the ability to divide and continue to do so. However, with limited nutrient supply, the growth slows down leading to a stationary phase. If we plot the parameter of growth against time, we get a typical **sigmoid** or **S-curve**, S-curve is characteristic of every living organism growing in a natural environment. It is typical for all cells, tissues and organs of a plant. However, S-curve of the individual cells/tissues/organs/organisms may not be synchronous, *i.e.*, not at the same phase of the curve. For example, one cell may enter stationary phase while other is just entering the lag phase. Individual leaves are initiated, and abscised (each with a S-curve), but the plant/branch on which they occur could be still in exponential (log phase) or other places of growth.

The exponential growth can be expressed as :

$$W_1 = W_{oe} r_t$$

W_1 = final stage (weight, height, number, etc.)

W_0 = initial size at the beginning of the period

r = growth rate

t = time of growth

e = base of natural logarithms

Here, r = is the relative growth rate and is also the measure of the ability of the plant to produce new plant material, referred to as **efficiency index**. Hence, final size of W_1 depends on initial size, W_0.

Quantitative comparisons between the growth of living system can be also made in two ways :

(*i*) Measurement and the comparison of total growth per unit is called absolute **growth rate**.

(*ii*) The growth of the given system per unit time expressed on a common basis, *e.g.*, per unit initial parameter is called **relative growth rate**. For example, two leaves A and B of different area sizes show exact absolute increase in area in the given time to give leaves A′ and B′. However, one of them has much higher relative growth rate.

Apical Dominance

In many herbaceous plants which produce aerial stems, growth in length takes place at the apex of the main axis of the plant. Although a lateral bud is present in the axil of every leaf, side branches do not often develop from these buds as long as the terminal bud retains its vigour and continues to grow. If the terminal bud is destroyed or is artificially removed, development of the lateral buds usually starts at once. This inhibiting effect of a terminal bud upon lateral bud development is called **apical dominance** and is much more pronounced in some species than in others.

The phenomenon of apical dominance is usually also in evidence in all woody plants on which true terminal buds form. The lateral buds on current shoots usually do not develop unless the terminal bud is destroyed or injured. Development of the lateral buds on older shoot segments is of more frequent occurrence, indicating that the inhibitory effect of the apical bud diminishes with greater distance of the lateral buds from the apex of the stem.

The controlling effect of the apical bud in apical dominance results from its auxin content. Several other growth regulating compounds have similar effects upon the growth of lateral buds, if applied to the cut surface of a detipped stem. For example, when potato tubers begin to sprout the apical buds grow rapidly but the lateral buds usually fail to elongate. If, however, the apical and lateral buds are cut out of the parent tuber both grow at similar rates. Treatment of the tubers with ethylene chlorohydrin which causes the destruction of auxin, results in the rapid growth of both lateral and apical buds. This supports that the growth of lateral buds is checked by auxins formed in the apical buds.

Conditions for Growth

Water, oxygen and nutrients are very essential for growth. The plant cells grow in size by cell enlargement which in turn requires water. Thus, plant growth and further development is intimately linked to water status/availability water also provides medium for enzymatic activity.

Oxygen helps in releasing metabolic energy, so essential for growth activities.

Nutrients (macro and micro essential elements) are required by plants for synthesis of protoplasm and act as source of energy.

Temperature. In addition, every plant organism has an optimum temperature range best suited for its growth. Any deviation from this range could be detrimental to its survival.

Environmental signals such as light and gravity also affect certain phases/stages of growth.

GROWTH REGULATORS

Plants produce some specific chemical substances, which are capable of moving from one organ to the other, and provide physiological control on growth. These substances, which are active in very small amounts are called **plant hormones** or **growth regulators**.

The genetic make up and **growth regulations** constitute important internal factors that regulate plant growth and development. The main naturally occurring plant growth regulators are **auxins**, **gibberellins**, **cytokinins**, **ethylene** and **abscisic acid**. The first four are called plant hormones. Abscisic acid (ABA) is naturally occurring growth inhibitor. A plant hormone is an organic compound synthesized in one tissue of plant and migrates to another tissue wherein very minute quantity affects the growth.

1. Auxins (*Gk. auxein = to grow*). Auxin is a group of hormone produced by the apical apices of stem and roots which migrate from the apex to the zone of elongation. The role of auxin is essential for the elongation process. Most of the knowledge about auxins comes from the work

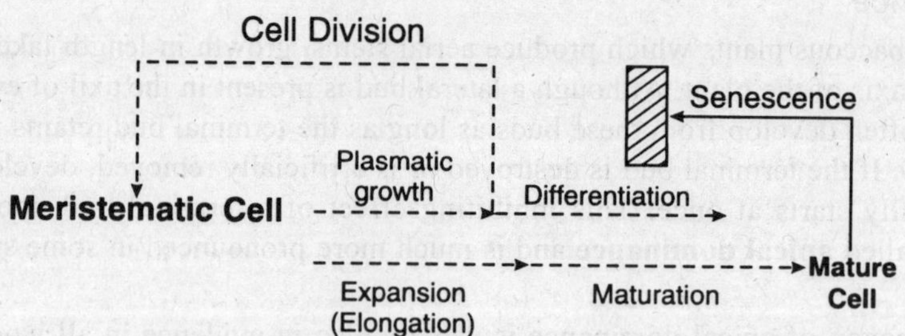

Fig. 6.7. Structures of some common synthetic auxins.

on oat (*Avena sativa*). If the growing tip of oat coleoptile is removed, the remaining portion of coleoptile will show a marked decrease in growth which will ultimately stop. This means growth is restricted to or dependent on upper region. If the removed tip is placed on agar block for a period of several hours and then removed and the agar block now transferred to the cut stem, it is observed that block partially substitutes for the tip and the growth is resumed. This may be concluded that substance synthesized in the tip migrates downwards (in agar block) and is responsible for elongation of tip. This substance was denoted as auxin. Went (1928) performed ***Avena*-curvature test.** He placed several freshly cut coleoptile tips on agar block, after few hours which is cut into several pieces. He placed the agar block eccentrically on the coleoptile cut portion for two hours in dark. The growth curvature occurred and the coleoptile bends on one side opposite to the agar block. It occurred due to auxin diffusion from block to coleoptile resulting in elongation and growth on that side, as a result coleoptile bends. The experiments performed by Went shared that auxin is synthesized in the coleoptile tip and is translocated downward.

Auxins are synthesized in the shoot tips, young leaf primordia, developing seeds and from there they migrate to the regions of elongation. The movement of auxin is polar and in stems, it takes place in a basipetal direction whereas in the roots, it is in an acropetal direction. Auxins cause elongation of the stems and roots.

Extensive work has been done on auxin and several related compounds have been isolated in pure forms. Among them the most common one is Indole 3-acetic acid (IAA) which is derivative of amino acid tryptophan.

The primary physiological effect of auxin is to promote the elongation of the cells which may be due to increasing osmotic pressure and permeability of cytoplasm to water and decreasing wall pressure. Some work done by Coartney *et al* (1967), Masuda *et al* (1967) and Nooden (1968) indicates that auxin is associated with nucleic acid metabolism. It affects the site of DNA responsible for increasing plasticity of walls and its extension.

The compounds which can be converted into auxins are called **auxin precursors**.

The compounds which inhibit the actions of auxins are called **anti-auxins**.

The auxins which are hard to extract and need the use of organic solvents are referred to as **bound auxins**. The free form of auxin is active, whereas the bound auxin is inactive in growth. A dynamic equilibrium exists in between these two forms.

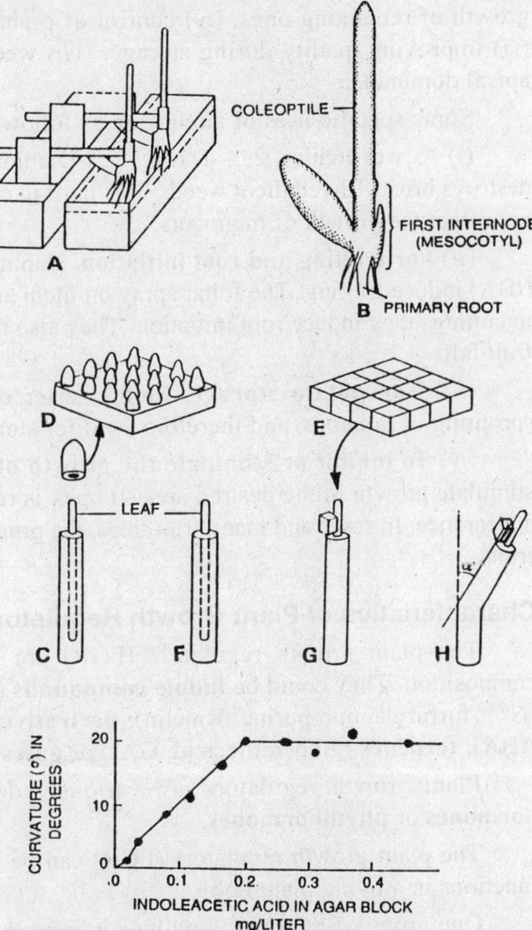

Fig. 6.8. Diagrammatic representation of the *Avena* curvature test.

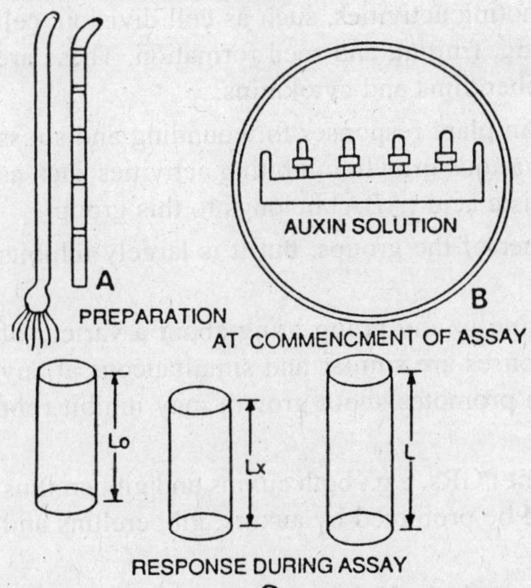

Fig. 6.9. Avena Section Test. LO = length of freshly cut segment. LX = Length of untreated segment after floating in water for length of test period ; L = length of treated segment after floating in test solution for length of test period.

The functions of auxins are generally tested with the help of bioassays. **Bioassay** means the testing of substance for its activity in causing a growth response in a living plant or its organ. This test is quantitative in nature, which measures the concentration required to produce the effect.

For example, *Avena* curvature test, and root growth inhibition test, are bioassays for examining auxin activity.

Uses of auxins. There are several horticultural and agricultural uses of auxins. They are as follows :

The applications of auxins have a wide range, such as (*i*) enhancing the formation of roots, (*ii*) development of parthenocarpic fruits (*iii*) thinning of flowers and fruits for healthy growth of remaining ones, (*iv*) control of preharvest fruit drop, (*v*) enhancement of flowering, (*vi*) improving quality during storage, (*vii*) weed control by herbicidal action, (*viii*) control of apical dominance.

Some specific uses of auxins are as follows :

(*i*) **As weedicides.** 2, 4–D (*i.e.*, 2, 4-Dichlorophenoxyacetic acid) is a selective weedicide. It destroys broad-leaved dicot weeds. It is used in cereal crops and in lawns to kill the weeds. It does not affect the growth of monocots.

(*ii*) **For fruiting and root initiation.** Naphthalene acetic acid (NAA) and indole butyric acid (IBA) induce fruiting. The foliar spray on litchi and pineapple induces flowering and when applied to cuttings they induce root initiation. They also retard abscission, and therefore prevent premature fruit fall.

(*iii*) **For potato storage.** Methyl ester of naphthalene acetic acid (NAA) inhibits the sprouting of potatoes, and therefore used for storage.

(*iv*) **To inhibit or stimulate the growth of lateral buds.** The pruning of plants is done to stimulate growth in the desired area. If apex is removed lateral buds develop, giving a branching appearance. In roses and many fruit trees, the pruning is done to improve the quality of flowers and fruits.

Characteristics of Plant Growth Regulators

The plant growth regulators (PGR) are small, simple molecules of diverse chemical composition. They could be **indole compounds** (indole-3-acetic acid IAA); **adenine derivatives** (N^6 – furfurylamino purine, Kinetin); **derivatives of carotenoids** and **fatty acids** (abscisic acid, ABA); **terpenes** (gibberellic acid, GA_3) or **gases** (ethylene, C_2H_4).

Plant growth regulators are variously described as **plant growth substances, plant hormones** or **phytohormones**.

The plant growth regulators (PGR) can be broadly divided into two groups based on their functions in a living plant body.

One group of PGRs are involved in growth promoting activities, such as cell division, cell enlargement, pattern formation, tropic growth, flowering, fruiting and seed formation. These are also called **plant growth promoters**, *e.g.*, Auxins, gibberellins and cytokinins.

The PGRs of other group play an important role in plant responses to wounding and stress of biotic and abiotic origin. They also are involved in various growth inhibiting activities such as promotion of dormancy and abscission. The PGR abscisic acid (ABA) belongs to this group.

The gaseous PGR, ethylene (C_2H_4), could fit either of the groups, but it is largely inhibits growth activities.

Functions of PGRs. Each of these PGRs belonging to any group bring about a variety of growth and developmental responses, and not all responses are similar and simultaneous at any given time, *e.g.*, a given concentration of auxin which promotes shoot growth may inhibit root growth.

Secondly, similar responses are affected by different PGRs, *e.g.*, both auxins and gibberellins can result in cell enlargement, while cell division could be promoted by auxins, gibberellins and cytokinins.

The PGRs are active at very low concentrations usually in 10^{-6} M range. However, site of production and sites of action (target cells) of PGRs could be same or different. Also they are synthesised at several sites and in different tissues within a plant body and not in any specialised gland or tissue, as in animals.

Moreover, each of the major classes of PGR bring about a variety of growth and differentiation responses.

The Discovery of Plant Growth Regulators

The discovery of each of the five major groups of PGRs have been accidental.

1. All this started with the observation of Charles Darwin and his son Francis Darwin when they observed that the coleoptiles of canary grass (*Phalaris canarensis*) responded to unilateral illumination by growing towards the light source (phototropism). After a series of experiments, it was concluded that the tip of coleoptile was the site of transmittable influence that caused the bending of the entire coleoptile.

   Thereafter, several experiments in different laboratories, culminated in the discovery of first PGR, the **auxin**, by F.W. Went (1928) in coleoptile tips of *Avena sativa* seedlings.
2. The *bakane* (foolish seedling) disease of rice seedlings which resulted in weak, elongated stems that produced little or no grains. Japanese plant pathologist indentified the causal organism, *Gibberella fujikoroi*, a fungus. In 1926 Kurosawa reported the appearance of symptoms of disease in uninfected rice seedlings. When they were treated with sterile filtrates of the fungus, the active substance was later identified as **gibberelic acid**.
3. F. Skoog (USA) and his co-workers were studying the nutritional requirements of tissue cultures derived from the internodal segments of tobacco stems. They observed that the callus proliferated only, it in addition to auxins nutrient medium was supplemented with one of the following extracts of vascular tissues or yeast extract or coconut milk or DNA. Skoog and Miller, soon identified and crystallized the cytokinesis-promoting active substance that they termed **kinetin**.
4. As the research in plant growth regulators was progressing, it became apparent that certain substances of plant origin interfered with auxin responses. Advent of paper chromatography technique as an analytical tool helped to identify the PGR, **abscisic acid** in 1967.
5. Businessmen involved in shipping and storing fruits have been aware that ripe and rotten fruits could accelerate the ripening of other fruits stored nearby. Cousins (1910) confirmed the release of a volatile substance from ripened oranges that hastened the ripening of unripened bananas stored nearby. R. Gane (1934) provided undisputable evidence that this volatile substance was **ethylene** (C_2H_4), a gaseous PGR using gas chromatography technique.

PHYSIOLOGICAL EFFECTS OF PLANT GROWTH REGULATORS

Auxins

Auxins (GK *Auxein* = to grow), isolated initially from human urine were the first plant PGR to be discovered.

However, IAA also occurs in human urine, especially in persons suffering from pellagra (niacin or nicotinic acid deficiency)

The term **auxin** is applied to the indole-3-acetic acid (IAA), and to other natural and synthetic compounds having certain growth regulating properties. They are generally produced by the growing apices of the stems and roots, from where they migrate to the regions of their action.

Auxins, like IAA and indole butyric acid, IBA, have been isolated from plants.

2, 4–D (2, 4 dichlorophenoxyacetic), NAA (naphthalene acetic acid) are the synthetic auxins.

All above mentioned auxins have been used extensively, particularly in agricultural practices.

Applications of auxins have a wide range. They have been used for more than 50 years in horticulture and agriculture. They help to initiate rooting in stem cuttings, an application widely used for plant propagation. Auxins promote flowering, *e.g.*, in pineapples. They help to prevent fruit and leaf drops at early stages but promote the abscission older/mature leaves and fruits.

In most higher plants, the growing apical bad inhibits the growth of lateral (axillary) buds, a phenomenon called **apical dominance**.

Removeal of shoot tips (decapitation) usually results in the growth of lateral buds. It is widely applied in tea plucking, hedge-making.

Auxins also induce parthenocarpy, *e.g.*, in tomatoes. They are widely used as herbicides. Auxin 2, 4–D, is widely does not affect mature monocotyledonous plants. It is also spread to kill dandelions/daisies in lawns by gardeners. Auxin also control xylem differentiation and help cell-division.

Gibberellins

This is another kind of promotery PGR. There are more than 100 gibberellins reported from different organism such as fungi and higher plants, and the number is still increasing.

They are denoted as GA_1, GA_2, GA_3 and so on. However, GA_3 was one of the first gibberellins to be discovered and remains the most intensively studied form.

All GA_5 are acidic. They produce a wide range of physiological responses in the plants. They specifically make genetic dwarfs, *e.g.*, in peas, grow tall. Their ability to cause increase in length of axis is used to make the stalks of graphs lengthy, thereby allowing the grapes to grow larger.

Gibberellins GA_4 and GA_7 cause apple fruits to elongate and improve shape. They also delay senescence, thus the fruits can be left on the tree longer to extend market period.

GA_3 is used to speed up malting process in brewing industry.

Spraying sugarcane crop with gibberellins increases the length of the stem. It helps increase yield an much as by 20 tons per acre.

Spraying juvenile conifers with GA_4 and GA_7 hastens the maturity period. Thus leading to early seed production. Gibberellins also promote bolting (*i.e.*, internode elongation prior to flowering) in beet, cabbages and many plants with rosette habit.

Cytokinins

These PGRs have specific effects on cytokinesis, and were discovered as **kinetin** (a modified form of adenine, a purine) from the herring sperm DNA. Kinetin does not occur naturally in plants. The searches for natural substances with cytokinin like activities led to the isolation of **Zeatin** from corn-kernels and coconnt milk, etc. Since the discovery of zeatin, several naturally occurring cytokinins, and some synthetic compounds, with cell division promoting activity, have been identified from plants.

Natural cytokinins are synthesised in regions where rapid cell division is occurring, *e.g.*, root apices, developing shoot buds, young fruits, etc.

They help in production of new leaves, chloroplasts in leaves, lateral shoot growth as well as adventitious shoot formation. Cytokinins overcome apical dominance. They promote nutrient mobilization that helps in delay of leaf senescence.

Ethylene

This is a simple gaseous PGR. It is aynthesised in large amounts by tissues undergoing senescence and ripening of fruits. Ethylent influences of plants include horizontal growth of seedlings, swelling of the axis, tighter apical look formation in dicot seedlings.

Ethylene promotes senescence and abscission of plant organs (especially leaves, flowers etc.)

Ethylene is highly effective in fruit ripening. It enhances the respiration rate during ripening of fruit. This rise in rate of respirsation is called **respiratory climacteric**.

Exposure of plants to ethylen causes drooping of leaves and flowers, a phenomenon known as **epinasty**.

Ethylene is a natural PGR and product of metabolism in plants.

Ethylene break seed and bud dormancy, initiate germination in peanut seeds, sprouting of potato tubers.

Ethylene promotes rapid internode/petiole elongation in deep water rice plants. It also helps leaves/upper parts of shoot to remain above water.

This also promotes root growth and root hair formation. Thus helping plant to have more absorption surface.

Ethylene promotes in pineapple family (Bromeliaceae), and is used for synchronizing fruit set.

Ethylene also induces flowering in mango. Ethylene regulates so many physiological processes.

It is one of the most widely used PGR in agriculture. The most widely used compound is **ethepon, 2-chloroethyl phosphonic acid (Ether).** Ethepon in aqueous solution is readily absorbed and transported within the plant and release ethylene slowly. Ethepon hastens fruit ripening in tomatoes and apples. It accelerates abscission in flowers and fruits (thinning of cotton, cherry, walnut, etc.). It promotes female flowers in cucumbers, and therefore, increases the yield.

Abscissic Acid

Abscissic acid regulates abscission and dormancy. However, like other PGRs it also has wide range of effects on plant growth and development. It acts as general plant growth inhibitor and also inhibits plant metabolism. Abscissic acid (ABA) inhibits closure of stomata in epidermis, and increases tolerance of plant to various kinds of stresses, and therefore, it is also called **stress hormone**.

ABA plays an important role in seed development, maturation and dormancy. By inducing dormancy, ABA helps seeds to withstand desiccation and other factors unfavourable for growth. In most situations ABA acts as an antagonist to gibberellins.

However, for any and every phase of growth, differentiation and development of plants, one or the other PGR has to play some role. Such roles may be complimentary or antagonistic. They could be individualistic or synergistic.

Abscission. The shedding of leaves, flowers and fruits from the living plant is known as abscission. These parts abscise in a region called the **abscission zone**, which is located near the base of petioles of leaves. In most plants leaf abscission is preceded by the differentiation of a distinct layer of cells, the **abscission layer**. During abscission the walls of the cells in the abscission layers are digested, which causes them to become soft and weak.

Auxins delay the onset of early stages of leaf abscission but promote the later stages.

Auxin level within leaves is very high in young leaves and progressively decreases in maturing leaves and in very low in senescent old leaves.

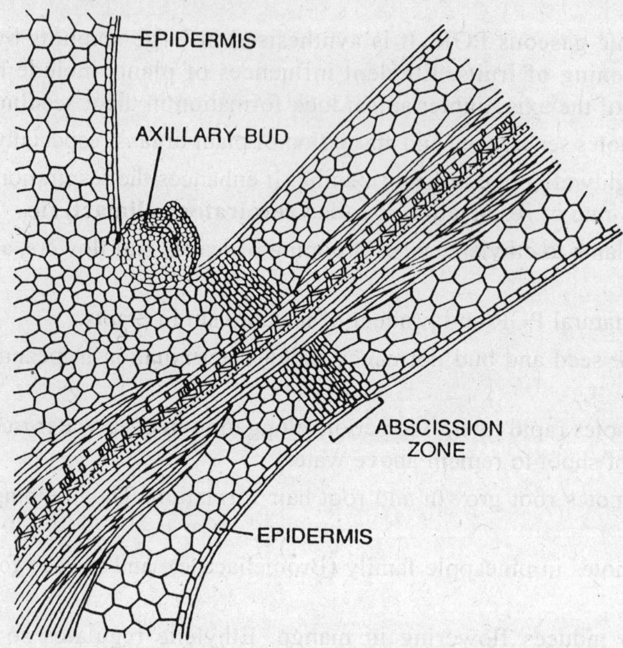

Fig. 6.10. Leaf abscission. Formation of abscission zone.

When applied externally to young leaves, IAA (indole acetic acid) inhibits leaf drop. However in later stages of leaf growth, if applied externally, it hastens the process.

It is also known that high levels of auxin trigger biosynthesis of ethylene within cells. Ethylene strongly hastens the formation of abscission zone.

There is an interaction between auxin and ethylene for an event, *i.e.*, the abscission zone formation.

In the similar way, there are so many events in the life of a plant where more than one PGRs interact to affect that event, *e.g.*, dormancy in seeds/buds, senescence, apical dominance, etc.

PRACTICALS

Experiment

Object : Demonstration of effect of auxin by *Avena* section test.

Requirements : Coleoptiles, distilled water, auxin solution, petri dishes.

Experiment : Take some germinating *Avena* seedlings. Cut the coleoptiles, 25-30 mm in length. Remove the tip of coleoptile and sections 3-5 mm in length of each coleoptile. Put all the sections for an hour in distilled water and then transfer them into other petri dishes containing auxin solution. Keep the experiment as such for a day or two.

Observation : The length of each section, measured after some time is found more than the previous one.

Explanation : This experiment explains the growth response of the solution which is found to be directly proportional to the logarithm of the concentration of auxin used.

2. Gibberellins. Gibberellin was first extracted from the fungus *Gibberella fujikuroi*, the causal organism of **'foolish seeding'** disease of rice. The infected plants were distinctly taller, seedless and pale in colour. These symptoms were produced due to a substance secreted by a fungus. Yabuta and Sumuki (1938) isolated Gibberellin A and B in form of crystals from the fungus but soon after it was isolated from the higher plants. Many types of gibberellins have been isolated among which GA_3 (gibberellic acid) is more well studied and most common in plants. There are 100 different types of gibberellins isolated so far and they are named as GA_1, GA_2, GA_3, to up to GA_{100} and so on. However, GA_3 was one of the first gibberellins to be discovered and remains the most intensively studied form. All GAs are acidic in nature. Gibberellins contain a gibbane ring system with specific biological properties.

Gibberellins are widely used in agriculture and horticulture. Some of important uses are as follows :

The most important effect of GA is the stem elongation, *i.e.*, GA, induces cell elongation. It has been confirmed on several plants such as pea, bean, tomato, pepper, sweet corn, summer squash, cucumber, lettuce, cabbage, etc. Gibberellins promote seed germination and break dormancy of buds and tubers. They promote flowering of long day plant under non-inductive conditions. They increase leaf and flower size in many plants. They also induce parthenocarpy (seedless fruit). GA increases the length of hypocotyl and cotyledonary leaf area. GA affects positively on fruit setting. In the germinating grains of cereals like barley, gibberellins stimulate the formation of *a*-amylase, responsible for the hydrolysis of starch. Bolting in cabbage is induced by gibberellin application. Here enormous elongation of the stem takes place, and flowering occurs.

3. Cytokinins. Skoog and his coworkers discovered this growth regulator. Its specific property is to promote cell division. Cytokinins were discovered as kinetin (a modified form of adenine) from sperm DNA. Kinetin does not occur naturally in plants. Cytokinins are synthesised in areas where cell division occurs, *e.g.*, in root apices, developing shoot buds, young fruits, etc.

Patan, Das and Skoog reported that kinetin promotes cell division in tobacco pith by activating DNA synthesis. (Cytokinin promotes cell division by activating DNA synthesis). It has antagonistic property with auxin. Kinetin promotes the growth of lateral buds by neutralizing the effect of natural auxin. Thimann (1960) reported that kinetin activates protein synthesis in buds by incorporation of labelled amino acids. Owing to its ability to bring about cytokinesis in cells, the

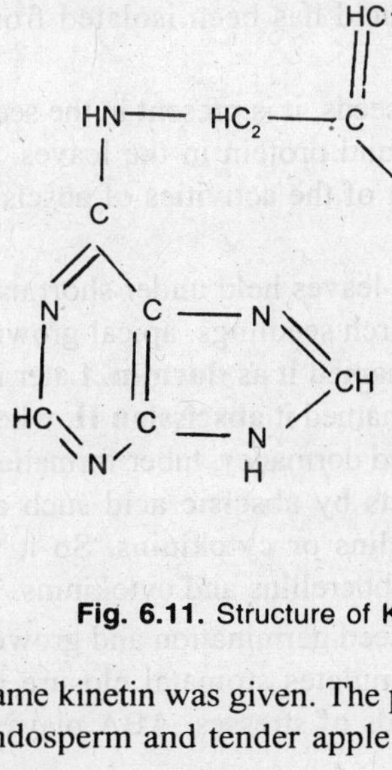

Fig. 6.11. Structure of Kinetin.

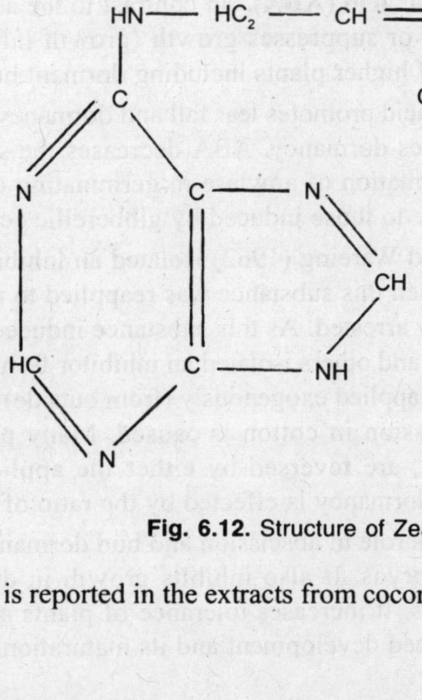

Fig. 6.12. Structure of Zeatin.

name kinetin was given. The presence of kinetin is reported in the extracts from coconut milk, maize endosperm and tender apple fruits.

Letham (1963) isolated another type of cytokinin from fractionation of sweet corn extract and termed it as **Zeatin**. The high activity of Zeatin is due to side chain of a very reactive allylic hydroxyl group.

Cytokinins reduce apical dominance, and cause delay in senescence. The common bioassays for cytokinins include promotion of cell division in tobacco pith culture, expansion of excised radish cotyledons, delay in senescence, etc.

Practicals

Two experiments on gibberellins and cytokinins are given here :

Experiment

Object : Demonstration of effect of gibberellic acid GA_3 on elongation.

Requirements : Prepared aqueous (1000.0) solutions of GA_3 having 0.01, 0.1, 1.0, 10.0 and 1000.0 mg/ml, young plants of dwarf and tall varieties of peas, distilled water.

Experiment : Take ten dwarf variety plants and ten tall variety plants. Apply 1 ml. of a GA_3 solution to a young leaf of each of the plants. Do this application again after one week. Distilled water is applied as a control in separate plants. Measure the height of each daily for fifteen days.

Observation and Explanation : Each time there is a clear elongation in the plant height. The growth response in dwarf variety increases with the increase of concentration of GA_3.

Experiment

Object : Demonstration of effect of kinetin of senescence.

Requirements : Solutions of kinetin containing 0, 0.001, 0.01 and 100.0 mg./litre of hormone, mature fresh leaves of radish, sterile distilled water, petri dish, filter paper, etc.

Experiment : Take mature fresh leaves of radish plant and wash them with sterile distilled water. Keep them in petri dishes on moist filter paper. Apply 0.5 ml of one of the kinetin solutions per leaf. Examine the leaves daily for a week.

Observation and Explanation : The leaves turn yellow after a week. This is due to disappearance of chlorophyll and degradation of proteins.

4. Abscisic acid (ABA). In contrast to the above discussed growth hormones, abscisic acid (ABA) retards or suppresses growth (growth inhibitor). Abscisic acid has been isolated from several parts of higher plants including dormant buds and seeds.

Abscisic acid promotes leaf fall and dormancy of buds. In some seeds, it is present in the seed coats and causes dormancy. ABA decreases the synthesis of RNA and protein in the leaves. It inhibits the formation of amylase in germinating cereal grains. Most of the activities of abscisic acid are reverse to those induced by gibberellic acid and cytokinins.

Eagles and Wareing (1962) isolated an inhibitor from the birch leaves held under short day conditions. When this substance was reapplied to the leaves of the birch seedlings, apical growth was completely arrested. As this substance induced dormancy, they named it as **dormin**. Later in 1965, Ohkuma and others isolated an inhibitor from cotton fruits and named it **abscission II**, when abscisic acid is applied exogenously (from outside) bud dormancy, seed dormancy, tuber formation and boll abscission in cotton is caused. Many physiological effects by abscisic acid such as dormancy, etc., are reversed by either the application of gibberellins or cytokinins. So it is proposed that dormancy is effected by the ratio of abscisic acid to gibberellins and cytokinins.

Besides its role in abscission and bud dormancy, ABA inhibits seed germination and growth of excised embryos. It also inhibits growth in duckweeds and stimulates stomatal closure in epidermal strips. It increases tolerance of plants against various kinds of stresses. ABA plays a major role in seed development and its maturation.

5. Ethylene. Ethylene is the only hormone which is found in the form of gas. It was previously used for early ripening of fruits. Now it is proved that plant cells produce ethylene when there is a stress on them. When ethylene is applied to a plant, loss of sensitivity of gravity (ageotropism) in roots and shoot **abscission** of leaves, ripening of fruits, adventitious root formation and hook formation in germinating seedlings takes place. When microorganisms invade a plant or cut is made in it, large quantity of ethylene is produced. This ethylene promotes the formation and secretion of peroxidase enzymes, *e.g.*, sweet potato tuber. This wound repair is done by ethylene. Ethylene is evolved by plants, specially during the ripening of fruits. This effect of ethylene has been utilized in citrus industry where oranges, lemons and grape fruits are sometimes picked while green and ripened in gas chambers containing ethylene (similar is the case now with many fruits, including bananas and mangoes.)

Ethylene is a natural product of metabolism in plants.

Exposure of plants to ethylene causes drooping of leaves and flowers. This phenomenon is called **epinasty**.

6. Vitamins. Vitamins (*vita*, life) are a heterogeneous group of organic products of plants, which are invaluable for normal growth and development of the body, maintenance of health, vigour, nervous stability, and proper function of the digestive system.

Vitamins are also essential in preventing or curing certain deficiency diseases such as scurvy, beriberi, rickets, malnutrition, loss of appetite, poor physical growth, eye infection, nervous break-down, etc., caused by the absence of vitamins in the food or their faulty absorption due to intestinal troubles.

Up till now several vitamins have been discovered and their value established. It is known that vitamins do not take any direct part in body-building nor they make a source of energy. However, they play a very important role in the metabolism of proteins, fats and carbohydrates, and their proper assimilation into the body. They also help proper utilisation of certain minerals in the body.

Some of the vitamins act as prosthetic groups or coenzymes and help in synthesis of certain enzymes, and control many important biochemical reactions in both plants and animals.

Vitamins are required in minute quantities for a particular effect, and are used up in the metabolic processes.

Vitamins are mostly synthesized by plants and stored up in their different organs. Certain vitamins are also formed by bacteria and fungi.

However, animals including human beings, cannot synthesize them in their body, and therefore, plants are the sources of vitamins for them.

Vitamins are of many types, such as vitamins A, B (B complex – B_1, B_2, B_6, B_{12}, etc.), C, D, E and K. Some common and important vitamins are as follows :

Vitamin A is growth-promoting and anti-infective vitamin, soluble in fats and oils, less soluble in water and fairly resistant to heat.

Deficiency of vitamin A results in eye-diseases, loss of weight, skin diseases, nervous weakness and respiratory diseases.

Main sources of this vitamin are carrot, green leafy vegetables, cereals, sprouting pulses, many fruits, such as tomato, mango, orange, banana, apple, papaya, etc., milk, butter, meat, liver of mammals, egg yolk, fish, cod liver oil, etc.

Carotene of plants is the source of this vitamin, and animals can synthesize vitamin A in their body by taking food containing carotene. Excess of vitamin A is stored in the liver and utilized when deficiency occurs.

Vitamin B consists of a group of closely allied vitamins, commonly called **vitamin B complex**. This is required for proper maintenance of health throughout life.

Deficiency of this results in digestive troubles, loss of appetite, diarrhoea, constipation, beriberi and neuritis. Some of the vitamins of this group are—thiamine, riboflavin, pyridoxine, niacin, etc.

Vitamin B complex is widely distributed in plants in almost all their parts. Rich sources are—dried yeast, whole grains, pulses, green vegetables, many fruits, *e.g.*, tomato, papaya, mango, banana, apple etc., nuts, milk, cheese, egg-yolk, meat, liver, fish, etc.

Vitamin C also known as ascorbic acid, soluble in water, sensitive to heat, and therefore, lost by cooking, destroyed when exposed to sunlight. Vitamin C is a general activator of metabolic processes. It cannot be stored in the body, and therefore, a daily supply of this vitamin is a necessity.

Vitamin C prevents scurvy, mental depression, swelling and bleeding of gums, degeneration of teeth, cold and sore mouth.

It is found in high concentrations in the fruits, such as orange, lemon, guava, tomato, pineapple, emblic myrobalan, papaya, etc. This is also found in green vegetables, sprouting pulses, cereals and milk.

Vitamin D also known as calciferol. This is an anti-ricket vitamin, soluble in fats and oils.

Its deficiency causes rickets and dental carries in children, and osteomalacia in adults.

It is essential for normal development of bones and teeth, and for general growth.

Vitamin D is commonly found in dried yeast, milk, butter, cheese, egg-yolk, fish and cod-liver oil. Vitamin D can also be produced in the human body by sunrays.

Vitamin E or anti-sterility vitamin. This is soluble in fats and oils, resistant to heat and light but destroyed by ultra-violet rays.

Its deficiency causes sterility in animals, degeneration of muscles and falling-out of hairs.

It is found in green vegetables, germinating grains, wheat embryo, milk, egg-yolk, meat, etc.

Vitamin K is fat-soluble.

Its deficiency causes lowering of pro-thrombin value of the blood. This means, this does not help proper blood clotting in wounds and cuts. This vitamin is found in green vegetables, fresh fruits, and in yeast.

Differentiation

The cells derived from root apical meristem (RAM) and shoot apical meristem (SAM) and cambium differentiate, mature to perform specific functions. This act leading to maturation is termed **differentiation**. They, undergo a few or major structural changes both in their cell walls and protoplasm. For example, to form, a trachery element the cells lose its protoplasm but develops a very strong, elastic, lignocellulosic secondary cell wall is best suited to carry water to long distances even under extreme tension. Similarly, cells designated to be mesophyll come to possess many chloroplasts so as to, perform photosynthesis. On one hand, a parenchyma in hydrophytes develop large schizogenous interspaces for mechanical support, buoyancy and aeration, but on the other hand, in a potato tuber (or perennating organs) develops more amyloplasts.

Dedifferentiation

In plants, the living differentiated cells can regain the capacity to divide mitotically under certain conditions. The sum of events, that bestow this capacity to divide once again, are termed

dedifferentiation. A dedifferentiated tissue can act as meristem (*e.g.*, interfascicular vascular cambium, wound meristem, cork cambium).

Redifferentiation

The product of dedifferentiated cells/tissue which lose the ability to divide are called redifferentiate cells/tissues and the event, **redifferentiation**.

However, the growth in plants is open, and even differentiation in plants is open, because, *e.g.*, the same apical meristem cells give rise to xylem phloem, fibres, etc., cells/tissues arising out of same meristem have different structures at maturity. The final structure at maturity of a cell/tissue arising out of same meristem is determined by the location of the cell within. For example, cells positioned distal to root apical meristem (RAM) differentiate as root cap cells, while those pushed to periphery mature as epidermis. However we do not know for as well as determined is called **commitment**. Terms **determination** and **commitment** are used as synonyms.

DEVELOPMENT

Development is a term that includes all changes that an organism goes through during its life cycle from germination of the seed through growth, maturation, flowering, seed formation and senescence. The term is also applied to formation of cells/tissues/organs. It can also applied for change in phases of life, *e.g.*, from vegetative to flowering, and from leaf initiation to leaf expansion. It is also used at sub-cellular level, *e.g.*, appearance of chloroplasts in cells exposed to light. The last phase of the development process in plants is **senescence** during which, events leading to ageing and death take place.

Diagrammatic representation of the sequence of processes of development of a cell are summarised as follows :

The plants follow different pathways in response to environment or phases of life to form different kinds of structures. This ability is called **plasticity**, *e.g.*, heterophylly in cotton, coriander and larkspur. In such plants the leaves of the juvenile plant are different in shape than those in mature plants. On other hand, difference in shapes of leaves produced in air and while in water in buttercup (*Ranunculus*) also represent heterophyllous development due to environment. This phenomenon of heterophylly is an example of plasticity.

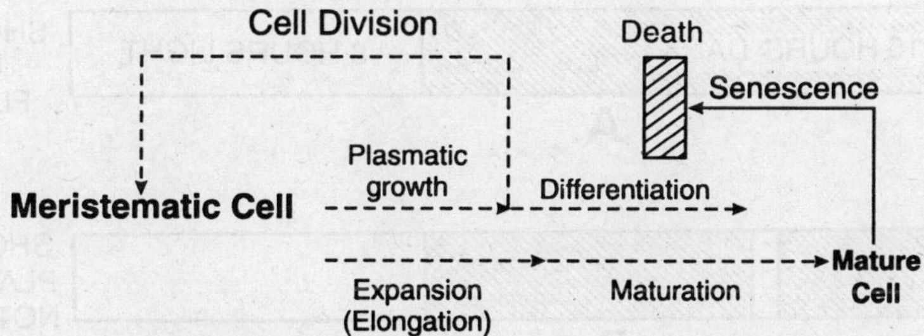

Fig. 6.13. Sequence of development process in a plant cell

Thus, growth differentiation and development are very closely related events in the life of a plant. However, development in considered as sum of growth and differentiation. The development in plants (*i.e.*, both growth and differentiation) is under control of **intrinsic** and **extrinsic** factors. The former includes both intracellular (genetic) or intercellular factors (chemicals such as plant growth regulators) while extrinsic factors include light, temperature, water, oxygen, nutrition, etc.

Table 6.1. Comparison of influences of PGRs on plant development
Plant growth regulators (PGRs)

	Auxin	Gibberellins	Cytokinins	ABA	Ethylene
Root development	×	×	×		×
Extension growth	×	×	×		×
Juvenility	×	×			
Flowering	×	×	×	×	×
Fruit development	×	×	×	×	×
Senescence	×	×	×	×	×
Dormancy		×	×	×	×

× = Involvement of PGR in the plant development

However, the role(s) of PGRs are only one kind of intrinsic control. Along with genomic control and extrinsic factors, they play an important role in plant growth and development. The extrinsic factors, such as temperature and light control plant growth and development via PGR. Such events are : vernalisation, flowering, dormancy, seed germination, plant movements, etc.

PHOTOPERIODISM

During the last few decades, it has been observed that plants respond to day length in their flowering. This has come to be known as the phenomenon of **photoperiodism**. The plants have been classified as **short day**, **long day**, **day neutral**, **short-long day** or **long-short day types**. Interestingly SD, LD or ND types have now been observed not only within the same genus but also among the ecotypes or variants within the same species.

The observation of Garner and Allard (1920) that a variety of tobacco, maryland mammoth flowered when the relative length of the day was shorter than the length of dark period led to the discovery of photoperiodism. They used the term photoperiodism to designate the response of organism to the relative length of day and light and photoperiod to designate the favourable length of day for each plant.

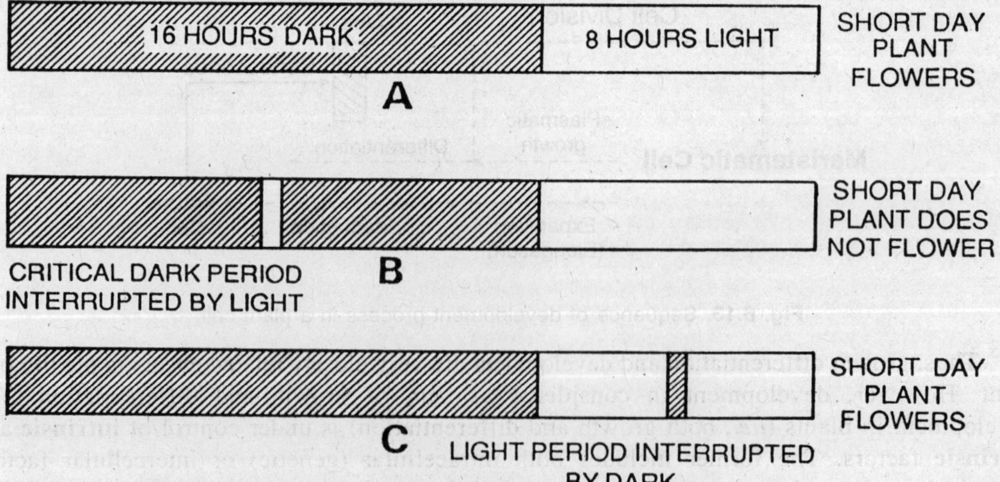

Fig. 6.14. Photoperiodism. Effect of a brief exposure of red light during dark and light periods on flowering in short day plant.

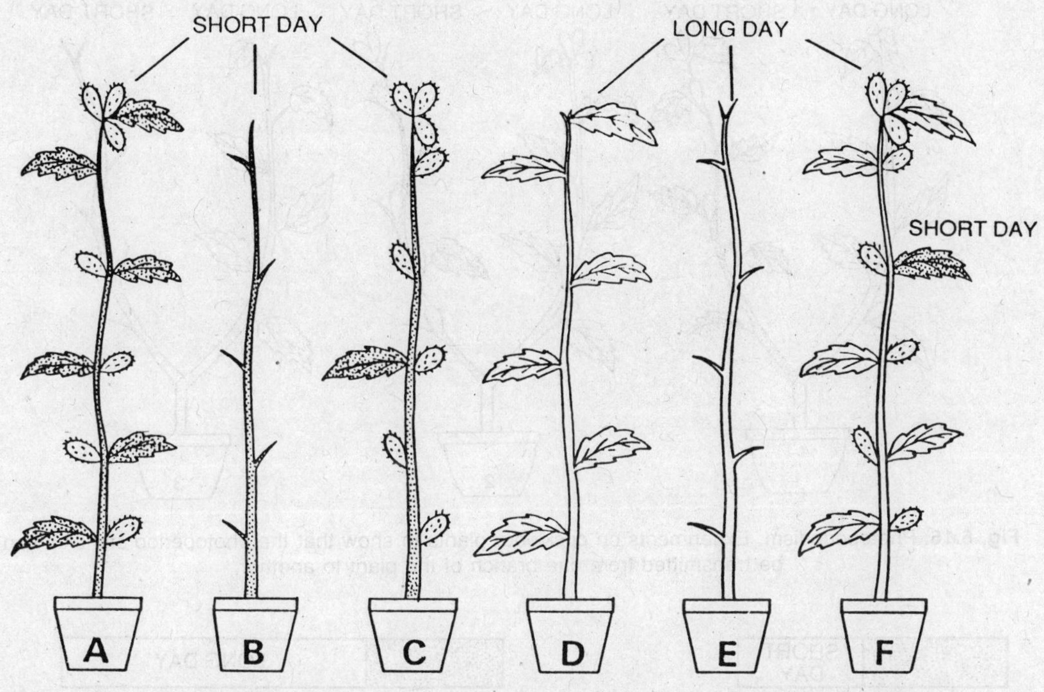

Fig. 6.15. Photoperiodism. Experiment on cocklebur plants to show that photoperiodic stimulus is perceived by the leaves. Flowering occurs even if a single leaf is exposed to appropriate photoperiod.

Short day plants. For flowering of short day plants, the day length must not exceed a certain critical value, the day length required is less than a certain critical length. Short day plants be more correctly called as long night plants, as a certain minimum of uninterrupted dark period in 24 hours is necessary for their flowering. In the case of the short day plants, the critical thing is the length of dark period. Even if the dark period is less than a critical length, flowering will not occur. Short day plants will not flower even if a flash of light is provided during the continuous dark period. Flowering is also checked even if very weak light is provided during the dark period. Some examples of short day plants are—soyabean, potato, sugarcane, cosmos, chrysanthemum, dahlia, tobacco, etc.

Long day plants. Long day plants require a photoperiod of more than a critical length which may vary from 4 to 18 hours. The best flowering of long day plants usually occurs in continuous light. For flowering, they require either no dark period or a very short dark period. Examples of long day plants are—spinach, radish, lettuce, alfalfa, opium poppy, maize, oats, wheat, etc.

Day neutral plants. Tomato, cucumber, cotton, pea, sunflower, etc., are examples of day neutral plants. They can flower even if the light period provided is from few hours to round the clock illumination.

In recent years, efforts have been made to define different states leading to flowering. These states are described as **induction, evocation** and **differentiation.** At one time, it was assumed that the plant responses could be qualified as **'qualitative'** (obligate) **or quantitative** (facultative). The distinction between the characteristics has been narrowed down, **if not** eliminated altogether. Furthermore, it has become clear that the intensity of flowering is a **quantitative** expression even in qualitative plants.

LONG DAY
SHORT DAY
LONG DAY
SHORT DAY
LONG DAY
SHORT DAY
1
2
3

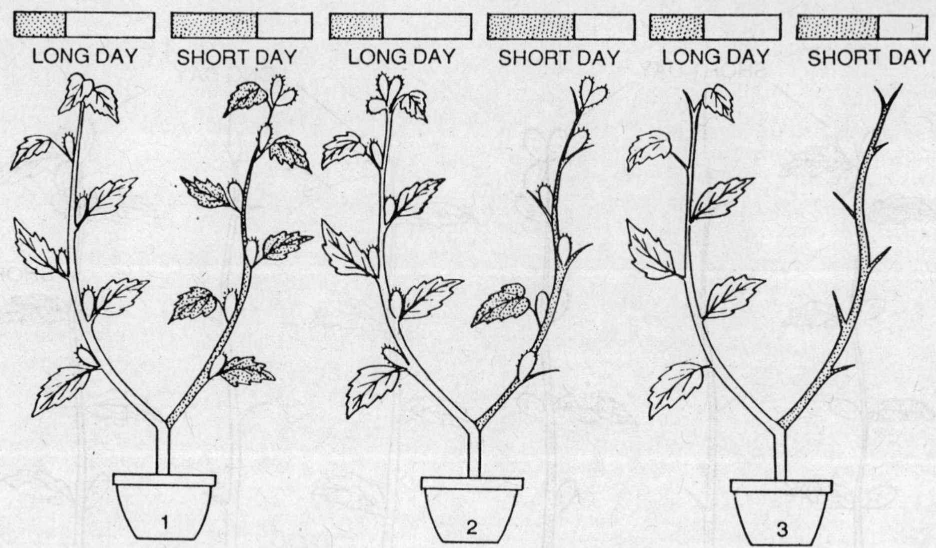

Fig. 6.16. Photoperiodism. Experiments on cocklebur plants to show that the photoperiod stimulus can be transmitted from one branch of the plant to another.

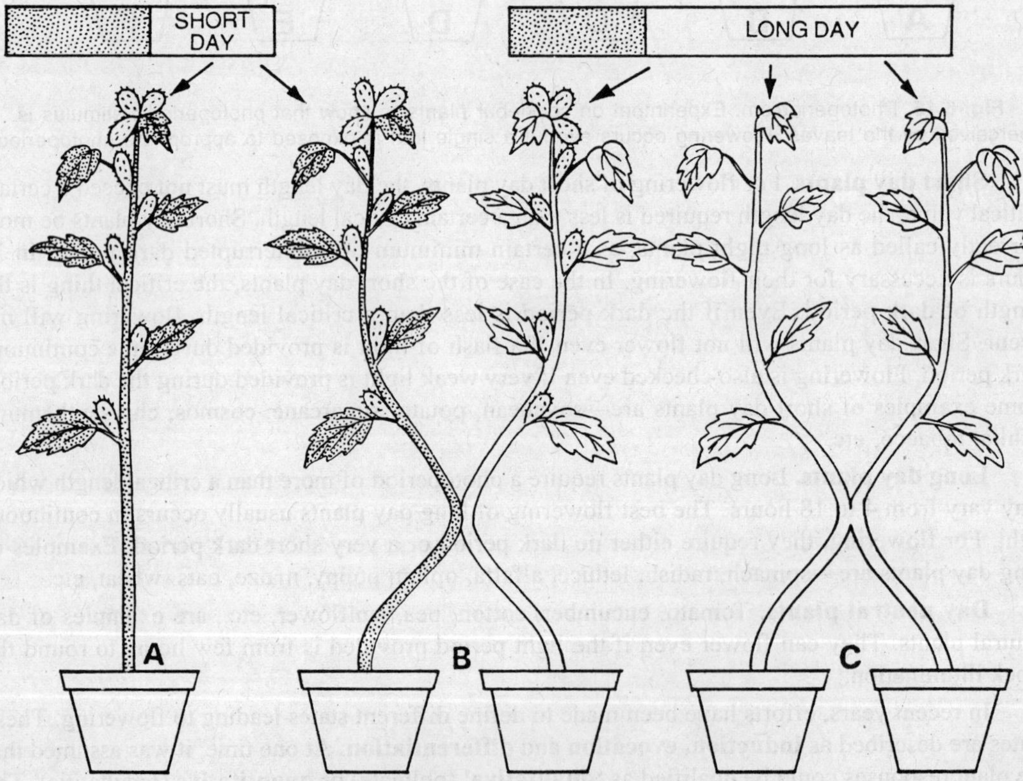

SHORT DAY
LONG DAY
A
B
C

Fig. 6.17. Photoperiodism. Grafting experiments in cocklebur plants to show the translocation of floral hormone from one plant to another through graft union.

The graft experiments between **'induced'** plants and **'non-induced'** under non-inductive conditions resulted in the development of the concept of a transmissible flowering stimulus. A unified concept of such a stimulus gained prominence because flowering could be induced between SD and ND, LD and ND or SD and LD. In addition, it was shown through graft experiments that the formation and transmission of an inhibitor could be responsible in controlling flowering. Either any flowering stimulus or hormone or an inhibitor has not been isolated, but the influence of many chemicals such as gibberellins, cytokinins, ethylene, etc., has been shown. In recent years, several phenolic compounds such as salicylic acid have been found to induce flowering.

It has been hypothesised that there is a hormonal substance called **florigen** which is responsible for flowering. But there is no experimental proof in support of this hypothesis. It is believed that this active principle is synthesised in the leaves, when the plants are exposed to the inductive photoperiod, and is translocated to the apical meristem for inducing flowering by conversion of vegetative meristem into reproductive meristem.

VERNALIZATION

In many plants, temperature has a profound effect on flowering. Some plants do not flower unless they are exposed to prolonged periods of cold temperatures. Biennials remain vegetative during the first season, and after being exposed to the cold temperatures of winter, they flower in spring. Majority of plants will not flower if they are not subjected to exposure and will grow vegetatively for an indefinite period. However, if they are subjected to cold treatment followed by correct photoperiod, they will flower. For example, the biennial henbane (*Hyoscyamus niger*) a long day plant, can be made to flower within one growing season, provided it is exposed to low temperatures followed by the correct photoperiod. The term **vernalization** has been coined to explain the effect of chilling treatment that prepares a plant to flower. Vernalization is defined as the method of inducing early flowering in plants by pretreatment of their seeds at very low temperature. Chourad (1960) has defined it as the acquisition of acceleration of the ability to flower by chilling treatment.

For vernalization, the seeds are allowed to germinate for some time and then are given cold treatment by keeping them at 0-5°C. The period of cold treatment varies from few days to many weeks from species to species. After the cold treatment, seedlings are allowed to dry for some time and then sown. They should not be sown immediately after the cold treatment.

Thus vernalization prepares the plant to flower whereas the photoperiodic stimulus not only prepares the plant for flowering but also induces it. The shoot tip is the part that perceives the stimulus of cold temperatures. The stimulus is then translocated to other parts of the plant. In some plants, application of gibberellin may replace the cold treatment.

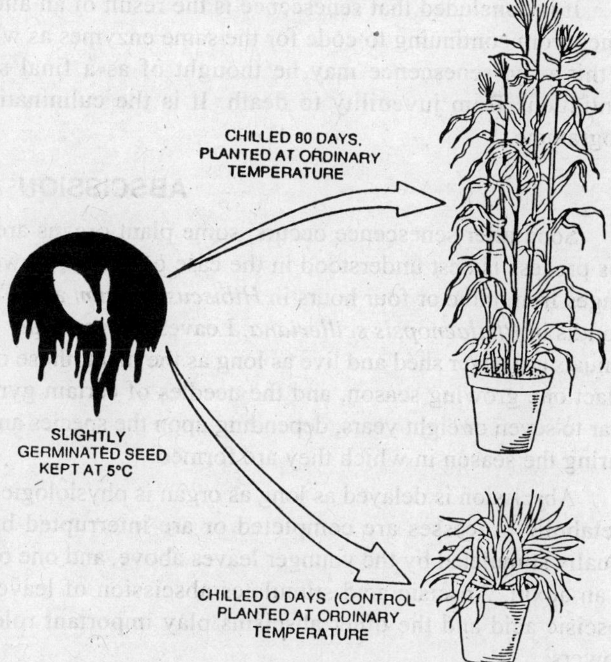

Fig. 6.18. Vernalization. Diagram to illustrate effects of vernalization.

Vernalization is defined as the method of inducing early flowering in plants by pretreatment of their seeds at very low temperatures. Chourad (1960) has defined it as the acquisition or acceleration of the ability to flower by chilling treatment. *Hyoscyamus niger* has two types, one is annual and the other is biennial. Both of them are long day plants and when exposed to short day will grow vegetatively. Even the biennial type can be made to flower if its ten days old seedlings are subjected to cold treatments. For vernalization the seeds are allowed to germinate for some time and then are given cold treatment by keeping them at 0-5°C. The period of cold treatment varies from few days to many weeks from species to species.

Temperature as a vernalization effect as well as during photoperiodic treatments has been found to be an important factor regulating flowering. There is evidence to suggest that vernalization effect could involve such substances as gibberellins. The interaction between photoperiod and temperature is considered important in the flowering of many crop plants.

SENESCENCE

Senescence has been defined as the deterioration that ends the functional life of an organism or an organ. Although meristems do not undergo senescence and can perhaps be considered potentially immortal, all of the cells produced from these undergo aging, which eventually leads to death. Senescence occurs in leaves, flowers, fruits, stems, and roots, but it commonly occurs at different times in these organs. In many perennial herbs, such as alfalfa, essentially the entire above ground system dies each year, but the crown and root systems remain largely viable. In deciduous woody perennials, the leaves die, but much of the root and stem tissues remain alive. Another pattern of senescence is seen in several herbaceous annual species such as beans, tomatoes and the cereal grains. Here there is a progressive senescence of the leaves from the older to the younger, followed by death of both the stem and roots after flowering.

Senescence; A scene in New York states, Catskill Mountain; The breakdown of chlorophyll allows yellow, red and orange light-gathering pigments to show.

It is concluded that senescence is the result of an altered hormone balance that prevents the genes from continuing to code for the same enzymes as when the cells were younger. Considered in this way, senescence may be thought of as a final stage of differentiation which itself is continuous from juvenility to death. It is the culmination of the morphogenetic information programme.

ABSCISSION

Soon after senescence occurs, some plant organs are lost by **abscission**. The physiology of this process is best understood in the case of leaves, flowers and fruits. The longevity of flowers ranges from three or four hours in *Hibiscus trionum* and *Calandrinia compressa* upto three or four months in *Phalaenopsis scilleriana*. Leaves are generally longer lived. Those of most herbaceous annuals are never shed and live as long as the plant, those of deciduous woody dicotyledons remain intact one growing season, and the needles of certain gymnosperms are retained from about one year to seven or eight years, depending upon the species and growth conditions. Most fruits are lost during the season in which they are formed.

Abscission is delayed as long as organ is physiologically active and occurs only if the normal metabolic processes are completed or are interrupted by injury or disease. Leaf abscission is usually stimulated by the younger leaves above, and one of the causative factors supplied by these is an auxin. Abscisic acid stimulates abscission of leaves in a number of plants. It seems, that abscisic acid and the other abscisins play important roles in the loss of leaves, of fruit and of flowers.

Abscission benefits plants in various ways. Injury or disease often causes the abscission of the affected organ. The plant thus rids itself of a useless part or of a potential infection source.

PHOTOMORPHOGENESIS

Growth and development of plants are influenced by several environmental factors including light. However, light causes several responses other than photosynthesis. These responses greatly influence the course of plant development and the final plant appearance. They are **photomorphogenetic responses**. For example, the seeds of many plants do not germinate unless they are exposed to light. Germination of seeds in light shows that the seedlings require light to grow. Phototropic responses of seedlings and of leaves of mature plants are also beneficial photomorphogenetic processes. Photomorphogenetic responses are also important to older plants. Many such responses respond to the relative lengths of day and night by forming reproductive structures or by forming dormant buds that can resist a cold winter (*i.e.*, phenomenon of photoperiodism and vernalization.)

Photoperiodism. The discovery of photoperiodism, *i.e.*, the growth response to the length of light and dark periods by Garner and Allard, completely changed the concept of flower initiation. It was demonstrated by experiments that the length of the dark period rather than the light period is the critical factor in the photoperiod response. As per the result of investigations, it became apparent that a majority of plants fell into one of three categories : (*i*) short day, (*ii*) long day and (*iii*) day-neutral plants. Short-day plants require a dark period exceeding some critical length to flower, and cannot flower under continuous illumination (light). Long-day plants are inhibited from flowering when the dark period exceeds some critical length, and they can flower under continuous illumination. Day-neutral plants can flower under any night length. In some plants, the leaves need only to be exposed to one light-dark cycle of the proper day length to cause flower initiation, whereas most plants require several or many such cycles.

It has been shown that photoperiodic responses may be altered by brief exposure to low light intensity. For example, interruption of the dark period in short-day plants has shown that it is red light (wave length 660 mμ) which is effective. Similarly it has been proved experimentally that, far-red light (wave length 730 mμ) reverses the red light effect, *i.e.*, if exposure to red light is followed by exposure to far-red, the result is as if there had been no exposure at all. If repeated alternating exposure to red and far-red are given, the final exposure determines the response.

Phytochrome

The above given observations led to the discovery of the pigment system isolated from the leaf called **phytochrome**. This pigment system specifically receives the external photoperiodic message. The phytochrome was discovered by H.A Borthwick and Sterling B. Hendricks for the first time in 1959. Phytochrome is a pigment present in small amounts in all plants. Under short-day conditions, an interruption of the dark period with light (specifically red light) causes an intramolecular shift in phytochrome that brings about flowering in long-day plants and prevents flowering in short-day plants.

Characteristics of phytochrome. The cytoplasmic pigment system phytochrome now identified as a protein and partially purified, which is sensitive to red light.

This pigment can exist in two forms, one that absorbs red light (Pr) and one that absorbs far-red light (Pfr). Red light converts Pr to Pfr, but this form is unstable and is gradually converted back to the Pr form.

$$\text{Pr 660} \xrightarrow{\text{red}} \text{Pfr 730}$$

$$\text{Pr 660} \xleftarrow[\text{or dark (slow)}]{\text{Far-red (rapid)}} \text{Pfr 730}$$

Phytochrome is active in the Pfr form and thus the responses are produced by red light. In relation to flowering, reactions promoted by Pfr interfere with the dark reaction that promotes flowering in short-day plants and inhibits the flowering of long-day plants. Since it is the length of the dark period that is critical, and therefore, it is supposed that the reversion of Pfr to Pr determines the critical length of the dark period. The reversion takes place only about four hours in the dark.

Apart from photoperiodic phenomena, many other light effects upon plants are exerted through the phytochrome system. The germination of lettuce seeds for example, required a light stimulus, and this is effected by red light and reversed by far-red light. Stem elongation and leaf expansion could be influenced by red light, and the effect of etiolation could be checked by exposure to red light.

Phytochrome is a protein with chromophoric (pigment) groups. The molecular weight of phytochrome is about 60,000 with three chromophores per molecule. The chromophore pigment is closely similar but not identical to the chromophore of c-phycocyanin, an algal (blue alga) chromoprotein.

It is believed that phytochrome is also involved in other photoperiodically controlled processes such as tuber and bulk formation, and dormancy.

About the functioning of the phytochrome, it has been suggested that it may operate by influencing membrane permeability in the cells, not only of the plasma membrane but also of those of the nucleus and the mitochondria, thus influencing the metabolism of cells.

Though the discovery of phytochrome has made a very significant contribution towards explaining the first step in the mechanism of flowering, yet much experimentation is required, so that the substances specifically causing the transition of the apical meristems from the vegetative to the flowering conditions could be isolated and their mode of action determined.

PLANT MOVEMENTS

Though ordinarily we do not associate plants with movements, they do exhibit shift in the positions of their several organs. These adjustments are necessary so that the plant organs may fulfil their function adequately. Though largely, the movements that we come across in plants are not locomotory; movement of locomotion involving the shift of the entire body are not wanting. The swimming movements of Volvocales and the male gametes of several plants are examples of locomotory movements.

A movement is made in response to a stimulus. This ability to respond is reflection of irritability, a property inherent in all organisms. Certain plant movements are rapid and thereafter clearly visible and certain other movements take place over a long period of time and therefore are imperceptible. The rapid folding up of leaves exhibited by 'touch-me-not' is an instance of the former kind.

A stimulus is essentially a change in environment. This change may be of the nature of a shock or chemical or atmospheric. All parts of plants are sensitive to stimuli. Specialized parts called perceptive regions receive the stimulus and transmit it to regions where responsive movements are effected and these latter areas are "responsive regions".

A stimulus does not produce a response immediately. It has to act for some time before the plant organ may respond. This period necessary for a stimulus to induce a response is called **presentation time**. A weak stimulus requires a long presentation time and a strong one a short presentation time. A stimulation repeated a number of times before the effect of the previous stimulation wears out, brings about the response. In other words, the effect of each stimulation adds up to the total effect of the previous stimulations. The behaviour is embodied in a law called the **law of summation of stimuli**. The time taken by a plant organ to manifest its response to a stimulus is called the "reaction time". This is equal to presentation time plus the time taken for the stimulus to be conducted from the region of perception of the region of response (transmission time) plus the time necessary for the internal changes in the cells of the responsive region.

The movements exhibited by the organs of plants are classified into three groups: (1) **growth movements** caused by changes in the position of the organs due to enlargement of cells, or increase in the number of cells, or both; (2) **turgor movements** caused by reversible changes in cell volume, and (3) **hydration movements** concerned usually with non-living plant tissues or organs and resulting from hydration or dehydration of the cell walls. Here we deal with some of the growth and turgor movements that occur most commonly in plants.

Growth Movements

Growth movements are divided into three sub-groups: (1) **Tropic movements** occur in response to always related to the direction of the stimulus. (2) **Nastic movements** in which the response has no relation to the direction of the stimulus. It is diffused. (3) **Nutation** is a movement which is usually independent of the environmental factors. It is characterized by spiral type of growth of the shoot apex.

Tropic Movements

Most commonly found in the radially symmetrical organs of the plant such as roots and stems; tropic movements are curvature movements caused due to stimulus coming from one side. These movements controlled by the direction of the stimulus occur in one direction due to differences in the rates of growth on the two sides of the moving organs which results in curvature. A tropic movement may be phototropism (in response to light), geotropism (in response to gravity), chemotropism (in response to chemicals), hydrotropism (in response to water) and haptotropism or thigmotropism (in response to touch).

(*i*) **Phototropism.** Phototropic or curvature movements that take place when plant is provided with artificial or natural light only from one direction. If curvature movement is towards the source of light it is known as **positive phototropism** and if away from the source of light is known as **negative phototropism**. Leaves of some plants can keep themselves perpendicular to the source of light and diaphototropic. One example of phototropic movement is the leaves of *Neptunia oleracea* which keeping their faces towards sun keep on moving throughout the day turning from east to west. The flowers of *Helianthus annuus* (sun flower) also move the same way always keeping their faces towards sun. Ovary stalk of *Arachis hypogea* shows different type of phototropic movements

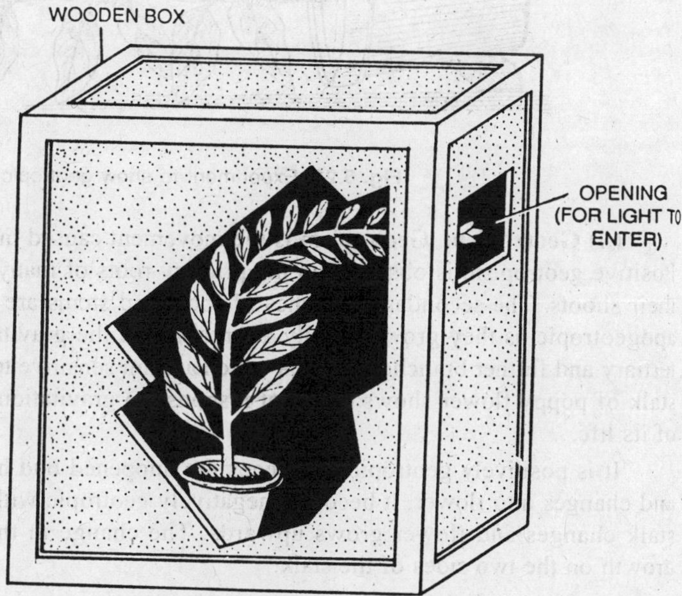

Fig. 6.19. Experiment to show phototropism.

during different phases of its life. It is positively phototropic when the ovary is unfertilized but after the ovary is fertilized, it becomes negatively phototropic with the result that it bends towards soil and moves the ovary inside the soil. The dispersal of spores in *Pilobolus* (a fungus) occurs when the sporangiophores have moved towards light.

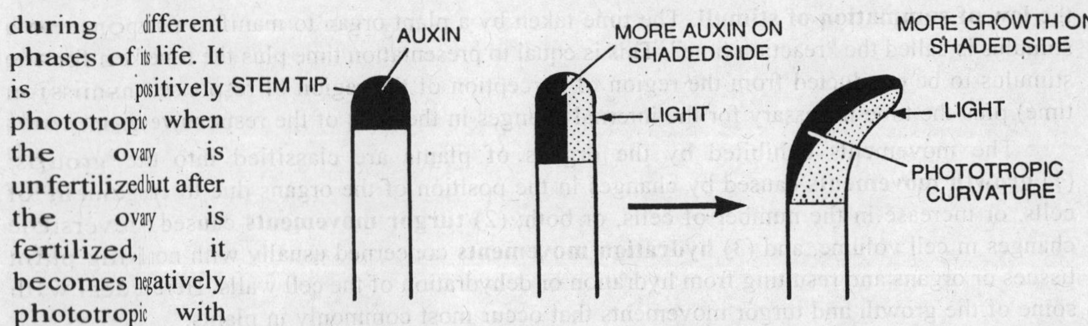

Fig. 6.20. Interaction of light and auxin in causing phototropic curvature in stem tip.

The phototropic movements are thought to be controlled by growth regulatory substances which collect on the side away from the light to increase its growth and result in curvature.

In the laboratory, phototropism can be demonstrated by keeping a potted plant inside a box with a small opening only on one side to provide light. After sometime it will be found that stem has bent towards the opening through which the light is coming.

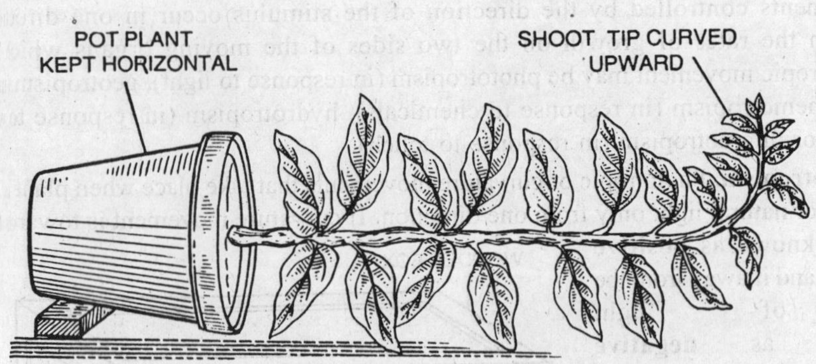

Fig. 6.21. Experiment to show geotropic curvature.

(*ii*) **Geotropism.** Geotropism is the movement caused in response to gravitational stimulus. Positive geotropism is observed in the primary roots of many plants and negative geotropism in their shoots. The secondary branches of roots and stems are diageotropic or plagiogeotropic or apogeotropic as they grow laterally at an angle to the gravitational force, called ageotropic are tertiary and further branches of roots which are not sensitive to gravitational stimulus. The flower stalk of poppy flower shows different response to gravitational stimulus during different phases of its life.

It is positively geotropic and keeps the unopened bud hanging down but as the bud opens and changes into flower, it becomes negatively geotropic with the result that the direction of the stalk changes and flower grows upwards. The change in the direction occurs due to unequal growth on the two sides of the stalk.

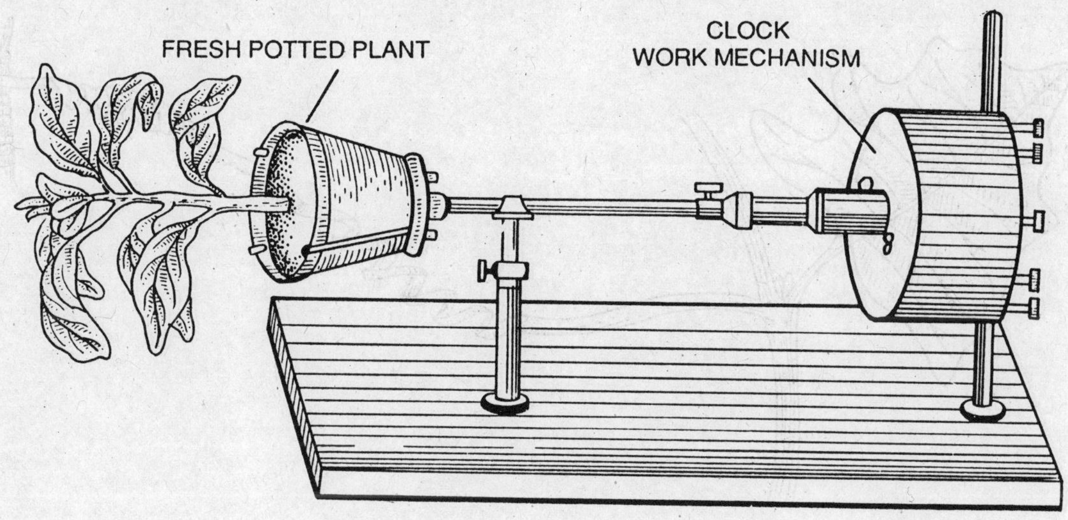

Fig. 6.22. Klinostat.

Depending on the contour, the rhizome of many plants also grows horizontally in the soil due to some reason.

The geotropism can be demonstrated in the laboratory with the help of an instrument known as klinostat which when fixed on it can keep a potted plant rotating with a definite speed. If two klinostats are taken with potted plants kept in horizontal positions and one is rotated while keeping the other one fixed and stationary, after some time it will be found that the shoot of the plant kept on the fixed klinostat bends upwards showing negative geotropism and root bends downwards showing positive geotropism. While both the roots and shoots of the plant fixed on the rotating klinostat do not show any bending because the gravitation stimulus in this case is not unilateral as it affects all the sides of the rotating organs equally.

Charles Darwin's experiment showed that if the shoot and root tips are cut off, there is no response to gravitational stimulus and the region of perception of the stimulus (the tip) is different from the region of response (the zone of cell elongation). To explain the reason of the curvature and bending of the roots and shoots, various theories have been put forward by Vernon, Nemec, Haberlandt, etc., in the part but the most convincing evidences are for the hormone theory proposed by Cholodny, Dolk, Bennet, Clark, Ball, etc. According to this theory, the hormones synthesized in the apical meristem uniformly distributed flow towards mature regions due to gravitational force when the plant is upright, but when it is kept in horizontal position, the concentration of flowing hormone becomes more on low side due to gravitational force and less on the other side. This unequal concentration of hormones on the two sides of the organ concerned results in an unequal growth rate of two sides affecting the tendency of the organ.

(*iii*) **Hydrotropism.** This is the movement of the plant organs to unilateral stimulus of water. This also results in curvature of the organ concerned due to unequal growth on its sides. Roots are positively hydrotropic. The demonstration of the hydrotropism is possible in the laboratory with the help of a shallow sieve in which moist saw dust is kept and some seeds are germinated. After some time, it will be observed that the radicles come out of holes due to positive geotropism but immediately after coming out start bending up and reenter the moist saw dust through other neighbouring holes of the sieve due to hydrotropism towards water.

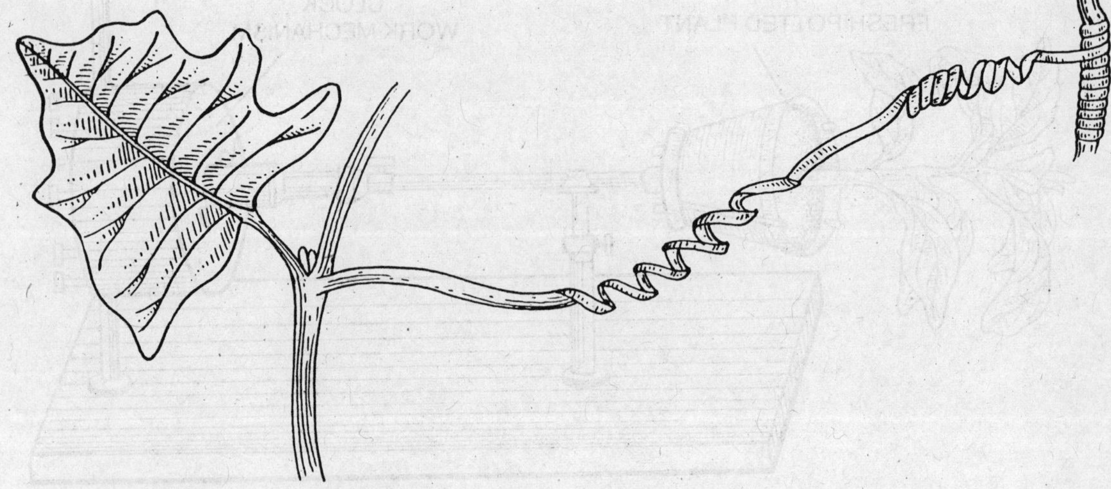

Fig. 6.23. Experiment to show hydrotropism.

(*iv*) **Chemotropism.** Chemotropism is the movement caused due to unilateral stimulus of some chemical. The pollen tube in many cases which moves through the style towards ovary due to chemical stimulus is a good example. Another example can be that of fungus mycelium which when grown in culture medium moves and grows towards the region of the medium where more nutrients are concentrated.

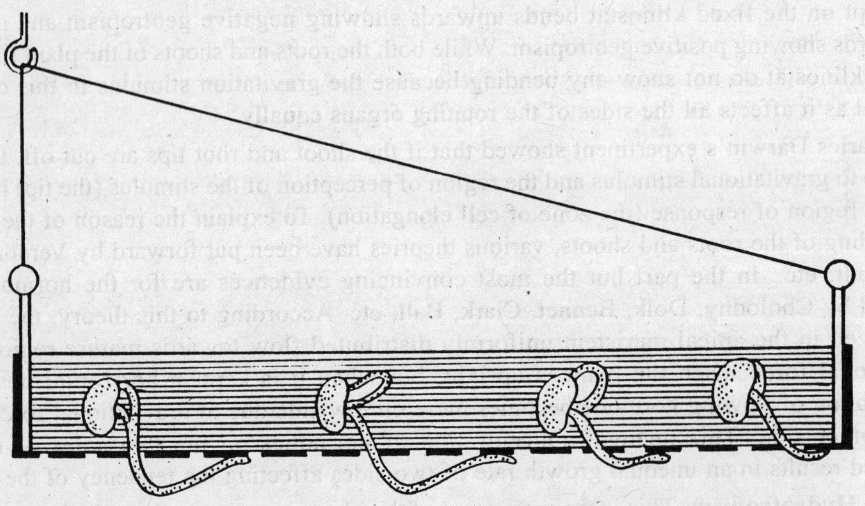

Fig. 6.24. Thigmotropic curvature of the tendril.

(*v*) **Thigmotropism.** Thigmotropism is the curvature movement of plant response to unilateral stimulus of touch and can be observed in so many twiners and climbers. Whenever the supporting organ touches a support, it makes a curvature movement to catch hold the object and tightens its hold. In tendril climbers, the tendrils first show nutation and as soon as its apex touches a support it makes thigmotropic movements to turn round it.

Turgor Movements

These movements are observed in case of sensitive plant like *Mimosa pudica*. Owing to its sensitivity to touch *Mimosa pudica* is called **"sensitive plant"** or **"touch me not"** It is a herbaceous plant with bipinnately compound leaves. The base of the petiole is swollen which is called pulvinus and similarly smaller pulvinules are present at the base of leaf-lets.

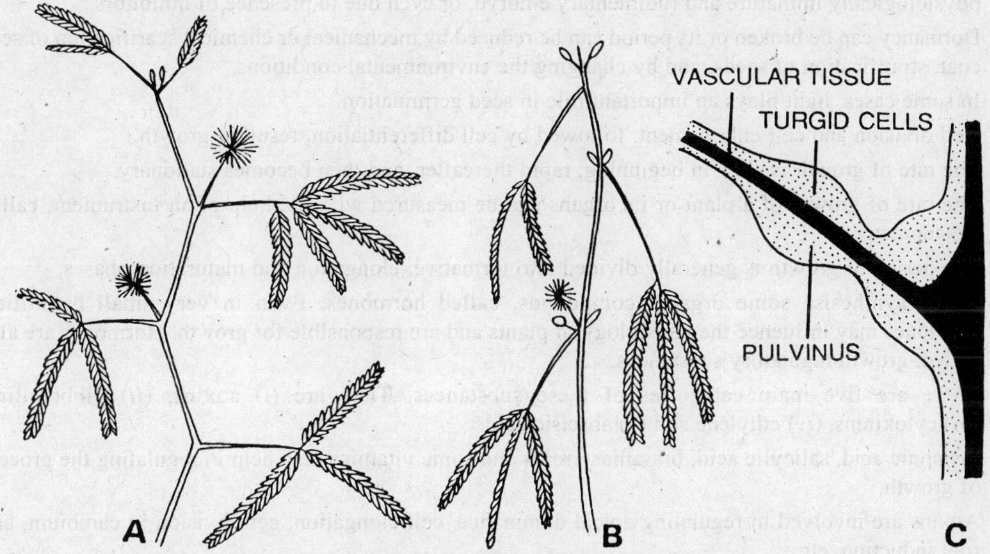

Fig. 6.25. Turgor movements in *Mimosa pudica*.

In case the terminal pinnule is struck a blow, cut or burnt or touched, the stimulus is conducted to its base and then to other pinnules. The stimulated pinnules get close up in pairs and drop down in succession from the tip backwards. The stimulus then passes to other pinnules where the same happening is repeated. If the stimulus is strong, the entire leaf becomes affected and the whole leaf bends downwards. A fast stimulus travels with a speed of 20 cm per second, thus the entire movement is very rapid, taking places in a few seconds.

The movement is caused by differential loss of turgor on the two sides of pulvinus. The two sides of pulvinus show different type of cellular organization of which lower half consists of thin walled loosely arranged parenchymatous cells with large intercellular spaces and upper half consists of a bit compactly arranged parenchymatous cells. As regards to sensitivity, the vascular trace passing through this region shows upper side stable and lower side sensitive.

During normal condition on both sides cells of pulvinus remain fully turgid and the leaf is held erect.

The stimulus reaches when in pulvinus, the osmotically active water enters into the cytoplasm of cells present on the lower half of pulvinus lowering the osmotic pressure of these cells effectively and forcing out water into the intercellular spaces. These flaccid cells now fail to hold pinnules erect and cause the dropping down of pulvinules and ultimately pulvinus. The whole leaf bends downwards and pulvinus and pulvinules act as hinges. The stimulus may pass from one leaf to another one. It is believed that the whole mechanism is operated under the influence of some hormone which produced in the stimulated part and travels to the pulvinus through the xylem. The reverse is the case true for restoring normal position.

SUMMARY

1. The irreversible increase in shape, size and weight of an organism or its part is called growth.
2. In plants, growth is generally confined to apical regions and in meristematic tissues.
3. Almost all the plants have a period of suspended growth and even a period of rest, is called dormancy.
4. The dormancy in seeds may be due to impermeable and mechanically resistant seed coat, physiologically immature and rudimentary embryo, or even due to presence of inhibitors.
5. Dormancy can be broken or its period can be reduced by mechanical or chemical scarification of seed coat, stratification of seeds, and by changing the environmental conditions.
6. In some cases, light plays an important role in seed germination.
7. Cell division and cell enlargement, followed by cell differentiation, result in growth.
8. The rate of growth is slow in beginning, rapid thereafter, and then becomes stationary.
9. The rate of growth of a plant or its organs can be measured with the help of an instrument, called auxanometer.
10. The period of growth is generally divided into formative, elongation and maturation phases.
11. Plants synthesise some organic compounds, called hormones. Even in very small quantities, hormones may influence the physiology of plants and are responsible for growth. Hormones are also called, growth regulatory substances.
12. There are five main categories of these substances. They are (*i*) auxins, (*ii*) gibberellins, (*iii*) cytokinins, (*iv*) ethylene and (*v*) abscisic acid.
13. Jasmonic acid, salicylic acid, brassinosteroids and some vitamins also help in regulating the process of growth.
14. Auxins are involved in regulating apical dominance, cell elongation, cell division in cambium, and root induction, etc.
15. Gibberellins are required for enhancing seed germination, cell elongation and flowering, overcoming dwarfism, and certain other general functions.
16. Cytokinins play main role in cell division, and help in delaying senescence and reducing apical dominance.
17. Ethylene and abscisic acid generally have a negative role and are responsible for immature falling of leaves, rapid fading of flowers, inhibiting germination, etc.
18. Sometimes these negative roles become beneficial for us, as inducing fruit ripening, breaking dormancy in potato, inhibiting the growth of duckweeds and stimulating the closure of stomata.
19. Vitamins are responsible for normal growth and development of plants.
20. Flowering of plants has a close relationship with the period of exposure to light. Some plants flower in longer periods of exposure to light, whereas others in shorter photoperiods. This phenomenon is called photoperiodism. Some flowering plants are neutral towards duration of light.
21. Low temperature, in particular, has profound effect on growth and flowering of plants. This phenomenon is known as vernalisation.
22. The period between complete maturity and final death of an organ or organism is called senescence.
23. Abscission is shedding of leaves, flowers or fruits by a plant.
24. The movement in plants is very slow process and can be observed at the interval of several days.
25. The streaming movement of protoplasm is spontaneous, whereas the movement in plastids is due to effect of light.
26. Induced movements are called taxis and they may be due to effect of chemical, light and heat.
27. In certain plants, the curvature is produced by diffused stimuli and affects the whole plant uniformity. This is called nastic movement or nastism.
28. In other plants, the stimuli do not affect the whole plant, and the curvature developed is called tropic movement or tropism. The stimuli that trigger movement in plants are light, heat, water, touch, chemical, gravity, etc.

29. Growth is open in higher plants. It is also called indeterminate.
30. Following cell division in root apical meristem and shoot apical meristem cells, the growth could be arithmetic or geometrical.
31. Growth rates could be absolute or relative
32. Growth is generally not sustained at higher rate throughout the life of cell/tissue/organ/organism.
33. There are three principal phases of growth, *i.e.*, lag, log and senescent phase.
34. When a cell loses the capactiy to divide it leads to differentiation.
35. Differentiation provides structures or is commensurate with the function the product has to perform.
36. General principles for differentiation for cell, tissues and organs are similar.
37. A differentiated cell may be differentiate and then redifferentiate.
38. Development is the sum of growth and differentiation.
39. Plant exhibits plasticity in development along with open differentiation.
40. Plants growth and development are under control of both intrinsic and extrinsic factors.
41. Intercellular intrinsic factors are the chemical substances, called plant growth regulators.
42. There are diverse groups of PGRs in plants, principally belonging to 5 groups : (*i*) auxins, (*ii*) gibberellins, (*iii*) cytokinins, (*iv*) abscisic acid, (*v*) ethylene.
43. The PGRs are produced in various parts of the plant; they control different differentiation and developmental events.
44. Any PGR has direct physiological effects on plant; similar affects are manifested by diverse PGRs.
45. PGRs may act synergistically or antagonistically.
46. Plant growth and development are also affected by light, temperature, nutrition, oxygen, gravity and such external factors.

TERMS TO REMEMBER

Abscission. The phenomenon of breaking of leaves and fruits from the plant.

Apical dominance. When the growth of apical bud at the apex of shoot inhibits the growth of lateral buds into branches, it is called apical dominance.

Aleurone layer. The outer layer of endosperm in cereal grains, (*e.g.*, maize, wheat, etc.) which produces hydrolysing enzymes, such as amylase and protease.

Autonomic movement. The self controlled growth movements of the plants.

Catabolism. The breaking down phase of metabolism. Such as digestion and respiration.

Cytokinin. A growth regulator concerned with stimulation of cell division and nucleic acid metabolism.

Day-neutral plants. The plants that flower after a period of vegetative growth indifferent to day length or the photoperiod.

Emergence. The appearance of root and shoot from seed.

Epigeal germination. During germination, the cotyledons are pushed above the ground level due to rapid elongation of hypocotyl.

Germination. The awakening of dormant embryo in the seed.

Growth. The phenomenon of growth represents an increase in mass or volume of an organism accompanied by an irreversible change in form and structure.

Hypogeal germination. In this type of germination, the cotyledons and endosperm remain below the soil and epicotyl elongates.

Imbibition. Initial intake of water by the seed.

Indole acetic acid (IAA). A type of auxin (a growth promoter).

Long-day plants. Such plants flower when the day length exceeds critical length.

Nastic movement. Non-directional growth movements.

NAA. Naphthalene acetic acid.

Paratonic movements. The growth movements induced by external stimuli.

Photoperiodism. Response of a plant to the changing relative lengths of the day and night.

Phototropism. Growth movements of plants induced by stimulus of light.

Phytochrome. A light sensitive protein pigment involved in several plant activities.

Response. The reaction of a plant to a stimulus.

Senescence. Gradual decline in the smooth functioning of the organisms ultimately leading to death.

Short-day plant. When the flowering of a plant is induced by uninterrupted dark period, longer than the critical length.

Vernalization. Low temperature treatment required by plants for flowering.

QUESTION BANK

Very short answer questions

1. Gibberellins promote the production of :
   (*a*) male flowers (*b*) female flowers (*c*) neutral flowers (*d*) abscission layer
2. Give any one function of phytochrome.
3. A student cultures a callus from the tobacco pith in a sterilised minimal nutritive medium but adds more cytokinins than auxins. What would develop first from the callus-the shoot buds or the roots.
4. Name the stress hormone in plants that functions during drought.
5. What is the full form of IBA ?
6. What is the full form of NAA ?
7. Name the two synthetic auxins used for inducing the rooting in woody plants.
8. What is vernalisation ?
9. Certain plants will flower only when they are exposed to low temperature for a few weeks. What do you call for this requirement ?
10. Name any two synthetic auxins in agriculture.
11. What induces parthenocarpy in grapes ?
12. Which is the only one gaseous natural plant growth regulator ?
13. What does an overripe apple release which affects other apples in the basket ?
14. Name the substances which cause seed dormancy.
15. Placing a ripe apple in a bag of green bananas will cause them to ripen quickly. This occurs because the apple:
    (*a*) absorbs cytokinin (*b*) forms digestive enzymes
    (*c*) lacks auxin (*d*) produces ethylene

ANSWERS

1. (*a*) ; **2.** It is required for breaking dormancy and initiation of germination in some light sensitive seeds, such as lettuce ; **3.** The shoot buds ; **4.** Abscisic acid (ABA) ; **5.** Indole butyric acid (IBA) ; **6.** Naphthalene acetic acid (NAA) ; **7.** Indole butyric acid (IBA) and Naphthalene acetic acid (NAA) ; **8.** This is low temperature treatment to plants for flowering ; **9.** Vernalisation; **10.** IBA and NAA ; **11.** Auxins ; **12.** Ethylene ; **13.** Ethylene ; **14.** They are germination inhibitors, which include organic acids, phenolics, alkaloids, etc. ; **15.** (*d*).

Short Answer Questions

1. Explain inhibitory effect of auxins with the help of one example.
2. Taking the examples of auxins and cytokinins together explain :
   (*i*) synergetic action in plants (*ii*) an antagonistic action in plants
3. What is bolting ? What conditions can induce bolting naturally and how can it be induced artificially ?

4. What is the effect of each of the following on germination of seeds :
   (*i*) Abscisic acid (*ii*) Far-red region of light
5. Write the full forms of two synthetic auxins NAA and IBA. What for are they used ?
6. What are nastic movements ? Give one example.
7. List any four uses of auxins.
8. What will you do to prevent leaf fall and fruit drop in plant ? Support your answer with reason.
9. Which among the following is a long day plant? Why is it so called ? Sugarbeet, sugarcane, tomato.
10. How does abscisic acid act antagonistically to auxins and gibberellins ?
11. What is the role of phytochrome in seed germination ?
12. Name two synthetic auxins. How are they used in agriculture ?
13. What is senescence ? What are its causes ?
14. In most plants the terminal bud suppresses the development of lateral buds. What is this phenomenon called ? Name the phytochrome that can promote the phenomenon.
15. Give at least four roles of gibberellins on plant growth.
16. Differentiate between quiescence and dormancy with reference to seed germination.
17. What is seed dormancy ? What causes it ? How is it broken naturally and how can it be broken artificially ?
18. Name the only gaseous natural plant hormone. Describe its three different actions in plants.
19. Write briefly about abscisic acid.
20. Describe the various steps involved in the process of seed germination.
21. "Growth is an important phenomenon of living". Justify the statement with reasons.
22. Explain different phases of growth with the help of a diagram.
23. What do you understand by photoperiodism and vernalisation ? Describe their significance.
24. Write an essay on growth regulators in plants.
25. How will you measure the rate of growth ? Describe an instrument used to measure the increase in height of an angiospermic plant.
26. What are the important characteristics of growth ? Describe in brief.
27. Describe an experiment that would demonstrate that growth stimulating hormone is produced at the tip of coleoptile.
28. Mention any two causes of seed dormancy. Add a note on the significance of this phenomenon.
29. Describe the phenomenon of photoperiodism.
30. Explain briefly the role of auxins on plant growth.
31. Describe the experiment conducted by F. Went regarding discovery of auxins.

Long Answer Questions

1. What do you understand by the spontaneous and induced movements in plants ? Illustrate your answer with suitable diagrams.
2. Write an essay on growth regulators in plants.
3. Write short notes on the following :
   (*a*) Senescence (*b*) Abscission (*c*) Quiescence (*d*) Vivipary
4. What do you understand by photoperiodism and vernalization ? Describe their significance.
5. Discuss the practical applications of growth regulators.
6. Describe the important factors that affect growth and development of plants.
7. Write a note on short day, long day and day neutral plants.
8. Write a note on seed germination and seed dormancy. What are the methods of breaking dormancy of seeds ?
9. Write an essay on vernalization.
10. Write the physiological effects of gibberellins and cytokinins.

MULTIPLE CHOICE QUESTIONS (Objective Type Questions)

1. The substance related phototropism in the shoot is :
(*a*) Ethanol (*b*) Cytokinin (*c*) Auxin (*d*) Gibberellin
2. In leaves of grasses bulliform cells show :
(*a*) Tropic movements (*b*) Growth movements
(*c*) Nastic movements (*d*) Turgor movements
3. Some flowers open during the day time and close at night. This is known as :
(*a*) Phototaxy (*b*) Phototropism (*c*) Photoperiodism (*d*) Photonasty
4. Gibberellins cause :
(*a*) Shortening of genetically tall plants (*b*) Elongation of genetically dwarf plants
(*c*) Promotion of rooting (*d*) Yellowing of young leaves
5. The hormone which has negative effect on apical dominance is :
(*a*) Cytokinin (*b*) Auxin (*c*) Gibberellin (*d*) Ethylene
6. A genetically dwarf plant can be converted into a tall plant by the use of :
(*a*) Kinetin (*b*) AA/X-rays (*c*) 2, 4–D (*d*) GA_3
7. Which one of the following hormones is not found in plants :
(*a*) 2, 4–D (*b*) GA_3 (*c*) Gibberellin (*d*) IAA
8. The chemical nature of Kinetin is :
(*a*) Butyric acid (*b*) Indole butyric acid
(*c*) 6-furfuryl amino purine (*d*) Indole acetic acid
9. Which of the following hormones owes its discovery to tissue culture technique ?
(*a*) Auxin (*b*) Abscisic acid (*c*) Gibberellin (*d*) Cytokinin
10. Germination of the seed is promoted by :
(*a*) Green light (*b*) Red light (*c*) Blue light (*d*) Infrared light
11. Which of the following induces cell division in a cell :
(*a*) Cytokinins (*b*) Gibberellins (*c*) Auxins (*d*) ABA
12. Movement of *Drosera* is :
(*a*) Thigmotropism (*b*) Thigmonastic (*c*) Thermonastic (*d*) Photonastic
13. Which one of the following induces flowering in long day plants?
(*a*) Gibberellins (*b*) Cytokinin (*c*) Auxins (*d*) Ethylene
14. Thigmotropism is best exemplified by :
(*a*) Root apex (*b*) Thorns (*c*) Tendrils (*d*) Lamina
15. The rate of growth of a plant stem can be measured by means of :
(*a*) Potometer (*b*) Auxanometer (*c*) Osmometer (*d*) Hydrometer
16. Touch induced closure of leaflets in *Mimosa pudica* is the case of :
(*a*) Thigmotropism (*b*) Phototropism (*c*) Thermotropism (*d*) Siesmonasty
17. Which hormones promote flowering in long day plants and also control sex expression ?
(*a*) Cytokinins (*b*) Gibberellins (*c*) Ethylene (*d*) Auxin
18. *Avena* coleoptile test was conducted by :
(*a*) Darwin (*b*) N. Smit (*c*) Paal (*d*) F.W. Went
19. The growth hormones responsible for bolting are :
(*a*) Auxins (*b*) Kinetin (*c*) Coumarins (*d*) Gibberellins
20. In short day plants, flowering is interrupted by :
(*a*) Dark condition is interrupted by white or red light
(*b*) Dark condition is interrupted by far-red light

(*c*) Dark condition is interrupted by red followed by far-red light

(*d*) In short day plants, flowering cannot be interrupted.

21. Growth regulator ethylene is used for :

(*a*) Retarding ripening of bananas (*b*) Speeding up ripening of fruits

(*c*) Slowing down ripening of apples (*d*) Both (*b*) and (*c*)

22. Coconut milk factor is :

(*a*) an auxin (*b*) a gibberellin (*c*) abscisic acid (*d*) cytokinin

23. Differentiation of shoot is controlled by :

(*a*) High auxin : cytokinin ratio (*b*) High cytokinin : auxin ratio

(*c*) High gibberellin : auxin ratio (*d*) High gibberellin : cytokinin ratio

24. Seed dormancy is due to the :

(*a*) Ethylene (*b*) Abscisic acid (*c*) IAA (*d*) Starch

25. By which of the following means the testa is made permeable to water :

(*a*) Scarification (*b*) Stratification (*c*) Vernalization (*d*) Denudation

26. One of these prevents elongation in plants :

(*a*) Auxin (*b*) Gibberellin (*c*) ABA (*d*) Cytokinin

27. Long day plant amongst the following is :

(*a*) *Glycine max* (*b*) Spinach (*c*) *Mirabilis jalapa* (*d*) Soya bean

28. Hormone responsible for preventing seeds during drought is :

(*a*) IAA (*b*) NAA (*c*) IBA (*d*) ABA

29. What will happen if terminal buds are removed from a plant ?

(*a*) The plant will die (*b*) The lateral buds will grow profusely

(*c*) The roots will die (*d*) The shoot will die

30. The closing and opening of leaves of *Mimosa pudica* is due to :

(*a*) Thermonastic movement (*b*) Hydrotropic movement

(*c*) Seismonastic movement (*d*) Chemonastic movement

31. ABA is involved in :

(*a*) Dormancy of seeds (*b*) Root elongation

(*c*) Shoot elongation (*d*) Increased cell division

32. A plant hormone used for inducing morphogenesis in tissue culture is :

(*a*) ABA (*b*) Ethylene (*c*) Gibberellin (*d*) Cytokinin

33. Gibberellins can be extracted from :

(*a*) Rice only (*b*) Fungus only

(*c*) Germinating seeds only (*d*) Many healthy plants

34. Thigmotropism is the response of the plant to :

(*a*) Gravity (*b*) Water (*c*) Light (*d*) Contact

35. Auxanometer is an instrument with which we can measure the :

(*a*) Rate of respiration (*b*) Rate of photosynthesis

(*c*) Geotropism (*d*) Rate of growth

36. During germination, stem grows upward and root grows downward because :

(*a*) it depends upon light (*b*) of auxin

(*c*) it does not depend on light (*d*) of epinasty and hyponasty

37. 2, 4–D is a :

(*a*) Flowering hormone (*b*) Rooting hormone

(*c*) Herbicide (*d*) Pesticide

38. To promote the growth of lateral branches of a plant :
(*a*) Axillary buds are removed (*b*) Apical bud is removed
(*c*) Auxin is applied to the apical bud (*d*) Auxin is applied to the decapitated shoot tip

39. The plants respond to photoperiods due to the presence of :
(*a*) Phytochromes (*b*) Stomata (*c*) Enzymes (*d*) Phytohormones

40. The hormone present in the liquid endosperm of coconut is :
(*a*) Ethylene (*b*) Gibberellin (*c*) Auxin (*d*) Cytokinin

41. Stress hormone is :
(*a*) Auxin (*b*) Gibberellin (*c*) ABA (*d*) Cytokinin

42. To remove seed dormancy by mechanical removing of seed-coat is called :
(*a*) Stratification (*b*) Vernalization (*c*) Scarification (*d*) Photoperiodism

43. Vernalization is done at :
(*a*) Low temperature (*b*) Low light intensity
(*c*) High temperature (*d*) High light intensity

44. Which one of the following statements is true for the phytochrome :
(*a*) Phytochrome is a phytohormone
(*b*) Phytochrome is a photosynthetic pigment
(*c*) Phytochrome is a pigment that controls growth, photomorphogenesis and development of many plants
(*d*) Phytochrome is a regulatory protein that controls several dark dependent development processes.

45. All the cytokinins are :
(*a*) Acidic (*b*) Aminopurines (*c*) Phenol (*d*) Glucosides

46. Certain chemical substances having profound effect on growth, are called :
(*a*) Compost (*b*) Catalytic agent (*c*) Enzymes (*d*) Phytohormones

47. The presence of auxins in a solution could be tested by :
(*a*) *Avena sativa* stem tip test (*b*) Carbon tetrachloride test
(*c*) Iodine test (*d*) Defoliation test

48. The photoperiodic stimulus is perceived by :
(*a*) Leaves (*b*) Flowers (*c*) Buds (*d*) Meristem

49. The activity of α–amylase in the endosperm of barley germinating seed is induced by :
(*a*) Ethylene (*b*) Cytokinin (*c*) IAA (*d*) Gibberellin

50. Leaf fall starts when the amount of :
(*a*) Auxin increases (*b*) Auxin decreases
(*c*) Abscisic acid decreases (*d*) Gibberellic acid decreases

51. The rate of growth of any organism follows :
(*a*) Hyperbolic curve (*b*) J-shaped curve (*c*) Sigmoid curve (*d*) Parabolic curve

52. Which of the following fungus is associated with the discovery of gibberellins :
(*a*) *Fusarium* (*b*) *Rhizopus* (*c*) *Aspergillus* (*d*) *Penicillium*

53. Induction of formation of interfascicular cambium is by :
(*a*) Kinetin (*b*) Gibberellin
(*c*) Auxin (*d*) Ethylene

54. The following is a naturally occurring growth inhibitor :
(*a*) IAA (*b*) ABA (*c*) NAA (*d*) GA_3

55. In plants, the induction of flowering, by low temperature treatment, is called :
(*a*) Pruning (*b*) Cryobiology (*c*) Vernalisation (*d*) Photoperiodism

ANSWERS

1. (*c*)	**2.** (*d*)	**3.** (*d*)	**4.** (*b*)	**5.** (*a*)	**6.** (*d*)	**7.** (*a*)	**8.** (*c*)	**9.** (*d*)	**10.** (*b*)
11. (*a*)	**12.** (*b*)	**13.** (*a*)	**14.** (*c*)	**15.** (*b*)	**16.** (*d*)	**17.** (*b*)	**18.** (*d*)	**19.** (*d*)	**20.** (*a*)
21. (*b*)	**22.** (*d*)	**23.** (*b*)	**24.** (*b*)	**25.** (*a*)	**26.** (*c*)	**27.** (*b*)	**28.** (*d*)	**29.** (*b*)	**30.** (*c*)
31. (*a*)	**32.** (*d*)	**33.** (*d*)	**34.** (*d*)	**35.** (*d*)	**36.** (*b*)	**37.** (*c*)	**38.** (*b*)	**39.** (*a*)	**40.** (*d*)
41. (*c*)	**42.** (*c*)	**43.** (*a*)	**44.** (*c*)	**45.** (*b*)	**46.** (*d*)	**47.** (*a*)	**48.** (*a*)	**49.** (*d*)	**50.** (*b*)
51. (*c*)	**52.** (*a*)	**53.** (*a*)	**54.** (*b*)	**55.** (*c*)					

7
CHAPTER

Lipid (Fat) Metabolism

What are lipids ? The term lipid is derived from the Greek word *lipos*, meaning fat. It is an operational term for a variety of substances that are not soluble in polar solvent, such as water but dissolve in non-polar solvents, such as benzene, ether and chloroform. This occurs because lipids contain relatively long or complex hydrocarbon chains that are non-polar and thus hydrophobic.

The main groups of compounds classified as lipids are : (*i*) **triglycerides,** (*ii*) **phospholipids,** (*iii*) **steroids and** (*iv*) **waxes.**

Triglycerides

One important group of stored lipids is **triglycerides,** a category that includes **fats** and **oils.** Triglycerides are composed of a single molecule of glycerol bound to three fatty acids (fig. 7.1).

Glycerol is a 3 carbon alcohol with 3 OH groups that serve as binding sites.

Fatty acids are long chain hydrocarbon molecules with a carboxyl group (COOH) at one end which is free to bind to one of the OH groups of the glycerol, thus forming a bond called ester bond.

The hydrocarbon portion of a fatty acid can vary in length from 4 – 24 carbons.

The fats may be saturated or unsaturated. If the carbons in the chain are single bonded, the fat is saturated; if there is at least one C = C double bond in the chain, it is unsaturated. The structure of fatty acids is responsible for the physical nature of fats and oils (liquid fats) which are greasy and insoluble.

In general, solid fats are saturated, and oils are unsaturated.

In most cells for long term, triglycerides are stored in concentrated form as droplets or globules.

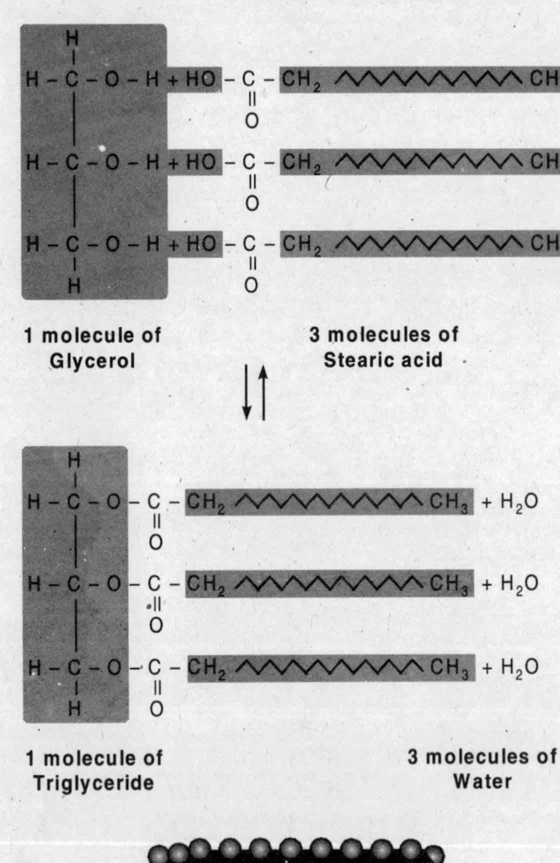

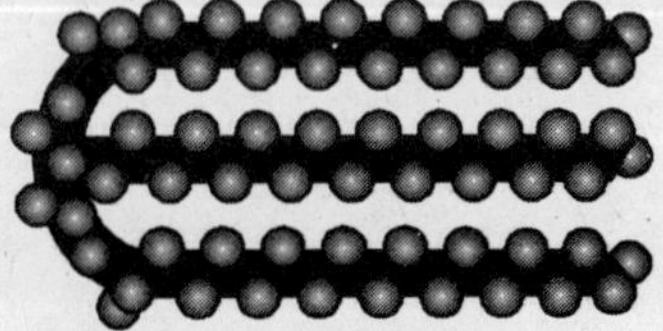

Fig. 7.1. Triglyceride molecule.

Phospholipids (Membrane Lipids)

A class of lipids that serve as major structural component of cell membranes is **phospholipids.** Although phospholipids are similar to triglycerides in containing glycerol and fatty acids, there are some significant differences. Phospholipids contain only two fatty acids attached to the glycerol, while the third glycerol binding site holds a phosphate group (fig. 7.2). This phosphate is in turn bonded to an alcohol. These lipids have both **hydrophilic** and **hydrophobic** regions due to a change on the phosphoric acid/alcohol 'head' of the molecule and lack of a charge on the long 'tail' of the molecule (formed by the fatty acids).

When exposed to an aqueous solution, the charged heads are attracted to the water phase and the non-polar tails are repelled from the water phase.

The way lipids naturally assume single and double layered (bilayer) configuration makes them a valuable component of the primary framework of cell membranes.

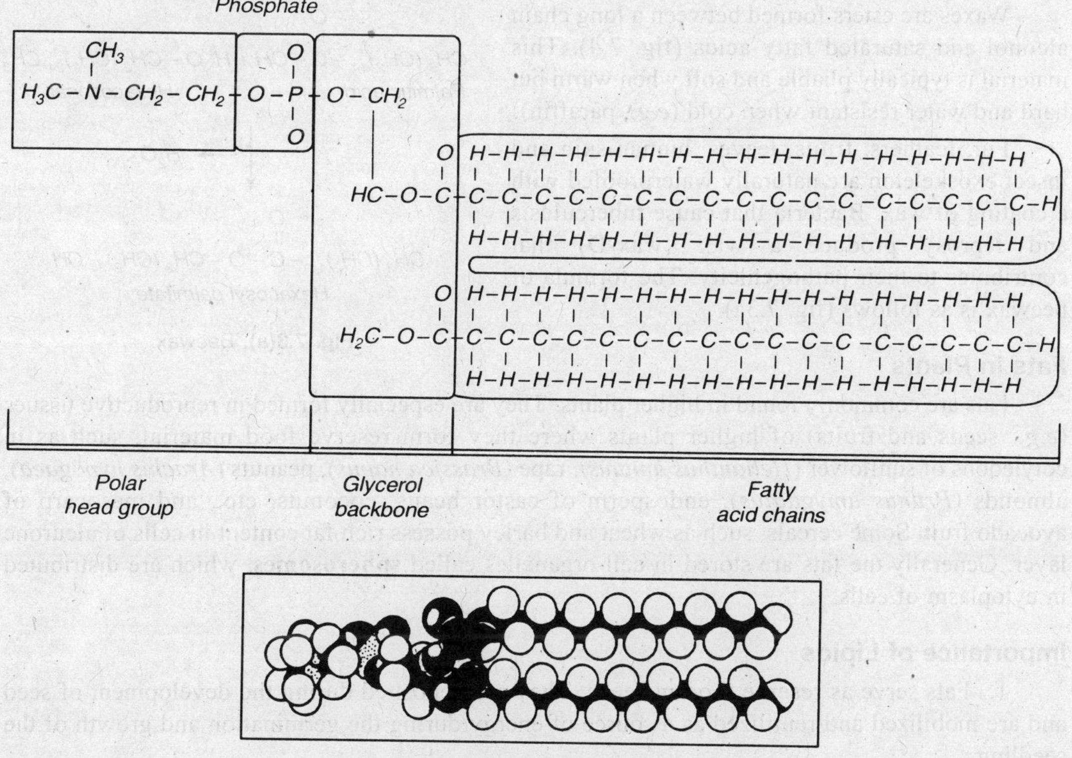

Fig. 7.2. Phospholipid molecule.

When two single layers of polar lipids come together to form a double layer, the outer hydrophilic face of each single layer will orient itself towards, the solution and the hydrophobic portion will become immersed in the core of the bilayer.

The structure of lipid bilayer helps the membrane in functions such as selective permeability and fluid nature.

Steroids

These are complex compounds commonly found in cell membranes and animal hormones. The best known of these is the sterol called **cholesterol** (fig. 7.3) which reinforces the structure of the cell membrane in animal cells and in an unusual group of cell-wall deficient bacteria called **mycoplasmas.** The cell membrane of fungi also contains a sterol called **ergosterol.**

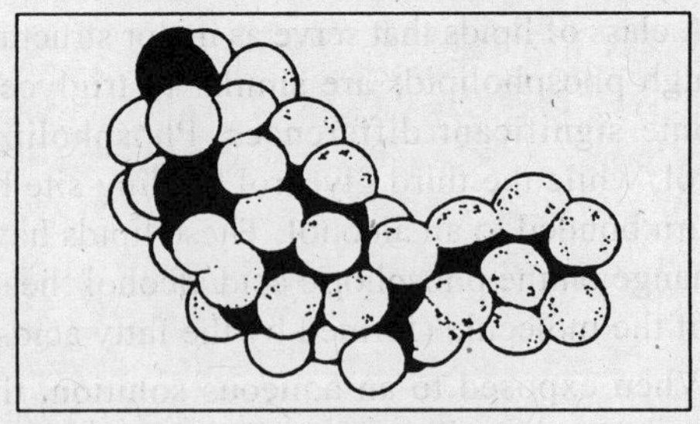

Fig. 7.3. Cholesterol molecule.

Wax

Waxes are esters formed between a long chain alcohol and saturated fatty acids (fig. 7.4). This material is typically pliable and soft when warm but hard and water resistant when cold (*e.g.*, paraffin).

Fur, feathers, fruits, leaves, human skin and insect exoskeleton are naturally waterproofed with a coating of wax. Bacteria that cause tuberculosis and leprosy produce a wax (wax-D) that contributes to their pathogenicity. The formula of beewax is as follows (fig. 7.3a).

$$\underset{\text{Palmitic acid}}{CH_3\,(CH_2)_{14}-\overset{\overset{O}{\|}}{C}-OH} + \underset{\text{Hexacosonal}}{H\;O-CH_2\,(CH_2)_{24}\,CH_3}$$

$$\downarrow \rightarrow H_2O$$

$$\underset{\text{Hexacosyl palmitate}}{CH_3\,(CH_2)_{14}-\overset{\overset{O}{\|}}{C}-O-CH_2\,(CH_2)_{24}\,CH_3}$$

Fig. 7.3(a). Beewax.

Fats in Plants

Fats are commonly found in higher plants. They are especially formed in reproductive tissues (*e.g.*, seeds and fruits) of higher plants where they form reserve food material, such as in cotyledons of sunflower (*Helianthus annuus),* rape (*Brassica napus*), peanuts (*Arachis hypogaea*), almonds (*Prunus amygdalus*), endosperm of castor beans, coconuts, etc., and mesocarp of avocado fruit. Some cereals, such as wheat and barley possess rich fat content in cells of aleurone layer. Generally the fats are stored in cell-organelles called **spherosomes**, which are distributed in cytoplasm of cells.

Importance of Lipids

1. Fats serve as reserve food in seeds. They are deposited during the development of seed and are mobilized and reutilized as a source of energy during the germination and growth of the seedling.

2. Oils are used by people for various purposes. The edible oils are obtained from the seeds of mustard, cotton, groundnut, soybean, sunflower, sesame and fruits of coconut.

3. Lipids provide the structural frame-work to the living tissues of plants and animals. Plasma membrane and other cell organelles, such as chloroplasts, mitochondria, endoplasmic reticulum, lysosomes and golgi complex contain the lipids in the form of layers. In the plants they are found in the form of small droplets dispersed in the cytoplasm.

4. Lipids serve as the main fuel reserve for metobolism and provide more energy than carbohydrate and proteins.

5. Linseed oil is primarily used in the manufacture of paints and varnishes. It is also used in the manufacture of linoleum and oil-cloth, printing and lithographic inks and soft soap.

6. Castor oil is used as a purgative. It is also used for making transparent soap, textile soap, typewriter inks, perfume aromatics, varnishes and paints.

7. The corn oil is used as an edible oil. It can be used for cooking and in bakeries.

8. The surface of leaves and fruits of different plants often possesses a covering of wax secreted by the plant. This wax is similar to fat in its composition. Jojoba wax, carnauba wax, candelilla wax and myrtle wax are more important waxes produced by plants which are generally used in making candles, soaps, varnishes, paints, ointments and several other useful products.

FATTY ACIDS

The acids which are combined with glycerol in the naturally occurring fats are called **fatty acids.** These acids contain even number of carbon atoms linked together in long chains which are in general unbranched. The chain contains from six to about twenty carbon atoms in plant triglycerides and the most frequent number being 16 or 18. The long chain molecule possesses the carboxyl group at one end, while the remainder part consists of carbon and hydrogen atoms only.

Classification of Fatty Acids

The fatty acids can be classified into four groups as follows :

1. **Saturated fatty acids**
2. **Unsaturated fatty acids**
3. **Branched chain fatty acids**
4. **Cyclic fatty acids.**

N.B. Only (1) saturated fatty acids and (2) unsaturated fatty acids are described.

1. Saturated fatty acids. Saturated fatty acids are those which contain only single bonds in their hydrocarbon chain. The general formula is :

$$R — COOH$$

where R is $CH_3(CH_2)n$; *n* varies from zero in acetic acid to 24 in cerotic acid and to 86 in mycolic acid.

The most commonly occurring saturated fatty acids of the higher plants are palmitic acid (C_{16}) and stearic acid (C_{18}). Other major saturated fatty acids found in plant lipids are shown in table.

Table 7.1. Major Saturated fatty acids and their sources

S. No.	Name of fatty acids	Structure	Source
1.	Butyric acid	$CH_3(CH_2)_2COOH$	Butter
2.	Caproic acid	$CH_3(CH_2)_4COOH$	Butter, palm oil, coconut oil
3.	Caprylic acid	$CH_3(CH_2)_6COOH$	Palm oil, coconut oil
4.	Capric acid	$CH_3(CH_2)_8COOH$	Palm oil, coconut oil
5.	Lauric acid	$CH_3(CH_2)_{10}COOH$	Plants of lauraceae, coconut oil, palm oil.
6.	Myristic acid	$CH_3(CH_2)_{12}COOH$	Seed fats of mace, butter, coconut oil
7.	Palmitic acid	$CH_3(CH_2)_{14}COOH$	Plant fats, palm oil, peanut oil.
8.	Stearic acid	$CH_3(CH_2)_{16}COOH$	Plant and animal fats

S. No.	Name of fatty acids	Structure	Source
9.	Arachidic acid	$CH_3(CH_2)_{18}COOH$	Peanut oil
10.	Behenic acid	$CH_3(CH_2)_{20}COOH$	Plant lipids
11.	Lignoceric acid	$CH_3(CH_2)_{22}COOH$	Plant lipids
12.	Cerotic acid	$CH_3(CH_2)_{24}COOH$	Beewax, wool

2. Unsaturated fatty acids. Unsaturated fatty acids are those which contain one or more double bonds in their hydrocarbon chain. The general formula is :

$$R - CH = CH\ (CH_2)_n - COOH$$

These fatty acids are also characterized by the presence of cis-trans isomerism at their double bonds. The most common examples of unsaturated fatty acids are mentioned in table. Oleic and linoleic unsaturated fatty acids are widely distributed in plants.

Table 7.2. Major Unsaturated fatty acids and their sources

S. No.	Name of fatty acids	Structure	Source
1.	Palmitoleic acid	$CH_3(CH_2)_5CH = CH(CH_2)_7COOH$	Sardine oil
2.	Oleic acid	$CH_3(CH_2)_7CH = CH(CH_2)_7COOH$	Olive oil, peanut oil, linseed oil
3.	Linoleic acid	$CH_3(CH_2)_4CH = CHCH_2CH =$ $CH(CH_2)_7COOH$	Olive oil, peanut oil, linseed oil, soyben oil
4.	γ-linolenic acid	$CH_3(CH_2)_4CH = CHCH_2$ $CH = CHCH_2CH = CH(CH_2)_4COOH$	linseed oil
5.	Parinaric acid	$CH_3CH_2CH = CHCH = CHCH =$ $CH(CH_2)_7COOH$	Plant lipids
6.	Erucic acid	$CH_3(CH_2)_7CH = CH(CH_2)_{11}COOH$	Rapeseed oil
7.	Arachidonic acid	$CH_3(CH_2)_4CH = CHCH_2CH =$ $CHCH_2CH = CHCH_2CH =$ $CH(CH_2)_3COOH$	Peanut oil

Properties of Fatty Acids and Fats

Physical Properties

1. Fats and fatty acids are soluble in organic solvents, such as petroleum ether, benzene and chloroform. They are insoluble in water.
2. Saturated fatty acids are solid at room temperature, while unsaturated fatty acids are liquid.
3. Unsaturated fatty acids show cis-trans isomerism due to presence of double bonds.
4. They are bad conductors of heat.
5. Saturated glycerides containing fats require high temperature for melting, whereas unsaturated glycerides containing fats require relatively lower temperature for its melting.

Chemical Properties

1. Hydrolysis. Fats undergo hydrolysis when treated with mineral acids, the alkalies or fat splitting enzyme *lipase* or hydrolases to yield glycerol and the constituent fatty acids.

$$\begin{array}{l} CH_2O\ \boxed{OC\cdot C_{15}H_{31}\ HO}\ H \\ \quad | \\ CHO\ \boxed{OC\cdot C_{15}H_{31}\ HO}\ H \\ \quad | \\ CH_2O\ \boxed{OC\cdot C_{15}H_{31}\ HO}\ H \end{array} \xrightarrow[\text{Lipase}]{\text{Enzyme}} \begin{array}{l} CH_2OH \\ \quad | \\ CHOH \\ \quad | \\ CH_2OH \end{array} + 3C_{15}H_{31}COOH$$

Fat (tripalmitin) Water Glycerol Palmitic acid

Hydrolysis by alkalies, such as NaOH or KOH leads to the formation of sodium or potassium salts of fatty acids. The salts are known as **soaps** and process of its formation is **saponification.**

2. Hydrogenation. Oils containing unsaturated fatty acids can be hydrogenated in presence of high temperature, pressure and finely divided nickel. By this process the oils are converted into solid fats (glycerides of saturated fatty acids). This reaction forms the basis of the industrial production of hydrogenated oil (vegetable ghee).

$$\begin{array}{l} CH_2OOC\,C_{17}H_{33} \\ | \\ CHOOC\,C_{17}H_{33} \\ | \\ CH_2OOC\,C_{17}H_{33} \end{array} \xrightarrow[\text{Ni } 180° - 200°C]{3H_2} \begin{array}{l} CH_2OOC\,C_{17}H_{35} \\ | \\ CHOOC\,C_{17}H_{35} \\ | \\ CH_2OOC\,C_{17}H_{35} \end{array}$$

Triolein (m.p.17°C) (oil) — Tristearin (m.p.60°C) (solid fat)

3. Hydrogenolysis. Oils and fats are converted into glycerol and a long chain aliphatic alcohol when excess of hydrogen is passed through them under pressure and in presence of copper-chromium catalyst. This splitting of fat by hydrogen is called **hydrogenolysis.**

4. Halogenation. When unsaturated fatty acids are treated with halogens, such as iodine and chlorine, they take up iodine or other halogens at their double bond site. This process of taking of iodine is called **halogenation** which is an indication of unsaturation. **Iodine number** is the percentage of iodine absorbed by a fat.

5. Rancidity. Oils and fats on long storage in contact with heat, light, air and moisture, develop an unpleasant odour. Such oils and fats are known as **rancid oils** and **fats.**

The rancidity develops due to certain chemical changes taking place in the fat. These changes include (*i*) enzymatic hydrolysis, (*ii*) air oxidation of unsaturated fatty acids, and (*iii*) β-oxidation of saturated fatty acids.

In all above mentioned changes, hydrolysis of fats is caused by enzyme *lipase* which is produced by microorganisms present in them. To check the rancidity, it becomes essential to protect oils and fats from air, light and moisture during storage.

6. Emulsification. The process of breaking of large-sized fat molecules into smaller ones is called **emulsification.** In animals, this process is brought about by bile juice liberated from liver. Other emulsifying agents are water, soaps, proteins and gums.

Breakdown (Degradation) of Fats

Fats are found in dynamic state in plants, *i.e.*, at one time they are synthesized; while at other time they break down to meet specific requirement of the cells.

Active breakdown of fats (insoluble) takes place as follows :

(*i*) During germination of fatty seeds so that the decomposition products may enter into glycolysis and Kreb's cycle to release energy and also to synthesise soluble **sucrose** through **glyoxylic acid cycle** which is then translocated to the growing regions of the young germinating seedling till it develops green leaves to manufacture its own food.

(*ii*) In plants, when carbohydrates reserve declines, the fats (and also proteins), may form the **respiratory substrates** which are broken down and oxidised to release energy.

The fats are first hydrolysed in the presence of the enzymes *lipases* to yield fatty acids and glycerol.

$$\begin{array}{l} CH_2OOC{\cdot}R \\ | \\ CHOOC{\cdot}R \\ | \\ CH_2OOC{\cdot}R \end{array} + 3H_2O \xrightarrow{\text{Lipase}} \begin{array}{l} CH_2OH \\ | \\ CHOH \\ | \\ CH_2OH \end{array} + \underset{\text{Long chain fatty acids}}{3R{\cdot}COOH}$$

Fat (Triglyceride) — Glycerol

OXIDATION OF GLYCEROL

The glycerol may react with ATP under the catalytic influence of *glycerol kinase* to form **glycerol-3-phosphate** which is then oxidised in the presence of *glycerol-3-phasphate dehydrogenase* and NAD^+ to produce **dihydroxyacetone phosphate** and enters into glycolysis.

$$\underset{\text{glycerol}}{\begin{array}{l} CH_2OH \\ | \\ CHOH \\ | \\ CH_2OH \end{array}} + ATP \longrightarrow \underset{\text{glycerol-3-phosphate}}{\begin{array}{l} CH_2OH \\ | \\ CHOH \\ | \\ CH_2OPO_3H_2 \end{array}} + ADP$$

$$\underset{\text{glycerol-3-phosphate}}{\begin{array}{l} CH_2OH \\ | \\ CHOH \\ | \\ CH_2OPO_3H_2 \end{array}} + NAD^+ \longrightarrow \underset{\text{dihydroxyacetone phosphate}}{\begin{array}{l} CH_2OH \\ | \\ C=O \\ | \\ CH_2OPO_3H_2 \end{array}} + NADH + H^+$$

$$\text{Dihydroxyacetone phosphate} \xrightleftharpoons{\text{Glycolytic pathway}} \text{Pyruvic acid} + 2ATP_S + NADH + H^+$$

The conversion of glycerol into pyruvic acid that takes place in cytoplasm yields 2ATPs by substrate level phosphorylation and 2NADH which on reoxidation by terminal electron transport chain *via* the external NADH dehydrogenase (located on the outer surface of the inner mitochondrial membrane in plants) further generate 4 ATP molecules (2 mol./NADH oxidised). If the pyruvic acid also undergoes complete oxidation into CO_2 and H_2O in Kreb's cycle (or TCA cycle), it will produce another 15ATP moleccules. Thus a total of 2 + 4 + 15 = 21ATPs are produced per glycerol moleule oxidised. However, 1ATP molecule is consumed in the *glycerol kinase* catalysed reaction. Hence, the net gain is 21 – 1 = 20 ATPs per glycerol molecule oxidised. Thus :

$$\text{Glycerol} + ATP \longrightarrow \text{Glycerol-3-phosphate} + ADP$$

$$\text{Glycerol-3-phosphate} + NAD^+ \longrightarrow \text{Dihydroxyacetone phosphate} + NADH + H^+$$

$$\text{Dihydroxyacetone phosphate} \underset{\text{isomerase}}{\overset{\text{Phosphotriose}}{\rightleftharpoons}} \text{3-phosphoglyceraldehyde}$$

$$\begin{array}{c} \textbf{Dihydroxyacetone phosphate} \\ + \\ \textbf{3-phosphoglyceraldehyde} \end{array} \xrightarrow{\text{aldolase}} \textbf{Fructose 1,6 bisphosphate}$$

$$\text{Fructose 1, 6 bisphosphate} \xrightarrow[\text{bis-phosphatase}]{\text{Fructolse 1, 6}} \text{Fructose-6-phosphate}$$

$$\text{Fructose-6-phosphate} \underset{\text{isomerase}}{\overset{\text{Phoasphohexose}}{\rightleftharpoons}} \text{Glucose-6-phosphate}$$

$$\text{Glucose-6-phosphate} \xrightleftharpoons{\text{Glucose-6-phosphatase}} \textbf{Glucose}$$

$$\text{Glucose-6-phosphate} \xrightleftharpoons{\text{Phosphoglucomutase}} \text{Glucose-1-phosphate}$$

$$\text{Glucose-1-phosphate} \longrightarrow \textbf{Oligo and polysaccharides}$$

OXIDATION (BREAKDOWN) OF FATTY ACIDS

Now, long chain fatty acids are broken down by **α-oxidation** and **β-oxidation.** The β-oxidation ultimately produces the **active 2-C units,** the **acetyl-CoA** ($CH_3CO \cdot CoA$).

α-Oxidation

Here, the long chain of fatty acid is gradually broken down until it is reduced to **12C-atoms.** Fatty acids with less than 13-C atoms are not affected by this process. One complete α-oxidation results in the elimination of one carbon atom in the form of CO_2 from the — COOH group of the fatty acid, whereas **α–C–atom**, *i.e.*, C-atom no., 2, adjacent to COOH is oxidised (α-oxidation).

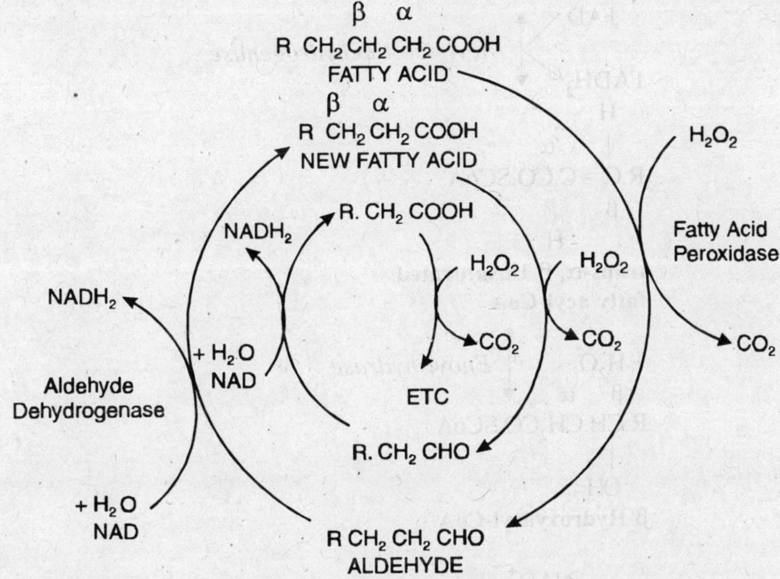

Fig. 7.4. α-oxidation of fatty acid.

Process of α-oxidation is as follows :

(*i*) The fatty acid is **oxidatively decarboxylated** in presence of enzyme *fatty acid peroxidase* and H_2O_2 to form an **aldehyde.** Here, CO_2 comes out from the carboxylic group and oxidation takes place at **α–C–atom** that converts into aldehyde group.

$$\underset{\text{Fatty acid}}{R{\cdot}CH_2\overset{\beta}{C}H_2\overset{\alpha}{C}H_2COOH} \xrightarrow[2(H_2O_2)]{\text{fatty acid peroxidase}} \underset{\text{Aldehyde}}{R{\cdot}CH_2CH_2CHO} + CO_2 + H_2O$$

(*ii*) Now, the aldehyde is further oxidised in the presence of enzyme *aldehyde dehydrogenase* to form the new fatty acid containing one carbon atom less than in the original fatty acid. NAD+ is reduced in the reaction.

$$\underset{\text{Aldehyde}}{R \cdot CH_2CH_2CHO} \xrightarrow[H_2O]{\text{Aldehyde dehydrogenase}} \underset{\text{New fatty acid}}{R \cdot CH_2CH_2COOH} + NADH + H^+$$

The new fatty acid will oxidise repeatedly, till it consists of 12 – C atoms by the same process of α-oxidation.

β-Oxidation

β-oxidation is the main process of fatty acid degradation in plants. This mechanism is well established for saturated fatty acids, while obscure for unsaturated fatty acids.

$$\overset{3}{\beta}\ \overset{2}{\alpha}\ 1$$
$$R.CH_2CH_2COOH$$
Fatty acid

↓ *Thiokinase* ATP, CoASH

$$R.\overset{\beta}{C}H_2\overset{\alpha}{C}H_2CO.SCoA + AMP + P - Pi$$
Fatty acyl-CoA

↓ FAD → $FADH_2$ — *Acyl–CoA dehydrogenase*

$$R.\underset{\beta}{\overset{H}{C}} = \underset{H}{\overset{\alpha}{C}}.CO.SCoA$$
trans-α, β Unsaturated fatty acyl-CoA

↓ $+ H_2O$ — *Enoyl hydrase*

$$R.\underset{OH}{\overset{\beta}{C}H}.\overset{\alpha}{C}H_2CO.SCoA$$
β-Hydroxyacyl-CoA

↓ NAD^+ → $NADH + H^+$ — *β-Hydroxyacyl-CoA dehydrogenase*

$$R.\underset{\|\atop O}{\overset{\beta}{C}}.\overset{\alpha}{C}H_2CO.SCoA$$
β-Keto fatty acyl-CoA

↓ +CoASH — *β-Ketoacyl thiolase*

$$R.\underset{\|\atop O}{C}.SCoA + CH_3CO.SCoA$$
Fatty acyl-CoA **Acetyl-CoA**

Fig. 7.5. β-oxidation of fatty acid.

β-oxidation takes place in mitochondria and also in glyoxysomes. It involves sequential removal of 2–C in the form of acetyl-CoA ($CH_3CO \cdot SCoA$) molecules from the carboxyl end of the fatty acid. This is called β-oxidation because β-C (*i.e.*, C atom No. 3) of the fatty acid is oxidised during this process. Various steps of this process are as follows :

(*i*) The first step involves the activation of fatty acid in the presence of ATP and enzyme *thiokinase.* CoASH is consumed and CoA derivative of fatty acid is produced.

$$\underset{\text{Fatty acid}}{R{\cdot}\overset{\overset{3}{\beta}}{C}H_2{\cdot}\overset{\overset{2}{\alpha}}{C}H_2\overset{1}{C}OOH} + CoASH + ATP \xrightleftharpoons{Mg^{++}} \underset{\text{Fatty acyl–CoA}}{RCH_2CH_2{,}\underset{\underset{O}{\|}}{C}{\cdot}SCoA} + AMP + \underset{\text{Pyrophosphate}}{P - Pi}$$

The AMP (adenosine Mono-phosphate) molecule thus produced reacts with another ATP molecule under the catalytic influence of enzyme *adenylate kinase* to form 2 ADP molecules.

$$AMP + ATP \rightleftharpoons ADP + ADP$$

(*ii*) In the second step of β-oxidation, two hydrogen atoms are removed between α and β-C atoms and a **trans α, β-unsaturated fatty acyl CoA** is formed. This reaction is catalysed by FAD-containing enzyme *acyl-CoA dehydrogenase.*

$$\underset{\text{Fatty acyl-CoA}}{R{\cdot}\overset{\beta}{C}H_2\overset{\alpha}{C}H_2\underset{\underset{O}{\|}}{C}{\cdot}SCoA} + FAD \rightleftharpoons \underset{\substack{\text{trans α, β-unsaturated}\\ \text{fatty acyl-CoA}}}{R{\cdot}\overset{\overset{H}{|}}{C} = \underset{\underset{H}{|}}{C}{\cdot}\underset{\underset{O}{\|}}{C}{\cdot}SCoA} + FADH_2$$

(*iii*) The third step involves the addition of a water molecule across the double bond to form corresponding **β-hydroxyacyle-CoA** in the presence of enzyme *enoyl hydrase.*

$$R{\cdot}\underset{\underset{\beta}{|}}{\overset{\overset{H}{|}}{C}} = \underset{\underset{H}{|}}{\overset{\alpha}{C}}{\cdot}\underset{\underset{O}{\|}}{C}{\cdot}SCoA + H_2O \rightleftharpoons \underset{\beta\text{-hydroxyacyl-CoA}}{R{\cdot}\underset{\underset{OH}{\|}}{\overset{\beta}{C}H}{\cdot}\overset{\alpha}{C}H_2{\cdot}\underset{\underset{O}{\|}}{C}{\cdot}SCoA} + NADH + H^+$$

(*iv*) In the fourth step β-hydroxyacyl-CoA is dehydrogenated in the presence of NAD-specific *β-hydroxyacyl-CoA dehydrogenase.* Two hydrogen atoms are removed from the β-C atom (β-oxidation) which now bears a carbonyl function and **β-keto fatty acyl CoA** is formed.

$$\underset{\beta\text{-hydroxyacyl-CoA}}{R{\cdot}\underset{\underset{OH}{|}}{\overset{\beta}{C}H}{\cdot}\overset{\alpha}{C}H_2\underset{\underset{O}{\|}}{C}{\cdot}SCoA} + NAD^+ \rightleftharpoons \underset{\beta\text{-keto fatty acyl-CoA}}{R{\cdot}\underset{\underset{O}{\|}}{\overset{\beta}{C}}{\cdot}\overset{\alpha}{C}H_2\underset{\underset{O}{\|}}{C}{\cdot}SCoA} + NADH + H^+$$

(*v*) The fifth and the last step involves the thioclastic cleavage of β-ketofatty acyl-CoA in the presence of the enzyme *β-ketoacyl thiolase* and results in the formation of an active 2-C unit acetyl-CoA and a fatty acyl-CoA molecule which is shorter by two-carbon atoms than when it entered the β-oxidation spiral.

$$R{\cdot}\underset{\underset{O}{\|}}{\overset{\beta}{C}}{\cdot}\overset{\alpha}{C}H_2\underset{\underset{O}{\|}}{C}{\cdot}SCoA + CoASH \rightleftharpoons \underset{\text{Fatty acyl-CoA}}{R{\cdot}\underset{\underset{O}{\|}}{C}{\cdot}SCoA} + \underset{\text{Acetyl CoA}}{CH_3CO{\cdot}SCoA}$$

The fatty acyl-CoA so produced again re-enters the β-oxidation spiral at step 2, by passing the first step as it is already activated, and losing a further 2-C unit. This sequence continues till whole molecule is degraded.

Each turn of the β-oxidation generates one $FADH_2$ (step 2), one NADH + H^+ (step 4) and one acetyl-CoA molecule (step 5). However, in the last turn of the spiral two acetyl-CoA molecules will be produced. Reoxidation of $FADH_2$ and NADH + H^+ by the electron transport chain will yield 2 and 3 ATP molecules respectively. Thus each turn of β-oxidation generates 5 ATP molecules. However, in the first turn 2 ATPs are consumed in the first step, thus in this turn there will be a net gain of only 3 ATP molecules.

Complete oxidation of one acetyl-CoA molecule in TCA cycle to CO_2 and H_2O will result in the production of 12 ATP molecules.

Thus huge amount of energy is generated in the form of ATP molecules by the mitochondrial oxidation of fatty acids through the β-oxidation spiral and TCA cycle. For instance, one molecule of palmitic acid (with 16 C atoms) on complete oxidation will produce 129 ATP molecules.

While complete oxidation of stearic acid (with 18 C atoms) through β-oxidation spiral and TCA cycle will result in production of 146 ATP molecules.

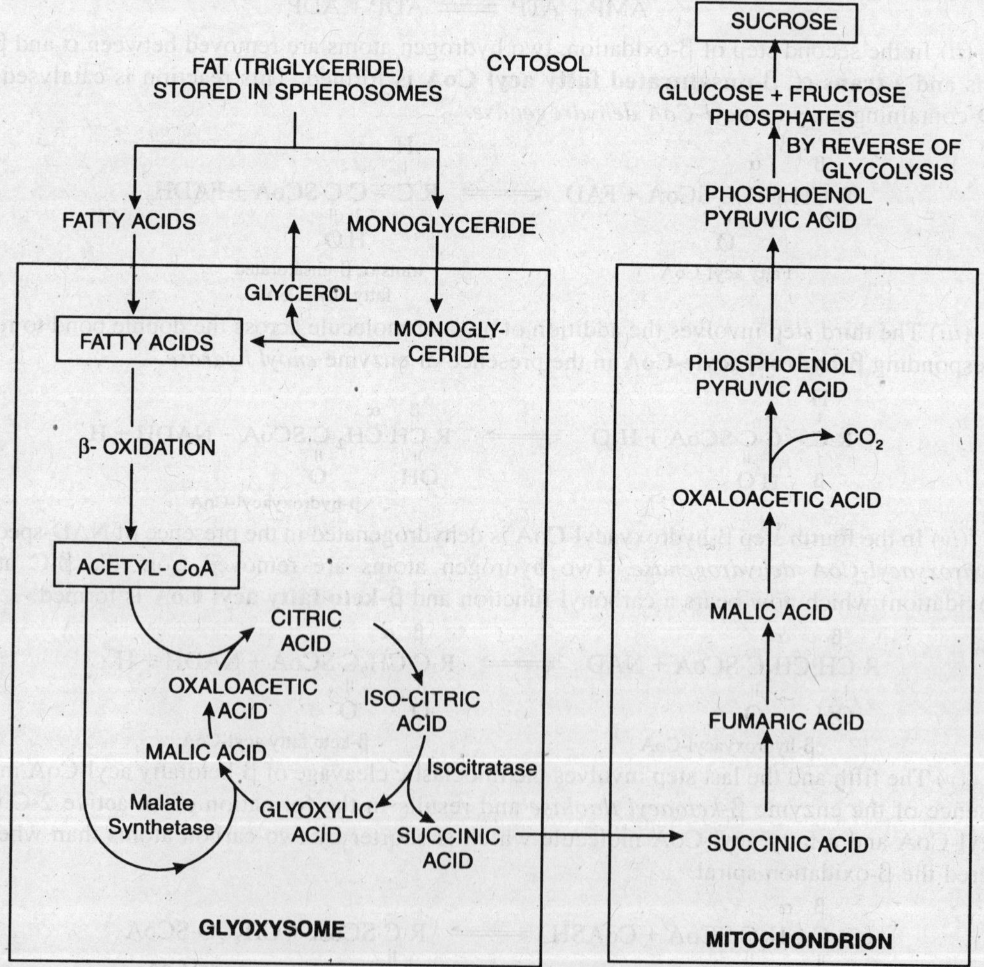

Fig. 7.6. Glyoxylate cycle.

Fate of acetyl–CoA ($CH_3CO{\cdot}CoA$). Acetyl-CoA units are end products of β-oxidation of fatty acids, which may enter (*i*) into Kreb's cycle (TCA cycle) and are oxidised to release energy as mentioned in preceding paragraphs, or (*ii*) in case of germination of fatty acids, they are converted into soluble sucrose through the **glyoxylic acid cycle.**

GLYOXYLATE CYCLE

(Glyoxylic Acid Cycle)

Korenberg and **Krebs** (1957) framed a cycle which is known as **Glyoxylic acid cycle** or **Glyoxylate cycle** through which the fats could be converted into sucrose (carbohydrates) during the germination of fatty seeds in plants.

The glyoxylate cycle is completed in **glyoxysomes, mitochondria** and **cytosol.** Various steps of this cycle occurring in higher plants especially during the germination of fatty seeds are shown in fig. 7.6. (Glyoxylate cycle) that occur in **glyoxysome** and **mitochondrion.**

BIOSYNTHESIS OF FATTY ACIDS

Long chain saturated fatty acids are synthesized in plants from active two carbon units, the acetyl-CoA ($CH_3CO \cdot CoA$). Though the reactions of β-oxidation of fatty acids are reversible yet the fatty acids are not formed simply by the reverse reactions of β-oxidation.

However, synthesis of fatty acids from $CH_3CO{\cdot}$ CoA takes place step by step. In each step the fatty acid chain is increased by two carbon atoms. Each step involves two reactions. The steps are as follows :

(*i*) In the first reaction which takes place in the presence of the enzyme *acetyl-CoA carboxylase,* acetyl–CoA combines with CO_2 to form **malonyl-CoA** (malonic acid is 3-C compound). ATP provides energy while Mn^{++} and biotin are required as co-factors.

$$\underset{\text{Acetyl CoA}}{CH_3CO{\cdot}CoA} + CO_2 + ATP \xrightarrow[\text{biotin } Mn^{++}]{\text{acetyl-CoA carboxylase}} \underset{\text{malonyl-CoA}}{H_2C\begin{matrix}\diagup COOH \\ \diagdown CO{\cdot}CoA\end{matrix}} + ADP + iP$$

(*ii*) Malonyl-CoA reacts with another molecule of $CH_3CO \cdot CoA$ (acetyl CoA) in the presence of the enzyme *fatty acid synthetase* and co-enzyme $NADPH_2$ to form coenzyme-A derivative of butric acid (butric acid contains 4C-atoms). One molecule of CO_2, H_2O and CoA are released while $NADPH_2$ is oxidised.

$$\underset{\text{malonyl-CoA}}{H_2C\begin{matrix}\diagup COOH \\ \diagdown CO{\cdot}CoA\end{matrix}} + CH_3CO{\cdot}CoA + 2NADPH_2 \longrightarrow \underset{\text{butyryl CoA}}{CH_3CH_2CH_2CO{\cdot}CoA} + CO_2 + H_2O + 2NADP$$

Butyryl CoA, in the next step combines with malonyl CoA to form CoA derivative of fatty acid containing 6-C atoms. This process is repeated again and again till coenzyme-A derivative of long chain fatty acid (which may contain 16-18 C atoms) is produced.

The enzyme *fatty acid synthetase* is not simple but a complex of many enzymes, called **multienzyme complex** and an acyl carrier protein called ACP, which is similar to CoA in having phosphopantetheine as the functional unit in their structure. However, the reaction (*ii*) described above only summarises a number of reactions involved in synthesis of fatty acids from acetyl-CoA and malonyl-CoA which can be grouped under three categories. They are as follows :

1. Initiation Reaction. In this reaction, acetyl CoA transfers its acetyl group to one of the —SH groups of multienzyme complex, *i.e., fatty acid synthetase.*

$$CH_3CO{\cdot}SCoA + \begin{matrix} HS \\ HS \end{matrix}\!\!>Enz \longrightarrow \begin{matrix} HS \\ CH_3CO{\cdot}S \end{matrix}\!\!>Enz + CoASH$$

2. Chain Elongation Reactions. Six different reactions involved are — (*i*) malonyl transfer, (*ii*) condensation, (*iii*) reduction, (*iv*) dehydration, (*v*) reduction and (*vi*) acyl transfer.

Chain elongation starts with the transfer of malonyl group from malonyl-CoA to second – SH group of the multienzyme complex. Thereafter, there is condensation of the latter so that a 4-C unit is produced. This unit by next three reactions, *i.e.*, reduction, dehydration and reduction converts into a saturated 4-C unit (*i.e.*, butyryl-CoA). In acyl transfer reaction the fatty acid residue is transferred back to the —SH group to which the acetyl group was transferred in initiation reaction. The cycle is repeated again and again with malonyl transfer, condensation, etc., till the fatty acid residue consists of up to 16 to 18 C atoms. Each such turn elongates fatty acid chain by 2-C atoms. Chain elongation reactions are as follows :

(*i*)
$$\begin{matrix} COOH \\ | \\ CH_2CO{\cdot}SCoA + HS \\ CH_3CO-S \end{matrix}\!\!>Enz \rightleftharpoons \begin{matrix} COOH \\ | \\ CH_2{\cdot}CO-S \\ CH_3CO-S \end{matrix}\!\!>Enz + CoASH$$

(*ii*)
$$\begin{matrix} COOH \\ | \\ CH_2{\cdot}CO-S \\ CH_3{\cdot}CO-S \end{matrix}\!\!>Enz \rightleftharpoons \begin{matrix} CH_3{\cdot}CO{\cdot}CH_2{\cdot}CO-S \\ HS \end{matrix}\!\!>Enz + CO_2$$

(*iii*)
$$\begin{matrix} CH_3{\cdot}CO{\cdot}CH_2{\cdot}CO-S \\ HS \end{matrix}\!\!>Enz \xrightleftharpoons{NADPH_2} \begin{matrix} CH_3{\cdot}\underset{\displaystyle OH}{\underset{|}{C}H}{\cdot}CH_2{\cdot}CO{\cdot}S \\ HS \end{matrix}\!\!>Enz$$

(*iv*)
$$\begin{matrix} CH_3\underset{\displaystyle OH}{\underset{|}{C}H}{\cdot}CH_2{\cdot}CO-S \\ HS \end{matrix}\!\!>Enz \rightleftharpoons \begin{matrix} CH_3CH=CH{\cdot}CO-S \\ HS \end{matrix}\!\!>Enz$$

(*v*)
$$\begin{matrix} CH_3CH=CH{\cdot}CO-S \\ SH \end{matrix}\!\!>Enz \xrightleftharpoons[(FMNH_2)]{NADPH_2} \begin{matrix} CH_3CH_2CH_2{\cdot}CO{\cdot}S \\ SH \end{matrix}\!\!>Enz$$

(*vi*)
$$\begin{matrix} CH_3CH_2{\cdot}CH_2{\cdot}CO-S \\ HS \end{matrix}\!\!>Enz \rightleftharpoons \begin{matrix} HS \\ CH_3CH_2CH_2CO-S \end{matrix}\!\!>Enz$$

3. Termination Reaction. When the fatty acid residue attains a desired length, the chain elongation stops at reaction (*v*) and the cycle is not repeated. The acyl group instead of being transferred to the —SH of the enzyme, is transferred to—SH group of Co-enzyme A (CoA SH) molecule. Thus, CoA derivate of the fatty acid is produced which can then be utilised in fat synthesis. The enzyme becomes free.

$$\begin{matrix} CH_3(CH_2-CH_2)_n-CO-S \\ HS \end{matrix}\!\!>Enz \rightleftharpoons CH_3(CH_2-CH_2)_n-CO{\cdot}SCoA + \begin{matrix} HS \\ HS \end{matrix}\!\!>Enz$$

STORAGE AND MOBILIZATION

Fats and oils are found in all groups of plants and in almost all living cells of the plant body. In angiosperms they are specially common in seeds and fruits. When oil is in abundance usually very little carbohydrate is present. Big aleurane grains often accompany fats and oils, as in castor seed. At ordinary temperature fats are solid and oils are liquid. They are not soluble in water or in alcohol (except castor oil); all of them are, however, soluble in ether, petroleum, chloroform, etc. They occur in the form of globules in the cytoplasm often saturating it. The fats and oils are formed from fatty acids and glycerine.

It is a well known fact that these substances enter into the composition of protoplasm; but they undergo many chemical changes before they are utilised by the protoplasm. They are decomposed by the enzyme *lipase* into fatty acids and glycerol. By this decomposition a large amount of heat is liberated.

When oily or fatty seeds germinate a considerable quantity of fatty acids may be detected in them, while glycerol disappears almost immediately.

Fatty acids and glycerol may also be translocated to the growing regions where they are utilised by the protoplasm. Both of them may readily pass through cell walls.

Oils and fats are deposited in most seeds to a greater or less extent. There is a special deposit of them in seeds such as groundnut, gingelly, coconut, castor, sunflower, safflower, palm, linseed, etc.

SUMMARY

1. Lipids are long chain of fatty acid bonding with glycerol.
2. Major types of lipids are—sterols, triglycerides, phospholipids and waxes.
3. Triglycerides are of two types—saturated and unsaturated.
4. Phospholipids are structural components of cell membrane. They have hydrophobic or hydrophilic ends.
5. The sterol called cholesterol reinforces the structure of the cell membrane in animals and wall less bacteria (mycoplasma).
6. Wax, an ester of long chain alcohol with saturated fatty acid, provides extra protection to cells by making them water proof.
7. The acids combined with glycerol in naturally occurring fats are called fatty acids.
8. Long chain fatty acids are broken down by α-oxidation and β-oxidation.
9. Long chain saturated fatty acids are synthesized in plants from active 2 carbon units, the acetyl-CoA ($CH_3CO{\cdot}CoA$).
10. Fats and oils are found in all groups of plants and in almost all living cells of plant body. In angiosperms, they are commonly found in seeds and fruits.

QUESTION BANK

Short and Long Answer Questions

1. Write short notes on : (*i*) steroids (*ii*) wax.
2. Describe structure of phospholipid. How are they arranged in the cell membrane ?
3. Explain the composition of triglyceride.
4. Mention the differences between saturated and unsaturated fats.
5. Write short notes on : (*i*) lipids, (*ii*) phospholipids and their occurrence.
6. Write short notes on : (*i*) phospholipids, (*ii*) oxidation of fatty acid, (*iii*) β-oxidation.
7. Differentiate between α-oxidation and β-oxidation of fats.
8. Describe classification of lipids.
9. Differentiate between saturated and unsaturated fatty acids.
10. Explain the chain elongation reactions in fat synthesis.
11. Describe the β-oxidation path way of fatty acids.

12. Give the structure and mechanism of fat synthesis in plants.
13. What are lipids ? Explain the distribution and degradation of lipids in plants.
14. Draw β-oxidation path way. No description is required.
15. What are plant lipids ? How are they biosynthesised ? Discuss mechanism of their biosynthesis.
16. Describe degradation of fats. What is the fate of end product of their oxidation ?
17. Enumerate the steps leading to the synthesis of fatty acids.
28. Give an account of fat synthesis in plants.

Multiple Choice Questions

1. Fats are found in abundance in plants especially in
   (*a*) reproductive tissues (*b*) vegetative tissus
   (*c*) meristematic tissues (*d*) all the above.
2. Which of the following is not a component of triglycerides ?
   (*a*) glycerol (*b*) long chain saturated fatty acid
   (*c*) long chain unsaturated fatty acid (*d*) long chain monohydric alcohol.
3. In cells, fats are stored in
   (*a*) spherosomes (*b*) lysosomes (*c*) cytosol (*d*) all the above.
4. β-oxidation of fatty acids takes place in
   (*a*) mitochondria (*b*) glyoxysomes (*c*) both (*a*) and (*b*) (*d*) cytosol.
5. The reactions of β-oxidation of fatty acids in plants are
   (*a*) reversible (*b*) irreversible (*c*) both (*a*) and (*b*) (*d*) non-enzymatic.
6. Which of the following is not a phospholipid ?
   (*a*) lecithin (*b*) cephalin (*c*) cardiolipin (*d*) none of the above.
7. Phospholipids make important structural component of
   (*a*) plasmamembrane (*b*) tonoplast (*c*) all cell membranes (*d*) none of the above.
8. Conversion of succinate into fumarate in glyoxylate cycle takes place in
   (*a*) glyoxysome (*b*) mitochondrion (*c*) cytosol (*d*) all the above.
9. Conversion of fats into carbohydrates through glyoxylate cycle is an example of
   (*a*) gluconeogenesis (*b*) glucogenesis (*c*) synthesis of sucrose (*d*) none of the above.
10. Glyoxylate cycle was discovered by
   (*a*) Hans A. Krebs (*b*) Korenberg and Krebs
   (*c*) Harden and Youngs (*d*) none of the above.
11. Glyoxylate cycle is completed in
   (*a*) glyoxysome (*b*) glyoxysome and mitochondrion
   (*c*) glyoxysome, mitochondrion and cytosol (*d*) none of the above.
12. The acetyl-CoA molecules produced at the end of β-oxidation spiral may be converted into sucrose through
   (*a*) glyoxylate cycle (*b*) glycolate cycle (*c*) TCA cycle (*d*) none of the above.
13. How many carbon atoms are removed in complete turn of α-oxidation spiral in plants ?
   (*a*) one (*b*) two (*c*) three (*d*) none of the above.
14. α-oxidation of fatty acids in plants is
   (*a*) one step process (*b*) two step process (*c*) three step process (*d*) none of the above.
15. In eukaryotic plants, complete aerobic oxidation of one molecule of glycerol results in net gain of
   (*a*) 5 ATP_S (*b*) 20 ATP_S (*c*) 15 ATP_S (*d*) none of the above.
16. Common 18–C unsaturated fatty acid, oleic acid contains.
   (*a*) one double bond (*b*) two double bonds (*c*) one triple bond (*d*) none of the above.

17. The phospholipids are broken down (hydrolysed) to their residues by the action of enzymes called
(*a*) phosphatases (*b*) phosphatidases (*c*) phosphorylases (*d*) none of the above.

18. Which of the following is not a saturated fatty acid ?
(*a*) lauric acid (*b*) myristic acid (*c*) palmitic acid (*d*) none of the above.

19. Palmitic acid contains
(*a*) 14-C atoms (*b*) 16-C atoms (*c*) 18-C atoms (*d*) 20-C atoms.

20. Complete oxidation of one palmitic acid molecule (16-C) through β-oxidation spiral and TCA cycle results in net gain of
(*a*) 127 ATP mols (*b*) 128 ATP mols (*c*) 129 ATP mols (*d*) 130 ATP mols.

21. In each turn of β-oxidation spiral, there is sequential removal of 2-C atoms in the form of
(*a*) two CO_2 mols (*b*) one CO_2 and one acetyl-CoA mol
(*c*) one acetyl-CoA mol (*d*) none of the above.

22. The enzyme fatty acid synthase is a complex of many enzymes (multienzymes complex) plus
(*a*) CoASH (*b*) ACP (*c*) FMN (*d*) none of the above.

23. Each turn of β-oxidation spiral generates two reduced coenzymes in the form of
(*a*) two NADH mols (*b*) two $FADH_2$ mols
(*c*) one NADH and one $FADH_2$ mols (*d*) none of the above.

24. Which of the following are called neutral lipids ?
(*a*) triglycerides and waxes (*b*) phospholipids and glycolipids
(*c*) sterols (*d*) none of the above.

25. The two key enzymes of glyoxylate cycle are
(*a*) malate synthetase and citrate synthase (*b*) malate synthetase and isocitratase
(*c*) citrate synthease and isocitratase (*d*) none of the above.

ANSWERS

1. (*a*) **2.** (*d*) **3.** (*a*) **4.** (*c*) **5.** (*a*) **6.** (*d*) **7.** (*c*) **8.** (*b*) **9.** (*a*) **10.** (*b*)
11. (*c*) **12.** (*a*) **13.** (*a*) **14.** (*b*) **15.** (*b*) **16.** (*a*) **17.** (*b*) **18.** (*d*) **19.** (*b*) **20.** (*c*)
21. (*c*) **22.** (*b*) **23.** (*c*) **24.** (*a*) **25.** (*b*)

UNIT - 3
Genetic Engineering

8
CHAPTER

Genetic Engineering, Cloning and Genomics

As we know, genes regulate growth and differentiation, and all fundamental life processes are also regulated by them. Ageing and cancer result from differential gene expression. Several other pathological disorders are also caused by some defective gene. Thus, it becomes quite important to understand the structure of genes and their organization and how their expression is regulated. However, in microorganisms, it is easy to study the organization and expression of genes. This is because several mutants, with altered characteristics can be isolated. In bacteria, conjugation and transduction experiments can also be carried out, where the functional characters could easily be related to specific genes and their arrangement.

In this chapter, we will learn about the tools and techniques of molecular biology for cleaving and rejoining DNA sequences from two or more different organisms.

Such genetically modified DNA fragments are known as recombinant DNA molecules. This process involves manipulation or engineering of the DNA (genes) involved in the process and thus the term **'genetic engineering'** has been coined.

The recombinant DNA molecules can be cloned and amplified virtually to an unlimited extent.

Application of DNA manipulation or genetic engineering ranges from cloning genes to cloning organisms including transgenic microbes, agriculturally important crops and farm animals.

Thus by the process of genetic engineering we can get genetically modified food products, human gene products, pharmacologically and therapeutically useful products.

Gene therapy or gene replacement can be used in the treatment of hereditary diseases which are otherwise incurable.

In modern days, DNA fingerprinting has been successfully used in forensic investigations.

Now human genome has been worked out, which reveals the blue print of our life itself.

The latest of this series is claim of first human clone, which was announced by Brigitte Boisselier, head of a company known Clonaid on December 26, 2002, a baby girl called Eve by scientists.

GENETIC ENGINEERING

Genetic engineering is considered as a kind of biotechnology.

This is a process in which the alteration of the genetic make up of cells is done by deliberate and artificial means.

This process involves transfer or replacement of genes to create recombinant DNA.

This can be done by cutting DNA molecules at specific sites to get fragments containing desirable and useful genes from one type of cell.

Thereafter, these genes can be inserted into a suitable carrier or vector.

Now, these recombinant DNA can be put into completely different cell of a bacterium or plant or animal cell.

By this technique they acquire useful characteristics, such as disease resistance or to make useful enzymes, hormones, vaccines, etc.

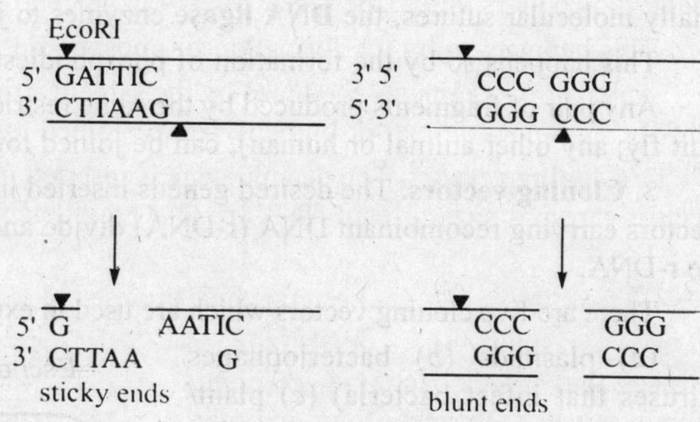

Fig. 8.1. Restriction enzymes.

Tools and Techniques of Genetic Engineering

Genetic engineering involves cutting and pasting fragments of DNA may be cut and pasted in the vector, a plasmid or a virus whose DNA then becomes the recombinant DNA.

Tools of genetic engineering. The three tools for recombinant DNA technology are as follows :

1. Cell culture containing cells with the desired DNA segment (gene).

2. DNA modifying enzymes such as : (*a*) Restriction endonuclease from various bacteria. It cleaves double stranded DNA but only at a limited number of sites (*b*) DNA ligase from *Escherichia coli* which joins DNA molecules. (*c*) DNA polymerase from *E. coli* which synthesises a new strand of DNA complementary to an existing DNA or RNA template.

DNA segment can be excised by "**molecular scissors**" or '**chemical scalpels**' which are known as **restriction enzymes** in biotechnological language.

Restriction enzymes, synthesized by microbes as a defense mechanism, are specific endonucleases, which can cleave double-stranded DNA.

However, they can do this only at limited number of sites depending on the number of recognition sequences in DNA.

With the use of **restriction enzymes** it is possible to cut a DNA sequence. For example, a restriction enzyme EcoR1 will cut DNA only if sequence $\begin{array}{c}\text{GAATC}\\\text{CTTAG}\end{array}$ is present in the double stranded DNA. Hundreds of such restriction enzymes with different sequence specificities have been isolated from several bacteria and are commercially available. These enzymes are known as **restriction enzymes** because they restrict infection by bacteria by certain bacteriophages, by degrading the viral DNA without affecting the bacterial DNA. Thus, such enzymes are helpful in breaking DNA molecule at a specific point. In recombinant DNA technology, they are used to cut DNA into specific fragments that contain specific genes.

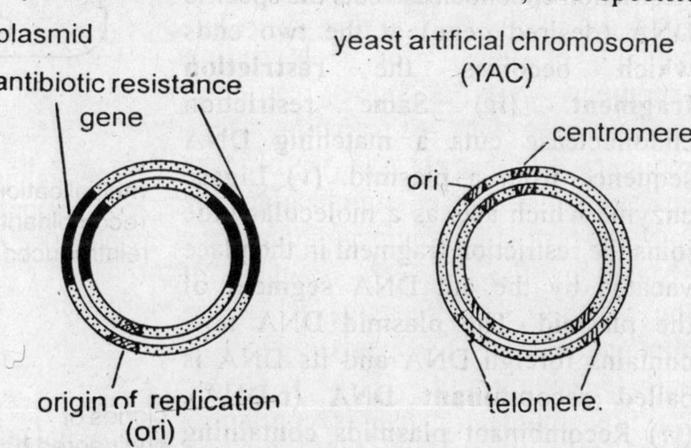

Fig. 8.2. Plasmid and yeast artificial chromosome

DNA synthesising enzymes. DNA synthesising enzymes, such as **DNA polymerase** or **reverse transcriptase** to make DNA complementary to an existing DNA or RNA template respectively called c DNA (complementary or copy DNA) are also required.

Because of the presence of an universal genetic code, the DNA polymerases (synthesising enzymes) can accurately transcribe a DNA sequence from one organism to another organism, and finally molecular sutures, the **DNA ligase** enzymes to join DNA fragments.

This happens so by the formation of phosphodiester bond, again.

Any pair of fragments produced by the same restriction enzyme, from any DNA source (*e.g.*, fruit fly, any other animal or human), can be joined together.

3. **Cloning vectors.** The desired gene is inserted in the vectors or gene carrier to form DNA. Vectors carrying recombinant DNA (r-DNA) divide and thus help in producing several copies of the r-DNA.

There are five cloning vectors which are used in experiments with *E. coli* animals/humans are:

(*a*) plasmids (*b*) bacteriophages (viruses that infect bacteria) (*c*) plant/animal viruses (*d*) transposons (jumping genes) and (*e*) artificial chromosomes of bacteria, yeast and mammals.

The most commonly used vectors are plasmids and viruses.

Recombinant DNA (r-DNA) technique. The sequence of steps in r-DNA technology are as follows :

(*i*) Cell culture of cells with the required DNA sequences is obtained through tissue culture technique. (*ii*) Specific restriction endonuclease is selected from the specific bacterium. (*iii*) Restriction endonuclease cuts the specific DNA (desired gene) at the two ends which becomes the **restriction fragment**. (*iv*) Same restriction endonuclease cuts a matching DNA sequence from a plasmid. (*v*) Ligase enzyme which acts as a molecular glue joins the restriction fragment in the place vacated by the cut DNA segment of the plasmid. The plasmid DNA now contains foreign DNA and its DNA is called **recombinant DNA** (r-DNA). (*vi*) Recombinant plasmids containing recombinant DNA are now inserted into a host cell, *e.g.*, *E. coli*. This is commonly done by a process called **electroporation** in which transient temporary pores are created in the cell membrane of the host for the uptake of foreign DNA molecules. (*vi*) Host bacterium divides to give multiple copies of the recombinant DNA which may be preserved in a DNA library which is also called genomic DNA library.

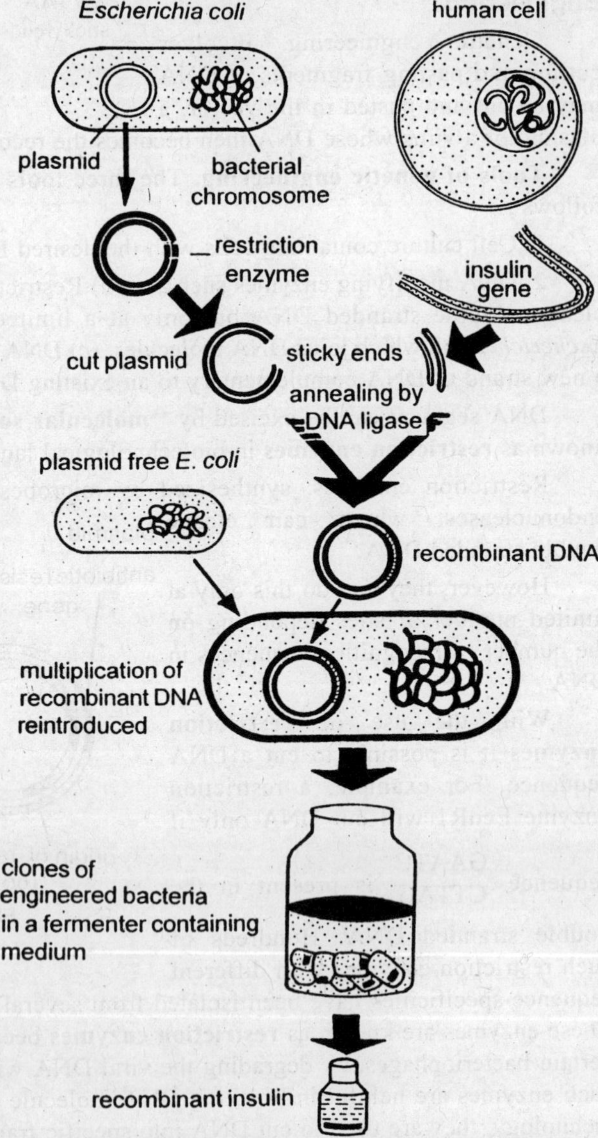

Fig. 8.3. Steps showing gene transfer for the production of human insulin.

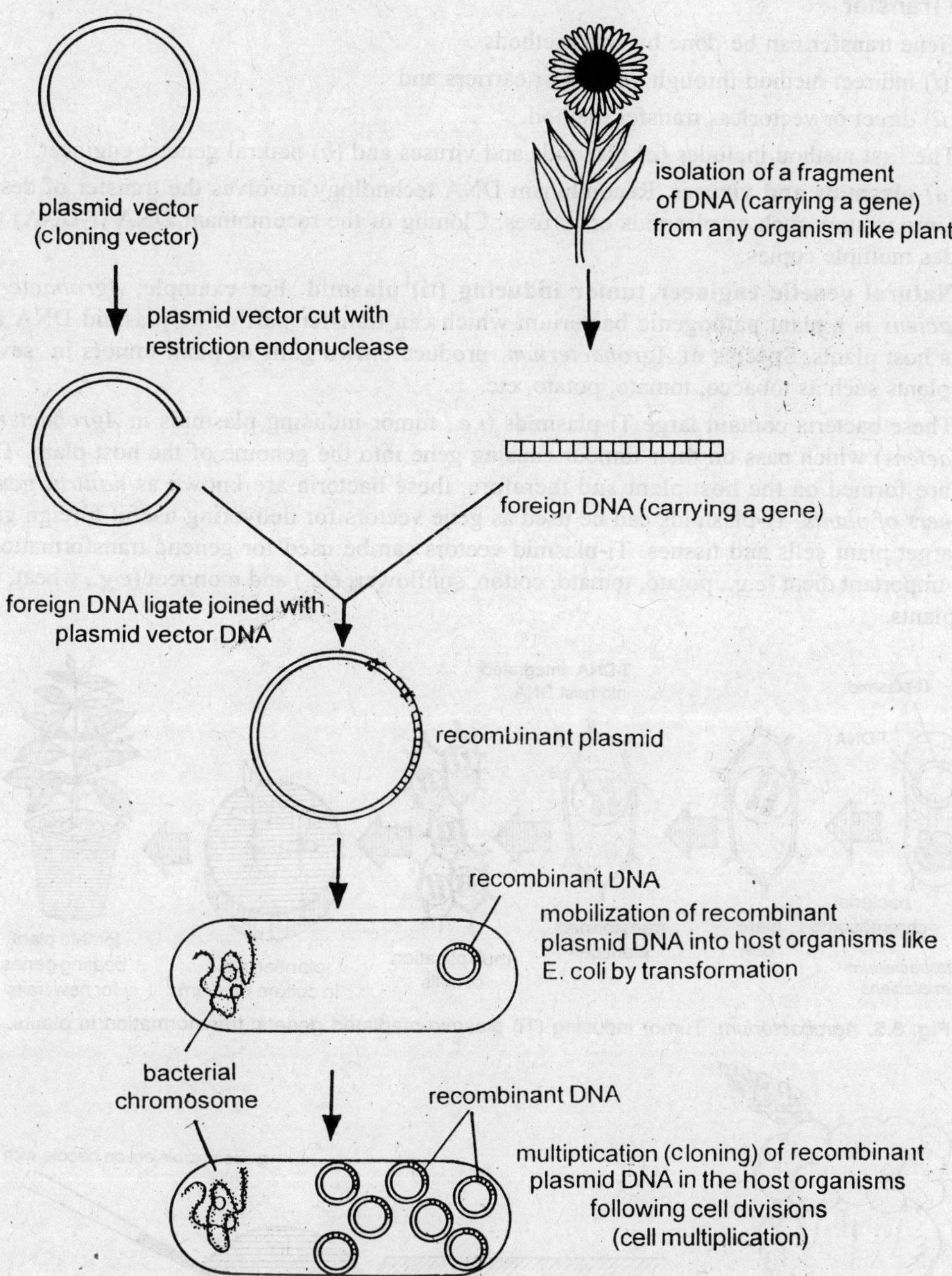

Fig. 8.4. Different steps involved in the cloning of a foreign DNA fragment into a plasmid vector (diagrammatic representation).

Genomic DNA library. A gene library contains several copies of many of the desired genes or DNA fragments. These may be present in bacterial clones with **recombination DNA** or by obtaining **complementary DNA** (c- DNA) by obtaining a DNA from an RNA template by using the enzyme reverse transcriptase. A collection of different c-DNA fragments is a c-DNA library. Gene libraries are maintained through special techniques.

Gene Transfer

Gene transfer can be done by two methods :

(*i*) indirect method through vectors or carriers and

(*ii*) direct or vectorless transfer method.

The first method includes (*a*) plasmids and viruses and (*b*) natural genetic engineer.

(*a*) **plasmids and viruses.** Recombinant DNA technology involves the transfer of desired genes into vectors such as plasmids or viruses. Cloning of the recombinant DNA (r-DNA) then provides multiple copies.

Natural genetic engineer tumor inducing (ti) plasmid. For example, *Agrobacterium tumefaciens* is a plant pathogenic bacterium which can transfer part of its plasmid DNA as it infects host plants. Species of *Agrobacterium* produce crown galls or plant tumors in several dicot plants such as tobacco, tomato, potato, etc.

These bacteria contain large Ti-plasmids (*i.e.,* tumor-inducing plasmids in *Agrobacterium tumefaciens*) which pass on their tumour causing gene into the genome of the host plant. Thus, galls are formed on the host plant and therefore, these bacteria are known as *natural genetic engineers of plants*. Ti-plasmids can be used as gene vectors for delivering useful foreign genes into target plant cells and tissues. Ti-plasmid vectors can be used for genetic transformation of many important dicot (*e.g.*, potato, tomato, cotton, sunflower, etc.) and monocot (*e.g.*, wheat, rice, etc.) plants.

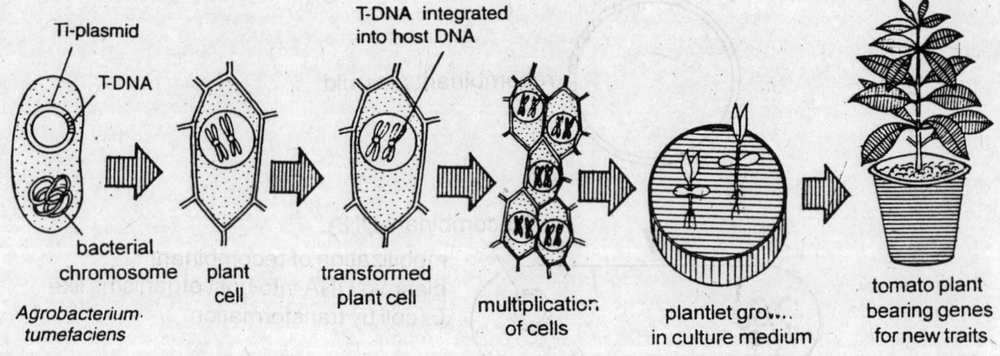

Fig. 8.5. *Agrobacterium.* Tumor inducing (Ti) plasmid-mediated genetic transformation in plants.

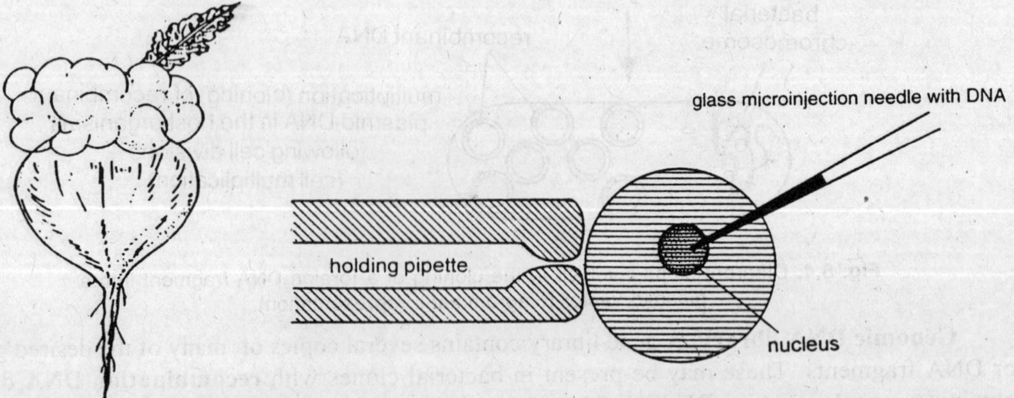

Fig. 8.6. Tumor inducing (Ti) plasmid. Crown caused on turnip by bacteria (*Agrobacterium tume-faciens*) carrying Ti plasmid.

Fig. 8.7. Microinjection. Micropipette is used for injecting DNA into the nucleus of a cell.

The **vectorless gene transfer** includes the following four important methods :

(*a*) Electroporation. In which temporary holes produced in plasma membrane of the cell mediate entry of foreign DNA.

(*b*) Chemical mediated gene transfer. In which certain chemicals (*e.g.*, polyethylene glycol) help in the uptake of foreign DNA into host cell.

(*c*) Microinjection. Introduction of foreign genes into plant or animal cells using micropipettes.

(*d*) Particle gun. In which Tungsten particles coated with foreign DNA are bombarded into target cells.

APPLICATIONS OF RECOMBINANT DNA TECHNOLOGY

The technique of recombinant DNA can be used as follows :

It can be used to elucidate molecular events in the biological process of cellular differentiation and ageing.

This can also be utilised in making gene maps with precision.

This can be used to spell out the complete nucleotide sequence of genome of various organisms including humans.

By this technique, useful chemical compounds can be produced cheaply and efficiently.

Vaccines for hepatitis B have been produced by recombinant DNA technology.

Application of Genetic Engineering

The most important application of genetic engineering is to unravel the structure and mechanism of expression of genes. For example the concept of exons and introns.

The operon concept of gene makes the genes active or inactive.

The introduction of specific genes into bacteria or yeast cells is helpful in synthesising human insulin, human growth hormones, etc.

It makes possible to introduce useful procaryotic genes into eucaryotes. For example, the possibility of the transfer of nitrogen fixing gene from bacteria to food crops develops there by enabling them to fix nitrogen. Thus, it reduces the need of nitrate fertilizers.

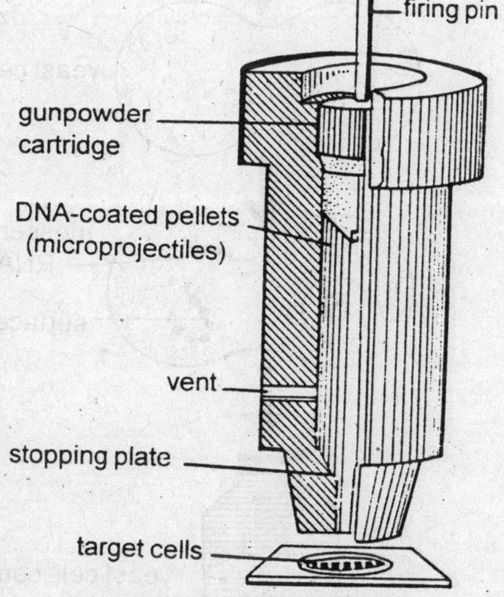

Fig. 8.8. Particle gene gun.

The introduction of genes coding for vitamins, antibiotics, hormones from higher animals to bacteria has helped to produce substances which are otherwise difficult to synthesise.

Genetic engineering has also helped man to use therapy for certain hereditary diseases, such as phenylketonuria and hemophilia.

However, there are some possibilities that the gene splicing are tempered with possible danger. For example, the introduction of new gene into bacteria may lead to the formation of new strain. The new strain may be virulent and may be responsible for epidemics.

A large number of drugs are becoming ineffective due to various microorganisms developing resistance due to uncontrolled recombinant DNA.

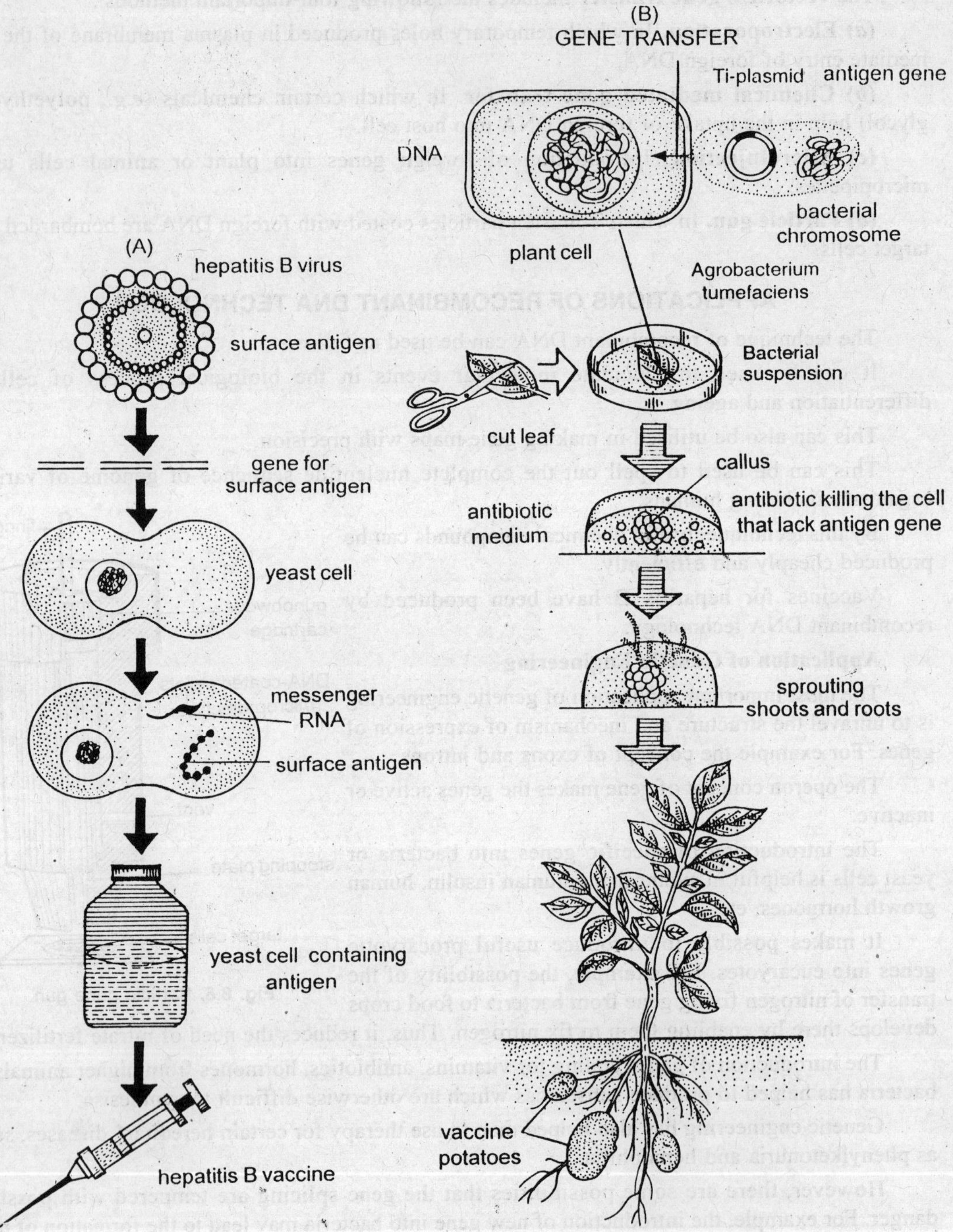

Fig. 8.9. Production of vaccines by rDNA technology. A, injectible hepatitis vaccine, B, edible vaccine.

The pathogenic microorganisms produced as a result of genetic engineering if used in biological warfare may be responsible for causing destruction of human race and other living beings.

Table 8.1. Application of Recombinant DNA products

Recombinant DNA products	Applications of r DNA products
1. Blood clotting factor VIII/IX	Replacement of clotting factor missing in patients of anaemophilia A / B.
2. Calcitonin	Treatment of rickets.
3. Chronic gonadotropin	Treatment of infertility.
4. Erythropiotin	Stimulation of the formation of RBCs for patients suffering from anemia during kidney dialysis or side effects of AIDS patients .
5. Human insulin	Treatment of diabetes.
6. Human growth hormone	Replacement of missing hormone in short stature people.
7. Interferon	Treatment of pathogenic viral infections, cancer.
8. Interleukins	Enhances action of immune system.
9 Platelet derived growth factor	Stimulation of healing of wounds.
10. Vaccines	Prevention of infectious diseases, such as hepatitis B, herpes, influenza, pertussis, meningitis, etc.

DIAGNOSIS OF HUMAN DISEASES

Recombinant DNA technology can be used in the diagnosis of several diseases.

Most of the tools and techniques are involved in the construction of probes, *i.e.*, short segments of single stranded DNA attached to a radioactive or fluorescent marker. Such probes are now used for identification of infectious agents, such as food poisoning *Salmonella*, pus forming *Staphylococcus*, hepatitis B virus, HIV, etc.

By testing DNA genetic disorder carrier parents, their genotype can be determined and predictions can be made about their afflicted children.

Gene Therapy

In humans, gene therapy is used to replace a faulty gene by a normal healthy functional gene.

The intensive investigation has been done in several human diseases, such as sickle cell anaemia caused by single mutations and on the other hand Severe Combined Immuno-Deficiency (SCID)

Example of SCID is mentioned here :

SCID is caused by a defect in the gene for the enzyme adenosine deaminase (ADA) in 25 per cent of the cases.

The patients having SCID do not have functioning T-lymphocytes and therefore, they cannot mount immune responses against invading pathogens.

The ideal approach for SCID treatment is to give the patient a functioning ADA that breaks down toxic biological products.

For this purpose, lymphocytes, a kind of white blood cells are extracted from the bone marrow of a SCID patient.

Now, a good copy of human gene encoding this enzyme (ADA) is introduced into these cells. This is done by exploiting the properties of special viruses called **retroviruses**, as vector.

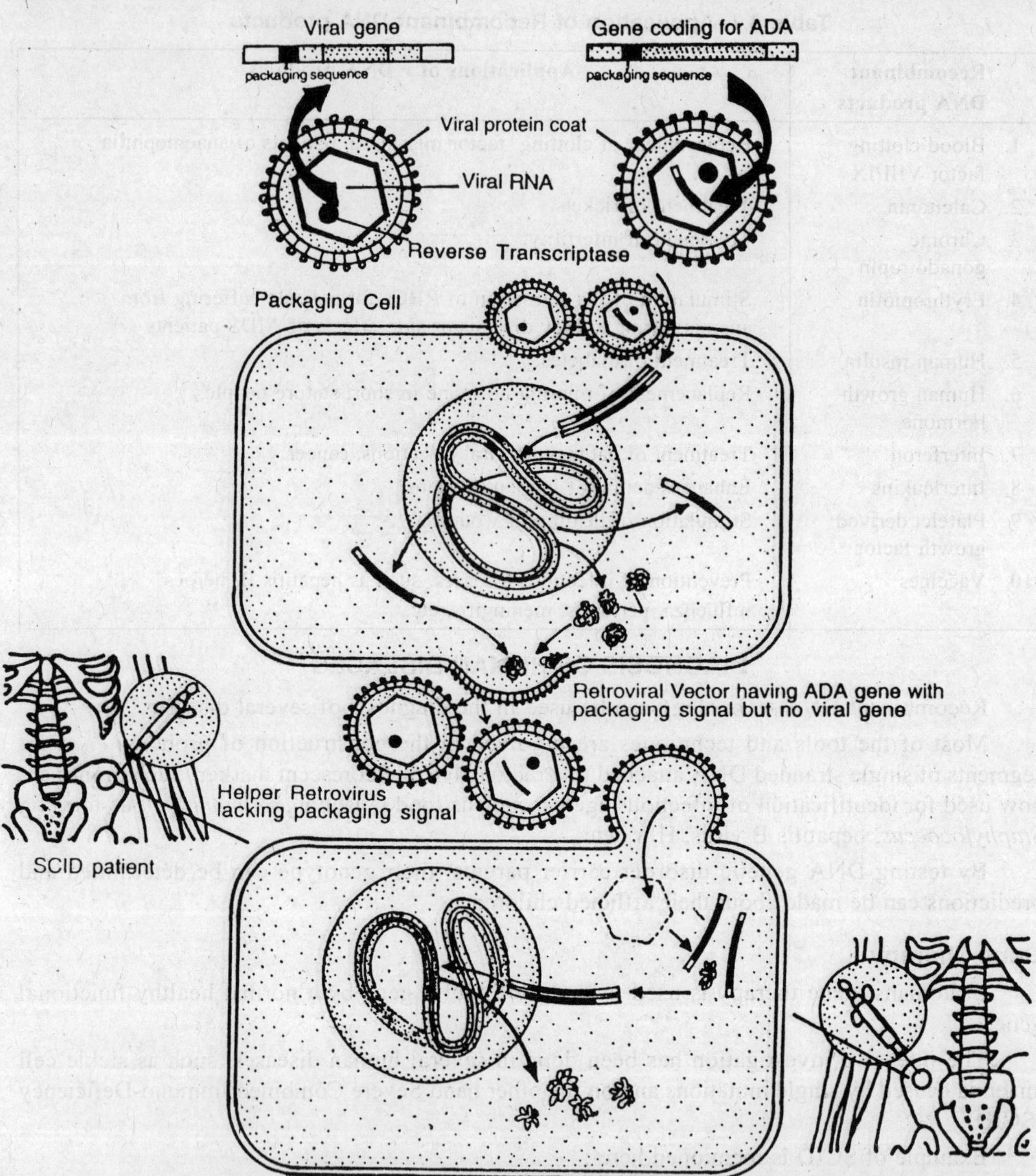

Fig. 8.10. Gene Therapy to treat Severe Combined Immuno-Deficiency (SCID).

This vector (retrovirus) is engineered to incorporate the ADA gene which has the packing sequence but no viral genes.

Now, co-infection of the bone marrow stem cells, removed from a SCID patient, with this vector and helper virus produces new virus particles with ADA gene inserted inside, but lacks the packing sequence.

This equips these cells with a good copy of the gene, in addition to the existing faulty copy of the gene.

Now, the treated cells are reinjected to the patient's bone marrow.

The lymphocytes subsequently produced by these bone marrow stem cells have a fully functional ADA gene. These cells can play their normal role in patient's immune system.

Such molecular surgical approach, is confronted by many practical challenges but even then this provides a ray of hope to the medical world.

CLONING

A **clone** may be defined as exact carbon copy or copies of a single parent. The word **clone** only has its meaning to living species.

The cloning of cells is a type of experiment of microsurgery in biological research programme. For example, using two microprobes, a biologist can carefully remove the nucleus from the egg cell of a frog with the same two probes and a micropipette, the biologist can also remove the nucleus from a normal body cell of another frog. The egg cell eventually develops into an exact copy of the frog from which the body cell was taken.

There are lots of clones in the world, for example, monozygotic identical twins are clones. They started off as a single cell in their mother's womb. The fertilized egg divided into two and each cell developed into a separate twin having the same genetic characteristics.

The most famous clones of the world are :

(*i*) **Dolly**, a sheep, recently died (*ii*) **Eve**, a human baby girl, born Thursday, December 26, 2002.

Dolly was produced from a single cell from her mother. She does not have a father, because no sperm was needed. She has exactly the same genetic characteristics as her mother, the single parent.

The birth of the first human clone, a baby girl called **Eve** by scientists was announced by Brigitte Boisselier, head of a company named Clonaid. According to Boisselier, the child is an exact genetic duplicate (clone) of her mother.

Microbial Cloning and Genetic Engineering

From very ancient times, it is known that a variety of microbes such as bacteria and yeast (fungus) are helpful in making a large number of products, *e.g.*, lactic acid, ethanol, vinegar, cheese, etc., as also of industrially useful substances like amylase, protease and life-saving antibiotics like penicillin and tetracycline. Cell proteins can also be produced by microbes (bacteria and fungi). Several improved strains of these microbes were developed by using mutants and other conventional techniques.

Today, gene cloning and genetic engineering techniques are generally used to improve the useful microbial strains for so many purposes.

(*a*) To produce useful compounds such as enzymes and vitamins on mass scale.

(*b*) They can be engineered to produce several pharmaceutically useful products such as human insulin (*e.g.*, in *Escherichia coli*), interferon, human growth hormone and viral vaccines.

(*c*) Several microbial strains are used in the various industries for performing many important functions such as the removal of undesired lignin (*e.g.*, *Trametes*).

(*d*) Microbes (*e.g.*, heavy-metal accumulating bacteria like *Pseudomonas*) can be used to purify polluted environment (called bioremediation).

(*e*) Certain microbes such as *Trichoderma* (fungus) and *Rhizobium* (bacterium) are used for bio-control (against plant diseases and pests) and as biofertilisers respectively.

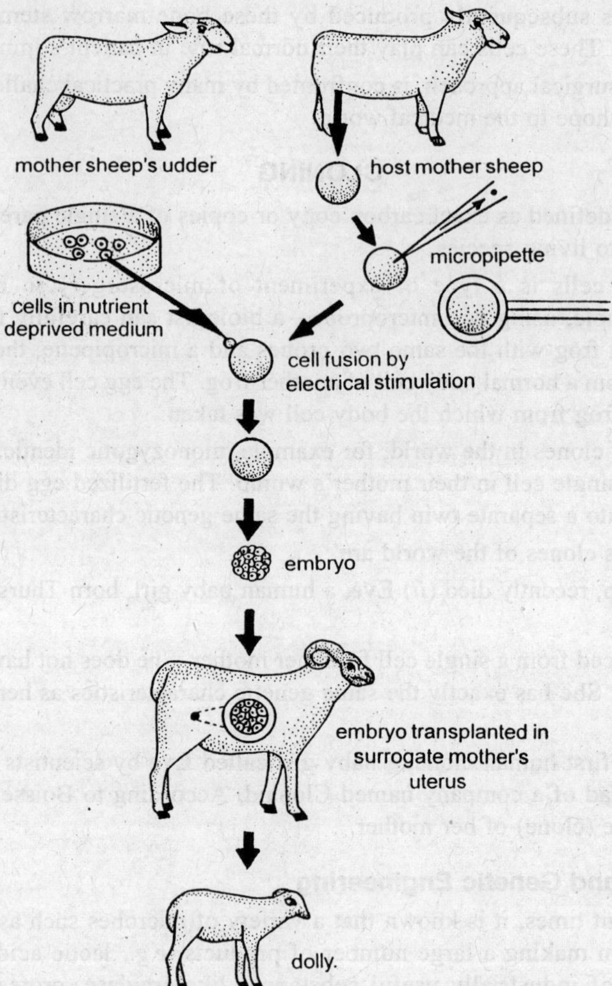

Fig. 8.11. Dolly. Steps in cloning of sheep.

Table. 8.2 Application of Genetically Engineered Microbes

Microbes	*Applications*
1. *Bacillus thuringiensis* (a soil bacterium)	Production of endotoxin (Bt toxin), a biodegradable insecticide for plant protection.
2. *Escherichia coli.* (a gut bacterium)	Production of human insulin, human growth factor interferons and interleukin.
3. *Pseudomonas fluorescence* (a bacterium)	Prevention of frost damage to the plants.
4. *Pseudomonas putida* (a bacterium)	Scavanging of oil spills. Digests hydrocarbons of crude oil.
5. *Rhizobium meliloti* (bacterium)	Nitrogen fixation by incorporating ''nif'' gene in cereal crops.
6. *Trichoderma* (a fungus)	Produces enzyme chitinase for biocontrol of fungal diseases in plants.

In the process of DNA cloning fragments of eukaryotic DNA are introduced into bacterial cells, which are then grown into individual colonies on a culture dish. As a cell proliferates to form a colony, the single DNA molecule taken in by the founder of the colony is repeatedly replicated by the enzymes present in the bacterial cells such that by the end of the growth period, the number of copies of the foreign sequence has been greatly amplified. Cloning, therefore, can be used as a technique to produce large quantities of a particular DNA sequence, but more important, it can be used as a technique to isolate, in an absolutely pure form, any specific DNA fragment among a large, heterogeneous population of DNA molecules.

Gene cloning. The object of the cloning technique is to introduce a copy of a eukaryotic DNA fragment into a recipient bacterial cell under conditions in which it will replicate autonomously from that of the host chromosome. In order to get this objective, the eukaryotic DNA must be linked to the DNA of a **vector** that normally replicates within a bacterial environment. The most common vector employed in cloning procedures is the bacterial **plasmid**. The use of plasmids is as follows: Plasmids are nonessential extrachromosomal DNA molecules present in many bacteria. Like the main bacterial chromosome, they are double-stranded circles, but they are much smaller and of variable genetic constitution. Since the plasmids and chromosomal DNAs have very different physical properties, they can be readily separated from one another.

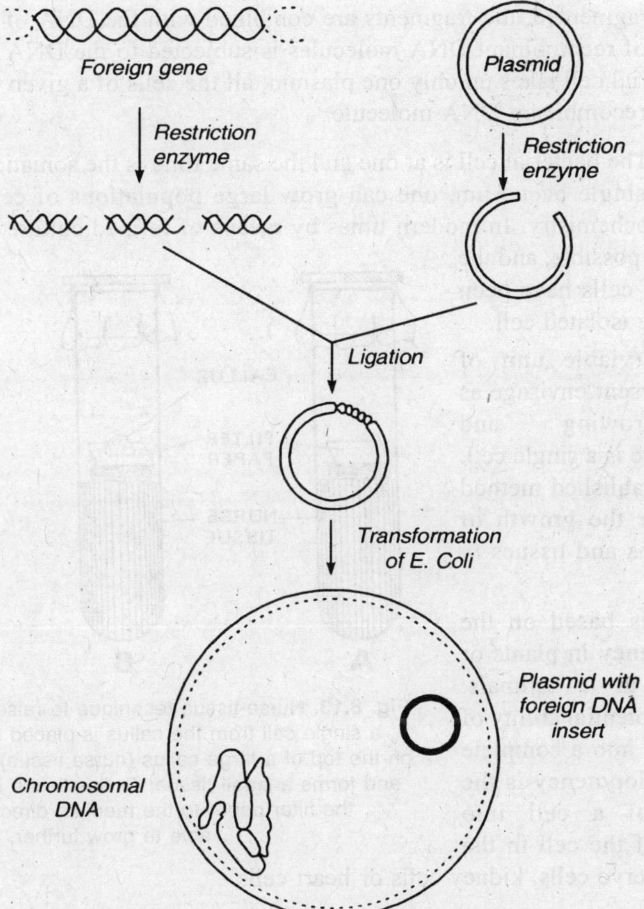

Fig. 8.12. Plasmid. Vectors and transformation.

The foreign DNA to be cloned is inserted into the plasmid to form the recombinant DNA molecule. This modified plasmid is then added to a bacterial culture as naked DNA, and in the presence of $CaCl_2$, a small percentage of the bacteria will pick up one of these molecules from the medium. Once a plasmid is taken up, it will replicate autonomously within the recipient and will be passed on to its progeny during cell division. A wide variety of plasmids have been isolated, many of which carry genes for resistance to antibiotics. If these plasmids are added to a bacterial culture and the cells are then treated with specific antibiotics, only these bacteria which have taken up the plasmid will be able to survive.

How is the DNA of the vector combined with that form the eukaryotic genome? Recombinant DNA molecules can be formed in several ways.

Recombinant DNAs can be constructed between DNA fragments having blunt ends. This can be done by enzymatically adding a string of A- or C- bearing nucleotides to one strand of the fragment and a complementary string of T- or G- bearing nucleotides to the opposite strand of the vector, mixing the two populations, and sealing the molecules with the ligase.

Cloning genomic fragments. When working with genomic DNA, one has to isolate a particular gene or genes, such as those which code for the histones, from among the thousands of unrelated sequences. The first step in this type of experiment is to fragment the DNA with restriction endonucleases. These enzymes recognize sequences from 4 to 6 nucleotides in length. Once the DNA is fragmented, the fragments are combined with the DNA of the vector, and the diverse population of recombinant DNA molecules is subjected to the DNA cloning procedure. Since a given bacterial cell takes up only one plasmid, all the cells of a given colony will contain copies of the same recombinant DNA molecule.

Cell cloning. The bacterial cell is at one and the same time is the somatic and the germ cell. Having selected a single bacterium, one can grow large populations of cells, all identical in morphology and biochemistry. In modern times by means of refined culture techniques the cell cloning has become possible, and the pure populations of cells have been grown from a single isolated cell.

The smallest viable unit of plant one can at present envisage as reproducing, growing and developing in culture is a single cell. Several workers established method which would allow the growth of isolated plant organs and tissues *in vitro*, in culture.

Cell cloning is based on the property of **totipotency** in plants or **pluriopotency** in animals. Totipotency is the potential ability of a plant cell to grow into a complete plant, whereas pluriopotency is the potential ability of a cell into develop any type of the cell in the animal body, *e.g.*, nerve cells, kidney cells or heart cells.

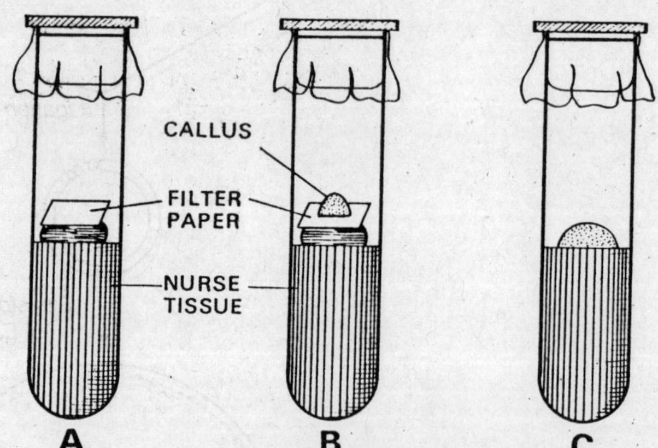

Fig. 8.13. Nurse-tissue technique to raise single-cell clones. A, a single cell from the callus is placed on filter paper lying on the top of a large callus (nurse tissue); B, the cell divides and forms a small tissue; C, this tissue is transferred from the filter paper to the medium directly where it is able to grow further.

Virtually all plants are totipotent but in animals only fertilized egg and stem cells in the embryonic blastocyst are totipotent.

Plant cloning. Cellular totipotency is the capacity of mature cells showing that when freed from the plant body, they have the ability to reorganize themselves into the new individuals. Steward and his co-workers, showed this phenomenon in the carrot cultures. Here the small pieces of mature carrot root, were grown in a liquid medium supplemented with coconut milk in special containers. These cultures were shaken generally which freed all the cell clusters into the medium. Some of them developed into rooting clumps. When these were transferred to the tubes containing a semi-solid medium, they gave rise to whole plant that flowered and set seeds.

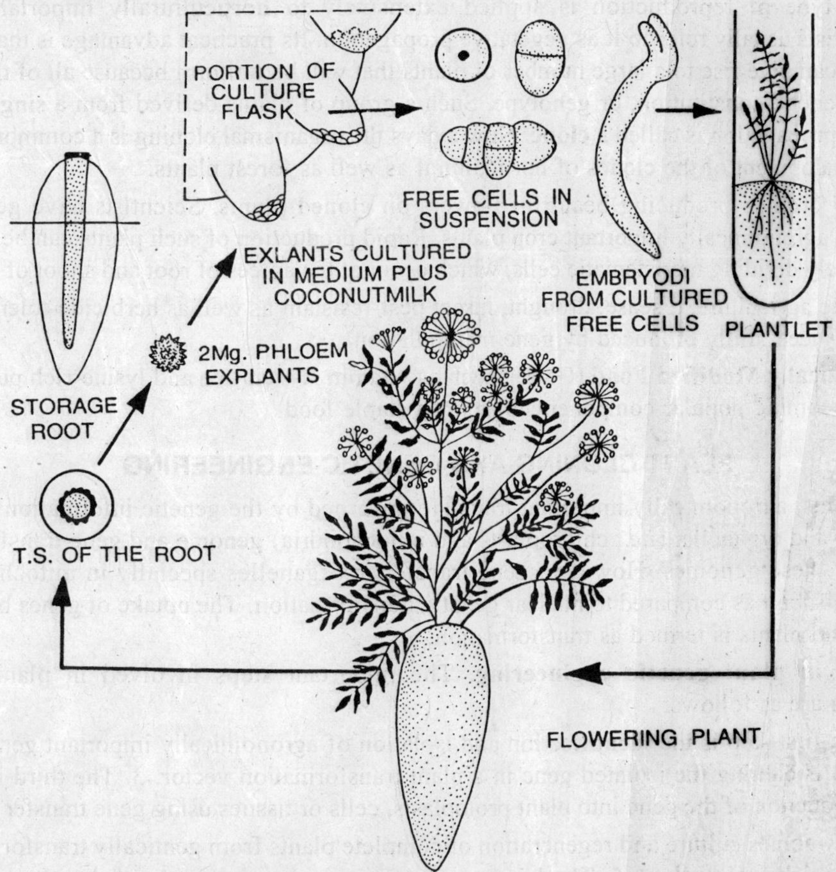

Fig. 8.14. An experiment to demonstrate the totipotency of plant cells. Slices of the carrot root were cut and small pieces of tissue were taken from the phloem region. These were inoculated into a liquid medium in special flasks, the tissue grew actively and single cells and small cell aggregates dissociated into the medium. Some of the cell clumps developed roots, and when transferred to a semisolid medium, these rooted nodules formed shoots. These plants could be transferred to soil.

Organismal cloning. It is characteristic of vegetative reproduction that a vegetative part, when removed from the plant and placed under suitable conditions, will replace the missing parts and produce a complete individual. It has been possible to grow several types of plant organs in sterile culture including roots, shoot apices, leaves, floral parts and fruits. The nutrient requirements for such organ culture vary considerably from species to species and according to the type of organ in question but certain general requirements can be recognised

Intact higher plants are autotrophic, that is, they are able to synthesize all the organic substances required for their own life from water, carbon dioxide, oxygen and mineral nutrients. But since most sterile cultures are unable to carry on photosynthesis, it is clear that they will require atleast a carbon source, usually supplied in the form of a sugar as sucrose or glucose. In addition, sterile cultures require the same mineral nutrients as the intact plant, including both macronutrients (*e.g.*, nitrogen, phosphorus, potassium and calcium) and micronutrients or trace elements (*e.g.*, Mg, Fe, Mn, Zn, etc.). In addition to the requirements for a carbon source and mineral nutrients, it is found that most isolated organs have also a requirement for certain special organic substances, such as vitamins, nicotinic acid, etc.

This type of reproduction is applied extensively to horticulturally important plants. Horticulturists usually refer to it as vegetative propagation. Its practical advantage is that a single individual can give rise to a large number of plants that will be uniform because all of them have the same genetic constitution, or genotype. Such a group of plants derived from a single one by vegetative propagation is called a **clone**. Now a days the organismal cloning is a common practice for the development of the clones of horticultural as well as forest plants.

Many orchids producing beautiful flowers on cloned plants. Scientists have genetically engineered agronomically important crop plants. Rapid production of such plants can be achieved using actively dividing meristematic cells, which occur at the apices of root and shoot of the plant.

For the agriculture, disease, drought, insect pest-resistant as well as herbicide-tolerant crops have been successfully produced by gene manipulation.

Genetically Modified Food (GMF), such as vitamin A-rich rice and lysine-rich pulse seeds are now becoming popular components of human staple food.

PLANT CLONING AND GENETIC ENGINEERING

In plants, agronomically important traits are governed by the genetic information stored in the nuclear and organellar (*i.e.*, chloroplasts and mitochondria) genome and gene transfer can be done in all these genomes. However, gene transfer in organelles specially in mitochondria is relatively difficult as compared to nuclear genetic transformation. The uptake of genes by cells in microbes and plants is termed as transformation.

Steps in plant genetic engineering. The important steps involved in plant genetic engineering are as follows:

1. The first step is the identification and isolation of agronomically important gene. 2. The second step is cloning the isolated gene in a plant transformation vector. 3. The third important step is introduction of the gene into plant protoplasts, cells or tissues using gene transfer methods.

4. Now comes culture and regeneration of complete plants from gentically transformed cells on suitable selection medium. 5. The last important step is demonstration of the integration and expression of foreign gene in the transgenic plants by using molecular techniques.

Applications of plant genetic engineering. Till now, the plant genetic engineering has been successful in producing disease and pest resistant as well as herbicide tolerant transgenic plants. For example, in tomato the slow ripening process of the fruits is developed; in soybean, corn and cotton herbicide tolerance is developed; in potato, tomato, tobacco and rice viral resistance is developed.

ANIMAL CLONING AND GENETIC ENGINEERING

Animal cloning. Animal cloning is more difficult than plant cloning because animal cells lose their totipotency on reaching the gastrula stage of animal development. However, animal tissue cultures from tumours and embryonic tissue cells have been successful. Standard techniques are available for isolating animal cells and tissues from different systems. Some more important

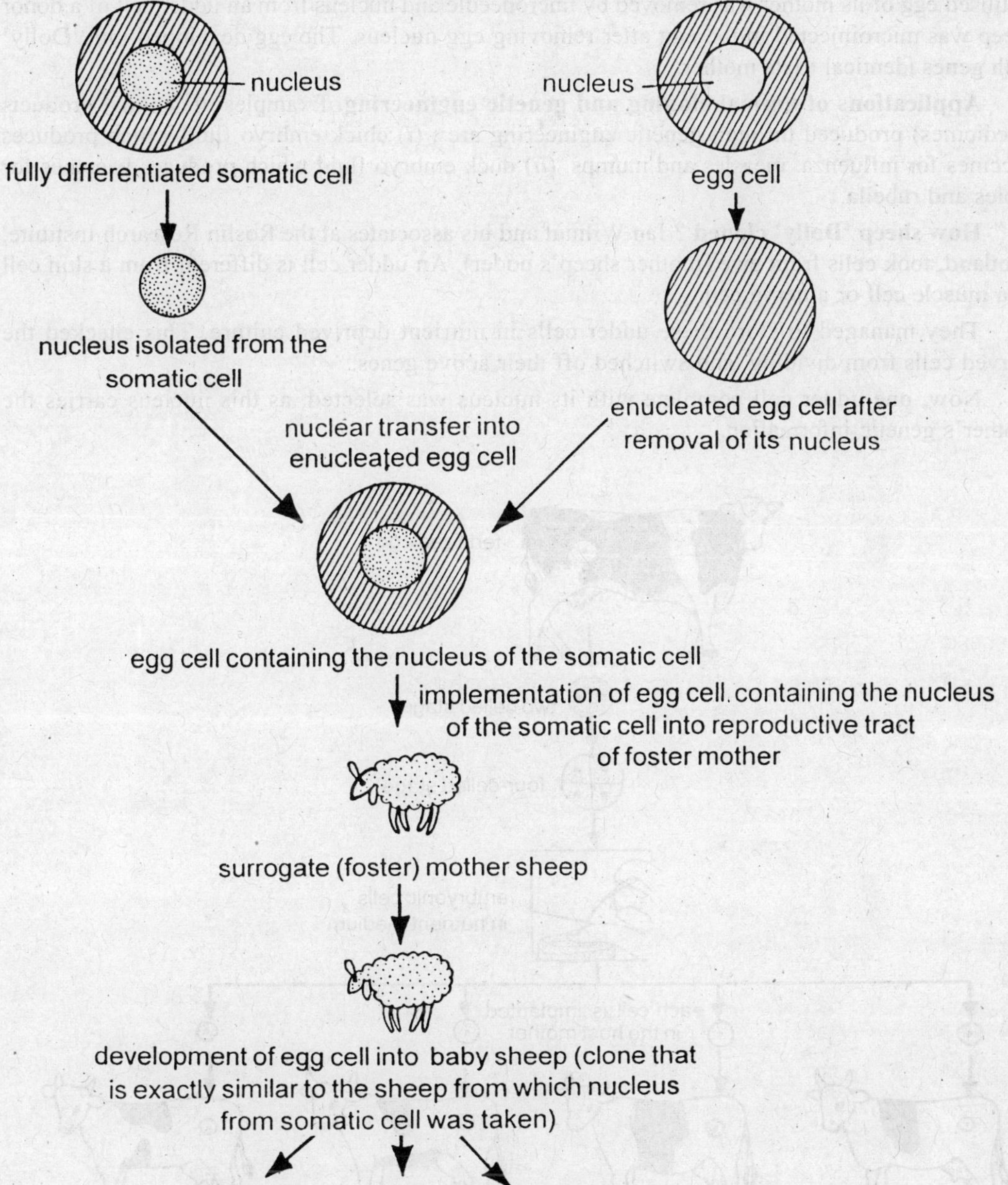

Fig. 8.15. Steps involved in the cloning of the sheep **Dolly.**

examples of animal cloning are tissue culture, somatic cell fusion, cell cloning and creating transgenics.

Gene transfer in animals. Gene transfer in animals is mostly through direct methods such as electroporation or microinjection or using particle gun. In creating 'Dolly' the cloned sheep, fertilised egg of its mother was removed by microneedle and nucleus from an udder cell of a donor sheep was microinjected in the egg after removing egg nucleus. The egg developed into 'Dolly' with genes identical to its mother.

Applications of animal cloning and genetic engineering. Examples of animal products (medicines) produced through genetic engineering are : (*i*) chick embryo fluid which produces vaccines for influenza, measles and mumps. (*ii*) duck embryo fluid which produces vaccines for rabies and rubella.

How sheep 'Dolly' cloned ? Ian Wilmut and his associates at the Roslin Research Institute, Scotland, took cells from ewe (mother sheep's udder). An udder cell is different from a skin cell or a muscle cell or a nerve cell.

They managed to store these udder cells in nutrient deprived culture. This checked the starved cells from dividing, and switched off their active genes.

Now, one udder cell complete with its nucleus was selected, as this nucleus carries the mother's genetic information.

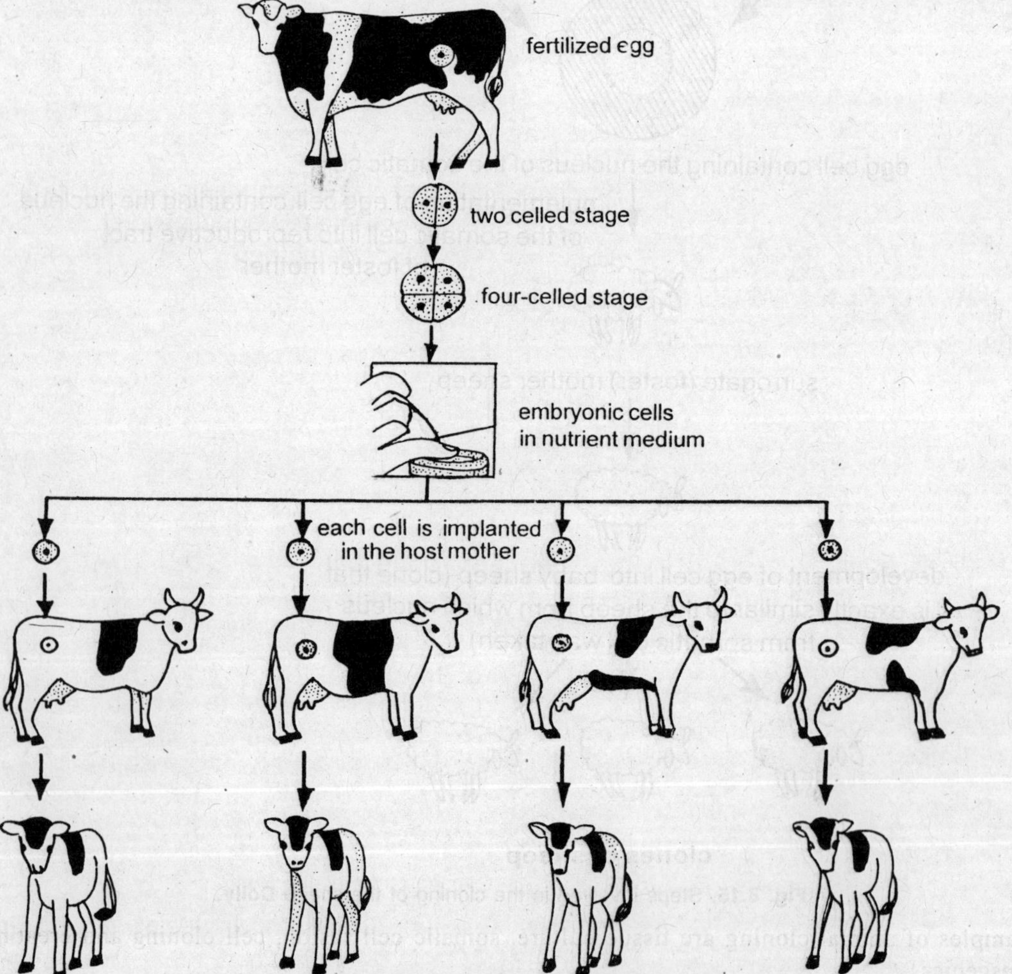

Fig. 8.16. Japanese method of cloning cattle.

Meanwhile, unfertilized egg cell was taken from a different ewe (host mother sheep). Its nucleus was sucked out leaving an empty cell containing all the necessary components to produce an embryo. This cell was now ready to receive udder cell nucleus.

They now fused udder cell nucleus with the empty egg cell by electrical stimulation. Then this egg cell had the mother's nucleus.

When a normal or altered egg is implanted in a different female is termed **'surrogate mother'**. This means, substitute mother.

Then the altered egg was cultured for six days. Out of many resulting embryos, one was implanted in the uterus of the surrogate mother, where it grew into a lamb.

Thus, **Dolly** was born genetically identical to mother sheep as her first cell nucleus came from mother's cell.

How calves are cloned in Japan ? Scientists from Japan have cloned cattle in a different way.

They have got success in growing as many as eight identical calves from one fertilized cell of their mother. The process is as follows :

When the mother cow has mated with the bull, she has a fertilized egg in her womb.

Now this zygote divides in two and then in four and then in eight. This embryo is carefully removed from the womb, and the embryonic cells are separated using as enzyme.

Each isolated cell is kept in a nutrient medium and later implanted in the womb of a different 'host mother' cow.

The host mother's womb must accept the cell and make it grow. Each cell may grow into a normal baby calf, if all goes well.

First human clone. The birth of the first human clone (December 26, 2002), a baby girl called **Eve** by scientists, was announced by Brigitte Boisselier, head of a company named Clonaid. According to Boisselier, the child is an exact genetic duplicate of her mother.

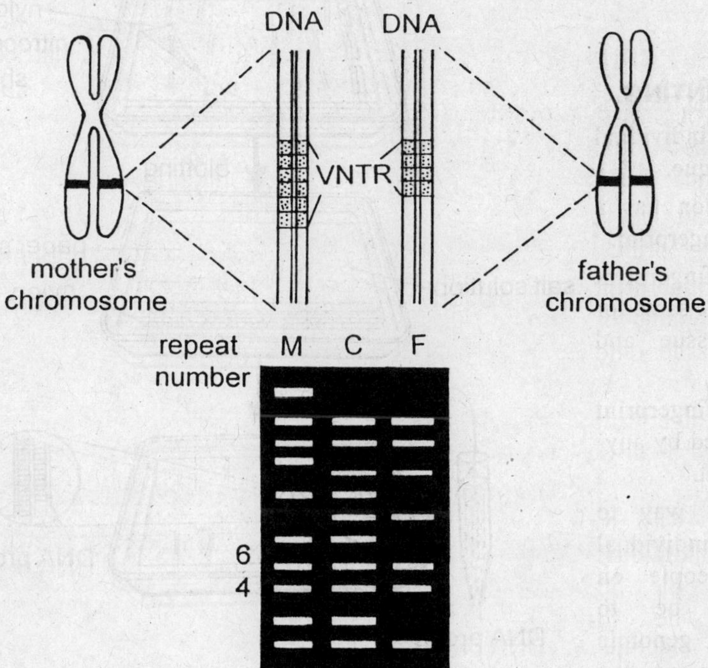

Fig. 8.17. Variable Number Tandem Repeats (VNTRs). M = mother C = child, F = father.

To clone, scientists slip the nucleus of an adult cell, like a skin cell, into an unfertilized egg from which its own genetic material has been removed.

Then, if stimulated to act like a fertilised egg, the newly altered genetic material can then direct the egg to divide and grow into an embryo, then a foetus, then a newborn, if all goes well. Yet, as scientists have discovered, only rarely does all go well.

In animal work so far only about 1 to 5 per cent of cloning attempts succeed, said Randall Prather, a cloning expert. That is, for every 100 eggs, one to five clones are born.

DNA FINGERPRINTING

Every individual organism is unique.

Each person has a unique DNA fingerprint.

A DNA fingerprint of a person is the same for every cell, tissue and organ.

DNA fingerprint cannot be altered by any known treatment.

The ideal way to distinguish an individual from other people on Earth would be to describe entire genomic DNA sequence of the person.

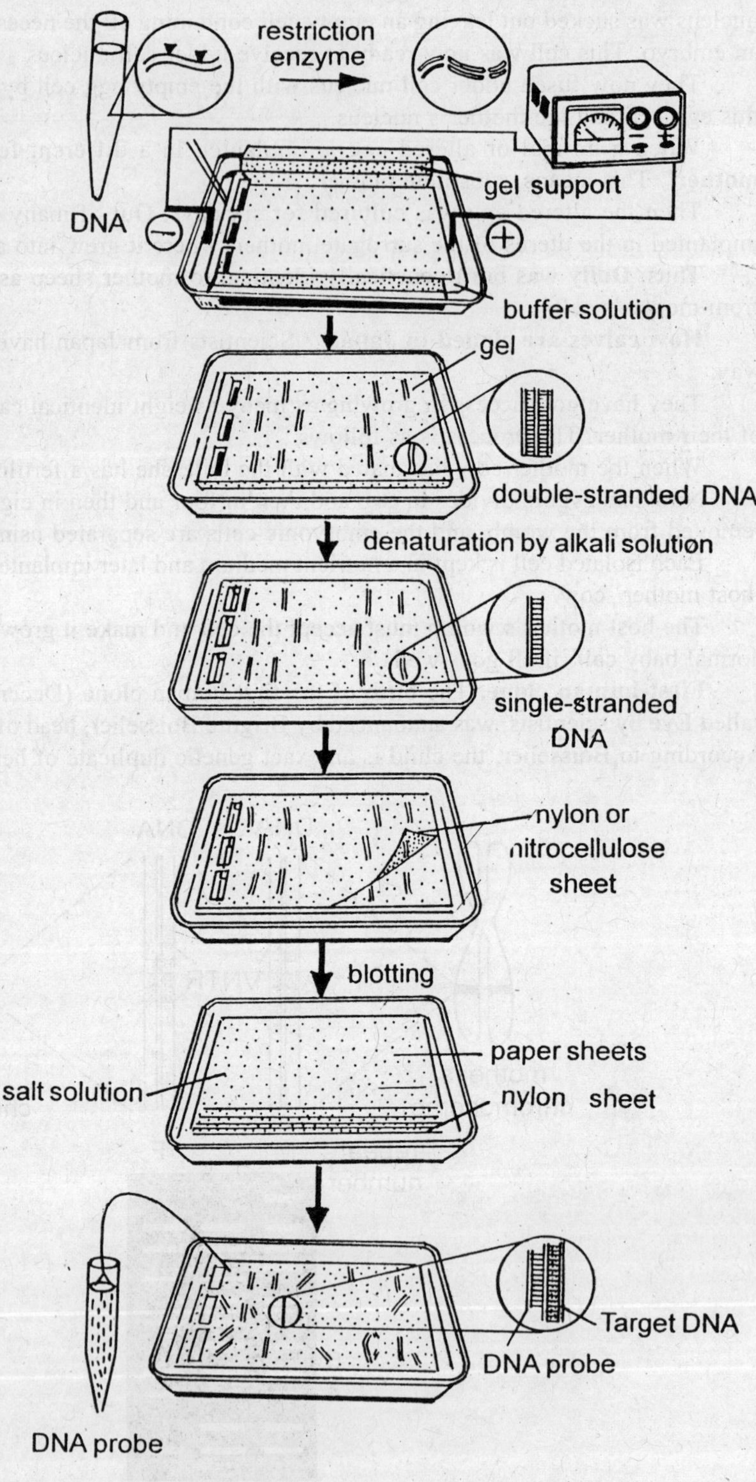

Fig. 8.18. Steps in technique of DNA fingerprinting.

The technique of DNA fingerprinting was used for the first time by English geneticist Alec Jeffreys, in 1984.

Important for DNA fingerprinting are short nucleotide repeats which vary in number from person to person and are inherited.

They are called **Variable Number Tandem Repeats** or **VNTRs.**

The VNTRs of two persons may be of the same length and sequence at certain sites, but vary at other sites.

For example, a child might inherit a chromosome with six tandem repeats from the mother and the same tandem repeated four times in the homologous chromosome inherited from the father.

However, the half of VNTR alleles of the child resemble that of the mother and half that of the father.

DNA Fingerprinting

This is also called **DNA typing** or **DNA profiling.** This is method of ascertaining relationships and the identity of a person by means of DNA fingerprint which is unique to each individual. The DNA fingerprint consists of the patterns of DNA fragment obtained in restriction analysis of certain highly variable repeated DNA fragments within the genome whose number and arrangement are unique for each person.

For example, two court cases are undermentioned : 1. Mr. X and Mr. Y both claim that Wilson is their son and want to take him away. 2. In another case a murder is committed and an innocent person is being implicated.

How can these cases be decided? Here DNA fingerprinting helps to solve these cases scientifically. In the first case the DNA of Mr. X and Mr. Y can be compared with that of Wilson and in the second case the blood stains or any hair found at the scene of crime can be tested for the identification of the murderer. Nowadays DNA fingerprinting has become a useful technique in forensic science to reach at correct decisions.

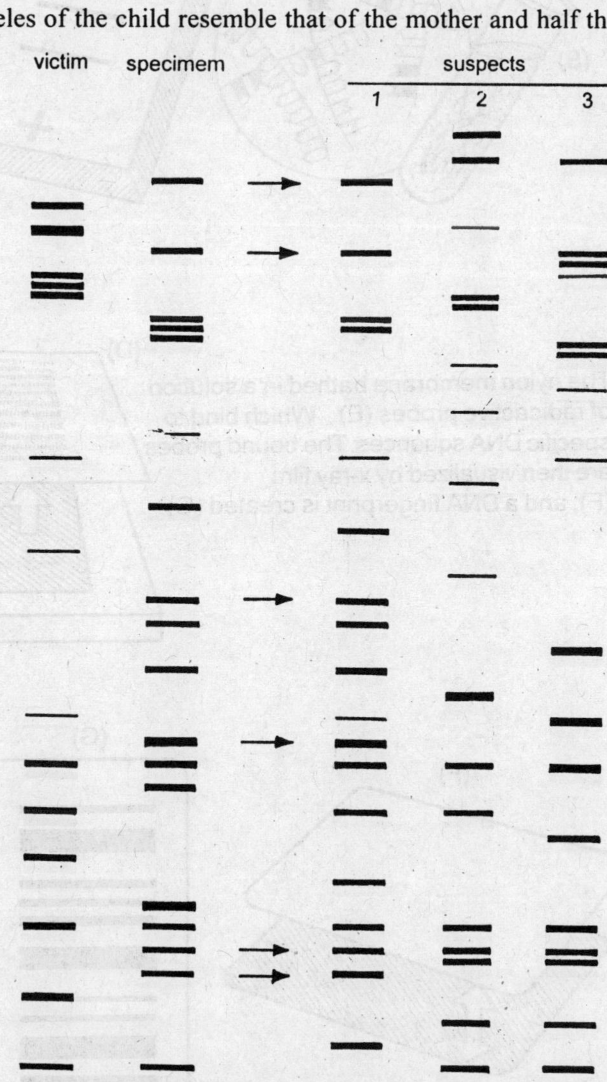

Fig. 8.19. DNA Fingerprints from a Murder Case (somewhat schematic). Here the DNA fingerprint of the specimen (evidence such as blood stain which does not belong to the victim but to the suspected murderer) obtained from the crime-scene, appears to match with that of suspect No. 1.

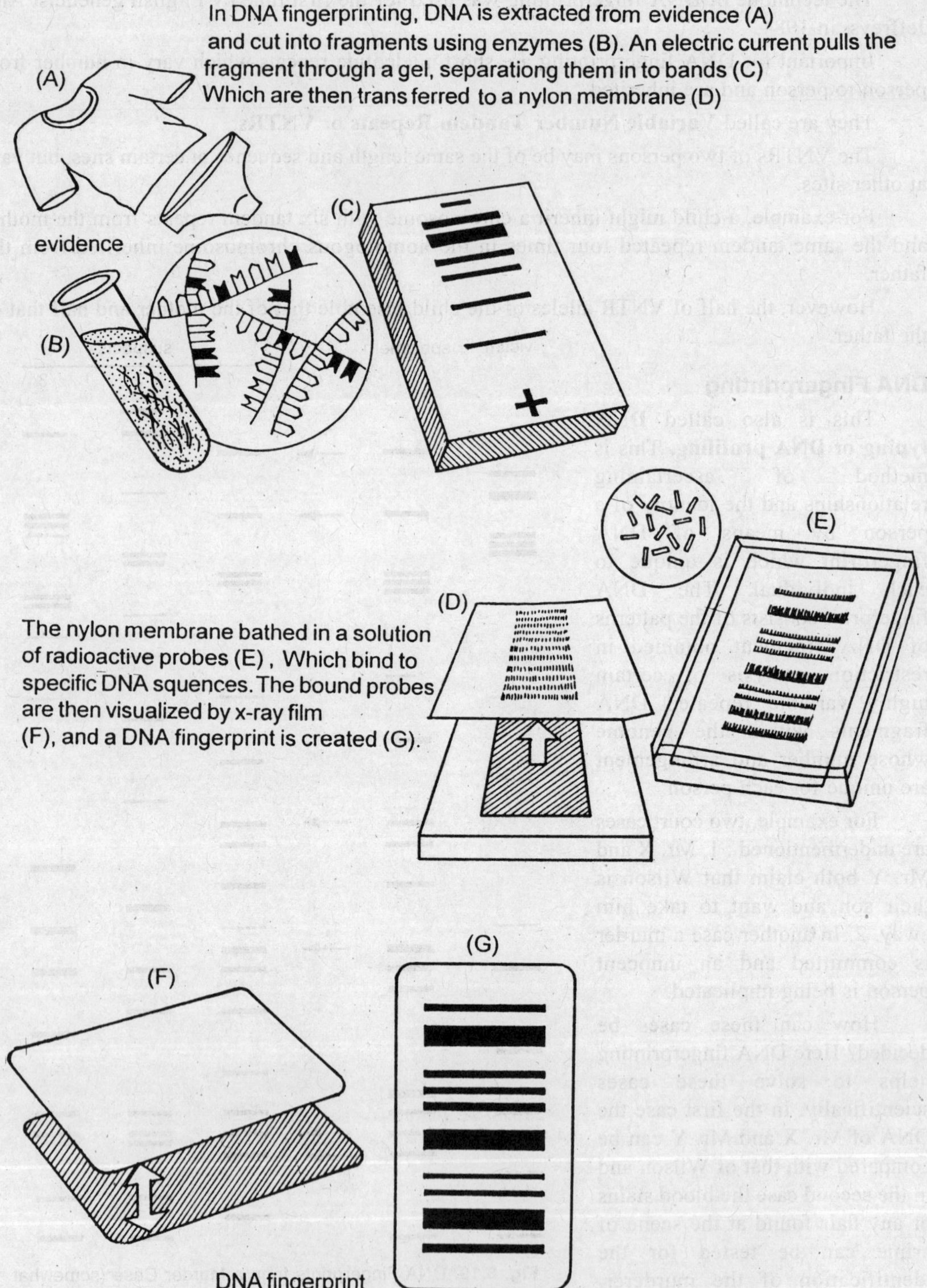

Fig. 8.20. DNA fingerprinting. How technique helps us to settle criminal (murder) cases.

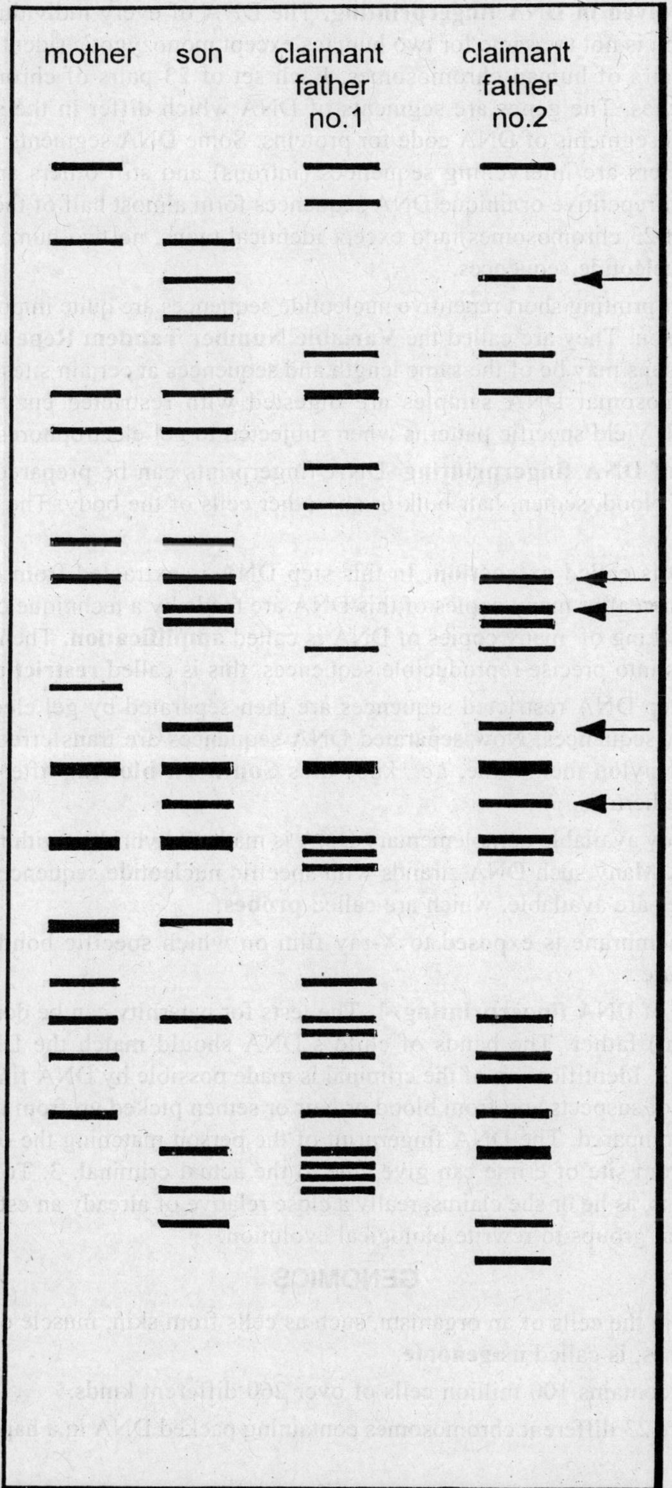

Fig. 8.21. DNA Fingerprints of mother under consideration, the child and two claimant "fathers" (somewhat schematic). Claimant No. 2 is proved to be the real (biological) father.

Principle involved in DNA fingerprinting. The DNA of every individual has distinctive characteristics which is not the same for two humans except monozygotic (identical) twins. There are twenty three pairs of human chromosomes. Each set of 23 pairs of chromosomes has 1.5 million pairs of genes. The genes are segments of DNA which differ in the sequence of their nucleotides. Not all segments of DNA code for proteins. Some DNA segments have a regulatory function, while others are intervening sequences (introns) and still others are repeated DNA sequences. The non-repetitive or unique DNA sequences form almost half of the haploid genome (DNA in one set of 23 chromosomes) and except identical twins, no two humans have genomes with same DNA nucleotide sequences.

In DNA finger printing short repetitive nucleotide sequences are quite important as they vary from person to person. They are called the **Variable Number Tandem Repeats (VNTRs).** The VNTRs of two persons may be of the same length and sequences at certain sites and vary at other sites. When chromosomal DNA samples are digested with restricted enzymes, DNA from different individuals yield specific patterns when subjected to gel electrophoresis.

Mechanism of DNA fingerprinting. DNA fingerprints can be prepared from extremely minute amounts of blood, semen, hair bulk or any other cells of the body. The main steps are as follows.

The first step is called **extraction.** In this step DNA is extracted from any of the above mentioned cells. Thereafter many copies of this DNA are made by a technique called Polymerase Chain Reaction. Making of many copies of DNA is called **amplification.** Then DNA is cut with restriction enzymes into precise reproducible sequences, this is called **restriction digestion.**

In the next step DNA restricted sequences are then separated by gel electrophoresis, *i.e.*, separation of DNA sequences. Now separated DNA sequences are transferred from gel onto a trinitrocellulose or nylon membrane, *i.e.*, known as **Southern blotting** after the name of the inventor E. M. Southern.

Then an already available complementary DNA is made to hybridise with the isolated DNA, *i.e.*, **hybridisation.** Many such DNA strands with specific nucleotide sequences complementary to VNTR sequences are available, which are called **probes.**

Lastly, the membrane is exposed to X-ray film on which specific bonds can be seen as depicted in the figure.

Applications of DNA fingerprinting. 1. The tests for paternity can be done for identifying the true (biological) father. The bands of child's DNA should match the DNA prints of the biological parents. 2. Identification of the criminal is made possible by DNA fingerprinting. Here DNA finger prints of suspects and from blood or hair or semen picked up from the scene of crime are prepared and compared. The DNA fingerprint of the person matching the one obtained from sample obtained from site of crime can give clue to the actual criminal. 3. To verify whether a hopeful immigrant is, as he or she claims, really a close relative of already an established resident. 4. To identify social groups to rewrite biological evolution.

GENOMICS

All the DNA in the cells of an organism, such as cells from skin, muscle or brain and every thing including genes, is called its **genome**.

Human body contains 100 million cells of over 260 different kinds.

In all, there are 23 different chromosomes containing packed DNA in a haploid set of human genome.

Additional DNA is in cell's mitochondria which is inherited from one's mother.

In human genome, according to estimates there are about three billion of biochemical letters, *i.e.*, nucleotide base pairs (in human genome), and 30,000 genes per cell.

Human Genome Project

This project began in 1990. The main aims of the project are as follows :

1. To develop the ways of mapping the human genome.

2. To store this information in databases and to develop tools for data analysis.

3. To address the ethical, legal and social issues that may arise from this ambitious human genome project.

Fig. 8.22. Human genome.

Genomic Revelations

The human gene count is approximately 30,000 to 40,000. This count is nearabout the same in mice.

Nine-tenths of human genes are identical to that of the mouse which is generally used for experimental purposes.

The fruit fly (*Drosophila melanogaster*) has about 13,000 genes per cell.

Escherichia coli has about 4,000 genes.

Humans are incredibly 99.9 per cent identical with *E. coli* at the DNA level. In other words, genomically humans and bacteria are similar.

However, most of human differences are shared among all ethnicities and races spread into the geographic regions around the planet.

Different human genes vary widely in the length often over thousands of base pairs.

White β -globin and insulin gene are less than 10 kilo base pairs.

The gene responsible for Duchenne Muscular Dystrophy on X chromosome is made up of 2400 kilo base pairs, most probably the longest gene known.

However, a lily has 18 times more DNA than a human, *i.e.,* 106 billion base pairs per cell.

Such large amounts of genes are split by long non-coding introns. Only less than 2 per cent of the genome is known to include the exons, the protein coding sequences. In fact, a lily produces fewer proteins than a human.

As already studied, clusters of VNTRs of variable lengths, arranged in tandem.

Approximately one million copies of short 5 to 8 base pair repeated sequence clustered around the centromeres and near the ends of the chromosomes represent, the **junk DNA**.

Prospects and Implications of Human Genome

It is expected and seems possible that we will soon get snapshots of more than 1200 genes that are responsible for common cardiovascular ailments, endocrine diseases like diabetes, neurological disorders like Alzheimer's disease, deadly cancers and many other such diseases.

Even single gene defects may account for several inherited diseases.

Efforts are in progress to determine genes that will revert cancerous cells to normal.

The human genome will definitely make a halthier living.

The human genome also holds the prospects of vast database of knowledge about designer drugs, genetically modified diets and genetic identity.

Human Genome Revelations

The year 2003 marks fifty years of the discovery of the double-helix of DNA, twenty five years of the world's first test-tube baby and now the complete mapping of the human genome.

The Human Genome Project that commenced in 1990 under a US-UK led public consortium of academic centres headed by Francis Collins, has now announced that the map, valued at nearly $800 million, is complete two years ahead of schedule.

Three draft maps of human genome have already made public, and this is the final one, as the scientists say. This however, does not mean the end of genome mapping as the work will shift to other species.

Also, variations in the human genome will have to be studied and then there is the whole task of analysing the tomes of data generated by the map. For, it is in data-crunching that applications can be formulated.

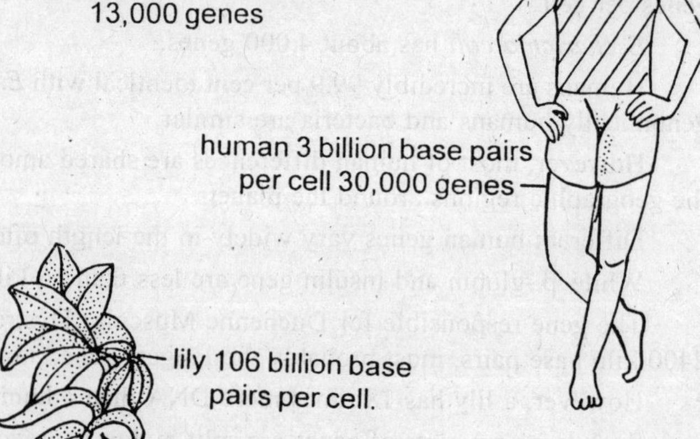

Fig. 8.23. Comparative genomes.

It all began with the discovery of the structure of DNA by James D. Watson and Francis Crick. They made their stunning discovery public in an article that first appeared in **Nature** magazine on April 25, 1953.

Since then, biology, has got transformed into a dynamic field of study, generating whole new subjects : **genetics, biotechnology, proteomics, pharmacogenomics, bio-informatics** and **chem-informatics**, among others.

The human genome map will serve as a ready-reckoner for human body functions. Its 35,000 - odd mapped genes are intended to unravel the nature and history of disease, *i.e.*, what makes us ill, and what makes us well ? The information generated will enable gene therapists formulate suitable treatments. It means analysing the 3.1 billion base or DNA letters responsible for heredity in humans.

The 'language of life' is a tool that has increasingly made designer drugs possible, especially to treat hereditary diseases. With the Canadians claiming they have now sequenced the SARS virus genome, hopes are up that a vaccine or cure could soon be found for the killer-pneumonia.

TRANSGENICS

Any organism containing a foreign gene (segment of DNA) from a different species is a **transgenic organism**. As genes can be transferred to the genome of another organism only through genetic engineering, transgenics can be defined as "genetically engineered organisms carrying gene/genes of another species in their genome".

Transgenic organisms are also called **Genetically Modified Organisms** (GMOs). The GMOs are created by incorporating into their genetic make up foreign gene/genes or extra copies of an endogenous gene or cloned or modified gene.

Transgenic crops with useful agronomic characters can be produced. For example, the gene coding for the insecticidal protein from bacterium *Bacillus thuringiensis* has been transferred to the cotton plant. This transgenic cotton plant known as **genetically modified cotton** Bt is resistant to boll worm.

Another example of transgenic plant is tomato. The transgenic genetically modified organism (GMO) tomato called Flavr Savr has a much longer and blended with flavour than conventional tomatoes, because of delayed ripening. This is achieved by reducing the amount of cell wall degrading enzyme 'polygalacturonase' responsible for fruit softening.

Production of high value pharmaceutically important compounds such as human insulin, interferons, hormones and blood clotting factors in transgenic plants or seeds. In fact transgenic plants would serve as minifactories to produce these and many other compounds in huge quantities.

Production of antibodies and vaccines in transgenic plants like banana and tomato. Children may get immunized against several deadly diseases like cholera by eating banana or raw tomatoes from vaccine producing transgenics.

Bioethics of Genetic Engineering

The ultimate success of genetic engineering depends upon consumer acceptance of the technology and the products. Certain products like human insulin produced in genetically engineered bacteria and other therapeutic products, such as oral vaccines may be readily accepted by public because of easier availability.

Similarly the transgenic plants that are not used for food are less likely to cause public concern.

On the other hand transgenic crop plants that are used for human consumption have led to lot of controversies about their safety because of the presence of certain unwanted properties.

Transgenic cattle that have received additional gene, for instance, to increase growth rate, can be marketed more quickly.

Transgenic animals have been genetically engineered with human genes in order to make products that can be used in human medicine.

For example, transgenic cattle (cow, sheep and goat) have been produced with the intention of increased milk yield, as well as, therapeutic human proteins. The cattle secrete there useful proteins in their milk from where it can be obtained easily.

Transgenic pigs have been given human genes so that their organs carry human antigens. These pig organs like heart, kidney, pancreas can be safely transplanted into humans.

However, the thorough assessment of the risks associated with transgenics for both plants and animals is important before they are released. Many scientists believe that genetic engineering is the technology which can solve various human problems especially hunger and disease.

GENE LIBRARIES AND GENE BANKS

Gene libraries

Each chromosome is like a volume of a book in a library.

Even the genome of a bacterium *Escherichia coli* contains 4,000 genes.

Gene cloning or DNA cloning is done to construct several copies of the desired genes in large amounts. This precise technique allows specific DNA sequences to be separated from other sequences permitting detailed analysis or manipulation.

Gene libraries can also be created using RNA.

The enzyme Reverse Transcriptase, a RNA-directed DNA polymerase, is used to convert the RNA into complementary DNA (*i.e.,* cDNA), which can then be converted in the same way as DNA cloning.

Such gene libraries are maintained through special techniques.

Gene Banks

A gene bank is a repository of clones of known DNA fragments, genes, gene maps, seeds, spores, frozen sperms or eggs or embryos.

These stored genes may be used in genetic engineering and breeding experiments where species have become extinct.

These gene banks will be used increasingly as the rate of extinction increases, depleting the Earth's biodiversity and genetic variety.

Now the genomic information is available for a dozen of species. The most remarkable is the prospects and contribution of Human Genome Project.

The Human Genome Project that commenced in 1990, has now announced that the map, is complete two years ahead of schedule in year 2003.

SUMMARY

1. Genetic engineering involves the isolation of specific genes, making of copies of genes and their transfer to target genomes.
2. The key to genetic engineering technology is a special class of enzymes known as restriction endonucleases, which recognise specific sites and cleave DNA molecules into sticky or blunt fragments.
3. The cleaved two strands of DNA can be joined back together with the aid of a sealing enzyme, ligase.
4. Plasmids and artificial chromosomes from bacteria and yeast are important tools for indirect gene transfer.
5. Bacteria, yeast and cultured plant and animal cells are commonly used as hosts for recombinant DNA experiments.
6. Agriculture is a major focus of genetic engineering activity. Successes have included incorporation of genes for herbicide resistance into crop plants, conferring natural resistance to insect pests.

7. Genetic technologies have made many advancements in the production of drugs, such as insulin and interferon.
8. Several recombinant vaccines directed against human diseases such as hepatitis and herpes, are now available as a therapeutic tool in the prevention of disease.
9. In gene therapy, a faulty gene is replaced with a normal healthy gene, to treat diseases where other medical treatments are not effective.
10. With cloning of 'Dolly' sheep the cloning of any mammalian species is now virtually possible.
11. The first human clone, a baby girl called Eve by scientists, was announced on December 26, 2002.
12. Many useful transgenic microbes, crop plants and farm animals have been cloned successfully.
13. Because the DNA of an individual or species is unique, the genetic analysis methods can be used to identify an organism from a small sample of its cells, that is, to create a DNA fingerprint. DNA fingerprinting is effectively applied in forensic science.
14. Genomic libraries can be created as sources of genes for cloning and their use.
15. The map of entire DNA sequence of a human besides many other comparative genomes, is the most significant achievement of Human Genome Project.
16. Besides many benefits, the ethical, social and legal implications of these potent gene technologies have led to considerable concern about the possibility of accidentally producing new pathogens responsible, for fatal diseases or developing 'genetic monsters'.

TERMS TO REMEMBER

Bacteriophage. A virus capable of infecting a bacterium and replicating in it.

Bioethics. Considering of what is good or bad from morality point of view in the application of biological innovations specially on humans.

Biotechnology. The use of living cells or microorganisms (*e.g.*, bacteria) in industry and technology to manufacture drugs and chemicals or to break down wastes, etc. In recent years, it specially refers to the use of genetically modified cells and microorganisms.

Callus. Mass of undifferentiated cells that initially arises from plant cell or tissue in artificial culture.

Cellular totipotency. Ability of a plant or animal cell to repeatedly divide and differentiate into a complete organism on a defined nutrient medium under suitable laboratory conditions.

Cloning. Producing a population of identical DNA molecules, cells or individuals derived from a single such unit.

DNA fingerprinting. Also called DNA typing or DNA profiling. A method of ascertaining relationships and the identity of a person by means of DNA fingerprint which is unique to each individual.

DNA library. A collection of cloned DNA fragments, usually representing an entire genome present in an organism. Also called genomic DNA library.

DNA ligase. An enzyme that can seal one DNA fragment with another DNA segment both having sticky ends. Ligase is the 'molecular glue'.

Electroporation. Creation of transient (temporary) pores in the cell membrane of a host allow the entry of foreign DNA.

Enucleation. Removal of the nucleus from the cell used, in cloning.

Gene cloning. Also called DNA cloning. Technique for obtaining clones or identical copies of a particular DNA molecule.

Gene therapy. Replacement and alteration of a faulty gene. Technique used for hereditary disease.

Genetic engineering. A technique for artificially and deliberately modifying DNA (genes) to suit human needs.

Genome. Total number of genes contained in one haploid set of chromosomes of a diploid organism.

Genome library. DNA library consisting of fragments of chromosomal DNA.

Microinjection. Introduction of foreign genes into animal or plant cell by injecting DNA directly.

Molecular scissors. Restriction endonuclease enzyme.

Particle gun. A technique for bombarding microprojectiles (gold or tungsten particles) coated with foreign DNA with great velocity into the target cells.

Plasmid. Extrachromosomal small circular double stranded DNA molecules in a bacterial cell which have sequences matching those of the required gene and can be similarly cut by the same restriction enzyme. Plasmids are autonomous and replicate separately from the host chromosome.

Probes. DNA strands with specific nucleotide sequences complementary to VNTR (Variable Number Tandem Repeats) sequences.

Recombinant DNA or r DNA. DNA produced by joining together genes from different sources.

Restriction endonuclease. Any such enzyme which very specifically recognises a particular DNA sequence, and cuts it. These enzymes are the 'molecular scissors'.

Southern blotting. A technique named after the inventor, Southern, involving hybridisation of DNA with probes.

Surrogate mother. A female who grows someone else's zygote/embryo in her own uterus for full gestation period and delivers the baby.

Syndrome. A group of symptoms appearing together for a particular condition.

Transgenic. Genetically modified organism (GMO) carrying foreign genes. The introduced foreign gene is called **transgene**.

Vector. An intermediate carrier (in genetic engineering), a phage, plasmid or virus DNA into which another DNA is inserted for introduction into bacterial or other cells for amplification (DNA cloning).

QUESTION BANK

N.B. Undermentioned questions are asked repeatedly in various Indian Universities and competitive examinations.

Very Short Answer Questions

1. What is 'genetic engineering' ?
2. Define cell cloning and gene cloning.
3. What is gene therapy ?
4. Give exact meaning of cloning.
5. What are cloning vectors ?
6. What is restriction fragment ?
7. What is electroporation ?
8. Define 'genomic DNA library'.
9. Give an example of 'natural genetic engineer'
10. Give the name of engineered microbe which produces human insulin.
11. Give one famous example of cloned sheep.
12. Give two examples of transgenic plants with their useful characters.
13. What is 'transgenics' ?
14. What is 'DNA' 'fingerprinting' ?
15. Expand VNTR and GMO.

ANSWERS

1. Genetic engineering is a technique for artificially and deliberately modifying DNA (genes) to suit human needs, **2.** Cell cloning is a technique by which genetically identical cells may be produced from a single cell; gene cloning is the technique to obtain clones or identical copies of a particular DNA molecule; **3.** To cut genes responsible for hereditary diseases and replace them with normal genes; **4.** Cloning means producing exact copy/copies of a single parent; **5.** The desired gene is inserted in the vectors or gene carrier to form DNA, *e.g.*, plasmids and viruses; **6.** Restriction endonuclease cuts the specific DNA (desired gene) at the two ends which becomes the restriction fragment; **7.** This is a process in which transient (temporary) pores are created in the cell membrane of the host for uptake of foreign DNA molecules; **8.** A gene library contains several copies of many of the desired genes or DNA fragments; **9.** *Agrobacterium tumefaciens*, a plant pathogenic bacterium which can transfer part of its plasmid DNA as it infects host plants; **10.** *Escherichia coli* ; **11.** Dolly; **12.** Tomato : increased shelf life of fruits (slow ripening); Rice : Vitamin A rich; **13.** 'Genetically engineered organisms carrying gene/genes of another species in their genome'; **14.** DNA typing or DNA profiling, DNA fingerprint is unique to each individual; **15.** Variable Number Tandem Repeats and Genetically Modified Organism.

Short Answer Questions

1. Name two of the common pharmaceutically important products obtained through genetic engineering.
2. What is meant by 'genetic engineering' ? List the various steps involved in this technology.
3. What are the proposed benefits of genetic engineering in crop improvement?
4. Explain the principles involved in DNA fingerprinting.
5. List two applications of DNA fingerprinting.
6. Gene therapy can cure the important genetic disorders in human. Comment.
7. Give one example each of transgenic plant and transgenic animal.
8. What are Variable Number Tandem Repeats or VNTRs ?
9. Name at least three diseases against which genetically engineered vaccines are now available.
10. What is a cloned animal ? Give one example.
11. What is Human Genome Project ?
12. Give an outline of the steps involved in DNA fingerprinting.
13. Name the various cloning vectors and explain how a plasmid can be used for genetic engineering.
14. Describe different methods to introduce foreign genes into animal cells.
15. List three applications of DNA fingerprinting.
16. Why is "Agrobacterium-mediated genetic transformation" in plants described as natural genetic engineering of plants ?
17. What is Southern blotting ? Give an example where it is applied.

Long Answer Questions

1. What is the Human Genome Project ? What has been revealed about our genome so far ?
2. What is genetic engineering ? Explain briefly the distinct steps common to all genetic engineering technology.
3. Differentiate between
   (*i*) Cell cloning and organismal cloning.
   (*ii*) Direct gene transfer and indirect gene transfer.
4. Write short notes on :
   (*i*) Recombinant DNA; (*ii*) Transgenics (*iii*) Genomics.
5. Write a comprehensive account of " Tools and Techniques of Genetic Engineering".
6. Write notes on :
   (*i*) Applications of Recombinant DNA Technology.
   (*ii*) Diagnosis of human diseases by recombinant DNA technology.
7. Write a comprehensive account of gene therapy, citing example of Severe Combined Immuno Deficiency (SCID).
8. Write an account of cloning, citing atleast three important examples.
9. How sheep 'Dolly' was cloned ? Give the process of cloning step wise.
10. Write short notes on :
    (*i*) How calves are cloned in Japan ? (*ii*) First human clone.

Multiple Choice Questions (Objective Type Questions)

Select the most appropriate alternative:

1. The technique in which a foreign DNA precipitated on the surface of tungsten or gold particles and shot into the target cells is known as
   (*a*) Microinjection (*b*) Chemical-mediated gene transfer
   (*c*) Particle gun (*d*) Electroporation.
2. The enzyme that cuts specifically recognition sites in the DNA is known as:
   (*a*) DNA ligase (*b*) DNA polymerase
   (*c*) Reverse transcriptase (*d*) Restriction endonuclease.

3. Restriction endonucleases
   (*a*) cleave DNA at highly specific recognition sequences,
   (*b*) are inserted into bacteria by bacteriophages
   (*c*) are made only by eukaryotic cells.
   (*d*) add methyl groups to specific DNA sequences.
4. Plasmids
   (*a*) are circular protein molecules. (*b*) are required by bacteria.
   (*c*) are tiny bacteria (*d*) confer resistance to antibiotics.
5. Genetic diagnosis by DNA testing
   (*a*) detects only mutants and normal alleles (*b*) can be done only on eggs or sperms
   (*c*) involves hybridization to ribosomal RNA.
   (*d*) utilizes restriction enzymes and a polymorphic site.
6. DNA can be introduced into any cell by
   (*a*) injection. (*b*) being complexed with calcium salts.
   (*c*) gel electrophoresis. (*d*) being placed along with the cells into a gene gun.
7. In DNA fingerprinting
   (*a*) The variability of repeated sequences between two restriction sites is evaluated.
   (*b*) a positive identification can be made.
   (*c*) The polymerase chain reaction amplifies fewer DNA.
   (*d*) multiple restriction enzyme digests/generate unique fragments.
8. Ability of a plant or animal cell to repeatedly divide and differentiate into a complete organism is :
   (*a*) Cloning (*b*) DNA fingerprinting
   (*c*) Cellular totipotency (*d*) Mitosis.
9. Restriction endonuclease is also known as :
   (*a*) Molecular glue (*b*) DNA ligase
   (*c*) DNA polymerase (*d*) Molecular scissors.
10. Extrachromosomal small circular double stranded DNA molecule in a bacterial cell is :
    (*a*) Plastid (*b*) Plasmid
    (*c*) Mitochondrion (*d*) Chloroplast.
11. Introduction of foreign genes into plant or animal cells using micropipettes is
    (*a*) Electroporation (*b*) Chemical-mediated gene transfer
    (*c*) Microinjection (*d*) Particle gun.
12. Natural genetic engineer is
    (*a*) *Agrobacterium* spp. (*b*) *Rhizobium* spp.
    (*c*) *Bacillus* spp. (*d*) Bacteriophage.
13. The vectorless gene transfer includes
    (*a*) Particle gun (*b*) Microinjection
    (*c*) Electroporation (*d*) All the above.
14. Three billion base pairs are found in
    (*a*) Lily (*b*) Human
    (*c*) Fruit fly (*d*) Yeast

ANSWERS

1. (*c*) **2.** (*d*) **3.** (*a*) **4.** (*a*) **5.** (*d*) **6.** (*d*) **7.** (*c*) **8.** (*c*) **9.** (*d*) **10.** (*b*)
11. (*c*) **12.** (*a*) **13.** (*d*) **14.** (*b*)

UNIT - 4
Biotechnology

9 CHAPTER

Plant Tissue Culture and Biotechnology

The technique of **plant cell culture** occupies a key role in the second green revolution in which gene modifications and biotechnology are being used to improve crop yield and quality. In recent years, plant protoplast, cell and tissue cultures have become an important tool for crop improvement, commercial production of natural compounds and many more in the development of forestry.

M.S. Swaminathan (1987), emphasized the significance of biotechnological application of *in vitro* cultured plant protoplasts/cells/tissues as mentioned here :

(*i*) Tissue culture applications in order to capitalise upon the totipotency of cells.

(*ii*) Cell and protoplast cultures coupled with DNA vectors to overcome problems caused by barriers to gene transfer through sexual means.

(*iii*) Culture of plant cells for the production of useful compounds.

(*iv*) Transfer of genes for nitrogen fixing (*nif*) ability to non-fixing species.

This technology has been employed to a wide range of crop plants and tree species to solve the problems. Now, with the better understanding of the technique of plant tissue culture and nutritional requirements of plant cell, it has become possible to develop newer technologies by culturing plant organs, such as an anther, ovary, ovule, petal, leaf and meristem, which leads to establishment of new research lines, *viz.*, haploids, virus-free plants, *in-vitro* fertilisation, embryo rescue and direct regeneration from leaf disc for genetic engineering.

Biotechnology is defined by different organisations in different ways. It has been broadly defined as "the development and utilisation of biological processes, forms and systems for obtaining maximum benefits to man and other forms of life". Thus, "biotechnology is the application of scientific and engineering principles to the processing of materials by biological agents to provide goods and service".

In recent years revolution in biology has occurred due to the potential of biotechnology. Techniques have been developed to produce rare and medicinally valuable molecules, to change hereditary traits of plants and animals, to diagnose diseases to produce useful chemicals, and to clean up and restore the environment. Thus, biotechnology has great impact in the field of health, agriculture and environmental protection.

In most of the developing countries, the recombinant DNA technology has become one of the major thrusts. In 1982, Government of India has set up an official agency, The National Biotechnology Board (NBTB) which started functioning under The Department of Science and Technology (CDST) in the Ministry of Science and Technology for planning, promotion and coordination of various biotechnological programmes.

PLANT TISSUE CULTURE

Plant tissue culture is the maintenance and growth of plant cells, tissues and organs on a suitable culture medium *in vitro*, *e.g.*, in a test tube or any other suitable vessel.

Plant tissue cultures are often classified according to the type of *in vitro* growth, such as callus and suspension cultures, or the explant used for culture initiation, *e.g.* embryo culture, anther culture, etc.

An **explant** is the part of a plant that is excised from its original location and used for initiating a culture.

Surface sterilisation and sterilisation. It is essential that the explants, glassware, culture containers or vessels, media and the instruments used for plant tissue culture must be free from microbes.

Hence the explants are treated with specific anti-microbial chemicals, and the process is called **surface sterilisation.** Suitable sized plant material (explant) is sterilised as follows :

(*i*) Bring the explant to be sterilised in well sterilised laboratory and prepare pieces for sterilisation.

(*ii*) Clean the working area and hands with alcohol, put on mask and cap, and light the spirit lamp.

(*iii*) Keep 3 or 4 petridishes in a line, add disinfectant (*e.g.*, mecuric chloride 0.01 to 0.1% aqueous solution, or 20% sodium hydrochloride) in first plate and autoclaved distilled water in subsequent plates.

(*iv*) Place plant pieces in first plate and immerse the material with the help of sterilised forceps for 5–10 minutes depending upon the disinfectant used.

(*v*) Transfer material from first to second petriplate, rinse gently and pass to third and fourth plates, one by one with thorough rinsing.

(*vi*) Finally drain the distilled water, and prepare suitable sized explants.

A quick dip in 70% ethanol (15–30 seconds), is always advantageous, before surface sterilisation with disinfectant.

The vessels, media and instruments are also suitably treated with steam, dry heat, alcohol or subjected to filtration to make them free from microbes.

Generally, autoclave is used to sterilise medium, glassware and tools for the purpose of plant tissue culture. Sterilisation of material is carried out by increasing moist heat (121°C) due to increased pressure inside the vessel (15–22 *psi, i.e.*, pounds per square inch) for 15 minutes for routine sterilisation. Moist heat kills the microorganisms and makes the material free from microbes.

Surface sterilisation of explants and their transfer to culture media must be done under aseptic conditions.

Nutrient media could be prepared in a separate room where sufficient space is available for keeping and weighing chemicals, and putting glassware.

Vitamins and growth hormones are carefully weighed.

Stock solutions of chemicals are kept in refrigerator to avoid contamination.

Culture Technique

As already mentioned, the technique of plant tissue culture enables us to study the cells, tissue or organs by isolating them from the plant body and growing aseptically, in suitable containers, on an artificial nutrient medium, under controlled environmental conditions. Thus (*i*) **nutrient medium**, (*ii*) **aseptic conditions** and (*iii*) **aeration of the tissue** are important aspects of the technique of *in vitro* culture.

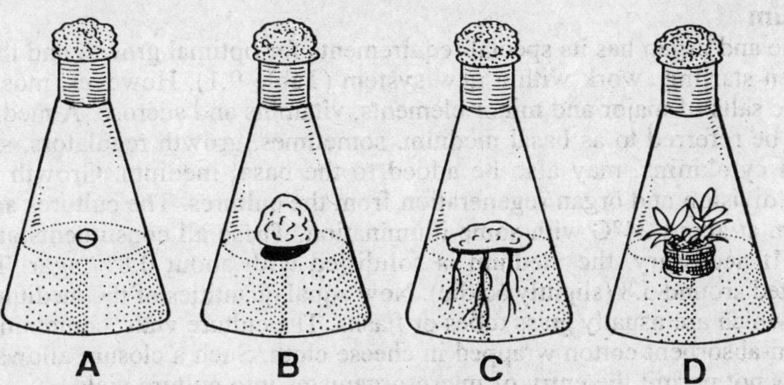

Fig. 9.1. An experiment to demonstrate the effect of growth regulators (auxins and cytokinins) on growth and differentiation in plant tissue culture. A, on the nutrient medium lacking a growth regulator the tobacco pith tissue showed very poor growth; B, in the presence of a cytokinin a vigorously growing callus was formed; C, when transferred to a medium containing 3 mg/l of indole acetic acid and 0.2 mg/l of kinetin this callus differentiated only roots; D, if a portion of the same callus was planted on a medium with higher concentration of kinetin (1 mg/1) and lower concentration of indole acetic acid (0.03 mg/1), it differentiated only shoots.

Table 9.1. Composition of Nutrient Medium (Murashige and Shoog, 1962)

	Chemical constituents		*Mg/l*
1.	**Macronutrients**		
	NH_4 No_3		1650
	KNO_3	...	1900
	$CaCl_2 . 2H_2O$	...	440
	$MgSO_4 . 6H_2O$	...	370
	KH_2PO_4	...	170
2.	**Iron**		
	Na_2 EDTA	...	37.3
	$FeSO_4 . 7 H_2O$	...	27.8
3.	**Micronutrients**		
	$MnSO_4 . 4H_2O$	...	22.3
	$ZnSO_4 . 4H_2O$	...	8.6
	$Na_2MO_4 . 2H_2O$	...	0.25
	$CuSO_4 . 5H_2O$	...	0.025
	$COCl_2 . 6H_2O$	...	0.025
	KI	...	0.83
	H_3BO_3	...	6.2
4.	**Vitamins**		
	Glycine	...	2.0
	Nicototinic acid	...	0.5
	Pyrodoxime-HCL	...	0.5
	Thiamine- HCL	...	0.1
5.	**Cytokinins**		
	Kinetin	...	0.04–10.0
	Myo-inositol	...	100.0
	Indole acetic acid (IAA)	...	1.0–30.0
	Sucrose (g)	...	30.0
	pH	...	5.7

Nutrient Medium

Every tissue and organ has its special requirements for optimal growth and these need to be worked out when starting work with a new system (Table 9.1). However, most of the media contain inorganic salts of major and minor elements, vitamins and sucrose. A medium with these ingredients will be referred to as basal medium. sometimes, growth regulators, such as auxins, gibberellins and cytokinins, may also be added to the basal medium. Growth regulators are required for cell division and organ regeneration from the cultures. The cultures are usually kept in a culture room at about 24°C with some illumination. These all constituents are dissolved in distilled water. If necessary, the medium is solidified with about 0.8% agar. The *pH* of the medium is adjusted around 5.8 (slightly acidic). Now equal quantities of the medium are dispersed in culture vials, which are usually glass tubes or flasks. The culture vials, containing medium, are plugged with non-absorbent cotton wrapped in cheese cloth. Such a closure allows the exchange of gases but does not permit the entry of micro-organisms into culture vials.

Aseptic Conditions

The sugar content of the nutrient media may support a luxuriant growth of many micro-organisms, like bacteria and fungi. It is, therefore, extremely important to maintain a completely aseptic environment inside the culture vials. Micro-organisms can contaminate the medium in at least three ways :

(*a*) The micro-organisms present in the medium right from the beginning many be destroyed by sterilizing the properly plugged culture vials. It can be done by maintaining the temperature at 120°C for about 15 minutes.

(*b*) The micro-organisms may also be carried along with tissue that is being cultured. To prevent this, the plant material from which the tissue is to be excised is surface sterilized. The material may be surface sterilized with saturated chlorine water and then thoroughly washed with sterilized distilled water and to remove all traces of chlorine. If the material is fairly hard, as are some fruits and seeds, it may be surface sterilized by rinsing in alcohol.

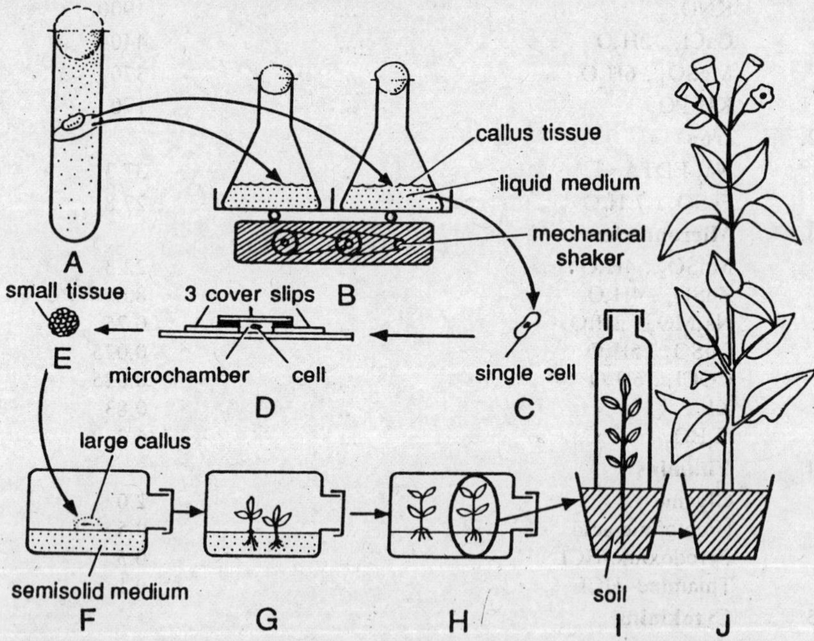

Fig. 9.2. Development of tobacco plants from single cells. A callus is raised from a small piece of tissue excised from the pith (*A*). By transferring it to a liquid medium and shaking the culture flasks ; (*B*), The callus is dissociated into a single cell; (*C*), is mechanically removed from the flask, and is placed on a slide in a drop of culture medium; (*D*), a microchamber is formed around the culture drop using the three-cover slips; (*D*), a small tissue ; (*E*) derived from the cell through repeated division is then transferred to a semi-solid medium where it grows into large callus (*E*) and eventually differentiated plants (*F, G*), when transferred to soil; (*H*), these plants grow to maturity, flower and set seeds (*I*).

(*c*) Finally, precautions must also be taken to prevent the entry of micro-organisms while the plug of a culture vial is removed to transfer the tissue to the nutrient medium (inoculation). For this, all operations from surface sterilization of the tissue up to inoculation are done in an aseptic environment.

Aeration

Proper aeration of the cultured tissue is also an important aspect of culture technique. If the tissue is grown on the surface of a semi-solid medium it acquires enough aeration without special device for liquid medium, special device "filter paper bridge" is used. Here, two legs remain dipping in the medium and the horizontal part carrying the tissue is raised above the level of the medium.

Cellular Totipotency

This is the capacity of mature cells showing that when freed from the plant body, they had the ability to reorganize themselves into the new individuals. Steward and his co-workers, showed this phenomenon in the carrot cultures. Here the small pieces of mature carrot root, were grown in a liquid medium supplemented with coconut milk, in special containers. These cultures were

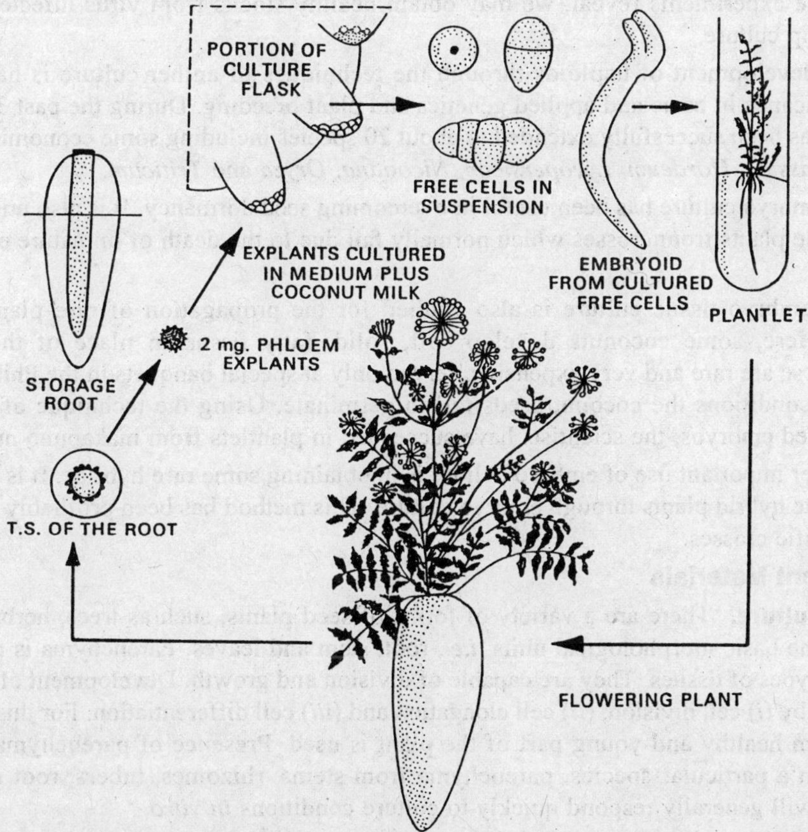

Fig. 9.3. An experiment to demonstrate the totipotency of plant cells. Slices of the carrot root were cut and small pieces of tissue were taken from the phloem region. These were inoculated into a liquid medium in special flasks, the tissue grew actively and single cells and small cell aggregates dissociated into the medium. Some of the cell clumps developed roots, and when transferred to a semisolid medium, these rooted nodules formed shoots. These plants could be transferred to soil.

shaken generally which freed all the cell clusters into the medium. Some of them developed into rooting clumps. When these were transferred to the tubes containing a semi-solid medium, they gave rise to whole plant that flowered and set seeds. On the basis of these experiments, it can be inferred that, at least theoretically, every living plant cell, irrespective of its age and location, is totipotent. However, this phenomenon cannot be compared with the mode of the development of the zygote, wherein the divisions give predictable manner. But in case of cultured cells, the isolated single cells of tobacco divide, quite irregularly to form a mass from which roots and shoot buds differentiate eventually.

However, Guha and Maheswari, while culturing mature anthers of *Datura innoxia* with an aim of understanding the physiology of meiosis, accidentally noticed that on basal nutrient medium containing kinetin, coconut milk, or grape juice numerous embryo-like structures appeared from the inside of anthers, which eventually, developed into plants. Later on, these workers confirmed the origin of the embryoids from pollen grain. As expected, the plantlets of pollen-origin were haploid.

Applications of Tissue Culture

(*a*) The technique provides a way for rapid multiplication of desirable and rare plants.

(*b*) As the experiments reveal, we may obtain healthy stocks from virus infected plants through shoot-tip culture.

(*c*) The development of haploids through the techniques of anther culture is having its potential significance in basic and applied genetics and plant breeding. During the past 30 years, the technique has been succesfully extended to about 20 species including some economic plants, *e.g. Atropa, Brassica, Hordeum, Lycopersicon, Nicotiana, Oryza* and *Triticum.*

(*d*) The embryo culture has been useful in overcoming seed dormancy. It is also utilized for producing viable plants from crosses which normally fail due to the death of immature embryos, *e.g.*. jute, rice.

(*e*) The embryo tissue culture is also applied for the propagation of rare plants, *e.g.*, "makapuno". Here, some coconuts develop soft, solid, fatty tissue in place of the liquid endosperm. These are rare and very expensive, served only at special banquets in the Philippines. Under normal conditions the coconut seeds fail to germinate. Using the technique of *in vitro* culture of excised embryos, the scientists have succeeded in plantlets from makapuno nuts.

(*f*) Another important use of embryo culture is in obtaining some rare hybrids. It is possible to raise complete hybrid plants through embryo culture. This method has been profitably used for many interspecific crosses.

Culture of Plant Materials

Explant culture. There are a variety of forms of seed plants, such as trees, herbs, grass, which exhibit the basic morphological units, *i.e.*, root, stem and leaves. Parenchyma is the most versatile of all types of tissues. They are capable of division and growth. Development of a tissue is characterised by (*i*) cell division, (*ii*) cell elongation and (*iii*) cell differentiation. For this reason, the explant from healthy and young part of the plant is used. Presence of parenchyma is first consideration in a particular species. parenchyma from stems, rhizomes, tubers, root is easily accessible and will generally respond quickly to culture conditions *in vitro.*

Callus formation and its culture. In nature, callus develops by infection of microorganisms from wounds due to stimulation by endogenous growth hormones, the auxins and cytokinins. However, it has been artificially developed by adopting tissue culture techniques. A callus is an amorphous mass of loosely arranged thin walled parenchyma cells developing from proliferating cells of the parent tissue. The unique feature of callus is that the abnormal growth has biological potential to develop normal root, shoots and embryoids ultimately forming a plant.

In callus culture, cell division in the explant forms a callus, an unorganised mass of cells. It is maintained on a medium gelled usually with agar. The medium ordinarily contains the auxin 2, 4 – D (2, 4 – dichlorophenoxyacetic acid), and often a cytokinin BAP (benzylaminopurine). When an explant is placed on such a medium, many of the cells become meristematic and begin to divide. In about 2 to 3 weeks, a callus mass is obtained.

Cell (suspension) culture. A suspension culture consists of single cells and small groups of cells suspended in liquid medium. Cell suspension is prepared by transferring a fragment of callus (about 500 mg.) to the liquid medium (500 ml.) and agitating them aseptically to make the cells free in medium. The medium ordinarily contains the auxin 2, 4 – D. Suspension cultures must be constantly agitated at 100 – 250 rpm (revolutions per minute).

Agitation serves three important purposes. They are as follows :

(*i*) aeration of culture; (*ii*) constant mixing of the medium and (*iii*) breakage of cell aggregates into smaller cell groups. Suspension cultures grow much faster than callus cultures.

It is difficult to have suspenstion of single cell. However, the suspension includes single cell, cell aggregates (varied number of cells), residual inoculum and dead cells. King (1980) has described that a good suspension consists of a high proportion of single cells than small cluster of cells.

Cell suspension cultures have many advantages over the callus cultures. They are as follows.

(*i*) The suspension can be pipetted.

(*ii*) They are less heterogeneous and cell differentiation is less pronounced.

(*iii*) They can be cultured in volumes upto 1500 litres.

(*iv*) They can be subjected to more stringent environmental controls.

(*v*) They can be manipulated for production of natural products by feeding precursors.

Subculturing. After some time, the undermentioned three things happen in all types of plant tissue cultures :

(*i*) Cell/tissue dry matter known as biomass increases, (*ii*) the level of nutrients in the medium decreases, and (*iii*) the medium volume declines due to evaporation.

Hence, if tissue cultures were kept in the same culture vessel, they will die in due course of time. Due to this reason, cells/tissues are regularly transferred into new culture vessels containing fresh media. This process is called **subculturing.**

Precaution is taken that during subculture, only a part of the culture from a vessel is transferred into the new culture vessel.

Table 9.2. Difference Between Callus and Suspension cultures.

Callus culture	Suspension culture
1. Her cell division in the explant forms a callus, which is an unorganised mass of cells.	1. Suspension culture consists of single cells and small groups of cells.
2. It is maintained on a medium gelled usually with agar.	2. Here, cells are suspended in liquid medium.
3. The medium ordinarily contains the auxin 2, 4-D, and often a cytokinin like BAP.	3. Usually the medium contains the auxin 2, 4-D.
4. There is no need of agitation.	4. They must be constantly agitated at 100-250 rpm (revolutions per minute).
5. In about 2-3 weeks, a callus mass is obtained	5. Suspension culture grow much faster than callus cultures.

Uses of callus and suspension cultures. The callus and suspension cultures can be used to achieve cell biomass production which may be used for biochemical isolation.

Also used for regeneration of plantlets, *i.e.*, newly regenerated plants through tissue culture. Production of transgenic plants and isolation of protoplasts.

Regeneration of Plantlets

Plants cells cultured *in vitro* can ultimately give rise to complete plants. Potentiality of a plant cell to regenerate the entire organism or plant is called **totipotency.** This potentiality has been exploited through the culture of protoplasts, cells, tissues and organs *in vitro*. Gottlieb Haberlandt in 1902, for the first time started the technique of plant tissue culture, when he attempted to culture isolated single cells from leaf mesophyll to determine their totipotency.

Regeneration describes the development of an organised structure, such as root, shoot or somatic embryo from cultured cells. However, plantlets can be obtained from cultured cells by two different ways:

(*i*) Where shoot regeneration is followed by rooting of the shoots and

(*ii*) Where regeneration of somatic embryos followed by their germination.

In cultured materials it has been possible to study such process as differentiation of a parenchyma cell into tracheid (*i.e.*, cytodifferentiation), organ formation (*i.e.*, organogenesis) and somatic embryo formation (*i.e.*, somatic embryogenesis). Besides the study of fundamental process of differentiation, the capacity of cell to form organs and embryos can be exploited to regenerate plantlets for clonal propagation.

Organogenesis. The process by which cells and tissues are forced to undergo changes which lead to the production of a structure, known as shoot or root primordium. This system is commonly produced in callus cultures. The earliest reports on controlled organogenesis *in vitro* were by White (1939) who obtained shoots on callus of a tobacco hybrid and by Nobecourt (1939), who observed root formation in carrot callus.

The earliest report on controlled **somatic embryogenesis** *in vitro* was with carrot reported simultaneously in 1958 by Reinert and Steward (see Fig 9.3). The process occurs naturally in a wide range of species from both reproductive and somatic tissues. Somatic embryos can be formed on callus, in cell suspensions and protoplasts cultures, or directly from cells of organised structures, such as stem segments or zygotic embryo.

Advantages of plantlet regeneration. There are several advantages of plantlet regeneration through plant biotechnological methods using organogenesis or embryogenesis, in comparison to conventional methods of propagation. The advantages include the efficiency of process, *i.e.*, formation of plantlet in fewer steps, with reduction in labour, time and cost, and the potential for the production of much higher number of plantlets, and their morphological and cytological uniformity. Till now about 150 species from angiosperms and gymnosperms have been reported to produce somatic embryos in culture.

Shoot and root regeneration. Shoot regeneration is promoted by a cytokinin, such as BAP (benzylaminopurine). While root regeneration is promoted by an auxin, such as NAA (naphthalene acetic acid). Thus, shoot and root regenerations are generally controlled by auxin-cytokinin balance. Usually, an excess of auxin promotes root generation, and that of cytokinin promotes shoot generation.

Callus cultures are first kept on a BAP (cytokinin) containing medium. After sometime shoots regenerate from callus cells. When the shoots become 2-3 cm long, they are excised and transferred to an auxin-containing medium. After sometime, roots regenerate from the lower ends of these shoots to give rise to complete plantlets.

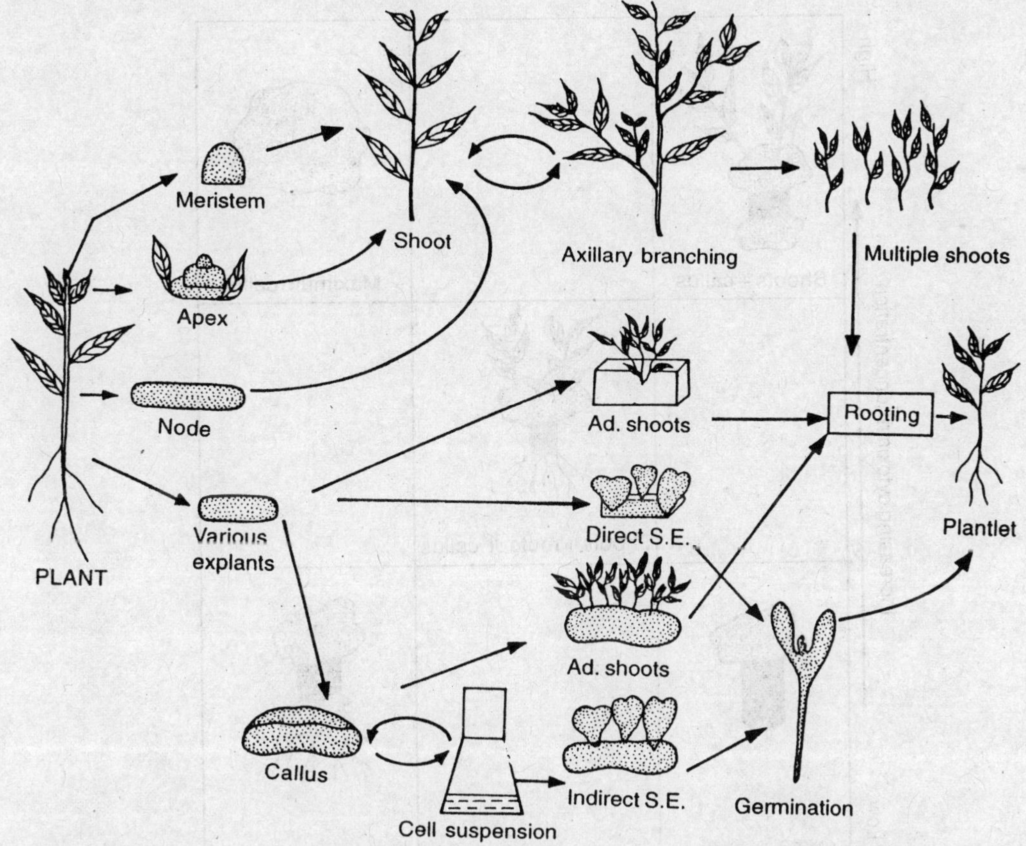

Fig. 9.4. Different methods of regeneration from explant and callus/cell cultures.

Somatic embryo regeneration. Plant cells are totipotent and can produce complete new plants under favourable conditions of nutrients and plant growth regulators. Steward and Reneirt, almost simultaneously reported for the first time, somatic embryo formation in carrot cell suspension cultures in 1958. These somatic embryos were similar to zygotic embryos in development and structure.

A somatic embryo develops from a somatic cell. The pattern of development of a somatic embryo is similar to that of a zygotic embryo.

Somatic embryo regeneration is induced usually by a relatively high concentration of an auxin like 2, 4-D (2, 4-dichorophenoxyacetic acid). The young embryos develop into mature embryos either in the same medium or on another medium. Mature somatic embryos germinate to yield complete plantlets.

During somatic embryo generation in cell suspension cultures, embryos of different sizes are produced. For any experimental or micropropagation method, embryos of uniform size are required. This can be achieved by sieving or fractionation of suspension with appropriate sieve size.

Somatic embryo regeneration is a versatile technique for micropropagation of plant species. A large number of herbaceous dicots and monocots have been regenerated through somatic embryogenesis.

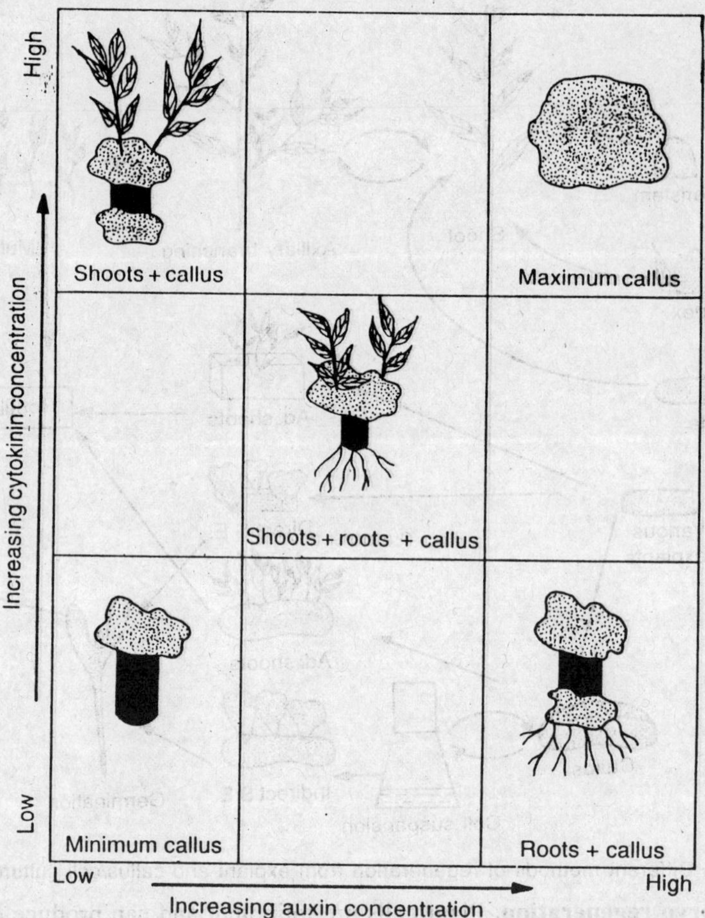

Fig. 9.5. Effect of a combination of auxin and cytokinin ratio.

Establishment of plantlets in the field. Plantlets are produced through rooting of isolated shoots or germination of somatic embryos. Now, the plantlets can be removed from culture vessels and established in the field. This stage is concerned with transfer of plantlets in pots, their **hardening** and establishment in soil. Hardening of plants imparts some tolerance to moisture stress and plants become autotrophic from heterotrophic condition. During hardening, plantlets are kept under a reduced light and high humidity for a suitable period of time. Hardening procedures make the plantlets capable of tolerating the relatively harsher environments outside the culture vessels.

Hardened plants are then transferred to glass or poly-houses with normal environmental conditions. Generally the poly-houses are erected by mounting polythene or polycarbonate sheets on metal frame support. Now, plants are irrigated frequently and their growth and variation monitored regularly. Plants are generally transferred to fields, *e.g.*, plantation crops, after 4-6 weeks of acclimatization.

In India, there are now many commercial companies which produce millions of plantlets through micropropagation.

Large glass houses and green houses are essential components of micropropagation industry. Hardening and acclimatization of delicate *in vitro* raised plantlets is carried out in these glass houses. Now a days chambers made of polycarbonate and polypropylene sheets are used for creating large working place. These houses are fitted with mist and fog generating units with cyclic auto-regulation. Light is provided through proper light sources.

Application of plant tissue culture. Tissue culture has been successfully employed for the multiplication of orchids and many other ornamental plants.

In the developed countries tissue culture is a routine method of multiplication while in developing countries, such as India, the techniques are largely used for producing plants for export markets.

With tree species, the technique of tissue culture remianed restricted for many years to the laboratory stage and have generally invited only academic interest. To bridge this gap between research and application, the Department of Biotechnology (DBT), Government of India, set two pilot-scale national facilities for large-scale production of elite planting material of forest trees through tissue culture, one at the Tata Energy Research Institute (TERI), New Delhi, and the other at the National Chemical Laboratory (NCL), Pune.

However, practical applications of plant tissue culture are mainly based on the ability of plant cells to give rise to complete plantlets. The use of plant cells to generate useful products is called **plant biotechnology.** In most of its activities, the useful product is plantlet that, in many cases may have been genetically altered. Usually, these plantlets are used for the following important purposes:

***(i)* Clonal propagation.** Clonal propagation by vegetative methods is a practice followed since man started cultivation of plants. The main objective of clonal propagation has been to reproduce plants of selected desirable qualities uniformly and in bulk. The traditional propagation methods, require long duration, whereas tissue culture helps in rapid plant multiplication.

A **clone** is a group of individuals or cells derived from a single parent individual or cell through a asexual reproduction. All the cells in callus or suspension culture are derived from a single explant by mitotic division.

Hence, all plantlets regenerated from a callus or suspension culture generally have the same genotype and constitute a clone.

These plantlets can be used for rapid clonal propagation of superior lines.

Selected examples of clonal multiplication of trees and horticulture plants are as follows : oil palm, citrus peach, prunus, poplar, etc. Improved cultivars developed by biotechnological methods are then clonally multiplied to replace inferior cultivated varieties, *e.g.*, in *Mentha* and other aromatic plants.

***(ii)* Somaclonal variation.** Genetic variation present among plant cells of a culture is called **somaclonal variation.**

In vitro cell and tissue cultures of plant give rise to genetic variation spontaneously. While spontaneous variation is not desired during propagule multiplication, it has been useful in providing some genetic variants among crop species. The observed frequencies of such somaclonal variants vary widely and are probably related to both time and culture. Several point mutations have been observed, the majority of analysed spontaneous genetic variants from somatic cultures appear to have resulted from induction of aneuploidy, loss of interchanges or intra-chromosomal segment duplication, deletions or structural arrangements.

Spontaneous or mutagen induced genetic variation in somatic cell culture coupled with *in vitro* selection techniques have been effective in isolating desired novel genetic variants with cellular level expression while in heterozygous condition. The somaclonal variation has been used to develop several useful varieties of crop plants.

***(iii)* Transgenic plants.** A gene that is transferred into an organism by genetic engineering is called **transgene**. An organism that contains and expresses a transgene is called **transgenic organism.**

The plants, in which a functional foreign gene has been incorporated by any biotechnological methods that generally not present in plant, are called **transgenic plants.** However, a number of transgenic plants carrying genes for traits of economic importance have either been released for commercial cultivation or are under field trials.

However, transgenes can be introducd into individual plant cells. The cells containing and expressing transgenes can be easily selected *in vitro*. Ultimately, plantlets can be regenerated from these cells. These plantlets yield highly valuable transgenic plants.

Some of the commercially grown transgenic crop plants in developed countries are : 'Flavr Savri' and 'Endless Summer' tomatoes, 'Freedom II' squash, 'High lauric' rapeseed and 'Roundup Ready' soybean. So far more than 60 transgenic dicot plants including herbs, shrubs and trees, and several monocots, such as maize, oat, rice, wheat, etc., have been produced. In future the number of these crops certainly will go up. These transgenic plants contain certain selected traits such as herbicide resistance, insect resistance, virus resistance, seed storage protein, modified ripening, modified seed oil, agglutinin, etc.

Meristem Culture

The meristem is a dome of actively dividing cells about 0·1 mm in diameter and 0·25 mm in length. Shoots of all flowering plants grow by virtue of their apical meristems. The totipotency of the plant cells forms the basis of meristem culture.

Here, one can use an explant containing pre-existing shoot meristems, and produce shoots from them. Such cultures are known as **meristem culture.** The explants generally used in meristem culture are shoot tips or nodal segments.

These explants may be cultured on a medium containing a cytokinin, usually BAP (benzylaminopurine). Cytokinins promote axillary branching by overcoming apical dominance. Hence, they support multiple shoot development from each explant.

When axillary branching takes place, individual shoots are cultured, whereas when axillary branching does not take place, the single shoot is cut into nodal segments, which are then cultured. Shoots of 2-3 cm are excised and rooted on a suitable medium.

The *in vitro* culture of meristem and shoot tip involves several phases such as initiation of culture and establishment of explant, growth and differentiation, proliferation of shoots and finally plantlet fomation by rooting of shoots.

The plantlets thus obtained are subjected to hardening and ultimately established in the field.

Uses of meristem cultures. Meristem cultures are used for rapid clonal multiplication. Development of virus free plants is one of the most significant application of meristem culture. Generally this technique has applications in diverse areas such as clonal multiplication of vegetatively propagated crop plants, virus elimination and germplasm preservation.

Embryo Culture

Interspecific crosses may fail due to several reasons, but when the development of embryo is arrested owing to the degeneration of the endosperm, or when the embryo aborts at an early stage of development, embryo culture is the only technique to recover hybrid plants. It is being used extensively in the extraction of haploid barley (*Hordeum vulgare*) from the crosses *H. vulgare* × *H. bulbosum*. Embryo culture is also a routine technique employed in orchid propagation and inbreeding of those species that show dormancy. Das and Burman (1992) developed the method of regeneration of tea shoots from embryo callus.

Excision of young embryos from developing seeds and their cultivation on a nutrient medium is called **embryo culture.** *In vitro* older embryos are more easily cultured than young embryos.

The objective of embryo culture is to allow the young embryos to complete development and ultimately, give rise to seedlings.

More recently, a number of hybrids have been successfully raised through embryo culture : *Hordeum vulgare* × *Secale cereale, H. vulgare* × *Agropyron repens, H. vulgare* × *Triticum aestivum,* interspecific hybrids in *Abelmoschus* and *Secale.*

In some interspecific crosses, the endosperm of developing hybrid seeds degenerates at an early stage. Here, young embryos also die on the degeneration of endosperm. Hence such interspecific crosses cannot be normally made. In such cases, young hybrid embryos are excised and cultured *in vitro* to obtain hybrid seedlings.

The orchid industry owes a great deal to the technique of embryo culture. The culture of orchid embryos was initiated at the Singapore Botanic Garden in 1928. Orchids lack stored food. Here, embryo culture allows development of seedlings from most of the embryos. In such cases, embryo culture is also used for rapid clonal propagation.

Embryo culture has helped in overcoming self-sterility of seeds, especially of crop plants propagated vegetatively, when the seeds do not germinate in nature. In a wild relative of commercial banana, *Musa balbisiana,* for example, the seeds do not germinate under natural conditions. However, if the embryos are excised and grown on a simple culture medium of mineral salts, seedlings are readily obtained.

Indian work. Sometimes some interspecific hybrids of *Oryza* show quite a low percentage of seed germination under normal conditons because the hybrid embryos succumb at an early stage of their development. The scientists raised seedllings of certain interspecific crosses in *Oryza* by culturing the excised embryos on Nitsch's medium . The embryo culture technique has also been successfully employed in *Colocasia esculenta* where the seeds fail to germinate in nature. They grew the excised embroys on synthetic media and obtained seedlings which on transplanting in soil gave rise to normal healthy plants.

Rangaswamy and Rangan (1963) have cultured the embroys of a stem parasite, *Cassytha filiformis,* in the absence of the host, on a modified White's medium supplemented with IAA. The embryos of *Cuscuta reflexa,* a total stem parasite, were also cultured. Somewhat similar results have been obtained with the embryos of *Dendrophthoe falcata*, a partial stem parasite. Like the embryos of stem parasites, those of some root parasites have also shown considerable capacity for proliferation. In cultures of seeds of *Orobanche aegyptiaca,* a total root parasite, the ovoid unorganized embryos produced a massive callus capable of continuous growth followed by differentiation into shoot tips.

Anther Culture and Haploid Production

Anther, a male reproductive organ, is diploid (2n) in chromosome numbers. As a result of microsporogenesis, tetrads of microspores or pollen grains are formed from a single spore mother cell (2n), which are haploid (n).

In nature, haploid plants originate from unfertilised egg cells. However, in laboratory, they can be produced from both male and female gametes. In many plant species, haploids are produced when their anthers are cultured on a suitable medium. This process is called **anther culture.**

The first successful pollen cultures were established by Tulecke in the 1950's using mature pollen grains of certain gymnosperms. In 1964, Guha and Maheshwari discovered that when excised anthers of *Datura innoxia* were cultured intact in a mineral salt medium in conjunction with coconut milk and other complex organic substances and growth hormones, embryo-like outgrowths appeared from the sides of the anther in about 6-7 weeks. In subsequent studies, they confirmed the haploid nature of the embryoids and their origin from microspores.

Investigations have shown that for some reason, anthers from flowers of the Solanaceae respond best to excision and culture by production of embryoids. Convincing demonstration of the direct transformation of microspores into embryoids has been provided by several workers in different species of the genera of Solanaceae, *i.e., Datura, Nicotiana, Atropa, Lycium, Petunia, Solanum, Capsicum* and *Hyoscyamus.* It has been observed that embryoids go through the globular, heart-shaped and torpedo stages typical of the ontogeny of normal diploid zygotic embryos before they finally elongate and form shoot and root meristems. Embryoids have also been obtained form cultured anthers of certain cereals like rice. All in all therefore, the capacity of microspores in cultured anthers to form haploid embryoids and plants can now scarcely be questioned.

In some other plants, haploids do not arise directly from the microspores, but do through the intervention of a callus. The microspores first develop into multicellular bodies which later give rise to an exceedingly dense callus. Subculture of the callus in an embryogenesis-inducing medium containing specific concentration of auxins and a cytokinin led to the initiation of haploid seedlings. There, thus seem to exist the following alternative pathways for haploidy in anther cultures :

1. Microspore → callus → embryogenesis → plantlet *e.g.,* Solanaceae
2. Microspore → callus → embryogenesis → plantlet *e.g., Oryza, Brassica, Hordeum, Coffea, Populus.*
3. Microspore → callus → plantlet, *e.g., Datura metelloides.*

In several species, the pollen grains can be isolated and cultured to obtain haploids. In many species of plants, haploids can also be produced by culturing unfertilised ovaries or ovules.

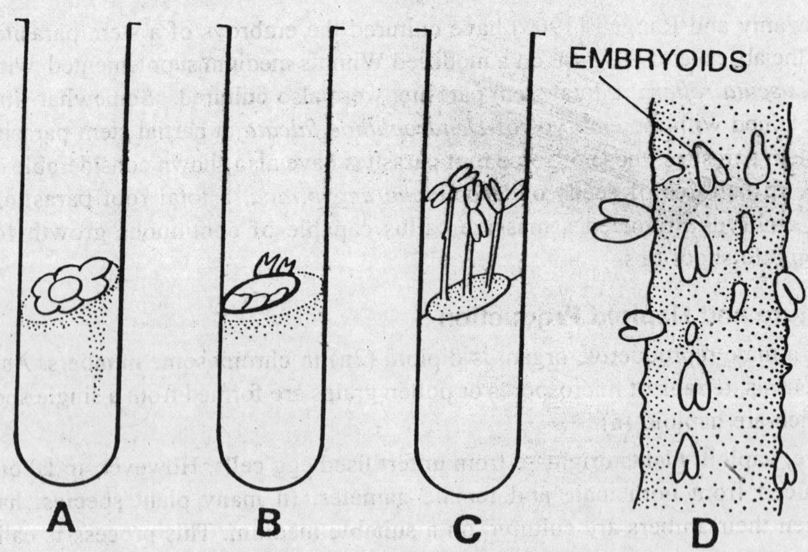

Fig. 9.6. Embryoid formation in tissue culture of butter cup. A, an unorganized callus ; B, six-week old culture showing numerous embryoids arising from the cells ; C, the embryoids have developed into plantlets and a fresh crop of embryoids is seen arising from the surface of the hypocotyledonary region of these plants; D, the embryoid-bearing portion enlarged from C.

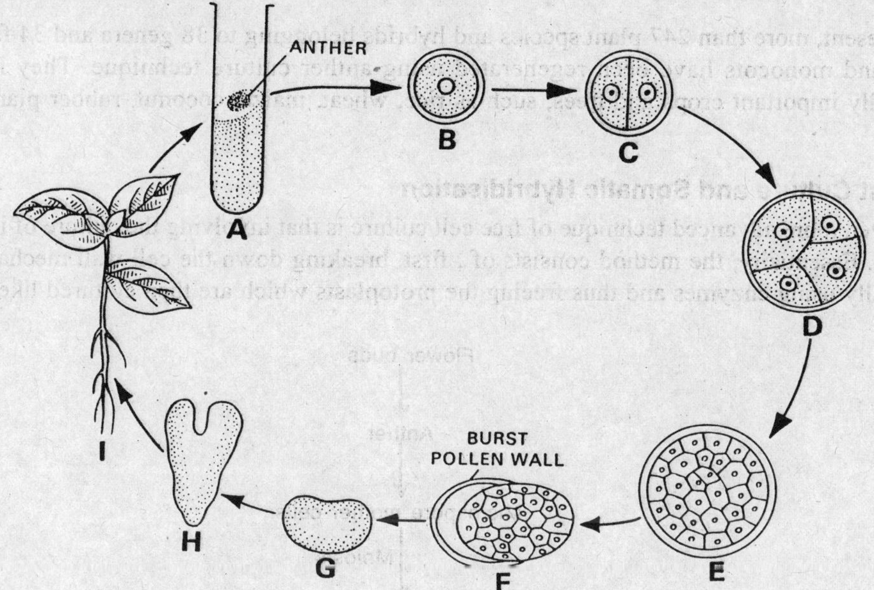

Fig. 9·7. Diagram showing the development of haploid plants from the pollen grains of tobacco. The anther of the right stage is excised from the flower bud and cultured on a suitable medium (A). Some of the pollen inside the anther undergoes repeated division (B-D), forming a multicellular tissue within the parent wall (E). Eventually, the pollen wall bursts, releasing the tissue mass directly develops into an embryoid and germinates to form a haploid plant.

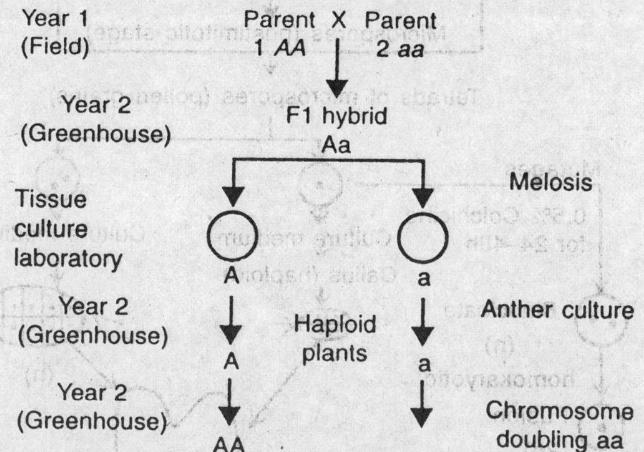

Fig. 9.8. Production of homozygous lines using anther culture.

Applications. Haploids, such produced are completely sterile and of no direct value. But they are important as they are used to produce homozygous lines in 2-3 years. This strategy can be of much value in breeding programmes. Anthers from F_1, plants, obtained by crossing two or more lines, are cultured to obtain haploid plants. The chromosome number of haploid plants may be doubled by using colchicine to obtain homozygous plants. The progeny from these plants are then subjected to selection to isolate superior homozygous lines.

Thus, haploid plants are very useful in (*i*) direct screening of recessive mutation in diploid or polyploid, screening of recessive mutation is not possible, and (*ii*) development of homozygous diploid plants following chromosome doubling of haploid plant cells.

In China, the most widely grown wheat is a doubled haploid produced through homozygous diploid lines. Anther culture of rice is also successfully grown. Haploid plants have been produced in tobacco, wheat and rice through pollen culture which are used for the development of disease resistant and superior diploid lines.

At present, more than 247 plant species and hybrids belonging to 38 genera and 34 families of dicots and monocots have been regenerated using anther culture technique. They include economically important crops and trees, such as rice, wheat, maize, coconut, rubber plantation, etc.

Protoplast Culture and Somatic Hybridisation

An even more advanced technique of free cell culture is that involving the culture of isolated protoplasts. Essentially, the method consists of , first, breaking down the cell wall mechanically or chemically using enzymes and thus freeing the protoplasts which are then cultured like whole

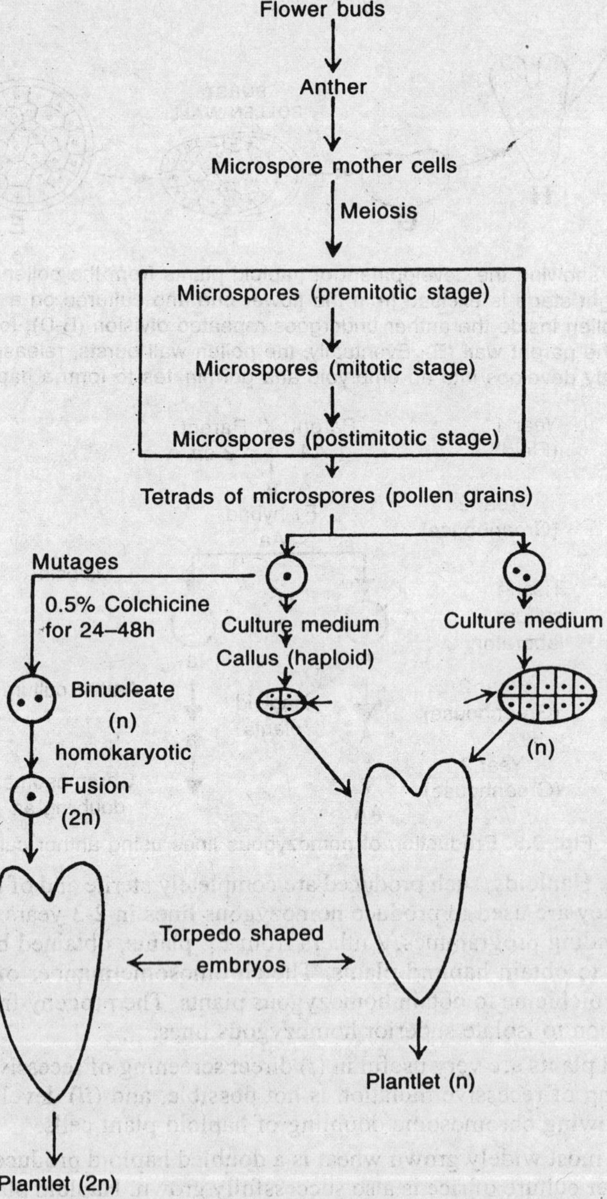

Fig. 9.9. Methoas of anther culture, and production of haploid and diploid plants *in vitro*.

cells using appropriate culture media. The cells are, in other words, rendered naked and then cultured.

When cell wall is mechanically or enzymatically removed, the isolated protoplast is known as 'naked plant cell' on which most of recent researches are based.

Plant cell wall acts as physical barrier and protects cytoplasm from microbial invasion and environmental stress. Cooking (1960), for the first time isolated the protoplasts of plant tissues by using cell wall degrading enzymes, *viz.*, cellulose, hemicellulose, pectinase and protease extracted from a saprophytic fungus *Trichoderma viride.* Later on protoplasts were cultured *in vitro*.

Protoplast culture and regeneration. The protoplasts regenerate a cell wall, undergo cell division and form callus. The callus can also be subcultured. Embryogenesis begins from callus when it is placed on nutrient medium lacking mannitol and auxin. The embryo develops into seedlings and finally mature plants.

Somatic hybridisation. Fusion between protoplasts of the selected parents is induced by a solution of polyethylene glycol (PEG), or by very brief high voltage electric current. Somatic bybridisation allows the production of hybrids between lines and species, that cannot be produced normally by means of sexual hybridisation.

Fusion of cytoplasm of two protoplasts results in coalescence of cytoplasm. The nuclei of two protoplasts may or may not fuse together even after fusion of cytoplasms. The binueleate cells are known as **heterokaryon.** When nuclei are fused the cells are known as **hybrid,** and when only cytoplasms fuse and genetic information from one of the two nuclei is lost is known as **cybrid**, *i.e.,*cytoplasmic hybrid.

However, production of cybrids which contain the mixture of cytoplasms but only one nuclear genome can help in transfer of cytoplasmic genetic information from one plant to another. Thus, information of cybrid can be applicable in plant breeding experiments. For example, in China, cybrid technology in rice is a great success. Such plants are very useful in producing hybrid seeds without emasculation. Today, cybrid technology has successfully been applied to carrot, mustard, citrus, tobacco and sugar beet.

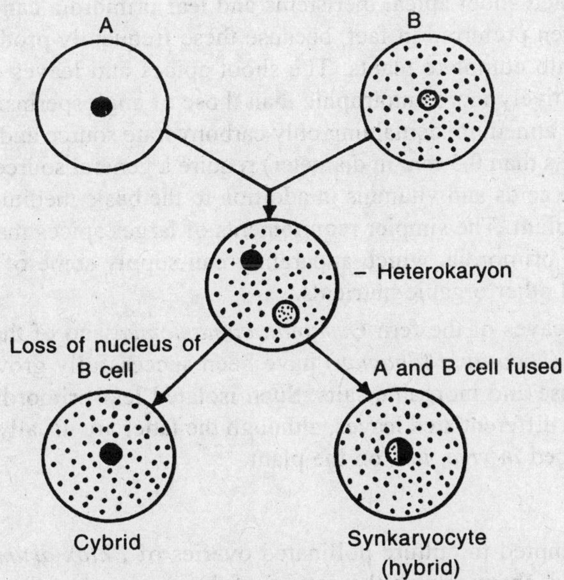

Fig. 9.10. Hybrid/cybrid production through protoplast fusion.

Somatic hybrids may also be used for gene transfer, transfer of cytoplasm and production of useful allopolypoloids.

Tissue and protoplast cultures have been used in genetic engineering for the transfer of DNA and extrachromosomal bodies — plasmids, mitochondria, chloroplasts, *nif* (nitrogen fixing) genes from the nitrogen fixing bacterium *Klebsiella pneumoniae* to a strain of the colon bacterium *Escherichia coli.* Isolated protoplasts have great advantage in all the afore mentioned uses. For transformation purposes cultured apical meristems are also usable because these can easily be regenerated into whole plants and also because intact DNA taken up by plants appears to be rapidly transported to meristematic regions, where growth and differentiation are centred.

Root culture. Excised roots were the first plant organs of higher plants to be successfully brought into sterile culture. In addition to the usual requirements for a carbon source and mineral nutrients most isolated roots grown in sterile culture required to be supplied with certain vitamins. This is because, in the intact plant, certain vitamins, are synthesized in the leaves and the roots are dependent upon shoots for the supply of these substances which they are unable to synthesize themselves. For example, tomato roots require only sucrose, mineral nutrients and thiamin, and given all these, will grow successfully in culture for many years. But excised roots of some monocotyledonous plants fail to grow even when supplied with a full complement of β-vitamins and other vitamins. In some of these (*e.g.*, rye), an exogenous auxin supplement to the nutrient medium allows growth to proceed. In order to maintain the culture, the excised roots must be regularly subcultured on to a fresh medium, by excising a piece of root bearing a lateral which then proceeds to grow rapidly and maintain the culture.

As a rule, excised root of most species produce only root tissues in culture. There are exceptions, however, where they regenerate shoot buds as well as further roots, *e.g.*, *Convolvulus, Taraxacum* and *Rumex*.

Generally excised root is cultured in liquid medium. The techniques of root culture give certain important informations such as, (*i*) nutritional requirements, (*ii*) infection by *Rhizobium* and nodulation and (*iii*) physiological activities.

Culture of Shoot Apices and Leaves

Like roots, isolated shoot apical meristems and leaf primordia can also be grown in sterile culture. These are often preferred in fact, because these frequently produce roots and can hence eventually develop into complete plants. The shoot apices and leaves of vascular cryptogams, such as ferns, are relatively more autotrophic than those of angiosperms. Thus, even a small fern apex can be grown on a medium containing only carbohydrate source and mineral nutrients. Small angiosperm apices (less than 0.5 mm in diameter) require a general source of organic nitrogen and certain specific amino acids and vitamins in addition to the basic medium, but larger apices will grow on a simple medium. The simpler requirements of larger apices may be due to the fact that they carry larger leaf primordia, which apparently can supply some of the requirements of the apex for vitamins and other organic nutrients.

Isolated young leaves of the fern *Osmunda cinnamomea* and of the sunflower (*Helianthus annuus*) and tobacco (*Nicotiana tabacum*) have been successfully grown on a simple medium containing only sucrose and inorganic salts. Such isolated leaf primordia continue to grow and develop into normally differentiated leaves, although the latter are usually very much smaller than normal leaves developed *in vivo, i.e.*, on the plant.

Culture of Ovaries

Rau (1956) attempted to culture pollinated ovaries of *Phlox drummondii* and studied the influence of extraneous chemicals on the pattern of development of the endosperm and embryo.

The addition of colchicine to the medium caused aberrant divisions, fusion and aggregation of endosperm nuclei and finally a degeneration of the endosperm. In 12 to 14-days old cultures the embryo also aborted and the seeds were malformed.

Nirmala Maheshwari and Lal (1961) excised the flowers of *Iberis amara* one day after pollination and planted them on agar nutrient media. Fruits of normal size were obtained in two weeks on a medium containing Knop's mineral salts + Nitsch's trace elements + sucrose + B vitamins. The seeds formed *in vitro* contained viable embryos. Removal of the sepals slowed down the growth of the ovaries showing that the calyx is by no means an unessential structure but has an important role in the physiology of the fruit.

Pollinated ovaries of a few other plants have also been reared into fruits, but when unpollinated ovaries are taken, the fruits are either not formed or are seedless.

Johri and Sehgal (1963) cultured the ovaries of *Anethum graveolens* at the zygote stage on White's basal medium containing yeast extract and /or casein hydrolysate thirteen to twenty weeks later, a polyembryonate mass developed by the proliferation of the zygotic proembryo and ruptured the pericarp. Further growth resulted in the production of multiple shoots some of which flowered in the test tube after about seven months. Thus this was possible to obtain several seedlings from a single fertilized ovule.

However, it is easier to maintain ovules in sterile culture when they are left *in situ* within the ovary. The physiological requirements of fertilized ovules do not appear to be species specific, since young ovules of widely different species have grown to mature seeds following transplantation on to the placenta of *Capsicum* fruits.

Culture of Ovules and Parts of Ovules

White (1932) was the pioneer to culture the ovules. White and La Rue cultured ovules of *Erythronium* and *Antirrhinum* on White basic medium containing indole acetic acid IAA.

Nirmala Maheshwari and Lal (1961) cultured the ovules of *Papaver somniferum* excised six days after pollination when they contained only a 2-celled proembryo and a free nuclear endosperm. These grew to maturity in twenty days and even germinated and produced seedlings in the culture tubes.

Several genera, such as *Citrus, Eugenia* and *Mangifera* show the occurrence of nucellar embryos. Since they have the same genetical composition as the maternal parent, they are of much importance for the clonal propagation of desirable varieties. Rangaswamy (1961) reported that if the nucellar tissue of *Citrus microcarpa* is grown on a suitable nutrient medium containing casein hydrolysate, it proliferates profusely and on subculturing produces embryo-like regenerants termed **"pseudobulbils"**, which can develop into seedlings so that an indefinite number of new plants can be obtained from a single nucellus.

A. N. Rao (1963) successfully germinated seeds of an interspecific hybrid of *Vanda* on a simple agar nutrient medium. Some of the fertilized ovules directly produced seedlings; others gave rise to a callus from which new plants arose subsequently.

Culture of Endosperms

It is somewhat more difficult to raise tissue cultures from the endosperm. However, Rangaswamy and Rao (1963) established a continuously growing tissue from the mature endosperm of *Santalum album* when they cultured it along with the embryo on White's medium containing yeast extract, kinetin and 2, 4-D. On the other hand, the endosperm failed to callus when it was grown without the embryo. Johri and Bhojwani (1965), also cultured the endosperm of *Exocarpus cupressiformis* on White's medium containing indole actic acid IAA, kinetin and casein hydrolysate. If merely the endosperm devoid of the embryo was grown, it did not proliferate. In their experiments, the shoots were differentiated and not the roots.

BIOTECHNOLOGY

Biotechnology is that portion of microbiological science which deals with the possible utilisation of micro-organisms in industrial processes, or in processes in which their activities may become of indusrial or technical significance. Thus, biotechnology may be defined as the use of microorganisms, animal or plant cells, or their components to generate products and services useful to human beings. However, the use of complete animals and plants is not generally included in biotechnology.

For several decades it has been known that numerous kinds of yeasts, moulds and other lower fungi, and several types of groups of bacteria have direct relation, to certain types of economic processes carried out in connection with industrial or factory operations, such as brewing, wine making, cheese making, etc. However, people did not know that microorganisms were involved in these processes. Microorganisms were used for the first time to produce some organic compounds like citric acid in the beginning of twentieth century. Thereafter, micro-organisms were employed to generate a variety of products, including antibiotics. In all such processes, only the natural capabilities of the organisms and cells are exploited. All such activities carried over by micro-organisms were grouped under **old biotechnology.**

In recent years revolution in biology has occurred due to the potential of biotechnology. Techniques have been developed to produce rare and medicinally valuable molecules, to change hereditary traits of plants and animals, to diagnose diseases, to produce useful chemicals, and to clean up and restore the environment. Thus, biotechnology has great impact in the fields of health, food, agriculture and environment protection.

During 1970, the technique of recombinant DNA technology was developed. Recombinant DNA technology is popularly known as genetic engineering. Recombinant DNA technology is responsible for the isolation of desired gene from any organism, and its transfer and expression into the organism of choice. In most of the developed countries, the recombinant DNA technology has become one of the major thrusts. The United Nations Industrial Development Organisation (UNIDO) has recognised the potential of genetic engineering and biotehnology for promoting the economic progress of the developing countries.

Transgenic micro-organisms are produced to obtain novel pharmaceutical proteins, such as human insulin is commercially produced from a transgenic *Escherichia coli* (bacterium) strain, that contains and expresses the human insulin gene. Proteins poduced by transgenes are called **recombinant proteins.**

Many valuable recombinant proteins, also have been produced by use of transgenic animal cell lines and transgenic plants. However, the recombinant proteins cannot be naturally produced by the concerned cells or organisms. The production technologies based on genetic engineering are grouped under **modern biotechnology.**

In India, in 1988, a full fledged Department of Biotechnology (DBT) has been set up, in the Ministry of Science and Technology for planning, promotion and coordination of various biotechnological programmes.

Recombinant DNA technology has made it easier to detect the genetic diseases and cure before the birth of a child. In genetic engineering programmes, it has become possible to map the whole genome of an organism. Gene bank and DNA clone bank have been constructed to make available different types of genes of its known function.

Table 9.3. Area of Biotechnology

Area of interest	Products
1. Recombinant DNA technology (genetic engineering)	Fine chemicals, enzymems, vaccines, growth hormones, antibiotics, interferon.
2. Treatment and utilisation of biomaterials (biomass)	Single cell protein, mycoprotein, alcohol and biofuels.
3. Plant and animal cell culture.	Fine chemicals (alkaloids, essential oils, dyes, steriods), somatic embryos, encapsulated seeds, interferon, monoclonal antibodies.
4. Nitrogen fixation	Microbial inoculants (biofertilizers).
5. Biofuels (bioenergy)	Hydrogen (via photolysis), alcoholos (from biomass), methane (biogas produced from wastes and aquatic weeds).
6. Enzymes (biocatylysts)	Fine chemicals, food processing, biosenser, chemotherapy.
7. Fermentation	Acids, enzymes alcohols, antibiotics, fine chemicals, vitamins, toxins (biopesticides).
8. Process engineering	Effluent, water recycling, product extraction, novel reactor, harvesting.

Genetically Modified Crops

The plants, in which a functional foreign gene has been incorporated by any biotechnological methods that generally not present in plant, are called **transgenic plants**. A transgenic crop that contains and expresses a **transgene** (*i.e.*, functional foreign gene). Generally, transgenic crops are called **genetically modified crops** or **GM crops**.

The techniques used for production of transgenic crops have two great advantages. They are as follows :

(*i*) Any gene (from any organism or chemically synthesised) can be used as a transgene.

(*ii*) The change in genotype can be controlled to some extent as only the transgene is added into the crop genome.

In contrast, breeding activities can use only those genes which are present in such species that can be hybridised with them. Moreover, changes occur in all those traits for which the parents used in hybridisation differ from each other.

However, when a transgene is introduced into the genome of an organism, it can achieve one of the following characteristics :

(*i*) Produces the desired protein.

(*ii*) Produces a protein which itself produces the desired phenotype.

(*iii*) Modifies an existing biosynthetic pathway, and therefore, a new end product is obtained.

Some examples are mentioned here :

For example, **hirudin** is a protein that prevents blood clotting. The gene encoding hirudin was chemically synthesised. Thereafter, this gene was transferred into *Brassica napus,* where hirudin accumulates in seeds. Now, the hirudin is purified and used medicinally. Here, the transgene product itself is the desired product.

The other example is of a soil bacterium *Bacillus thuringiensis* which produces a crystal (Cry) protein. The Cry protein is toxic to larvae of certain insects. There are several different types of Cry proteins, and each of them is toxic to a different group of insects. The gene encoding Cry protein is *cry* gene, which has been isolated and transferred into several crops. A crop, that

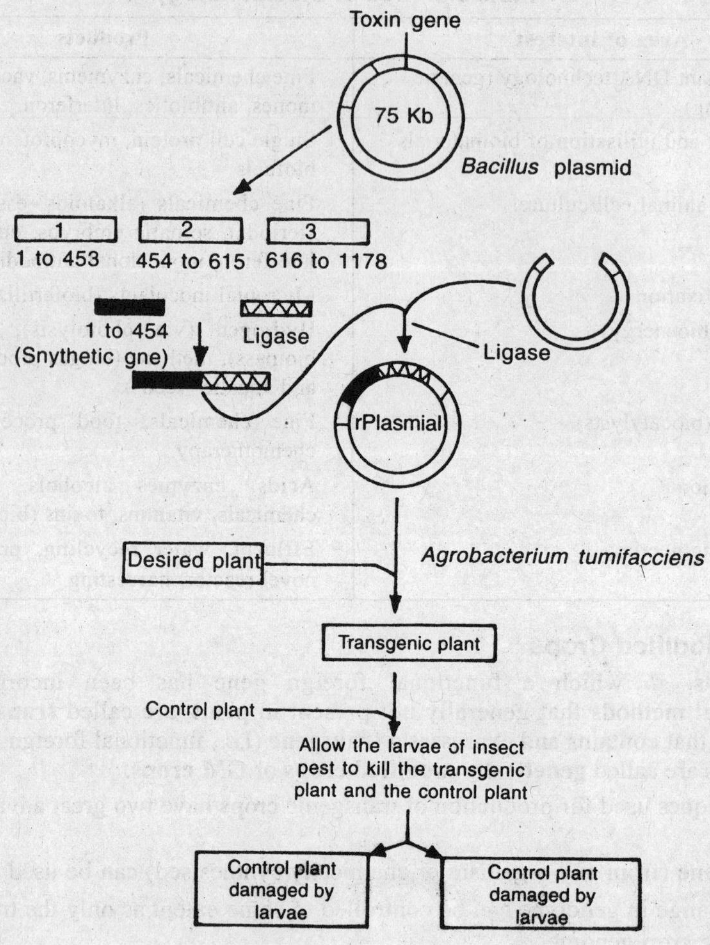

Fig. 9.11. Contruction of a transgenic plant and expression of **Bt** gene against insect larvae.

expresses a *cry* gene is usually resistant to the group of insects for which the concerned Cry protein is toxic. This is a case where the transgene product is directly responsible for the production of the phenotype of interest. Here this is noteworthy that the symbols for a gene (*cry*) and for its protein (Cry) product are the same. However, transgene symbol which has small letters is written in italics (*cry*), whereas the first letter of protein symbol is capital and written in Roman (Cry).

Insect-resistant transgenic plants. The *Bt* gene of a bacterium, *Bacillus thruingiensis* has been found to encode the toxins called endotoxins which have cidal effect of certain insect pests. These toxins are of different types, such as beta-endotoxin and delta-endotoxin. Preparations of *Bt* gene in powder form have been made available in market for commercial use.

The other approach has been isolation of the toxin gene *Bt 2* from *Bacillus thruingiensis* and its introduction into Ti-DNA plasmid of *Agrobacterium tumefaciens*. Thus, Ti-plasmid mediated transformation of several plants has been done, *e.g.*, tobacco, cotton, tomato, corn, etc.

Tomato variety 'Flavr Savr' is an example where expression of a native tomato gene has been blocked. Expression of native gene can be blocked by several methods. For example, fruit softening is promoted by the enzyme **polygalacturonase**, which is responsible for degrading pectin. Production of polygalacturonase was blocked in the transgenic tomato variety 'Flavr Savr'. Hence, fruits of this tomato variety remian fresh and retain their flavour for a longer period in comparison of the fruits of normal tomato varieties. The fruits of this transgenic variety have a superior taste and increased total soluble solids.

Genetically modified crops (GM crops) are already in cultivation in advanced countries, such as U.S.A. and many European countries.

However, in India, some insect resistant cotton varieties expressing *cry* genes have reached onto farmers for cultivation.

It is thought that transgenic crops may be harmful to the environment because of the following reasons :

(*i*) The transgene may be transferred through pollen from GM crops to their wild relatives and such a gene transfer may make the weeds more persistent and damaging.

In such cases, transgenic crops should not be grown in close vicinity of their wild relatives.

(*ii*) The transgenic crops may themselves become persistent weeds.

(*iii*) In view of this, such crops may damage the environment in some mysterious manner.

Investigations are on to check such menace.

Genetically Modified Food

The food prepared from the produce of genetically modified crops (GM crops) is called **genetically modified food** (GM food).

The GM food differs from the food prepared from the produce of conventionally developed used during gene transfer by genetic engineering or recombinant technology.

(*iii*) GM food contains the antibiotic resistance gene itself.

It has been argued that the above mentioned features of GM foods may be harmful and problem making if such foods are consumed. These problems may be as follows :

(*i*) The transgene product (GM food) may cause toxicity and produce allergies.

(*ii*) The enzyme produced by the antibiotic resistance gene may cause allergies, as it is a foreign protein.

(*iii*) The bacteria present in the intestine of the humans may take up the antibiotic resistance gene that is present in the GM food. These bacteria will become resistant to the concerned antibiotic and become unmanageable.

The biotechnologists involved in the production of transgenic crops are aware of above men tioned aspects, and effortrs are being made to use other genes in place of antibiotic resistance genes.

Ban on genetic food. It is a growing concern all over the world that the genetic food may pose risks to human health, ecology and the environment. However, it has forced the governments of many countries to rethink on introduction of such crop. For the first time the European Commission's Scientific Advisors have recommended that a genetically engineered potato be withheld from the market because they cannot guarantee its safety. The United States, the world's biggest producers of genetically modified foods, has also threatened New Zealand to ban his genetically engineered foods.

Sustainable Agriculture

In modern days, in agricultural practices non-renewable resources are utilised which cause pollution. However, such practices cannot be continued indefinitely. This means, they are not sustainable.

Sustainable development may be defined in several ways. **Sustainable agriculture** primarily have renewable resources, which cause minimum pollution and maintain the optimum yield level.

Any such development which reduces the use of non-renewable resources, and level of pollution, will definitely enhance the sustainability of agriculture.

Biotechnology contributes in several ways for enhancement of the sustainability of agriculture. They are as follows :

Biofertilizers. The term 'biofertilizers' denotes all the 'nutrient inputs of biological origin for plant growth' (Subba Rao, 1982). However, micro-organisms employed to enhance the availability of nutrients like nitrogen, and phosphorus to crops are called biofertilizers.

As we know, nitrogen is available in atmosphere in high amount in the form of gas. It is converted into combined form of organic compounds by some prokaryotic micro-organisms through biological reactions. The phenomenon of fixation of atmospheric nitrogen by biological means is known as '**diazotrophy**' or 'biological nitrogen fixation' and these prokaryotes as 'diazotrophs' or 'nitrogen fixers' (nif). They may be in free living or in symbiotic forms.

Examples of nitrogen-fixing micro-organisms are bacteria and cyanobacteria (blue-green algae). Some of these micro-organisms are free-living, while others form symbiotic association with plant roots. **Rhizobia** form root nodules in leguminous crops, while cyanobacteria form symbiotic association with the pteridophyte *Azolla*.

On the other hand, insoluble forms of soil phosphorus are converted into soluble forms by certain micro-organisms. This makes the phosphorus available to the plants.

Phosphate is made soluble by some bacteria and by some fungi that form association with roots of higher plants. The fungus and plant root association is called *mycorrhiza*. Here the fungi absorb their food from the roots and in response are beneficial to the plants. The mycorrhiza may be external or internal. The external mycorrhiza also called 'ectophytic mycorrhiza' are confined to the outer region of the roots, whereas the internal mycorrhiza are found deeply in the root cells. These fungi solubilise phosphorus, produce plant growth promoting substances and protect host plants from soil pathogens.

Advantages. Biofertilizers make a low cost and easy technique and can be used by small farmers.

It is free from pollution hazards and increases soil fertility. Cyanobacteria secrete growth promoting substances, amino acids, proteins, vitamins, etc. They add sufficient amount of organic matter in soil.

Rhizobial biofertilizer can fix 50-150 kg N/ha/annum.

Azolla supplies N, increases organic matter and fertility in soil and shows tolerance against heavy metals.

The biofertilizers increase physico-chemical properties of soil, such as soil structure, texture, water holding capacity, etc.

The mycorrhizal biofertilizers make the host plants available with certain elements, increase longevity and surface area of roots, reduce plant response to soil stresses, and increase resistance in plants. In general, plant growth, survival and yield are increased.

However, extensive efforts are made to enhance the effectiveness and the contribution of biofertilizers to agricultural production.

Biopesticides. Biopesticides are those biological agents which are used to control weeds, insects and pathogens. There is a vast majority of micro-organisms, such as viruses, bacteria, fungi, protozoa and mycoplasma known to kill the insect pests. The suitable preparations of such micro-organisms for control of insects are called 'microbial insecticides'. The microbial insecticides are non-hazardous, non-phytotoxic and selective in their action. Pathogenic micro-organisms which kill insects are viruses (DNA containing viruses), bacteria (*e.g.*, *Bacillus thuringiensis*), and fungi (*e.g.*, *Aspergillus, Fusarium,* etc.). Now a days, some of the biopesticides are being used even at commercial scale.

For example, *Bacillus thuringiensis* is a widely distributed soil bacterium, and can be isolated from soil, litters and dead insects. It is a spore forming bacterium and produces several toxins. Spores of this bacterium produce the insecticidal Cry protein. Therefore, spores of this bacterium kill larvae of certain insects. After ingestation of spores, larvae are damaged, as the rod-shaped bacterial cell secretes at the opposite end, a single large crystal (Cry) in the cell. This crystal is toxic and proteinaceous in nature. The commercial preparations of *B. thuringiensis* contain a mixture of spores, Cry protein (toxin) and an inert carrier.

Bacillus thuringiensis, was the first biopesticide to be used on a commercial scale. Certain other bacteria and fungi are also used for control of some weeds and diseases of various crop plants.

Microbial pesticides are produced by many multi-national companies by using viruses, bacteria and fungi. *B. thuringiensis* preparations have been produced in U.S.A., France, Russia and U.K. in the form of wettable powder and water suspensions.

A number of **viruses** have been discovered which belong to groups Baculoviruses and cytoplasmic polyhedrosis viruses (CPV). Preparations of viruses or their products have been developed as effective biopesticides and being successfully used for the control of insect pests in agriculture and horticulture.

Recent studies on the use of **mycopesticides** for the control of insect pests are of much value. Mode of action of these fungi is different from viruses and bacteria. The infective conidia, spores, etc., of the antagonistic fungi reach the haemocoel of the insect either through integument or mouth. They multiply in haemocoel followed by secretion of mycotoxins which result in death of insect hosts.

The use of biopesticides may reduce the application of synthetic chemicals for control of diseases, insect pests and weeds. The synthetic insecticides, generally affect non-target organisms, and many beneficial organisms for agriculture, are killed. In turn, they cast hazardous effects on human health, and therefore, use of biopesticides has been suggested.

Disease-resistant Varieties

Genetic engineering has been also used in the development of such crop varieties which are resistant to certain diseases. Usually, plant diseases are caused by fungi, bacteria, viruses and nematodes.

The most successful approach for the production of virus resistant plants is the transfer of the virus coat protein gene into the plants. The genetic material of viruses is found enclosed in a protein coat.

The gene that encodes coat protein is isolated from the genome of the virus that causes concerned disease. Now this gene is transferred and expressed in the host of the concerned virus.

Expression of the coat protein produces resistance in the host to this virus. This approach has been used in producing a virus-resistant variety of squash. Such disease-resistant varieties are used to minimise the use of chemicals which are generally used for control of crop diseases. This approach also reduces pollution. Such varieties are successful in reducing yield losses due to various crop diseases, thus they enhance agricultural production.

Single Cell Protein (SCP)

The dried cells of micro-organisms, such as algae, bacteria, actinomycetes and fungi, used as food or feed are collectively known as **microbial protein.** Since the time immemorial a number of micro-organisms have been used as part of human diet. Micro-organisms are widely used for preparation of a variety of fermented foods, such as cheese, butter, leavened bread, *idlis* and several other bakery products. Some other micro-organisms have long been used as human food, *e.g.*, the blue green alga (cyanobacteria), *Spirulina*, and the fungi commonly called edible mushrooms.

The term 'microbial protein' was replaced by a new term 'single cell protein' (SCP) during first International Conference on 'microbial protein' held in 1967, at Masachusetts, U.S.A. In recent years, NBRI, Lucknow and CFTRI, Mysore, have established centres for mass production of SCP from *Spirulina* (cyanobacteria).

Substrates used for production of SCP. A variety of substrates are used for SCP production. Algae which contain chlorophylls, do not require organic wastes. They use free energy from sunlight and carbondioxide from air, while bacteria and fungi require organic wastes, as they do not contain chlorophylls, the major components of substrates are the raw materials which contain sugars, starch, lignocellulose from woody plants and herbs having residue with nitrogen and phosphorus contents and other raw materials.

Nutritional value of SCP. SCP is rich in high quality protien and poor in fats. They are ideal for human food. SCP provides a valuable protein-rich supplement in human diet.

Now a days, many pilot plants for the production of *Spirulina* powder have been established in Japan, U.S.A. and European countries. In India, food grade *Spirulina* at two main centres, one at MCRC, Chennai and the other at Central Food Technology and Research Institute (CFTRI), Mysore. The products are marketed in India and abroad.

The use of *spirutina* (SCP) should help bridge the gap between the requirement and the supply of proteins in the human diet. *Spirulina* (SCP) is a rich source of protein, amino acids, vitamins, minerals, crude fibres, etc., it is used as supplemented food in diets of under-nourished children, adults and old aged people in developing countries. *Spirulina* is also popular as health food.

SCP as therapeutic and natural medicine. *Spirulina* posseses many medicinal properties. It has been recommended by medicinal experts for reducing body weight, cholesterol and for better health. It lowers sugar level in blood of diabetics. It is a good source of β-carotenes, and helps in monitoring healthy eyes and skin.

Biopatent

Dictionary meaning of **patent** is, 'an official right to be the only person to make, use or sell a product or an invention'. Thus, a patent is the right granted by a government to prevent others from commercial use of his invention.

A patent is granted for :

(*i*) an invention, including a product,

(*ii*) an improvement in an earlier invention,

(*iii*) the process of generating a product, and

(*iv*) a concept or design.

Initially, patents were granted for industrial inventions by a particular company, such as patent medicines, etc.

But, now a days, patents are also being granted for biological entities and for products derived from them, such patents are called **biopatents**, *e.g.*, *neem* and its products; *haldi* and its products.

However, industrialised countries, such as U.S.A., Japan and European Union Countries, are awarding Biopatents.

Biopatents are awarded for the following :

(*i*) strains of micro-organisms,
(*ii*) cell lines,
(*iii*) genetically modified strains of plants and animals,
(*iv*) DNA sequences,
(*v*) the proteins enclosed by DNA sequences
(*vi*) various biotechnological products
(*vii*) production processes
(*viii*) products, and
(*ix*) product applications.

On the basis of ethical and political reasons, such biopatents have been opposed from time to time by different societies of the world. However, arguments in favour of biopatents are given primarily of increased economic growth.

Many biotechnological patents are quite broad in their coverage. For example, one patent covers 'all transgenic plants of family Brassicaceae/mustard family. Such broad patents are unacceptable and not fair, as they would enable financially powerful corporations to have their monopoly control over biotechnological processes. Such powerful corporations try to control the direction of whole agricultural research, including plant breeding. Such a position seems to be a threat to the food security of the world.

Biopiracy

When big organisations and multinational companies exploit patent biological resources or bioresources of other nations without proper authorisation from the countries concerned; such exploitation is called **biopiracy.**

The advanced or industrialised nations are generally rich in technology and financial resources. However, they are poor in biodiversity and traditional knowledge related to bioresources. While developing nations are poor in technology and financial resources, but quite rich in biodiversity and traditional knowledge related to bioresources.

Biological resources or **bioresources** are those organisms which can be used to derive commercial benefits from them.

Traditional knowledge related to bioresources is the knowledge developed by various communities from time immemorial, regarding the utilisation of the bioresources, *e.g.*, use of plants and other organisms in healing art. Such traditional knowledge of a particular nation can be exploited to develop modern commercial processes. Here, the traditional knowledge is primarily used in the direction to be followed which saves a lot of time, and bioresources are easily commercialised.

Institutions and multinational companies of industrialised advanced nations are collecting and exploiting the bioresources, as follows :

(*i*) They collect and patent the genetic resources themselves. For example, a patent granted in U.S.A. covers entire 'basmati' rice germplasm indigenous to our country.
(*ii*) The bioresources are analysed for identification of valuable biomolecules. A **biomolecule** is a compound produced by a living organism.
(*iii*) Useful genes are isolated from the bioresources and patented, and thereafter, used to generate useful commercial products.

(*iv*) Sometimes, even traditional knowledge itself of other countries may be patented.

For example, a plant, *Pentadiplandra brazzeana* of West Africa, produces a protein called **brazzein.** This protein is approximately two thousand times as sweet as sugar. Moreover, this is a low calorie sweetner.

Local people of West Africa have known and used the super-sweet berries of this plant for centuries. However, the protein brazzein was patented in U.S.A. , where the gene encoding this protein was also isolated, sequenced and patented.

It is proposed to transfer the brazzein gene into maize and express it in maize grains. These grains (kernels) will be used for the extraction of brazzein, which can give a serious jolt to the countries exporting large quantities of sugar.

Bioresources of third world countries have always been commercially exploited by the industrialised nations without an adequate compensation. This exploitation has increased a lot with the development of biotechnological techniques. Some developing nations coming forward and raising voice to make laws to prevent unauthorised exploitation of bioresources and traditional knowledge.

Biowar

This word denotes, the use of harmful bacteria as weapons of war. The biological weapons are generally used against humans, and their crops and animals. A **bioweapon** is a device that carries and delivers to the target organisms, a pathogen or a toxin derived from it.

The **bioweapon agent**, is kept in a suitable container so that it remains acitive and virulent during delivery. The container with bioweapons could be delivered to the target by several ways, including missiles and aircrafts.

For example, anthrax is an acute infectious disease caused by the spore-forming bacterium *Bacillus anthracis*. Spores of *B. anthracis* can be produced and stored in a dry form keeping them viable for several decades in storage or after release. A cloud of anthrax spores, if released at a strategic location to be inhaled by the individuals under attack may act as an agent of effective weapon of **biowar**. For example, the anthrax bacteria were sent through letters after September 2001, in U.S.A.

An attack with bioweapons using antibiotic-resistant strains would initiate the incidence and spread of communicable diseases, such as anthrax and plague, on either an endemic or epidemic scale.

Bioweapons are low cost weapons, and cause far more casualities than chemical or conventional weapons. Bioweapon agents are microscopic and invisible with naked eyes, and therefore, difficult to detect.

Such type of biowar and use of bioweapons against civilised human society is a major threat for all inhabitants of this planet, the earth.

The possible defences against bioweapons include use of gas mask, vaccination, administration of specific antibiotics, and decontamination. However, biologists should play an important role in creating awareness about the impact of misuse of biology on the human society and the whole Biokingdom.

Bioethics

Ethics includes 'moral principles' that control or influence a person's behaviour. This is connected with beliefs and principles about what is right or wrong, morally correct or acceptable. This includes a set of standards by which a community regulates its behaviour and decides as to which activity is legitimate and which is not.

Thus, **bioethics** makes a set of standards which is used to regulate our activities in relation to the whole biokingdom.

Now a days, biotechnology, particularly recombinant DNA technology, is used for exploitation of the biological world by various ways. Biotechnology has been used in various ways, from 'unnatural' to 'detrimental' to 'biodiversity'. The major bioethical ways concerning to biotechnology, are as follows :

1. Use of animals in biotechnology is cruelty towards animals which causes great suffering to them.

2. When animals are used for production of certain pharmaceutical proteins, they are treated as a 'factory' or 'machine'.

3. Introduction of a transgene from one species into another species threats integrity of species.

4. Transfer of human genes into animals or *vice-versa* is great ethic threat for humanness.

5. Biotechnology is only used for fulfilment of the motive of selfishness by humans. This is used only for the benefit of human beings.

6. However, biotechnology poses unforeseen risks to the environment and biodiversity.

Besides ethical arguments, techniques of biotechnology are used in the production of things on a much larger scale and at a much faster rate. Each society has to evaluate bioethical issues and take right decision about their application.

SUMMARY

1. Plant cells and plant organs can be cultured *in vitro* (*e.g.*, in test tube or any other container) on a suitable medium (*e.g.*, agar medium), is called plant tissue culture.
2. A plant part excised from its original location and used for initiating a culture is known as explant.
3. The explants, media, vessels and instruments used in tissue culture must be sterilised to make them free from any microbes.
4. Culture medium should be provided with nutrition that is required for growth and development of explants.
5. Culture of explants on an agar medium provided with 2, 4-D yields callus cultures.
6. Plant cells can also be grown in a liquid medium as a suspension of single cells as well as groups.
7. A constant agitation of suspension cultures is essential.
8. Complete plantlets can be obtained from agar and suspension cultures.
9. Explants that contain pre-existing shoot meristems and produce shoots from them are called meristem culture.
10. Excision of young embryos from developing seeds and their cultivation on a nutrient medium is called embryo culture.
11. Haploids may be produced in many plant species when anthers are cultured on a suitable medium, this is called anther culture.
12. Chromosome numbers of haploid plants can be doubled (diploid, 2n) to produce homozygous lines within 2 to 3 years.
13. A hybrid produced by fusion of somatic cells of two species is called somatic hybrid, and process is known as somatic hybridisation.
14. Cell walls can be digested with a combination of pectinase and cellulase.
15. The plant cells lacking cell walls are called protoplasts.
16. The fusion of protoplasts can produce somatic hybrids.
17. Biotechnology may be defined as the use of microorganisms, animal or plant cells, or their components to generate products and services useful to humans.

18. Recombinant DNA technology or genetic engineering permits the isolation of desired gene from any organism, and its transfer and expression into the organism of choice. Generally this is used to produce transgenic organisms that possess novel specialties.
19. The most important application of plant tissue culture is production of transgenic crops.
20. A transgenic crop contains one or more genes transferred by recombinant DNA technology (genetic engineering).
21. Transgenic crops resistant to insects and pests, with changed quality, capable of producing a medicinal protein, are already cultivated.
22. Food obtained from transgenic crops is called genetically modified food (GMF).
23. However, such foods may be toxic, may cause allergies and may affect useful microbial populations of alimentary canal. Efforts are in progress to check this obnoxity.
24. Modern agricultural practices utilise non-renewable resources and cause pollution. These practices are not sustainable. However, sustainable agriculture would use renewable resources, would cause minimum pollution and maintain optimum yield level.
25. Biotechnology contributes to sustainable agriculture by developing biofertilisers, biopesticides, disease and insect resistant varieties and single-cell proteins.
26. The microorganisms that are employed to enhance the availability of nutrients, such as nitrogen (N) and phosphorus (P) to crops are called biofertilisers.
27. Biopesticides consist of viruses, bacteria, protozoans, fungi and mites that help in controlling diseases, insects and weeds.
28. The use of biofertilisers, biopesticides and resistant varieties helps to save non-renewable resources and checks pollution.
29. Single-cell protein is microbial biomass rich in high quality protein. They serve as valuable food and feed supplements.
30. Patents that are granted for biological entities and for products derived from them are called biopatents. Developments in biotechnology have accelerated biopatenting.
31. Biopiracy is unauthorised use of bioresources and traditional knowledge related to bioresources for commercial benefits.
32. Such developments of biotechnology favour rich industrialised nations, at the cost of developing nations.
33. Now, some nations are developing comprehensive laws to prevent unauthorised exploitation of their bioresources and traditional knowledge.
34. Biowar or biological war is the use of biological weapons, or disease causing agents, such as viruses, bacteria and fungi to destroy crops, domesticated animals and human beings.
35. Bioethics is a set of standards by which a community regulates its behaviour and decides as to which activity is legitimate and which is not. Each society has to evaluate such issues and take right decision regarding them. However, biotechnology may pose unforeseen risks to biodiversity.

TERMS TO REMEMBER

Anther culture. Haploids are produced, when anthers are cultured on a suitable medium.

Bioethics. A set of standards which may be used to regulate own activities in relation to the biological world.

Biofertilizers. Micro-organisms employed to enhance the availability of nutrients.

Biomolecule. A compound produced by a living organism.

Biopatents. The patents granted for biological entities and products derived from them.

Biopesticides. Biological agents used for control of weeds, insects and pathogens.

Biopiracy. Exploitation of bioresources of other nations without proper authorisation from the countries concerned.

Bioresources. All these organisms that can be used to derive commercial benefits.

Biotechnology. Use of organisms, and their cells or components to generate products and services useful to human beings.

Biowar. Use of biological weapons against humans, or their crops and animals.

Bioweapon. A device that carries and delivers to the target organisms a pathological agent or a toxin derived from it.

Bioweapon agent. A biological agent or toxin.

Brazzein. A plant, *Pentadiplandra brazzeana* produces a protein called brazzein, which is about 2000 times as sweet as sugar. A low-calorie sweetener.

Callus culture. Cell division in the explant forms a callus, an unorganised mass of cells is called callus culture.

Clone. A group of individuals or cells derived from a single parent individual or cell through asexual reproduction.

Embryo culture. Excision of young embryos from developing seeds and their cultivation on a nutrient medium.

Explant. A plant part, excised from its original location and used for initiating a culture.

G.M. crop. Genetically modified crop or transgenic crop that contains and expresses a transgene.

G.M. food. The food prepared from genetically modified (transgenic) crops.

Hardening. The procedure done, when plantlets are transferred from culture vessels and established in the field.

Hirudin. A protein that prevents blood clotting.

Meristem culture. Use of an explant that contains pre-existing shoot meristems and produce shoots from them.

Mycorrhiza. The fungus and plant root association.

Plant tissue culture. Maintenance and growth of plant cells, tissues and organs on a suitable culture medium *in vitro*.

Protoplasts. The plant cell lacking cell wall.

Recombinant proteins. Proteins produced by transgenes.

Regeneration. Development of an organised structure (*e.g.*, root, shoot, embryo) from cultured cells.

Rhizobia. The bacteria which form root nodules in legume crops.

Somatic hybrid. A hybrid produced by fusion of somatic cells of two varieties or species.

Subculturing. When tissues/cells are regularly transferred into new culture vessels containing fresh media, the process is called subculturing.

Suspension culture. It consists of single cells and small groups of cells suspended in a liquid medium.

Sustainable agriculture. Primarily use renewable resources, cause minimum pollution and obtain optimum yield level.

Totipotency. The ability of plant cells to regenerate into complete plants is called totipotency.

Transgenic crop. A crop that contains and expresses a transgene.

QUESTION BANK

Very Short Answer Questions

1. Surface sterilisation
2. Subculture
3. Somatic embryo
4. Protoplast
5. Somatic hybrid
6. Biotechnology
7. Transgene

8. Transgenic organism
9. Sustainable agriculture
10. Sterilisation
11. What is plant tissue culture ?
12. What is an explant ?
13. What is callus culture ?
14. What is a suspension culture ?
15. What is totipotency ?
16. What is regeneration ?
17. What is plant biotechnology ?
18. What are protoplasts ?
19. What is modern biotechnology ?
20. What is a transgenic crop or/GM crop ?
21. What is GM food ?
22. What are biopesticides ?

ANSWERS

1. When the explants are treated with specific anti-microbial chemicals, the procedure is called surface sterilisation; **2.** The regular transference of tissues, into new culture vessels containing fresh media, is called subculturing ; **3.** A somatic embryo develops from a somatic cell, which is comparable to a zygotic embryo; **4.** Plant cells lacking cell wall are called protoplasts ; **5.** A hybrid produced by fusion of somatic cells of two species is called somatic hybrid ; **6.** Biotechnology may be defined as the use of microorganisms, animals or plant cells, or their components to generate products and services useful to human beings ; **7.** A gene that is transferred into an organism by genetic engineering is called transgene ; **8.** An organism that contains and expresses a transgene is called transgenic organism ; **9.** Sustainable agriculture primarily uses renewable resources, which cause minimum pollution and obtain optimum yield level ; **10.** When media and instruments are treated with steam, dry heat or alcohol, or subjected to filtration to make them free from microbes is called sterilisation ; **11.** It is maintenance and growth of plant cells, tissues and organs on a suitable culture medium *in vitro* ; **12.** It is a plant part that is excised from its original location and used for initiating a culture ; **13.** In callus culture, cell division in the explant forms a callus which is an unorganised mass of cells ; **14.** It consists of single cells and small groups of cells suspended in a liquid medium ; **15.** The ability of plant cells to regenerate into complete plants is called totipotency ; **16.** The development of an organised structure, like root, shoot or somatic embryo from cultured cells, **17.** It is the use of plant cells to generate useful products and/or services ; **18.** Plant cells lacking cell wall are called protoplasts ; **19.** The production technologies based on genetic engineering are often referred to as modern biotechnology; **20.** A crop that contains and expresses a transgene ; **21.** The food prepared from the produce of transgenic crops (genetically modified GM) ; **22.** Biopesticides are those biological agents that are used for control of weeds, insects and pathogens.

Short Answer Questions

1. Write a note on single-cell protein (SCP).
2. Explain the term 'biopatent'.
3. Explain the term 'biowar'.
4. What is 'biopiracy' ?
5. What are 'biowar agents' ?
6. Explain the term 'bioethics'.
7. What is 'genetically modified food' ? Explain.
8. What is 'sustainable agriculture' ? Explain.
9. What are 'bioweapons' ? Explain.
10. Write a note an 'suspension cultures'.

11. What is 'subculturing' ? Explain the process.
12. What are uses of somatic hybrids ?
13. What is Hirudin ? How is it obtained ?
14. What is anther culture ? Explain.
15. What is old biotechnology ? Write a note.
16. Explain the terms 'callus and 'suspension cultures'.
17. What is 'meristem culture' ? Explain.
18. What is 'biopiracy' ? Write a note.
19. Write a note on 'embryo culture'.
20. What are biofertilizers ? Explain.
21. Describe the importance of mycorrhiza, cyanobacteria and bacteria in biofertilizers.
22. Why are biofertilizers preferred over chemical fertilizers ?
23. What are green manures ? What are their uses ?
24. What are biofertilizers ? How do mycorrhiza act as an entity ?
25. Explain the meaning of the following and their relevance to human welfare in not more than 100 words.
26. Why are biofertilizers or biopesticides preferred to chemical fertilizers or pesticides ?
27. What are advantages and disadvantages of biofertilizers ?

Long Answer Questions

1. Discuss the role of plant tissue culture in increasing food production.
2. Elaborate as to how biotechnology can be helpful in achieving sustainable agriculture.
3. 'Biotechnology can greatly promote human welfare, but it can also be misused to increase human sufferings', comment on the statement with the help of suitable examples.
4. Bring out the salient features through which biotechnology can lead to higher food production.
5. Give a schematic representation of initiation of callus and suspension cultures ?
6. How would you differentiate between callus and suspension cultures ?
7. Give a schematic representation of micropropagation.
8. What are biopatents ? Give their advantages and disadvantages.
9. How does GM food differ from conventional food ? What harm is expected from it ?
10. What can a transgene do on entering the genome of another organisation ?
11. Define biofertilizers. Describe any four sources of biofertilizers.
12. Dicuss the biological methods for the control of insect pests.

MULTIPLE CHOICE QUESTIONS (Objective Type Questions)

1. The tool of genetic engineering which makes it successful is
   (*a*) DNA polymerase (*b*) Permease
   (*c*) Restriction endonuclease (*d*) Topoisomerase
2. Restriction endonucleases are :
   (*a*) Synthesised by bacteria
   (*b*) Present in mammalian cells for degradation of DNA
   (*c*) Used for *in vitro* DNA synthesis
   (*d*) Used in genetic engineering
3. Restriction endonucleases are used in genetic engineering, because :
   (*a*) They can degrade harmful proteins
   (*b*) They can join DNA fragments

(*c*) They can cut DNA at specific base sequences
(*d*) They can cut DNA at variable sites.

4. Structure involved in genetic engineering is :
(*a*) Plasmid (*b*) Codon (*c*) Plastid (*d*) Scissors

5. One of the following is used to join the segments of DNA during genetic engineering :
(*a*) Lipase (*b*) Ligase (*c*) Gyrase (*d*) Helicase

6. Genetic engineering is possible because :
(*a*) Phenomenon of transduction in bacteria is well understood
(*b*) We can see DNA by electron microscope
(*c*) We can cut DNA at specific sites by endonucleases like DNAase I
(*d*) Restriction endonucleases purified from bacteria can be used *in vitro*.

7. The transgenic animals are those which have :
(*a*) Foreign DNA in some of their cells (*b*) Foreign DNA in all of their cells
(*c*) Foreign RNA in all of their cells (*d*) Both (*a*) and (*c*)

8. Wine and beer are produced directly by fermentation. Brandy and whisky require both fermentation and distillation because :
(*a*) Fermentation is inhibited at an alcoholic level of 10-18%
(*b*) Distillation prolongs storage
(*c*) Distillation improves quality
(*d*) Distillation purifies the beverage

9. Which of the following organelles is related with genetic engineering :
(*a*) Mitochondria (*b*) Golgi apparatus (*c*) Plasmids (*d*) Lysosomes

10. Genetically engineered bacteria have been used in commercial production of :
(*a*) Thyroxine (*b*) Testosterone (*c*) Human insulin (*d*) Melatonin

11. The main technique involved in agricultural biotechnology is called :
(*a*) Tissue culture (*b*) Transformation (*c*) Plant breeding (*d*) NA replication

12. When the genotype of an organism is improved by the addition of foreign genes, the process is called :
(*a*) Biotechnology (*b*) Tissue culture (*c*) Genetic engineering (*d*) Genetic diversity

13. Genetic engineering would not have been possible if which of the following were not known :
(*a*) DNA polymerase (*b*) RNA synthetase
(*c*) DNA ligase (*d*) Reverse transcriptase

14. Which of the following is a genetic vector :
(*a*) Plasmid (*b*) Phage (*c*) Cosmid (*d*) All of these

15. Genetic engineering is :
(*a*) Plastic surgery (*b*) Addition or removal of genes
(*c*) Study of extranuclear genes (*d*) All the above

16. Plasmids occur in :
(*a*) Viruses (*b*) Chromosomes (*c*) Bacteria (*d*) Chloroplasts

17. 'Molecular scissors' used in genetic engineering is :
(*a*) DNA polymerase (*b*) DNA ligase
(*c*) Restriction endonuclease (*d*) Helicase

18. Which of the following correctly defines a transgenic animal ?
(*a*) An animal which has foreign DNA in all its cells because of an injection of DNA into the nuclei of the zygote from which it is developed
(*b*) An animal which has foreign DNA in all its cells because of an injection of DNA into the nuclei of some of the cells in adulthood.

(*c*) An animal which has foreign DNA and RNA in some of its cells because of an injection of DNA and RNA into the nucleus of the zyyote from which it is developed.

(*d*) An animal which has foreign DNA in some of its cells because of an injection of DNA, into the nuclei of some of the cells of the blastocyst.

19. A chemical substance derived from a living source that has the capacity to inhibit growth or destroy microbes is called :

(*a*) Vaccine (*b*) Toxoid (*c*) Toxin (*d*) Antibiotic

20. Biopiracy is related to which of the following :

(*a*) Traditional knowledge

(*b*) Biomolecules and regarding bioresources, genes isolated from bioresources

(*c*) Bioresources

(*d*) All of the above

21. Which of the following is used in biowar ?

(*a*) A pathogen

(*b*) Toxin from a pathogen

(*c*) A delivery system for the bioweapon agent

(*d*) All of the above

22. Which of the following can be controlled by using biopesticides ?

(*a*) Insects (*b*) Diseases (*c*) Weeds (*d*) All of them

23. Which of the following is included in biopesticides ?

(*a*) Viruses and bacteria

(*b*) Viruses, bacteria and fungi

(*c*) Viruses, bacteria, fungi, protozoa and mites

(*d*) Viruses, bacteria, fungi and protozoa

24. Biofertilizers include :

(*a*) Blue-green algae, rhizobia, other nitrogen-fixing bacteria and mycorrhiza fungi

(*b*) Blue-green algae, rhizobia and other nitrogen-fixing bacteria

(*c*) Rhizobia, other nitrogen-fixing bacteria and mycorrhiza fungi

(*d*) Blue-green algae, rhizobia and mycorrhiza fungi.

25. Embryo culture is used for :

(*a*) establishing suspension culture (*b*) recovery of interspecific hybrids

(*c*) somatic hybridization (*d*) haploid production

ANSWERS

1. (*c*) **2.** (*d*) **3.** (*c*) **4.** (*a*) **5.** (*b*) **6.** (*d*) **7.** (*b*) **8.** (*a*) **9.** (*c*) **10.** (*c*)
11. (*a*) **12.** (*a*) **13.** (*c*) **14.** (*a*) **15.** (*c*) **16.** (*c*) **17.** (*c*) **18.** (*a*) **19.** (*d*) **20.** (*d*)
21. (*d*) **22.** (*d*) **23.** (*c*) **24.** (*a*) **25.** (*b*)

COURSE – II

ECOLOGY AND UTILIZATION OF PLANTS

UNIT - 5
ECOLOGY

10
CHAPTER

Organisms and the Environment

The world 'environment' commonly means 'a surrounding'. The life containing and life supporting environment is restricted to a very irregular laye~, five to twenty ten thick, around the globe. Life could orignate and flourish on earth only because the environment of this planet provides the necessary conditions for it.

Every organism lives amongst living (biotic) and non-living (abiotic) objects, events, and influences, the aggregate of which constitutes the environment for their sustenance. Close interaction between the organism and the environment of which it is a part, is essential to satisfy the basic necessities of life.

Thus, the environment can be defined as the sum total of all the physical and biotic conditions which influence the response of organisms.

Ecology is the branch of biology that deals with the study of interaction between organism and their environment. For the first time, the term ecology was coined by a German biologist, Ernst Haeckel in 1869. The origin of this word, was from the Greek word Oikos meaning 'home' and logos meaning 'to study'.

Like other fields of learning, the developments in the field of ecology have been gradual. This science has progressed from natural history and biography to ecosystem ecology. In modern days, global ecology with an emphasis on climate change, biodiversity conservation and ecological sustainability has been emphasised.

What do we study in ecology ? In ecology, we study the distribution and abundance of organism, their interactions among themselves and with the physical environment. The physical environment consists of (*i*) lithosphere (*ii*) hydrosphere and the (*iii*) atmosphere.

In this chapter, the effects of environment factors on organisms, the significance of the range of tolerance, and adaptation of organisms to varying physical environmental conditons have been discussed.

LEVELS OF ORGANISATON

Biology is a field of enquiry from genes to the ecosystems, which form a spectrum of several levels of organisation.

These levels, in the ascending order of complexity are arranged into a graded series, they are : genes → cells → organs → individual organism → population → biological community → ecosystem → land scape → biome → biosphere.

Here, one level of organisation integrates with the other level. However, there are no sharp lines in the functional sense among various levels. The ecological levels of organisation are as follows :

Organisms. They make the basic unit of study in ecology. At each level, the biological unit has a specific structure and function. At this level, the form, physiology, behaviour, distribution and adptations in relation to the environmental conditions are studied.

The organisms of the smilar type have the potential for interbreeding, and produce fertile offspring, which are called **species**.

The organism performs all the life processes independently. However, parts of organism cannot exist independently of one another.

An organism is fully adapted to its environment. It has a definite life span including definite series of stages like birth, hatching, growth, maturity, senescence, aging and death.

Competition, mutualism and predation are various types of interaction between organisms.

Population. In ecology, a population is a group of individuals of the same species, inhabiting the same area, and functioning as a unit of biotic community.

For example, all individuals of the common grass, *Cynodon*, in a given area constitute its population. Similarly, the individuals of elephants or tigers in an area constitute their population.

The interaction between populations are generally studied. These interactions may be a predator and its prey, or a parasite with its host. Competition, mutualism, commensalism, parasitism, and predation are various types of interactions.

Biological community. Biotic community organisation results from intedependence and interactions amongst population of different species in a habitat. This is an assemblage of populations of plants, animals, bacteria and fungi that live in an area and interact with each other.

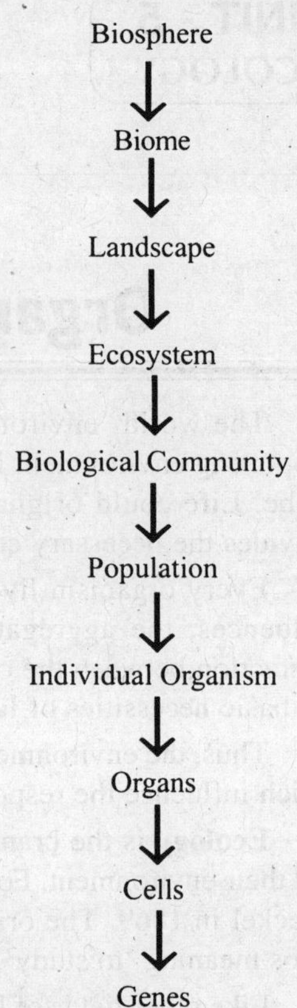

Fig. 10.1. Levels of organisation in ascending series.

A biotic community is a higher ecological category next to population. These are three types of biotic community, they are : animals, plants and decomposers (*i.e.*, bacteria and fungi). A biotic community has a distinet species composition and structure.

Ecosystem. The ecosystems are parts of nature where living organisms interact amongst themselves and with their physical environment. An ecosystem in composed of a biotic community, integrated with its physical environment through the exchange of energy and recycling of the nutrients. The term ecosystem was coined by Sir Arthur Tansley in 1935.

Ecosystems can be recognised as self-regulating and self-sustaining units of landscape, *e.g.*, a pond or a forest. An ecosystem has two basic components : (*i*) abiotic (non-living) and (*ii*) biotic (living organisms). Abiotic components comprise inorganic materials, such as carbon, nitrogen, oxygen. CO_2, water etc., while biotic components include producers, consumers and decomposers.

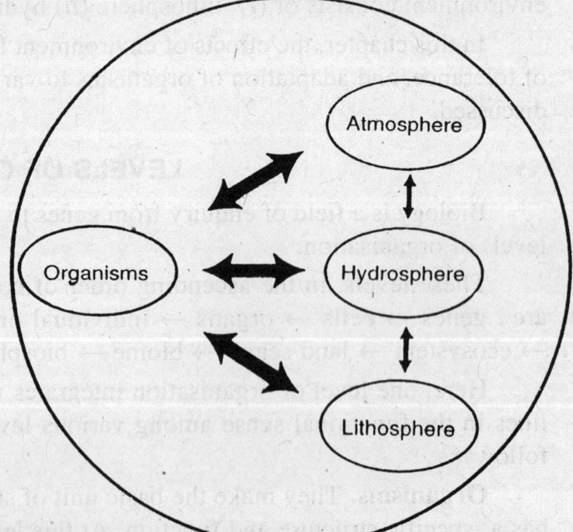

Fig. 10.2. The interaction of organisms with atmosphere.

Landscape. A landscape is a unit of land with a natural boundary having a mosaic of patches, which generally represent different ecosystems.

Biome. This is a large regional unit characterised by a major vegetation type and associated fauna found in a specific climate zone. The biome includes all associated developing and modified communities occurring within the same climatic region, *e.g.*, forest biomes, grassland and savanna biomes, desert biome, etc.

On a global scale, all the earth's terrestrial biomes and aquatic systems constitute the biosphere.

Biosphere. The entire inhabited part of the earth and its atmosphere including the living components is called the biosphere. The global environment consists of three main sub-divisions : (*i*) the hydrosphere which includes all the water components, (*ii*) the lithosphere comprises the solid components of the earth's crust, and (*iii*) the atmosphere formed of the gaseous envelope of the earth. The biosphere consists of the lower atmosphere, the land and the oceans, rivers and lakes, where living beings are found.

ENVIRONMENT, HABITAT AND NICHE

Environment

The word 'environment' commonly means 'a surrounding', but in the context of life it means much more. Life could originate and flourish on earth only because this environment of this planet provides the necessary conditions for it. All organisms, *i.e.* microbes, plants and animals including man, depend upon the environment for their sustenance.

Thus, environment is the sum total of all biotic (living) and abiotic (non-living) factors that surround and potentially influence an organism. Every organism lives in the midst of various living and non-living objects, events and influences. Some components of the environment act as resource, which others act as regulatory factor.

The environment and the organisms are both products of a continuing process of universal change. Each plant or animal species of today is the outcome of millions of years of biological evolution, and any form of life once lost cannot be recreated. Thus, all existing living forms including man are related somehow to one another through common ancestry.

The different components, of the environments are interlinked and interdependent.

Spatial and Time Scales of Environment

Interaction of organisms with their environment takes place at several spatial and time scale.

For example, a single bacterium in soil, interacts with air and water within a fraction of a cubic centimeter space.

On the other land, a tree interacts with a large volume of air, water and soil at a large spatial scale.

The environment varies from place to place due to variations in climate, soil type and topography.

The activities of the organisms influence the hydrosphere, the lower atmosphere and the near surface part of the lithosphere. These activities happen due to exchanges of matter and energy.

On the other land, organisms have to cope with the external environment over a range of time scales. The time scales may be of few minutes, days, seasons or even much longer periods of geological time scale.

For example, phytoplankton populations may change within a few days with the change in light conditions in aquatic systems.

However, the variations in lithosphere occur very slowly over a long period of time.

The ecological factors that influence theorganisms have been classified into two major groups. They are :

(*i*) **Direct factors.** For example : light, temperature, humidity, soil moisture, soil nutrients, soil air, etc.

(*ii*) **Indirect factors.** For example : soil structure, precipitation, wind, soil organisms, altitude, slope, etc.

Climate

Climate is the referred to as the average weather of an area, including general patterns of atmospheric conditions seasonal variations and weather extremes averaged over a long period.

While, the short-term properties of the atmosphere, such as temperature, pressure, humidity, rainfall, sunshine, cloud cover and wind, at a given place and time, as a whole called weather.

The weather reflects the hourly, daily or weekly changes in the above mentioned properties, while the climate exhibits the longer periods, such as season or years together.

Temperature and the rainfall are two most important factors which determine the climate of an area. Thus the climate includes the undermentioned main factors : (*a*) Light, (*b*) Temperature (*c*) Rainfall and atmospheric humidity and (*d*) Air and atmosphere.

Global variations of temperature and rainfall result from : (*i*) differential input of solar radiation in different regions, and (*ii*) from the redistribution of heat energy by winds and ocean currents.

Variations in temperature, rainfall and humidity in different regions of the world form global climate patterns, which govern all life that prevails on earth.

Climatic zones. On the basis of variation in mean temperature along latitude, the main climatic regions are follows :

They are distributed in four categories.

(*i*) Tropical region (0°– 20° latitude)

(*ii*) Sub-tropical region (20°– 40° latitude)

(*iii*) Temperate region (40°– 60° latitude)

(*iv*) Arctic and Antarctic regions (60°– 80° latitude)

The special characteristic of climatic zones are as follows :

(*i*) The mean temperature declines from tropical to arctic region.

(*ii*) In the similar way, as we move from the lower altitude to the higher altitude, in the mountains the mean temperature decreases. Thus, a mountain located in a tropical region will sucessively have tropical → subtropical → temperate and → alpine zones, with increasing alitutude.

(*iii*) Similarly, in temperate zone, the high altitudes will have alpine climatic conditions.

(*iv*) Within each temperature-based climatic zone, the annual rainfall or snowfall varies considerably, and these two factors, *i.e.* temperature and rainfall, determine the vegetation and soil types.

Generally, climate in of two types : (*i*) Microclimate and (*ii*) Macroclimate.

(*i*) **Microclimate.** The microclimate represents the climatic conditions that prevail at local scale or in areas of limited size, *e.g.* the immediate surroundings of plants and animals. This is owing to the uneven topography, plants cover, etc. The microclimates are recognisable only in a very small area and they are sufficient to induce marked local variations in type of composition of vegetation. Thus, microclimate is of much ecological value in the study of plant and animal behaviour.

Microclimate generally differs from the prevailing regional climatic conditions. For example, in a forest, dense foliage reduces the amount of light reaching the ground.

Microclimate also results in changed air temperature profile. In day-time, the air temperature inside the forest is lower than the outside. The interior of a forest in generally more humid than a non-forested area in close vicinity.

(*ii*) **Macroclimate.** In one macroclimatic region there may exist a number of microclimates differing in several respects from the general climate. Forest is an example of macroclimate, where dense forest tall trees are exposed to high intensity of light and temperature, but shrubs growing in the shade of trees do not get the same climatic conditions. Thus, shrubs grow in the environment of dim light, high moisture and low temperature, and again at ground level the conditions are changed where only some peculiar shade loving herbs can thrive.

Habitat and Niche

Habitat. This is a specific place or locality where an organism lives. It is a physical entity that comprises of the sum total of the abiotic factors to which a species or a group of species is exposed. Habitat usually refers to a relatively large area, such as a pond, a forest, an estuary or an ocean.

Habitats are characterised by conspicuous physical features, which include the dominant forms of plant and animal life. The habitat may be referred to the place occupied by an entire biological community. For example, a large number of species of plants and animals are formed in a forest habitat.

The habitats can be further subdivided into parts, which exhibit distinct properties and give shelter to different types of organisms, such habitats are called **microhabitats**. For example, the muddy bottom and a surface water of a pond tree canopy and forest floor are microhabitats.

Some animals are very specific to their habitat, while others have wider choices. For example, *Fubifex*, an annelid lives in flowing fresh water which is rich both in organic content and oxygen on the other hand, the *Chironomous* larvae (insect larvae) can survive in water with very low oxygen content. Certain fishes can live only in fresh water, while others e.g. Hilsa can live in both fresh water and saline water.

Ecological niche. The ecological niche of an organism represents the range of conditions that if can tolerate, the resources it utilises, and its functional role in ecological system. Each species of a community lives in a very specific part of a habitat and performs certain functions. The habitat together with the functions forms the niche of the species.

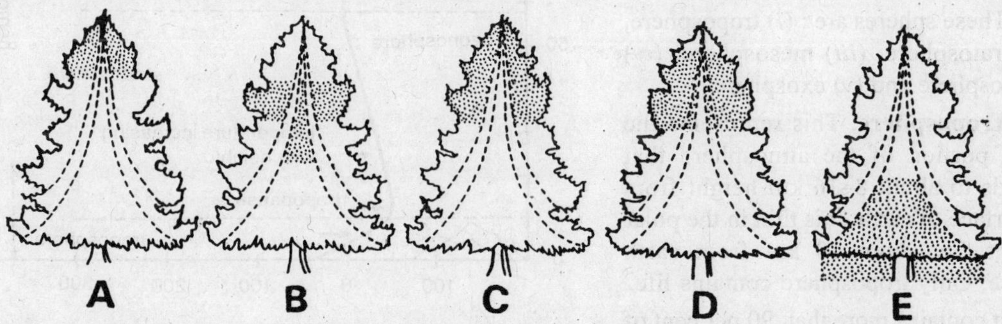

Fig. 10.3. Several species of birds often use the same tree for feeding and nesting, but each species uses a specific zone.

In other words, the concept of ecological niche includes factors, such as food relations, predators, tolerance levels and other habits as well as abiotic factors. Each species within a community has a separate ecological niche. No two species within a given community can have exactly the same niche and live permanently together. In such case, there would be a direct competition with each other till one eliminates the other.

ENVIRONMENTAL FACTORS

As we know, the global environment consists of three main sub-divisions :

1. The Hydrosphere. It includes all the water (liquid) component of the oceans, seas, rivers and other inland water bodies.

2. The Lithosphere. This comprises of solid components of the earth's crust, i.e., rocks, soil and minerals.

3. The Atmosphere. This is formed of the gaseous cover which envelops the hydrosphere and the lithosphere.

The entire inhabited part of the earth and its atmosphere including the living and the non-living components forms the **biosphere**. Thus, biosphere is the combination of all the ecosystem or indefinite units of the earth environment's.

The physical environment consists of factors, such as air, light, heat, water, soil and wind. These abiotic factors determine the success of an organism through their effect on structure, life-history, physiology and behaviour.

Growth and reproduction of an organism are also affected by biotic factors, which include all other organisms in the habitat.

Role of various factors in biotic and abiotic process have been discussed here.

The Atmosphere

The atmosphere is divided into a series of **atmospheric layers** or **spheres**, due to variations in temperature and pressure at various attitudes.

These spheres are : (*i*) troposphere, (*ii*) stratosphere, (*iii*) mesosphere, (*iv*) thermosphere and (*v*) exosphere.

Troposphere. This represents the linear portion of the atmosphere that extends to about 8–16 km height from the surface of earth. It is thin in the polar regions *i.e.* about 10 km from earth surface. Only troposphere contains life.

It contains more than 90 per cent of gases in the atmosphere. Nitrogen and

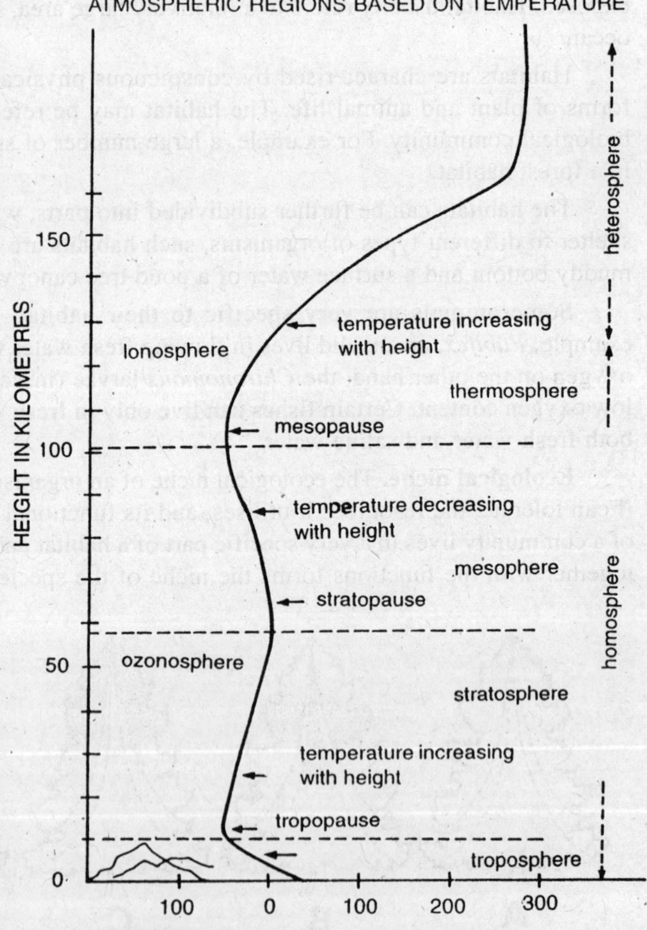

Fig. 10.4. Different layers of the atmosphere.

oxygen are the most aboundant gases in the troposphere, constituting 78.0 per cent and 20.9 per cent of the volume of total gases, respectively.

Generally temperature decreases with increasing height upto **tropopause**, *i.e.*, top of troposphere. The average temperature near the soil surface in about 15°C. The temperature lowers down to –57°C at the tropopause, which denote the transition to the stratosphere.

Stratosphere. The second layer of air mass above tropopause is called stratosphere. It extends upto 30 – 50 km. There is little mixing of gases between the troposphere and stratosphere. However, a thin layer of ozone is present in the stratosphere at the height of 15 to 30 km. The uppermost layer of stratosphere is **stratosphere**, which is the transition layer between the stratosphere and the mesosphere. In this zone, the temperature shows an increase in temperature from a minimum of about – 60°C to maximum of 5°C. The increase in temperature is due to ozone formation under the influence of UV (ultraviolet) rays of solar radiation.

Mesosphere. The third layer of atmosphere next to stratopause in called mesosphere. It extends upto an altitude of 80 km and shows a decrease of temperature with height. The mesosphere is characterised by low atmospheric pressure and low temperature. The temperature begins to drop down from stratopause, goes on decreasing with the increase in height, and seconds a minimum of about – 95°C at a level of 80 km or above. The upper limit of the mesosphere is called mesopause.

Thermosphere. Next to mesosphere is thermosphere, which extends upto 500 km above the earth surface. Thermosphere is characterised by steady temperature increase with the height from mesopause. The thermosphere includes the regions in which ultraviolet radiations and cosmic rays cause ionisation of molecules like oxygen and nitric oxide. This region is called **ionosphere**. In ionosphere, molecules of gases are so widely spaced that high frequency audible sound is not carried by the atmosphere.

Exosphere. The rest of the region of the atmosphere above the thermosphere is called **exosphere** or outer space, which lacks atoms, except those of hydrogen and helium. This extends upto 32190 km above the earth surface. Exosphere has a very high temperature due to solar radiation. The earth's magnetic field becomes more important than gravity in distribution in of atomic particles in the exosphere.

Homosphere extends upto 100 km from earth's surface. It is homogeneous due to air circulation in it.

Hetrosphere is the outer region of thermosphere, which bears gases mixed with radioactive substances.

Air

Air Compositon. Nitrogen and oxygen are major gases in the troposphere, where nitrogen is 78 per cent while oxygen 20.9 per cent. The remaining 1 per cent consists of Argon, water vapour, carbon dioxide, ozone and other gases.

Table 10.1. Air composition

Constituent	*Percentage of volume*
Nitrogen	78 per cent
Oxygen	20.9 per cent
Carbon dioxide	0.03 per cent
Argon	0.23 per cent

Though water vapour, carbon dioxide and ozone (O_3) occur in minute quantities in the atmosphere, yet they are essential for maintaining life on earth.

Ozone occurs only in minute traces in the atmosphere and has a peak concentration around 25 km above the surface of the earth. At no level is its concentration more than a millionth of the main atmosphere. If the entire ozone content is compressed at normal temperature and pressure, it will constitute a layer only 3 mm thick. But it has played a vital role in the evolution of life. The life sustaining role comes from the fact that radiations below 300 mm are biologically harmful, the UV radiation. Any large scale depletion of the ozone content can have devastatory effect on all biological systems, including plants.

Water vapour regulates the hydrological cycle, which sustain life in both terrestrial and aquatic systems. This also absorbs infrared radiation from the earth.

Carbon dioxide, water vapour and ozone play an important role in maintaining the heat balance of the earth. The composition of the atmosphere is a product of the activities of living organisms.

Gases in water. In the aquatic systems, oxygen (O_2), carbon dioxide (CO_2), and other gases are paritally dissolved in water.

In water bodies, oxygen may be a limiting factor for the growth of phytoplankton, and other aquatic organisms, generally in deep lakes, which receive heavy loads of organic materials.

The oxygen supply in water in regulated through diffussion from the air and from photosynthetic activity.

Carbondioxide is highly soluble in water and may be found in variable amounts. CO_2 combines with water and results in carbonic acid (H_2CO_3). The carbonic acid reacts with available limestone and form carbonates (CO_3^{-2}) and bicarbonates (HCO_3^{-1}).

Light

Light is the most essential abiotic factor without which no life can sustain on earth. It is non-lethal limiting factor both at the maximum and minimum levels. The main natural sources of light are-sunlight, moonlight, star light and light producing or luminescent organisms.

Light affects many physiological activities of the plants, *e.g.* photosynthesis, respiration, opening and closing of stomata, growth and flowering of plants, etc. Similarly, light as a complex physical factor, affects the colour vision, eye-size, skin pigmentation, reproduction and biological periodicity of the animals.

Electromagnetic spectrum. Sun is the ultimate source of energy for most organisms on earth, directly or indirectly. The radiant energy coming from the sun in the form of visible spectrum is called light or luminous energy. When the visible sublight in passed through a prism, it is disposed into a series of wavelengths exihibiting seven different colours, *i.e.* violet, indigo, blue, green, yellow, orange and red (VIBGYOR).

Solar radiation, before entering the atmosphere, at 83 km above the earth surface carries energy at a constant rate of 2 cal cm^{-2} min^{-1} in known as solar constant.

Solar spectrum comprises short wave radiation, light and long wave radiation. The short wave radiations include cosmic rays, X-rays and ultraviolet (UV)rays, which have wavelengths shorter than 0.4 to 0.7 mm. This is also known as photosynthetically active radiation (PAR). However, the infrared wavelengths are longer than 0.740 mm.

The ultraviolet (UV) radiation wavelengths are 0.1 mm to 0.4 mm. They are strongly absorbed by ozone (O3) layer present in the stratosphere, and only a small fraction of UV radiation reaches the surface of the earth.

As regards to their wavelength, these are three types of ultraviolet radiation. These are : (*i*) **UV–C** (0.100 to 0.280 mm), (*ii*) **UV–B** (0.280 to 0.320 mm) and (*iii*) **UV–A** (0.320 to 0.400 mm). Out of these three radiation types, **UV–C** radiation type is lethal, and UV–B is harmful to the organisms.

Earth is a very small object in solar system and if receives only 1–50 millionth part of the total solar radiation reaching the universe. The journey of solar energy from sun to earth surface is very much moderated while if passes through stratosphere and the atmospheric envelope surrounding the earth. In the strasphere lies the ozone layer which absorbs the harmful solar radiations of short wavelengths. At any place on the earth surface the energy flux varies with season, time of day and atmospheric humidity and drought.

From the ecological point of view, the energy and its utilisation is most important in sustenance of life. Many of the environmental phenonmena, such as movement of wind currents, rainfall, chemical actions and organic behaviour are controlled by energy.

Effects of light on plants. Light affects many physiological activities of the plants the quality, the intensity and duration of light are important to organisms. Here quality of light means, its wavelength, the intensity means, energy measured in calories and the duration means, the length of day.

The intensity of light varies with latitude and time of the day.

Light affects the processes of (*i*) photosynthesis, (*ii*) respiration, (*iii*) opening and closing of stomata, (*iv*) growth and flowering of plants, (*v*) movements, (*vi*) germination of seeds and (*vii*) reproduction. Light is also an important factor in distribution of plants. Some plants grow and flourish in full sunlight, while others prefer shade and called shade loving plants.

The quality of light plays an important role in flower induction, seed germination and plant movements.

The duration of light regulates the phenological processes, such as flowering and fruiting in plants. The timing of seasonal activites of plants in relation to change in environmental comditions is called **phenology**.

For more details you can consult chapters 3, 4 and 13, *i.e.*, Photosynthesis, Respiration, Plant Growth and Movements, respectively.

On the other hand, in many animals, migration, hibernation and reproduction are controlled by the relative lengths of day and night. The photoperiod is known to affect breeding behaviour of

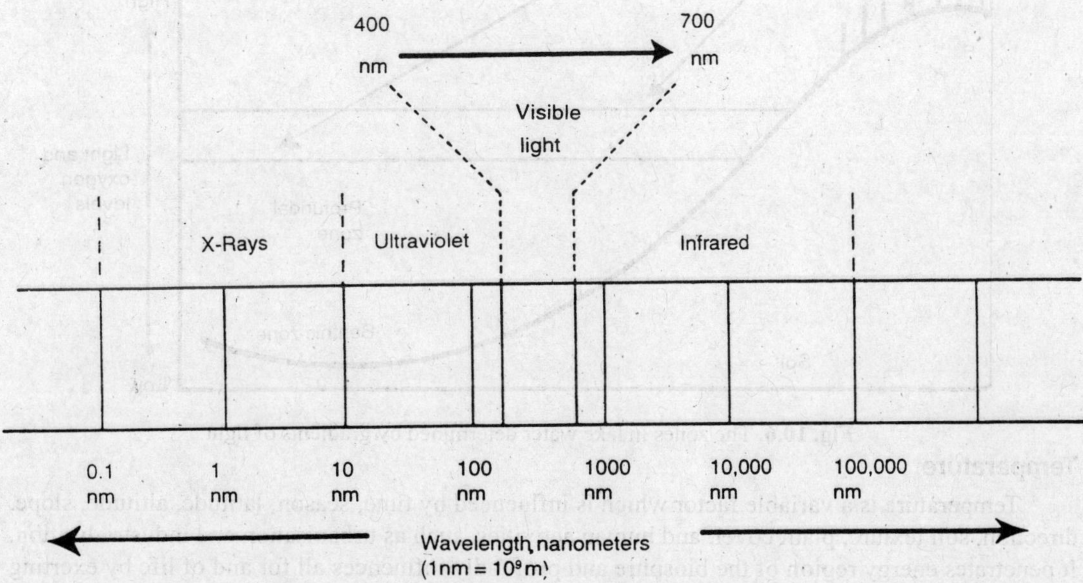

Fig. 10.5. The electromagnetic spectrum of solar radiation.

of insects, certain birds and mammals. The animals are divided into three categories : (*i*) short-day animals, *e.g.* sheep, deer, etc. (*ii*) long-day animals, *e.g.*, turkeys, starlings, etc, and (*iii*) day-neutral animals, rabbit, ground squirrel, etc.

Effects of light on aquatic systems. Life activity under water is often controlled by the availability of light.

Light makes a limiting factor for plants in deep waters, *e.g.*, in oceans and deep lakes.

In aquatic systems, the pressure of light determines where producers and consumers are to live in water. For example, ther phytoplanktons (phyto=plants, plankton = small) live in illuminated surface layer of water, whereas benthic organisms are found in the sediments of a lake.

According to penetration of light, a deep lake can be divided into three zones :

1. The littoral zone. This is shallow water around the edge of the lake that supports rooted vegetation. The light penetrates through shallow water.

2. The limnetic zone. The open water zone found beyond the littoral zone in called limnetic zone. In this zone, phytoplankton grows in abundance,. Here light can penetrate, 20 to 40 m depth depending upon clarity of water.

3. The profundal zone. The dark zone, where light does not reach, is called **profundal zone**. The sediments at the bottom of lakes and ponds form the **benthic** region which makes a habitat for benthic organisms, such as snails, slugs and microorganisms.

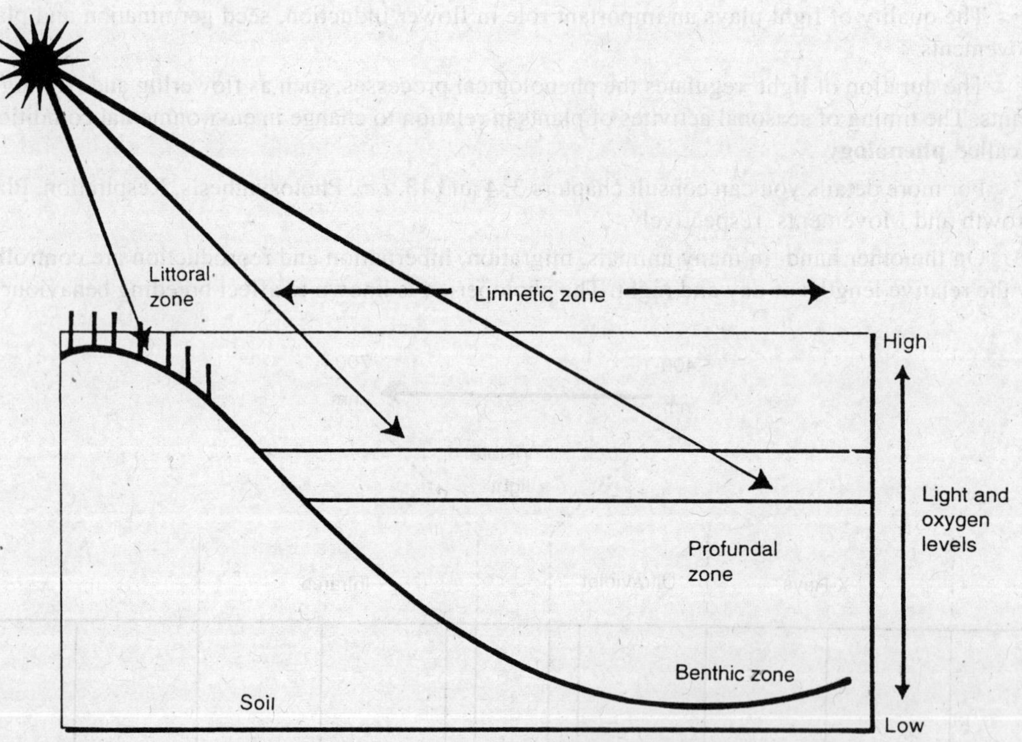

Fig. 10.6. The zones in lake water determined by gradients of light.

Temperature

Temperature is a variable factor which is influenced by time, season, latitude, altitude, slope, direction, soil texture, plant cover, and human activities, such as urbanisation and industrialisation. It penetrates energy region of the biosphre and profundly influences all for and of life by exerting its action through increasing and decrcasing some of the vital activities such as the metabolism, reproductive behaviour, enbryonic development and growth.

Temperature is a measure of intensity of heat. In terms of standard unit its is commonly expressed as degrees either in Fahrehenit or Celcius scale. Heat is a form of energy, called thermal energy. Thermal energy is exchanged between animal and environment by radiation, conduction, convection and evaporation. These four basic needs of transfer of heat energy occur within the organisms and in the interface between the organisms and their environment.

Temperature is the degree of hotness or coldness of a substance. The vertical temperature gradient over earth's surface is called **lapse rate**, the value of which is 6.5°C per 1000 m elevation.

Temperature has an emphatic effect on the climatic conditions, growth responses of plants and activities of organisms.

Temperature also acts as a stimulus for plants, determining the timing of their development. for example, thermoperodism (*i.e.*, a day-night temperature differential) is essential for optimum growth of some plants.

Exposure to cold, (*i.e.*, vermalisation) enhances seed germination, as well as induces flouering in some plants.

Thermoregulation and homeostasis. Temperature also controls the rate of processes inside an organism, and other activities.

In cold-blooded animals, also known as **ectotherm**, the body temperaure tends to match with the temperature of the environment, in which they live.

Many active ectotherms, *e.g.*, frogs and snakes, control their body temperature by moving around or by seeking shade. While some ectotherms are nocturnal in habit and they feed during the night.

On the other hand, the **endotherms**, *i.e.*, warm-blooded animals, such as birds and mammals, regulate their body temperature by physiological processes and maintain a more or less constant temperature within their bodies even when the temperature of the environment fluctuates.

The warm-blooded animals (endotherms) have certain physiological mechanisms to keep their body temperature constant or within tolerance limits.

The maintenance of relatively constant internal environment is called **homeostasis**. To maintain a constant body temperature in the case of endotherm or warm-blooded animals provides them a metabolic advantage over other organisms.

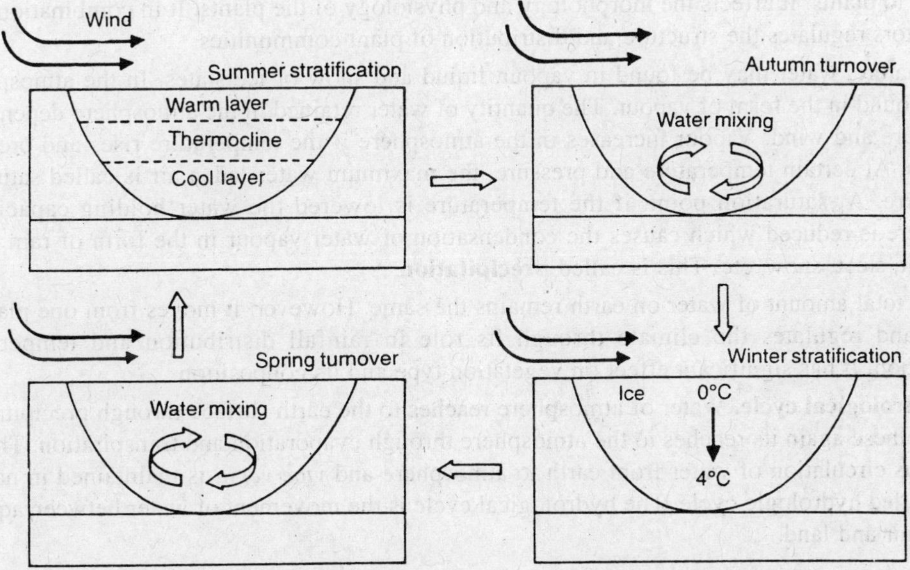

Fig. 10.7. Thermal stratification in temperate lakes.

This is an established fact that biochemical pathways and enzymes often function at their maximum at a temperature about 37°C in endotherms, and threfore, these animals may remain active even under cold conditions by exhibiting different types of adaptations to minimise the loss & heat.

Thermal stratification in lakes. Differences in temperature of at different depths in deep lakes, result in thermal stratification.

During summer, temperature is higher in the surface water, which is separated from the deeper water mass by a **thermocline**, *i.e.*, a zone of gradual change in temperature. During summer, surface water bears a higher temperature, which is separated by deep cold water by thermocline zone. A thermocline often creates two different layers : (*i*) epilimnion, and (*ii*) hypolimnion.

1. Epilimnion. The epilimnion represents the upper layers of water. During winter, in a temperate lake water is at a freezing temperature on the surface. The surface water is cool during autumn and warm in spring. This results in a free mixing of autumn and **spring turnover**. This turnover redistributes oxygen and nutrients, resulting in a bloom of shytoplankton growth.

2. Hypolimnion. The hypolimnion represents the lower layer of water. During winter, the temperature of lower layer remains about 4°C in temperate lakes.

Due to stratified conditions, both in winter and summer, growth of phytoplankton in low due to low nutrients and oxygen supply.

In terrestrial environment temperature fluctuations are varied and marked. Lowest temperature recorded for any land mass is –70°C (in Siberia). Higher temperature, may often reach upto 85°C in deserts at noon. In Rajasthan, the highest temperature exceeds 50°C. Water in hotsprings and geysers may be as high as 100°C.

The temperature varies from place to place and likewise the vegetations of different areas also differ considerably. Desert plants grow in extreme heat; aquatic plants grow in low temperature range, and grasses prefer to grow in the area of moderate temperature.

Water

Water is one of the most important climatic factors. It affects the vital processes of all the living beings. This is the sole agent that sets in motion the nutrients of the soil and makes them available to plants. It affects the morphology and physiology of the plants. It in combination with other factors regulates the structure and distribution of plant communities.

In nature, water may be found in vapour, liqiud and snow or ice states. In the atmosphere, water is found in the form of vapour. The quantity of water retained in the atmosphere depends on temperature and wind. Vapour increases in the atmosphere if the temperature rises and pressure decreases. At certain temperature and pressure, the maximum water-laden air is called saturated atmosphere. At saturation point, if the temperature is lowered the water holding capacity of atmosphere is reduced which causes the condensation of water vapour in the form of rain drop, dew, front, sleet, snow, etc. This is called **precipitation**.

The total amount of water on earth remains the same. However, it moves from one place to another and regulates the climate through its role in rainfall distribution and temperature modification. It has significant effect on vegetation type and its conposition.

Hydrological cycle. Water of atmosphere reaches to the earth surface through precipitation, and from these again its reaches to the atmosphere through evaporation and transpiration. Thus, a continuous circulation of water from earth to atmosphere and *vice versa* is maintained in nature. This is called hydrologic cycle. The hydrological cycle is the movement of water between aquatic systems, air and land.

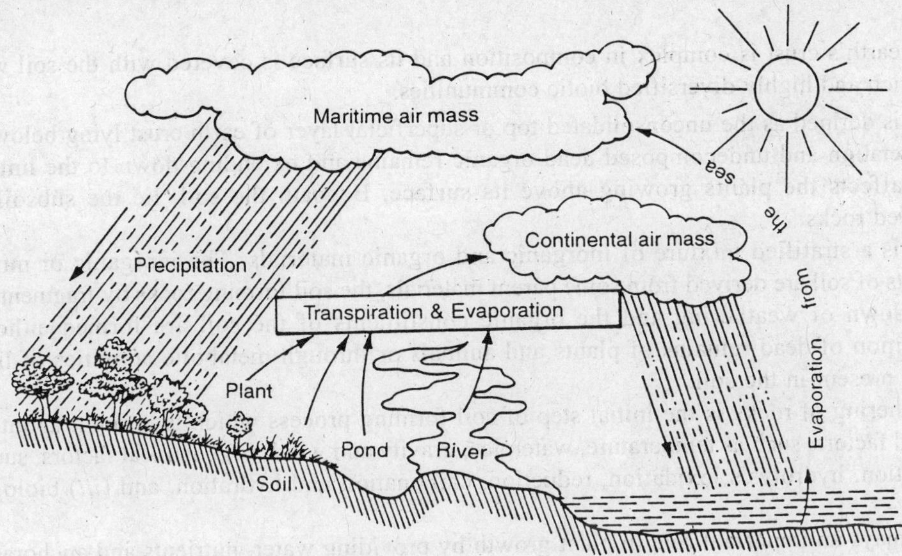

Fig. 10.8. Hydrological cycle in nature.

Water vapour, an important constituent of the atmosphere, can condense to form clouds, and eventually rainfall. Plants play an important role in the hydrological cycle through the process of transpiration. In tropical forests as much as 75 per cent of annual rainfall is returned to the atmosphere by plants, which recharges the atmospheric water vapour.

Rainfall is uneven on earth's surface. For example, deserts recieve very low rainfall, (*i.e.*, less than 100 mm year^{-1}) each year, while same areas receive high rainfall, *e.g.*, in Cherrapunji, average rainfall is more than 11,000 mm per year.

Atmospheric humidity. The water vapour present in unit volume of air is called **absolute humidity**. This is expressed in terms of percentage of water vapour present in unit volume of air at certain temperature. The amount of water vapour required to saturate the same unit volume of air under constant physical conditions is called **relative humidity**. This is affected by light intensity, temperature, wind velocity, vegetation and soil water. Relative humidity (RH) is measured by an instrument called hygrometer. It plays an important role both in life of plants and animals.

Plant-water relations. For land plants, the main source of water to soil is rainfall and melting of snow.

The potential energy of water is referred to as water potential. This is the source with which soils hold water and it is quantified in terms of pressure.

At sea level pressure exerted by the atmosphere is one bar which equals about 0.1 megapascals (MPa).

The upper limit of water availability of a soil is called **field capacity**. The water potential of a soil at field capacity is about – 0.01 MPa.

The lower limit of water availability of a soil is called wiliting point. The water potential of a soil at wilting point is about – 1.5 MPa.

While the **osmotic potential** in the roots of plants is responsible for the entrance of water in the roots from the soil.

Plants have adapted to variations in water availability by habitat selection (*e.g.* xerophytes, mesophytes, hydrophytes, etc.) phenological adjustments, or by having higher water use efficiency.

Soil

The earth's crust is complex in composition and its surface is covered with the soil which supports rich and highly diversified biotic communities.

Soil is defined as the unconsolidated top or superficial layer of earth crust lying below any aerial vegetation and undecomposed dead organic remains and extending down to the limits to which if affects the plants growing above its surface. Beneath the soil, lie the subsoil and unweathered rocks.

Soil is a stratified mixture of inorganic and organic materials. The inorganic or mineral constituents of soil are derived from some parent materials, the soil forming rocks by fragmentation or break-down of weathering and the organic constituents of the soil are formed either by decomposition of dead remains of plants and animals or through metabolic activities of living organisms present in the soil.

Weathering of rocks is the initial step of soil forming process which is brought about by : (*i*) physical factors, such as temperature, water, ice, gravity and wind, (*ii*) chemical factors, such as solvent action, hydrolysis, oxidation, reduction, carbonation and hydration, and (*iii*) biological factors.

Soil plays an important role in plant growth by providing water, nutrients and anchorage.

Soils support the growth of crops, grassland and forests on which we depend for food, fibre, wood and building materials.

Edaphic factors are those which are dependent on the soil, such as soil constitution, soil water, soil air, soil organisms, etc.

Soil at different places vary considerably in their structure, components and properties. These differences in the soils are often largely responsible for differences in vegetation within the same climatic region, and consequently of great significance in the distribution of plant communities.

The mineral composition of soils depends on the minerals in the parent material and the extent of weathering.

Soil profile. The vertical layered structure of soil is called the **soil profile**. It develops due to weathering process, accumulation of organic matter and the leaching of mineral matter.

These are four main horizons in a soil profile : the O-horizon is the organic layer composed of lead organic residues, found above the A-horizon.

The A-horizon, also called the **top soil**, is the uppermost mineral layer, which also contains roots and partially decomposed organic matter. The A-horizon designates the top stratum which is subjected to marked leaching.

The B-horizon is the subsoil, lying under A-horizon has little organic matter, very few plant roots and a sparse microflora and fauna. In it, iron and aluminium compounds are often accumulated. A and B-horizons collectively represent the true soil.

The C-horizon represent, the less weathered parent maeatrial. In this layer the organic matters are present in small amounts and little or no life is present.

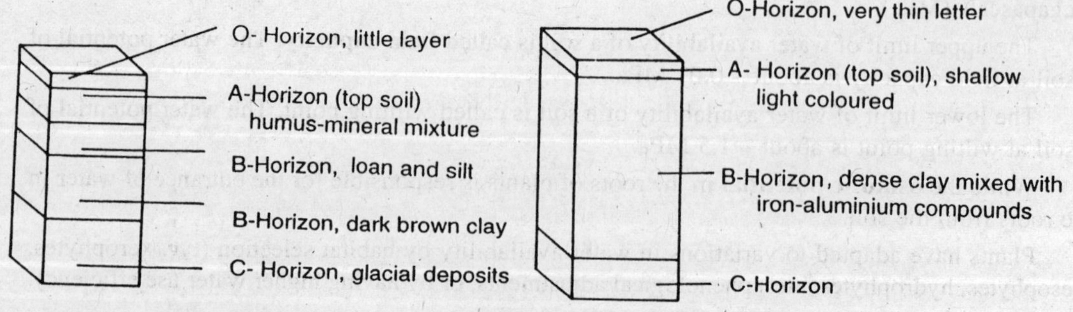

Fig. 10.9. Soil profile. A, temperate deciduous forests; B, tropical rain forests.

The development of soil profile is mainly governed by the climate and the kind of vegetation.

Here, in the following figures, the soil profiles of a temperate deciduous forest and tropical rain forest are given :

Characteristics of soil profile. Soils in forest, grasslands and deserts differ markedly in colour, clay, organic matter content and depth.

In desert soils, the top soil (A-horizon) has little or no humus, while subsoils (B-horizon) make a mixture of sand, clay, minerals and salts.

However, in the grasslands the roots of native plants penetrate deep in the ground, forming a dense sod that holds moisture and prevents erosion. This adds large amount of organic material to soil each year.

While in temperate forest soils, the top soil (A-horizon) is a rich mixture of humus and inorganic soil components as exhibited in Fig. 16.9 A.

In tropical rain forest, dense clay subsoil (B-horizon), heavy rainfall, and high temperatures result in nutrient-poor and shallow soils, as exhibited in Fig. 16.9 B.

Properties of soils. Soils are made up of four components : The two solid substances are : (*i*) small mineral particles and (*ii*) organic matter. The other components are : (*iii*) air (gaseous), and (*iv*) water (liquid).

(*i*) The small mineral particles come from weathering of rocks. They include about 25 elements and of which 16 elements are essential. These are divided into two categories, *i.e.*, macronutrients, such as C, H, O, N, P, S, Ca^{++}, K^{+}, Mg^{++} and micronutirents, such as Fe, Mn, B, Mo, Cu, Zn, and Cl.

(*ii*) The organic matter is of plant and animal origin, both from dead and living organisms.

In an average soil, about 45 percent of the volume is filled with mineral particles and about 5 percent ley organic matter.

The remaining 50 percent of the volume is filled with soil (*iii*) air and (*iv*) water. The amount of air or water varies greatly, depending on how wet soil is at the time.

The proportion of mineral particles of different sizes in soil represents its texture. According to varying soil particles they are divided into following categories. The particles vary in sizes, such as **gravel** (more than 2 mm); **coarse sand** (between 0.2 to 2 mm); **fine sand** (0.02 to 0.2 mm); **silt** (0.002 to 0.02); **clay** (less than 0.002 mm).

Fine texture soils have a predominance of clay and silt particles; while the coarse textured soils have a predominance of sand particles.

Sandy soils. It is formed of 85 percent sand and 15 percent clay and silt, also known as **light soil**. This soil is not rich in nutrients and is less fertile. It is characterised by xerophytic vegetation.

Clayey soil. It is formed of 50 per cent clay and 50 per cent silt or sand or both also known as **cold** or heavy soil. It has high water holding capacity and less water logging.

Silt soil. It has 90 percent silt and 10 percent sand. It has good soil porosity and water holding capacity. This is poor in nutrient supply.

Loamy soil. It is composed of 70 per cent sand and 30 per cent clay or silt or both. This is considered best soil for plant growth. It has good water holding capacity, water infiltration and adequate aeration. Roots penetrate deep in loamy soils.

Soil moisture. Plants absorb a small quantity of rain water and dew directly, but they take a large quantity of water from the soil. Water held in the soil is of following types :

(*i*) **Gravitational water.** This is free water which percolates downwardly through the pore spaces between soil particles and is accumulated in the pore spaces in the form of **ground water**.

(*ii*) **Capillary water.** The amount of water present around the soil particles and held by surface tension and attraction force of water moelcules. Capillary water remains readily available to the roots upto a certain soil moisture tension.

(*iii*) **Hygroscopic water.** Water, which is adsorbed on the soil particles and held on the surface of particles by forces of attraction and cohesion of its molecules. This is the moisture, remaining in air-dry soil, and cannot be used by the plants.

Soil organic matter. The organic matter stored in soil may be divided into two categories:

(*i*) **Detritus or litter.** This is freshly dead and partially decomposed plant and animal material.

(*ii*) **Humus.** This is colloidal, amorphous and dark coloured substance that constitutes the organic component of soil. It is formed by the decomposition of plant and animal remains and has a complex and variable chemical composition. Being a colloid, it can hold water and therefore improves the water retaining properties of soil; it also enhances soil fertilily. **Acidic humus (mor)** is found in region of coniferous forest, where the decay is brought about mainly by fungi. **Alkaline humus (mull)** is typically found in grassland and deciduous forest : it supports an abundance of micro-organisms and small annelids, *e.g.*, earthworms.

A part from the dead organic matter, soil harbours very large number of organisms which include detritus feeders, such as termites, earthworms, carpenter ants and crabs; bacteria and fungi.

The soil organisms act as decomposers and play an important role in decomposition of plant and animal residues.

Fire

Fire is a biological factor rather than a physical factor as it is mostly caused by man's activity. Lightning initiated fires have destroyed plants and animals since their early appearance on earth. A large body of information has been developed on the effects of fire on grassland and forests as well as on the use of fire in land management. Native people in various lands have used fire to enhance hunting, improve visibility and provide forage, etc.

The branch of ecology which deals with the effects of the fire an ecosystem is called **'fire ecology'** or **'ecopyrology'** plants having ability to withstand fire with little or no damage are referred to as **pyrophytes**. Many pyrophytes are known to occur in Siwalik hills of India. The examples of pyrophytes are *Cochlospermum religiosa*, *Combretum nanum*, *Grewia sapida*, etc. These small plants are supposed to have become permanently dwarf by annual forest fires. Pyrophytes are generally woody plants with thick bark.

Fires caused by man's activity are responsible for complete destruction of vegetation at certain places resulting in temporary or permanent alterations in the characters of vegetations. Fire has marked effects on the physical environment through removal of the plant cover, burning of litter mass present on the soil surface, and loss of nutrients due to volatilisation.

There is mortality among a variety of animal groups due to occasional or recurrent fires.

Generally in fires the aerial parts of plants are destroyed completely but their roots, rhizomes or other underground parts may sometimes remain unaffected which under favourable conditions may grow and produce new plants.

Fire generally makes the area suitable for the growth of grasses and thus imporves the quality of forage. Post-burn plants are preferred by herbivores. Animals grazing on burnt grasslands are found to gain in weight more rapidly than these grazing on unburnt grasslands. Fire removes harmful plant parasites and pests.

Fire is an important factor in temperate forest and grassland regions, as well as in tropical areas having dry seasons.

Biotic Factors

The biotic factors include the influence of living organisms,both plants and animals upon the vegetation. The living species may be the producers, (*e.g.*, the green plants) of food or consumers such as herbivores predators, parasites or omnivores.

Any activity of the living organism, which may cause marked effects on vegetation in any way is referred to as **biotic effect**. The biotic effect may be both direct and indirect. It may be beneficial to the plants in some respects but detrimental in other respects. The species may have positive, negative or neutral effect on population growth and well-being of other species.

The plants live together in a community and influence one another. In the forest there are many plant communities, such as trees, shrubs, herbs, mosses and lichens. These communities interact with one another and adjust according to environmental conditions. Trees cast their shadow on many shade-loving plants, which grow around or beneath them. The micro-organisms, such as bacteria, algae and fungi affect the life of plants of a given area in many ways.

Many animals use plants as their food and for shelter as well. Besides animals, the man is most significant agent for modifying the vegetation.

Range of tolerance

The environmental factors vary in their effects, and biological species have a range of tolerance for these factors. Any of these factors which is present in least amount may become limiting. The examples of such limiting factors are as follows :

(*i*) Low temperature limits plant growth at higher elevations.

(*ii*) Water availability limits plant growth in deserts.

(*iii*) Whereas low phosphorus availability limits the growth of phytoplankton in deep lakes.

However, this is noticeable that not only too little amount of something is a limiting factor, but also too much may be a limiting factor for the growth and distribution of an organism.

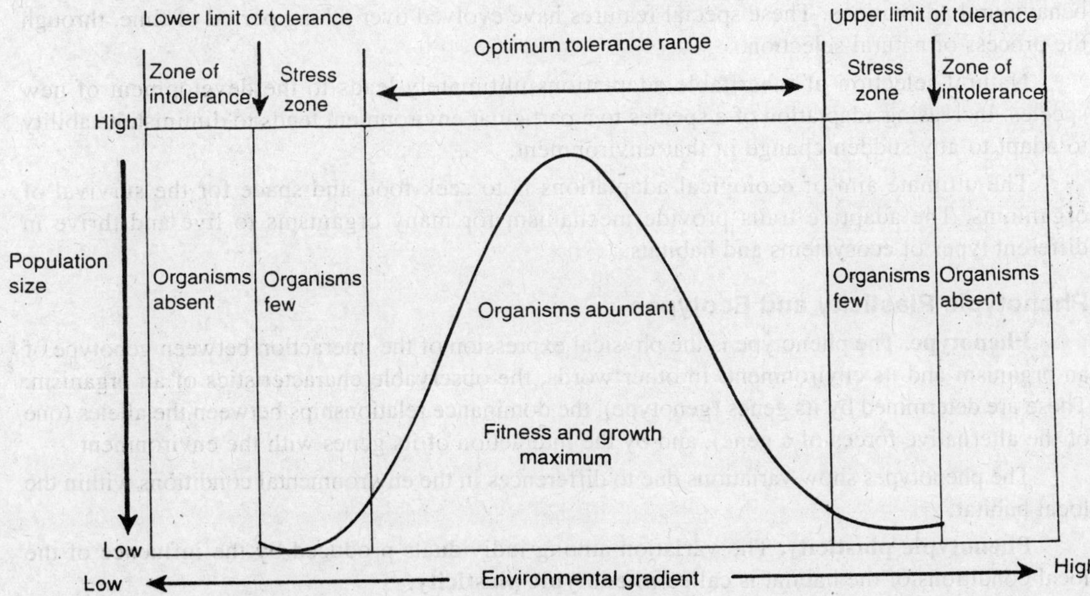

Fig. 10.10. The response of an organism to a range or gradient of an environmental factor, *i.e.*, temperature, light, nutrient.

The organisms show a bell shaped curve (Fig. 10.8) to a range or gradient of a single environmental factor, such as temperature, sunlight or nutrient concentration.

In this curve, the organisms are abundant in the central **optimum** range showing greatest fitness, growth and survival.

On the other hand, in the zone of **stress**, only a few organisms survive and are not able to reproduce. While in the zone of **intolerance** the organisms are absent.

Thus, the range between 'too little' and 'too much' shows the range of tolerance that determines the existence and abundance of an organism.

If the range of tolerance is wide then the organisms are widely distributed, on the other hand when this range is narrow, the distribution of organisms is restricted.

Acclimatisation

The gradual physiological adjustment to slowly changing new environmental conditions is known as **acclimatisation**.

The limits of tolerance and optimal range of distibution may vary in different seasons, in some species of animals. If some environmental factor shifts beyond the limit of tolerance of an organism; in such conditions, the organisms can come to the resting stage or migrate or they can acclimatise them in new environmental conditions.

ECOLOGICAL ADAPTATIONS

The term **adaptation** may be defined : Any change in the structure or functioning of an organism that makes it better suited to its environment. However, the special charcteristics of plants and animals that enable them to be successful under prevailing set of environmental conditions are called **adaptations**.

The organisms in natural world have various types of morphological, physiological and behavioural adaptations. These special features have evolved over a long period of time, through the process of natural selection.

Natural selection of inheritable adaptations ultimately leads to the development of new species. Increasing adaptation of a species to a particular environment tends to diminish its ability to adapt to any sudden change in that environment.

The ultimate aim of ecological adaptations is to seek food and space for the survival of organioms. The adaptive traits provide mechanism for many organisms to live and thrive in different types of ecosystems and habitats.

Phenotypic Plasticity and Ecotypes

Phenotype. The phenotype is the physical expression of the interaction between genotype of an organism and its environment. In other words, the observable characteristics of an organism. These are determined by its genes (genotype), the dominance relationships between the alleles (one of the alternative forces of a gene), and by the interaction of its genes with the environment.

The phenotypes show variations due to differences in the environmental conditions within the local habitat.

Phenotypic plasticity. The variation among individuals produced by the influence of the local conditions of the habitat is called **phenotypic plasticity**.

Ecotypes. Commonly, species having a wide range of distribution evolve genetically adapted local populations, called **ecotypes**. The ecotypes differ from each other on the basis of morphological and physiological characters. Though ecotypes of a species differ genetically yet they are interfertile. Such ecotypes are commonly found in plants and sessile animals.

Strategies of Adaptations in Plants

The living organisms react with their environments and they bear full impression of the environments in which they grow. In order to withstand adverse conditions of the environment and utilise to their maximum benefit the nutrients and other conditions prevailing therein, the organisms develop certain morphological, anatomical, physiological and reproductive features. Any feature of an organism or its part which enables it to exist under condition of its habitat is called **adaptation**. Every organism develops certain adaptations and so does the population or a community.

Plants have special traits that help them to enlarge their tolerance lmits to : (*i*) light regimes, (*ii*) dry conditions, (*iii*) high temperature, (*iv*) water saturated condition and (*v*) saline environments.

In plants, flowers evolve special structures to ensure pollination by insects or other animals.

Plants also deveop various mechanisms to deal with stress conditions of the environment.

Adaptation to light. Plant communities, as well as individual plants, adapt to different light intensites. These are two categories of such plants :

(*i*) **Sciophytes.** They are also known as shade loving plants. The shade loving or shade adapted plants generally have low photosynthetic, respiratory and metabolic activities. For example, several ferns and other herbaceous plants growing on the ground under dense canopy of trees in forests, are shade tolerant plants.

(*ii*) **Heliophytes.** Such plants are adapted to high intensity of light. They are higher temperature optima for photosynthesis, as well as have high rates of respiration. For example, many tropical plants, particularly grasses which grow in not and arid climates possissing C_4 pathway of photosynthesis.

Adaptations to Water Scarcity and Heat

Plants of hot deserts are adapted to survive in dry conditions of soil and high temperature. Places where available water is not present in adequate quantity are termed **xeric** habitats.The plants growing in xeric conditions are xerophytes. Xeric habitats may be of following types :

(*i*) Habitats physically dry; where water retaining capacity of the soil is very low and the climate is dry, *e.g.*, desert, rock surface, waste land, etc.

(*ii*) Habitats physiologically dry; places water is present in excess amount but it is not such as can be absorbed by the plants easily. Such habitats may be either too salty or too acidic, too hot or too cold.

(*iii*) Habitats dry physically, as well as physiologically, *e.g.*, slopes of mountains.

Drought evading plants. The plants which evade dry conditions are known as **ephemerals**. They are short lived plants. During critical dry periods, they survive in the form of seeds and fruits which have hard and resistant seedcoats and pericarps respectively. For example, in desert areas of Rajasthan, many annual plants germinate from seeds, complete their life cycle quickly during the rainy season and survive in the form of seeds in the following dry season. Thus, plants remain unaffected by extreme dry conditions. Such plants (ephemerals) are quite common in the semiarid zones where rainy season is of short duration. The common examples are : *Astragalus*, *Artemesia* members Zygophyllaceae, Boraginaceae, grasses, etc.

Deep rooted plants. Some plants have deep tap roots, which can reach even upto water table, in arid climates, hence they are capable of absorbing water from deep soil. Some examples of deep rooted plants are : *Prosopis* (mosquito), palms and some species of *Acacia*.

Plants having deciduous leaves. Many shrubs and trees of dry climate, such as *Acacia nilotica* and *Capparis decidua* have deciduous leaves. But in the majority of the plants the leaves are generally reduced to seeds, as in *Casuarina*, *Ruscus*, *Asparagus*, etc.

Leaves with sunken stomata. For example, stomata are found is sunken pits in the common *Nerium* plant, reducing the rate of transpiration.

In many xerophytes, the leathery leaf surface and waxy cuticle also check the transpiration.

Sometimes the leaves may be reduced to spines, *e.g.* in *Ulex*, *Opuntia*, *Enphorbia*, *Capparis*, *Acacia*, etc.

Succulence. Succulents are those plants in which some organs become smaller and fleshy due to active accumulation of water, or the bulk of the plant body is composed of water storing tissues. Water stored in these tissues is consumed up during the period of extreme arid conditons when the soil becomes depleted of available water. In cacti leaves are reduced to spines, while stems are modified into fleshy and spongy structures.

In leaf succulents, the leaves swell remarkably and become very fleshy owing to storage of excess amount of water and latex in them. Plants with succulent leaves generally develop very reduced stems. The examples of leaf succulents are : *Sedum*, *Aloe*, *Mesembryanthenum*, *Kleinia* and several members of family Chenopodiaceae.

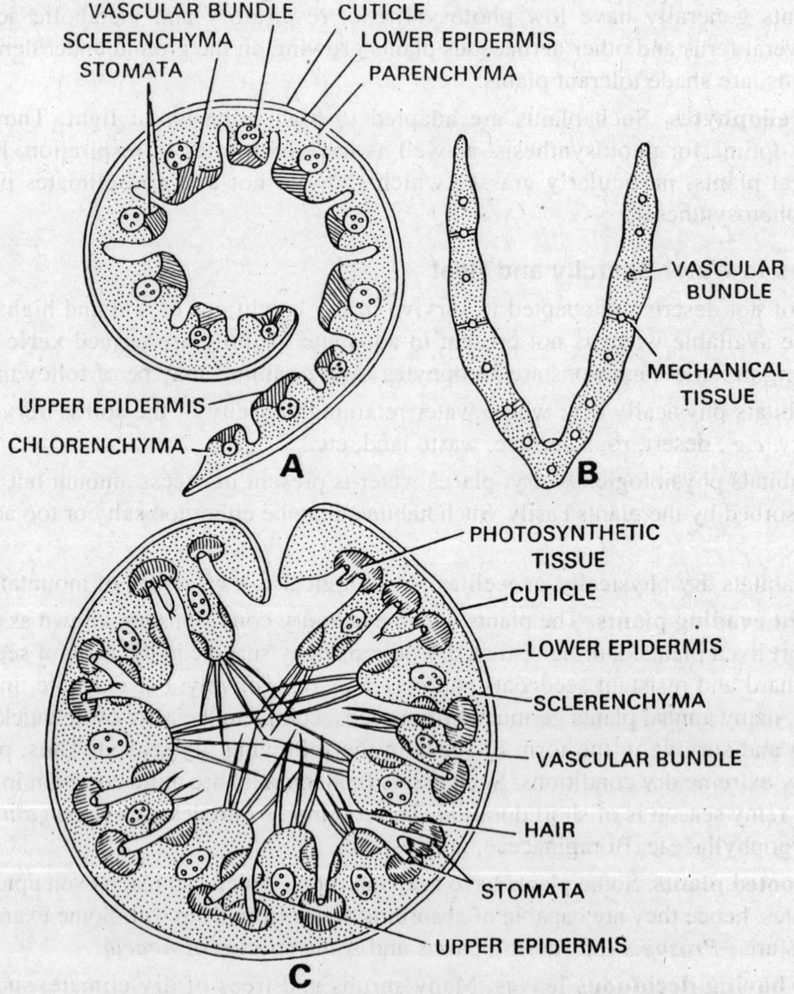

Fig. 10.11. Rolling of xerophytic leaves. A. Cross-section of *Agropyron* leaf; B. C.S. of *Poa pratensis* leaf; C.C.S. of *Ammophila arenaria* leaf.

Trichophylly. In some xerophytes, especially these growing well exposed to strong wind, the undersurfaces of the leaves are covered with thick mats of hairs which protect the stomatal guard cells and check the transpiration. These xerophytes which have hairy covering on the leaves and stems are known as trichophyllous plants, *e.g.*, *Zizyphus*, *Calotropis*, *Banksia*, *Nerium*, etc.

Rolling of leaves. The leaves of many xerophytic grasses, roll tightly under dry conditions. In these grasses, the stomata are combined to the ventral surface of the leaf, so that when the leaf edges roll inward, the stomata are effectively shut away from to outside air. As the stomata are located on the inner surface of the leaf, the air enclosed by the rolled leaf soon becomes saturated with water and the outward diffusion steps, *e.g.*, *Ammophila*, *Erica*, *Spartina*, etc.

Other significant adaptations. Many tropical plants, especially grasses which grow in hot and arid climates, have C_4 pathway of photosynthesis. The plants having C_4 pathway of photosynthesis perform better in low soil water environments. Thus, such plants use less water to achieve higher rates of photosynthesis, at higher temperatures.

Many other xerophytes, such as cacti and succulents close their stomata during the day and open them in the night, hence reducing transpiration. Crassulacean acid metabolism (CAM) occurs in succulents and other plants that normally grow in dry conditions.

Many xerophytes accumulate proline (an amino acid) in the cells which maintain osmotic and water potential in their leaves.

On the other hand, the heat shock proteins (chaperonins) provide physiological adaptations to plants to high temperatures. Such protein help other protein to maintain their structure and avoid denaturisation at high temperature.

Characteristics of xerophytes. The peculiar characteristics of xerophytes are as follows :

1. They grow in deserts or in very dry places; they can withstand a prolonged period of drought uninjured, for this purpose they have certain peculiar adaptations.

2. The xerophytic plants have to guard against excessive evaporation of water; this they do by reducing evaporating surfaces.

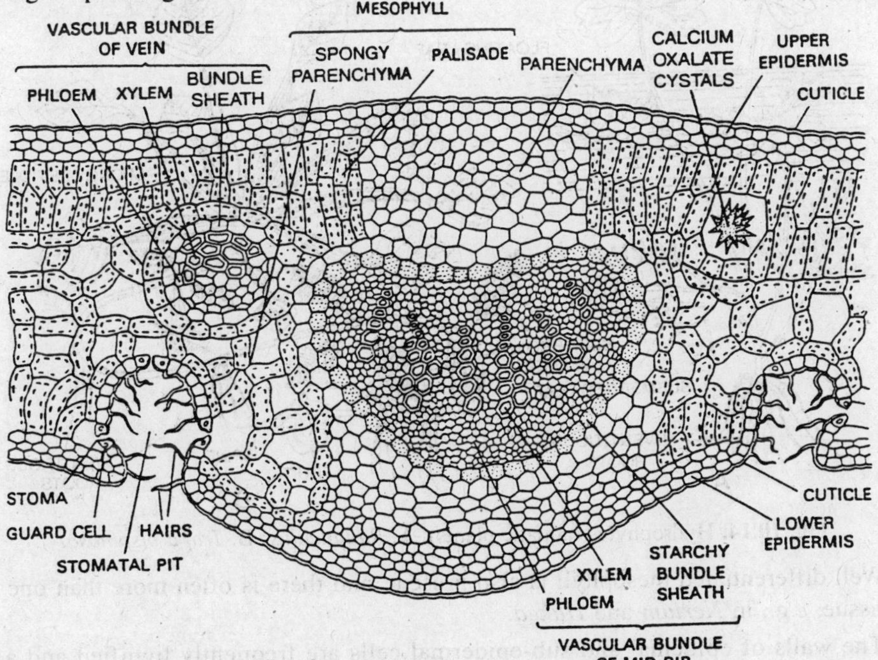

Fig. 10.12. C.S. of *Nerium* leaf showing xerophytic characters..

3. Plants produce a long top root which goes deep into the sub-soil in search of moisture.

4. To retain the water absorbed by the roots, the leaves and stems of some plants become very thick and fleshy, *e.g.*, *Aloe*, *Agave*, etc. Water tissue develops in them for storing upwater; this is further facilitated by the abundance of mucilage contained in them.

5. Multiple epidermis sometimes develops in the leaf, *e.g.*, in *Nerium*.

6. Modification of the stem into phylloclade for storing water and food at the same time performing functions of leaves is characteristic of many desert plants, *e.g.*, *Opuntia* and other cacti.

In xerophytes, certain structural features are also common. Leaves are thick and leathery, with a well developed cuticle and abundant hairs.

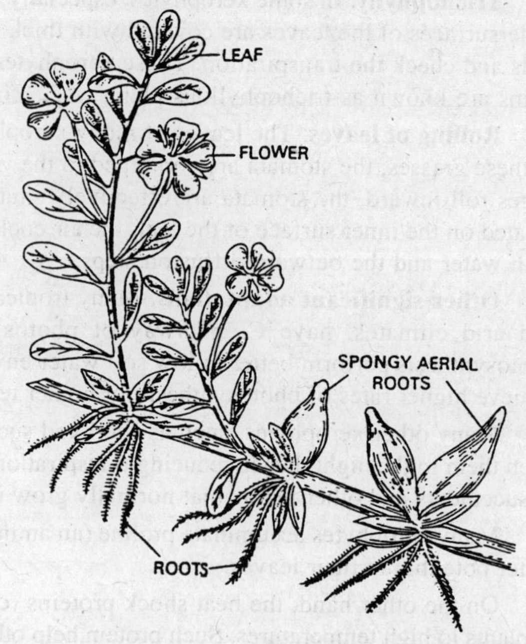

Fig. 10.13. *Jussiaea,* a typical hydrophyte.

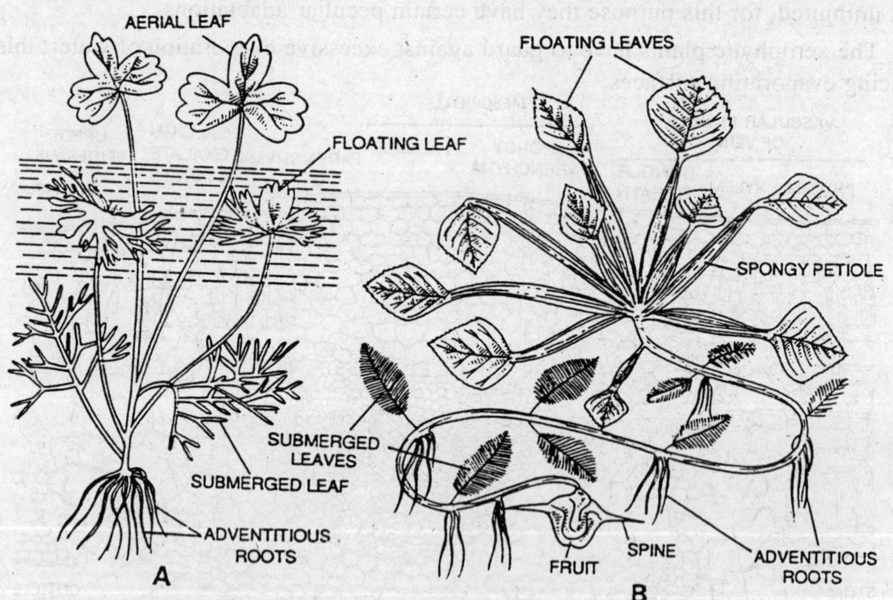

Fig. 10.14. Hydrophytes (aquatic plants). A, *Ranunculus;* B. *Trapa bispinosa.*

8. Well differentiated mesophyll is also present, and there is often more than one layer of palisade tissue, *e.g.*, in *Nerium* and *Hakea*.

9. The walls of epidemal and sub-epidermal cells are frequently lignified and a distinct hypodermis may be present.

10. They have well developed vascular system and often an abundance of sclernchyma, either in the form of sclereids or fibres, *e.g.*, in *Hakea* and *Ammophila*.

11. The leaf is sometimes cylindrical or related. This organisation is to protect the stomata, which may occur in furrows.

Adaptations in aquatic environments. Plants which remain permanently immersed in water are called **hydrophytes**. Such plants grow in water or very wet places. They may be submerged or partially submerged, floating or amphibious. Their structural adaptations are mainly due to the high water content and the deficient supply of oxygen. The various adaptations are as follows:

Epidermis. In aquatic plants, the epidermis is not protective but absorbs gases and nutrients directly from the water. The epidermis in the typical hydrophyte has an extremely thin cuticle, and the thin cellulose walls permit ready absorption from the surrounding water. Commonly chloroplasts are found in epidermal cells of leaves, especially when the leaves are very thin; these chloroplasts utilise the weak light under water for photosynthesis.

In submerged plants, stomata are not present, and exchange of gases takes place directly through cell walls. The floating leaves of aquatic plants have abundant stomata on the upper surface.

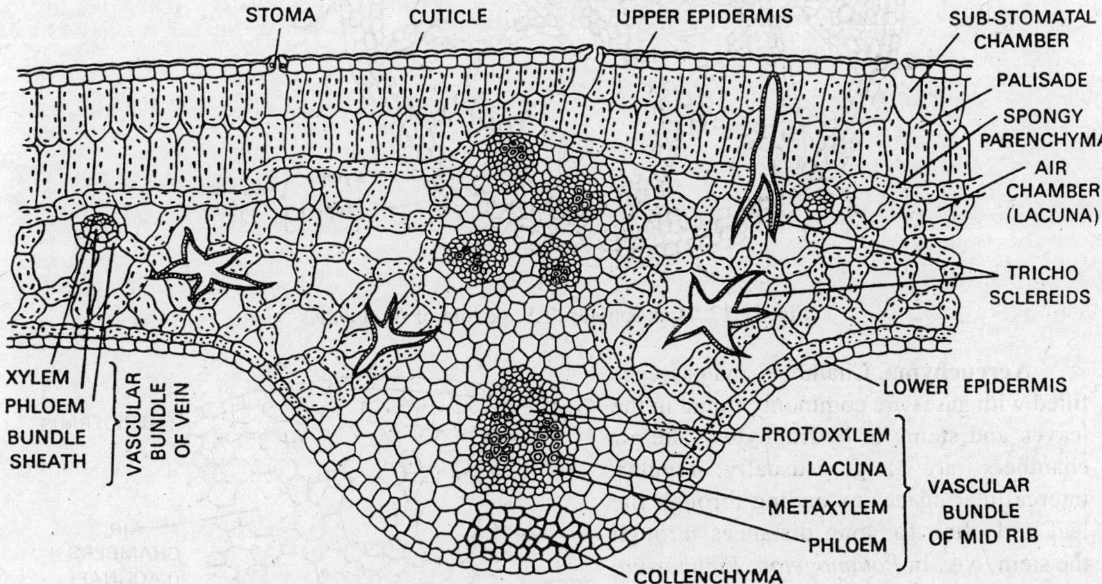

Fig. 10.15. Cross-section of Nymphaea leaf (a floating hydrophyte).

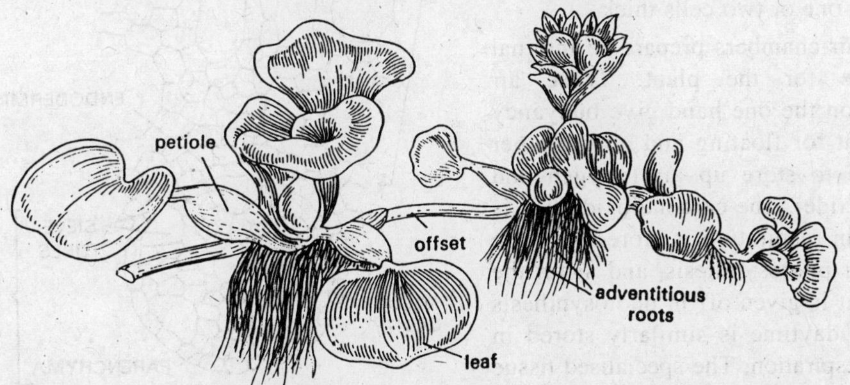

Fig. 10.16. Hydrophytes. *Eichhornia* (a floating hydrophyte).

Reduction of absorbing tissue. The root-system in hydrophytes is feebly developed and root hairs and root caps are absent. In some floating plants, such as *Utrieularia*, *Ceratophyllum*, etc., no roots are developed, and in submerged plants, such as *Vallisneria*, *Hydrilla*, etc., water dissolved mineral salts and gases are absorbed by their whole surface. An aquatic plant is in reality, submerged in or floating upon a nutrient solution. In hydrophytes the root-system is functioning chiefly as holdfasts or anchors and a large part of the absorption takes place through leaves and stems.

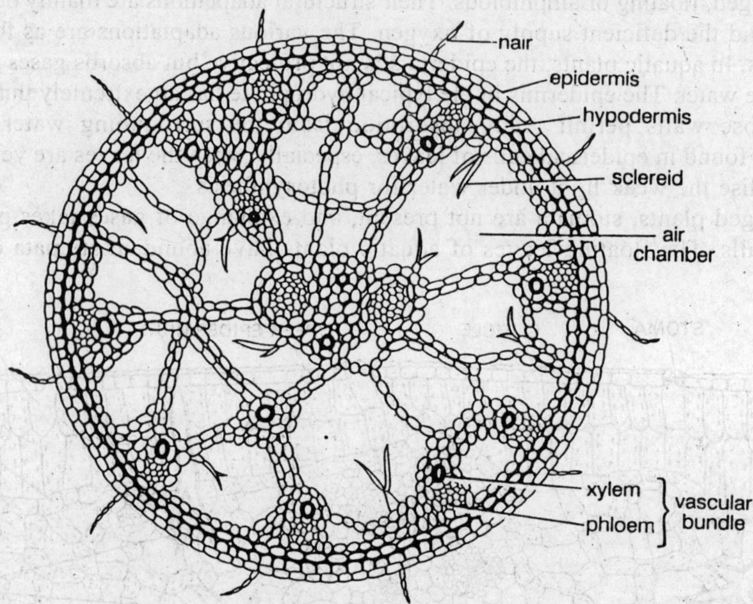

Fig. 16.17. Hydrophytes. T.S. petiole of *Nymphaea*.

Aerenchyma. Chambers and passeges filled with gases are commonly found in the leaves and stems of hydrophytes. The air chambers are large, usually regular, intercellular spaces extending through the leaf and often for long distances through the stem, *e.g.*, in *Potamogeton*, *Eichhornia*, *Pontederia*, etc. The spaces are usually separated by partitions of photosynthetic tissue only one or two cells thick.

The air chambers prepare an internal atmosphere for the plant. These air chambers on the one hand give buoyancy to the plant for floating and on the other they serve to store up air (oxygen and carbon dioxide). The carbondioxide that is given off in respiration is stored in these cavities for photosynthesis, and again the oxygen that is given off in photosynthesis during the daytime is similarly stored in them for respiration. The specialised tissue frequently found in aquatic plants that

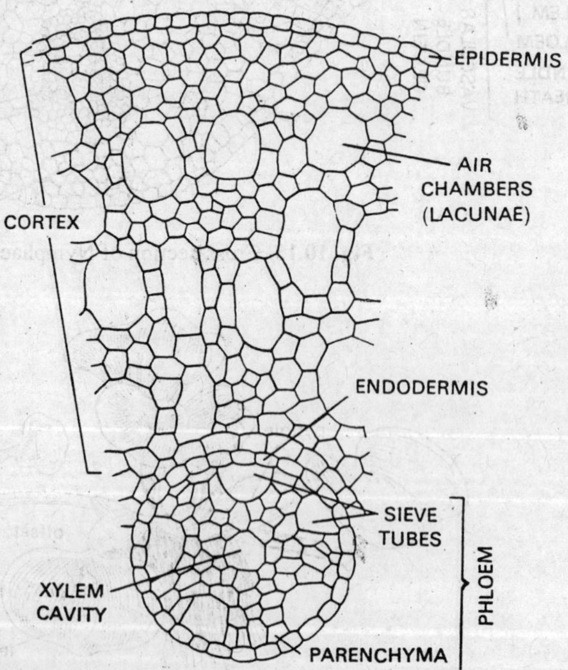

Fig. 10.18. Hydrophytes. T.S. root of *Potamogeton*.

gives buoyancy to the plant parts on which it occurs is **aernechyma**. This tissue helps to transport oxygen produced during photo-synthesis and permits its free diffusion to other parts, including roots located in anaerobic soil.

Presewa of inflated peteioles in Eichhornia (water hyacinth) Keeps the plants floating on the surface of water.

Roots are poorly developed or absent in free floating hydrophytes, such as *Wolffia*, *Solvinia* and *Cerato-phyllum*, or poorly developed, as in *Hydrilla*. *Nymphora* and *Nelumbo* are examples of rooted hydrophytes which grow in fresh water pond.

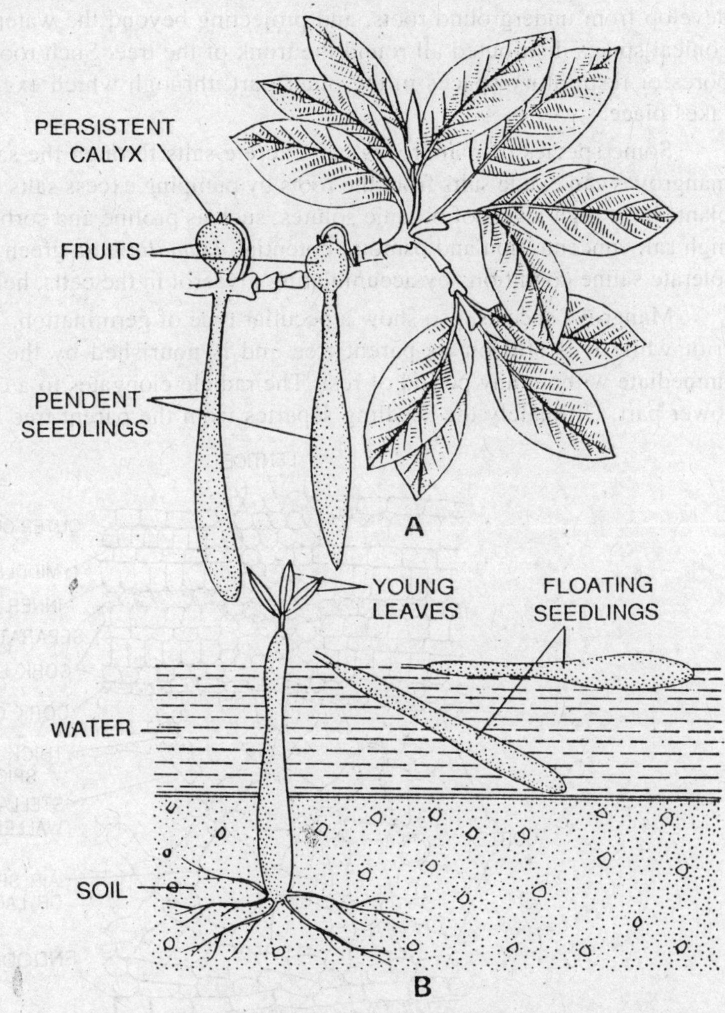

Fig. 10.19. Vivipary. Germination of the seed inside the fruit while the latter still remains attached to the plant.

Examples of aquatic plants are :

Submerged – *Vallisneria*, *Potamogeton*, *Hydrilla*, etc.

Floating – *Ceratophyllum*, *Pistia*, *Utricularia*, *Trapa*, *Eichhornia*, *Neptunia*, etc.

Plants with floating leaves – *Nymphaea*, *Nelumbo*, *Victoria*, *Euryale*, *Limnanthemum*, etc.

Amphibious plants – *Ranunculus aquatilis*, *Sagittaria*, *Typha*, *Alisma*, *Cardenthera*, *Myriophyllum* etc.

Adaptations in saline environments. Some plants grow and complete their life cycle in the habitats with a high salt content. They are called salt plants or **halophytes**. Halophytes occur in tidal marshes and coastal dunes, mangroves and saline soils. They show some special characters. The halophytic plants, under hot and dry conditions, may become succulent and dilute the ion cocnentration of salts with, water they store in cells by stems and leaves. Leaves may be modified into or provided with spines. Typical examples of such plants are : *Suaeda maritima*, *Salsola*, *Acanthus ilicifolius*. *Chenopodium* and some species of family Asclepiadaceae.

Halophytes growing in marshy places near the seashore, as in Sundarbans (West Bengal), from a special vegetation known as the **mangrove**. Mangrove plant produce a large number of stilt roots from the main stem and the branches. In many cases, in addition to the still roots, special roots called **respiratory roots** or **pneumatophores**, are also produced in large numbers. Such roots develop from underground roots, and projecting beyond the water level, they look like so many conical spikes distributed all round the trunk of the tree. Such roots are provided with numerous pores or respiratory spaces in the outer part, thruogh which exchange of gases for respiration takes place.

Some species of mangroves can excrete salts through the salt glands on the leaves. Some mangroves can exude salts from the roots by pumping excess salts back into soil. Many mangrove plants have high levels of organic solutes, such as proline and sorbitol, to cope with conditions of high salt concentration and osmotic potential. *Dumaliella*, a green alga found in saline lakes, can tolerate saline conditions by accumulating glycerol in the cells, helping in osmoregulation.

Mangrove species also show a peculiar type of germination. The seed germinates inside the fruit while it is still on its parent tree and is nourished by the same. Germination is almost immediate without any period of rest. The radicle elongates to a certain length and swells at the lower part. Ultimately the seedling separtes from the parent tree and falls vertically down. The

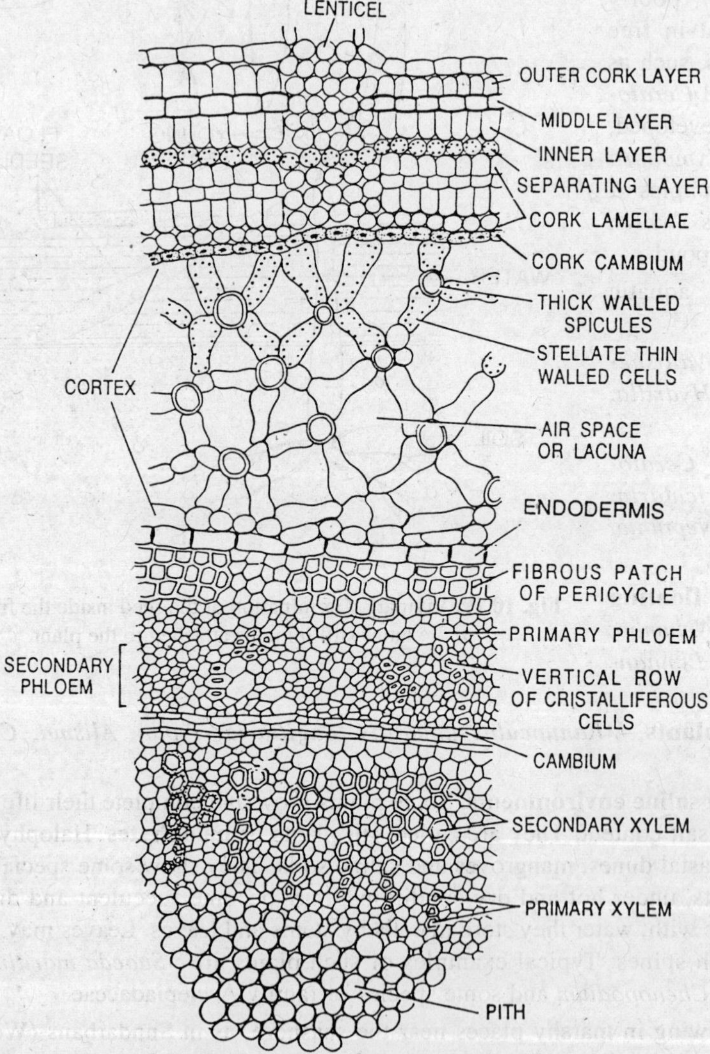

Fig. 10.20. *Rhizophora mucronata.* T.S. of a portion of subterranean stilt root.

radicle presses into the soft mud, keeping the plumule and cotyledons clear above the saline water. This kind of germination of the seed inside the plant is known as **vivipary**. Typical examples of mangrove plants are : *Rhizophora*, *Avicennia*, *Sonneratia*, *Ceriops*, *Heritiera*, etc.

Avicennia and *Rhizophora* are the domiant species in mangrove forests. The pneumatophores are capable to take up oxygen from the atmosphere and transport it to the main roots. The stilt roots give support to the mangrove plants in wet and muddy substratum, while vivipary permits plants to escape the effect of salinity on seed germination.

Adaptations to oligotrophic soils. The oligotrophic soils are those which contain lone amounts of nutrients. Such soils are generally found in geologically stable areas, such as soils of tropical rain forest region. Due to intense weathering and high rates of leaching, such soils have a poor nutrient retention capacity.

However, in nutrient poor soils, nutrient accumulation in vegetation is high.

Mycorrhizae. Many plants that grow in nutrient-poor soils have mycorrhizae or fungus root which have mutualistic association of roots with fungi. These are two types of mycorrhizae

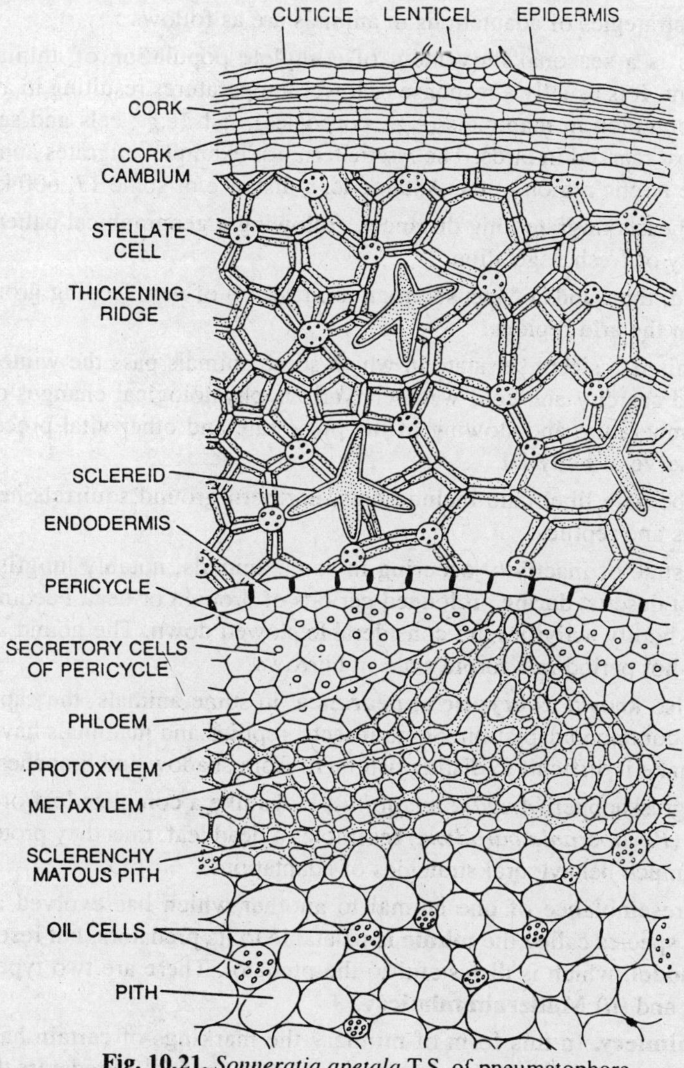

Fig. 10.21. *Sonneratia apetala* T.S. of pneumatophore.

(*i*) **endomycorrhizae** and (*ii*) **ecotomycorrhizae**. In endophytic mycorrhizae, the fungal hyphae are found within the tissues of roots in several vascular plants. While in ectomycorrhizae, the fungal mycelium forms a mantle of mat outside the root of vascular plant. Ectomycorrhizae generally occur in roots of shrubs and trees of temperate regions :

Mycorrhizae help in efficient absorption of nutrients, such as phosphorus and other nutrients, thus fulfil the scarcity of nutrient-poor soils.

Strategies of Adaptations in Animals

Several strategies of adaptations are found in animals. Many animals adjust themselves in different environmental conditions to survive and flourish.

For example, carnivores and herbivores have adaptations to eat a specific kind of food. On the other hand, some animals have adaptation to avoid being eaten by the predators.

Some animals have behavioural adaptations to attract a mate. The moles of some animals have bright colouration., particularly the plumage of the birds, which are helpful to attract their mates.

Some important strategies of adaptations in animals are as follows :

Migration. This is a seasonal movement of complete population of animals to a more favourable environment. It is usually a response to lower temperatures resulting in a reduced food supply. Migration is common in mammals (*e.g.*, porpoises), fish (*e.g.*, eels and salmon) and in some insects but is most marked in birds. The Arctic tern, for example, migrates annually from its freeding ground in the Arctic region to the Antaractic, a distance of some 17, 600 km.

In Africa, wild beasts migrate long distances, following a geographical pattern of seasonal rainfall and availability of fresh vegetation.

On the other land, thousands of locusts migrate in search of new feeding grounds from the food depleted areas, in the arid regions.

Hibernation. This is a sleeplike state in which some animals pass the winter months as a way of surviving food scarcity and cold weather. Various physiological changes occur, such as lowering the body temperature and slowing of the pulse rate and other vital processes, and the animal lives on its reserve of body fat.

Animals that hibernate libermate include bats, northern ground squirrels hedgehogs, and many fish, amphibians and reptiles.

Aestivation. A state of inactivity occuring in some animals, notably lungfish and ground squirrels in south-west deserts, during prolonged periods of drought or heat. Feeding, respiration, movement, and other bodily activities are considerable slowed down. The gound squirels, avoid heat by spending dry-hot period in a torpid state in burrows.

Camouflage. Also known as **cryptic appearance**; in some animals, the capacity to blend with surrounding is a common adaptation. Some insects, reptiles and mammals have markings on their bodies, which make it difficult to distinguish them from shadow and branches.

For example, a grasshopper (*Arantia rectifolia*), looks like a complete leaf or a part of leaf, while praying mantis (*Phyllocrania paradoxa*) resembles a dead leaf, thus they protect themselves from the enemies by much behavioural strategies of adaptations.

Mimicry. The resemblance of one animal to another which has evolved as a means of protection. Here, one species called the **mimic** is palatable to its predators, but resembles another species, called the **model**, which is distasteful to the predator. There are two types of minicry : (*i*) **Batesian minicry** and (*ii*) **Mullerain mimicry**.

(*i*) **Batesian mimicry.** In this form of mimicry the markings of certain harmless insects closely resemble the warning colouration of another insect (the model). Predators that have learnt to avoid the model will also avoid good minics of it. This phenomenon is often found among butterflies. For examples, monarch butterfly and the mimics vicerory butterfly. Here, monarch butterfly contains toxins in its body and mimicked by vicreroy butterfly which has no toxins.

(*ii*) **Mullerian mimicry.** This form of mimicry involves the mutual resemblance of a group of animals, all harmful, such as the wasp, bee and hornet, so that predator, having experienced one will subsquently avoid all of them. The other examples are-monoarch butterfly and the mimic queen butterfly. Here the mimic shares the same defensive mechanisms as the model.

Echolocation. A method used by some animals (such as bats, dolphins and certain birds) to detect objects in the dark. The animal emits a series of high-pitched sounds that echo back from the object and are detected by the ear or some other sensory receptor. From the direction of the echo and from the time between emission and reception of the sounds the object is located, often very accurately.

For example, horseshoe bat produces high frequency sound, and detects the pressure of echoes produced from the objects.

Warning colouration (aposematic colouration). The conspicous markings of an animal that make it easily recognisable and warm would-be predators that it is a poisonous, foul-tasting, or dangerous species. For example, the brightly colored and highly poisonous dart frogs (*Phylobates bicolour*), of the tropical rain forests of South America are easily recognised and avoided by the predators. The other example is the yellow and black striped abdomen of the wasp warns of its sting.

Adaptations to water scarcity. Two main types of adaptations are found in animals of arid regions, *i.e.*, (*i*) lowering of water loss as much as possible and (*ii*) adapting to arid conditions. The examples are as follows :

The kangaroo rat conserves water by excreting solid urine, and can live throughout life without drinking water.

On the other hand, the camels show unique adjustments to desert conditions. They are very economical in water use they can tolerate the wide fluctuations in body temperature and can maintain blood stream moisture even during extreme heat stress.

Adaptations to cold. Some sessile animals, such as barnacles and molluses which live in very cold inter-tidal zones of northern shores, and several insects and spiders resist the effect of cold spells by a process known as **cold hardening**.

The organisms that tolerate freezing conditions passes ice nucleating proteins which induce ice formation in the extracellular spaces at low rule zero temperatures.

While some free-avoiding animals can tolerate environmental temperature below 0°C by accumulating glycerol or antifreeze proteins that lower freezing point of their body fluids. Presence of such antifreeze substances allows the fish in Antarctica region to remain active in sea water.

SUMMARY

1. Ecology is the study of reciprocal relationships between organisms and their environment.
2. The levels of biological organisation of interest in ecology are : Organism-Population-Community-Ecosystem-Biosphere.
3. A community contains many organisms of different species.
4. An ecosystem includes biological community, integrated with its physical environmet.
5. Biosphere includes all ecosystems on the earth.
6. The prevaling regional environment has large-scale effets on climate, biota and vegetation.
7. Microclimate represents the climatic conditions closer to living organisms.
8. Habitat is the place where an organisms lives.

9. Ecological niche is the functional role of an organisms.

10. Light and temperature affect climate, activities of plants and animals, and metabolic process of organisms.

11. Hydrological cycle is regulated by water vapour in atmosphere, evaporation and precipitation.

12. Soil plays an important role in plant growth by providing water, nutrients and anchorage, and a habitat for diverse population of soil organisms.

13. Moisture and temperature are generally responsible for limiting the activities of terrestrial organisms.

14. However, aquatic organisms are mainly affected by oxygen and light.

15. Organisms with broad range of tolerance have wide distribution.

16. Organisms with narrow range of tolerance have restricted distribution.

17. Desert plants adapt to water scarcity by mining water with deep tap root system, storing water in succulent leaves and stems, and having low transpiration.

18. Plants adapt to high concentration of salt in soil or water by salt excretion, or by accumulating organic solutes in stems and roots.

19. The oligotrophic soils are nutrient-poor soils. In such soils, nutrient retention is high in vege tation and mycorrhizae help in mineral nutrition of plants.

20. A large number of animal adaptations, are physiological and behavioral that include migration, regulation of body temperature, and mimicry to avoid predators.

21. In *Batesian* mimicry, the mimic is defenseless.

22. In *Mullerian* mimicry, the mimic shares the same defensive mechanisms as the model.

23. Some organisms avoid predation by cryptic appearance or warning colouration.

24. In cold environments, organisms adapt by super cooling their body fluids, or by lowering the freezing point of their body fluids by anitfreeze compounds.

TERMS TO REMEMBER

Acclimatisation. The gradual physiological adjustment to slowly changing new environmental conditions.

Adaptations. The special characteristics of organisms that enable them to be successful under prevailing set of environmental conditions.

Aestivation. Bounding the dry-hot period in an inactive state.

Biological community. This is an assemblage of populations of plants, animals, bacteria and fungi that live in an area and interact with each other.

Biomes. A large regional unit characterised by a major vegetation type and associated fauna found in a specific climatic zone.

Benthic region. The sediments at the bottome of lakes and ponds.

Batesian mimicry. Here, the mimic is defenseless, but has anti-predatory marks, like the model which has a defense against predators.

Climate. The average weather of an area.

Comouflage. The capacity to blend with surroudings.

Detrius. Freshly dead and partially decomposed plant and animal material.

Ecology. The branch of biology that deals with the study of interactions between organisms and their environment.

Ernst Haeckel. German biologist, coined word 'ecology' in 1869, (*oikos* = home; *logos* = to study).

Ecosystem. This is composed of a biological community integrated with its physical environment through the exchange of energy and recycling of the nutrients.

Environment. The sum total of all biotic and abiotic factors that surround and influence organisms.

Ecotypes. The species having a wide range of distribution evolve genetically adapted local population.

Ecological niche. The ecological niche of an organisms represents the range of conditions that if can tolerate the resources, its utilities and its functional role in the ecological system.

Ectotherms. Cold-blood animals, the body temperature tends to match with the environmental temperature is which they live.

Endotherms. Warm-blooded animals, regulate their body temperature by physiological means.

Ephemerals. The plants which evade dry conditions.

Endomycorrhizae. The fungal hyphae live inside roots.

Ectomycorrhizae. The fungal mycelium forms a mat outside the root.

Epilimmion. Upper layer of water in a simple body of water.

Field capacity. The upper limit of water availability.

Habitat. The place where an organism lives.

Homostasis. The maintenance of relatively constant internal environment under varying external environment.

Hypolimnion. Linear layer of water in a single body of water.

Hydrological cycle. This is the movement of water between aquatic systems, air and land.

Humus. Colloidal, amorphous and dark coloured substance.

Heliophytes. The plants adapted to high intensity of light.

Hydrophytes. Plants which remain permanently immersed in water.

Halophytes. Plants of saline environments which are adapted to grow in high concentrations of salt in soil or water.

Hibernation. Spending winter in dormant conditon.

Landscape. A unit of land with a natural boundary having a mosaic of patches.

Littoral zone. The shallow water zone around the edge of the lake which supports vegetation.

Limnetic zone. The open water zone beyond the littoral zone, where phytoplankton grows in abundance.

Lapse rate. The vertical temperature gradient over earth's surface, the value being 6.5°C per 1000 m elevation.

Litter. See detritus.

Mullerian mimicry. The process when the mimic shares the same defensive mechanism as the model.

Migration. Long-distance or short-distance movement of animals from one region to another.

Mimicry. Two species resemble each other closely, one species called the mimic which resembles another species called the model. The former is palatable, while the latter distasteful to the predator.

Mycorrhizae. Mutualistic association of roots with fungi (fungus root).

Mesosphere. Beyond stratosphere is extends up to an altitude of 80 km and shows a decrease of temperature with height.

Microclimate. The climatic conditions that prevail at a local scale, or in areas of limited size.

Oligotrophic soils. The soils, which contain low amounts of nutrients.

Phenotype. The physical expression of the interaction between genotype of an organism and its environment.

Profundal zone. The dark zone in water, where light does not reach.

Phenology. The timing of seasonal activities of plants in relation to change in environmental conditions.

Photosynthetically active radiaton (PAR). Light or visible spectrum having wavelengths of 0.4 to 0.7 mm.

Population. A group of individuals of a plant or an animal species, inhabiting a given area.

Species. Similar organisms having the potential for interbreeding and producing fertile offspring.

Stratopause. The transition layer between stratosphere and the mesosphere.

Solar constant. Solar radiation, befoer entering the atmosphere, carries energy at a constant rate of 2 cal cm^{-2} min^{-1}.

Soil. The uppermost weathers layer of earth's crust and is composed of minerals and partly decomposed organic matter.

Soil profile. The vertical layered structure of soil.

Sciophytes. Shade tolerant plants

Succulents. The plants having fleshy leaves and stems to store water.

Troposphere. The lower portion of the atmosphere, extending to about 8 to 16 km height from the earth surface.

Tropopause. Top of troposphere, where temperature remains –57°C.

Warning colouration. Concealing form and colouration enables a species to avoid its naturale predator.

Wilting point. The lower limit of water availability in soil.

Weather. The short term properties of the atmosphere, at a given place and time.

QUESTION BANK

Very Short Answer Questions

1. Define 'ecology'.
2. What is 'biological community'?
3. Which are the main climatic zones ?
4. What is PAR ?
5. What does 'phenology' refer to ?
6. What is 'biome'?
7. What is environment ?
8. What do you understand by the term 'microclimate'?
9. What is 'troposphere' ?
10. Give the ecological levels of organisation in ascending or desending order.
11. What is 'littoral zone' of a lake ?
12. Define 'hydrological cycle'?
13. What are lithophytes ?
14. What are eutrophic soils ?
15. Define hibernation.
16. What are microothermal organious ?
17. Where the mangrove vegetation in found ?
18. What is warning colouration ?
19. What are 'ecotypes'?
20. What is profundal zone of water ?

ANSWERS

1. Ecology is the branch of biology that deals with the study of interactions between organisms and their environment; **2.** The biological community is an assemblage of population of plants, animals, bacteria and fungi that live in an area and interact with each other; **3.** Depending on variation of mean temperature, the main zones are : tropical, subtropical, temperate, Arctic and Antarctic; **4.** Light or visible spectrum, which has wavelength 0.4 to 0.7 mm is called 'photosynthetically active radiation, *i.e.*, PAR; **5.** Phenology refers to the time of seasonal activities of plants with respect to the change in environmental conditions; **6.** A biome is a large regional unit characterised by a major vegetation type and associated fauna found in a specific climate zone; **7.** Environment is the sum total of all biotic and abiotic factors that surround and potentially influence an organism; **8.** The microclimate repersents the climatic conditions that prevail at a local scale; **9.** The lower portion of the atmosphere, extending to about 8-16 km hight from the earth surface; **10.** Biome-ecosystem-community-population-species organisms; **11.** The littoral zone is the shallow water zone around the edge of the lake which supports rooted vegetation; **12.** The hydrological cycle is the movement of water between aquatic system, air and land; **13.** Plants found on rocky surface; **14.** The soils which have almost optimum concentration of minerals in soil solution; **15.** Hibernation is the period of dormancy during winter; **16.** Organisms living at moderate summer temperature and low winter temperature; **17.** Mangroves are found in marshy condition of tropical deltas and along ocean edges; **18.** Concealing form and colouration enables a species to avoid its natural predator; **19.** The species having a wide range of distribution evolve genetically adapted local population, called ecotypes; **20.** The dark zone where light does not reach is called profoundal zone.

Short Answer Questions

1. What is the difference between climate and weather ?
2. Distinguish between comouflage and mimicry.
3. How do plants adapt to oligotrophic soils.
4. What do you understand by ecological niche ?
5. What is solar constant ?
6. Some freeze avoiding animals can tolerate temperatures below 0°C. How ?
7. What type of plants are called ephemerals ? Cite an appropriate example.
8. What are the types of ultraviolet radiation, on what basis their division is being done ?
9. How does Dunaliella tolerate saline conditions?
10. Classify the vegetation of world into major groups.
11. Give a brief account of wilting coefficient.
12. What is the effect of temperature on metabolism of plants ?
13. Describe environmental factors and their importance to plants and animals.
14. How do plants adapt to water scarcity and saline environments ?
15. Define the following terms :
    (*a*) Migration (*b*) Stratosphere (*c*) Community (*d*) Biosphere
16. How do kangaroo rats and camels adapt to arid conditions ?
17. How do plants adapt in saline environment and oligotrophic soils ?
18. Distinguish between :
    (*a*) Batesian mimicry and Mullerian mimicry
    (*b*) Endomycorrhizae and Ectomycorrhizae
19. Differentiate between ectotherms and endotherms.
20. What do you understand by ecological factors?
21. Give a comprehensive account of soil moisutre.
22. What is the importance of soil in determining the vegetation of a place ?
23. Explain the importance of climatic factors in determining the nature of plant community.
24. How do sessile animals like barnacles and molluscs combat very cold spells in the intertidal zones of northern shores.

Long Answer Questions

1. Explain the following terms :
   (*a*) Mimicry (*b*) Acclimatisation (*c*) Ectotherms (*d*) Endotherms
2. Explain how tolerance to environmental factors determines distribution of species.
3. What are the different of the different types of adaptation in animals ? Explain with suitable examples.
4. Give a brief account of the different layers of atmosphere.
5. Give a comprehensive account of the effect of light on aquatic systems.
6. Define soil profile. How does it develop ? What are the main horizons in a soil profile ?
7. What are biotic factors ? Explain the effect of humans on vegetation.

8. Describe the effect of soil composition and soil texture on the growth and distribution of plants.
9. What is soil ? What is the role of microorganisms in soil formation ?
10. Describe in brief the main edaphic factors that affect the vegetation of a given area.
11. Give the importance of soil reaction, soil organisms and soil cover.
12. Explain the effect of light on the growth and distribution of plants.

MULTIPLE CHOICE QUESTIONS (OBJECTIVE TYPE QUESTIONS)

1. A zone of gradual change in temperature is :
(*a*) Thermostat (*b*) Thermal (*c*) Thermocline (*d*) Thermal tolerance
2. The branch of biology that deals with the study of interaction between organisms and environ ment is called :
(*a*) Physiology (*b*) Morphology (*c*) Ecology (*d*) Methology
3. Who coined the word 'ecology':
(*a*) Strasburger (*b*) P. Odum (*c*) Ernst Haeckel (*d*) Roxburgh
4. Whose organisms occur with other individuals of the same species form a part of :
(*a*) Biome (*b*) Biosphere (*c*) Population (*d*) Environment
5. Organisms interact with physical environment that comprises of :
(*a*) Atmosphere (*b*) Hydrosphere (*c*) Lithosphere (*d*) All of these
6. Warm-blooded animals are called :
(*a*) Ectotherms (*b*) Endotherms (*c*) Homeostasis (*d*) None of these
7. An association of different interelated poplation belonging to different species in common environment which can survive in nature in called :
(*a*) Family (*b*) Hard (*c*) Population (*d*) Biotic community
8. A direct food relation between two species of animals in which one animal kills and feeds on another is referred to as :
(*a*) Scavanging (*b*) Predation (*c*) Parasitism (*d*) Symbiosis
9. Which are of the following is protective device :
(*a*) Commensalism (*b*) Mutualism (*c*) Camouflage (*d*) Competition
10. Synecology is the study of :
(*a*) Biosphere (*b*) Individuals
(*c*) Environment (*d*) Community in relation to envirionment
11. Plants growing on burnt soil are called :
(*a*) Heliophilous (*b*) Hetrophyllous (*c*) Hydrophylous (*d*) Pyrophilous
12. When one population is harmed and the other remains unaffected the relationship is called :
(*a*) Amensalism (*b*) Commensalism (*c*) Parasitism (*d*) Mutualism
13. The relationship between a climber and the host correspond to :
(*a*) Mutualism (*b*) Parasitism (*c*) Commensalism (*d*) Neutralism
14. The mangrove plants are generally characterised by :
(*a*) Microphylly (*b*) Pneumatophores and vivipary
(*c*) Absence of aernchyma (*d*) Poor water storage
15. Plant groups growing in zones where high temperatuer alternates with low temperature are called :
(*a*) Megatherms (*b*) Mesotherms (*c*) Microtherms (*d*) Hekistotherms
16. Plants growing on sand and gravel are called :
(*a*) Eremophytes (*b*) Prammophytes (*c*) Prilophytes (*d*) Oxylophytes
17. The mesophytes are characterised by :
(*a*) Uniform mesophyll (*b*) Absence of cuticle
(*c*) Well developed root-shoot systems (*d*) All of these
18. Plants growing on cold soil are called :
(*a*) Psychrophytes (*b*) Premmophytes (*c*) Sciophytes (*d*) Oxylophytes
19. The biome is :
(*a*) That part of earth and its atmosphere, which inhabits living organisms.

(*b*) A complex of community interacting with one another
(*c*) The flora in an ocean
(*d*) The flora on land

20. Ecosystem has two components :
(*a*) Plants and animals' (*b*) Needs and trees (*c*) Biotic and abiotic (*d*) Frogs and lizards

21. Which one of the following animals can live from birth to death without even drinking water ?
(*a*) Kangaroo rat (*b*) Kangaroo (*c*) Camel (*d*) Desert cat

22. The study of periodic biological events such as flowering, breeding, migration etc., in relation to climatic and other environmental factors :
(*a*) Polynology (*b*) Pomology (*c*) Anthology (*d*) Phenology

23. Hydrilla is a :
(*a*) Free floating hydrophyte (*b*) Floating and anchored hydrophyte
(*c*) Submerged hydrophyte (*d*) Suspended hydrophyte

24. Among the following which plant is completely devoid of roots :
(*a*) Azolla (*b*) hydrilla (*c*) Ceratophyllum (*d*) Vallisneria

25. Among the following characters which one does not characterise a hydrophyte ?
(*a*) Poor development of roots
(*b*) Leaves having stomata only on the upper surface or none
(*c*) Plentiful xylem and sclerenchyma
(*d*) Abundant air spaces and air chambers

26. Aerenchyma is very extensive in almost all the vegetative organs of :
(*a*) Epiphytes (*b*) Mesophytes (*c*) Xerophytes (*d*) Hydrophytes

27. Xerophytes have deep roots in soil :
(*a*) Due to light (*b*) To draw water from deep water beds
(*c*) To give mechanical support (*d*) None of these

28. Mechanical tissues are very poorly developed in :
(*a*) Mesophytes (*b*) Hydrophytes (*c*) Halophytes (*d*) Lithophytes

29. Which of the following is the most characteristic feature of a xeric environment ?
(*a*) Low atmospheric humidity (*b*) High rate of vaporisation
(*c*) Extremes of temperature (*d*) The precipitation

30. Humus is an example of :
(*a*) Crystalloids (*b*) Organic colloids (*c*) Soil structure (*d*) None of these

31. The substance having organic and inorganic constituent is :
(*a*) Water (*b*) Soil (*c*) Air (*d*) All of these

32. Who proposed the term 'ecosystem'?
(*a*) Gardner (*b*) Tansley (*c*) Tansley (*d*) Clements

33. Who coined the term 'biocoemosis'
(*a*) Huxley (*b*) Tensley (*c*) Haeckel (*d*) Carl Mobius

34. The role of an organism in the ecological system is known as :
(*a*) Habitat (*b*) Herbivary (*c*) Niche (*d*) Interaction

35. Adaptation to low temperature and freezing in animals occurs due to the production of :
(*a*) Antifreeze proteins (*b*) Chaperonins (*c*) Proline (*d*) Analine

ANSWERS

1. (*a*)	**2.** (*c*)	**3.** (*c*)	**4.** (*c*)	**5.** (*d*)	**6.** (*b*)	**7.** (*d*)	**8.** (*b*)	**9.** (*c*)	**10.** (*d*)
11. (*d*)	**12.** (*a*)	**13.** (*c*)	**14.** (*b*)	**15.** (*b*)	**16.** (*a*)	**17.** (*c*)	**18.** (*a*)	**19.** (*b*)	**20.** (*c*)
21. (*a*)	**22.** (*d*)	**23.** (*c*)	**24.** (*c*)	**25.** (*c*)	**26.** (*d*)	**27.** (*b*)	**28.** (*b*)	**29.** (*c*)	**30.** (*b*)
31. (*b*)	**32.** (*c*)	**33.** (*d*)	**34.** (*c*)	**35.** (*a*)					

11
CHAPTER

Population, Biotic Community and Succession

The organization of an ecosystem starts with the **organism**. This organism may be simple and one-celled or complex and multicellular. A group of organisms of the same species living in a particular place at the same time makes a **population**. For instance, a population may be a single flock of animals or all the members of a plant species growing on a mountain range, or a mass of slime mold. It is not possible to study all the members of a species, and therefore, the ecologist must decide on the size of the sample group to be studied. It may be the antelope of a sanctuary or the rats of a particular area.

Social interaction in a marine population of guppies exhibiting sexual dimorphism : The male (left) is normally much brighter in color than the female.

All population of organisms that live together in a specific area and depend upon one another for food and shelter make up a **biological (biotic) community**. Different types of environments, support different populations. A desert community consists of populations of plants and animals different from those in a tropical rain forest.

However, the organisms in a habitat are not there simply by chance; their presence and suitability to a particular environment shows a long-drawn struggle for existence.

In this chapter population and community levels of organisation, population characteristics and growth, species interaction in the community and change in community, are dealt with.

POPULATION

The term **population** has been interpreted in several ways, they are as follows :

1. In human demography, a population is a set of humans in a given area, such as village, town, city, state or country or even the whole world.

2. In genetics, a population is a group of interbreeding individuals of the same species, which is isolated from other groups.

3. In ecology, a population is a group of individuals of the same species, inhabiting the same area, and functioning as a unit of biotic community.

For example, all individuals of the common grass snails or pine trees, of particular habitats, constitute their population.

Organisms in a population undergo the same life cycle. They experience similar ecological process at a particular stage of the life cycle.

Populations can be defined at various spatial scales. For example, local population can occupy very small habitat patches, like a temporary pond of water. A set of local populations connected by dispersing individuals is called a **metapopulation**.

Populations can also be considered at the scale of regions, islands, continents or oceans. Even the entire species can be treated as a population.

Characteristics of Population

A population has group characterstics, such as density, natality (birth rate), mortality (death rate), dispersal, age distribution (*i.e.*, ratio of one age group to the other), biotic potential and growth forms.

Populations are groups of individuals of the same species living in a particular place at the same time. They may differ in their number, patterns of living together, behaviour and organization. All of these factors are important when considering the interactions of a population with its environment and other populations. The description of a population involves several parameters such as (*i*) the number and kind of individuals of a species, (*ii*) a given space or an area, and (*iii*) time.

Density. The number of individuals of a species living in a particular area is the **density** of that population. Density is dependent upon many factors, such as danger of predators and the availability of food, water, light, heat and nesting places. The density of a population increases when conditions are favourable to the populations and decreases when they are not favourable. Population density is determined by counting all the individuals and dividing it by the number of units of space. For terrestrial organisms usually two dimensional units m^2 (a quadrate of $1m \times 1m$) or soil surface are taken. For aquatic organisms and organisms suspended in a medium, space is expressed in three dimensions (m^3). If the total number of individuals is taken as N and the number of units of space as S, the population density *D* can be calculated as

$$D = \frac{N}{S}$$

For example, 50 individuals of a tree species may occur per hectare, or 70 individuals of the grass *Cynodon* per square metre. Density of plants is determined by counting individuals in sample units of predetermined sizes. Species density varies from time to time and from one area to another. For human population, counting is being done every ten years.

Distribution. The arrangement of the individuals of a population within a particular space is called the distribution. While population density gives us knowledge that how many individuals exist in an area, whereas the distribution tells us how those individuals are arranged in that area. The patterns of distribution are **random, uniform** and **clumped**. In random distribution organisms occur in no particular order. The organisms that are distributed in an uniform manner are spread evenly over an area, whereas the clumped organisms are concentrated at points throughout an area. Clusters may offer the population protection from enemies, wind and cold.

Factors Affecting Population

Populations can change in many ways. They can get larger or smaller. The areas they occupy can increase or decrease. The ratio of males to females can vary. Changes in a population occur in response to some change in the factors that influence the population. A population tends to increase as individuals are added by **natality**, *i.e.*, birth, hatching or germination, and **immigration**, *i.e.*, moving in from outside. On the other hand, a population tends to decrease by **mortality** or death, and **emigration** or moving out. However, in the case of land plants incapable of locomotion, active immigration or emigration do not occur. An occasional addition to a plant population may occur through arrival of seeds from a long distance dispersed by air, water or animals.

Natality. The increase in number of individuals in a population under given environmental conditions is called **natality**. For example, birth, hatching, germination and vegetative propagation are main causes which are responsible for increase of number of individuals.

When the increase in individuals is expressed per unit time, it is called, **natality rate**.

Natality is calculated as follows :

$$\text{Birth rate or natality (B)} = \frac{\text{number of births per unit time}}{\text{average population}}$$

The maximum number of births produced per individual under ideal conditions of environment is called **potential natality**.

Natality varies from organism to organism. It depends upon the population density and environmental factors. It is a general rule that if the population density is low, the birth rate is also low, because the chances of mating between males and females are low. On the other hand, if population density is unusually high, the birth rate may also be low due to poor nutrition, or physiological or other probelms related to crowding.

The maximum or absolute natality is observed when the species exists under ideal ecological and genetic conditions. The actual number of births occurring under the existing environmental conditions is much less as compared to absolute natality. It is referred to as **ecological natality** or **realised natality**. It is not constant for population and may vary with the size of population as well as with the time.

Mortality. The loss of individuals due to death in a population under given environmental conditons is called **mortality**.

Mortality refers to the number of deaths per unit time.

Mortality rate = $\frac{D}{t}$ where D is the number of deaths in the time t.

Mortality can be expressed as follows :

(*i*) **Minimum or specific or potential mortality.** It represents the minimum or theoretical loss of individuals under ideal or non-limiting conditions. Thus even under the best conditions individuals of a population would die of old age determined by their physiological longevity. So it is constant for a population.

(*ii*) **Ecological or realised mortality.** It refers to the death of individuals of a population under existing environmental conditions. Since it varies with environmental conditions, it is never constant. The maximum mortality occurs at the egg, larval, seedling and old age.

Mortality is affected by a number of factors, such as density, competition, disease, predation and environment. Death rates vary among the species and are correlated with birth rates. When the rate of natality is equal to the rate of mortality, the population is stationary.

A birth-death ratio $\left(\frac{\text{Births}}{\text{Deaths}} \times 100\right)$ is called **vital index**.

For a population, the survival of individuals is more important than the death. The number of births in relation to the carrying capacity of the habitat is a fundamental factor influencing the moratality rate. When more youngs are born than the habitat can support, the surplus must either die or leave the area. As the number of survivors is more important than the number of dying individuals, mortality is better expressed as survival or as **life expectancy**. The life expectancy refers to the average number of years the members of population have left to live.

Dispersal. The majority of organisms disperse at one time or the other during their life cycles. The individuals move into (immigration) and move out of the population (emigration), and such movements affect the size of population.

Movement of individiuals into and out of the population is called **population dispersal**. It plays important role in the geographical distribution of organisms even to the areas previously occupied by the members of population. Dispersal of oraganisms occurs for various reasons such as food, protection, prevention from overcrowding, action of wind and water, environmental factors such as light, temperature, breeding behaviour, physiological reasons or for interchange of genetic materials between the populations.

Population dispersal occurs in nature as follows :

(*i*) **Emigration.** It is one way movement of individuals out of a population. This movement is permanent and causes spread of a species to new areas. Emigration under natural conditions occurs when there is overcrowding in the population and is generally regarded as an adaptive behaviour that regulates the population on a particular site and prevents over-exploitation of the habitat. This type of dispersal offers new opportunity to the indiviudals of a population to interbreed with those of the other population leading to more genetic heterozygosity and adaptability.

(*ii*) **Immigration.** This is one way movement of individuals into a population. It leads to rise in density of population. It may result in decreased mortality among the immigrants or decreased reproductive capacity of the individuals.

(*iii*) **Migration.** Migration is two way mass movement of the entire population. It involves a periodic departure and return of the individuals of a population and occurs only in mobile organisms during unfavourable periods. It is shown by many birds, fishcs and certain animals. In most cases, migration of population occurs for food, shelter or reproduction.

Age distribution. Age distribution is an important characteristic of population which influences natality and mortality. Mortality, usually varies with age, as chances of death are more in early and later periods of life span. Similarly, natality is restricted to certain age groups, *e.g.*, in middle age-groups in higher animals.

The individuals of a population can be divided in three ecological stages. They are : (*i*) **pre-reproductive**, (*ii*) **reproductive** and (*iii*) **post-reproductive**. The individuals of pre-reproductive group are young, those of reproductive group are mature and those in post-reproductive group are old.

The distribution of ages may be constant or variable. It is directly related to the growth rate of the population. Depending upon the proportion of the three age groups, populations can be said to be **growing**, **mature** or **stable**, and **diminishing**. Rapidly increasing population contains a large proportion of young individuals, a stable population shows even distribution of individuals in reproductive age group, and declining or diminishing population contains a large propotions of old individuals.

Age pyramids. Age pyramid a model in which the numbers or proportions of individuals in various age groups at any given time are geometrically presented. In an age pyramid, the number of pre-reproductive individuals is shown at the base, that of reproductive age group in the middle and the number of post-reproductive individuals at the top. The age pyramid indicates whether a population is expanding or stable or diminishing. The hypothetical age pyramids are as follows :

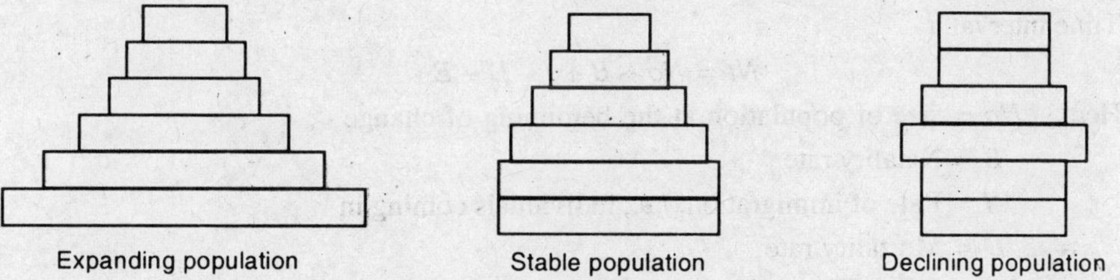

Fig. 11·1. Age pyramids of hypothetical populations. A. pyramid with broad base: B. bell-shaped pyramid; C. urn-shaped pyramid.

(*i*) **Pyramid with broad base.** This pyramid shows a high percentage of young individuals and an exponential growth of population due to high birth rate, *e.g.*, yeast, housefly, etc.

(*ii*) **Bell-shaped pyramid.** This type of age pyramid shows a stable population having, more or less equal number of young and middle-aged individuals and post-reproductive individuals being smallest in number.

(*iii*) **Pyramid with narrow base.** This is an urn-shaped pyramid which shows increased numbers of middle-aged and old organisms as compared to young ones in the population. This shows diminishing of population.

Biotic potential and environmental resistance. Every kind of organism produces more offspring than are necessary for that species to survive. The maximum reproduction rate of a population, under ideal conditions, is called **biotic potential**. This factor assumes that all of the young will live long enough to reproduce. To reach its biotic potential a population must have all of the food or raw materials, water and space it needs to survive. It also needs an ideal climate and the absence of competitors and disease. For instance, some bacteria divide about every 20 minutes. In 24 hours, the number of individuals produced would be enough to cover the entire surface of the earth to a depth of over 20 cm. However, populations do not reach their biotic potential. Many factors influence the growth rate of population. **Environmental resistance** means the collection of factors that reduce the growth rate of a population.

Biotic and abiotic factors influencing population growth. Biotic factors that influence population growth are living organisms or their characteristics. For example, the biotic factors that influence a population are the presence of predators or hunters, the food supply, the density of the population itself, the biotic potential of the species and disease. Abiotic factors are nonliving physical elements in the environment, such as temperature, water, humidity, light, soil, slope, air and wind.

Density dependent and density independent factors. Density dependent factors influence a population differently if the population is crowded than if it is not crowded. The effects of predators, disease and competition vary with the cocentration of the population. For example, a disease spreads more rapidly through a crowded population severely. The biotic factors influence all population than through a sparse one. The loss of a food source affects a crowded population, regardless of their density, and are called **density-independent**. For example, a forest fire affects the food supply and shelter of all organisms in the areas whether they are crowded or not crowded.

Population Growth Forms

The growth is one of the dynamic features of species population. Population size increases in a characteristic way. The growth of a population can be measured as increase in its size over a period of time. The change in population size may be expressed as follows :

Population having initial size, *No*

Population increases to size, *Nt*, and

Time interval, *t*

$$Nt = No + B + I - D - E$$

Here, No = size of population at the beginning of change

B = Natality rate

I = Rate of immigration, *i.e.*, individuals coming in

D = Mortality rate

E = Rate of emigration, *i.e.*, individual moving out.

When the number of individuals of population is plotted on the y-axis, and the time on the x-axis, a curve is obtained that indicates the trend in the growth of population size in a given time. This curve is called **population growth curve**. There are two types of growth curves :

(*i*) **J-shaped curve.** In the case of J-shaped growth form, the population grows exponentially, and after attaining the peak value, the population may abruptly crash. This increase in population is continued till large amount of food materials exist in the habitat. After some time, due to increase in population size, food supply in the habitat becomes limited which ultimately results in decrease in population size. For example, many insect populations show explosive increase in numbers during the rainy season, followd by their disappearance at the end of the season.

The following equation exhibits J-shaped growth :

$$\frac{dN}{dt} = rN$$

Here *dN/dt* represents rate of change in population size, *r* is biotic potential and *N* stands for population size.

2. **S-shaped or sigmoid curve.** When a few organisms are introduced in an area, the population increase is very slow in the beginning, *i.e.*, **positive acceleration phase** or **lag phase**, in the middle phase, the population increase becomes very rapid, *i.e.*, logarithmic phase, and finally in the last phase the population increase is slowed down, *i.e.*, **negative acceleration phase**, until an equilibrium is attained around which the population size fluctuates according to variability of environment. The level beyond which no major increase can occur is referred to as **saturation level** or **carrying capacity** (K). In the last phase the new organisms are almost equal to the number of dying individuals and thus there is no more increase in population size.

The S-shaped sigmoid growth form is represented by the following equation :

$$\frac{dN}{dt} = rN\left(\frac{K-N}{K}\right) = rN\left(1-\frac{N}{K}\right)$$

where, *dN/dt* is the rate of change in population size,
r is biotic potential
N is population size,
K – *N*/*K* or 1 – (*N*/*K*) is for environmental resistance.

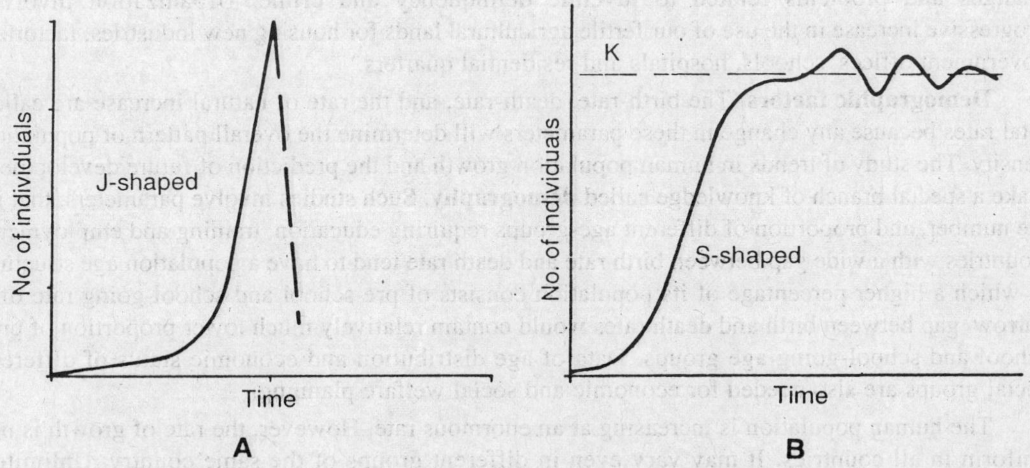

Fig. 11·2. Population growth curves. A. J-shaped; B. S-shaped (sigmoid) curve. K stands for carrying capacity.

Human Population

The distribution of human population in various regions of the world has been and is quite uneven in terms of its density mainly because of geographical, socio-economic and demographic factors. Only about thirty per cent of the world's land area is inhabited by man.

The pattern of human population and population growth depend upon several complex factors, such as—(1) geographical factors like climate, location of water, soil, energy, mineral resources, transportation etc., (2) socio-economic factors, and (3) demographic factors.

Geographical factors. The north and south poles are free of human habitation mainly because ʻhey are extremely cold and also agriculturally unproductive. Human settlements occur in places where adequate sources of water are available. The fertility of the soil for farming is another important determinant of population distribution. For instance, the main basis of the highly dense populations in the Indus valley and Indo-Gangetic plains is the alluvial nature of the soils in these regions.

In certain places both iron ore and coal and other fossil energy sources are located close together. Industrial cities based around steel plants have come up in Jamshedpur, Bokaro, Durgapur, Bhilai and Rourkela.

Means of transportation have played an important role in the redistribution of population. This made it possible for the first time for persons to live far away from the source of produce or of goods required for living. Transportation by water is cheaper than by land or air and this is probably the main basis for the coastal location of most big cities such as Kolkata, Mumbai, London, New York and Tokyo.

Socio-economic factors. The size of population, and birth and death rates have great significance on the standard of living of the people, their aspirations, and their economic and social development. The birth rate of human population is regulated more by socio-economic factors than biological factors. These factors may be social status of women, age of women at marriage, family structure, education, acceptability of family planning practices, and religious beliefs.

Prosperity in the cities is the basic cause of continued urban growth. Some important consequences of increasing urbanization are—overcrowding, leading to problems of sanitation and sewage disposal; transportation and associated traffic problems; environmental pollution generated by industrial activities and automobiles; noise pollution and various socio-economic and cultural changes and problems related to juvenile delinquency and crime. Uroanization involves progressive increase in the use of our fertile agricultural lands for housing new industries, factories, government offices, schools, hospitals and residential quarters.

Demographic factors. The birth-rate, death-rate, and the rate of natural increase are called vital rates because any change in these parameters will determine the overall pattern of population density. The study of trends in human population growth and the prediction of future development make a special branch of knowledge called **demography**. Such studies involve parameters, that is, the number, and proportion of different age-groups requiring education, training and employment. Countries with a wide gap between birth rate and death rate tend to have a population age structure in which a higher percentage of its population consists of pre-school and school-going rate of a narrow gap between birth and death rates would contain relatively much lower proportion of pre-school and school-going age groups. Data of age distribution and economic status of different social groups are also needed for economic and social welfare planning.

The human population is increasing at an enormous rate. However, the rate of growth is not uniform in all countries. It may vary even in different groups of the same country. Unlimited population growth leads to over-crowding, has adverse environmental implications, especially in urban areas. It tends to reduce food, water, fuel, land and other natural resources. It is a fact that the growth of human population is more stable in the developed countries, *i.e.*, both the birth rates and death rates are low. When the birth rate of a population is high, the population increases, and on the other hand when the death rate is high the population decreases. In the United States of America, France and Germany, the population growth is more stable than in the developing countries like Pakistan, Bangladesh, India and China. Usually, the relationship between population growth rate and the level of industrial development and education is inversely proportional. Such relationship is also found among different groups of the people in the same country.

SPECIES, ECOLOGICAL INTERDEPENDENCE AND INTERACTIONS

The Individual Organism

The individual organism is a distinct package carrying out life processes within its body as a entity separate from those being preformed in the bodies of others. The organisms may be unicellular or multicellular. The example of individual organisms are potato plants, bean plants, cats, rats, tigers and others. Such organisms can be counted, measured and studied. However, in may cases they remain organically connected with one another such as in grasses or in colonies of corals and sponges. The individual organisms always develop from pre-existing ones by means of reproduction which may be vegetative, asexual or sexual.

Species

A large variety and numerous individual organisms are found in nature. The organisms do not occur in isolation. Such organisms may be of a kind, as cats, rats, tiger, bean plants, humans, etc. The individual organisms aggregate and form population. A geographically localised group of individuals of the same kind form a **species**. The individual organisms of a species may differ in structure and function depending upon their sex, stage of development or specialisation for specific functions. For example, in many insects the adult, larval and pupal stages are considerably different. In frogs, toads and salamanders the adults and larval stages are quite different in structure and function. On the other hand, the sporophyte and gametophyte of a fern or moss are quite different in their structure and function.

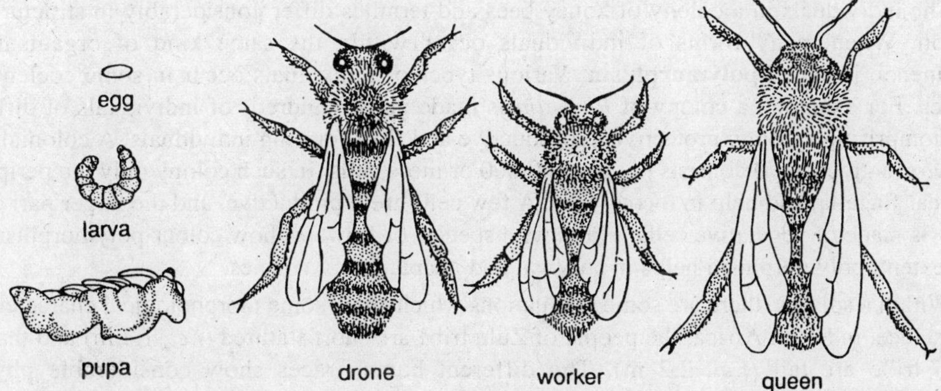

Fig. 11.3. Polymorphism in adults of honey bee. Stages in the development of honey bee.

On the other hand, in many organisms the males and females are different in size, form and colour. When two forms occur among the organisms of the same kind the phenomenon is called **dimorphism**. The horse and the mare, the dog and the bitch, the lion and the lioness, man and woman, peacock and peahen are not alike. They exhibit **sexual dimorphism**. Plants such as the

Fig. 11.4. Some breeds of dog showing variation within a species.

date palm and Cycas have male and female individuals and bear different types of flowers and sporophylls.

The individuals in a colony of honey bees and termites differ considerably in structure and function. When many forms of individuals occur within the same kind of organism, the phenomenon is called **polymorphism**. Various types of individuals occur in some coelenterate colonies. For example, a colony of *Physalia* is made up of hundreds of individuals of different kinds comprising feeding, protective, reproductive and float forming individuals. A colonial form of *Volvox* (a green alga) consists of about 60,000 or more cells. In such colony only the peripheral cells bear flagella, and help in locomotion. A few cells are reproductive, and the larger part of the colony is made of vegetative cells. Many wild species of flowers show colour polymorphism. To some extent, polymorphism helps in survival and adaptation of species.

Within a species, there are some populations which show some morphological characterstics. For instance, in South Africa, the people of Zulu tribe are short statured (*i.e.*, 1.2 m) and those of Watusi tribe are tall (*i.e.*, 2.2 m). The different human races show considerable physical differences. The negroids, caucasoids and mongoloids, can be recognized by their peculiar faces. Similarly, the various breeds of dogs, cows, sheep, roses and other cultivated plants look quite different in size, colour and appearance (Fig. 17.4). For example, there are several breeds of dogs, such as Keeshond, Eskimo, Cairn terrier, Indian greyhound, Welsh terrier, Chow, Dachshund, Dalmatian, Irish terrier, Yorkshire terrier, Buldog and Irish wolfhound, which are quite different in size, colour and apperarance. However, all these different looking individuals of the same species may interbreed among themselves. While the populations or individuals belonging to different species may appear quite similar but do not breed in natural conditions.

Thus, a species can be defined as a *group of closely related, structurally and functionally similar organisms which interbreed with one another in nature, but not with organisms of other groups.*

The inability to interbreed or reproductive isolation makes the most important distinction between different species. Thus, reproductive isolation is the main factor that maintains the distinctness of different species. But there are several examples where members of two species are alike and still do not interbreed. However, members of two different species may breed under captivity or under artificial conditions. As we know, the horses and donkeys are of distinct species but they interbreed and produce sterile offspring, the **mule** (*i.e.*, progeny of male donkey and mare) or **hinny** (*i.e.*, progeny of stallion and female donkey). The artificially produced hybrids between bull frogs leopard frogs die before attaining the reproductive age. There are some examples of species that interbreed under captivity to produce fertile offspring, they are—the mallard and pintail duck, the tiger and the lion, the polar bear and the Alaskan brown bear, the platy and swordtail fishes. The interbreeding between members of two different species can occur only when the isolating condition is removed artificially. There is great diversity in the living world. To study the diversity, the scientists have developed several hierarchical terms–many similar species make a **genus** ; many related genera constitute a **family**, and several related families form the **order**.

Origin of Species

It has already been mentioned in the preceding paragraphs that prevention of interbreeding between members of different species is essential for the maintenance of identity of the species. It is an established fact that life arises only from pre-existing life. It is also thought, that new species is derived from the par-existing ancestral species. Now the question arises, how does a new species originate. Here, we will consider the origin of a terrestrial species which has a wide range of distribution. For example, due to development of a physical barrier, such as a high mountain range or a large body of water, it becomes impossible to maintain interbreeding between the members of the groups thus separated. Sometimes the long distance between populations may act as a reproductive barrier between them. In the initial stages, interbreeding between the individuals of these two populations will be possible if the isolation factor is artificially removed. However, the absence of a reproductive contact for a long time will lead to a loss of capacity for interbreeding. In the due course of time, the individuals of these populations undergo physical and physiological changes depending upon their environment, so that, they become morphologically different from each other. They would gradually go away from each other in structure and function and ways of living, and become so different as to give rise to a new species.

Interactions Between Members of a Species

The individuals of a species interact with each other in several ways, such as cooperation, communication and competition.

Cooperative interaction. This type of interaction is essential for reproduction and prepetuation of the species. Organized social interactions offer the advantages of cooperation in reproduction, caring for the young, obtaining food and in defence. A herd of musk oxen form a circle for protection. Social organization may be very simple or very complex. The simplest social behaviour is herding, schooling or flocking. The individuals are less likely to be captured by a predator while in a school than if alone. Social organization also aids in reproduction and rearing. For the seal, meeting at the breeding area makes contact between male and female easier. The baboons and monkeys live together for such advantages as grooming each other and caring for the young. Herding, rank establishment and territory and other interactions are restricted to certain animals only.

Group formation. Family life or group formation in different animals varies. It is found only in few types of animals. Foxes, wolves and swans are **monogamous**, *i.e.*, once a male and female come together the partnership lasts throughout life. However, in sparrow the partnership breaks down at the end of a breeding season, and the male and female birds find new partners in the subsequent years. Seals, walruses and deers are **polygamous**. A male keeps a harem of many females. Rats, mice and some other animals do not have any fixed partners.

Many animals, such as geese, ducks, parrots, antelopes, deer, zebras, elephants and monkeys live in groups or herds. Members of each herd cooperate for mutual protection and for finding food and water for the herd. For example, it has been observed in the Bandipur Santuary of Karnataka that when a group of cheetals is attacked by a predator, such as the wild dog, panther or tiger, the stag with the best antlers is closely surrounded by other individuals so that it is saved from destruction. This is a unique example of the self-sacrifice of a few individual deer allowing themselves to be killed in the interest of the species.

Usually a community has a few populations that are more important in determining the nature of community than the others. These **dominant species** strongly influence the food supply and the envrionment of the other species. In a forest, the large trees are usually dominant species. Large trees dominate forests because their canopy of leaves over the other populations determines how much light is available to other plants.

On the other hand, within the herd there is a social organisation and ranking of individuals. Usually the females are subordinate to males and the young are subordinate to the adult. In the red deer the female is the leader of the herd. **Dominance** or ranking of individuals is generally settled by battles. Dominance is also established by an aggressive posture or display of feathers by birds. In most species, the individuals of the highest rank has the first choice of mate and food followed by individuals of subsequent ranks.

The honey bees, ants and termites form well organised societies having division of labour among its members showing **polymorphism**. In these social insects (honey bees) the individuals of different kinds (*i.e.*, polymorphic) live together in a colony and work together for the benefit of the group. Here the insect societies are formed of different castes such as workers, males (drones) and queen. All of them are specialized for different kinds of work. The workers collect and store food. They also build houses of complicated design, and also do other special tasks. The males and the queen are mainly involved in reproduction and act as the progenitors of future colony.

In the termites incomplete metamorphosis is found, which shows that the insects hatched from eggs are very similar to adults and develop into adults through growth and moulting. Most of these function as worker termites, and have perpetual adolescent phase and do not develop into adults. Only the **king** and the **queen** reach the adulthood and produce future colony through sexual reproduction.

The insect societies are quite different from the human society. They differ both in origin and organisation. Human societies show more psychosocial evolution than biological evolution. The members are made of numerous family units composed of males and females, adults and juveniles. However, the insect societies are the products of biological evolution based on learnt behaviour. All the members of the colony are derived from common parents.

Home range and territory. Individuals, pairs or groups of vertebrates and higher invertebrates, commonly restrict their activities to a definite area. This area over which the animals wander in search of food and mates, is called their **home range**. Variations in home range are associated with the species, sex, age, season and with such ecological condition as availability of food, shelter and intraspecific struggle. Home range provides an animal with its necessities of life, such as food, cover, shelter and to rear young ones.

A number of breeding groups, that is pairs and herds inhabit a home range. Each breeding group, or an individual occupies a particular **territory**. The home range of different groups can overlap but not the territory. Each territory centres round the individuals or group's home, nests, holes, burrow or other types of shelter. The special odours of the urine or faeces delineate the territory of animals like dogs and tigers. The territory is defended by its members. The territorial concept is most developed in birds where the male first establishes a territory. It drives off all other males of the same species from this territory and sometimes also females other than its mate.

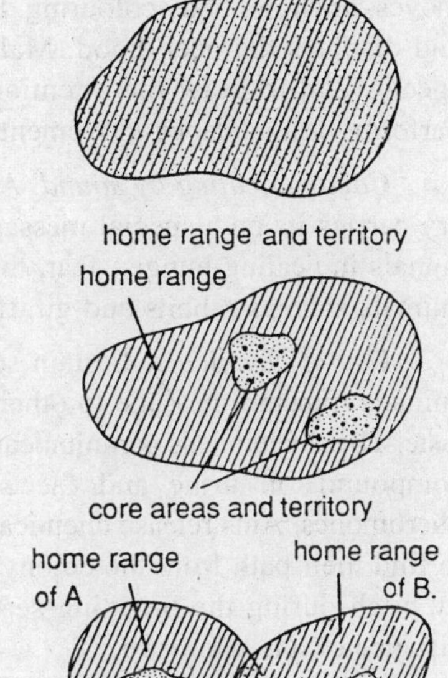

Fig. 11.5. Home ranges and territories of two species.

Communication. The survival and well-being of an animal population is largely dependent upon the communication among the members of the population. Even simple organisms have developed methods of communication, such as sight, sound, chemicals or contact. Through these methods of communication, members of a population indicate dominance, submission, the source of food, the presence of danger, hunger, the urge to mate, and the limits of territory.

Visual displays. Animals may use visual displays to communicate. This is the way where the animal stands,

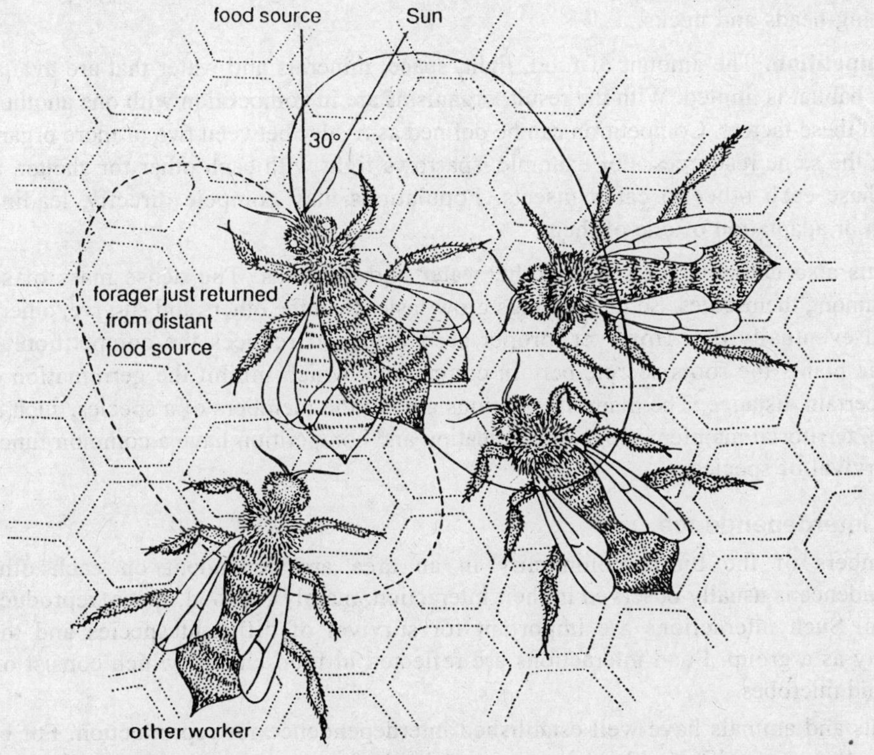

Fig. 11.6. Honey bees communicate through special dance patterns.

moves, or shows body colouring. For example, an aggressive cobra arches its back, bares its fangs, and displays the neck hood. Male birds are often brightly coloured to attract females of their species and to draw the attention of enemies away from the female and young. Honey bees perform various dance movements to indicate the distance and direction of the food source.

Communicating by sound. Animals may also communicate by making sounds. Each call or cry carries its own special message. Some animals have as many as two dozen different sound signals indicating hunger, fear, threat, annoyance, desire for mating and parental concern. Even animals such as rabbits and giraffes, utter sounds when they are intensely afraid.

Use of chemicals. Certain special chemical compounds such as **pheromones**, secreted by animals transmit message to other members of the species. Pheromones are detected by smell or taste. These chemicals communicate territorial rights, fear, or a desire to mate. The territory making compounds in urine and faeces as already mentioned in preceding paragraphs are also pheromones. Ants release chemicals containing pheromones, which enable the members of colony to find their path from the colony to the source of food and back. The sex pheromones are used by bitch during the breeding season to attract the male dog. A dog can smell a bitch from a distance of one kilometre.

Chemicals can also be produced and secreted by **specialized glands** or **tissues** and may be included in the urine or saliva. The male lion marks his territory by spraying urine at points along the outer edge of his territory. Among social animals, the dominant males have large **scent-marking glands** as compared to the others on the group and use them more often.

Physical contact. To convey concern, hunger, play or aggression, animals may use physical contact. A young bird pecks at the parent's bill to communicate hunger. A pet cat will rub up against a person to communicate pleasure. Giraffes use their long necks in struggles for dominance by knocking heads and necks.

Competition. The amount of food, light, space, minerals and water that are available in a particular habitat is limited. With the result, organisms are in competition with one another for one or more of these factors. Competition can be defined as rivalry between two or more organisms for obtaining the same resources. For example, sparrows fight with each other for shelter, and wall lizards chase each other to catch insects. Populations may compete directly, leading to the extinction or adaptation of one of them.

Plants also compete for space, light, water and minerals. The dense mass of seedlings compete among themselves. Some seedlings grow faster than the others and survive, others remain weak and eventually die. However, proper seed dispersal reduces the competition. In some xerophytic plants the roots secrete certain compounds which inhibit the germination of seeds within a certain distance. The many interactions among the members of a species, such as group formation, territorial maintenance, communication and competition have a common function and that is survival of species.

Species Interdependence

Members of the biotic community in an area are dependent on each other. The interdependence is usually observed in their interaction, mainly for food, space, reproduction and protection. Such interactions are important for survival of different species and the biotic community as a group. Food interactions are reflected in food chains, which consist of plants, animals and microbes.

Plants and animals have well established interdependence for reproduction. For example, insects are flower-specific and have structures suitable for nectar sucking and pollination.

Contrary to it, some flowers are insect-specific. For example, snapdragon flower possesses lip-like petals, which facilitate insect entry and landing. Here, both flowers and insects have developed and evolved in a way leading to their interdependence. The nectar of flowers is main attraction for insects. Insects visit the flowers in search of nectar and pollination is effected. The fluctuations in insect pollination would affect pollination that leads to plant reproduction.

Birds and mammals help in the dispersal of seeds and fruits. Interdependence may also be shown by certain other specific examples. For example, cuckoo lays eggs in the nest of other birds, and gall wasps bury their eggs deep into the tissues of plants that ensure hatching and protection.

Camouflage. Some members of biotic community, particularly animals, sometimes camouflage to avoid detection. For example, butterflies and moths have colours which match the colours of flowers, and therefore, their detection becomes difficult. The camouflage can be observed in praying mantis and leaf insect, which mimic the shapes of leaves and branches.

Mimicry. For example, orchids mimick the butterfly, as their colours are concerned. Generally, weaker members of animals, mimic the strong, fast moving species and which have a fewer natural enemies. For example, non-poisonous snakes mimic the poisonous snakes to avoid their enemies. Usually, birds mimic voice of other animals.

Interaction Between Different Species

In the great web of life, which exists in nature, living organisms not only live in an environment, but are also themselves a part of the dynamic environment for other organisms. The relationship between one species and another within a community has evolved through their interactions, based on the requirement and the mode of nutrition and shelter and also on the habits of species. The relationships between members of different populations are termed **interspecific relations**.

The interactions between populations of species in a community are broadly divided into two categories : (*i*) **positive** (beneficial) and (*ii*) **negative** (inhibition) **interactions**. This depends upon the nature of effect on the interacting organisms of different species.

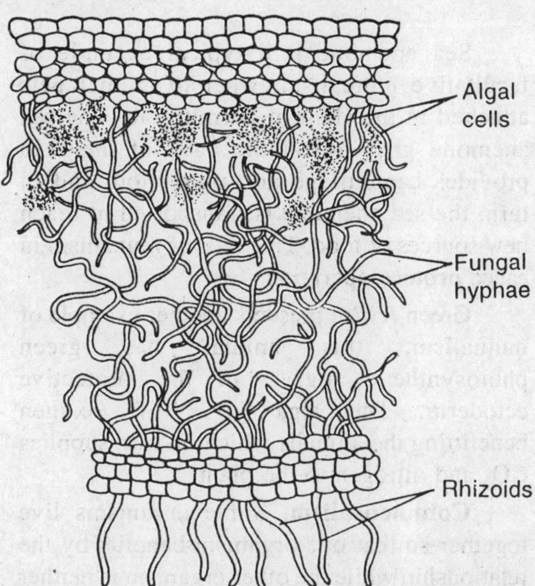

Fig. 11.7. V.S. of Lichen thallus showing association of alga and fungus.

Positive interactions

Symbiosis or mutualism. When two species live together in a close association that is helpful to both species, the relationship is known as **symbiosis**. The oxpecker bird and the rhinoceros exhibit this relationship. The oxpecker receives protection and obtains food from the ticks and other pests infesting the rhino's skin. The rhino receives cleaning and warning of approaching dangers. Algae and fungi live together in symbiotic relationship in **lichens**, whereas the fungi live on the roots of higher plants, and the association is known as **mycorrhiza**. In lichens, the algae are able to produce food by photosynthesis, and the fungi obtain water and minerals and provide attachment for the lichen. The bacterium *Rhizobium leguminosarum* found in the root nodules of leguminous plants fixes atmospheric nitrogen in the soil in the form of nitrates which is used by the plant, and in return plant supplies water, minerals and organic food to the bacterium. The cyanobaterium *Anabaena* also lives in symbiotic association with water fern *Azolla*. Bacteria in the gut of some domestic animals help in cellulose digestion.

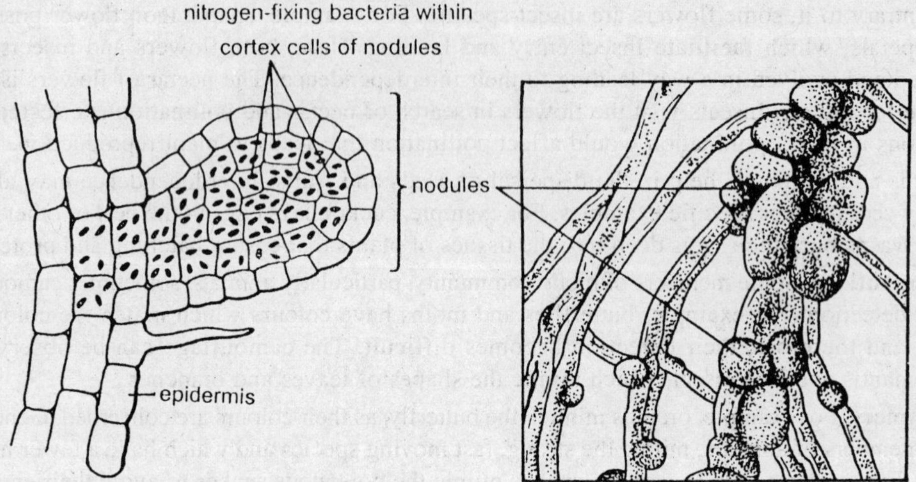

Fig. 11.8. Nitrogen-fixing bacteria in root nodules of a legume.

Sea anemone is a typical example of facultative mutualism, wherein animal gets attached to the shell of hermit crab. The sea anemone growing on the back of the crab provides camouflage and protection, and in turn, the sea anemone is transported to reach new sources of food. This type of mutualism in called **protoco-operation**.

Fig. 11.9. Crab and sea anemone mutualism.

Green *Hydra* presents another example of mutualism, this animal has green photosynthetic alga in the protective ectoderm. The alga gives off oxygen benefitting the animals which, in turn supplies CO_2 and nitrogen to the plant.

Commensalism. Some organisms live together so that one organism benefits by the relationship while the other organism is neither helped nor harmed. This type of relationship is known as **commensalism**. An example of this association is the relationship between the shark and the small remora fish. The remoras may attach themselves to the shark as it swims through the water. When the shark finds food, the remoras eat some of the food not consumed by the shark. The shark is not harmed by the remora, while the

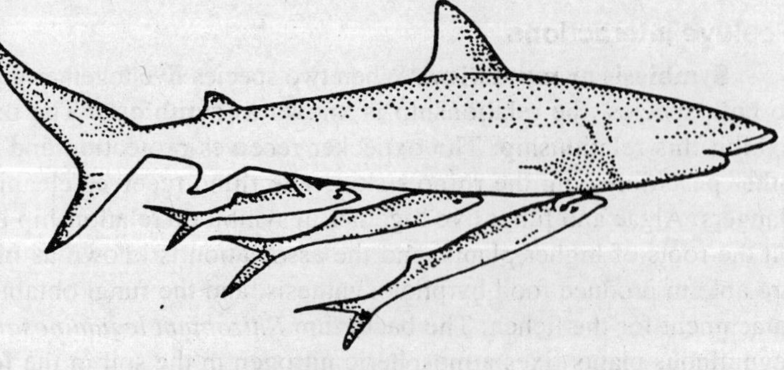

Fig. 11.10. Commensalism of shark and remora fish.

remora is helped. The attachment of the sedentary sea anemone to the body of a hermit crab and barnacles to a whale are other examples of commensalism.

Some epiphytes, such as orchids, mosses, ferns, etc., are the best examples. Epiphytes depend upon the other trees for support and nutrients. They manufacture their own food but do not help the supporting plant any way.

Several woody climbers (lianas) take the support of the trees for exposing their canopy above ground without doing any harm to the supporting trees.

Negative Interactions

Certain interactions between different species give rise to negative effect on either or both species. Parasitism and predation are interaction where one species gains and the other suffers. While in the interaction called competition both species are harmed.

Parasitism. This is a relationship in which one organism, the **parasite** spends much or all of its life living in or another organism, the **host**. The parasite is dependent upon the host for food. The parasite benefits from the relationship and the host is always harmed. Parasites may bring about the death of their host, but most often only weaken their host. Human parasites may be external (**ectoparasites**) such as body lice, ticks, mites and leeches, or internal (**endoparasites**) such as tapeworms, some types of roundworms, malarial parasite, microfilaria and guineaworm. The dodder (*Cuscuta*), broomrape (*Orobanche*), mistletoe (*Viscum*), *Dendrophthoe*, *Striga* and *Rafflesia* are parasitic plants. Certain bacteria are parasitic on human beings and animals and cause fatal diseases such as *Vibrio coma* (cholera), *Mycobacterium tuberculosis* (tuberculosis), *Salmonella typhosa* (typhoid fever), *Mycobacterium leprae* (leprosy), etc. A large number of fungi are parasitic on several crop plants causing diseases like rusts, smuts, blights, mildews and wilts.

A parasite usually parasitises a host which is larger in body size than it, and ordinarily it does not kill the host, at least not until it has completed its reproductive cycle.

Predation. This is commonly associated with the idea of strong attacking the weak such as the tiger pouncing upon the deer, the hawk upon the sparrow, the frog upon the insects and so on. A species such as the frog may be both a prey and a predator. The relationship between a snake and a rat is more than that between a prey and a predator as the snake also seeks shelter in the rat holes. Thus "predation represents a direct and often complex interaction of two or more species, of the eaters and being eaten." This is a negative interaction which results in negative effects on the growth and survival of one of the two populations.

In this type of association and interaction one species (**predator**) kills and feeds on second species (**prey**). Predation is important process in the community dynamism. Predator is always stronger than pery. From population ecology point of view predation is the action and reaction in the transfer of energy from one trophic level to the other. It represents a direct and complex interaction between two or more species of eaters and eaten.

Parasites and predators have some points of difference, they are as follows :

Parasites, like the predators, limit the population of the host species, but they are generally host-specific, and do not have choice or alternatives like predators.

Parasites are smaller in size and have higher biotic or reproductive potential composed to the predators.

Parasites have poor means of dispersal and require specialised structures to reach or invade the host. While the predators are quite mobile and swift, and capable of capturing the prey.

However, newly acquired predators and parasites are more damaging than the older ones, as the latter are familiar and the species getting affected have adjusted by then.

Competition. The amount of food, light, space, minerals and water that are available in a particular habitat is limited. As a result, organisms are in competition with one another for one or more of these factors. Competiton occurs not only among individuals of a given species but also between members of two or more species. For instance, carnivorous animals such as tigers and leopards, compete for the prey. The members of kingdom Plantae, such as trees, shrubs and herbs in a forest are to compete for sunlight, nutrients and water and biological agents for pollination and dispersal.

Populations may compete directly, leading to the extinction or adaptation of one of them. Many animals establish territories within which they live and which they will defend against other of their species who try to intrude. By staying in their own territories, competition and combat are lessened.

This is of two types :

(*i*) **Interspecific competition** occurs between the individuals of the same species and their requirements are common and,.

(*ii*) **Interspecific competition** occurs between individuals of two different species occurring in a habitat.

Generally the intraspecific competition is more intense than interspecific competition. Requirements of individuals of same species are similar, and therefore, competition is more intense.

Scavanging. This is a type of direct food relationship where animals such as a vulture or hyena, or a jackal feed on other animals which have died naturally or have been killed by another animal. For instance, the omnivorous animals such as the common crow consume many types of foods including dead animals.

Other relationships. Many insects, reptiles, birds and mammals get shelter and protection on trees and shrubs. Many animals protect them by a highly interesting device, called **mimicry**. This phenomenon is seen in many insects where they develop a superficial resemblance in shape and colour to specific plant parts on which they live. The examples are—the stick insect (*Carausius morosus*) mimics thin dry branches; the dead leaf butterfly (*Kalima parolecta*) resembles a dry leaf, whereas the praying mantis (*Mantis religiosa*) and *Rhyllium frondosum* resemble with the green leaves. By their mimicry, they conceal themselves from predators and other foes.

For the survival of a community, the availability of the other nonliving components is also important, and sometimes one of the components plays a dominant role in determining the character of the community. For example, a community having fruit bearing trees may consist mainly of fruit eating animals, such as bats and insects. On the other hand, in a grassland there will be only seed eating birds, mice, voles and predatory birds living on others lower down the food chain. In a marshy land, frogs, toads, fish, aquatic insects and water birds which feed on aquatic insects are maximum. Aquatic plants adapted to different intensities of light may be found either at the surface or at different depths in water. Thus the environmental factors act as the determinants of the types of individuals in a community.

BIOTIC COMMUNITY ORGANISATION

All the living things make the biotic system of the earth. Neither organisms nor species populations exist by themselves in nature, but are always part of an assemblage of different species populations living together in the same area. *Any assemblage of populations of living organisms in the prescribed area or habitat is termed biotic community*. For example, the different species of organisms occurring in a pond constitute the **pond community**. Some other communities such as swamps, deserts, large lakes, grass lands and many others of different dimensions. Similarly, the interdependent species in any environment, such as a forest, a lawn or a desert track, constitute biotic communities. A community may be composed primarily of animals, or primarily of plants, but most communities consist of both animals and plants. The factors that influence the living orgnisms are known as **biotic factors**. In the natural world there is interdependence of one form of life on another. A population of single species is not a viable entity by itself. It will not be able to survive for long. For example, the sparrows require worms, insects or seeds or some other kind of food. Thus the most appropriate definition of biotic community— "a biotic community is a naturally occurring assemblage of plants and animals that live in the same environment, are mutually sustaining and interdependent, and are constantly fixing and dissipating energy."

However, the different species in a biotic community share a common environment and their relationships are based on direct or indirect functional interactions. The nature of relationship is determined by the requirements of the members of the community.

Biotic community organisation results from interdependence and interaction among populations of different species in a habitat. Large number of biotic communities found in nature due to two specific reasons. They are as follows :

(*i*) Existence of diverse habitats with characterstic environmental conditions and

(*ii*) Co-occurrence of different species whose tolerance ranges overlap with the environmental condition obtained in that habitat. When similar habitat conditions are repeated at another location, the same biotic community gets established there.

Characteristics of Community

Each biotic community consists of very diverse organisms belonging to different kingdoms of living things. The number of species and abundance of population in communities also vary greatly. As mentioned earlier, the organisms in a community depend upon each other as well as upon the non-living environment for food, shelter and reproduction.

Species composition. The kinds of plants and other organisms present in a community indicate its **species composition**, which differs from one community to another. Sometimes, in the same community, there may be seasonal variation in plant species.

Each species of community has got definite range of tolerance towards the physical and biological enviromental conditions of the habitat. The range of environment a species can tolerate is called its **ecological amplitude**. The nature of community of a particular habitat is determined by the species, and physical and biotic influences prevailing in the locale of community.

Dominance. A biotic community may have major categories of growth forms, such as trees, shrubs, herbs and mosses. Out of hundreds of species present in the community, relatively only a few exert a major controlling influence due to their large size, numbers of activities. The phenomenon is called **dominance**. "Dominant species are those which are highly successful ecologically and which determine to a considerable extent the conditions under which the associated species must grow."

The dominance in the community may be the result of co-action between two or more species. Different communities are generally recognised and named on the basis of dominant species occurring in them. For example, a forest community in which pine trees are dominant is called pine forest. Grassland represents a community which has grass species dominating over the other herbs. Sometimes, communities are named after environmental factors, such as desert community, marine community, mangrove vegetation, etc.

Physiognomy. General appearance of vegetation is referred to as **physiognomy**. It constitutes general stature, shape and life-forms of the species comprising the vegetation and actually the classification of vegetation types has been done on the basis of physiognomy. The species of a community can be grouped into several life-forms on the basis of general appearance and growth. The physiognomy is the total effect created by the combination of vertical structure and architecture of dominant species of vegetation. For example, the high physiognomy of a forest differs distinctly from a low physiognomy of a grassland. However, several communities though possess similar physiognomy, yet differ markedly on the basis of species composition and dominants, *e.g.*, different types of forests.

Stratification. Every biotic community has a vertical layering or **stratification** of organisms or environmental conditions. A number of examples can be cited to support the concept of community stratification from different habitats. In **grassland community** three strata, namely (1) **subterranean**, (2) **floor** and (3) **herbaceous** may be recognized. The subterranean stratum contains the roots of the principal vegetation and provides permanent shelter to bacteria, fungi, protozoans, nematodes, earthworms, ringworms and several other invertebrates. The floor stratum consists of basal parts of the vegetation, including the rhizomes of grass plants. In this stratum, generally the insects, spiders, reptiles and rodents are peresent. The herbaceous stratum of grassland community is represented by upper parts of grasses and herbs. Several types of insects, birds and grazing mammals are included in this stratum.

In a **forest community**, five vertical strata are present. They are (1) **overstorey stratum**, (2) **understorey stratum**, (3) **transgressive stratum**, (4) **seedling stratum** and (5) **subterranean stratum**. A typical forest may have an overstorey stratum comprised of trees that are more than forty feet in height. These trees make a canopy. Just beneath this canopy there is an understorey stratum that extends from twenty feet in height to a short distance below the overstorey. A transgressive stratum extends from four feet to twenty feet or more and comprises of shorter shade loving species. The seedling stratum begins at the soil level and extends to the lower limit of the transgressive strautm. The subterranean stratum in the forest community is moist and contains a large amount of humus. It is very prominent upto a depth of two to three meters. Each strautm has its own group of animals. However, most animals can move from one stratum to another in search of food or in response to many biotic factors, such as insects, snails, birds and squirrels. The animals found above the soil include turtles, snakes, some birds, insects and a variety of mammals, such as rabbits, deer, wolves and foxes. The animals living in humus soil are beetles, fly larvae, spiders, annelids, antipedes, mites, protozoans, nematodes and some springtails.

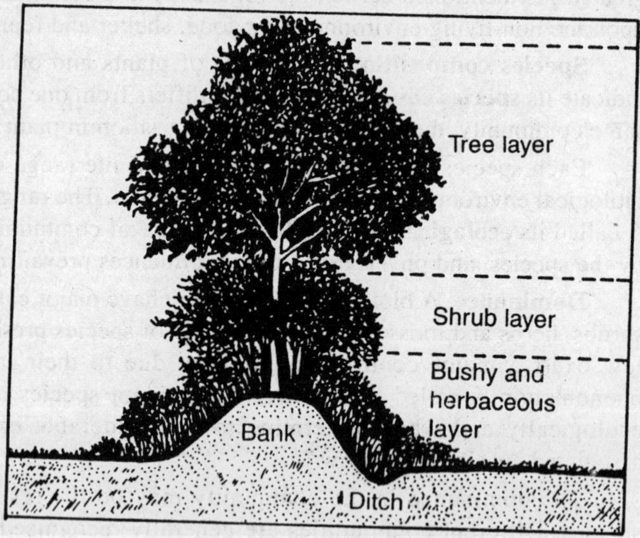

Fig. 11.11. Stratification.

In **pond community**, vertical stratification is very little. However, in deep ponds and lakes three strata—(1) **littoral zone**, (2) **limnetic zone** and (3) **profundal zone** can be recognized. The littoral zone comprises shallow water region and is occupied by rooted plants. The limnetic zone occupies the depth upto which the light penetrates and inhabited by planktons, nektons and neustons. In profundal zone, there is no penetration of light, and therefore, it has no photosynthetic organisms.

Species diversity. The biotic community is a natural assemblage of a large number of plant and animal species in an area. However, in any particular habitat there is no considerable variation in environmental condition, the plants growing together in a community show unique uniformity in their behaviour. Vegetation, therefore, is reflection of a climate and, in general, widely separated areas having similar climate have similar aspects of landscape.

Some communities, for example, tropical rain forest and coral reef community, show high species diversity with many different kinds of species living at each trophic level.

Some community areas have limits but more often the community boundaries are hard to define. A clearly distinguished area or a type of area with uniform habitat conditions and supporting characteristic type of vegetation is termed **biotype**.

Ecotone and edge effect. The zone where two or more different communities meet and integrate, is called **transition zone** or **ecotone**. This zone of integration may be narrow or wide, local (*e.g.*, a zone between field and a forest) or regional (*e.g.*, the transition between forest and grass land). Ecotone contains few species from both communities. The total number of species is often greater in the ecotone than in the adjoining communities. The ecotone or transition zone exhibits a shift in dominance of the conspicuous species of both sides. It may also include a number of highly adaptable species that tend to colonize such transitional areas. Beacause of this, the variety (*i.e.*, species diversity) and density of life is often greatest in such areas. This potential for the ecotone to act as a habitat for species found in neither major community is called **edge effect**. Thus the tendency of increased variety and density of some organisms at the community border is known as **edge effect**. The organisms that occur primarily, or most abundantly, or spend the greatest amount of their time in junctions between communities are called **edge species**. A common example of the edge effect in action can be seen in those species of owl that live in or near ecotones between forests and grasslands. They depend on forest trees for nesting and do their hunting in the grassland, where they depend on field rodents for food.

In man-made communities such as agricultural fields, the ecotone between the field and the forest act as refuge for species formerly found in the ploughed area, as well as for other plants such as weeds. Ecotones of this type are also the prime habitat of may species of insects, game birds, and mammals.

Keystone species. The species, which have much greater influence on community characteristics, relative to their low abundance or biomass, are known as **keystone species**. Such species play a vital role in controlling the relative abundance of other species.

When keystone species is removed, it causes serious disruption in the functioning of the community. For example, in the tropical rain forests, the different species of figs are the keystone species as they produce large quantity of fruits. During the time of food scarcity, these fruits are consumed by monkeys, bats, birds, etc. Thus, by protecting the fig trees, the animals dependent on them are also conserved.

Link species. However, only a few species act as keystone species, while others act as **link species**. For example, mycorrhizal fungi in soil are critical link species as they establish essential links in the absorption of nutrients from the soil and other organic substances.

Some critical link species are responsible for providing food to the network species, while others act as pollinators of flowers, and some act as agents for dispersal of seeds and fruits. Tropical rain forests are rich in critical link species due to high degree of animal dependent pollination and dispersal.

Analysis of Plant Communities

Generally, analysis of community characters is being done for two specific purposes :

(*i*) to record variation within and between communities and

(*ii*) for naming and calssifying communities.

Analytical characteristics are those features of community which can be observed or measured directly in each aspect. It involves measurements of various charactrs in sample plots, commonly known as **quadrats**. Measurements made in sample plots (quadrats) are scientifically processed to reflect the characteristics of the entire community.

Two sets of characters, *viz.*, (*i*) **analytical** and (*ii*) **synthetic** are studied in a community at the same time.

1. Analytic characters. They are directly observed or measured in sample plots. They include kinds and number of species, distribution of individuals, number of individuals, height of plants, etc.

2. Synthetic characters. They are derived from the measurements of analytic characters and utilise data obtained in the analysis of a number of stands.

Analytical characters are of two types :

(*i*) **Qualitative.** They are based on non-quantitative observations, *e.g.*, species composition and stratification of vegetation. They are expressed only in qualitative way.

(*ii*) **Quantitative.** They are expressed in quantitative terms. They are measured. The major quantitative characters include frequency, diversity, cover, biomass, leaf size, abundance, dominace, etc. They are as follows :

Frequency. This is based on precentage of sample plots in which a species is present, indicating its dispersion in space. This frequency of each species is calculated as follows :

$$\text{Frequency percentage} = \frac{\text{number of sampling units in which that species occurred}}{\text{total number of sampling units studied}} \times 100$$

Diversity. This is denoted by number of individuals per unit area, indicating the relative abundance of a species.

Cover and basal area. This is percentage land area occupied by a species, indicating the influence zone of a species. Although sometimes used in general sense for the area occupied by a plant, (which may be the herbage cover or the cover of basal area), it is generally used for above ground parts. Thus, cover or herbage cover signifies primarily the area of the ground occupied by the above ground parts of plants, such as leaves, stems and inflorescences as viewed from above.

However, basal area refers to the gound actually penetrated by the stems and is readily seen when the leaves and stems are clipped at the ground surface. It is one of the chief characterstics to determine dominance. It is measured either at 2.5 cm above ground or actually on the ground level.

Biomass. This expresses quantity of living materials per unit area, indicating the growth of a species. Thus, **biomass** is the standing crop expressed in terms of weight (*i.e.*, organism mass) of the living matter present. The amount of living material, present in a component population at any time, is known as the **standing crop**, which may be expressed in terms of weight per unit area.

Leaf area. The precentages of species having different leaf sizes, indicating the adaptation of the vegetation to the prevailing environment. As the leaves are essential part and are very much affected by climate condition, their shapes and sizes have been taken as important criteria in determination of quantitative characters.

Density. Density represents the numerical strength of a species in the community. The number of individuals of that species in any unit area is its **density**. This gives an idea of degree of competition. It is calculated as follows :

$$\text{Density} = \frac{\text{Total number of individuals of the species in all the sampling units}}{\text{Total number of sampling units studied}}$$

The value thus obtained is then expressed as number of individuals per unit area.

Abundance. This is the number of individuals of any species per sampling unit of occurrence. It is calculated as follows :

$$\text{Abundance} = \frac{\text{Total number of individuals of the species in all the sampling units}}{\text{Number of sampling units studied}}$$

Synthetic characters. These are determined after computing the data on the quantitative and qualititative characters of the community. For comparing the vegetation of different areas, community comparison needs the calculation of their synthetic characters. These are determined in terms of presence and constance, fidelity, etc.

Presence and constance. It expresses the extent of occurrence of the individuals of a particular species in the community.

Fidelity. This is the degree with which a species is restricted in distribution to one kind of community. Such species are sometimes known as **indicators**.

Dominance. Here, the dominance is expressed in synthetic form. On the basis of density, frequency and dominance (cover) values, there has been proposed idea of **Importance Value Index** (IVI). IVI of a species in the community give the idea of its relative importance. For IVI, values of **Relative density**, **Relative frequency** and **Relative dominance** (cover basis) are obtained as follows :

$$\text{Relative density} = \frac{\text{Density of the species} \times 100}{\text{Total density of all the species}}$$

$$\text{Relative frequency} = \frac{\text{Frequency of the species} \times 100}{\text{Total frequency of all the species}}$$

$$\text{Relative dominance (cover)} = \frac{\text{Dominance (cover) of the species} \times 100}{\text{Total dominance (cover) of all the species}}$$

Now for IVI, three values are added. IVI values of different species are then arranged in decreasing order.

Other synthetic characters. In addition to above mentioned characters, there are some other synthetic characters. They are quite useful in comparative studies on communities. Such characters include, **interspecific association** and **association index, index of similarity, dominance index, species diversity** and **diversity index**, etc.

BIOTIC SUCCESSION

Abiotic factors such as water, solar energy and minerals determine which populations, especially plant populations, can survive in an area. In general, climate and soil are the main factors that control where plants can survive on land. It is noticeable that plants serve as food for animals and other consumers. The result is that plant distribution will determine which animal populations will be present . On the other hand, it is true that the nonliving environment influences populations, and it is also true that populations influence the nonliving environment.

Abiotic factor and biotic factors interact to produce changes in the ecosystem. As conditions change with time, some populations gradually become less capable of surviving while others become more capable. The ecologists have studied that there is a successive replacement of one community by another. This sequence of replacement of one community of organisms with another is called **ecological** or **biotic succession**. Succession is a process with certain types of organisms always present in a particular stage of development in a given area.

The plants that invade the base land initially, are called **pioneer species**. The accumulation of pioneer species becomes the **pioneer community**. Generally, the pioneer species show high rate of growth but of short life span.

In time, a pioneer community is replaced by another community with combination of different species. This second community is replaced by third community, and process goes on. Different communities, are represented by combinations of lichens, mosses, herbs, shrubs and trees replacing one another during succession, they are called **seral stages** or **seral communities**.

The plant species which are established later are called late succcessional species.

For any particular kind of climate and soil, succession leads to a certain community structrure that is relatively stable and undergoes little further change. The stable commuinity structure of an ecosystem is called the **climax community**. A specific sequence of development of a community is related to a particualr set of physical and chemical conditions. This is known as a **sere** and is composed of a number of biotic successions replacing each other in course of time.

The biotic succession may be of two types—(1) **primary succession** and (2) **secondary succession**. When the succession proceeds from a bare and sterile area or from an area which has not been changed physically by organism, is known as **primary succession**. For example, succession on a bare rock **(xerarch)**, succession in a pond **(hydrarch)** succession in a newly exposed sea floor, lava sediments or sand dunes.

The succession occurring in terrestrial areas with low moisture, such as bare rock, is known as **xerarch**. The succession, that occurs in water bodies, such as ponds and lakes is called **hydrarch** succession.

These two types of succession are described here :

Succession on a Bare Rock (Xerarch)

For example, in the succession from bare rock, there must first be the long process of rebuilding the soil. Organisms such as lichens can survive on bare rock. Gradually, the effects of the lichens together with wind, rain, ice and temperature changes begin to break rocks into soil. The lichens also trap soil that is transported by wind and water. As soil formation takes place, other plants such as mosses, grasses and flowering herbs are able to survive and reproduce. After hundreds of years such succession from bare rocks may result in a climax forest in the area.

Fig. 11.12. Ecological succession on a bare rock (xerarch).

The pioneer colonisers on the bare rock are crustose lichens which occur on the rock surface in the form of membranous crusts. Their spongy nature enables them absorb excess amount of water and minerals. The lichens secrete carbonic acid (H_2CO_3) in excess, which reacts with the rocky materials and loosens the rock particles. The corroded rock particles together with decaying lichens make the first thin layer of the soil on the rock surface. After accumulation of little soil and humus the rock surface previously occupied by crustose lichens, now gets covered with foliose and fruticose lichens. Now because of enough moisture mosses also invade, and a lichen-moss carpet is formed which provides suitable substratum for the germination of seeds of herbaceous plants which are dispersed in it.

The seeds of higher plants germinate and grow successfully in pockets of newly formed soil on the rock. When the soil increases in thickness, the herbaceous vegetation, which consists mainly of annual and perennial herbs, develops very quickly. The roots of these plants penetrate deep in the crevices of the rock, where they secrete acids and accelerate the process of rock disintegration. Decaying roots, stems, leaves and other parts of the plants become deposited on the soil surface in the form of humus. This further increases the water holding capacity of the soil. These reactions make the habitat more suited for woody plants than to the existing herbs. Now herbaceous species make way for the invasion of shrubs followed by the trees. Decaying roots, leaves, twigs, etc., of these shrubs also enrich the soil with humus. These important reactions bring about such conditions in the habitat, that are most suited to the trees rather than to shrubs or herbs. Now the trees invade the area which has been occupied previously by shrubs. The first trees growing in such areas are dwarf and widely spaced. With the increase in the water holding capacity of the soil these trees gradually disappear and subsequently the mesophytic large trees are developed. Passing through seral stages in course of time, **climax community** gets established, which is generally dominated by trees.

Dominance of species. This refers to the fact that not all the species in the community are equally importnat in determining the nature of the community out of hndreds of species present in the community, relatively only a few exert a major controlling influence due to their large size.

The stability of a community depends upon the large number of species of a community, the larger the number of speices that interact, the more stable that community will be. For example, if a biotic community is represented by a single species of eucalyptus or casuarina (*i.e.*, monoculture), it may be completely wiped out if it is infested by fungal, viral or bacterial disease. However, when a mixture of many species of trees is raised up, only one species would be destroyed by a particular disease, and the rest species will survive.

Various studies and records have revealed that development of a climax forest from bare rocks or sand dunes in primary succession takes about 1000 years or more, whereas a secondary succession requires a shorter period of time. A destroyed grassland recovers in 50 to 100 years, while a destroyed forest in more than 200 years. When a man destroys a forest by felling the trees, or denudes a grassland by overgrazing his cattle, the process of community succession reverses.

Succession in Aquatic Environment (Hydrarch)

Hydrarch succession occurs in the aquatic environment. Such succession is recognisable only if the colonisation of plant communities takes place in small shallow ponds, lakes, etc., where wave action speeds up the process by allowing the erosion of soil towards edge regions. Water bodies are prove to sitting as a result of soil erosion from surrounding areas. On the other hand, blockage of rivers by landslides and construction of dams lead to formation of new lakes, where hydrarch succession proceeds in due to invasion of aquatic species. Thus, the filling process speeds up quickly and consequently the body of water disappears within few years time.

In a new and virgin pond hydrarch starts with the colonisation of phytoplankton and zooplankton which constitute the pioneer community. The simple forms of life, such as bacteria, algae and many other aquatic forms of plants (phytoplankton) and minute animals (zooplankton) floating in water are the pioneer colonisers. All these organisms add large amount of organic matter and nutrients due to their various life activities and after their death, they settle in the bottom of pond to form a layer of silt.

The phytoplankton stage is flollowed by submerged plant stage. The submerged plants, with their roots anchored in the mud are next to colonise the pond. When these plants die their remains are deposited at the bottom of the ponds or lakes. This process gradually raises the bottom of the ponds and lakes up. As this process progresses, the body of water becomes more and more shallow, consequently the habitat becomes less suited for the submerged plants but more favourable for other plants.

Now the **floating plant species** invade the pond. In the beginning the submerged and floating plants grow intermingled but in the course of time the submerged plants are replaced completely. The most tolerant species in the area are able to reproduce and perpetuate. Due to continuous interaction between plant communities and aquatic environment, the habitat becomes changed chemically as well as physically. More water and air borne soil and dead remains of plants are deposited at the bottom. Thus, the substratum rises up in vertical direction. Important floating plants that replace the submerged vegetation are : *Nelumbo*, *Trapa*, *Nymphaea*, *Pistia*, *Limnanthemum*, etc.

With the continued siltation, the pond bottom is gradually raised and water layer becomes shallow and rich in nutrients. The habitat is changed so much that it becomes less suiting to the floating plants, some other plants, such as reeds, are able to colonise the pond. This is accompanied by the invasion of dragon flies, crustaceans and more rooted species, such as *Typha*, *Phragmites*, etc. The foliage leaves of such plants are exposed much above the surface of water and roots are generally found either in mud or submerged in water. The foliage leaves form a cover over a submerged and floating plants and thus they cut off light from the plants underneath them. Under such condition neither submerged nor floating plants can survive. Further deposition of soil and plant debris at the bottom reduces the depth of water and makes the habitat less suitable for the pre-existing plants. When the bottom reaches very close to the water surface many secondary species of plants, such as *Polygonum*, *Sagittaria*, etc., make their appearance. Thus, the species composition of the pond keeps changing with time.

The filling process finally results in a marshy soil which may be too dry for the plants of pre-existing community. Now the plants well adapted to this new habitat begin to appear in the pre-existing community in mixed state. As most of the plant species are sedges and rushes, and therefore, this is called **sedge marsh stage**. With increased setting of silt and deposition of dead organic matter derived from pre-existing community of plants, the pond becomes shallower until it gets transformed into a terrestrial habitat.

Gradually grasses, bushes and other mesophytes start appearing and after sometime the sedge and marshy vegetation is totally replaced by them.

In the beginning some shrubs and later medium sized trees form open vegetaion or **woodland stage**. These plants produce more shade and absorb and transpire very large quantity of water. Ultimately these terrestrial species colonise the pond area and a **climax community** is established. In the climax forest, all types of plants are met with. Trees are dominant and they have control over the entire vegetation. During hydrarch succession, the colonisation by land plants usually progresses from margins toward the centre of the pond area. At the climax stage, a complete harmony is established among plant community and habitat.

Characteristics of Community that change during Succession

The characteristics of the community and their species that appear during primary or secondary succession, vary with their habitats and its conditions.

The seral stages differ from climax stages with regard to their structure and function. The average size of individual plants or animals increases and the organisation of community becomes more complex in comparison to seral community.

Now the food webs become more complex during various successional stages. At the same time, the efficiency of energy use and nutrient conservation increase as the community progresses towards the climax stage.

The composition of species at the climax stage is determined by the climate of particular region, as well as local edaphic factors or soil conditions.

Topography of the region and water availabilty are also responsible for species composition. Where heavy rains are available the forests appear.

The climax stage shows the highest level of vegetaion and the associated fauna that can be supported under the surroundings of environment.

SUMMARY

1. Population is defined as a group of individuals of the same species that inhabit the same area.
2. The members of the population are capable of interbreeding among themselves.
3. The characteristic parameters of population are : density, natality, mortality, age distribution, growth form, biotic potential, etc.
4. Under favourable conditions, the magnitude of population increases.
5. Two patterns of growth form occur, which are known as J-shaped and S-shaped.
6. The size of population is determined by the balance between number of individuals added usually by natality and immigration.
7. The highest magnitude of the population that can be supported in a given habitat is called its carrying capacity.
8. Population of different species that occur in a particular habitat make the biotic community.
9. The characteristic features of biotic community are : dominance, stratification and species interactions.
10. Community analysis may be done on the basis of qualitative, quantitative and synthetic characters.
11. The characters of quantitative analysis include frequency, density, cover, biomass and leaf size of species present in the community.
12. The transition zone between two communities is known as ecotone.
13. The members of the community are interdependent for food, reproduction and protection.
14. The insects and moths play an important role in pollination of plants.
15. The interaction in between members of the biotic community may have positive or negative effects.
16. Organisms living together when benefit each other, the process is called mutualism.
17. When one organism may benefit the other without affecting it, the process is commensalism.
18. In certain cases one organism may adversely affect the other, they are called predator and prey respectively.
19. The relationship in whih the smaller organism, called parasite adversely affects the larger host, is called parasitism.
20. The biotic community is always dynamic, and undergoes changes with the passage of time; such changes are sequentially ordered and constitute succession.
21. The process of succession involves replacement of one community by the other.
22. Succession begins with invasion of the barren area by the pioneer species and thereafter other suitable species develop.
23. Ultimately succession leads to a climax community. The climax community is stable till the environment does not change.

TERMS TO REMEMBER

Analytic characters. The characters, which are directly observed or measured in sample plots.

Biomass. Quantity of living materials per unit area, indicating the growth of a species.

Biotic community. Poupulations of a variety of plants, animals and micro-organisms living together within a habitat.

Biotic potential. The inherent maximum capacity of an organism to reproduce or increase in number.

Carrying capacity. The maximum number of individuals of a population that can be sustained indefinitely in a given habitat in carrying capacity (K).

Commensalism. A relationship between two species where one species in benefitted, while the other neither gets any benefit nor adversely affected.

Climax community. The terminal stage of succession.

Competition. Interaction between two species, where both suffer adverse effects.

Cover. Percentage of land area occupied by a species, indicating the influence zone of a species.

Density. It is expressed as the total number of individuals present per unit area or volume at a given time.

Dominance. Where one or few species in a community are in abundance to dominate and influence other species in terms of numbers and biomass production.

Diversity. Number of individuals per unit area, indicating the relative abundance of a species.

Ecological succession. The succession replacement of communitites in an area over a period of time.

Ecotone. The transition zone between two communities.

Ectoparasites. The parasites which remian outside the host.

Edge species. The organisms which spend greatest amount of their time in junctions between communities.

Edge effect. The tendency of increased variety and density of some organisms at the community border.

Environmental resistance. The environmental check on population size, or its biotic potential.

Frequency. It is based on precentage of plots in which a species is present, indicating its dispersion in space.

Hydrarch. Succession in aquatic environment.

Keystone species. The species having much greater influence on community characteristics, relative to their low abundance or biomass.

Leaf size. Percentages of species having different leaf sizes, which indicate adaptation of vegetation to the prevailing environment.

Mortality. The loss of individuals due to death in a population under given environmental condition.

Metapopulation. A set of local populations connected by dispersing individuals.

Mutualism. An association of two species where both species are benefitted.

Natality. The increase in number of individuals in a population under given environmental conditions.

Parasitism. In the interaction, the parasite lives on the host, i.e., host parasite relationship.

Pioneer species. The plants that invade the base land initially during succession.

Physiognomy. External look or appearance of the community.

Pioneer community. The assemblage of pioneer species.

Population. A group of individuals of the same species, inhabiting the same area, and functioning as a unit of biotic community.

Population growth forms. Population characteristic patterns of growth with time.

Symbiosis. The condition in which there is a close physical association between the individuals (symbionts) of a pair of species, benefitting each other.

Xerarch. Succession on a bare rock.

QUESTION BANK

Very Short Answer Questions

1. Select the statement which best explains commensalism :
   (*a*) One organism is benefitted
   (*b*) Both the organisms are benefitted
   (*c*) One organism is benefitted; other is not affected
   (*d*) One organism is benefitted; other is affected
2. Community is defined as aggregation of :
   (*a*) individuals of the same kind (*b*) individuals of different kinds
   (*c*) individuals of a population (*d*) population of different species.
3. Parasite can be explained as an organism which depends on others :
   (*a*) for food (*b*) for shelter
   (*c*) for both food and shelter (*d*) for reproduction
4. Define symbiosis, and name the organisms that form lichens.
5. Define the term succession.
6. What is 'hydrarch succession' ?
7. What is 'xerarch succession' ?
8. Define the term 'climax community'.
9. What are 'link species' ? Give one example.
10. What is 'mutualism' ?
11. What is 'biotic potential' ?
12. What is 'ecotone' ? Give one example.

ANSWERS

1. (*c*) ; **2.** (*d*) ; **3.** (*c*) ; **4.** The condition in which there is a close physical association between the individuals of a pair of species; algae and fungi ; **5.** The successive replacement of communites in an area over a period of time ; **6.** The succession occurring in water bodies like ponds and lakes ; **7.** The succession that occurs in terrestrial areas with low moisture ; **8.** The terminal stage of succession, which is stable and does not show changes in species composition ; **9.** Link species are those which establish esential links in the absorption of nutrients from the soil and organic residues, *e.g.*, mycorrhizae ; **10.** An association of two species, in which both species are benefitted ; **11.** The inherent maximum capacity of an organism to reproduce or increase in number ; **12.** The transition zone between two communities; *e.g.*, the ecotone between the grassland and the forest.

Short Answer Questions

1. Write a short note on 'natality'.
2. What is 'environmental resistance' ? Support your answer with suitable example.
3. Write a note on 'population growth forms'.
4. What is 'parasitism' ? Give suitable examples.
5. What do you understand 'edge effect' and 'edge species' ? Write a note.
6. Write a note on 'camouflage' and 'mimicry'. Give suitable examples.
7. Write a note on 'commensalism' with one suitable example.
8. What do you understand by 'predation' ? Write a note.
9. Distinguish between population and community.
10. Explain J-shaped pattern of population growth.
11. Give any example of mimicry in organisms.
12. How is a biotic community named ? Give examples.

13. Compare J-shaped pattern with S-shaped pattern of population growth.

14. What are the important analytic characteristics in community analysis ?

15. Write some special characteristics of a parasite.

16. Explain how predation is beneficial in the long run.

17. Define biotic community. How is it different from the term community used in human population ?

18. What do you understand by the term 'physiognomy' ? Support your answer with suitable examples.

19. What is 'stratification' ? Support your answer with suitable examples.

Long Answer Questions

1. Describe the process of succession on a bare rock.
2. How does succession differ in terrestrial and aquatic systems ?
3. Explain the differences between the seral stage and the climax community during succession.
4. Explain species interaction under :
   (*a*) Positive interaction (*b*) Negative interaction
5. Describe the different community characteristics.
6. How do you distinguish the following ?
   (*a*) Ecotones and edge effect
   (*b*) Kestone species and Critical link species
   (*c*) Ectoparasite and endoparasite
   (*d*) Primary succession and secondary succession
   (*e*) Pioneer community and Climax community
7. Describe the process of succession in lakes.

MULTIPLE CHOICE QUESTIONS (OBJECTIVE TYPE QUESTIONS)

1. A mutually beneficial association which is neceassary for the survival of both the partners is called :
   (*a*) Symbiosis (*b*) Commensalism (*c*) Parasitism (*d*) Mutualism
2. A parasite living in the tissues of a host is :
   (*a*) Ectophyte (*b*) Endophyte (*c*) Epiphyte (*d*) None of these
3. A saprophyte which can act as parasite is :
   (*a*) Facultative saprophyte (*b*) Obligate saprophyte
   (*c*) Facultatative parasite (*d*) Obligate parasite
4. In which of the following association, one of the species gets benefit and the other is unaffected ?
   (*a*) Commensalism (*b*) Mutualism (*c*) Symbiosis (*d*) Parasitism
5. Food derived by killing organism is known as :
   (*a*) Commensalism (*b*) Symbiosis (*c*) Parasitism (*d*) Predation
6. Obligate parasites are the organisms that :
   (*a*) Live on dead, decaying organic matter
   (*b*) Live on living organism only
   (*c*) are essentially saprophytes but can also become parasites
   (*d*) are essentially parasites but can also become saprophytes
7. Relationship between Remora and Shark is :
   (*a*) Symbiosis (*b*) Commensalism (*c*) Parasitism (*d*) Protocooperation
8. Mycorrhiza show :
   (*a*) Amensalism (*b*) Parasitism (*c*) Commensalism (*d*) Mutualism
9. Lichen is the asociation of :
   (*a*) Alga and alga (*b*) Alga and fungus
   (*c*) Alga and roots of higher plant (*d*) Fungus and fungus

10. The tiny free-floating animals in surface water constitute :
(*a*) Phytoplankton (*b*) Symbiosis (*c*) Benthos (*d*) Zooplankton

11. A bird introduced from another country becomes a serious pest due to :
(*a*) Better adaptation to new area (*b*) Increased sexual reproduction
(*c*) Better nesting habitats (*d*) Absence of natural competition

12. Relationship between sea anemone and hermit crab is called :
(*a*) Symbiosis (*b*) Mutualism (*c*) Commensalism (*d*) None of them

13. Interaction in which one partner is benefitted and other is unaffected is called :
(*a*) Amensalism (*b*) Scavanging (*c*) Commensalism (*d*) Symbiosis

14. Association of animals in which both the partners are benefitted :
(*a*) Colony (*b*) Commensalism (*c*) Mutualism (*d*) Amensalism

15. Individuals of one kind (species) occupying a particular geographic area at a given time form :
(*a*) Community (*b*) Population (*c*) Species (*d*) Biome

16. Which of the following utilises inorganic materials :
(*a*) Autotrophs (*b*) Decomposers (*c*) Saprophytes (*d*) Heterotrophs

17. The independent evolution of flowering plants and pollinating insects together is known as :
(*a*) Mutualism (*b*) Coevolution (*c*) Commensalism (*d*) Cooperation

18. Climax community is :
(*a*) Unstable community (*b*) First stable community
(*c*) Intermediate community (*d*) Last stable community

19. The effect of individuals of the same species upon each other is called as :
(*a*) Reaction (*b*) Competition (*c*) Coaction (*d*) Ecesis

20. The invasion of a new community in a bare area is initiated by :
(*a*) Aggregation (*b*) Ecesis (*c*) Migration (*d*) All the above

21. Primary succession on rocks starts with :
(*a*) Herbs (*b*) Shrubs (*c*) Trees (*d*) Lichens

22. The ultimate seral community appears in an area as a result of :
(*a*) Reaction (*b*) Invasion (*c*) Nudation (*d*) Ecesis

23. In a 'hydrarch', the submerged stage in followed by :
(*a*) Reed-swamps (*b*) Sedges (*c*) Floating plants (*d*) Phytoplanktons

24. Plant succession occurring in sand dune is :
(*a*) Psammosere (*b*) Halosere (*c*) Xerosere (*d*) Hydrosere

25. Primary succession is development of biotic communities on :
(*a*) Newly exposed habitat (*b*) Cleared forest area
(*c*) Freshly harvested crop field (*d*) Pond filled after a dry season

ANSWERS

1. (*d*) **2.** (*b*) **3.** (*a*) **4.** (*a*) **5.** (*d*) **6.** (*b*) **7.** (*b*) **8.** (*d*) **9.** (*b*) **10.** (*d*)
11. (*d*) **12.** (*b*) **13.** (*c*) **14.** (*c*) **15.** (*b*) **16.** (*a*) **17.** (*b*) **18.** (*d*) **19.** (*a*) **20.** (*b*)
21. (*d*) **22.** (*a*) **23.** (*c*) **24.** (*a*) **25.** (*a*)

12
CHAPTER

Ecosystem: Structure and Function

The populations of organisms interact with each other in biological communities, and no biotic community can live in isolation. It lives in an environment which supplies its material and energy requirements and provides other living conditions. A biological community interacting with the non-living environment is called an **ecosystem**. An ecosystem can be natural or artificial, temporary or permanent. A pond, a lake, a tract of forest, a coral reef, a large grassland, a village, an aquarium, a field of sugarcane, a manned spaceship, or a laboratory culture can all be regarded as ecosystems. Thus, an ecosystem may be defined as *a dynamic system which includes both organisms (biotic component) and abiotic environment influencing the properties of each other and both necessary for the maintenance of life*.

The concept of ecosystem was first put forth by Sir Arthur Tansley in 1935. Ecosystems can be recognised as self-regulating and self-sustaining units of landscape. Ecosystem is the major ecological unit. In nature, two major categories of ecosystems may be distinguished: (*i*) **terrestrial** and (*ii*) **aquatic**. Forests, grasslands and deserts are main examples of terrestrial ecosystems, while ponds, lakes, streams or salt water (marine estuaries) represent aquatic ecosystems.

Ecosystem has both **structure** and **function**. The structure is related to species diversity. The more complex is the structure the greater is the diversity of the species in the ecosystem. The function of ecosystem is related to flow of energy and cycling of materials through structural components of the ecosystem.

According to E.P. Odum, the ecosystem is the basic fundamental unit of organisms and their environmental interacting with each other and with their own components. All the ecosystems of the earth are connected to one another, *e.g.*, river ecosystem is connected with the ecosystem of oceans, and a small ecosystem of dead logs is a part of large ecosystem of a forest.

However, human activities may modify natural ecosystems into man -made or anthropogenic ecosystems. For example, natural forests have been cut and the land converted to tree plantations or agricultural systems. Sometimes dam construction involves submergence of forests and conversion to water reservoirs.

COMPONENTS OF ECOSYSTEM

Organisation of the Ecosystem

The living parts **(biotic component)** and the non-living parts **(abiotic component)** of ecosystems are very closely interconnected. As we have studied that the energy flows in the form of food from one organism to another organism within a community. This is being done through feeding relationships such as grazing, predation and parasitism. The biotic community is only one part of ecosystem. Another important part of ecosystem is the **abiotic** portion. The abiotic part of the ecosystem includes chemical elements such as oxygen, carbon and nitrogen. It also includes physical elements such as oxygen, carbon and nitrogen. It also includes physical elements such as wind, fire and climate. Together, the biotic and abiotic parts determine the characteristics of the ecosystem.

The study of the functional aspect of an ecosystem consists of the volume and the rate at which various materials circulate and the rate at which energy flows through it. Energy moves through the food chain in one direction only. The movement of matter, is **cyclic** or circular. Energy flows through the ecosystem and is finally lost to space in the form of heat. Matter may be changed from one form to another, but it always remains in the ecosystem. The study of the functional aspect of an ecosystem also includes the processes by which living organisms change the non-living or abiotic environment. The processes by which abiotic environment affects living organisms and the events by which population levels of organisms are regulated.

Biotic Components

All organisms require energy for their life processes, and this energy enters the world of living things through the process of photosynthesis carried on by green plants. Green plants produce carbohydrates by photosynthesis and also synthesise proteins and fats. The green plants are called **primary producers**. Producers are autotrophic, *i.e.*, self-nourishing. A variety of photosynthetic bacteria, chemosynthetic bacteria and photosynthetic protozoans also produce organic substances in terrestrial and aquatic habitats, in very small amounts. In terrestrial ecosystems, the autotrophs are usually rooted plants, such as herbs, shrubs and trees, while in deep aquatic ecosystems, floating plants called **phytoplankton** are the major autotrophs. The dominant producers of shallow water are called **macrophytes**, which are generally rooted plants.

All other populations in a community are dependent upon green plants and are referred to as **consumers**. Under this category are included all different kinds of animals that are found in an ecosystem. Consumers are also known as **phagotrophs** (*phago* = to eat). They are **hetrotrophic** organisms, mostly animals which generally ingest or swallow their food. There are different categories of consumers, such as primary, secondary, tertiary, and scavangers, etc.

Primary consumers are purely herbivorous animals that are dependent for their food on producers or green plants, *e.g.*, insects, rodents, rabbit, deer, cow, buffalo, goat, etc. The herbivores serve as the chief food source for carnivores.

Secondary consumers are carnivores. Generally, they are flesh eating animals that are adapted to consume herbivores, *e.g.*, fox, wolves, cats, snakes, etc.

Tertiary consumers are the top carnivores which prey upon other carnivores and herbivores, *e.g.*, lion, tiger, hawk, vulture, etc.

Besides different categories of consumers, the parasites, scavangers and saprobes are also included among consumers. The parasitic plants and animals utilize the living tissues of different plants and animals. The scavangers and saprobes utilize dead remains of animals and plants as their food.

The non-green organisms like the fungi and some bacteria, which are incapable of producing their food, live on dead and decaying plants or animal parts, and are consumers of a special kind, called the **decomposers**. These simplify step by step the organic constituents of each dead body. Their activities make chemical substances available for other living beings, the producers. Decomposers are also known as **saprotrophs** (*sapro* = to decompose). Generally, they are hetrotrophic organisms, consisting mostly of bacteria and fungi which live on dead organic matter. However, decomposers or saprotrophs release different enzymes from their bodies into the dead and decaying plant and animal remains. The extracellular digestion of the dead remains leads to the release of simpler inorganic substances which are then used by the decomposers. The decomposers play an important role in maintaining the dynamic nature of ecosystems.

Food-getting interactions are made up of several **feeding**, or trophic levels. The first trophic level is made up of populations of green plants and other autotrophs which use energy from sunlight to make food through the process of photosynthesis. These organisms which are mostly green plants, are called **primary producers**. The organisms depending for their food on primary producers are **consumers**. The herbivores such as cattle, deer and goat are **primary consumers** or the **first order consumers**. Animal populations that feed upon other consumers are **secondary consumers** or **second order consumers**, such as lion, tiger, leopard, etc. Secondary consumers that only eat other animals are **carnivores**. Within a food chain, primary and secondary carnivores can be found. **Primary carnivores** eat animals and are in turn eaten by other animals, **secondary carnivores**. **Omnivores** are consumers that eat both plants and animals. For example, humans are omnivores.

Abiotic Components

The living parts and the non-living parts of ecosystems are very closely interconnected. The biotic community is only one part of the ecosystem. Another important part of the ecosystem is the **abiotic** portion. The abiotic part of the ecosystem includes chemical elements such as oxygen, carbon and nitrogen. Among the abiotic factors the most important are **climatic** and **edaphic** factors. The climatic factors consist of temperature, humidity, rain and snowfall. The edaphic factors comprise the soil and substratum.

Temperature. A certain temperature is essential for all the vital functions of the plant and for its growth. All living organisms can survive only in a narrow range of temperatue which allows their metabolism. For instance, plants normally prefer a temperature varying from 20°C to 40°C. Most flowering plants are killed at a temperature below 0°C and above 45°C, while seeds remain injured at a temperature far beyond these limits. However, many living organisms develop physiological and behavioural adaptations to withstand extremes of temperature, such as the polar bear can live in very cold regions and hibernates during winter. Some birds and mammals migrate to warmer places in winter to avoid extreme cold. Some desert animals live inside burrows to avoid the intense heat of the desert.

Water. Without water the life cannot exist. The unique chemical properties of water make it the supporter and carrier of life, and all biological activities depend on it. A given molecule of water does not necessarily stay in the same place forever. It may evaporate from a lake into the atmosphere or it may precipitate from the atmosphere as rain or snow. It may be taken up by plant roots and become part of plant tissue. It may be frozen into a glacier or run off the land to meet the ocean. This movement of water throughout the biosphere is called the **hydrologic cycle**.

Plants, in relation to the factor of water are divided into three groups, *i.e.*, **hydrophytes**, **mesophytes** and **xerophytes**. Hydrophytes are adapted to extremely wet situations. The leaves of such plants are fine and thin and possess little cutin. Mesophytes are plants adapted to inhabit regions with a moderate water supply. Examples are cultivated crops, garden vegetables, trees, ferns and mosses. Xerophytes are plants adapted to inhabit in arid or semiarid regions, such as cactus, yucca, etc.

Whether an animal lives in the water, or on land, the water balance in its body must be maintained. The amphibians and earthworms always live in a moist environment. Most animals obtain water by drinking, Many animals get most of the water they need from their food. Some carnivorous animals obtain their water from the blood of their prey. The termites live without the intake of water. They satisfy their water requirements from the water of metabolism. The camel can live for long periods without drinking water partly because it has a flexible water temperature.

The animals that cannot actively swim live under stones or in burrows and crevices, to avoid the strong water currents. The plants found in fast flowing water possess finely divided or thin ribbon-shaped leaves. Sea anemones and limpets, living in the intertidal zone of sea have sessile habit, whereas *Neries* and some tube worms live in the burrows in soil. The sea weeds such as *Fucus*, *Laminaria*, *Macrocystis* and *Nereocystis* are found in attached condition to face the waves and water currents.

Humidity. The humidity of the atmosphere directly regulates the rate of transpiration and perspiration from the plants and animals respectively. Various plants and animals develop several adaptations to face dry conditions.

Light. Light is an important ecological factor which is essential for photosynthetic organisms for the preparation of food, on which the whole living world depends. The reproductive cycle of many species of plants is affected by duration of light or length of day. The response of plants to day length is called **photoperiodism**. Light has been involved in the evolution of **photoreceptors** in the various animal groups. These photoreceptors, such as the pigment spots of many invertebrates, play a prominent role in the procurement of food, the avoidance of enemies, and the recognition of mates during the breeding season. Animals such as true frogs and chameleons have the ability to change their colour. They do so by photic stimulation through the eyes. Relative lengths of day and night make an important factor in periodical animal migration. Some animals, such as the deer, mate in autumn when the days are short. Poultrymen have been able to increase egg production by exposing hens to longer light periods through the use of artificial light.

Light plays an important role in determining the periods of activity of animals. Some animals are active only at dawn, others are active through the morning hours, and still others are active only during brightest hours of the day. At night diurnal forms withdraw from the scene and nocturnal animals become active. Many nocturnal animals have eyes specialized for nocturnal vision. The leaves of many leguminous plants fold up or droop at night.

Wind. Air gusts or wind affect the plants in several ways. It enhances transpiration. In the forests it has been noticed that some plants can resist the action of wind far better than others. On seashore the coconut-palms can withstand them well because their leaves are cut into narrow segments with stout mid-rib. Where the wind gusts are quite strong, only the plants with tough roots and stems can survive. Wind is useful in disseminating seeds and fruits, particularly those provided with some kind of appendages.

Soil pH. The living organisms require an optimal pH range for their survival. Certain plants and animals thrive best in acidic conditions, other prefer neutral or alkaline conditions. pH of the soil and water is responsible to a great extent for the distribution of organisms.

Mineral elements. Soil is defined 'as mineral material that may exist in solid or unbroken form, such as boulders and gravels or as finely divided particles of mineral matter referred to as sands, silts or clays, depending upon the texture'. Often these finely divided soils will contain considerable amounts of organic matter that form a very rich productive humus. A soil is described as 'a complex physical and biological system providing support, water, nutrient and oxygen for the plant'. The major elements present in the soil are aluminium, silicon, calcium, magnesium, sodium, potassium and iron. The minor elements are -cobalt, boron, iodine, zinc, arsenic, etc. The concentration and availability of essential mineral elements control the distribution of all living organisms. Plants found in nitrogen deficient soil develop special adaptations to obtain it, such as harbouring nitrogen-fixing bacteria, and the insectivorous habit of plants.

The salinity of soil or water has great influence on the distribution of organisms. Special salt secreting glands develop in plants and animals to eliminate excessive salt. The fresh water species cannot survive in a marine habitat. Animals living in the estuaries near the seashore have special physiological adaptations for facing fluctuations in salinity accompanying tidal rhythms. The pneumatophores and vivipary are special adaptations in the mangrove vegetation.

Topography. The topography of an area is represented by physical features, such as hills, plains or slopes. It influences distribution of organisms in the direct way. Sometimes, it indirectly affects the factors, such as wind, water current, light or wave action. The centre and edge of a pond, top-side and under-side of a rock or boulder, north and south faces of a ridge are generally occupied by different species of organisms.

Background. The background of the habitat is responsible for the distribution of animals by enabling them to paint against the colour, pattern and general texture, such as desert animals like lion and the camel are sand-coloured. The jelly fishes and sea cucumbers are glassy in appearance. The chameleon changes its colour according to its background.

STRUCTURE AND FUNCTION OF ECOSYSTEM

Structure

As described earlier, biotic and abiotic components are physically organised to provide a characteristic structure of the ecosystem. Important structural features are:

(*i*) **Species composition** and (*ii*) **Stratification**

For example, some ecosystems, such as tropical rain forests, show a canopy of tall plants and a large number of biological species. On the other hand, a desert ecosytem shows a low discontinuous layer of herbs or small shrubs which consist of less number of species and extensive bare patches of soil.

In the forest, tallest trees, influence the under storey plants and ground vegetation. The plants form more or less, distinct strata or storeys on vertical as well as in horizontal planes, which is characteristically known as **stratification**. The individuals of different layers represent different species of plants.

Trophic levels. There is another way to depict the structure of ecosystem through food relationships of producers and consumers. Trophic or food structure of ecosystem is based on the existence of several trophic levels.

The energy in a community is passed from one organism to another in feeding patterns. These patterns of energy transference together form a **food chain**. Each step in the food chain represents a **trophic level**. The plants represent the **first trophic level**, the herbivores make the **second trophic level**, the primary carnivores constitute the **third trophic level**, and the secondary carnivores, such as large fish, man, etc., constitute the **fourth trophic level** of an ecosystem. Thus energy from the sun enters the living world through photosynthetic organisms and passes open from one organism to another in the form of food.

Trophic structure may be described in terms of living material, called **standing crop**, found in different trophic levels at a given time. The standing crop is commonly expressed as the **biomass** of organisms per unit area. The biomass is defined as the total weight of dry matter (dry weight) present in the ecosystem at any one time.

Nutrients are required for the proper growth and development of living organisms. They flow from abiotic to biotic components and back to the non-living component again in a more or less cyclic manner. This is known as biogeochemical cycle or inorganic-organic cycle. The amounts of nutrients, such as nitrogen, phosphorus and calcium present in the soil at any given time, is called the **standing state**. The standing states of nutrients differ from one ecosystem to another, or with seasons even in the same ecosystem.

Function of Ecosystem

An ecosystem is a discrete structural, functional and life sustaining environmental system. The environmental system consists of biotic and abiotic components in a habitat. The synthesis and perpetuation processes involve energy exchange and this energy comes from the sun in the form of light or solar energy. Thus, in any ecosystem, there are three functional components:

(*i*) Inorganic constituents (*i.e.*, air, water and minerals)

(*ii*) Organisms (*i.e.*, plants, animals and microbes), and

(*iii*) Energy input which enters from outside (*i.e.*, sun).

These three interact and form an environmental system. Inorganic constituents are synthesized into organic structures by the green plants (primary producers) through photosynthesis and solar energy is utilized in the process. Green plants become the source of energy of renewals (herbivores) which, in turn become source of energy for the flesh eating animals (carnivores). Animals, of all types grow and add organic matter to their body weight and their source of energy is complex organic compound taken as food. They are known as secondary producers. All the living organisms, whether plants or animals in an ecosystem have a definite life span after which they die.

The dead organic remains of plants and animals provide food for saprophytic microbes, such as bacteria, fungi, and many other animals. The saprobes ultimately decompose the organic structure and break the complex molecules and liberate the inorganic compounds into their environment. These organisms are known as decomposers.

During the process of decomposition of organic molecules the energy which kept the inorganic components bound together in the form of organic molecules gets liberated into the environment as heat energy. Thus, in an ecosystem energy from the sun, the input is fixed by the plants and transferred to animal components.

The ecosystems operating in different habitats, such as deserts, forests, grasslands and seas are interdependent on one another. The energy and nutrients of one way find their way into another, so that, ultimately all parts of the earth are interrelated, each comprising a part of the total system that keeps the biosphere functioning.

Thus, principal functions of ecosystems are as follows:

(*i*) Reception of radiant energy of sun.

(*ii*) Manufacture of organic materials from inorganic ones by producers.

(*iii*) Consumption of producers by consumers and further elaboration of consumed materials

(*iv*) After the death of producers and consumers, complex organic compounds are degraded and finally converted by decomposers and converters into such forms as are suitable for reutilisation by producers.

PRODUCTIVITY

The relationship between the amount of energy accumulated and the amount of energy utilised within one trophic level of food chain has an important bearing on how much energy one trophic level passes on to the next trophic level in the food chain. The ratio of output of energy to input of energy is referred to as **ecological efficiency**.

Productivity

The rate of biomass production is called **productivity**. The portion of fixed energy, a trophic level passes on to the next trophic level is called **production**. Productivity in ecosystems is of two kinds, i.e., primary and secondary. Green plants fix solar energy and accumulate it in organic forms as chemical energy. As this is the first and basic form of energy storage, the rate at which the energy accumulates in the green plants or producers is known as **primary productivity**. Productivity is a rate function, and is expressed in terms of dry matter produced or energy captured per unit area of land, per unit time. It is more often expressed as energy in calories /cm^2/yr or dry organic matter in g/m^2/ yr (g/m^2 $\times$ 8.92 = lb/acre). Hence, the productivity of different ecosystems can be easily compared.

Primary productivity has two aspects (*i*) gross and (*ii*) net. The total solar energy trapped in the food material by photosynthesis is referred to as **gross primary productivity** (GPP).

However, a good fraction of gross primary productivity is utilised in respiration of green plants. The amount of energy-bound organic matter created per unit area and time that is left after respiration is **net primary productivity** (NPP).

Net productivity of energy = Gross productivity — Energy lost in respiration.

The rates at which the heterotrophic organisms resynthesise the energy-yielding substances is called **secondary productivity**. Here, the net primary productivity (NPP) results in the accumulation of plant biomass, which serves the food of herbivores and decomposers. It is notable that the food of consumers has been produced by the primary producers, and secondary productivity depicts only the utilisation of this food for the production of consumer biomass. Secondary productivity is the productivity of animals and saprobes in ecosystem.

There are three fundamental concepts of productivity:

1. Standing crop
2. Materials removed and
3. Production rate.

1. Standing crop. This is abundance of organisms existing in the area at any one time. It may be expressed in terms of number of individuals, as biomass of organisms, as energy content or in some other suitable terms. Measurement of standing crop reveals the concentration of individuals in various populations of the ecosystem.

2. Materials removed. The second concept of productivity is the materials removed from the area per unit time. It includes the yield to man, organisms removed from the ecosystem by migration, and the material withdrawn as organic deposit.

3. Production rate. The third concept of productivity is the production rate, at which the growth processes are going forward within the area. The amount of material formed by each link in the food chain per unit of time per unit area or volume is the production rate.

Environmental factors affecting the production processes in an ecosystem are as follows:

1. Solar radiation and **temperature**

2. Moisture, *i.e.*, leaf water potential, soil moisture, fluctuation of precipitation, and transpiration

3. Mineral nutrition, *i.e.*, uptake of minerals from the soil, rhizosphere effects, fire effects, salinity, heavy metals and nitrogen metabolism.

4. Biotic activities, *i.e.*, grazing, above ground herbivores, below ground herbivores, predators and parasites and diseases of primary producers.

5. Impact of human populations, *i.e.*, populations of different sorts, ionising radiations, such as atomic explosions, etc.

6. In aquatic systems, productivity is generally limited by **light**, which decreases with increasing water depth. In deep oceans **nutrients** often become limiting for productivity. **Nitrogen** is most important nutrient limiting productivity in marine ecosystems.

The largeness of primary productivity depends on the photosynthetic capacity of producers and the existing environmental conditions, such as solar radiation, temperature and soil moisture.

In tropical conditions, primary productivity may remain continuous throughout the year, provided adequate soil moisture remain available.

While in temperate regions, primary productivity is limited by cold climate and a short snow-free growing period during the year.

Primary productivity of the major ecosystems of the world is as follows:

Table 12.1. Geographical area, mean plant biomass and net productivity in major world ecosystems

Major world ecosystems	*Geographical area* 10^6 km^2	*Mean plant biomass* ($t\ ha^{-1}$)	*Mean net primary productivity* ($t\ ha^{-1}\ year^{-1}$)
1. Tropical rain forest	17	440	20
2. Tropical deciduous forest	8	360	15
3. Temperate deciduous forest	7	300	12
4. Temperate coniferous forest	12	200	8
5. Savanna	15	40	9
6. Temperate grassland	9	20	5
7. Desert shrub	18	10	0.7
t = ton = 1000 kg ; ha = 10,000 m^2			

Decomposition

Decomposers and **transformers** are the living components of the ecosystem and they are **fungi** and **bacteria**. Decomposers attack the dead remains of producers and consumers and degrade the complex organic substances into simpler compounds. The simple organic matters are then attacked by another kind of bacteria, and transformers which change these organic compounds into the inorganic forms that are suitable for reuse by producers or green plants.The decomposers and transformers play very important role in maintaining the dynamic nature of ecosystems.

Decomposition concerns breakdown of complex organic matter by decomposers to inorganic raw materials like carbon dioxide, water and various nutrients. The upper layer of soil is the main site for decomposition processes in the ecosystem.

Dead plant parts and animal remains are called **detritus**.

Dried plant parts, such as leaves, stems, bark, flowers, etc., and dead remains of animals, including faecal matter, drop over the soil which constitute the **aboveground detritus**, which is also known as **litter fall**.

Contrary to it, the **belowground detritus** is composed of dead roots and also known as root **detritus**.

As the time passes, decomposing detritus loses weight, until it disappears completely.

Sometimes, when decomposition is retarded or stopped, large amount of partially decomposed organic matter is accumulated in the ecosystem.

Processes of Decomposition

Generally, there are three main processes of decomposition. They are: (*i*) fragmentation of detritus, (*ii*) leaching and (*iii*) catabolism.

1. Fragmentation of detritus. This process takes place, mainly due to the action of detritus feeding invertebrates, also known as **detrivores**, where detritus is broken into smaller particles. Here, the detritus is pulverised, when passes through the digestive tracts (alimentary canals) of animals. Due to this process, the surface area of detritus particles is greatly increased.

2. Leaching. During this process, water percolating through soil removes soluble substances, such as sugars and several other nutrients, from the fragmented detritus due to leaching action.

3. Catabolism. During process of catabolism, the extracellular enzymes released by bacteria and fungi carry out enzymatic conversion of the decomposing detritus to simpler compounds and inorganic substances.

It is notable that all the above mentioned processes of decomposition operate simultaneously on the detritus.

Besides above mentioned decomposition processes, **humification** and **mineralisation** occur during decomposition in the soil.

Humification. During this process of decomposition a dark coloured amorphous substance is formed and accumulated, which is called **humus**. This is highly resistant to microbial action and undergoes quite slow decomposition. Humus serves as a reservoir of nutrients.

Mineralisation. A process, which results in the release of inorganic substances, such as CO_2, H_2O and other nutrients like NH_4^+, Ca^{++}, Mg^{++}, K^+, etc., in the soil.

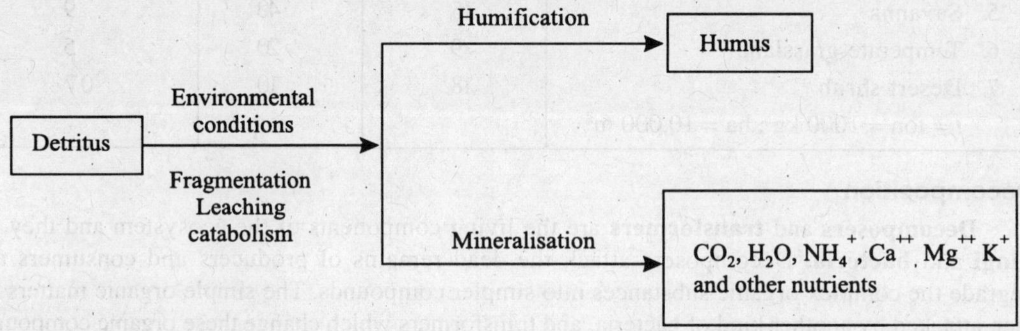

Nutrient immobilisation. During process of decomposition, under certain specific conditions, soil nutrients get tied up with the biomass of microbes and become temporarily unavailable to other organisms. This type of incorporation of nutrients in living microbes is called **nutrient immobilisation.**

Such nutrients remain immobilised for variable periods and get mineralised later, after the death of microbes.

The immobilisation of nutrients prevent them from being washed out from the ecosystem.

Factors Affecting Decomposition

The rate of decomposition of detritus is primarily regulated by factors categorised under two important heads, (*i*) climatic conditions and (*ii*) chemical quality of detritus.

1. Climatic conditions. Here, key role is played by temperature and soil moisture, which have their effect on the activities of soil microbes.

Detritus decomposes very rapidly, within a few weeks or months, in a climate where higher temperature prevails, *i.e.*, more than 25°C, and moist conditions remain constant for long time, *e.g.*, in humid tropical regions.

However, low temperature, *i.e.*, less than 10°C, sharply reduces decomposition rate even if moisture is in abundance. For example, in regions of high altitude, complete decomposition of detritus takes place in several years or decades.

Decomposition rate is lowered in prolonged dryness of soil, even if the temperature remains high, as found in tropical deserts.

2. Chemical quality of detritus. The accumulation of certain substances is responsible for promotion or retardation of rate of decomposition. The chemical quality of detritus is determined by the relative proportions of water soluble substances, which include sugars, polyphenols, lignin and nitrogen.

Even, within same climatic conditions decomposition rate becomes slower, if the detritus is rich in lignin and chitin.

The nitrogen-rich detritus, on the other hand, having low amounts of lignin, decomposes relatively rapidly.

However, the overall rate of decomposition of detritus in natural conditions depends upon the combined effect of environmental conditions and quality of detritus.

BIOSPHERE

The entire global environment is basically made of **abiotic** (non-living) and **biotic** (living) components. These components together constitute the **biosphere** or **ecosphere**. *Biosphere is, therefore, defined as that part of the globe where life exists*. The abiotic global environment is composed of the **atmosphere** (air), the **lithosphere** (earth) and the **hydrosphere** (water), and the biotic component is made of various forms of life inhabiting in the abiotic environment. However, so far it is known, life exists only on the planet we live, *i.e.*, the earth. Life exists only on earth because it possesses all the physical conditions essential for sustaining life; earth too, possesses only certain areas having optimal conditions for the sustenance of life, and also all types of environmental conditions are not favourable to all kinds of organisms, *e.g.*, terrestrial animals like lion or tiger, etc., cannot live in water and aquatic fishes do not survive on land. In fact, living organisms cannot survive without non-living environment around them and the maintenance of non-living environment also depends on its living counter-parts.

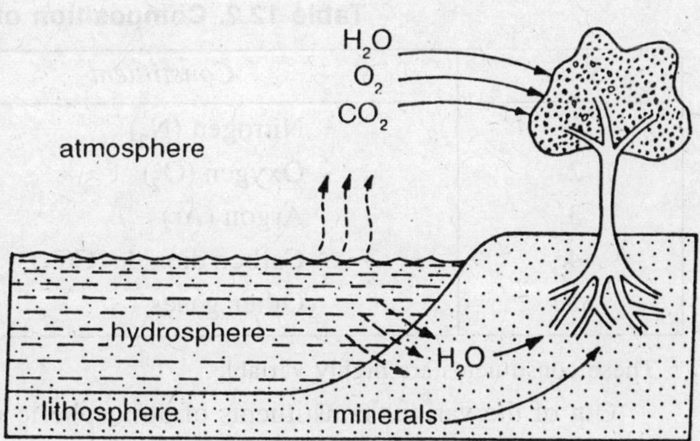

Fig. 12.1. Diagram illustrating the supply of inorganic substances for the maintenance of living matter from the abiotic components of biosphere.

Biosphere is made of different types of ecosystem. It is, therefore, a biochemical system capable of capturing, converting, storing and utilising the energy of sun.

Living organisms require inorganic metabolites like water, minerals, and oxygen, nitrogen and carbon dioxide, etc., for building and maintenance of lives. Living organisms obtain all such inorganic substances from the abiotic counter-parts of the biosphere (Fig. 12.1). Thus, these components are described in detail in the following pages.

1. Hydrosphere. It includes water in the oceans, lakes, ponds, etc., and covers about 73% area of the earth's surface. Water is the major inogranic nutrient needed by all living organisms, hence, water is essential to all life. First life originated in water. The means of obtaining and conserving water have shaped the nature of terrestrial life; means of living within the water have the overwhelming influence on aquatic life. Water is one of the main agents in pedogenesis and is also the medium for several different ecosystems.

Water continuously circulates between atmosphere and the earth's surface; this cycle is referred to as the **water cycle**. The energy for driving the cycle and, thus, ensuring a constant supply of fresh water on land comes from the sun. Solar heat evaporates water from the ocean which is the great reservoir of water. A lesser amount of water is also evaporated from the surface of the land from plants, a process known as **evapotranspiration**. All this vaporized water forms clouds which moved by winds, may pass over land where they are cooled enough to precipitate the water as rain or snow. Some of the precipitated water soaks into ground, some runs off the surface into stream and goes directly back to the seas. The ground water is returned to the surface by springs, by pumps and the transpiration—the movement of water in plants from roots to leaves. Water inevitably ends up back in the sea, but it may become incorporated into the bodies of several different organisms, one after another, *en route*.

2. Atmosphere. It is a thick gaseous mantle which envelopes the hydrosphere and earth's surface, called the **air**. It supplies oxygen, nitrogen, carbon dioxide and other gases and is subjected to physical cycles by the sun and spin of earth. Cooled polar air sinks, while warm equatorial air rises and the axial rotation of the earth causes to shift air masses laterally resulting into the formation of global air currents. The composition of atmosphere is somewhat uniform upto a height of about 80 kms (Table 12.1). Higher levels have lighter gases. The atmosphere is denser near the earth and rarer away from the earth. The atmospheric pressure is more near the earth than away from it. The atmosphere has a layer of ozone at a height of about 32–48 km; this layer acts as a barrier that checks ultraviolet rays of sun from reaching the earth which could be lethal for living organisms.

Table 12.2. Composition of dry atmosphere.

S.No.	*Constituent*	*Percentage*
1.	Nitrogen (N_2)	78.084
2.	Oxygen (O_2)	20.947
3.	Argon (Ar)	0.934
4.	Carbon dioxide (CO_2)	0.0314*
5.	Other gases	0.003

* These constituents are highly variable.

Out of the various constituents of atmospheric gases, oxygen, nitrogen and carbon dioxide serve as metabolites for living organisms and, hence, they are essentially needed by them for their normal functioning. However, all these undergo characteristic cycles in nature in which organisms play an important role and, hence, their normal percentage is maintained in atmosphere.

3. Lithosphere. The solid component of earth is called **lithosphere**. It is multilayered and includes following three main layers: **crust, mantle** and **core** (outer and inner). The core is the central fluid or vapourized sphere having diameter of about 2500 km from the **centre** and is possibly composed of nickel-iron. The **mantle** extends about 2900 km above the core. This is in a molten state. The **crust** is the outermost solid zone of the earth and it is about 8 to 40 km above mantle. The crust is very complex and its surface is covered with the **soil** supporting rich and varied biotic communities. The soil provides food, shelter, anchorage and concealment from predators to living organisms.

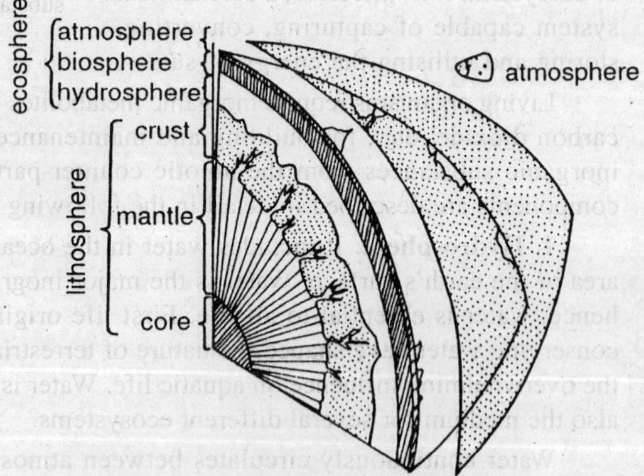

Fig. 12.2. Cross section of earth showing different layers.

The soil is the most characteristic feature of terrestrial environment. It is the top layer of earth's crust and is the mixture of weathered rock materials (*i.e.*, minerals) and organic detritus, both of which are formed through the physical, chemical and biological processes occurring slowly and slowly for a long period at the earth's surface. Besides being the source for the entry of nutrients and water in plants, the soil is the medium for the detritus food chain : nutrients released in detritus are decomposed by various soil microbes like bacteria, algae, fungi, protozoa, etc., bound in or on soil particles and taken back into plants through their roots. Soil (mud) is also the main source of nutrients for all aquatic plants; rooted, submerged or free-floating. In addition, soil is the means to support for all terrestrial organisms: plants are anchored to the soil by their root systems; animals walk upon it and are supported by it, as many animals like nematodes, polychaetes, arachnids, insects, rodents, etc., live under the soil. However, the chemical composition of earth's crust has been shown in Table 12.2. In addition, soil also contains organic matters like **humus**, etc.

Table 12.3. Chemical composition of earth's crust.

S.No.	*Constituent*	*Percentage*
1.	Oxygen (O)	46.6
2.	Silicon (Si)	27.7
3.	Aluminium (Al)	8.1
4.	Iron (Fe)	5.0
5.	Calcium (Ca)	3.6
6.	Sodium (Na)	2.8
7.	Potassium (K)	2.6
8.	Magnesium (Mg)	2.1
9.	Compounds of elements like B, Mn, Cu, Zn, Mo, Co, I, F, etc.	1.4

Thus, these three ecological components and the biotic component of the world constitute the **biosphere**. All these four components, thus, represent the four major global components of the world ecosystem. These four spheres continuously exchange matter with one another in a cyclical manner. Thus, biosphere is that part of the earth in which life exists. More specifically, the sum of those portions of the hydrosphere, lithosphere and atmosphere into which life penetrates, is the biosphere. However, together with the geological, chemical features of the totality of our habitats, these (air, water, earth and organisms) are sometimes grouped under the term **ecosphere**. Biosphere or ecosphere may be thought as a biochemical system capable of capturing, converting, storing and utilizing the energy of the sun. Approximately three hundred thousand species of green plants and microorganisms are recognized as **primary producers** which utilize inorganic elements and compounds to synthesize the organic minerals of life. Their productivity is consumed by more than a million other species of organisms which convert this organic store-house into animal form, adding to the beauty, and value of the biosphere as well as its complexity. Still other species, primarily bacteria and fungi, accomplish the recycling process by returning plant and animal wastes and residues to inorganic form so the process may be renewed. In this cyclic process of life, many elements are shared form a common global pool and are converted from inorganic to organic form and back again. Examples of such elements which commonly shuttle among the air, water, earth and organism are carbon, oxygen, hydrogen, nitrogen, phosphorus, sulphur, sodium, potassium, calcium, magnesium, iron, manganese, cobalt, copper, and zinc.

In physical terms, the biosphere is a relatively thin and incomplete envelope covering most of the world. It represents a mosaic of different biotic communities from simple to complex, aquatic to terrestrial, and tropical to polar. It does not exist in the extremities of the polar regions, the highest mountains, the deepest ocean troughs, the most extreme deserts, or the most highly polluted areas of land and water. Its total thickness, including all portions of the earth where living organisms can exist, is less than 26 kilometres. Its zone of active biological production, in terms of photosynthesis, is much narrower, and varies from a few centimetres to over 100 metres. This zone would, for instance, be only a few centimetres in muddy or turbid water, whereas in very clear ocean water, it could be more than 100 metres in thickness. On land, the zone of biological production might be only a few millimetres in a desert or rock environment, whereas it might again be more than 100 metres in a sequoia or tropical rain forest. Living organisms can exist,of course, beyond the range of active biological production; some insects or birds may be airborne to altitudes above 20,000 feet, and viable spores, seeds of plants, and microorganisms may be found in the atmosphere and mountain tops above 25,000 feet. In the ocean depths, many animals can exist well below a thousand feet—one hydra-like animal has been photographed at a depth of 15,900 feet in the south Atlantic. However, in both extreme altitude and extreme depth, the organisms depend upon the much thinner zone of active biological production, that portion of the system which converts the energy of sunlight into the chemical and physical energy of living organisms.

Spaceship Earth

The earth is the third planet of our solar system which revolves round the sun. In the initial stages of earth's formation, it probably lost some materials in form of light gases. But, as the conditions became suitable for the origin of life on it, the material loss had been reduced and at the same time there is a little inflow of material, from outside the earth. Therefore, the earth together with its atmosphere can be considered a **closed system**. The amount of materials like oxygen, carbon, nitrogen, water and minerals, essential for sustaining life, is finite. The supply of these materials in forms that can be utilised by the living organisms is exhaustible.

On the other hand, earth is an **open system** as far as energy is concerned. The sole source of its energy is the sun. Earth receives solar energy in form of heat, light, X-rays, ultraviolet rays, electric waves and in other forms; some forms of energy which are harmful to life are filtered by the atmosphere. In fact, a very small amount of energy received from the sun is utilised by the living organisms for their energy need and an enormous amount of it is radiated by the earth in its outer space.

Concepts of Matter and Energy

In fact, matter and energy determine the environment, *e.g.*, all objects are formed of matter and any phenomenon or process or activity involves **energy**. Thus, matter and energy are essential for the existence of all living and non-living objects, processes of life and the physical, chemical and biological events taking place in the environment. Therefore, what we see, feel, hear, taste, touch or use are all manifestations of matter and energy.

Matter has mass and volume. It exists in solid, liquid or gaseous states. These states of matter are inter-convertible, *e.g.*, liquid water can be changed into solid ice and gaseous vapours but cannot be destroyed. Matter, in normal conditions, neither can be created nor destroyed. But, it can be broken down into its constituent components, *e.g.*, water can be broken down into its elements of hydrogen and oxygen by electrolysis. Destruction of matter, as it occurs during nuclear fission, generates a large amount of energy.

It is a well known fact that an individual organism is a well organised orderly system. Every organised system tends towards disorganisation due to entropy. So, to reduce entropy energy is required. Therefore, we need energy to be living, so even if we sleep energy is needed, pumping of heart, ventilation of lungs, digestion of food, etc., all need energy. In the same way all non-living affairs, *e.g.*, falling of rain, the blowing of breeze, the running of a car or scooter, etc., involve energy. Objects emit energy as light, heat and sound in the non-living environment and our senses perceive these forms of energy to make us aware of the changes happening in our surroundings. Energy may be in form of kinetic or potential; either directly involved in doing an activity, or may be stored in form of energy-rich chemical bonds, *e.g.*, ATP in living beings. One form of energy can by changed into other form, *e.g.*, electrical energy can be changed into mechanical energy. Thus, energy cannot be created or destroyed but can be changed from one form to other. However, under specific conditions, energy can be changed into matter and vice-versa. Therefore, both these are so closely associated to each other that it is difficult to isolate them. Hence, it is true that matter and energy determine the environment.

TRAPPING OF SOLAR ENERGY BY LIVING ORGANISMS

Amongst the living organisms green plants are only equipped to trap solar energy, the process is called **photosynthesis**. This process is complex one and essentially involves two stages. First of all, **chlorophyll** present in green plants absorbs energy from sunlight. This solar energy is finally converted and captured in the form of chemical energy as a substance called ATP. This acts as transporter and temporary store of energy in the cell. For long-term storage of energy compounds more stable than ATP are needed. Hence, in the second stage which goes without light, carbon dioxide is taken into the plant cell and reduced to organic compounds. The energy needed for this reaction is supplied by ATP. The organic compounds, thus, formed, are various types of sugars. These are used for building new tissues, for food storage, or for respiration to generate energy essential to keep the plant cells alive and functioning. The most important by-product of the process of photosynthesis is oxygen, which is liberated in the atmosphere. However, plants constitute the source of energy for animals.

ENERGY FLOW IN ECOSYSTEM

Energy has been defined as the capacity to do work. Energy exists in two forms: (*i*) **potential** and (*ii*) **kinetic**. Potentical energy is the energy at rest, *i.e.*, stored energy, capable of performing work. Kinetic energy is the energy of motion, *i.e.*, free energy. It results in work performance at the expense of potential energy. Conversion of potential energy into kinetic energy involves the imparting of motion.

The source of energy required by all living organisms is the chemical energy of their food. The chemical energy is obtained by the conversion of the radiant energy of sun. The radiant energy is in the form of electromagnetic waves which are released from the sun during the transmutation of hydrogen to helium. The chemical energy stored in the food of living organisms is converted into potential energy by the arrangement of the constituent atoms of food in a particular manner.

In any ecosystem there should be unidirectional flow of energy. This energy flow is based on two important laws of thermodynamics, which are as follows:

1. **First law of thermodynamics.** It states that the amount of energy in the universe is constant. It may change from one form to another, but it can neither be created nor destroyed. Light energy can be neither created nor destroyed as it passes through the atmosphere. It may, however, be transformed into another type of energy, such as chemical energy or heat energy. There forms of energy can be transformed into electro magnetic radiation.

2. **Second law of thermodynamics.** It states that non-random energy, *i.e.*, mechanical, chemical and radiant energy, cannot be changed, without some degradation into heat energy. The change of energy from one form to another takes place in such a way that a part of energy assumes

waste form, *i.e.*, heat energy. Thus, after transformation the capacity of energy to perform work is decreased, and energy flows from higher to lower level.

Thus, in ecosystem, the transfer of food energy from one organism to another leads to degradation and loss of a major fraction of food energy as heat due to metabolic activities, with only a small fraction being stored in living tissues or biomass. Energy in food is in concentrated form, while heat energy is highly dispersed. All changes in energy forms can be accounted for in any system.

Main source of energy is sun. Approximately 57 percent of sun energy is absorbed in the atmosphere and scattered in the space. Some 35 per cent is expanded to heat, water and land areas and to evaporate water of the approximately 88 per cent of light energy striking plant surface, 10 percent to 15 per cent is reflected, 5 per cent is transmitted and 80 per cent to 85 per cent is absorbed, and an average of only 2 per cent (0.5 to 3.5 per cent) of the total light energy striking on a leaf is used in photosynthesis and rest is transformed into heat energy.

However, only the visible light, *i.e.*, the **photosynthetically active radiation** (PAR), which carries about 50 per cent of the energy of total incident solar radiation, is available to primary producers for absorption.

By multiplying the estimated productivity with the **calorific value**, *i.e.*, energy content per unit weight, of biomass, the energy captured can be determined. The calorific value is determined by burning known weight of dry biomass in a bomb calorimeter in the presence of oxygen, and measuring the heat evolved.

Energy flow in ecosystems. Living organisms can use energy in several forms: **radiant** and **fixed energy**. Radiant energy is in the form of electromagnetic waves such as light. Fixed energy

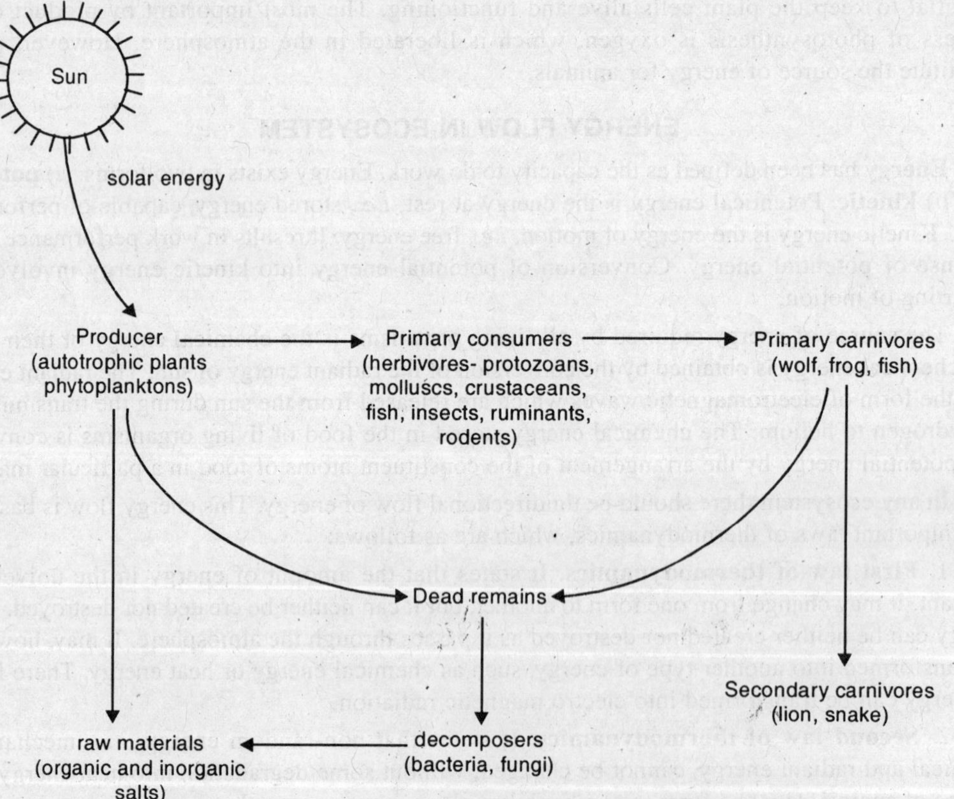

Fig. 12.3. Flow of energy at different levels of ecosystem.

is potential chemical energy bound in various organic substances which can be broken down in order to release their energy content.

Organisms that can fix energy from inorganic sources into organic molecules are called **autotrophs**. Organisms that cannot obtain energy from abiotic source but depend on energy-rich organic molecules synthesised by autotrophy are called **heterotrophs**. Those which obtain energy from living organisms are called consumers and those which obtain energy from dead organisms are called **decomposers**.

When the light energy falls on the green surfaces of plants, a part of it is transformed into chemical energy which is stored in various organic products in the plants. When the **herbivores** consume plants as food and convert chemical energy accumulated in plant products into kinetic energy, degradation of energy will occur through its conversion into heat. When herbivores are consumed by **carnivores** of the first order, *i.e.*, **secondary consumers**, further degradation will occur. Similarly, when primary carnivores are consumed by **top carnivores**, again energy will be degraded.

However, under favourable environmental conditions, only about 1-5 per cent energy of incident radiation, or 2–10 per cent of photosynthetically active radiation, as actually captured by the photosynthetic process, *i.e.*, gross primary productivity, and the remaining portion is dispersed.

As the simultaneously occurring respiratory processes are energy consuming and use up the part of photosynthetic gain, the net capture of energy, i.e., net primary productivity is reduced to only 0.8 – 4 per cent of the incident total radiation, or 1.6 to 8 per cent of photosynthetically active process (PAR). Only the energy captured in net productivity is used by other trophic levels.

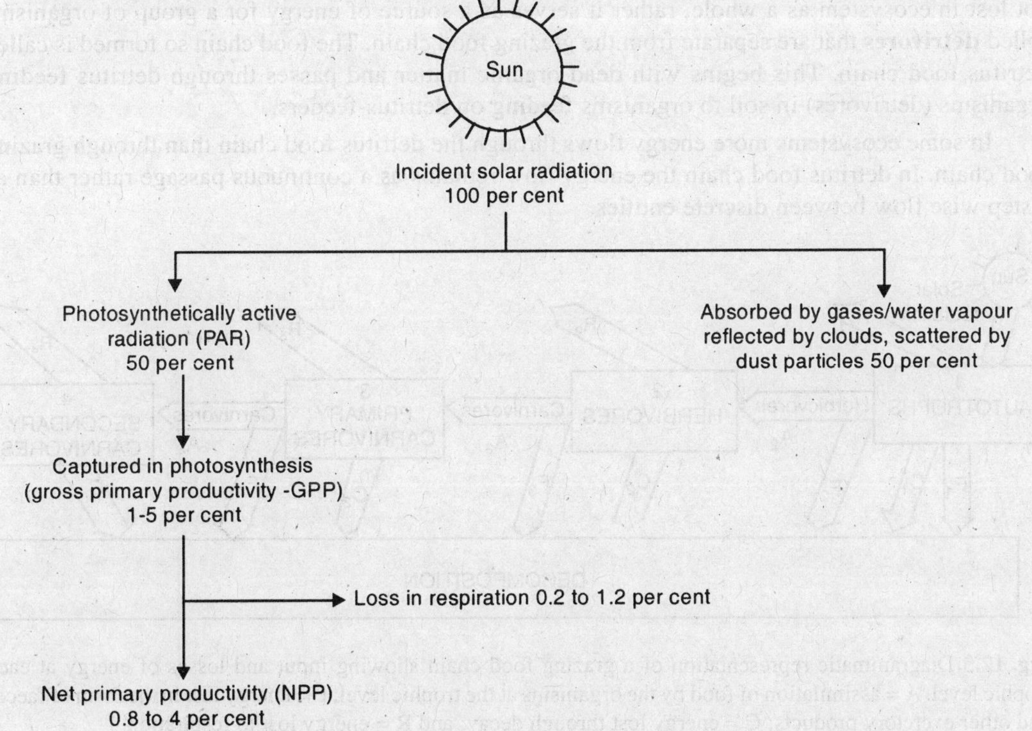

Fig. 12.4. Fate of solar radiation.

Trophic level. The producers and consumers in ecosystem can be arranged into several feeding groups each known as **trophic level** or **feeding level**. In any ecosystem, **producers** represent the first trophic level, **herbivores** represent the second trophic level, **primary carnivores** represent the third trophic level and **top carnivores** represent the last trophic level.

Food Chain and Food Web

In the ecosystem, green plants alone are able to trap in solar energy and convert it into chemical energy. The chemical energy is locked up in the various organic compounds, such as carbohydrates, fats and proteins present in the green plants. All trophic levels in 'an ecosystem are connected by transfer of food or energy. The transfer of energy from one trophic level, *e.g.*, producers, to the next trophic level, *e.g.*, consumers, is called **food chain**. Thus, a food chain may be defined as the transfer of energy and nutrients through a succession of organisms through repeated process of eating and being eaten. Mainly, food chains are of two types: (*i*) **grazing food chain** and (*ii*) **detritus food chain**.

1. Grazing food chain. The grazing food chain starts from green plants and goes from herbivores (primary consumers) to primary carnivores (secondary consumers) and then to secondary carnivores (tertiary consumers) and so on. Cattle grazing in grasslands, deer grazing in forest, and insects feeding on crops and trees, are most common biotic constituents of the grazing food chain.

Autotroph ⟶ Herbivore ⟶ Primary carnivore ⟶ Secondary crnivore, etc.

2. Detritus food chain. The dead organic remains including metabolic wastes and exudates derived from grazing food chain are generally called **detritus**. The energy contained in detritus is not lost in ecosystem as a whole, rather it serves as a source of energy for a group of organisms called **detrivores** that are separate from the grazing food chain. The food chain so formed is called detritus food chain. This begins with dead organic matter and passes through detritus feeding organisms (detrivores) in soil to organisms feeding on detritus-feeders.

In some ecosystems more energy flows through the detritus food chain than through grazing food chain. In detritus food chain the energy flow remains as a continuous passage rather than as a step wise flow between discrete entities.

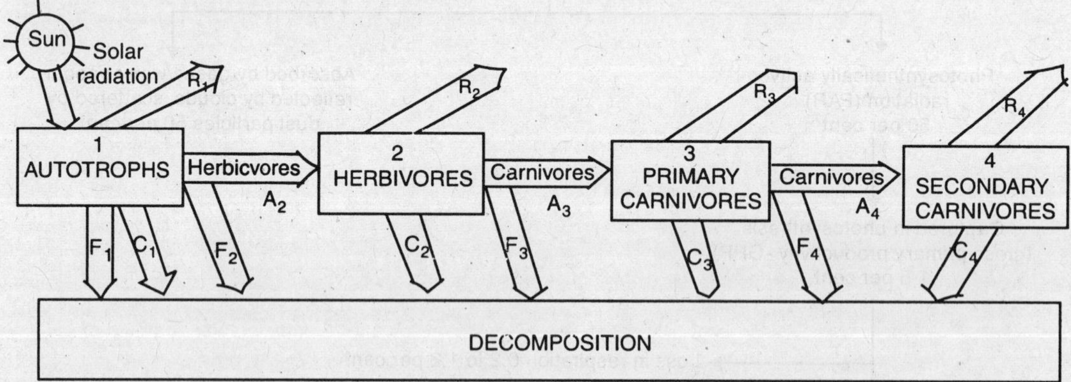

Fig. 12.5. Diagrammatic representation of a grazing food chain showing input and losses of energy at each trophic level. A = assimilation of food by the organisms at the trophic level; F = energy lost in the form of faeces and other excretory products; C = energy lost through decay; and R = energy lost to respiration.

In detritus food chain, there are many organisms which include algae, fungi, bacteria, slime molds, actinomycetes, protozoa, etc. Detritus organisms ingest pieces of partially decomposed organic matter, digest them partially and after extracting some of the chemical energy in the food to run their metabolism excrete the remainder in the form of slightly simpler organic molecules. The waste from one organism can be immediately utilised by a second which repeats the process. Gradually the complex organic molecules present in the organic wastes or dead tissues are broken down to much simpler compounds, sometimes to carbon dioxide and water, and whatever is left that is humus. In a normal environment the humus is quite stable and forms an essential part of the soil.

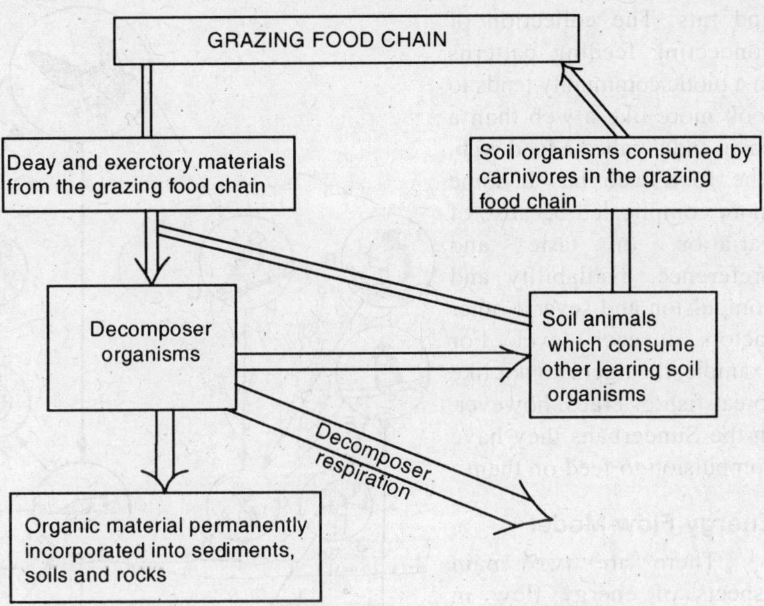

Fig. 12.6. Diagrammatic representation of detritus food chain showing energy transfers between it and the grazing food chain.

3. Links Between Grazing and Detritus Food Chains

Just as energy enters the detritus food chain from the grazing food chain as leaf litter, or dead organisms, energy can re-enter the grazing food chain from the detritus food chain if organism from the latter are consumed by grazers. Considering the diversity of the detritus community, it is not surprising that many of these serve as prey for carnivores in the grazing food chain. A common example is robin eating earthworms. In addition, many insects, among them beetles and flies, spend their larval period in the detritus food chain and their adulthood in the grazing food chain. But through such links through which energy passes from detritus food chain basic into grazing food chain, very little amount of energy is passed in comparison to the amount of energy which flows from grazing food chain to detritus food chain.

Food web. In nature simple food chains occur rarely. The same organism may operate in the ecosystem at more than one trophic level, *i.e.*, it may derive its food from more than one source. Even the same organism may be eaten by several organisms of a higher trophic level or an organism may feed upon several different organisms of a lower trophic level. Usually the kind of food changes with the age of the organism and the food availability. Thus, in a given ecosystem various food chains are linked together and intersect each other to form a complex network called food web.

Communities are not made up of unrelated chains. A consumer often has several major sources of food. A predator such as a lion or a tiger may feed on several different species of herbivores. Ticks and mites, leeches and blood sucking mosquitoes and other insects are dependent on herbivores and even on carnivores. On the other hand, a population of rabbits may be the prey of several different predators. Thus different organisms at each level have food relationship with more than one organism at the lower levels, such as a rat feeds on several kinds of grains and fruits; a snake eats the rat, and a falcon eats the snake. The snakes eat both frogs and rats. The collection of connecting feeding patterns in a biotic community tends to look more like a web than a chain and is called a **food web**. The food web has become more complicated because of variation in taste and preference, availability and compulsion and several other factors at each level. For example, the tigers do not like to eat fish or crabs, however, in the Sunderbans they have compulsion to feed on them.

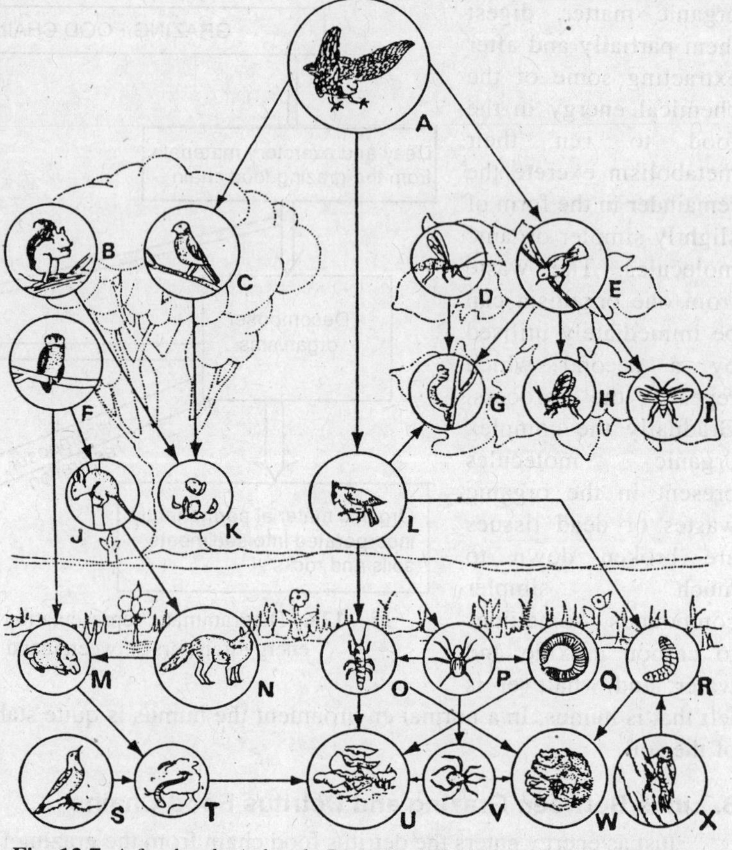

Fig. 12.7. A food web. A, hawk, B, squirrel; C, fruit eating bird ; D, insect ; E, L and S, birds feeding on insects, worms etc.; F, owl ; G, caterpillar; H and I, insects ; J, rodent ; K, fruits and nuts ; M, shrew ; N, fox ; O, insect ; P, spider; Q, millipede ; R, grule ; T, earthworm; U, mushroom; V, mite ; W, bracket fungus ; X, woodpecker.

Energy Flow Model

There are two main aspects of energy flow in ecosystem : 1. The energy flows one way, *i.e.*, from producers through herbivores to carnivores. This cannot be transferred in reverse direction. 2. The amount of energy flow decreases with successive trophic levels.

Producers capture only a small fraction of solar energy, *i.e.*, only 1-5 per cent of total solar radiation, and the bulk of unutilised energy is dispersed mostly as heat.

However, part of energy captured in gross production of producers is used for maintenance of their standing crop (respiration) and for providing food to **herbivores**, *i.e.*, **herbivory**.

The unutilised net primary production is ultimately converted to detritus, which serves as source of energy to decomposers.

Thus, energy actually used by the herbivore trophic level is only a small fraction of the energy captured at the producer level.

On an average, in different ecosystems, the herbivore assimilation or productivity approximately comes to 10 per cent of gross productivity of producers.

The energy assimilated by herbivores is used in respiration and a fraction of unassimilated energy is transferred to decomposers, *e.g.*, faecal matter.

The remaining herbivore level energy is either utilised by the carnivores, or gets transferred to decomposers after the death of herbivores.

The energy available at carnivore trophic level is again partitioned, leaving a very small fraction to support the next trophic level.

The species which occupy each trophic level in the food chain form a link, called the **food - link**. The links of one food chain are a flower, a butterfly, a dragonfly, a frog, a snake and a hawk. Here the flower nectar is eaten by butterfly, the butterfly is eaten by dragon fly, and so on through each successive link of the food chain. The chain ends with the hawk. When a hawk dies, its cells and tissues are digested and used as food by bacteria and other organisms of decay. In this way, the organic materials of the food chain are returned to the soil and recycled through another chain. However, the organisms that trap solar energy and act as producers also use up some energy for their own life processes. Animals being more active than plants use much of the energy acquired before they are consumed by the next trophic level. The studies have shown that only about ten per cent of the food energy in green plants is utilized by themselves at the primary consumer level. Much of the other 90 per cent of the energy goes to heat and movement. The actual percentages of energy from food at one level converted to energy from food at the next level be more than 10 per cent or less than 10 per cent. This general pattern forms an **energy pyramid** as each successive trophic level has less energy available to it. So, at each stage of energy transfer, a considerable energy is lost from the food chain. The decomposition of dead organisms also releases chemical energy. The chemical energy is transferred to living world through food. Food means "materials containing energy that organisms can use". Since food is the means of transfer of both matter and energy in the living world hence the study of food relation becomes necessary to understand the community and environment.

Ecological Pyramids

Studies of various food chains and food webs show that starting from the producers there is a regular decrease in the characteristics (number, energy and biomass) of the organisms occupying a trophic level, whether in a food chain or a level as a whole. If shown graphically these relationships will show a figure of a pyramid, known as **ecological pyramid**. Generally the pyramids are of three types — 1. pyramid of numbers; 2. pyramid of energy and 3. pyramid of biomass.

Pyramid of numbers. In a food chain the members at the successive higher levels are smaller in number. For example in a pond the lowest trophic level is represented by algae and diatoms, which are largest in number. The second trophic level is represented by herbivorous zooplankton, such as copepods, *Ranatra* etc., which are less abundant in number. While the third and fourth trophic levels are occupied by smaller and larger fish, respectively. There is a considerable reduction in the number of individuals from the base to the top of the pyramid. Another example may be quoted of a forest.

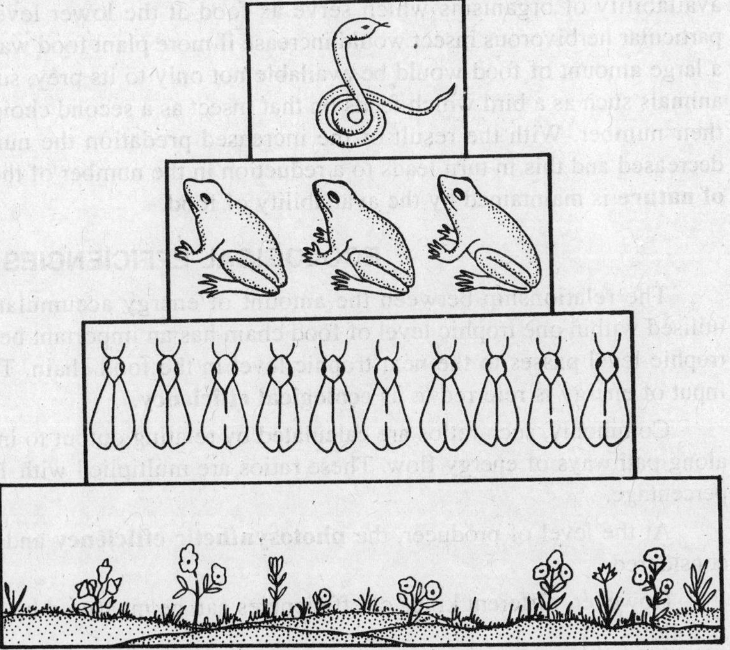

Fig. 12.8. Pyramid of number in food chain.

In a forest the small herbivorous insects are more abundant than the insectivorous birds. Similarly the preying birds, such as hawks are fewer than insectivorous birds.

Pyramid of energy. The amount of energy and matter transferred through food to successive higher levels become less and less. Greater amount of energy is available at the producer level than at the primary consumer level (herbivores). The energy production of the primary consumer is greater than that of the secondary consumers (*i.e.*, primary carnivores) which form next link in the energy chain. The energy at the tertiary consumer level (secondary carnivores) is produced in minimum amount. The actual fractions or percentages of energy from food at one level converted to energy from food at the next level may be more than 10 per cent or less than 10 per cent. This general pattern forms an **energy pyramid**, because each successive trophic level has less energy available to it.

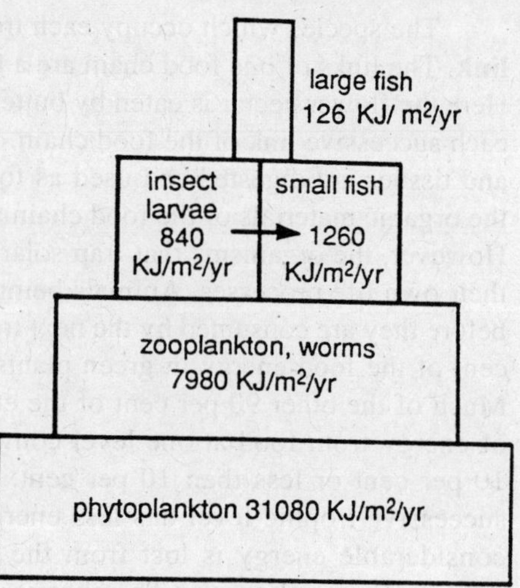

Fig. 12.9. Pyramid of energy.

Pyramid of biomass. The biomass is defined as the total weight of dry matter (dry weight) present in the ecosystem at any one time. The biomass can be measured graphically. This graph represents the shape of a pyramid which is known as **pyramid of biomass**. This pyramid shows the total mass of organisms and gives a rough picture of the overall effect of food chain relationships for the ecological group as a whole.

In studies of energy flow, the number of organisms at any trophic level depends upon the availability of organisms which serve as food at the lower level. For example the number of particular herbivorous insect would increase if more plant food was available to them. In this way, a large amount of food would be available not only to its prey, such as the frog but also to other animals such as a bird which feeds on that insect as a second choice which leads to an increase in their number. With the result of the increased predation the number of herbivorous insects is decreased and this in turn leads to a reduction in the number of their predators. Thus the **balance of nature** is maintained by the availability of food.

ECOLOGICAL EFFICIENCIES

The relationship between the amount of energy accumulated, and the amount of energy utilised within one trophic level of food chain has an important bearing on how much energy one trophic level passes to the next trophic level in the food chain. The ratio of output of energy to input of energy is referred to as **ecological efficiency**.

Commonly, such ratios are calculated by relating output to input of energy at various points along pathways of energy flow. These ratios are multiplied with 100 to express the efficiency as percentage.

At the level of producer, the **photosynthetic efficiency** and **net production efficiency** are considered.

However, different kinds of efficiencies can be measured by different parameters, which are as follows:

(*i*) **Ingestion.** This indicates quantity of food or energy taken by trophic level. This is referred as **exploitation efficiency**. The ability of herbivores and carnivores to use the food energy ingested, varies from one species to another.

(*ii*) **Assimilation.** This indicates the amount of food absorbed and fixed.

(*iii*) **Respiration.** This indicates the energy lost in metabolism.

The photosynthetic efficiency measures the ability to utilise incident solar radiation. This efficiency is expressed in relation to photosynthetically active radiation (PAR).

However **net production efficiency** varies widely in different species. For example, tree species having larger nonphotosynthetic biomass, *i.e.*, huge trunks and thick branches, depict lesser net production efficieny.

The major ecological efficiencies may be summarised as follows:

Percentage of photosynthetic efficiency

$$= \frac{\text{Gross primary productivity} \times 100}{\text{Incident total solar radiation}}$$

Percentage of net production efficiency

$$= \frac{\text{Net primary productivity} \times 100}{\text{Gross primary productivity}}$$

Percentage of assimilation efficiency

$$= \frac{\text{Food energy assimilated} \times 100}{\text{Food energy ingested}}$$

Percentage of ecological efficiency

$$= \frac{\text{Energy of biomass production at a trophic level} \times 100}{\text{Energy in biomass production at previous trophic level}}$$

NUTRIENT CYCLING-BIOGEOCHEMICAL CYCLES

Out of over 105 elements discovered so far, only about 20 of them have been found to be essential for plant growth and metabolism. These essential elements are divided into two broad categories, (*i*) macronutrients and (*ii*) micronutrients. The macronutrients include C, H, O, N, P, S, K, Ca, Mg, and Si, while micronutrients include Mn, Cu, Mo, Zn, B and Cl. These materials flow from abiotic to biotic compounds and back to the non-living component in a more or less cyclic manner, and referred to as **nutrient cycling**.

Plants can photosynthesise simple compounds, such as carbohydrates, using C, H, and O obtained from carbon dioxide and water. On the other hand, for the synthesis of more complex substances, such as proteins, they require supply of additional essential elements or nutrients. Some nutrients, such as nitrogen and phosphorus are required in large amounts (macronutrients), while others are needed in trace amounts, *e.g.*, zinc, molybdenum, copper, etc. Nutrient cycles involve storage and transfer of nutrients through various components of the ecosystem, and therefore, they are repeatedly used.

The nutrient cycle is also known as the **biogeochemical cycle** or **inorganic-organic cycle**. The flow of the elements through the ecosystem must be cyclic with matter being consistently reused. As the flow involves not only living organisms, but also a series of chemical reactions in the abiotic environments, such cycles are called biogeochemical cycles. Nutrient cycles, are of two types: (*i*) **gaseous** and (*ii*) **sedimentary**.

1. Gaseous. The reservoir of gaseous type of nutrient cycle is generally located in the atmosphere or hydrosphere, *e.g.*, nitrogen (reservoir in atmosphere).

2. Sedimentary. However, in the sedimentary type, the reservoir exists in the earth's crust, *e.g.*, phosphorus (reservoir in lithosphere).

Nutrient cycles are generally considered under three aspects, which are as follows:

Input of Nutrients

Ecosystem receives nutrients from external sources and stores them for further use through biological processes, *e.g.*, nutrients in dissoloved state are received from rainfall, *i.e.*, **wet deposition**. When nutrients are gained from dust fall, it is called **dry deposition**.

Symbiotic biological fixation of nitrogen in soil is also an input of nutrients.

Another example of input is weathering of soil parent materials, which releases available nutrients from their fixed state.

Output of Nutrients

Here, nutrients are moved out of an ecosystem and many become input to another ecosystem. For example, lot of loss of nutrients like calcium and magnesium occurs through run off water or through soil erosion.

The other example is considerable loss of nitrogen in gaseous form by the process of denitrification in soil.

Harvesting of agricultural crops, and transportation of logs from forests, represent loss of nutrients from these ecosystems.

Internal Nutrient Cycling

Nutrients in varying amounts are absorbed by the plants from the soil.

Nutrients are continuously regenerated and stored in soil due to process of decomposition of dead organic matter.

Both processes, *i.e.*, regeneration and absorption of nutrients occur simultaneously.

The transfer of nutrients from the soil to the plants by absorption is called **uptake** process.

The absorbed nutrients are used metabolically for the use of plant growth. At the time of death, when plant debris again goes to soil, nutrients are recycled and brought back to soil through **litter fall** of organisms.

Both aboveground and root detritus are decomposed, with the result, nutrients are regenerated.

However, when the uptake of nutrients exceeds the amount recycled, a fraction of the uptake is retained in the **standing crop**. This type of **retention** of balance nutrients in the standing crop increases the nutrient content of the ecosystem. This means:

Retention = uptake–recycle.

Rates of nutrient uptake, retention and recycle vary in different ecosystems, and may be calculated by various chemical methods.

ECOSYSTEM NITROGEN CYCLE

Of all the elements that plants absorb from the soil, nitrogen is most important for plant growth. This is required in greatest quantity. Nitrogen is required for the synthesis of amino acids, proteins, enzymes, chlorophylls, nucleic acids, etc. Green plants obtain nitrogen from the soil solution in the form of ammonium, nitrate and nitrite ions, but the main source of all these nitrogen compounds is the atmospheric nitrogen. The ultimate source of nitrogen to the ecosystems is molecular nitrogen in the atmosphere, which cannot be directly metabolised by plants or animals. Thus, all the nitrogen accumulated in biomass, detritus and humus in various ecosystems, is derived from the atmosphere by biological activities. Nitrogen cycle consists of the following steps:

(*i*) Nitrogen fixation, (*ii*) nitrogen assimilation, (*iii*) ammonification, (*iv*) nitrification, (*v*) denitrification and (*vi*) sedimentation. The details are as follows:

1. **Nitrogen fixation.** Conversion of free nitrogen of atmosphere into the biologically acceptable form or nitrogenous compounds is referred to as **nitrogen fixation**. This process is of two types: (*i*) physicochemical or non-biological nitrogen fixation, and (*ii*) biological nitrogen fixation.

In **physico-chemical process**, the atmospheric nitrogen combines with oxygen (as ozone) during lightning or electrical discharges in the clouds and produces different nitrogen oxides. The equations are as follows:

$$N_2 + 2(O) \xrightarrow[\text{discharge}]{\text{electric}} 2NO$$

$$2NO + 2(O) \longrightarrow 2NO_2$$

$$2NO_2 + (O) \longrightarrow N_2O_5$$

These nitrogen oxides get dissolved in rain water, and on reaching earth surface they react with mineral compounds to form nitrates and other nitrogenous compounds:

$$N_2O_5 + H_2O \longrightarrow 2HNO_2$$

$$2HNO_2 + CaCO_2 \longrightarrow Ca(NO_3)_2 + CO_2 + H_2O$$

During combustion of various types, some nitrogenous compounds are formed, which are washed down along with rain water.

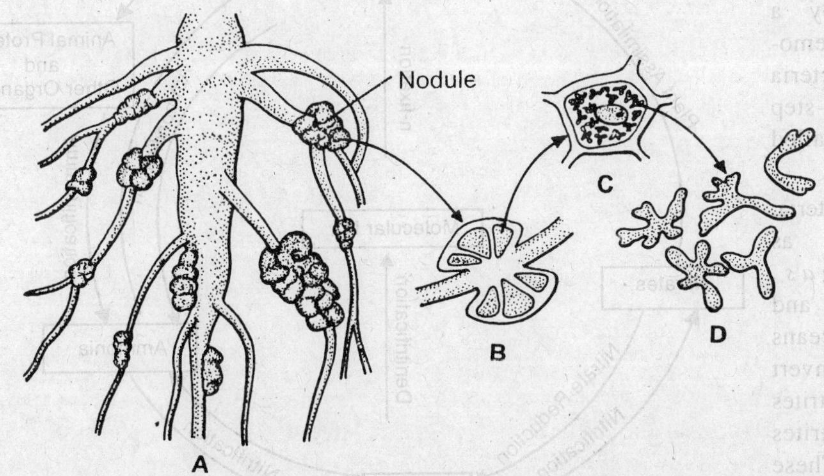

Fig. 12.10. Root nodules of a legume plant. A, legume root with nodules; B, T.S. of root nodule; C, single cell of nodule; D, nitrogen fixing bacterium, *Rhizobium*.

Biological nitrogen fixation is carried out by certain prokaryotes. The cyanobacteria (blue-green algae) fix significant amounts of nitrogen in the oceans, lakes and soils. Symbiotic bacteria (*Rhizobium*) inhabiting the root rodules of legumes and symbiotic cyanobacteria, such as *Nostoc*, *Anabaena*, etc., found in free state, or in thalli of *Anthoceros* (*bryophyte*), *Azolla* (water fern), coralloid roots of *Cycas* (gymnosperm) fix atmospheric nitrogen.

Certain **free living** nitrogen fixing bacteria, such as *Azotobacter*, *Clostridium*, *Beijerinckia*, etc., also fix free nitrogen of atmosphere in the soil. Frankia, an actinomycetous fungus found in the roots of higher plants, such as *Alnus* and *Casuarina*, also fix nitrogen.

Nitrogen fixing organisms combine the gaseous nitrogen of atmosphere with hydrogen obtained from respiratory pathway to form ammonia, which then reacts with organic acids to form amino acids.

Biological nitrogen fixation is the major source of fixed nitrogen upto 140 – 700 mg/m^2/year as against 35 mg/m^2/year by electrical discharge and photochemical fixation.

2. Nitrogen assimilation. Inorganic nitrogen in the form of nitrates, nitrites and ammonia is absorbed by the green plants and converted into nitrogenous organic compounds. Nitrates are first converted into ammonia which combines to organic acids to form amino acids. Aminoacids are used in the synthesis of proteins, enzymes, chlorophylls, nucleic acids, etc.

Animals derive their nitrogen requirement from the plant proteins. Plant proteins are not directly utilised by the animals. They are first broken down into aminoacids during digestion and then the aminoacids are absorbed and manipulated into animal proteins, nucleic acids, etc.

3. Ammonification. The dead organic remains of plants and animals and excreta of animals are acted upon by a number of microorganisms, especially actinomycetes and bacilli, such as Bacillus ramosus, B. vulgaris, etc. These organisms utilise organic compounds in their metabolism and release ammonia. This process is called **ammonification**. After meeting their own metabolic requirement, these microbes release the excess ammonia in the soil.

4. Nitrification. In next step of ammonia formation, ammonia is converted into nitrate by a group of chemo-autotrophic bacteria through a two-step process called **nitrification**.

Certain bacteria, such as *Nitrosomonas*, *Nitrococcus* and *Nitrospira* in oceans and soils convert ammonia into nitrites and then nitrites into nitrates. These bacteria primarily use the energy of dead organic matter in their metabolism. The equation is as follows:

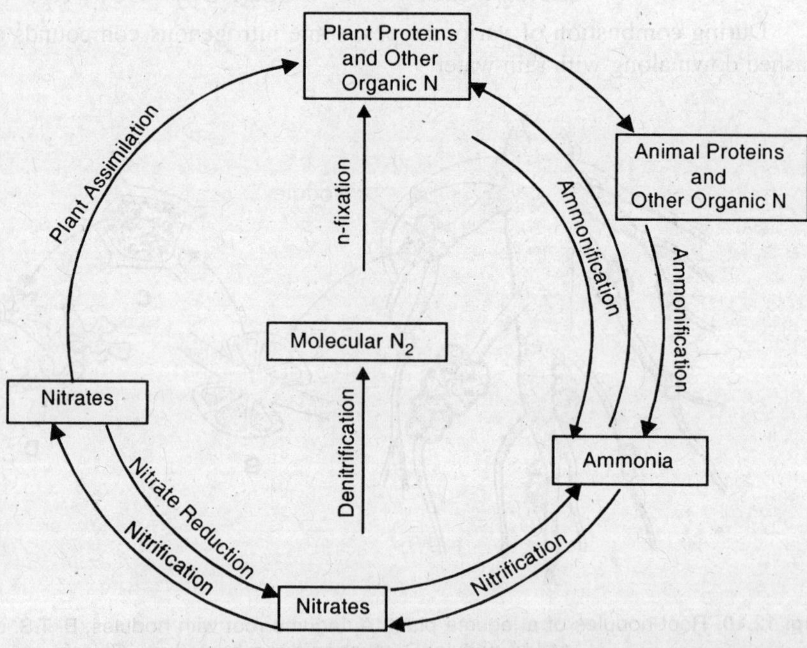

Fig. 12.11. Nitrogen cycle in ecosystem.

$$2NH_4 + 2O_2 \longrightarrow NO_2^- + 2H_2O + \text{energy}$$

Conversion of nitrites to nitrates is brought about by several microbes, such as *Penicillium* (a fungus), *Nitrobacter*, etc.

$$2NO_2^- + O_2 \xrightarrow{\textit{Nitrobacter}} 2NO_3^- + \text{energy}$$

Some nitrates are also made available through weathering of nitrate containing rocks.

5. Denitrification. Ammonia and nitrates are converted into free nitrogen by certain microbes. This process is referred to as **denitrification**. *Pseudomonas*, the most common denitrifying bacterium, thrives best under poorly aerated and detritus-rich conditions. Denitrifying bacteria transform nitrate nitrogen to nitrous and nitric oxides, and ultimately to gaseous nitrogen, which goes to atmosphere.

Denitrification by denitrifying bacteria.

$$2NO_3^- \longrightarrow 2NO_2^- \longrightarrow 2NO \longrightarrow N_2$$

6. Sedimentation. Nitrates of the soil are washed away to the sea or leached deep into the earth along with percolating water. Nitrates thus lost from the soil surface are locked up in the rocks. This process is called **sedimentation** of nitrogen. Nitrogen of rock is released only when the rocks are exposed and weathered.

Thus a large part of nitrogen is fixed up and stored up in plants, animals and microbes. Most higher plants absorb nitrate from the soil; the absorbed nitrate is ultimately converted to organic nitrogen. A fraction of nitrogen incorporated in plant tissues is used by consumers, and ultimately all dead remains convert into detritus and used by decomposers. Thus, complex nitrogen cycle completes.

Carbon Cycle

The carbon is the basic constituent of all organic compounds found in the protoplasm, such as carbohydrates, proteins, fats and nucleic acids. In the non-living world, there are three main sources of carbon; 1. the carbon dioxide in air and found dissolved in water, 2. carbonates of earth's crust derived from rocks give rise to carbon dioxide on chemical changes, and 3. the fossil fuels like coal and petroleum give rise to carbon dioxide on burning.

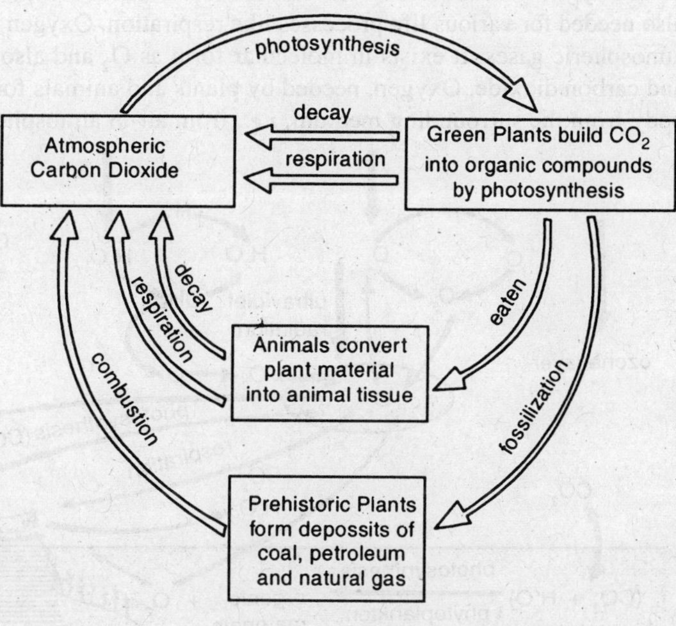

Fig. 12.12. Carbon Cycle.

The basic source of carbon for the living organisms is carbon dioxide. Atmosphere, having a concentration of 0.031 per cent of carbon dioxide, acts as its reservoir. Carbon dioxide from the atmosphere enters in the living world by the process of photosynthesis in green plants (producers). The amount of carbon fixed by photosynthesizers is nearly 7×10^{13} kg per year. As we know, during photosynthesis simple sugar like glucose ($C_6H_{12}O_6$) is synthesized which is subsequently converted into other organic compounds like polysaccharides (sucrose, starch, glucose, etc.), proteins and lipids. These are passed from plants (producers) to the herbivorous and carnivorous animals (consumers).

Available data suggest that one hectare of a healthy forest can produce about 10 tonnes of oxygen and absorb 30 tonnes of carbon dioxide in a year.

The respiratory activity of plants and animals return back fixed carbon as carbon dioxide into the atmosphere. The excretory wastes of living organisms having accumulated carbon compounds and they after their death are decomposed by microorganisms (decomposers) in the soil to release carbon dioxide back into the environment. Burning of wood and fossil fuels, dissolution of carbonate rocks and volcanic activities, all release carbon dioxide into the environment for its recycling (Fig. 12.12).

Under certain conditions, large number of organisms got buried deep into the layers of the earth and they gradually transformed into coal, petroleum and natural gas under high pressure, and temperature. Or, if the process of decay is extremely slow then a sufficient amount of carbon goes out of the cycle. Such events take millions of years. But, when such fuels are burnt, carbon dioxide goes back to the cycle. Similarly, carbon is also locked up in hard calcareous shells of animals and also in their skeletal structures which the decomposers cannot act upon. Therefore, the chief pathway of carbon cycle is from the atmosphere or hydrosphere into the living organisms and back. The carbon cycle is essentially a self-regulating feedback system but may be disturbed by man by the over-use of fossil fuels.

Oxygen Cycle

Oxygen, like carbon, is another essential element for the construction of living body and it is also needed for various life processes like respiration. Oxygen constitutes nearly 21 per cent of the atmospheric gases. It exists in molecular form as O_2 and also as a part of compounds like water and carbon dioxide. Oxygen, needed by plants and animals for respiration, enters directly in their body from the surrounding medium, *i.e.*, from air in atmosphere or from water.

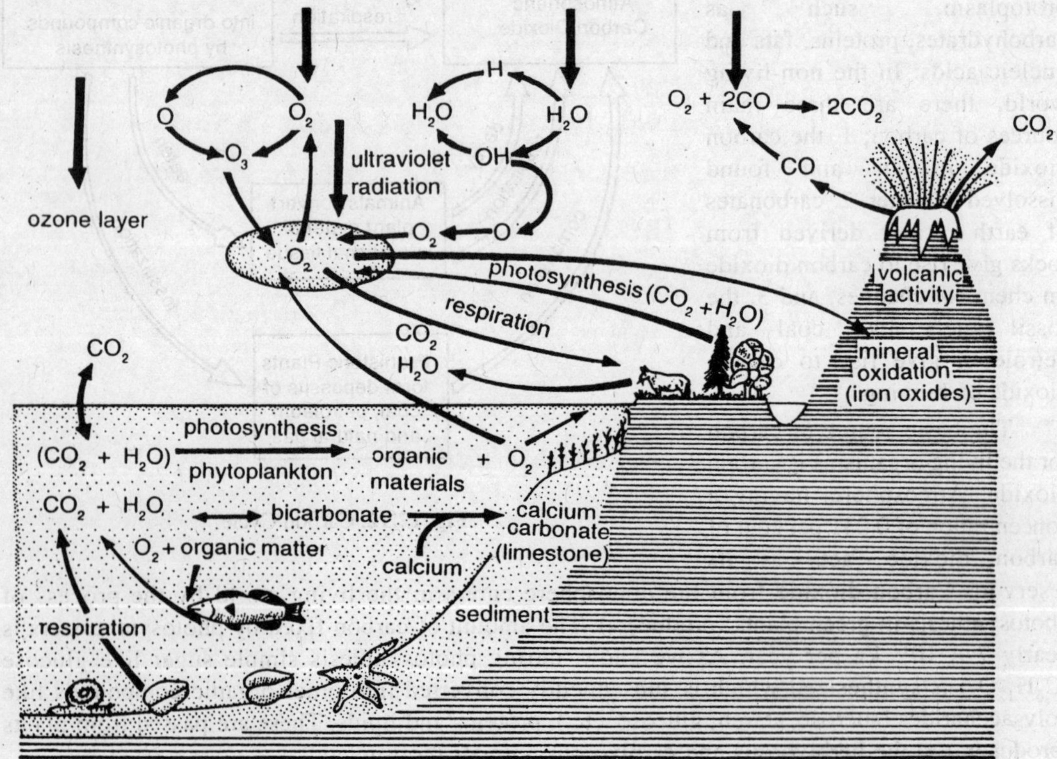

Fig. 12.13. Oxygen cycle.

However, cycling of oxygen is very complex (Fig. 12.13). As a constituent of CO_2, it circulates freely throughout the biosphere. Some carbon dioxide combines with calcium to form carbonates. Oxygen combines with nitrogen compounds to form nitrates, with iron to ferric oxides, and with many other materials to form various other oxides. In these states oxygen is temporarily withdrawn from circulation. In photosynthesis the oxygen freed is split from the water molecule. This oxygen is then reconstituted into water during plant and animal respiration. Some part of atmospheric oxygen that reaches the higher levels is reduced to ozone (O_3) by high energy ultraviolet radiations. Thus, in this way oxygen level in the atmosphere is maintained at a constant level.

The earth's atmosphere, after its formation, has undergone a slow but steady change due to various human activities. Human activity, however, has not had much effect on the oxygen content of the atmosphere because it is always replenished by photosynthesizers. But the condition is altogether different for other atmospheric gases, collectively these are called **greenhouse gases** and the effect they produce is called the **greenhouse effect**.

The greenhouse gases allow the penetration of incoming solar radiations having short wave lengths (0.15 to 4 m m) but they prevent the escaping of long wave length radiations into the outer space and reflect back on the earth causing rise in temperature of the earth. This is referred to as greenhouse effect. For example, carbon dioxide (a greenhouse gas) tends to prevent the long wave radiations (*i.e.*, infra-red heat radiation) from earth from escaping into outer space and reflect it back to the earth. This has caused an increase in earth's temperature. Since, carbon dioxide content of the atmosphere is increasing with an alarming rate due to excessive use of fossil fuels liberating carbon dioxide into the earth's atmosphere, it has a greenhouse effect (Fig. 12.14).

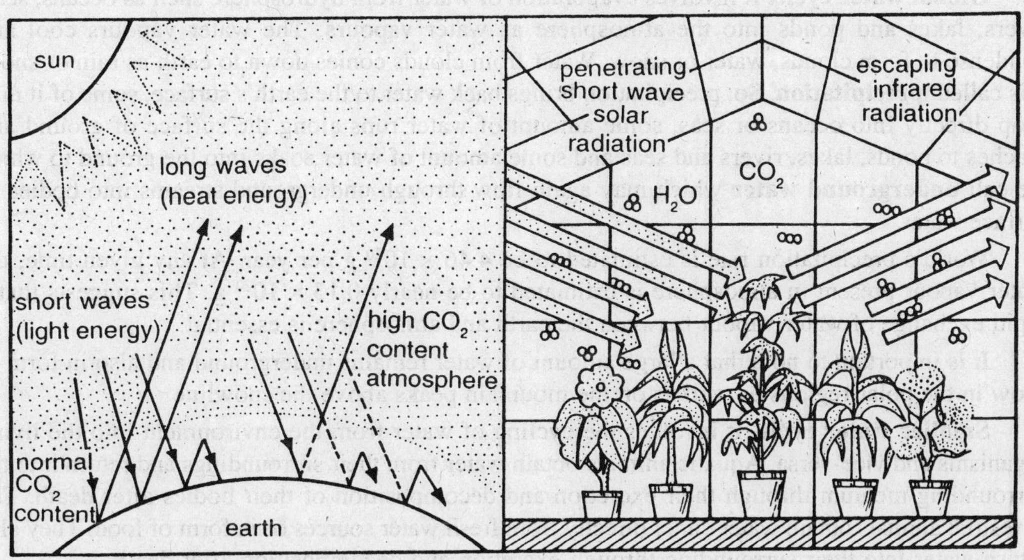

Fig. 12.14. Greenhouse effect caused due to high concentration of carbon dioxide.

Water Cycle

Water cycle basically forms a link between gaseous and sedimentary biogeochemical cycles. Hence, it needs a discussion at this stage before discussing the sedimentary cycles.

Water is indispensable for life constituting more than 80 per cent of living cells. The water cycle is represented by two overlapping cycles in nature. (*i*) **Global water cycle**, being controlled by physical processes involving no life, and (*ii*) **Smaller water cycle**, being controlled by biological processes involving life.

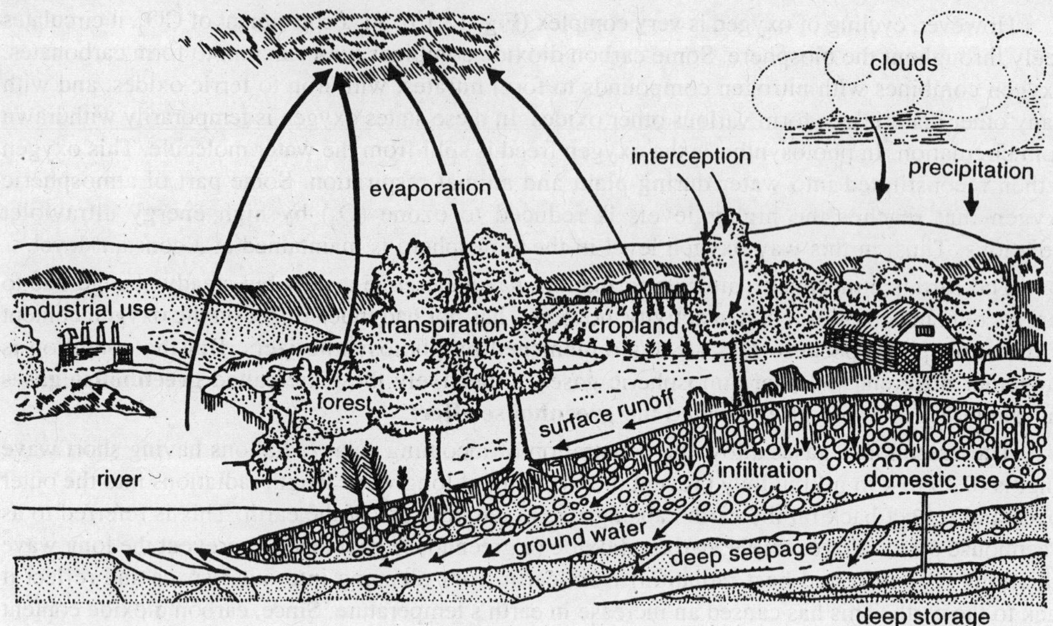

Fig. 12.15. Water cycle.

Global water cycle. It involves evaporation of water from hydrosphere such as oceans, seas, rivers, lakes and ponds into the atmosphere as water vapours. The water vapours cool and condense to form clouds, water or snow. Water from clouds comes down to earth as rain or snow, it is called **precipitation**. So, precipitation brings back water to the earth's surface, some of it may drop directly into oceans or seas, some amount of water runs along the surface of ground and reaches to ponds, lakes, rivers and seas and some amount of water soaks into the ground to which we call **underground water** which may again flow through underground streams into bodies of surface water.

World's precipitation rate is estimated to be 4.46×10^{20} g per year. At any given time, the water vapour present in atmosphere is estimated to be nearly 0.13×10^{20} g. This suggests that a rapid exchange of water vapour between the earth and atmosphere is essential.

It is important to note that a large amount of water remains underground and also in form of snow in the polar regions as well as on the mountain peaks above the snowline.

Smaller water cycle. It involves the cycling of water from the environment into the living organisms and vice-versa. Aquatic animals obtain water from their surroundings and return it to the surrounding medium through their excretion and decomposition of their bodies after death. The terrestrial animals obtain water either directly from fresh water sources or in form of food. They also return water into their surrounding through excretion and decomposition after death.

The plants absorb water from the soil by their roots. These retain some amount of water in their bodies and the rest is given out in the environment as water vapours during the process of transpiration and amount of water left in their bodies returns back to the environment after the death and decomposition of their bodies. Evaporation of water from terrestrial plants cools the surrounding air and, thus, it helps in maintaining a microclimate around them.

The smaller water cycle is complicated one and differs considerably from the global water cycle but both of these operate simultaneously in nature to keep a water balance between the living and non-living world.

ECOSYSTEM PHOSPHORUS CYCLE

Plants and animals obtain phosphorus from the environment. Phosphorus is a component of nucleic acids, ADP, ATP, NADP, phospholipids, etc. It occurs in the soil as rock phosphate, apatite or calcium phosphate, fluorapatite Ca_{10} Fe_2 $(PO_4)_6$ ironphosphate or aluminium phosphate. Soils derived from the rock beds rich in phosphates are rich in phosphorus.

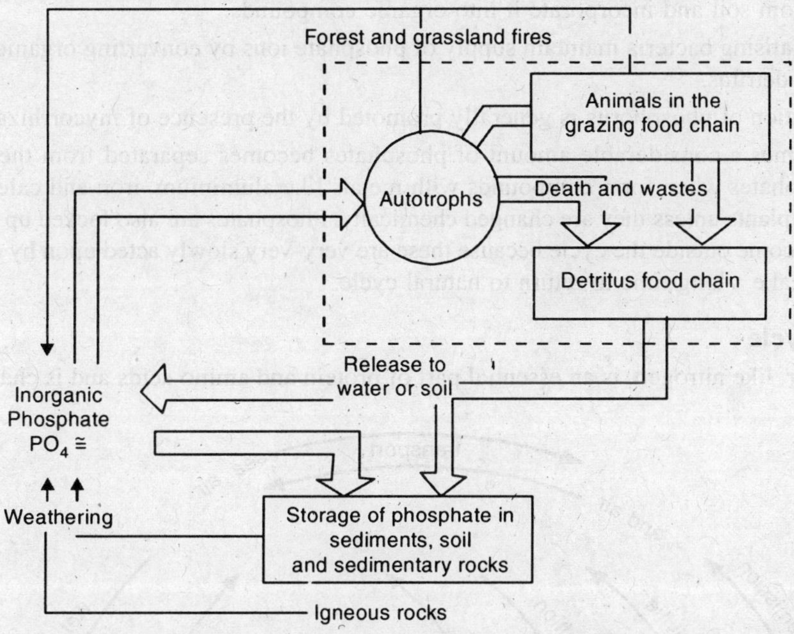

Fig. 12.16. Phosphorus cycle in ecosystem.

Phosphorus occurs in the soil in five forms:

P_1 (stable organic), P_2 (labile organic), P_3 (labile inorganic), P_4 (soluble) and P_5 (mineral form) and of these forms P_3 and P_4 are in equilibrium and entry of phosphorus in the green plants is considered to occur via labile inorganic pool.

The dissolved phosphorus is absorbed by plants and converted to organic form. From plants it travels to various trophic levels in the form of organic phosphates. When the plants and animals die, the decomposers attack them and liberate phosphorus to the environment. Thus this process proceeds in cyclic way.

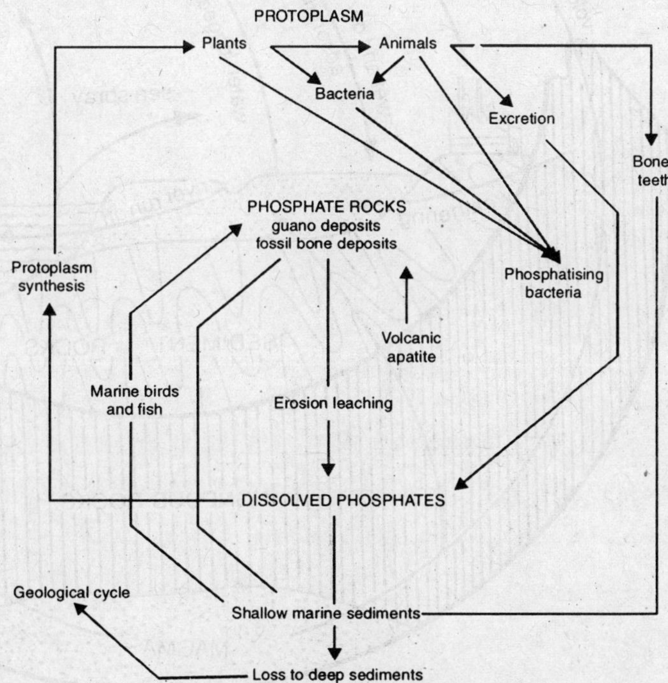

Fig. 12.17. The phosphorus cycle.

Phosphorus along with many other mineral elements reaches the oceans and settles down as sediment. A good proportion of phosphorus leaches down to deep layers of soil. In this way, major proportion of phosphate becomes lost to this cycle by physical processes, such as sedimentation and leaching. Biological processes such as formation of teeth and bones also keep phosphorus locked up for some time.

Phosphorus occurs in soil solution mainly as orthophosphate (PO_4^{3-}). Plants absorb phosphate from soil and incorporate it into organic compounds.

Phosphatising bacteria maintain supply of phosphate ions by converting organic phosphorus contained in detritus.

Absorption of phosphorus is generally promoted by the presence of mycorrhizae.

Sometimes a considerable amount of phosphates becomes separated from the phosphorus cycle ; phosphates when form compounds with metals like aluminium, iron and calcium become unusable for plants unless they are changed chemically. Phosphates are also locked up in bones and teeth, and become outside the cycle because these are very-very slowly acted upon by decomposers and, hence, take a long time to return to natural cycle.

Sulphur Cycle

Sulphur, like nitrogen, is an essential part of protein and amino acids and is characteristic of

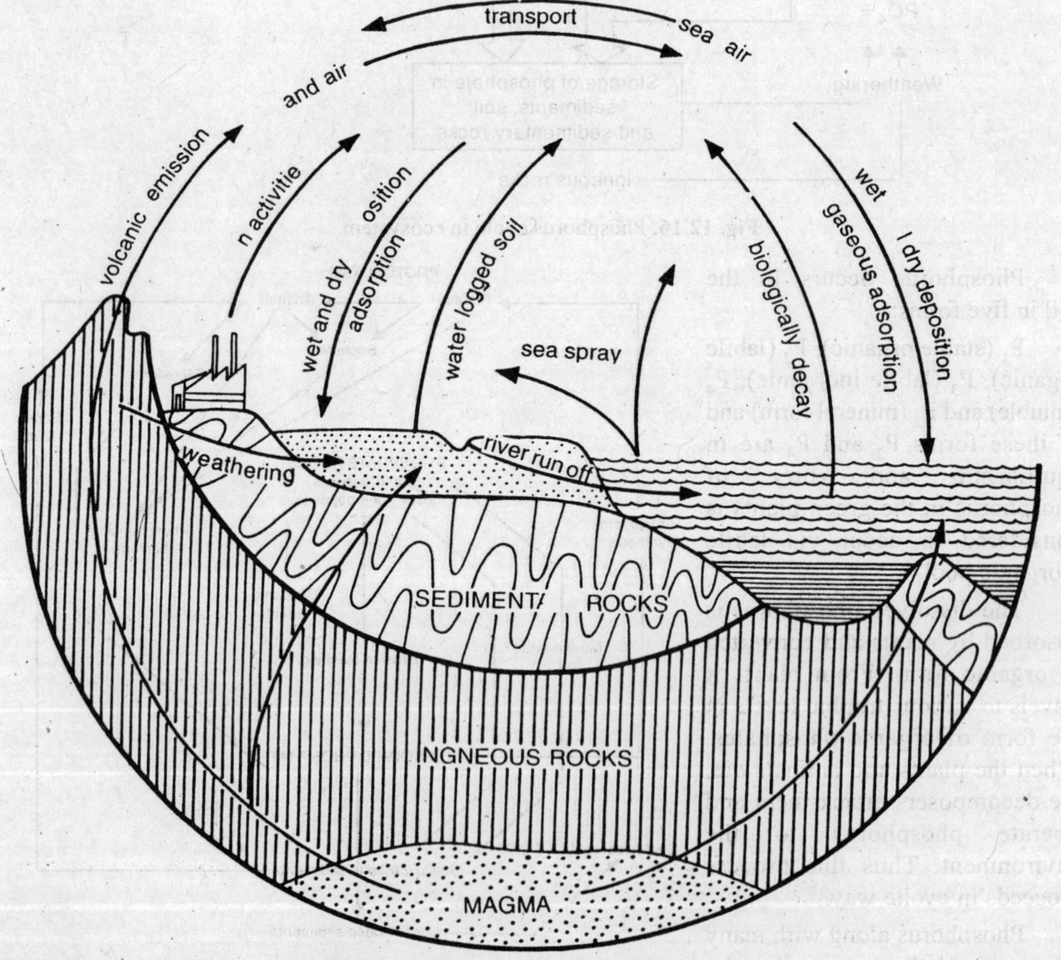

Fig. 12.18. The sulphur cycle.

organic compounds. The main source of sulphur for living organisms is the inorganic sulphur in form of sulphates.

The photosynthesizers (producers) obtain sulphur in form of sulphates from the soil or from water in case of aquatic plants and make use in their protein synthesis. This organic sulphur reaches to animals through food. After death, the animals and plants are decomposed in the soil by microorganisms like *Aspergillus*, *Neurospora* and *Escherichia* under aerobic conditions as sulphates ; these sulphates are mineralised in the soil. But under anaerobic conditions such as in marshes, it is reduced to sulphides. Sulphides, thus, produced are converted into soluble sulphates by sulphur bacteria, *viz.*, *Thiobacillus* into sulphates.

Sulphur is added in the atmospheric air as sulphur dioxide from various industries. The fossil fuels also add sulphur dioxide to the environment when they are burnt. During rains, sulphur dioxide of the atmosphere forms H_2SO_4 (sulphuric acid) which later on decomposed to give rise to SO_4 ions and then incorporated into living body.

Sulphur is removed from the organic phase in form of elemental sulphur which is insoluble and accumulates into reservoir pool in deep sediments.

Other Minerals and their Cycling

Calcium, magnesium, iron and other minerals (micronutrients), are also cycled in nature like other cycles, as they are also required by living organisms.These are slowly released from the rocks either by water or wind actions and finally reach to the soil and water. Plants obtain these through their roots and animals get them in form of food. Animals like molluscs and corals, deposit a good amount of calcium in their shells and skeletons which is locked for a considerable period and, hence, not available for cycling. The sedimentary elements in the sea return very slowly to the cycle through marine food chains and through coastal sea water sprays. The process may take hundred to millions of years.

Flow of Energy

As referred to, sun is the only source of energy to our planet earth. The solar energy is released in form of invisible ultra-violet light and infra-red light, and visible light. Most of the ultra-violet light is absorbed by **ozone layer** in the upper atmosphere and most of the infra-red light having longer wavelengths are absorbed by atmospheric gases and some are reflected back in the environment. Visible light constitutes nearly 50 per cent of solar energy; a part of it is reflected by clouds and dust particles, and a part is absorbed by the atmosphere (Fig. 12.19). In fact, sun radiates energy of nearly 10^{26} Joules/second but the earth gets only a fraction of it. On the top of the atmosphere, the earth receives energy equal to 173×10^{15} Joules/second ; of this about one-third is reflected back into the space by the atmosphere.

However, whatever amount of energy from the sun reaches the earth in form of light and heat regulates the climatic phenomena on this planet. The green plants trap only a very small fraction (nearly one per cent) of available solar energy during the process of photosynthesis. It has already been mentioned that green plants trap light energy with the help of the pigment chlorophyll and convert it in chemical energy. Thus, the energy trapped by green plants, *i.e.*, **autotrophs** is utilised then in synthesizing organic compounds during the process of photosynthesis. The green plants, thus, are the only organisms which can trap and fix solar energy and can manufacture food used by them as well as by other organisms ; hence, these are called **producers**.

The rate at which organic compounds are synthesized in a green plant or a population of green plants is called **gross primary productivity** (GPP). GPP gives us an idea of the proportion of assimilation of light energy into living matter of the plant. The organic compounds synthesized

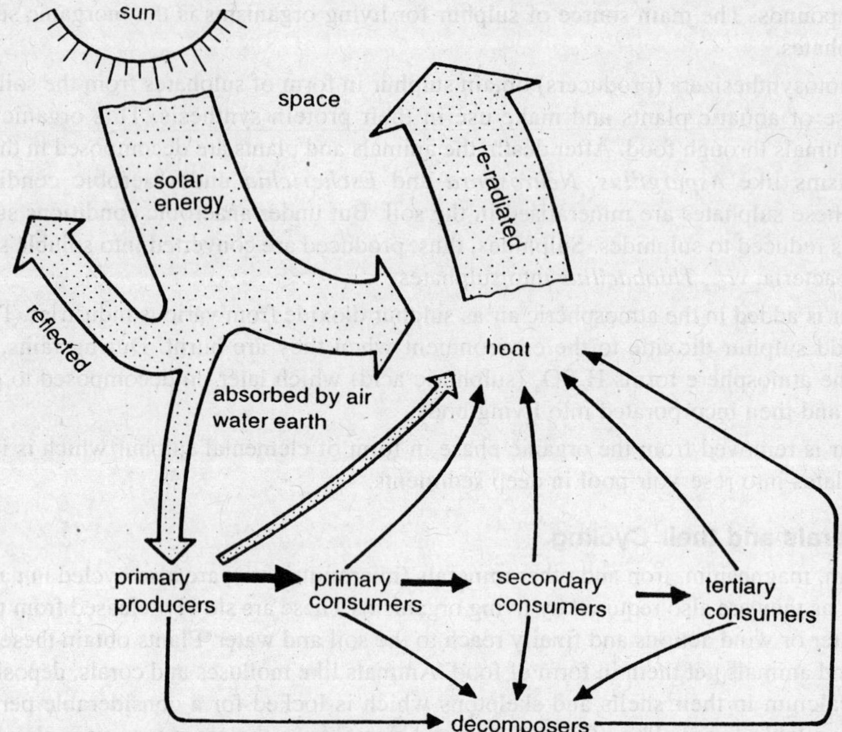

Fig. 12.19. Flow of energy and cycling of materials.

by plants are used as a source of energy for their body building, storage and various catabolic activities. The total stored energy in form of organic matter (which is not used by the producer itself) is called **net primary productivity** (NPP). Exactly in the manner as a single plant contributes to the gross primary productivity and net primary productivity, all autotrophs in a community contribute to **community productivity**.

It has been mentioned earlier that all heterotrophs depend upon autotrophs (producers) directly or indirectly for food. So, by consuming autotrophs, heterotrophs consume organic matter from them. Energy, thus, enters the ecosystem through photosynthesis and passes through the different trophic levels. In other words, flow of energy takes place from the sun to autotrophs to heterotrophs. The food consumed by heterotrophs is utilized in their maintenance and for storage too. If there is more storage than consumption, the biomass of the population would be higher at the end of a time period than at the beginning. So, this rate of increase in the biomass of heterotrophs is referred to as **secondary productivity**.

There occurs a continuous loss of energy all along the path when it is transferred from producer to a consumer and from consumer to the next consumer at the higher trophic level. For this, Lindemann (1942) formulated a concept called **10 per cent law**. To explain, when plants (producers) are eaten by a herbivore (primary consumer), about 10 per cent of energy in the food is fixed in herbivore's body and when a carnivore (secondary consumer) eats up the herbivore, only about 10 per cent of energy in food is fixed in carnivore's body. Therefore, the second level consumer gain in energy is only 0.01 per cent of the net productivity of a plant (producer). This clearly shows that there is a loss of energy at each trophic level. So, the longer the food chain, the minimum is the energy available to the last member of the chain. It is, therefore, there are usually four or five steps in natural food chains.

The flow of energy is, however, unidirectional.

Ecological Imbalance

Ecological studies suggest that there exists an **ecological** balance between various constituents of abiotic and biotic components in the **biosphere**. For example, the various biogeochemical cycles operate in nature without any interruption and the abiotic as well as biotic components including producers, consumers and decomposers are significant. If producers be destroyed, carbon from atmospheric carbon dioxide cannot move to organic compounds in living organism and if consumers and decomposers be destroyed the organic compounds will accumulate resulting into locking up of carbon and further not available for cycling. However, when all organisms of an area are studied, it becomes evident that in spite of different types of competitions and struggles between them, they cooperate each other on many ways or they seem to be interdependent. This dependence is called ecological balance. It is very necessary for the continuity of ecosystems and, hence, the biosphere.

However, man being a dominant organism of most ecosystems controls and modifies environment more extensively than any other organism. In fact, there have been significant changes in the natural environments due to man's intervention and his rapid progress in colonization, urbanization, industrialization, agriculture, transportation and technology. Thus, deforestation has provided land for agriculture and rural inhabitation, rural land has been converted into urban settlements and open spaces are fast vanishing. All these have resulted into a mass-scale destruction of flora and fauna which has become detrimental to ecological balance. If, this balance is disturbed for long, it is bound to create a number of crisis to the biological world including our lives. Hence, considering this importance, a number of ways are being adopted by government and private agencies so that ecological balance could be maintained.

MAJOR BIOMES OF THE WORLD

Natural ecosystems of plants and animals extend over large areas of the world. Each of such major terrestrial ecosystem or distinct terrestrial areas with their natural groupings of climax plants and associated animals make **biomes**. Thus a biome is a group of ecosystems similar in types of dominant vegetation. A biome is the largest terrestrial biotic community. The abiotic factors such as temperature, sunlight, soil barriers, latitude, altitude and availability of water are important in determining the nature and extent of biomes. Across the surface of the earth, examples of the same biome a desert, for example can be found wherever the climate and soil conditions are similar. Similarly there are distinct major ecosystems or life supporting zones in marine and fresh water environments. However, the characteristic of the aquatic systems are different from the terrestrial biomes. Ecosystems of the same biome from anywhere on earth are similar in many ways.

The major biomes of the world are (*i*) Tundra, (*ii*) Northern conifer forest or Taiga, (*iii*) Temperate deciduous forest, (*iv*) Tropical rain forest, (*v*) Chapparal, (*vi*) Tropical Savannah, (*vii*) Grassland, (*viii*) Desert.

Tundra. The literal meaning of word **Tundra** is north of the timberline. The tundra extends above 60°N latitude. It is almost treeless plain in the far northern parts of Asia, Europe and North America. A tundra consists of plains characterised by snow, ice and frozen soil most of the year. The permanent frozen soil of tundra is called **permafrost**. Winters are very long on the tundra with little daylight. In contrast summers are short but there are many daylight hours. Precipitation is low, amounting to only 25 cm or less per year, because cold air can hold relatively little moisture. The ground is soggy in the summer because moisture cannot soak into the permanently frozen ground. Ponds, small lakes and marshes are abundant due to the nearly flat terrain.

There are no upright trees on the tundra. Only trees such as dwarf willows and birches, which grow low to the ground, can escape the drying effect of the wind which upright trees would experience. This biome consists mainly of mosses, grasses, sedges, lichens and some shrubs. Seasonal thawing of the frozen soil occurs only upto a few centimetres depth, which permits the growth of shallow rooted plants.

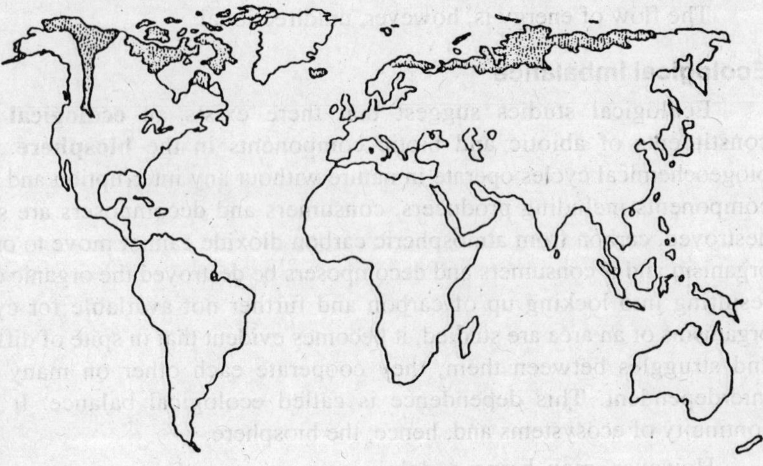

Fig. 12.20. Tundra biome.

Caribon, arctic hare and musk ox are important herbivores of tundra biome. Some important carnivores that prey on the herbivores are the arctic fox, arctic wolf, bobcat and snowy owl. Polar bears live along coastal areas, and prey on seals. Because of the severe winters, many of the animals are migratory and move from one region to another with the change in seasons. Many shorebirds and water fowls, such as ducks and geese, nest on the tundra during the summer but migrate south for the winter. The tundra make a very delicate ecosystem, and may be recovered from any disturbance very slowly.

Northern conifer forest or taiga. The northern coniferous forest or **taiga** is a 1300-1450 km wide band south of the tundra. This extends as an east-west band across North America, Europe and Asia. This area also has long, cold winters, but summer temperatures may reach 10-12°C, and the summer and the growing season are longer than in the tundra. Precipitation is higher than in the tundra, ranging from 10 to 35 cm annually. The moisture is the combined result of summer rains and winter snows. Lakes, ponds and bogs are abundant. The duration of growing period of plants is only about 150 days. Since five physical conditions are variable, the organisms are resistant to fluctuations of temperature.

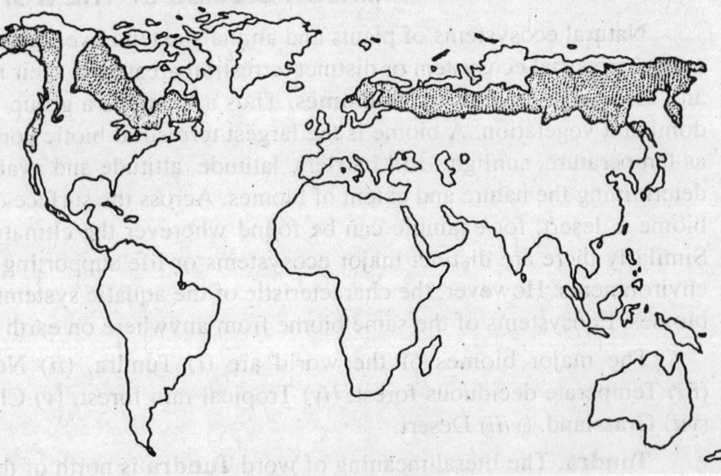

Fig. 12.21. Taiga biome.

The taiga makes really a northern forest of coniferous trees such as spruce, fir, pine, cedar and hemlock. In disturbed areas, deciduous trees such as birch, willow and poplar are abundant. In certain areas the trees are so dense that little light may reach the floor of the forest. Vines, maple and spring wild flowers are common. Mosses and ferns also grow in moist areas.

The common smaller mammals are herbivores, such as squirrels, snowshoe hare, and predatory martens. Important migratory herbivores include moose, elk, deer and caribon. Moose and caribon migrate to the taiga for winters and to the tundra for summers. Important predators are the timber wolf, grizzly bear, black bear, bobcat and wolverine. Many insects are found during the warmer months. Migratory shore birds and waterfowls are abundant during summer months.

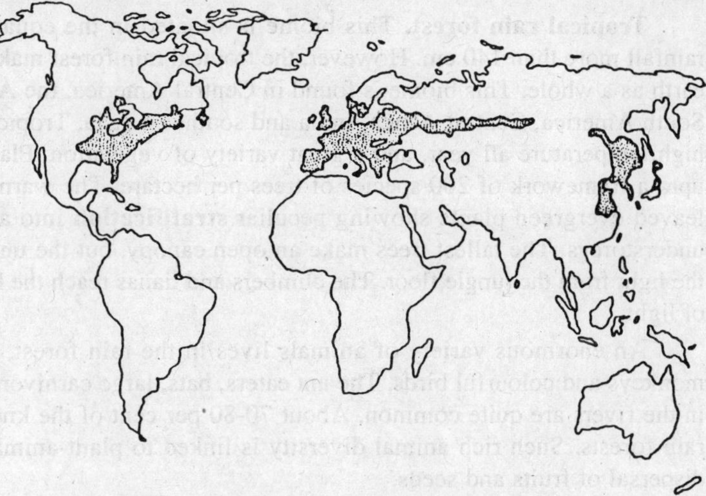

Fig. 18.22. Temperate deciduous forest biome.

Temperate deciduous forests. The deciduous forests are found in the temperate regions of north central Europe, east Asia and the eastern United States, that is, south of the taiga in the Northern Hemisphere. Such forests occur in regions having hot summers, cold winter, rich soil and abundant rain. Annual rainfall is typically around 100 cm per year.

Common deciduous trees are the hardwoods such as beech, maple, oak, hickory and walnut. They are broad-leaved trees. The trees shed their leaves in the late fall so the biome has an entirely different appearance in the winter than in the summer. The fallen leaves provide food for a large variety of consumer and decomposer populations, such as millipedes, snails and fungi living in or on the soil. The temperate deciduous forest produces flowers, fruits and seeds of many types which provide a variety of food for animals.

The common herbivores of this biome are deer, chipmunks, squirrels, rabbits and beavers. Tree-dwelling birds are abundant in number and diversity. Important predators are—black bears, bobcats, and foxes. Predatory birds are also found, such as hawks, owls and eagles. The cold-blooded or ectothermic animals, such as snakes, lizards, frogs, and salamanders are also common.

The temperate deciduous forest makes a very complex biome. Many changes take place during the year, and a large variety of species inhabit the soil, trees and air.

Fig. 12.23. Stratification in a tropical rain forest strata : 1. Tallest trees; II. dense canopy of trees ; III. short trees ; IV. shrubs ; V. herbs and decaying.

Tropical rain forest. This biome is situated in the equatorial regions having the annual rainfall more than 140 cm. However, the tropical rain forest makes an important biome across the earth as a whole. This biome is found in Central America, the Amazon Basin, Orinocon Basin of South America, Central Africa, India and southeast Asia. Tropical rain forests have high rainfall, high temperature all year, and a great variety of vegetation. Plant life is highly diverse reaching upto a framework of 200 species of trees per hectare. The warm, humid climate supports broad-leaved evergreen plants showing peculiar **stratification** into an upper storey and two or three understoreys. The tallest trees make an open canopy, but the understoreyed plants block most of the light from the jungle floor. The climbers and lianas reach the highest level of the trees in search of light.

An enormous variety of animals lives in the rain forest, such as insects, lizards, snakes, monkeys and colourful birds. The ant eaters, bats, large carnivorous animals, and a variety of fish in the rivers are quite common. About 70-80 per cent of the known insects are found in tropical rain forests. Such rich animal diversity is linked to plant-animal interaction for pollination and dispersal of fruits and seeds.

Chapparal. This biome is also known as mediterranian scrub forest. This is marked by limited winter rain followed by drought in the rest of the year. The temperature is moderate under the influence of cool, moist air of the oceans. The biome extends along the mediterranian, Pacific coast of North America, Chile, South Africa and South Australia. This biome has broad-leaved evergreen vegetation. The vegetation is generally made up of fire resistant resinous plants and drought-adapted animals. Bush fires are very common in this biome.

Tropical savannah. The savannahs are warm climate plants characterised by coarse grass and scattered trees on the margins of tropics having seasonal rainfall. Primarily they are situated in South America, Africa and Australia. However, there is no savannah **vegetation** in India. The average total rainfall in such regions is 100 to 150 cm. There is alternation of wet and dry seasons. Plants and animals are drought tolerant and do not have much diversity. The animal life of tropical savannah biome consists of hoofed herbivorous species, such as giraffe, zebra, elephant, rhinoceros and several kinds of antelope. Kangaroos are found in the savannahs of Australia.

Grassland. Some grasslands occur in temperate areas of the earth and some occur in tropical regions. Temperate grasslands usually possess deep, rich soil. They have hot summers cold winters and irregular rainfall. Often they are characterised by high winds. The main grasslands include the **prairies** of Canada and U.S.A., the **pampas** of South America, the **steppes** of Europe and Asia, and the **veldts** of Africa.

The dominant plant species comprise short and tall grasses. In tall-grasses prairies in the United States, important grasses are tall bluestem, Indian grass and slough grass. Short-grass prairies generally have blue grama grass, mesquite grass and bluegrass. Many grasses have long, well-developed root systems which enable

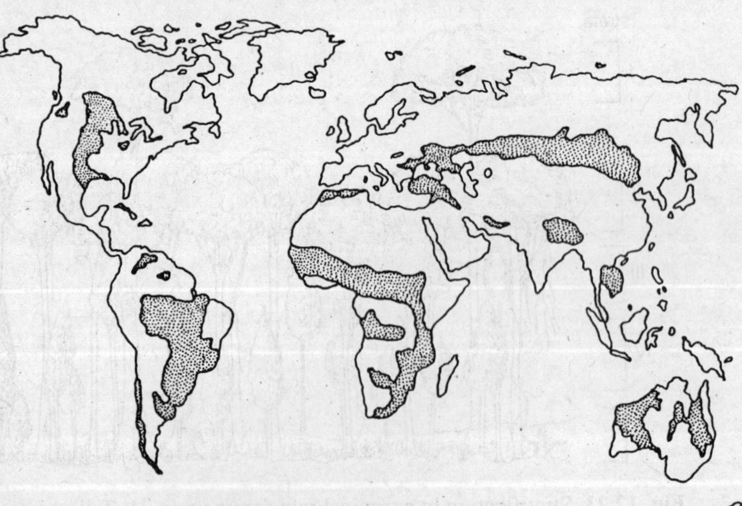

Fig. 12.24. Grassland biome.

them to survive limited rainfall and the effects of fire. The main animals of this biome are—the prong-horned antelopes, bison, wild horse, jack rabbit, ground squirrel and prairie dogs. Larks, the burrowing owl and badgers are also found. Important grassland predators include coyotes, foxes, hawks and snakes.

Desert. The desert biome is characterised by its very low rainfall, which is usually 25 cm per year or less. Most of this limited moisture comes as short, hard showers. Primarily the deserts of the world are located in the south-west U.S.A., Mexico, Chile, Peru, North Africa (Sahara desert), Asia (Tibet, Gobi, Thar) and central western Australia. Deserts generally have hot days and cold nights, and they often have high winds. The reason for the difference of temperature between day and night is due to the lack of water vapour in the air. Deserts are characterised by scanty flora and fauna. Desert organisms must meet some initial requirements if they are to survive. The plants must be able to obtain and conserve water. In order to meet these requirements, many adaptations have been made by desert plants. Such adaptations are—reduced leaf surface area, which reduces evaporation from the plants, loss of leaves during long dry spell; small hairs on the leaf surfaces, and the ability to store large amount of water. The example of important desert plants are—yuccas, acacias, euphorbias, cacti, many other succulents and hardy grasses. Many of the small plants are annuals.

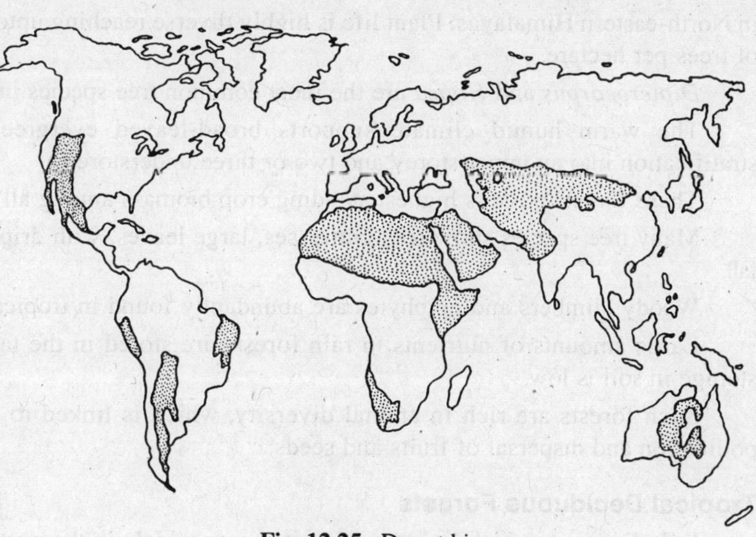

Fig. 12.25. Desert biome.

Animals also must meet the requirements of heat, cold and limited water. Many desert animals are nocturnal in habit, and are active mainly at night. Many reptiles and small mammals burrow to get away from the intense heat of midday. The other common desert animals are the herbivorous kangaroo, rat, ground squirrel, and jack rabbit. The important predators are—coyotes, badgers, kit fox, eagles, hawks, falcons and owls. Ants, locusts, wasps, scorpions, spiders, insect-eating birds, such as swifts and swallows, seed-eating quails, doves and various cats are other common desert animals.

Altitudinal Biomes

The biomes can also be noted along the slopes of high mountain ranges, such as Himalayas (Asia), Andes (South America) and the Rockies. In such mountain ranges the gradual change in the biome types follows the altitude rather than latitude. The biomes extending thousand of kilometres in horizontal or latitudinal sequence may be seen only in few thousand metres in altitudinal sequence. Temperature and precipitation, *i.e.*, rain fall and snowfall have much effect on these biomes. In the tropical mountains the sequence of biomes from base to snowline is as follows—tropical forest (known as Terai forest in India), deciduous forest, coniferous forest and tundra. In the higher latitude, where the mountain is situated, its community and other conditions will be more of the northern type. Thus the number of habitats that cover its slopes is much reduced.

Major biomes of India may be divided in categories mentioned below:

FOREST BIOMES OF INDIA

Major forest biomes in India are commonly divided in four categories; they are as follows :

(*i*) Tropical rain forest biome,

(*ii*) Tropical deciduous forest biome,

(*iii*) Temperate broad-leaf forest biome and

(*iv*) Temperate needle-leaf or coniferous forest biome.

Tropical Rain Forests

The tropical rain forest makes an important biome across the earth as a whole. Tropical rain forests have high rainfall, high temperature all the year, and a great variety of vegetation.

In India, tropical rain forests are discontinuously distributed mainly along Western Ghats and in North-eastern Himalayas. Plant life is highly diverse reaching upto a framework of 200 species of trees per hectare.

Dipterocarpus and *Hopea* are the most common tree species in Indian rain forests.

The warm humid climate supports broad-leaved evergreen plants showing peculiar stratification into an upper storey and two or three understoreys.

These forests possess highest standing crop biomass among all biomes.

Many tree species show root buttresses, large leaves, with drip tips and round the year leaf fall.

Woody climbers and epiphytes are abundantly found in tropical rain forests.

Large amounts of nutrients in rain forests are stored in the tall vegetation, while nutrient storage in soil is low.

Such forests are rich in animal diversity, which is linked to plant-animal interaction for pollination and dispersal of fruits and seeds.

Tropical Deciduous Forests

In India, the tropical deciduous forests occur widely in the northern and southern parts both in hilly areas and plains. Towards the north-west zone, they make scrub thorn forests. Sal (*Shorea robusta*) and teak (*Tectona grandis*) trees are the dominant tree species in tropical deciduous forest of India.

Other common species in deciduous forests are tendu (*Diospyros melanoxylon*), chiraunji (*Buchanania lanzan*) and khair (*Acacia catechu*), which are of great economic importance.

The trees of deciduous forests are generally short statured, only 10 to 20 metres high.

During rainy season, such forests are lush green, while the leaves of most trees fall before advent of summer and the whole forest becomes leafless along with dried herbs. The forests are rich in nutrients due to scarce leaching.

Temperate Broad-leaf Forests

In India, temperate broad-leaf forests mainly occur between 1500 and 2400 m altitude in the western Himalayas, *i.e.*, Himachal Pradesh and Jammu and Kashmir.

Several species of oak (*Quercus*) are dominating in the temperate broad-leaf forests. The main species of oak found in western Himalayas are: banj oak (*Q. leucotrichophora*), khrsn oak (*Q. semecarpifolia*), tilonaj oak (*Q. floribunda*) and rianj oak (*Q. lanuginose*). Though species of oak in western Himalayas are evergreen, yet their leaves drop during summer, but the trees never become naked.

The trees of oak of different species are 25 to 30 m in height. The trees are thickly populated, and tree canopy is dense. The epiphytes are common on oak trees.

Temperate Coniferous Forests

They are also known as needle-leaf forests. In the Himalayas, the temperate coniferous or needle-leaf forests are generally distributed over 1700 to 3000 m altitude, a.s.l. Such forests mainly consist of gymnosperms, *e.g.*, blue pine (Pinus wallichiana), deodar (*Cedrus deodera*), spruce (*Picea smithiana*) and silver fir (*Abies pindrow*). The wood of these trees is of great economic value, and they yield resins, etc. The trees are quite tall (30 – 35 m) and possess evergreen needle-like leaves. The canopy is generally cone-shaped.

AQUATIC ECOSYSTEMS

Aquatic Ecosystems

Many kinds of aquatic ecosystems are found. They differ widely with regard to abiotic factors. The biotic communities living in such ecosystems also vary. The aquatic ecosystems range from open ocean to temporary puddles, ponds and lakes. These possess the conditions of salinity, depth and fluctuations of temperature. The depth of the water is important for vegetation, since more light penetrates near the surface of the water than farther down in deep water. The fresh water and marine environments have many ecosystems with overlapping boundaries which cannot be clearly demarcated.

The adaptations of aquatic organisms have great diversity. For instance, fishes live exclusively in water. On the other hand, the frogs, crocodiles, hippopotamus and aquatic birds are amphibious in nature. The animals like coelenterates and echinoderms live only in saline water, while hilsa fish can live in fresh water as well as in sea.

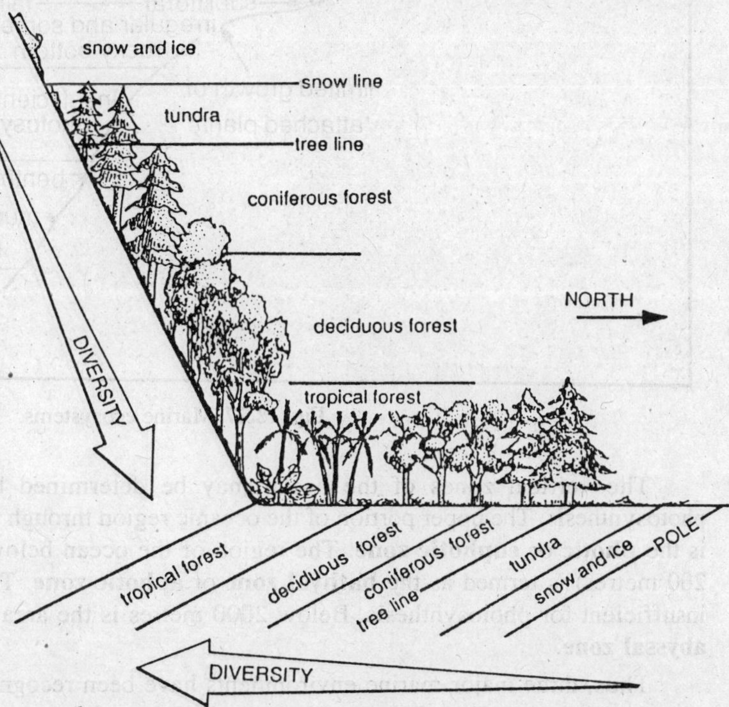

Fig. 12.26. Altitudinal and latitudinal biomes showing correlation.

Marine Environment

Marine biomes are the biomes of the oceans. The marine environment is characterised by its high concentration of salt and mineral ions. The salt concentration in the open sea is about 3.5 per cent. The most dominant ions are sodium and chloride followed by sulphur, magnesium and calcium. The sea is not a single environment. It is subdivided into distinct ecosystems both in abiotic and biotic composition. More important determinants of marine ecosystems are—depth of the water, currents, tide, light penetration and temperature. Ocean coasts are fascinating areas in which the communities are made up of populations which at high tide are under water but at low tide actually on land.

The sea is divided into regions or zones and each region or zone is characterised by different physical conditions and by different plant and animal inhabitants. From the coastline of the sea a gradually sloping region extends to an average of 160 km in the sea. This zone of depth of 8-200 metres is known as the **continental shelf**. Beyond this zone there is **continental slope**, extending from the outer edge of the shelf into deeper water. The continental slopes do possess the ridges, basins and trenches formed of mud and sand, deposited by slow-flowing rivers. The sea is divided into two major regions or zones, *i.e.*, **neritic** and **oceanic**. The neritic zone, adjacent to the shore, is a shallow-water zone which lies above the continental shelf. The oceanic region is the deep water region beyond the neritic zone and beyond the continental shelf.

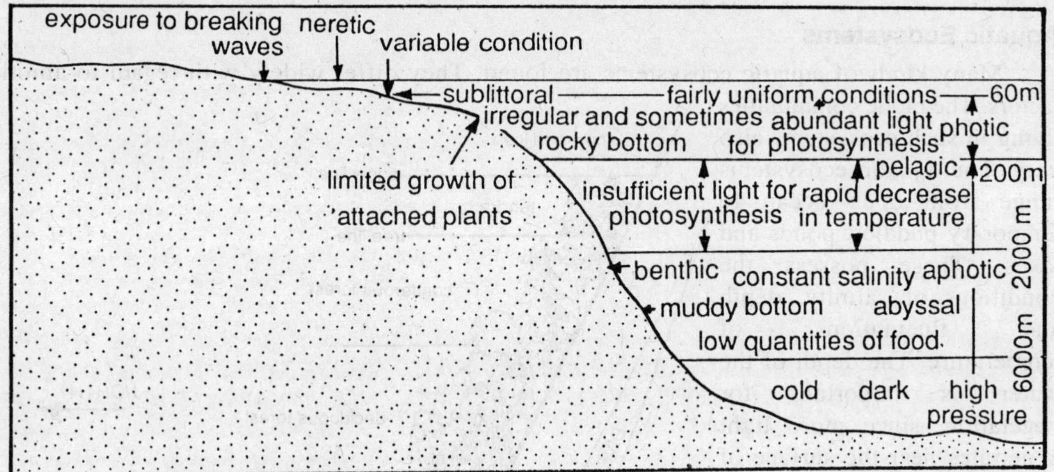

Fig. 12.27. Marine ecosystems.

The vertical zones of the ocean may be determined by the availability of light for photosynthesis. The upper portion of the oceanic region through which light effectively penetrates is the **photic** or **euphotic zone**. The region of the ocean below the euphotic zone (*i.e.*, below 200 metres) is termed as the **bathyal zone** or **aphotic zone**. This zone gets less light which is insufficient for photosynthesis. Below 2000 metres is the area of constant darkness, called the **abyssal zone**.

Thus, three major marine environments have been recognised (*i*) **the littoral zone**, which consists of the sea floor from the shore to the edge of the continental shelf; (*ii*) **the benthonic zone**, comprising of the sea floor along with the continental slope and the aphotic and abyssal zones and (*iii*) **the pelagic zone**, which constitutes the water of the ocean basin.

Life in the sea is abundant and diverse and extends to all its depths. The total mass of marine organisms is greater than the combined masses of terrestrial fresh water organisms.

Marine organisms can be classified as **plankton**, **nekton**, **benthos** and **neuston**.

1. Plankton. The plankton consists of floating organisms. Most of these minute organisms are confined to the euphotic zone. The plankton includes photosynthesizing organisms including diatoms are termed **phytoplankton**, and heterotrophic organisms are **zooplankton**. Zooplankton includes many of the protozoans, especially the dinoflagellates, radiolarians, foraminifera and a large number of small crustaceans.

2. Nekton. They comprise of actively moving organisms with well-developed locomotory organs. They include actively swimming organisms found in the surface or **deep** waters. They feed on planktons or smaller nektons.

3. Benthos. The benthonic organisms are found along the floor of sea-bed. They include creeping, crawling and sessile organisms. Most of these organisms are scavangers.

The open sea. The open sea has **photic**, **aphotic** and **abyssal** regions. The photic region contains largest number of organisms including phytoplanktons and zooplanktons. The nekton comprises of large and small fishes, squids and whales. The nektons are met within the both photic and aphotic zones.

The producers are not found in abyssal zones, and they are inhabited by predators. Some of these predators have stout jaws and sharp teeth (shark fish). Some animals are luminescent and attract other organisms. The examples of such animals are devil fish and hatchet fish.

The benthonic fauna consists of less producers, and more consumers, decomposers and scavangers. Among the benthic animals are star fish, snails, clams, crabs, worms, brancles, sponges, sea anemones, corals, oysters, tunicates and others.

Coastal region. The continental shelf adjacent to coastal region can be divided into **littoral** (substratum), **neretic** (water) and **intertidal zones**. Since there is no wave action in neretic zone, and therefore, it harbours a large variety of planktons and other organisms. The organisms found in the littoral zones face the strong wave action, and therefore they adhere strongly to the substratum, such as many large and small sea weeds. The organisms found in the intertidal zones face rhythmic alteration between aquatic and sub-terrestrial conditions. In the estuaries the physical conditions are different as the fresh water from the rivers mixes with the sea water. The estuaries make the most fertile ecosystems of the world.

Fresh Water Environment

Such biomes occur in ponds, lakes, streams and rivers.

Lakes and ponds. The lakes and ponds have stagnant fresh water. They occur practically in every biome. They vary in size from less than a hectare to hundreds of hectares. The life span of ponds ranges from a few weeks or months in the case of small seasonal ponds to several hundred years for large lakes. Lakes also differ in their origin, and may be of direct or indirect result of **glaciation**. Certain ponds or lakes are natural or man-made. The depth of ponds and lakes varies from a few centimetres to over a hundred metres. The shallow lakes, also known as **eutrophic lakes** have a rich accumulation of organic products. The deep lakes, having steep and rocky side are called **oligotrophic lakes**. Such lakes are poor in circulating nutrients, such as phosphates. Some lakes contain saline or brackish water, such as Sambar lake of Rajasthan.

The **autotrophic** and **heterotrophic** components are found in perfect harmony in a pond or lake. The fresh water pond autotrophs include microscopic phytoplankton such as *Chlamydomonas*, *Volvox*, *Euglena* and other such organisms. Other free floating plants are found on the surface of water such as *Azolla*, *Lemna*, *Wolffia* and *Pistia*. Macrophytes like *Ceratophyllum*, and *Utricularia* are free-floating plants found beneath the surface of water. *Vallisneria*, *Nelumbo*, and *Typha* are the common rooted plants.

The heterotrophs or first order consumers include ciliates, flagellates, rotifers and small crustaceans, such as *Daphnia*, *Cypris*, *Cyclops* and eggs and larvae of various aquatic animals. Other common consumers are leeches, molluscs, fishes, frogs and larger carnivorous fishes.

Streams and rivers. The flowing fresh water makes the streams, rivers and brooks. Such bodies of flowing fresh water differ widely in their volume of water, speed of flow, oxygen content, temperature and other chemical and physical conditions. The flora and fauna depend on the source of river and the environment of land through which the stream or river passes.

The rivers which originate from melting glaciers or springs are penennial and have cool water with swift currents. Floating populations of plankton are generally not found in such waters. When the water reaches in the plains, the current becomes slower and the river-body wider. Sediments deposit in the river-bed. The total volume of water is also increased. Since the water surface remains exposed to sun light the temperature of water rises up. The rate of photosynthesis and other biological activities also become higher. In the lower reaches, the water current diminishes further. The water becomes muddy, and the light becomes poor at the river bed. Both phytoplankton and zooplankton develop in abundance. The nature and composition of fauna becomes like that of lakes. Many mammals, birds and reptiles obtain their food from the water of river or stream.

MAN-MADE ECOSYSTEMS — AGROECOSYSTEM

The man-made ecosystems are artifical and also known as **agroecosystem**. Such ecosystems are variable in their duration and stability. The terrestrial ecosystems developed by man may be villages, cities, parks, plantations, orchards, gardens and crop fields. The man-made aquatic ecosystems are dams, reservoirs, lakes, ponds, canals, fishery tanks and aquaria.

The important man-made modifications in the biotic community are brought by cultivation of plants, domestication of animals and use of fire. The human civilisation has been established and supported by agriculture. Most of the human population depends upon agriculture for their food. The agriculture is the process of cultivating soil for raising the economically important plants. In broad sense, the term agriculture is used for both animal husbandry and crop production. Man has converted very large areas of forests and grasslands by cutting them into croplands for growing cereals, pulses, oilseeds and fodder crops. Thus, the cropland and pastoral land are essentially grasslands with economic plants of choice.

The **agroecology** is the science of the study of the relation between agricultural crops and environment. The biotic communities and the physical conditions of the different croplands vary widely. These croplands possess some common features and are called **agroecosystems**. The physical condition of the cropland depends upon latitude, climatic conditions and edaphic factors. The physical deficiencies of cropland are removed by proper irrigation and use of fertilisers, biofertilisers and manures. The climate, season and choice of farmer, are the main determinants of the type of crop.

Generally one crop **(monoculture)** is grown in a field and there is a strong selective pressure to reduce diversity. However, some weeds grow in the crop field, but their number is quite less as compared to a natural grassland. The animal population of cultivable land consists of earthworms, nematodes, insects, domestic animals, pests, pollinators, birds and rodents.

The man-made ecosystems are simple and efficient because of less diversity of organisms. They do not have the diversity of natural ecosystems. A diversified ecosystem (natural ecosystem) is more stable because some species of organisms would survive even if they are affected by certain pests and pathogens. A monoculture would be destroyed completely by drought, floods, pests or pathogens, whereas a natural diversified ecosystem has capacity for many adjustments.

BOUNDARIES OF ECOSYSTEMS

An ecosystem is considered as an area where the inputs and outputs can be studied across its boundaries, and it is treated as a separate entity. The ecosystems are not strictly separate and isolated. Their boundaries are indistinct and overlapping. For example, a pond may be a separate entity from forest or grassland, but birds of their regions can feed on fishes and crabs of the pond. The movement of energy and materials can be achieved by biological, climatic, and geological processes. The top soil from Himalayas is brought in the deltas of the rivers such as Ganga, Jamuna and Brahmputra. Siberian cranes migrate to Ghana bird sanctuary for overwintering. Thousands of the ecosystems are interrelated with each other, and a single global ecosystem is formed, which is known as **biosphere**.

SUMMARY

1. Ecosystems are self-regulating and self-sustaining units of landscape.
2. The term ecosystem was coined by Sir Arthur Tansley in 1935.
3. Ecosystem is comprised of two basic components: (*i*) abiotic (inorganic materials, air, soil, water, etc.) and (*ii*) biotic (producer, consumer and decomposer organisms).
4. Biotic and abiotic components are physically organised to provide a characteristic structure, such as species composition, stratification, etc., of the ecosystem.
5. Ecosystem structure can also be depicted through food relationship of producers and consumers, which is called trophic structure.
6. Productivity and decomposition are two important functions in ecosystem.
7. Productivity is involved in production of food.
8. While decomposition leads to the breakdown of organic matter.
9. Primary productivity is the rate of capture of solar energy or production of biomass by producers.
10. The rate of total production of organic material or capture of energy is known as gross primary productivity.
11. Net primary productivity represents the balance after meeting the cost of respiration.
12. Secondary productivity is the rate at which food energy is assimilated at the trophic levels of consumers.
13. Decomposition is the process of breakdown of complex organic compounds in detritus to carbon dioxide, water and inorganic nutrients by decomposers.
14. There are three main processes of decomposition, that is, (*i*) fragmentation of detritus, (*ii*) leaching and (*iii*) catabolism, which operate simultaneously on the detritus.
15. Energy flow and nutrient cycling are important functions in the ecosystem.
16. Energy flow is unidirectional, which involves (*i*) the conversion of light energy into food and (*ii*) transfer of energy from one organism to another.
17. The trophic levels are connected by the transfer of food or energy along the food chain.
18. Ecological efficiencies show the efficiency with which the organisms exploit their food and convert the food into biomass make available to the next higher trophic level.
19. Nutrient cycles involve strorage and transfer of nutrients through various components of the ecosystem, and the nutrients are repeatedly used.
20. Nutrient cycles are of two kinds : (*i*) gaseous and (*ii*) sedimentary.
21. The reservoir of gaseous type of nutrient cycle is generally located in the atmosphere or the hydrosphere; while the reservoir of sedimentary type exists in the earth's crust.
22. Ecosystem nutrient cycles possess three aspects: (*i*) input, (*ii*) output and (*iii*) internal cycling of nutrients.
23. Nitrogen, oxygen, carbon dioxide are good examples of gaseous nutrient cycle, while Phosphorus, sulphur are examples of sedimentary types.
24. The biomes are the life forms of the climax vegetation.
25. Forest, grassland, savanna, and desert biomes occupy most of the land area under natural vegetation.
26. Forest biomes are generally dominated by trees having a closed canopy.

27. Grasslands are dominated by herbaceous species.

28. Savanna is characterised by herbs interspersed with trees.

29. Deserts show scarce vegetation. They have low herb or shrub cover with mostly base ground.

TERMS TO REMEMBER

Abiotic components. They include inorganic materials and dead organic matter.

Aboveground detritus or litter fall. Dried plant parts, and dead remains of animals, including faecal matter.

Ammonification. Formation of ammonia from organic nitrogenous compounds.

Associative symbiosis. A type of combination of bacterium and root of higher plant, where nitrogen-fixing microbe does not penetrate deep into host tissues.

Biotic components. They include producers, consumers and decomposers.

Biogeochemical cycle. Nutrient cycle, generally considered in a regional or global context.

Belowground detritus. Dead organic material primarily composed of dead roots.

Biomes. The regional biotic units, recognised by the life form of climax vegetation.

Carnivore. The secondary consumer, an animal that eats flesh of other animals.

Consumers or phagotrophs. The heterotrophic organisms, mostly animals which generally ingest their food.

Decomposers or saprotrophs. The heterotrophic organisms, consisting mostly of bacteria and fungi which live on dead organic matter or detritus.

Detritus. Dead plant parts and animals remains.

Ecosystems. Parts of nature where living organisms interact amongst themselves and with their physical environment.

Food chain. The transfer of energy from one trophic level to the next trophic level.

Food web. Interconnected matrix of food chains is called food web.

Grassland. Such ecosystems have treeless herbaceous plant cover, dominated by a wide variety of grass species (family Poaceae).

Gross primary productivity. The rate of total capture of energy or the rate of total production of organic material (biomass).

Macrophytes. Rooted plants, in shallow waters.

Nitrification. Ammonia is converted into nitrate by a group of chemo-autotrophic bacteria through a two-step process.

Nutrient cycle. They involve storage and transfer of nutrients through various components of the ecosystem so that the nutrients are repeatedly used.

Net primary productivity. The balance energy or biomass remaining after meeting the cost of respiration of producers.

Phytoplankton. In deep aquatic ecosystems, floating plants called phytoplankton are the major autotrophs.

Productivity. The rate of biomass production.

Producers. They are autotrophic, generally chlorophyll-bearing organisms which produce their own food by photosynthesis.

Primary productivity. This refers to the rate at which sunlight is captured by producers for the synthesis of energy-rich organic compounds through photosynthesis.

Savanna. A well developed grass cover, interspersed with scattered shrubs or small trees.

Secondary productivity. At the trophic level of consumers, the rate at which food energy is assimilated.

Standing crop. Trophic structure may be described in terms of the amount of living material, called standing crop, present in different trophic levels at a given time.

QUESTION BANK

Very Short Answer Questions

1. What man made ecosystems are called?
2. Who coined the term 'ecosystem'?
3. Expand PAR.
4. What are biomes?
5. Which are the types of food chains in all ecosystems?
6. Cite three examples of N_2- fixing cyanobacteria.
7. What is detritus ?
8. Several higher plant species other than legumes also produce nodules. Name them.
9. Name two most common trees found in Indian tropical rain forest.
10. Name any two dominant deciduous tree species.
11. What are coral reefs?
12. Define leaching.
13. Give the name for a network formed by interlinking of a number of food chains.
14. Name two components of ecosystem.
15. Give the term for the formation of soil.
16. Give the terms for the plants found on saline soils and sandy soils
17. Name the term used to indicate the soil factors affecting the organisms.
18. Give the term for the grasslands with scattered trees.
19. Name the most productive agro-ecosystem.
20. Name two kinds of cycles of materials.
21. Name the process by which CO_2 enters into living world.
22. Name a denitrifying bacterium.
23. Name greenhouse gas.
24. Give the alternative term for 'biosphere'.
25. Give the term for the amount of living matter in a component population of a specific trophic level at a given time.

ANSWERS

1. Anthropogenic ecosystem; **2.** Sir Arthur Tanslay, 1935; **3.** Photosynthetically active radiation; **4.** Biomes are major ecosystem delimited by particular climates or geographical areas; **5.** Grazing food chain and detritus food chain; **6.** Anabaena, Aulosira and Nostoc; **7.** The dead parts of plants and animal remains; **8.** Casuarina and Alnus, which produce nodules harbouring Frankia, a nitrogen-fixing actinomycete; **9.** Dipterocarpus and Hopea; **10.** Sal and Teak; **11.** Rocks formed in the coastal seas by calcareous secretions of Cnidarian polyps; **12.** Loss of soluble substances from A-horizon of soil as these are carried by downward moving gravitational water; **13.** Food web; **14.** Abiotic component and biotic component; **15.** Pedogenesis; **16.** Halophytes and psammophytes; **17.** Edaphic; **18.** Savannah; **19.** Sugarcane crop; **20.** Gaseous and sedimentary cycles; **21.** Photosynthesis; **22.** Pseudomonas; **23.** Carbon dioxide; **24.** Ecosphere; **25.** Standing crop.

Short Answer Questions

1. What are the processes involved in decomposition of detritus?
2. What are the two basic components of ecosystem? Define them.
3. Define food chain and food web.
4. What do you understand by ecological efficiency?
5. Enlist basic requirements of an ecosystem.

6. Define ecological pyramids. Give their different types.
7. Define biogeochemical cycles. Name two types of them.
8. Differentiated between grazing food chain and detritus food chain.
9. Give the main characteristics of Tundra biome.
10. Differentiate between gaseous and sedimentary types of nutrient cycling.
11. What is ecological efficiency? Give its significance.
12. What is a grassland? How does it differ from savannah?
13. Differentiate between abiotic and biotic components.
14. Briefly describe the processes and products of decomposition.
15. What are greenhouse gases? Name two gases exerting greenhouse effect?
16. Differentiate between food chain and food web.
17. Distinguish between Net primary productivity and Gross primary productivity.
18. What is xerosere? Name its two types.
19. What is greenhouse effect?
20. List the processes involved in nitrogen cycle.
21. Distinguish between wet deposition and dry deposition.
22. With the help of labelled diagram only, give nitrogen cycle.
23. Explain phosphorus cycle by line diagram.
24. Give an account of biotic components of ecosystem.
25. Give an account of factors affecting the rate of decomposition.
26. Give an account of energy flow in ecosystem.
27. What is biosphere? What are the main sub-divisions of biosphere?
28. Explain ten per cent law of flow of energy.
29. Give the major characteristics of the desert biome.
30. Explain the concept of food chain.
31. What is biome? Enumerate the factors that determine the characteristics of a biome.
32. What is primary productivity? Give the range of primary productivity in different ecosystems of the world.
33. Explain three biotic components of any natural ecosystem.
34. Explain the term 'stratification' in relation to a tropical rain forest.
35. Describe the various kinds of food chains in a food web.
36. Give an account of keystone species.
37. Define an ecosystem. Mention some important instances of an ecosystem.
38. Name the two fundamental trophic levels and describe their general characteristics.
39. Describe in brief the biotic components found in an ecosystem.
40. Give a schematic representation of nitrogen cycle of a terrestrial ecosystem. Name any two microorganisms involved in it.

Long Answer Questions

1. Briefly describe the tropical forest biomes.
2. Outline the salient features of the ecosystem nitrogen cycling.
3. What are ecological pyramids? Mention their various types.
4. Give an account of major forest biomes found in India.
5. Explain the pond ecosystem as a classical example of ecosystem.
6. What is meant by nutrient cycling? Explain nitrogen cycle.
7. What are sedimentary cycles? Depict phosphorus cycle diagrammatically.

8. Describe the marine ecosystems.
9. Give the major characteristics of the desert.
10. Define the term 'biome' and list the factors that determine characteristics of a biome.
11. Discuss briefly the abiotic components of an ecosystem.

MULTIPLE CHOICE QUESTIONS (OBJECTIVE TYPE QUESTIONS)

1. Succulent plants grow in :
(*a*) Tundra (*b*) Deserts
(*c*) Tropical rain forests (*d*) Temperate deciduous forests
2. Total amount of living material at the various trophic levels of a food chain is dipicted by pyramids of :
(*a*) Number (*b*) Energy (*c*) Biomass (*d*) All of these
3. In an upright pyramid of biomass the herbivores occupy the position :
(*a*) 1st (*b*) 2nd (*c*) 3rd (*d*) 4th
4. In submerged plants, the stomata lie on :
(*a*) Upper surface (*b*) Lower surface
(*c*) None on any surface (*d*) Equally on both surfaces
5. Pyramid of energy in a pond ecosystem is always :
(*a*) Upright (*b*) Inverted (*c*) Both (*a*) and (*b*) (*d*) None of these
6. Acacia arabica is a :
(*a*) Halophyte (*b*) Mesophyte (*c*) Xerophyte (*d*) Hydrophyte
7. N_2 content is kept constant in the biosphere due to :
(*a*) N_2 fixation (*b*) Industrial pollution (*c*) N_2 cycle (*d*) Absorption of N_2
8. Zone of atmosphere that lies near the ground is :
(*a*) Troposphere (*b*) Stratosphere (*c*) Homosphere (*d*) All of these
9. Greenhouse effect is due to :
(*a*) CFCs
(*b*) O_3
(*c*) Gases that check reflected infra-red radiations to pass through
(*d*) All of these
10. Mangrove of marshy Sunderbans is characterised by :
(*a*) Pneumatophores (*b*) Prop roots (*c*) Vivipary (*d*) All of these
11. The food chain in which micro-organisms breakdown the energy rich compounds, is called :
(*a*) Ecosystem (*b*) Detritus food chain (*c*) Parastic food chain (*d*) Predator food chain
12. The relationship in an ecosystem can be depicted by :
(*a*) Pyramid of energy (*b*) Pyramid of biomass (*c*) Pyramid of numbers (*d*) All of these
13. The biome which is characterised by broad-leaved vegetation, fire resistant resinous plants and drought evading plants, is known as :
(*a*) Savannah (*b*) Chapparal (*c*) Steppes (*d*) Deciduous forests
14. The niche of the population is the :
(*a*) Place where it lives (*b*) Geographical area that it covers
(*c*) Set of conditions that interacts (*d*) Geographical and resources it uses
15. In a lake, the upper warmer oxygen rich water zone is called :
(*a*) Epilimnion (*b*) Hypolimnion (*c*) Limenetic zone (*d*) Profundal zone
16. Which part of the world has a high density of organisms :
(*a*) Grasslands (*b*) Savannahs (*c*) Deciduous forests (*d*) Tropical rain forests

17. An ecological pyramid of biomass represents :
(*a*) Population in each food web (*b*) Tissue organisation at each trophic level
(*c*) Energy flow through each trophic level (*d*) All of these

18. Which of the following is correctly matched :
(*a*) Butterfly .... camouflage (*b*) Garden lizard .... mimicry
(*c*) House lizard .... autotomy (*d*) None of these

19. Organisms found at the bottom of the sea are :
(*a*) Planktons (*b*) Nekton (*c*) Benthos (*d*) Phytons

20. In the submerged hydrophytes exchange of gases occurs through :
(*a*) Hydathodes (*b*) Lenticels (*c*) Stomata (*d*) General surface

21. The plants growing in deserts, to tolerate water stress have :
(*a*) No stomata (*b*) Long roots reaching water table
(*c*) Stem modified into leaf-like form (*d*) Stipular spines

22. Food chain refers to :
(*a*) Number of humans forming a chain for food
(*b*) Animals gathered near a source of food
(*c*) Transfer of energy from producers to consumers
(*d*) None of these.

23. Deciduous forests have :
(*a*) Variety of grasses (*b*) Broad-leaved trees (*c*) Narrow-leaved trees (*d*) Variety of crocodiles

24. Transition zone between vegetational types is :
(*a*) Ecotone (*b*) Ecotype (*c*) Ecocline (*d*) Ecosystem

25. Which of the following can be recognised as a major environment in oceanic basin :
(*a*) Littoral zone (*b*) Benthonic zone (*c*) Pellagic zone (*d*) All of these

26. When the number of organisms at successive trophic levels are plotted, they assume the shape of a pyramid. This is called the pyramid of :
(*a*) Number (*b*) Energy (*c*) Biomass (*d*) None of these

27. How much energy is consumed from one trophic level to another trophic level ?
(*a*) 5% (*b*) 10% (*c*) 15% (*d*) 20%

28. Desert can be converted into green land by planting :
(*a*) Oxylophytes (*b*) Psammophytes (*c*) Halophytes (*d*) Tropical trees

29. Two similar halophytes are called :
(*a*) Isotype (*b*) Neotype (*c*) Synotype (*d*) Mesotype

30. Halophytes are peculiar in having :
(*a*) Pneumatophoric stems (*b*) Velamen tissue
(*c*) Vivipary (*d*) Saline soil

31. Deer in a forest ecosystem acts as a :
(*a*) Primary consumer (*b*) Secondary consumer
(*c*) Decomposer (*d*) None of these

32. Pyramid of energy is always :
(*a*) Inverted (*b*) Upright (*c*) Spindle-shaped (*d*) Cup-shaped

33. In India, tropical rain forests are found in :
(*a*) Darjeeling (*b*) Karnataka (*c*) Andaman (*d*) J&K

34. Halophytes grow on physiologically dry soil due to presaence of :
(*a*) Excess salts in water (*b*) Excess salts in plants
(*c*) Excess humidity outside (*d*) Dry soil

35. The free floating organisms of an open sea and sea shore are collectively called :
(*a*) Planktons (*b*) Nektons (*c*) Benthonic (*d*) Sea anemones

36. In an ecosystem, bacteria are considered as :
(*a*) Microconsumers (*b*) Macroconsumers
(*c*) Primary consumers (*d*) Secondary consumers

37. Field capacity is :
(*a*) Higher for sandy soils than clay soils (*b*) Amount of water in soil available to plants
(*c*) Same as permanent wilting point (*d*) Amount of water left in soil after drainage

38. Soil salinity is measured by :
(*a*) Porometer (*b*) Potometer (*c*) Conductivity meter (*d*) Calorimeter

39. The factors governing the structure of soil are called :
(*a*) Topographic (*b*) Edaphic (*c*) Biotic (*d*) Temperature

40. Concept of ecological pyramids was developed by :
(*a*) Reiter (*b*) Odum (*c*) Darwin (*d*) Elton

41. In an ecosystem, which one of the following is natural controlling force :
(*a*) Soil moisture (*b*) Temperature (*c*) Food (*d*) Predators

42. The maximum energy amongst escosystems is contributed by :
(*a*) Crops (*b*) Forests (*c*) Coal (*d*) Fuel gas

43. The most recently discovered ecosystem is :
(*a*) Vent (*b*) Crater (*c*) Tundra (*d*) Floating iceberg

44. Inverted pyramid of biomass can be traced in one of the following ecosysteme :
(*a*) Rain forest (*b*) Desert (*c*) Ocean (*d*) Tundra

45. The study of energy transfer in different trophic levels of an ecosystem is called :
(*a*) Bioenergetics (*b*) Biogeocoenosis (*c*) Holocoenosis (*d*) Biosystem

46. The 10% law for energy transfer in food chains was given by :
(*a*) Stanley (*b*) Tranley (*c*) Lindemann (*d*) Weismann

47. Frog that feeds on insects is :
(*a*) Primary consumer (*b*) Secondary consumer (*c*) Tertiary consumer (*d*) Decomposer

48. Atmospheric humidity is measured by :
(*a*) Auxanometer (*b*) Hygrometer (*c*) Photometer (*d*) Potometer

49. In a comparative study of grassland ecosystem and pond ecosystem, it may be observed that :
(*a*) Primary and secondary consumers are similar
(*b*) Both biotic and abiotic components are different
(*c*) Biotic components are almost similar
(*d*) Abiotic components are almost similar

50. Phytoplanktons of a pond ecosystem act as :
(*a*) Producers (*b*) Primary consumers
(*b*) Secondary consumers (*d*) Decomposers

51. The lentic ecosystem includes :
(*a*) Gravitational water (*b*) Standing water (*c*) Rain water (*d*) Running water

52. A rat feeding on potato tuber is a :
(*a*) Producer (*b*) Primary consumer (*c*) Carnivore (*d*) Decomposer

53. Which of the following habitats show highest density of living species ?
(*a*) Grassland (*b*) Deserts
(*c*) Tropical rain forests (*d*) Temperate forests

54. This is an inverted pyramid :
(*a*) Pyramid of number in a grassland (*b*) Pyramid of energy in pond system
(*c*) Pyramid of biomass in a grassland (*d*) Pyramid of biomass in pond ecosystem

55. Food chains are met only in the :
(*a*) Sea (*b*) Forests (*c*) Deserts (*d*) All of these

56. The zooplankton forms are :
(*a*) Primary consumers (*b*) Secondary consumers
(*c*) Carnivores (*d*) Primary producers

57. The driving force of the ecosystem is :
(*a*) Carbohydrates in plants (*b*) Biomass
(*c*) Producers (*d*) Solar energy

58. The snakes are included under :
(*a*) Saprophytes (*b*) Primary consumers
(*c*) Secondary consumers (*d*) None of these

59. Transfer of energy from organisms to organisms in a natural community establishes :
(*a*) Food chains (*b*) Biological control (*c*) Natural barriers (*d*) All of these

60. Green plants constitute :
(*a*) 1st trophic level (*b*) 2nd trophic level (*c*) 3rd trophic level (*d*) 4th trophic level

61. In an ecosystem, one of the following occurs as abiotic compounds :
(*a*) Flow of energy (*b*) Cycling of materials (*c*) Consumers (*d*) Both (*a*) and (*b*)

62. Which of the following habitat is most unsuitable for primary productivity ?
(*a*) Meadow (*b*) Forested river bank (*c*) Cave (*d*) Pond

63. Ecosystem has two components :
(*a*) Plants and animals (*b*) Weeds and trees (*c*) Biotic and abiotic (*d*) Frogs and men

64. The pyramid of number in a grassland ecosystem is :
(*a*) Linear (*b*) Upright (*c*) Inverted (*d*) Irregular

65. Biological equilibrium is found between :
(*a*) Producers and consumers (*b*) Producers and decomposers
(*c*) Producers and light (*d*) Producers, consumers and decomposers

66. The number of individuals of a species in a particular ecosystem at a given time remains constant due to :
(*a*) Human beings (*b*) Parasites (*c*) Predators (*d*) Food availability

67. Ecological pyramids are of :
(*a*) Two types (*b*) Three types (*c*) Four types (*d*) Five types

68. A lake ecosystem is :
(*a*) Artificial (*b*) Abiotic (*c*) Natural (*d*) Hydrological

69. Food level in an ecosystem is called :
(*a*) Trophic level (*b*) Consumer level (*c*) Producer level (*d*) Herbivore level

70. Savannahs are :
(*a*) Tropical rain forests (*b*) Dense forests with close canopy
(*c*) Grasslands with scattered trees (*c*) Desert scrub

71. The maximum biomass occurs in :
(*a*) Tropical forests (*b*) Temperate forests (*c*) Taiga (*d*) Alpine vegetation

72. Which is correct :
(*a*) Least productive ecosystem is desert and deep lake
(*b*) Sugarcane is the most productive agroecosystem

(c) Most productive ecosystem is coral reef
(a) All of the above

73. Stratification is more common in :
(a) Deciduous forest (b) Tropical rain forest (c) Temperate forest (d) Tropical savannah

74. Hygrometer is used to measure the :
(a) Temperature (b) Wind velocity (c) Relative humidity (d) Turbidity

75. Which of the following is an abiotic component of the ecosystem ?
(a) Bacteria (b) Humus (c) Plants (d) Fungi

76. What fraction of assimilated energy is used in respiration by the herbivore ?
(a) 20 per cent (b) 30 per cent (c) 40 per cent (d) 60 per cent

77. Which of the following is a free living nitrogen-fixing bacteria present in soil ?
(a) Azotobacter (b) Nitrosomonas (c) Rhizobium (d) Pseudomonas

78. Which of the following represents the sedimentary type of nutrient cycle ?
(a) Nitrogen (b) Carbon (c) Phosphorus (d) Oxygen

79. Which of the following processes helps in nutrient conservation ?
(a) Mineralisation (b) Immobilisation (c) Leaching (d) Nitrification

80. Decomposers are :
(a) Animalia and Monera (b) Protista and Monera
(c) Fungi and Plantae (d) Bacteria and Fungi

81. Transition zone between vegetational types is :
(a) Ecotone (b) Ecotype (c) Ecoline (d) Ecosystem

82. Bacterial role in carbon cycle is :
(a) Chemosynthesis (b) Photosynthesis
(c) Breakdown of organic matter (d) Assimilation

83. The rate at which light energy is changed into chemical energy of organic molecules is :
(a) Net primary productivity (b) Gross primary productivity
(c) Net secondary productivity (d) Gross secondary productivity

84. If there was no CO_2 in the earth's atmosphere, the temperature of earth's surface would be :
(a) Same as present (b) Less than present
(c) Higher than present (d) Depends upon amount of O_2

85. If we completely remove the decomposers from an ecosystem, its functioning will be adversely affected because :
(a) Energy flow will be blocked (b) Herbivores will not receive solar energy
(c) Mineral movement will be blocked (d) The rate of decomposition will be high

86. In a biotic community (ecosystem), the primary consumers are :
(a) Carnivores (b) Omnivores (c) Detritivores (d) Herbivores

87. What is true about photolithotrops ?
(a) Obtain energy from radiations and hydrogen from organic compounds
(b) Obtain energy from radiations and hydrogen from inorganic compounds
(c) Obtain energy from organic compounds
(d) Obtain energy from inorganic compounds

88. The frog that feeds on an insect is a :
(a) Tertiary consumer (b) Decomposer (c) Primary consumer (d) Secondary consumer

89. Relative Biological Effectiveness (RBE) is caused due to :

(*a*) Soil (*b*) Water (*c*) Radiation (*d*) Pollution

90. Pneumatophores are characteristics of :
(*a*) Halophytes (*b*) Oxylophytes (*c*) Fresh water plants (*d*) Psammophytes

91. Forest cover of India according to 'State of Forest Report 1997' is :
(*a*) 11.5% (*b*) 17.55% (*c*) 18.27% (*d*) 19.27%

92. Which of the following communities is more vulnerable to invasion by outside animals and plants ?
(*a*) Mangrove (*b*) Tropical evergreen
(*c*) Temperate forests (*d*) Oceanic island communities

93. In a terrestrial ecosystem (*e.g.*, forest) maximum energy is in which trophic level :
(*a*) T_1 (*b*) T_2 (*c*) T_3 (*d*) T_4

94. Plants such as Prosopis, Acacia and Capparis are found in tropical :
(*a*) Grasslands (*b*) Thorny forests (*c*) Deciduous forests (*d*) Evergreen forests

95. Which of the following ecosystems has the highest gross primary productivity ?
(*a*) Grassland (*b*) Coral reef (*c*) Mangroves (*d*) Tropical rain forest

96. The 'niche' of a species is meant for :
(*a*) Specific space where an organism lives
(*b*) Habitat and specific functions of a species
(*c*) Specific species function and its competitive power
(*d*) None of these

97. Benthonic animals are those, which are :
(*a*) Deep dwellers in sea (*b*) Free floating
(*c*) Submerged in sea (*d*) Float on sea surface

98. In any food chain, the largest population is that of :
(*a*) Primary consumers (*b*) Tertiary consumers (*c*) Producers (*d*) Decomposers

99. Which of the following is the most stable ecosystem ?
(*a*) Forest (*b*) Desert (*c*) Mountain (*d*) Ocean

100. In desert grasslands, which type of animals are relatively more abundant ?
(*a*) Diurnal (*b*) Arboreal (*c*) Aquatic (*d*) Fossorial

ANSWERS

1. (*b*)	**2.** (*c*)	**3.** (*b*)	**4.** (*c*)	**5.** (*a*)	**6.** (*c*)	**7.** (*c*)	**8.** (*a*)	**9.** (*d*)	**10.** (*d*)
11. (*b*)	**12.** (*c*)	**13.** (*b*)	**14.** (*d*)	**15.** (*a*)	**16.** (*d*)	**17.** (*c*)	**18.** (*c*)	**19.** (*c*)	**20.** (*d*)
21. (*c*)	**22.** (*c*)	**23.** (*b*)	**24.** (*a*)	**25.** (*c*)	**26.** (*a*)	**27.** (*b*)	**28.** (*b*)	**29.** (*a*)	**30.** (*c*)
31. (*a*)	**32.** (*b*)	**33.** (*c*)	**34.** (*a*)	**35.** (*a*)	**36.** (*a*)	**37.** (*d*)	**38.** (*c*)	**39.** (*b*)	**40.** (*d*)
41. (*d*)	**42.** (*b*)	**43.** (*b*)	**44.** (*c*)	**45.** (*a*)	**46.** (*c*)	**47.** (*b*)	**48.** (*b*)	**49.** (*d*)	**50.** (*a*)
51. (*b*)	**52.** (*b*)	**53.** (*c*)	**54.** (*d*)	**55.** (*d*)	**56.** (*a*)	**57.** (*d*)	**58.** (*c*)	**59.** (*a*)	**60.** (*a*)
61. (*d*)	**62.** (*c*)	**63.** (*c*)	**64.** (*b*)	**65.** (*d*)	**66.** (*c*)	**67.** (*b*)	**68.** (*c*)	**69.** (*a*)	**70.** (*c*)
71. (*a*)	**72.** (*d*)	**73.** (*b*)	**74.** (*c*)	**75.** (*b*)	**76.** (*b*)	**77.** (*a*)	**78.** (*c*)	**79.** (*a*)	**80.** (*d*)
81. (*a*)	**82.** (*c*)	**83.** (*b*)	**84.** (*b*)	**85.** (*c*)	**86.** (*d*)	**87.** (*b*)	**88.** (*d*)	**89.** (*a*)	**90.** (*a*)
91. (*a*)	**92.** (*a*)	**93.** (*a*)	**94.** (*b*)	**95.** (*d*)	**96.** (*c*)	**97.** (*a*)	**98.** (*d*)	**99.** (*d*)	**100.** (*d*)

13
CHAPTER

Natural Resources and Vegetation Types of India

Man lives in nature and depends on the resources of nature. The progress of human beings depends upon the exploitation of different natural resources. The utilisation of soil, water, coal, electricity, oil, gas and nuclear energy is important for the development of a nation. Any component of the natural environment that can be utilised by man to promote his welfare is considered as a **natural resource**.

Food, shelter and clothing are the primary requirements of the man. Today, man's dependence on the environment is greater than that of other organisms because he is more than a mere biological creature. Modern man has developed an obsession for greater comfort and security. To meet his requirements, he consumes larger amounts of material and energy than any other organism. He has developed a new kind of environment–the **socio-cultural environment**–within the natural environment. Such environment consists of things developed by man through his tools, skills, efforts and social institutions. It depends on and draws sustenance from the natural environment.

Human civilisation has evolved gradually and passed through several distinct stages. The hunting, fishing and food-gathering stages of the primitive man were followed by the agricultural and the industrial stages. With these events the biosphere became transformed into a human-dominated environment or **noosphere** (*noo* = mind; *sphere* = domain).

Due to increase in human population, India is facing an ecological crisis and is depleting her natural resources day by day. Now the shortage of natural resources is a matter of international concern. There is increasing deficiency of energy, metals, coal, non-fuel and non-metallic materials. In the subsequent paragraphs, the natural resources, causes of their degradation and conservation will be discussed.

CLASSIFICATION OF NATURAL RESOURCES

Natural Resources

The word 'resource' means a source of supply or support generally held in reserve.

In other words, natural resources are the components of the atmosphere, hydrosphere and lithosphere, which can be drawn upon for supporting life.

The 'natural resources' include energy, air, water, land, minerals, plants and animals.

For human-beings resources are those materials and sources of energy which are needed for their survival and prosperity.

The nature of resources varies from society to society depending on a number of factors, such as culture, the level of development and the nature of work of that particular society. For example, uranium, gold or silver are of no use for the Onge tribals of Andamans.

Natural resources vary greatly in their location, quantity and quality. For example, a particular type of forest may occur only in certain countries.

Classification

Natural resources can be classified in different ways. A most convenient classification of resources is based upon their exhaustibility and renewability. They are most commonly classified as (*i*) **inexhaustible**, and (*ii*) **exhaustible** on the basis of their abundance and availability.

Inexhaustible Resources

Inexhaustible resources are available in unlimited quantities on the earth.

Some inexhaustible resources remain unaffected by human activities, while others may show changes in their quality, but their quantity is not affected.

Inexhaustible resources like solar energy, wind power, tide power, rainfall and even atomic energy cannot be exhausted significantly at global level due to human activities.

However, they may sometimes be affected at local level by human activities, *e.g.*, pollution may change the quality of air.

Exhaustible Resources

A large number of natural resources are exhaustible, *i.e.*, they have limited supply on the earth and can be exhausted if used indiscriminately. For example, the availability of water in lakes and rivers and water stream energy can be diminished or degraded if used improperly.

Exhaustible resources are classified as (*i*) **renewable**, and (*ii*) **non-renewable**.

Renewable resources. Resources that have the inherent capacity to reappear or replenish themselves by quick recycling, reproduction and replacement within a reasonable time and maintain themselves. The growth and reproduction of such resources can be successfully managed so that these resources are continuously regenerated. Soil, water and living organisms are the main renewable resources. The rate at which their renewal occurs varies.

However, if the consumption of renewable resources continues to exceed their rate of renewal, not only their quality becomes affected, they may even get totally exhausted.

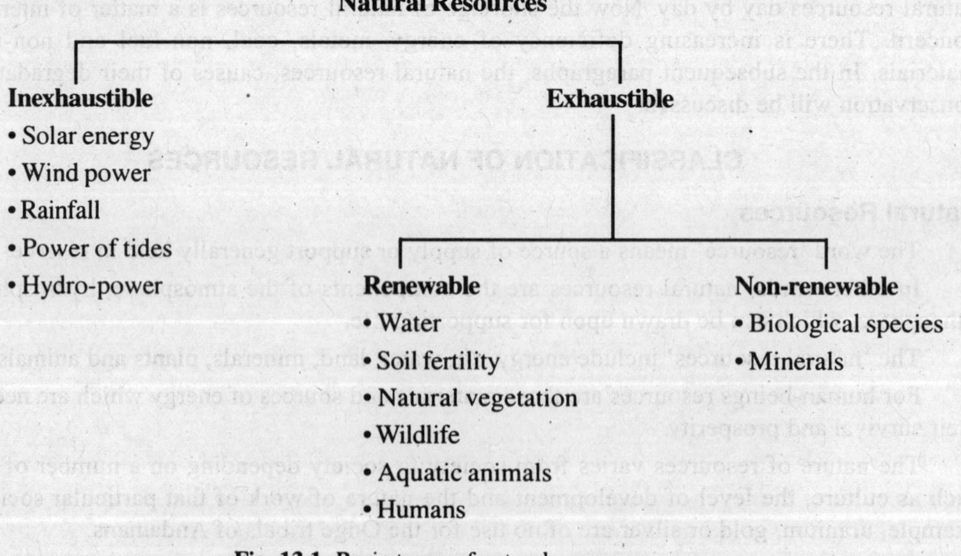

Fig. 13.1. Basic types of natural resources

Some important ecosystems and their renewable resources are :

(*i*) **Forests.** They yield timber and other plant products.

(*ii*) **Rangelands.** They sustain grazing animals for milk, meat and wool production.

(*iii*) **Wild life.** They maintain food chain.

(*iv*) **Agricultural systems.** They yield food, fibre, spices, etc.

(*v*) **Marine and fresh water systems.** They provide various products from plants and animals.

(*vi*) **Soil and water.** They are other renewable resources of great importance.

Non-renewable resources. Resources that lack the ability for recycling and replacement are called non-renewable resources. Substances with a very long recycling time are also regarded as non-renewable resources. For example, fossil fuels, like coal and petroleum, and minerals are non-renewable resources. Biological species, which have evolved in nature during the course of millions of years, are considered non-renewable. Once a biological species becomes extinct, it cannot be recreated again by man.

Resources such as underground water, forests and wildlife are renewable resources but can become non-renewable if used too rapidly by improper management.

Principal Natural Resources, that contribute significantly to human welfare are :

(*i*) Soil, (*ii*) Water, (*iii*) Land, (*iv*) Energy, (*v*) Marine, and (*vi*) Mineral.

SOIL RESOURCES

Land is a major constituent of the lithosphere and the source of many minerals essential to man and other organisms. It forms about one-fifth of the earth's surface, covering about 13, 393 million hectares. About 36.6 per cent of the land area is occupied by human dwellings and factories, roads and railways, deserts and dunes, glaciers, polar ice marshes, rocks and mountains. About 30 per cent of the total land mass is under forests, approximately 22 per cent is occupied by meadows and pastures and only 11 per cent is arable land (*i.e.*, suitable for growing crops).

The surface layer of the land is called soil. About four-fifths of the land area is covered by soil. Soil is composed of inorganic particles, organic matter, air, water and a variety of organisms. It takes hundreds of years for the development of soil horizons having different physico-chemical properties.

The physical properties of soil depend on its texture, structure, bulk, density, porosity and water retentivity. The salt content, pH, and organic and inorganic nutrients like nitrogen, phosphorus and potassium determine its chemical properties. The topography, climate and biotic factors control the conditions of the soil.

Human activities often create worldwide problems. Soil erosion, salinisation, water-logging, and acidification and alkalisation of soil by erroneous human activities are the great dangers to the land resources.

Construction of dams, roads, and railways, urban encroachment and industrialisation and mining activities are responsible for causing depletion of productive land.

Soil Erosion

The word 'erosion' literally means 'to wearing away'. In soil erosion, fertile soil surfaces are detached and removed from their original places and are deposited at some other places. According to Fox (1950), the soil erosion covers a wide range of physical and chemical actions, such as removal of soluble matters, chemical changes, disintegration by frost or by rapid changes of temperature, attrition by dust charged wind, scouring by silt laden currents, alternate impact and succession by storm waves, land slides and so on. Thus, soil erosion is the removal of top soil by movement of water and air.

Human activities accelerate soil erosion by removing natural plant cover. From croplands in India, millions of tons of top soil are eroded into sea each year. Deforestation, undesirable forest biota, and mechanical practices by man are important factors which cause soil erosion. Erosion causes a significant loss of soil fertility by transporting organic matter and nutrients that are essential part of the soil.

The eroded soil, which gets into streams, rivers and lakes in the form of sediments, affects water quality and the habitats of aquatic organisms. There are several serious effects of soil erosion, which are as follows :

1. Due to uprooting of trees, scarcity of timber and fuelwood.
2. Loss of soil stability and fertility.
3. Shortage of fodder.
4. Destruction of cropable land in plains.
5. Formation of sand dunes.
6. Greater frequency of floods and threat to communication channels.
7. Silting of river beds, lakes and dams.
8. Higher temperature and lower rainfall.

However, abundant plant cover significantly reduces soil erosion.

Depletion of Soil Fertility

If the rate of removal or loss of components is greater than the rate of addition, the soil will naturally less fertile. Leaching is an important factor which makes the soil poor in its resources. In this process minerals and organic substances are removed from the top layer of soil by rain water. Biological agencies are also active in causing loss of resources from the soil. Cultivation of crops regularly year after year makes the soil less productive. In cultivation there are little chances for the compensation of lost nutrients of the soil by way of death and decay of the vegetation which grows on it.

On the other hand, when natural vegetation is removed to develop agricultural systems, as in many parts of India, not only the nutrients stored in vegetation are removed, the organic matter and nutrients accumulated in the soil are also lost. From agricultural systems, nutrients are exported through crop harvest, thus depleting soil fertility.

Leguminous plants, however, compensate the loss of nitrogenous compounds because bacteria (rhizobia, etc.) inhabiting in their root nodules fix considerable amount of free atmospheric nitrogen into its compounds, such as ammonia, nitrites, nitrates, etc.

Soil Conservation

The main aims of soil conservation are : (*i*) to protect the soil from erosion, and (*ii*) to maintain the productive capacity of the soil.

Both engineering and biological methods have been used to check the soil erosion but it is still without a plausible check. In India, we are aware of advancing deserts of Rajasthan, and erosional losses, floods, etc., in other parts of the country. The problem has received the attention of forest ecologists, soil scientists and engineers these days. With the result several soil and crop management practices have been evolved which can minimise erosion and reduce nutrient depletion of agricultural soils. Such practices include (*i*) conservation tillage, (*ii*) organic farming, (*iii*) crop rotation (especially cereals with legumes), (*iv*) contour ploughing, and (*v*) strip-cropping terraces, etc.

(*i*) **Tillage conservation.** Many researches support the view that in dry areas, shallow ploughing gives comparatively good crop yields. Shallow ploughing removes the weeds and enables the soil to absorb water. Deep ploughing often leads to soil erosion but in the areas where rainfall is sufficiently high, deep ploughing (upto 15 to 30 cm deep) is effective in removing weeds and increasing crop yields.

In contrast to conventional tillage, conservational tillage incorporates residues from previous crops into the soil, thus increasing the organic matter which in turn improves soil moisture and nutrients. There are two kinds of conservation tillage : (i) reduced tillage, and (ii) no-tillage.

The efforts to improve erosion-affected soils are summarised in two steps. They are :

(*i*) stabilising the soil to prevent further erosion, and

(*ii*) restoration of soil fertility.

For soil stabilization, seeding of base ground with plants that can survive adverse conditions is required. For this purpose usually drought resistant grasses are grown. Such plants eventually establish vegetation cover on the soil, preventing further erosion.

(*ii*) **Organic farming.** With increasing addition of detritus, the soil organic matter, nutrient and moisture levels improve. Application of biofertilizers is also useful for enhancement of soil fertility. Various organic farming measures that provide increased organic input to soil have long-term beneficial effects on soil fertility.

Humus is formed by the accumulation of partially decayed and partially synthesized organic materials. Humus helps in making the soil granular. It causes air spaces to be formed in the clay soil and increases the water holding capacity of sandy soils. Humus is rich in nutrients and therefore, enhances plant growth. Being black humus absorbs heat and warms up the soil.

WATER RESOURCES

Water is the major constituent of the hydrosphere and covers four-fifths of the earth's surface. In fact, water is all around us. Every cubic millimetre of air, even over dry deserts, has water vapour. Water is present in the soil and also hidden in underground pools.

The total volume of water in the hydrosphere is 1.4 billion cubic kilometres (km^3), about 97.5 per cent of the earth's water is found in oceans, which is strongly saline. The rest 2.5 per cent is fresh water, and all of this is not available for direct human use. Most of the fresh water is frozen as polar or glacial ice (*i.e.*, 1.97 per cent). Remaining fresh water occurs as ground water (*i.e.*, 0.5 per cent), and water in lakes and rivers (0.02 per cent), soil (0.01 per cent) and atmosphere (0.001 per cent). Thus, only a small fraction of fresh water is available for human consumption.

About 84 per cent of the total global evaporation occurs from ocean surface and 16 per cent from land surface.

At any given time the amount of moisture in the air is only enough to meet a total rainfall requirement of ten days. Thus, there is a fast movement of water from ocean and land into the atmosphere, and an average stay time of water in the air is only about ten days.

About 77 per cent of the total rainfall on earth is received on the sea surface (as against 84 per cent evaporation from ocean surface) and 23 per cent on land (as against 16 per cent from land surface). Thus, there is net gain of 7 per cent rainfall water on land, and this excess is returned to the oceans by surface run off through rivers and sub-surface water flows.

However, on global basis, the hydrological cycle is perfectly balanced as the total annual evaporation matches with annual rainfall.

Global water is unequally distributed. Precipitation is seasonal and, therefore, the amount of water in inland bodies is variable. Irregularity in the duration and intensity of rainfall often causes floods or droughts. Scarcity of fresh water results in serious regional disparities. Arid regions suffer perennially from water shortage.

Table 13.1. Global Distribution of Fresh Water

	Water : Cubic km (km^3)
1. Water in snow caps, ice sheets, glaciers, etc.	24, 000, 000
2. Surface ponds, lakes and reservoirs	280, 000
3. Water in streams and rivers	1, 200
4. Water present as soil moisture	85, 000
5. Ground water	60, 000, 000
Total amount of fresh water on our planet	84, 366, 200

The global distribution of fresh water on our planet is only about 84.4 million cubic km. The global distribution of fresh water on earth's crust including ground water and water present as vapours in the atmosphere is given in table 13.1.

Table 13.2. Annual Water Budget of our Planet

	Water in cubic km (km^3)
1. Evaporation from sea surface	452, 600
2. Evaporation from land surface	72, 500
3. Precipitation on ocean surface	411, 600
4. Precipitation on land surface	113, 500
5. Surface and ground water run-off	41, 000
Total evaporation from land and sea surface	525, 100
Total precipitation on land and sea surface	525, 100

The excess water received by land surface, about 41, 000 cubic km has to flow back to sea, it cannot be retained on earth's surface ordinarily.

Water Use

On global basis, the water use has increased 4 to 8 per cent per year since 1950, and the consumption rate varies among different countries.

Agriculture sector is the biggest consumer of freshwater. Worldwide, about 70 per cent of total water use is accounted by agriculture.

Following agriculture, power generation (6.2 per cent), and industries (5.7 per cent) are the biggest consumers of fresh water.

Domestic requirement and livestock management taken together consume only 4.3 per cent of the total water drawn.

However, only about 1.1 per cent of fresh water is used for domestic and municipal supplies. The rest of water is consumed by various industries, such as cement, mining, pharmaceutical, detergent and leather industry, etc.

Fisheries, hydro-electric power generation, recreational activities, etc., also require a huge quantity of water, much of which flows down to the sea.

Problems Related with Water Resources

Today in almost all spheres of human activity more water is drawn than what is actually needed. Due to carelessness much of it is wasted and allowed to flow out in an impure state. At many places enough clean water is no longer available. The rapid rise in demand for fresh water is naturally a result of an equally rapid growth in the number of consumers. But there are many other causes as well which have contributed significantly to the wastage and degeneration of fresh water.

For example, about 40 per cent of the world's population lives in arid or semi-arid regions. These people spend lot of energy, time and effort in obtaining water for their domestic and agricultural uses.

To meet the needs of huge population, surface waters, *i.e.*, ponds, lakes, rivers, streams, etc., are overdrawn. Due to over use of surface water, the nearby wetlands may dry up.

When more ground water is removed for human use than can be recycled by rainfall or snow-melt, the ground water may also dry out.

However, everywhere we tend to use more water than is actually needed. A little care can result in substantial savings and longer life for our resource base or else, the water thus saved can be diverted to regions where there is shortage.

Excessive irrigation in arid and semi-arid regions can cause salt accumulation in the soil, due to which crop productivity will decline.

On the other hand, the continuous depletion of ground water along the coastal regions often leads to the movement of saline sea water into fresh water wells, spoiling the quality of water. Estuaries become more saline and consequently less productive when surface waters are over drawn.

Addition of wastes and sewage causes the water body to become exceedingly rich in plant nutrients. Blooms of algae and other organisms appear and make the water useless for human purposes. The entire biomass may suddenly die and start decomposing causing several other problems. Organically rich waters also support a population of many pathogenic organisms and vectors which transfer diseases from one individual to another.

Heavy rainfall results in rapid run-off from areas having exposed soil, particularly on mountain slopes. This causes soil erosion which results in heavy floods. Uncontrolled soil erosion is the main cause of sedimentation of water ways which causes harm to fisheries.

Now a days, pressure of demand on underground water resourses has gone up considerably. Every year more and more water is drawn up from sub-surface layers whereas recharging of underground waters has been slowed down. Massive deforestation has caused disappearance of plant cover over a large area of land surface. In absence of plant cover most of the rain water flows down quickly in streams and rivers. Little of it percolates down to sub-surface layers to recharge the ground water stock. With rapidly flowing waters considerable amount of top soil is lost with torrential flow of water. In a number of localities where underground water is regularly drawn, the water table is receding deeper, thus output exceeds the input.

Conservation and Management of Water

As we know. water is the basic need of living organisms. An acute crisis of water is expected to follow in some regions of the world, particularly in our country. The shortage of water shall make many localities barren, dry and devoid of life. Fertile land will convert into deserts. Conservation of fresh water is therefore, an absolute necessity of today. A number of steps are being undertaken to minimise wastage of fresh water resources and to make more efficient use of the available water. Main approaches for conservation of water are as follows :

(*i*) Reducing agricultural water wastage by increasing efficiency of irrigation.

(*ii*) Reducing water wastage in industry by recycling the used water.

(*iii*) Reducing domestic water wastage by constructing waste water treatment plants and recycling the treated water.

(*iv*) Rainwater harvesting by using practices to store rainwater and recharge groundwater.

(*v*) Afforestation and protection of water sheds to improve water economy.

(*vi*) Reducing pollution and recycling of water.

(*vii*) By development of an efficient distribution system of water.

(*viii*) By enhancement of surface water storage capacity.

(*ix*) By improvement of underground water storage capacity.

(*x*) By adopting water economy.

Main approaches for management of water are as follows :

(*i*) Construction of dams and reservoirs to regulate year-round supply of water. They are also helpful in controlling flood and generating electricity.

(*ii*) Desalinisation of sea water and saline groundwater, making it fit for drinking and agricultural purposes. Desalinisation plants are already under operation in Middle East countries.

(*iii*) Diversion of water and bodies (*e.g.*, through canals) to increase the natural supply of water to a particular area.

(*iv*) Regular dredging and desiltation of water bodies.

Swajaldhara. Reforms in the Rural Drinking Water Sector were adopted in 1999 and also a few projects were launched as pilots. They were intended to be implemented during the Ninth Five Year Plan and with the experience gained there on, the reform initiatives were to be firmed up and scaled up during the Tenth Plan period for adopting the demand responsive strategy and also for institutionalising community participation for the sustainability of drinking water supply system and sources in rural areas. On 15.12.2002, the reform initiative in the Rural Drinking Water Sector were scaled up throughout the country by launching the **Swajaldhara** by the Hon'ble Prime Minister of India.

LAND RESOURCES

Land is a major constituent of the lithosphere and is the source of many materials essential to man and other organisms. Earth's one fourth area is formed by land which is largely covered with natural forests, grasslands, wetlands and man made urban and rural settlements along with agriculture. About 36.6 per cent of the land area is occupied by human dwellings and factories, roads and railways, deserts and dunes, glaciers, polar ice marshes, rocks and mountains. About 30 per cent of the total land mass is under forests, approximately 22 per cent is occupied by meadows and pastures and only 11 per cent for agriculture. Low-lying areas covered with shallow water are called wetlands. The wetlands make the transitional zones between terrestrial and aquatic areas.

FORESTS

The forest is 'a plant community predominantly of trees and other woody vegetation, usually with a closed canopy'. The word **forest** has been derived from the Latin '*foris*' meaning outside, the reference being to a village boundary or fence, and it must have included all uncultivated and uninhabited land. Today a forest is any land managed for the diverse purpose of forestry, whether covered with trees, shrubs, climbers, lianes, etc., or not. The Indian word 'jungle' has been adopted in the English language to describe a collection of trees, shrubs, etc., that are not grown in a regular manner. Forests are a very striking feature of the land surface. They vary greatly in composition and density, and stand in marked contrast with meadows and pastures. Certain forests are evergreen, like the Deodar forests of Kashmir and Himachal Pradesh, while others are deciduous, becoming leafless either before the advent of winter when vegetative activity almost ceases, such as the oak forests of the Himalayas, or else just before the onset of intense dry summer, to reduce transpiration to the minimum, like the teak forests of Madhya Pradesh.

Importance of Forests

The forests of a country make a natural asset of immense value. Unlike its mineral resources, including fossil fuels, which in course of time will either get exhausted or their utilisation will become uneconomic due to increased costs for obtaining and processing them, the forests, if of adequate extent ideally dispersed, scientifically managed, and judiciously utilised can be kept perpetually productive and useful, conferring many benefits, direct and indirect, on the people. Directly, forests meet the needs of timber wood, fuel, bamboos and a variety of other products, including fodder which are indispensable requirements of the people living in close proximity of the forests. The forests provide facility for the grazing of livestock, a yield and variety of products, of commercial and industrial value such as structural timber, charcoal, raw materials for making paper, newsprint, panel products, **bidi** leaves, gums, resin, tanstuff and a number of other economic products including medicinal drugs. Indirectly, forests preserve the physical features, check soil runoff, mitigate floods and make the stream flow perennially and thus help agriculture. Forests make the climate equable and have a considerable hygienic and strategic value. They also provide shelter to the wildlife.

In developing countries the heaviest demand on forests is for fuel wood. More than 1500 million people depend on wood for cooking and other purposes. The world consumption of wood for fuel is estimated to be more than 1,000 million m^3. This is well over 80 per cent of the total use. About 58 per cent of the total energy used in Africa and 42 per cent in South East Asia is obtained from fuel wood.

Bamboos are tall, perennial, arborescent grasses. More than 100 species of them are found in the Indian forests. Bamboos are characterised by woody stems, called culms. The strength of culms, their straightness, hardness, and the facility with which they can be split make them suitable for a variety of purposes. Indeed bamboos are the poor man's timber, used as rafters, scaffolding, roofing, walling, flooring, matting, basketry, carthoods, cordage, etc. Tender shoots are eaten and pickled. Bamboos are the only indigenous plant of wide occurrence in the Indian forests which has a comparatively long fibre suitable for making paper and rayon. Canes, that come from family Palmaceae, and occur in the forests of Andamans, Assam, Kerala, Tamil Nadu, Maharashtra, Bihar, Orissa and Uttar Pradesh, are used as plaiting material, as ropes and cables of suspension bridges in the forests, for furniture, walking sticks, umbrella handles, and sports goods.

Essential oils are obtained from a variety of forest plants. These are used in the manufacture of soaps, cosmetics, pharmaceuticals, confectionery, tobacco flavouring and incense. Important from the commercial point of view are the oils of *Eucalyptus*, used in medicine, rusa grass, khus and sandalwood. Tannins, dyes, gums, resins and oleoresins are other important forest products. Babul bark (*Acacia nilotica*) is an important tanning material used in northern India, the main centre being Kanpur. Gums and resins are exuded by plants. The gum is used in the textile trade, cosmetics, dentrifices, cigar and food industry. The chief producing centres of gum are Madhya Pradesh, Uttar Pradesh, Rajasthan and Gujarat. The most important resin is that obtained by tapping the chir pine (*Pinus roxburghii*). It is a flourishing forest industry. Natural camphor is also a forest product. Drugs, spices, poisons and insecticides are other important forest products. They are derived from trees, shrubs, climbers, herbs, ferns, mosses, lichens and consist of fruits, flowers, leaves, bark, stem or root. **Sarpagandha**, a famous drug is obtained from the roots of *Rauwolfia serpentina*, whereas quinine, a drug used against malaria fever, is obtained from the *bark of Cinchona*. **Kala zira** is the seed of *Carum carvi*, a much valued spice. Several forest fruits, flowers and even leaves and tuberous roots are eatern, such as **bel**, **ber**, **phalsa**, **jamun**, **khirni** and **tendu**. The parts of many plants are used as pickles or vegetables. Such important trees and shrubs are—**amla**, **anar**, **imli**, **karaunda**, **kokam**, **kachnar** and **kaith**. Mushrooms and morels are found in wild state in Kashmir and Himachal Pradesh. They are edible and used as delicacy. Palmyra palm yields a fermented drink **nira**. **Mahua** corollas are fermented to get a liquor. **Phulwara** butter is obtained from the seeds of Indian butter tree (*Diploknema butyracea*). **Shahtoot** fruit is eaten or made into a sherbet. **Bhilama** fruits are salted and pickled.

The leaves of **tendu** (*Diospyros melanoxylon*) are used as wrappers of tobacco to make **bidis**. Leaves of the woody climber *Bauhinia vahlii* found in deciduous forests are collected in large quantities and made into plates and leaf cups used for feasts and other festive occasions.

Miscellaneous forest products include leaves of certain species used for various purposes, soap substitutes, such as **ritha** (seeds of *Sapindus emarginatus*), pods of *Acacia concinna* called **shikakai**, sola pith, ornamental seeds, such as **gumchi** and **rudraksha** (the drupes of *Elaeocarpus ganitrus*).

Lac is the most important animal product obtained from forests. It is a resinous secretion, formed as a protective covering on the bodies of larvae of a minute insect *Laccifer lacca*. The lac insects live on the hosts, such as **palas** (*Butea monosperma*) and **ber**. India produces about 85 per cent of the world production. Lac has been mentioned in Vedas, and known from ancient times in India. Other animal products from forests are honey and wax. **Tussar** or **kosa** silk moths, horns and hides of dead animals are also the products of forests. Antlers of deer and ivory are also animal forest products.

The forests play an important role in the life of tribal people living in forests. The forests provide them food, fruits, meat, medicines, hides, skins and other products of commercial use.

Forests make large biotic communities, and they give shelter to diverse species of plants, animals and micro-organisms.

Forests protect the environment. A forest and its environment are always interacting. The most important environmental factors are micro-climate soil characteristics, moisture availability and action of animals and insects. Micro-climate is governed by solar radiation, rainfall, wind, humidity and temperature of the air and the soil. In a dense forest temperature range is narrow, and although the air temperature in and outside a forest is more or less the same, the humidity is greater inside. Soil temperature is also influenced by the forests, which in turn affects the biological activity in the top layer of the soil. Heavy winds may cause a serious erosion if the soil is dry and not covered with vegetation. Extensive forests condense low clouds and thus to some extent increase precipitation. When the tree canopy is open there is little humus on the forest floor and, therefore, rain strikes the bare soil with full fury and brings about soil erosion, which leads to denudation. In rocky tracts, roots of trees penetrate deep into the soil and thus bind it together.

The need for adopting soil conservation measures is the practice of forestry. A well managed and properly stocked forest reduces soil runoff. Rain water is absorbed by the humus and then it seeps into the sub-soil and finally moves underground, thus keeping the streams flowing perennially. Forests thus mitigate floods which can otherwise do immense damage to agricultural land in the lower region. The forest soil absorbs water during rains, and does not allow it to evaporate or run off swiftly. This ensures a perennial supply of water in springs and wells. The hill slopes with forest cover conserve water and gradually release it into streams, rivers and sub-soil springs. Forests help to increase the humidity of the atmosphere by drawing subsoil water and sending it out in the air by means of transpiration. The increased humidity of the forest helps plants and animals for their survival in a warm season by making it cool and pleasant. Forests reduce atmospheric pollution by collecting the suspended particulate matter and by absorbing carbon dioxide from atmosphere.

Forests have a great aesthetic value. All people appreciate the beauty and tranquillity of forests. In ancient times, to arrest devastation of forests, which was adversely affecting the life of people, some wise ancestors and **rishis** declared cutting of trees a sin and planting and protecting them an act of piety. Several useful species were thus saved from extinction, such as the banyan, the **pipal** and **bel**. Forests are of great educative value. The forests are supposed to be nature's laboratories which have contributed to the study and research in various branches of natural sciences. Indian culture was born in **ashramas** situated in deep forests, where the **rishis** sang— "May the gods, the water, the plants and the forest trees, accept our prayers, and may their blessings protect us for ever, and ever".

The forests are storehouse of biodiversity and provide important environmental services to mankind. These services originate from the undermentioned key functions of forests :

(*i*) **Productive functions.** They include production of wood, fruits and a wide variety of compounds, such as resins, alkaloids, essential oils, latex, pharmaceuticals, etc.

(*ii*) **Protective functions.** They include conservation of soil and water; prevention of drought, shelter against wind cold, radiation, noise, sights and smells, etc.

(*iii*) **Regulative functions.** They include absorption, storage and release of gases (CO_2, O_2), water, mineral elements and radiant energy.

Forests effectively regulate floods and drought and the global biogeochemical cycles, particularly carbon cycle.

The regulative functions also improve atmospheric and temperature conditions. They enhance the economic and environmental value of the landscape.

Forest Area in India

At the beginning of the twentieth century, about 30 per cent of land in India was covered with forests.

However, by the end of the twentieth century, the forest cover was reduced to 19.4 per cent.

This is less than the optimum 33 per cent area recommended by the National Forest Policy (1988) for the plains, and 67 per cent for the hills.

Of the existing forests, less than two-third are dense forests, and the rest are open degraded forests.

Today, per capita forest area avaiable in India is 0.06 ha, comparatively, which is much below the average for the world, *i.e.*, 0.64 ha per capita.

Table 19.3. Forest Cover in India, 1999

Class	*Area* (km^2)	*Percentage-Geographic area*
Dense forest	3,77,358	11.5
Open degraded forest	2,55,064	7.8
Mangrove	4,871	0.1
Scrub forest	5,896	1.6
Non-forest (other land use)	25,98,074	79.0
Total	32,87,263	100.0

Forests in India

Palaeobotanical studies have shown that there were dense forests in India in the Permian period, 250 million years ago. Man was evolved in the beginning of the Pleistocene Age, only a million years ago. At this time India had thick forests except in Rajasthan and parts of Punjab which lay buried under a swamp. Archaeological evidence shows that the Rajasthan swamps existed till as late as 4000 B.C., when Mohenjodaro culture flourished. During **Ramayana era** there were dense forests in Naimisharanya, Chitrakoot, Dandakaranya and Panchwati which abounded in wild life. The chronicles of Chinese pilgrims mention dense Indian forests in the birth place of Lord Krishna. Records relating to the invasion of Alexander in 327 B.C. mention the existence of almost dark and deep forests along the Indus. Later in Kautilya's time, protection of forests, planting of new species of trees and preservation of wild life were considered desirable. By the time of Samrat Ashoka, the inroads were into the forests. Ashoka also encouraged the cultivation of exotic, medicinal plants. Emperor Jahangir introduced the famous **chinar** tree in Kashmir valley. Over-population and unplanned felling of trees have drastically reduced the forest area and the quality of forests in India today.

The area under forests in India is about 75 million ha and consists of about 23 per cent of the total land area. The National Forest Policy lays down that India should maintain one-third of its land under forests. The forests of India may be classified into three classes : (*i*) reserved forests, (*ii*) protected forests, and (*iii*) unclassed forests, forming 47.1, 30.3 and 22.6 per cent respectively of the total forest area. The forests in India are largely owned (about 96 per cent) or managed by the states.

Forest types. The distribution of natural forests in India is governed principally by rainfall which may vary from 12 cm to 125 cm annually in different parts of the country resulting in a wide variety of vegetation, from xerophytic scrub forests in the arid zones in Rajasthan to evergreen rain forests in the tropics as in the Western Ghats, Assam, North Bengal, Andamans and conifers in the temperate region. In between the extremes, there are many intermediate types and sub-types. The major forest types are : (*i*) moist tropical forests, (*ii*) dry tropical forests, (*iii*) montane sub-tropical forests, (*iv*) montane temperate forests, (*v*) sub-alpine forests, and (*vi*) alpine scrub forests.

Moist tropical forests occur in Western Ghats, Andamans and Assam. They consist of many evergreen species, such as dipterocarps, hopea, irul, toon, terminalias and several others.

Dry tropical forests occur in areas with rainfall 75 cm to 125 cm containing thorn forests. Tropical thorn forests are found in the desert and arid zones of Rajasthan, Punjab, Haryana, Gujarat, Western U.P. and Madhya Pradesh. They consist mainly of acacias.

Montane sub-tropical forests occur at elevations of 750 to 1800 m in the Himalayas and also at the similar elevations in the Nilgiris in South India. The forest flora is represented by pines.

Montane temperate forests occur at elevation of 1800 to 2700 m and with a rainfall of about 125 cm to 200 cm. Deodar, blue pine, spruce and fir and the important conifers in the Western Himalayas mixed with oaks, maples, walnut and *Ulmus*.

Sub-alpine forests occur between 2900 to 3500 m a.s.l. The vegetation consists of a dense growth of small crooked trees or large shrubs with patches of conifers like *Abies* and blue pine. Birch and *Rhododendron* are the common broadleaved trees.

Alpine scrub forests occur above 3500 m along the entire length of the Himalayas. The vegetation consists of low evergreen forests mostly of *Rhododendron* with birch and other deciduous trees forming a dense growth with short branchy stems.

Deforestation. Deforestation is a serious threat to the economy, quality of life and future of the environment of this country. The forests in India have declined from about 7,000 million hectares in 1900 to 2890 million hectares in 1975. It is estimated that by the year 2000 only 2370

million hectares will be left. Tropical rain forests have been reduced from 1,600 million hectares to 938 million hectares.

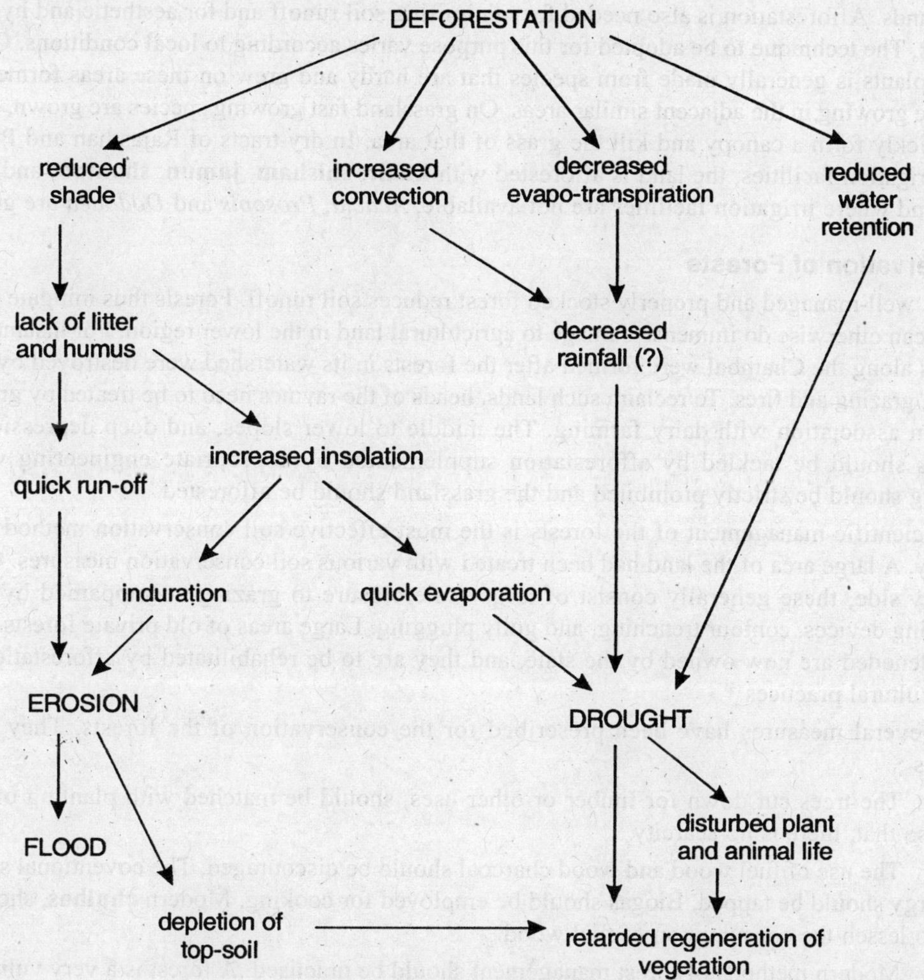

Fig. 13.2. Deforestation.

In India, the main causes of deforestation are : explosion of human and livestock population, increased requirement of timber and fuel wood, expansion of croplands for farming and enhanced grazing by cattle. Construction of roads in the hilly terrain is another important factor for deforestation. The total length of such roads is about 30,000 km in ecologically fragile areas. Development of industries, extension work for mining, quarrying, irrigation and agriculture are other causes of deforestation. Increased demand for fuel wood, timber, wooden crates, paper, newsprint and paper board led to felling of trees on large scale.

Deforestation is the main cause of soil erosion. The floods, drought and loss of precious wild life are also due to deforestation. Thus the economy of the country and quality of life of weaker sections have been much deteriorated. In this country, particularly in the Himalayan region, shortage of fire wood and water has been causing great misery to the womenfolk. They have to spend the whole day in the collection of fuel wood for cooking the meals of the day. It is thought that time is not far when the cost of food would be lesser than the cost of fuel to cook it.

Afforestation. As there is a general shortage of forests and as wood is a bulky commodity which cannot be transported to a long distance, it would be desirable to develop forests on all the wastelands. Afforestation is also needed for minimising soil runoff and for aesthetic and hygienic reasons. The technique to be adopted for this purpose varies according to local conditions. Choice of the plants is generally made from species that are hardy and grew on these areas formerly or they are growing in the adjacent similar areas. On grassland fast growing species are grown, which can quickly form a canopy and kill the grass of that area. In dry tracts of Rajasthan and Punjab, with irrigation facilities, the land is afforested with **khair**, **shisham**, **jamun**, **shahtoot** and **safed siris**, and where irrigation facilities are not available, *Acacia*, *Prosopis* and *Dodonea* are grown.

Conservation of Forests

A well-managed and properly stocked forest reduces soil runoff. Forests thus mitigate floods which can otherwise do immense damage to agricultural land in the lower region. For instance, the ravines along the Chambal were formed after the forests in its watershed were destroyed by over-felling, grazing and fires. To reclaim such lands, heads of the ravines need to be treated by growing grass in association with dairy farming. The middle to lower slopes, and deep depressions of ravines should be tackled by afforestation supplemented by appropriate engineering works. Grazing should be strictly prohibited and the grassland should be afforested.

Scientific management of the forests is the most effective soil conservation method in the country. A large area of the land had been treated with various soil conservation measures. On the forestry side, these generally consist of large-scale closure to grazing accompanied by water disposing devices, contour trenching, and gully plugging. Large areas of old private forests which were denuded are now owned by the state, and they are to be rehabilitated by afforestation and other cultural practices.

Several measures have been prescribed for the conservation of the forests. They are as follows ;

1. The trees cut down for timber or other uses, should be matched with planting of more trees, so that, there is no scarcity.

2. The use of fuel wood and wood charcoal should be discouraged. The coventional sources of energy should be tapped. Biogas should be employed for cooking. Modern **chulhas**, should be used to lessen the consumption of fuel wood.

3. Modern methods of forest management should be practised. A forest is a very vulnerable asset. The slight neglect or unscientific working can rapidly deplete and may even annihilate it after which reforesting of the area may not be economically feasible.

4. The silvicultural practices should be adopted. Cultivation of forest crops consists of regeneration, that is, renewal of forest crop, and tending, that is carrying on operations from time to time for the benefit of the growing forest crop during its lifetime. The silvicultural practices include use of irrigation, fertilisers, bacterial and mycorrhizal inoculations, disease and pest management, control of weeds, cleaning, thinning, breeding of elite trees and use of tissue culture techniques.

5. The Social Forestry Programme (1976) should be encouraged. Under this programme, the so called waste lands should be utilised to produce fire wood, fodder and small timber for the use of agricultural implements. This programme includes raising, planting and protecting trees for various purposes. This programme is a boon for rural community.

6. The annual deforestation must be followed by annual reforestation of the deforested areas. Thus the forest is conserved. There should be intensification of afforestation or reforestation rates to nearly three times of the existing rate.

7. The agroforestry programme should be encouraged. Under this programme the attempt should be made to use the same land for farming, forestry and animal husbandry.

8. The urban forestry programme should also be encouraged. Under this programme, the small gardens and house compounds should be well maintained by planting flower and fruit trees. The trees of aesthetic value should also be planted along the road side. The ornamental trees should be grown in the parks and other wasteland. Such forestry programme will give nice look to the town, and mitigate the foul environment.

According to the opinion of experts in this field, it is believed that economy of this country would be sound if at least one-third of country's land comes under forest. To achieve this, there should be prompt action at the levels of individuals, community and the government. In India, several such movements have been taken by the people. For instance, *Chipko* movement of Tehri-Garwal region, and the public agitation against the construction of the hydroelectric project in silent valley, are current examples. India became independent in 1947, and thereafter, attention of the government was drawn to the need for remodelling the management of Indian forests so that they could play an increasingly useful role in promoting national welfare. As a first step towards making the people forest-conscious, a nation-wide celebration of an annual tree-planting festival, called **Van-Mahotsava** was conceived. All the waste-lands, road sides, railway track sides, parks and barren lands should be planted to regenerate forests.

Forest conservation and management programmes may be summarised as follows :

(*i*) They should ensure sustainable supply of tree products and services to people and industry.

(*ii*) They should ensure the maintenance of long-term ecological balance through protection, restoration and conservation of forest cover.

Extensive programmes of **afforestation** are needed to save the diminishing forest cover. The goals can be achieved by undermentioned practices :

(*i*) **Conservation forestry.** Also known as **protection forestry**. This involves protection of degraded forests to allow recoupment of their flora and fauna.

In such forestry practices, forest areas designated as national parks and sanctuaries are protected from human interference.

On the other hand, well-stocked forests are managed scientifically for producing timber and other forest products without causing any negative environmental impact on the forest.

(*ii*) **Production forestry.** Also known as **commercial forestry**. This aims to fulfil the commercial demand, without causing denudation of natural forests, through intensive plantation in available land. Social forestry and agro-forestry programmes are also included in this category.

Social forestry. This programme aims to plant trees and shrubs on all unused and fallow land to provide fuel wood, fodder, etc., thus, reducing pressure on existing forests. For example, unused farmland, usar land, community land, road and rail sides, etc., are planted with suitable tree species.

Agro-forestry. This includes a variety of land uses, where woody species are grown in combination with herbaceous crops, either at the same time or in time sequence.

Taungya system involves growing agricultural crops between rows of planted trees, such as sal, teak, poplars, etc.

Jhum system is well known shifting cultivation. This is a traditional agro-forestry system usually practiced in the north-eastern region of our country. This system involves felling and burning of forests, followed by cultivation of crops for few years and then gain abondoning cultivation to allow forest regrowth.

FORESTS AND WILDLIFE LAWS

National Forest Policy (1988). The National Forest Policy, 1988, emphasizes on scientific forestry research toward attainment of policy objectives which aim at restoring ecological balance and conservation of country's natural heritage by evolving socially acceptable, technically viable and ecologically sustainable management practices. Some priority areas of research broadly indicated in the policy are :

(*i*) Increasing forest productivity per unit area and per unit time.

(*ii*) Conservation eco-restoration and management of existing forest resources.

(*iii*) Social forestry in rural/tribal development.

(*iv*) Efficient utilisation of forest produce and substitutes for wood

(*v*) Formulate a meaningful curriculum and training model for forestry education to ensure professional competence.

(*vi*) Creating awakening amongst masses for increasing tree cover and adopting agroforestry models using suitable extension methodologies.

To enhance the effectiveness of management, conservation, sustainable development and utilisation of forest resources it becomes expedient to build up technical skills and improve the quality of human resources, research capability and support, forestry extension, information dissemination system and public education.

The principal aim of Forest Policy is to ensure environmental stability and maintenance of ecological balance, including atmospheric equilibrium, which are vital for all life forms, human, animal and plant.

Wildlife (Protection) Act 1972 (Amended 1991). This Act aims to protect wild animals, birds and plants. The main objectives of this Act are :

(*i*) Restriction and prohibition on hunting of animals.

(*ii*) Protection of specified plants ;

(*iii*) Setting up and managing sanctuaries and national parks;

(*iv*) Empowering zoo authority with control of zoos and captive breeding; and

(*v*) Control of trade and commerce in wildlife, wildlife products and trophies.

Forest Act 1927. This Act aims to consolidate the law relating to forests. The basic objectives of this act are :

(*i*) Setting up and managing reserved forests, protected forests and village forests ;

(*ii*) Protection of non-government forests and forest land;

(*iii*) Control of movement of forest produce ; and

(*iv*) Control of cattle grazing.

WWF – INDIA

The Word Wide Fund for Nature-India (WWF – India) has been working to promote harmony between humankind and nature for almost three decades. Today it is recognised as a premier NGO (non-government organisation) in the country dealing with conservation and development issues.

At a time when the Web of Life has come under increasing threats, WWF – India's attempts have been to find and implement solutions so that human beings can live in harmony with nature, and leave for future generations a world rich in natural resources and natural wonders.

Formerly known as the World Wildlife Fund, WWF – India was established as a Charitable Trust in 1969. With its network of State/Divisional and Field offices spread across the country to implement its programmes. WWF – India is the largest and one of the most experienced conservation organisations in the country. The Secretariat functions from New Delhi. The organisation is part of the WWF – India family with 27 independent National Organisations. The coordinating body, the WWF – International, is located at Gland in Switzerland.

WWF – India Mission

'The promotion of nature conservation and environmental protection as the basis for sustainable and equitable development'

To achieve this Mission WWF – India undertakes the following activities :

(*i*) **Tiger and Wildlife conservation**
(*ii*) **Freshwater and Wetland Conservation Programme**
(*iii*) **Educational Awareness**
(*iv*) **Centre for Environmental Law**
(*v*) **Oceans and Coasts Programme**
(*vi*) **Library and Documentation Services**
(*vii*) **Communication and Fund Raising**
(*viii*) **Indira Gandhi Conservation Monitoring Centre (IGCMC)**

WWF – India, through its biodiversity conservation, ecorestoration and environment education programmes has been extending its support to the conservation efforts and trying to maintain the ecological balance.

Table 13.4. Environmental Calendar

Month	*Date*	*Event*
January	14 – 30	Animal Welfare Fortnight
February	2	World Wetlands Day
February	28	National Science Day
March	21	World Forestry Day
March	22	World Water Day
April	22	World Earth Day
April	28	World Heritage Day
June	5	World Environment Day
June	17	World Day to Combat Desertification
July	1	Vanmahotsava Day
July	11	World Population Day
September	16	World Ozone Day
October	1 – 7	Worldlife Week
October	2	International National Disaster Prevention Day
October	5	World Habitat Day
November	1	World Ecology Day
November	10	Forest Martyr's Day
November December	19 to 18	National Environment Month
December	29	International Day for Biological Diversity

GRASSLANDS

The **grasslands**, also known as **rangeland**, provide forage and habitat to domestic animals and wildlife. Natural grasslands occur where rainfall is intermediate between that of desert lands and forest lands. In the temperate zone this generally means an annual precipitation between 10 and 30 inches, depending on temperature, seasonal distribution of the rainfall, and the water-holding capacity of the soil. Tropical grasslands may receive up to 60 inches concentrated in a wet season that alternates with a prolonged dry season. Grasslands are one of the most important of terrestrial ecosystem types.

A well-developed grassland community contains species with different temperature adaptations, one group growing in the cool part of the season (*i.e.*, spring and winter) and another in the hot part (summer). Forbs (*i.e.*, nongrassy herbs) are often important components, and woody plants (trees and shrubs) also occur in grasslands either as scattered individuals (savanna) or in belts or groups along streams and rivers.

In India, the area under various kinds of grass cover, including fallow and waste lands is about 18 per cent of the total land area. While forest cover is about 19 per cent of total land. This means, about 37 per cent of land can be said to be available for grazing.

Uses. In rural areas of our country, dried hay obtained from grasslands, particularly from tall grasses, is used as fuel or thatching material.

Grass cover is extremely effective in binding soil particles with the help of highly branched fibrous root system, thus significantly reducing soil erosion.

The annual average production of dry grass or hay in India is about 250 million tons.

Large herbivores are a characteristic feature of grasslands.

When man uses grasslands as natural pastures he usually replaces the native grazers with its domestic animals, *i.e.*, cattle, sheep and goats.

Since grasslands are adapted to heavy energy flow along the grazing food chain, such a switch is ecologically sound.

Degradation of Grassland

However, man has had a persistent history of misuse of grassland resources by virtue of allowing overgrazing and overplowing. The result is that many grasslands are now man-made deserts. The conversion of grassland (or forest) to desert is called **desertification**.

Degradation of destruction of grassland is mainly related to overpopulation. To enhance food production, grasslands possessing fertile soils are ploughed and converted to agricultural lands.

In India, grassland areas are frequently overgrazed. For example, the number of animals grazing in the arid and semi-arid regions has been found to be 2 – 10 times greater than the capacity of the grassland to feed the animals.

The lack of plant cover due to overgrazing causes soil erosion. When overgrazing occurs in combination with drought, the deserts are resulted.

Grassland Management

The important measures of grassland management are :

(*i*) **Protection from grazing.** This allows recovery of severely damaged vegetation.

(*ii*) **Use of rotational grazing.** While some areas are closed to grazing, allowing the plant cover to recover, grazing is permitted in other selected areas.

(*iii*) **Removal of unwanted plants.** This includes removal of woody bushes or shrubs and weeds, which usually adversely affect the productivity of grasses.

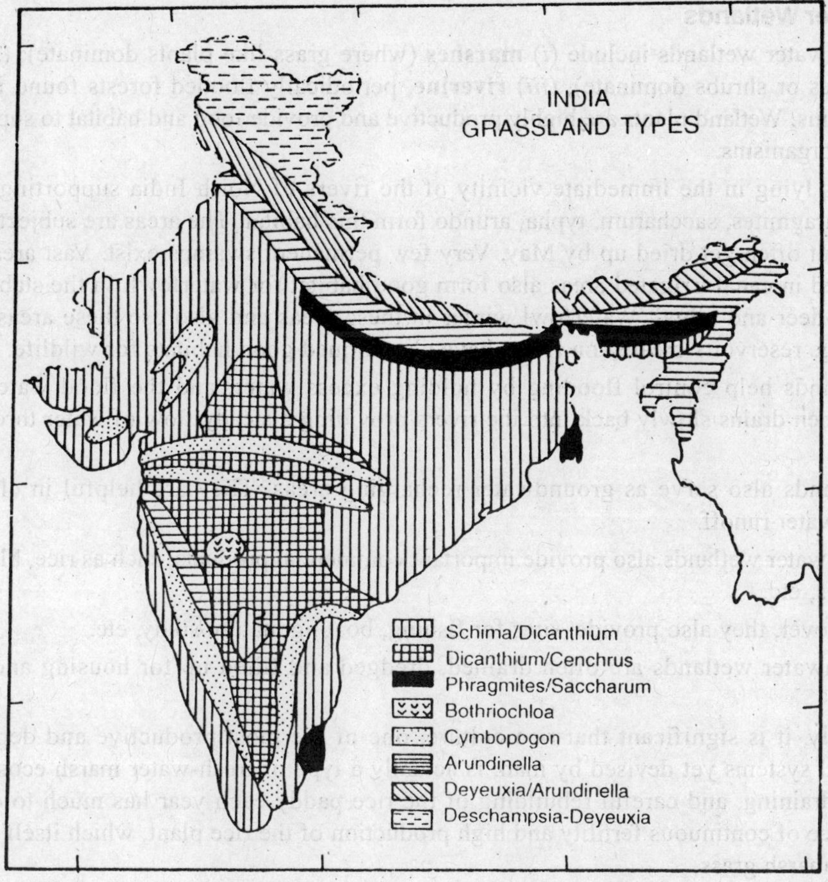

Fig. 13.3. Major grassland types of India.

(*iv*) **Conservation of soil and water.** This may be done by reducing loss of soil and water from the grassland.

(*v*) **Use of controlled burning.** This promotes recycling of nutrients stored in dried mulch and to reduce woody species invasion.

Grassland Types of India

The tropical grasslands of India may be classified a. follows :

(*i*) **Xerophilous grasslands.** They are found in dry regions of North West India under semiarid conditions. The common examples are : *Andropogon contortus*, *Cenchrus ciliaris*, *C. barbatus*, etc.

(*ii*) **Mesophilous grasslands.** Also known as 'savannahs' they are extensive grassy plains typically occurring in moist deciduous forests of Uttaranchal. The dominant species are : *Saccharum munja*, *Saccharum narenga*, etc.

(*iii*) **Hygrophilous grasslands.** These are called wet 'savannahs' adapted to wet soil.

WETLANDS

Wetlands are low-lying areas, usually covered by shallow water and have characteristic soil and water tolerant vegetation. Wetlands may be either freshwater or salt water.

Wetlands occupy almost 6 per cent of the world's land surface and provide crucial environmental services. Now a days the wetlands are increasingly threatened by agriculture, pollution and construction of dams, etc.

Freshwater Wetlands

Freshwater wetlands include (*i*) **marshes** (where grass-like plants dominate), (*ii*) **swamps** (where trees or shrubs dominate), (*iii*) **riverine**, periodically flooded forests found in lowlands along streams. Wetland plants are highly productive and provide food and habitat to support a wide variety of organisms.

Areas lying in the immediate vicinity of the rivers of North India supporting tamarisks, acacias, phragmites, saccharum, typha, arundo form this habitat. The areas are subject to summer flooding but often get dried up by May. Very few permanent swamps exist. Vast areas rendered waterlogged in canal irrigated areas also form good habitat for waterfowl. In the stabilised areas occur hog deer and nilgai. Waterfowl winter in these areas and also use these areas as staging points; large reservoirs created on rivers have also formed good habitats for wildlife.

Wetlands help control flooding by holding excess water, and the flood water stored in wetlands then drains slowly back into the rivers providing a steady flow of water throughout the year.

Wetlands also serve as groundwater recharging areas. They are helpful in cleaning and purifying water runoff.

Freshwater wetlands also provide important commercial products, such as rice, black berries, blue berries, etc.

Moreover, they also provide, sites for fishing, boating, nature study, etc.

Freshwater wetlands are often drained, dredged and filled up for housing and industrial purposes.

Finally, it is significant that rice culture, one of the most productive and dependable of agricultural systems yet devised by man, is actually a type of fresh-water marsh ecosystem. The flooding, draining, and careful rebuilding of the rice paddy each year has much to do with the maintenance of continuous fertility and high production of the rice plant, which itself is a kind of cultivated marsh grass.

Saltwater Wetlands

Coastal wetlands are also known as saltwater wetlands. They include highly productive estuaries which provide food and habitat for a large number of marine organisms. Between the seas and the continents lie a belt of diverse ecosystems that are not just transition zones but have ecological characteristics of their own. Whereas physical factors such as salinity and temperature are much more variable near shore than in the sea itself, food conditions are so much better that the region is packed with life.

Along the shore live thousands of adapted species that are not to be found in the open sea, on land, or in freshwater. The word 'estuary' (Latin, *aestus* = tide) refers to a river mouth or coastal bay where the salinity is intermediate between the sea and freshwater and where tidal action is an important physical regulator. Estuaries and inshore marine waters are amongst the most naturally fertile in the world.

Tidal action promotes a rapid circulation of nutrients and food, and aids in the rapid removal of the waste products of metabolism. A diversity of plant species and life forms provides a continuous photosynthetic carpet despite variable physical conditions.

Mangrove swamps are coastal wetlands in tropical regions which contain certain trees and shrubs growing best in the intertidal zone. As mangroves expand into the ocean, other plants colonise the soil left behind. Mangrove roots provide habitat for oysters, crabs and other marine organisms. However, the coastal wetlands are also being destroyed for coastal development and agricultural lands.

Organisms have evolved many adaptations to cope. with tidal cycles, thereby enabling them to exploit the many advantages of living in an estuary. Some animals such as fiddler crabs, have internal 'biological crops' that help to time their feeding activities to the most favourable part of the tidal cycle.

Wetland Conservation

Conservationists and trained marine environmental engineers have become alarmed by the needless destruction of coastal resources. Ultimately it may be necessary to set up some kind of conservation plan so that the use of such areas by man can be placed on a sound ecological basis. Some wetland conservation programmes are as follows :

(*i*) Preparation of wetland inventories

(*ii*) Identification of wetlands of critical importance for their protection.

(*iii*) Checking waste disposal in wetlands

(*iv*) Reduction of excessive inflow of nutrients and silt into wetlands from surrounding uplands by keeping them under plant cover.

ENERGY RESOURCES

The energy crisis is a global problem today. The survival of the man will be difficult if the energy problem is not solved on the priority basis. Future energy needs of rapidly expanding human population will demand the exploitation of most energy resources. The energy consumption is maximum in developed countries as compared to developing nations.

The nineteenth century industrial societies used fossil fuels, and the daily per capita energy utilisation jumped to 70,000 kcal. Today we require energy for agriculture, industry, transport, communication, comfort and defence. The per capita energy consumption per day in the USA has reached about 2,50,000 kcal. People in other countries use far less energy. It is as low as 10,000 kcal in several developing countries. However, it is increasing fast with rapid industrialisation.

Today the world's energy resources have reached a critical stage. The reasons are many. First, the world's almost total dependence on the fossil fuels-coal, petroleum and natural gas has caused such depletion that these fuels may last only another few centuries.

Today, 30 per cent of the world's population, living in industrialised countries, consumes about 80 per cent of the global energy. On the other hand the poorer nations rely more on firewood for cooking and heating and animal power for transport.

(*i*) **Non-renewable energy resources.** Such resources include various fossil fuels and nuclear energy. Fossil fuels include petroleum products, natural gas and coal. While nuclear energy is mainly obtained from the nuclear fission of the uranium.

The world reserves of fossil fuels and uranium are limited and will eventually be depleted. It is estimated that the stock of the mineral oil will be depleted in twenty first century, if it continues to be used at present rate.

However burning fossil fuels for energy has negative environmental consequences, such as global warming, air pollution, acid rain and oil spills. Now, this has become necessary to minimise use of non-renewable energy resources, and to replace them with renewable resources.

(*ii*) **Renewable energy resources.** Such energy resources are regenerated by natural processes so that they could be used indefinitely.

Renewable energy generally causes much less negative environmental impact than fossil fuels or nuclear energy. However, the generation of renewable energy is often more expensive than energy produced by fossil fuels or nuclear energy. With the advancement of technology, the costs of renewable energy is expected to decrease.

The most important renewable energy is solar energy, while other resources of renewable energy are hydropower, wind geothermal energy, ocean waves and tidal energy.

Solar Energy

There are two types of solar energy :

(*i*) Direct solar energy, and

(*ii*) Indirect solar energy.

Direct solar energy is the radiant energy, while **indirect solar energy** is obtained from materials that have previously incorporated the sun's radiant energy.

Solar energy can be used for direct heating and alternatively the heat converted into electricity, *i.e.*, thermal electric generation. Solar thermal pumps have been developed and are being tested under actual field conditions. Photovoltaic cells convert direct solar energy into electricity. However, a back-up system is required to store and generate electricity when solar power is not operative at night or during cloudy days. Solar photovoltaic panels, cookers, heaters and solar battery driven cars are being made and are in use.

Biomass energy. Among various energy resources, where solar energy is utilised indirectly, biomass energy is most important one. This is detained from those materials whose origin can be traced to photosynthesis, *e.g.*, live plant material and dried residues, freshwater and marine algae, agricultural and forest residues, such as straw, husks, corn cobs, bark, sawdust, roots, animal wastes, etc. Biomass also includes biodegradable organic wastes from industries, such as sugar mills, breweries, etc. About half of the world's population depends upon biomass as their main source of energy for domestic use.

In India, fuel wood is still a major source of energy for domestic purposes, at least in rural areas. Biomass fuel, may be solid, liquid or gas. This is burned to release its energy. Solid biomass includes wood, charcoal, animal dung and peat.

Biomass can be converted to liquid fuels, such as methanol (methyl alcohol) and ethanol (ethyl alcohol), which can be used in internal combustion engines of automobiles. Gasoline mixed with 10–20 per cent ethanol can be used in conventional gasoline engines.

Biogas. Biomass in the form of animal waste and hydrophytes, such as *Eichhornia* can be converted into biogas in biogas digesters by using the process of anaerobic microbial decomposition. This is a clean anaerobic fuel whose combustion produces less pollutants than other combustible energy resources. The biogas is composed of 60 per cent methane and 40 per cent carbon dioxide.

Energy Plantations

Production of biomass for energy requires sufficient area of land and water. Fuel wood has been the primary energy source for mankind from the beginning of civilisation and still continues to be the main source of energy in the developing countries, particularly in rural areas. Plant based energy is obtained through energy plantations which can produce biomass from slected species of trees in the shortest possible time at a low cost. These plants yield solid, liquid or gaseous fuel through burning, gasification, digestion, etc. Such plantations can be raised in both hills and plains particularly on marginal land. It has been calculated that about four tons of dry matter can be obtained in 0.25 ha land, which is sufficient for an average family.

Some important plants for energy plantation, which are extensively grown in India are *Leucaena leucocephala*, *Casuarina*, poplar tree and *Eucalyptus*.

Petroplants. Also called **hydrocarbon plants** are known to yield liquid hydrocarbons the substitute for liquid fuels. The hydrocarbons present in such plants can be converted into petroleum hydrocarbons of high molecular weight. *Jatropha curcas* tree yields 2 kg of seed oil per plant/year. The oil obtained from this plant is expected to replace the conventional diesel fuel. These plants can easily be grown in arid regions and other waste lands.

Other Renewable Resources of Energy

Among other renewable resources of energy some are important which are mentioned here

(*i*) **Hydropower.** Hydroelectric power is the most important and widely used resource of renewable energy. This is generated from water. Water falling from a height turns turbines at the bottom of dams to generate electricity. Hydropower produces approximately one-fourth of the worlds electricity, and is usually cheaper than electricity produced by thermal power plants. India has big potential for hydroelectric power generation. The energy obtained from hydropower is widely used in agriculture, industry, transport, domestic sectors, etc.

However, building a dam to hold the water results in several environmental problems, such as submergence of plant and animal habitats and displacement of people.

(*ii*) **Wind energy.** The wind can also be exploited as a source of energy in different ways. When fans are rotated by the action of wind, its energy can be used for generation of electricity. National Aeronautics Limitd, Bangalore is engaged in research and development of power generation through wind mills. In many states wind mills have been set up for irrgation purposes. However, harvesting wind energy is possible only in the areas that receive fairly continual winds, such as islands, coastal areas and mountain passes. Wind energy may be converted into mechanical and electrical energy. Now a days, wind energy is being used for pumping water is rural areas. According to an estimate, about 20,000 mw electricity can be generated in India from wind alone.

(*iii*) **Tidal energy.** The difference in the level of water between high tide and low tide can be used to generate electricity.

(*iv*) **Geothermal energy.** Heated groundwater flowing upward as hot water or steam, or as hot springs, can be used to turn turbines and generate electricity is geothermal power plants.

(*v*) **Ocean waves energy.** Ocean waves, produced by winds, possess the potential to turn a turbine and generate electricity.

The present critical energy position demands an organised effort at all levels from individual practices to international action. Considerable amount of energy can be saved by reducing wastage and using energy efficient devices. Some of such measures are :

(*i*) Reduction in man's dependence on fossil fuels and development of newer alternative sources of energy.

(*ii*) Preparation of smokeless and efficient chulhas (wood stoves).

(*iii*) Improvement and expansion of sources of solar energy. Solar photovoltaic panels, cookers, heaters, and solar-battery driven cars need to be improved technically and made cost-effective.

(*iv*) The use of biogas plants must be encouraged so that agricultural and animal wastes can be used to produce both energy and fertiliser.

(*v*) Improvement in design, manufacture and maintenance of biogas plants, engines and pumps.

(*vi*) Generation of hydroelectric, wind and tidal power has to be explored more extensively.

(*vii*) Programmes for growing fuel wood trees and shrubs under the control and maintenance of local communities have to be implemented in the rural areas of the developing countries.

MARINE RESOURCES

The oceans cover nearly three-fourth of earth's surface. Ocean has been the source of many needs of man from the time immemorial. The rapid growth of human population and the advancement of industrialisation have exerted great pressure on the resources and the environment of oceans. Land resources are depleting at tremendous rate and in view of this man has started thinking to exploit the oceanic resources.

Marine resources can be broadly divided into two major categories : (*i*) **Living resources**, such as algae and the animals of the sea, (*ii*) **Non-living resources**, *e.g.*, various kinds of minerals.

Alga Resources

Marine algae vary greatly in form and range from one-called microscopic phytoplanktons to giant helps, which attain a length of 100 to 150 metres, *e.g.*, *Nereocystis*, *Macrocystis*, etc.

Green, blue, red and brown algae are commonly found in oceans. From time immemorial, algae have been widely used as human food. The algae are used as direct source of food by several sea animals and fishes. Marine planktonic diatoms are of fundamental biological importance since all life of sea is dependent upon them. Since the prehistoric times several sea weeds have been used as direct source of food to human beings. The sea weeds form the most important part of the diet of Japan and China. **Suimono** is a Japanese preparation of dried fish and several sea weeds. **Mitsu** is another Japanese preparation which contains sea weeds, fruits, sugars and dried kidney beans.

In many countries, animals are still regularly fed on fresh or processed sea weed (*e.g.*, *Laminaria* and *Fucus*).

Marine algae have been used as a manure in many countries because of their high nutrient content. Balanced fertiliser can be made by mixing sheep manure, fish and shells with seaweeds. Seaweeds are a store-house of the important potash, ionic sulphate, microelements and growth substances, besides having every other element and radical required by plants. Seaweed manure seems to increase disease resistanc to plants.

Agar extraction. Mainly red algae, *e.g.*, Gracilaria, Gelidium, etc., are used for the extraction of commercially important agar. Japan produces the largest quantity of agar, and exports to most of the countries of the world. The agar is used in several ways. It is employed in the preparation of ice creams, jellies, desserts, pharmaceuticals, photography, metal plating, lithography, metallurgy, and also in the manufacture of explosives, detergents, pesticides, in sizing the textiles and clearing many liquids. It is also used in preparing shaving creams, cosmetics and shoe polishes. The agar has constantly been used in biological laboratories for media preparation for fungal and bacteriological cultures.

Animal Resources

Fish, molluscs, crustaceans and many mammals found in sea waters are used as human food. Oceans provide an excellent habitat for above mentioned organisms.

Fish. The bulk of fish and other aquatic animals come from marine habitats. Marine fish provide considerable amount of food throughout the world. Fish are also used for the manufacture of several edible products, such as fish glue, fish meal, fish oil, fish proteins and vitamins.

Economically important fish are generally divided into two categories : (*i*) **demersal fish** and (*ii*) **pelagic fish**. The demersal fish are found at the sea-bottom, while pelagic fish floating free in the water column.

Fishing activity near the shore-line is carried on by small boats and vessels. This is called **coastal fishing**. On the other hand a large part of the total annual catch of fish comes from deep waters, and called **deep sea fishing**. This type of fishing is done by large boats or ships. In developed countries, where fishing is an organised industry, large boats and vessels are used for the purpose.

Molluscs. They are soft-bodied and shelled animals, (Latin, molluscus = soft bodied). Important molluscs from commercial point of view are the mussel, oyster, clam, etc. Many types of molluscs are used as food. However, pearl oyster is commercially important. The pearl worn as jewellery is secreted by certain oysters (bivalves). The inner calcareous surface of several bivalve shells is iridescent and pearl-like (nacreous) in appearance.

Crustaceans. They include crabs and prawns. They are the dominant arthropods of the sea. Prawns, lobsters and crabs are important economically. They make valuable items of food. India ranks first among the prawn producing countries of the world.

Phytoplankton-feeding, tiny crustaceans form an important link in the food chain of oceans.

Mammals. Whales, dolphins and porpoises are the economically important mammals found in the sea. The blue whales are the largest animals, some 30 m in length. Whales may be filter-feeders, living on planktons, or may feed on fish. Whales provide many valuable primary products, such as meat, ivory, skin, and frozen glands. They also produce many secondary products, such as oils, liver oil, etc. Fresh meet of all cetaceans, *i.e.*, whales, dolphins, porpoises, etc., has been used for human consumption.

Minerals in Sea

The sea is a store house of many valuable minerals. Most abundant elements in sea water are sodium, chlorine, magnesium, and bromine, that are commercially extracted from sea water. Phosphorite nodules are also found in sea, which can meet the shortage of phosphate fertilizers.

The sea weeds make an unique supplement for a well balanced diet. Potassium, sodium and chloride are found in the ionic form in sea weeds. The significance of iodine, as a constituent of food, is that besides being present in organic combination it is also available in part in the readily available form of the precursor of thyroxine and hence this source of iodine surpasses mineral iodine in drinking water and iodised table salt. The another micronutrients besides iodine which are important in human metabolism are iron, copper, manganese and zinc and all of them present as the trace elements of sea weeds. The highest copper content is found is *Sarconema* and *Acanthophora*. Today, iodine is produced from brown algae (seaweeds) in Japan, France, Norway and Jawa.

MINERAL RESOURCES

Some mineral elements are essential for the formation and functioning of the living body of all organisms, including man. Most of the essential elements enter animals through food chains and food webs. However, minerals are essential to our industrialised society and daily life. Mineral resources are non-renewable. The geographical distribution of essential minerals is unequal. Industry, transport, agriculture and defence preparation are making ever-increasing demands on the limited mineral deposits of the world. Depletion of almost all known and easily accessible deposits are anticipated within a few decades. Moreover, there may be shortage of some crucial elements, such as mercury, silver, gold, platinum, copper, tungsten within next 20 to 100 years. The limited resources of phosphorus which is an essential component of fertilisers may also cause retardation of agricultural growth.

Minerals can be divided into two groups : (*i*) metallic, *e.g.*, iron, copper, gold, silver, platinum, etc., and (*ii*) non-metallic, *e.g.*, sand, stone, salt, phosphates, asbestos, sulphur, corundum, etc.

Mining. This includes extraction, processing and disposal of minerals. However, mining has adverse effects on environment. This not only disturbs and damages the land, but also pollutes the soil, water and air.

The land that has been destroyed due to mining is known as **mine spoil**. Such destroyed lands can be reclaimed to a semi-natural condition by revegetation to prevent further degradation.

Conservation of Minerals

Efforts are being made to check the wasteful and injudicious use of minerals by recycling and adopting more efficient technologies, exploiting untapped deposits and by deep sea-mining. In process of recycling, used and discarded items are collected, remelted and reprocessed into new products, such as iron scraps, aluminium cans, etc.

The problem of high mineral consumption can be minimised by scientific and technological developments. Search for new production processes, designing smaller equipment, use of new raw materials and restrained use of materials have to be intensified. Finding new uses for glass, plastics, ceramics and synthetic fibres and using them as substitutes for exhaustible minerals is in progress.

Some minerals present in products can be recycled, e.g., gold, silver, lead, copper, nickel, steel, zinc, aluminium, etc. However, minerals in several other products are lost through normal use, such as paints containing lead, zinc, chromium, etc.

Table 13.5. Some Important Minerals and Their Uses

Minerals	*Uses*
1. Metallic	
Uranium	Nuclear bombs, electricity, tinting glass
Iron	Steel
Manganese	Alloy steel, disinfectants
Chromium	Metallurgy, refractory, chemicals
Nickel	Over 3000 alloys, coins, metal plating
Copper	Electrical products, alloys, gold jewellery, silver, etc.
Lead	Batteries, gasoline, paints, pipes, etc.
Tin	Tin plate, solder, chemicals, cans, containers, alloys.
Zinc	Galvanising, solder, die-casting, chemicals, brass, electrodes
Aluminum	Air crafts, rockets, building materials, electrical wiring, utensils
Gold	Monetary purposes, jewellery, dentistry, alloys
Silver	Jewellery, photography, utensils, alloys
Potassium	Fertiliser, glass, photography
Platinum	Jewellery, equipments, industrial catalyst
Radium	Medical and industrial uses, radiography
2. Non-metallic	
Phosphorus	Fertilisers, chemicals, medicine, detergents
Sulphur	Fertilisers, acid, iron and steel industries, insecticides, medicines
Corundum	Abrasives
3. Metallic liquid	
Mercury	Thermometers, dentistry, electric switches, etc.

Recycling and reusing not only renew the mineral resources, but also help in : (*i*) saving land from disruption of mining, (*ii*) reducing the amount of solid waste that is disposeable, and (*iii*) reducing pollution and consumption of energy.

However, to maintain the extended supply of minerals for a longer time, consumers must decrease their mineral consumption.

Durable and repairable products should be encouraged to be used again. Manufacturing industries should use the waste products of one manufacturing process as the raw materials for mother industry.

ENVIRONMENTAL ETHICS AND RESOURCE USE

Every morning the sun rises, and enlightens the whole land. The darkness disappears, and all becomes lighted. Besides light, we get energy from sun. All greens absorb light and food is produced for us. This great phenomenon of nature is **photosynthesis**. Only the plants are capable of this phenomenon. The plants give us food; they give us life. The plants are producers while rest consumers. The plants are biggest creation of god for us ; they are ideal donors. They only give, and from the time immemorial we worship them. The tree is eternal in its origin, and provides us the eternal peace. Beyond the kingdom of plants there is absolute darkness. Hundreds of plants are being considered sacred by the great people of this country, dealing with cultural heritage, festivals and religious ceremonies, starting from birth till death.

Lord Krishna says (Gita X, 26), "of all trees (I am) *Asvattha* (peepal tree)".

"They speak of the imperishable *asvattham* (peepal tree) as having its root above and branches below. Its leaves are the *Vedas* and he who knows this is the knower of the *Vedas*". (Gita XV, 1).

"Its branches extend below and above, nourished by the modes, with sense objects for its twigs and below in the world of men stretch forth the roots resulting in action". (Gita XV, 2).

"Its real form is not thus perceived here, nor its end nor beginning nor its foundation. Having cut off this firm-rooted *asvattham* (peepal tree) with the strong sword of non-attachment". (Gita XV, 3).

It is said in *Taittiriya Upnishad*, I, 10

"I am the originator of the world tree"

The cosmic tree is supreme–to know the cosmic tree is true knowledge, and the knower is complete in wisdom.

Plato says, "As regards the most lordly part of our soul, we must conceive of it in this wise ; we declare that God has given to each of us, as his demon, that kind of soul which is housed in the top of our body and which raises us-seeing that we are not an earthly but a heavenly plant-up from earth towards our kindred in heaven". (*Timaeus*, 90–A).

Dr. Sarvapallai Radhakrishnan says in his famous work '*The Bhagavadgita*' : 'As the tree originates in god, it is said to have its roots 'above' ; as it extends into the world, its branches are said to go downwards'.

He further says; "According to ancient belief, the vedic sacrificial cult is said to sustain the world and so the hymns are said to be the leaves which keep the tree with its trunk and branches alive".

"India is a unique country with a great cultural diversity associated with all kinds of climates and rich flora and fauna. The human societies in our country have evolved within magnificent environments and reverence to nature is inherent in our cultural ethos". I quote the lines: "The vast sky; and so the vast land; The vast ocean and the great Mountains; The great people and the majestic animals; The beautiful birds and the colourful butterflies; All the insects and all the moths; The creator has created them. Why ? Because they are need of creation; They are components of creation". One can enjoy nature by giving up greed.

Here I quote 'the philosophy I ever dreamt': "Where I find, no concrete walls, and no cement pillars; No chimneys and no factories; Neither transport nor poisonous gases, nothing like pollution, Only nature, the vast nature; Nature from horizon to horizon; great Nature in full grandeur ; Only huts and hamlets and their inhabitants; the innocense and serenity; the philosophy I ever dreamt; This is life; This is God; The 'beauty' and 'love' at the same time; great elements *khsiti* (soil), *Jal* (water), *Pavak* (fire) *gagan* (ether) and *Samira* (air), are components of great Nature; Where air develops from ether and water from air; where fire bakes our bread and earth makes base to stand; Then survival of all creatures and survival of whole 'global family'; The philosophy I ever dreamt".

Through history, the Indian people have not been exploiters but utilisers of nature. We belive in 'Live and let live'. We must recognise our responsibility to conserve earth resources for future generations. We must grow trees to maximum to protect the environment from poisonous gases. "In this hour of peril; Every hand is on trigger, and waits for order to fire; where does this race end, they do not know; Humanity is on brink of disaster; War mongers are roaming to convert the world in flames and ashes; Whole bio-kingdom is under threat; Count down has begun; By order or by mistake; All have to face doom's day; Oh contractors of humanity; Thou trust in God; Then why thou do not know; what His sons and daughters need; I tell thee, they need 'plain drinking water' ; Neither *agni* nor fire.

To conclude, I may say that the plant is the biggest creation of God; the gift of nature for us; static *Brahmarishi*, the ideal of donors, what he takes, takes from the mother, mother the land, mother the sole donor; sucks the nectar from the mother and gives to us. Grow more plants; protect the earth; protect the environment.

I worship the plants; I bow my head to plant kingdom.

SUMMARY

1. Any component of the natural environment that can be utilised by man to promote his welfare is known as a natural resource.
2. The natural resource can be a substance, an energy unit or a natural process, or phenomenon.
3. Natural resources can be divided into two categories: inexhaustible and exhaustible.
4. Inexhaustible resources, such as solar energy, wind power, are available in unlimited quantities on the earth.
5. Exhaustible resources have finite supply and can be either renewable or non-renewable.
6. The growth and reproduction of renewable resources, such as products from forests, grasslands, etc., can be managed so that these resources are continuously regenerated.
7. Non-renewable resources, such as biological species cannot be regenerated or reconstructed once they are used up.
8. Amongst the natural resocurces of earth, soil, water, land, energy and mineral resources contribute a lot to human welfare.
9. Soil resource regulates the yield of biotic products in terrestrial ecosystems.
10. Erosion by water and air makes a significant loss of soil.
11. World's fresh water reserves consist of 2.5 per cent of the total water, while the rest of water is saline.
12. The water reserves play significant role in global hydrological cycle.
13. The world's use of water has been increasing, and agriculture and industries account for most of the water consumption.
14. Thus, water conservation practices are significantly important to ensure water availability.

15. Amongst land resources the most important are; forests, grasslands and wetlands.
16. World's forest cover is shrinking rapidly, especially in the developing countries.
17. Deforestation is the main cause for the extinction of plant, animal and microbial species, and may induce regional and global climatic changes.
18. Massive afforestation programmes are required to save forest cover.
19. On the other hand, grasslands provide forage and habitat to domestic animals and wildlife. The grass cover also effectively binds the soil with its highly branched fibrous root system.
20. Wetlands of freshwater or salt water provide important environmental services, such as high productivity, control of floods, and serve as groundwater recharging areas.
21. Energy resources are non-renewable, such as fossil fuel, nuclear energy or renewable, such as hydropower, wind, geothermal energy, ocean waves, tidal energy, etc.
22. Renewable energy generally causes much less negative environmental impact than fossil fuels or nuclear energy.
23. Amongst various solar energy resources, biomass energy is most important.
24. Various algae and animals form important marine resources.
25. However, minerals are non-renewable resources.
26. India is a unique country with a great cultural diversity, associated with all kinds of climates and rich flora and fauna.
27. The human societies in this great country have evolved within magnificent environments, and reverence to nature is inherent in our culture.
28. *Isha-Upanishad* states : 'The whole universe together with its creatures belongs to the Lord. Let no one species encroach over the rights and privileges of other species. One can enjoy nature by giving up greed'.

TERMS TO REMEMBER

Afforestation. Extensive planting of trees.

Agro-forestry. Growing of woody species in combination with herbaceous crops.

Biogas. Biomass, particularly animal waste can be converted into biogas in biogas digesters, by the process of anaerobic microbial decomposition.

Biomass energy. This may be obtained from those materials whose origin can be traced to photosynthesis.

Conservation forestry. Involved protection of degraded forests to allow recoupment of their flora and fauna.

Commercial forestry. To fulfil the commercial demand, without causing denudation of natural forests, through intensive plantation in available land.

Deforestation. Destruction of forests by cutting or felling of trees.

Desertification. The conversion of grassland or forest to desert.

Direct solar energy. The radiant energy from sun.

Energy plantations. Plant species, showing high calorific value and growth rate.

Exhaustible resources. They have finite supply on the earth and can be exhausted.

Forest Act. This act aims to consolidate the law relating to forests.

Geothermal energy. Where hot springs, can be used to turn turbines to generate electricity.

Grasslands. Also called rangeland, which provides forage and habitat to domestic animals and wildlife.

Hydropower. Water falling from a height turns turbines at the bottom of dams to generate electricity.

Indirect solar energy. Energy obtained from materials that have previously incorporated the sun's radiant energy.

Inexhaustible resources. Resources available in unlimited quantities on the earth, e.g., solar energy, wind power, etc.

Jhum. The shifting cultivation, a traditional agro-forestry system.

Mineral conservation. Recycling and reuse of minerals.

Mine spoil. The land that has been destroyed due to mining.

Mangrove swamps. Coastal wetlands in tropics having certain trees and shrubs growing best in the intertidal zone.

Marshes. Where grass-like plants dominate in wetlands.

Natural resource. Any component of the natural environment utilised by man for his welfare.

Non-renewable resources. Such resources cannot be regained or reconstructed once they are used up.

Non-renewable energy. Resources include various fossil fuels and nuclear energy.

National Forest Policy. To ensure environmental stability and maintenance of ecological balance.

Ocean waves energy. Ocean waves produced by winds are capable to turn a turbine and generate electricity.

Production forestry. See commercial forestry.

Protection forestry. See conservation forestry.

Renewable energy. Resources are regenerated by natural processes so that they can be used indefinitely.

Riverine forests. Forests found in lowlands along streams.

Renewable resources. Such resources can be successfully managed so that they are continuously regenerated.

Soil conservation. Soil and crop management practices which can minimise erosion and deplete nutrient depletion of agricultural soils.

Social forestry. Planting of trees and shrubs on all unused and fallow land to provide fuel wood, fodder, etc.

Swamps. Where trees or shrubs dominate in wetlands.

Tidal energy. The difference in the level of water between high tide and low tide can be used to generate electricity.

Taungya system. This involves growing agricultural crops between rows of planted trees.

Wetlands. Low-lying areas covered with shallow water.

Wind energy. When fans are rotated by the action of wind, its energy can be used to generate electricity.

Wildlife Laws. Legal provisions to safeguard the national interests related to wildlife.

QUESTION BANK

Very Short Answer Questions

1. What do you understand by natural resources ?
2. Why marine algae are used as manure ?
3. What is India's share in total global fish catch ?
4. In which part of atmosphere ozone layer is found ?
5. What is the percentage of forest of our country, and what should it be ?
6. What is total land area under forests in the world ?
7. How will you classify land vegetation ?
8. What are wetlands ?
9. What is weathering ?
10. Give few examples of inexhaustible natural resources.
11. Give some examples of renewable exhaustible natural resources.
12. Give two examples of non-renewable exhaustible natural resources.
13. What is per capita forest area available in India ?

14. What is social forestry ?

15. What is desertification ?

16. Give a single main cause of soil erosion and flood.

17. Name the sea weeds that are processed and fed to animals.

18. Which algae are used for extraction of agar.

ANSWERS

1. Any component of the natural environment that can be utilised by man to promote his welfare; **2.** Due to their high nutrient content; **3.** 2.5 million tons; **4.** Stratosphere; **5.** 19.4 per cent and 33.0 per cent; **6.** 30.0 per cent; **7.** Forests, grasslands, shrubs and crops; **8.** Low lying areas covered with shallow water; **9.** This is destructive process of transformation of solid rocks into soil; **10.** Solar energy, wind power, rainfall, power of tides and hydro-power; **11.** Water, soil fertility, natural vegetation, wildlife, aquatic animals and humans; **12.** Biological species and minerals; **13.** 0.06 ha per person; **14.** Planting of trees and shrubs on all unused and fallow land to provide fuel wood, fodder, etc; **15.** The conversion of grassland or forest to desert; **16.** Deforestation; **17.** Laminaria and Fucus; **18.** Gracilaria and Gelidium of a group called the red algae.

Short Answer Questions

1. What is desertification ? How is it caused ?

2. What is soil erosion ? How is it accelerated ?

3. What are the advantages of recycling and reusing ?

4. What are the kinds of solar energy ?

5. What are the forms of forests ?

6. Give a short account of fresh water wetlands.

7. What are the advantages of energy plantation ?

8. What is humus ? Give its two main functions.

9. How is land vegetation classified ?

10. Write a short note on agroforestry.

11. How is soil fertility depleted ?

12. How can you say that global hydrological cycle is perfectly balanced ?

13. What are the steps involved to improve erosion affected soil ?

14. List some legal provisions for safeguard of wildlife.

15. Explain why soil resource is important for human welfare.

16. What are the methods of conserving water resources ?

17. What is the importance of forest resources ?

18. How do we manage our grassland ?

19. Why are wetlands considered ecologically important ?

20. Explain the significance of biomass energy.

21. Describe major marine resources.

22. What do you mean by field capacity ?

23. What is agar ? From which organisms is it obtained ?

24. What is the importance of humus in the soil ?

25. Explain the value of minerals as natural resources.

26. What are the main sources of energy in the world ?

27. What role do the animals play in the soil ?

28. Give important marine resources.

29. Name the various regions of the atmosphere, and give the position of each of them.

30. What are alternative sources of energy ?

31. How can mineral resources be conserved ?
32. Give an account of soil profile.
33. Give a comprehensive account of commercial and social forestry.
34. How is water present in the earth ?

Long Answer Questions

1. Distinguish between the following :
   (*a*) Inexhaustible and exhaustible resources (*b*) Renewable and non-renewable resources
   (*c*) Afforestation and agro-forestry
2. Distinguish between the following :
   (*a*) Freshwater wetlands and marine wetlands (*b*) Grasslands and wetlands
   (*c*) Deforestation and desertification
3. What are the main approaches for conservation of water ?
4. What are the frequently employed measures of grassland management ?
5. Discuss the methods of soil conservation.
6. Give the list of different categories of natural resources. Illustrate your answer with suitable examples.
7. Explain the role of plants as essential resources.
8. Describe the major marine resources.

MULTIPLE CHOICE QUESTIONS (OBJECTIVE QUESTIONS)

1. Tick the correct answer. Soil erosion can be prevented by :
   (*a*) Overgrazing (*b*) Removal of vegetation
   (*c*) Afforestation (*d*) Deforestation
2. Mild grazing in grassland by herbivores :
   (*a*) Retards growth of grasses (*b*) Arrests growth of grasses
   (*c*) Stimulates growth of grasses (*d*) Destroys vegetation
3. Deforestation generally decreases :
   (*a*) Rainfall (*b*) Soil erosion (*c*) Drought (*d*) Global warming
4. Forest area in India is about :
   (*a*) 9 per cent of geographical area (*b*) 19 per cent of geographical area
   (*c*) 29 per cent of geographical area (*d*) 37 per cent of geographical area
5. Which one of the following represents a renewable source of energy :
   (*a*) Petroleum (*b*) Coal (*c*) Nuclear fuel (*d*) Trees
6. Which of the following represents the regulative function of forest :
   (*a*) Storage and release of gases (*b*) Production of wood
   (*c*) Production of essential oils (*d*) Conservation of soil and water
7. In India per capita forest area is :
   (*a*) 0.06 ha (*b*) 0.60 ha (*c*) 1.0 ha (*d*) 1.6 ha
8. Extensive planting of trees to increase forest cover is called :
   (*a*) Afforestation (*b*) Agro-forestry (*c*) Deforestation (*d*) Social forestry
9. Wetlands occupy :
   (*a*) 6 per cent of world's land (*b*) 10 per cent of world's land
   (*c*) 12 per cent of world's land (*d*) 14 per cent of world's land
10. Which of the following functions performed by a forest help most in controlling drought :
    (*a*) Forests prevent soil erosion (*b*) Forests bring rainfall
    (*c*) Forests have lot of water plants (*d*) Forests act as water sheds

11. The value of a resource depends upon :
(*a*) Quantity of resource (*b*) Quality of resource
(*c*) Type of society and its needs (*d*) All the above

12. All the following are proper plans for increasing agriculture production except :
(*a*) Expansion of irrigation facilities
(*b*) Increase in application of fertilisers
(*c*) Mechanisation of various aspects of farming
(*d*) Cutting of forests to add land for cultivation

13. Amount of water a soil can hold against the pull of gravity is known as :
(*a*) Hygroscopic water (*b*) Storage capacity (*c*) Gravitational water (*d*) Field capacity

14. Edaphic factors are related to :
(*a*) Man (*b*) Soil (*c*) Animals (*d*) Temperature

15. Quantity of water the soil can hold against the force of gravity is known :
(*a*) Storage water (*b*) Gravitational water (*c*) Hygroscopic water (*d*) Field capacity

16. American water plant which has become a troublesome water weed in India is :
(*a*) Cyperus rotundus (*b*) Trapa latifolia (*c*) Eichhornia crassipes (*d*) Trapa bispinosa

17. The least porous soil among the following is :
(*a*) Sandy soil (*b*) Clayey soil (*c*) Loamy soil (*d*) Gravel soil

18. Waterlogging occurs in :
(*a*) Sandy soil (*b*) Gravel soil (*c*) Loamy soil (*d*) Clay soil

19. Humus is an example of :
(*a*) Soil structure (*b*) Crystalloids (*c*) Organic colloids (*d*) None of these

20. Alluvial soil is mostly found in which region :
(*a*) Southern India (*b*) Northern India
(*c*) Gangetic Jamuna Plains (*d*) Eastern India

21. Water remaining in the soil after a plant has permanently wilted is :
(*a*) Hygroscopic water (*b*) Ground water (*c*) Gravitational water (*d*) Capillary water

22. Desertification is due to :
(*a*) Deforestation (*b*) Soil-erosion due to water
(*c*) Soil-erosion due to wind (*d*) Very high temperature and low rainfall.

23. Geothermal energy is :
(*a*) Non-renewable, conventional source of energy
(*b*) Non-renewable, non-conventional source of energy
(*c*) Renewable, non-conventional source of energy
(*d*) Renewable conventional source of energy

24. Deforestation has an alarming effect on :
(*a*) Soil erosion (*b*) Weed control
(*c*) Sunlight (*d*) Increase in grazing area

25. The best soil for healthy and vigrous growth of most plants is :
(*a*) Sandy soil (*b*) Loam soil (*c*) Clay soil (*d*) None of these

26. Who will maintain the organic matter of the soil :
(*a*) Fungi and semiparasites (*b*) Cyanobacteria and parasitic animals
(*c*) Protozoans and slime molds (*d*) Bacteria and Fungi

27. Terai forest is :
(*a*) Tropical forest (*b*) Coniferous forest
(*c*) Deciduous forest (*d*) Temperate deciduous forest

28. Soil erosion can be prevented by :
(*a*) Deforestation (*b*) Afforestation (*c*) Overgrazing (*d*) All of these

29. Soil water available to plants is :
(*a*) Surface water (*b*) Hygroscopic water (*c*) Capillary water (*d*) Gravitational water

30. Water holding capacity is maximum in :
(*a*) Silt soil (*b*) Sandy soil (*c*) Clayey soil (*d*) Loamy soil

31. Source of energy which does not produce CO_2 is :
(*a*) Oil (*b*) Coal (*c*) Nuclear energy (*d*) Organic compounds

32. A non-renewable resource is :
(*a*) Coal and minerals (*b*) Plants and coal
(*c*) Water and natural gas (*d*) Energy and water

33. Non-conventional source of energy is :
(*a*) Coal (*b*) Wood (*c*) Petroleum (*d*) Biogas/solar energy

34. Bulk of carbon fixation by photosynthesis occurs in :
(*a*) Oceans (*b*) Croplands (*c*) Tropical rain forests (*d*) (*b*) and (*c*)

35. Plants die in water-logged soil because of :
(*a*) Nutrient leaching (*b*) Dilution of soil nutrients
(*c*) Stoppage of root respiration (*d*) Dilution of cell sap

36. Main reason of animal extinction is :
(*a*) Felling of trees (*b*) Habitat destruction (*c*) Predation (*d*) Pollution

37. Soil best suited for plant growth is :
(*a*) Loamy (*b*) Clayey (*c*) Sandy (*d*) Gravelly

38. Soil fertility is reduced by :
(*a*) Nitrogen-fixing bacteria (*b*) Decaying organic matter
(*c*) Crop rotation (*d*) Intensive agriculture

39. Black soil is due to the presence in it of :
(*a*) Organic matter (*b*) More moisture (*c*) Minerals (*d*) Air

40. The renewable source of energy is :
(*a*) Coal (*b*) Petroleum (*c*) Biomass (*d*) Kerosene

41. Black soil is productive because of abundant :
(*a*) Gravel and calcium (*b*) Clay and humus (*c*) Silt and earthworms (*d*) Sand and zinc

42. A non-renewable resource is :
(*a*) Water (*b*) Forest (*c*) Wild life (*d*) Coal/Petroleum

43. Fertility of soil is determined by its ability to :
(*a*) Hold organic matter (*b*) Retain water (*c*) Retain nutrients (*d*) Support life

44. Of the total amount of water, the water found in the seas and oceans forms about :
(*a*) 60 per cent (*b*) 70 per cent (*c*) 80 per cent (*d*) 95 per cent

45. Soil is composed of :
(*a*) Minerals, water and air (*b*) Minerals, organic matter and water
(*c*) Minerals, organic matter, air and water (*d*) Organic matter and water.

ANSWERS

1. (*c*)	**2.** (*c*)	**3.** (*a*)	**4.** (*b*)	**5.** (*d*)	**6.** (*a*)	**7.** (*a*)	**8.** (*a*)	**9.** (*a*)	**10.** (*b*)
11. (*c*)	**12.** (*d*)	**13.** (*d*)	**14.** (*b*)	**15.** (*d*)	**16.** (*c*)	**17.** (*b*)	**18.** (*d*)	**19.** (*c*)	**20.** (*c*)
21. (*a*)	**22.** (*a*)	**23.** (*b*)	**24.** (*a*)	**25.** (*b*)	**26.** (*d*)	**27.** (*a*)	**28.** (*b*)	**29.** (*c*)	**30.** (*c*)
31. (*c*)	**32.** (*a*)	**33.** (*d*)	**34.** (*a*)	**35.** (*c*)	**36.** (*b*)	**37.** (*a*)	**38.** (*d*)	**39.** (*a*)	**40.** (*c*)
41. (*b*)	**42.** (*d*)	**43.** (*d*)	**44.** (*d*)	**45.** (*c*)					

14
CHAPTER

Biodiversity and Biogeographical Regions of India

'Biological diversity' encompasses all species of plants, animals, and microorganisms and the ecosystems and ecological processes of which they are parts. It is an umbrella term for the degree of nature's variety including both the number and frequency of ecosystems, species, or genes in a given assemblage. Thus, the term '**biodiversity**' refers to the totality of 'genes, species and ecosystems' of a region.

It is usually considered at three different levels; (*i*) genetic diversity, (*ii*) species diversity and (*iii*) ecosystem diversity.

Genetic diversity is the sum total of genetic information, contained in the genes of individuals of plants, animals and microorganisms that inhabit the earth.

Species diversity refers to the variety of living organisms on earth and has been variously estimated to be between 5 and 50 million or more, though only about 1.4 million have actually been described.

Ecosystem diversity relates to the variety of habitats, biotic communities, and ecological processes in the biosphere, as well as the tremendous diversity within ecosystems in terms of habitat differences and the variety of ecological processes.

Biodiversity is defined as "the variability among living organisms from all sources, including terrestrial, marine and other aquatic ecosystems and the ecological complexes of which they are a part; this includes diversity within species, between species and of ecosystems."

Conservation and sustainable use of biodiversity is fundamental to ecologically sustainable development. Biodiversity is part of our daily lives and livelihood, and constitutes resources upon which families, communities, nations and future generations depend.

Every country has the responsibility to conserve, restore and sustainably use the biological diversity within its jurisdiction. Biological diversity is fundamental to the fulfilment of human needs. An environment rich in biological diversity offers the broadest array of options for sustainable economic ability, for sustaining human welfare and for adapting to change. Loss of biodiversity has serious economic and social costs for any country.

The experience of the past few decades has shown that as industrialisation and economic development in the classical sense takes place, patterns of consumption, production and needs, change, straining, altering and even destroying ecosystems.

India, a megabiodiversity country, while following the path of development, has been sensitive to needs of conservation and hence is still rich in biological resources. Ethos of conservation and harmonious living with nature is very much ingrained in the lifestyles of India's people.

MAGNITUDE OF BIODIVERSITY

The foundation for assessing the importance of biodiversity is an inventory of how many species exist and which species exist where. At the global level the plants and vertebrates are relatively well known. Erwin (1982), for example, suggests as many as 30 million species in total, with most undescribed species living in tropical forests. The known and described number of species of all organisms on the earth is between 1.7 and 1.8 million which is fewer than 15 per cent of the actual number'. About 61 per cent of the known species are insects. Only 4650 species of mammals are known to biological science. A large number of plant species (2,70,000) and vertebrates are known. But the fact remains that basic knowledge of the organisms that make up most ecosystems, especially in the tropics is inadequate. Information about bacteria,viruses, protists and Archaea is only fragmentary.

For convenience, many assume that about 10 million species exist, though the final figure is likely to be 30-50 million.

Given these limitations, the following table 14.1, represents a summary of the current state of knowledge (from Wilson, 1988):

Table 14.1. How many species exist

Group	*Number of described species*
Bacteria and blue-green algae: (*i.e.*, cyanobacteria)	4760
Fungi:	46,983
Algae:	26,900
Bryophytes (mosses and liverworms):	17,000 (WCMC, 1988)
Gymnosperms (conifers):	750 (Raven *et al.*, 1986)
Angiosperms (flowering plants):	250,000 (Raven *et al.*, 1986)
Protozoans:	30,800
Sponges:	5,000
Corals and Jellyfish:	9,000
Roundworms and earthworms:	24,000
Crustaceans:	38,000
Insects:	751,000
Other arthropods and minor invertebrates:	132,461
Molluscs:	50,000
Starfish:	6,100
Fishes (Teleosts):	19,056
Amphibians:	4,184
Reptiles:	6,300
Birds:	9,198 (Clements, 1981)
Mammals:	4,170 (Honacki *et al.*, 1982)
Total	**1,435,662**

However, new species are being discovered faster than ever before due to the efforts of Global Biodiversity Information Facility and the Species, 2000. The approximate number of species is being mentioned in the table 14.2.

Table 14.2. Approximate Numbers of Species.

Group	*Number of described species*
Higher plants	2,70,000
Algae	40,000
Fungi	72,000
Bacteria and cyanobacteria	4,000
Viruses	1,550
Mammals	4,650
Birds	9,700
Reptiles	7,150
Fish	26,959
Amphibians	4,780
Insects	10,25,000
Crustaceans	43,000
Molluscs	70,000
Nematodes and worms	25,000
Protozoa	40,000
Others	1,10,000

The number of plant and animal species in different groups recorded in India are given in table 14.3.

Table 14.3. Number of Species of Different Taxonomic Groups

Plants		*Animals*	
Group	***No. of species***	***Group***	***No. of species***
Angiosperms:	17,500	Mammalia:	390
Gymnosperms:	64	Aves:	1232
Pteridophytes:	1100	Reptilia:	456
Bryophytes:	2850	Amphibia:	209
Lichens:	200	Pisces:	2546
Fungi:	14,500	Protochordata:	119
Algae:	6,500	Other invertebrates:	8329
Bacteria:	850	Arthropoda:	68389
		Mollusca:	5070
		Protozoa:	2577

India has 47,000 species of flowering and non-flowering plants representing about 12 per cent of the recorded world's flora. Out of 47,000 species of plants, 5,150 are endemic and 2532 species are found in the Himalayas and adjoining regions and 1,782 in the peninsular India. India is also rich in the number of endemic faunal species it possesses.

Table 14.4. Comparative statement of recorded number of animal species in India and the world

Taxa	INDIA			World	Percentage of India to the world
	Species	Endemic species	Threatened species		
Protista	2577			31,259	8.24
Mollusca	5070	967		66,535	7.62
Arthropoda	68389	16214 (insects)		987,949	6.90
Other Invertebrates	8329		22	87,121	9.56
Protochordata	119			2106	5.65
Pisces	2546		4	21,723	11.72
Amphibia	209	110	3	5150	4.06
Reptilia	456	214	16	5817	7.84
Aves	1232	69	73	9026	13.66
Mammalia	390	38	75	4629	8.42

Source: MoEF 1999; Baillie 1996.

LEVELS OF BIODIVERSITY

Biodiversity is usually considered at three different hierarchical levels:

(*i*) Genetic diversity,

(*ii*) Species diversity, and

(*iii*) Community and ecosystem diversity.

Though these levels of biodiversity are interrelated yet distinct and can be studied separately to understand the interrelationships that sustain life on earth.

Genetic Diversity

Genetic diversity is the sum total of genetic information, contained in the genes of individuals of plants, animals, and microorganisms that inhabit the earth. Each species, stores an immense amount of genetic information.

For example, the number of genes is about 450-700 in Mycoplasma; 4000 in *Escherichia coli*; 13,000 in *Drosophila melanogaster*; 32,000-50,000 in *Oryza sativa* and 35,000 to 45,000 in *Homo sapiens*.

The characteristics of genetic diversity are as follows:

Genetic diversity refers to the variation of genes within species. The differences could be in alleles, *i.e.*, different variants of same genes; in entire genes, *i.e.*, the traits that determine particular characteristics or in chromosomal structures.

The genetic diversity enables a population to adapt to its environment and to respond to natural selection. On the other hand, if a species has more genetic diversity, it can adapt better to the changed environmental conditions.

Lower diversity in a species leads to uniformity, for example, in large monocultures of genetically similar crop plants. This may be an advantage when increased crop production is considered, whereas this may be a problem when a pest or pathogen attacks the crop field, and the whole crop is threatened.

The amount of genetic variation is the basis of **speciation**, *i.e.*, evolution of new species. It has a key role to play in the maintenance of diversity at species and community levels.

The total genetic diversity of a community will be greater if there are many species, as compared to an area with few species. Genetic diversity within a species often increases with environmental variability.

Species Diversity

Species diversity refers to the variety of living organisms on earth and has been variously estimated to be between 5 and 50 million or more, though only about 1.4 million have actually been described.

Species are distinct units of diversity, and each unit plays a specific role in an ecosystem. Thus, loss of species has consequences for the whole ecosystem.

Species diversity refers to the variety of species within a region.

Species richness or the number of species per unit area, denotes the measure of species diversity. The number of species increases with the area of the site, that is greater the species richness, greater is the species diversity. On the other hand, number of individuals among the species may also vary, resulting into differences in **evenness**, or **equitability** and consequently in diversity.

Sample area 1

Sample area 2

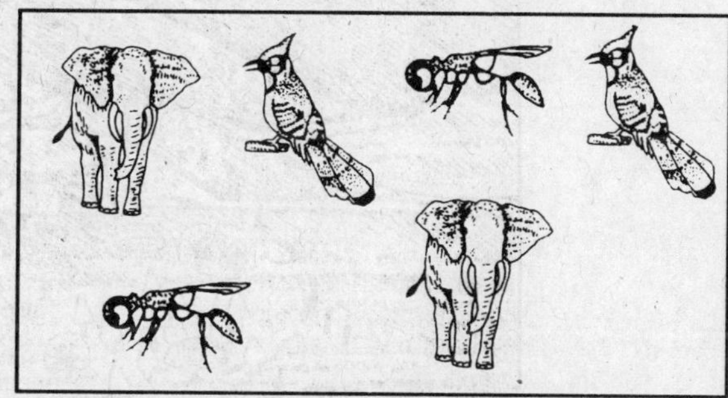

Sample area 3

Fig. 14.1. Different sample areas species richness (sample 1), species evenness (sample 2) and diversity due to taxonomically unrelated species (sample 3).

Here the example of three sample areas may be quoted:

1. The sample area 1, consists of three species of birds.Two species are represented by one individual each, while the third species has four individuals.

2. The sample area 2, also consists of three species, here, each species is represented by two individuals. This sample area exhibits greater evenness, and there are equal chances for a species being represented in a sample. The sample area 2 is more diverse than sample area 1.

3. The sample area 3, depicts the presence of an insect, a mammal and a bird. This sample area is considered most diverse, as it comprises of taxonomically unrelated species.

This example shows, there are equal number of species in each sample area, *i.e.*, samples 1, 2 and 3, but having varying number of individuals per species.

However, in nature, both the number and kind of species, as well as the number of individuals per species vary,which result in greater diversity.

Community and Ecosystem Diversity

Ecosystem diversity is related to the variety of habitats, biotic communities, and ecological processes in the biosphere, as well as the tremendous diversity within ecosystems in terms of habitat differences and the variety of ecological processes.

Diversity at the level of community and ecosystem has been divided into three categories:

1. Alpha diversity (= within community diversity). This refers to the diversity of organisms sharing the same community or habitat.

A combination of species richness and equitability or evenness is used to represent diversity within a community or habitat. Species frequently change when habitat or community changes.

2. Beta diversity (= between-community diversity). This refers the rate of replacement of species along a gradient of habitats or communities. In beta diversity, there are differences in species composition of communities along environmental gradients, *e.g.*, altitudinal gradient, moisture gradient, etc. Greater the dissimilarity between communities, higher is the beta diversity.

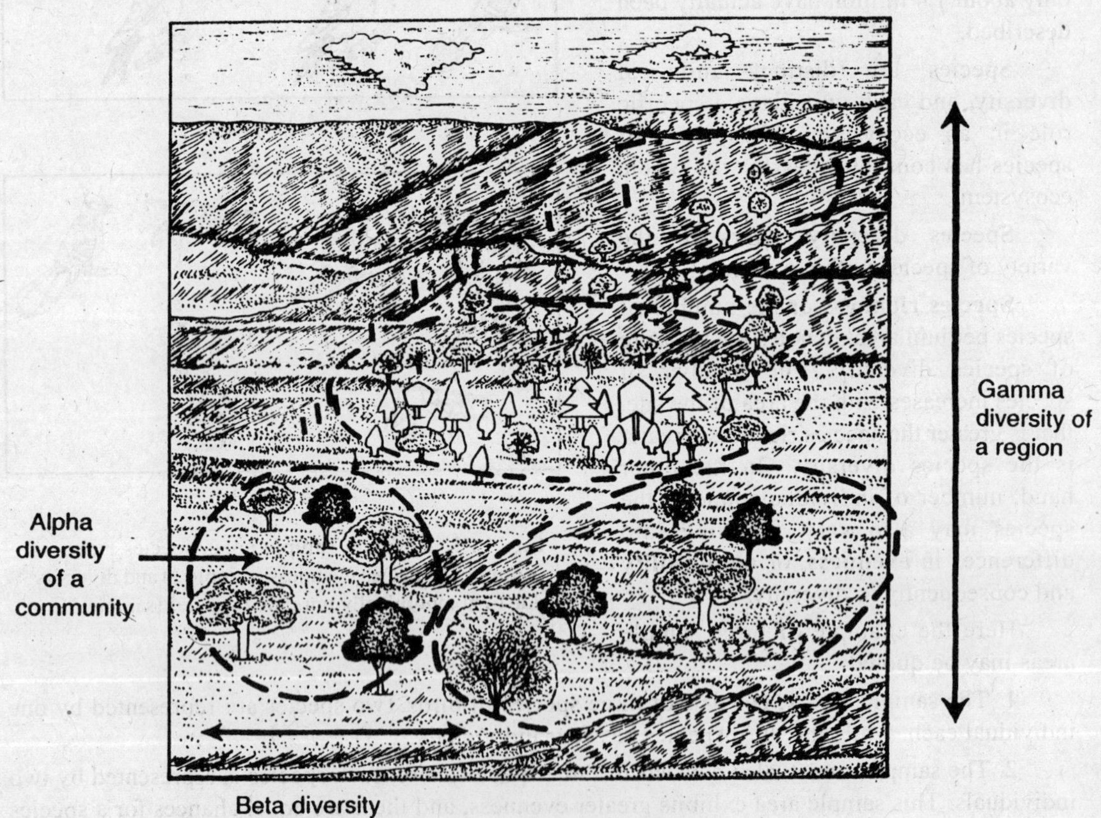

Fig. 14.2. Alpha, beta and gamma diversity.

3. Gamma diversity. This refers to the diversity of the habitats over the total landscape or geographical area.

Characteristics of ecosystem diversity. In ecosystem diversity there are a number of ecological niches, trophic levels and various ecological processes that sustain energy flow, food needs and the recycling of nutrients.

Ecosystem diversity has a focus on various biotic interactions and the role and function of keystone species.

However, the diverse communities are functionally more productive and stable, even under environmental stresses, such as dry or drought conditions.

STATUS OF BIODIVERSITY IN INDIA

India occupies only 2.4 per cent of the world's land area but its contribution to the world's biodiversity is approximately 8 per cent of the total number of species, which is estimated to be 1.75 million. Of these, 126,188 have been described in India. The species recorded includes flowering plants (angiosperms), mammals, fish, birds, reptiles and amphibians, constitute 17.3 per cent of the total wheareas nearly 60 per cent of India's bio-wealth is contributed by fungi and insects.

Based on the available data, India ranks tenth in the world and fourth in Asia in plant diversity, and ranks tenth in the number of endemic species of higher vertebrates in the world. There are ten biogeographical zones in India. They can be classified as follows:

1. Trans-Himalayas. An extension of the Tibetan Plateau, harbouring high-altitude cold desert in Laddakh (J and K) and Lahaul Spiti (H.P.) comprising 5.7 per cent of the country's landmass.

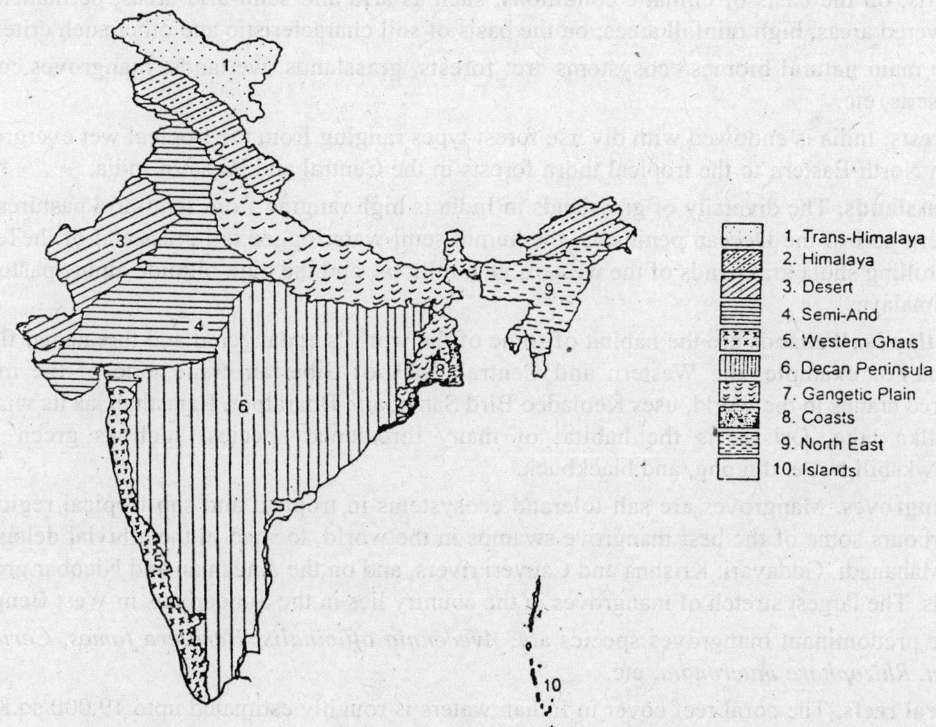

Fig. 14.2(*a*). Biogeographical regions of India (source WII Dehradun).

2. Himalayas. The entire mountain chain running from north-western to north-eastern India, comprising a diverse range of biotic provinces and biomes, 7.2 per cent of the country's landmass.

3. Desert. The extremely arid area west of the Aravalli hill range, comprising both the salty desert of Gujarat and the sand desert of Rajasthan. 6.9 per cent of the country's landmass.

4. Semi-arid. The zone between the desert and the Deccan plateau, including the Aravalli hill range. 15.6 per cent of the country's landmass.

5. Western ghats. The hill ranges and plains running along the western coast line, south of the Tapti river, covering an extremely diverse range of biotic provinces and biomes. 5.8 per cent of the country's landmass.

6. Deccan peninsula. The largest of the zones, covering much of the southern and south-central plateau with a predominantly deciduous vegetation. 4.3 per cent of the country's landmass.

7. Gangetic plain. Defined by the Gangas river system, these plains are relatively homogenous. 11 per cent of the country's landmass.

8. North-east India. The plàins and non-Himalayan hill ranges of north eastern India, with a wide variation of vegetation. 5.2 per cent of the country's landmass.

9. Islands. The Andaman and Nicobar Islands in the Bay of Bengal, with a highly diverse set of biomes. 0.03 per cent of the country's landmass.

10. Coasts. A large coastline distributed both to the west and east, with distinct differences between the two; Lakshadeep Islands are included in this with the percent area being negligible.

Apart from the above mentioned biogeographic classification, ecosystems can also be demarcated on the basis of purely geographical or geological features, such as mountains, islands, valleys, plateau, oceans, etc., on the basis of vegetative cover like forests, grasslands, mangroves and deserts; on the basis of climatic conditions, such as arid and semi-arid areas, permanently snow-covered areas, high rainfall areas; on the basis of soil characteristic and other such criteria.

The main natural biomes/ecosystems are: forests, grasslands, wetlands, mangroves,coral reefs, deserts, etc.

Forests. India is endowed with diverse forest types ranging from the tropical wet evergreen forests in North-Eastern to the tropical thorn forests in the Central and Western India.

Grasslands. The diversity of grasslands in India is high ranging from semi-arid pastures of the western part of the Deccan peninsula, the humid semi-waterlogged tall grassland of theTerai belt, the rolling shola grasslands of the western ghat hilltops, and the high-altitude alpine pastures of the Himalayas.

Wetlands. Wetlands are the habitat of some of the world's endangered and threatened flora and fauna.For example, the Western and Central flock of Siberian crane, one of the most endangered cranes in the world, uses Keoladeo Bird Sanctuary, Bharatpur, Rajasthan, as its winter site. Chilka lake, Orissa, is the habitat of many threatened species, such as green sea turtle,Hawksbill turtle, dugong, and blackbuck.

Mangroves. Mangroves are salt-tolerand ecosystems in tropical and sub-tropical regions. India harbours some of the best mangrove swamps in the world, located in the alluvial deltas of Ganga, Mahanadi, Godavari, Krishna and Cauveri rivers, and on the Andaman and Nicobar group of Islands. The largest stretch of mangroves in the country lies in the Sunderbans in West Bengal.

The predominant mangroves species are; *Avecennia officinalis*, *Heritiera fomes*, *Ceriops decandra*, *Rhizophora mucronata*, etc.

Coral reefs. The coral reef cover in Indian waters is roughly estimated upto 19,000 sq.Km. Indian coral reefs are as follows:

Palk Bay and Gulf of Mannar; Gulf of Kachchh; Andaman and Nicobar Islands; Lakshadeep Islands and Central West Coast.

Deserts. Three kinds of deserts are noticeable in India:

1. The sand desert of western Rajasthan and neighbouring areas.
2. The vast salt desert of Gujarat
3. The high-altitude cold desert of Jammu and Kashmir and Himachal Pradesh.

Desert fox, desert cat, houbara bustard and some sandgrouse species are restricted to Thar Desert, Rajasthan.

In the Rann of Kachchh, lies the nesting ground of Flamingoes and the only known population of Asiatic wild ass.

The distinctive animals of cold desert are:

Kiang (relative of wild ass of Kachchh), snow leopard, Yak, Tibetan antelope, Ibex, Blue sheep, Tibetan gazelle, Woolly hare, etc.

GRADIENTS OF BIODIVERSITY

Biodiversity varies with change in latitude or altitude.When one moves from high to low latitudes (*i.e.*, from the poles to the equator), the biodiversiy increases.

In the temperate region, the climate is severe having short growing period for plants. On the other hand, in tropical rain forest the conditions are favourable for growth of plants throughout the year. Favourable conditions are helpful for speciation, and large number of species develop and grow.

The mean number of species of vascular plants per 0.1 ha sample area in tropical rain forests varies from 118-236, while in temperate zones, this range is 21-48 species.

Such a correlation between diversity and latitude is also found for a variety of several other taxa, such as ants, birds, butterflies, moths, etc.

In the similar way, there is a decrease in species diversity from lower to higher altitudes on a mountain. Increase of 1000 metres in altitude results in a temperature drop of about 6.5° C. Such

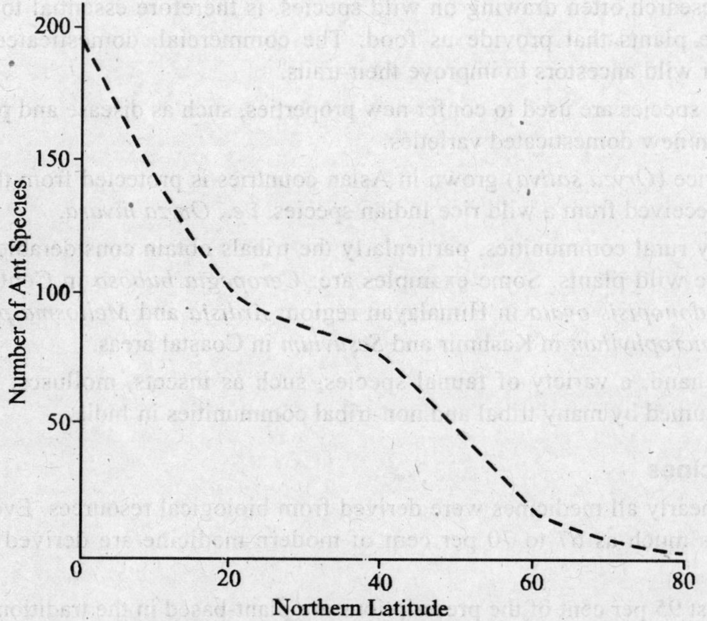

Fig. 14.3. Latitudinal gradient of biodiversity.

temperature drop and seasonal variability at higher altitudes are responsible for lowering the diversity.

The **latitudinal** and **altitudinal** gradients of species diversity are two master gradients. To some extent regional and taxa-related gradients also affect biodiversity. It is expected that prevalence of more complex and heterogeneous environment makes the flora and fauna more diverse and complex.

USES OF BIODIVERSITY

Humans derive many direct and indirect benifits from biological diversity. All our food comes from wild species brought into domestication. Most of our medicines, pharmaceutical, fibres, rubber and timber come from biological resources. The biodiversity also provides many ecological services free of charge that are responsible for maintaining ecosystems. Our water is supplied by one of nature's most important processes, called the 'hydrological cycle'. Forested watersheds provide clear,high-quality water for domestic or agricultural use,while healthy rivers provide water, transport and fish.

Source of Food and Improved Varieties

Biodiversity has direct consumptive value in agriculture. In modern agriculture, biodiversity is used as follows:

(*i*) As a source of new crops

(*ii*) As a source material for breeding improved varieties, and

(*iii*) As a source new biodegradable pesticides.

Approximately 80,000 edible plants have been used at one time or another in human history, of which only about 150 have even been cultivated on a large scale. Today a mere 10 to 20 species provide 80 to 90 per cent requirements of the world. Over half of human nutrition is provided by the three major carbohydrate crops, *i.e.*, rice, wheat and maize. Fats, oils, fibres, etc., are other uses for which many new species are to be investigated.

Continuous research,often drawing on wild species, is therefore essential to maintain the productivity of the plants that provide us food. The commercial, domesticated species are crossbred with their wild ancestors to improve their traits.

Genes of wild species are used to confer new properties, such as disease and pest resistance or increased yield in new domesticated varieties.

For example rice (*Oryza sativa*) grown in Asian countries is protected from the four major diseases by genes received from a wild rice Indian species, *i.e.*, *Oryza nivara*.

In India, many rural communities, particularly the tribals obtain considerable part of their daily food from the wild plants. Some examples are: *Ceropegia bubosa* in Central India and Western Ghats; *Codonopisis ovata* in Himalayan region; *Ardisia* and *Meliosma pinnata* in the North-east; *Cicer microphyllum* in Kashmir and *Sesuvium* in Coastal areas.

On the other hand, a variety of faunal species, such as insects, molluscs, spiders, wild herbivores are consumed by many tribal and non-tribal communities in India.

Drugs and Medicines

At one time, nearly all medicines were derived from biological resources. Even today they remain vital and as much as 67 to 70 per cent of modern medicine are derived from natural products.

In India, almost 95 per cent of the prescriptions are plant-based in the traditional systems of Ayurveda, Unani and Sidha. Many indigenous medicines also utilise medicines and their parts or extracts as remedies for various diseases.

Examples of plant-derived substances developed into valuable drugs are: Morphine obtained from *Papaver somniferum* and used as an analgesic; Quinine obtained from *Cinchona ledgeriana*, used for treatment of Malaria, and Taxol, an anticancer drug obtained from the bark of yew tree (*Taxas baccata*, *T.brevifolia*).

Botanochemicals. Many plants are also used for the manufacture of hundreds of synthetic products, which are called 'botanochemicals'.

Aesthetic and Cultural Benefits

Biodiversity has also great aesthetic value. Diverse habitats and species have non-consumptive use-value. Examples of aesthetic benefits include ecotourism, bird watching, wildlife, pet keeping, gardening, recreation and scientific research.

Many plants, animals and their parts are used in rituals all over the country. In a majority of Indian villages and towns, plants like *Ocimum sanctum* (Tulsi), *Ficus religiosa* (Pipal), *F. bengalensis* (Bargad), etc., are considered sacred and worshipped by the people. Flowers of *Hibiscus*, *Datura* and *Euphorbia*, leaves of *Aegle marmelos* (Bel), *Eragrostis cynasuroides* (Kusa grass), rice, til, chenopods sacred, odorous roots of *Dolomiaea macrocephala* (Dhup) are used in rituals. Besides, sacred values are attached to entire ecosystems, for example, patches of forests were believed to be the abode of demigods and are used only for prayers and rituals. Many 'sacred groves' exist in different parts of our country.

Several birds, snakes and other animals are treated sacred and worshipped.

Ecosystem Services

The indirect use-value of biodiversity includes ecosystem process of biological diversity, which provides valuable ecological services to the biosphere.

For example, the ecosystem's ability to absorb pollution, maintain soil fertility and microclimates, recharge groundwater and other invaluable services. These services include maintenance of gaseous composition of the atmosphere, climate control by forests and oceanic systems, natural pest control, pollination of plants by insects and birds, formation and protection of soil, conservation and purification of water, and nutrient cycling, etc.

The above mentioned ecosystem services have been valued in the range of 16 to 54 trillion US dollars per year.

THREATS TO BIODIVERSITY

Major biodiversity threats are as follows:

(*i*) Habitat destruction and fragmentation
(*ii*) Extention of agriculture
(*iii*) Filling up of wetlands
(*iv*) Conversion of rich biodiversity site for luman settlement and industrial development
(*v*) Destruction of coastal areas,
(*vi*) Uncontrolled commercial exploitation
(*vii*) Disturbance and pollution, and
(*viii*) Introduction of non-native (exotic) species

Habitat Loss and Fragmentation

The destruction of habitats is the primary reason for the loss of biodiversity. This erosion of biodiversity is largely due to habitat loss caused by the expansion of various development projects, such as mines, dams, roads and canal construction. When people cut down trees, fill a wetland, plough a grassland or burn a forest the natural habitat of a species is changed or destroyed. Such changes can kill or force out many plants, animals and microorganisms. These changes also disrupt complex interactions among the species.

Habitat loss leads to the fragmentation of continuous stretches of land and consequently fragments wildlife populations inhabiting them. These small populations are increasingly vulnerable to inbreeding depression, high infant mortality, susceptibility to environmental stochasticity and in the long run possibly to extinction.

Apart from the primary loss of habitats, there are many other problems contributing to the loss and endangered status of several plant and animal species. Habitat degradation can lead to declines in primary food species for wildlife.

Poaching is another threat as one of the primary reasons for extinction of species such as the tiger.

Population pressures, such as collection of fuelwood and fodder, and grazing in forests by local communities also take their toll on the forests and consequently its biodiversity.

Other minor factors include fire, and natural calamities like droughts, diseases, cyclones and land slides.

It is estimated that, after independence, the country has lost 4,696 million ha of forest land to non-forestry purposes.

Disturbance and Pollution

Communities are affected by natural disturbances, such as fire, tree fall, landslides and defoliation by insects.

Anthropogenic disturbances differ from natural disturbances in intensity, rate and spatial extent. For example, man by using fire more frequently may change species richness of a community.

The new anthropogenic impacts may be large number of synthetic compounds, massive releases of radiations by nuclear projects, and spill over of oil in sea. These impacts may lead to a change in the habitat quality.

Over exploitation of natural resourses, such as fishing by boats, mechanical catching of other animal species, etc., are serious threats to wildlife. Disturbance in migratory routes of several fishes, due to construction of dams, etc., hence they are not able to reach their spawning grounds and face extinction.

Pollution is also responsible for reduction and elimination of populations of sensitive species. For example, pesticide linked decline of fish-eating birds and falcons.

Lead poisoning of water is also a cause of mortality of many species, such as ducks, swans and cranes, as they take in the spent shotgun pellets that fell into lakes and marshes. Eutrophication, *i.e.*, nutrient enrichment of water bodies is also responsible for reduction of species diversity.

Introduction of Exotic Species

New species that enter a geographical region are called **exotic** or alien species. The invasion of such species may cause disappearance of native species through changed biotic interactions. The invasion of exotic species is considered second one to habitat destruction as a major threat to extinction of species.

Exotic species have more effective impact in island ecosystems, which give shelter to the world's threatened biological diversity. Examples are as follows:

(*i*) An exotic species *Lantana camara* has invaded many forest lands in different parts of our country, and strongly competes with the native species.

(*ii*) On the other hand water hyacinth (*Eichhornia crassipes*), an exotic aquatic species, clogs rivers, lakes and village ponds in different parts of India and several other tropical countries, which threaten, the survival of several native aquatic species.

(*iii*) Nile perch, an exotic predatory fish has been introduced into Lake Victoria of South Africa, where this has threatened the entire ecosystem of the lake, and eliminated several native species of small fish, *e.g.*, Cichild fish species.

(*iv*) *Periplaneta americana* (American cockroach) has threatened the existence of native cockroach, *i.e.*, *Blatta orientalis*.

(*v*) *Parthenium hysterophorus* has threatened several native herbs and shrubs of this country.

(*vi*) Goats and rabbits introduced in the islands of Indian and Pacific oceans are destroying the habitats of several plants, birds and reptiles.

Extinction of Species

Extinction has been a fact of life since life first emerged. The present few million species are the modern-day survivers of the estimated several billion species that have ever existed. All past extinctions have occurred by natural processes, but today humans are the main cause of extinction.

The rapid destruction of the world's most diverse ecosystems, especially in the tropics,has led most experts to conclude that perhaps a quarter of the earth's total biological diversity is at serious risk of extinction during the next 20-30 years. By many indications, the world is already experiencing extinction rates of greater scale and impact than at any previous time in the earth's history.

More species than even before are threatened with extinction, with thousands, mostly insects, disappearing each year. There are three main types of extinction processes:

1. Natural extinction. With the change in environmental conditions, some species become extinct, while others, which are more adapted to changed conditions,take their place.This type of extinction of species which took place in the geological past at a very slow rate, is called **natural extinction**.

2. Mass extinction. As we know, there have been several periods in the geological history of earth when large number of species disappeared because of catastrophes. This kind of extinction is called '**mass extinction**'. Such mass extinctions occurred in millions of years.

3. Anthropogenic extinction. All past extinctions occurred either by natural or mass extinctions, but today humans are the main cause of extinctions. Today an alarming number of species are disappearing from the surface of earth due to irresponsible human activities. This man-made extinction of species represents a very severe depletion of biodiversity, particularly because this occurs within a short period of time.

It has been estimated at World Conservation Monitoring Centre (WCMC), that 533 species of animals (mostly vertebrates) and 384 species of plants (mostly angiosperms) have become extinct since 1600 AD. More species have gone extinct from the islands than from the mainland or oceans. The current rate of extinction of species is 1000 to 10,000 times higher than the background rate of extinction. Some interesting observations about the current rate extinction of species are as follows:

1. From ten high-diversity localities in tropical forests which cover about 300,000 sq. km, some17,000 endemic plant species and 350,000 endemic animal species are threatened and could be lost in near future.

2. The tropical forests alone are losing about 14,000 to 40,000 species per year,which comes about 2-5 species per hour.

3. It is estimated that by the end of twenty first century, if the current rate continues, the earth will lose up to fifty per cent of the species.

Susceptibility to Extinction

The characteristics of species that make them susceptible for extinction are as follows:

1. Large body size. Many animals are large-sized, *e.g.*, Bengal tiger, lion and elephant.

2. Small population size and low reproductive rate. This characteristic of species is responsible for susceptibility to extinction, *e.g.*, Blue whale and Giant panda.

3. Feeding at high trophic levels in the food chain. This characteristic also makes the animals susceptible to extinction. The examples are: Bengal tiger and Bald eagle.

4. Fixed migratory routes and habitat. Such routes cannot be changed, if any interruption occurs that leads towards extinction. The examples are; Blue whale and Whooping crane.

5. Localised and narrow range of distribution. This characteristic also makes the species susceptible to extinction, *e.g.*, woodland caribon and many island species.

THE IUCN RED LIST CATEGORIES

The World Conservation Union (WCU), formerly known as International Union for the Conservation of Nature and Natural Resources (IUCN), has recognised eight Red List Categories of species. They are as follows:

1. Extinct. A taxon is **Extinct** when there is no reasonable doubt that the last individual has died.

2. Extinct in the Wild. A taxon is **Extinct in the Wild** when exhaustive surveys, in known and/or expected habitats, have failed to record an individual.

3. Critically Endangered. A taxon is **Critically Endangered** when it is facing an extremely high risk of extinction in the wild in the immediate future.

4. Endangered. A taxon is **Endangered** when it is not Critically Endangered, but is facing a very high risk of extinction in the wild in the near future.

5. Vulnerable. A taxon is **Vulnerable** when it is not Critically Endangered or Endangered, but is facing a high risk of extinction in the wild in the medium-term future.

6. Lower Risk. A taxon is **Lower Risk** when it has been evaluated and does not satisfy the criteria for Critically Endangered, Endangered or Vulnerable.

7. Data Deficient. A taxon is **Data Deficient** when there is inadequate information to make a direct, or indirect, assessment of its risk of extinction.

8. Not Evaluated. A taxon is **Not Evaluated** when it has not yet been assessed against the above criteria.

The IUCN Red List is a catalogue of those taxa which are facing the risk of extinction.

The uses of the Red List are:

(*i*) To develop awareness about the importance of threatened biological diversity.

(*ii*) Identification and documentation of endangered species.

(*iii*) Providing a global index of the decline of biodiversity.

(*iv*) Defining conservation priorities at the local level and guiding conservation action.

The species which are threatened with extinction are included in **Vulnerable, Endangered** or **Critically Endangered** categories.

Rare species. Taxa with small world populations that are not at present 'Endangered' or 'Vulnerable' but are at risk.

In practice, 'Endangered' and 'Vulnerable' categories may include, temporarily, taxa whose populations are beginning to recover as a result of remedial action, but whose recovery is insufficient to justify that transfer to another category. These taxa are usually localised within restricted geographical areas or habitats or are thinly scattered over a more extensive range.

Indeterminate. Taxa known to be 'Endangered', 'Vulnerable', or 'Rare', but where there is not enough information to say which of the three categories is appropriate.

The IUCN Red List System was initiated in 1963, and since then evaluation of the conservation status of species and subspecies is continuing on a global scale.

The 2000 IUCN Red List is the most comprehensive inventory of the global conservation status both of plant and animal species.

The 2000 IUCN Red List contains assessments of more than 18,000 species, of which 11,000 are threatened.

Status of Threatened Species. According to 2000 IUCN Red List, there are 11,046 species listed as threatened, *i.e.*, Critically Endangered, Endangered, or Vulnerable. Of these 5,485 are animals and 5,611 are plants. 1939 species are listed as Critically Endangered, of which 925 are animals and 1,014 are plants.

The percentage of threatened species of Angiosperms and four verteberate groups of animals are as follows:

Table 14.5. Percentage of Threatened Species

Categories	*Angiosperms*	*Amphibians*	*Reptiles*	*Birds*	*Mammals*
1. Critically Endangered	16%	14%	15%	9.0%	10%
2. Endangered	19%	22%	21%	17%	19%
3. Vulnerable	51%	48%	43%	48%	34%
4. Lower Risk	14%	16%	21%	16%	37%

According to 2000 IUCN Red List, in India 44 species of plants are critically endangered; 113 endangered and 87 vulnerable. While amongst animals 18 species are critically endangered, 54 endangered and 143 species, vulnerable.

Table 14.6. The Number of Plant and Animal Species of Various Threat Categories in India

Categories	*Plants*	*Animals*
1. Critically Endangered	44	18
2. Endangered	113	54
3. Vulnerable	87	143
4. Lower Risk	73	109
5. Data Deficient	14	31

Table 14.7. Examples of Threatened Species in India

Categories	*Plants*	*Animals*
1. Critically Endangered	*Berberis nilghiriensis*	**Sus salvanius** (Pigmy hog)
2. Endangered	*Bentinckia nicobarica*	**Ailurus fulgens** (Red Panda)
3. Vulnerable	*Cupressus cashmeriana*	**Antilope cerbicapra** (Black buck)

CONSERVATION OF BIODIVERSITY

The complex threats to biological diversity call for a wide range of responses across a large number of private and public sectors. All are necessary, with the mix of responses adjusted to the local conditions. As government policies are often responsible for depleting biological resources, it stands to reason that policy changes are often a necessary first step toward conservation.

National policies dealing directly with wildlands management or forestry, or influencing biodiversity indirectly through land tenure, rural development, family planning, and subsidies for food, pesticides, or energy can have significant impacts on the conservation of biodiversity.

Protecting species can best be done through protecting habitats. Most national governments have established legal means for protecting habitats that are important for conserving biological resources. These can include: national parks and other categories of reserves; local laws protecting particular forests, reefs or wetlands; planning restrictions on certain types of land; and customary laws protecting sacred groves or other special sites. The responsibility for such management is often spread widely among public and private institutions.

The Ministry of Environment and Forests (MOEF) is the main agency in the Government of India for planning, promotion, coordination and overseeing the implementation of the environmental and forestry programmes. The Mandates of the MOEF included survey of flora, fauna, forests and wildlife, and conservation of natural resources. A number of institutions affiliated with the Ministry are involved in the work related to various aspects of biological diversity.

Today, the ecosystems are undergoing change due to pollution, invasive species, anthropogenic over exploitation and climate change. Biodiversity at all the three levels (*i.e.*, genetic, species and ecosystem or community biodiversity) is important and needs to be conserved. The most effective and efficient mechanism for conserving biodiversity is to prevent further destruction or degradation of habitats by us.

People form the foundation for the sustainable use of biological resources. Local communities need to be more or involved in the management of biological resources, and to benefit from their sustainable use.

There are two basic strategies of biodiversity conservation, (*i*) *in situ* (on site) and (*ii*) *ex situ* (off site).

In situ Conservation (Within natural habitat) Strategies

Protected areas. The areas of land and/or sea especially for protection and maintenance of biodiversity, and of natural and associated cultural resources. These areas are managed through legal or other effective means, *e.g.*, National Parks and Wildlife Sanctuaries. The earliest national parks are: The Yellowstone National Park in USA and the Royal National Park near Sydney, Australia. These parks were chosen because of their scenic beauty and recreational values.

Today, many such protected areas throughout the world protect rare species. World Conservation Monitoring Centre (WCMC) has recognised 37,000 protected areas around the world.

In India, some important measures are taken. They are as follows:

Approximately 4.7 per cent of the total geographical area of the country has been earmarked for extensive *in situ* conservation of habitats and ecosystems. A protected area network of 89 National Parks and 492 Wildlife Sanctuaries have been created (MOEF, 2002). The results of this network have been significant in restoring viable population of large mammals, such as tiger, lion, rhinoceros, crocodiles, elephants, etc.

The Jim Corbett National Park, Nainital, Uttaranchal, was the first National Park, in India.

The Indian Council of Forestry Research (ICFRE) has identified 309 forest preservation plots of representative forest types for conservation of viable and representative areas of biodiversity. 187 of these plots are in natural forests and 112 in plantations covering a total area of 8,500 hectares.

Eco-development. A programme entitled 'eco-development' for *in situ* conservation of biological diversity involving local communities has been initiated in recent years. The concept of 'eco-development' includes the ecological and economic parameters for sustained conservation of ecosystems by involving the local communities with the maintenance of earmarked regions surrounding protected areas. The economic needs of the local communities are taken care of under this programme through provision of alternative sources of income and a steady availability of forest and related produce.

The main benefits of protected areas are:

1. To maintain viable populations of all native species and subspecies.
2. To maintain the number and distribution of communities and habitats. Conservation of the genetic diversity of all the existing species.
3. To prevent human caused introductions of alien species.
4. To make it possible for species and habitats and shift in response to environmental changes.

Biosphere reserve programme. Biosphere reserves are a special category of protected areas of land and/or coastal environments, wherein people are an integral component of the system.

The biosphere reserves are representative examples of natural biomes and contain unique biological communities.

The concept of Biosphere Reserves was launched in 1975 as a, part of UNESCO's 'Man and Biosphere Programme, dealing with the conservation of ecosystems and the genetic resources contained therein.

Till May 2002, there were 408 biosphere reserves dispersed in 94 countries.

In India, thirteen biodiversity rich areas have been designated as Biosphere Reserves applying the diversity and genetic integrity of plants, animals and microorganisms. (See map and table 14.8).

In India, Biosphere Reserves are also notified as National Parks.

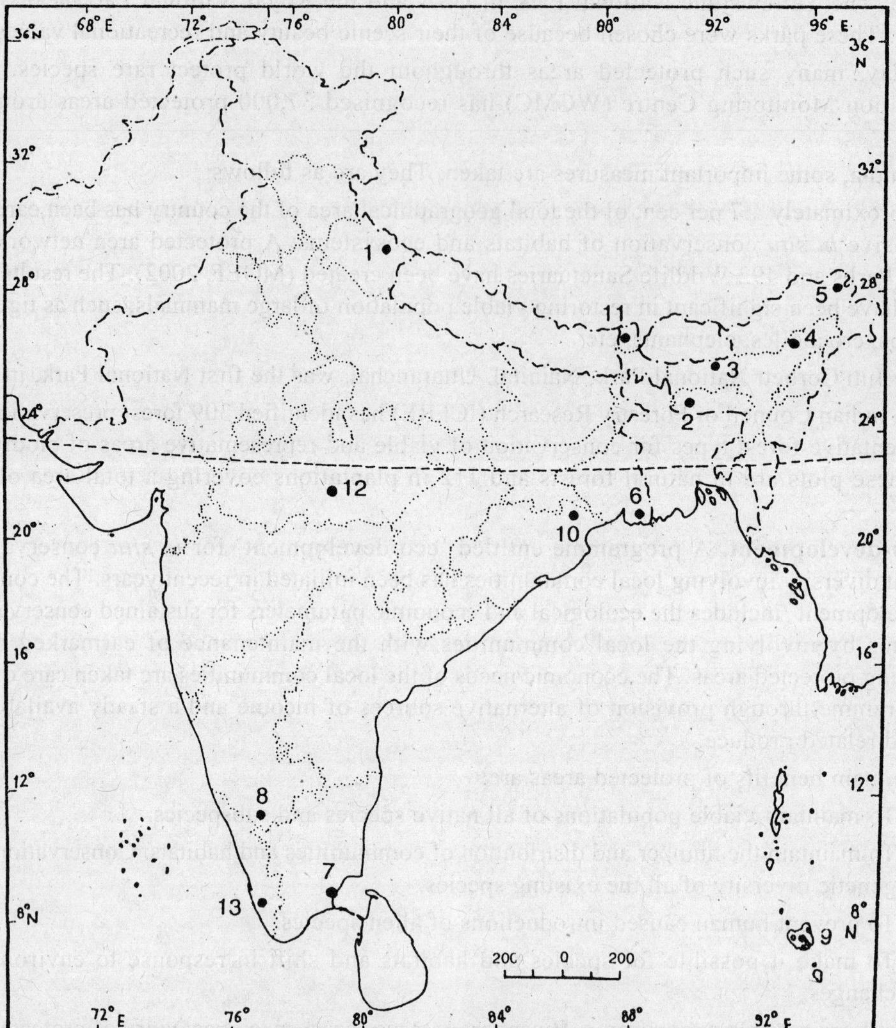

Fig. 14.4. The Biosphere Reserves in India.

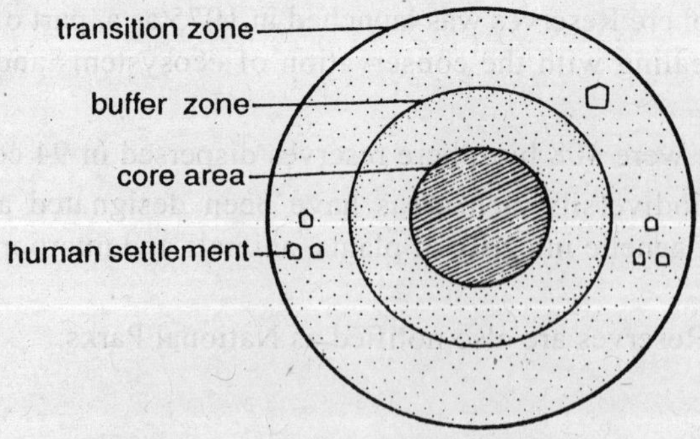

Fig. 14.5. Zonation in a terrestrial biosphere reserve.

Table 14.8. Biosphere Reserves

Name of the site	*Location (State)*
1. Nanda Devi	Part of Chamoli, Pithoragarh, Almora Districts (Uttaranchal)
2. Nokrek	Part of Gora Hills (Meghalaya)
3. Manas	Part of Kokrajhar, Bongaigaon, Barpeta, Nalbari Kamprup and Darang District (Assam)
4. Dibru-Saikhowa	Part of Dibrugarh and Tinsukia District (Assam)
5. Dehang Debang	Part of Siang and Debang Valley in Arunachal Pradesh
6. Sunderbans	Part of delta of Ganga and Brahamputra river system (West Bengal)
7. Gulf of Mannar	Indian part of Gulf of Mannar between India and Sri Lanka (Tamil Nadu)
8. Nilgiri	Part of Wynad, Nagarhole, Bandipur and Madhumalai, Nilambur, Silent Valley and Siruvani Hills (Tamil Nadu)
9. Great Nicobar	Southern most islands of Andaman and Nicobar (A and N islands)
10. Similipal	Part of Mayurbhanj District, (Orissa)
11. Khanghendzonga	Part of Kanchanjanga Hills and Sikkim (Kanchanjanga)
12. Pachmarhi	Parts of Betul, Hoshangabad and Chindwara Districts of Madhya Pradesh
13. Agasthyamalai	Part of Kerala State

Zonatism of a terrestrial biosphere reserve. A terrestrial biosphere reserve consists of **core**, **buffer** and **transition** zones.

(*i*) The **natural** or **core zone** comprises an undisturbed and legally protected ecosystem.

(*ii*) The **buffer zone** surrounds the core area, and is managed to accommodate a greater variety of resource use strategies, and research and educational activities.

(*iii*) The **transition zone**, the outermost part of the Biosphere Reserve. This is an area of active cooperation between reserve management and the local people, wherein activities like settlements, cropping, forestry, recreation and other economic uses continue in harmony with people and conservation goals.

The main functions of biosphere reserves are:

(*i*) **Conservation.** To conserve the ecosystems, a biosphere reserve programme is being implemented, for example, conservation of landscapes, species and genetic resources. It also encourages traditional resource use.

(*ii*) **Eco-development.** The concept of eco-development integrates the ecological and economic parameters for sustained conservation of ecosystems by involving the local people with the maintenance of earmarked regions. Biosphere reserves are also used to promote economic development which is culturally, socially and ecologically sustainable.

(*iii*) **Scientific research programme.** Programmes have also been launched for scientific management and wise use of fragile ecosystem. Specific programmes for management and conservation of wetlands, mangroves and coral reef systems are also being implemented. Under this programme, 21 wetlands,15 mangrove areas and 4 coral reef areas have been identified for management. National and sub-national level committees oversee and guide these programmes to ensure strong policy and strategic support.

Sacred forests and sacred lakes. In India and some other Asian countries, a traditional strategy for the protection of biodiversity has been in practice in the form of **sacred forests** or **groves**. These forest patches of varying dimensions are protected by local people due to their religious sanctity. Generally, they are most undisturbed forests without any human impact. In India, sacred forests are located in several parts, such as Karnataka, Maharashtra, Kerala, Meghalaya, Uttaranchal, Uttar Pradesh, etc., and serve as refuge for a number of rare and endangered taxa.

Similarly several water bodies are declared sacred by the people, *e.g.*, Khecheopalri lake in Sikkim. Such water bodies protect aquatic flora and fauna.

Ramsar sites. Six internationally significant wetlands of India have been declared as **Ramsar Sites** under the Ramsar Convention. To focus attention on urban wetlands threatened by pollution and other *anthropogenic* activities, state Governments were requested to identify lakes that could be include the National Lake Conservation Plan (NLCP).

World heritage sites. Under the World Heritage Convention, five natural sites have been declared as 'World Heritage Sites'.

Table 14.9. World Heritage Sites

Site	*Location*
1. Kaziranga National Park	Assam
2. Keoladeo Ghana National Park	Rajasthan
3. Manas Wildlife Sanctuary	Assam
4. Nanda Devi National Park	Uttaranchal
5. Sunderban National Park	West Bengal

Five natural World Heritage Sites are as follows:

1. The Tura Range in Gora Hills of Meghalaya is a gene sanctuary for preserving the rich native diversity of wild *Citrus* and *Musa* species.

2. Sanctuaries for rhodendrous and orchids have been established in Sikkim.

3. **Project tiger.** A potential example of an highly endangered species in the Indian Tiger (*Panthera tigris*). It is estimated that India had about 40,000 tigers in 1900, and the number declined to a mere about 1,800 in 1972. Hence project tiger was launched in 1973.

At present these are 25 Tiger Reserves spreading over in 14 states and covering an area of about 33875 sq. km and the tiger population has more than doubled now due to total ban on hunting and trading tiger products at national and international levels.

4. **Project elephant.** This project was launched in 1991-92 to assist states having free ranging population of wild elephants to ensure long term survival of elephants in their natural habitats.

5. **Rhinos.** Rhinos have been given special atention in selected sanctuaries and national parks in the North East and North West India.

All these programmes, though focussed on a single species, have a wider impact as they conserve habitats and a variety of other species in those habitats.

Ex-situ Conservation (outside natural habitats) Strategies

The *ex situ* conservation strategies include: botanical gardens, zoological gardens, conservation stands and gene, pollen, seed, seedling, tissue culture and DNA banks.

Seed gene banks make the easiest way to store germplasm of wild and cultivated plants at low temperature.

While in **field gene banks**, preservation of genetic resources is being done under normal growing conditions.

Table 14.10. In Situ and Ex-situ Biodiversity Conservation

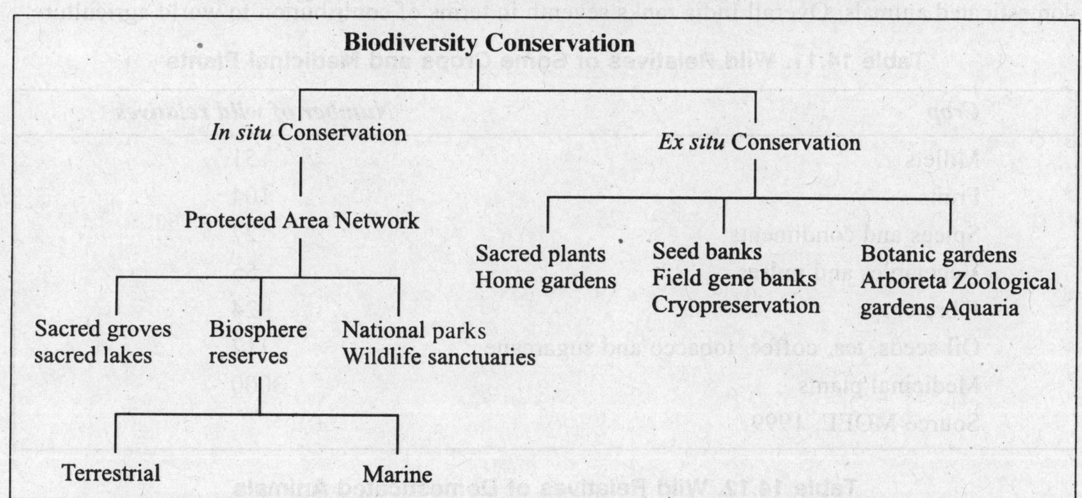

Cryopreservation. This type of *in vitro* conservation is done in liquid nitrogen at a temperature of –196ºC. This is particularly useful for conserving vegetatively propagated crops, *e.g.*, potato.

Cryopreservation is the storage of material at ultra-low temperature (*i.e.*, –196º C) either by very rapid cooling, as used for storing seeds, or by gradual cooling and simultaneous dehydration, as being done in tissue culture.

In cryopreservation, the material can be stored for a considerable long period of time in compact low maintenance refrigeration units.

Botanical gardens. According to currently available survey, Central Government and State Governments together run and manage 33 Botanical Gardens, while Universities have their own botanic gardens.

An scheme entitled Assistance to Botanic Gardens provides one-time assistance to botainc gardens to strengthen and institute measure for *ex situ* conservation of threatened and endangered species in their respective regions.

There are more than 1500 botanic gardens and arboreta in the world containing more than 80,000 species. Many of these botanic gardens now have seed banks, tissue culture facilities and other *ex situ* technologies.

Zoological gardens (zoos). In India, there are 275 zoos, deer parks, safari parks, aquaria, etc. A Central Zoo Authority was set up to secure better management of zoos.

There are more than 800 professionally managed zoos around the world with about 3000 species of mammals, birds, reptiles and amphibians.

Many of these zoos have well developed captive breeding programmes.

Conservation of wild species. The conservation of wild relatives of crop plants, animals or cultures of microorganisms provides breeders and genetic engineers with a ready source of genetic material.

India has 47,000 species of flowering and non-flowering plants representing about 12 per cent of the recorded world's flora. Out of 47,000 species of plants, 5150 are endemic and 2532 species are found in the Himalayas and adjoining regions, and 1782 in peninsular India.

India is also rich in the number of endemic faunal species it possesses, while its record in agro-biodiversity is very impressive as well.

There are 166 crop species and 320 wild relatives along with numerous wild relatives of domesticated animals. Overall India ranks seventh in terms of contribution to world agriculture.

Table 14.11. Wild Relatives of Some Crops and Medicinal Plants

Crop	*Number of wild relatives*
Millets	51
Fruits	104
Spices and condiments	27
Vegetables and pulses	55
Fibre crops	24
Oil seeds, tea, coffee, tobacco and sugarcane	12
Medicinal plants	3000
Source MOEE, 1999	

Table 14.12. Wild Relatives of Domesticated Animals

Group	*Numbers*
Cattle	27
Sheep	40
Goats	22
Camels	8
Horses	6
Donkeys	2
Poultry	18
Buffalo	8
Source MOEF, 1999	

HOT SPOTS OF BIODIVERSITY

Norman Myers developed the **hot spots** concept in 1988 to designate priority areas for in situ conservation.

Biodiversity hot spots are areas that are unusually rich in species, most of which are endemic, and are under a constant threat of being overexploited.

The main criteria for determining a hot spot are:

(*i*) Number of endemic species (*i.e.*, the species confined to that area, and not found elsewhere).

(*ii*) Degree of threat, which is measured in terms of habitat loss.

Among the 25 hot spots in the world, two are found in India.

These are two distinct areas:

(*i*) The Eastern Himalayas, and

(*ii*) The Western Ghats.

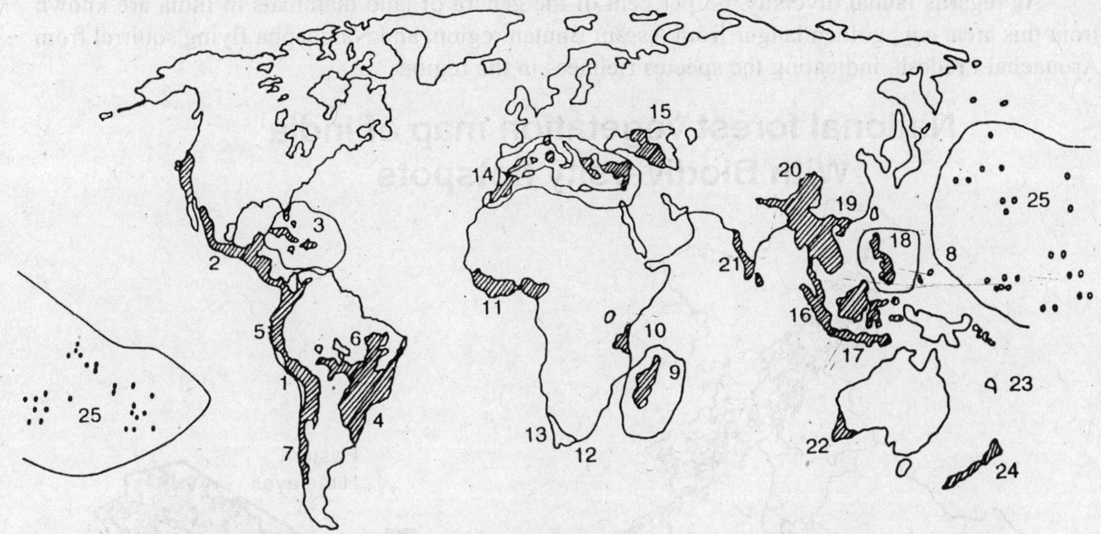

1. Tropical Andes, 2. Mesoamerica, 3. Caribbean, 4. Brazil's Atlantic Forests, 5. Choco/Darien/Western Ecuador, 6.Brazil's Cerrado, 7. Central Chile, 8. California Floristic Province, 9. Madagascar, 10. Eastern Arc & Coastal Forests of Tanzania/Kenya, 11. West African Forests, 12. Cape Floristic Province, 13. Succulent Karoo, 14. Mediterranean Basin, 15. Causcasus, 16. Sundland, 17. Wallacea, 18. Philippines, 19. Indo-burna, 20, South-Central China, 21. Western Ghats Sri Lanka, 22. Southwest Australia, 23. New Caledonia, 24. New Zealand, 25. Polynesia/Micronesia.

Fig. 14.6. The terrestrial biodiversity worldwide hot spots.

These hot spots extend into the neighbouring countries also. These areas are rich in angiosperms (flowering plants), also in reptiles, amphibians, swallow-tailed butterflies and some mammals. Most of the species are endemic.

Together these 25 sites of the world contain approximately 49,955 endemic plant species or 20 per cent of the world's recorded plant species, in only 746,400 sq km or 0.5 per cent of the earth's land surface.

1. **Eastern Himalayas.** Phytogeographically, the Eastern Himalayas forms a distinct floral region and comprises Nepal, Bhutan, neighbouring states of east and north-east India, and a part of southwestern China.

The temperate forests are found at altitudes of 1780 to 3500 metres a.s.l. In the whole of Eastern Himalayas, there are an estimated 9000 plant species, with 3500, *i.e.*, 39 per cent of them are endemic.

In India's sector of the area, there occur 5800 plant species, approximately 2000, *i.e.*, 36 per cent of them are endemic.

At least 55 flowering plants endemic to this area are recognised as rare, *e.g.*, the pitcher plant (*Nepenthes khasiana*). The species of numerous primitive angiosperm families such as Magnoliaceae and Winteraceae, and primitive genera of plants, like *Magnolia* and *Betula* are found in Eastern Himalayas.

The area is also rich in wild relatives of plants of economic significance, such as rice, banana, citrus, ginger, chilli, jute and sugarcane. The rigion is regarded as the centre of origin and diversification of commercially important palms, like coconut, arecanut, palmyra palm, sugar palm and wild date-palm.

As regards faunal diversity, 63 per cent of the genera of land mammals in India are known from this area, *e.g.*, golden langur from Assam Bhutan region, and Namdapha flying squirrel from Arunachal Pradesh, indicating the species richness in the region.

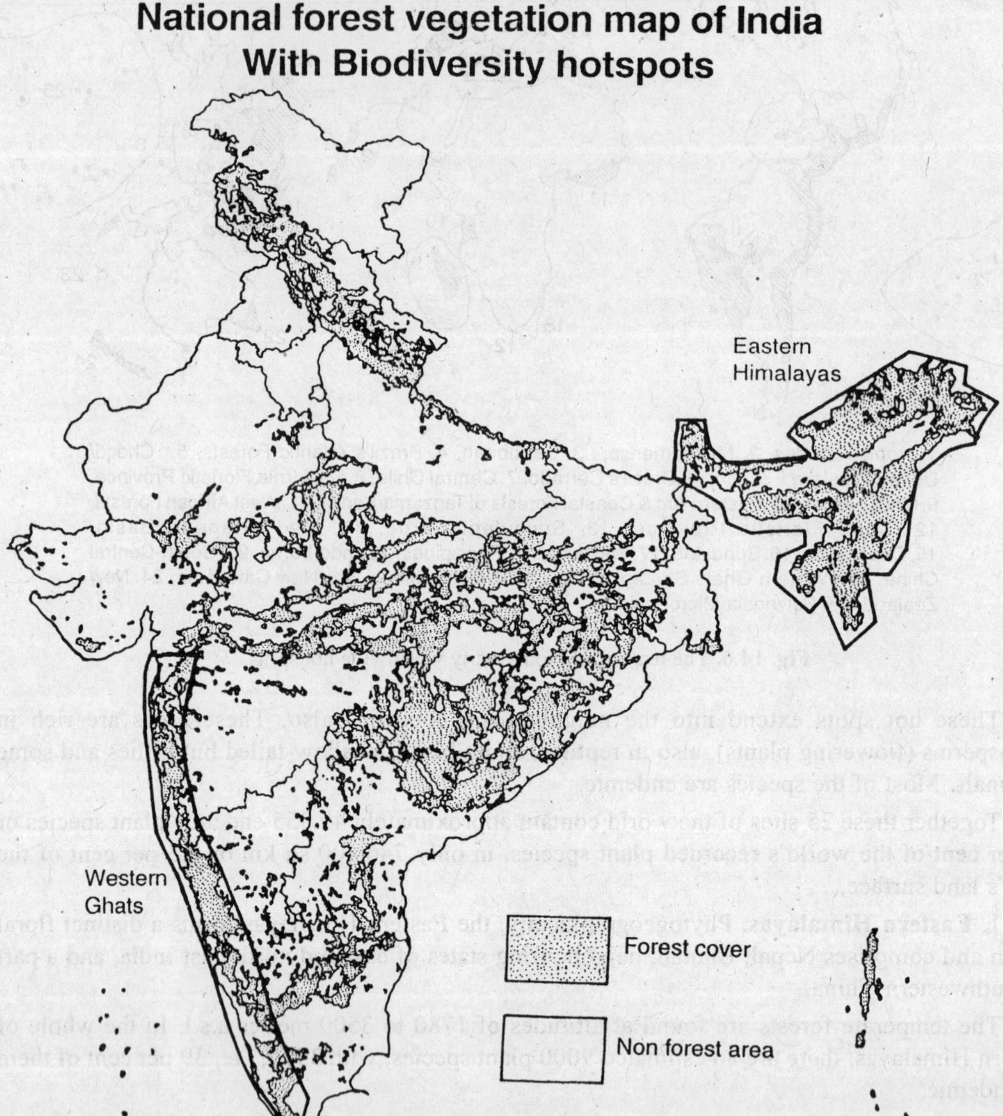

Fig. 14.7. National forest vegetation map of India with biodiversity hotspots.

2. **The Western Ghats.** The Western Ghats region is considered as one of the most important biogeographic zones of India, as it is one of the richest centres of endemic species. This region lies parallel to the western coast of Indian peninsula for almost 1600 km in Maharashtra, Karnataka, Tamil Nadu and Kerala. The forest at low elevation (500 m a.s.l.) are mostly evergreen, while those found at 500 to 1500 metres a.s.l. are generally semi-evergreen forests.

Due to varied topography and micro-climate, some areas within the region are considered to be active zones of speciation.

The region has 490 arborescent taxa, of which 308 species are endemic. About 1500 endemic species of dicotyledonous plants are reported from the Western Ghats. 245 species of orchids (monocotyledons) belonging to 75 genera are found in this region, of which 112 species in ten genera are endemic to this region.

As regards the fauna, as many as 315 species of vertebrates belonging to 22 genera are endemic, which include 12 species of mammals, 13 species of birds, 89 species of reptiles, 87 species of amphibians and 104 species of fish.

Nearly 235 species of endemic flowering plants are considered endangered.

Rare fauna of the region includes: Lion Tailed Macaque, Nilgiri Langur, Nilgiri Tahr, Flying Squirrel, and Malabar Gray Hornbill.

In the Wester Ghats, the Agasthyamalai hills, the Silent Valley, and the new Amambalam Reserve, are the main centres of diversity.

BIODIVERSITY CONSERVATION IN INDIA

India is one of 12 megadiversity countries of the world. The innumerable life forms harboured by the forests, deserts, mountains, other land, air and oceans provide food, fodder, fuel, medicine, textiles, etc. There are innumerable species, the potential of which is not as yet known. It would therefore be prudent to not only conserve the species we already have information about, but also species we have not get identified and described from economic point of view.

Taxus baccata, a tree found in the sub-Himalayan regions, once believed to be of no value is now considered to be effective in the treatment of certain types of cancer. The diversity of genes, species and ecosystem is a valuable resource that can be tapped as human needs and demands change, the still more basic reasons for conservation are the moral, cultural and religious values.

Major problems with biodiversity conservation in India are:

(*i*) Low priority for conservation of living natural resources.

(*ii*) Exploitation of living natural resources for monetary gain.

(*iii*) Values and knowledge about the species and ecosystem inadequately known.

(*iv*) Unplanned urbanisation and uncontrolled industrialisation.

Biodiversity conservation in India is also impeded by a lack of knowledge of the magnitude, patterns, causes and rates of deforestation and biodiversity loss at the ecosystem and landscape level. Poaching and trade in wildlife species are among the most important concerns in the management of protected areas today but information on poaching, trade and trade routes is sketchy and current wildlife protection and law enforcement measures are inadequate and inefficient.

Existing Policy of Biodiversity Conservation

1. A protected area network of National Parks and Wildlife Sanctuaries have been created.

2. The Indian Council of Forestry Research and Education (ICFRE) has identified forest preservation plots of representative forest types for conservation of viable and representative areas of biodiversity.

3. A programme entitled **'ecodevelopment'** for *in situ* conservation of biological diversity involving local communities has been initiated.

4. To conserve the respective ecosystems, a **Biosphere Reserve Programme** is being implemented.

5. Programmes have also been launched for scientific management and wise use of fragile ecosystem.

6. Specific programmes for management and conservation of wetlands, mangroves and coral reef systems are also being implemented.

7. Six internationally significant wetlands of India have been declared as '**Ramsar Sites**' under the Ramsar Convention.

8. Wildlife Protection Act is in the final stage of revision and provisions have been made for conservation reserves and community reserves to allow restrictive use to make it more people oriented. There will also be State Biodiversity Boards to control access to domestic consumers.

9. Under the World Heritage Convention, five natural sites in India, have been declared as **World Heritage Sites.**

10. Project Tiger and Project Elephant have been launched to protect the wildlife. Rhinos have been given special attention in selected sanctuaries and national parks.

11. The Ministry of Environment and Forests (MOEF) constituted the National Afforestation and Ecodevelopment Board (NAEB) in August 1992, which has evolved specific schemes for promoting afforestation and management strategies and ecodevelopment packages for promoting biomass production through a participatory planning process of Joint Forest Management (JFM) and microplanning.

Certain *Ex situ* conservation strategies are as follows:

Central Government and State Government together run and manage Botanical Gardens, zoos, deer parks, safari parks, aquaria, etc. A Central Zoo Authority has been set up to secure better management of zoos. A scheme entitled Assistance to Botanic Gardens provides asstistance to botanic gardens.

State Governments have already received the funds for the preparation of management action plans for Pongdam in Himachal Pradesh, Wullar lake in Kashmir, Loktak in Manipur, Rudrasagar in Tripura and Kolleru in Andhra Pradesh.

The National Bureau of Plant, Animal and Fish Genetic Resources has a number of programmes to collect and conserve the germplasm of plants and animals in seed gene banks, and field gene banks *in vitro* conservation.

The diverse food and medicinal plants are also being conserved successfully by the tribal people. The women particularly have an important role in the conservation of agrobiodiversity. The contribution of natural and agricultural biodiversity in terms of crops, live stock, fisheries, etc., is very substantial in terms of commercial value.

Biodiversity Contribution To Indian Economy

Biodiversity products have obtained a commercial value and have been increasingly exchanged in the markets having a monetary value, from which their share in the national economy can be judged.

The contribution of natural and agricultural biodiversity has a major contribution to the Indian GDP (gross domestic product).

The large economic implication of biodiversity in its wild and domesticated forms is the rice improvement programme. Rice accounts for 22 per cent of the total cropped area and 39 per cent of the total area under cereals, which reflects its importance in the country's struggle to attain self sufficiency in food.

When the rice crop was doomed in the 1970s, one single gene from the wild strain of rice, namely *Oryza nivara* from Uttar Pradesh, showed resistance to this virus and proved vital in the fight against the virus.

With respect to the commercial value of the plant species of medicinal value, the world trade is of several billion dollars and this is growing. The export market for medicinal plants has also increased. India's foreign exchange reserves from horticultural products are from high yielding varieties. Increased production of oilseeds also helped in saving large amounts of foreign exchange spent on edible oil import.

International Efforts For Conserving Biodiversity

It is apparent that action plans and strategies, when designed appropriately and implemented, can make important contributions to conservation. A collaborate effort of the World Resources Institute, IUCN (International Union for Conservation of Nature and Natural Resources), and UNEP (United Nations Environment Programme), working with other institutions, is leading to the preparation of a Global Strategy for the Conservation of Biodiversity, as a companion to the new version of the **World Conservation Strategy** now being prepared. The aim of the Strategy is to provide a comprehensive framework to stimulate urgent, positive, innovative and coordinated action to stem the loss and degradation of the world's biological resources and enhance the contribution of these resources to human well-being. The strategy will be developed by and for national governments, NGOs (Non-government organisations), resource managers, scientists, international institutions, multilateral banks, and bilateral aid agencies. The development of the Global Strategy will be centred around a series of regional workshops in Asia, Africa, Europe, Latin America and North America. The Global strategy will include considerations of a variety of factors influencing biological resource conservation, such as international financing, international cooperation, research, education, training, public awareness, and ecological restoration.

The Earth Summit held in 1992 at Rio de Janeiro, Brazil resulted into a Convention on Biodiversity, which came into force on December, 29, 1993. The Convention has three main objectives:

(*i*) Conservation of biological diversity
(*ii*) Sustainable use of biodiversity, and
(*iii*) Fair and equitable sharing of benefits that arise by the utilisation of genetic resources.

The World Conservation Strategy (WTS) and the World Wide Fund for Nature (WWF) support projects worldwide to promote conservation of biological diversity.

WILDLIFE CONSERVATION

When we hear the term "wildlife", it generally refers to large ferocious animals living in jungles and forests such as tigers, lions, elephants, wolves, etc. But in fact, "wildlife" implies to any living organism in its natural habitat which includes all plants, animals and microorganisms except cultivated plants and domesticated animals. From ecological view point, wildlife is a renewable resource.

Necessity for Wildlife Conservation

The conservation of wildlife is required for the following benefits:

1. The wildlife helps us in maintaining the '**balance of nature**'. Once this equilibrium is disturbed it leads to many problems. The destruction of carnivores or insectivores often leads to the increase of herbivores which in turn affects the forest vegetation or crops.

2. The wildlife can be used **commercially** to earn money. It can increase our earning of foreign exchange, if linked with tourism.

3. The preservation of wildlife helps many naturalists and behaviour biologists to study morphology, anatomy, physiology, ecology, behaviour biology of the wild animals under their natural surroundings.

4. The wildlife provides best means of **sports** and **recreation**.

5. The wildlife of India is our **cultural asset** and has deep-rooted effect on Indian art, sculpture, literature and religion.

Reasons for Depletion of Wildlife

Many wild animals have become extinct due to various human and natural activities:

1. Absence of cover or shelter to wild animals.

2. Due to deforestation for cultivation or for urbanisation, reduction in area for free movement of wild animals which retard reproductive capacity of certain wild animals like deer, bison, rhino, tiger, etc.

3. Destruction of wild plants of forests for timber, charcoal and fire wood often deprives wild animals their most palatable food and affects their survival.

4. Noise pollution by different transporting media (trucks, buses, rails, aeroplanes, etc.) and polluting river water have adversely affected wild animals.

5. Various natural calamities such as floods, droughts, fires, epidemics,etc., have also caused great destruction of wild animals.

6. Hunting methods of all kinds and for any purpose (*i.e.*, for food, recreation, hide, fur, plumage, musk, tusk, horn, etc.) have caused destruction of wildlife.

SOME IMPORTANT WILDLIFE OF INDIA

Indian subcontinent is unique in having immense natural beauty in its different regions (like Himalayas, Gangetic Plain and Plateau of Deccan, etc.) and also in possessing a rich and diverse wildlife fauna. However, some of the mammalian wildlife representatives of our country are:

1. **The Majestic Elephants** (*Elephas maximus*). It is the largest terrestrial mammal and is confined to the Terai and foot-hills because of its dependence on succulent grass, bamboo and plenty of water. It is gregarious and its herds move constantly in search of new feeding grounds; it cannot tolerate high temperature. It is an intelligent animal and can be domesticated easily. Its lifespan is about 70 years. Its population is hunted out for tusk and, therefore, its number is decreasing slowly.

2. **The great one-horned Rhinoceros** (*Rhinoceros unicornis*). It is found in the grasslands and jungle areas of the foot hills of the Himalayas and also in the plains of West Bengal and Assam. It prefers swamps and open savannah, covered with tall elephant grass. It has a single horn near the tip of its snout; the horn is made of keratinized skin and matted hair and it is said to have aphrodisiac and other medicinal values. It is, therefore, hunted for this purpose and is at the verge of extinction.

3. **The Sloth Bear** (*Melursus ursinus*). It is a stout carnivore which is found in the hilly forests like Mount Abu and Erinpura ranges. It lives mostly on fruits, honey and insects, it sucks up white ants from their hills which are broken by its claws with great speed. It is rapidly vanishing due to ruthless hunting and it needs conservation. **Brown Bears** (*Ursus arctus*) are found on mountains upto 3,000 metres height and **Black Bears** (*Selenarctos tibetanus*) at 1500–3000 metres height.

4. **The lion** (*Panthera leo*). It is a rare mammal and was found in Thar desert, in arid plains of Sind, Rajasthan and Punjab but now it is restricted to Gir forest of Gujarat. It is somewhat nocturnal and referred to as the **King of Animals**. Male possesses mane which is not found in female. This animal is said to possess enormous strength. The lion is the emblem of the Government of India. Its extermination is well known.

5. **Indian Panther** (*Panthera paradus fusca*). It is smaller than tiger and is spotted. It can live in all types of forests. Once it was found all over India but now its number has been drastically reduced. Now, it is under protection in Mount Abu sanctuary. It is commonly called the **Leopard**. The **snow leopard** (*Panthera unica*) is found in the high Himalayas from Kashmir to Sikkim near snow line. It has a creamy grey coat with large black rings. The leopards, especially the snow leopards are greatly hunted for their fur. It is an endangered species.

6. **The Indian Tiger** (*Panthera tigris*). It is gracious carnivore and distributed in Uttar Pradesh from the Himalayas to the Vindhya forests in the South. It lives in a variety of habitats, from thorn forests to the dense terai forests. It is carnivorous and nocturnal. Its number, once estimated to be nearly 25,000 in 1948, has gone considerably down nearly 1540 in 1970. Realizing this fact, Government of India has launched a project commonly called **Project Tiger** to save the tigers from extinction. This project was launched in 1972; due to this effort the population of tiger has improved considerably.

The **White Tiger** is a mutant variety of Tiger which was first noticed in the forests of Gidhaur in Monghyr District of Bihar State.

7. **The Cheetah** (*Acinonyx jubatus*). It was found in Central India,Deccan region and in Thar desert upto Jaipur in Rajasthan. Now the Cheetah, one of the most beautiful, agile and fastest animals, is regarded to be extinct from India.

8. **The Clouded Leopard** (*Neofelis nebulosa*). It is found in the evergreen forests of Assam. It is black in colour and lives on trees, *i.e.*, arboreal in habit. It feeds on birds and small animals.

9. **Striped Hyaena** (*Hyaena striata*). It is found generally roaming near jungles throughout our country. It is terrestrial and carnivorous; it feeds on goat, sheep, etc.

10. **Wolf** (*Canis lupus*). It is found in the plains throughout our country. It preys on goat, sheep and deer.

11. **Wild Ass** (*Equus hemionus khur*). It had a fairly wide distribution in the dry regions of North-West India (Jaisalmer and Bikaner), but its distribution has become restricted to the southern part of the Thar desert. Its number is estimated to be nearly 870. Its population is severely reduced due to hunting, catching for breeding mules and surra disease, etc. Since, it is the only true **wild ass** of the world, hence, it needs protection and conservation.

12. **Wild Boar** (*Sus cristatus*). The wild boar or pig has a long mobile snout with terminal nostrils. The canines of both jaws grow continuously and form long triangular and up curved tusks for defence and digging roots; the tusks are better developed in males. It is omnivorous, it feeds on carrion, snakes, insects, roots, tubers and cultivated crops. It is gregarious living in families in grass or bush of marshy places. It is now rare animal and is mostly found in small numbers along the Aravalli ranges in the eastern part of Thar desert where it inhabits rocky slopes.

13. **Wild goat and sheep.** These are restricted only to the Northern part of our country. The **Nilgiri Tahr** (*Hemitragus hylocrius*) is a variety of wild sheep; the only variety found in south (in Nilgiris, Tamil Nadu). It lives in groups and grazes on grasses. This species is greatly endangered due to habitat disturbance, predation by black panther, wild dogs and hunting by poachers. The **Himalayan Ibex** is a variety of wild goat, *Capara siberica*.

14. **Nilgai or Blue Bull** (*Boselaphus tragocamelus*). It lives in plains and Rajasthan deserts in groups of 4 to 10 in number. It prefers to live near cultivated lands and is also found in desert grasslands.Males have small cone-shaped curved horns. Both sexes have a short dark mane on the neck. It is hunted under Crop-Protection Act, because herds of Nilgai inflict severe damage to crops. In Thar desert it is regarded sacred like 'cow', so is spared from hunting.

15. **Wild-Buffalo** (*Bas bubalis*). It is large-sized, robust animal having streamlined body. It prefers tall grassy forests close to marshy areas in the vicinity of rivers and lakes. It was once distributed over the grass jungles and rivers in forests of Gangetic plain and in the foot-hills of Assam, Madhya Pradesh and Nepal. Its number has also decreased rapidly due to great hunting.

16. **Indian Bison** (*Bos gaurus*). It is related to bull and is found in Andhra Pradesh, Orissa, Madhya Pradesh and Bihar. Due to various reasons, it is now confined only to Bandipur of Karnataka. Its number has also reduced down to a considerable extent.

17. **Indian Gazelle** (*Gazella bennetii*). It is also called **Chinkara** and is mostly found through out desert region of Southern part of Uttar Pradesh and Rajasthan. It lives in herds and adapted to live in extremely arid conditions. It prefers to live in open grass-lands with thick bushes.

18. **Four-horned Antelope** (*Tetracerus quadricornis*). It is also called **Chausingha** and has two pairs of horns. It is found in the foot-hills of Himalayas and in Bundelkhand region. It generally leads solitary life.

19. **The Black Buck** (*Antelope cervicapra*). It has four subspecies in India. It is a graceful and fast running animal, found in the grassy plains of India and avoids forests. Males have spirally ringed, unbranched long and permanent horns. It has been hunted out from the most parts of India and now it is confined only in certain localities in Rajasthan.

20. **The Musk Deer** (*Moschus moschiferus*). It is found in Himachal Pradesh, Northern Uttar Pradesh, Nepal and Sikkim. It is a shy animal and lives a solitary life. Male possesses a musk gland for which it is hunted. The secretion of this gland is a valuable product used as perfume fixative, and as an important ingredient of several Aurvedic and Homeopathic drugs.

21. **The Barking Deer** (*Cervulus muntjac*). It is found in the foot-hills and plain. It has suborbital glands below eyes, antlers and non-protruding canines. It is diurnal and lives in pairs.

22. **The Dancing Deer** (*Cervus unicolor*). It is also called **Sambhar**. It is a rare beautiful deer, lives in herds and found in the forests all over the country. It has very long antlers. It comes out on plains only at mid-night. It is mostly hunted for its skin and its flesh is used as food by many people.

23. **The Spotted Deer** (*Axis axis*). It is also called **Cheetal**. It is beautiful deer with white spots on its brown coat. It has long slender antlers, each having three branches which shed annually. It is mostly found in the forests of Indian plains and is also reported from Assam, Punjab and Rajasthan. It is also hunted greatly for skin and meat.

24. **The Hog Deer** (*Axis porcinus*). It is smaller and stout in build and has hog-like appearance. It lives either solitary or in pair and prefers to live in grass patches at the borders of the forests. It is diurnal and comes out to feed only during morning and evening.

25. **The Nilgiri Langur** (*Cercopithecus johni*).

26. **The Lion-tailed Monkey** (*Macaca silensus*).

27. **The Hanuman Monkey** (*Semnopithecus entellus*).

28. **The Rhesus Monkey** (*Macaca mulatta*).

29. **The Indian Giant Squirrel** (*Rotufa indica*).

30. **The Porcupine** (*Hystrix lencura*).

31. **The Pangolin** (*Manis crassicaudata*).

The above mentioned list is related to mammals only. Likewise, a large number of birds, reptiles, etc., are also among wildlife of India.

PROTECTED SPECIES OF INDIAN WILDLIFE

The Government of India has declared followings species as protected:

1. Tiger,
2. White tiger,
3. Lion,
4. Panther,
5. Snow Leopard,
6. Clouded Leopard,
7. Leopard cat,
8. Kashmir stag,
9. Brown Antlered Deer,
10. Black Buck,
11. Chital,
12. Indian swamp deer,
13. Four-horned Antelope,
14. Golden Cat,
15. Indian Bison,
16. Indian Rhinoceros,
17. Wild Ass,
18. Barking Deer,
19. Slender Loris,
20. Pigmy hog,
21. Musk deer,
22. Indian Gazelle,
23. Gibbons,
24. Nilgiri Langur,
25. Slow Loris,
26. Golden Langur,
27. Lion-tailed Monkey,
28. Great Indian Bustard,
29. Mountain Quail
30. Grey Jungle Fowl,
31. White Winged Wood Duck,
32. Pea Fowl,
33. Horned Owl,
34. Indian Gharials,
35. Crocodiles.

Measures for Wildlife Conservation in India

In India, following measures have been undertaken for wildlife conservation:

(*i*) **Protection of natural habitats.** Jungles and forests are the natural habitats of many wild animals. Hence, these natural habitats should not be destroyed and for this law should be framed to protect such areas from any damage and hunting should be prohibited. However, realizing the importance of natural resources, several national and international organisations have been established to look after such problems.

An international organisation **IUCN** was established in Switzerland in 1948. This organisation has framed many projects to overcome such problems for which this organisation arranges a meeting once a year in any country. Its 10th general meeting was held in New Delhi in 1969. For the normal and effective functioning of this organisation, an additional organisation referred to as **WWF** (World Wildlife Fund) has been established. The **WWF** is related to the collection of funds from its member countries, of course, India is also a member, which are 24 in number. This fund is utilized to run several projects dealing with the conservation of natural resources. The **Tiger Project** in India has been financed by **WWF** and working efficiently since 1972.

(*ii*) **Protection by Law.** India was probably the first country to enact a Wildlife Protection Act. The **Wild birds and Animals Protection Act** was passed in 1887 and repealed in 1912. For game protection in the states the **Forest Act XVI** was enacted in 1927. Then **Indian Board of Wildlife** was established in 1952 which is followed by setting up of Wildlife Boards in different states. In 1972, new **Wildlife Protection Act** was passed. Under this Act, possession, trapping, shooting of wild animals alive or dead; serving their meat for eating, their transport and export and all controlled and watched by special staff (Chief Wildlife Warden and anthorized officers). This Act prohibits hunting of females and young ones under this Act, threatened species are absolutely protected and the rest and provided graded protection according to the condition of their population size. However, in India every year we observe a **Wildlife Week** from 1st October to 8th October.

(*iii*) **Establishment of sanctuaries and national parks.** In India, nearly 200 sanctuaries and national parks have been established for wildlife management. These are meant for providing natural habitat to wild animals.

A **sanctuary** is a place or area with natural environment having optimum living conditions and protection for wild animals; in fact, in a sanctuary wild indigenous animals are kept for protection. Shooting and hunting is prohibited in a sanctuary. A sanctuary is established by **State Forest Department** by Notification and it can be abolished by similar procedures. Contrary to a sanctuary, **a national park** is a permanent place or area established by **Central Legislation** for preserving flora, fauna, landscapes and historic objects of a particular area. However, shooting and hunting is prohibited in such areas also.

Conservation Strategies

For the judicious use of resources, the scientists from nearly 100 countries of the world have formulated some conservation strategies. Few of the steps proposed to save the existing species of the wildlife are given below:

1. Every effort should be made to preserve the endangered species; these species should be given priority over a vulnerable one, a vulnerable species over a rare one and a rare species over other categories.

2. Proper management and planning should be made to preserve wildlife in their natural habitat (*in situ*) and in zoos and national parks (*ex situ*).

3. A wide range of varieties of livestock, animals for aquaculture and their wild varieties should be preserved, and priority should be given to those that are most threatened and needed for national and international breeding programme.

4. Each country should locate the habitats of wild relatives of the economically valuable and useful animals and ensure their preservation in protected areas.

5. The feeding, breeding, nursery and resting areas of species should be safeguarded.

6. In case of migratory animals, a network of protected areas should be established to preserve their habitats. For those species which migrate from one national jurisdiction to another, bilateral or multilateral agreements should be made to meet out the required network.

7. Unique ecosystems should be protected as a matter of priority.

8. The national protection programmes should be coordinated with the international programmes like UNESCO's Man and Biosphere Project 8 and national parks and protected areas of International Union for Conservation of Nature and Natural Resources. This would safeguard the genetic diversity and their continuing evolution.

9. The productive capacities of species should be determined, so that, their utilisation should not exceed those capacities.

10. Proper legislative and administrative measures should be taken to regulate the International trade of wild animals.

Protected Areas in India

As referred to, Indian subcontinent is very rich in possessing natural beauty. Its biogeographical range consists of the coldest Ladakh and Spiti, the hot deserts of Thar, the temperate forest in Himalayas to the lush green tropical rain forests of the low lands; Indian subcontinent also possesses large fresh water bodies such as Wular and the Manasbal lakes in Kashmir, the Chilka in Orissa and Kolleru lake in Andhra Pradesh and the ruggest and rich coastline and coral reefs of the Deccan.

Government of India passed the Wildlife Protection Act in 1972, under which national parks and sanctuaries could be created.

However, out of 434 national parks and sanctuaries in India, 17 have been selected as "Project Tiger" areas.

During the past twenty years UNESCO has evolved **Biosphere Reserves** under its Man and Biosphere (MAB) Programme. Accordingly, India has located 13 areas as Biosphere Reserves; from these Nilgiri Biosphere Reserve, including parts of Karnataka, Kerala and Tamil Nadu was declared in 1986 and the Nanda Devi Biosphere Reserve in 1988. Probably two more Biosphere Reserves, one at Uttaranchal (including the Valley of Flowers in North-Western Himalayas) and another at Nokrek (North-Eastern Himalaya), has also been included in the list of Biosphere Reserve.

In a Biosphere Reserve, the land is designated into different zones; these are the **core zone** (where no human activity is allowed), the **buffer zone** (where limited human activity is allowed) and the **manipulation zone** (where a large number of human activity may go on). In Biosphere Reserves, wild population as well as traditional life styles are protected.

Some Important Wildlife Sanctuaries

1. **Annamalai Sanctuary.** This sanctuary is situated in the southern part of Coimbatore District in Tamil Nadu and extends in a vast area of 958 sq km. This sanctuary has rich fauna of animals like elephant, gaur, sambhar, spotted deer, barking deer, Nilgiri tahr, lion-tailed monkeys, tiger, panther, sloth bear, langur, porcupine and pangolin.

2. **Kaziranga Wildlife Sanctuary.** This sanctuary is situated on the south bank of the river Brahmaputra in Sibsagar district of Jorhat subdivision of Assam. It consists of 430 sq km of forest grasslands and swamps and supports a fauna of nearly 700 rhinoeros. Besides, it also has elephant, wild buffalo, bison, tiger, leopard, sloth bear, sambhar, swamp deer, hog deer, barking deer, wild boar, gibbon, and birds like pelicon, stork and ring-tailed fishing eagles. This is, in fact, a **Rhino Sanctuary**.

3. **Jaldapara Wildlife Sanctuary.** This sanctuary is situated in Jalpaiguri district of West Bengal and is extended in a 65 sq km stretch of grassland. It is known for rhinoceros and, hence, referred to as a **Rhino Sanctuary**. Besides, it also has gaur, elephant, tiger, leopard, deer, and a variety of birds and reptilian fauna.

4. **Manas Wildlife Sanctuary and Tiger Reserve.** This sanctuary is situated in Kamrup district in Assam and is extended in an area of 540 sq km. River Manas passes through it. It contains the following wild animals: tiger, panther, wild dog, wild boar, rhinoceros, gaur, wild buffalo, sambhar, swamp deer and golden langur.

5. **Periyar Wildlife Sanctuary.** This sanctuary is situated in Kerala and has an area of 777 sq km. It has the fauna of wild elephants, gaur, leopards, sloth bear, sambhar, barking deer, wild dogs, wild boars, black Nilgiri Langur and some birds.

6. **Mudumalai Wildlife Sanctuary.** This sanctuary is situated in north-western part of Nilgiris in Tamil Nadu and extends in an area of 22 sq km. It has diverse fauna of wild elephant, gaur, sambhar, chital, barking deer, mouse deer, four-horned antelope, tiger, panther, bonnet monkey, langur, giant squirrel, flying squirrel, wild dog, jackal, wild cat, sloth bear, porcupine, pangolin, flying lizard, monitor lizard, rat snake, python and various birds.

7. **Bandipur Wildlife Sanctuary.** This sanctuary is situated in south of Mysore city of Karnataka state and extends in an area of 874 sq km. Its fauna includes elephant, leopard, sloth bear, wild dog, chital, panther, barking deer, porcupine and langur.

8. **Sesan Gir Sanctuary.** This sanctuary is situated near Ahmedabad in Gujarat state and extends in a vast area of 1295 sq km. It is the largest sanctuary of our country. It is known for Asiatic lions and other fauna includes spotted dear, blue bull, four-horned antelope, chinkara, striped hyaena, wild boar, porcupine, langur, python, crocodiles and some birds.

9. **Dachigam Wildlife Sanctuary.** This sanctuary is situated in Kashmir and extends in an area of 89 sq km. Its fauna includes Kashmir stag or hungal, musk deer, leopard, black bear, brown bear and baboon.

10. **Bharatpur Bird Sanctuary.** It is also named as Koeldeo Ghana Sanctuary. This sanctuary is situated at Bharatpur of Rajasthan state and extends in an area of 29 sq. km. It is a'sanctuary meant for various bird fauna.

In addition to these, the mention of the following sanctuaries may also be made:

(*i*) **Point Calimera Wildlife Sanctuary** in Tamil Nadu and includes the fauna of Chital, wild boar and black buck.

(*ii*) **Sariska Sanctuary** near Alwar in Rajasthan and includes tiger, leopard, spotted deer, four-horned antelope, etc.

(*iii*) **Chilka Lake Sanctuary** in Orissa and has variety of water birds.

(*iv*) **Shivpuri Sanctuary** in Madhya Pradesh and is known for its tiger fauna.

(*v*) **Gautam Buddha Sanctuary** in Gaya of Bihar state and has tiger, leopard, chital, sambhar, etc.

(*vi*) **Bhimbadh Sanctuary** of Monghyr district in Bihar state and has wolf, tiger, leopard, etc.

(*vii*) **Valmikinagar Sanctuary** in Champaran of Bihar state and has tiger, leopard, sambhar, chital, etc.

Some Important National Parks

1. **Simlipal National Park.** This national park is situated in Mayurbhanj district in Orissa state and is extended in an area of 2750 sq km. It is covered over by dense sal forests and is chosen for **project tiger**. Its fauna includes tiger, elephant, deer, pea fowl, talking mynas, chital, sambhar, panther, gaur, hyaena and sloth bear. It is the largest national park of our country.

2. **Corbett National Park.** This national park is situated between Nainital and Garhwal districts of Uttar Pradesh and extends in an area of 525 sq km. It supports rich and diverse fauna of tiger, panther, sloth bear, elephant, blue bull, swamp deer, barking deer, Indian antelope, porcupine, and many birds and reptiles. This national park is one of the most prestigious wildlife parks of our country.

3. **Hailey National Park.** This national park is situated in Ramnagar in the valley of Nainital in Uttar Pradesh and covers an area of 200 sq km. It supports the fauna of elephant, sambhar, chital, langur, wild fowls, pea fowls, etc.

4. **Palamu or Betla National Park.** This national park is situated in Daltonganj district of Bihar state and covers an area of 345 sq km. It supports the fauna of tiger, panther, sloth bear, elephant, chital, gaur, nilgai, chinkara, mouse deer, etc.

5. **Hazaribagh National Park.** This national park is situated in Hazaribagh district of Bihar state and covers an area of 184 sq km. It supports the fauna of wild boar, sambhar, nilgai, tiger, leopard, sloth bear, hyaena and gaur.

6. **Kanha National Park.** This national park is situated in Mandala district of Madhya Pradesh and extends in an area of 160 sq km. It supports the fauna of tiger, chital, deer, black buck, sambhar, blue bull, bison, panther, wild dogs, hyaena, etc.

7. **Tandoba National Park.** This national park is situated in Chandrapur district of Maharashtra state and covers an area of 116 sq km. It supports the fauna of tiger, sambhar, sloth bear, bison, chital, chinkara, barking deer, blue bull, four-horned deer, langur, pea fowl, and few crocodiles.

Zoological Gardens

The **zoological gardens** are somewhat different from sanctuaries and wildlife national parks as the animals are kept in cages in zoological gardens for show. The animals, thus, do not get natural habitat in zoological gardens. There are various zoological gardens in our country but the zoological garden (Zoo) of Delhi is worth mentioning

SUMMARY

1. The totality of genes, species and ecosystems of a region, as a whole are known as biodiversity.
2. Biodiversity has many medicinal and economic use to its credit.
3. Genes of wild species are used to confer new properties, such as disease resistance or improved yield in domesticated species of animals and plants.
4. Biodiversity provides valuable indirect services through natural ecosystems.
5. The total number of species found on earth is estimated to range from 5 to 50 million.
6. However, only about 1.8 million species have so far been described.
7. Biodiversity has three levels to its credit: (*i*) genetic, (*ii*) species, and (*iii*) community and ecosystem.
8. Species are distinct units of diversity and each species is responsible for a specific role in an ecosystem.
9. However, the diversity within a species often increases with the variability of environment.
10. Species diversity refers to the variety of species within a region.
11. Within community,diversity is known as alpha diversity.
12. Between communities, diversity is called beta diversity.
13. Overall diversity of a region is called gamma diversity.
14. Biodiversity increases from poles towards equator, and from high elevations to low elevations.
15. Loss of habitat, fragmentation, disturbance and introduction of alien species seem to be greatest threats to biodiversity.
16. Invasive species are second only to habitat destruction as a major cause of extinction of species.
17. Exotic species especially are having large impacts in island ecosystems, which harbour much of the worlds threatened biodiversity.
18. In the name of modernity, humans have caused the extinction of many species and the rate of species loss is increasing day by day.
19. According to approximate estimation about 14,000—40,000 species are being lost every year from the tropical forests alone.
20. There are three main categories for species threatened with extinction, they are: (*i*) Vulnerable (*ii*) Endangered, and (*iii*) Critically Endangered.
21. The 2000 IUCN (International Union for the Conservation of Nature and Natural Resources) Red List is the world's most comprehensive inventory of the global conservation status of threatened plant and animal species.
22. The IUCN is important to ensure the conservation of landscapes, ecosystems, species and genetic resources.
23. Conservation strategies include *in situ* (on site) and *ex situ* (off site) approaches.
24. Protected Area Network, includes: national parks, wildlife sanctuaries, and biosphere reserves, etc.
25. Sacred groves are traditionally protected areas, which serve as refuge for a number of rare and endangered species.
26. The hot spots are the richest and most threatened reservoirs of plant and animal life on earth.

27. Twenty-five terrestrial hot spots for conservation of biodiversity have been identified worldwide.

28. Amongst the 25 hot spots of the world, two are found in India, *i.e.*, (*i*) Western Ghats, and (*ii*) Eastern Himalayas.

29. The World Conservation Union and World Wildlife Fund (WWF) support projects worldwide to promote conservation and development of Biosphere Reserves.

TERMS TO REMEMBER

Alpha diversity. Means within community diversity, refers to the diversity of organisms sharing the same community/habitat.

Anthropogenic extinction. Many species are disappearing from the face of the earth due to human activities.

Biodiversity. The totality of genes, species and ecosystems of a region.

Beta diversity. Means between community diversity.

Botanochemicals. Plants are used for the manufacture of many synthetic products.

Biosphere reserves. They are a special category of protected areas of land and coastal environments, wherein people are an integral component of the system.

Buffer zone. Surrounds the core area, and is managed to accommodate a great variety of resource use strategies, and research and educational activities.

Core zone. An undisturbed and legally protected ecosystem.

Earth Summit. Held in 1992 at Rio de Janeiro resulted into a convention on biodiversity.

Genetic diversity. Each species, stores an immense amount of genetic information.

Gamma diversity. Diversity of the habitats over the total landscape or geographical area.

Hot spots. Norman Myers, in 1988 developed this concept to designate priority areas for *in situ* conservation.

Mass extinction. When large number of species become extinct.

Natural zone. See core zone.

Natural extinction. When some species disappear and others, which are more adapted to changed conditions, take their place.

Protected areas. Areas of land and/or sea, especially dedicated to the protection and maintenance of biological diversity, and of natural and associated cultural resources.

Speciation. Evolution of new species.

Species diversity. The variety of species within a region.

Species richness. The number of species per unit area.

Sacred forest. Forest patches protected by people due to religious sanctity.

Transition zone. The outermost part of the Biosphere Reserve, is an area of active cooperation between reserve management and the local people.

QUESTION BANK

Very Short Answer Questions (1 mark each)

1. What is biodiversity?

2. How many species of mammals are known to science?

3. Give hierarchical levels of biodiversity.

4. Name three zones of a biosphere reserve.

5. Expand BNHS and WPSI.

6. The number of genes in *Escherichia coli* is ________.

7. What is genetic diversity?

8. What do you understand by 'species diversity'?

9. What is alpha diversity?
10. Define gamma diversity.
11. What are sacred forests?
12. Who developed hot spots concept?
13. How many terrestrial hot spots for conservation of biodiversity are identified worldwide?
14. Which day is celebrated as 'World environment day'?
15. Which State of India has richest flora and fauna?
16. Name the national parks famous for rhinoceros, and lion respectively.
17. What are the threats to biodiversity?
18. Where the Wildlife Institute of India is situated?

ANSWERS

1. The totality of genes, species, and ecosystems of a region are referred to as biodiversity; **2.** 4650 species of mammals; **3.** There are three hierarchical levels of biodiversity, they are: (*i*) general diversity, (*ii*) species diversity and (*iii*) community and ecosystem diversity; **4.** They are: core, buffer and manipulative zones; **5.** Bombay Natural History Society and Wildlife Preservation Society of India; **6.** 4000; **7.** Genetic diversity refers to the variation of genes within species; **8.** Species diversity refers to the variety of species within a region; **9.** Diversity within-community; **10.** Diversity of the habitats over the total landscape is called gamma diversity; **11.** They are forest patches protected by tribals due to religious sanctity accorded to them; **12.** Norman Myers developed the hot spots concept in1988; **13.** Twenty five hot spots; **14.** 5th, June; **15.** Madhya Pradesh; **16.** Kaziranga National Park, Assam and Gir National Park, Gujarat; **17.** Increased human population, resources consumption, pollution; **18.** Dehradun, Uttaranchal

Short Answer Questions

1. Define genetic diversity, species diversity and community or ecosystem diversity.
2. How do forests help in maintenance of climate?
3. How biodiversity is used in agriculture?
4. How does latitude or altitude effect biodiversity?
5. Describe the role of the wildlife in modern agriculture.
6. Give the list of threats to biodiversity.
7. Define wildlife. Which animal is used as symbol for WWF?
8. Define extinction. List the causes of extinction.
9. What are threatened species? Give three categories of threatened species.
10. Name the international organisations involved in the conservation of wildlife.
11. How conservation of wildlife is defined? What are three objectives of conservation of wildlife.
12. Give an account of biogeographical regions of India.
13. What efforts have been taken in India to conserve wildlife?
14. What are the key criteria for determining a hot spot?
15. Name three endangered animal species of India.
16. Why would it be unethical to destroy wild species?
17. What are the patterns of biodiversity.
18. Explain what is meant by species diversity.
19. What is biodiversity? Why has it become important recently?
20. What kinds of threats to the biodiversity may lead to its loss?
21. How is diversity at all levels generally conserved?
22. Broadly classify the extinction processes.
23. Why should killing wild animals be regarded unethical?

24. Distinguish between endangered, critically endangered and vulnerable species.
25. What do you understand by management of wild life?
26. Which organisations are working for the safety of wildlife in India?
27. What is the social forestry?
28. What are the specific objectives of conservation of wildlife?

Long Answer Questions

1. What are the three major threat categories of species? Describe with examples.
2. Write an explanatory note on the efforts for conservation of biodiversity in India.
3. How is biodiversity distributed along major environmental gradients?
4. Write short notes on the following:
   (*a*) Ex-situ conservation
   (*b*) Hot spots of biodiversity
   (*c*) Biosphere reserves
   (*d*) IUCN Red List
   (*e*) Protected areas
5. What is the economic importance of a forest?
6. What is deforestation? Give its causes, and explain them.
7. What are exotic species? How do they threaten biodiversity?
8. Define conservation of wildlife. What are the various steps taken to conserve wildlife?
9. How is extinction process distinguished? Explain its types.
10. What are the effects of deforestation?
11. What is wildlife? How does it help in the maintenance of environment?
12. Give three main causes of extinction of a species by human intervention.
13. Give the *in situ* and ex situ approaches of conserving biodiversity in India.
14. Write a note on forest and grassland management.

MULTIPLE CHOICE QUESTIONS (OBJECTIVE TYPE QUESTIONS)

1. The Earth Summit held at Rio de Janeiro in 1992 resulted into:
   (*a*) Compilation of Red List
   (*b*) Establishment of Biosphere reserves
   (*c*) Convention of Biodiversity
   (*d*) IUCN
2. Approximate percentage of endemic flowering plants in India is:
   (*a*) 23 (*b*) 33 (*c*) 53 (*d*) 63
3. The species become extinct most easily by:
   (*a*) Urbanisation (*b*) Heavy rains (*c*) Deforestation (*d*) Sliding of hills
4. Bandipur in Karnataka is the site of project:
   (*a*) Elephant (*b*) Hangul (*c*) Tiger (*d*) Peacock
5. Red Data Book deals with:
   (*a*) Endemic plant
   (*b*) Plants that are extinct
   (*c*) Animals on verge of extinction
   (*d*) Plants showing photoperiodism
6. Lions in India are now found in:
   (*a*) Jim Corbett N.P.
   (*b*) Forests of M.P.
   (*c*) Forests of Western Ghats
   (*d*) Gir forests
7. Saffron is obtained from:
   (*a*) Styles and stigmas of *Crocus*
   (*b*) Stamens of *Hibiscus*
   (*c*) Petals of *Hibiscus*
   (*d*) Stamens of *Crocus*
8. Wildlife of Nature is renamed as:
   (*a*) World Wildlife Conservation of Nature
   (*b*) World Wildlife of Nature
   (*c*) World Wild Fauna
   (*d*) World Wild Flora

9. Commercial latex is obtained from:
(*a*) Euphorbiaclae (*b*) Acacia arabica (*c*) Tectona grandis (*d*) Manihot esculenta
10. Which animal is the symbol of the World Wildlife Fund?
(*a*) Tiger (*b*) Red Panda (*c*) Horn bill (*d*) White bear
11. In a national park, protection is provided to:
(*a*) Flora only (*b*) Fauna only (*c*) Both flora and fauna (*d*) Entire ecosystem
12. Which one of the following is First National Park in India?
(*a*) Kanha NP (*b*) Periyar NP (*c*) Corbett NP (*d*) Bandipur NP
13. Forest Research Institute is situated at:
(*a*) Nainital (*b*) Madras (*c*) Calcutta (*d*) Dehradun
14. Animals and plants are best protected in:
(*a*) Zoos (*b*) Botanical gardens (*c*) Sanctuaries (*d*) National Parks
15. The species of animals protected in Kaziranga sanctuary is :
(*a*) Macaca mullata (*b*) Panthera tigris (*c*) P. leo (*d*) Rhinoceros unicornis
16. Red Data Books are produced by:
(*a*) IUCN (*b*) WWF (*c*) IBWL (*d*) None of these
17. The national bird of India is:
(*a*) Horn bill (*b*) Black swan (*c*) House sparrow (*d*) Peacock
18. Nandan Kanan Zoo is known for the:
(*a*) Nilgiri Tahr (*b*) Hippopotamus (*c*) White tiger (*d*) Whale
19. Dudhwa National Park is located in:
(*a*) Himachal Pradesh (*b*) Uttar Pradesh (*c*) Madhya Pradesh (*d*) Arunachal Pradesh
20. The biological name of the domestic cat is:
(*a*) Panthera domestica (*b*) Panthera indica (*c*) Felis leo (*d*) Felis domestica
21. Kanha National Park is famous for:
(*a*) Birds (*b*) Rhinocers (*c*) Tigers (*d*) Crocodiles
22. 'Project tiger' in India was started in the year:
(*a*) 1973 (*b*) 1981 (*c*) 1985 (*d*) 1986
23. The lion-tailed monkeys Macaca silenus are found only in:
(*a*) Kaziranga
(*b*) Eastern Ghats and Chennai
(*c*) Western Ghats including Travancore and Mysore
(*d*) Himalayas
24. Now a days, biological reserves are commonly destroyed by:
(*a*) Pollution (*b*) Population (*c*) Rains (*d*) None of these
25. *Idri Idri* is found in:
(*a*) Australia (*b*) Madagascar (*c*) Mauritius (*d*) Tasmania
26. What is major cause of diminishing wildlife numbers?
(*a*) Cannabalism (*b*) Felling trees
(*c*) Paucity of drinking water (*d*) Habitat destruction
27. The animal extinct from India is:
(*a*) Panthera leo (*b*) Antelope (*c*) Rhinoceros (*d*) Acinomyx
28. At present the most significant cause of dwindling biodiversity is:
(*a*) Biological magnification of DDT (*b*) Global warming
(*c*) Destruction of habitat (*d*) Destruction of ozone layer

29. A high density of a protected animal in a National Park can result into:
(*a*) Emigration (*b*) Predation
(*c*) Mutualism (*d*) Intraspecific competition

30. Best method to conserve genetic material of wildlife is:
(*a*) Cold storage (*b*) Tissue culture
(*c*) Seed storage (*d*) Growing in natural habitats

31. The Ranthambore National Park is located in:
(*a*) Rajasthan (*b*) Gujarat (*c*) U.P. (*d*) Maharashtra

32. MAB stands for:
(*a*) Man and Biology Programme (*b*) Mammals and Biology Programme
(*c*) Mammals and Biosphere (*d*) Man and Biosphere Programme

33. Which one of the following fish is introduced in India by foreigners:
(*a*) Labeo rohita (*b*) Pomphret (*c*) Mystus singhala (*d*) Clarius betrachus

34. The breeding place of Flamingo (Hansawar) in India is most likely in:
(*a*) Sambhar lake (*b*) Chilka lake (*c*) Rann of Kachchh (*d*) Ghana vihar

35. Which monkey has prehensile tail?
(*a*) Spider monkey (*b*) Semnopithecus (*c*) Rhesus monkey (*d*) Bounet monkey

ANSWERS

1. (*c*)	**2.** (*b*)	**3.** (*c*)	**4.** (*c*)	**5.** (*c*)	**6.** (*d*)	**7.** (*a*)	**8.** (*d*)	**9.** (*a*)	**10.** (*b*)
11. (*d*)	**12.** (*c*)	**13.** (*d*)	**14.** (*d*)	**15.** (*d*)	**16.** (*a*)	**17.** (*d*)	**18.** (*c*)	**19.** (*b*)	**20.** (*d*)
21. (*c*)	**22.** (*a*)	**23.** (*c*)	**24.** (*a*)	**25.** (*b*)	**26.** (*d*)	**27.** (*d*)	**28.** (*c*)	**29.** (*d*)	**30.** (*d*)
31. (*a*)	**32.** (*d*)	**33.** (*c*)	**34.** (*b*)	**35.** (*a*)					

15

CHAPTER

Environmental Changes

POLLUTION

Pollution of the environment is one of the most horrible **ecological crisis** to which we are subjected today. The air, land (soil) and water are the three basic amenities for living organisms. Sometimes in the past, these amenities were pure, virgin, undisturbed, uncontaminated and basically most hospitable for living organisms. But, the situation is just reverse today because our environment has become foul, contaminated, undesirable and, therefore, harmful for the health of living organisms including man. The various factors responsible for it are numerous but the apparent ones are the explosive population, rapid industrialization, urbanization leading to maximum destruction of natural resources, etc., are of utmost importance. Thus, pollution is an undesirable change in the physico-chemical and biological characteristics of our basic amenities causing harmful effect on our life or that of other desirable species and cultural assets. However, **Southwick** (1976) has defined pollution as **the unfavourable alteration of our environment, largely as a result of human activities.**

Pollution adversely affects biological species, including humans. It damages our industrial processes, living conditions and cultural assets. Other significant changes brought about through human activities are seen in the lower atmosphere. Such changes occur due to the increase in concentration of carbon dioxide and other greenhouse gases, and the depletion of ozone layer of stratosphere. These environmental changes, occur on a global scale, which influence the air, water, land resources, biological diversity and human health.

The loss of forests, fertility of soil, productivity and energy crisis have created many problems. The pollution created by industry, technology and over-consumption by the affluent society lead to the rapid depletion of basic natural resources. Many human problems are also due to mismanagement which is created by humans themselves.

Environmental implications due to pollution are in various aspects. These produce serious problem for human beings to maintain its existence, protection, survival and for the improvement of general standard. The basic needs of human beings have disturbed natural resources and finally led to a situation which has threatened to be disastrous. In the recent years every one has started thinking over the problem of over-population and its consequences, which is primarily concerned with the environmental pollution and every effort should be made to focus public attention to save mankind from self destruction, and steps should be taken at national and international levels so that the consequences may not become worse.

The ecological state of biosphere is becoming more and more disbalanced day by day due to technical and industrial advancements as well as population explosion. Vast changes are taking place in the environment due to interaction between human society and environment itself. Man is exploiting the natural resources for its own interest and many such instances are there as clearly indicate that man has disturbed the natural balance for the sake of small benefits and has changed the environment of many places to such an extent that they are not fit for inhabitation by living beings.

Environmental pollution is the unfavourable alteration of our surroundings, wholly or largely as byproducts of man's action through direct or indirect effects of changes in energy patterns, radiation levels, chemical and physical constitutions and abundance of organisms. These changes may affect man directly or through his supplies of water and of agricultural and other biological products, his physical objects or possessions, or his opportunities for recreation and appreciation in nature.

"The environment is very important in the aspects of peace because when we destroy our resources, they become scarce and we fight over that."

Wangart Maathai, 2004 *Nobel Peace Prize Winner*

POLLUTANTS

The materials which cause pollution of environment are called **pollutants**. The pollutants are, all man made because these are the residues of various materials we make, use and throw away. Certain common pollutants of well-developed and developing countries are as under : **1. Deposited matter** such as soot, smoke, tar, dust and grit. **2. Gases** like sulphur dioxide, carbon monoxide, carbon dioxide, nitrogen oxide, hydrogen sulphide, ammonia, fluorine, chlorine, etc. **3. Chemical compounds** such as aldehydes, arsines, hydrogen fluorides, phosgenes, detergents, etc. **4. Metals** like lead, iron, zinc, mercury, etc. **5. Economic poisons** like herbicides, fungicides, pesticides, nematocides, insecticides, rodenticides and other biocides. **6. Fertilizers**, **7. Sewage**, **8. Radioactive substances**, **9. Noise** and **heat**, etc.

On the other hand, from the viewpoint of ecosystem, pollutants can be divided into two main categories, *i.e.* (*i*) **non-biodegradable** and (*ii*) **biodegradable** pollutants.

1. Non-biodegradable pollutants. Such pollutants are not acted upon by microbes but can be oxidised and dissociated automatically :

For example, chlorinated hydrocarbon pesticides (*i.e.*, dichloro diphenyl trichloroethane (DDT), benzene hexachloride (BHC), waste plastic bottles and bags, polyethylene bags, used soft drink cans, and other plastic materials. They are either not degraded or degraded very slowly in the nature.

However, non-biodegradable pollutants are difficult to manage, and in most cases there is no process of treatment to handle this man-made menace in the ecosystem.

2. Biodegradable pollutants. These include market garbage, livestock wastes, municipal sewage, etc. Such pollutants can be decomposed efficiently by the decomposers.

The biodegradable pollutants are easily manageable by natural processes, or in engineered systems, such as the waste treatment plants.

Pollutants are also classified into the following two categories on the basis of their forms they exist in the environment after their release :

A. Primary pollutants. These are the pollutants which exist as such after being added or released into the environment, *e.g.*, oxides of sulphur, nitrogen, etc.

B. Secondary pollutants. These are the pollutants formed by the reactions of primary pollutants, particularly in the environment in presence of sunlight, *e.g.*, hydrocarbons and nitrogen oxides react in atmosphere in presence of sunlight to form a group of nitrous compounds like **peroxyacetyl nitrate (PAN)**.

An air pollutant may be a **gas** or **particulate matter**, *i.e.*, suspended aecrosols composed of solids and liquids.

Concentrations of atmospheric pollutants depend mainly on the total mass emitted into the atmosphere, and the atmospheric conditions that affect their fate and transport.

Most of the air we breathe constitutes of O_2 and N_2. About 1 per cent is composed of other constituents, such as CO_2 (.03 per cent) and water vapour.

A small part of this 1 per cent may be air pollutants including gases and particulate matter. This small concentration is sufficient to harm life and property.

Natural sources of air pollution consist of pollen, dust, smoke from forest fires, volcanic ash, which are emitted into the atmosphere.

On the other hand, anthropogenic air pollutants enter the atmosphere from (*i*) **fixed** and (*ii*) **mobile** sources.

Fixed sources are those which include large factories electrical power plants, mineral smelters and different small-scale industries.

Mobile sources include all sorts of transport vehicles moving by road, rail or air.

However, air pollutants can be divided into two broad categories : (*i*) **primary** and (*ii*) **secondary** air pollutants.

Primary air pollutants enter the atmosphere directly from various sources.

While **secondary air pollutants** are formed during chemical reactions between primary air pollutants and other atmospheric constituents, such as water vapour. Generally these reactions take place in the presence of sunlight.

Primary Air Pollutants and their Effects

The most important primary air pollutants are particulate matter, carbon monoxide (CO), hydrocarbons, sulphur dioxide (SO_2) and nitrogen oxides (NO_x).

Particulate matter. Solid and liquid aerosols suspended in the atmosphere are referred to as particulate matters. They arise from grinding, erosion, spraying, etc. Aerosols are the chemicals which are released in the air with force in the form of a mist on vapour. The examples are soot, smoke, dust, asbestos, fibres, pesticides, some metals, including mercury, lead, copper and iron, and also biological agents like tiny dust mites, and flower pollen.

Atmospheric particles having diameter approximately about 10 μm, generally settle out in a day or so, while particles with diameter 1 μm or less can remain suspended in air for weeks.

Suspended particulate matter in the lower atmosphere (troposphere) causes human respiratory ailments, like asthma and chronic bronchitis, etc.

Particulate matter, when accumulated in the upper atmosphere (stratosphere) alters the radiation and thermal budgets of the atmosphere, lowering the temperature at the surface of the earth.

The aerosols containing fluorocarbons deplete the ozone layer in the stratosphere and thus permit more harmful ultraviolet radiations to reach the earth surface. Ozone acts as a preventive covering against ultraviolet rays.

From the emission of supersonic aeroplanes nitrogen and sulphur oxides are released which also cause thinning of ozone layer.

The effects of particulate matter include corrosion of metals, erosion and soiling of buildings, sculptures, painted surfaces, and soiling of clothings, damage of electrical equipments, etc.

Many toxic particles including metal dusts, asbestos, aromatic hydrocarbons have been discovered in a polluted urban atmosphere. Lead from vehicle exhausts, resulting from the use of tetraethyl lead as an anti-knock additive to petrol may build up to dangerous levels in urban areas. In small dose, it shortens life span and causes deterioration of nervous system.

Carbon monoxide (CO). It is one of the important gaseous air pollutants. It originates from incomplete combustion of fossil fuels. Nearly 50 per cent of all CO emission originates from automobiles. Carbon monoxide is also present in cigarette smoke. It is short-lived in the atmosphere and oxidises to CO_2. Carbon monoxide is highly poisonous to humans and other animals. The toxic effects of CO on human beings or animals arise from its reversible combination with haemoglobin in the blood. Haemoglobin has much greater affinity with CO, and it lessens the oxygen carrying capacity of blood. However, when inhaled it reduces the oxygen carrying capacity of blood and also reduces the dissociation of oxyhaemoglobin.

Hydrocarbons. They are also known as **volatile organic carbons**. These compounds are composed of hydrogen (H) and carbon (C). Natural sources of hydrocarbons are largely biological. Hydrocarbons are produced naturally during decomposition of organic matter and by certain types of plants, such as pine trees. Worldwide methane production mainly from anaerobic decay of organic matter probably amounts to one billion metric tons annually. Methane (CH_4), the most abundant hydrocarbon in the atmosphere, is evolved from soil in flooded rice fields and swamps. Hydrocarbons are also generated during the burning of fossil fuels, *i.e.*, coal and petroleum.

Benzene and its derivatives, such as formaldehyde, carcinogenic or cancer causing substances.

In urban areas, ethylene is known to inhibit the plant growth. Many aromatic compounds cause irritation and injury in the mucous membrane.

Hydrocarbons are of particular concern because of their involvement in the production of photochemical oxidants which cause irritation and other effects.

Formaldehyde emitted from indoor sources, such as newly manufactured carpeting, causes indoor pollution.

Sulphur dioxide (SO_2). The most important air pollutant is SO_2. About one-third of SO_2 present in the atmosphere is believed to be produced by activities of human beings. SO_2 is the major constituent in the emission when sulphur containing coal is burnt. SO_2 present in atmosphere comes largely from coal and petroleum combustion. Biologically produced H_2S when oxidised produces SO_2. Ore smelters and oil refineries also emit large amounts of SO_2.

SO_2 can damage building materials and buildings, mainly through its conversion into the highly reactive H_2SO_4.

It accelerates corrosion of metals, especially iron, steel and zinc.

A high concentration of SO_2 causes severe respiratory problems in human beings.

Exposure to high SO_2 concentration is harmful to plants. High concentration of SO_2 over short period of time can produce leaf necrosis and browning of pine leaves.

Nitrogen oxides (NO_x). Nitrogen oxides are formed mainly from N_2 and O_2 during combustion of fossil fuels in automobile engines. The biological production of NO and NO_2 amounts to about one billion metric tons annually, while man's combustion processes produce 48 million metric tons NO_2 annually.

Nitrogen oxides produce fading of textile dyes; deterioration of cotton and nylon and corrosion of metals due to production of particulate nitrates.

Nitrogen oxides cause the reddish-brown haze, known as **brown air**, in traffic congested city air, which contributes to heart and lung problems and may be carcinogenic for human beings.

NO_2 causes leaf injury and reduction of growth in several NO_2 sensitive plants.

Nitrogen oxides also contribute to acid rain because they combine with water droplets to produce nitric acid and other acids.

Besides, it is considered to be a major factor in causing eye irritation.

Secondary Air Pollutants and their Effects

Photochemical smog. This is a secondary air pollutant, and generally formed in traffic congested metropolitan cities where warm conditions and intense solar radiation are present.

Photochemical smog mainly consist of ozone (O_3), peroxyacetyl nitrate (PAN) and Nitrogen oxides (NO_x).

It is often called **brown air** where solar radiation is intense. When lesser solar radiation prevails, smog formation is incomplete and the air is referred to as **grey air**.

Automobile exhaust contains hydrocarbons and nitrogen oxides, which play an important role in ozone (O_3) and peroxyacetyl nitrate (PAN) formation in environment of metropolitan cities.

Photochemical reactions involved in smog formation are as follows :

Reaction that occurs inside engine of automobile :

$$N_2 + O_2 \longrightarrow 2NO$$

Reactions that occur in atmosphere :

$$2NO + O_2 \longrightarrow 2NO_2$$

$$NO_2 \xrightarrow{\text{ultraviolet radiation}} NO + O$$

$$O + O_2 \longrightarrow O_3$$

$$NO + O_3 \longrightarrow NO_2 + O_2$$

$$HC + NO + O_2 \longrightarrow NO_2 + \text{peroxyacetylnitrate (PAN)}$$

The effects of photochemical smog are as follows :

(*i*) Smog ozone causes damage to plants as well as animal life. In plants, generally the leaves are affected.

(*ii*) Ozone (O_3) is responsible for causing lung diseases in humans.

(*iii*) Ozone (O_3) is also an effective oxidant, corrodes the building surfaces of historical importance and damages marble surfaces and statues.

(*iv*) Several species of plants are susceptible to PAN in smog. With the result, chloroplasts are damaged and photosynthetic efficiency is reduced.

(*v*) Smog also inhibits electron transport system in plants and interferes with enzyme systems that play important role in cellular metabolism.

(*vi*) PAN causes acute irritation of eyes.

Acid rain. Acid rain refers to several ways in which acids from the atmosphere are deposited on the earth. The term acid rain was first used by Robert Angus in 1872. Literally it means presence of excessive acids in rain water. In fact, acid rain is a mixture of acids, such as H_2SO_4, HNO_3 and HCl. These acids may travel long distances in the atmosphere and undergo several physical and chemical transformations to produce more hazardous substances, which may fall on earth with rain.

Acid deposition includes wet and dry deposition :

Wet deposition refers to acidic water received through rain, fog, snow, dew, mist, etc.

Dry deposition refers to the wind blown acidic gases and particles in the atmosphere, which settle down on the earth.

About half of the acidity is transferred to earth through dry deposition. Dry deposited gases and particles can also be washed from trees and other surfaces by rain fall.

In the atmosphere the following reactions take place in case of sulphur :

$$S + O_2 \longrightarrow SO_2$$

$$2SO_2 + O_2 \longrightarrow 2SO_3$$

SO_3 in humid atmosphere forms droplets of H_2SO_4

$$SO_3 + H_2O \longrightarrow H_2SO_4$$

In case of nitrogen following reactions take place :

$$NO + O_3 \longrightarrow NO_2 + O_2$$

$$NO_2 + O_3 \longrightarrow NO_3 + O_2$$

$$NO_3 + NO_2 \longrightarrow N_2O_5$$

N_2O_5 in humid atmosphere gives rise to nitric acid

$$N_2O_5 + H_2O \longrightarrow 2HNO_3 \text{ (nitric acid)}$$

HNO_3 and H_2SO_4 combine with HCl to generate precipitation, which is referred to as acid rain.

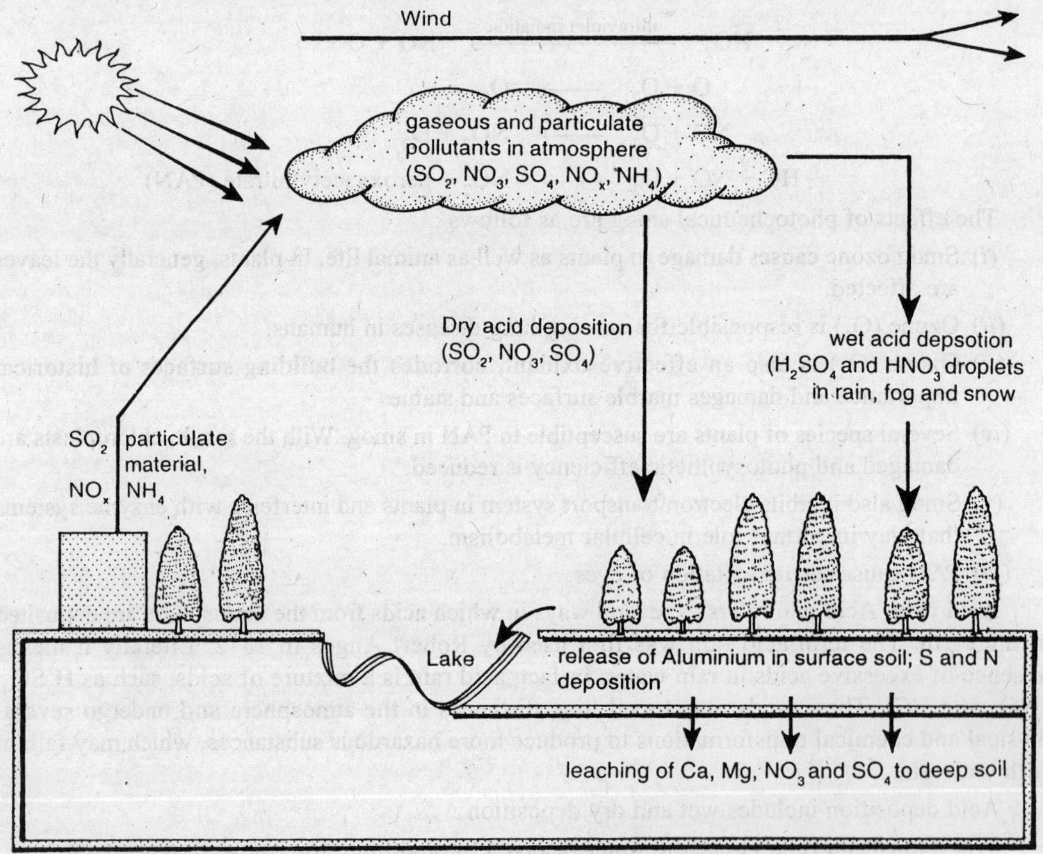

Fig. 15.1. Dry or wet acid deposition (acid rain).

Nitrogen oxides (NO_x), volatile organic carbons (VOCs) and sulphur dioxide (SO_2) are produced during the combustion of coal and petroleum in industries and automobiles respectively.

Lightning in sky also produces NO_x naturally.

The above mentioned gases are highly reactive in air. They rapidly oxidise to acids, sulphuric acid (H_2SO_4) or nitric acid (HNO_3), which dissolve in water and are washed out to the ground as **acid rain**.

In normal cases, the rain water is slightly acidic (pH 5.6 to 6.5) as water and carbon dioxide combine in air to form a weak acid.

While pH of acid rain in less than 5.6 and sometimes as low as 4 or lesser.

Adverse effects of acid rain are as follows :

(*i*) Acid rain causes extensive damage to buildings and structural materials such as marble, lime, stone, mortar and slate. Our heritage monuments, such as Taj Mahal at Agra, are threatened by the corrosive action of acid deposition.

(*ii*) Acid rain adversely affects terrestrial and aquatic vegetation.

(*iii*) The fish and molluscs of rivers and lakes cannot tolerate water having pH below 5.0.

(*iv*) Low pH conditions also damage the soil microbial community.

Control of Particulate Matter

Generally, two important devices are used to remove particulate air pollutants. These devices are:

(*i*) **Arresters.** Such devices are used to separate particulate matters from contaminated air.

(*ii*) **Scrubbers.** These devices are used to clean air for both dusts and gases by passing the contaminated air through a dry or wet packing material.

Particulate matter arresters are of different kinds. The most common are **cyclonic separators** and **trajectory separators**. They are generally used to separate out particulate matters from industrial emissions with minimum moisture content. These separators work on the principle of dust separation by centrifugal force and efficiently work for coarser dust particles.

Filters are usually used for the collection of extremely fine particulate matters. Different types of filters are used for different size and quality of the particulate matters.

Electrostatic precipitator (ESP) is considered the most effective device to remove particulate matters. This type of precipitator works on the principle of electrical charging of the dust particles and collecting it on a differently charged platform.

Scrubbers are also used for dust separation. They are either dry or wet types. However, scrubber is the least used device for separation of particulate matters from the emissions. This device is suitable for the removal of gaseous pollutants.

Control of Gaseous Pollutants

There are some well known techniques that are used to control gaseous pollutants. These techniques are as follows :

(*i*) **Combustion.** In this process, oxidisable gaseous pollutants are completely burnt at a high temperature.

Petro-chemical, fertiliser, paints and varnish industries use this type of combustion device to control gaseous pollutants.

(*ii*) **Absorption.** In this process, gaseous pollutants are absorbed in suitable absorbent materials.

(*iii*) **Adsorption.** This technique is applied to control toxic gases, vapours and inflammable compounds. Such air pollutants are adsorbed on large solid surfaces.

Control of Automobile Exhaust

An efficient type of automobile engine can reduce the unburnt hydrocarbons (HC_s) in auto-emissions.

Catalytic converter filters in the vehicle can convert nitrogen oxides (NO_x) to nitrogen, reducing the hazards of nitrogen oxides.

The automobile fuels of good quality can also reduce the toxic contaminants in exhaust. Lead-free petrol to some extent reduces the load of lead in the exhaust.

Automobile engines operated with **compressed natural gas** (CNG) reduce significantly toxic contaminants in exhaust. The buses with CNG fuel are now running on roads of metropolitan cities like Delhi and Mumbai, which have reduced toxic contamination in exhaust.

Kinds of Pollution

Pollution can be classified as follows :

Generally pollution is classified on the basis of part of environment where it occurs, such as atmosphere, hydrosphere and lithosphere, where it can be classified as (*i*) **air pollution**, (*ii*) **water pollution** and (*iii*) **soil pollution**.

In terms of origin, pollution may be classified into two important categories :

1. Natural. This originates from natural processes, such as volcanic eruptions, forest fires, natural organic and inorganic decays; tons of toxic gases released from volcanic eruptions and forest fires and particulate matter, dust and pollen, etc., naturally present in atmosphere.

2. Anthropogenic. This is called man-made pollution, *e.g.*, industrial pollution, agricultural pollution, etc.

According to the physical nature of the pollutants, the pollution may be categorised as follows :

(*i*) **Gaseous pollution**, where pollution is produced by toxic gases.

(*ii*) **Dust pollution**, is produced by particulate matter in atmosphere.

(*iii*) **Thermal pollution** is produced by higher temperature.

(*iv*) **Noise pollution**, where furious sound is produced by man-made instruments.

(*v*) **Radioactive pollution**, generally produced by nuclear weapons, etc.

AIR POLLUTION

Degradation of air quality and natural atmospheric conditions constitute **air pollution**. When, due to some natural processes or human activities, the amount of solid wastes or concentration of gases other than oxygen increase in the air which are harmful to man and his environment, then the air is said to be polluted and this process is referred to as **air** or **atmospheric pollution**. Air pollution is one of the most dangerous and common kind of environmental pollution that is reported in most industrial towns and metropolitan cities of our country and abroad such as Delhi, Mumbai, Kolkata, Kanpur, Chennai, Hyderabad, Bhopal, Jaipur, Ahmedabad, Nagpur, Firozabad, Agra and also in London, New York, Tokyo, Pittsburg, etc. The importance of air pollution can be estimated by a calculated fact that air constitutes nearly 80 per cent of man's daily intake by weight. A man breathes about 22,000 times a day inhaling about 16 kg of air.

Sources of Air Pollution

Pollution of air started from the very moment when the primitive man knew to make fire, since then it has increased and still increasing every moment. However, the major causes of air pollution are **increase in population, deforestation, burning of fossil fuels and fires, emissions from vehicles, industrialization, agricultural activities, wars** and **natural causes**.

Increase in population. We are worried about our fast increasing population. It is believed that the present world population is likely to be doubled by the year 2000 reaching about six billion. If this rate remains the same then the population would be doubled after every 35 years. The increase in population has resulted several serious problems including the worsening of the condition of environment. An estimate suggests that man alone has dumped 36,00,00,00,000 tons of carbon dioxide in the atmosphere. If its concentration goes on increasing in the atmosphere, in addition to increase in atmospheric temperature it also pollutes the environment making it unsuitable for animal life.

Deforestation. Since, plants purify the air by taking in carbon dioxide for their use and liberating oxygen which is used up by animals during respiratory activity liberating carbon dioxide used by the plants, hence, plants maintain a balance of these gases in nature. Deforestation, i.e., cutting of plants, trees and cleaning of the jungles and forests indiscriminately by man for his own need has disturbed the balance referred to. This has caused in an increase in carbon dioxide and decrease in oxygen concentration in the atmosphere.

Burning of fossil fuels and fibres. Wood, charcoal and fossil fuels are the conventional sources of energy. Burning of these sources causes air pollution. We see smoke coming out from every house every morning and evening. In fact, the by-products of burning are poisonous gases like carbon dioxide, carbon monoxide, sulphur dioxide, etc. The kind, quality and concentration of these gases depend on the type of fuel being used. In the surroundings of factories, the concentration of sulphur dioxide in air is much higher. Sometimes, sulphur dioxide (SO_2) mixes in the air with small particles of metals near the factories and gets oxidised into sulphur trioxide (SO_3). These gases are harmful and they react with water to form sulphuric acid (H_2SO_4) or sulphurous acid (H_2SO_3) and come down to the earth with rain water; it is called **acid rain** or **acid precipitation**. Thus, various poisonous gases coming out from the burning of fossil fuels and fibres, pollute the air making it foul or unfit for breathing.

Emissions from vehicles. Nearly 80 per cent of total air pollution is caused by automobile, locomotives, aircrafts, exhausts, *i.e.*, emissions from vehicles in cities. These vehicles release poisonous fumes by incomplete combustion of petrol and diesel which react with oxides of nitrogen in presence of sunlight in atmosphere to form photochemical smog; it is very toxic. The important pollutants released from vehicles are carbon monoxide—77.2 per cent, nitrogen oxide—7.7 per cent and hydrocarbons—13.7 per cent; in addition it also contains leaded gas and particulate lead.

In a survey conducted at Kolkata, it was estimated that carbon monoxide concentration of 35 ppm (parts per million) due to auto exhaust during the peak hours at selected points was the highest incidence of air pollution in India. Particulates measured about 530 microgram per cubic metre. In a similar survey conducted in New York city suggests that a man who remains outdoors for 24 hours, inhales fumes and smoke equivalent to the smoke of 38 cigarettes,

Industrialization. Industries such as metallurgical plants and smelters, chemical plants, petroleum refineries, pulp and paper mills, sugar mills, cotton mills, and synthetic rubber manufacturing plants are responsible for about one-fifth of air pollution. They release various types of inorganic and organic gases in smoke they produce. The most common pollutants in this case are carbon dioxide, carbon monoxide, sulphur dioxide, hydrogen sulphide, nitrogen oxide, etc. in addition to small particles of dust, carbon, metals, other solids, liquids and radioactive materials which get mixed into the smoke, and pollute the air. All such gases and suspended particles in them are injurious to human health.

Accidental release of poisonous gases, due to carelessness of authorities and other faults, from the industries also pollute the air badly and sometimes fatal instantaneously. An example is **Bhopal gas tragedy** in which methyl isocynate **(MIC)** gas was released accidentally killing thousands of persons while they were asleep.

Cotton dusts, released from the industries located at Ahmedabad and Surat, are one of the most important air pollutants in these cities. The pesticide industries are also concern of threat to the environment. Food processing factories and tanneries produce repulsive odours.

Smog. Smog is yet another problem to which industrial areas are subjected. It envelops the entire areas like a thick blanket. In addition to reducing visibility, it harms animals and even plants and trees. The harmful particles suspended in air block the stomatal openings of plant and tree leaves rendering then to wilt. These suspended particles come down to the earth with rain and enter the root. Sometimes they are harmful to the extent that they kill the entire vegetation of the area.

Tokyo is the most polluted city of the world; it is estimated that nearly 34 tonnes of carbon particles mixed with other waste particles fall per square km every day in this city. The air in Tokyo is polluted to the extent that the traffic policemen have to breathe pure oxygen from oxygen booths. New York is, however, next polluted city of the world where nearly 17 tonnes of carbon and other waste particles fall per square km/day.

Agricultural activities. Different types of biocides used in agricultural practices like pesticides, insecticides, larvicides, herbicides, etc., also cause air pollution because some amount of these poisonous substances is carried away by wind, during their spray over crops, to distant places making the air foul for human health as well as for other animals.

Wars. Various types of explosives used during wars also pollute the air. Radioactive rays coming out of nuclear explosions pollute the air extensively and bring sufferings to mankind to a very large extent. The effects of nuclear explosions on Hiroshima and Nagasaki during the World War II are well known to us.

Natural causes. The natural sources of air pollution are volcanic eruptions releasing poisonous gases, forest fires, natural organic and inorganic decays, marsh gases, pollen grains of flowers and fungal spores. All these are produced naturally and released in the air, making it foul and injurious for human health.

Air Pollution in India

Nearly 80 per cent of industries, the major source of air pollution in India, are located in few big cities; these cities are often referred to as the industrial centres. The major pollutants in the atmosphere of these cities are smoke dust, carbon dioxide, sulphur dioxide, carbon monoxide, hydrogen fluorides, hydrocarbons, nitrogen oxides, fly ash, soot and rarely radioactive substances. A survey conducted regarding sulphur dioxide, nitrous oxide and suspended particulate matter (SPM) in the air of certain important industrial cities of India suggests the data given in the following table :

Table 15.1. Air quality over some Indian cities (annual average of 24 hours). All values are in microgram/cubic meter.

Cities	*SPM*	*SO_2*	*NO_2*
Agra	308.0	15.50	–
Ahmedabad	249.47	41.57	25.67
Baroda	316.9	14.2	37.9
Chennai	143.34	13.32	12.6
Cochin	114.5	33.35	7.7
Delhi	400.88	31.14	31.06
Hyderabad	248.66	21.8	9.54
Jaipur	301.74	7.98	11.94

Cities	*SPM*	*SO_2*	*NO_2*
Kanpur	326.34	15.4	14.28
Kolkata	430.58	53.26	26.76
Mumbai	190.6	38.34	19.88
Nagpur	209.78	10.62	14.96
Vishakhapatnum	129.08	24.35	8.5

From the above table, it is evident that Kolkata is the most polluted city of our country.

Common Air Pollutants

The most common air pollutants are : **Oxides of carbon** such as carbon dioxide and carbon monoxide primarily produced by cigarette smoking, automobiles, burning of fuels in houses, industrial plants, power houses and carbon dioxide is also produced by respiratory activity of animals and plants. **Oxides of sulphur** such as sulphur dioxide and sulphur trioxide produced while melting metallic ores having sulphur impurities, by burning coal and fossil fuels. **Oxides of nitrogen** such as nitrogen monoxide and nitrous oxide released by motor vehicles and particularly fertilizer industries. **Photochemical oxidants**, formed by photochemical reactions between primary pollutants like oxides of nitrogen and hydrocarbons; the oxides of nitrogen react with unburnt hydrocarbons derived from incomplete combustion of automobile fuel in presence of sunlight, *e.g.*, peroxyacetyl nitrate (PAN), ozone, aldehydes and other organic compounds in the air. These are called secondary pollutants. **Aerosols** such as fluorocarbons (carbon compounds having fluorine) released by jet aeroplane emission causing thinning of ozone layer which forms a protective shield against harmful ultraviolet radiations reaching the earth. **Hydrocarbons**, these are unburnt discharges from incomplete combustion of automobile fuels. These form very toxic PAN after reacting with oxides of nitrogen in atmosphere. **Smog**, formed by condensation of smoke and exhausts from industries and automobiles, contains several gaseous and non-gaseous air pollutants. **Radioactive substances** released during nuclear explosions used in wars and experimentations.

Effects of Air Pollution

Air pollution effects our health, our vegetation, our livestock and our belongings in different ways. However, few important of them are discussed below.

1. Effects on human health. Severe air pollution affects our health causing many fatal disceases and disorders. Some of the effects caused by inhaling polluted air are listed below.

(*i*) Flower's pollen grains, fungal spores and suspended fine dust particles responsible for causing various allergic reactions.

(*ii*) Sulphur dioxide enters the soft tissues causing drying of the mouth, scratchy throat and smarting eyes.

(*iii*) Oxides of carbon, sulphur and nitrogen diffuse into the blood stream to combine with haemoglobin causing reduction in its oxygen carrying capacity. Carbon monoxide severely damages cardiovascular system and disturbs psychomotor functions.

(*iv*) Hydrocarbons and many other air pollutants are responsible for causing cancer.

(*v*) Air pollutants cause a number of respiratory diseases like emphysema, chronic bronchitis and lung cancer.

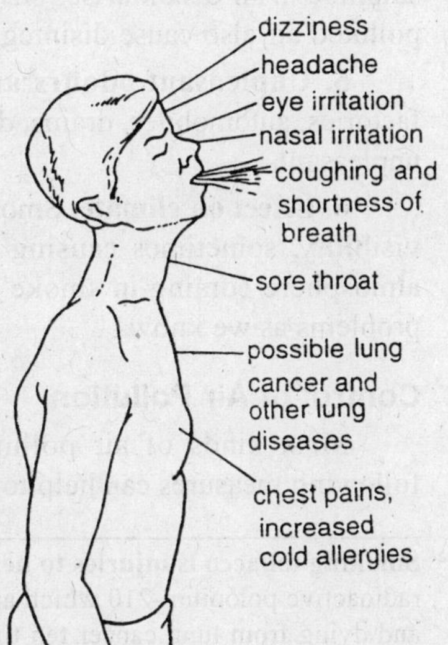

Fig. 15.2. Diseases in man caused due to air pollution.

(*vi*) Fine dust particles and cotton dusts cause bronchitis and asthma.

(*vii*) Cough, sore throat, wheezes, eye irritations, nasal irritations, headache, dizziness, etc., are commonly caused if a person inhales polluted air.

(*viii*) Black lung disease is caused among coal miners who inhale mine dust for many years.

(*ix*) Asbestosis is caused among pipe fitters and insulation workers exposed to airborne asbestos fibres.

2. Effects on vegetation. Air pollution has serious harmful effects on vegetation causing severe economic loss to us. Some of such effects are listed below.

(*i*) Sulphur dioxide causes chlorosis, *i.e.*, loss of chlorophyll and, therefore, it has deleterious effects on trees, plants and vegetation. Its effect completely kills the vegetation.

(*ii*) Oxides of nitrogen and fluorides reduce crop yields.

(*iii*) Fluorides destroy the leaf tissues causing chlorosis and necrosis of leaf apex and margins.

(*iv*) Photochemical smog bleaches and blazes foliage of economically important plants and crops.

(*v*) Hydrocarbons like ethylene cause immature leaf fall, shedding of floral buds, curling and discolouration of petals.

(*vi*) Suspended particles in air and smoke when cover the leaf surface, hamper the photosynthetic activity of the plant causing heavy reduction in yield.

(*vii*) Ozone damages cereals, fruits, cotton, crop, etc. It also causes immature yellowing and shedding of leaves.

3. Effects on livestock. Pollutants present in air also cause widespread damage to our livestock. The general effects of such pollutants on livestock are the same as in the case of human beings. In addition, various fluorine compounds which fall on foliage of plants are eaten by livestock causing abnormal calcification of bones and teeth in them. It is called fluoride toxicity, *i.e.*, fluorosis. It may cause lameness, loss of weight and frequent diarrhoea in animals.

4. Effects on metals and buildings. Acid rains and photochemical smog affect metals and buildings; they also pollute the soil and water. A large number of historical monuments, like Taj Mahal at Agra are affected by air pollution particularly created by Mathura refinery. Hydrogen sulphide in air decolourises silver and lead paints, ozone oxides, rubber goods. Acidic products in polluted air also cause disintegration of textiles, papers and marbles.

5. Unpleasant odours and asthetic insult. Smoke and foul smelling odours emitted by factories, automobiles, drains, dumped garbages, etc., make the urban life most discomfortable and unpleasant.

6. Effect on climate. Smoke, smog and dust make the atmosphere less transparent and reduce visibility, sometimes causing severe accidents on roads and rails. The various gases in the atmosphere coming in smoke from factories imbalance the composition of air causing severe problems as we know.

Control of Air Pollution

Most kinds of air pollutions can be controlled by modern technology. However, the following measures can help to a great extent in keeping the environment free from air pollution :

Smoking tobacco is injuries to health because its smoke contains nearly seven polycyclic hydrocarbons and radioactive polonium-210 which are said to be carcinogenic. An average smoker has the risk of developing and dying from lung cancer ten times more than a non-smoker. The risk of lung diseases is six times and that of heart diseases twice as compared to a non-smoker. A number of other diseases are also said to be linked to smoking.

1. The use of crude fuels should be avoided and the use of high quality fuels should be recommended. This will considerably reduce sulphur contents and hydrocarbons in the atmosphere.

2. Use of automobiles should be minimised which will reduce the nitrogen contents in the atmosphere.

3. The industrial smokes must be filtered before releasing into the atmosphere.

4. The population growth which is supposed to be a major cause of pollution should be brought under control.

5. The gaseous pollutants from industries and power plants can be removed by chemical means.

6. Nuclear explosions should be avoided.

7. Tree plantation should be increased which will reduce carbon dioxide contents of the environment and make it pure by releasing oxygen.

8. Legal control of air pollution is also helpful in making the environment free from air pollutants to some extent. We have **Air Pollution Boards** at central and state level. Their power is to issue and cancel licences to polluting industries, enforce emission standards and frame rules and regulations for the control of air pollution.

In fact, some pollutant gases which are more soluble in a particular liquid than in air can be separated by dissolving them before being released in the atmosphere. It is done by the technique of electrostatic precipitation device. For this, cyclone collectors consisting of a spirally made

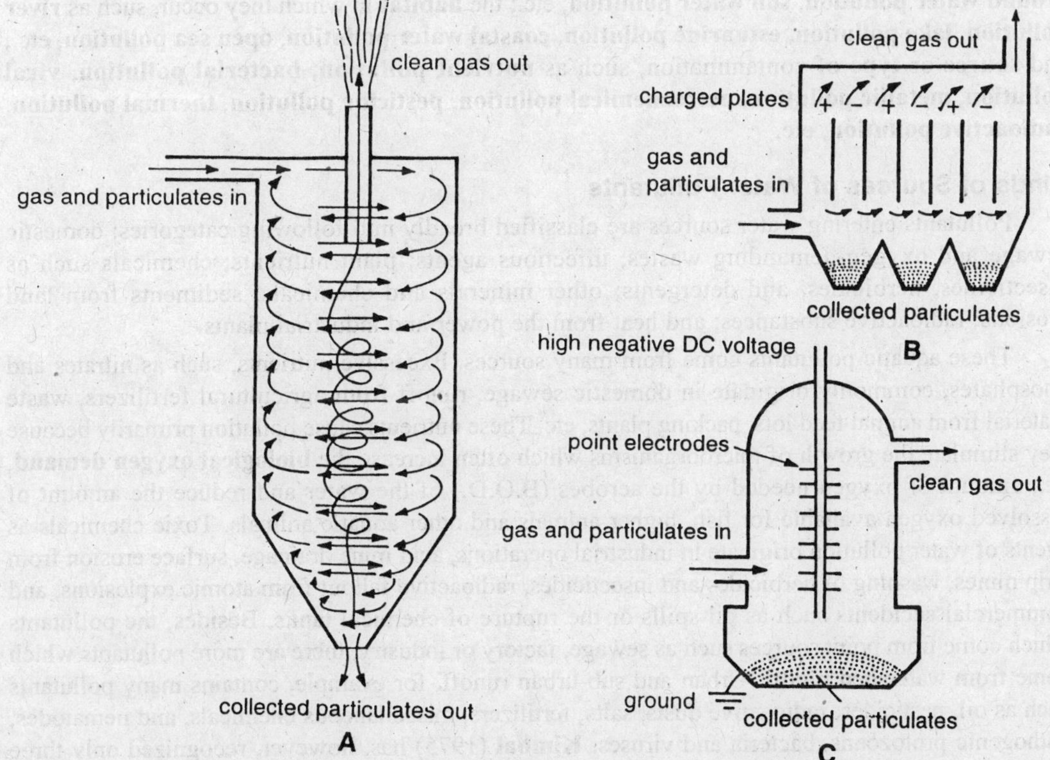

Fig. 15.3. Common types of electrostatic precipitators. A-Cyclone collector; B, Plate-type precipitator; C, Tube type precipitator.

Air pollution inhibits the growth of lichens on trees, hence, lichens can serve as pollution indicators; clean air is indicated by heavy growth of lichens on trees.

chamber or any type of precipitator is taken. In this, the gas to be cleaned, is passed and whirled round through a tight circular spiral. The pollutant particles are centrifuged outward and collected which is removed periodically. For this purpose different types of collectors used are shown in Fig. 15.3.

WATER POLLUTION

'**Water is life**' is a common saying because all organisms including man need water for various purposes; it is needed for various metabolic activities and some organisms use it as habitat as well; man needs water for several purposes like bathing, washing, cooking, irrigation, flushing away wastes, generating electricity, manufacturing different products in industries, navigation, etc. However, during all these processes the wastes and undesirable substances are continuously added to the water sources. A general survey has revealed that nearly 70% streams and rivers in our country contain polluted water. Rivers of our country like Ganga, Jamuna, Hooghly, Godawari, Cauvery, etc., which were once very pure in their waters, are polluted to the extent of 70%. Realizing the importance, the Government has now started several projects to clean the waters of these rivers in phases to make them free from pollutants. The term **water pollution**, therefore, refers to any type of aquatic contamination which affects the living organisms.

Types of Water pollution

It may be classified by the **medium** in which they occur, such as **surface water pollution**, **ground water pollution**, **soil water pollution**, etc.; the **habitat** in which they occur, such as **river pollution**, **lake pollution**, **estuarine pollution**, **coastal water pollution**, **open sea pollution**, etc.; and **source** or **type** of contamination, such as **nutrient pollution**, **bacterial pollution**, **viral pollution**, **metallic pollution**, **petrochemical pollution**, **pesticide pollution**, **thermal pollution**, **radioactive pollution**, etc.

Kinds of Sources of Water Pollutants

Pollutants entering water sources are classified broadly into following categories: domestic sewage and oxygen-demanding wastes; infectious agents; plant nutrients; chemicals such as insecticides, herbicides, and detergents; other minerals and chemicals; sediments from land erosions; radioactive substances; and heat from the power and industrial plants.

These aquatic pollutants come from many sources. Excessive nutrients, such as nitrates and phosphates, commonly originate in domestic sewage, runoff from agricultural fertilizers, waste material from animal feed lots, packing plants, etc. These nutrients cause pollution primarily because they stimulate the growth of microorganisms which often increase the **biological oxygen demand**, *i.e.*, amount of oxygen needed by the aerobes (B.O.D.) of the water and reduce the amount of dissolved oxygen available for fish, higher animals and other aquatic animals. Toxic chemicals as agents of water pollution originate in industrial operations, acid mine drainage, surface erosion from strip mines, washing of herbicides and insecticides, radioactive fallout from atomic explosions, and commercial accidents such as oil spills or the rupture of chemical tanks. Besides, the pollutants which come from point sources such as sewage, factory or industry, there are more pollutants which come from watershed runoff. Urban and sub-urban runoff, for example, contains many pollutants such as oil, pesticides, radioactive dusts, salts, fertilizers, miscellaneous chemicals, and nematodes, pathogenic protozoans, bacteria and viruses. **Kimbal** (1975) has, however, recognized only three major sources of pollution of water : **domestic**, **industrial** and **agricultural**.

Effects of Water Pollution

Each type of water pollution affects the abiotic and biotic factors of different aquatic systems in different degrees and its ultimate effect on man remains quite in medical, asthetic, and economical sense. Some of the well known effects of water pollution are following :

1. Sewage pollution. Contamination of fresh-waters and shallow offshore seas by sewage is a common occurrence. Sewage generally includes biodegradable pollutants such as human faecal matter, animal wastes and certain dissolved organic compounds (*e.g.*, carbohydrates, proteins, urea, etc.) and inorganic salts such as nitrates and phosphates of detergents and sodium, potassium, calcium and chloride ions. Under natural processes most of the biodegradable pollutants of sewage are rapidly decomposed, but when they accumulate in large quantities, they create problem, *i.e.*, when their input in the environment exceeds the decomposition capacity of the latter. Most cities of developed countries like USA, Britain, etc., and some cities of developing countries like India have evolved certain engineering systems, such as, **septic tanks, oxidation ponds, filter beds, waste-water treatment plants and municipal sewage treatment plants** for the removal of many harmful bacteria and other microbes, organic wastes and other pollutants from the sewage, before it is tipped into river or sea.

The sewage treatment is usually performed in following three stages : **(1) primary treatment**, which removes large objects and suspended undissolved solids of raw sewage and converts them into a biologically inactive and aesthetically inoffensive state, called the **sludge**, a valuable fertilizer. **(2) secondary treatment**, which supplies aeration and bacteriological action to decompose organic compounds into harmless substances such as carbon dioxide, sulphate and water. During later stages of secondary treatment, whole waste water is chlorinated (*i.e.*, treated with chlorine) to reduce its contents of bacteria. **(3) tertiary treatment**, which removes nitrates and phosphates and releases pure water. These three stages of sewage treatment have become increasingly expensive and only in most advanced countries all the three treatments of sewage are done. However, most Indian cities either lack any sewage or waste-water treatment plant or have inadequate sewage treatment facilities. Consequently, normally and especially during heavy downpour and floods raw sewage or incompletely treated sewage is dumped into rivers which causes severe water pollution problems.

2. Industrial pollution. Most of the Indian rivers and fresh-water streams are seriously polluted by industrial wastes or effluents (Table 15.2 and 15.3) which come along waste waters of different industries such as petro-chemical complexes, fertilizer factories, oil refineries, pulp, paper, textile, sugar, steel mills, tanneries, distilleries, drugs, fibres, rubber, plastic, etc. All these

Table 15.2. Some important Indian rivers and their major sources of pollution

Name of the river	*Sources of pollution*
1. Yamuna near Delhi	1. D.D.T. factory, sewage, Indraprastha Power Station, Delhi.
2. Ganga at Kanpur (U.P.)	2. June, chemical, metal and surgical industries, tanneries, textile mills and a great bulk of domestic sewage of highly organic nature.
3. Gomati near Lucknow (U.P.)	3. Paper and pulp mills; sewage.
4. Damodar between Bokaro and Panchet (Bihar)	4. Fertilizers, flyash from steel mills, suspended coal particles from washeries, and thermal power station.
5. Hoogly at Kolkata	5. Power stations, paper and pulp, jute, textile, chemical mills, metal steel, varnishes, rayon and soap and a huge bulk of sewage.
6. Sone at Dalminagar (Bihar)	6. Cement, pulp and paper mills.
7. Godavari	7. Paper mills.
8. Cauvery (Tamil Nadu)	8. Sewage, tanneries, distillaries, paper and rayon mills.

chemicals of industrial wastes are toxic to animals and may cause death or damage to different body systems of the aquatic fauna and flora.

3. Thermal pollution. Various industrial processes may utilize water for cooling and resultant warmed water has often been discharged into streams or lakes. Coal or oil-fired generators and atomic energy plants generate large amounts of waste heat which iscarried away as hot water and cause **thermal pollution**. The thermal pollution produces distinct changes in aquatic biota. However, the thermal pollution can exert a disruptive effect on aquatic ecosystem.

4. Effects on aquatic ecosystem. Organic and inorganic wastes decrease the **dissolved oxygen** (DO) content of water bodies.

Water having DO content less than 8.0 mg L^{-1} is considered containinated. Heavily polluted waters have DO content less than 4.0 mg L^{-1}.

DO content water is important for the survival of aquatic organisms.

The turbidity, photosynthetic activity, oxygen consumption by organisms and decomposition of organic matter are the factors that determine the amount of DO present in water.

Higher amounts of organic waste increase the rate of oxygen consumption, thus causes drop in DO content of water.

The demand of oxygen is directly related to increasing input of organic wastes and is referred to as **biochemical oxygen demand** (BOD) of water.

BOD is a measure of oxygen required by aerobic decomposers for the biochemical degradation of organic materials (biodegradable materials) in water.

Higher the BOD, lower would be dissolved oxygen (DO).

Chemical oxygen demand (COD) is also a measure of pollution load in water. This is measure of oxygen equivalent of the requirement for oxidation of total organic matter, (i.e., biodegradable and non-biodegradable) present in water.

Thus, contamination of water bodies by pollutants reduces DO content, and therefore, sensitive organisms, such as plankton, mollusc and fish are eliminated.

In such state, only a few tolerant species, like annelid *Tubifex* and some insect larvae, survive in highly polluted, low DO water, and they are referred to as **indicator species** for polluted waters. Biocide residues, PCBs and heavy metals like Hg, Cd, Pb, Cu, As, etc., can eliminate different species of organisms.

Higher the temperature of water, lower is the rate of dissolution of oxygen in water.

However, hot wastewaters discharged from industries, when added to water bodies, cause lowering of DO content of water.

5. Biological magnification. The phenomenon through which certain pollutants are accumulated in tissues in increasing concentration along the food chain, is refered to as **biological magnification**. Such pollutants are non-biodegradable, such as DDT. Once they are absorbed by an organism, they cannot be metabolised and broken down or excreted out.

Such pollutants (non-biodegradable), generally get accumulated in fat containing tissues of the organism.

For example, DDT an insecticide, is sprayed on water bodies to check the growth of mosquitoes.

BOD is measured by keeping a sample of water containing known amount of oxygen for five days at 20°C in the dark. At the end of this period, the oxygen content is again measured. A high BOD indicates intense level of microbial pollution.

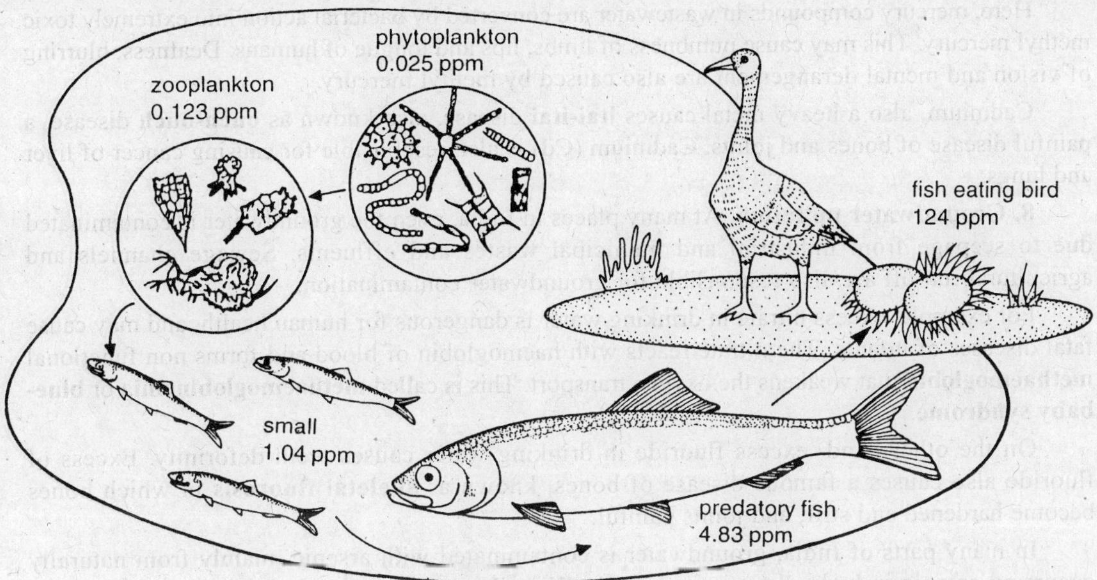

Fig. 15.4. Process of biological magnification. DDT concentrations increase in organisms along food chain.

In an island of USA, after regular DDT spraying for few years, the population of fish-eating birds decreased. Later, this was found that the concentration of DDT had increased about 800 times in the phytoplankton relative to the concentration in water.

However, zooplankton contained about five times greater DDT than phytoplankton.

In different fish, the DDT concentration increased 9 to 40 times, relative to the concentrations in zooplankton.

Birds showed about 25 times greater DDT concentration relative to that in fish.

Many other persistent pesticides and radionuclides also show biological magnification.

6. Eutrophication. Here, instead of inorganic nutrient input with the inflow of waste water, decomposition of organic wastes, increases the nutrient content of the water bodies.

The excess of nutrients causes profuse growth of algae, especially the blue-green algae (cyanobacteria). Such profuse growth of algae is called **algal bloom**.

Such algal blooms completely cover the water surface.

The blue-green algae release toxins in water, and sometimes cause deficiency of oxygen in water. Therefore, in bloom-infected water body the growth of other fresh water is checked due to toxins, and aquatic animals may die because of toxic effect or lack of oxygen.

The process of enrichment of nutrients of water, and then loss of species diversity is known as **eutrophication**.

7. Effects on human health. Domestic sewage contains pathogens, such as virus, bacteria, parasitic protozoans, and several types of worms.

Contaminated water with above mentioned pathogens causes several water borne diseases, like jaundice, cholera, typhoid, amoebiasis, etc. The contamination makes the water unfit for drinking purpose. Such water cannot be used even for bathing, swimming and irrigation purposes.

On the other hand, the heavy metal contamination of water causes serious ailments of human beings. Mercury poisoning, also known as **Minamata disease** is caused by consumption of fish captured from Hg-contaminated Minamata Bay in Japan. This disease was detected for first time in 1952.

Here, mercury compounds in wastewater are converted by bacterial action into extremely toxic methyl mercury. This may cause numbness of limbs, lips and tongue of humans. Deafness, blurring of vision and mental derangement are also caused by methyl mercury.

Cadmium, also a heavy metal causes **itai-itai** disease, also known as **ouch-ouch** disease, a painful disease of bones and joints. Cadmium (Cd) is also responsible for causing cancer of liver and lungs.

8. Groundwater pollution. At many places in India, even the groundwater is contaminated due to seepage from industrial and municipal wastes and effluents. Sewage channels and agricultural run off are also responsible for groundwater contamination.

For example, excess nitrate in drinking water is dangerous for human health, and may cause fatal diseases of infants. The nitrate reacts with haemoglobin of blood and forms non functional **methaemoglobin** that weakens the oxygen transport. This is called **methaemoglobinemia** or **blue-baby syndrome**.

On the other hand, excess fluoride in drinking water causes teeth deformity. Excess of fluoride also causes a famous disease of bones, known as **skeletal fluorosis** in which bones become hardened and stiff, and joints painful.

In many parts of India, groundwater is contaminated with arsenic, mainly from naturally occurring arsenic in bedrocks, particularly in West Bengal. Excess of arsenic in water causes **black-food disease**. Arsenic also causes diarrhoea, peripheral neutritis, hyperkeratosis, and also skin and lung cancers.

Improving Water Quality

The industrial and municipal waste waters are treated in **Effluent Treatment Plant (ETP)** prior to disposal in water bodies.

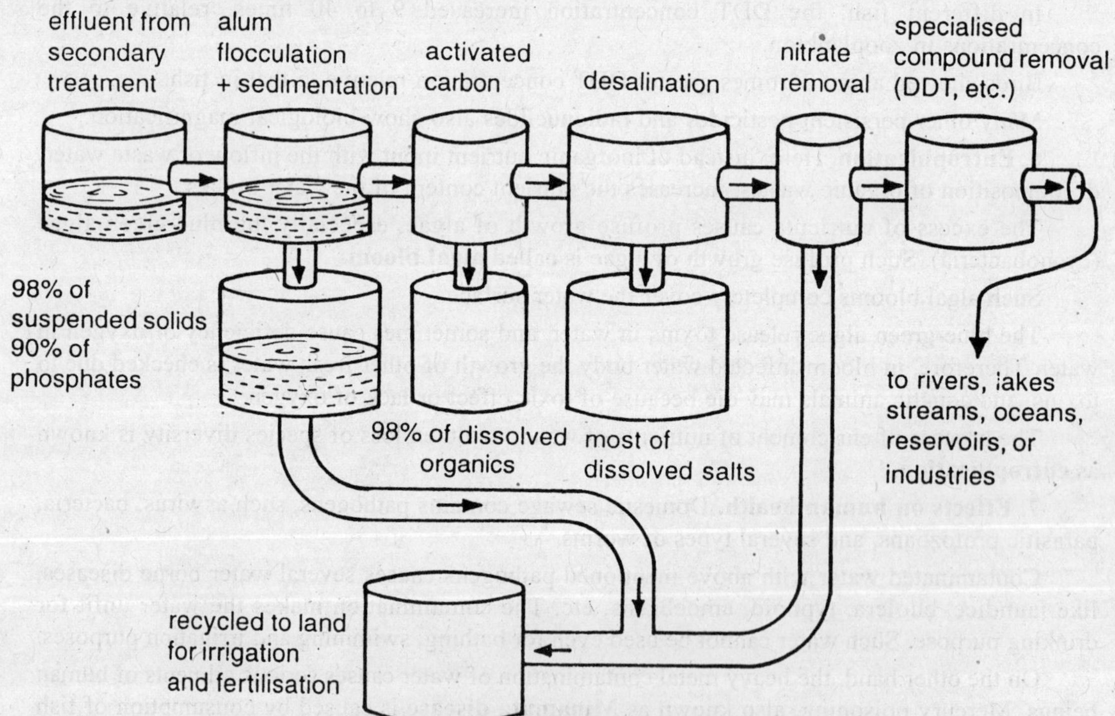

Fig. 15.5. Treatment of wastewaters in Effluent Treatment Plant (ETP).

Generally the undermentioned treatments are given in ETP :

(*i*) **Primary treatment.** This is a physical process, when the separation of large debris is done, which is followed by sedimentation in tanks or clarifiers of improved type.

(*ii*) **Secondary treatment.** This is a biological process and carried out by the activity of microorganisms. In this treatment, the waste water is pumped in shallow **stabilisation** or **oxidation** ponds, where the microorganisms (microbes) oxidise its organic matter.

This process results in release of carbon dioxide and formation of **sludge** also known as **biosolid**. The sludge is continuously aerated to accelerate its oxidation.

Algae grown in the upper lighted zone of the wastewater provide aeration by generating oxygen.

(*iii*) **Tertiary treatment.** This is physicochemical process, which removes turbidity in wastewater caused by nutrients, such as nitrogen, phosphorus, etc., dissolved organic matter, heavy metals or pathogens.

Table 15.3. Various organic and inorganic pollutants of water and their sources

Type of Industry	*Organic pollutants*	*Organic pollutants*
Mining	**Mine wastes:** Chlorides, various metals, ferrous sulphate, sulphuric acid hydrogen sulphide, ferric hydroxide, surface wash offs, suspended solids, chlorides and heavy metals.	—
Iron and steel	Suspended, solids, iron cyanide thiocyanate, sulphides, oxides, of copper, chromium, cadmium, and mercury.	Oil, phenol and naptha
Chemical plants	Various acids and alkalies, sulphates, nitrates of metals, phosphorus, fluorine, silica, and suspended particles.	Aromatic compound solvents, organic acids, nitro compound dyes, etc.
Pharmaceuticals	—	Proteins, carbohydrates, organic solvents, intermediate products, drugs and antibiotics.
Soap and detergent	Tertiary ammonia compounds, alkalies.	Fats and fatty acids, glycerol, polyphosphates, sulphonated hydrocarbons.
Food processing	—	Highly putrescible organic matter and pathogens
Paper and pulp	Sulphides, bleaching liquors	Cellulose fibres, bark, wood, sugars and organic acids.

This step involves chemical oxidation of wastewater by strong oxidising agents like chlorine gas, perchlorate salts, ozone gas and ultra-violet (UV) radiation.

After tertiary treatment, the wastewater is discharged into natural waters and may be used for irrigation.

Control of water pollution. It is very difficult to check the pollution of water but the intensity of pollution can be minimised to a great extent by the following methods :

1. Adequate sewage treatment. Raw sewage should not be dumped into rivers or oceans. Before its release into them, sewage should be properly treated in sewage treatment plants.

2. Treatment of industrial effluents. The industrial effluents should be cleaned before they are discharged into rivers.

3. Recycling. The best method of prevention and control of water pollution is the recycling of various kinds of wastes. Dung of cows and buffaloes can be used for the production of **gobar gas**, a cheap source of fuel and also of manure.

Ganga Action Plan

As referred to, it is one of the largest projects adopted recently by the Government of India, with a cost of nearly 292 crores of rupees, to clean Ganga water.

Ganga, the sacred river of India, is one of first ten mighty rivers of the world. It travels nearly 2,525 km through U.P., Bihar and West Bengal supporting about 45 million people. Once, the purest water of Ganga is now polluted at a number of places by the pollutants of domestic and industrial wastes (see Table 15.2). An estimate suggests that nearly 1,200 million litres of waste water are released everyday in it from nearly 27 major cities situated at its banks. At Varanasi, a city in U.P., it receives the largest amount of biological and chemical filths from millions of people taking bath every year. Besides this, nearly 60 million litres of untreated sewage are dumped into it every day. It also receives ashes of about 40,000 human bodies, about 10,000 incompletely burnt dead human bodies and those of the dead bodies of animals are dumped annually into it at Varanasi. A survey conducted by CBPCWP (Central Board for the Prevention and Control of Water Pollution) suggests that B.O.D. in the river reaches upto 9.7 mg/litre as against the acceptable limit of 3 mg/litre.

However, realising the consequences of the pollution of Ganga water, Government of India has recently prepared a project to clean its water; this project is called **Ganga Action Plan**. This project is to be implemented in phases. This Action Plan covers nearly 27 cities (6 in U.P., 4 in Bihar and 17 in West Bengal). Government has established "**Ganga Project Directorate**" in 1985 as a wing of the **Department of Environment** to coordinate the activities taken by the Action Plan and another government agency called **Central Ganga Authority** was also established in the same year to look into the implementation of the Action Plan. Now, its first phase is on with the following tasks :

1. Renovation of existing trunk sewers to prevent the outflow of sewage into the Ganga.
2. Construction of interceptors, drains and sewers to divert the flow of sewage and other liquid wastes.
3. Renovation and installation of sewage treatment plants with special emphasis on recovery of resources like methane gas for the generation of power and treated water for irrigation and agriculture.
4. Low cost sanitation schemes in the areas adjoining the river.
5. Biological conservation measures.
6. River front developments including afforestation on the banks.
7. Prevention of industrial pollution particularly from main pollutors, in cooperation with regulatory agencies and other schemes.
8. Regular monitoring of water quality.

SOIL POLLUTION

Land and soil are very essential like air and water, either directly or indirectly for the survival and maintenance of living organisms including man. Soil supports vegetation on which every living animal depends. However, like air and water soil may also get polluted. There are many natural and synthetic materials that can adversely affect the physical, chemical and biological properties of soil and seriously affect its productivity. Therefore, any substance that reduces the productivity of soil is called soil pollutant and such soil is referred to as polluted.

Soil gets polluted either directly by dumping and disposal of solid wastes or indirectly by air pollution. The major soil pollutants are different types of biocides used in agricultural practices, fertilizers, industrial wastes such as chemicals, flyash and cinders which are residues of combustion of solid fuels, garbage, discarded food, paper, plastics, rubbers, cloth, leather, asphalt, tin, iron, lead, copper, building materials, dead animals like cattle, dogs, cats, birds, and discarded goods like refrigerators and autos, etc. Huge quantity of unwanted materials causes serious disposal problems. From these, the organic wastes decompose and befoul the air; others do not decompose, hence, cause a problem for their disposal.

Biocides such as pesticides, insecticides, larvaecides, herbicides reduce the population of undesired species as also of the desired microorganisms in the soil. Therefore, their uncontrolled use affects the structure and fertility of the soil. Several pesticides or their residues are absorbed by the plants, that affect the entire food chains and food webs.

The excretory products of humans and livestock, and digested sewage sludge used as manure pollute the soil. In underdeveloped and developing countries the poor sanitary condition aggravates soil pollution. Several disease germs contained in such wastes contaminate the soil and vegetables causing severe health hazards for human beings and domestic animals.

Salination of soil also affects its productivity and degrades the quality of land. Intensive farming with poor drainage also results in salination of the soil. An estimate suggests that there are nearly six million hectares of saline land in our country. In Punjab alone nearly 6000—8000 hectares of farm land are becoming saline and unfit for agriculture. About one-sixth of the arid and semi-arid lands of the world has high salinity.

Control of Soil Pollution

Several modes adopted to control soil pollution are given below :

1. Transfer stations should be established at various places in a city for bulk transfer of refuge to ultimate discharge sites.
2. Pneumatic pipes should be used for collecting and disposing wastes.
3. Materials such as paper, plastics and glass should be recycled.
4. Animal refuge should be used for production of biogas for various domestic as well as industrial use.

Fluorosis. Fluoride containing minerals are most widely distributed in our country. Most underground and surface waters are contaminated with fluoride ions. High amounts of fluoride ions are reported in drinking water nearly in 13 states of our country. A human body can tolerate fluoride to the extent of 1.5 ppm (mg/litre of water). When its concentration reaches beyond the tolerance level, it causes fluorisis resulting into various neuro-muscular disorders, gastro-intestinal problems, allergies, dental disorders and skeletal disorders leading to crippling of people. So, it is very essential to check and minimise fluorine content in drinking and cooking waters within the range of tolerance. Industrial fluorisis has also been reported to affect agriculture and horticulture. Hence, defluoridation technology must be adopted in India, unfortunately this technology has not yet been introduced in our country.

5. Use of chemical fertilizers should be minimised by applying biofertilizers and manures.
6. Use of biocides should be minimised by adopting biological control methods.
7. Soil erosion can be checked by planting more trees and the process of deforestation should be checked.

Degradation of Land

In addition to pollution, soil and land face several other problems. These include erosion, deforestation and overgrazing, flooding and water logging, salination, desertification and ill-planned urbanization. It is estimated that if the present rate of soil depletion and land degradation continues, about one-third of arable land will be destroyed by the end of the present century.

Erosion. Removal of top, fertile layer of soil is called soil erosion. The agencies responsible for it are water, wind, ocean waves, glaciers, rains, over-grazing, over-cropping, and improper tilling. In fact, erosion occurs both in dry and wet regions and results in floods. In India, soil erosion is a serious problem in northern and north-eastern regions. Soil erosion is a global problem especially in countries like Nepal, India, China, Spain, Australia, U.S.A., USSR and Central Asia.

Deforestation and over-grazing. In India, these are the main causes of soil erosion. An estimate suggests that nearly 40,000 hectares of land are affected by wind and water erosion. The quality of top soil loss in India is about 18.5% of the total global loss. India has highest number of livestock in the world, the pressure of grazing, therefore, is intensively higher.

Roots of grasses, plants and trees are excellent binding material and keep the soil intact. Disruption of grass cover by plough, intensive grazing, deforestation, etc., loosen the soil making it suitable for being washed away by wind or rain.

Shifting cultivation. It is also one of the causes of land degradation. Many tribal communities, particularly in the north-eastern India, follow the practice of cutting trees, burning them and raising crops on resulting ashes called **jhuming**. This practice is most prevalent in Jhum forests covering an area of about 30,000,000 km^2. It degrades forests and disturbs soil stability.

Desertification. It results from erosion of top soils, shifting of sand dunes by wind and over-grazing in lands sparsely covered by grass. Many deserts are believed to be man-made.

In addition to above causes, various developmental activities like rapid urbanisation and human settlements, construction of dams, canals, mining, roads, railways, airports and establishment of industries have considerably reduced the areas of fertile and productive crop lands, woodlands and grasslands. All these have, thus, aggravated the problem of land degradation.

Control of land degradation. Degradation of land can be controlled by adopting following measures :

1. By restoring forests and grass cover, soil erosion and flood can be checked.
2. Shifting cultivation can be replaced by crop rotation practices. Mixed cropping or developing plantation crops would improve soil fertility and support a larger population.
3. Adequate drainage should be made to prevent salinity of soil.
4. Salt affected lands can be recovered by bleaching them with more water.
5. Sufficient plantation of appropriate plant species can prevent shifting sand; mulching or covering the area with artificial protective covering also prevents shifting sand.

India is basically an agricultural country. Out of India's total land mass of 305 million hectares nearly half is waste land, hence, nearly 143 million hectares are used for agricultural purposes. From remaining, nearly 18 million hectares are under urban productive use, 21 million hectares are rocky or snowcapped, 17 million hectares are culturable waste lands, 23 million hectares are fallows, 83 million hectares are forests and pasture lands.

Due to rapid increase in population the demand of land for agriculture, settlement and industry is increasing rapidly but, on the other hand good land is shrinking considerahly because of soil degradation problems. Obviously, the only solution of the problem appears to develop the waste lands so that they may be used by man for their various needs.

The waste lands are grouped into two categories, *viz.*, **culturable** and **unculturable waste lands**. The culturable waste lands are those which can be brought to use by their development and reclamation. Examples of such lands are ravinous and gullied lands, surface water-logged land and marsh, saline lands, forest lands, strip lands, industrial waste lands, shifting cultivation areas, etc. The unculturable waste lands are those which cannot be used. Examples of such lands are barren rocky areas, steep slopes, snow-capped mountains, glaciers, etc. So, the development and reclamation of culturable waste lands can increase the availability of productive land.

No doubt, reclamation of waste lands involves high expenditure, expertise hands and manpower. Even then, proper steps within the financial means must be undertaken by the Government agencies immediately which will definitely solve the crisis of productive land to some extent.

Water Management in India

Implementation of environmentally unsound developmental programmes, practices and defective land use in mountains have created severe problems in the plains of countries like ours, Pakistan and Bangladesh. Nearly 400 million people are ravaged by recurring floods and their consequences. Even today, the canal irrigated land's productivity in our country is very poor in comparison to other countries due to salinity and water logging. So, proper drainage development and desalination methods should be taken to overcome such problems.

In our country, however, we have not yet thoroughly tapped underground water. It should be tapped because of the following advantages :

1. It involves very low cost, *i.e.*, installation of tube-wells.
2. It takes very short time.
3. A large amount of underground water is available even in desert areas.
4. The underground water is especially suitable in canal command areas because it helps in lowering the water tables.
5. Irrigation by underground water is economical because it does not involve storage and transportation.
6. It is easier to rouse the interest of the farmers.
7. The underground water is not vulnerable to evaporation and seepage losses.

RADIOACTIVE POLLUTION

It is a phenomenon of physical pollution to all life-supporting system like air, water and land. **Radioactivity is spontaneous disintegration of a radioactive substance which results in the emission of radioactive rays.** In fact, in this phenomenon, the nuclei of radioactive substances spontaneously disintegrate to emit protons (α-particles), electrons (β-particles) and gamma rays (short wave electromagnetic waves). The emitted radiations of radioactive elements affect the biotic community in various ways. Though the biotic community including human beings has been exposed to low levels of radiations from natural sources for several millenniam, but the danger of radiations exposure has increased in the recents years.

Sources of environmental radiations are both **natural**, comprising cosmic rays that reach the surface of the earth from space and terrestrial radiations from radioactive element's nuclei present in the earth's crust (*e.g.*, radium-224, Uranium-235 and 238, thorium-232, radon-222, potassium-40 and carbon-14), and **man-made** (*e.g.*, mining and refining of plutonium and thorium, production and explosion of nuclear weapons, nuclear power plants and fuels and preparation of radioactive isotopes).

Nuclear Weapons

Amongst the nuclear weapons, atomic bombs were first used in World War II on the twin cities of Japan, Nagasaki and Hiroshima in 1945. Since then the craze for possessing nuclear weapons has caused the nuclear race by the big powers of the world.

Tests of nuclear arms comprise the use of uranium-235 and plutonium-239 for fission and hydrogen or lithium as fusion material. Nuclear explosions are uncontrolled chain reactions, they give rise to very large neutron flux conditions that make other materials radioactive in the surroundings. Such materials include iodine-131, strontium-90, cesium-137. These materials are transformed into gases and fine particles which are thrown high up into the air like a mushroom cloud. The radioactive particles are blown away by winds and, thus, get distributed in wide areas. When rain drops containing these particles fall on the ground, the radioactivity is transferred to the soil. Thus, soil and water get polluted by radioactive substances. These substances, from the soil, enter into the food chain affecting different forms of life including man.

Radioactive isotopes like C^{14}, I^{125}, P^{32} and their compounds are widely used in scientific research. Waste waters containing these radioactive substances reach the rivers and lakes through sewer. These radioactive isotopes enter the food chain and finally reach to man (final victim of radioactive pollution) through fish. Man also receives radiations voluntarily from diagnostic X-rays and radiation therapy for cancer. People living in nearby areas and also those who work in nuclear reactors, fuel processors and power plants are vulnerable to radiation exposure.

The radiation exposure sources may be summarised as under :

1. Natural

A. Cosmic rays

B. Environmental

(*i*) Terrestrial

(*ii*) Atmospheric

(*iii*) Internal

2. Man-made

A. Diagnostic exposures, *i.e.*, medical and dental (X-rays and radiation therapy)

B. Nuclear explosions, leakages and exposures (Radioactive fallout, nuclear reactor wastes, etc.)

C. Occupational exposures

D. Miscellaneous exposures (TV. luminous dials, luminous markers, isotope tagged products, etc.)

The nuclei of radioactive elements possess a constant rate of disintegration. Half-life is the time taken for half of the atoms to disintegrate. This varies from a fraction of second to thousands of years. The nuclei of elements having long half-life are usually the main sources of environmental concern.

Effects of Radioactive Pollution

Atomspheric and climatic conditions such as wind, temperature, rainfall and also half-life, energy releasing capacity and rate of diffusion of the radioactive pollutants determine their effects. However, some generalized effects are listed below :

1. In lower doses, it can affect seriously and impair their functions.
2. In higher doses it can cause instantaneous death.
3. Long and repeated exposures can cause cancer and leukemia, and induce mutations.
4. The mutant genes having deleterious effects can persist and passed on to the progeny where they show their effects.

Control of Radioactive Pollution

In fact, there is no cure of radiation damage, hence, every precaution should be taken to prevent radioactive pollution, such as :

1. Leakages of radioactive substances from nuclear reactors, careless handling, transport and use of radioactive fuels, fission products and radioisotopes have to be totally stopped.
2. Safety measures against accidents in the power plants, fuel processors, nuclear reactors should be strictly enforced to avoid any chance of leakage.
3. The disposal of waste must be very safe.
4. Regular monitoring through frequent sampling and quantitative analysis has to be ensured in the risk areas.
5. Use of nuclear weapons should be completely banned.

NOISE POLLUTION

Noise is defined as sound without value or any sound that is undesirable by the recipient. In other words, noise can be defined as **wrong sound, in the wrong place, at the wrong time**. Noise pollution, thus, refers to the unwanted sound dumped into the atmosphere leading to health hazards. Noise is a physical form of pollution and has no persistent effects on the life-supporting systems but it has direct effects on the recipients.

Sources of Noise Pollution

Noise is either **natural** such as thunder or **man-made**. The main sources of man-made noise in developed urban areas are automobiles such as motors, trucks, buses, scootors, fire engines, police cars, ambulances, etc., factories, industries, trains, air-planes, and accessory noise producers such as horns, sirens, musical instruments, TV, radio, transistors, barking of dogs, shouting, loud-speakers, etc. There has been a considerable increase in noise from man-made sources during the last 100 years which is now doubling after every 10 years.

Effects of Noise Pollution

The various effects of noise pollution on human beings are summarised as under :

1. Auditory effects. The most acute and immediate effect of noise pollution is impairing of hearing leading to auditory fatigue and may finally lead to deafness.

The unit of noise is deciBels (dB). Modern conversation has a noise value of 60 deciBels, any value greater than 80 dB causes noise pollution. Noise becomes unbearable at 140 dB. A man subjected for long periods continuously with 85 dB sound noise, his hearing ability is affected. A jet plane when takes off produces 120 dB and big cities where cars and trucks are continuously running produce a noise of 90 dB.

2. Non-auditory effects. These effects are very alarming as they may cause severe diseases. These effects can be briefed as under :

(*i*) Interference with speech communication.

(*ii*) Annoyance leading to ill-temper, bickering, mental disorientation and violent behaviour.

(*iii*) Loss of working efficiency due to physiological disorders. These may lead to cause neurosis, anxiety, insomnia, hypertension, cardio-vascular diseases, hepatic stress, giddiness, nausea, fatigue, visual disturbance, change in skin temperature and blood circulation, quickening of human foetus's heart rate and malformation of its nervous system.

Control of Noise Pollution

The different methods to control noise pollution are as under :

1. Control of noise at source may be achieved by designing and fabricating silencing devices and their use in air-craft engines, trucks, cars, auto-scooters and industrial machines.
2. Noise at source can be minimised by proper lubrication and better maintenance of machines.
3. Noise producing parts of machines may be insulated in sound proof materials.
4. Noise producing industries should be installed away from dwelling areas.
5. Use of loudspeakers and aimplifiers should be restricted at a fixed intensity and hours of the day.
6. Stuffing of cotton balls in the ears, covering our ears with hands under noise conditions, being away from the sources of noise and using ear plugs to reduce noise pollution to the intensity of about 40–50 dB are personal protection measures.

The Central Pollution Control Board has recommended zone wise ambient noise levels. They are as follows :

Table 15.4. Zone-wise permissible Ambient Noise Levels

Zones	*Day : 6.00 to 21.00 hr*	*Night : 21.00 to 6.00 hr*
1. Industry	75 dB	70 dB
2. Commercial	65 dB	55 dB
3. Residential	55 dB	45 dB
4. Silent	50 dB	40 dB

ENVIRONMENTAL LAWS FOR CONTROLLING POLLUTION

The Insecticide Act, 1968

This act deals with the regulation of import, manufacture, sale, transport, distribution and use of insecticides with a view to prevent risk to human health and other organisms.

The Water (Prevention and Control of Pollution) Act, 1974

An Act to provide for the prevention and control of water pollution and the maintaining or restoring of wholesomeness of water, for the establishment, with a view to carrying out the purposes aforesaid, of Boards for the prevention and control of water pollution, for conferring on and assigning to such boards powers and functions relating thereto and for matters connected therewith.

In brief, this Act deals with the preservation of water quality and the control of water pollution with a concern for the detrimental effects of water pollutants on human health and also on the biological world.

The Air (Prevention and Control of Pollution) Act, 1981

An Act to provide for the prevention, control and abatement of air pollution, for the establishment, with a view to carrying out the aforesaid purposes, of Boards, for conferring on and assigning to such Boards powers and functions relating thereto and for matters connected therewith.

In brief, this Act deals with the preservation of air quality and the control of air pollution with a concern for the detrimental effects of air pollutants on human health and also on the biological world.

In 1987, important amendments to the Air Act 1981 were made and noise was recognised as an air pollutant.

The Environment (protection) Act, 1986

(*a*) In this Act, unless the context otherwise requires; 'environment' includes water, air and land and the inter-relationship which exists among and between water, air and land, and human beings, other living creatures, plants, microorganisms and property;

(*b*) 'environmental pollutant' means any solid, liquid or gaseous substance present in such concentration as may be, or tend to be, injurious to environment;

(*c*) 'environmental pollutant' means the presence in the environment of any environmental pollutant;

(*d*) 'handling', in relation to any substance, means the manufacture, processing, treatment, package, storage, transportation, use, collection, destruction, conversion, offering for sale, transfer or like of such substance;

(*e*) 'hazardous substance' means any substance or preparation which, by reason of its chemical or physicochemical properties or handling, is liable to cause harm to human beings, other living creatures, plant, micro-organism, property or the environment;

(*f*) 'occupier', in relation to any factory or premises, means a person who has, control over the affairs of the factory or the premises and includes in relation to any substance, the person in possession of the substance;

(*g*) 'prescribed' means prescribed by rules made under this Act.

In brief, this Act deals with the protection of air, water and soil quality, and the control of environmental pollutants, including noise under its purview.

GLOBAL ENVIRONMENTAL CHANGE

An outcome of the atmospheric pollution is warming of the earth's environment. This is a serious problem and in twenty-first century this may be a major global problem. The temperature of the earth has increased to some extent during the past fifty years has been exhibited in the following fig. 15.6. This increase in temperature of atmosphere is dangerous for living organisms. The human activities are mainly responsible for global warming. Though there are several causes for global warming, however, some important ones are mentioned here: They include deforestation, industrialisation, energy production from fossil fuels, such as diesel, petroleum, coal, etc., and urbanisation. The conversion of a forest to a grazing land or a cropland through deforestation causes loss of carbon stored in soil and vegetation to the atmosphere, and affects the global carbon cycle.

Biomass burning associated with agricultural practices releases a large amount of CO_2 into the atmosphere. Besides, due to domestic and industrial coal burning, huge amount of carbon dioxide is also released in the atmosphere.

Besides carbondioxide, some other gases like methane (CH_4), nitrous oxide (N_2O) and chlorofluorocarbons (CFCs) are also responsible for increasing the temperature of the earth and its environment.

These gases CO_2, CH_4, N_2O and CFCs are **radiatively active gases**, also called **greenhouse gases** as they can absorb long wave infrared radiation.

The increased amounts of greenhouse gases in the atmosphere affect the global climate, and this phenomenon is called **global change**.

Greenhouse Gases (Greenhouse Effect)

The gaseous mantle around the globe allows a considerable portion of solar radiations to enter right upto the surface of earth which absorbs it and radiates back infra-red and heat waves. This heat is transferred to layers above, as warm layer rises and in turn passes on to higher and higher layers. The outgoing longwave infrared radiation is absorbed by the greenhouse gases normally present in the atmosphere. The atmosphere radiates part of this energy back to the earth. This downward flux of radiation, known as **greenhouse flux**, keeps the earth warm. Thus, the greenhouse gases of the atmosphere form a blanket over the earth that controls the escape of heat from the surface of earth to outer space so as to keep it warm and cosy. This phenomenon is called **greenhouse effect**.

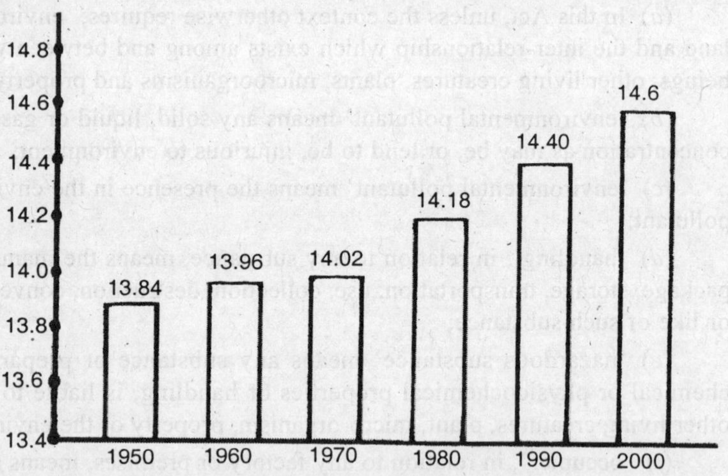

Fig. 15.6. Average temperature of earth recorded in different years.

The name, greenhouse effect is derived from the fact that inside a glass enclosed **green house**, temperature is warmer than outside. The mean annual temperature of the earth is about 15°C. If there would have no greenhouse gases in the atmosphere, the earth's mean temperature would go sharply down to about—20°C.

This capacity of the atmosphere to keep the earth warm depends upon the concentration of the greenhouse gases. The excessive increase in concentrations of greenhouse gases in the atmosphere would retain more and more of the infrared radiation, resulting in enhanced **green house effect**. The consequent increase in the global mean temperature is referred to as **global warming**.

The rise in temperature of earth's atmosphere caused due to greenhouse effect depends on the amount of CO_2 present in the atmosphere. At normal CO_2 concentration (0.03 per cent) in the atmosphere the surface temperature of the earth remains constant due to energy balance of the sun rays which strike on the earth, heat it and then radiate back into the space. This is called **energy budget**. But when there is increase in CO_2 concentration in the atmosphere, the thick layer of CO_2 prevents the heat from being radiated out into space. This layer of CO_2 thus functions as a glass pane of green house which allows the sunlight to filter through it but prevents heat from being radiated back into the outer space.

However, an international panel on climate change periodically makes an assessment of the atmospheric abundance of greenhouse gases and its possible impact on climate and other such issues.

The trends in the increase in concentrations of greenhouse gases since pre-industrial times, are as follows :

Carbon dioxide. CO_2 is the most abundant greenhouse gas in the atmosphere. This is about sixty per cent of the greenhouse gases. The level of the concentration of carbon dioxide in the

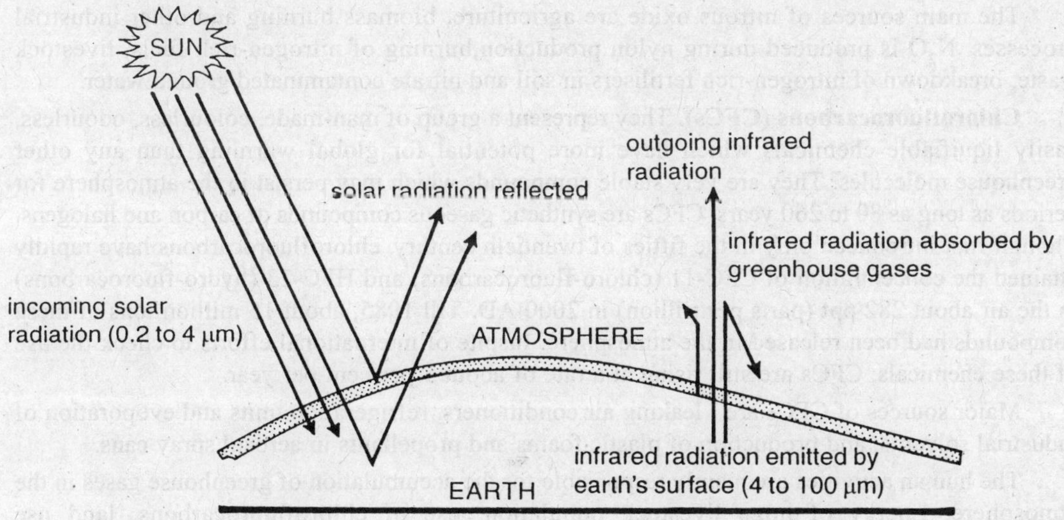

Fig. 15.7. The greenhouse effect.

atmosphere has increased from the pre-industrial level of 280 ppm (parts per million) to about 368 ppm in 2000 AD. Here, pre-industrial time is considered 1750 AD. This increase in concentration is mainly due to fossil fuel burning, deforestation and change in land use.

Methane. CH_4 is another greenhouse gas which is produced when organic matter decays under anaerobic conditions. Methane concentration in atmosphere was about 750 ppb (parts per billion) in pre-industrial level, *i.e.*, 1750 AD., which had become 1750 ppb in 2000 AD. Methane is generally produced as a result of incomplete decomposition by a group of bacteria called **methanogens**, under anaerobic conditions. The concentration of this gas

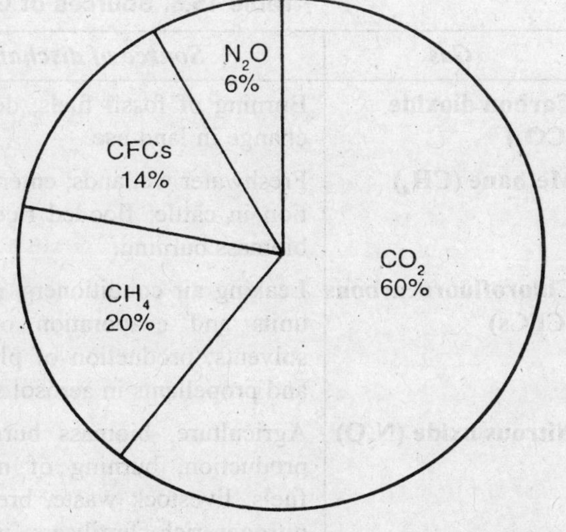

Fig. 15.8. Contribution of percentage of greenhouse gases to global warming.

is rising at a rate of about one per cent per year. Between the years 1980-90 about 15 per cent of the total warming has been attributed to this gas alone.

The major sources of methane are : freshwater wetlands, enteric fermentation in cattle, and flooded rice fields. Biomass burning is also a source of methane. In the atmosphere, methane undergoes oxidation to carbon dioxide and water, both of which tend to emphasize the greenhouse effect.

Nitrous oxide. N_2O is another troublesome greenhouse gas. Other oxides of nitrogen are reacted upon or cleared rapidly, while nitrous oxide undergoes decomposition very slowly, and therefore, it tends to accumulate in the atmosphere. The concentration of nitrous oxide in the atmosphere has increased from about 270 ppb in pre-industrial time (*i.e.*, 1750 AD.) to about 316 ppb (parts per billion) in year 2000 AD.

The main sources of nitrous oxide are agriculture, biomass burning and other industrial processes. N_2O is produced during nylon production,burning of nitrogen-rich fuels, livestock waste, breakdown of nitrogen-rich fertilisers in soil and nitrate contaminated ground water.

Chlorofluorocarbons (CFCs). They represent a group of man-made, colourless, odourless, easily liquifiable chemicals which have more potential for global warming than any other greenhouse molecules. They are very stable compounds which may persist in the atmosphere for periods as long as 80 to 260 years. CFCs are synthetic gaseous compounds of carbon and halogens. Though first introduced only in the fifties of twentieth century, chlorofluorocarbons have rapidly attained the concentration of CFC-11 (chloro-fluorocarbons) and HFC-23 (hydro-fluorocarbons) in the air about 282 ppt (parts per trillion) in 2000 AD. Till 1985, about 15 million tons of these compounds had been released in the atmosphere. Inspite of international efforts to check the use of these chemicals, CFCs are still rising at a rate of about 5 per cent per year.

Major sources of CFCs are : leaking air conditioners, refrigeration units and evoporation of industrial solvents; and production of plastic foams and propellants in aerosol spray cans.

The human activities are mainly responsible for the accumulation of greenhouse gases in the atmosphere. Energy of huge livestock population, use of chlorofluorocarbons, land use modification and industrial production are some of the aspects of human activity which are responsible for accumulation of greenhouse gases.

Table 15.5. Sources of Greenhouse Gases

Gas	*Source of discharge*	*Percentage of global warming*
Carbon dioxide (CO_2)	Burning of fossil fuels; deforestation; change in land use.	60 per cent
Methane (CH_4)	Freshwater wetlands; enteric fermentation in cattle; flooded rice fields and biomass burning.	20 per cent
Chlorofluorocarbons (CFCs)	Leaking air conditioners, refrigeration units and evaporation of industrial solvents; production of plastic foams and propellants in aerosol spray cans.	14 per cent
Nitrous oxide (N_2O)	Agriculture, biomass burning; nylon production, burning of nitrogen-rich fuels, livestock waste, breakdown of nitrogen-rich fertilisers in soil and nitrate contaminated ground water.	6 per cent

Percentage of greenhouse gases. It is estimated that carbon dioxide contributes about 60 per cent of the total global warming. The share of methane is 20 per cent, while that of chlorofluorocarbons is 14 per cent. However, a smaller contribution to global warming is made by nitrous oxide, *i.e.*, 6 per cent.

EFFECTS OF GREENHOUSE GASES

The increasing abundance of greenhouse gases in the atmosphere has the following three main effects :

(*i*) CO_2 fertilisation effect on plants

(*ii*) Global warming and

(*iii*) Depletion of ozone (O_3) layer in the stratosphere.

Carbondioxide (CO_2) Fertilisation Effect on Plants

The data produced in USA have shown that atmospheric carbondioxide concentration has been rapidly rising since 1959 as shown in the graph. If such rising trend continues, by the end of twenty first century the atmospheric concentration of CO_2 shall increase to a level between 540 and 970 ppm.

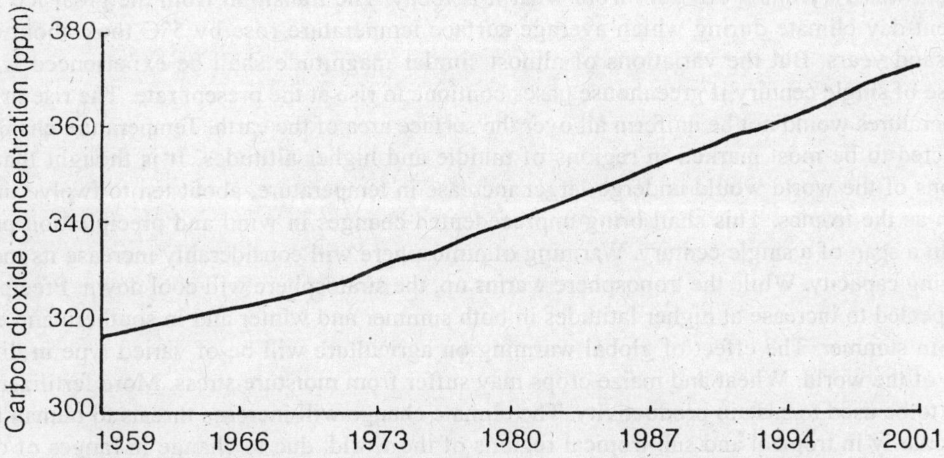

Fig. 15.9. Increase in CO_2 concentration in atmosphere from 1959 to 2001.

When the CO_2 concentration of the atmosphere is more or less doubled, the growth of many plants, *i.e.*, C_3 plants in particular, under favourable conditions of water, nutrients, light and temperature, could increase by about thirty per cent on average, in the short term of few years or so.

The response of plants to elevated concentrations of CO_2 is called **carbon dioxide fertilisation effect**.

Due to increased carbondioxide concentration, the rate of photosynthesis also increases, and the stomatal conductance decreases due to partial closure of stomata. Hence, the transpiration rate reduces, and water-use efficiency increases.

Such effect allows many species to grow successfully in regions of water scarcity.

Under higher atmospheric carbon dioxide concentrations, plants allocate, a greater proportion of photosynthate to roots. However, greater root production increases development of mycorrhiza and fixation of nitrogen in root nodules, thus, makes possible the plants to grow in soils which are poor in nutrition.

However, in natural conditions the beneficial effects of increased carbon dioxide may not be there because of negative effects of global warming.

Possible Effects of Global Warming

Global records of earth's surface temperatures indicate that a warming of about 0.6° (*i.e.*, 0.3° – 0.7°C) has occurred during the last century alone. Results from recent climatic models suggest that mean global temperatures would rise by 2° – 6°C during the twenty first century, if we assume that carbon dioxide concentration increases to 600 ppm (parts per million). The predicted global warming in near future has the potential to affect the (*i*) weather and climate, (*ii*) sea level change, (*iii*) distribution and phenology of species and (*iv*) food production and the fishery resources in the oceans.

(*i*) **Effects on weather and climate.** As already mentioned in the preceding paragraph, the global mean temperature has increased about 0.6°C in the twentieth century. It is thought, that mean global temperature may increase by 1.4 to 5.8°C by the end of twenty first century, (*i.e.*, 2100 AD).

The projected change in mean surface temperatures may appear insignificant because variatious of this magnitude are experienced in course of seasonal or even daily weather. The world climate was very much different from what it is today. The transition from the great Ice Age to present-day climate during which average surface temperature rose by 5°C took about twelve thousand years. But the variations of almost similar magnitude shall be experienced within a course of single century if greenhouse gases continue to rise at the present rate. The rise in global temperatures would not be uniform all over the surface area of the earth. Temperature changes are expected to be most marked in regions of middle and higher altitudes. It is thought that polar regions of the world would undergo larger increase in temperature, about ten to twelve times as much as the tropics. This shall bring unprecedented changes in wind and precipitation patterns within a span of a single century. Warming of atmosphere will considerably increase its moisture carrying capacity. While the troposphere warms up, the stratosphere will cool down. Precipitation is expected to increase at higher latitudes in both summer and winter and in southern and eastern Asia in summer. The effect of global warming on agriculture will be of varied type in different parts of the world. Wheat and maize crops may suffer from moisture stress. More fertilizers shall have to be used to sustain productivity. The climate change will increase threats to human health, particularly in tropical and sub-tropical regions of the world, due to change in ranges of disease vectors, water-born pathogens, etc. Global warming shall bring with it an entirely new environment in which life though not impossible yet its existence shall be tougher to maintain.

(*ii*) **Sea level change.** Though there are so many uncertainties regarding the consequences of global warming, one of the obvious results of the general heating up of earth's surface shall be a rather rapid rise in mean sea level. Evidences suggest that in Ice Age, about 12,000 years back it was nearly 100 m lower than the present-day level. It is expected that global rise in temperature shall further enhance the rate of already rising sea level in two ways : (*i*) Large deposits of ice present on earth's surface shall melt, which will add more water to the oceans. (*ii*) Rise in temperature shall also cause thermal expansion of the upper layers of the water.

Sea level has been raised by 1 to 2 mm per year during the twentieth century. It is expected that by the end of twenty first century (*i.e.*, 2100 AD.), the global mean sea level can increase upto 0.88 m over the 1990 level. A rise of even half a meter in sea level would largely effect human population, one-third of which lives within 60 km of a coastline.

Thus, many of the world's important cities and coastal areas will come under the threat of flood, and several low-lying islands will be submerged.

The flooding of coastal salt marshes and estuaries will deprive many important birds and fish their breeding grounds and their extinction will be imminent.

Thus, sea-level rise is predicted to have negative impacts on human settlements, such as tourism, fresh-water supplies, fisheries, wetlands, etc.

(*iii*) **Effects on distribution and phenology of species.** As the climatic belts shift away from equator towards poles, vegetation will also shift in the same direction to stay in favourable climatic conditions. Those species, which will not do so shall die. There will be losses of genetic resources on large scale. Only hardy and resistant forms shall come up and survive. For example, with a global temperature rise by 2 to 5°C during the twenty first century, the temperate region vegetation may shift 250-600 km poleward.

As trees are more sensitive to temperature stress, a rapid rise in temperature may cause large scale death of trees and there replacement by scrub vegetation.

As temperature changes will affect wind and precipitation patterns also water could play an important role in altering the biotic communities. It has been suggested that some rise in precipitation, however, shall be balanced by an enhanced evaporation and this could lead to water deficit and moisture stress in many regions of the world. Insects and pests may increase as warmer conditions could be more favourable to their growth and coupled with higher humidity pathogens shall multiply.

(*iv*) **Food production.** Increased temperature will cause eruption of plant diseases and pests. Besides, there will be explosive growth of weeds and increased basal rate of respiration of plants. Thus, all these factors will be responsible for decrease of crop production.

However, small temperature increase may slightly enhance crop productivity in temperate regions, but larger temperature changes will reduce crop productivity there also.

Even a small rise in temperature will have detrimental effect on crop-productivity, both in tropical and sub-tropical regions.

It is expected that rice yield alone, in southeast Asia will go down by 5 per cent for each 1°C rise in temperature. However, in spite of beneficial carbon dioxide fertilisation effect, the overall world crop productivity will decline considerably due to projected global warming, and the food supply of the whole world shall be affected.

As the climatic belts will shift away from equator towards poles, vegetation will shift in the same direction to stay in favourable climatic conditions. Those species which shall be unable to do so, they will die.

There will be losses of genetic resources on large scale. Only hardy and resistant species shall come up and survive. An altogether changed biotic spectrum shall replace the earlier ones and almost all important biomes shall be affected.

Alterations in cropping pattern shall occur and pest resistant varieties more suitably to warmer conditions shall have to be developed. In brief greenhouse warming (global warming) shall bring with if an entirely new environment in which life though not impossible yet its existence shall be tougher to meet with.

Approaches to Deal with Global Warming

Some strategies which could reduce the global warming are as follows :

(*i*) Reduction of the greenhouse gas emissions by the use of limited fossil fuels, by developing alternative renewable sources of energy, such as solar energy, wind energy, etc.

(*ii*) By increasing the vegetation cover, particularly the forests, for photosynthetic utilisation of corbon dioxide.

(*iii*) By using minimum nitrogen fertilisers in agriculture, so that, the N_2O emissions may be reduced.

(*iv*) By developing substitutes for chlorofluorocarbons (CFCs).

Besides, above mentioned strategies, adaptations to address localised impacts of climate change shall be important.

Stratospheric Ozone Depletion

Sun emits radiations at all wavelengths. Of these, radiations important to our planet are : ultraviolet (UV) radiations, light rays, and infrared rays or heat waves. Radiations of visible spectrum and infrared rays carry little energy which does not harm living beings. However, the energy content of ultraviolet radiations is larger than the limits of tolerance of a living cell, and therefore, is harmful or even lethal to a living system.

Stratospheric O_3 layer. High up in the stratosphere, that is, about 20 to 26 km above the sea level, short wavelength ultraviolet radiations are absorbed by molecular oxygen which splits up into its constituent atoms. These atoms combine with molecular oxygen to produce ozone.

$$O_2 \xrightarrow{\text{UV rays}} O + O$$

$$O_2 + O \xrightarrow{\text{UV rays}} O_3$$

Thus, ozone is a result of photochemical reactions in which the starting molecule is oxygen. Along with this reaction another photochemical reaction which causes breakdown of ozone molecules due to absorption of radiations also occur :

$$O_3 \longrightarrow O_2 + O$$

$$O_3 + O \longrightarrow 2O_2$$

The two reactions, *i.e.*, the formation and destruction of ozone molecule normally balances each other and ultimately result in effective absorption of short wavelength ultraviolet radiations in the stratospheric region. Life underneath is thus protected from the biocidal solar radiations.

Most of the ozone present in the atmosphere, almost 90 per cent is concentrated in the stratosphere at an altitude of about 20 to 26 km above the surface of earth. The thickness of the vertical column of stratospheric ozone layer, condensed to standard temperature and pressure, averages 0.29 cm above the equator and may exceed 0.40 cm above the poles at the end of the winter season. Though maximum amount of ozone production occurs over equator, yet its concentration is lowest here, as most of the ozone formed is displaced towards the poles with massive movement of atmospheric air.

Ozone shield. The stratospheric ozone layer acts as the ozone shield, which protects the earth biota from harmful effects of strong UV radiation. Absorption of UV radiation by ozone layer increases more rapidly with its thickness. Thus, maximum amount of UV radiation passing through the atmosphere reaches the earth surface in the tropics, *i.e.*, near the equator, and this amount decreases towards the poles.

The concentration of ozone in the stratosphere changes with seasons. In spring season (February to April), this is highest, while during fall season (July to October), this is lowest.

Ozone hole. During the period 1956-1970, the spring season O_3 layer thickness above Antarctica varied from 280 to 325 Dobson Unit (DU), *i.e.*, 1 DU = 1 ppb (part per billion). The thickness was sharply reduced to 225 DU in 1979 and to 136 DU in 1985. Later the O_3 layer thickness continued to decline to about 94 DU in 1994.

The decline in spring time ozone layer thickness is termed **ozone hole**. The ozone hole was first discovered in 1985 over Antarctica. The presence of ozone hole was also discovered above

Arctic in 1990. The global average total column ozone amount for the period 1997-2001 was about 3 per cent below the pre-1980 average values.

Ozone hole is formed by intense UV activity on CFCs. It starts appearing at the end of winter, shows maximum display in spring and disappears in summer. This is because in winter the air is heavy and industrial gases like CFCs get concentrated over an area in Antarctica where the maximum damage of ozone layer has been noticed. In Antarctica CFCs bearing winds in winter become frozen to form ice crystals. During spring time when ice melts the trapped CFCs and ClO are freed at once which in turn release nascent chlorine to attack ozone. Thus, ozone hole becomes large in the spring season.

CFCs, CH_4 and N_2O escape into the stratosphere and cause destruction of O_3 there. Most damaging effect is of CFCs (chlorofluorocarbons) which produce active chlorine (*i.e.*, Cl and ClO radicals) in the presence of UV radiation. These radicals catalytically destroy ozone and convert it into oxygen. CH_4 and N_2O also cause ozone destruction through a series of reactions.

For making above mentioned discoveries related to O_3 destruction, Sherwood Rowland and Mario Molina, along with Paul Crutzen, were awarded Nobal Prize for chemistry in 1995.

Effect of ozone depletion. Ozone layer shields us from biocidal UV radiations. Depletion of stratospheric ozone concentration shall result in an increased penetration of UV–B radiations. A higher loss of stratospheric ozone could cause entry of not only UV–B radiations but also UV–C radiations. This could turn the geological clock back to the era when the ozone umbrella was very thin. Terrestrial life shall be drastically affected so also will be aquatic life which occurs in shallow waters.

The thinning of the O_3 layer results in an increase in the UV–B radiation reaching the earth surface. A 5 per cent loss of ozone results in a 10 per cent increase in UV–B radiation.

In humans the increased UV–B radiation increases the chances of cataract and skin cancer. This also diminishes the functioning of immune system.

Increased levels of UV–B radiation affect photosynthesis, and damage nucleic acids in living organisms. UV–B radiation inhibits photosynthesis in most phytoplankton as it penetrates through the clear open sea waters. Thus, the whole food chain of the organisms would be affected that depend on phytoplankton for thier food.

INTERNATIONAL CONCERN FOR MITIGATING GLOBAL CHANGE

Global warming is the global concern. The long term challenge of stabilising the atmospheric concentrations of greenhouse gases requires that global emissions be significantly lowered than what they are today. In this context, the representatives of many countries assembled from time to time to pass some resolutions, which can be implemented.

Montreal Protocol. In 1987, seventy seven industrially developed countries signed the Montreal Protocol. This conference was held at Montreal. This was a landmark international agreement to protect the stratospheric ozone by agreeing to limit the production and use of ozone-depleting substances. It was decided in the conference to limit the use of CFCs and cut level of CFCs production to 50 per cent by 1999. To-date, more than 175 countries have signed the Montreal Protocol. However, India and China did not sign the Protocol. Developed countries have also agreed to establish a multilateral fund under Montreal Protocol to provide financial support to developing countries for that purpose.

London Protocol. In 1990, London Protocol was signed. In London Protocol, India, China and several other countries demanded more time to reduce CFCs consumption. However, the developed countries agreed to stop CFCs consumption completely by 2000 AD, and for developing nations the time limit was extended to 2010 AD.

Toronto Conference. In 1988, the world conference on the changing atmosphere was held in Toronto (Canada). There it was decided to reduce 20 per cent carbon dioxide emissions by the year 2005.

Earth Summit. In 1992, The First Earth Summit was held in Rio de Janeiro, Brazil. The Earth Summit, was held under the banner of UNCED (*i.e.*, The United Nations Conference on Environment and Development). This summit, established the principles for reducing greenhouse gas emission.

Kyoto Protocol. The Kyoto Protocol, approved by a follow-up conference held in Kyoto, Japan, during December 1997, has specified the commuitments of different countries to mitigate climate change. This protocol has established the rules to reduce overall greenhouse gas emissions to a level at least 5 per cent below the 1990 level by 2008-2012.

At present, on the basis of Montreal Protocol, India has established **ozone cell** under the Ministry of Forests and Environment with the help of UNDP (United Nations Development Programme). By now, India has developed R-22, and R-134 as CFCs substitutes which are presently at trial stage.

SUMMARY

1. Pollution may be defined as an undesirable change in physical, chemical or biological characteristics of air, water and land, resulting in air pollution, water pollution and soil pollution.
2. There are five types of primary air pollutants : (*i*) particulate matter, (*ii*) carbon monoxide (CO), (*iii*) hydrocarbons (HCs), (*iv*) sulphur dioxide (SO_2) and (*v*) nitrogen oxides (NO_2).
3. Secondary air pollutants are formed during chemical reactions between primary air pollutants and other atmospheric constituents, such as water vapour.
4. Different types of control measures are adopted to combat pollution from gaseous and particulate matters in industrial emissions and automobile exhausts.
5. Types of water pollutants may be (*i*) biological, such as pathogens like viruses, bacteria, protozoa, blue-green algae and helminths, (*ii*) chemical, *e.g.*, organic chemicals like biocides, polychlorinated biphenyls (PCBs) and (*iii*) inorganic chemicals, such as phosphates, nitrates, fluoride, etc., and also heavy metals, such as arsenic, lead, cadmium, mercury, etc., and (*iv*) physical, such as hot water from industries, oil spills from oil carriers, etc.
6. Water pollutants adversely affect the physical, chemical and biological health of the aquatic ecosystems, quality of ground water and human health.
7. Most of the industrial and municipal waste waters are treated in Effluent Treatment Plants (ETPs) for safe disposal.
8. Soil pollution and land degradation usually result from different human activities, such as waste dumping, agricultural practices, deforestation, mining operations and urbanisation.
9. Control measures for soil pollution are : safe disposal and management of soil wastes from industries and human habitations.
10. Noise pollution : man-made noise that originates from industry, transport, sound amplifiers pollute our environment.
11. Several laws have been enacted in India for controlling air water, soil and noise pollution.
12. Major issues of global environmental change are global warming and ozone depletion in stratosphere.
13. Naturally occurring greenhouse gases, such as carbon dioxide, methane and nitrous oxide, in the atmosphere absorb long-wave radiations and help in maintaining the average temperature of earth around 15°C.
14. The increase in concentrations of greenhouse gases is casusing enhanced green house effect, increased global temperature, sea level rise, and change in rainfall pattern.
15. The ozone layer of stratosphere plays an important role by protecting the living organisms from the harmful effects of ultraviolet radiations.

16. Ozone hole is referred to as thinning of stratospheric ozone layer during spring. For the first time, ozone hole was discovered over Antarctica, but it also occurs over the Arctic region.

17. The man-made chemicals such as chlorofluorocarbons (CFCs) are mainly responsible for ozone depletion.

TERMS TO REMEMBER

Acid rain. The ways, in which acids from the atmosphere are deposited on the earth.

Air pollution. Degradation of air quality and natural atmospheric condition.

Algal bloom. Profuse growth of algae.

Anthropogenic pollution. Man-made pollution, *e.g.*, industrial pollution, agricultural pollution.

Biodegradable pollutants. The pollutants, which can be decomposed efficiently by the decomposers, *e.g.*, market garbage, livestock wastes, etc.

Brown air. Nitrogen oxides cause reddish brown haze in traffic congested city air, called brown air.

Biochemical oxygen demand (BOD). A measure of oxygen required by aerobic decomposers for the biochemical degradation of organic materials (*i.e.*, biodegradable materials) in water.

Biological magnification. The phenomenon through which certain pollutants get accumulated in tissues in increasing concentrations along the food chain.

Carbon dioxide fertilisation effect. The response of plants to elevated concentrations of CO_2.

Chlorofluorocarbons (CFCs). They are non-toxic and non-flammable, highly stable and synthetic gaseous compounds of carbon and halogens.

Chemical oxygen demand (COD). The measure of oxygen equivalent of the requirement for oxidation of total organic matter present in water.

Decibel (dB). The unit of sound level.

Earth Summit. Held at Rio de Janeiro, Brazil in 1992.

Eutrophication. The process of nutrient enrichment of water, and consequent loss of species diversity, is called eutrophication.

ETP. Effluent Treatement Plant. The industrial and municipal waste waters are treated in this plant prior to disposal in water bodies.

Grey air. In areas of lesser solar radiation, smog formation is incomplete and the air is referred to as 'grey air'.

Greenhouse gases. These gases (CO_2, CH_4, N_2O and CFCs) are radiatively active gases, because they can absorb, long wave infrared radiation.

Global change. The increased amounts of greenhouse gases in the atmosphere that affect the global climate, the phenomenon is global change.

Greenhouse flux. Downward flux of radiation.

Greenhouse effect. The atmospheric greenhouse gases forming a blanket over the earth, control the escape of heat from the earth's surface to outer space so as to keep it warm and hospitable, the phenomenon is greenhouse effect.

Global warming. The increase in the global mean temperature.

Hydrocarbons (HCs) or volatile organic carbons. They are compounds composed of hydrogen and carbon.

Incineration. Burning the solid wastes in presence of oxygen.

Indicator species. Tolerant or resistant species that survive in highly polluted water.

Kyoto Protocol. Conference held in Kyoto, Japan, in 1997.

Montreal Protocol. A landmark international agreement to protect the stratospheric ozone, in Montreal, in 1987.

Noise pollution. The loud disturbing sound dumped into the ambient atmosphere without regard to the adverse effects it may have.

Non-biodegradable. The pollutants, which are not decomposed by decomposers.

Oil spills. The accidental discharge of petroleum in oceans or estuaries.

Ozone hole. The decline in spring-time ozone layer thickness is termed ozone hole.

Photochemical smog. Mainly formed in traffic congested metropolitan cities where warm conditions and intense solar radiation are present. This is composed mainly of ozone (O_3), peroxyacetyl nitrate (PAN) and NO_2.

Physical pollution. Pollution caused by changes in physical properties of water, *e.g.*, change in temperature.

Smog. Combination of smoke ang fog.

Synergism. Increase in toxicity from primary pollutants to secondary pollutants.

QUESTION BANK

Very Short Answer Questions

1. Define pollution.
2. Classify pollution, on the basis of part of environment.
3. Name types of pollution in terms of origin.
4. Give at least four types of pollution in terms of physical nature of pollutants.
5. Give the important categories of the pollutants from the ecosystem viewpoint.
6. Give four examples of biodegradable pollutants.
7. Expand PAN.
8. Expand BOD.
9. What is COD ?
10. Expand DDT.
11. Expand BHC.
12. Give four examples of non-biodegradable materials.
13. What are hydrocarbons ?
14. Give five examples of primary air pollutants.
15. Give two examples of secondary air pollutants.
16. What is grey air ?
17. What is acid rain ?
18. What is combustion ?
19. Classify water pollutants.
20. What is algal bloom ?
21. What is Minamata Disease, where this was discovered ?
22. Name greenhouse gases.
23. When and where, ozone hole was discovered ?
24. When Montreal Protocol was signed ?
25. When and where Earth Summit was held ?
26. What are aerosols ?
27. Define greenhouse effect.
28. Which aerosol pollutant is present in jet aeroplane emission ?
29. Name main components of photochemical smog.
30. What is jhuming ?
31. Expand ODS.
32. Which one of the following is not a greenhouse gas ?

    CO_2, CH_4, O_3, CFCs

ANSWERS

1. Pollution is an undesirable change in physical, chemical or biological characteristics of air, land or water, caused by excssive accumulation of pollutants; **2.** Air pollution, water pollution and soil pollution; **3.** Natural or anthropogenic; **4.** Gaseous pollution, dust pollution, thermal pollution and noise pollution; **5.** Biodegradable and non-biodegradable pollutants; **6.** Market garbage, livestock wastes, municipal sewage, faecal matter; **7.** Peroxyacetyl nitrate; **8.** Biochemical oxygen demand; **9.** Chemical oxygen demand; **10.** Dichloro diphenyl trichloro ethane; **11.** Benzene hexachloride; **12.** DDT, BHC, waste plastic bottles, polyethylene bags; **13.** Compounds composed of hydrogen and carbon, also called volatile organic carbon; **14.** Particulate matter, carbon monoxide (CO), hydrocarbons (HCs), sulphur dioxide (SO_2) and nitrogen oxides (NO_x); 15. Photochemical smog and acid rain; **16.** Where smog formation is incomplete, the air is referred to as grey air; **17.** Acid rain refers to several ways in which acids from the atmosphere are deposited on the earth; **18.** A process, in which oxidisable gaseous pollutants are completely burnt at a high temperature; **19.** (*i*) Biological, (*ii*) chemical and (*iii*) physical; **20.** Profuse growth of algae; **21.** Mercury poisoning due to consumption of fish captured from Hg-contaminated Minamata Bay in Japan, detected in 1952; **22.** CO_2, CH_4, N_2O and CFCs; **23.** The ozone hole was first discovered in 1985 over Antarctica; **24.** In 1987; **25.** Earth Summit was held at Rio de Janeiro, Brazil in 1992; **26.** Chlorofluorocarbons compressed in the form of droplets released by jet aeroplanes; **27.** Increase in environmental temperature due to increased concentration of green house gases (CO_2, CH_4, N_2O and CFCs); **28.** Chlorofluorocarbons (CFCs); **29.** PAN and O_3; **30.** Raising crops on the resulting ash due to burning of trees; **31.** Ozone Depleting substances; **32.** O_3.

Short Answer Questions

1. What is photochemical smog composed of ? How does this affect the plants ?
2. Define pollution. Compare the biodegradable and non-biodegradable pollutants.
3. What do you understand by fixed and mobile sources of air pollution ?
4. Distinguish between the primary and secondary air pollutants.
5. What is particulate matter ? How do particulate matters harm human health ?
6. What is photochemical smog ? How does smog affect the biological world ?
7. What is acid rain ? What are its effects on plants ?
8. Distinguish between point and non-point sources of water pollution.
9. State the consequences of stratospheric ozone depletion.
10. What do you mean by 'biochemical oxygen demand' (BOD) ?
11. What is Minimata disease ? Where and when this was detected ?
12. Name three types of particulate pollutants of air.
13. What is thermal pollution ?
14. Give two harmful effects of ozone.
15. What causes enhanced greenhouse effect ?
16. Mention the effects of excess fluoride in drinking water.
17. What is global change ?
18. What is greenhouse effect ?
19. What are the effects of increased greenhouse gases in the atmosphere ?
20. What are the harmful effects of PAN ?
21. Explain the phenomenon of biological magnification.
22. What measures do you suggest to control pollution from automobile exhausts ?
23. Why are industrial effluents more difficult to manage than municipal sewage ? Name a disease that is caused by heavy metal contamination.
24. How can pollution by domestic sewage be controlled ?
25. Describe the ways by which soil gets polluted.
26. What is noise ? Describe briefly the effects of noise on human health.

27. Discuss the causes and effects of global warming.
28. What are the effects of ultraviolet radiations on humans ?
29. What is the environmental significance of the increasing Antarctica ozone hole ?
30. How are gaseous pollutants controlled ?
31. How can automobile exhaust be controlled ?
32. Give the steps involved in management of solid wastes.
33. What are pollutants ? How are they classified?
34. How is soil salinity caused ?
35. What do you understand by ocean water pollution ?

Long Answer Questions

1. What is meant by the ozone shield ? How do the CFCs and ozone depleting substances affect ozone shield ?
2. Explain the following :
(*i*) Greenhouse effect
(*ii*) CO_2 fertilisation effect
(*iii*) Ozone hole.
3. What are the control measures used for controlling soil pollution.
4. Give an account of the possible effects of global warming.
5. What is 'ozone shield' ? How does O_3 depleting substances affect it ?
6. Give the details of various sources of air pollution.
7. Write a comprehensive note on control measures used for controlling soil pollution.
8. What is environmental pollution ? How this is being affected by human activities ?
9. Mention the steps which can reduce noise pollution.
10. How can water pollution be controlled ?
11. What is radioactivity ? Discuss the man-made radioactive pollution.

MULTIPLE CHOICE QUESTIONS (OBJECTIVE TYPE QUESTIONS)

1. Ozone layer of upper atmosphere is being destroyed by :
(*a*) Sulphur dioxide (*b*) Photochemical oxidants
(*c*) Chloroflurocarbons (*d*) Smog
2. Acid rain is caused by increase in the atmospheric concentration of :
(*a*) Ozone and dust (*b*) SO_2 and NO_2 (*c*) SO_3 and CO (*d*) CO_2 and CO
3. Ozone depletion in the stratosphere will cause:
(*a*) Increased incidence of skin cancer (*b*) Forest fires
(*c*) Global warming (*d*) None of these.
4. Attack of asthma in certain persons may be due to :
(*a*) Eating of some seasonal vegetables
(*b*) Inhalation of certain air-borne pollens
(*c*) Exposure to cold temperature
(*d*) Absence of CO_2 in the air due to increased rate of photosynthesis.
5. Eutrophication causes decrease in :
(*a*) Dissolved hydrogen (*b*) Dissolved salts (*c*) Dissolved oxygen (*d*) All of these
6. Positive pollution of soil is due to :
(*a*) Reduction in soil productivity (*b*) Addition of wastes on soil
(*c*) Excess use of fertilisers (*d*) All of these

7. If the fertilisers are added to fresh water :
   (*a*) Plants will die (*b*) Eutrophication will occur
   (*c*) Fish population will decrease (*d*) Overall animal population will increase
8. Photochemical smog formed in congested metropolitan cities mainly consists of :
   (*a*) Oxides of nitrogen (*b*) Suspended particulate matter
   (*c*) Carbon dioxide and carbon monoxide (*d*) Oxides of sulphur
9. Biological Oxygen Demand (BOD) is a measure of :
   (*a*) Amount of oxygen needed by green plants during night
   (*b*) Amount of carbon monoxide inseparately combined with haemoglobin
   (*c*) Industrial wastes poured into water bodies
   (*d*) Extent to which water is polluted with organic compounds
10. A person is suffering from impaired nervous system and madness after prolonged consumption of polluted water. Which metal is responsible :
   (*a*) Hg (*b*) Ca (*c*) Mn (*d*) Pb
11. The major source of methane in India is :
   (*a*) Rice fields (*b*) Wheat fields
   (*c*) Sugarcane plantation (*d*) Fruit orchard
12. Natural sink of stratospheric ozone layer is :
   (*a*) SO_2 (*b*) HFCs
   (*c*) Ferons (*d*) Sulphur flux of oceans
13. Which of the following has been declared a killer disease under Factory Act ?
   (*a*) Asbestosis (*b*) Tuberculosis (*c*) Silicosis (*d*) Shigellosis
14. Green house effect is caused by :
   (*a*) Green plants (*b*) Infra red rays (*c*) UV-rays (*d*) X-rays
15. Peeling of ozone hole is due to :
   (*a*) CFCs (*b*) PAN (*c*) CO_2 (*d*) Coal mining
16. Acid rains are caused due to :
   (*a*) Production of SO_2 and NO_2 (*b*) Phosphates
   (*c*) SO_2 and CO (*d*) All of these
17. Ozone hole refers to :
   (*a*) Reduction in thickness of ozone layer in trophosphere
   (*b*) Reduction of thickness of ozone layer in stratosphere
   (*c*) Hole in ozone layer in stratosphere
   (*d*) Increased concentration of ozone
18. Ozone umbrella is affected by :
   (*a*) CFM (*b*) CO_2 (*c*) Methane (*d*) PHN
19. Greenhouse effect is because of :
   (*a*) CFC
   (*b*) O_3
   (c) Those gases which fail to allow reflected infra-red radiations through them
   (*d*) All of these
20. DDT is considered harmful because :
   (*a*) It is organochloride compound (*b*) It is pesticide
   (*c*) It is non-biodegradable (*d*) It is easily degradable
21. The pollutants released by jet planes are :
   (*a*) Fogs (*b*) Smogs (*c*) Colloids (*d*) Aerosols

22. Greenhouse effect refers to :
(*a*) Ozone hole moon light (*b*) Global warming caused by increasing CO_2
(*c*) Increased chlorofluorocarbons (*d*) Plants grown in green houses

23. The compound associated with Bhopal gas tragedy is :
(*a*) DDT (*b*) Atomic radiations (*c*) methyl isocyanate (*d*) Malathion

24. Sudden mass scale death of fishes from O_2 depletion is more likely in case of :
(*a*) Oligotrophic lake (*b*) Mesotrophic lake (*c*) Eutrophic lake (*d*) Oxbow lake

25. The World Environment Day is on :
(*a*) 7th, August (*b*) 5th, June (*c*) 10th, April (*d*) 28th, February

26. Greenhouse effect is due to :
(*a*) Higher CO_2 concentration
(*b*) Absorption of infra-red gases and dust particles
(*c*) Both (*a*) and (*b*)
(*d*) Rarification of atmosphere

27. Gas leaked in Bhopal tragedy was :
(*a*) Potassium isocyanide (*b*) Sodium isocyanate
(*c*) Ethyl isothiocyanide (*d*) Methyl isocyanate

28. Which of the following is not a pollutant ?
(*a*) SO_2 (*b*) CO_2 (*c*) CO (*d*) NO_2

29. Which of the following on inhalation dissolves in the haemoglobin more rapidly than oxygen ?
(*a*) SO_2 (*b*) O_3 (*c*) CO (*d*) N_2O

30. The most harmful environmental pollutants are :
(*a*) Human organic wastes (*b*) Wastes from fields
(*c*) Non-biodegradable (*d*) Natural nutrients in excess

31. Acid rain is caused by :
(*a*) SO_2 (*b*) NO_2
(*c*) Both (*a*) and (*b*) (*d*) None of these

32. Smog is a combination of :
(*a*) Fire and water (*b*) Smoke and fog (*c*) Water and smoke (*d*) Air and water

33. As it travels along the food chain, the concentration of DDT :
(*a*) Increases (*b*) Stays constant (*c*) Decreases (*d*) Fluctuates randomly

34. Result of ozone hole is :
(*a*) Greenhouse effect (*b*) Global warming
(*c*) Acid rain (*d*) Increased UV radiations

35. The term biomagnification refers to :
(*a*) Increase in concentration of non-degradable pollutant through a food chain
(*b*) Increase in population size
(*c*) Growth of organisms due to food consumption
(*d*) Blowing up of environment issues by man

36. Acid rain is due to :
(*a*) Excess CO_2 (*b*) Excess O_2
(*c*) A natural phenomenon (*d*) Excess of SO_2

37. DDT-spraying on the crop results in the pollution of :
(*a*) Soil and water (*b*) Air and soil (*c*) Crops and air *d*) Air and water

38. Which one is not dangerous for life ?
(*a*) Biopollutants (*b*) Ionosphere (*c*) Nuclear blast (*d*) Deforestation

39. Acid rain is due to :
(*a*) Sulphur dioxide pollution (*b*) Carbon monoxide pollution
(*c*) Pesticide pollution (*d*) Dust particles

40. Plants are known as purifiers of air due to process of :
(*a*) Respiration (*b*) Transpiration (*c*) Photosynthesis (*d*) Desiccation

41. Carbon monoxide is a major pollutant of :
(*a*) Water (*b*) Air (*c*) Noise (*d*) Soil

42. Air pollution effects are usually found on :
(*a*) Roots (*b*) Stems (*c*) Flowers (*d*) Leaves

43. One of the important effect of SO_2 and transformation products on plants is :
(*a*) Plasmolysis (*b*) Destruction of chlorophyll
(*c*) Destruction of Golgi bodies (*d*) Destruction of cell wall

44. Major pollutant in the jet plane emission is :
(*a*) Carbon tetrachloride (*b*) SO_2 (*c*) CO (*d*) Fluorocarbons

45. Pollutants having most lasting effects are :
(*a*) Pesticides (*b*) Smoke from chimneys
(*c*) SO_2 (*d*) CO

46. Thermal pollution is more common near :
(*a*) Hot water springs (*b*) Coal based power plants
(*c*) Tropical regions (*d*) Temperate regions

47. Pollution indicator plants :
(*a*) Purify atmosphere (*b*) Resist pollution
(*c*) Are sensitive to pollutants (*d*) Can indicate onset of rain

48. Sound becomes a hazardous noise pollution if its level is above :
(*a*) 30 dB (*b*) 80 dB (*c*) 120 dB (*d*) 150 dB

49. Pollution of water is caused by :
(*a*) Industrial effluents (*b*) Sewage (*c*) Form run off (*d*) All of these

50. Photochemical smog always contains :
(*a*) PAN and O_3 (*b*) CH_4 and O_3 (*c*) CO and O_3 (*d*) PAN and CH_4

51. Greenhouse effect is due to the presence of :
(*a*) NO_2, CO_2, SO_2, CFC (*b*) CO, CO_2 H_2
(*c*) CO, O_2, H_2, CH_4 (*d*) CO_2 and CH_4 only

52. Acid rains are caused due to :
(*a*) SO_3 only (*b*) CO_2 only (*c*) NO_2 and SO_2 (*d*) SO_3 and CO_2

53. The depletion of ozone layer is due to :
(*a*) Chlorofluorocarbons (*b*) Oxides of nitrogen (*c*) Hydrocarbons (*d*) All of these

54. The indiscriminate use of fertilizers causes :
(*a*) Air pollution (*b*) Water pollution (*c*) Soil pollution (*d*) None of these.

55. In big cities, air pollution is mainly due to :
(*a*) Burning of fossil fuels (*b*) Thermal power plants
(*c*) Sewage (*d*) Suspended particles

56. As compared to tap water, BOD of a water polluted with sewage waste would be :
(*a*) High (*b*) Low (*c*) Normal (*d*) Nil

57. Which of the following does not occur when the sewage is discharged into the rivers ?
(*a*) Eutrophication (*b*) Depletion of O_2 occurs
(*c*) Increasing O_2 (*d*) Cyanophycean bloom occurs

58. Increase in the concentration of pollutants in higher trophic levels is known as :
(*a*) Biomagnification (*b*) Biodegradation (*c*) Eutrofication (*d*) Recycling

59. Greenhouse effect is increasing due to :
(*a*) Increasing CO_2 concentration (*b*) Increasing SO_2 concentration
(*c*) Ozone hole (*d*) Increasing CFCs concentration

60. Major cause of air pollution in big cities is due to :
(*a*) Domestic exhaust (*b*) Burning of fossil fuels
(*c*) Thermal power plants (*d*) Automobile exhaust

61. Effect of pollution is on :
(*a*) Cell division (*b*) Crop rotation (*c*) Mutations (*d*) Ecological balance

62. The UV-rays in atmosphere are checked by :
(*a*) O_3 (*b*) O_2 (*c*) H_2 (*d*) Cl_2

63. Which of the following is indicatar of SO_2 pollution ?
(*a*) Moss (*b*) Lichens (*c*) Algae (*d*) Pteridophytes

64. One of these contributes maximum air pollution :
(*a*) India (*b*) USA (*c*) Russia (*d*) Britain

65. Which of the following will not be found in polluted water ?
(*a*) Blue-green algae (*b*) Larvae of stone fly (*c*) Water hyacinth (*d*) Sewage fungi

66. Greenhouse effect is related to :
(*a*) Global warming (*b*) Terrace gardening
(*c*) Kitchen gardens (*d*) Increased growth of algae

67. The lake, in which fishes will die in bulk, is :
(*a*) Oligotrophic lake (*b*) Eutrophic lake
(*c*) Organotrophic lake (*d*) None of these

68. Taj Mahal is threatened due to effect of :
(*a*) Chlorine (*b*) SO_2 (*c*) O_2 (*d*) Hydrogen.

69. The major contribution of greenhouse gases to the atmosphere is :
(*a*) Russia (*b*) Germany (*c*) Brazil (*d*) USA

70. In the coming time, skin-related disorders might be more common due to :
(*a*) Use of detergents (*b*) Water pollution
(*c*) Depletion of ozone layer (*d*) Air pollution

71. The CO_2 content in the atmospheric air is approximately :
(*a*) 0.034 % (*b*) 0.34 % (*c*) 3.34 % (*d*) 6.5 %

72. Phosphate pollution is mainly caused by :
(*a*) Sewage and phosphate rocks (*b*) Phosphate rocks only
(*c*) Sewage and agricultural fertilisers (*d*) Agricultural fertilisers only.

73. Formation of ozone hole is maximum over :
(*a*) Antarctica (*b*) Africa (*c*) Europe (*d*) India

74. A sewage treatment process, in which a portion of the decomposer bacteria present in the waste is recycled into the beginning of process, is called :
(*a*) Cyclic treatment (*b*) Primary treatment
(*c*) Activated sludge treatment (*d*) Tertiary treatment

75. Which important greenhouse gas other than methane is being produced from the agricultural fields ?
(*a*) Arsine (*b*) SO_2 (*c*) Ammonia (*d*) Nitrous oxide

76. If there was no CO_2 in the earth's atmosphere, the temperature of earth's surface would be :

(*a*) Dependent on amount of O_2 in the atmosphere

(*b*) Higher than the present

(*c*) Less than the present

(*d*) The same

77. Greenhouse effect with respect to global climate refers to :

(*a*) Cooling of earth (*b*) Warming of earth

(*c*) Increased rainfall and greenery (*d*) Desertification

78. Which of the following is secondary air pollutant ?

(*a*) SO_2 (*b*) CO_2 (*c*) PAN (*d*) Aerosol

79. In the year 1984, the Bhopal gas tragedy was caused by the leakage of :

(*a*) CO (*b*) SO_2 (*c*) NO_2 (*d*) Methyl isocyanate

80. What is the intensity of sound in normal conversation ?

(*a*) 10–20 decibel (*b*) 30–60 decibel (*c*) 70–90 decibel (*d*) 120–150 decibel

81. What is BOD ?

(*a*) Amount of O_2 utilised by organisms in water

(*b*) Total amount of O_2 present in water

(*c*) Amount of O_2 utilised by microorganisms for decomposition

(*d*) All of these

82. Effective Biological Resistance (EBR) is associated with :

(*a*) High temperture (*b*) Low temperature (*c*) Radiation (*d*) Pollution

83. Which of the following is a pollution-related occupational health hazard ?

(*a*) Fluorosis (*b*) Silicosis (*c*) Pneumoconiosis (*d*) Asthma

84. At present the most significant cause of dwindling biodiveristy is :

(*a*) Biological magnification of DDT (*b*) Global warming

(*c*) Deterioration of ozone layer (*d*) Destruction of habitat

85. Fluoride pollution mainly affects :

(*a*) Brain (*b*) Teeth (*c*) Heart (*b*) Kidney

86. *Escherichia coli* is used as an indicator organism to determine pollution of water with :

(*a*) Industrial effluents (*b*) Heavy metals

(*c*) Pollen of aquatic plants (*d*) Faecal matter

ANSWERS

1. (*c*)	**2.** (*b*)	**3.** (*a*)	**4.** (*b*)	**5.** (*c*)	**6.** (*d*)	**7.** (*b*)	**8.** (*c*)	**9.** (*d*)	**10.** (*d*)
11. (*a*)	**12.** (*c*)	**13.** (*a*)	**14.** (*b*)	**15.** (*a*)	**16.** (*a*)	**17.** (*b*)	**18.** (*b*)	**19.** (*c*)	**20.** (*c*)
21. (*d*)	**22.** (*b*)	**23.** (*c*)	**24.** (*c*)	**25.** (*b*)	**26.** (*c*)	**27.** (*d*)	**28.** (*c*)	**29.** (*c*)	**30.** (*c*)
31. (*c*)	**32.** (*b*)	**33.** (*a*)	**34.** (*d*)	**35.** (*a*)	**36.** (*d*)	**37.** (*a*)	**38.** (*b*)	**39.** (*a*)	**40.** (*c*)
41. (*b*)	**42.** (*d*)	**43.** (*b*)	**44.** (*d*)	**45.** (*a*)	**46.** (*b*)	**47.** (*b*)	**48.** (*b*)	**49.** (*d*)	**50.** (*a*)
51. (*a*)	**52.** (*c*)	**53.** (*a*)	**54.** (*b*)	**55.** (*a*)	**56.** (*a*)	**57.** (*c*)	**58.** (*a*)	**59.** (*a*)	**60.** (*d*)
61. (*d*)	**62.** (*a*)	**63.** (*b*)	**64.** (*b*)	**65.** (*b*)	**66.** (*a*)	**67.** (*b*)	**68.** (*d*)	**69.** (*d*)	**70.** (*c*)
71. (*a*)	**72.** (*c*)	**73.** (*a*)	**74.** (*c*)	**75.** (*d*)	**76.** (*c*)	**77.** (*b*)	**78.** (*c*)	**79.** (*d*)	**80.** (*b*)
81. (*c*)	**82.** (*b*)	**83.** (*c*)	**84.** (*d*)	**85.** (*b*)	**86.** (*d*)				

UNIT - 6
Utilization of Plants Economic Botany

16
CHAPTER

Introduction

The economic botany deals with application of botanical knowledge to the well-being of mankind. The primary necessities of man are threefold—*food, clothing and shelter*. The most essential need of man is food. The food primarily comes from plants in the forms of cereals (*e.g.*, rice, wheat, maize, oat, barley and rye), millets (*e.g.*, sorghum, pearl millet, finger millet, etc.), pulses, vegetables and fruits. For clothing again plants are indispensable. The plants that yield fibres are second only to food plants. From the times immemorial the man needed some form of clothing, and this need was fulfilled only by fibre-yielding plants. In this respect cotton occupies an all important position, supplemented by jute and some other fibres for coarse clothing. To meet an everincreasing demand for food and clothing an application of the knowledge of botany has been considered to be of great importance for better utilization of plant products. From the time immemorial shelter from the weather inclemencies and as a protection against natural enemy has been felt too much. In this respect the forest products have been of service to mankind from the very beginning of his history. The most familiar and most important, of these products is wood, which is used in all types of construction work. Other products like bamboo, reed, cane, thatch grass, etc., are of equal value for house making. With the advancement of civilization the need of man is also ever on the increase. To meet his requirements he has tried to tap plants as sources for his comforts and varied use by exploiting his scientific knowledge and has been successful in this direction to a great extent. Other useful plant products include—wood (for furniture, boat building, bridge building, railway sleepers, fuel, etc.), fibres (for gunny bags, rope, cordage, matting, canvas, carpet, etc.), drugs, tannins, dyes, paper, sugar, starch, gum, resin, rubber, vegetable fats, fatty oils, essential oils, tea, coffee, cocoa, tobacco, spices, condiments, cork, etc. Utility of bacteria, yeasts, other edible and antibiotics producing fungi, algae and lichens cannot be denied. It becomes quite evident that a knowledge of botany and its proper application led to the well-being of humanity in several ways. The more important plants and their products have been grouped in several important categories. In the present text the plants and their products have been grouped as follows:

FOOD PLANTS

Cereals and millets. *Cereals*: The cereals are the most important sources of plant food for man. They constitute the most important group in the food plants of India. The cereals are the members of family Gramineae. There are six true cereals rice, wheat, maize, barley, oat and rye. They contain a high percentage of carbohydrates, together with a considerable amount of proteins and some fats. Even vitamins are present.

Millets: They are also known as small grains. They are considered to have been cultivated in India from prehistoric times. Some of the commonly grown millets in India are sorghum, pearl millet and finger millet.

Legumes and nuts. *Legumes*: The legumes or pulses are next in importance to cereals as sources of food. They belong to the family Leguminosae. They contain more protein material than any other vegetable products. Carbohydrates and fats are also present in the legumes. The pulses form an important item in India where the majority of the population consists of vegetarians. The important Indian pulses are gram, black gram, green gram, pea, pigeon pea, lentil, etc.

Nuts: A nut is a one-celled, one-seeded dry fruit with a hard pericarp. Only a few nuts of commerce fulfil this description, *e.g.*, chestnut. The others included among nuts may be peanuts, almonds, coconuts, cashewnuts, walnuts, pistachio nuts, etc. These "so called" nuts make a valuable food material. The food value is due to a high protein and fat content.

Vegetables. The term vegetable is usually applied to edible plants which store up reserve food in roots, stems, leaves and fruits and which are eaten cooked, or raw as salad. The vegetables rank next to cereals as sources of carbohydrate food. The nutritive value of vegetables is tremendous, because of the presence of indispensable mineral salts and vitamins.

Fruits. Morphologically a fruit is the seed-bearing portion of the plant, and consists of the ripened ovary and its contents. Simple fruits are derived from a single ovary, and compound fruits from more than one. The aggregate fruits are formed from numerous carpels of the same flower,

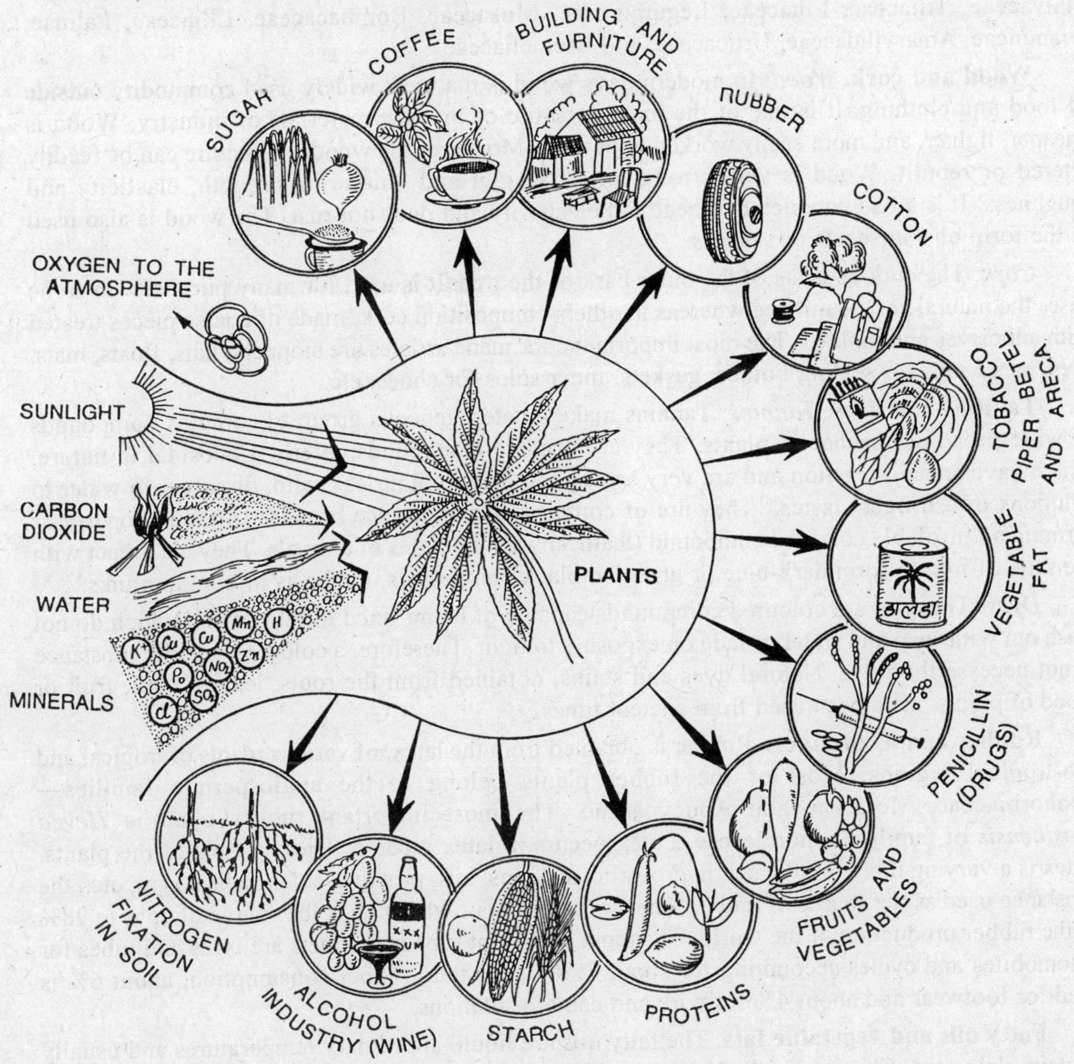

Fig. 16.1. Plants and human welfare

while composite fruits develop from ovaries of different flowers. In economic botany only those fruits are considered which are usually eaten without cooking. For convenience the fruits have been divided into two groups *tropical fruits* (*e.g.*, mango, citrus fruits, litchi, banana, guava, sugar apple, fig, papaya, pine-apple, etc.) and *temperate fruits* (*e.g.*, apple, pear, plum, peach, strawberries, grape, etc.).

PLANTS AND PLANT PRODUCTS OF INDUSTRIAL VALUE

Fibres and fibre-yielding plants. The fibre-yielding plants rank second only to food plants in their usefulness to human kind. The utilization of fibres is directly related to the advancement of civilization. As civilization advanced, the use of vegetable fibres increased greatly. They are of enormous value in our daily life. About two thousand species of the plants yield various fibres all over the world. The chief fibres of commercial importance have been classified in six groups— textile fibres (*e.g.*, cotton, flax, etc.), brush fibres, filling fibres, rough weaving fibres, natural fabrics and paper-making fibres. The most important fibre-yielding plants belong to the families— Malvaceae, Tiliaceae, Linaceae, Leguminosae, Musaceae, Bombacaceae, Liliaceae, Palmae, Gramineae, Amaryllidaceae, Urticaceae, and Bromeliaceae.

Wood and cork. *Wood*: In modern days wood is the most widely used commodity outside of food and clothing. It is one of the most versatile of the raw materials of industry. Wood is cheaper, lighter, and more easily worked with tools. Moreover the wooden structure can be readily altered or rebuilt. Wood is very strong for its weight and unique in strength, elasticity and toughness. It is a bad conductor of heat and electricity and does not rust. The wood is also used in the form of thin sheets or *veneers*.

Cork: The cork consists of the outer bark of the tree. It is used for many purposes. In some cases the natural cork is utilized whereas in others composition cork, made of coarse pieces treated with adhesives and molded. The most important cork made articles are stoppers, hats, floats, mats, tiles, metal tops for sealing bottles, gaskets, inner soles for shoes, etc.

Tannins and dyes. *Tannins*: Tannins make a heterogeneous group of complex compounds of widespread occurrence in plants. They are organic compounds chiefly glucosidal in nature, which have an acid reaction and are very astringent. It is a colourless solid, dissolves in water to solutions of astringent tastes. They are of commercial importance because of their property of forming an insoluble colloidal compound (leather) with the hides of animals. They also react with the salts of iron to form dark-blue or greenish-black compounds, the basis of common inks.

Dyes: The dyes are coloured compounds capable of being fixed to fabrics and which do not wash out with soap and water or fade on exposure to light. Therefore, a coloured organic substance is not necessarily a dye. Natural dyes and stains, obtained from the roots, leaves, bark, fruit or wood of plants, have been used from ancient times.

Rubber and its products. Rubber is obtained from the latex, of various plants of tropical and sub-tropical regions. Most of the rubber plants belong to the angiospermic families— Euphorbiaceae, Moraceae and Apocynaceae. The most important rubber plant is *Hevea brasiliensis* of family Euphorbiaceae. Latex occurs in latex tubes or latex vessels of the plants. Latex is a varying mixture of water, hydrocarbons, resins, oils, proteins, acids, salts, sugar, etc., the substance used as the source of rubber. The Hevea or Para rubber tree is the source of 95% to 98% of the rubber production of the world. The most important rubber products are tyres and tubes for automobiles and cycles accounting for about 75% of the total rubber consumption, about 6% is used for footwear and about 4% for wire and cable insulations.

Fatty oils and vegetable fats. The fatty oils are liquid at ordinary temperatures and usually contain oleic acid. Chemically they consist of glycerin in combination with a fatty acid.

The fats are solid at ordinary temperatures and contain stearic or palmitic acid. When a fat is boiled with an alkali, it decomposes and the fatty acid unites with alkali to form soap.

Usually the fatty oils are stored up in seeds of the plants belonging to many angiospermic families. Sometimes, to a less extent they are also stored in fruits, stems and other plant organs. Several fatty oils are edible and used as cooking media. These edible oils contain both solid and liquid fats.

Sugars and starches. *Sugars*: The glucose manufactured by the green plant in photosynthesis, is almost universally present in plant cells. This basic material of metabolism, the glucose, has the formula $C_6H_{12}O_6$. Fruit sugar or fructose is another product of photosynthesis which has the same formula. The more complex sugars are built up from these simple sugars. The most important complex sugar is sucrose or cane sugar, which has the formula $C_{12}H_{22}O_{11}$. This sugar is accumulated in abundance in sugarcane and sugar beets. Today, the sugar industry is the second largest in India, next only to textiles.

Starch: Starch occurs in all green plants. It is a complex carbohydrate with the formula $(C_6H_{10}O_5)\,n$. Like the sugars, they are derived also from glucose and constitute the first visible product of photosynthesis. This is the commonest type of reserve food in green plants and plays the most important role in their metabolism. Commercial sources of starch are wheat, barley, maize, potatoes and arrowroot. In these plants starch occurs in the form of starch grains which vary in shape and size.

Pulp and paper. The paper is a cellulose product. Cellulose is the most complex carbohydrate and is universally present in the cell walls of plants. One of the important uses of cellulose is in the manufacture of paper. Paper can be made from any natural fibrous material.

Gums and resins. *Gums*: The true gums are formed as the result of disintegration of internal tissues, for the most part from the decomposition of cellulose. The gums are insoluble in alcohol but soluble in water, readily swell up in it, and form a viscous mass. They are colloidal in nature. They contain a large amount of sugar and are closely allied to the pectins. The gums are formed in various kinds of plants. *Acacia senegal* yields the best gum-arabic of commerce. The gums exude naturally from the stems, or in response to wounding. The commercial gums are the dried exudations of the plants.

Resins: They represent oxidation products of various essential oils and are very complex and varied in their chemical composition. They are mostly found in the stems of the trees, and occur in abundance in special canals or ducts. They are yellowish solids, insoluble in water but soluble in alcohol, turpentine and spirit. It normally oozes out through the bark and hardens on exposure to the air.

MEDICINAL PLANTS AND DRUGS

Medicinal plants. The branch of medical science, which deals with the drug plants, is known as *pharmacognosy*. This branch of medical science is concerned with the history, commerce, collection, selection, identification and preservation of crude drugs and raw materials. Most of the drugs are obtained from wild plants growing in all parts of the world, and especially in tropical regions. The medicinal value of drug plants is due to the presence in the plant tissues of some chemical substances that produce a definite physiological action on the human body. The most important of these substances are alkaloids. Some of these chemicals are powerful poisons and therefore the drugs should be prepared and prescribed only by expert physicians.

Fumitories and masticatories. From the time immemorial the human beings have smoked or chewed various substances for pleasure. For example, the bulk of tobacco is used for smoking in the form of cigarette, *bidi*, cigar, cheroot and *chuttas* and in pipe and *hookah*. Large quantities of tobacco are also consumed for chewing.

On the other hand betel leaves and arecanuts are widely used as a masticatory. The basic preparation for chewing purposes consists of betel leaf smeared with hydrated lime, and catechu to which scrapings of arecanut are added; flavourings such as coconut shavings, clove, cardamom, fennel, powdered liquorice, nutmeg, and also tobacco are used according to one's taste. Chewing of betel leaves with various adjuncts is an ancient practice in India and other countries of East Asia.

FOOD ADJUNCTS

Spices and condiments. The spices cannot be grouped as foods, for they contain less nutritive value. They stimulate the appetite and increase the secretion and flow of gastric juices. They give a good flavour and aroma to food, and add greatly to the pleasure of eating. For this reason they are commonly known as "*food adjuncts*". The aromatic value of the spices is due to the presence of the essential oils. Sometimes the term "spice" is restricted to hard parts of plants, which are generally used in a pulverized state. Condiments are spices or other flavouring substances which possess a sharp taste, and are commonly added to food after it has been cooked.

Vitamins. The vitamins are absolutely essential for well-being of both plants and animals. They are formed by plants. They are necessary for normal metabolism, growth, development and reproduction. Vitamins are also essential for the prevention of several human diseases. Fruits, vegetables and seeds of various plants are good sources of vitamins. Seaweeds contain nearly all the known vitamins. The yeasts also contain several important vitamins.

Non-alcoholic beverages. The beverages containing caffeine are used all over the world for their stimulating and refreshing qualities. Caffeine is an alkaloid, which has definite medicinal values and acts as a diuretic and nerve stimulant. The most important non-alcoholic beverages are—tea, coffee and cocoa. Caffeine is harmful in large quantities, it is present in these beverages in very small amounts, not exceeding two per cent. The other beverages which do not contain alcohol are commonly known as *soft drinks*. They contain high sugar content and make a good source of energy. The fruit juices are the simplest soft drinks.

LOWER PLANTS IN ECONOMIC BOTANY

Fungi. Some fungi have been esteemed as articles of food from early times. Edible fungi have delicate flavours and are often valued delicacies. The important edible fungi are—*Agaricus* spp., *Cantharellus* spp., *Volvaria* spp., *Lycoperdon* spp., *Morchella* spp., *Tuber* spp. and many others.

The fungi, like yeasts are responsible for fermentation. The baking industry depends solely on the enormous budding of the yeast cells.

Penicillium is best known to the non-botanist because it is the source from which the antibiotic penicillin is extracted. The famous drug, ergotine is obtained from *Claviceps purpurea*, the causal organism of plant disease, ergot of rye.

Lichens. As food lichens have been used from the earliest times as animal feed and also as food for human consumption in times of scarcity. Lichens were once employed for dyeing wool and silk and colouring materials derived from them were highly valued before the advent of synthetic dyes. Lichens were formerly used as sources of fermentable sugars for the production of ethyl alcohol. Several lichens possess medicinal properties.

Algae. From the earliest times several sea weeds have been used as direct source of food to human beings. High mineral content up to 5% of the wet material, in which all the mineral elements important in human and animal physiology are found, makes sea weeds a unique supplement for a well balanced diet. The sea weeds (marine algae) are the richest source of vitamins. The vitamins A, B and E are found in abundance in sea weeds. Several important foodstuffs prepared by algae are—suimono, mitsu, dulse, seatron, laver, kompu, carrageen, etc.

17
CHAPTER

Food Plants

CEREALS

The cereals are the most important source of plant food for man. They constitute the most important group in the food plants of India. The cereals are the members of family Gramineae, and possess the characteristic fruit, the caryopsis. In this fruit the wall of the seed becomes fused with the ovary wall to form the husk. The term 'grain' is applied to this type of fruit. There are six true cereals—rice, wheat, maize, barley, oats and rye. Of which rice, wheat and maize, are most important, and they have played an important part in the development of civilization. Sometimes the millets and sorghums are erroneously referred to as cereals. Cereals contain a high percentage of carbohydrates, together with a considerable amount of proteins and some fats. Even vitamins are present.

RICE

Oryza sativa Linn; family—Gramineae (Poaceae); Eng.—rice; Hindi—*chaval*, *dhan*; Sanskrit—*dhanya*, *vrihi*, *nivara*, *syali*; Bengali—*chal*; Marathi—*tandula*, *dhan*, *bhat*; Gujarati—*choka*, *dangar*; Telugu—*vadlu*, *varidhanyamu*, *biyyamu*; Tamil—*nellu*, *arisi*; Kannada—*akki*, *bhatta*, *nelu*; Malayalam—*nellu ari*.

History. The earliest mention of rice cultivation is connected with China. According to Stanislas Julien, a ceremony was established in that country about 2800 B.C., by the Emperor Chin-nung, in which the sowing of five kinds of grain is the chief observance. The reigning Emperor himself has to sow the rice, but he may delegate the sowing of the other four kinds to the princes of his family. However De Candolle, affirms that rice cultivation in India, though subsequent to that of China, in point of date of first record, has been a valued crop since classic periods. One of its Sanskrit names *Dhanya* means the supporter or nourisher of mankind. By the Hindus it is regarded as the emblem of wealth or fortune.

According to Dutt's *Materia Medica of the Hindus*, there are three principal classes of rice—*Sali*, or that reaped in the cold season, *Vrihi*, or that ripens in the rainy season, and *Shashtika*, or that grown in the hot weather in low lands. It may be said that rice enters extensively into the ordinances of Hinduism. Several kinds of rice are used as votive offerings at many religious ceremonies. The Sanskrit word *syala* denotes the custom of the bride's brother scattering fried grains at the marriage ceremony. In the *Ordinances of Manu* the word *food* is given as synonymous with *rice*. Wild rice in Sanskrit works is spoken of as *nivara*, and various Sanskrit names are given to the cultivated rices, such as *dhanya*, *vrihi*, *syali*, *jiva-sadhana*, *tanonu* and *shashtika*. Most of these names are traceable to roots which denote life, existence and subsistence. As a possible historic fact, which would have a bearing on the cultivation of rice in Upper India, it may be said the word *vrihi*, is met with in the *Atharva Veda*.

The most general vernacular name for wild rice is, however, *uri* and the plant occurs mainly in Tamil Nadu, Orissa, Bengal and Kerala. The area of wild rice, corresponding to that in which *uri* and other such names are used leads to the concept that in India rice cultivation may have spread from there all over the rest of the peninsula and ultimately across the Himalayas.

The chief wild habitat of the plant is from south India. The cultivation of rice appears to have spread from there eastward to China, perhaps 3,000 years before the Christian era, and at perhaps a slightly more recent date, westward and northward, throughout India to Persia and Arabia, and ultimately to Egypt and Europe.

The Chinese were the first to cultivate rice (approximately 4,000 B.C.). In classical Chinese the words for agriculture and rice culture are synonymous. It is believed that the cultivation of rice in India dates back to ancient periods even earlier than 3,000 B.C. (Parthasarathy, 1954). The cultivation of rice was introduced into India before the time of Greeks, and then reached Syria and Northern Africa. The rice was first cultivated in Europe in 1468 in Italy. In America this was first grown in 1694 in South Carolina.

Rice is predominantly a crop of Asia, 95% of the world's area being concentrated in south-eastern Asia extending from West Pakistan to Japan. India has the largest area, 79 million acres in the world under rice cultivation, and production is about 28 million tons per year. India and Indo-China are supposed to be the centres of origin of rice (Ramiah, Ghose and Vachhani, 1952).

The traditional use of rice in the religious ceremonies of the Hindus, associated with birth, marriage and death is a testimony to its great antiquity as also to the intimate place it has in the life of the people. The crop is cultivated in almost all the states of India, extending from the delta region of the south to the higher altitudes of 3,000 to 5,000 feet above sea level in the Kashmir valley in the north. It is mostly cultivated in the valley of the Ganga and its delta and the low-lying coastal areas in southern and north-eastern India.

Of all the principal crops grown in the country rice has the largest area under it. Andhra Pradesh, Assam, Bihar, Maharashtra, Madhya Pradesh, Tamil Nadu, Orissa, Uttar Pradesh and West Bengal are the main producing states; together they account for more than 90% of the total area under the crop.

Cultivation. The rice grows best on damp soils underlaid with a semi-impervious subsoil in places where it can be flooded. A considerable interest has been aroused in India during last two decades in the Japanese method of rice cultivation. This combines improved cultural practices with proper manuring and plant protection measures. The increased acre yields obtained by this method have naturally led to its popularity. The results of breeding for high yield have also been very satisfactory, and many improved varieties have been evolved which give an enhanced yield. It has also been realized that the newer intensive methods of cultivation and manuring demand varieties which are capable of giving higher yields. With this aim in view, there has been started an intensive hybridization programme between the *japonica* and *indica* varieties of rice. The *japonica* varieties are high yielding, while the *indica* varieties have a better tillering capacity and are resistant to disease and adverse climatic conditions.

The fields are ploughed and the rice is transplanted from seed beds when 9 or 10 in. high. The young plants are covered with water, and the water is kept in circulation. As soon as ripening starts, the water is drawn off and the field is allowed to dry out. Rice is harvested and the stalks are stacked up to dry.

Investigations have been carried out to determine the role of blue-green algae in rice nutrition and their effect on the fertility of soil. The results indicate that the fixation of nitrogen is increased by the activity of blue-green algae.

It has been established that incorporation of green matter in paddy soil improves yield. It provides the cheapest way of manuring rice fields wherever facilities for raising green-manure crops are available. *Sesbanea aculeata* (*dhaincha*) has been found to be the most suitable crop for

Fig. 17.1. Rice *(Oryza sativa)* variety 'IET 4106' takes about 102 days for maturity with a yield potential of 40–45 q/ha.

rice areas as it can stand the soil and water conditions prevailing in these areas. In Kashmir valley, green manuring of rice crop with *Lens esculentus* has been found to be very effective. In U.P. *Crotalaria juncea* is becoming popular.

Milling. The grains of rice are removed by threshing. The grains are to be husked when they are to be used, and are then pounded in a mortar (*okhli*) with a wooden mallet (*musal*) and winnowed. The resulting grain is very nutritious for it contains protein and fat as well as starch.

In the commercial preparation the impurities are removed and the paddy is passed between millstones to break up the husk. This chaff is moved by blowers. Now the grain is pounded in huge mortars and a portion of the bran layer and embryo is removed. The waste is called rice bran. The white rice is then scoured by friction and polished. In polishing the embryo which contains proteins, vitamins and oils is knocked out due to lack of strong link between the embryo and the endosperm. During this process the outer, more nutritive parts of the grain are removed.

Uses. The chief use of rice is as food, and more people use it than any other cereal. The rice is generally eaten with pulses (legumes) or some other food rich in proteins. A diet of rice and soyabeans makes the food of millions. The rice straw is used for making straw boards, paper and mats. Rice bran oil is used for making soaps and cosmetics. Rice starch is much used in European countries. In several tropical countries intoxicating beverages are prepared from rice. Important beverage of Japan, *Sake* is prepared by fermenting rice. In India also in some parts of Orissa and Andhra Pradesh intoxicating beverages are prepared from rice.

This starch-rich food is eaten after boiling. It may also be converted into parched rice, beaten rice or rice flakes, and puffed rice; mixed with black gram (*urd*) powder, used for fermented preparations like *idli* and *dosa*. Broken grains obtained during milling, used as human and cattle food, for making alcoholic beverages, and as a source of starch and rice flour. Paddy husk is used as fuel; the bran is also used as cattle feed. The fatty oil obtained from bran is used for edible purposes.

WHEAT

Triticum aestivum Linn.; family—Gramineae (Poaceae); Hindi—*Gehu*; Sanskrit—*Godhuma*; Bengali—*Giun, gom, gam*; Marathi—*Gahum, gahung*; Gujarati—*Ghavum, gawn, govum*; Telgu—*Goodhumalu*; Tamil—*Godumai, godumbayarisi*; Kannada—*Godhi*; Malayalam—*Gendum, kotanpam, godamba.*

History. According to M.A. de Candolle, the cultivation of wheat is prehistoric in Old World. Very ancient Egyptian monuments show the establishment of this cultivation. However, when Egyptians or Greeks speak of its origin, they attribute it to mythical personages—Isis, Ceres, Triptolemus, etc. A small-grained wheat has been found at the earliest lake dwellings of Western Switzerland, the inhabitants of which were at least contemporary with the Trojan war, and perhaps earlier. The same form of wheat, *i.e., Triticum vulgare antiquorum* Heer, was found by Unger in a brick of the pyramid of Dashur in Egypt, to which he assigns a date of 3359 B.C. Another small-grained wheat, *T. vulgare compactum muticum* Heer, was less common in Switzerland in the earliest stone age, while a third intermediate form was cultivated in Hungary at the same period. However, De Candolle believes that the wheat cultivation in the temperate parts of Europe, Asia and Africa is probably older than the most ancient known languages. The Chinese certainly grew it in 2700 B.C., and considered it a gift direct from heaven. In the annual ceremony of sowing five kinds of seeds, instituted by the Emperor Shen-nung or Chin-nung, wheat was one of the species employed.

De Candolle has written in his famous work the *Origin of Cultivated Plants* that the original home of wheat in very early prehistoric times was in Mesopotamia. According to Berosus, the earliest of all Western Historians, and a Chaldean priest who wrote some twenty-three centuries ago, it occurred wild there. According to Strabo, who was born 50 B.C., a grain very similar to wheat grew wild upon the banks of the Indus.

According to De Candolle there is strong evidence in favour of India being the home of some of the forms of wheat as can be shown for any other part of the globe. India possesses perhaps as comprehensive a series of time honoured forms of wheat as can be shown for any other country. India has its hard wheats and soft wheats ; its starch wheats and speltwheats ; its bearded and beardless wheats. It has been established that most of these have been grown for countless ages on the same fields as they are to be found at the present day. It has also been proved that wheat cultivation in India is as ancient as in Europe or any other part of the world. The wheat grains discovered as a result of Indus valley excavations at Mohenjo-Daro indicate that north-western India was one of the ancestral lands of this cereal (Pal, B.P., 1954). The carbonised grains resemble those of *Triticum sphaerococcum* Percival, an endemic species which is still found in a few places in India.

According to Vavilov, the wheats have had a multiple origin, the soft wheats have come from the mountains of Afghanistan and the southwestern Himalayas ; the durum wheats from Abyssinia, Algeria and Greece; and einkorn from Asia Minor. The archaeological evidences establish that the wheat has been cultivated for at least 6000 years.

However, it must be kept in mind, while discussing the question of the origin of this ancient, cultivated cereal, that it is very nearly impossible, whatever the original home of wheat may have been, to determine with accuracy the character of the first parent from which it was derived.

India accounts for about 3.5 per cent of the global wheat production. The area under wheat constitutes roughly 14.0 per cent of the total area under cereals, and 10.0 per cent of the total area under food grains in the country. The crop is grown mainly in the northern and central parts. It is not much significance in the south.

It is cultivated as a food crop mainly in Uttar Pradesh, Punjab, Madhya Pradesh, Maharashtra, Bihar and Rajasthan.

Wheat is the second staple food crop of India and occupies about 29 million acres of land. It is consumed mainly by the people in the north. The wheat grains discovered as a result of Indus Valley excavations at Mohenjo-Daro indicate that north-western India was one of the ancestral lands of this cereal (Pal, B.P., 1954). The carbonised grains resemble those of *Triticum sphaerococcum* Percival an endemic species which is still found in a few places in India. The wheat growing area in India can be divided into three zones—the Gangetic alluvium, the Indus valley, and the black soil tract of Peninsular India. Wheat is also cultivated in Gujarat. A special feature of Indian wheat is the relatively short season in which it completes its life cycle. While the crop stands in the field for 9 to 10 months in some western countries, in India it is ready for harvest in four to six months after sowing. However, in the hilly tracts of Northern India the growing season for wheat is about nine months but the area under the crop in this region is very small.

Of the 18 recognized species under the genus *Triticum*, five are cultivated in India (Mirchandani and Khan, 1953). Of these *Triticum aestivum* Linn., commonly known as bread wheat, occupies the largest area and is mainly grown in the northern regions. Next in importance is the macaroni wheat, *T. durum* Desf., which is the predominant species in Central and Western

Fig. 17.2. *Triticum aestivum* : Wheat **(Gehu)** 'Moti'. A trigenic variety of wheat

India. The area under *T. dicoccum* Schrank is grown in South India. *T. turgidum* Linn., the rivet wheat, and *T. sphaerococcum* Percival, the dwarf wheat are even less important.

Wheat was the first crop to attract the attention of Indian breeders. The main object was to evolve high-yielding varieties with good grain quality. A notable achievement of the last few years has been the evolution of a new series of improved wheats. Their special characters include adaptability to certain regions, high yield, good grain quality, tolerance to rusts, and resistance to loose smut.

Cultivation, harvesting and threshing. In northern India sowing of wheat is done in October-November. Wheat may be sown broadcast, either by hand or by sowing machines.

Germination begins immediately and the first leaves appear within a fortnight. Wheat is properly manured and irrigated. The crop is harvested by cutting the plants with a sickle close to the ground, in March–April. Threshing is the next process, and this involves the separation of the grain from the spike. Threshing is generally done under the feet of bullocks or by threshing machines. After threshing the wheat is winnowed and sifted. In the Punjab, recently the *combines* are introduced for the purpose. The combines reap, clean, thresh, winnow and sift the grains ; wheat must be stored in firmly built structures, and it must be well ventilated.

Milling. In India in most of the motor driven flour mills there remains a fixed lower stone upon which a movable upper one revolves. The grains are dropped in the openings in the upper stone and gradually worked out between the stones which possess grinding surfaces cut in radiating lines. In this process whole grain is used.

The roller process of milling is an advanced and perfect process. The first step in this process comprises cleaning and scouring. After making the grains clean they are thoroughly washed and scoured. The next step is tempering. In this process a little water is added, which toughens the bran and prevents its breaking up, so that it will flake out all in one piece. In the last the conditioned and tempered wheat is submitted to breaking, grinding and rolling. The grain is first cracked or crushed gradually through a series of four to six pairs of chilled iron break rolls. The surface of the break rolls is made rough by sharp lengthwise foldings. The break flour is separated out by sieves, while the main portion goes to the second break. This process is repeated until five sets of rollers have been utilised. All bran is removed during this process and the purified material is passed to smooth rollers for final granulation. The final product is the flour.

Uses. There are three main kinds of flour—*suji*, *maida*, and *ata* which are used for various purposes. The flour is used chiefly for making 'bread' and 'chapatis'. The flour is also used for making biscuits, cakes, pastry and similar articles. Wheat flakes are used as breakfast food.

Wheat is also used in the manufacture of beer and other alcoholic beverages. Wheat straw is used for seating chairs, stuffing mattresses, etc. It makes a good food for livestock. Wheat straw is also used as fodder.

This makes a staple food in most parts of the world. Properties of gluten in the grains are such that it produces bread-stuffs generally superior to those from any other cereal grains. By products of wheat milling, such as bran, germ and middlings constitute valuable feed for stock, readily eaten; supplementary feeds are provided to supply protein and minerals in which the straw is deficient. The straw is used as bedding for cattle; it is also used for padding, as in mattresses, for packing fragile goods, for thatching and many other purposes. It may be used also for production of furfuryl alcohol. Straw-pulp is utilized for the manufacture of paper, straw-board, and building-board.

Non-feed industrial uses of wheat include the manufacture of starch, industrial alcohol, malted wheat, and core-binder flour; only small quantities of wheat are used for starch and gluten manufacture. Grain is regarded as a stand-by for alcohol production. Low-grade flours are utilized in the preparation of pastes for wall papering and ply-wood adhesives, and in iron foundries as a core binder.

Wheat products include peeled wheat ; *Bulgur*, a parboiled wheat product; *Wurld wheat*, similar to *Bulgur*, but of lighter colour ; *Instant* or *agglomerated* flour; Farina or *semolina*; *Wheat flakes*; *Shredded wheat* ; *Puffed wheat* ; *Grape-nuts*, prepared from toasted slices of malted bread; *Gluten*, used in special breads; and *wheat germ*, rich in vitamin E.

MAIZE

Zea mays Linn., family—Gramineae (Poaceae) ; Eng.—Maize, corn, Indian corn; Hindi—*Makai*, *Makka*, *Bhutta* ; Bengali—*Janar*, *bhutta*, *jonar*; Marathi—*Maka*, *makai*, *buta*; Gujarati—

Makkari, *makkai*; Telugu—*Mokka-janna*, *makka jonnalu*; Tamil—*Makka-cholam*; Kannada—*Mekkejola*, *musukojola*, *goinjol*; Malayalam—*Cholam*; Oriya—*Maka*, *buta*; Assam—*Gomdhan*, *makoi*; Manipur—*Chujak*, *nahom*.

History. It is now universally admitted that Maize or Indian corn is a native of America. This species probably originated in a wild state in the tropical South America. In Andes its cultivation goes back to prehistoric time. It has been established by evidence that the plant must have been grown for many centuries prior to the period of the Inca civilization. The maize was grown by the Indians in new Mexico as early as 2000 B.C. De Candolle on this subject writes "No one denies that maize was unknown in Europe at the time of Roman Empire but it has been said that it was brought from the east in the Middle ages". The various names which it bears in Europe, Egypt and

Fig. 17.3. 'Protina Opaque 2 Composite', a protein-rich variety of maize *(Zea mays)* (developed by Govind Ballabh Pant University of Agriculture and Technology Pantnagar).

Asia only show that in each country it was supposed to come from some not very distant region. Thus it is Turkish-wheat, Indian Corn, Roman Corn, Sicilian Corn, Barbary and Guinea Corn, etc. The Turks call it Egyptian Corn and the Egyptians speak of it as Syrian grain. Its most general name in India may be rendered Mecca Corn (*Makkai*). However, there is no authentic Sanskrit name for the plant, nor is the grain in any way associated with the Hindus. Crawfurd also writes, — "Maize is, beyond all question, a native of America, and before the discovery of the New World was wholly unknown to the old." Royle says that the Portuguese very probably introduced the richest products of America into India, such as maize, capsicum, guava, custard-apple and pine-apple. Abul Fazl writes in *Ain-i-Akbari* about *kewarah*, where he says that its leaves are like those of "maize". Rox-burgh wrote about the beginning of last (nineteenth) century, that Indian-Corn was "cultivated in various parts of India in gardens, and only as a delicacy; but not any where on the continent of India, so far as I can learn, as an extensive crop." Buchanan-Hamilton (1819) wrote about Kangra—"The poor people live much on Maize." Graham (1839) wrote about Western India that maize was commonly cultivated. Dalzell and Gibson (1861) wrote that maize was extensively grown in the early part of the rains, especially near large towns. It is thus very probably that in the Upper India maize was much more extensively grown at the beginning of the last (nineteenth) century. It is a field crop upon which at least the bulk of the aboriginal tribes of the hilly tracts of India are very largely dependent for subsistence.

According to De Candolle, the Maize was brought to China after the discovery of America (1492). The Portuguese came to Java in 1496, that is to say four years after the discovery of America and to China in 1516, Magellan's voyage from South America to Philippines took place in 1520. It is thought, that the seeds of maize may have been taken to China by navigators from America or from Europe. Maize had reached Europe a short time before the dates mentioned for India and China. Dodoens (1583), Camerarius (1588) and Matthiole, adopted the name Mays, which they knew to be American.

Conclusively this can be said that Maize was first introduced into Europe by Columbus, and into Asia by the earlier Portuguese explorers. Maize had acclimatized wherever the climate permitted, and has now spread all over the world.

Maize also known as 'Indian corn' or simply as 'corn', is a food crop of considerable importance in many parts of the world, especially in the U.S.A., which produces nearly 57 per cent or the total maize grown in the world as compared to about 1.4 per cent produced in India. The other important maize producing countries are—China (4.4 per cent), the Russia (3.3 per cent), Manchuria (2.9 per cent), Yugoslavia (2.4 per cent) and Mexico (2.3 per cent).

A tall annual cultivated grass ; it has been introduced in this country from tropical America about the beginning of the seventeenth century. The area under this crop measures 9.6 million acres with an annual total production of three million tons. Maize is mainly grown as 'kharif crop'. Although it is grown throughout India, the chief concentration is in the north-eastern parts. The crop is of special importance in the hilly and submontane regions where it makes the staple food of people. It is grown as a food crop mainly in Uttar Pradesh, Punjab, Madhya Pradesh, Bihar, Andhra Pradesh, Jammu and Kashmir.

Cultivation and harvesting. Maize is a summer annual. It thrives best in fertile, well irrigated, medium, heavy loamy soil. It is also commonly grown in the coarse gravel soils of hilly tracts. In India the crop is generally sown in June–July and harvested in September–October. The maize stems are cut close to the ground with the help of corn knife or sickle. The stalks are stacked to allow the grain to ripe further. After a month of this curing process, the ears (cobs) are husked by hand or by machine. Maize must be stored in well ventilated bins so that the excess of moisture is evaporated.

The table shows that maize contains all the essential nutrients of food when compared to Wheat and other foodgrains.

Uses. The chief use is as a food for man and livestock. The grain is very nutritious, with a high percentage of carbohydrates, fats and proteins. Not only is the grain valuable as a stock feed, but the plant as a whole is an important fodder crop. The immature cobs are largely eaten after roasting. The grains are also used in making corn starch and industrial alcohol. The glucose is also manufactured from the grain. The corn oil is prepared which is used for soap making, lubrication and as salad oil. Corn flakes make a good breakfast food. The fibres in the stalks are utilized for making paper and yarn. Zein, the protein which occurs in maize grain, is utilized for making artificial fibres with good tensile strength and wool-like qualities.

Table 17.1

Cereals	*Proteins (Gms)*	*Fat (Gms)*	*Calories*	*Calcium Mgs*	*Iron Mgs*	*Vit 'A' I.U.*	*Vit 'B' Rib-oflavin Mgs*	*Vit 'C' Mgs*
Maize (Tender)	4.7	0.9	125	9	1.1	54	0.17	6
Maize (Dry)	11.1	3.6	342	10	2.0	1502	0.10	-
Ragi	7.3	1.3	328	344	17.4	70	0.10	-
Rice (milled)	6.8	0.5	345	10	3.1	-	0.03	-
Jowar	10.4	1.9	349	25	5.8	79	0.28	-
Wheat (whole)	11.8	1.5	46	41	4.9	108	0.12	-

It is of special importance in the hilly and sub-montane regions of the country where it forms the staple diet of the people, particularly in the winter months. In the northern parts, it is also extensively grown as a fodder.

Maize Products	**Uses**
1. Maize grits	Can be used directly for maufacture of breakfast foods like corn flakes. This can also be used in the manufacture of starch.
2. Maize rice	Can be used in replacement of rice or in conjunction with rice. This can also be used in the manufacture of starch.
3. Poultry grits	Can be used in replacement of rice and other cereals in breweries and distilleries.
4. Brewer's grits	Can be used in replacement of rice and other cereals in breweries and distilleries.
5. Maize flakes	Can be used to prepare various types of dishes in place of beaten rice. This can also be used in the preparation of soup and in breweries and distilleries in replacement of rice and other cereals.
6. Maize soji **7. Pre-cooked Maize soji**	Can be used in the preparation of sweet and savoury dishes like Kesaribhath, Kharabhath, Idli, Dosa, Vermicelli, etc. Pre-cooked Maize Soji can be used in the manufacture of Balahar and snack foods.
8. Maize maida	Can be used in the preparation of Bread, Biscuit, Vermicelli, Nippottu, Obbattu, Bajji, Pakoda, Kodubale, etc. This can also be

	used by mixing with Wheat, Maida and Bengal gram flour for preparing the above said dishes. Also used in gum manufacturing industry.
9. Maize flour **10. Pre-cooked Maize flour**	Can be used in the preparation of Ragi balls by mixing with ragi flour. Can also be used in the preparation of sweet and savoury dishes like Pakoda, Bajji, Kodubale, Chapaties, etc. As raw materials in the manufacture of binders (Gum) for being used in the foundry and paper industry. Pre-cooked flour can be used in the preparation of Soup, Curry, Sambar, Majjigehuli (khadi) and in the manufacture of cereal and other foods for babies.
11. Maize oil	Used as edible oil. This oil has got certain advantages like easy blood circulation, reducing body fat, etc. This can also be used in soap manufacturing industry.
12. Maize cake **13. Animal meal** **14. Bran**	Can be used in Cattle and Poultry Feed industry.

Other uses. However, due to low gluten content, it does not form dough with elastic properties, but dough with good elastic properties can be made from flour made from dehulled grain. Maize starch is extensively used as a sizing material in the textiles and paper industries. In the food industry it is used in the preparation of pies, puddings, salad dressings and confections. Maize starch is used for the production of dextrose and corn syrup; also employed as a diluent for pharmaceutical preparations, dusting material to prevent articles like surgeons gloves, from sticking together, ingredient of oil-well drilling muds, and as a depressant in ore-floatation process. In cosmetics, maize starch form an ingredient of various forms of toilet powders. Dried germs yield a semi-drying oil, known as maize oil or corn oil used as salad or cooking oil; it may also be used with linseed oil for paints. Zein, recovered from maize gluten used as a binder for cork particles in forming composition cork. A textile fibre has been produced from zein under the trade name *Vicara*; it combines the virtues of cotton and wool. Cobs are rich in pentosans and used for furfural production. They may also be used for making building-boards which are water and fire-resistant. Maize silk (styles) is astringent, diuretic and chloretic. Some of the maize products include dextrose, maize starch, syrup, corn flakes and popcorns.

VEGETABLES

The term vegetable is usually applied to edible plants which store up reserve food in roots, stems, leaves and fruits and which are eaten cooked, or raw as salad. The vegetables rank next to cereals as sources of carbohydrate food. The nutritive value of vegetables is tremendous, because of the presence of indispensable mineral salts and vitamins. India grows a large variety of vegetables belonging to the tropical, sub-tropical and temperate groups. However their present availability is only 1.3 ounces per adult per day. On the other hand a balanced diet requires ten ounces of vegetables per adult per day. A good number of the vegetables in India are introductions from foreign countries. For convenience the vegetables may be classified in three categories (*i*) underground vegetables, (*ii*) herbage vegetables, and (*iii*) fruit vegetables.

UNDERGROUND VEGETABLES

In these vegetables the food is stored in underground parts. The storage organs may be true roots or modified stems, such as rootstock, tubers, corms and bulbs. From earliest time roots and tubers have furnished food for man. The underground vegetables may be classified in two groups—(*i*) roots and (*ii*) underground stems. Underground stem vegetable **potato** is described.

UNDERGROUND STEM VEGETABLE

POTATO

Solanum tuberosum Linn.; Eng. Potato; Hindi—*Alu*; Family—Solanaceae.

Bengali and Hindi—*Alu*; Marathi and Gujarati—*Batata*; Telugu—*Bangaladumpa, uralagadda*; Tamil—*Uralakilangu, wallaraikilangu*; Kannada—*Batate, alu-gidde*; Malayalam—*Urulan kizhangu*; Punjab, Kumaon, Bihar, Orissa and Assam—*Alu*.

History. It has been firmly established that at the time of the discovery of America, the cultivation of the potato (*S. tuberosum* Linn.) was practised in the temperate regions extending from Chile to New Grenada. In Europe the potato was first introduced, at some period between 1535 and 1585, into Spain when its cultivation spread into Portugal, Italy, France, Belgium, and then into Germany. The introduction into Ireland in 1585 or 1586, was independent but after the era of its first cultivation in Spain.

The date of the introduction of the potato into India is uncertain. According to Roxburgh, it has been cultivated in India before the beginning of the eighteenth century (*i.e.*, middle or end of seventeenth century). The probability is that the cultivation of potato was introduced into India from Spain, sometimes in the seventeenth century.

This is one of the most important food plants of the world. It is native of South America now grown, chiefly in Uttar Pradesh, Assam, Punjab, West Bengal and Bihar.

Potato is also a cash crop of great importance. With proper care, it can be stored for several months and transported without much damage from the producing centres to the consuming area. The crop is mainly suited to colder latitudes and temperate zones. India accounts for only 0.7 per cent of the world production. The important potato producing countries are the Russia (32.0 percent), Poland (14.6 per cent), West Germany (10.5 per cent), France (5.7 per cent), the U.S.A. (4.6 per cent), United Kingdom (3.9 per cent), Czechoslovakia (3.0 per cent) and China (0.8 per cent).

Fig. 17.4. Potato, germinating tubers. Healthy sprouts emerge from healthy tubers.

The potato was introduced into India in the seventeenth century. The tubers are consumed in India as a vegetable rather than as a staple food. The per capita annual consumption of potatoes is estimated at nine lbs., while in some of the Western countries the figure is as high as 440 lbs. (Pushkernath, 1955). With an area of 702,000 acres the annual production is estimated at 1,674,000 tons (Maheshwari. 1959). About 90% of the crop is grown in the north Indian plains in winter and the rest in hills in the summer. In the Nilgiris, the crop is cultivated all the year round. In the plains, two successive crops are sometimes raised on the same land. The commonly grown varieties are 'Phulwa', 'Darjeeling Red Round', 'Up-to-date', and 'Gola'. Several newly evolved high-yielding hybrids, such as 208, 209, 2236 and No. 45, have been found to be very suitable for cultivation in north Indian plains.

The plant is a spreading annual. It has pinnately compound leaves, fibrous roots, and numerous rhizomes which are swollen at the tip to form the tubers.

They can be cultivated in many soils and many climates. The best environment is a cool moist climate, and a rich light soil. They are usually propagated vegetatively by means of tubers, or parts of tubers, the so-called 'seed potatoes'. The essential parts of the seed potatoes used for propagation are the eyes. The eyes are really groups of buds situated in the axils of aborted leaves. There is usually a central bud in each eye, surrounded by smaller lateral ones. The eyes are more numerous toward the apex of the tuber. Pieces of the tubers are cut at right angles to the main axis. The larger the piece, the more vigorous is the vegetative growth, and there is greater yield. In any case, at least one eye must be present. The tubers require a rest period of several weeks for maturation. Potatoes contain about 78% water, 18% carbohydrates, 2% proteins, 9.1% fat and 1% potash.

Uses. The tubers are used as vegetable. In European countries they make a universal table food. Small tubers are utilized for the production of starch and industrial alcohol. The potatoes are also fed to livestock.

Potato chips made from fresh potatoes are famous in the South while "*katri* and *patri*" are quite common in Maharashtra and Gujarat.

Katri and patri. The fresh potatoes are made into potato chips and shavings after boiling the potatoes. These are dried in the sun. These dried chips are deep fried as and when required.

Surplus and cull potatoes are used as feed for livestock and also as raw material for manufacture of starch, ethyl alcohol, and some other industrial products. Potato is among richest foods in potassium. Biological value of potato protein is 68 and protein efficiency ratio 1.9 and it decreases on storage, more so at room temperature than at 4°C. Cooking improves digestibility of potato protein. Tuberin has a high nutritive value, being superior to wheat protein. Potatoes contain phosphorylase, amylase and several other enzymes including proteolytic enzymes. Potato flesh, even at its natural high moisture content, is superior to 72 per cent extraction of wheat flour as a source of vitamin B1, riboflavin and nicotinic acid. Concentration of thiamine is lower and that of riboflavin and folic acid is higher at the periphery than in the interior of tuber. Potato leaf haulms are useful source of a-cellulose. Potatoes are used as antiscorbutic, aperient, diuretic, and galactagogue. Extract of leaves is used as antispasmodic in cough. Tuber is ground and made into a paste for application to burns with much benefit.

Potato products include potato chips, frozen French fries, potato granules, potato flakes, potato flour, spaghetti-like potato, canned potatoes, potato starch and fermentation products.

Potatoes are used for the preparation of nutrient media, *e.g.*, potato dextrose agar media (PDA), for culture of microorganism and for potato broth, used in microbiological work.

Variety: *Kufri Alankar*—Flourishes well in the plains of Northern India. Plants are tall, erect and vigorous. Flowers white in colour. Tubers are oval, white with medium deep to fleet eyes. Flesh is white with floury texture on cooking. It takes 80–90 days to mature. In the plains, the average yield is about 250–300 quintals per hectare. Early variety.

Variety : *Kufri Jyoti*—It is most suitable for hills and plains of Northern India. The plants are tall, erect, vigorous and medium compact and the flowers are white and abundant. The tubers are oval, white with fleet eyes. Flesh is dull white. It takes about 100 days in the plains and 120 days in hills to mature. Average yield 150 quintals per hectare. Early variety.

Variety : *Kufri Sheetman*—It is best suited for the plains of Punjab, western Rajasthan, Haryana and Tarai region of Uttar Pradesh. The plants are tall, vigorous, erect, open and the flowers have blue purple colour with white tips. The tubers are round oval slightly flattened with fleet eyes. The flesh is dull white and texture floury on cooking. It takes about 100–110 days to mature in the plains and 140–150 days in hills. It yields 230 quintals per hectare on an average. Medium variety.

SUGAR

SUGARCANE

Saccharum officinarum Linn.; Eng. Sugarcane; Hindi—*Ganna*, *Ikh*; Sanskrit—*Ikshu*, *khanda*, *sarkara* ; Bengali—*Paunda*, *Pundia* ; Telugu—*Cheruku* ; Tamil—*Poovan karumbu* ; Kannada—*Patta patti kabbu* ; Malayalam—*Karimbu*; Family—Gramineae/Poaceae.

History. Charaka wrote the oldest treatise extant on Hindu medicine, at about the beginning of the Christian era. The glossary of Sanskrit names in this work denotes an intimate knowledge with the various forms and preparations of sugar which is considerately more than 2000 years old. The term *Sarkara* appears to have given origin not only to the Arabic, Persian, Greek and Latin classic names, but to the extensive assortment of words in the modern languages of India and Europe which are every nearly the direct equivalents of the English word sugar. However, there are several sources of information that carry the Hindu knowledge in cane-sugar considerably further back than can perhaps be shown for any other country. The sugar-cane plant and the cane-sugar are frequently mentioned in the *Institutes of Manu*. The more generally accepted view is that these *Institutes* were begun two or three centuries before the advent of Christ. It may thus be safely accepted that the reference to a field of sugar-cane, in *Manusmriti*, is more than 2000 years old. It may thus be accepted that sugar-cane was very generally cultivated in India, during the time the *Institutes of Manu* were written. It is thus probable that sugar-cane has been cultivated in India for at least 3,000 years.

The culms of the wild species such as *Saccharum arundinaceum* (of Deccan) and *S. spontaneum* (of Kar-Nicobar) yield a certain amount of saccharine fluid, and it may be possible that one of these wild forms might have given origin to the sugar-cane, the cultivation of 3,000 years having destroyed its original specific characteristics.

In the *Ain-i-Akbari* (Abul Fazl, 1590 A.D.) a description of the methods of cultivation, manufacture of all forms of sugar, and the distillation of spirits from it, is being given in detail.

The Spaniards carried the cultivation and manufacture of sugar-cane to the Canary Islands in the fifteenth century, but prior to that in 1420 the Portuguese had conveyed it from Sicily to Madeira and to St. Thomas Island. In 1506 it was taken from the Canary Islands to San Domingo. The Dutch first established sugar works in Brazil in 1580, and then they carried the art of sugar manufacture to the West Indies in 1655. Sugar was manufactured by the English in Barbadoes in 1643 and in Jamaica in 1664. Sugar-cane was first introduced into the United States of America in Louisiana in 1741.

It is grown chiefly in Uttar Pradesh, Bihar, South India (especially Coimbatore) and the Punjab. The Indo-Gangetic belt previously grew mostly thin canes (*Saccharum barberi* Jeswiet). These are popularly known as Indian canes, have been extensively cultivated in the country for a long time. They are characterized by high fibre and an intermediate sucrose content. In the peninsular belt the thick canes (*Saccharum officinarum* Linn.) are the chief types. They have all the desirable agronomic and milling qualities. They are characterized by a thin rind and are low in fibre and high in sucrose content. The greatest landmark in the sugar industry is the evolution of improved canes in India. Ever since its inception in 1912 the Sugarcane Breeding Institute at Coimbatore has been engaged in the evolution of improved varieties for the sub-tropical regions of India and since 1926 for the tropical regions also. Its programme of hybridization has included various species of *Saccharum* and related genera. The improved Coimbatore canes now occupy 93% of the sugarcane area in India. They have increased the yields by at least 50%. Some of the improved types are: Co. 213, Co. 290, Co. 312, Co. 313, Co. 419, Co. 421, Co. 453, Co. 527, Co. 658, Co. 997, and Co. 1148. Of these, Co. 419 yields from 40 to 90 tons of cane per acre and contains 13 to 16% sugar.

The sugar industry in India is the second largest industry in the country, next only in importance to cotton textiles. The chief source of sugar is the sugarcane. This plant is a rapid growing perennial grass, which reaches a height in cultivation of 8 to 12 feet or more and a diameter of about 2 inches. It grows in clumps. The stem is solid with a tough rind and numerous fibrous strands and contains about 10% of juice, the sugar content of which varies greatly.

The commercial sugarcane is not known in a wild state. By 327 B.C. the sugarcane had become an important crop in India. The name "sugar" comes from the Sanskrit "sarkara", meaning gravel, and refers to the crude sugar.

Breeding for effective utilization of sunlight throughout the year. Sugarcane being a C. 4 plant has a very efficient photosynthetic system. Many of the tropical sugarcane varieties take 14 to 18 months from planting to harvest. This long duration also leads to sugar factories remaining idle for considerable periods. Hence, a breeding programme was initiated at the Sugarcane Breeding Institute, Coimbatore for selection of early ripening canes. Some of the cultures give good sugar recovery and yield in about 8 months. The production system with early ripening canes can be made even more efficient with the help of chemical ripeners. Such strains can enable either taking 3 crops (one main and two ratoon) of sugar cane in 2 years or promoting rice-sugarcane (with sugarcane planting in early autumn) and wheat-sugarcane (with sugarcane planting in late spring) rotations. By properly distributing the areas under spring and autumn planting the sugarcane crushing period can be considerably extended. In trials carried out at the Sugarcane Breeding Station, Coimbatore one variety recorded in 1977 about 19% sucrose with 88% purity in 240 days. During 1978 also, trials with short duration varieties gave over 18% sucrose in juice at 7 months, whereas the mid season variety (Co. 6304) used as control recorded only 12.5% sucrose at that time. (M.S. Swaminathan, 1978).

Fig. 17.4. *Sugarcane.* Sugarcane is cultivated practically in every state in India, but Uttar Pradesh has the largest area under the crop and accounts for about 50 per cent of the total area under sugarcane in the country. Andhra Pradesh, Bihar, Haryana, Punjab, Tamil Nadu and Karnataka are other major cane growing states which account for about 40 per cent of the total cane area in the country. Crop harvest is seen in the picture in Meerut district.

Cultivation. The sugarcane is propagated by cuttings of various sizes made from upper joints of old canes. These cuttings known as *seed*, are placed in trenches and nearly covered with soil. They begin to sprout in two weeks. The crop is harvested from 10 to 20 months after sprouting. The stems are cut close to the ground for the lower end of the cane is richest in sugar. The rhizomes will normally give rise to two or three more crops, known as *ratoons* before another planting is necessary. The canes are brought to the sugar mills by rail road or any available means of transportation.

Table 17.2. Sugarcane Varieties in Cultivation

Variety	*Parentage*	*Characteristics*	*Areas where cultivated*
Co. 312	Co. 213 x Co. 244	Grown well in all types of soil but lodges in rich loose soil; good ratooner; drought and frost resistant; suitable for Sugar and *Gur.*	Uttar Pradesh, Haryana, Madhya Pradesh and Rajasthan.
Co. 313	Co. 213 x Co. 244	Suited to light loamy soils; requires irrigation or well distributed rainfall; gives *gur* of good quality.	Uttar Pradesh and Assam.
Co. 419	POJ 2878 x Co. 290	Suited to all kinds of soils, but at its best in soils of medium depth and in soils with fairly high pH; good ratooner; good variety from factory point of view; *gur* hard, crystalline.	Andhra Pradesh, Maharashtra, Karnataka, Tamil Nadu, Madhya Pradesh, Orissa, Gujarat and Assam.
Co. 453	Black Cheribon x Co. 285	More suited to heavy soils; moderate tillerer and good ratooner; *gur* not of good quality.	Punjab and Haryana.
Co. 527	Co. 349 x Co. 312	Suited to well-drained rich loam; good tillerer, good for *rab* and sugar: *gur* golden yellow; highly crystalline and hard.	West Bengal, Eastern Uttar Pradesh and Andhra Pradesh.
Co. 785	Co. 443 x Co. 453	Satisfactory under flooded condition; tillering very good.	Kerala.
Co. 1148	Co. 301 x P. 4383	A good ratooner; *gur* outturn high.	Uttar Pradesh, Punjab and Haryana.
Co. 1158	Co. 421 x Co. 419	Mid season maturing cane with good yield of canes and sugar. Canes uniform and erect making harvest easy.	In large-scale cultivation in U.P.
Co. S. 510	Co. 453 x Co. 557	Adapted to sandy and loamy soils; tolerant to drought; tillering heavy, good ratooner; heavy yielder with high sucrose content.	Uttar Pradesh.

Variety	*Parentage*	*Characteristics*	*Areas where cultivated*
B.O. 17	Co. 331 x Co. 326	Suited for loamy soils, tolerant to water logged conditions and *usar* and saline soils; moderate tillering and good ratooning; yields *gur* of good keeping quality.	Uttar Pradesh and Bihar.
Co. L. 9	Co. 312 x Co. 285	Fairly frost resistant; good ratooner; produces *gur* of good quality.	Punjab and Haryana.
Co. L. 29	Co. 312 x Co. 285	Good tillerer and good ratooner; frost resistant; gives *gur* of very good quality.	Punjab and Haryana.

Sucrose or cane sugar. Cane sugar has been known from early days and its crude form was separated from cane juice. India has been the chief producer of cane-sugar from time immemorial. Later on, other countries like Java, Cuba, Philippines, etc., also started producing sugar and they contributed towards the development of better techniques for the production of refined sugar. The main sources of sugar are sugarcane and beet. In cold countries like Russia where beet is in abundance, it is extracted from beet. In tropical countries like India it is mostly obtained from sugarcane. Indian cane contains about 12–13% of sugar.

MANUFACTURE OF CANE-SUGAR

The process of manufacture of cane-sugar as followed in India involves the following steps:

Extraction of the juice. The canes are thoroughly cleaned and cut into small pieces. These cane pieces are dropped over a continuous belt of steel plates called the cane-carrier, moving over roller chains. From the cane-carrier, these are conveyed to a crusher fitted with a set of revolving knives, and cut into very small pieces. These small pieces in the form of a compact blanket are made to pass through two roller crushers and four sets of mills. After major quantity of the juice has been extracted by the crushers and the first two mills, cold or hot water is sprinkled over the *bagasse* whereby residual juice gets diluted and can be easily extracted by further milling. About 90-95% of the juice is usually extracted. The juice extracted is weighed or measured as desired and sent for clarification.

In case sugar is to be extracted from sugar-beet, it is sliced and put in hot water when the sugar present in the beet diffuses out. A number of tanks are used and worked on the principle of counter currents—concentrated solution comes in contact with fresh slices and the exhausted slices come in contact with fresh water and lose last traces of sugar in them. The solution obtained is treated like cane-juice.

Clarification of juice. The raw juice is dark opaque liquid containing about 15% sucrose and small quantities of glucose, fructose, vegetable proteins, mineral salts, organic acids, colouring matter, gums and fine particles of bagasse suspended in it. Allowed to remain untreated for some time, it begins to ferment and the sucrose present changes to mixture of glucose and fructose. In order to avoid this, the juice is not allowed to stand untreated for a long time.

There are two main processes for clarification of juice:

(*i*) *Sulphitation*. Adding milk of lime and treating with sulphur dioxide.

(*ii*) *Carbonation*. Adding milk of lime and treating with carbon dioxide.

Both processes have advantages of their own. In fact combination of both is made use of to get better results.

(*a*) *Defecation*. The juice is strained to remove suspended matter and treated with 2-3% lime till pH value reaches 7.2 in tanks heated with steam coils.

Heating of the juice helps the coagulation of albuminoids by lime. The vegetable proteins are thus coagulated and the organic acids neutralized. These form a scum on the surface of the juice when the hot juice is transferred to settling tanks called subsiders. The scum at the surface and the mud settling at the bottom are mechanically removed by passing the whole through a filter press and the clear solution sent on to the conical tanks for carbonation or sulphitation.

(*b*) *Carbonation or Sulphitation*. A current of carbon dioxide is passed through the defecated juice, which contains unreacted lime and calcium sucrosate (*carbonation*). This removes the excess of lime as calcium carbonate and decomposes calcium sucrosate.

Very often defecated juice is treated with sulphur dioxide (*sulphitation*) instead of carbon dioxide which serves the same purpose and bleaches the juice in addition, and produces a juice with much lighter colour. It prevents formation of brown mass by oxidation, and brings about coagulation of gums and albuminoids more effectively. During the process of sulphitation the solution is maintained neutral. In some cases 'sulphitation' follows 'carbonation'. The juice is filtered again to remove the precipitates.

At present all big sugar factories in India are following double carbonation and double sulphitation process for the manufacture of cane-sugar.

Concentration and crystallisation. The clear juice is concentrated in a multiple effect evaporator. Juice in the first pan is heated by exhaust steam from the engines of the factory. The concentrated juice from the first pan is taken to the second pan and heated there at a lower pressure, by steam from the first evaporator. The concentrated solution from the second pan is taken to the third pan and heated there at a still lower pressure by exhaust steam from the second evaporator.

To this concentrated juice sulphur dioxide is again applied. All through this process a strict control is maintained over the acidity of the solution otherwise there will be losses due to inversion, destruction or even discolouration may be there.

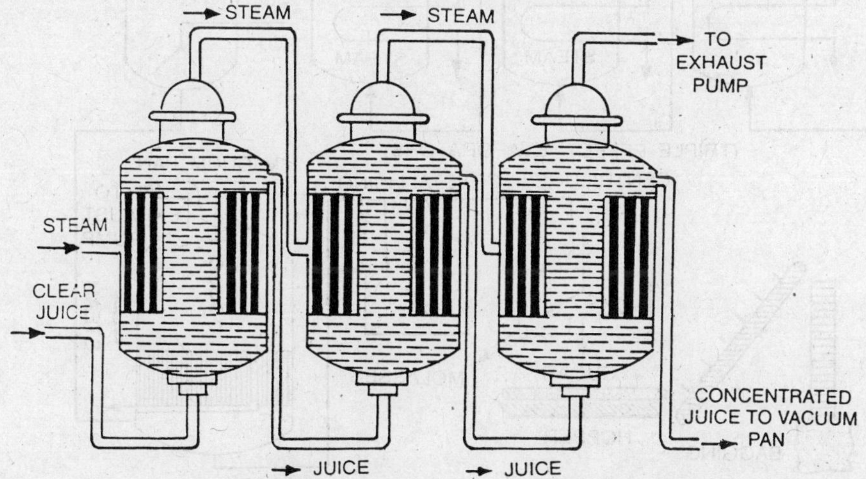

Fig. 17.6. Sugar manufacture. Multiple effect evaporator.

The clear syrupy juice is just boiled in a vacuum pan till formation of sugar crystals begins. The contents of the vacuum pan (*massecuite*) are taken into the crystallizing tank and allowed to cool slowly when the tiny crystals of sugar grow in size.

Separation of crystals. The crystals alongwith the mother liquor (molasses) are whirled in centrifugal machines wherein the molasses is removed. A little molasses adhering to the crystals is removed by spraying cold water and whirling in the centrifuge again. The crystals are given a

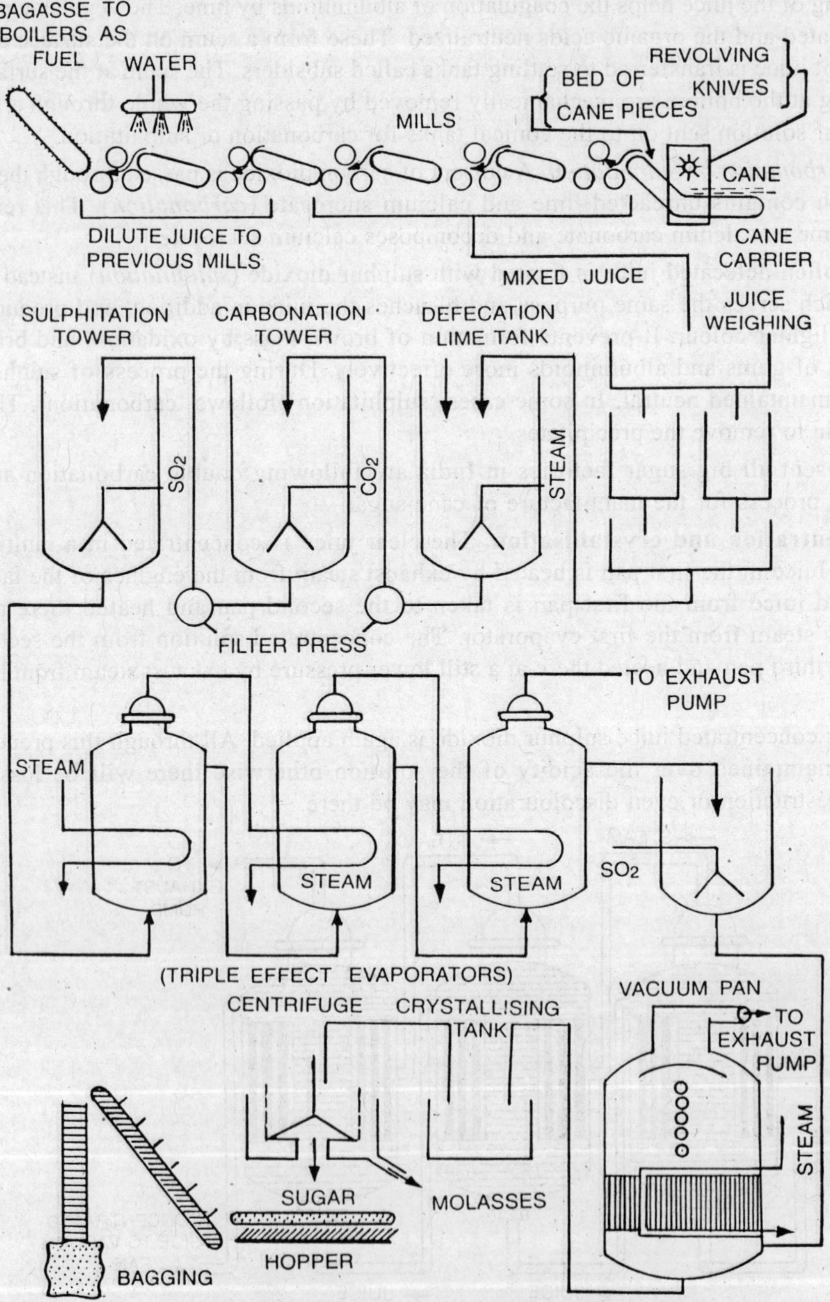

Fig. 17.7. The sugar industry. Flow sheet of the double carbonation and double sulphitation process for the manufacture of cane-sugar.

little blue colour and dried by dropping through a long pipe through which hot air is passing up and bagged.

Refining of sugar. Sugar in India is generally put in the market as obtained in the above process. In U.S.A. and other foreign countries it is further refined by dissolving the crystals in hot water and decolourising the solution with animal charcoal or coconut charcoal. It is followed by filtration, concentration under reduced pressure and crystallization as usual.

Recovery of sugar molasses. The molasses still contains sufficient amount of sugar. It is boiled over again to get a fresh crop of crystals. To recover sugar from the final molasses, the latter is treated chemically. It is diluted and treated with hot concentrated solution of strontium hydroxide. Sucrose present forms a precipitate of strontium sucrosate. This is separated, suspended in water and treated with carbon dioxide when strontium carbonate separates as a white precipitate and sucrose present in the solution is decolourised, concentrated and crystallized as above.

In India sugar is not extracted chemically from molasses although still about 30 to 35% sugar is there. It is now being sent to distilleries where it is fermented and used in the manufacture of alcohol.

Sugar Industry in India. India is the accredited birth place of sugar cane and sugar. However, the Industry progressed considerably after 1931 when it was granted protection. Today, the industry is the second largest in India, next only to textiles. The industry employs about 3 lakh workers, both skilled and unskilled. The industry is highly concentrated in Uttar Pradesh and Bihar.

ESTIMATED CAPACITY AND PRODUCTION OF SUGAR IN INDIA

The present output has outstripped annual home consumption considerably, thus presenting a serious problem of surplus stocks. The industry is experiencing difficulty in promoting exports as well, because of the high cost of domestic sugar. This is due to several weaknesses of the industry *viz.* (*i*) low yield per acre of cane in India which is just 15 tonnes as against 62 and 56 tonnes in Hawaii and Java, (*ii*) high price of cane which constitutes 60% of the cost of production of sugar and low sucrose content, (*iii*) uneconomic size of many factories and (*iv*) short crushing season which lasts only for 145 days in a year in India.

By-Products. There are three by-products of sugar industry, *viz.*, bagasse, molasses and press-mud. Bagasse was being used mainly as fuel for steam raising. Now it is used also for the manufacture of paper pulp, cardboard and insulation board. Molasses has demand in the preparation of acetic acid, aconitic acid, power alcohol, chemicals, chloroform, plastics, etc. Pressmud finds use in the manufacture of carbon paper and polish.

Uses of sugar. The sugar is used as an article of food and as sweetening agent in sweets and drinks. It is used in food preservation and in the preparation of oxalic acids in the laboratory. It is used also for the manufacture of octa-acetate.

Sugarcane juice is used mainly for three products used in foods and beverages, for sweetening purposes, *viz.*, *Gur* and jaggery, vacuum pan sugar, and open pan sugar or *khandsari*. A good quantity of sugar is also used for chewing. Byproducts obtained during the manufacture of *gur* and sugar are molasses. *Gur* is used for direct consumption by human beings and livestock. Sugar is used for sweetening milk, tea and coffee, and for making sweets.

18
CHAPTER

Fibres and Fibre Yielding Plants

It goes without saying that the fibre yielding plants rank second only to food plants in their usefulness to humankind. The utilization of fibres is directly related to the advancement of civilization. From the earliest times the man required three things for his living—food, clothing and shelter. Though other things were at hand to fulfil these demands yet they were insufficient and incomplete too without the genuine aid of the plants. As far as the clothing was concerned the skins and hides were not sufficient. He needed somewhat more and that gap was fulfilled by the plant fibre. As there was advancement of civilization the needs of man multiplied several fold and the use of the plant fibres increased greatly. Till now more than two thousand species of plants have been worked out which yield fibres. However, the commercial fibre yielding plants are few in number. The fibres from other plants are used by primitive tribal people of the world.

Economically the fibres may be grouped as follows:

Textile fibres. Fibres utilized for the manufacture of fabrics, netting and cordage, etc.

Brush fibres. Fibres utilized in the manufacture of brushes and brooms.

Rough weaving fibres. Fibres utilized in making baskets, chairs, mats, etc.

Filling fibres. Fibres used for stuffing mattresses, cushions, pillows, etc.

Natural fibres. Natural fibres used as cloth without weaving.

Paper-making fibres. Wood fibres, etc., utilized for paper-making.

As regards their structure, the sclerenchyma cells are grouped into fibres. Fibres are described as long cells with thick walls, correspondingly small cavities, and usually pointed ends. The walls of fibres usually consist of lining as well as cellulose. The fibres may occur singly or in small groups. The fibres may be found in any part of the plant—stems, leaves, roots, fruits and even seeds.

The plants belonging to several families of angiosperms yield fibres. They are chiefly tropical plants. The important families of the fibre yielding plants are—Malvaceae, Linaceae, Tiliaceae, Gramineae, Musaceae, Bombacaceae, Leguminosae, Palmae, Liliaceae, Urticaceae, Amaryllidaceae, Moraceae, Asclepiadaceae, and Bromeliaceae.

TEXTILE FIBRES

These fibres are supposed to be most important because they are connected with the textile industry. This industry is concerned with the manufacture of fabrics, netting and cordage. For making fabrics and netting, the flexible fibres are twisted together into yarn and thereafter woven, spun and knitted. For making cordage the individual fibres are twisted together rather than woven. The textile fibres are always long and possess high tensile strength, together with cohesiveness and pliability. They possess a fine, uniform, lustrous staple. The chief textiles are included in three categories—(*i*) surface fibres, (*ii*) soft fibres and (*iii*) hard fibres.

Surface fibres, *e.g.*, Cottons.

Soft fibres, *e.g.*, Flax, hemp, jute and ramie.

Hard fibres, *e.g.*, Sisal, coconut, pineapple, abaca, etc.

SURFACE FIBRES

COTTON

Gossypium Linn.; Eng.—Cotton; Hindi—*Kapas*; Bengali, Hindi, Gujarati, Punjabi and Marathi—*Kapas*, *rui*, *tula*; Kannada—*Hathi*; Telugu—*Patti*, *karpasamu*; Tamil and Malayalam—*Paruthi*, *panji*; Oriya—*Karpaso*, *kopa*; Family—Malvaceae.

History. The history of cotton (*Gossypium* sp.) is most interesting, and perhaps no more remarkable example of a sudden development exists in the whole history of Economic Products, than in the case of cotton. Only a few hundred years ago cotton and its products were practically unknown to the civilized nations of the West. However, evidences exist amongst certain of the aboriginal tribes of India which indicate a much more ancient knowledge of the plant. For example, the Khonds grow a cotton-bush in the place selected for a new settlement, and in after years the village plant is tended as sacred and carefully watered. This custom is probably of very ancient origin, and may denote a superstitious regard for the plant, probably derived from the knowledge and appreciation of its valuable properties.

The Sanskrit word translated 'cotton' is perhaps first mentioned in the *Institutes of Manu*, where it is stated that, "the sacrificial thread of a Brahman must be made of cotton, so as to be put on over his head in three strings." The word used in that passage and translated cotton is *Karpasi* from which has been derived, the *Kapas*.

However, it is certain that at the time of *Institutes of Manu* spinning and weaving were not only known, but the art of starching or weighing a textile was also practised, and the admissible amount prescribed. A still earlier notice of the process of starching probably occurs in the *Rigveda*. The picture thus raised of the character of the textile industries of India, some two thousand years ago. *Institutes of Manu* also deal with the regulations concerning washerman. We are thus led to believe that the arts of spinning, weaving and washing were perfectly understood in the East, at a time when the textile industries of the West, through ignorance of cotton, were in a much more backward condition.

Katan is an old Arabic word for flax, but may also have been used for cotton, since it apparently gave origin to the European words such as Cotton (*Eng.*) Cotone (*Ital.*), Cotn (*Fr.*), etc. In Greek the word *Karpason* is generally translated cotton. It is significant to mention that, with writers at the beginning of the Christian era, it was customary to speak of cotton and cotton goods by names not in use prior to that date. Herodotus is perhaps the first who seems to refer to the fibre. But perhaps the first unmistakable reference to cotton is by Theophrastus, subsequent to the expedition of Alexander into India. He mentions, that the plant was cultivated, at least in North-Western India, at that early date.

The muslins of Dacca seem also to have been known. They are described as superior to all others, and are said to have been called *gangitiki* by the Greeks, a name indicative of their manufacture on the banks of the Ganges.

The knowledge of the manufacture of cotton appears to have extended into Arabia and Persia about the beginning of the Christian era. From India, there is no doubt, the cultivation of cotton spread into Persia, Arabia and Egypt, from where it probably extended into Central and Western Africa. From Persia the culture migrated into Syria and Asia Minor, also into Turkey in Europe and thereafter into other parts of southern Europe. The earliest account of its cultivation on the

European shores, is to be found in the works of Eben el Awan of Seville who lived in the twelfth century.

In the new world, we find that cotton has probably been used from the earliest times. It is connected with some of the most ancient beliefs of the aboriginal peoples of South America. Columbus found cotton in use among the natives of Hispaniola, but only in the most primitive way. Cortez found the manufacture in a much more advanced condition in Mexico.

With the rise of Muhammadanism the knowledge of spinning and weaving cotton probably spread to Europe, consequently the delicate cotton fabrics of the East were first imitated in Italy and Spain during the twelfth or thirteenth century. From these localities the art of cotton manufacture became diffused and extensively established all along the southern shores of the Mediterranean, but it was confined to that area till sixteenth century. From the low countries the industry passed over to England in the seventeenth century. Towards the end of the seventeenth century cotton printing was established in England. Up to the commencement of the nineteenth century America contributed but a very insignificant portion of the cotton was consumed. At the close of the war in 1815, the production of cotton in America received a fresh impetus, and the subsequent progress was rapid and continuous. This great increase in the importance of America, as a cotton-growing country early led the East India Company to commence operation for the purpose of improving the quality and increasing the quantity of Indian cotton for export to England. The imports into Great Britain from India during the years 1800 to 1809 averaged 12,700 bales per annum.

This is the world's greatest industrial crop, the chief fibre plant. It was known to the ancient world long before written records were made. References to it are to be found in the works of the Greek and Roman writers. Cotton has been in the use in India since 1800 B.C., and from 1500 B.C. to A.D. 1500 India was the centre of the industry. The Hindus were the first people to weave cloth. Cotton was introduced into Europe by the Mohammedans. This was first grown in the United States soon after the first settlements were made. The first cotton mill was established in 1787.

All the cultivated cottons fall under four species, two belonging to the Old World and two to the New World. These are *Gossypium arboreum*, *G. herbaceum*; *G. hirsutum*, and *G. barbadense*.

G. arboreum. This is most widespread of all the species of Old World cottons, being distributed throughout the rain-fed savannah areas from Africa, through Arabia and India, to China, Japan and E. Indies. Its origin is obscure, but it is obviously Asian, since the area of its greatest variability is found around the Bay of Bengal. This species comprises a large number of varieties and races including many of the cultivated cottons in and around India.

The staple is coarse and very short, only 3/8 to 3/4 inch in length, but it is strong.

G. herbaceum. This is also an Old World species. It occurs in Africa, Middle East countries, Central Asia and Western India. It has been grown in India from time immemorial. Commercially the cottons belonging to this species constitute a fairly large percentage of medium staple cotton grown in India. The major part of the cotton crop in Maharashtra state comprises this species. This species comprises a large number of cultivated races.

Fig. 18.1. Staple cotton. **'Kanchana'** variety.

It is utilized for low-quality fabrics, carpets, and blankets and is especially suitable for blending with wool.

G. hirsutum. This is a New world species. The centre of variability for this species is Central America. It comprises a large number of varieties or races. Only three races—*punctatum*, *marie galante* and *latifoliam*—extend beyond Central America. The last one is of great importance agriculturally, as it comprises the Upland cotton, which has spread over vast areas in America, Asia and Africa. The types of cotton belonging to this race constitute the bulk of long staple cottons grown at present in India. The fibres are white with a considerable range in staple length, from 5/8 to $1^3/_8$ inch.

G. barbadense. This is a New World species. It includes perennial shrubs or small trees, 3–15 feet high, or annual shrub moderately high.

The centre of origin of this group is tropical South America particularly its north-western parts.

Two distinct types of cotton belongs to:

Sea-island Cotton. This is one of the most important selections which yields lint of perhaps the highest quality which the genus *Gossypium* has so far proved capable of yielding. Its fine, strong, light cream-coloured fibres are more regular in the number and uniformity of the twists and have silkier appearance than those of other cottons. Sea-island cotton was brought to the United States from the West Indies in 1785. The finest types yielded staples 2 in. or more in length, surpassing all the others in strength and firmness. Another form of sea-island cotton is grown along the coast in Georgia and Florida and in the West Indies and South America. This has a staple form 1½ to 1¾ in. in length.

Egyptian Cotton. The second important line of annual types derived from the perennial stock is the Egyptian cotton. Egyptian cotton is derived from a hybrid stock between a perennial G. barbadense and the annual sea island cotton. It is ecologically quite distinct from sea island cotton and is well adapted for irrigated conditions in the subtropics. Its practical value lies in its greater earliness. Some of the important commercial types at present in cultivation are Karnak, Menoui, Ashmuni, and Giza. The lint length of these cotton ranges from $1^1/_8$ in. (Ashmuni) to 1½ in (Karnak). Considerable quantities of these Egyptian types are imported every year into India for supplementing indigenous long staple cotton supplies. Besides Egypt, significant quantities of this cotton are produced in Sudan, U.S.A., Peru, French North Africa and Russia.

Because of its staple's length, strength and firmness this cotton is used for thread, underwear, hosiery, tire fabrics and fine dress goods.

Cultivation. Cotton is essentially a tropical crop, but its cultivation is carried on successfully over many parts of the world, far removed from the tropics. The limits of cultivation may be said to be the 40th latitude on both sides of the Equator. It is grown either at sea level or at moderate elevations not exceeding 3,000 feet. Cultivation is confined largely to flat open country and rough hilly tracts, where the minimum temperature does not fall below 70°F. Higher temperatures are very favourable, and the upper limit may go up even to 105°F in the picking season. The crop thrives well in moderate rainfall. Rainfall exceeding 35 inches is supposed to be harmful to the crop. The lower limit for a purely rain-fed crop is 20 inches. On black cotton soils, hardly any rainfall is needed over most of the growing period provided good showers have been received before the crop was sown and a satisfactory start has been made. Cotton is grown both as a dry crop and as an irrigated crop. If the rainfall is distributed over both the monsoons, the extra-ordinary fertility of the black cotton soil allows a wide variety of crops to be grown, and also taking of two crops in the year—one in the north-east monsoon period, and the other in the south-west monsoon period. On the other hand, if the rainfall is low and is confined to the north-east monsoon period, the only one crop is grown in the year. There is considerable mixed cropping practice with cotton. Pulses such as *arhar*, blackgram and green-gram and other crops such as groundnut and the castor, into the mixture. The 'New World' cottons are, however, grown pure—whether as dry crops or as irrigated crops.

THE COTTON INDUSTRY

Several operations are necessary in order to prepare the raw cotton fibre, as it comes from the field, for use in the textile industry. In brief these operations are as follows—ginning; baling; transporting to the mills; picking, a process in which a machine removes any foreign matter and delivers the cotton in a uniform layer; lapping, an operation whereby three layers are combined into one; carding, combing, and drawing during which the short fibres are extracted and the others are straightened and evenly distributed; and finally twisting the fibres into thread.

Harvest and yield. The cotton crop is usually harvested in three or four pickings, taken at suitable intervals. Picking is carried out by hand, mostly by women, the amount of cotton collected ranging from 20 to 50 lb. per day per person. Cotton should be picked only when the bolls are fully mature, fully open and the floss has puffed up consequent on exposure to sun.

The yield per acre is low in India as compared to yield in other countries.

Cotton marketing. The bulk of cotton produced in India is sold as *Kapas* or unginned cotton. *Kapas* is transported to the local markets or ginneries mainly in carts or, sometimes, on pack animals. The cultivator sells his cotton in the village market. The purchasers are village merchants, or agents of ginneries, spinning mills or exporting firms. In some states, co-operative societies organized by cultivators have taken up the purchase and sale of the cotton.

Ginning. Kapas or seed cotton collected from the field contains both lint and seed. For use in industry, cotton should be cleaned and the lint separated from the seed. A small amount of seed cotton is ginned in villages by the use of *charkha* gin. The bulk of it, however, is ginned in factories by power-driven machinery. The yield and quality of lint depend on the type of cotton and the machinery used for ginning.

Baling. Cotton is packed for trade purposes both in loose and compressed bales. Loose packing is adopted for inland transit of ginned cotton to a pressing factory, while compressed packing is adopted for transporting ginned cotton to the market and for storing in the godowns. Each loose bale contains 200 to 300 lb. of cotton. The usual weight of compressed bale is 392 lb. net with density of 40 lb. per cubic foot.

DEVELOPMENT OF FIBRE IN COTTON

A cotton fibre is a delicate tubular prolongation of the peripheral wall of the seed. The fibres are formed by the outer epidermis of the outer integuments. These hairs are single celled and thick walled and attain a length of up to 45 mm These fibres also show characteristic twists and are called *lint fibres*. Intermingled with the lint hairs are some *fuzz hairs*. The maximum length attained by the latter is 10 mm They are thick walled with the narrow lumen and lack the twists.

USES OF COTTON

The bulk of cotton production is consumed in the manufacture of woven goods, alone or in combination with other fibres. The principal types of woven fabrics are—print cloth, yarn fabrics, sheetings, fine cotton goods, napped fabrics, duck, tyre fabrics and towels. Products in the form of yarn and cord include unwoven tyre cord, thread, cordage and twine and crochet yarns. Unspun cotton finds use in mattresses, pads and upholsteries. Cotton constitutes one of the basic raw materials for cellulose industries including plastics, rayon and explosives. Sterilized absorbent cotton finds use in medical and surgical practice.

Yarns of varying size and fineness are needed in the production of fabrics. Coarse yarns are spun from short staple cottons and fine one from medium and long staple types. Long and uniform staples are utilized for yarns of high counts required for fine fabrics.

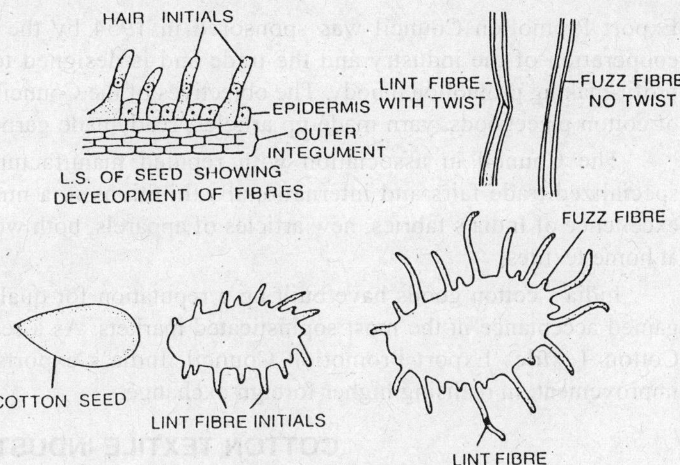

Fig. 18.2. Development of fibre in cotton.

Cotton waste is a by-product of the spinning and weaving mills and consists principally of short fibres rejected by combing and carding machines, floor sweepings, odds and ends from weaving and various scraps. The amount of waste given by cotton is an important factor in its quality evaluation. Cotton waste of good grade is employed in making cotton blankets, sheets, towels and flannelettes. Cylindrical strips from carding machine, which are constituted of fibres of good strength, are used for warps, twines, ropes and nets; they are also useful for wadding, padding for upholstery, bed quilts, etc. Strips from Egyptian cottons are mixed with wool for making mixed woollen goods. Floor sweeping and fibres unfit for spinning are bleached and used for gun-cotton, cellulose and artificial silk. Short remnants and thread waste that can not be respun are used as wiping and polishing material.

The *stalks* of plant contain a fibre that can be used in paper making or for fuel, and the *roots* possess a crude drug. The seeds are of the greatest importance and every portion is utilized. The *hulls* are used for stock feed; as fertilizer; for lining oil wells; as a source of xylose, a sugar that can be converted into alcohol and for many other purposes. The *kernels* yield one of the most important fatty oils, cottonseed oil and an oil cake and meal which are used for fertilizer, stock feed, flour, and as a dyestuff.

THE COTTON TEXTILES EXPORT PROMOTION COUNCIL OF INDIA—REPORT

As an exporter of quality textiles India's reputation dates back to the pre-Christian era. India's textile goods were shipped to distant corners of the world and they drew appreciation of high and mighty. To this day, these fabrics are known for the novelty and creativeness. The textile industry of India in those days was a cottage industry and the cloth was woven on handlooms. The skill and craftsmanship were handed over from generation to generation and this imparted a rare aesthetic quality to product. The glory of the beautiful handwoven cloth suffered a sort of eclipse in the wake of industrial revolution in the middle of the nineteenth century which created a vogue for mass produced textile products manufactured by mechanised mills. The beginning was made in India when a modern cotton mill was set up in 1854 in Bombay. The industry passed through many vicissitudes and pressure of competition from abroad in the process of its development until 1947 when India gained independence. The conditions created by the Second World War and the demand from many countries for their supplies of cotton textiles from India created an outlook for exports among the coteries of cotton mill industrialists. Cotton textiles are India's major source of foreign exchange. In 1974 cotton textiles brought in foreign exchange worth more than rupees 3,000 million.

A great need was felt to systematise the efforts to boost exports of textiles and the Government of India decided to sponsor an organisation towards the end. The Cotton Textile

Export Promotion Council was sponsored in 1954 by the Government of India with active cooperation of the industry and the trade and is designed to function as an autonomous non-profit-making promotional body. The objectives of the Council extend to the promotion of exports of cotton piecegoods, yarn made-up articles, readymade garments and hosiery articles.

The Council in association with reputed manufacturers and exporters participates in specialized trade fairs and international exhibitions in a number of countries to highlight the excellence of India's fabrics, new articles of apparels, both woven and knitted and the new range at home textiles.

India's cotton goods have built up a reputation for quality in all markets of the world and gained acceptance in the most sophisticated markets. As a result of the functions carried out by Cotton Textiles Export Promotion Council, India's exports have been helped by qualitative improvement, in realising higher foreign exchange.

COTTON TEXTILE INDUSTRY

Cotton Textile industry is one of the oldest and largest industries of India, which has made rapid strides during the current century. The industry is well organised in our country either in terms of value of annual output or number of labour employed. India has the world's highest acreage under cotton, but in terms of production among fifty-five major countries, it ranks only fourth after the U.S.A., Russia and China. Cotton is grown in our country over an area of about eight million hectares and average production during the last few seasons had been around eighty seven lakh bales. At present the industry provides direct employment to nearly twelve lakh workers. It also provides indirect employment to many millions like the cotton growers, processors, handloom and powerloom weavers who alone are estimated over three million and innumerable cloth dealers and shopkeepers. The industry contributes in ever increasing measures to the Central and State Governments by way of taxes and duties. A large domestic market and plentiful supply of domestic cotton were the two basic factors responsible for the growth and development of cotton textiles into a major key industry. The first factor still holds good. With an annual growth of 2.5 per cent in population the domestic demand for textiles is rising. But the second factor has been weakening and has tended to undermine the strength and stability of the industry. It may be said that the future of this premier industry primarily depends on our ability to increase the cotton production efficiently and economically. In this industry, concentration of management and ownership is mostly in the hands of the private sector.

Location

Though cotton textile industry being the largest industry in India has spread practically over all parts of the country, it is mostly localised in the states of Maharashtra and Gujarat. This is obvious from that these two states claim more than half of the spindlage and about 2/3 of the loomage. At the end of 1983 there were 805 mills in India. Out of this, 525 were spinning units and 280 were composite mills. Tamil Nadu, with 303, had the largest number of textile mills. But out of these only 24 were composite mills and the remaining made yarn. In Maharashtra, there were 108, out of these, 80 were composite mills. There were 89 composite units in Gujarat, out of a total of 114 mills. These three States account for the block of the production of yarn and growth in the organised sector. Gujarat and Maharashtra put together account for two-third of the mill loomage.

Reasons for this location are (*i*) availability of raw material both in good qualities and quantities; (*ii*) cheap and developed transport facilities including ports like Bombay and Kandla; (*iii*) availability of commercial and financial facilities like export and import houses and banking institutions; (*iv*) suitable climate and (*v*) enterprising capital. In recent years, the cotton textile industry has also spread to a number of other states like Madhya Pradesh, Bihar, Kerala, Andhra Pradesh, Tamil Nadu and Uttar Pradesh.

Size

On the whole, Indian market favour the organisation of the small or moderate sized mills. Cotton mills in Tamil Nadu, Kerala, Karnataka, West Bengal, Orissa, Assam and Bihar are small in size. Size of the cotton mill industry in these states ranges from ten to fifteen thousand spindles and 200 to 400 looms. In Andhra Pradesh, Uttar Pradesh, Madhya Pradesh and in Maharashtra the units are considered as medium sized. The size of model units in these regions ranges from fifteen to thirty thousand spindles and 400 to 600 looms. Big sized units with a range from 30 to 60 thousand spindles and 750 to 1500 looms are localised at Mumbai.

The Textile Policy

The textile policy announced by the Government in August 1978 emphasises the need for encouragement of handloom sector. The policy stated that the government would freeze the loomage of mills and powerlooms and the responsibility of meeting the additional requirements arising from the increase in population and also any improvement in the per capita consumption will be entrusted to the handloom sector. This policy is aimed at in view of the fact that this would create additional jobs in the rural areas. A number of handloom development centres have been proposed to be started. As far as the policy on controlled cloth is concerned, it has been stated that mill made controlled cloth would be limited to 400 million sq. metres and would be produced by both the National Textile Corporation and the private sector mills. The policy further laid that, "it is intended that the handloom sector should to the maximum extent possible and in the shortest period of time, meet the requirements of cloth for the weaker sections of the population. The manufacture of controlled cloth in the organised sector would be phased out over a period of time consistent with the growth of production of the required variety of cloth in the handloom sector.

Again, the Government of India announced its textile policy on the sixth June 1985, indicating significant changes in the Government's approach to the growth and development of the textile industry. It aims at free and unrestricted growth of the industry by removing sector-specific and fibre specific rigidities in production and capacity creation, prevention of unregulated growth of power looms, treating mill and powerloom sectors at par, providing protection to handloom sector, restructuring of the industry in terms of stages of manufacturing process and rationalisation of excise duty structure. It also emphasises rehabilitation of viable sick units, modernisation and upgradation of technology, as well as improvement in product quality and reduction in cost of production.

SOFT OR BAST FIBRES

JUTE

Corchorus Linn.; Eng. Jute; Hindi—*Patsan*; Sanskrit—*Kalasaka*; Hindi—*Narcha*; Bengali—*Chinalitapat, narcha, nalita, titapat*; Assam—*Titàmara*; Family—Tiliaceae.

History. J. F. Duthie found *C. capsularis* on the banks of the Gumpti near Judalpur in a wild state. However, Roxburgh reported the non-jute yielding species of *Corchorus* as natives of India. A special enquiry was, in 1873-74 in Madras of wild or cultivated jute found in that Presidency, and *C. olitorius* was reported to have been discovered both wild and cultivated. However, De Candolle was not convinced that the species existed in a truly wild state in the north of Calcutta. It might be found possible to produce forms of Corchorus from some of the truly wild species which would closely approximate to *C. capsularis* and *C. olitorius*. There are no Sanskrit names for these plants. There are neither Arabic nor Persian names for the species of *Corchorus*, known to the people of India. Mr. Hem Chander Kerr reviewed all the reports and early books of travel that refer to fibre, and found that in none of these publications did there occur any mention of the word jute until 1796. In several works *pat* was mentioned as a fibre in India as a form of hemp. By the beginning of the nineteenth century the word pat was completely superseded by jute in all commercial correspondence. With the facts of this kind it is believed that jute plant is a comparatively modern development from some wild stock which was unknown to the Sanskrit writers. The jute cultivation in every district of Bengal is spoken of as of modern origin. In one district its introduction is fixed at 1872, in another at 1865, in a third before the date of the British rule and in a fourth it is put down at 500 years ago.

Fig. 18.3. A Jute plant.

The commercial fibre jute is obtained from either one or both of the following species of *Corchorus*, *viz*., *C. capsularis* Linn., grown in West Bengal and Bangladesh, and *C. olitorius* Linn., raised in the vicinity of Calcutta. Little or no jute is produced in the other states of India.

Cultivation. In India, the crop is sown between March and May, and harvested between July and September, It is grown mainly in West Bengal, Assam and Bihar, which together account for 90 per cent of the total area sown. Orissa, Uttar Pradesh and Tripura are the other states, which contribute the remaining 10 per cent area.

Jute is a bast fibre obtained from the secondary phloem. The bast fibre is obtained from *C. capsularis*, a species with round pods which is grown in lowland areas subject to inundation. Fibre from *C. olitorius*, an upland species with long pods, is but little inferior.

In India, the time for harvesting the crop depends entirely upon the date of sowing; the season commences with the earliest crops about the end of June, and extends to the beginning of October. The crop is harvested within three or four months after planting, while the flowers are still in bloom.

Separation of fibre by retting. The fibre is separated from the stems by a process of retting in pools of stagnant water. The crop is stacked in bundles for two or three days, to give time for the decay of the leaves, which are said to discolour the fibre in the retting process. The period of retting depends upon the nature of the water, the kind of fibre, and condition of the atmosphere. It varies from two to twenty–five days. The operator has therefore to visit the tank daily, and ascertain, by means of his nail, if the fibre has begun to separate from the stem. This period must not be exceeded, otherwise the fibre becomes rotten and almost useless for commercial purposes. The bundles are made to sink in the water by placing on the top of them sods and mud. When the proper stage has been reached, the retting is rapidly completed. The cultivator, standing up to the waist in the foetid water proceeds to remove small portion of the bark from the ends next the roots. Grasping them together, he strips off the whole with a little management from end to end without breaking either stem or fibre. Thereafter the washing is done to remove the impurities. It is now wrung out so as to remove as much water as possible, and then hung up on lines prepared on the spot, to dry in the sun.

The very long pale-yellow fibres from six to ten feet in length, are quite stiff, as they are considerably lignified; they possess a silky lustre. They are very abundant, but are not particularly strong, and they tend to deteriorate rapidly when exposed to moisture, to which they are quite susceptible. Besides these disadvantages the jute is cheap and easily spun makes it valuable.

Utilization. Jute is used chiefly for rough weaving. The thick cloth made from jute fibre is used for making gunny-bags. Another type of fine cloth prepared from jute fibre is chiefly used as a cloth to sleep on. Another type of coarse cloth is largely used for making the sails of country boats, and also for bags to hold large seeds or fruits.

Jute is extensively used in the manufacture of carpets, curtains, shirtings, and is also mixed with silk or used for imitating silk fabrics. The fibre is also used for making twine and ropes. Short fibres and pieces from the lower ends of the stalks constitute jute butts, which are used to some extent in paper making.

India not only grows most of the jute, but it is the largest manufacturer and exporter of jute products.

JUTE INDUSTRY

In the industrial economy of India, Jute industry occupies an important place both in terms of value of annual foreign exchange and the number of people employed. This industry accounts for Rs. 300 crores by way of capital employed, provides employment to nearly four lakh workers and exports more than 75 per cent of the production. The 75 per cent of the production accounts for nearly 5 per cent of total foreign earnings of all commodities. This industry has an installed capacity of 13.0 lakh tonnes spreading over 44,516 looms and 5.92 lakh spindles. Besides, India is the largest producer of jute goods in the world. Apart from all these, about three million families of farmers derive a substantial part of their income from jute cultivation. In 1885, India's first jute factory was set up at Serampore in Bengal, followed by several others. However, after country's partition, nearly 75 per cent of India's jute growing areas went to East Bengal (Bangladesh) whereas most of the jute mills remained on the Indian territory, which gave a severe blow to the industry. In spite of this the industry made serious effort to increase jute production through both extensive and intensive cultivation. As days passed on Bangladesh became most powerful competitor which has built considerable jute manufacturing capacity over the years by putting up more productive modern machines as against India's old factories.

Fig. 18.4. The jute industry. A gunny bag, a jute product.

Location

Jute industry has been located in five states in India, *i.e.*, West Bengal, Andhra Pradesh, Bihar, Uttar Pradesh and Madhya Pradesh. Though the units engaged in the industry are located in all these five states, the main concentration of the industry is only in the State of West Bengal particularly about sixty miles long and two miles broad along both the banks of Hoogly, above and below Calcutta. The reasons for nearly ninety per cent of concentration of this industry around Calcutta are (*i*) raw jute is cheap and cannot afford much transport cost, as such the mills were forced to establish around Calcutta because of the availability of raw material in abundance in that area; (*ii*) availability of cheap power especially coal in West Bengal; (*iii*) transports, port, financial and commercial development of export market in Calcutta. The Central Government as well as some State Governments are encouraging the farming of jute with a view to feed our jute industry and to establish new units in various locations throughout the country. For example, Tamil Nadu proposes to launch a scheme for cultivation of jute with the ultimate aim of starting a jute mill.

Size

Market and managerial integration determine, to a large extent, the optimum size of units in the jute industry. The loomage capacity of a vast number of Indian mills ranges from 1,400 to 5,000 though mills outside West Bengal have less than 500 looms each. The reason for such large size in West Bengal is mainly influenced by location advantage. It is because of availability of adequate supplies of raw jute. Marketing and managerial integration has accelerated the situation and created an atmosphere which made the units big. It is not possible to say that a unit working with a particular loomage capacity is at optimum level because the problem of optimum size for the Indian jute industry becomes more complicated. It is complicated because of scarcity of raw material, the growing adoption of substitutes in other countries and increasing rate of development of technology and modernisation in the competing countries. If we say today that a particular unit is at optimum level it may not hold good tomorrow due to the dynamic conditions arising in day-to-day life of jute industry.

Competition from substitutes. The Indian jute goods are becoming dearer day by day because of increase in the cost of production due to increase in the price of raw material. Many foreign countries have developed the cheap substitutes of the jute goods. In recent years, the synthetic materials and other alternatives such as paper have emerged as major competitors of jute goods in the field of packing. Unless and until we don't reduce our cost of production and offer the jute goods at cheap rate, there will be no bright future for our jute industry.

Cost of production. In the last few decades the cost of production has gone up by about 75 per cent. Among the various components of cost, there has been a steep rise in wages while the increase in the cost of raw material is of lesser magnitude. The cost of other components such as dyes, chemicals and power charges have also increased.

Competition. The main competitor of India in this industry is Bangladesh. Bangladesh produces raw jute of fine quality and manufactures a variety of jute goods. Bangladesh has been underquoting Indian jute goods in foreign markets. In most of cases the foreign buyers have been placing orders first with Bangladesh; and India gets only residual orders. It is being stated that the jute goods of Bangladesh are cheaper than Indian jute products.

Modernisation. The survival of Industry is mostly dependent on its ability to accelerate the pace of modernisation. The Indian jute industry lags behind Bangladesh both in capital equipment and techniques. The Planning Commission has pointed out that the existing machinery has become to a large extent worn out and obsolete.

19 CHAPTER

Vegetable and Fatty Oils

The fatty oils are liquid at ordinary temperatures and usually contain oleic acid. They are sometimes called 'fixed oils' because, unlike the essential oils they do not evaporate. Chemically, the fatty oils consist of glycerine in combination with a fatty acid.

However, the fats are solid at ordinary temperatures and contain stearic or palmitic acid. When a fat is boiled with an alkali, it decomposes and the fatty acid unites with alkali to form soap.

Usually the fatty oils are stored up in seeds of the plants belonging to many families. Sometimes, to a less extent they are also stored in fruits, stems and other plant organs. Several fatty oils are edible and used as cooking media. The edible oils contain both solid and liquid fats.

Extraction of fatty oils. The seed coats are removed and the material is being converted into a fine meal. The oils may be removed by solvents or by hydraulic pressure. The hydraulic pressure method is commonly used for extracting the edible oils. The residue or oil cake is rich in protein and makes a good cattle feed. The pressure causes the cell walls to break and the fatty oils escape. These oils are filtered and purified. The refined oil is used for food purpose, and the lower grades of oil are used in industries.

Classification of fatty oils. There are three categories of fatty oils–(1) drying and semi-drying oils, (2) nondrying oils, and (3) vegetable fats. The refined grades of drying and semi-drying oils are used as edible oils, whereas the inferior grades are used in making soaps, varnishes, paints, candles and other such articles. The nondrying oils are edible and can be used for soap and lubricants. The fats are solid or semisolid at ordinary temperatures. They are edible and are also used in the soap and candle industries.

DRYING AND SEMI-DRYING OILS

SARSON OIL

Brassica campestris Linn.; var. ***sarson*** Prain; Eng. Yellow sarson, Indian colza; Hindi—*Sarson*; Family—Cruciferae (Brassicaceae).

A herb. It is grown as an oil-seed crop mainly in Uttar Pradesh, the Punjab, Bihar and Assam. The oil content is 30 to 45 percent, and the oil is extracted by expression or solvents.

Uses. The oil is used for cooking and burning purposes. The oil-cake is used as cattle feed.

NUTS WITH HIGH FAT CONTENT

COCONUT

Cocos nucifera Linn.; Family—Palmae/Palmacea/Arecaceae; Eng.—Coconut; Hindi—*Nariyal*; Sanskrit—*Narikela*; Bengali—*Dab*, *narikel*; Marathi—*Narel*, *naral*; Telugu—*Kobbarichettu*, *narikelamu*, *tenkaya*; Tamil—*Tennaimaram*, *tenaki*; Kannada—*Tengu*; Malayalam—*Thenna*, *thenga*, *marikelam*.

Fig. 19.1. *Cocos nucifera*;
Coconut (**Nariyal**). A coconut grove.

History. De Candolle argues in favour of an American as well as those of an Asiatic origin for coconut tree (*Cocos nucifera* Linn.) and concludes that it most probably belongs to the Indian Archipelago. According to him its introduction into India, Sri Lanka and China does not date further back than three thousand years, but the transport by sea to the coasts of America and Africa took place perhaps in a more remote epoch.

A tall palm; cultivated chiefly in Kerala, Tamil Nadu and Karnataka. The coconut has been called a *Kalpavriksha* because it provides a large range of useful products, among which ripe coconut, copra, coconut oil and coir are important. The value of this tree has been recognized from very ancient times. Indeed, the region from South Kanara to Cape Comorin is known as '*Kerala*', meaning the land of the coconut. In Malayalam or Sanskrit "*Keralam*" means land of coconut. The Kerala state in

Fig. 19.2. Germinating coconut.
I, II and III stages

Malayalam is known as "*Keralam*". In these parts, the coconut plays a vital part in the life and economy of the people. The coconut is mainly grown for nuts which it yields almost at monthly interval all through its life of about eighty years.

Varieties and cultivation. Of the varieties of coconuts grown in India, the West Coast variety, is the one that is extensively cultivated. It is a long-lived, hardy, multipurpose palm, yielding a large number of medium-sized nuts in almost every leaf axil. This variety has been cultivated from very ancient times and may be considered as indigenous to the county. Among other varieties are the Dwarf, New Guinea, Cochin-China, Java, Siam, Laccadive ordinary and Laccadive small. The New Guinea variety yields about 65 nuts a year, the Cochin-China about 90, the Java about 95, the Laccadive ordinary about 127 and the Laccadive small 160.

The primary requirements for the successful cultivation of coconut are a tropical climate, abundant rainfall and a well-drained soil. Seashores, river banks and hill-slopes afford ideal situations. In India, the principal areas of cultivation are Kerala, Karnataka, Tamil Nadu, Andhra Pradesh, Maharashtra, West Bengal and Orissa; 85 per cent of the total output comes from Kerala State.

Uses. The tender nuts are in demand for their liquid which provides a very refreshing and delicious drink. It is said to have laxative and diuretic properties. The soft kernel inside the tender nut is also scraped and eaten. The mature nuts are used mainly to make copra (dried meat or kernel) and for edible purposes as fresh kernel in food preparations.

The copra is crushed to obtain coconut oil which is in great demand for edible purposes. It is used as an adjunct for milk-fat, as a filler and thickener for milk and as an ingredient of artificial milk for infants. The cake left after the extraction of oil is largely used to feed cattle and poultry. It is occasionally used as food by poor people. The coconut cake helps to increase the milk content when fed to cows and the butter made from such milk has a firmer texture and better flavour than when other cakes are used.

Fig. 19.3. *Cocos nucifera.* VHC-1 coconut hybrid.

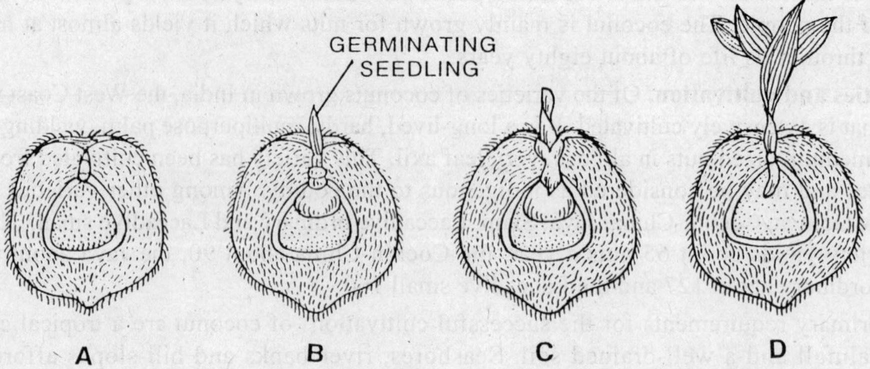

Fig. 19.4. *Cocos nucifera* (coconut). A–D, successive stages (line diagram) showing germination of coconut (seedling stage).

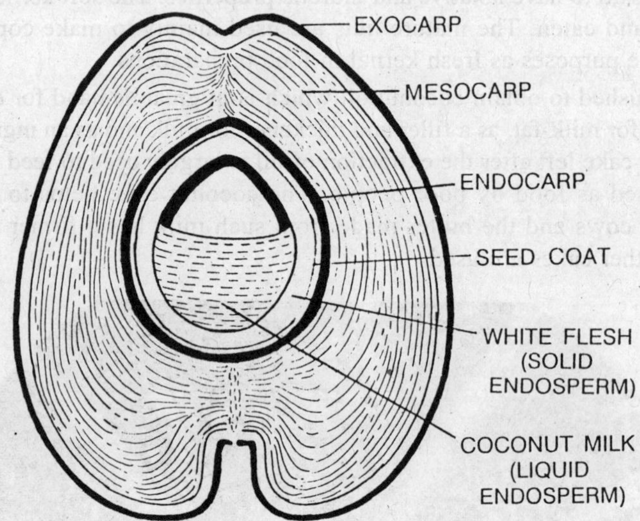

Fig. 19.5. *Cocos nucifera* (coconut). L.S. of fruit with seed.

The coconut inflorescences (spadices) are tapped to extract the juice which is converted into sweet or fermented toddy. Vinegar, jaggery, sugar or arrack (alcohol) are also made from the juice.

Shell of coconut is used for handicrafts; mesocarp (fibre) is used for coir manufacturing. The Kerala State is second to Philippines in the production of coir materials. The dry woven leaves are used for thatching the house and for temporary sheds and pandals in Southern States especially in Kerala.

GROUNDNUT

Arachis hypogaea Linn,: Eng. Peanut, groundnut, Hindi—*Mungphali*; Family—Leguminosae (Papilionaceae).

Bengali—*Chini badam*; Marathi—*Bhui mug*; Telugu—*Verusenagalu*; Tamil—*Verkadalai*; Kannada—*Nela-gadale*; Malayalam—

Nelakadala.

History. The peanut (*Arachis hypogea* Linn.) is a native of South America, now generally cultivated throughout India. This was not known in the Old World before the discovery of America. According to Dymock this plant reached India through China. It does not appear to have been cultivated for more than 150 years. This was brought to western India from Africa.

It is native of Brazil. It is widely grown in South India, Maharashtra and Uttar Pradesh. North Gujarat is famous for peanut cultivation. The plant is a bushy or creeping annual with the peculiar habit of ripening its fruit underground. A sandy soil is best for its cultivation. The soil must be friable so that the ripening fruit can be buried, and it must be well fertilized.

Fig. 19.6. Groundnut. High yielding variety.

Uses. The peanuts (seeds) are used for roasting or salting and for the preparation of peanut butter. Peanuts are a very nutritious food. One lb. of peanuts yields 2700 cal. The filtered refined oil is used for cooking and in making margarine. Peanut oil is an important food oil. The oilcake is used as fodder. The protein in peanuts is used in the manufacture of ardil, a synthetic fibre. The vegetable *ghee* is made from the peanut oil after hydrogenation.

The kernels are also used in various foods and confectionery. They are ground and made into peanut butter. Peanut flour is prepared by grinding the finest grades of peanut cake; it is used for supplementing the white flour. Cake is used as feed for cattle and other farm animals; also used as manure. Cake has high nutritive value. Seedcoats are mixed with groundnut husk and the product is called groundnut bran. Some commercial products are groundnut milk, peanut ice-cream and peanut massage oil for infantile paralysis. Hulls are used as filler for fertilizers, or ground into meal for insulation blocks, floor sweeping compounds, bedding the stables, etc. Peanut oil also finds some use as a lubricant, and blends with mineral oil have been developed.

20
CHAPTER

Wood and Cork

From the time immemorial foo7d, clothing and shelter have been the three great necessities of mankind. The wood is the most familiar and most important forest products. The wood has contributed a lot to the advancement of civilization. Today the wood is the most widely used commodity other than food and clothing. It is one of the most important and versatile of the raw materials of the industry. It has so many advantages over the metals.

It is cheap, light and may easily be worked with tools. It is very strong for its weight and it is embodiment of strength, toughness and elasticity. It is a bad conductor of heat and electricity. Wood is also used in the form of thin sheets or veneers. The wood is definitely superior to metals in several respects.

DIAGNOSTIC FEATURES OF WOODS

On the basis of several important diagnostic features the commercial woods may be identified to some extent. The more important diagnostic features are mentioned here:

Porous and nonporous woods. The presence or absence, and the nature and arrangement of pores, serve as a ready means of classifying woods. The coniferous woods do not possess pores, and are known as *nonporous* woods, whereas the angiospermic woods possess numerous pores and are termed as *porous* woods. On the basis of the distribution of pores, the woods may be of two types—*ring porous* and *diffuse porous* woods. In ring porous woods (*e.g.*, ash, elm, oak, etc.) the pores are found to be arranged in concentric circles, the outer and inner portions of which differ with regard to the number and size of the pores. In diffuse porous woods (*e.g.*, beech maple, walnut, etc.) the pores are small and nearly of the same size and are found to be scattered uniformly throughout the wood.

Early wood and late wood. In temperate regions, every year new wood is formed in a limited growing season, with the result definite growth layers develop, which show two distinct areas within each layer. The wood thus formed in the spring is called the *spring wood* or early wood, and that formed in winter is called the *autumn wood* or late wood. There is a sharp contrast between the late autumn wood and the early spring wood, and this makes the successive rings distinct. The growth ring of a single year is called an annual ring and the number of these annual rings gives an indication of the age of tree. Annual rings of successive years may vary greatly in width. Wide rings are formed under favourable conditions of growth of the tree, and narrow ones are formed when conditions are unfavourable.

Sapwood and heartwood. The outer region of the wood which is of lighter colour is known as the *sapwood*, and this alone is used for conduction of water and salt solutions. The cells of this region are alive and physiologically active. In old trees the central region of the secondary wood is filled up with tannin and other substances which make it hard and durable. This region is known as the *heartwood*. It looks black owing to the presence of tannins, oils, gums, resins, etc., in it. The vessels often become plugged with tyloses. The function of the heartwood is no longer conduction of water, but it simply gives mechanical support to the stem. The heartwood usually takes good polish and is used for cabinet work, furniture and other high grade wood-working industries.

Texture, grain and figure. *Texture* refers to the relative size and quality of the various woods, while *grain* refers to their structural arrangement. *Figure* is applied to the design or pattern which appears on the surface of wood.

Rays. The rays are made of parenchyma cells that are oriented at right angles to the main axis of the stem. They vary greatly in width, height and arrangement.

MECHANICAL PROPERTIES OF WOOD

Wood possesses some important mechanical properties which either alone or in combination, determine its usefulness and suitability for various purposes. These properties may differ in different species. The mechanical properties enable the wood to resist various external forces which tend to change its shape and size and produce deformations. The important mechanical properties of wood are mentioned here.

Strength

The strength is restricted to the ability to resist certain definite forces which may be termed—crushing strength, tensile strength, shearing strength and cross-breaking strength. They are as follows:

Crushing strength. It is the resistance offered to forces that tend to crush wood.

Tensile strength. It is the resistance to forces that tend to pull wood apart.

Shearing strength. It is resistance to those forces which tend to make the fibres slide past one another.

Cross-breaking strength. This is the resistance to forces which cause the beams to break, and all the above-mentioned forces are involved.

Stiffness. It is the measure of the ability of wood to resist forces that tend to change its shape.

Toughness. It is referred to the ability of wood to absorb a large amount of energy, and so resist repeated, sudden sharp blows or shock.

Hardness. It is the measure of the power of wood to resist indentations, abrasion and wear.

Cleavability. It is an expression of the ease with which wood can be split.

STRENGTH AND OTHER PROPERTIES OF WOOD

The strength of wood is the most important property in determining the value of any species for structural purposes. It is a very variable property, and is influenced by the density of the wood, the moisture content, the presence of defects and many other factors. Suitability figures for eight different properties have been calculated by taking into consideration the various strength functions of both green and seasoned timber as shown in the table.

Suitability	*Strength and other properties of wood*
1. **Weight**	Specific gravity based on weight oven dry and the volume of seasoned timber (approximately 12%) moisture content.
2. **Strength as a beam**	Modulus of rupture and fibre stress as the elastic limit in static bending. Fibre stress at the elastic limit in impact bending.
3. **Stiffness as a beam**	Modulus of elasticity in static bending. Modulus of elasticity in impact bending.

Suitability	*Strength and other properties of wood*
4. **Suitability as a post**	Maximum crushing strength and strength at the elastic limit in compression parallel to grain. Modulus of elasticity in static bending.
5. **Shock-resisting ability**	Work to maximum load and total work in static bending.
6. **Retention of shape**	Shrinkage green to oven dry in volume and in the radial and tangential directions.
7. **Shear**	Shearing strength in the radial and tangential directions.
8. **Hardness**	Fibre stress at the elastic limit in compression perpendicular to the grain. Radial, tangential and end hardness.

SEASONING OF WOOD

Wood always contains moisture, the amount of which varies from 40 per cent to 100 per cent or more of the dry weight. The water found within the wood is known as *hygroscopic water*. The amount of hygroscopic water constitutes from 20 to 35 per cent of dry weight. This property of wood is known as *hygroscopicity*. The moisture content of wood has an important bearing on its weight, density, and often on its strength. It is only the loss of hygroscopic water which is responsible for the increase in strength, that accompanies seasoning, as the drying out of wood is called. The loss of hygroscopic water causes the shrinkage of wood. This tendency of wood to shrink as it dries is one of the great drawbacks to its use. As the result of uneven shrinkage, warping and other defects may develop, which tend to counteract any increase in strength. However, dried or seasoned wood is for the most part stronger, harder, stiffer and more durable than unseasoned wood. Artificial methods, of wood seasoning are employed. There are two chief types of artificial wood seasoning—1. *air seasoning* and 2. *kiln drying*.

Air seasoning. Here, the moisture is removed by exposure to air without resorting to artificial heat. It is carried out in open until the wood ceases to lose weight. The ultimate water content varies from 12 to 30 per cent. The main objects of air-seasoning are to reduce weight, the amount of shrinkage and other defects. This is a long-time process.

Kiln drying. Here, heat is applied to wood in an enclosed space. In this process, the moisture is removed more rapidly and more completely. The final moisture content varies from 4 to 12 percent. If green wood is kiln-dried, warping and other defects may be prevented. This process is completed in short duration.

PRINCIPAL WOODS OF INDIA

Babul

Acacia nilotica (Linn.) Del.; Syn. *A. arabica* Willd.; Trade name—Babul; Hindi—*Babul, kikar*; Bengali—*Babul*; Hindi—Punjab and Uttar Pradesh—*Kikar*; Gujarati—*Baval*; Telugu—*Nallatumma*; Tamil—*Karuvelei*; Kannada—*Jaali, gobbli*; Malayalam—*Karivelan*; Family—Mimosaceae.

The sapwood is yellowish-white and usually wide. The heart-wood when freshly cut, is a pinkish or old rose colour, but it darkens on exposure to a dull red or reddish brown. It is often mottled with darker streaks. The wood is dull and without taste or smell, hard, fairly close-textured and with straight or slightly twisted grain.

There are several varieties of *babul*. The two best known are the *telia* and *kauria* varieties, the wood of the former having a better reputation than the latter.

Seasoning. Babul is a wood which can be air-seasoned with fairly good results if care is taken. One year is sufficient to air-season babul boards up to 2 in. in thickness. It can be kiln-seasoned without difficulty.

Strength. Babul is an extremely strong, hard and tough wood. It is nearly twice as hard as teak and has a very good shock resisting ability.

Durability. The sapwood is not durable. The heartwood is durable in most situations but not to the same degree as teak and sal.

Working qualities. An easy wood to convert and to saw when green, but it becomes harder and tougher when seasoned. It works well by hand, and machines and finishes to a good surface. It takes a fair polish but requires careful filling.

Uses. It is popular for parts of carts (body work, spokes, naves, axles, felloes, yokes and shafts), and for agricultural implements such as ploughs, harrows and clod-crushers. It is a useful tool handle wood but not for all types of handle, and it can be used for tent pegs. It can be

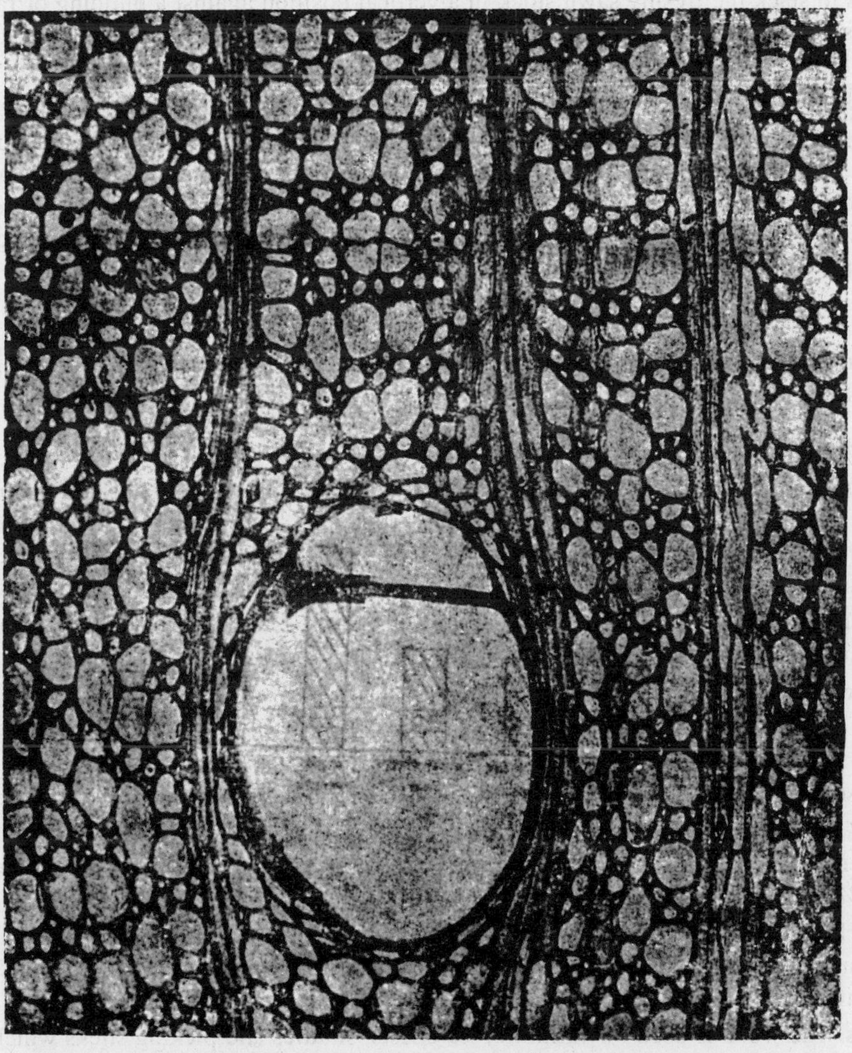

Fig. 20.1. *Bombax ceiba* (**Semul**). Micrograph X110, of wood.

described as one of the best Indian utility woods where hardness and toughness are required. It is a good turnery wood.

Sources of supply. Babul is usually available in small logs only, but in some districts larger logs are available. It is found throughout the drier regions of North, Central and South India.

Simul

Bombax ceiba L.; Syn. *Salmalia insiginis* (Wall.) Schott & Endl.; Hindi—*Semul*; Sanskrit—*Salmali*, *rakta-pushpa*; Bengali—*Simul*, *roktosimul*; Marathi—*Saur*, *simlo*; Gujarati—*Sawar*, *simalo*; Telugu—*Booruga*, *konda-buruga*; Tamil—*Mullialava*; Kannada—*Booruga*; Malayalam—*Pula-maram*; Oriya—*Bour*; Garo—*Panchu*; Mundari—*Edelsong*; Family—Bombacaceae.

A large tree found in the Western Ghats, Assam and the Andaman Islands. The wood is very soft creamy white or pale pink timber, very light in weight and with large open pores.

Seasoning. Quick drying can be achieved by exposing freshly converted timber directly to the drying action of the sun and wind or by kiln-seasoning. If air-seasoning is to be done, the best method is to stack the freshly sawn stock in a vertical position against a building or railing with the planks facing the sun. The boards should be turned from time to time so that both surfaces are exposed, kiln-seasoning is the surest and best method of drying *simul*.

Uses. The wood is used for matches, packing cases, penholders, veneers and ply-wood.

Wood is most widely used in match industry, especially for match boxes. Suitable for shingles, canoes, toys, scabbards, cooperage, bush handles, tea-chest plywood and picture frames. Also used for cushioning mine-props and for inside portions of opium chests. Floss used for stuffing life belts, mattresses, cushions and pillows, upholstry and quilts. Tree yields a gum called *mocharus*.

Albizia

There are several common commercial *Albizia* woods. The best known is perhaps ***Albizia lebbeck*** (Linn.) Benth.; the others *A. odoratissima*, *A. procera*, and *A. stipulata*.

Red Cedar

Toona ciliata Roem.; Eng. Cedrela tree, Red cedar; Hindi—*Toon*; Sanskrit—*Nandivriksha*, *tunna*; Hindi and Bengali—*Tun*, *mahanim*; Marathi—*Kuruk*; Tamil—*Santhanavembu*, *tunumaram*; Telugu—*Nandichettu*; Kannada—*Mandurike*; Malayalam—*Malarveppu*; Assam—*Poma*; Family—Meliaceae.

A tree cultivated in the Punjab, Assam, Bihar, the Western Ghats, Maharashtra, Madhya Pradesh and the Nilgiris.

Description of wood. The wood is a pinkish brick red, when freshly cut, toning to a light brownish red on exposure. The wood is a light weight, and usually straight grained timber. The wood has a distinct cedar like smell.

Seasoning. *Toon* is a fairly easy wood to air-season if care is taken with the stacking. Kiln-seasoning can also be done.

Strength. *Toon* is a timber of only moderate strength qualities. Hill *toon* is strongest and has about 80% the strength of teak.

Working qualities. *Toon* is one of the easiest of Indian woods to saw and work. If properly finished and filled, it takes a very good polish. It makes up into a good type of plywood. It makes high quality congealated wood.

Uses. The wood is used for furniture, tea-chests, shuttles and picking sticks which are used in the textile industry, and cigar boxes.

Walnut

Juglans regia Linn.; Eng. English walnut, Persian walnut; Hindi—*Akhrot*; Family—Juglandaceae. Found in Kashmir, Himachal Pradesh, the Khasia hills and in the hills of Uttar Pradesh.

Indian walnut varies considerably in colour, sometimes being a dull grey, while other wood may be a dark brown with even darker markings. It is relatively light wood for its strength and it works very easily and finishes to a fine surface. Its most important quality, however, is that, when once dried it does not shrink, swell or split. This, combined with its lightness, strength and good working qualities gives it pre-eminence as a wood for rifle parts, gun stocks, high class cabinet making, and delicate carvings.

Seasoning. Walnut wood seasons slowly and springs considerably while drying out, but apart from this, it is a model wood in so far as seasoning is concerned, both in air-drying and kiln-drying. Green conversion, followed by stacking under cover with a good air circulation is the method of seasoning recommended.

Strength. For its weight, walnut is a relatively strong wood. It is about 85% the weight of teak and in shock resistance is about equal to teak.

Working qualities. An extremely easy and pleasant wood to saw and work to a fine finish. It takes an excellent polish and needs very little filling.

Uses. The wood is used for musical instruments and cabinet work. Used extensively in Kashmir and North India for carving. Well known as an excellent veneer and plywood timber. It is also used for gun stocks.

White Willow

Salix alba Linn.; Eng. White willow; Hindi—*Bis*; Kashmir—*Vivir*; Punjab—*Bis*, *madnu*, *bhushan*; Family—Salicaceae.

A small tree, found in the North-Western Himalayas and in the valley of Kashmir. The twigs are used for making baskets. Cricket bats are made of the wood.

Used also for house building, match-boxes and splints, shoes, tool-handles, agricultural implements, boats, boxes, combs and toothpicks; suitable for paper pulp and charcoal making.

Mulberry

***Morus alba*,** Linn; Eng. White mulberry; Hindi—*Tut*; Family—Moraceae. A tree, native of China. In India, it is grown in Punjab, Uttar Pradesh, Kashmir and the North-Western Himalayas. It is also cultivated in Karnataka state for silk worm cultivation.

Description of the wood. The sapwood is white and sharply delimited from the heartwood, which is a bright yellowish brown when freshly cut. No odour or taste. Straight-grained and of rather open medium coarse texture.

Seasoning. The best results have been obtained by storing the logs for some months before final conversion. The wood can be kiln-seasoned without difficulty. In practice, the most of the mulberry used in Punjab is used in a green state by the sports goods manufacturers. The timber is converted green, steam bent, and the dried out in a bent form tightly held in a clamp.

Strength. It is approximately the same in weight as teak, and in shock resistance, shear and hardness it is considerably higher than teak.

Uses. The primary use is for sports goods, for which purpose it is eminently suitable. It is used mainly for the manufacture of hockey sticks. It is also used for tennis rackets, badminton and squash rackets, presses, cricket stumps, etc.

Fig. 20.2. *Cedrus deodara* (**Deodar**). A tree at Forest Division, Chakrata, Uttarakhand

Also suitable for house-building, agricultural implements, furniture, spokes, poles, shafts, bent parts of carriages and carts and for turnery. Bark is used for paper making; also yields a textile fibre.

Deodar

Cedrus deodara Loud.; Eng. Deodar; Hindi—*Deodar*, *Diar*; Family—Pinaceae.

A tall evergreen tree found in the North-Western Himalayas from Kashmir to Garhwal.

Description of the wood. It is of light yellow-brown colour and possesses distinctive odour. It is a medium weight wood which is very sturdy in use and durable. It is usually even-grained and of medium to fine texture, but the presence of large knots is a common feature.

Seasoning. It is an easy wood to air-season. It can also be kiln-seasoned.

Strength. Deodar is the strongest of the Indian conifers. Its weight is 20% less than teak and its strength is also about 20% less.

Working qualities. It is an easy timber to saw and work to a smooth finish.

Uses. The timber is used for construction work and for railway sleepers. It is also suitable for beams, floor boards, ports, window frames, light furniture and shingles.

This is strongest of Indian coniferous woods, also used for doors, furniture, packing cases, masts, spars and also for wooden bridges. Wood yields an oleoresin and a dark coloured oil used for ulcers and skin diseases.

Himalayan Spruce

Picea smithiana (Wall.) Boiss. ; Eng. Himalayan spruce; Hindi—*Bajur*; North West Himalayan region—*Rai*, *rau*, *re*, *kachal*, *salla*, *tos*; Jaunsar, Garhwal and Kumaon—*Roi*, *rha*, *ragha*, *morinda*; Trade—Spruce; Family—Pinaceae.

A tree commonly found in the Western Himalayas. The wood is used for construction work, railway sleepers, cabinet making, packing cases and wood pulp.

One of the most useful timber trees of the Western Himalayas. Wood is used for planking, general fittings and joinery, rough furniture, tea-boxes and packing cases. This is one of the best light boxwood of India. Wood is suitable also for match boxes, battery separators, for fence posts, transmission poles, piles and newsprint manufacture.

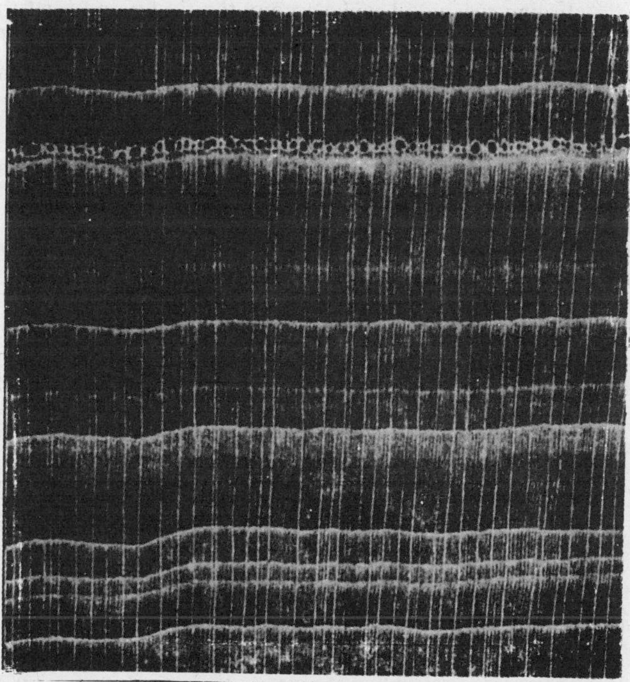

Fig. 20.3. Deodar wood (non-porous), sectional view of micrograph X 10.

Pine Wood

Pinus roxburghii Sar.; Eng. Long-leaved pine; Hindi—*Chir*, *Salla*; Family—Pinaceae.

An evergreen coniferous tree. It is found in the Western and Eastern Himalayas. The wood is used for construction work, railway sleepers, packing cases, furniture and in match industry.

Pinus merkusii Jungh & de Vries. A tree. The wood is very resinous with reddish-brown heartwood and yellow sapwood. It is used for building purposes and general carpentry.

Pinus wallichiana A. B. Jackson; Eng. Indian blue-pine; Hindi—*Kail*, *Chil*. A coniferous tree. The wood is used for pencils, penholders, splints, match boxes, construction work, railway sleepers, and furniture. Also used for shingles, packing cases, constructional work and house fitments. It is a good wood for pattern making, cores for laminboards, drawing boards, and planetables.

Indian Rosewood

Dalbergia latifolia Roxb.; Eng. Indian rosewood; Verna. *Kali Shisham*; Sanskrit—*Shishapa*; Bengali—*Sitsal, swetasal*; Marathi—*Shisham, siswa*; Gujarati—*Shissam, kalaruk*; Telugu—*Cittegi, irgudu*; Tamil—*Itti, karundorviral*; Kannada—*Bite, todagatti*; Malayalam—*Itti, colavitti*; Oriya—*Sisua*; Trade—Indian Rosewood; Family—Papilionaceae.

A large deciduous or nearly evergreen tree with cylindrical, fairly straight bole and full rounded crown, found in the sub-Himalayan tract of Eastern Uttar Pradesh, Bihar, Orissa and central, western and southern India. It attains its maximum development in the southern region of western ghats, where trees 130 ft. high up to 20 ft. in girth and 70 ft. clear bole are sometimes found. The minimum exploitable size is 6 ft. girth.

The sapwood is narrow and pale yellowish white in colour, often with a purple tinge. The heartwood ranges in colour from golden brown through shades of light rose, purple with darker streaks, to deep purple with rather distant, nearly black lines, darkening with age. It is fragrant, heavy, narrowly interlocked grained and medium coarse-textured. The timber is stronger and much harder than teak.

Indian rosewood ranks among the first wood for furniture and cabinet work. It is a valuable decorative wood suitable for carving and ornamental ply-boards and veneers. It is specially useful for pattern-making, calico printing blocks, mathematical instruments and screws. It is also used for gun carriage wheels, ammunition boxes and army wagons, pulleys, handles, shelves, decorative carriage parts, temple chariots, boat knees, agricultural implements, combs, razor handles and brush backs. Carefully selected and manufactured Indian rosewood ply-boards satisfy aircraft specifications.

Sissoo

Dalbergia sissoo Roxb.; Eng. Sissoo; Verna. *Shisham*; Sanskrit—*Shinshapa, aguru*; Hindi—*Shisham, sissu, sissai*; Bengali—*Shisu*; Gujarati—*Sisam, tanach*; Telugu—*Errasissu, sinsupa*; Tamil—*Sisu itti*; Kannada—*Agaru, biridi*; Malayalam—*Iruvil*; Oriya—*Sisu, simsapa*; Punjab—*Tali, shisham*; Marathi—*Sissu*; Trade—Sissoo, Shisham; Family—Papilionaceae.

A deciduous tree, often with crooked trunk and light crown. Normally the tree attains a height of about 100 ft., a girth up to 8 ft. and a clear bole up to 35 ft. It occurs throughout the sub-Himalayan tract from Ravi to Assam, ascending up to 5,000 ft. It is extensively cultivated in Punjab. Uttar Pradesh, Bengal and Assam. No other timber tree, except teak is cultivated to a greater extent.

The sapwood is white to pale brownish white in colour and the heartwood, golden brown to dark brown with deep brown streaks becoming dull on exposure. It is heavy, narrowly interlocked grained and medium textured. It can be air or kiln-seasoned. However, kiln-seasoning enhances the value of timber as the colour darkness and the difference between lighter and darker bands gets intensified. The wood can be converted into ornamental veneers or commercial plywood of good quality.

Sissoo (*Shisham*), is a high class furniture and cabinet wood widely used throughout north India. On account of its great strength, elasticity and durability, it is highly valued as constructional and general utility timber and is used for all the purposes for which Indian rosewood is employed in the south. It is esteemed also for railway sleepers, musical instruments, *Charpai* legs, hammer handles, shoe heels, *hookah* tubes and tobacco pipes. Carefully selected and manufactured plywood logs satisfy the specifications prescribed for aircraft, and for this purpose, the wood from trees, growing on canal banks and in plantations is considered to be the best. Sissoo wood is suitable for making laminated skiis. It can be worked into decorative ornamental carvings.

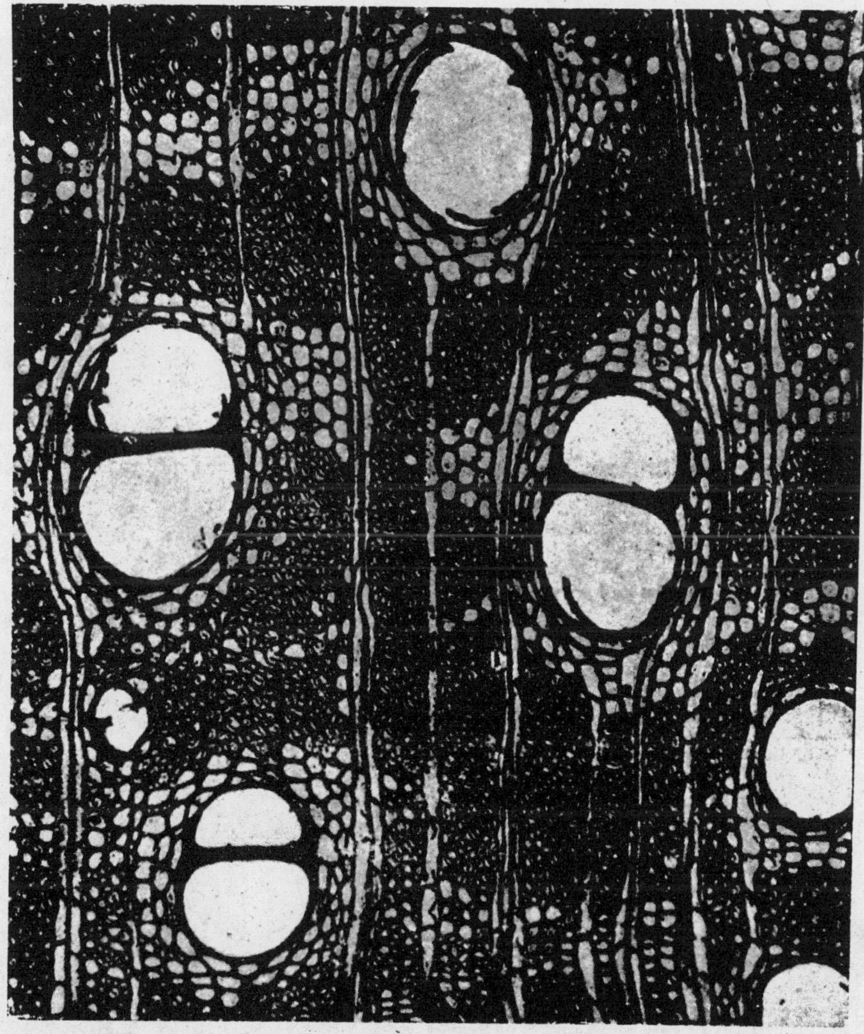

Fig. 20.4. *Dalbergia sissoo* (porous wood) micrograph of wood x 110.

Andaman Redwood

Pterocarpus indicus Willd; Eng. Andaman redwood; Verna. *Padauk*. It is a large and lofty tree found in the Andaman islands. Sapwood is small. Heartwood is dark-red, close-grained, moderately hard with a slight aromatic scent. It is durable and not attacked by white ants. When thoroughly seasoned it is unaffected by alternate dryness and moisture of the atmosphere. It seasons well, works well and takes a very fine polish. It is used for furniture, carts, guncarriages, and other purposes, and is said to be the most useful wood in the Andamans, and where it grows to an enormous size.

Indian Kino Tree

Pterocarpus marsupium Roxb. Sanskrit—*Pitasara*; Hindi—*Bijasal*; Bengali—*Pitshal*; Marathi—*Asan*, *bibla*; Gujarati—*Biyo*; Telugu—*Yegi*, *peddagi*; Tamil—*Vengai*; Kannada—*Honne*, *bange*; Malayalam—*Venga*; Oriya—*Byasa*; Trade—Bijasal; Family—*Leguminosae*.

Fig. 20.5. *Pterocarpus marsupium.* Indian Kino Tree. The trees in Karnataka State.

The wood is used chiefly for building purposes, such as doors, window frames, rafters, beams and posts. It is also used in railway carriages, wagons, carts, boats, ships, electric poles, pit-props in mines, agricultural implements, drums, tool handles, camp furniture, mathematical instruments, picture frames, combs, and parts of textile looms. It is a very important timber in peninsular India.

Red Sandalwood

Pterocarpus santalinus Linn.; Eng. Red sandalwood; Verna. *Rukhto-chandan*; Sanskrit—*Raktachandana*; Hindi and Bengali—*Raktachandan, lalchandan*; Marathi—*Tambada chandana*; Gujarati—*Ratanjali*; Telugu—*Agarugandhamu, raktagandhamu*; Tamil—*Atti, sivappu chandanam*; Kannada—*Agaru, honne, kempugandha*; Malayalam—*Patrangam*; Oriya—*Raktachandan*; Trade—Red Sanders.

A small tree of South India, chiefly found in Cuddapah, North Arcot, and the southern portion of the Karnul district. It favours a dry, rather rocky soil, and a hot fairly dry climate. Sapwood is white. Heartwood is purplish-black dark orange-red when fresh cut, extremely hard, the shavings giving a blood red-orange colour. It is used for building and for turning, and is said to be much prized because it is not subject to the attacks of white ants. It is much used for carvings in Andhra Pradesh especially at Tirupati.

Wood is highly prized for house posts; also used for agricultural implements, poles, shafts and bent rims of carts, picture frames, boxes and other joinery work. In Japan, it is used for a musical instrument called **Shamisen**. Wood is ground and used for dyeing wool, cotton and leather, and staining other woods; *santalin* is the colouring principle.

Ebony

Diospyros ebenum Koenig.; Eng. Ebony; Verna. *Abnus*; Hindi—*Ebans, abnus*; Telugu—*Nallavalludu*; Tamil—*Tumbi, karunkali*; Kannada—*Karemara*; Malayalam—*Karu, vauari*; Oriya—*Kendhu*; Trade—Ebony.

It is a moderate to large sized evergreen tree with a dense crown, sometimes attaining a height up to 80 ft. and a girth of 7 ft. The girth of heartwood is not more than 4 ft. It is found in the dry evergreen forests of Deccan, Karnataka, Coimbatore, Malabar, Cochin and Travancore. It prefers well-drained rocky soil and grows well on sandy loam. It is found sparsely scattered in the forests in mixture with other trees.

This is perhaps the best ebony yielding tree and the only one which yields jet black heartwood without streaks or markings, the true ebony of commerce, *D. ebenum* is not very commonly found in India, nor do the trees, where they occur, attain any large size. Only limited supply of ebony is available in the area of its occurrence.

The sapwood is light yellowish grey, often streaked with black. The heartwood is jet black, rarely with a few dark or light brown or golden streaks. It has a metallic lustre when smoothed. It is heavy, straight-grained or somewhat irregular or wavy-grained, fine and even-textured.

Ebony is difficult to season. It is best to convert the green logs to the smallest size permissible and to store them under cover, well protected from hot winds.

The heartwood is resistant to attack by insects and fungi. It is very durable. The value of this wood depends on its intense blackness, working quality and the high finish it takes. It is difficult to work, especially when dry, but finishes to a glossy surface. It requires little hand finishing and takes a high lasting polish. It is one of the most valuable woods in the fancy wood market. It is used for ornamental carving and turnery, and for special purposes in decoration. It is used for veneers, inlaying, musical instruments, sports goods, mathematical instruments, piano keys and caskets.

Diospyros virginianum Linn. is widely used for weaving shuttles and is the standard wood for the heads of golf clubs.

Saffron Teak

Adina cordifolia Hook. f.; Eng. Saffron teak; Verna, *Haldu*, Hindi—*Haldu*; Bengali—*Petpuria, dakom*; Marathi—*Heddi*; Telugu—*Pasupukadamba*; Tamil and Malayalam—*Manjakadamba*; Kannada—*Arsintega, yettega*; Trade—Haldu; Family—Rubiaceae;

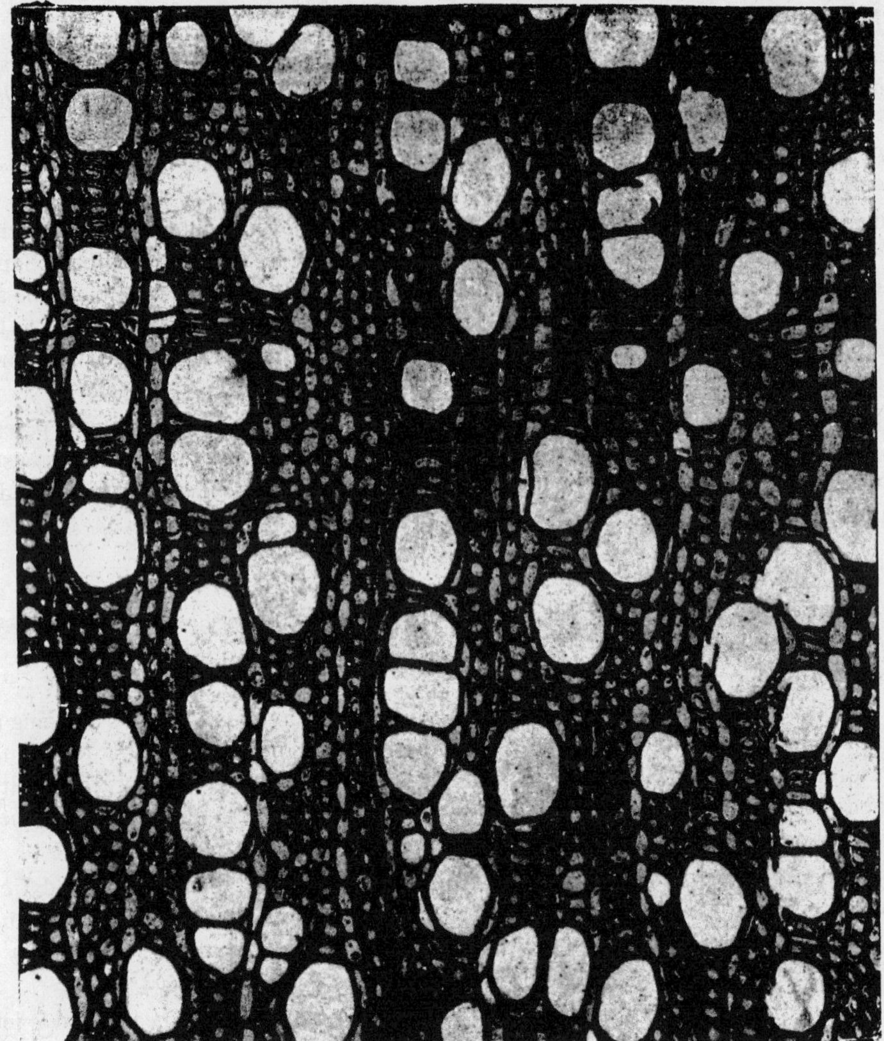

Fig. 20.6. *Adina cordifolia* (**Haldu**). Micrograph x 110, of porous wood.

This makes an important Indian timber, used for a great variety of purposes; Pearson and Brown (1932) state that it is one of the best turnery wood in India and that selected pieces are attractively figured. The wood is commonly used for carvings, construction work, flooring and for railway carriages.

Timber is easy to saw, seasons well and takes good polish; durable under cover, resistant against subterranean termites, but highly susceptible to drywood termites. Haldu is accepted as grade I commercial and moisture-proof plywood timber.

Wrapping, writing and printing paper is manufactured from the wood pulp.

Teak

Tectona grandis Linn.; Eng. Teak; Hindi—*Sagaun*; Sanskrit—*Saka*; Bengali—*Segun*; Marathi—*Sag, sagwan*; Gujarati—*Saga*; Telugu—*Adaviteeku, teeku*; Tamil—*Tekkumaram, tekku*; Kannada—*Jadi, sagwani*; Malayalam—*Thekku, tekha*; Oriya—*Singuru*; Assam—*Chingjagu*; Lepcha—*Ripnyok*; Family—Verbenaceae.

A large, deciduous tree, indigenous to both peninsulas of India. In Western India it does not extend far beyond the Mhye. In Central India it attains its northern most point in the Jhansi district, and from that point the line of its northern limit continues in a south-east direction to the Mahanadi river in Orissa. It is found wild in Assam. It is, however, cultivated throughout Bengal, Assam and Sikkim, and in North-West India without difficulty as far as Saharanpur. The forests richest in large timber on the west side of the Peninsula are the Travancore, Anamally, Wynad, South-west Karnataka and North Kanara forests. In the centre of the Peninsula the Godavary forests are most compact and valuable.

Structure and utility of wood. Sapwood white and small; the heartwood when cut green, has a pleasant and strong aromatic fragrance and a beautiful dark golden yellow colour, which on seasoning soon darkens into brown, mottled with darker streaks. The timber retains its fragrance to a great age, the characteristic odour being apparent whenever a fresh cut is made. It is moderately hard, exceedingly durable and strong, does not split, crack, warp, shrink or alter its shape when once seasoned; it works easily, takes a good polish. Teak owes its chief value to its great durability, which is ascribed, probably with justice, to the circumstance that it contains a large quantity of fluid resinous matter which fills up the pores and resists the action of water. (At the Karli caves near Poona the teak-wood-work, two thousand years old, seems perfectly good at the present day).

The many uses of teak are well known. In India it is highly prized for construction, ship building, and for making sleepers and furniture.

Wood is very durable and resistant to fungi. It is used for poles, beams, trusses, columns, roofs, doors, window frames, flooring, planking, panelling, stair cases, and other constructional work. It is one of the best timbers for furniture and cabinet making, wagons and railway carriages. Due to its better shape-retention ability, teak is popular in marine constructions and is a class by itself for boat and ship–building, particularly for decking. On account of its resistance to chemicals, teak articles are used in chemical industries and for making laboratory bench-tops; suitable for casks and vats for shipping corrosive liquids and for storing vegetable oils, fruit syrups, chutneys, etc. Teak is employed for sound-boards of musical instruments, keys, etc., and for different grades of plywood. Wood waste in the form of wood-shavings and sawdust is used for chip-boards, fibre-boards and plastic-boards.

Sal

Shorea robusta; Eng. Sal tree; Hindi—*Sal*; Bengali and Hindi—*Sal*, *sakhu*; Marathi and Gujarati—*Ral*; Telugu—*Gugal*; Tamil—*Kungiliyam*; Kannada—*Kabba*; Malayalam—*Maramaram*; Oriya—*Sal, sagua*; Punjab and Haryana—*Sal*; Lepcha—*Taksal-Kung*; Assam—*Sal*, *bolsal*; Trade—Sal; Family—Dipterocarpaceae.

A large gregarious tree, often covering certain interrupted tracts without the existence of connecting patches. It occurs along the base of Tropical Himalayas from the Sutlej to Assam, in the eastern districts of Central India, and on the Western Bengal Hills.

Structure and utility of wood. The *sal*, one of the most valuable timber trees in India, has a distinct sapwood which is small in amount, whitish and not durable. The heartwood is brown in colour, finely streaked with darker lines, coarse grained, hard, strong and tough, with a remarkably fibrous and cross-grained, structure. The fibres of successive concentric strata do not run parallel, but at oblique angles, to each other, so that when the wood is dressed the fibres appear interlaced. It does not season well, but warps and splits in drying, and even when thoroughly seasoned, absorbs moisture with avidity in wet weather. During the process of seasoning, it dries when first cut, and evaporation goes on afterwards with extreme slowness. *Sal*, when once thoroughly seasoned, stands almost without a rival, as a timber, for strength, elasticity, and durability, which

Fig. 20.7. Scientifically managed Sal Forest, at Dehradun.

qualities it retains without being sensibly affected, for an immense length of time. Average weight of the seasoned *sal* about 35 lb. per cubic foot (Brandis; Gamble).

The timber is the one most extensively used in Northern India. It is in constant request for piles, beams, planking and railing of bridges, doors, and window posts of houses, for the bodies of carts, and above all, for railway sleepers. In Assam it is favourite wood for boat-building, and in the hills of Northern Bengal where it is found, perhaps, of the largest size available, the trunks are hollowed out into canoes. Owing to the fact that when unseasoned it is not floatable, difficulty is experienced in most sal forests in getting the timber out of the forests in log. This is, however, overcome by floating the logs either with the assistance of the boats or with floats of light wood or bamboos (Gamble).

Sal wood ranks with teak and deodar as one of the best sleeper woods in India; also used in form of bellies and poles. After treatment, the poles are suitable for overhead electric, telegraph and telephone lines. As domestic timber it is used for beams, scantlings, rafters and floors; also used for piles, mine work and pit props, bridges, dug-out boats, carriages and wagon building, spokes, fellows and hubs of wheels, agricultural implements, tool-handles, tent pegs, liquid storage vats, and beer and oil casks.

WOODS FOR VARIOUS USES

Agricultural implements. Agricultural implements is a term of rather wide application but it refers in the main to such appliances as ploughs, harrows, rollers and clod crushers. A strong, hard tough timber is required for this type of work. *Acacia nilotica* (*Babul*), *Anogeissus latifolia* (*axle wood*), *Syzygium cumini* (*Jaman*), *Grewia tiliifolia* (*Dhaman*), *Shorea robusta* (*Sal*), *Xylia xylocarpa* (*Irul*), etc., are used for agricultural implements of different kinds.

Axe helves and tool handles. Woods for helves and tool handles must be strong and tough and must also possess great shock resisting abilities. *Dalbergia* spp. (*Shisham*), *Grewia* spp. (*Dhaman*), *Diospyros* spp. (ebony), etc., are recommended for this purpose.

Boat and ship-building. Timber used in all small boats and large ships is subjected to very great strains, and is often employed under circumstances which tax its durability to the utmost. For this reason any timber used for ship and boat-building should be strong, elastic, durable and free from defects. *Tectona grandis* (Teak—*Sagaun*) is the best ship-building timber in the world, due to its relatively small co-efficient of expansion and contraction and to its durability. The other recommended timbers for this purpose are—*Acacia nilotica* (*Babul*), *Dalbergia* spp., *Dipterocarpus* spp., *Shorea* spp., *Xylia* spp., *Grewia* spp., *Morus* spp., *Terminalia* spp., *Bombax* spp., etc.

Boot lasts and shoe heels. There is a considerable demand in India for boot and shoe lasts, and recently demand for shoe heels, and especially ladies shoe heels, has grown enormously. For boot lasts and shoe heels, a tough wood which is not too hard is required. In addition, the wood must be able to stand repeated nailing. The important woods used for the purpose are—*Dalbergia sissoo* (*Shisham*), *Gardenia* spp., *Mitragyna parviflora* (*Kaim*), *Mangifera indica* (*Am*), etc.

Cart and carriage building. The various parts of a cart or carriage are subjected to different kinds of stresses and strains, and require different qualities of wood for real efficiency. The important woods recommended for various parts of the carts are—*Dalbergia* spp., *Shorea* spp., *Dipterocarpus* spp., *Syzygium* spp., *Acacia* spp., *Grewia* spp., *Anogeissus* spp., *Terminalia* spp., and several others.

Construction and general joinery work. Constructional woods are those timbers used for superstructures, which include all parts of houses and buildings, bridges and similar structures not actually in contact with water or the earth. For this purpose a timber should be strong and durable. Lightness of wood is sometimes an asset if strength and durability are not sacrificed. Floor and wall planking should be non-warping and non–shrinking and the wood for interior work and panels should be ornamental. There are three very important woods which stand out above all others as building timbers. These are—*Tectona grandis* (Teak—*Sagaun*), *Shorea robusta* (*Sal*) and *Cedrus deodara* (*Deodar*). The other important timbers used for this purpose are—*Acacia* spp., *Toona ciliata* (*toon*), *Dalbergia* spp., *Mangifera indica* (*Am*), *Pinus* spp., *Terminalia* spp., *Xylia* spp.

Cooperage. Cooperage or barrel-making consists of two types, "*tight cooperage*", *i.e.*, barrels used for liquids, and "loose or slack cooperage" used for dry goods like cement. The oaks, dhaman are used for tight cooperage for beer and liquor casks. *Bombax ceiba* (*Simul*), mango spruce, etc., are used for slack cooperage.

Electric poles. The qualities necessary to make a good pole are that it should be straight, *i.e.*, without crooks and bends, that it should not split or crack excessively, and that it should have the required strength for the work it has to do. The woods used in India for use as electrical transmission poles are—*Pinus roxburghii* (*Chir*), *Shorea robusta* (*Sal*), *Tectona grandis* (*Teak*), *Cedrus deodara* (*Deodar*), *Terminalia tomentosa* (Laurel), etc.

Engraving and printing blocks. *Dalbergia sissoo* (*Shisham*) in North India, and *Dalbergia latifolia* (Rosewood) in South India, are the most popular woods for calico-printing blocks. Sissoo (*Shisham*) is considered to be excellent for the purpose. *Toona ciliata* (toon) and *Tectona grandis* (teak) are also sometimes used. *Tamarindus* (*Imli*) is also said to be good for printing by the Indian method.

Furniture, cabinet making and panelling. For high class furniture, cabinet-making and decorative panel work, there are several very ornamental and excellent woods in India. The chief characteristics required for these uses are non-liability to crack and split, retention of shape, ease of working and good colour figure and grain. The most commonly used and recommended timbers for the purpose are *Albizia* spp., *Toona ciliata* (toona), *Dalbergia* spp., *Juglans regia* (walnut), *Tectona grandis* (teak), *Terminalia tomentosa* (Laurel), etc.

Artocarpus spp., is wildly grown and commonly used for making furniture in South India especially in Kerala it is a valuable wood. It is also preferred in temple works in Kerala.

Match splints and boxes. A good match wood for splint manufacture must be soft, straight-grained, white and cheap. India possesses many woods which are suitable for box making and splints. The woods are—*Bombax ceiba* (*simul*), *Anthocephalus indicus* (kadam), *Salix* spp., etc.

Mathematical instruments. The better class of mathematical instruments such as set squares, rulers, etc., are usually made of *Buxus wallichiana* (box wood), *Juglans regia* (walnut) and *Aesculus indica* (horse chestnut), but cheaper instruments intended for school use are often made of *Adina cordifolia* (*Haldu*), *Toona ciliata* (Toon), *Gardenia* spp., etc. For set square and ruler edgings, good quality hardwoods such as rosewood sissoo and ebony are most frequently used.

Musical instruments. In India, many woods are used for making musical instruments as tanpura, sitars, violins, etc., *teak*, *toon*, *sissoo*, *mulberry*, *haldu* being among the commonest. For *sitars*, teak is used for the long neck and *deodar* or *shisham* for keys. Toon is almost universally used for the bodies. For banjos, teak is most commonly used. For drums, *mulberry*, *sissoo*, *siris* etc, are used. *Canarium euphyllum* (*Dhup*) is used for guitars.

Packing cases and boxes. Wood for these purposes must be light, easily worked and cheap. They must have good nail-holding power and should preferably be of whitish colour. Tea is now generally packed in plywood chests, which are strong, light and cheap. *Hollock* (*Terminalia myriocarpa*) and *hollong* are used for making good plywood tea chests. The commonest Indian woods for packing cases and boxes are—*Dalbergia* spp., toon, teak, mango, siris, kadam, *Terminalia chebula*, *Pinus* spp., etc., Cigar boxes are used made in South India of *Toona ciliata*, *Melia azedarach*, etc. *Terminalia chebula* is a popular wood in South India for coffee boxes, while poplar (*Populus* spp.) is much used in Kashmir for fruit crates.

Pencils and pen-holders. Amongst the timbers used in India for pencil making may be mentioned Cypress, blue pine or *kail*, *simul*, *toon*, *Salix tetrasperma*, *Melia* spp., etc. Pen-holders may be made of *Haldu*, *Gardenia* spp., *Kaim*, etc., Spruce and fir are used in North India for cheap pen-holders.

Picture framing. No very special qualities are required for picture framing, so long as the wood used is well seasoned and not liable to warp. There are two distinct types of picture framing, one in which the wood is used in its natural state, either polished or waxed, and the other in which the framing is painted or covered with a plaster composition which is moulded to the form required and covered with gilt or other colouring media. For the former type of framing and steady ornamental wood is suitable, the most commonly used being teak, sissoo, rosewood, ebony and *haldu*. For cheaper frames, in plain moulded form or painted, light conifer woods such as fir and spruce are the best. School slate frames can be made from a variety of woods, *e.g. Trewia nudiflora* (*gutel*), *Terameles nudiflora* (maina), *Terminalia chebula*, *Mangifera indica* (*Am*), spruce and fir, etc.

Railway carriage and wagon building. The qualities required of a wood to be in railway carriage and wagon work are that it should be sufficiently strong and durable. It should be free from bad seasoning defects, and it should be available in sufficient quantities. Teak complies with this specification better than any other timber, and for this reason teak is the main timber used by all the railway wagon and carriage works in India. The other important timbers used for this purpose are—*Shorea robusta*, *Pterocarpus* spp., *Adina cordifolia*, *Cedrus deodara*, *Dalbergia latifolia*, *Toona ciliata*, *Acacia nilotica*, *Dipterocarpus* spp., *Pinus* spp., and several others.

Railway sleepers. The woods most commonly used are—sal, deodar, teak, *pyinkado*, and *bijasal*.

Rifle parts. Walnut (*Juglans regia*) is the chief timber used for the manufacture of rifle work all over the world, as it stands up exceptionally well when worked to a fine finish on high speed cutting and drilling machines. The wood is also very steady and is not prone to excessive shrinking, swelling, warping or splitting, once it is properly seasoned.

Shuttles. The subject of shuttles is a very important one in India. All cotton, jute, wool and other textile mills use wooden shuttles in large quantities. The important wood which may be used for making shuttles are—*Diospyros melanoxylon* (ebony), *Gardenia latifola*, *Acacia nilotica*, *Mitragyna parviflora*, *Dalbergia latifolia*, *Dalbergia sissoo*, etc.

Sporting requisites. Billiard cue shafts in India are made of *Diospyros melanoxylon* (*tendu*), *Grewia tilaefolia* (*dhaman*), *Polyalthia fragrans*. For the butts of cues ebony, *Hardwickia pinnata* and *Dysoxylum glandulosum* are used.

More than 200 manufacturers in India are involved in the production and export of world-class sports goods. These goods are of such superior quality that over 70 countries import from India. Professional players use them in national, international and even Olympic tournaments. There are several good reasons for the tremendous popularity of Indian sports goods abroad: excellent quality, superb finish, carefully selected raw material, enhanced durability. And to top it all master craftsmanship from India, the land of traditional skills.

Cricket bats are made of *Salix* spp. (williow). *Populus* spp. has also been used for cheap types of bat.

Indian cricket bats have scored many a century for the world champions. Made from strong Kashmir willows (*Salix* spp.), the bats are designed for force and flexibility.

Golf clubs are made of *Terminalia* spp., *Grewia* spp., *Anogeissus* spp., and *Pyinma*.

Hockey sticks are made of *Morus alba* (mulberry). Hockey sticks from India are known the world over for their balance and strong drive.

The timbers employed for the skis are *Dalbergia sissoo* (*shisham*) and *Anogeissus latifolia* (axle wood).

Stumps and bales are made of *Morus* spp. (mulberry) and *Grewia tiliaefolia* (dhaman).

In India, tennis and badminton rackets are made of mahogany and maple wood. The racket frames are made of *Prunus padus*, *Dalbergia sissoo*, *Toona ciliata*, *Melia azedarach*, etc.

Turnery, carving, combs, toys, etc. Very close-grained woods are acquired for high class turnery and carving. Walnut is much in demand in North India and Kashmir for this purpose. The sandalwood carvings of Mysore are equally well known. Sissoo and ebony are also used high class carving and inlay work in North India.

Dalbergia latifolia is commonly used in Kerala for carvings of idols, animals, birds, etc.

Wood carving is an ancient craft in which the West Bengal excels. The items include are dolls and toys, owl, horse, etc. Recently many piece goods including stands for table lamps are being manufactured. Shisham wood is used in abundance.

Erythrina spp., *Gyrocarpus* spp., *Bauhinia malabarica* are used for making toys in Karnataka. *Hymenodictyon excelsum* (*Kuthan*) is the best toy wood in U.P.

Indian combs such as those used by Sikhs in the hair are made of *Buxus* spp. (boxwood), *Adina cordifolia* (*Haldu*), ebony and sandal-wood.

VENEERS

The veneers are thin slices or sheets of wood of uniform thickness. The common thickness is about 1/20 in. to 3/8 in. In making veneers or logs pieces of woods are peeled and boiled and then cut with sharp knife. Usually the veneers are cut by **rotary process** which involves turning a log on a lathe against a stationary knife, with the formation of a continuous sheet of veneer. The other process is known as **slicing process**. Here the logs are quartersawed and therefore show a more attractive figure. The logs are first quartered and thereafter sliced with a stationary knife, resulting in the formation of separate sheets. Freshly cut veneers are usually wet and must be thoroughly dried.

Uses. The veneers are used in the manufacture of boxes, baskets, door panels, cooperage, trunks, mirrors, musical instruments and several other articles.

PLYWOOD

To overcome some of the natural defects in wood is the manufacture of so-called "*built up*" stock or plywood. The plywood is made by gluing together from three to nine thin veneers. The grain of each successive layer is at an angle to the next, so that the strength is redistributed. With the result the finished product becomes very strong and stable and does not warp or twist like ordinary wood. The plywood does not split when the nails and screws are driven close to its edge.

The manufacture of a five-ply plywood panel consists of preparing the face, back, cross band, and core sheets; applying the adhesive; pressing the glued stock into a panel; and drying and finishing the product. The soft woods are generally used for the purpose, as they can be glued more easily. Several resins and gums are used in the manufacture of plywood.

Uses. The plywood is extensively used in the home for doors, walls, flooring, cabinets, shelves, partitions, furniture and interior trim. A large quantity of plywood is utilized in making concrete forms, boats, prefabricated houses, airplanes, railroad cars and the body making of station wagons.

LAMINBOARDS

Laminboards are usually made up with cheap inferior woods for the cores, and better quality decorative good for the faces. Such timber as pine, spruce, fir, *Bombax ceiba*, *Ailanthus grandis*, *Tetrameles nudiflora* and *Kydia calycina* are very suitable for core work, while better class timbers such as teak, rosewood, sissoo and toon make excellent face veneers. Face veneers are usually sliced.

Uses. The laminboards are commonly used for making doors and cabinets.

CORK

Most commercial cork is obtained from the cork oak (*Quercus suber*). Cork or "corkwood" consists of the outer bark of the tree. This can be removed without injury to the tree. At the age of about 20 years, when the tree is about 40 cm in circumference, this outer layer, known as virgin cork, is removed by stripping. The operation consists of making vertical and horizontal cuts with knives or saws, and then prying off large pieces of the bark. The underlying exposed phelloderm and cortical cells die, and a new cork cambium layer is formed several millimetres deeper in the cortex. By this cork cambium (*phellogen*) new cork is formed more rapidly than by the first layer, and after 9 to 10 years the new cork layer has attained sufficient thickness to be commercially valuable and is in turn removed. This cork is of better quality than the nearly useless virgin cork, but is not so good as that obtained at the third and subsequent stripings which take place at intervals of about 9 or 10 years until the tree is 150 years old. Usually the trees live for 100 to 500 years, and have an average yield of 40 to 500 lb. per tree. After stripping the pieces of cork are dried for several days. They are first boiled in large copper vats. This process removes the sap and tannic acid, increases the volume and elasticity and flattens pieces. The rough edges are then trimmed off and the flat pieces are sorted and baled.

Cork is exceedingly buoyant due to its composition of dead watertight cells. It is highly compressible. It is durable, a bad conductor of heat and electricity, and is resistant to passage of moisture and liquids. It absorbs sound and vibrations and also possesses some frictional characteristics.

Uses of Cork

It is used for many purposes and for the manufacture of several products. In some cases the natural cork is used whereas in other cases composition cork, made of finely ground pieces treated with adhesives and molded.

The natural cork is used to make stoppers, hats, helmets, cigarette tips, handles of golf clubs, penholders, fishing rods, floats, life preservers, life jackets, base balls, mats, tiles and several other such products. Cork board is used as an insulating material for cold storage plants, houses and refrigerators.

Composition cork is utilized for the lining of crown caps, the metal tops for sealing bottles; gaskets; toes, counters and innersoles for shoes.

LINOLEUM

A type of floor covering is made from cork or wood flour, linseed oil, resins, such as rosin or kauri gum, pigments and burlap. The oil is boiled and allowed to solidify by dripping on pieces of cloth. The solidified oil is ground up and melted with the resins. This mixture is cooled and hardened, and after 'curing' for several days it is mixed with the cork, which has been ground as fine dust, and with the dry colour pigments. It is then pressed into burlap cloth with hydraulic presses. It is then seasoned in ovens and finished by giving it a protective surface of nitrocellulose lacquer. The tiles, known as linotiles are made from ground cork and linseed oil, but thicker than linoleum.

FUELWOOD

Fuel is one of the great necessities of modern life and it is indispensable, both in the home and in industry, as a source of heat and power. Any material that burns readily in air can be utilised, and a great variety of plant products are so used. The most important of these are wood and coal.

Though much less wood is used for fuel at the present time than formerly, the total consumption is probably greater than for any other purpose. However, the extensive use of wood for fuel does not represent a proportional drain on the forest resources, for much waste material is used as well as wood unfit for other purposes.

Wood is 99 per cent combustible when perfectly dry, and thus makes an excellent fuel. It is also a flaming fuel and well adapted for heating large surfaces. Hard woods have the greatest fuel value. The woods of *Fagus*, *Betula*, *Pinus*, *Acacia*, *Mangifera*, *Quercus*, etc., are of much value for fuel purposes.

Coal comprises the fossilized remains of plants that lived in former geological periods. The original plant tissue has been more fully decomposed and converted into carbon. Coal possesses a great heating power. Anthracite or hard coals are the oldest and contain about 95 per cent carbon.

It is generally supposed that petroleum had an organic origin and was formed under pressure from the plant and animal life of former shallow seas.

Fuelwood in India

According to an ADE publication of 1984, 'A Perspective on Demand for Energy in India up to 2004–05', the population of India in 2004–05 would be 1046 million. The estimated energy requirement for this population would be 312 million tonnes of firewood, 17 million tonnes of kerosene, 94 million tonnes of agro-waste, 207 million tonnes of cattle dung, 176 million cu. metres of biogas and 100 twh of electricity. The Working Group on Energy Policy (1979) estimated the consumption of fuelwood, agriculture waste and animal dung to be 133.1, 41.0 and 73.0 million tonnes respectively, whereas the total production of fuelwood was estimated at 49 million tonnes by the Fuelwood Committee, Planning Commission (1982). Therefore, research in increasing the production of fuelwood assumes great significance. A number of steps towards the objectives, *e.g.*, high density planting, introduction of fast growing species, genetic improvement for higher biomas, production and afforestation of stress sites, etc., have been advanced by the Council.

21
CHAPTER

Bamboos in India

INTRODUCTION

Bamboos are integral not only to the culture of India but that of the entire Southeast Asia. The wide range of their uses qualifies this group of species to be the one foremost associated with the mainstay of rural life and culture in our country. Its versatility has led to the coinage of such terms as "bamboo culture", "green gold", "poor man's timber", "friend of the people" and "cradle to coffin timber". In fact, bamboo can be used as a substitute of timber in many respects.

Bamboos belong to the grass family *Poaceae* (Gramineae). In India, there are 125 indigenous as well as exotic species of bamboos belonging to 23 genera. Bamboo forests occupy an area of 10.03 million ha, roughly 12.8% of the total forest area in the country. They are found in almost all the states of the country, from the tropical to the temperate regions and the alluvial plains to the high mountains, the only exception where they do not occur naturally is Kashmir. The principal bamboo genera occurring in India are *Arundinaria*, *Bambusa*, *Chimonobambusa*, *Dendrocalamus*, *Dinochloa*, *Gigantochloa*, *Indocalamus*, *Ochlandra*, *Drepanostachyum*, *Phyllostachys*, *Pleioblastus*, *Pseudoxytenanthera*, *Schizostachyum*, and *Thamnocalamus*. More than 50 per cent of the bamboo species occur in Eastern India – Arunachal Pradesh, Assam, Manipur, Meghalaya, Mizoram, Nagaland, Sikkim, Tripura and West Bengal. Other areas rich in bamboos are the Andamans, Bastar region of Madhya Pradesh and the Western Ghats.

Bamboos in India.

HABITAT AND DISTRIBUTION

Bamboos grow from sea level, in the tropics, to 4000 metres above sea level, in the temperate region. Although, bamboos are more common to the tropics, they also occur naturally in subtropical and temperate zones of the world, except in Europe.

The geographical distribution of bamboos is governed largely by the conditions of rainfall, temperature, altitude and soil. Most of the bamboos require a temperature from 8 degree celcius to 36 degree celcius, a minimum annual rainfall of 1000 mm and high atmospheric humidity for good growth.

They form an important constituent of many deciduous and evergreen forests. In nature bamboos seldom occur as pure crops, but are generally found as an understorey in tropical evergreen, tropical moist deciduous, tropical dry deciduous, montane sub-tropical, montane wet temperate and Himalayan wet temperate forests. Generally, the genera *Bambusa* and *Dendrocalamus* are found under tropical conditions, whereas *Arundinaria* and its allies occur in the temperate region and are by far, most common on high elevations in the western and eastern Himalayas. *Dendrocalamus strictus* is predominant in dry deciduous forests, while *Bambusa bambos* flourishes best in moist deciduous forests. The most important bamboo of the semi-evergreen forests of the Andamans in *Gigantochloa rostrata*. In the eastern region, *i.e.*, West Bengal, Assam and north-eastern Himalayas, the commercially important bamboos are *Bambusa tulda*, *Dendrocalamus hamiltonii* and *Melocanna baccifera*.

Bamboos are usually found in moist valleys, sheltered depressions, along steams and the lower hill slopes, but occasionally also occur on higher slopes and hill tops. They usually occur mixed with or under tree species, except in case when they form secondary brakes, and then, they are more or less pure. Though many species form brakes, those of *Arundinaria*, *Ochlandra*, *Melocanna* and *Bambusa bambos* are typically more extensive. Where shifting cultivation is practised, the bamboo brakes are more common.

The distribution and site factors of some commercially important bamboos occurring in India are:

Table 21.1

Sl. No.	*Species*	*Nature of soil*	*Topography status*	*Moisture*	*Distribution*
1	*2*	*3*	*4*	*5*	*6*
1.	***Drepanostachyum falcata***	Rich sandy loam to clayey loam often mixed with stones.	Hill slopes and high terraces.	Moist situation.	Western Himalayas, at an elevation of 1300–2200 m.
2.	***Arundinaria racemosa***	–do–	–do–	–do–	Eastern Himalayas, at an elevation of 2200–3050 m.
3.	***Indocalamus wightianus***	Fertile loam	High hills	–do–	Evergreen Sholas of southern India, especially Nilgiris.
4.	***Bambusa bambos***	Flat alluvial and rich moist soil; prefers moderately fine clayey soils, and tolerates slight water-logging; avoids coarse-grained soils and open ridges.	Fluorishes best along river banks, in valleys, and ravines.	Moist localities.	All over the country except in the dry regions and desert areas; mostly in the Indian peninsula; cultivated in sub-himalayan tract and Indo-gangetic plains; scarce in Assam and Bengal.
5.	***Bambusa tulda***	Alluvial soils along water courses.	Flat land.	Prefers moist areas.	Hill tracts of Assam, West Bengal, central India and the Deccan

Sl. No.	Species	Nature of soil	Topography status	Moisture	Distribution
1	*2*	*3*	*4*	*5*	*6*
					plateau; in moist mixed deciduous forests.
6.	***Schizostachyum pergracile***	Well-drained fresh loam.	Low hill country.	Between dry and moist sites.	Assam, central India and the Deccan plateau; in dry and moist deciduous forests.
7.	***Dendrocalamus brandisii***	Black loam and loam with lateritic nodules and other good soils.	Along the rivers.	Moist localities.	Keral, Karnataka and Tamil Nadu; in moist west coast climate.
8.	***Dendrocalamus hamiltonii***	Rich loamy soil	Valley and along streams	Moist places	Central and Eastern Himalayas; upto an elevation of 1000 m; in moist deciduous and semi-evergreen forests.
9.	***Dendrocalamus longispathus***	Fertile loam	Most along streams	Moist	West Bengal, Assam, Kerala and Tamil Nadu; as an associate of moist mixed forests.
10.	***Dendrocalamus strictus***	Poor, coarse grained, well-drained dry soils and stony soils on hill slopes; avoids poorly drained and moist heavy clay soils.	Hill slopes, alluvial plains and ravines.	Dry areas	Most widely distributed all over the country except northern parts of Bengal, Assam and other very moist areas; occurs in dry, open type deciduous forests; common associate of dry sal and dry teak forests. Also occurs on hills, and ascends to 1000 m.
11.	***Melocanna baccifera***	Fertile loam	Hill slopes and abandoned clearings by shifting cultivation.	Moist areas.	Moist and wet hill tracts of Bengal, Assam and other parts of the country.
12.	***Ochlandra travancorica***	Rich loam soil	Banks of perennial or semi-perennial steams, lower slopes and flat lands.	Moist areas; prefers running water	Western ghats; as under-growth in evergreen and semi-evergreen forests; ascending upto 1600 m.

Sl. No.	*Species*	*Nature of soil*	*Topography status*	*Moisture*	*Distribution*
1	*2*	*3*	*4*	*5*	*6*
13.	***Psudoxytenanthera ritcheyi***	Shallow rocky soils.	Hill tops and ridges.	Dry areas	Western ghats and hills of Indian peninsula; as undergrowth in mixed deciduous forests.
14.	***Gigantochloa rostrata***	Fertile loam	Along streams and lower slips.	Moist areas	Orissa, Assam, central India and Deccan plateau; gregarious in semi-deciduous forests of Andaman and Nicobar islands.

UTILIZATION

Properties. Bamboo is considered to be the most efficient natural resource as far as strength vis-a-vis cost. Strength of the bamboo depends upon the species, climatic factors and its moisture content. Bamboo is strong, though not stronger than timbers like Sal and Teak. Seasoning is a essential pre-requisite for maintaining the appearance and strength of bamboo products and structures. Unseasoned bamboo is highly susceptible to insects and fungal attack, decay and mechanical degradation.

Seasoning of bamboos has to be done properly and carefully as rapid drying results in splitting. To prevent cracks during seasoning green bamboos may be split into halves. The bamboos are best dried by air seasoning under overhead cover for 2–3 months. Kiln drying under controlled conditions can be done successfully in about 2–3 weeks.

Bamboo can be cut and split easily with hand saw. Strips of any measurement can be made with knife. Immature bamboos are soft and pliable and can be moulded to desired shape. Bamboo takes polish and paint well.

Uses. The strength of the culms, their straightness and lightness, combined with hardness, range in size, hollowness, long fibre and easy working qualities, make the bamboos suitable for a variety of purposes. Whether it is the tender shoots, used for pickles and curries, the slivers, used for tooth picks and meat barbecues, ribs for fans or slats for sun screens, or the dried culm as the sturdy lathi and for construction of the so called bamboo houses, the versatility of bamboo is legendary. The thousand and one uses for bamboos have long been known and established. Some of the traditional uses are: Agricultural implements, anchors, arrows, back scratchers, baskets, beds, blinds, boats, bottles, bows, bridges, brooms, brushes, buildings, caps, cart-yokes, caulking material, chairs, chicks, chopsticks, coffins, combs, containers, cooking utensils, cordages, dust-pans, fans, fences, fish-traps, fishing-nets, fishing rods, flag-poles, floats for timber,

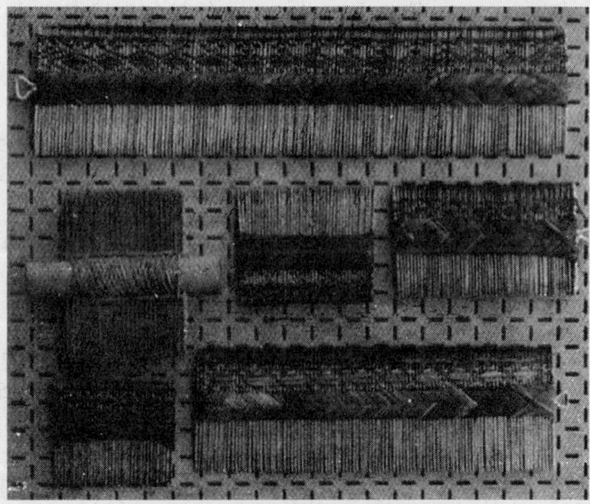

Weaving combs of bamboo.

flutes, flower-pots, food, food-baskets, fuel, furniture, hats, handicrafts, haystack stabilisers, hedges, hookah-pipes, joss-sticks, kites, ladders, ladles, lamps, lance staves, lanterns, lining of hats and sandals, masts, match-sticks, mats, musical instruments, nails, net floats, ornaments, props, paper, pens, polo mallets, rafts, rayon pulp, roofing, sails, scaffoldings, scoops, seed-drills, shoes, shuttles, slats, sports goods, sprayers, stools, sticks, tables, thatching, tobacco pipes, toys, tool handles, traps, trays, tubes, umbrella-handles, walking-sticks, walls, vessels and wrappers.

Weaving Fish trap and containers of bamboo

Structural utilisation of bamboos

Bamboo has been traditionally in use in building construction from times immemorial. It is used in foundation, frames, floors, walls, partitions, ceilings, door and windows, roofs and for reinforcement in cement concrete. Although, there are more than 125 species of bamboo reported to be occurring in India, 20 important species have been tested so far for their physical and mechanical properties. The mean values and ranges of specific gravity, ultimate bending and compressive strength (MOR and MCS) of bamboo in green condition are given below.

	Mean	***Ranges***
Specific gravity	0.639	0.515–0.817
Ultimate bending strength (MOR) (kg/sq cm)	559	172–986
Compressive strength (MCS) (kg/sq cm)	401	262–539

The classification of timber species for structural use in building is done on the basis of modulus of rupture (MOR) and modulus of elasticity (MOE) while for other structural uses like piles and mine use, timbers have been classified on the basis of modulus of rupture (MOR), modulus of elasticity (MOE) and maximum crushing stress (MCS). Similarly, bamboos have been classified on the basis of MOR, MOE and MCS in green condition in three groups as given below. The limits of these properties for the three groups have been fixed keeping in view the corresponding limits for wood and spread of ultimate bending and compressive strength of bamboos.

	MOR (kg/sq cm)	***MOE (1000 kg/sq cm)***	***MCS (kg/sq cm)***
Group I	> 700	> 90	> 350
Group II	500–700	> 60	> 300
Group III	300–500	> 30	> 250

Suitable species for constructional use are given below alongwith some traditional timber species falling in these groups.

Table 21.2

Group I	*Group II*	*Group III*
Bambusa multiplex	*Bambusa balcooa*	*Bambusa ventricosa*
(Syn. *B. nana*)	*Bambusa pallida*	*Bambusa vulgaris*
Dendrocalamus strictus	*Bambusa nutans*	*Dendrocalamus longispathus*
Oxytenanthera abyssinica	*Bambusa tulda*	*Bambusa bambos*
	Bambusa auriculata	(Syn. *B. arundinacea*)
	Bambusa burmanica	
	Schizostachyum pergracile	
	Melocanna baccifera	
	Thrysostachys oliveri	
	Timber species	
Shorea robusta	*Tectona grandis*	*Pinus roxburghii*
Albizia odoratissima	*Lagerstroemia lanceolata*	*Pinus wallichiana*
Mesua ferrea	*Terminelia paniculata*	
Hopea sp.		

For facilitating proper designing in bamboo safe working stresses of these species have also been worked out.

Uses deserving specific mention:

1. **Bamboo parquet:** (Block flooring). The term parquet is applied to flooring in which strips, 3.81 cm × 6.71 cm, are cut and laid out in geometric patterns.

2. **Laminated bamboo:** The bamboo culms are cracked, spread out and flattened into sheets, with suitable binding and filling material. Then the sheets are combined, lapped, arranged, glued, treated and pressed to the desired form. They are then cut and trimmed to the desired size and shape and finally given finishing touches, according to one's fancy.

3. **Bamboo-reinforced concrete:** Bamboo reinforcement was first used in China in 1919 in concrete piles in railways. It is also used for construction purposes.

4. **Artificially-shaped bamboo:** Bamboo can be made to grow into square, rectangular and triangular shapes for low cost housing, handicrafts, decoration items and cottage industries. Japan has successfully grown artificially-shaped bamboos. Production is simple, like moulding hollow blocks, but needs practice.

5. **Medicine:** Banslochan or Tabashir is a siliceous secretion found in the internodes of the culms of various species of bamboos. It occurs in fragments or masses, up to 2 cm thick. It is used as a cooling tonic and aphrodisiac; also in the treatment of asthma, cough and other debilitating diseases.

6. **Charcoal:** It is superior to many other charcoals for use in electric batteries; in India, used for pharmaceutical purposes. Jewellers prefer bamboo charcoal.

7. **Deodorizer:** The dried, mature bamboo leaves are used for deodorizing fish-oil in Asian countries.

8. **Soil conservation:** Many bamboos are quite useful for soil conservation due to their intricate rhizome system and roots.

SOME BAMBOOS AND THEIR USES

1. ***Bambusa polymorpha*** (Assam—*Betua*; Beng—*Jamabetua*)

This handsome species is distributed in Northeastern India. The clean culms attain a height of 20 to 27 m and a diameter of 15 cm. It is one of the best for construction, thatching and roofing; also used for paper, agarbati sticks, fibre boards, etc. tender shoots are of best quality for edible purposes.

2. ***Bambusa striata*** = *B. vulgaris* var. *striata* (Bengali—*Basini Bans*; Marathi—*Kalaka*; Tamil—*Ponmungil*)

This is a moderate sized bamboo 6 m to 10 m in height and 5–8 cms in diameter, with yellow or green striped culms. Commonly cultivated all over India upto an altitude of 1000 m. It forms loose clumps. It is widely used for furniture, toys, cages, etc., also used for construction. Commonly used as an ornamental.

3. ***Bambusa tulda*** (Bengali—*Tulda*; Hindi—*Peka*; Santal—Mak)

This bamboo is a native of eastern India and grows gregariously between 400 and 600 m above m.s.l. It is also cultivated in the plains and foothills of north India. It is one of the most useful bamboos. Due to its strong culms; it is used for building purposes, scaffolding and roofing; also used for general purposes, specially for mats and baskets, often for paper pulp.

4. ***Dendrocalamus giganteus***

This is the largest of the bamboos, cultivated often in northeastern parts and occasionally in other parts of our country. Indigenous to Myanmar (Burma). It is 24–35 m high and 20 30 cm in diameter. It is used for building purposes, water buckets, boxes, masts of boats, flower vases, etc. Tender shoot is of good quality for edible purposes. It is better than *D. strictus* for paper making.

5. ***Dendrocalamus hamiltonii*** (Assam—*Kokua*; Bengali—*Pecha*; Hindi—*Kaghzi bans*)

This species is found mainly in northwest and northeastern Himalayas mostly between an altitude of 600–1200 m. It is a large bamboo; with culms often over hanging, and up to 25 m long and 10–18 cm diameter, thin-walled. The most important use is for paper manufacture; also used for construction, baskets, mats, water and milk vessels, rafts, etc.

6. ***Dendrocalamus strictus*** (Hindi—*Narbans*; Marathi—*Nanvel*; Oriya—*Salia bhanso*; Tamil and Malayalam—*Kalmungil*; Telugu—*Sadanapa veduru*)

Distributed in deciduous forests almost all over India especially in drier zones usually upto an altitude 1000 m; it is commonly cultivated throughout India in the plains and foothills. The culms are 8–16 m high and 2.5–8 cm in diameter, almost solid or internodes are thick-walled. It is the most universally used Indian bamboo for house construction, baskets, mats, furniture, agriculture implements, tool handles, chicks, etc. The important raw material for paper and rayon in India.

7. ***Drepanostachyum falcatum*** = *Arundinaria falcata* (Hindi—*Ringal*)

This is a low level bamboo of the western Himalaya, ranging from Jammu to Nepal, at elevations between 1200–2200 m. It grows gregariously, forming an undergrowth in the forests of deodar, oak, etc., usually in moist shady locations. Its rhizome spreads widely and give out loosely a large number of culms, sometimes more than a hundred. The culms are 1.8 to 2.0 m high. It is used for basket work, hookah pipes, etc. It is especially prized for fishing rods.

8. ***Gigantochloa rostrata*** = *Oxytenanthera nigrociliata* (Bengali—*Kalia*; Oriya—*Bolangi*; Garo—Washut; Bastar—Pani bans)

This is a tufted bamboo, with culms 5–8 m high and 2.5–5 cm in diameter. It is distributed mainly in eastern India and to a limited extent in Madhya Pradesh, Maharashtra and Karnatak. It is usually gregarious. It is used for building huts, basket work and paper manufacture.

9. ***Indocalamus wightianus*** = *Arundinaria wightiana* (Tamil—*Chevani*)

This bamboo occurs in the Nilgiris, Palghat and Tinnevelly, in South India, at an elevation of 1800–2600 m. It forms a gregarious undergrowth in the evergreen sholas. Culms are 2 to 3 m high and about 1.2 cm diameter. It flowers annually. The split culms are used for making mats, baskets and fences.

10. ***Melocanna baccifera*** (Bengali—*Muli*; Assam—*Tarai*)

This species is the most common bamboo of northeastern India in the plains and lower hills. The culms do not form clumps, are straight thin-walled and attain a height of 10–20 m and a diameter of 4–7 cm. It is strong and extensively used for construction, baskets, mats, chicks, food grain containers, umbrella sticks, hats, toys, etc., an important material for superior paper pulp.

11. ***Ochlandra scriptoria*** (Malayalam—*Ammei*, *Ottal*)

A shrubby bamboo, with culms upto 5 m in height and 2.5 m in diameter. It occurs gregariously mostly along river banks and wetter parts of Western Ghats at low elevations. It is cultivated along the margins of ponds and paddy fields as a soil binder; also used for making mats, baskets and paper.

12. ***Ochlandra travancorica*** (Malayalam—*Eetta*; Tamil—*Eeral*, *Odai*)

This shrubby bamboo occurs mostly in Kerala, and in Tamil Nadu in hills of Kannyakumari and Tinnevelly districts upto an altitude of 1500 m. The culms grow up to 6 m in length and 5 cm in diameter, with rough thin-walled internodes upto 1.5 m in length. This is one of the main raw materials for paper pulp; also commonly used for mats, baskets and for making temporary huts. One of the best soil binders along water courses and other moist slopy areas.

13. ***Pseudoxytenanthera ritcheyi*** = *Oxytenanthera ritcheyi* (Kannada—*Chiwa*, *Choua*; Marathi—*Huda*, *Manga*)

This is an erect, thick-walled bamboo, occurring from the Konkan to the Anamalai hills on the western ghats. It is often cultivated. The culms are upto 6 m in height and 2.5 cm in diameter. They are strong and are used for tent poles, walking stricks, baskets and umbrella handles.

14. ***Schizostachyum polymorphum*** = *Pseudostachyum polymorphum* (Assam—*Bajal*, *Tolli*)

This is a thin-walled, shrubby bamboo, found along river banks and valleys in Northeastern states. It is in considerable demand among tea planters for baskets; also used for mats.

15. ***Thamnocalamus spathiflorus*** = *Arundinaria spathiflora* (Hindi—*Nigal*, *Ringal*)

This is a small bamboo, 3.5–6.0 m high, growing gregariously as an undergrowth in the fir, spruce and deodar forests of the western Himalayas from 2100–2700 m. The culms are used for hookah pipes, mats, baskets, fishing rod, etc.

22

CHAPTER

Spices and Condiments

The spices and condiments have played a prominent part in all the civilizations of antiquity, in ancient India and China. The majority of spices originated in the Asiatic tropics and were among the first objects of commerce between the East and the West. The quest for condiments and spices was once a powerful force in world history. The search for a sea route to the spice Islands of the East led to the voyages of Columbus, Magellan and many others.

Arabs brought the products of southern India and the spice Islands by caravan to Arabia, and then to Europe. Later other countries took over the spice trade. In the sixteenth century the Portuguese monopolized the trade for two hundred years. Later the British Empire owned most of the spice trade of the world.

The spices cannot be grouped as foods, for they contain less nutritive value. However, they give a good flavour and aroma to food, and add greatly to the pleasure of eating. They stimulate the appetite and increase the secretion and flow of gastric juices. For this reason they are commonly known as "*food adjuncts*". The aromatic value of the spices is due to the presence of the essential oils.

In general, all aromatic vegetable products that are used for flavouring foods and drinks are known as spices. Sometimes the term "*spice*" is restricted to hard or hardened parts of plants, which are generally used in a pulverized state. *Condiments* are spices or other flavouring substances which possess a sharp taste, and are commonly added to food after it has been cooked.

The majority of spices are obtained from the tropics, chiefly from Asia. In India, the major spices are pepper, chillies, cardamom, ginger and turmeric. Besides several others are also quite common, such as: coriander, cinnamon, clove, nutmeg, etc. The pepper plant (*Piper nigrum* L.) is indigenous to the forests of Kerala and has been cultivated in India from time immemorial. Large-scale cultivation of the crop is confined to the Western Ghat of Peninsular India. Chillies (*Capsicum* spp.) are an important condiment crop in India. They are indigenous to South America, and are said to have been introduced into India by the Portuguese. The principal cardamom (*Elettaria cardamomum*) producing countries are India, Ceylon and Indo-China. The big cardamom (*Amomum* spp.) are grown in Sikkim, Nepal, Cambodia, Java and some other tropical countries. Ginger of commerce is a dry product prepared from the rhizomes of the plant *Zinjiber officinale*. The most important producers of commercial dry ginger are India, Jamaica and Sierra Leone. Turmeric (*Curcuma longa* Linn.), is used as an important condiment almost in every Indian house, and has been cultivated in India from very ancient times. Coriander (*Coriandrum sativum* Linn.) is an important condiment used widely in all parts of India, Russia, Central Europe, Asia Minor and Morocco. The cinnamon of commerce is the bark of tree *Cinnamomum zeylanicum* Breyn., which is indigenous to Sri Lanka, Southern India, Myanmar and the Malayan Peninsula. The clove of commerce is the dried unopened flower bud of the clove tree, *Syzyzium aromaticum*. A native of Moluccas, it has spread to a number of countries, *viz.*, Zanzibar, the Spice Islands, the West Indies, the Mascarene Islands, Sumatra, Java, Sri Lanka etc. The cultivation of clove in India is limited to South India. The nutmeg of commerce is the de-arilled dried seed of *Myristica fragrans*

Houtt. A native of eastern island of the Moluccas, it is also indigenous to Banda island, Amboyna, Gilolo and Western New Guinea. It was introduced in India towards the end of the eighteenth century. Its cultivation is confined to the south, in the Nilgiris, the Tenkasi Hills and the coastal areas of Kerala. The important minor spices grown in India are ajowan, aniseed, caraway, celery, coriander, cumin, dill seed, fennel, fenugreek, garlic, onion, saffron and vanilla, etc. Several other spice yielding plants of India and abroad have been discussed in the present text.

India is one of the largest suppliers of spices and condiments. India alone contributes 25-30 per cent of the total world trade.

Most of the important spices are grown in India. The classification of *spices* and *condiments* in the present text is based on the morphology of the plant organ which is used for the purpose. The general categories of the plants yielding *spices* and *condiments* are as follows:

SPICES AND CONDIMENTS OBTAINED FROM ROOTS

Asafetida

Ferula Linn.; English—Asafetida; Hindi—*Hing*; Kashmiri—*Anjudan*; Gujarati—*Hing*; Mumbai—*Hing*; Tamil—*Kyam*, *perungayam*; Sanskrit—*Hingu*; Family—Umbelliferae (Apiaceae).

This is a genus of perennial herbs commonly distributed from the Mediterranean region to central Asia. Some of the species are important as sources of oleogum - resins, used as condiment and in medicine. Three species are found in India. The Main bulk of asafetida and other ferula gum - resins are being imported in India from Afghanistan.

Asafetida is the dried latex obtained mainly from living rootstocks or taproots of several species of *Ferula*, such as : *F. foetida* Regel, *F. alliacea* Boiss., *F. rubricaulis* Boiss., *F. assafoetida* Linn. and *F. narthex* Boiss., found in Central Asia. *F. narthex* and *F. assafoetida* are found in Himachal Pradesh and Kashmir.

The plants from which the asafetida of commerce is obtained bear massive carrot-shaped roots (5-6 inches in diameter at the crown when they are 4-5 year old). In March-April when the plants are about to flower, the upper part of the root is laid bare and the stem cut off close to the crown. The exposed surface is covered by a dome-shaped structure made of twigs and earth. A milky juice (latex) exudes from the cut surface. After some days the exudate is scraped off and a fresh slice of the root cut when more latex exudes. The collection of the resin and the slicing of the root are repeated until exudation stops (about three months after the first cut). Sometimes the resin is being collected from successive incisions made at the junction of the stem and the tap root. With three incisions, some plants yield 1 kg. or more of gum-resin.

Asafetida (asafoetida) occurs in commerce in three forms—(*i*) *tears*, (*ii*) *mass* and (*iii*) *paste*. The *tears* make the purest form of the resin. They are rounded or flattened, 5-30 mm. in diameter and greyish or dull yellow in colour. *Mass asafetida* is the common commercial form. It consists of tears united into a more or less uniform mass usually mixed with fragments of root, earth, etc. The paste also contains foreign matter.

Asafetida is adulterated with gum arabic, other gum-resins, gypsum, chalk, wheat or barley flour, etc.

Asafetida contains 40-64% resin, 25% gum, 10-17% volatile oil and 1.5 to 10% ash. The resin portion chiefly contains asaresinotannol, free or combined with ferulic acid.

Uses. Asafetida is bitter and acrid in taste and emits a strong and peculiar odour. In India, it is used for flavouring curries, sauces and pickles. It is also used in Ayurvedic medicines. It stimulates the intestinal and respiratory tracts and the nervous system. It is useful in asthma, whooping cough and chronic bronchitis. It is used as an enema for intestinal flatulence. It is also used in epileptic and hysterical affections and in cholera. It is also used in veterinary medicine

Lovage

Levisticum officinale Koth., English—Lovage; Family—Umbelliferae (Apiaceae).

This is a perennial plant, introduced from Europe. It is grown as a garden plant. It thrives best in well-drained soil. It is propagated either by seeds or roots. The roots are dug out in October of second or third year after setting the plants. The freshly dug roots are washed, cut into slices and dried.

Uses. Generally the roots are used for flavouring foods. Sometimes the seeds are used for flavouring confectionery. The leaves and stems are also eaten as salad. The dried roots; on steam distillation yield volatile oil known as 'lovage oil'. The oil is used for flavouring foods.

SPICES AND CONDIMENTS OBTAINED FROM UNDERGROUND STEMS

Onion

Allium cepa Linn.; English—Onion; Hindi—*Piyaz*; Tamil—*Vella-vengayam, irulli*; Telugu—*Vulli-gaddalu*; Kannada—*Vengayam, nirulli*; Malayalam—*Bawang*. They are cultivated all over India. Onion seed will not keep for certain more than one year. The selected bulbs be planted, and

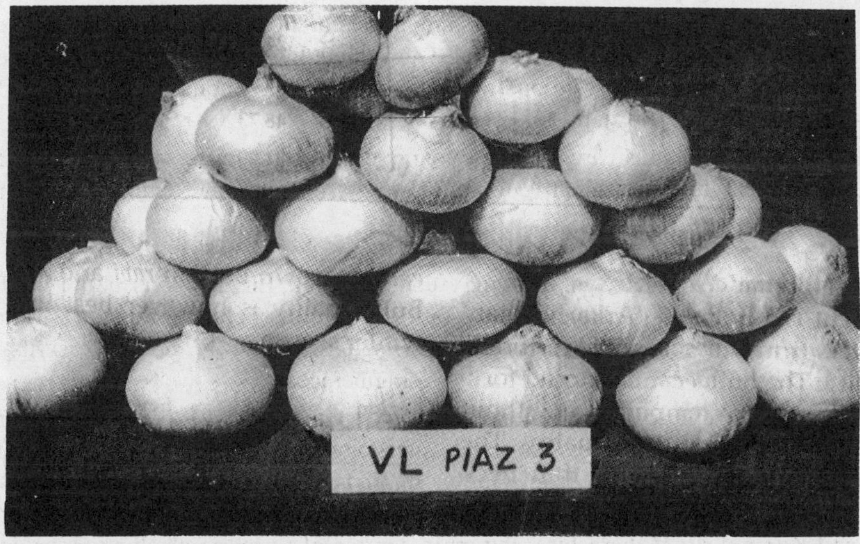

Fig. 22.1. Onion. A hybrid VL Piaz 3. It is less-prone to bolting during storage.

seed obtained from these. If planted in the cold season, they will produce seed in the beginning of the hot season, and if carefully preserved, after being well ripened and dry, the seed obtained in this way will be found to yield a good crop in the following cold season, from October to February.

Uses. The bulbs are used as food and condiment.

Garlic

Allium sativum Linn.; English—Garlic; Hindi—*Lasan, lahsun*; Bengali—*Lashan*; Sanskrit—*Maha-ushadha*; Tamil—*Vallaipundu*; Telugu—*Vellulli fella-gadda*; Kannada—*Belluli*; Family—Liliaceae. Cultivated all over the country. It is propagated by planting out the cloves

Fig. 22.2. Garlic bulbs.

singly, in October, in drills about 6-7 inches apart and 2-3 inches deep. The crop is taken up in the hot weather, and after being dried in the sun the bulbs are stored for further use.

Numerous bulbs remain enclosed in a common membranous covering. Stem simple, about 2 feet high; scape is smooth and shining, solid terminated by a membranous pointed spathe enclosing a mass of flowers and solid bulbils prolonged into leafy points; leaves long, flat, acute, sheathing the lower half of the stem; flowers small, white.

Uses. The bulbs (rhizomes) are used as a condiment and flavouring substance. Garlic powder is used as a condiment. It is a carminative, diuretic, stomachic, alterative, emmenagogue, and tonic.

Turmeric

Curcuma domestica Valet; Syn. *C. longa* Auct. non Linn.; English—Turmeric; Hindi—*Haldi*; Bengali—*Halud*; Punjabi—*Haldar*, *halja*; Tamil—*Manjal*; Telugu—*Pasupu*; Malayalam—*Mannal*, *marinalu*; Kannada—*Arishina*; Marathi—*Halede*; Gujarati—*Halada*; Sanskrit—*Haridra*, *nisa*; Persian—*Zard-chobah*, *dar-zard*; Family—Zingiberaceae.

Turmeric has been cultivated in India from very ancient times. It prefers sandy and clayey loams for its cultivation. The crop cannot stand water-logging or alkalinity in the soils. The largest supplies of turmeric are obtained from Guntur district of Andhra Pradesh. Orissa is the next important growing area for turmeric where production is concentrated in the districts of Ganjam, Phulbani and Koraput. In Maharashtra, the main centres of turmeric production are in Gujarat, Thane and Khandesh districts. Tiruchirapally, Salem and Coimbatore districts of Tamil Nadu are also important turmeric growing areas. Other important states for this crop are - Uttar Pradesh, Madhya Pradesh, Karnataka, West Bengal, Rajasthan and the Punjab. The total average under turmeric in India has been estimated variously from 60,000 to 100,000 acres, and the production is nearly 100,000 tonnes of rhizomes per annum. A large part is consumed within the country, but a portion is exported to the U.K., Pakistan, Sri Lanka and U.S.A.

Cultivation. The crop is propagated vegetatively by means of corms (swollen under ground stems). The preparatory cultivation requires deep ploughing several times (six time or so). It needs a fine tilth and heavy application of manure. The field is prepared into narrow beds with facilities for irrigation and drainage. The corms are planted from April to July in these beds, 6-9 inches apart in the furrows and 16 inches apart between the furrows. After planting the beds are levelled and covered with a mulch of leaves. The corms develop aerial shoots above the ground in a month or so. One or two weedings are given, and the beds are earthed up. By November, the leafy growth is complete and the corms beneath the soil begin to thicken and to develop uniformly deep colour. In the month of February the leaves turn yellow and dry up, which is an indication of maturity of the underground corms. Harvesting is done in March which continues till the end of April.

A great care has to be taken in harvesting operations. The whole clump consisting of both primary and secondary branches known as "*fingers*", and the main thickened portion, the "*bulb*", are to be lifted up without injuring the corms. The leaves and roots are then cut off, and the bulb and fingers separated from each other. Very small quantities are sold in this raw form. In India, the bulb of the crop is marketed as dry cured turmeric (*haldi*).

Curing. The raw produce of turmeric has to be cured properly before the commercial dry turmeric can be obtained. Curing consists mainly of three phases - 1. boiling, 2. drying and 3. polishing. After harvesting, the raw green turmeric is heaped up covered with turmeric leaves, and kept in this condition for some time. The entire produce is then transferred to an earthen pot or larger iron pan which is filled up with water, the water level is kept 2-3 inches above the level of the turmeric. After covering the bulk with dry turmeric leaves, it is boiled over fire. As soon as the rhizomes become soft to touch, they are removed from the vessel, thinly spread out and dried in the sun. After 5-7 days of drying, the produce becomes fit for sale and storage. When it is quite dry, it is cleaned to remove roots and other parts, and then rubbed well between hands. The dry turmeric is also polished by special appliances. Now the produce is sorted out into *bulbs*, *fingers* and *splits*, and graded into large and small sizes according to need.

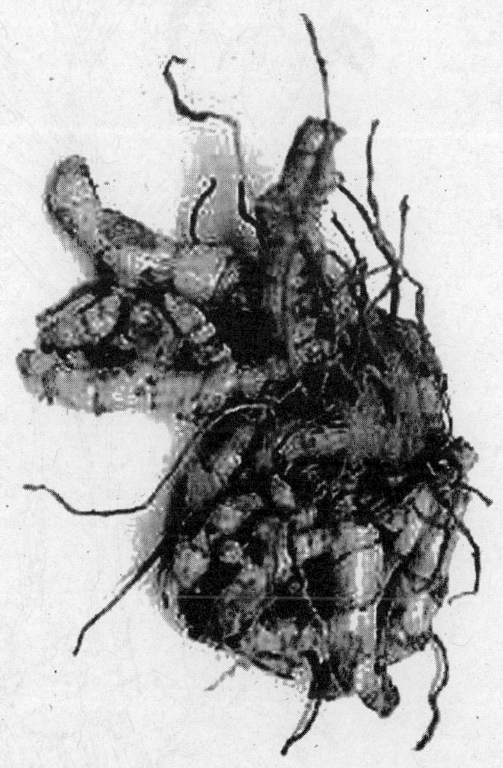

Fig. 22.3. Turmeric. Roots and rhizomes.

The plant of turmeric is a robust perennial with a short stem and tufted leaves. The pale-yellow flowers are found in dense spikes, topped by a tuft of pinkish bracts. The rhizomes, which yield the colourful condiment, are short and thick with blunt tubers. They are cleaned, washed and dried in the sun, it is very aromatic, with a musky odour and yellow colour. It has a pungent bitter taste.

Uses. It is used as a condiment. It is used to flavour and colour pickles, and foodstuffs. It is one of the principal ingredients of curry.

Ginger

Zingiber officinale Rosc.; English—Ginger; Hindi—Plant = *adrak*, Dried rhizome = *sonth*, fresh rhizome = *adrak*; Bengali—Plant = *ada*, dried rhizome = *sont*, fresh root = *adrak*, *ada*; Assamese—Plant = *ada*; Nepalese—Dried rhizome = *sunt*; Oriya—Plant = *ada*; Punjabi—Plant = *ada*, *adrak*, dried rhizome = *zangzabil*, *sunt*, fresh rhizome = *zunjbel*, *adrak*; Marathi—Plant = *ale*, Gujarati—Dried rhizome = *sunt*, fresh rhizomes *adu*, *adhu*; Tamil—Dried rhizome = *shukku*, fresh rhizome = *inji*; Telugu—Plant = *allam*, dried rhizome = *sonti*, *sonthi*, *allam*, fresh rhizomes = *allam*; Kannanda—Dried rhizomes = *vana sunthi*, fresh rhizome = *hasisunthi*; Malayalam—Dried rhizome = *Chukka*, fresh rhizome = *inchi*; Sanskrit—Plant = *adraka*, *sringavera*, dried rhizome = *vishva-bhishagam*, *nagara*, *sunti*, *mahaushadha*, fresh rhizome = *ardrakam*; Persian—Dried rhizome = *zanjabile-khushk*, fresh rhizome = *zanjabile-tar*; Family—Zingiberaceae.

Ginger of commerce is a dry product prepared from the green underground stems or rhizomes. The most important producers of commercial dry ginger are India, Jamaica and Sierra Leone. The

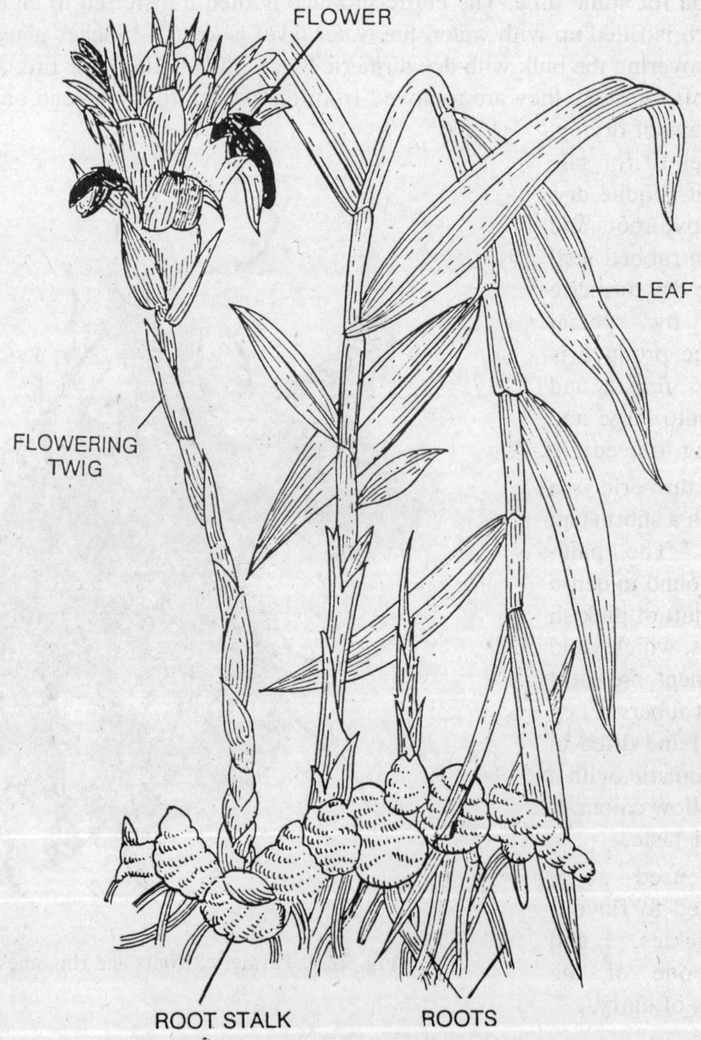

Fig. 22.4. *Zingiber officinale.* Ginger (*Adrak*). Rhizomes are used as spice and condiment.

most important ginger growing area in India is Kerala State. The preparation of the dry ginger (sonth) of commerce is also exclusively confined to this area. In Kerala, the main producing centres lie in the Kottayam division. The other main producing areas in the West Coast are in the Malabar district.

Ginger is grown generally in areas of heavy rainfall. It requires a rich and well drained soil for its cultivation. The most suited soils are sandy or clayey loams and laterites of the Malabar Coast. Ginger is cultivated usually in small home gardens.

Cultivation. Though a perennial plant, yet it is cultivated as an annual crop. It takes 9-10 months to mature, and every year a fresh crop is taken. The plant is propagated vegetatively by means of rhizomes. The field is well ploughed, and long and narrow beds are prepared. The beds are heavily manured with cattle-manure. Seed ginger, consisting of small portions of the rhizomes, collected from the previous crop, is then planted in these beds in small pits at a distance of about 6-10 inches each way. Planting is done in May-June, and the beds are covered with mulch leaves to protect the seed from following rains. In about 2-3 weeks, the beds shoot up. The plant grows to a height of about two feet, and develop several lateral shoots in each clump. The rhizomes develop and become ready for harvest by December-January. As the crop matures, the aerial shoots begin to dry up and lodge on the ground. The leaves also become yellow. Harvesting is done by digging the whole crop or only portions of it.

After digging out, the rhizomes are thoroughly washed with water to remove the soil and dirt sticking to them. These underground stems are kept in the sun for a day or so, and then taken to the market for sale as green ginger (*adrak*). The preparation of dry ginger (*sonth*) is a local craft of Kerala State. For purpose of export only dry ginger is used.

In India, about 48,000 acres are under the crop, and the estimated production of cured ginger is 24,000 tonnes of which about 18,000 tonnes are produced on the west coast alone. About 25 per cent of the produce on the west coast is sold as the green ginger (*adrak*) and (*sonth*) and exported to other countries. The chief countries which import ginger from India are the U.K., U.S.A., France, Germany and Arabia.

Chemical composition. Moisture 10.5%; volatile oil 1.8%; oleoresin 6.5%; water extract 19.6%; cold alcohol extract 6.0%; starch 53.0%; crude fibre 7.17%; crude protein 12.4%; total ash 6.64%.

Uses. Ginger is used more as a condiment than as a spice. The aroma of ginger is due to essential oil, whereas the pungent taste is due to the presence of carminative and digestive stimulant. The essential oil, obtained from the rhizomes, is used for flavouring food stuffs. Ginger is exceedingly popular for flavouring beverages, such as ginger bear and ginger malt.

Zerumbet Ginger

Zingiber zerumbet (Linn.) Rosc. ex Smith; English—Zerumbet ginger; Hindi—*Mahabari bach*, *nar kachur*; Punjabi—*Kachur*, *nar kachur*; Malayalam—*Kathu-inshi-kua*; Sanskrit—*Sthula granthi*; Family—Zingiberaceae.

A perennial herb. Found throughout India.

Uses. The rhizomes are used as spice and condiment. The rhizomes has a slightly aromatic odour and possesses similar properties to those of *Zingiber officinale*, but in a minor degree.

SPICES AND CONDIMENTS OBTAINED FROM BARK

Cassia China

Cinnamomum aromaticum Nees.; English—Cassia China; Family—Lauraceae.

Cassia China is obtained from *C. aromaticum* Nees cultivated in China. The bark is the regular cassia China of commerce. Because of its brittle nature, much of the bark is broken, thus yielding the '*broken grade*'. The main grades are: 1. Whole Quills or China Rolls, 2. Selected Brokens or Canton Rolls, and 3. Extra selected brokens. Trees growing at higher altitude (180-300 metres) yield better quality bark with higher volatrle oil content. The bark has been known from the earliest times. The dried immature fruits constitute the well known '*Cassia buds*', known in India as *Kala Nagkesar*. This species of Cassia is not grown in India but considerable quantities of it are imported into the country.

When the trees are about six years old, the first cut of bark is obtained. The season for barking commences in March and continues till the end of May, after which the bark loses its aroma. The bark, after its removal and while it is still moist with sap, is then laid with the convex side downwards. The bark is left to dry for about 24 hours, and tied up in bundles of about 45 cm. diameter and sent for marketing.

Uses. It is used as a spice or condiment in curries and similar preparations. The important constituent of commercial importance is the volatile oil which finds numerous uses in the various culinary preparations, perfumery and cosmetics as well as pharmaceutical preparations.

Indian Cassia

Cinnamomum tamala Nees. & Eberm.; English—Indian Cassia; Hindi, Bengali and Punjabi—*Tejpat, tejput*; Gujarati—*Tamalapatra*; Marathi—*Darchini*; Sanskrit—*Tamalaka, tejpatra*; Tamil—*Talishapattiri*; Telugu—*Talishapatri*; Family—Lauraceae.

A moderate-sized evergreen tree. Distributed in tropical and subtropical Himalayas (915 to 2440 metres altitude), Khasi and Jaintia hills (915-1220 metres altitude).

The bark of the tree, known in trade as Indian Cassia or Indian Cassia Lignea, is collected from trees growing at the foot of the Sikkim Himalayas. Regular plantations of *C. tamala* are grown in Khasi and Jaintia Hills, Gara Hills, Mikir Hills, Manipur and Arunachal.

Uses. The bark is aromatic. It is coarser than the bark of true cinnamon and is one of the common adulterants of true cinnamon. The essential oil from the bark is pale yellow and contains 70–85 per cent cinnamic aldehyde.

Cinnamon

Cinnamomum zeylanicum Bl.; English—Cinnamon; Hindi—*Dalchini, qalmi-dar-chini*; Bengali—*Dalchini*; Punjabi—*Dar-chini, kirfa*; Mumbai—*Taj, dalchini, tikhi*; Marathi—*Darachini, dalchini*; Tamil—*Karruwa, lavangap-pattai*; Telugu—*Sanalinga, lavanga-patta*; Malayalam—*Lavanga-patta*; Kannada—*Lavanga-patta, dalachini*; Sanskrit—*Gudatvak*; Persian—*Tali khahe, dar-chini*; Family—Lauraceae.

Cinnamon is one of the oldest known spices. The true cinnamon, also called as Ceylon Cinnamon is the dried bark of *Cinnamomum zeylanicum* Bl.

It is native of Sri Lanka. In India, it is grown in the Nilgiris, South Kanara, Malabar, Assam and Kumaon. It is an evergreen tree. The tree attains a height of 8 to 12 metres, but in cultivation it is coppiced or cut back to a height of about 2 metres. Its highly aromatic leaves are 12 to 17 cm. long, dark glossy green above and lighter beneath. The flowers are small, yellow and inconspicuous developing into dark purple ovoid and one seeded berries, about 1.5 to 2.5 cm. long.

Soil and climate. It is a hardy plant which can very well grow in almost all types of soils under a wide variety of tropical conditions. The quality of the bark is highly influenced by the soil and ecological factors. In Sri Lanka which is the major cinnamon growing country in the world the plants are cultivated on white sandy soil. Sandy or siliceous soils with an admixture of humus are considered to be ideal for cinnamon cultivation. Sheltered situations upto an altitude of 800 to 1000 metres receiving an annual rainfall of 200 to 250 cm. are considered to be good for the crop.

Propagation. The easiest and most widely adopted method of cinnamon propagation is by seed. It can also be propagated by planting, cutting and layers.

Cinnamon fruits ripen in July-August. The fruits fall down when fully ripe. The fleshy berries are left in heaps in shade to soften and rot. The mass is then trempled. The pulp free seeds are washed and dried in shade. They are sown without much delay as they have a short period of viability. The nursery is raised in a suitable spot in soil, rich in organic matter. The place is dug well twice or thrice. The soil is broken to powder and made loose, making it altogether free from stones, root bits, etc. The seed beds are made 1 metre wide and of suitable length with adequate provision for drainage. The seeds are sown in lines 12 cm. apart and are covered with a layer of soil to a thickness of about 2.5 cm. Germination of seed occurs in about 20 days. The beds have to be provided with artificial shade and watered regularly. The shade should continue untill the plants are about 12 cm. in height and then removed gradually. Transplanting is best done when the plants are about 10 to 12 month old.

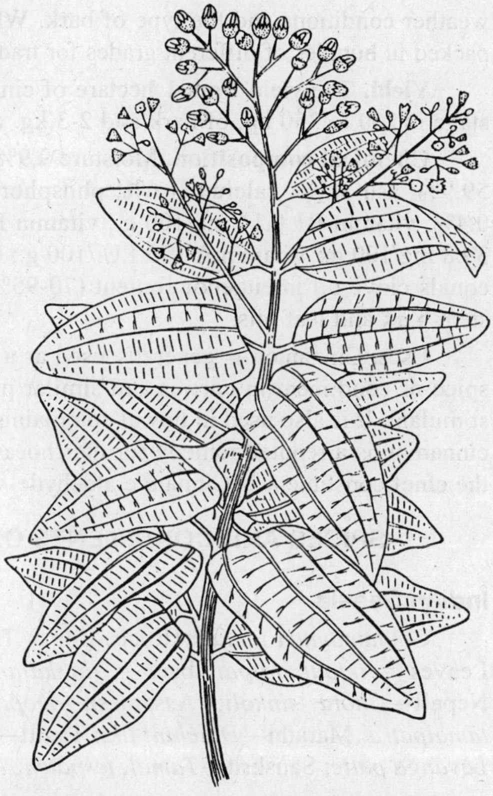

Planting. Seedlings are transplanted when they are about 12 months old at a distance of about 2 metres between plants and rows. The planting is done in ideal weather conditions, preferably in June-July.

Fig. 22.5. Spices and condiments. *Cinnamomum zeylanicum* Blume; Eng. Cinnamon (Lauraceae). A flowering twig.

Harvesting and Curing. There are two regular cutting seasons in South India, which more or less synchronize with two monsoons. The appropriate time for cutting the shoots for peeling is determined with reference to the circulation of sap between the wood and the corky layer. The sap flow and the time for peeling are judged by the peelers by making a test cut on the stem with a sharp knife. If the bark separates readily, the cutting is commenced immediately. The shoots should be of at least two years growth with the bark having a uniform brown colour and should have attained a length of 1 to 1.25 m. and a thickness of not less than 1.25 cm. Shoots satisfying these requirements are cut and bundled after the leaves and the terminal shoots have been removed.

Cutting is followed by scraping and peeling operations. The peeling is a specialized operation peculiar to this industry which requires some skill and considerable experience. It is done by using a specially made knife which has a small and round pointed end with a projection on one side to facilitate ripping of the bark. The peeler takes the cut shoots one by one and after scraping outer skin, first makes a longitudinal slit from one end to the other. Then working the knife between the bark is ripped quickly. The twigs cut in the morning are peeled on the same day. The peeled slips are gathered and kept overnight under shade for what is called '*fermentation*'. However, while no real fermentation process develops during the course of a night, a little softening of the bark does occur with the result that peels become more easy and pliable for ensuring piping operation. The bark as it dries contracts and assumes the shape of a pipe otherwise known as quill. The smaller quills or pipes are inserted into larger ones to form compound quills. They are then arranged systematically end to end in length of about 90 cm. The quills are rolled by hand when they are soft and fresh and after rolling they are dried on mats in shade. The drying lasts for 2-5 days, depending upon the weather conditions and the type of bark. When the drying is complete the bark is collected and packed in bundles of different grades for trade.

Yield. The yield from 1 hectare of cinnamon plantation after attaining the proper yielding stage is 200 to 350 kg. of bark and 2-3 kg. of leaf oil per year on an average.

Chemical composition. Moisture 9.9%; protein 4.6% fat 2.2%; fibre 20.3%; carbohydrates 59.9%; ash 3.5%; calcium 1.6%; phosphorus 0.05%; iron 0.004%; sodium 0.01%; potassium 0.4%; vitamin B1 0.14 mg/100 g.; vitamin B2 0.21 mg/100 g.; niacin 1.9 mg/100 g.; vitamin C 39.8 mg/100 g.; Vitamin A 175 I.U./100 g.; calorific value 355 calories/100 g.; cinnamon leaf oil equals clove oil in eugenol content (70-95%), root bark yields 3% oil which differs from both stem bark and leaf oils.

Uses. Cinnamon is generally used as a spice or ingredients of curry powder. It is used as a spice or condiment in curries and similar preparations. It is also used in medicine as a cordial stimulant. It is also used in bowel complaints such as dyspepsia, diarrhoea and vomiting. Powder cinnamon is a reputed remedy for diarrhoea and dysentery. The bark also yields an oil of which the chief constituent is Cinnamic aldehyde.

SPICES AND CONDIMENTS OBTAINED FROM LEAVES AND TWIGS

Indian Cassia

Cinnamomum tamala (Buch.-Ham.) T. Nees & Eberm.; English—Indian Cassia; Hindi—Leaves = *tejpat*, *tajpat*, bark = *taj kalami*, *dalchini*, *talispatri*, *silkanti*; Bengali—*Tejpat*; Nepali—*Chota sinkoli*; Assamese—*Dopatti*; Mumbai—*Darchini*, *tamala*; Gujarati—*Taj*, *tamalpatra*; Marathi—*Dalchini tiki*; Tamil—*Talisha-pattiri*; Telugu—*Tallisha-patri*; Kannada—*Lavanga patte*; Sanskrit—*Tamal*, *tespatra*; Arabic—*Zarnab*; Family—Lauraceae.

Fig. 22.6. *Cinnamomum tamala* (Tejpat). A plant with inflorescence and leaves.

A moderate-sized evergreen tree, found on the Himalayas, between 3,000 and 7,000 feet, in Bengal, the Khasia hills and Assam.

A cinnamon known as Indian Cassia is obtained from this plant. It is a coarser and sold in larger pieces than the true cinnamon or bark of *C. zeylanicum* Bl., for which it is often used as an adulterant. The bark of the root is quite as good as the true cinnamon bark. In Manipur, the natives collect the root-bark instead of the stem-bark.

Cassia buds. These are the immature fruits of tree yielding Indian Cassia. They are something like cloves, and consist of the unexpanded flower-heads. They possess properties similar to those of the bark. Cassia buds were once employed in preparing the speed wine called *Hippocras*.

Cassia bark. The bark has a general resemblance to cinnamon, but is in simple quills, not inserted one within the other. The quills, are less straight, even and regular, and of dark-brown colour. It breaks with a short fracture.

Uses. The leaves are known as *tejpat*, and bark as *taj*. The bark and dried leaves are used to flavour dishes. The leaves are commonly used as a condiment. They are famous for their pungency and added in cooked rice (*pulao*) and curries.

Leaves are also used medicinally. They are stimulant, carminative, used in rheumatism, colic, diarrhoea, and in scorpion sting. Bark is aromatic and used in gonorrhoea.

Leaves yield an essential oil, which contains d–phellandrene and 78% eugenol.

Sweet Marjoram

Majorana hortensis Moench; English—Sweet marjoram; Hindi—*Murwa*; Bengali—*Murru*; Kannada—*Maruga*; Malayalam—*Maruvamu*; Tamil—*Marru*; in Kumaon it is called Bantulsi and in Dakshin Bharat Murwa; Family—Labiatae (Lamiaceae).

This is an aromatic, branched perennial, 1-2 feet high. It is native of mediterranean region, North Africa and Asia Minor. The herb is commonly grown in Indian gardens. It is well suited for hilly places in India. Leaves oblong-ovate; flowers small, whitish or purplish, in terminal clusters; seeds minute, oval, dark brown.

It is treated as an annual under cultivation. It thrives well in well-drained fertile garden loam. It is propagated by seeds and cuttings. Seeds are sown in pots and seedlings, when large enough, are transplanted in the field 8-10" apart in rows. The crop becomes ready for harvesting in three and half months. The tops are cut when the plants are coming into flower and dried in the shade. The volatile oil content of leaves is maximum when the plant is harvested before seed formation.

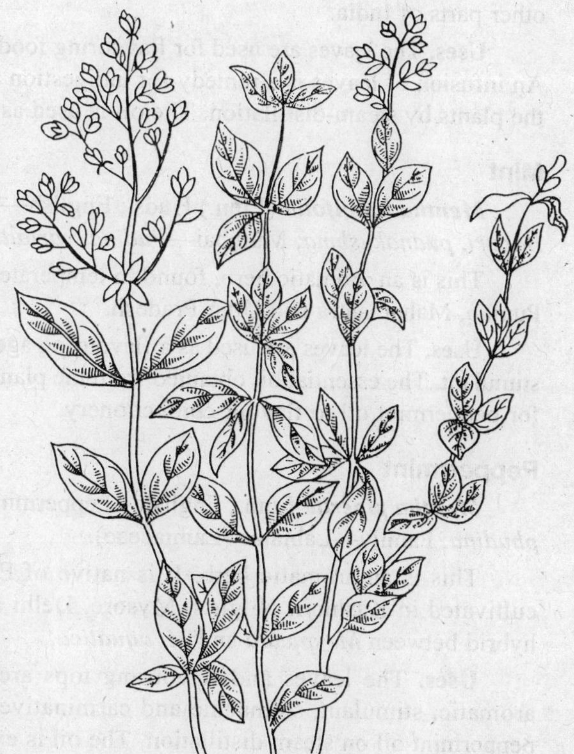

Fig. 22.7. Spices and condiments. *Majorana hortensis* Moench., Eng. Marjoram (Labiatae, Lamiaceae). The dried leaves of the herb are used as spice and flavouring agent.

Sweet marjoram possesses a strong, spicy and pleasant odour and flavour. The dry herb contains - water, 7.61%; protein 14.31%; fixed oil 5.6%; volatile oil 1.72%; pentosans 7.68%; fibre 22.06%; and ash 9.69%. Tannin and ursolic acid are also present.

Steam distillation of the leaves and flowering heads yields a volatile oil, known in the trade as *Oil of Sweet Marjoram*. The oil is colourless or pale-yellow to yellow-green with a peculiar odour.

Uses. The leaves of the plant are used fresh or dried and highly esteemed as a condiment for seasoning food. Fresh leaves are employed as garnish and incorporated in salad. They are used also for flavouring vinegar. The aromatic seeds are used in confectionery. It is a carminative, expectorant and tonic. Leaves and seeds are astringent. An infusion of the plant is used as stimulant, emmenagogue and gulactagogue. The oil is used in high grade flavour preparations and perfumes, and in soap and liqueur industries. It is used as an external application for sprains, bruises, paralytic limbs and toothache.

Field mint or Corn mint

Mentha arvensis Linn.; English—Field mint or corn mint; Hindi—*Podina*, *Pudina*; Bengali—*Mar*.; Gujarati and Telugu—*Podina*, *Pudina*; Kannada—*Chetni maragu*; Family—Labiatae (Lamiaceae).

An erect herb, 10-60 cm. in height found throughout temperate North Asia. Leaves 2.5-5 cm. long, shortly petioled or sessile, oblong ovate or lanceolate, serrate, cuneate at the base, hairy or glabrous; flowers in axillary, capitate whorls, borne on axils of leaves on upper stem.

Mentha arvensis grows wild in Kashmir at 1,500-3,000 m. It has also been reported from other parts of India.

Uses. The leaves are used for flavouring food stuffs. It is used as a stimulant and carminative. An infusion of leaves is a remedy for indigestion and rheumatism. A volatile oil is obtained from the plants by steam-distillation. The oil is used as a flavouring ingredient.

Mint

Mentha longifolia (Linn.) Huds.; English—Mint; Hindi—*Pudina*, *jungli pudina*; Punjabi—*Baburi*, *pudnakushma*; Mumbai—*Pudina*, *vartalau*; Family—Labiatae (Lamiaceae).

This is an aromatic herb, found in temperate Himalayas and in Kashmir, Garhwal, Kumaon, Punjab, Maharashtra and Uttar Pradesh.

Uses. The leaves are used as a flavouring agent. The dried leaves are used as carminative and stimulant. The essential oil obtained from the plant on steam distillation can be used as a substitute for peppermint oil to flavour confectionery.

Peppetrmint

Mentha piperita Linn.; English—Peppermint; Hindi—*Vilaiti pudina*, *paparaminta*, *gamathi phudina*; Family—Labiatae (Lamiaceae).

This is an aromatic herb. It is native of Europe. It is grown in Indian gardens and also cultivated in Kashmir, Nilgiris, Mysore, Delhi and Dehradun. *M. piperita* is considered to be hybrid between *M. spicata* and *M. aquatica*.

Uses. The leaves and flavouring tops are used for flavouring food stuffs. The herb is aromatic, stimulant, stomachic and carminative, and used medicinally. The herb yields a true peppermint oil on steam distillation. The oil is extensively used for flavouring food stuffs and in pharmacy. Peppermint oil is one of the most popular and widely used essential oils. It is employed for flavouring chewing gums, candies, ice creams, confectionery, mouth washes, pharmaceuticals, dental preparations and alcoholic liqueurs. It is widely employed for nausea, flatulence and vomiting. It has a refreshing odour and a persistent cooling taste. It may be taken with sugar or in the form of tablets or lozenges.

Spearmint

Mentha spicata Linn.; English—Spearmint; Hindi—*Pahari pudina*, *pudina*; Family—Labiatae (Lamiaceae).

A perennial aromatic herb, 1 to 3 feet high, with creeping rhizomes. It is native of temperate Europe and Asia. In India, it is cultivated in the Punjab, Uttar Pradesh and Maharashtra.

Uses. Both fresh and dried leaves are used for mint sauce and jelly and to flavour soups, stews, sauces and beverages. The leaves are the source of spearmint oil, which is used for flavouring food products. Spearmint oil is used in U.S.A. for flavouring chewing gums, confectionery and pharmaceuticals. In India, green leaves are used for making chutney and for flavouring culinary preparations, vinegars, jellies and iced drinks. The herb is considered stimulant and carminative. A soothing tea is brewed from the leaves.

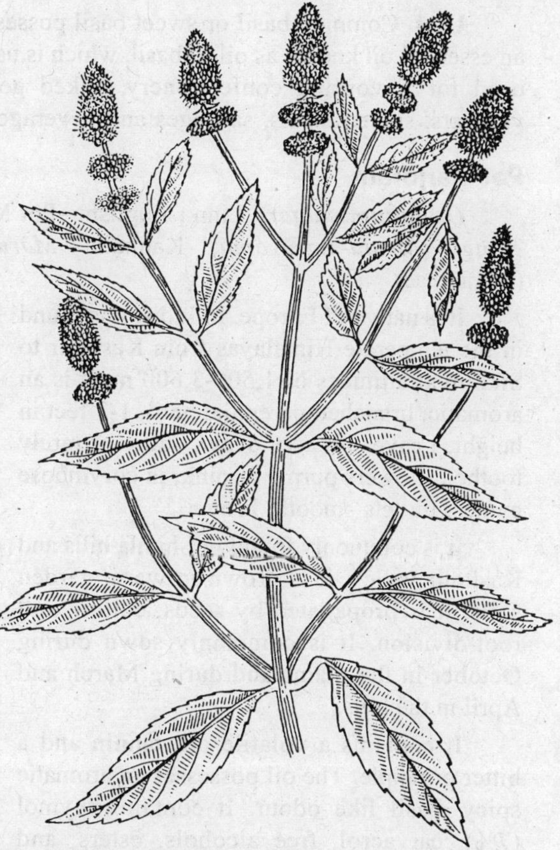

Fig. 22.8. Spices and condiments. *Mentha piperita* Linn.; Eng. Peppermint (Labiatae, Lamiaceae). A flowering twig. The leaves and flowering tops are used for flavouring purposes.

Curry Leaf Tree

Murraya koenigii (Linn.) Spreng.; English—Curry leaf tree; Hindi—*Kathnim*, *mitha nim*; Bengali—*Barsanga*; Marathi—*Karhinimb*; Gujarati—*Goranimb*; Telugu—*Karepaku*; Tamil—*Kattuveppilei*; Oriya—*Barsan*; Family—Rutaceae.

A large aromatic shrub or a small tree, upto 6 m. in height and 15-40 cm. in diameter, found throughout India and the Andaman Islands uptoan altitude of 1,500 m. Bark dark-brown or almost black; leaves imparipinnate; leaflets 9-25, ovate, lanceolate, crenate-dentate, obtuse or acute, glabrous above, pubescent beneath, gland dotted, strongly aromatic; flowers in terminal corymbose cymes, white, fragrant, berries sub-globose or ellipsoid, purplish black when ripe, 2-seeded. The plant is much cultivated for its aromatic leaves. Propagation is by seeds.

Uses. The leaves of the plant are extensively employed as flavouring agent in curries and chutneys. The leaves, root and bark are used also as tonic, stomachic and carminative.

Common Basil

Ocimum basilicum Linn.; English—Common basil, sweet basil; Hindi—*Babui-tulsi*, *Kali-tulsi*, *sabzah*, *marua*; Bengali—*Babui tulsi*, *khub kalam*, *debunsha*; Oriya—*Dhala tulasi*; Punjabi—*Baburi tulsi*, *rehan*; Marathi—*Sabza*; Tamil—*Tirmutpatchie*; Telugu—*Bhu-tulasi*, *rudra jada*, *vibudi-patri*; Kannada—*Kam-kasturi*; Malayalam—*Tiru-nitru*; Sanskrit—*Munjariki*. *Varvara*; Persian—*Firanj-Mushk*; Family—Labiatae (Lamiaceae).

An erect, herbaceous, glabrous or pubescent annual, 1-3 feet high, native of Central Asia and North-West Asia, growing throughout India. Leaves ovate - lanceolate, gland dotted; flowers white or pale purple. It is cultivated throughout tropical India from the Punjab to West Bengal.

Uses. Common basil or sweet basil possesses a clove like scent and a saline taste. It yields an essential oil known as oil of basil, which is used as a flavouring agent. Oil of basil is commonly used for flavouring confectionery, baked goods, sauces, ketchups, tomato pastes, pickles, vinegars, spiced meats, sausages and beverages.

Pot Marjoram

Origanum vulgare Linn.; English—Pot Marjoram or Common Marjoram; Hindi—*Sathra*; Telugu—*Mridu-maruvamu*; Kannada—*Maruga*; Punjabi—*Mirzanjosh*; Family—Labiatae (Lamiaceae).

It is native of Europe. In India, it is found in the temperate Himalayas from Kashmir to Sikkim at altitudes of 1,500-3,600 m. It is an aromatic, branched perennial herb, 1-3 feet in height. Leaves broadly ovate, entire or rarely toothed; flowers purple or pink, in corymbose cymes; nutlets smooth, brown.

It is commonly found in Shimla hills and Kashmir valley. It is grown in warm garden soils. It is propagated by seeds, cutting and root-division. It is commonly sown during October in the plains and during March and April in the hills.

It contains a volatile oil, tannin and a bitter principle. The oil possesses an aromatic spicy, basil like odour. It contains thymol (7%), carvacrol, free alcohols, esters, and bicyclic sesquiterpene.

Chemical composition of the leaves. Moisture 8.0%; protein 11.7%; fat 6.4%; crude fibre 11.0%; carbohydrates 53.9%; total ash 9.0%; calcium 1.7%; phosphorus 0.20%; iron 0.05%; sodium 0.02%; potassium 1.7%; vitamin A 1010 international units (I.U.)/100 g.; vitamin B_1 0.34 mg/100 g.; vitamin B_2 0.41 mg/100 g.; niacin 6.2 mg/100 g. and vitamin C 12 mg/100 g; calorific value 360 calories/100 g.

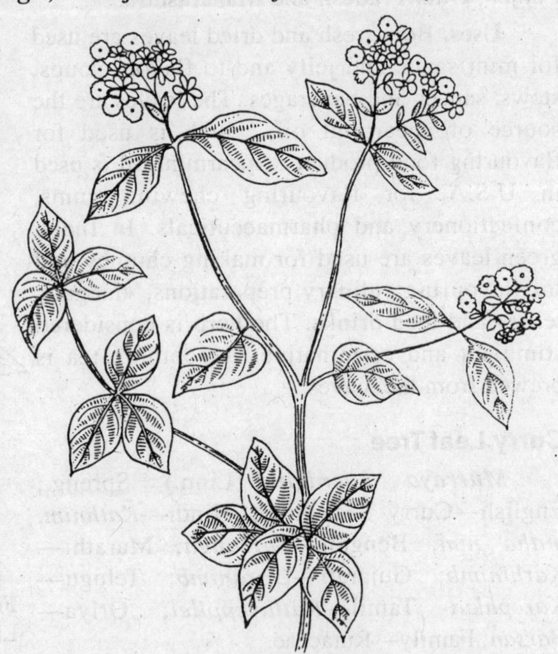

Fig. 22.9. Spices and condiments. *Origanum vulgare* Linn.; Eng. Pot marjoram (Labiatae, Lamiaceae). The dried leaves of the herb make the Origanum, Oregano, Mexican Origanum or Pot marjoram, which is used as a condiment. The taste of the herb is spicy, warm, pungent and bitter. The herb contains a volatile oil, tannin and bitter principle.

Uses. The leaves and tops of the herb cut prior to blooming are used to flavour food stuffs especially meats, egg preparations and salads. The volatile oil from herb is aromatic, carminative, stimulant, tonic, stomachic, diaphoretic, emmenagogue, rubifacient, given in colic, diarrhoea and hysteria, also applied in toothache.

Rosemary

Rosmarinus officinalis Linn.; English—***Rosemary***; Hindi—*Rusmary*; Family—***Labiatae*** (***Lamiaceae***).

A small shrub, native of Mediterranean region. It is found wild on dry rocky hill in the Mediterranean region. In India, it is cultivated to some extent in the temperate Himalayas and Nilgiris hills in dry to moderately moist climate.

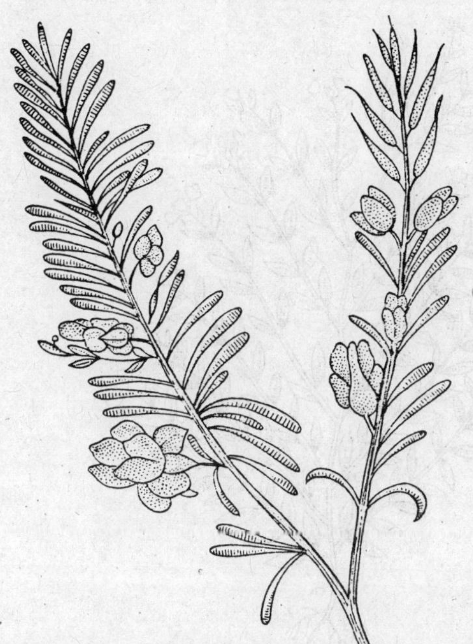

Fig. 22.10. Spices and condiments. *Rosmarinus officinalis* Linn; Eng. Rosemary (Labiatae, Lamiaceae). A flowering twig. The dried leaves of *R. officinalis* Linn., make the rosemary of commerce.

A small shrub. It is native of Mediterranean region, cultivated as a spice and for medicinal purposes. Stems shrubby, white woolly; flavouring branches tomentose-pubescent, leaves aromatic, petiolate, entire, oblong, base narrowed; flowers, purple, blue or white, in racemes. The greyish-green hairy leaves are very aromatic.

Uses. The leaves and tender twigs are used for flavouring food products. Sage has been extensively employed in the food industry as a standard spice in making stuffing for fowl, meats and sausage. Dried and powdered aromatic leaves are mixed with cooked vegetables and sprinkled on cheese dishes, cooked meats and other curries. Fresh sage leaves are used in salads, vegetable sandwiches.

Chemical composition of dried leaves. Moisture 5.7%; Protein 10.2%; fat 14.1%; crude fibre 16.0%; carbohydrates 46.3%; total ash 7.7%; Calcium 1.8%;

Leaves of rosemary possess an agreeable aromatic odour and a pungent, bitter and camphoraceous taste. It yields a volatile oil known as oil of rosemary. The volatile oil is obtained by steam distillation of leaves, flavouring tops and twigs. The finest produce is being obtained from the dried leaves freed of stalks. The oil is pale-yellow or colourless liquid with the characteristic odour of the leaves and a warm camphor, borneol and bornyl acetate.

Uses. The leaves are employed as a condiment dried and powdered. They are added to cooked meats, fish, eggs, soups, stews, sauces; dressings, preserves and jams. They are mixed with sage in pork and veal stuffings and added to biscuits. Carminative, diaphoretic, astringent, diuretic, emmenagogue, stimulant and stomachic oil or rosemary is used for flavouring of meats, sausages, soups, table sauces and other food products. The oil is mainly used in perfumery. It is also used in medicinal preparations.

Garden Sage

Salvia officinalis Linn.; English—Sage, Garden sage; Hindi—Salvia sefakuss; Family—Labiatae (Lamiaceae).

Fig. 22.11. Spices and condiments. *Salvia officinalis* Linn.; Eng. Sage, Garden sage (Labiatae, Lamiaceae). A flowering twig. The dried leaves of *S. officinalis* Linn., make the "sage" of commerce.

Phosphorus 0.09%; Iron 0.03%, Sodium 0.01%; Potassium 1.0%; vitamin B_1 0.75 mg/100 g.; vitamin B_2 0.34 mg/100 g.; vitamin C 39.8 mg/100 g.; niacin 5.7 mg/100 g.; vitamin A 2395 International Units (I.U.)/100 g.; calorific value 415 calories/100 g.

Thyme

Thymus vulgaris Linn.; English—Thyme; Family—Labiatae (Lamiaceae).

A flowering plant. It is a perennial garden plant. It is native of Mediterranean region. In India it is found in Himalayas from Kashmir to Kumaon.

The dried leaves and tender tops make "thyme" of commerce. A volatile oil of pleasant odour is obtained on steam distillation of herb.

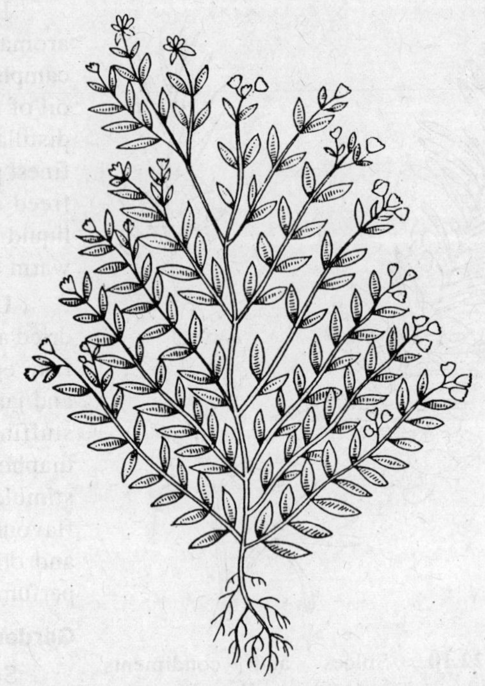

Fig. 22.12. Spices and condiments. *Thymus vulgaris* Linn.; Eng. Thyme (Labiatae, Lamiaceae). A flowering plant. The dried leaves and tender tops of *T. vulgaris* Linn., make "thyme" of commercial value. It is a perennial garden plant.

Chemical composition of dried leaves. Moisture 7.1%; Protein 6.8%; fat 4.6%; crude fibre 24.3%; carbohydrates 44.0%; mineral matter 13.2%; Calcium 2.1%; Phosphorus 0.20%; Iron 0.14%, Sodium 0.08%; Potassium 0.9%; vitamin B_1 51 mg/100 g.; vitamin B_2 0.4 mg/100 g.; vitamin C 12.0 mg/100 g.; niacin 4.9 mg/100 g.; vitamin A 175 International Units (I.U.)/100 g.; calorific value 340 calories/100 g.

Uses. The dried leaves are used for dressing and seasoning meat preparations, egg curries, pork sausage, etc.

SPICES AND CONDIMENTS OBTAINED FROM FLOWER BUDS, FLOWERS AND INFLORESCENCE

Saffron

Crocus sativus Linn.; English—Saffron; Hindi—*Kesar*, *zafran*; Bengali—*Jafran*; Mumbai—*Safran*, *kessar*, *kecara*; Marathi—*Kecara*; Gujarati—*Keshar*; Tamil—*Kungumapu*; Telugu—*Kumkum apave*; Kashmiri—*Kong*; Sanskrit—*Kunkuma*, *kashmirajanma*, *kumkuma*; Arabic & Persian—*Zaafaran*; Family—Iridaceae.

A rhizomatous herb. It is cultivated in Kashmir at Pampur near Srinagar. To a small extent it is cultivated also at Bhersar and Chaubattia in Uttar Pradesh. The dried stigmas and tops of the styles make the saffron of commerce. The product is obtained from the stigmas of the flowers, 4,000 of which are required to produce 25 gms. of saffron.

Uses. The saffron is used as a spice and dyestuff. It possesses a pleasant aroma. Saffron is an ingredient of many Indian dishes, particularly cooked rice and sweet rice (Kesaria chaval) and pulao.

Clove

Syzygium aromaticum (Linn.) Merr. & Perry; Syn. *Eugenia aromatica* O. Kuntze; *E. caryophyllata* Thunb.; English—Clove; Hindi—*Laung*; Bengali—*Lavange, langa*; Punjabi—*Laung, karanfal*; Kashmiri—*Raung*; Marathi & Gujarati—*Lawanga, lavinga*; Tamil—*Kiramber, ilavang ap-pu*; Telugu—*Lavangalu*; Malayalam—*Chanki*; Sanskrit—*Lavanga*; Persian—*Mekhak*; Family—Myrtaceae.

Fig. 22.13. Spices and condiments. *Syzygium aromaticum* (Linn.) Merr. & Perry, syn. *Eugenia aromatica* O. Kuntze, *E. caryophyllata* Thunb; Eng. Clove (Myrtaceae). A, a flowering twig; B, dried unopened flower bud.

Clove is one of the most ancient and valuable spices of the orient and holds a unique position in the international spice trade. Native to Moluccas, the so called 'Spice Islands' in the East Indian Archipelago, this spice was first introduced in India around 1800 A.D. by the East India company. The company's spice garden in Courtallam in Tamil Nadu was then established to cultivate clove and nutmeg as the principal spice crops. Induced by the success of the cultivation in Courtallam, cultivation of clove was extended during the period after 1850 A.D. to Nilgiris (Burliar) in Tamil Nadu, Southern regions of Travancore and also to Cochin State on the slope of Western Ghats. The Important clove growing regions in India now are Nilgiris, Tenkasi hills, and Kanyakumari districts of Tamil Nadu and Kottayam and Quilon districts of Kerala.

Although clove has been under cultivation in India for over about 170 years, its development has been very slow owing probably to its long pre-bearing period and lack of knowledge regarding the method and economics of its cultivation.

Clove is the dried unopened flower bud of *Syzygium aromaticum*, a medium statured, cone-shaped ever-green tree belonging to the Family—Myrtaceae. Clove tree attains a height of 10 to 12 metres. The stem is usually forked near its base with two or three main branches. Smaller branches are slender, rather brittle and covered with grey bark. The leaves appearing in pairs, are lanceolate, acute at both ends and are of dark shining green colour. The aromatic nature of the leaves is due to numerous oil glands found on their under-surfaces. The flower buds are greenish when fresh and are borne on ends, which are picked green and dried in the sun till they become dark brown, form the 'clove' of commerce. The buds have slightly cylindrical base and are surmounted by the plump ball like unopened corolla which is surmounted by the four toothed calyx. If the bud is left unpicked, the flower develops after fertilization into a fleshy, purple and one-seeded oval fruit as 'Mother of clove'. The fruit is about 2.5 cm. long and 1.25 cm. in width. The seed is oblong, rather soft in texture and grooved on one side. The leaves, unripe fruit and broken clove, including the stalk are all aromatic and yield an essential oil.

Soil and climate. Deep and rich loams with high humus content are best suited for clove cultivation. In India, clove has developed well in the open sandy loams and the laterite soils of South Kerala region. But the best growth is seen in black loams of the semi forest regions. Clove abhors water logging and, therefore, perfect drainage is essential.

Clove is strictly a tropical plant and it requires a warm humid climate. Although there has been a general belief that clove requires proximity to sea for the proper development and cropping, experience in India has shown that the trees do well in the submontane regions and have been found to perform better than those in other areas. Humid atmospheric condition and an annual rainfall of 150 to 250 cm. are the other ideal requirements of the crop. Clove thrives altitude of 800 to 900 metres.

Fig. 22.14. *Syzygium aromaticum*-clove (**laung**). The dried flower buds are used as spice.

Propagation. Clove is propagated through seed. Usually the seeds become available for sowing from August to October. The seeds lose their viability within one week after harvest under normal conditions and hence it is necessary to sow them immediately after collection from the tree. The seeds can be sown with or without the fruit coat. Raised nursery beds are prepared in a shady place and the seeds are sown in rows adopting a spacing of about 12 cms. The seeds begin to germinate in four to five weeks after sowing. The seedlings are very slender and delicate and grow very slowly. Watering is necessary throughout the nursery period. The seedlings after about six months of nursery life are transferred to baskets made of bamboo or mud pots and nurtured properly under the shade till they attain an age of 12 to 18 months. Clove can also be propagated vegetatively by grafting on its own stock.

Clove can conveniently be grown mixed with other commercial crops like arecanut, coconut, nutmeg, etc. The shade cast by these crops will provide enough protection to clove from the sun.

Harvesting and curing. Clove tree begins to yield from the seventh or eighth years after planting. The full bearing stage is attained after about 15 to 20 years. The flowering season is September-October in the plains and December-January in high altitudes. The buds are ready for harvest in about four months.

Just before flowering there is fresh flush of young leaves and soon after this, the flower buds begin to appear. The optimum stage for picking clove buds is indicated by the change in the colour from green to slightly pinkish tinge. The unopened clove buds are carefully picked with hand when they turn pink in colour. It is necessary to pick the buds before they open, otherwise, the value of spice will be lost to a considerable extent.

The harvested buds are spread evenly to dry in the sun either on grass-mats or on cement drying floor. During nights the buds should be stored under cover, lest they re-absorb moisture. Normally it is possible to dry the cloves in four to five days under direct sun and in about four hours when they are heated in zinc trays over a regulated fire. Fully dried buds develop the characteristic dark brown colour and crisp. If the produce is uniformly good, approximately 8,000 to 10,000 cloves would weigh one kilogram. Clove is graded according to its appearance and impurity content. Good quality clove should be brownish black in colour, with full and plump crown, somewhat rough to the touch and without wrinkles and it should not contain more than 16% moisture and 5% foreign matter. Also it should have fine aroma and flavour and should readily exude oil when the stem is pressed with the finger nail.

Yield. There is considerable variation in the yield of clove. Under favourable conditions well grown trees may yield as much 4 to 8 kg. of cloves. It is, however, common to meet with low yielding tree in clove plantations, sometimes in large numbers, resulting in low and uneconomic production, particularly in certain years. The average yield from a bearing tree in a well maintained plantation, in India is reported to be about 2.5 kg. considering that the percentage of bearing tree will be around 60, one hectare of plantation containing about 250 trees will yield about 375 kg. of cloves.

Uses. Clove is very aromatic and fine flavoured and imparts warming qualities. In all Indian homes, it is used as a culinary spice as the flavour blends well with both sweet and savoury dishes. Clove is used for flavouring pickle, curries, ketchup and sauces. It is highly valued in medicine as a carminative, aromatic and stimulant. Clove has stimulating properties and is one of the ingredients of betel chewing. In Jawa, clove is used in preparation of a special brand of cigarette for smoking. The essential oil which is obtained by distilling clove with water or steam, has even more uses. It is used medicinally in several ways. The chief constituent of the oil eugenol, is extracted and used as an imitation carnation in perfumes.

Clove requirement in India. The requirement of clove in India is being met through imports. We have been importing large quantities of clove in the past. The average annual imports of cloves was about 2,940 tonnes valuable at about Rs. 183 lakhs in the 1950's. However, owing to the restrictions imposed from time to time, it has been possible to cut down the imports considerably and in the 1960's the annual import was of the order of 500 tonnes valued at about Rs. 47 lakhs.

The chief clove producing countries are Zanzibar, which grow 90% of the total output, Indonesia, Mauritius, and the West Indies.

Chemical composition. Moisture 5.4%; protein 6.3%; volatile oil 13.2%; fat (non volatile ether extract) 15.5%; crude fibre 11.1%; carbohydrates 57.7%; mineral matter 5.0%; ash 0.24%; calcium 0.7%; phosphorus 0.11%; iron 0.01%; sodium 0.25%; potassium 1.2%; vitamin B_1 0.11 mg/100 g.; vitamin B_2 0.04 mg/100 g.; niacin 1.5 mg/100 g.; vitamin C 80.9 mg/100 g. and vitamin A 175 International Units (I.U.), calorific value 430 calories per 100 g.

SPICES AND CONDIMENTS OBTAINED FROM FRUITS

Celery

Apium graveolens Linn. var. *dulce* DC.; English—Celery, celery seed, garden celery; Hindi—*Ajmud*, *bari-ajmud*, *karafs*; Bengali—*Chanu*, *randhuni*; Mumbai—*Bori ajamoda*, *ajmud*; Punjabi—*Bhut jhata*; Arabic—*Karafs*; Persian—*Karasb*; Family—Umbelliferae (Apiaceae).

It is native of England and other parts of Europe. In India, it is cultivated in the North-Western Himalayas and in the hills of Uttar Pradesh, the Punjab and South India. In Indian plains, it is cultivated during the cold weather, chiefly as a garden crop in the vicinity of towns. The fruits are small and dark brown with peculiar flavour.

Uses. The herb is eaten as salad, or made into soup. The seed (fruit) is eaten as spice. The seed tied into a piece of cloth are used to flavour soup. The soil extracted from the seeds is stimulant and used chiefly for flavouring soups and curries. The seeds are also cordial, tonic, carminative, diuretic, emmenagogue and antiseptic.

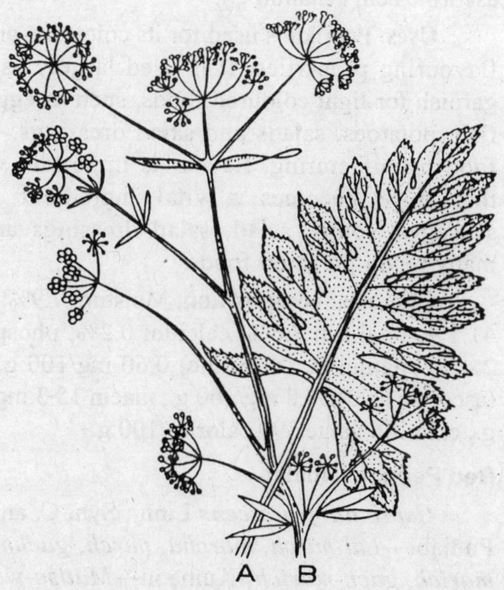

Fig. 22.15. Spices and condiments. *Apium graveolens* Linn. var. *dulce* DC.; Eng. Celery, Celery seed, Garden celery (Umbelliferae, Apiaceae). A, leaf; B, floral axis with typical umbels.

Chemical composition. Moisture 5.1%; protein 18.1%; fat (ether extract) 22.8%; crude fibre 2.9%; carbohydrates 40.9%; total ash 10.2%; calcium 1.8%; phosphorus 0.55%; iron 0.45%; sodium 0.17%; potassium 1.4%; iron 0.53%; vitamin B_1 (thiamine) 0.41 mg/100 g.; vitamin B_2 (riboflavin) 0.49 mg/100 g.; vitamin C (ascorbic acid) 17.2 mg/100 g.; niacin 4.4 mg/100 g.; vitamin A 650 international unit (I.U.); calorific value 450 calories/100 g.

Sweet Pepper

Capsicum annuum Linn.; English—Sweet pepper, paprika, Hungarian paprika, Spanish pimento; Family—Solanaceae.

This is a non-pungent variety of chilli. The red paprikas are famous for their brilliant red colour and mild flavour. Several varieties of paprika have been successfully grown at the IARI, New Delhi and Central Food Technology Research Institute (CFTRI), Mysore. Since several varieties of *Capsicum annuum* Linn. are used to produce paprika, pods in one growing area may differ in shape and appearance of those from another. Some pods are round in shape with pointed end whereas, the others are elongated. They are medium to small and quite fleshy. They are produced on small, bushy plants. When ripe, the pods are picked and dried in the sun. It is processed into powder where it is grown. The paprika pods are supposed to be one of the richest sources of ascorbic acid (vitamin C).

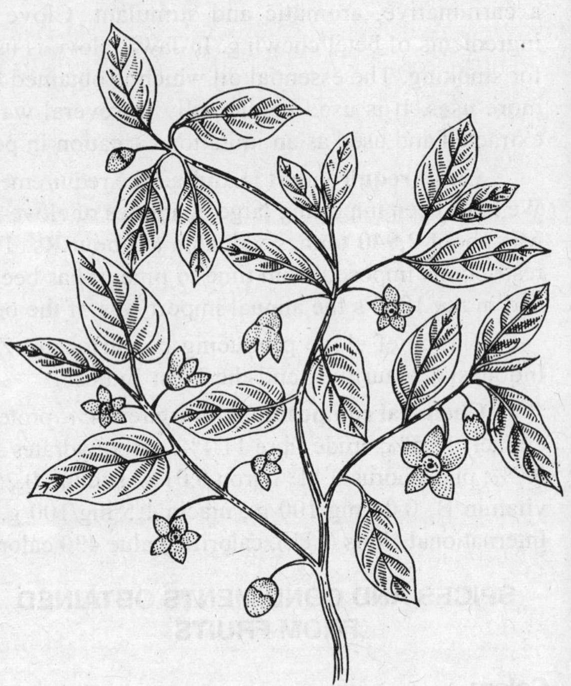

Fig. 22.16. Spices and condiments. *Capsicum annuum* Linn.; Eng. Paprika, Hungarian Paprika, Sweet pepper, Spanish pimento (Solanaceae). A flowering twig. This is a mild or non-pungent variety of chilli.

Uses. Paprika is used for its colouring and flavouring properties. It is used largely as a garnish for light coloured foods, such as eggs, fish, potatoes, salads and salad dressings. In food manufacturing, flavour is important, as this spice becomes a vital ingredient of sausages, soups, salad, salad dressings and many ready prepared foods.

Chemical composition. Moisture 7.9%; protein 13.8%; fat 10.4%; fibre 19.2%; carbohydrates 41.1%; total ash 7.6%; calcium 0.2%; phosphorus 0.3%; iron 0.23%; sodium 0.02%; potassium 2.4%; vitamin B_1 (thiamine) 0.60 mg/100 g.; vitamin B_2 (riboflavin) 1.36 mg/100 g.; vitamin C (ascorbic acid) 58.8 mg/100 g.; niacin 15.3 mg/100 g.; vitamin A 4915 International Units (I.U.)/100 g.; calorific value 390 calories/100 g.

Red Pepper, Chillies

Capsicum frutescens Linn.; Syn. C. annuum Linn.; English—Red pepper; chillies; Hindi & Punjabi—*Lal mirch, marcha, mirch, gachmirch, mattisa, wangru*; Bengali—*Lal-mirch, lanka-marich, gach-marich*; Kumaon—*Matisa-wangru*; Kashmiri—*Mirtz-a-vangun, mirch-wangum*; Gujarati—*Lal-mirch, marchu*; Marathi—*Mirsinga*; Tamil—*Milagay, mollagu*; Telugu—*Mirapakaya*; Malayalam—*Kappal-melaka*; Kannada—*Mena-sina-kayi*; Sanskrit—*Marichi-phalam*; Persian—*Filfile-surkh, pilpile-surkh*; Family—Solanaceae.

It is native of the West Indies and tropical America, most probably of Brazil. Commonly cultivated for its fruit throughout the plains of India, and on the lower hills such as Kashmir, and in the Chenab valley upto altitude 6,500 feet. When grown on the hills it is said to be very pungent. There are several varieties, differing chiefly in the length, shape, and colour of the fruit, some being round, others oblong, obtuse, pointed or bifid, smooth or rugose, and red, white, yellow or varigated. The very pungent principles are present in the flesh and rind as well as seeds. The ripe fruits are dried in the sun and used whole or powdered.

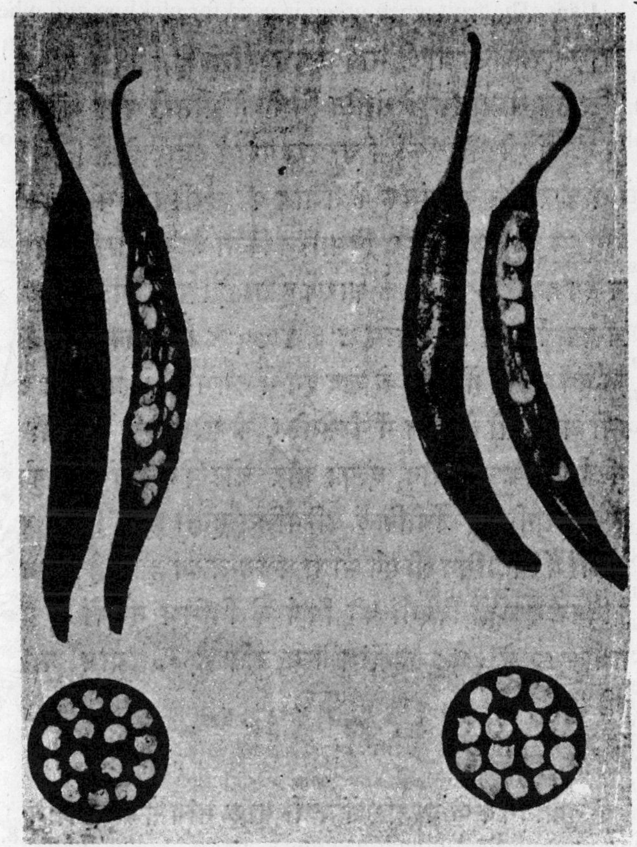

Fig. 22.17. *Capsicum frutescens* (red pepper - **Lal mirch**). Fruits and seeds.

It is a herb, 2 or 3 feet in height, with entire ovate leaves, white flowers with a rotate corolla, and many-seeded fruits, which are technically berries.

Cultivation. Chillies can be cultivated as a field crop over a wide range of tropical and sub-tropical conditions. It grows well up to an altitude of 5,000 feet a.s. 1., in tracts receiving an annual rainfall of about 25 to 50 inches. Heavy rains during the growing period lead to rotting of leaves and fruits. The crop can be grown both in the rainy season commencing from about June and in the hot weather about February. In areas of heavy rainfall the crop is taken after the monsoon period. In tracts of low rainfall, the crop is grown under irrigation.

The soils best suited for the cultivation are of the heavy loam type, from grit, gravel, and stones and well drained.

The seed is sown in the end of June or in the beginning of July in well prepared nursery beds. When the seedlings are about 40-45 days old, they are transplanted. At the nursery stage frequent irrigation is required. The flowering of the plants begin after about a month of transplanting. In another one month, the first crop of green chillies is ready for picking. From about the month of November the fruits begin to ripen and change in colour, first to a deep orange and then to red. The fruits are gathered as they become red. About the middle of December the picking is in full swing. The plants continue to bear the fruit to the end of February. After being gathered, the fruits are dried in the sun. If the colour is not fully developed, they are heaped up until the colour develops fully, and then dried in the sun. They are usually trampled lightly when they are drying, so that they may be flattened and their packing in bags for storage and transport may become easy. Process of drying usually takes about a fortnight. Well irrigated crops yield 1,500 lb. to 2,500 lb. per acre.

Uses. The fruit when green is used for pickling and when ripe is mixed with tomatoes to make sauces. It is also dried and ground for use. The consumption of chillies is very great, and both rich and poor daily use them. The chillies form the principal ingredient in all chutneys and curries. They are ground into a paste, between two stones (Silbatta), with a little mustard oil, ginger and salt, they form the only seasoning which the millions of poor can obtain to eat with rice.

Caraway

***Carum carvi* Linn.**; English—Caraway; Hindi—*Shiajira*, *zira*; Bengali—*Jira*; Punjabi—*Zira siyah*; Kashmiri—*Gunyun*; Ladakh—*Unbu*; Tamil—*Shimai-shombu*; Telugu—*Shimai-sapu*; Sanskrit—*Sushavi*; Persian—*Karoya*; Family—Umbelliferae (Apiaceae).

A herb; native of Europe and West Asia, now cultivated in Bihar, Orissa, the Punjab, Bengal and Andhra Pradesh, as a cold season crop. It is frequently cultivated on the hills, as a summer crop in Kashmir and Garhwal at an altitude of between 9,000 and 12,000 feet. It is a perennial with thick roots, compound leaves with linear segments and small white flowers. The fruits are brown in colour, slightly curved and tapering. The spice seems to have come into use in Europe about the thirteenth century, and it was known in England at the close of the fourteenth century.

Fig. 22.18. Spices and condiments. *Carum carvi* Linn.; Eng. Caraway (Umbelliferae, Apiaceae). A flowering twig.

Uses. The fruit (seed) is used parched and powdered, or raw and entire. In the former case it is employed to flavour curries; in the latter it is put in cakes. It is used in confectionery and in flavouring drinks. It also produces a spirit cordial. It is carminative and largely used in curry powder.

A valuable essential oil is obtained from the "seeds" called caraway oil. The oil is a mixture of a hydrocarbon $C_{10}H_{16}$ (*carvene*) and an oxygenated oil, $C_{10}H_{14}O$ (*carvol*). The oil possesses a strong odour and flavour of the fruit. It is used medicinally and to flavour curries.

Chemical composition of fruit of *Carum carvi* Linn. Moisture 4.5%; protein 7.6%; fat 8.8%; fibre 25.2%; carbohydrates 50.2%; total ash 3.7%; calcium 1.0%; phosphorus 0.11%; sodium 0.02%; potassium 1.9%; iron 0.09%; vitamin B_1 (thiamine) 0.38 mg/100 g.; vitamin B_2 (riboflavin) 0.38 mg/100 g.; vitamin C (ascorbic acid) 12.0 mg/100 g.; niacin 8.1 mg/100 g.; vitamin A 580 International Units (I.U.); calorific value 465 calories/100 g. of spice.

Coriander

Coriandrum sativum Linn.; English—Coriander; Hindi—*Dhaniya*; Bengali—*Dhane*; Sanskrit—*Dhanyaka*; Tamil—*Kotamalli*; Telugu—*Danyalu, kotimiri*; Kannada—*Kotambari*; Nepali—*Danya*; Persian—*Kushniz*; Family—Umbelliferae (Apiaceae).

It is a native of the Mediterranean region. It is grown extensively in India, Russia, Central Europe, Asia Minor and Morocco. In India, it is cultivated in all the States. The more important

States for its cultivation are - Andhra Pradesh, Assam, Maharashtra, Tamil Nadu, Uttar Pradesh, Karnataka, Himachal Pradesh and Madhya Pradesh.

The plant is generally 2-3 feet in height, with white or pinkish flowers. The lower leaves have broad segments, while the upper are very narrow. The fruits are small, oval and aromatic. Technically the fruit is known as "cremocarp". Each fruit consists of two one-seeded carpels, or mericarps with numerous oil ducts (vittae).

Cultivation. It is cultivated mostly as rain-fed crop, either pure or mixed with other crops. It is also grown as an irrigated crop in some areas. The sowing time differs in different parts of India. In Northern India, the seeds are sown during winter; in Maharashtra during rainy season; in Tamil Nadu during autumn and in Karnataka and some parts of Tamil Nadu from May to August and from October to January.

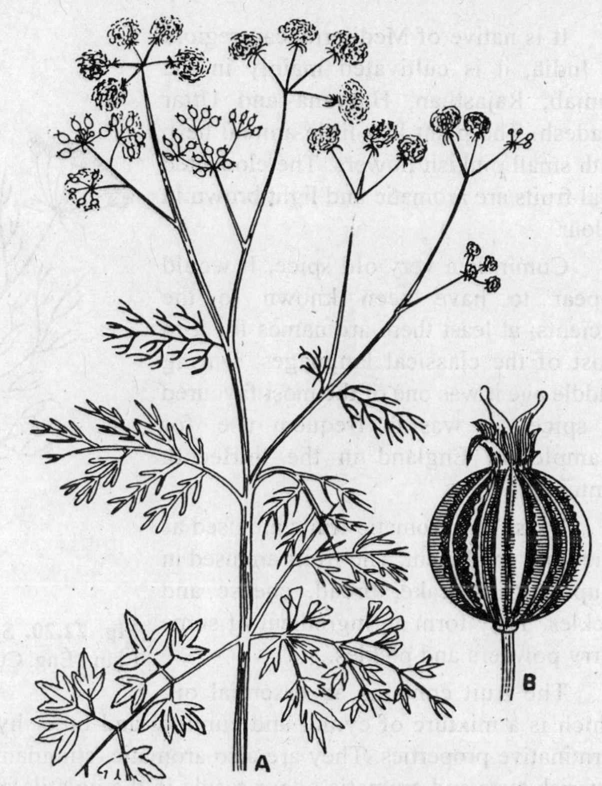

Fig. 22.19. Spices and condiments. *Coriandrum sativum* Linn.; Eng. Coriander (Umbelliferae, Apiaceae). A, a flowering twig; B, a fruit. India produces, on an average, 80,000 tons per year from about 2.5 lakh hectares.

The seed germinates in 10-25 days after sowing. Two or three weedings are necessary for the crop. Irrigation is done when it is needed. Proper manuring increases the yield. The crop is ready in three to three and half months after sowing. The plants are pulled out along with the roots and dried. Therefore, the fruits are threshed out, further dried in the sun, winnowed and stored. The yields vary from place to place. It ranges from 300 kg. to 900 kg. per acre.

Uses. The fruits and leaves are aromatic and used as flavouring materials. The fruits are used extensively in the preparation of curry powder, pickles, sauces, soups, sausages and seasonings. They are also employed for flavouring pastries, cookies, buns and cakes and tobacco products. It is also used for flavouring liqueurs, particularly gin. The fruits are used both as spice and condiment. It is an important condiment used widely in all parts of India. A major part of the produce is used within the country. The fruits are also used as stimulant, carminative, stomachic, diuretic, antibilous, refrigerant, aphrodisiac and tonic. Oil of Coriander is used in medicine and for flavouring beverages. The Coriander leaves make one of the richest sources of vitamin C (*i.e.*, 250 mg/100 g.) and vitamin A (*i.e.*, 5200 I.U./100 g.).

Cumin

Cuminum cyminum Linn.; English—*Cumin*; Hindi—*Zira*; Bengali—*Jira*; Gujarati—*Jiru*, *jira-utmi*; Marathi—*Jire*; Tamil—*Shirangam*; Telugu—*Jiraka*, *Jilakarra*; Kannada—*Jiringe*, *jirage*; Malayalam—*Jirakam*; Sanskrit—*Jiraka*, *jirana*; Persian—*Zira*; Arabic—*Kamun*; Family—Umbelliferae (Apiaceae).

It is native of Mediterranean region. In India, it is cultivated mainly in the Punjab, Rajasthan, Haryana and Uttar Pradesh. The plant is a little annual herb with small pinkish flowers. The elongated oval fruits are aromatic and light brown in colour.

Cumin is a very old spice. It would appear to have been known to the ancients; at least there are names for it in most of the classical languages. During middle age it was one of the most favoured of spices. It was in frequent use, for example, in England in the thirteenth century.

Uses. The aromatic fruits are used as spice and condiment. The fruits are used in soup, curries, cake, bread, cheese and pickles. They form an ingredient of some curry powders and pickles.

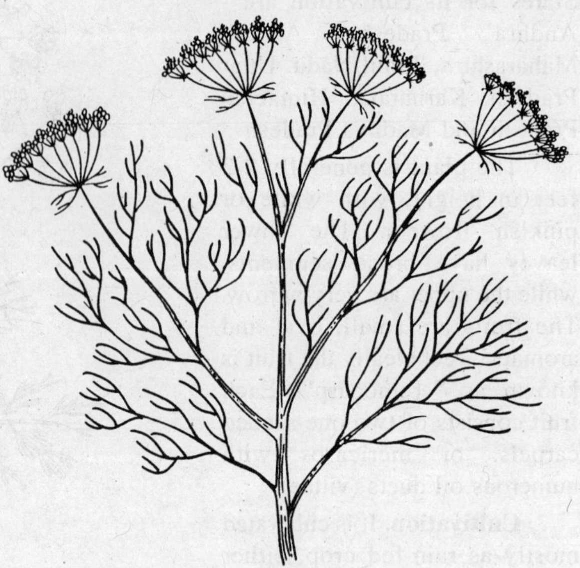

Fig. 22.20. Spices and condiments. *Cuminum cyminum* Linn.; Eng. Cumin (Umbelliferae, Apiaceae). A flowering

The fruit contains an essential oil, which is a mixture of cymol and cuminol and other hydrocarbons. Both fruit and oil possess carminative properties. They are also aromatic, stimulant, stomachic and astringent. Their warm bitterish taste and aromatic odour reside in the volatile oil.

Chemical composition. Moisture 6.2%; protein 17.7%; fat 23.8%; crude fibre 9.1%; carbohydrates 35.5%; mineral matter 7.7%; calcium 0.9%; phosphorus 0.45%; sodium 0.16%; potassium 2.1%; iron 0.048%; vitamin B_1 (thiamine) 0.73 mg/100 g.; vitamin B_2 (riboflavin) 0.38 mg/100 g.; vitamin C (ascorbic acid) 17.2 mg/100 g.; niacin 2.5 mg/100 g.; vitamin A 175 International Units (I.U.); calorific value 460 calories/100 g.

Fennel

Foeniculum vulgare Mill.; English—Fennel; Hindi—*Saunf*; Bengali—*Mauri*, *pan-muhori*; Mumbai—*Bari-shopha*, *panmohuri*; Gujarati—*variari*, *Variyali*; Marathi—*Badishep*; Kannada—*Badisopu*; Tamil—*Sohikira*; Telugu—*Pedda-jila-kurra*; Sanskrit—*Madhurika*; Family—Umbelliferae (Apiaceae).

It is a stout, glabrous, aromatic herb, 5-6 feet high; leaves pinnately decompound; flowers small, yellow, in compound terminal umbels; fruit oblong, ellipsoid or cylindrical, 6-7 mm. in length, straight or slightly curved, greenish or yellowish brown; mericarp 5-ridged with prominent vittae.

Fennel is a very old flavouring substance. It was known to the ancient Hindus, Chinese and Egyptians as a culinary spice. It is native of Mediterranean region. In India, it is cultivated mainly in the Punjab, Uttar Pradesh, Assam, Maharashtra and Baroda. It is cultivated mostly as a garden or homeyard crop throughout India at all altitudes upto 6,000 feet. It requires a fairly mild climate and is cultivated as a cold weather crop in parts of Northern India.

Cultivation. It grows well in any good soil, but thrives best in rich, well-drained loam or black sandy soil containing sufficient lime. It is propagated mainly by seeds. Seeds are sown broadcast by hand or by shallow drills. The time of sowing is October-November in the plains of India and March-April on the hills. When 3-4 inches high, the plants are thinned out to about 1 feet apart. Occasional weeding and irrigation once a week during dry weather is being done.

Harvesting. The crop is harvested before the fruits are fully ripe. The stems are cut with a sickle and spread out to dry in the sun. When dry (4-5 days after cutting) the fruits are threshed out and cleaned by winnowing. The yield ranges from 2800 to 1120 lb. per acre.

Uses. The leaves are used in fish sauce and for garnishing. Leaf stalks are used in salad. The leaf stalks are also used as vegetable. The leaves also possess diuretic properties. The roots are used as purgative and they possess an aromatic odour and taste.

Dried fruits of fennel have a fragrant odour and a pleasant aromatic taste. They are used for flavouring curries, soups, meat dishes, sauces, bread rolls, pastries and confectionery. They are also used for flavouring liqueurs and in the manufacturing of pickles.

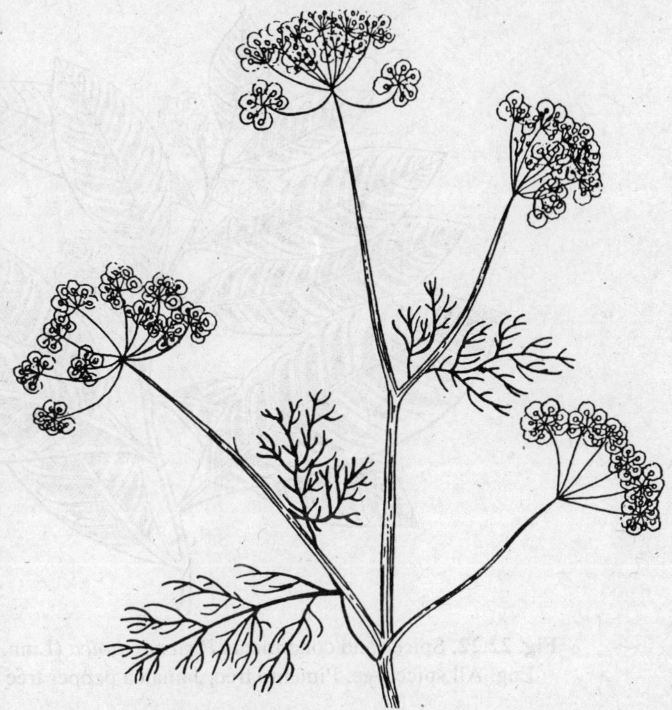

Fig. 22.21. Spices and condiments. *Foeniculum vulgare* Mill.; Eng. Fennel (Umbelliferae, Apiaceae). A flowering twig.

The fruits are aromatic, stimulant and carminative. They are useful in the diseases of chest, spleen and kidney. It is used as one of the condiments in liquorice powder. A hot infusion of the fruits is used to increase lacteal secretion and to stimulate sweating. Fennel fruits from Lucknow are considered to be the best and are priced higher than those from other areas.

Fennel oil. The fruit of fennel also contains a volatile oil. The oil is obtained by steam distillation of crushed fruits. It is a colourless or pale yellow liquid with a characteristic taste and odour. The main constituent of the fennel oil is anethole. Oil of good quality contains 50-60% anethole. Fennel oil is largely used as a flavouring agent in culinary preparations, confectionery, cordials and liqueurs. It is a good aromatic and widely carminative. It is used in colic and flatulence of infants. It makes a good vermicide against hookworm. It is used in the preparation of fennel water. It is used in perfumery and in scenting soaps.

Chemical composition. Moisture 6.3%; protein 9.5%; fat 1.0%; crude fibre 18.5%; carbohydrates 42.3%; mineral matter 13.4%; calcium 1.3%; phosphorus 0.48%; sodium 0.09%; potassium 1.7%; iron 0.01%; vitamin B_1 (thiamine) 0.41 mg/100 g.; vitamin B_2 (riboflavin) 0.36 mg/100 g.; vitamin C (ascorbic acid) 12.0 mg/100 g.; niacin 6.0 mg/100 g.; vitamin A 1040 International Units (I.U.); calorific value 370 calories/100 g.

Allspice

Pimenta dioica (Linn) Merrill syn. *P. officinalis* Lindl.; English—Allspice tree, Jamaica Pepper tree; Family—Myrtaceae.

It is a bushy evergreen tree, 6-9 m. high. It is native of West Indies and tropical America. It is grown in gardens in India. Leaves oblong-lanceolate, coriaceous; flowers white, in terminal and

Fig. 22.22. Spices and condiments *Pimenta dioica* (Linn.) Merrill, syn. *P. officinalis* Lindl.; Eng. All spice tree, Pimento tree, Jamaica pepper tree (Myrtaceae). A flowering twig.

axillary trichotomous paniculate cymes; fruit a globose berry, about the size of a pea seed, black or purple, two seeded; seeds deep brown, reniform.

The unripe berries, when rapidly dried, form the Allspice, Jamaica Pepper or Pimento of commerce. In India, the plant is cultivated in gardens, especially in Bengal, Bihar, Orissa and Mysore. It grows well in Mysore and fruits heavily. The plants are propagated by seeds. Seeds are sown under shade in freshly prepared beds or in pots. The beds should be well watered before sowing, and the seeds should be thinly broadcast and lightly covered with coir waste. Seedlings are potted in large containers at the 2- and 4- leaf stage and are ready for planting in the field when they are 25-35 cm. in height. The tree begins fruiting when about seven years old and continues so upto twenty years. The berries are picked when mature, but still green. The berries are quickly dried in the sun for 4-10 days; They become wrinkled like black pepper and turn reddish brown in colour and the aroma becomes more pronounced. Average yield of a tree is about 35-45 kg. of berries. Dried berries are nearly globular, 4-7 mm. in diameter, reddish brown and with a wrinkled surface. The berries possess an aromatic taste and flavour resembling a mixture of cinnamon, cloves and nutmeg, and hence the name Allspice. Allspice owes its characteristic odour to the presence of an essential oil (3.3-4.5%), found mainly in the pericarp. Allspice is available whole or ground.

Uses. The berries are used as a condiment, as a flavouring agent in sauces, soups, pickles, canned meats, sausages, etc., It is an important ingredient of whole mixed pickling spice, and spice mixtures, *viz.*, curry powders, mincemeat spice, poultry dressing, pastry spice, etc., They are used for flavouring liqueurs. Pimenta berry oil is used for flavouring condiments and food products. The oil is used also as a carminative and stimulant.

Chemical composition of ground berries. Moisture 8.8%; protein 6.0%; fat 6.6%; fibre 21.6%; carbohydrates 52.8%; total ash 4.2%; calcium 0.8%; potassium 1.1%; iron 7.5%; vitamin B_1 (thiamine) 0.1 mg/100 g.; vitamin B_2 (riboflavin) 0.06 mg/100 g.; vitamin C (ascorbic acid) 39.2 mg/100 g.; niacin 2.9 mg/100 g.; vitamin A 1445 International Units (I.U.); calorific value 380 calories/100 g.

Anise

Pimpinella anisum Linn.; English—Aniseed, Hindi—*Saunf*, *badian*; Bengali—*Muhuri*, *mitha-jira*; Marathi—*Somp*; Gujarati—*Anisa*; Telugu—*Kuppi*; Tamil—*Shombu*; Kannada—*Shombu*; Oriya—*Sop*; Family—Umbelliferae (Apiaceae).

It is native of the Eastern Mediterranean region. It is widely cultivated in Europe, Russia, North Africa, Mexico and South America. In India, it is grown to a small extent. It is grown in Uttar Pradesh, the Punjab and Orissa. It is stated that a commodity commonly available in Indian market under the name *Lucknow saunf* is called Indian Aniseed.

The plant is an annual about 2 feet in height with simple based leaves and pinnately dissected steam leaves. The small fruits are greyish brown and covered with short hairs.

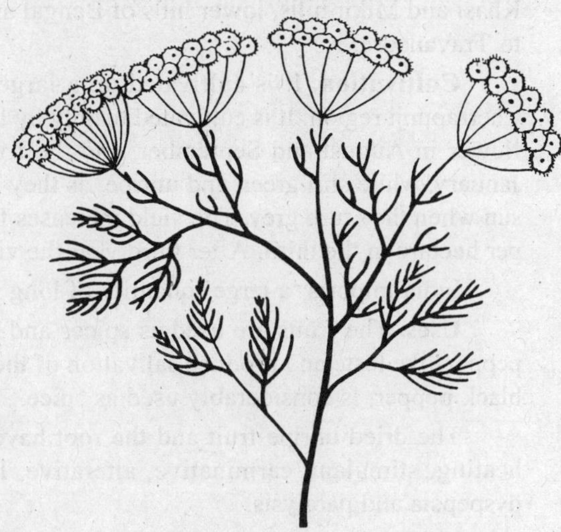

Fig. 22.23. Spices and condiments. *Pimpinella anisum* Linn.; Eng. Anise, Aniseed (Umbelliferae, Apiaceae). A flowering twig.

The plant is well grown in light, fertile or moderately rich, well drained sandy loam. It is propagated by seeds. Seeds are sown broadcast or in drills, 20-40 cm apart. It is commonly sown from the middle of October to the end of November in the plains, and from the beginning of April to the end of May on hills. When the seedlings are 5-8 cm. in height, they are thinned to stand 10-15 cm. apart in a row. The plants require, occasional weeding and watering once a week. The crop is ready to harvest in three and half months after planting when the fruits turn greyish green. The plants are tied in bundles and then stacked with the fruiting heads towards the centre. Fruits ripen in 4-5 days and are then threshed out, cleaned and bagged for the market. A yield of 445-665 kg. of fruit per hectare is expected.

Aniseed on steam distillation yields an essential oil, known as oil of anise, anise oil is a colourless or pale yellow liquid having the characteristic odour and taste of the fruit.

Uses. The fruits are used as condiment. Now, oil of anise replaces the fruits for medicinal and flavouring purposes. Aniseed possesses a sweet aromatic taste and emits when crushed, a characteristic agreeable odour and is used for flavouring food, confectionery, bakery products, beverages and other liqueurs. The fruits are expectorant, stimulating, carminative, diuretic and diaphoretic.

Fresh leaves of the plant are used as a garnish and for flavouring salads.

Chemical composition of fruits. Moisture 9-13%; protein 18%; fatty oil 8-23%; essential oil 2-7%; sugars 3-5%; starch 5%; nitrogen free extract 22-28%; fibre 12-25%; ash 6-10%; choline is also present.

Indian Long Pepper

Piper Longum Linn.; English—India long pepper; Hindi—*Pipal*, *pipulmul*, *pipli*, *gaz pipal*; Bengali—*Pipul*, *pipli*; Nepali—*Popal pipal*; Punjabi—*Pipal*, *pipla mul*; Mumbai—*Pipli*, *pipla mul*; Marathi—*Pimpli*; Gujarati—*Pipli*, *piper*; Tamil—*Tippili*, *pipili*, *pippallu*; Telugu—*Pipili*, *pippali katte*; Kannada—*Yippali*; Malayalam—*Lada*, *mulagu*; Sanskrit—*Pippali*, *upukulya*, *videhi*, *ushuna*; Persian—*Pilpil*, *filfil-i-daraz*; Family—Piperaceae.

It is a native of India, Sri Lanka and Philippines Islands. It is a slender aromatic climber with perennial woody roots occurring in the hotter parts of India, from central Himalayas to Assam,

Khasi and Mikir hills, lower hills of Bengal and evergreen forests of Western Ghats from Konkan to Travancore.

Cultivation. It is cultivated on a large scale in limestone soil, 450 to 600 m. below the Cherrapunji region. It is cultivated mainly by layering of mature branches or by suckers. The plants flower in August and September, and the fruits mature in January. The spikes are harvested in January, while still green and unripe, as they are most pungent at this stage. They are dried in the sun when they turn grey. The yield increases from 560 kg. per hectare in the first year to 1,680 kg. per hectare in the third. After third year the vines become less productive and should be replaced.

India imports 'a large' quantity of long pepper from Malaysia and Singapore.

Uses. The fruits are used as spices and also in pickles and preserves. They have a pungent pepper-like taste and produce salivation of the mouth. Long pepper, though considered inferior to black pepper, is considerably used as spice.

The dried unripe fruit and the root have long been used in medicine. They are considered heating, stimulant, carminative, alterative, laxative, and useful in cough, hoarseness, asthma, dyspepsia and paralysis.

The work on the fruit of *Piper longum* has shown the presence of the alkaloids—*piperine* and *piplartine*.

Black Pepper

Piper nigrum Linn.; English—Black pepper; Hindi—*Kalimirch*, *golmirch*, *choca mirch*; Bengali—*Kalamorich*, *golmorich*; Marathi—*Kalimirch*, *mire*; Gujarati—*Kalamari*, *kalomirich*; Tamil—*Milagu*; Telugu—*Miriyala tige*; Kannada—*Kare menasu*; Malayalam—*Kurumulaku*, *nallamulaku*; Punjabi—*Golmirich*; Kashmiri—*Martz*; Sanskrit—*Maricha*, *ushana*, *hapusha*; Family—Piperaceae.

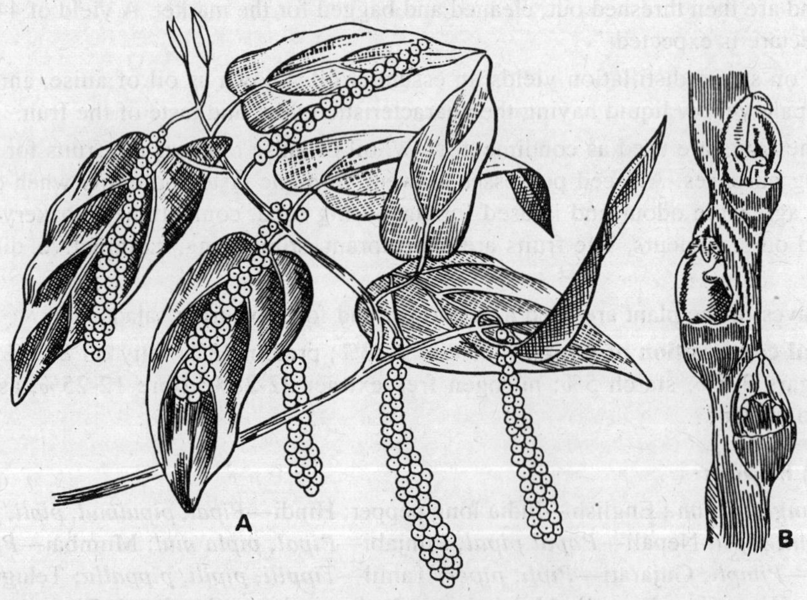

Fig. 22.24. *Piper nigrum* (Black pepper—**Kali mirch** or **golmirch**).
The dried berries are used as black pepper.
A, a branch with fruits; B, Part of a spike.

A branching, climbing perennial shrub, mostly found cultivated in the hot and moist parts of India, Ceylon and other tropical countries. Branches stout, trailing and rooting at the nodes; leaves entire, 12.5-17.5 by 5.0-12.5 cm., very variable in breadth, sometimes glaucous beneath, base acute rounded or cordate, equal or unequal; flowers minute in spikes, usually dioecious; fruiting spikes very variable in length and robustness, rachis glabrous; fruits ovoid or globose, bright red when ripe; seeds usually globose, testa thin, albumin hard.

Pepper is one of the most ancient crops cultivated in India and has probably originated in the hills of south-western India, where it is met with in a wild state in the rain forests from North Kanara to Kanyakumari. The pepper vine in its wild state is mostly dioecious and consequently rarely sets fruit. Under cultivated condition, however, the fruiting is very much better, since most of the cultivated types are bisexual. Almost all the types cultivated at present are selections from wild plants. Some crosses have already been evolved at the Pepper Research Station, Panniyur (Kerala). Of these, the hybrid, *Panniyur I*, a progeny from a cross between *Taliparamba No. 1*, noted for its high yield. *Panniyur I* is a good hardy climber and has long spikes with close set, large fruits.

Cultivation. In India, pepper is cultivated mostly as a mixed crop in homestead gardens. Under this system, the vines are trained on to existing trees like jack, mango, coconut or arecanut. The vines receive intensive care and continue to be productive upto 60 or even 100 years. The vines are also grown as in inter - or subsidiary crop along with other plantation crops like coffee, cardamom, arecanut, coconut and orange. This practice is more common in North Kanara, Coorg and parts of Kerala. As a pure crop, pepper is grown on a plantation scale only in small areas in the sub-montane tracts and lower altitudes in South Kanara, Malabar, Wynad and Travancore. India has been a leading producer and exporter of pepper. India, Indonesia, Ceylon, Cambodia and Sarawak, Brazil and Malagasy are main producers of pepper. In India Kerala on the west coast produces nearly 95% of the total output. In India, pepper is grown mainly in Kerala, Tamil Nadu and Mysore and to a less extent in Assam.

Climate and soil. The pepper vine is found growing upto an altitude of 1,500 m., but thrives best at about 500 m. Pepper requires a warm and humid climate. It thrives in places where the annual rainfall is well over 200 cm. Pepper plants are pollinated by rain-water and consequently satisfactory yields are obtained when frequent showers of rain occur during blossoming period.

Pepper thrives best on virgin soil rich in humus and other plant nutrients. Naturally well drained, red laterite soils or alluvial soils rich in humus seem to be highly suitable for this crop. In Kerala, major part of the pepper cultivation is on laterite or sandy loam along the alluvial banks of the rivers.

Propagation. The pepper vine can be propagated either vegetatively or by seeds. Vegetative propagation is universally adopted. However, it is believed that the plants raised from seeds live longer, and give heavier crops in later years than plants raised vegetatively.

There are two types of vegetative shoots preferred for planting, *viz.*, the runners which are usually confined to the bottom portion of the vines, and the terminal shoots towards the top region. The runners form the most common material for propagation. In some areas, particularly in North Kanara, the terminal shoots of the top fruiting branches are preferred. Plants raised from such cuttings are said to yield even in the second year of planting and produce a satisfactory crop from the fourth year onwards while vines raised from runner shoots give satisfactory yields only from the sixth or seventh year.

Planting. The planting of the pepper vine is done either in July or August. The pits for planting are made preferably on the north and north-eastern sides of the standard so that the severe western sun is avoided.

Manuring. Pepper is an exhausting crop and the soil needs high level of nitrogenous manuring.

Harvesting and yield. An adult vine yields annually 0.5 kg. dried fruits under Indian conditions. The yield goes on increasing as the vines get older upto about the tenth year. Usually there are two crops in a year. One in August-September and the other in March-April. When the fruits are ready for harvest, the whole spikes are removed from the vines with the aid of a ladder. They are dried in the sun for 3-4 days, and the fruits separated from the stalks by beating the dried spikes with a stick or by trampling under foot. When completely dry, the outer skin of the fruits become black and wrinkled. It is estimated that 100 kg. of green pepper roughly yield 33 kg. of dried pepper.

In India, the average yield of commercial black pepper has been found to vary from 110 to 335 kg. per hectare.

Black Pepper. Black pepper consists of the dried, fully developed unripe fruits. It is nearly globular in shape, about 4-5 mm in diameter with the characteristic coat with deep set wrinkles. The freshly harvested spikes are spread on cement floors or mats and dried in the sun for about a week. During drying the green or red fruits gradually change in colour to dark brown or almost black and their skin becomes tough and wrinkled. The fruits are detached from the stalks by beating the heaped up material with sticks or treading upon it barefooted. The outturn of dried black pepper obtained from green fruits varies from 26-29% depending upon the type.

White Pepper. It consists of dried ripe fruits, freed of their pericarp. In preparing white pepper the separation is done through the zone of fibro-vascular bundles, *i.e.*, only a part of the mesocarp is removed. In India, the production is limited to very small quantities, by some households in Kerala, for medicinal and domestic usage. For this purpose, the fully ripe fruits are rubbed with hand to remove the soft rind, and dried. The white fruits obtained are then washed and finally dried.

Uses. Pepper fruits are used as spice or condiment. In Kerala, fresh green pepper is sometimes used for preparing pickles. Black and white pepper make the major condiments employed for seasoning freshly cooked and prepared foods. In U.S.A. and other European countries they are used mainly for preserving meat. The whole fruits are added to pickles, certain types of sausages, etc., but the bulk of the product is generally ground before use. Black pepper is mostly used for its characteristic aroma and pungent taste. White pepper is less pungent. The aromatic odour of pepper is due to a volatile oil, while the pungent taste is caused by an *oleoresin*. An alkaloid is also present. Pepper stimulates the flow of saliva and the gastric juices and has a cooling effect. In modern Indian medicine, it is much employed as an aromatic stimulant in cholera, weakness following fevers, coma, etc., as a stomachic in dyspepsia, as an antiperiodic in malarial fever.

Chemical composition of different type of black pepper (*Piper nigrum* Linn.) varies. Moisture 8.7-14.1%; total nitrogen 1.55-2.60%; nitrogen in non-volatile ether extract 2.70-4.22%; volatile ether extract 0.3-4.2%; non-volatile ether extract 3.9-11.5%; alcohol extract 4.4-12.0%; starch 28.0-49.0%; ash 3.6-5.7%; crude fibre 8.7-18.0%; crude piperine 2.8-9.0%; piperine 1.7-7.4%; Alkaloid piperine makes the biting taste.

Ammi

Trachyspermum ammi (Linn.) Sprague; English—Ammi, Lovage; Hindi—*Ajwain, ajowan*; Bengali—*Jowan, juvani*; Gujarati—*Ajamo*; Marathi—*Owa*; Kashmiri—*Jawind*; Tamil—*Aman, oman*; Telugu—*Omami, omamu*; Kannada—*Omu, oma*; Mumbai—*Ajwan owa*; Sanskrit—*Yamani*; Persian—*Zinian*; Family—Umbelliferae (Apiaceae).

This is an aromatic herb. Cultivated extensively in India for its fruits (seeds); from the Punjab and Bengal to the South Deccan. It is first mentioned in Europe as brought from Egypt about 1549 A.D.

Uses. The fruits are used as spice. They yield an oil of distillation with water, which is used medicinally. The fruits (seeds) are much valued for their antispasmodic, stimulant, tonic and carminative properties. They are considered to combine the stimulant quality of capsicum or mustard with the bitter property of chiretta, and the antispasmodic virtues of asafoetida.

Vanilla

Vanilla planifolia Andr.; Syn. *V. fragrans* (Salibs) Ames; English—Vanilla; Family—Orchidaceae.

It is climbing vine with fleshy adventitious roots, large succulent leaves, and greenish-yellow flowers. The fruits are long, thin, pod like capsules, known as vanilla beans. It is a tropical species and thrives in hot climate with frequent rains. It is propagated by means of cuttings. It is native of tropical America. In India, it is cultivated in the Nilgiris, Wynad, Coorg and Mysore.

The flavour and aroma are not present in the pods until they have been cured. The unripe fruits are subjected to a sweating process. They are exposed to the sun in the morning and are then protected by blankets in the afternoon, while at night they are placed in airtight boxes. During curing process a glucoside is changed by enzyme action into a crystalline substance, vanillin, which possesses the characteristic odour.

Chemical composition. Moisture 25.85-30.93%; protein 2.56-4.87%; fatty oil 1.68-6.74%; volatile oil 0.0-0.64%; nitrogen free extract 30.35-32.90%; fibre 15.27-19.6%; carbohydrates 7.1-9.1%; ash 4.5-4.7%; vanilline 1.48-2.90%; resins 1.5-2.6%; calcium 19.7%; phosphorus 9.5%; sodium 6.7%; potassium 16.2%; iron 0.3%.

Fig. 22.25. Spices and condiments. *Vanilla planifolia* Andr.; Eng. Vanilla (Orchidaceae). A, a flowering twig; B, a seed. The vanilla pods or sticks of commerce are the cured fruits, or beans of *V. planifolia* Andr. (an orchid).

Uses. The pods are used for flavouring confectionery and in perfumery. Vanilla is used in flavouring chocolate, ice creams, candy, beverages, cakes and other food products.

SPICES AND CONDIMENTS OBTAINED FROM SEEDS

Bengal Cardamom

Amomum aromaticum Roxb; English—Bengal cardamom, Nepal cardamom; Hindi—*Bari elaichi*; Marathi—*Veldode*; Family—Zingiberaceae.

It is a perennial herb, grown in North Bengal and Khasia hills. The fruit ripens in September; the capsules are then carefully gathered by the natives and sold to the druggists.

Uses. The fruits and seeds are used as a condiment. It is one of the chief ingredients of 'garam masala'.

Fig. 22.26. *Amomum aromaticum* (**Bari elaichi**). The fruits and the seeds are used as spice.

White Mustard

Brassica hirta Moench; Syn. *B. alba* (Linn.) Boiss.; English—White mustard; Hindi—*Sufed rai*, *safed rayan*; Bengali—*Dhop-rai*; Marathi—*Pandhora-mohare*; Gujarati—*Ujlo-rai*; Tamil—*Vellai-Kadugu*; Telugu—*Tella-avalu*; Malayalam—*Vella-Katuka*; Kannada—*Bili-sasave*; Sanskrit—*Siddhartha*; Persian—*Sipandane-supid*; Family—Cruciferae (Brassicaceae).

This is considered native of Europe and Western Asia. It is a herb, 2-6 feet high, with yellow flowers, hairy lobed leaves and a long-beaked pod with bristles. The small round seeds are yellow on the outside and white within. The seeds contain a glucoside, *sinalbin*. When ground seeds, are treated with water, the glucoside breaks down through enzyme action, and yields a sulphur compound with a characteristic sharp and pungent taste.

Uses. It is used as a condiment in food stuffs and pickles. The powdered seeds made into a paste with cold water act as a stimulant. It is used also in medicine. The seeds yield fatty oil.

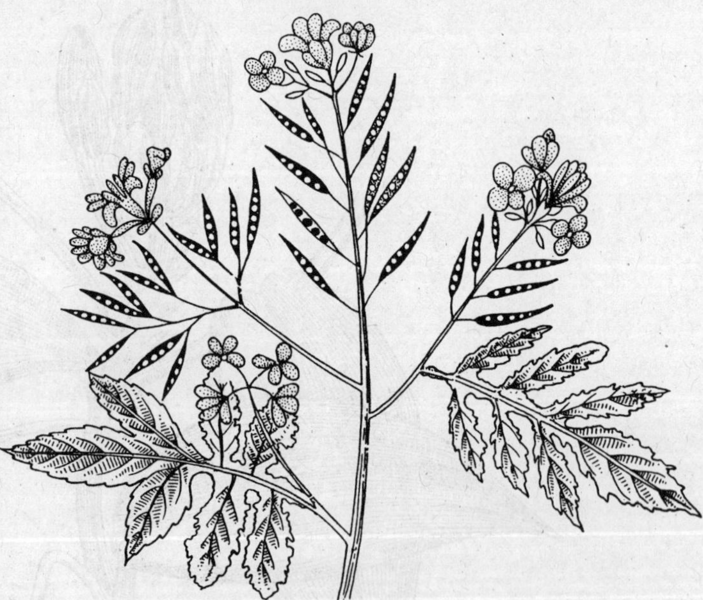

Fig. 22.27. Spices and condiments. *Brassica* spp.; Eng. Mustard (Cruciferae, Brassicaceae). *Brassica hirta* Moench. (White mustard); *B. juncea* (Linn.) Czern. & Coss., (Indian mustard); *B. nigra* Koch (black mustard).

Indian Mustard

Brassica juncea (Linn.) Czern. & Coss; English—Indian Mustard; Hindi—*Rai, sarson, sarson-lahi, gohna-sarson, barirai, barlai, badshahi-rai, shahzada-rai, khas-rai*; Bengali—*Rai sarisha*; Kashmiri—*Asur*; Gujarati—*Rai*; Mumbai—*Rai, sarson, rajika*; Marathi—*Mohari, rayan*; Sanskrit—*Rajika*; Family—Cruciferae (Brassicaceae).

A tall, erect annual herb, 3-5 feet high, with bright green foliage; racemes terminal, with a rough reticulate testa. It is cultivated chiefly in the Punjab, West Bengal and Uttar Pradesh. It is sown in October-November and cut in March-April.

Uses. The seeds are used as a condiment in pickles. The seed oil is used for pickles and cooking purposes.

Black Mustard

Brassica nigra Koch; English—Black mustard; Hindi—*Kali rai, rai, tira, tara mira, lahi, jag rai, makra rai*; Bengali—*Rai sarisha*; Gujarati—*Rai, kalirai*; Mumbai—*Rai, sarson*; Tamil—*Kadagho*; Telugu—*Avalo*; Kannada—*Bile sasive*; Sanskrit—*Rajika*; Persian—*Jarshaf*; Family—Cruciferae (Brassicaceae).

It is a native of Eurasia. In India, it is cultivated mainly in the Punjab, Uttar Pradesh and Tamil Nadu. The plant is smaller than the white mustard. It has smooth pods with dark brown seeds, which are yellow inside. The seeds contain a glucoside, sinigrin. This glucoside yields on decomposition a volatile oil containing sulphur, which is responsible for the aromatic odour and pungency.

Uses. Ground black mustard seeds are used as a condiment in pickles. The seed oil is used for cooking purposes.

Chemical composition of Brassica seeds. Moisture 7.6%; nitrogen substances 29.1%; nitrogen free extract 19.2%; ether extract 28.2%; crude fibre 11%; ash 5%; seed oil 27-33%.

Lesser Cardamom

Elettaria cardamomum Maton; English—Lesser cardamom, cardamom; Hindi—*Chhoti elaichi*; Bengali—*Ilachi*; Gujarati—*elachi*; Punjabi—*Illachi*; Mumbai—*Malabari elachi*; Marathi—*Velloda*; Tamil—*Ellakay, aila-cheddi, elakaya*; Telugu—*Ellakay, elaki chettu, ela-kaya*; Kannada—*Yalakki, yelaki, yerakki*; Sanskrit—*Prith weeka, chundruvala, ela, bahoola*; Family—Zingiberaceae.

This is a tall herbaceous perennial, with branching subterranean rootstock, from which arise a number of upright leafy shoots, 5-18 feet high, bearing alternate, elliptical or lanceolate sheathing leaves, 1-3 feet long. Flowers borne in panicles 2-4 feet long, arising from the base of vegetative shoots; panicles upright throughout their length

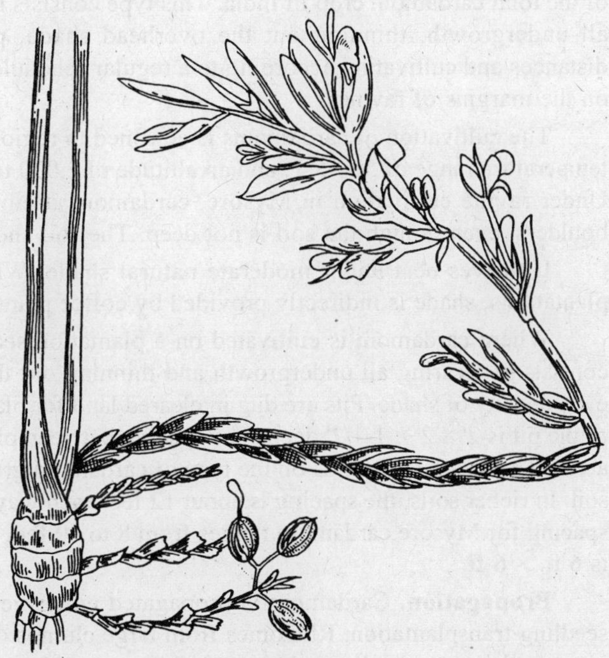

Fig. 22.28. Spices and condiments. *Elettaria cardamomum.* Maton.; Eng. Lesser cardamom, Cardamom. (Zingiberaceae). A plant with inflorescence and fruits.

or upright at first and ultimately pendent or prostrate; flowers about 1.5 inches long, white or pale green in colour with a central lip streaked with violet, borne in a close series on the rachis; bisexual. Fruits trilocular capsules, fusiform to ovoid, pale green to yellow in colour, containing 15-20 hard, brownish black, angled and rugose seeds, covered by a thin mucilaginous membrane.

It is native of the moist evergreen forests of South India, growing wild in the Western Ghats, between 2,500 and 5,000 feet. It is found wherever the overhead canopy has been thinned. It is also found along stream banks, where the overhead shade is less dense.

There are two varieties: 1. *E. cardamomum* var. *major* Thw. comprising all the wild races and 2. *E. cardamomum* var. *minor* Watt. (syn. *E. cardamomum* var. *minuscula* Burkill) comprising all the cultivated races, particularly those included under the names Malabar and Mysore cardamoms. Variety major is the more primitive variety from which the cultivated variety minor is derived. The minor variety is commonly grown in India.

Malabar cardamom is characterised by a short leafy shoot, rarely exceeding 9 feet in height, leaves 1-1.5 feet long, hairy on the under-surface; panicles 2-3 feet long and prostrate; fruits globose, rounded or ovoid and lightly ribbed. It is chiefly cultivated in Mysore and Coorg, and to some extent in Travancore.

Mysore cardamom is characterised by robust growth, leafy stem, upto 17 feet high, and large leaves with smooth underside; panicles erect; fruit fusiform, three-angled and ribbed. It is cultivated mainly in Travancore, Annamalai and Nelliampathy hills.

Distribution. It is usually cultivated in those regions which form the natural habitat of the species. They are chiefly cultivated in Kerala, Mysore, Maharashtra, Assam, Sikkim and Tamil Nadu.

Cultivation. The mode of cultivation varies considerably in different regions. The Mysore system of cultivation is the most important for commercial production and accounts for about 90% of the total cardamom crop in India. This type consists in clearing selected areas of jungle land of all undergrowth, thinning out the overhead shade, planting cardamom seedlings at regular distances and cultivating according to a regular schedule. In Mysore, cardamom is planted mostly on the margins of ravines.

The cultivation of cardamoms is confined to regions with an annual rainfall of 60-130 in., a temperature range of 50-95°F. and an altitude of 2,000 to 5,000 feet. It grows on a variety of soils. Under ravine cultivation in Mysore, cardamom attains good growth in pockets of soil among boulders, even though the soil is not deep. The soil should be moist and well drained.

It thrives best under moderate natural shade. When inter-planted in coffee and arecanut plantations, shade is indirectly provided by coffee plants and areca palms.

Where cardamom is cultivated on a plantation scale in virgin forest areas, the initial work consists in clearing all undergrowth and thinning out the overhead canopy in order to obtain an even density of shade. Pits are dug in cleared land for planting cardamom seedlings. The usual size of the pit is 2' × 2' × 1-1/2' deep. The spacing between pits varies from 5 feet to 10 feet in different areas. The spacing depends on the type of cardamom grown and also on the natural fertility of the soil. In richer soils, the spacing is about 12 feet each way however, it is much less in poor soil. The spacing for Mysore cardamom ranges from 8 to 10 feet each way, while for Malabar cardamom it is 6 ft. × 6 ft.

Propagation. Cardamom is propagated either vegetatively by division of rhizomes or by seedling transplantation. Rhizomes from large clumps of growing plants are taken out, separated into small clumps, each consisting of atleast one old and one young shoot and planted in prepared pits. The method, however, is unsatisfactory for planting large areas as the number of rhizomes obtainable is limited.

Propagation by seedlings is common adopted in North Kanara, Mysore, Coorg, Tamil Nadu and Travancore-Cochin. Seedlings which come up spontaneously in jungle clearings and those which occur in old cardamom plantings are collected and planted in prepared nursery beds until they are required for planting in permanent positions. The seeds may be sown broadcast in favourable positions on cleared land and transplanted into permanent positions when they are sufficiently developed. The sowing time is usually August-October in Coorg, Mysore and North Kanara, and February-March in Tamil Nadu.

Seeds germinate in 6-7 weeks after sowing. Seed beds are prepared on raised ground and seeds are sown broadcast and covered with soft earth. Nurseries are generally opened near stream with abundant water. The beds are not shaded from the sun until after the seedlings appear. Low pandals are put up, 4-5 feet above the beds to provide shade for seedlings. Seedlings when 3-4 months old are transplanted 6-18 in. apart in nursery beds. Seedlings are usually kept in the nursery for about two years. The time of planting out seedlings is June-July, before the onset of monsoon rains. Generally two seedlings are planted in each pit, one of which is later used for filling up gaps in the plantation. Frequent weedings are necessary in the first two years to prevent young cardamom plants from being choked up.

Harvesting. Cardamom comes into bearing in three years after planting, which may be the fourth or fifth year after sowing. Flowering commences in April-May and continues till July-August; flowers may be seen almost throughout the year. The peak flowering period is May-June and as flowers appear for long, fruits ripen irregularly at intervals necessitating several pickings. Fruits are gathered at intervals of 30-40 days. Only those fruits which are just to ripe are picked. This is necessary to prevent capsules from splitting on the drying floors. If picked in the under-ripe stage, the fruits shrink on drying and have a shrivelled appearance. Individual fruits are picked with the peduncle and each clump has to be periodically visited to gather the fruits. The first crop obtained in the third or fourth year is usually small, higher and sustained yields are obtained in subsequent years upto the tenth or fifteenth year, depending upon the type cultivated, after which the plants become exhausted.

Drying and curing. The fruits are dried after harvesting either in the sun on barbecues of beaten earth or by artificial heat in drying houses. Artificial drying takes about 48 hours and cardamoms so dried retain their green colour, which is much liked in European and American markets. Sun drying, on the other hand, takes 3-5 days, but the mucilaginous coats on the seeds remain intact in sun-dried capsules and the seeds possess a characteristic sweetish aroma. A certain amount of bleaching also takes place during sun-drying and the capsules attain a straw colour. Dried capsules are hand-rubbed and winnowed to remove stalks, calyces and foreign matter, and then bagged.

The average yield of dry capsules from a well maintained cardamom estate is 100-150 lb. per acre.

Uses. Cardamom is used as a spice and masticatory, and in medicine. The seeds possess a pleasant aroma and a characteristic, warm, slightly pungent taste. It is used for flavouring curries, cakes, bread and for other culinary purposes. It is also used for flavouring liqueurs. In the Arab countries, cardamom is used for flavouring coffee and tea. In medicine, it is used as an aromatic, stimulant, carminative and flavouring agent.

The cardamom capsules contain—Moisture 20%; protein 10.2%; ether extract 2.2% mineral matter 5.4% crude fibre 20.1% carbohydrate 42.1% calcium 0.13% phosphorous 0.16% iron 5.0 mg/100 g. The seeds of cardamom contain 2-8% volatile oil. Seeds of green cardamom yield appreciably more oil than those of bleached cardamom.

Chemical composition of seeds. Moisture 8.3%; volatile oil 8.3%; total ash 3.7%; non-volatile ether extract 2.9%; crude fibre 9.2%; crude protein 10.3%; calcium 0.3%; phosphorus 0.21%;

sodium 0.01%; potassium 1.2%; iron 0.012%; vitamin B_1 (thiamine) 0.18 mg/100 g.; vitamin B_2 (riboflavin) 0.23 mg/100 g.; vitamin C (ascorbic acid) 12.0 mg/100 g.; niacin 2.3 mg/100 g.; vitamin A 175 International Units (I.U.) per 100 g. of seeds.

Cardamom oil of commerce is obtained by the distillation of the whole fruits of *E. cardamomum*. The cardamom oil is a colourless or pale yellow liquid with a penetrating, camphoraceous odour and a strong pungent taste. The main constituents of the oil are - cineol, terpineol, terpinene, limonene, sabinene and terpineol in the form of formic and acetic esters. Cardamom oil is used in flavouring beverages.

The chief importing countries of Indian cardamom are - Arabia, Sweden, U.K., U.S.A., Germany and Middle East countries.

Nutmeg and Mace

Myristica fragrans Houtt.; English—Nutmeg tree; Hindi—Jaiphal (fruit kernel), Japatri (aril); Bengali—Jaiphal (nut), jotri (aril); Punjabi—Japhal (nut), jauntari (aril); Marathi—Jaiphala (fruit), jayapatri (aril); Gujarati—Jayephal (nut), javantari, japatri (aril); Tamil—Jadikkay (nut), jadi-pattiri (aril); Telugu—Zevangam, jajikaya (nut), japatri (aril); Malayalam—Jatikka (nut), jati-pattiri (aril); Sanskrit—Jajiphalam (nut), jajipatri (aril); Family—Myristicaceae.

It is native of the Moluccas Islands. In India, it is grown in the Nilgiris, Kerala, Mysore, Andhra Pradesh, Assam and West Bengal.

This is a dioecious or rarely monoecious evergreen aromatic tree, usually 30-40 feet high. The bark is greyish black and longitudinally fissured in old trees. The leaves are elliptic or oblong-lanceolate, coriaceous. The flowers are arranged in umbellate cymes. The flowers are creamy-yellow and fragrant. The fruits are yellow, broadly pyriform or globose, 6-9 cm. long, glabrous often drooping. The pericarp is fleshy, 1.25 cm. thick, splitting into two halves at maturity. The seeds are broadly ovoid, arillate, albuminous, with a shell like purplish brown testa. The aril is red and fleshy.

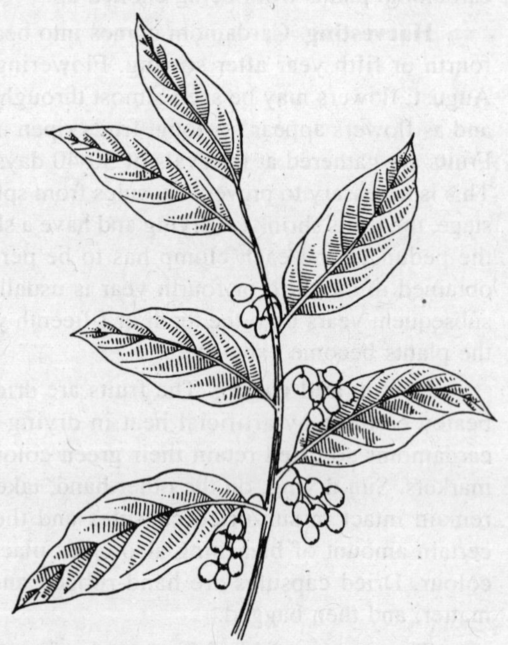

Fig. 22.29. Spices and condiments. *Myristica fragrans* Houtt.; Eng. Nutmeg (kernel) and mace (aril) (Myristicaceae). A flowering twig.

Cultivation and propagation. It requires a hot and moist climate with a rainfall of 150-300 cm. per annum. It grows best at low elevations in alluvium formed of deep friable loam with good drainage, well sheltered from high winds.

The plant is propagated from seeds. Fresh seeds with shells, collected from mature fruits and dried for a day, are sown either directly at the site or in nursery beds, sheltered from wind and strong sun. Seeds are sown 30 cm. apart, 2.5 cm. below the soil in nursery beds and take 2-3 months to germinate. The seedlings are transplanted when 60-90 cm. high to their permanent in the field during wet weather. The spacing between trees varies from 6-7.5 m., depending upon the gradient and fertility of the soil; wider spacing is adopted in rich level lands. Young plants require shade. This is provided by growing bananas in advance of planting. It may also be cultivated as a mixed crop in orchard, and in coffee, coconut, arecanut and rubber plantations.

In plantations raised from seeds there are usually more male than female trees and the sex is not determined until the plants are 6-7 years old and begin to bear flowers. In order to ensure a sufficient number of female trees in the plantation, two seedlings are planted close together and the one which has proved to be female is retained.

Harvesting and yield. Under favourable conditions, the tree starts to bear fruit from the seventh year or earlier and the yield increases upto the fifteenth year or beyond. Trees continue to bear for many years and 70-80 years old trees give good and sustained yields. Fruits are borne more or less throughout the year but the main harvesting season is June-October. The fruits are picked up every morning from the ground or are gathered from the trees by the use of a long stick to which a hook is attached. Seeds that drop out of split fruits are also gathered. The average annual yield at Burliar is 1,250-4,000 fruits per tree.

Preparation of nutmeg and mace. The harvested ripe fruit with the valves split, discloses the seed with a shell-like testa covered by a scarlet fibrous aril. After collection, the pericarp is removed and the seed separated from the aril and dried. Drying is complete when the kernel rattles in the shell. The shells are cracked off with wooden hammers and the kernels removed and sorted. Dried kernels make the nutmeg of commerce.

Fig. 22.30. Spices and condiments. *Myristica fragrans* Houtt; Eng. Nutmeg (kernel) and mace (aril) (Myristicaceae). A twig with an opened fruit.

Mace is the dried fibrous aril covering the testa. It is obtained by separating the arils and drying in the sun. Mace of superior quality is produced by drying in specially constructed ovens. During drying the arils at first become dark red and brittle, and in about six weeks they become bright amber coloured.

Nutmeg. The commercial nutmeg is ovoid, 2.0-3.5 cm. long × 1.5-2.8 cm. diameter, greyish brown in colour with minute reddish brown spots and lines and reticulately furrowed. East Indian nutmeg is of three grades - 1. Banda nutmeg is supposed to be the best which contains upto 8% essential oil; 2. Siauw nutmeg is of second grade containing 6.5% essential oil and 3. Penag nutmeg, which is suitable only for distillation purposes.

Composition. The nutmeg contains - Moisture 14.3%; protein 7.5%; ether extract 36.4%; fibre 11.6%; carbohydrates 28.5%; mineral matter 1.7%; calcium 0.12%; phosphorus 0.24%; iron 4.6 mg/100 g.; Nutmeg contains a volatile oil (6-16%), starch (14.6-24.2%); pentosans (2.25%); furfural (1.5%), and pectin (0.5-0.6%). The flavour and therapeutic action are due to the volatile oil.

Nutmeg oil. The percentage of volatile oil varies from 6-16% according to the origin and quality of the spice. Commercial oil is derived from broken and wormy nutmegs. The material is comminuted, pressed to remove fixed oil, and immediately subjected to steam distillation.

The nutmeg oil is a mobile, colourless or pale yellow liquid with a characteristic odour. The major constituents of the oil are *d*-pinene and *d*-camphene.

Nutmeg butter. Commercial nutmeg butter, a highly aromatic fat, is obtained from undersized, damaged or worm-eaten kernels which are unfit for sale as spice. The material is ground and cooked or steamed before pressing.

The nutmeg butter is a soft solid, yellow or yellowish red in colour, with the odour and taste of nutmeg.

Mace. Commercial mace consists of flattened lobed pieces, 2.5 cm. or more in length, somewhat less in breadth and 1 mm. thick. When soaked in water the lobes swell up and regain their original form. It is dull yellowish red in colour, translucent and brittle. It is nutmeg-like in odour and taste. There are three grades of mace.

1. *Banda mace* is supposed to be finest — it is bright orange in colour and possesses fine aroma; 2. *Java Estate mace* is golden yellow in colour with brilliant crimson streaks; 3. *Siauw mace* is of light colour and contains less volatile oil.

Mumbai mace is derived from *Myristica malabarica*, it is dark red in colour and consists of narrow pieces, divided into numerous lobes twisted together at the apex. It is almost devoid of aroma, and is useless as a spice.

Composition. Mace contains — Moisture 15.9%; protein 6.5%; ether extract 24.4%; fibre 3.8%; carbohydrates 47.8%; mineral matter 1.6%; calcium 0.18%; phosphorus 0.10%; iron 12.6 mg/100 g. It contains a volatile oil (4-15%; average 10%), amylodextrin (25%), reducing sugars, pectin and resinous colouring matter. The chief constituents are volatile oils and amylodextrin.

Uses. Both nutmeg and mace are used as condiment and in medicine. They are commonly used more as a drug than as condiment. Nutmeg is stimulant, carminative, astringent and aphrodisiac. It is used in tonics and forms a constituent of preparations prescribed for dysentery, flatulence, nausea, vomiting, malaria, rheumatism, sciatica and leprosy.

Nutmeg oil and mace are used for flavouring food products and liqueurs. Nutmeg butter is used as mild external stimulant in ointments, hair lotions and plasters, and forms a useful application in case of rheumatism.

Black Cumin

Nigella sativa Linn; English—Black cumin, small fennel; Hindi—*Kalonji*, *kalajira*; Bengali—*Mugrela*, *kala-jira*; Kashmiri—*Tukm-i-gandna*; Mumbai—*Kalonji*, *kalenjire*; Tamil—*Karunshirogam*; Telugu—*Nalla-jilakra*; Kannada—*Karijirigi*, *kare-jirage*; Malayalam—*Karun-chirakam*; Sanskrit—*Krishna-jirata*; Persian—*Siyah-danah*; Myanmar—*Samon-ne*; Family—Ranunculaceae.

A herb. It is a native of Southern Europe. In India, it is cultivated in the Punjab, Bengal, Assam and Bihar for its seeds. Its Sanskrit names indicate, its introduction at a very early period.

Uses. The seeds, possess a strong, pungent, aromatic taste, and therefore, are much used in curries, pickles and other dishes. They are also frequently sprinkled over the surface of bread along with sesamum seed. French cooks employ the seeds of this plant under the name of *Quatre epices* or *toute epices*, and they were formerly used as a substitute for pepper.

Poppy Seeds

Papaver somniferum Linn. var. album; English—Poppy seeds; Hindi—*Post*; Bengali—*Pasto*; Marathi—*Posta*; Gujarati—*Posta*; Family—Papaveraceae.

It is native of West Asia. In India, it is cultivated in the East Punjab, Uttar Pradesh, Rajasthan and Madhya Pradesh. The plant is an erect, rarely branched, usually glaucous annual herb, 60-120 cm. in height. Leaves ovate-oblong or linear-oblong, amplexicaul, lobed, dentate or serrate; flowers large, usually bluish white with a purplish base or white, purple or variegated; capsules large, 2.5 cm. diameter, globose, stalked; seeds white or black, very minute, reniform.

As we know poppy is cultivated also for its seeds. In India, var. *album*, with white seeds has been cultivated for many years for the production of seeds (*post*) under licence in Dehradun and Tehri Garhwal districts of Uttar Pradesh, and in Jalandhar, Kapurthala, Hoshiarpur and Patiala districts of the Punjab. In India, best seeds are obtained when the capsules have not been incised for extraction of opium. The crop yields 220-275 kg. of seeds per hectare.

Uses. Poppy seeds are considered nutritive and are used in breads, curries, sweets and confectionery. Poppy seeds contain upto 50% of an edible oil. The oil is odourless and possesses a pleasant almond - like taste. In India the oil is extracted by cold-pressing the seed in small presses. The oil is widely used for culinary purposes. It is free from narcotic properties and used as a salad oil. The cake or the meal left after extraction of the oil from the seeds is sweet and nutritious and is eaten by poor people.

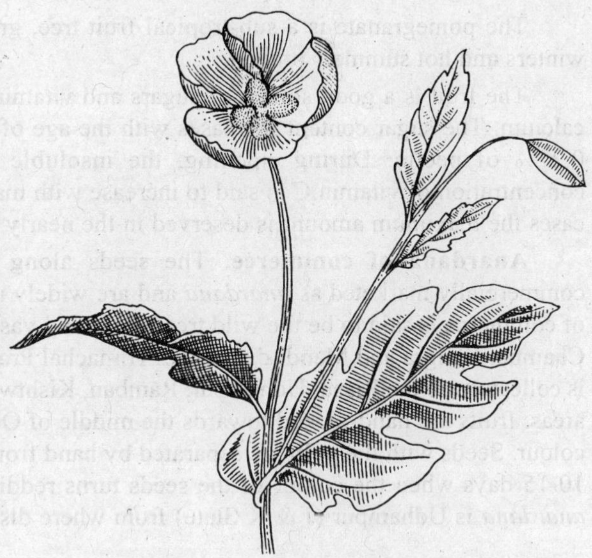

Fig. 22.31. Spices and condiments. *Papaver somniferum* Linn. var. *album*; Eng. Poppy seeds (Papaveraceae). A flowering twig. The poppy seeds are white and very small.

Chemical composition of seeds. Moisture 4.3-5.2%; protein 22.3-24.4%; ether extract 46.5-49.1%; nitrogen free extract 11.7-14.3%; crude fibre 4.8-5.8%; ash 5.6-6.0%; calcium 1.03-1.45%; phosphorus 0.79-0.89%; iron 8.5-11.1 mg/100 g.; thiamine 740-1181 ug/100 g.; riboflavin 756-1203 ug/100 g.; nicotinic acid 800-1280 ug/100 g; iodine 6 ug/kg.; manganese 29 mg/kg.; copper 22.9 mg/kg.; magnesium 15.6 g/kg.; zinc 130 mg/kg.; lecithin 2.80%; Oxalic acid 1.62%; pentosans 3.0-3.6%; traces of narcotin, an amorphous alkaloid and enzymes—diastase, emulsin.

Pomegranate

Punica granatum Linn.; English—Pomegranate; Hindi—*Anardana*; Assamese—*Dalim*; Bengali—*Dalimb*; Gujarati—*Dalamb*; Kannada—*Dalimbari*; Kashmiri—*Daan*; Malayalam—*Mathalam pazham*; Marathi—*Dalimb*; Tamil—*Mathalam pazham*; Telugu—*Dannima pandu*; Punjabi—*Anardana*; Sanskrit—*Dadima*; Family—Punicaceae.

A shrub or small tree, 5-10 m. high, native of Iran, Afghanistan and Baluchistan, found growing wild in the warm valleys and outer hills of the Himalayas between 900 and 1,800 m. and cultivated throughout India. Leaves 2.0-8.0 cm. long, oblong or obvate, shining above; flowers usually scarlet red; fruits globose, crowned by persistent calyx, with a coriaceous woody rind and an interior septate with membranous walls, containing numerous seeds; seeds angular with a fleshy testa which is red, pink or whitish.

Though mostly found cultivated in many parts of India, the tree is also very common and gregarious in the gravel and boulder deposits of dry ravines in the outer Himalayas, upto about 1,800 m. In Jammu, it is found gregariously on dry limestone soils in the upper extremities of sub-tropical forests; it is also found in Chamba, Kangra and Mandi districts of Himachal Pradesh and valleys below 1,800 m. in Jaunsar and Tehri - Garhwal districts of Uttar Pradesh. As a cultivated crop, maximum area is said to be devoted to it in Maharashtra, particularly in Pune, Sholapur and Satara districts. In Gujarat, its cultivation is concentrated in Dholka taluka. It is cultivated in the districts of Almora, Aligarh, Tehri - Garhwal, Meerut and Farrukhabad in Uttar Pradesh; Uthukuli, Michaelpatti, Vellodu and Dindigul in Tamil Nadu state; Penukonda and Madakasira in Andhra Pradesh; Tumkur, Kolar, Bangalore and Mysore districts in Karnataka state.

Among the numerous types grown in India, the *Bedana* and *Kandhari* are considered the best.

The pomegranate is a sub-tropical fruit tree, growing best in semiarid climate where cool winters and hot summers prevail.

The fruit is a good source of sugars and vitamin C, and a fair source of iron but is poor in calcium. The sugar content increases with the age of the fruit and of the tree. The fruit contains 0.27% of pectin. During ripening, the insoluble pectin changes into soluble pectin. The concentration of vitamin C is said to increase with maturity and ripening of the fruit, and in most cases the maximum amount is deserved in the nearly ripe fruit.

Anardana of commerce. The seeds along with the fleshy portions are dried and commercially marketed as *anardana* and are widely used as condiment. The source of *anardana* of commerce is said to be the wild trees, covering vast tracts of hill slopes in Jammu, and parts of Chamba, Kangra and Mandi districts of Himachal Pradesh. The main areas from where *anardana* is collected are the Riasu, Udhampur, Ramban, Kishtwar and Bhadarwah Forest Divisions. In these areas, fruits are hand-picked towards the middle of October, when they are ripe and brown red in colour. Seeds with the pulp are separated by hand from the rind and are dried in the sun for about 10-15 days when the colour of the seeds turns reddish brown. The main assembling centre for *anardana* is Udhampur (J & K State) from where distribution is done all over the country.

Fenugreek

Trigonella foenum-graecum Linn.; English—Fenugreek; Hindi—*Methi*, *muthi*; Bengali—*Methi*, *methika*; Punjabi—*Methi*, *methri*; Gujarati—*Methi*, *methini*; Tamil—*Vendayam*; Telugu—*Menti kura*; Kannada—*Menthya*; Malayalam—*Ventayam*; Sanskrit—*Methi*, *methika*; Persian—*Shanbalid*; Family—Papilionaceae.

It is a native of Southern Europe and Asia. It is grown mainly in Northern India. It is an annual legume with white flowers and long slender pods with a pronounced beak. The crop is sown in October-November. It is ready to harvest in April.

Uses. The seeds are chiefly used as a condiment to flavour curries made of rice, pulse, flour and meat, or as a relish with unleavened bread. The aromatic leaves are used as vegetable. The seeds are carminative, tonic and aphrodisiac.

SPICES AND CONDIMENTS OBTAINED FROM LICHENS

Parmelia

Parmelia abessinica Kremp.; Telugu—*Rathipoovvu*; Family—Parmeliaceae.

This is a crustaceous lichen found in rocky areas of Bellary, Anantpur and Guddapah districts of Andhra Pradesh. The lichen is available in large quantities in the market and used as food material and condiment. It contains atranorin (1.1%), lecanoric acid (3.3%), salazinic acid (0.1-0.5%) and isolichenin (3.4%).

Parmelia tinctorum Despr. is a foliose lichen commonly found in the plains of India and in the Himalayas on tree bark and rocks. It is commonly used in certain parts of South India as a component of various food preparations. It contains—crude protein, 13.8; ether extract, 6.0; isolichenin, 25.0; crude fibre, 13.4; and ash, 12.6%; calcium 1,728 mg.; ascorbic acid, 5.0 mg.; and riboflavin 100 mg./100 g.

Uses. The above mentioned lichens are used as spice and condiment.

23
CHAPTER

The Medicinal Plants—Pharmacognosy

The branch of medical science which deals with the drug plants is known as *pharmacognosy*. This science is concerned with the history, commerce, collection, selection, identification and preservation of crude drugs and raw materials. The study of the action of the drugs is known as *pharmacology*. There are several thousand drug yielding plants all over the world. Most of the plants are known and utilized by herb doctors and *ayurevdic vaids*. Only a few drug plants are cultivated. Most of the supply of drugs is obtained from wild plants growing in all parts of the world and especially in tropical regions. These drug-yielding wild plants are popularly known as *jari-butis* in India. The drug plants are collected and prepared in crude indigenous way. The medicinal value of drug plants is due to the presence of some chemical substances in the plant tissues which produce a definite physiological action on the human body. The most important chemical substances are alkaloids, carbon compounds, hydrogen, oxygen, nitrogen, glucosides, essential oils, fatty oils, resins, mucilages, tannins, gum, etc. Some of these substances are powerful poisons, and, therefore, the preparation and administering of drugs should be done by competent physicians.

CLASSIFICATION OF DRUGS

The classification of drugs in the present text is based on the morphology of the plant organ from which the drug is obtained. Here the drug plants have been considered and arranged on a morphological basis. The general categories of the drug plants are as follows:

1. Drugs obtained from roots.
2. Drugs obtained from underground stems.
3. Drugs obtained from bark.
4. Drugs obtained from stem and wood.
5. Drugs obtained from leaves.
6. Drugs obtained from flowers.
7. Drugs obtained from fruits.
8. Drugs obtained from seeds.
9. Drugs obtained from all parts of plant.

DRUGS OBTAINED FROM ROOTS

Aconite

Aconitum heterophyllum Wall., of family Ranuncualceae; Eng. Aconite; Hindi—*Atis*; Sanskrit—*Ativisha*.

This herb is commonly found in the subalpine and alpine zones of the Himalayas from the Indus to Kumaon, from 6,000 to 15,000 feet.

Uses. The drug commonly known as aconite is obtained from the roots. The roots are used as astringent, tonic, antiperiodic in diarrhoea, dyspepsia and cough. It contains the alkaloids, atisine, dihydroatisine, heteratisine and hetisine. It is used externally for rheumatism and internally to relieve pain and fever.

Total alkaloidal content is 0.79 per cent. Plant is considered a valuable febrifuge and bitter tonic. Roots are used for hysteria, throat infections, dyspepsia and vomiting, abdominal pain and diabetes

Indian Belladona

Atropa acuminata Royle ex Lindley, of family Solanaceae; Eng. Indian belladonna; Hindi—*Sag-angur*.

This is tall erect herb, 3-5 feet high. It is found in Kashmir and Himachal Pradesh.

Uses. The roots are used in the manufacture of tinctures and plasters. It is used as an anodyne of rheumatism, neuralgia and local inflammations.

Belladona

Atropa belladonna Linn., of family Solanaceae; Eng. Belladona; Hindi—*Sag angur*.

This is also a tall erect herb. It is native of Europe, now cultivated in Kashmir.

Uses. The roots are used as sedative, stimulant and antispasmodic. It is used externally to relieve pain and internally to check excessive sweat, cough, etc.

Dried roots and leaves are used as sedative, antispasmodic, and anodyne; also used in ophthalmology to dilate pupils. Leaves are stimulant and narcotic, contains atropine and hyoscyamine, extremely toxic alkaloids.

Liquorice

Glycyrrhiza glabra Linn., of family Papilionaceae, Eng. Liquorice, sweetwood; Hindi—*Mulhatti*; Sanskrit—*Madhuka*; Bengali—*Jashtimadhu*; Marathi—*Jestha madha*; Gujarati—*Jethi madha*; Telugu—*Atimadhuramu*; Tamil—*Atimadhuram*; Kannda—*Atimadhura*; Malayalam—*Iratimadhuram*.

This is a perennial herb; native of the Mediterranean region but grown in Himachal Pradesh, Jammu and Kashmir.

Uses. The root contains glycyrrhizic acid. It is sweet demulcent, emollient, pectoral, alterative; laxative and expectorant; it is given in cough, bronchitis and other chest complaints, catarrhal condition of bowels, urinary passages, asthma, hoarseness of voice, etc. Its infusion is given in sore throat. It is also used in preparation of nauseous drugs.

Liquorice is also chewed with betel leaves. Principal constituents of liquorice to which it owes its characteristic sweet taste is *glycyrrhizin*, 2 to 14 per cent.

Chitraka

Plumbago zeylanica; Hindi—*Chitraka*; Family Plumbaginaceae. The plants grow as perennial herbs in the plains of West Bengal and South India. The root has been reported to be a powerful poison. It yields *plumbagin* (chitraka) a crystalline substance. It is when given internally or applied to the *osuteri* causes abortion. Plumbagin has been reported to be a powerful irritant to the smooth muscles and uterus.

Serpentine

Rauvolfia serpentina Benth., of family Apocynaceae; Eng. Serpentine; Hindi—*Sarpagandha*; Sanskrit—*Sarpagandha*; Bengali—*Chandra*; Marathi—*Harkaya*; Telugu—*Paataalagani*;

Tamil—*Chivan amelpodi*; Kannda—*Sarpagandhi*; Malayalam—*Chuvannavilpori*; Oriya—*Patalgarur*; Mundari—*Simjenga*; Assam—*Arachontita*; Khasi Hills—*Todong-pait-parao*; Trade—*Rauvolfia*.

Habitat. A large climbing or twining shrub, found in the tropical Himalayas and plains near the foot of the hills from Sirhind and Moradabad to Sikkim. It occurs also in Assam (up to 4000') and in South India Peninsula along the Ghats to Travancore and Kerala. The genus is distributed in the tropics, *i.e.*, Central and South America, Africa, India, Sri Lanka, Burma, China and Japan. The largest number of species are found in Africa and South America. It is also distributed to Malaysia and Java.

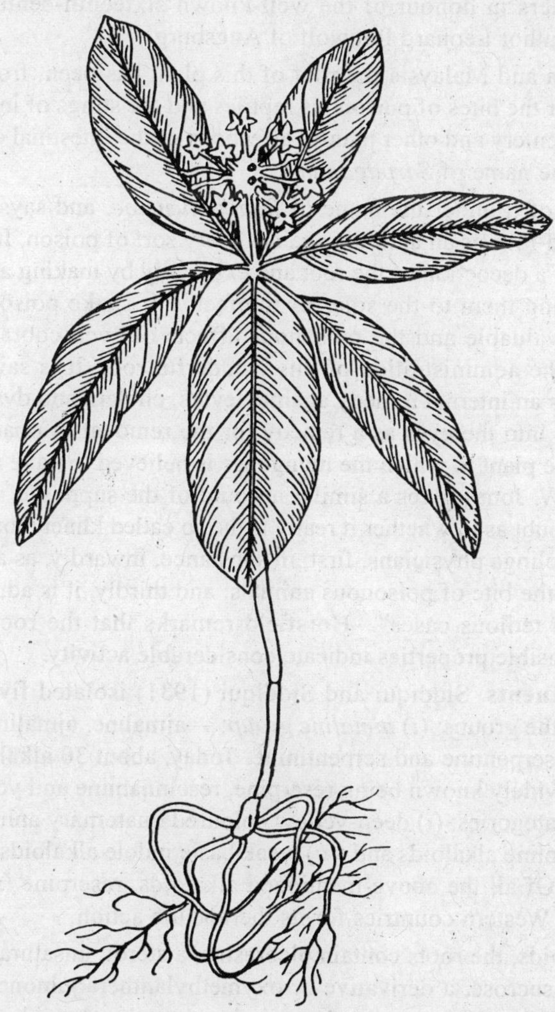

Fig. 23.1. *Rauvolfia serpentina* (*Sarpa-gandha*). The roots are used as medicine. It has been used in indigenous medicine for hundreds of years, has acquired great importance in modern therapeutics following recent scientific research. It is being extensively used in reducing blood pressure in hyperplesis and as a sedative in the treatment of insomnia, hypochondria, mental disorders and certain forms of insanity. It has been recently used to advantage in numerous other diseases like intractable skin disorders such as psoriasis, excessive sweating and itching, gynaccological ailments such as menopausal syndrome; in old-age afflictions such as toxic goitre; and in conditions such as angina pectoris and rapid or irregular heart action. As high blood pressure now is one of the commonest constitutional diseases, the drug is considered a great boon and is assured of an ever increasing demand.

History. The story of *Rauvolfia serpentina* is a most fascinating one. The plant is mentioned in ancient literature including the works of Charaka (1000-800 B.C.) where it is described under its Sanskrit name of Sarpagandha as a useful antidote against snake-bite and insect stings. There are many folk-lores about rauvolfia. One such is that a mongoose would first chew upon its leaves to gain strength before engaging a cobra in combat. According to another, its leaves when freshly ground and applied to the toes, could serve as an antidote for snake poison. A third one has it that the mentally deranged person is relieved of his madness if he eats the pieces of the root. Because of this supposed curative effect in case of insanity the plant has also been known in India as '*Pagal-ki-dava*'. Its generic name was coined towards the end of the seventeenth century by the French Botanist, Plumiers in honour of the well-known sixteenth-century German physician, botanist, traveller and author Leonard Rauwolf of Augsburg.

Medicine. In India and Malaysia the root of this plant has been, from ancient times, much valued as an antidote for the bites of poisonous reptiles and the stings of insects, also as febrifuge, and as a remedy for dysentery and other painful affections of the intestinal canal. In Sanskrit works it is mentioned under the name of *Śarpagandha*.

Rumphius speaks of it under the name of *Radix musteloe*, and says that in his time it was widely used in India and Java as an antidote against every sort of poison. It was administered both internally in the form of a decoction of the root and externally by making a plaster of the roots and fresh leaves and applying them to the soles of the feet. For snake poison, he continues, it was esteemed as specially valuable and the poisonous effects of even cobra's bite were viewed as rendered harmless by the administration of this wonderful root. It is said by him to have been universally employed as an internal remedy against fevers, cholera and dysentery, and the juice of the leaves was instilled into the eyes as a remedy for the removal of opacities of the cornea. He states also that this is the plant to which the mongoose is believed to have recourse when bitten by poisonous snakes. Sir W. Jones gives a similar account of the supposed medicinal virtues of the plant, but expresses a doubt as to whether it really is the so called khneumon plant. Roxburgh states that it is used by the "Telinga physicians, first, in substance, inwardly, as a febrifuge; secondly, in the same manner, after the bite of poisonous animals; and thirdly, it is administered, in substance to promote delivery in tedious cases". Horsfield remarks that the root yields a strong bitter infusion and that its sensible properties indicate considerable activity.

Chemical constituents. Siddiqui and Siddiqui (1931) isolated five crystalline alkaloids. These are divided into the groups: (*i*) *ajmaline group*; —ajmaline, ajmalinine and ajmalicine; and (*ii*) *serpentine group*—serpentine and serpentinine. Today, about 30 alkaloids are known to exist in this plant, the more widely known being reserpine, rescinnamine and yohimbine. Now they can be grouped into three categories: (*i*) deep-yellow coloured quaternary anhydronium bases, (*ii*) the intermediate strong indoline alkaloids and (*iii*) weak basic indole alkaloids; the last mentioned two groups are colourless. Of all the above-mentioned alkaloids, reserpine (serpasil) has gained the greatest prominence in Western countries for its therapeutic action.

Besides the alkaloids, the roots contain oleoresins, a sterol, unsaturated alcohols, oleic acid, fumaric acid, glucose, sucrose, a derivative or oxymethylantheraquinone, a strongly fluorescent substance and mineral salts. Of these, the oleoresin is said to be physiologically active and shows typical hypnotic and sedative action of the drug.

Uses as a drug. Vast developments have been made in the chemistry of its different alkaloids and their role in hypertension, neuropsychiatric and other pharmacodynamics. The two main pharmacological actions of reserpine, *viz.*, lowering the blood pressure and sedative influence, have led to successful trials of the drug in hypertension, neuropsychiatric, gynaecological and geriatrics. With a moderate hypotensive action it controls mainly subjective symptom of high blood pressure when given alone. The drug is also reported to be useful in certain gynaecological conditions, such as menstrual, moliminia, frigidity and women complaining of menopausal syndrome. In the clinical use of reserpine in geriatrics it is reported to be of value in hypertension, nervousness, insomnia, etc.

Some important preparations are as follows:

Rauvolfia serpentina (tablets)—Gluconate Limited Calcutta.

Ralfen (alcoholic preparation) (tablets)—Bengal Chemical and Pharmaceutical Works, Calcutta.

Serpina or S 117—Himalaya Drug Company, Dehradun.

Raudixin (tablets)—Sarabhai Chemicals, Ahmedabad.

Adelphene-CIBA India Limited, Mumbai.

Serpasil—CIBA Pharma Limited, Basle, Switzerland.

Ashwagandha

Withania somnifera Dunal.; Hindi—*Ashwagandha* of family Solanaceae; Sanskrit—*Ashwagandha*; Bengali—*Ashvagandha*; Marathi—*Askandha*; Gujarati—*Ghodakun*; Telugu—*Pulivendram*, *panneru*; Tamil—*Amukkura*; *aswagandhi*; Kannada—*Panneru*, *aswagandhi*; Punjab—*Asgand*; Rajasthan—*Chirpotan*; Trade—Aswagandha.

The wild plants are generally erect branching shrubs up to one metre in height. The cultivated plants are morphologically distinct from wild plants. The cultivated plants are "different from the wild ones not only in their therapeutic properties but in all morphological characters like roots, stems, leaves, flowers, pollen grains, mature seeds and the enlarged calyx." (Kaul, K.N., 1957).

Except for a limited collection of roots from *W. somnifera* plants growing wild, the supplies of Ashwagandha roots are obtained exclusively from cultivated sources.

The entire plants are uprooted for the collection of roots. The roots are separated from the overground parts as soon as possible after collection by cutting the stems 1-2 cm. above the crown. The entire root system along with the remains of aerial stems is either cut transversely into smaller pieces or dried in the whole in the sun. The roots undergo further cleaning, trimming and grading before they are despatched to the various drug markets of India.

Uses. This is used as a tonic in geriatrics, being efficacious in relieving hand and limb tremors of elderly people. It is prescribed for all kinds of weaknesses and is supposed to promote strength and vigour, being regarded as an aphrodisiac and rejuvenator. It is used for the treatment of rheumatic pain, inflammation of joints and certain paralytic conditions. Several preparations of this drug are used for all types of nervous disorders and as sedatives in the treatment of insanity and in hypertension. The leaves of the plant are used as a local application for all types of skin lesions, ulcers, boils, and swelling to reduce pus formation, inflammation to promote healing processes.

Roots have long been in use for hiccup, cough, dropsy, rheumatism, and female disorders. The pharmacological activity is ascribed to the presence of several alkaloids. The leaves are used as a febrifuge and applied to lesions, painful swellings and sore eyes. Withaferin A is the most important withanolide to which the curative properties of leaves are attributed.

Asafetida

Ferula asafoetida Linn., of family Umbelliferae (Apiaceae); Eng. Asafetida; Hindi—*Hing*; Sanskrit—*Balhika*, *hingu*; Hindi, Bengali, Marathi, Gujarati and Kannada—*Hing*; Telugu—*Inguva*; Tamil and Malayalam—*Perungayam*; Oriya—*Hengu*; Kashmir—*Yang*; Mumbai—*Hing*.

This is an unpleasant smelling perennial herb, commonly grown in the Punjab and Kashmir.

Uses. The oleo-resin gum that exudes from the root-stock is carminative, antispasmodic, nervine, stimulant, digestive, sedative, expectorant, diuretic, anthelmintic and emmenagogue. It is prescribed in flatulent colic, dyspepsia, asthma, hysteria, convulsions, cholera, chronic constipation, chronic bronchitis, whooping cough, cough, spasmodic disorders of the bowels and angina pectoris. The fried gum is given either as a solution or emulsion or pills, preferably as pills.

DRUGS OBTAINED FROM UNDERGROUND STEM

Turmeric

Curcuma longa Linn., of family Zingiberaceae; Eng. Turmeric; Hindi—*Haldi*; Sanskrit—*Haridra*; Hindi, Bengali, Marathi and Gujarati—*Haldi, Halada*; Tamil—*Manjal*; Telugu—*Pasupa*; Kannada—*Arishina*.

A herb, cultivated commonly in Maharashtra, Bengal, Tamil Nadu, Andhra Pradesh and Orissa for its rhizomes.

Uses. The rhizome is aromatic, stimulant, antiperiodic, tonic, alterative and carminative. It is given in diarrhoea, intermittent fevers, dropsy, jaundice, liver disorders and urinary troubles. The fresh juice of rhizome is used as an anthelmintic. It is commonly used for cleaning foul ulcers. Along with lime or alum it makes a good dressing for sprains, bruises, wounds, etc. Its decotion is an effective eye wash in ophthalmia. The fresh juice is also used as antiparasitic for many skin affections.

Fig. 23.2. The rhizomes/corms of turmeric – a medicinal plant.

Curcuma paper is used as an indicator of chemicals. Rhizomes yield an oil used as a carminative, stomachic and tonic.

Mango Ginger

Curcuma amada Roxb., of family Zingiberaceae; Eng. Mango ginger; Hindi—*Amahaldi*; Sanskrit—*Karpura-haridra*; Bengali—*Amada*; Marathi—*Amba haladi*, Telugu—*Mamidiallam*; Tamil—*Mangai inji*.

This herb is grown in Bengal, Andhra Pradesh and Tamil Nadu. Also found wild in parts of Bengal, Konkan and Tamil Nadu.

Uses. The rhizome is carminative, stomachic and cooling. It is applied over contusions and sprains.

Garlic

Allium sativum Linn., of family Liliaceae; Eng. Garlic; Verna. *Lahsuna*; Sanskrit—*Arishta, lashuna*; Hindi, Gujarati, Bengali and Marathi—*Lasun, lasan*; Telugu and Malayalam—*Velluli, tellagadda*; Tamil—*Vellaipundu*; Kannada—*Bellulli*; Punjab—*Thom*.

This is a strong smelling, glabrous, bulbous-rooted, perennial about a foot in height. The bulbs are short, compressed, breaking up into 10-12 bulbils or cloves, surrounded by a few, dry membranous scales. It is cultivated throughout the country.

Uses. Garlic is of great medicinal value. It is given in fevers, coughs, flatulence, disorders of the nervous system, pulmonary phthisis, whooping cough and dilated bronchitis. It has diuretic properties. A decoction of garlic made with milk and water is given in small doses in hysteria, flatulence, sciatica, etc. In the form of syrup, it is a valuable remedy for asthma, and disorders of the chest and lungs. It is used as an anthelmintic. Externally it is used as rubefacient, vesicant and disinfectant. Its juice is introduced in the ear for relief of earache. The oil in which garlic has been fried is a useful liniment for rheumatic pains, nervous diseases like scabies and maggot infested wounds.

Fig. 23.3. Garlic bulbs.

Ginger

Zingiber officinale Rosc., of family Zingiberaceae; Eng. Ginger; Hindi—*Adrak*; Sanskrit—*Ardraka*; Bengali—*Ada*; Marathi—*Ale*; Telugu—*Allamu*, *sonthi* (dry); Tamil—*Allam*; Kannada—*Hasisunti*; Malayalam—*Andrakam*, *inchi*.

This is a perennial herb; the root stock is horizontal, tuberous and aromatic; native of South-East Asia but now cultivated mainly in Kerala, Uttar Pradesh, West Bengal, Maharashtra, Himachal Pradesh and Andhra Pradesh for the rhizomes.

Uses. The rhizome is used as a stimulant, carminative and flavouring agent. It is given in dyspepsia and flatulent colic. It is administered as an adjunct to many tonic and stimulating remedies. It makes a valuable drug for disorders of the digestive system, rheumatism, piles, pulmonary and catarrhal diseases, dropsy, febrile diseases, neuralgia, etc. The juice of fresh rhizome, with honey is a remedy for coughs and asthma. A paste of ginger is a local stimulant and rubefacient in headache and toothache.

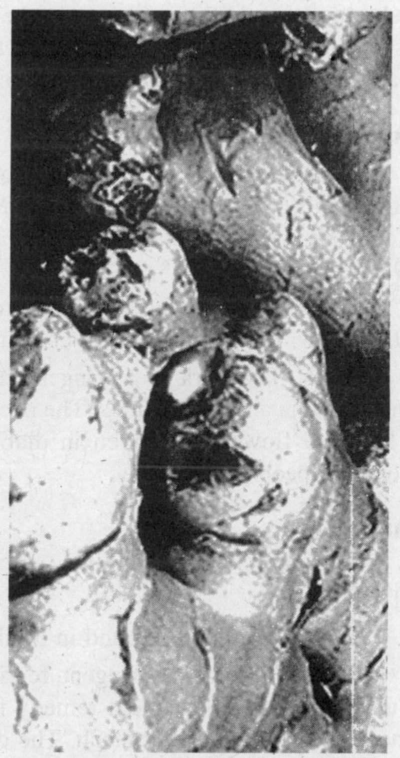

Fig. 23.4. Ginger rhizomes

DRUGS OBTAINED FROM BARKS

Cinnamon

Cinnamomum zeylanicum Bl., of family Lauraceae; Eng. Cinnamon; Hindi—*Dalchini*.

This is native of Sri Lanka but grown in the Nilgiris, South Kanara, Malabar, Assam and Kumaon.

Uses. The bark is aromatic, cordial, astringent, carminative and stimulant; it is given in diarrhoea, nausea, gastric irritation, flatulence, vomiting, spasmodic affections of the bowels, toothache, paralysis of the tongue and labour caused by defective uterine contractions. It is administered in the form of a powder, infusion or decoction. A paste made of the bark is applied to the temples to get relief from neuralgic pains and severe headaches. The oil obtained from the stem bark is used as a carminative, antiseptic and astringent.

Fig. 23.5. Cinnamon. The bark quills.

Quinine

Cinchona calisaya Wedd., *C. ledgerina* Moens. ex Trimen; *C. officinalis* Linn; *C. robusta* How., *C. succirubra* Pavon ex Klotzsch. of family Rubiaceae; Eng. Quinine; Hindi—*Kunain*.

The quinine tree is native to the Andes of South America. In India it is found in West Bengal, the Khasia hills, Nilgiris, South India, Sikkim and Madhya Pradesh.

Uses. Quinine is one of the most important drugs known and it is the only adequate cure for malaria. Quinine is obtained from the hard thick bark of above mentioned species of *Cinchona*. The most important constituent of Cinchona bark is quinine, a very bitter, white, granular substance. In addition to its use in the treatment of Malaria, it is a valuable tonic and antiseptic. It is also used in the treatment of other fevers. Some 29 other alkaloids have been isolated from the bark, including cinchonidine, cinchonine, and quinidine, all of which are useful in medicine.

Asoka Tree

Saraca indica Linn., of family Caesalpiniaceae; Eng. Asoka tree; Hindi—*Ashok*; Sanskrit, Hindi and Bengali—*Asok*, *ashoka*; Marathi—*Ashoka*; Gujarati—*Ashopalava*; Telugu—*Asoka*; Tamil—*Asogam*; Kannada—*Asokadamara*; Malayalam—*Asokam*; Oriya—*Oshoko*; Khasi Hills—*Dieng-sohkyrkha.*

This is a spreading evergreen tree; native of India; commonly found in Central and East Himalayas, Bengal and West Peninsula.

Uses. The bark is astringent. Its decoction is given in uterine disorders, especially in menorrhagia and leucorrhoea. The decoction is also an efficacious remedy for piles and dysentery. The dried flowers are given in diabetes. The dried flowers of it are fried and used even by common people in Kerala.

Arjun Tree

Terminalia arjuna (Roxb. ex DC.) Wight and Arn., of family Combretaceae; Eng. Arjun tree; Hindi—*Arjuna.*

A tree, commonly found in Madhya Pradesh, Bihar and the Western Peninsula.

Uses. The bark is astringent, febrifuge, cooling, cardiac stimulant, cholagogal, lithontriptic and vulnerary. It is an excellent remedy for heart diseases. The decoction with milk is given every morning on an empty stomach. The decoction is also used for cleaning sores and ulcers.

DRUGS OBTAINED FROM STEMS AND WOODS

Ephedrine

Ephedra gerardiana Wall., of family Gnetaceae or Ephedraceae; Eng. Ephedrine; Hindi—*Khanda*.

A small shrub, found in the drier regions of the temperate and Alpine Himalayas from Kashmir to Sikkim at 7,000 to 16,000 feet elevation.

Uses. The stems are the source of famous drug ephedrine. The liquid extract of stem is used for controlling asthma. The tincture of the plant is cardiac and circulatory stimulant. The decoction of stems and roots is used as a remedy for rheumatism and syphilis in Russia. Juice of berries is used in affections of the respiratory passages. The stems contain the alkaloids—ephedrine and pseudo-ephedrine.

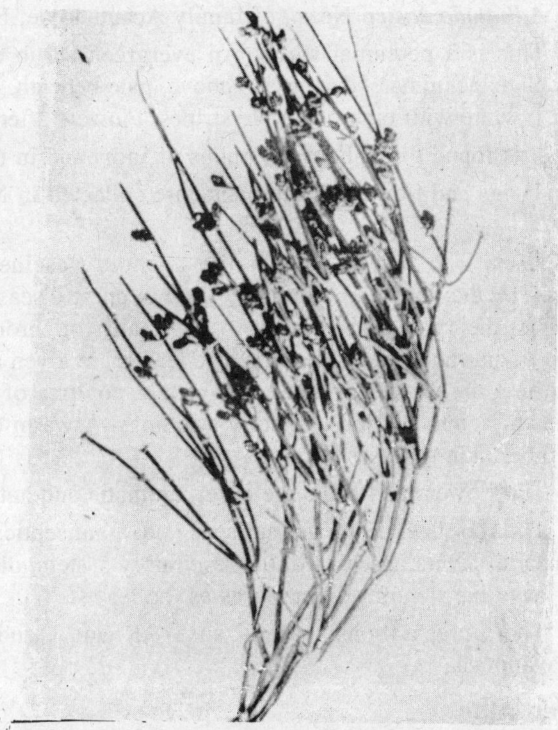

Fig. 23.6. *Ephedra*. The stems are the source of famous drug '*ephedrine*'.

White Sandal Wood

Santalum album Linn., of family Santalaceae; Eng. White sandalwood; Hindi—*Safed chandan*.

A tree, found in the Western Peninsula, Karnataka, Tamil Nadu, and forests of Kerala.

Uses. The wood is cooling, astringent, bitter, sedative, cardiac, tonic and diuretic. It is given in fevers and thirst. The powder of wood is given in gonorrhoea, fevers and bilious disorders. A paste of wood is a cooling dressing in inflammatory eruptive skin diseases. The paste is also applied to the temple in headache and fever. The essential oil of wood is stimulant, antiseptic demulcent and diuretic. It is used in gonorrhoea, chronic catarrh of the bladder, urethral haemorrhage, chronic bronchitis and intermittent fevers. The oil is applied over scabies and other skin diseases.

Catechu

Acacia catechu Willd., of family Mimosaceae; Eng.. Catechu tree; Verna. *Kattha*, *Khair*.

Uses. It is a valuable astringent. It is generally used for disorders in which a mild, non-irritating and powerful astringent is required, such as chronic diarrhoea, dysentery, bleeding piles, uterine haemorrhage, leucorrhoea, etc. It is very efficacious for mercurial salivation bleeding or ulcerated or spongy gums, hypertrophy of the tonsils, enlarged uvula, aphthous ulceration of the mouth, etc. It may be used as lozenge made with the addition of gum arabic and sugar. A powder of catechu is used for plugging the cavity of an aching tooth to relieve pain. A tincture of catechu is useful for bed sores and painful mammary glands. A mixture of catechu and myrh, known as *Kathbol*, is used as a tonic, as a galactagogue for nursing mothers. Khersal or catechuic acid, found in small fragments in cavities of the wood, is a useful drug for chest diseases. The juice of the fresh bark is given with asafoetida in haemoptysis.

DRUGS OBTAINED FROM LEAVES

Vasaka

Adhatoda vasica Nees., of family Acanthaceae; Eng. Malabo nut; Verna—*Arusi, rusa.*

This is a perennial shrub. An evergreen shrub about 4-8 feet in height. Leaves elliptic, lanceolate, acuminate, dark green above, pale beneath. Flowers in short dense–axillary peduncled spikes, white with pink or purple stripes. Flowers after rainy season.

It is found throughout the plains of India and in the sub–Himalayan tract up to 4,000 feet.

Twigs and flowers of the plants are collected in March-April. The dried drug is sold in the market.

Uses. The leaves contain the alkaloid vascine; they are a powerful expectorant and antispasmodic; they are commonly used in chest diseases, especially phthisis. A decoction of the leaves or their powder is also given, especially for chronic bronchitis and asthma. The leaf juice is given in diarrhoea and dysentery. The leaves are given as a febrifuge in malarial fevers. For relief of asthma the dried leaves are smoked. A poultice of the leaves is applied over fresh wounds, rheumatic joints and inflammatory swellings. A warm decoction of the leaves is used for scabies and other skin diseases.

The flowers and fruits are bitter, aromatic and antispasmodic.

The root is expectorant, antispasmodic, antiseptic, antiperiodic and anthelmintic. It is given in malarial fevers, diseases of the respiratory system, diphtheria and gonorrhoea. The root and the bark have the same medicinal uses as the leaves.

The preparations are—Vansa Khand and Chyavanprash.

Indian Aloe

Aloe barbedensis Mill., of family Liliaceae; Eng. Indian Aloe; Verna. *Gheekunvar, ghrita kumari.*

The plant is about 2-3 feet high. The leaves are fleshly tapering to a blunt point, smooth, pale green, having horny prickles on their margins. The plant is found throughout our country.

Uses. After removing the skin of leaves they are given in fevers, enlargement of the liver, spleen and other glands, skin diseases, gonorrhoea, constipation, menstrual suppression, piles, jaundice and rheumatic affections. In the diseases of liver and spleen the pulp of one leaf is administered with black salt and ginger. The juice of roasted leaf is being given with honey for cough and cold. The pulp of one leaf is being given daily in abdominal tumors, dropsy, carbuncles, piles, sciatica, rheumatism and retention of urine in fevers. A salad of leaves is eaten in indigestion, constipation and flatulence. The leaf juice is given as a remedy for intestinal worms in children. A poultice of the leaves is applied to tumours, cysts, inflamed parts and scalds.

Fig. 23.7. *Aloe barbedensis*. A common medicinal plant with succulent leaves. The pulp of leaves is given in fever and liver enlargement.

Life Plant

Bryophyllum pinnatum Kurz., of family Crassulaceae; Eng. Life plant; Hindi—*Ghamari.*

This is a succulent perennial plant. Distributed throughout India.

Uses. The leaf juice is given in diarrhoea, dysentery, calculous affections and cholera. The leaves are styptic, astringent and antiseptic; they are slightly toasted before they are applied to wounds, bruises, boils, cuts, ulcers, bites of venomous insects; by its application swelling is prevented, incised wounds heal rapidly and irritation is allayed.

Fig. 23.8. *Bryophyllum.* A common medicinal plant. The leaves possess antiseptic properties.

Indian Pennywort

Centella asiatica (L.) Urban., of family Umbelliferae (Apiaceae); Eng. Indian pennywort; Hindi—*Brahmi.*

This is a common trailing herb, rooting at the nodes. Commonly found throughout India.

Uses. The weed is alterative, tonic, diuretic, antiphlogistic, blood purifier and local stimulant. It is a remedy for skin diseases, like chronic eczema, chronic ulcers etc., enlargement of glands, chronic rheumatism, chronic nervous diseases, madness, cholera, amenorrhoea and piles. The leaves are a household remedy in early stages of dysentery of children. The powder of leaves is given with milk in small doses in mental weakness and to improve memory. The fresh juice of leaves mixed with milk is given as an alterative in gonorrhoea, jaundice and fevers; this is also useful for children in skin diseases and for improving the blood and nervous system. An ointment made of leaf juice and lanoline is of great value in elephantiasis.

Fig. 23.9. A flowering twig of *Tulsi*. The leaves on steam distillation yield a bright yellow volatile oil possessing a pleasant odour. The oil is reported to possess antibacterial and insecticidal properties. It has marked insecticidal activity against mosquitoes. Besides the volatile oil the plant is reported to contain alkaloids, glycosides, saponins and tannins.

Holy Basil

Ocimum sanctum Linn., of family Labiatae; Eng. Holy basil; Hindi—*Tulsi.*

This is an erect softy hairy aromatic herb or undershrub. Found throughout India.

Uses. The leaves are expectorant, stomachic, anticatarrhal, diaphoretic and aromatic; their decoction or infusion is given in malaria, gastric diseases of children and liver disorders; as a prophylactic against malaria fresh leaves are taken with black pepper in the morning. The leaf juice is given in chronic fever, haemorrhage dysentery and dyspepsia; it is also used to check vomiting and as an anthelmintic.

Betel Pepper

Piper betle Linn., of family Piperaceae; Eng. Betel pepper; Hindi—*Pan*.

This is a climbing shrub cultivated for its leaves.

Uses. The leaves are astringent, aromatic, antiseptic, carminative, aphrodisiac, stimulant, expectorant and sialogogue. The leaf juice with honey is given to children in colic, indigestion, diarrhoea and laryngitis. The leaves are chewed to remove foul odour from the mouth. The leaf juice is used as eye drops in ophthalmia, and other painful eye diseases and night blindness. Oil coated leaves are a useful dressing for blistered surfaces and ulcers. The essential oil extracted from the leaves, is given in catarrhal and pulmonary diseases.

Fig. 23.10. *Piper betle*. The leaves are of medicinal value.

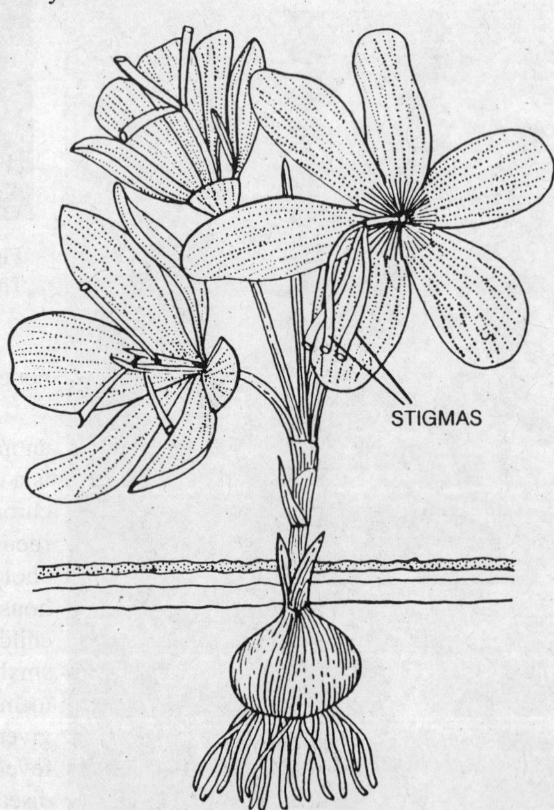

Fig. 23.11. *Crocus sativus* (Saffron). The stigmas and tops of the styles make the saffron of commerce.

Tylophora

Tylophora asthmatica; Syn. *Tylo-phora indica*; Hindi—*Jangli Pikavan* or *Antamul*; Family—Asclepiadaceae.

It is a trailing creeper, growing abundantly in Northern India, Bengal, Assam; the Western Ghats, Burma and Sri Lanka. The leaves are thick and deep green, 5 to 12 cm long and 2 to 6 cm wide. In dry state, they are rather, thick and of a pale-yellow colour.

Uses. The chewing and swallowing of leaves by the patient of asthma have shown wonderful results. The leaf was also tried at Vallabhbhai Patel Chest Institute, Delhi on the patients suffering from bronchial asthma, the results were quite satisfactory.

DRUGS OBTAINED FROM FLOWERS

Saffron

Crocus sativus Linn., of family Iridaceae; Eng. Saffron; Hindi—*Kesar*.

This is a herb, cultivated in Kashmir (Pampur), Bhersar and Chaubattia in Uttar Pradesh.

Uses. The dried stigmas and tops of the styles make the saffron of commerce. It is of medicinal value. It contains the bitter principle picrocrocin. Saffron is mild stimulant, stomachic, carminative, antispasmodic, nerve sedative, mild narcotic, diuretic and emmenagogue. It is a remedy for promoting menstruation. It is also used in fevers, hysteria, flatulent colic, asthma, leucorrhoea, piles and catarrhal diseases of children.

Iron Wood

Mesua ferrea Linn., of family Guttiferae; Eng. Ironwood; Hindi—*Nagkesar*.

A tree, found in the Eastern Himalayas, Assam, West Bengal, the Western Ghats, Travancore and the Andaman Islands.

Uses. The flowers are astringent and stomachic. The dried flowers are given in vomiting, dysentery, coughs, thirst, irritability of the stomach, excessive perspiration and bleeding pile.

Violet

Viola odorata Linn., of family Violaceae; Eng. Violet; Hindi—*Banafsha*.

A herb; native of Europe but now cultivated in Kashmir.

Uses. The flowers are demulcent astringent, diuretic, emollient, diaphoretic and laxative. They are given in biliousness and lung troubles. The flowers are used for the treatment of coughs, sore throat, kidney diseases, liver disorders and infantile affections.

DRUGS OBTAINED FROM FRUITS

Bael

Aegle marmelos Corr., of family Rutaceae; Eng. Bael tree; Verna. Bael, *Siriphal*.

This is a tree.

Uses. The fruit is of great medicinal value. The ripe fruit is aromatic, astringent, cooling and laxative. The unripe or half-ripe fruit is astringent, stomachic, antiscorbutic and digestive. It is best given in sub-acute or chronic cases of diarrhoea and dysentery and in irritation of the elementary canal; it is a useful adjunct in after treatment of bacillary dysentery. The ripe fruit is prescribed in intestinal disorders and certain forms of dyspepsia characterized by alternate constipation and diarrhoea. A sherbet of the ripe fruit is given for chronic constipation and dyspepsia.

Coriander

Coriandrum sativum Linn., of family Umbelliferae (Apiaceae); Eng. Coriander: Hindi—*Dhaniya*.

This is an aromatic herb; native of the Mediterranean region now cultivated in Madhya Pradesh, Maharashtra, Karnataka, Bihar and throughout India.

Uses. The fruit is aromatic, stimulant, carminative antispasmodis, correctivem diuretic, aphrodisiac, and refrigerant. A decoction of the dried fruits is given in flatulent colic, rehumatism and neuralgia. The volatile oil distilled from the fruits is given in flatulent colic, rheumatism and neuralgia. The watery paste of seeds is used for the cure of ulcers of the mouth and throat.

Cumin

Cuminum cyminum Linn., of family Umbelliferae (Apiaceae); Eng. Cumin; Hindi—*Zira*.

This is an annula aromatic 1-3 feet high herb. It is native of Mediterranean region but now commonly grown in the Punjab and Uttar Pradesh fro aromatic fruits.

North Gujarat especially Unja is well known for its cultivation and market.

Uses. The fruit is a good source of thymol; it is a stomachic, diuretic, carminative, stimulant, astringent and emmenagouge. It is prescribed in dyspepsia, diarrhoea, and hoarseness of voice. The fruit is given with lime juice to pregnant women for checking nausea; for promoting secretion of milk, cumin seeds are given after childbirth.

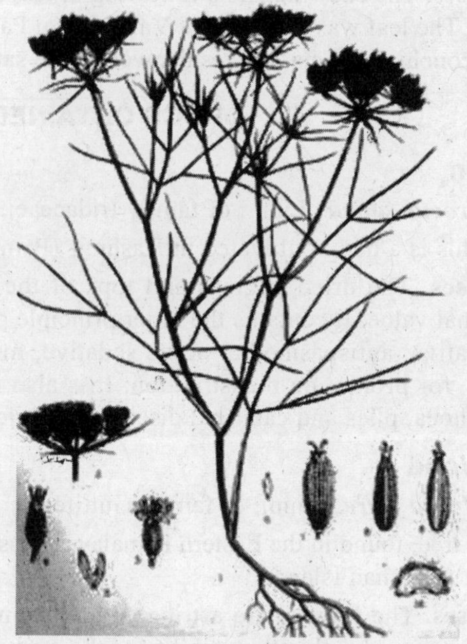

Fig. 23.12. *Cumin.* An aromatic herb.

Emblic

Emblica officinalis Gaertn., of family Euphorbiaceae; Eng. Emblic; Hindi—*Amla*.

A tree found throughout India.

Uses. The fruit is of great medicinal value. It is one of the richest sources of vitamin C. The fresh fruit is refrigerant, tonic, antiscorbutic, diuretic and laxative. It is used in fevers, vomiting, indigestion, habitual constipation and other digestion troubles. The dried fruit is a good astringent, refrigerant, stomachic, antiscorbutic and blood purifier. It is given in diarrhoea, dysentery and haemorrhage. The infusion of seeds is a useful eye-wash in ophthalmic diseases.

Fig. 23.13. *Emblica officinalis* (*Amla*). The fruit is one of the richest sources of vitamin C.

Fennel

Foeniculum vulgare Mill., of family Umbelliferae (Apiaceae, Fennel; Hindi—*Saunf*.

This is an aromatic herb. It is native of Mediterranean region, now cultivated mainly in the Punjab, Assam, Maharashtra and Baroda.

Uses. The fruit is sweet, laxative, aphrodisiac, stomachic, stimulant, aromatic, corrective of flatulence, appetizer, anthelmintic, carminative, lactagogue, diuretic, stimulant and antidotal. It is administered in eye diseases, burning sensation, fever, thirst, wounds, dysentery, diseases of the chest, spleen and kidney, headache, amenorrhoea and bellyache of children; it also promotes female monthly regularity. In dysentery a decoction of fruit is given; in fever a decoction of fruit is given with sugar. The oil extracted from the seed is an anthelmintic against hookworms. The leaves are diuretic, digestive, stimulant and aromatic; they are useful in cough, flatulence, colic and thirst. They are also eaten to improve the eyesight.

Opium

Papaver somniferum Linn., of family Papaveraceae; Eng. Opium; Hindi—*Afim*.

A herb; native of W. Asia, now grown in Uttar Pradesh, in the East Punjab, Rajasthan and Madhya Pradesh.

An inspissated juice obtained by scratching the unripe capsules of *Papaver somniferum* Linn.; and allowing the milky sap, which exudes thereby, to dry spontaneously. The bulk of the medicinal article is obtained from Asia Minor, at the present day, as it has been for nearly the past 2,000 years. Indian Opium is, however, the most important form commercially, being that which is smoked, eaten or drunk, in various preparations by the inhabitants of Asiatic countries, chiefly the Chinese.

Fig. 23.14. *Papaver somniferum* (*Afim*)—The latex is obtained by scrapping the capsules. The latex, obtained from the immature fruits; is the source of opium, which is used to induce sleep, relieve pain, and relax spasms.

In India the opium-poppy is grown here and there throughout the country, but its cultivation is mainly confined to three centres, which afford the following opiums—'Patna Opium' in Bihar, 'Benares Opium' in Uttar Pradesh, and 'Malwa Opium' in Madhya Pradesh and certain parts of Rajasthan.

Medicinal properties of opium. Opium is well known to the medical profession. The medicinal drug of Europe is mainly derived from Asia Minor and is the produce of *P. somniferum* var. *glabrum*.

In the words of Pharmacopaeia of India it may be said that opium in its primary effects medicinally is stimulant, and its secondary action narcotic, anodyne, and antispasmodic. "It operates chiefly on the cerebrospinal system, and, through the nerves arising therefrom, it affects more or less every organ of the body. It tends to diminish every secretion excepting that of the skin, which increases under its use. In overdoses it is a powerful poison." In inflammation, especially of serious membranes, it has been employed extensively either alone, or in combination with calomel, antimony, and other remedies. As a general rule it is less applicable to inflammatory and other diseases, in which the tendency to death is by coma or by apnoea, then to those which produce death by asthenia. In the latter, as in peritonitis, it holds the first place as a remedial agent. In fevers, especially in the advanced stages, it is of the highest value, either alone or in combination with antimony, camphor, allaying vascular and nervous excitement, and procuring sleep; but as a general rule, it is inadmissible when contraction of the pupil is strongly marked. In painful spasmodic affections, opium in large and repeated doses often affords immediate relief. In various morbid states of the abdominal viscera, *e.g.*, simple or cancerous ulceration of the stomach, chronic gastritis, gastrodynia, nervous and sympathetic vomiting, diarrhoea, dysentery, colica pictonum, strangulated hernia, visceral destructions, it is often given with the best results. In diseases of the genito-urinary system, *e.g.*, cystitis, cistirrhoea, spasmodic stricture of the urethra, also in menorrhagia, dysmenorrohea, irritable states of the uterus, metritis, it is a remedy of the highest value. Tetanus, acute rheumatism, and delirium tremens are amongst the other diseases in which opium has been employed as a sheet anchor. In cholera it has been used but with doubtful results. In cancerous and gangrenous ulceration, opium, by allaying constitutional disturbances, often exercises a favourable influence on the local symptoms. As an external application, opium proves valuable in various rheumatic, neuralgic, and other painful affections; also in ophthalmia and other diseases of the eye".

"*Dose.* From a quarter of a grain to two grains or more according to circumstances, in many painful affections the sole criterion for the regulation of the dose is the amount of relief afforded".

Long Pepper

Piper longum Linn., of family Piperaceae, Eng. Long pepper; Hindi—*Pipar*, *Piplamul.*

This is a slender creeping undershrub. It is native of India and cultivated in the Western Ghats, Karnataka and Tamil Nadu for the fruits.

Uses. The sun-dried unripe fruit (berries) or the spike of the small flowers is known as long pepper. It is a cardiac, stimulant carminative, alterative tonic, laxative, digestive, emollient, stomachic and antiseptic. It is given with honey for indigestion, dyspepsia, cough, bronchitis, asthma, hysteria, cholera, fever, leprosy, piles, gout, etc. To prevent fever after childbirth it is an efficacious remedy. The dried root (*Piplamul*) possesses the same medicinal qualities as the berries but in an inferior degree.

Black and White Pepper

Piper nigrum Linn., of family Piperaceae; Eng. Black pepper; Hindi—*Kali mirch*; *gol mirch.*

This is a stout glabrous creeper; native of the Indo-Malayan region, also cultivated in the Western Ghats, Karnataka, Maharashtra, Assam and Kerala.

Uses. The unripe dried berries are the black pepper of commerce; and the ripe fruit with the outer covering removed is the white pepper of commerce. The berries contain the alkaloid piperine, The fruits are aromatic, stimulant, carminative, digestive, stomachic, nervine tonic, diuretic emmenagogue and antiperiodic. It is given in dyspepsia, flatulence, diarrhoea, cholera, piles, disorders of the urinary system, cough, coma, gonorrhoea and malarial fever. It is given either as an infusion or powder. White pepper is more aromatic and less pungent than black pepper. It is given in dyspepsia and constipation.

Belleric Myrobalan

Terminalia bellirica Roxb., of family Combretaceae; Eng. Belleric myrobalan; Hindi—*Bahera.*

A common tree grown as an avenue.

Uses. The ripe dry fruit known as belleric myrobalan (bahera) is astringent, bitter, tonic and laxative. It is given in piles, dropsy, diarrhoea, leprosy, biliousness, dyspepsia and headache. Half ripe fruit is given as purgative and fully ripe as an astringent.

Chebulic Myrobalan

Terminalia chebula Retz., of family Combretaceae; Eng. Chebulic myrobalan; Hindi—*Harara*, *Haritaki.*

A tree. Commonly found in Northern India. Also found in South India in the Deccan tablelands at 1,000–3,000 feet and up to 6,000 feet in Travancore; higher forests of the Mumbai Ghats, Satpuras, Belgaum and Kanara.

Uses. The fruit is the chebulic myrobalan of commerce. It is an efficacious laxative, astringent, stomachic, tonic and alterative. The fruit pulp is given in chronic diarrhoea, dysentery, flatulence, asthma, urinary disorders, vomiting, enlarged spleen and liver, etc. It is used externally as a local application to chronic ulcers and wounds and as a gargle in stomatitis. Fine powder of fruit is used as a dentrifice and considered useful in various teeth, bleeding and ulcerations of the gums.

Ammi

Trachyspermum ammi (L.) Sprague of family Umbelliferae. Eng. Ammi; Hindi *Ajwain*.

A herb, cultivated throughout India.

Uses. The fruits are carminative, stimulant, antispasmodic and tonic. They are given in colic, flatulence, dyspepsia, indigestion, diarrhoea, cholera, tympanites, hysteria, etc. The fruit yields an essential oil containing thymol.

Solanum

Solanum viarum Dunal; Syn. *Solanum khasianum* var. *Chatterjeeanum*, Sengupta; family Solanaceae.

Solanum viarum, is a wild perennial herb. It grows well in the hilly areas particularly of Khasia and Jaintia hills (Assam) in North East and Nilgiri hills of South India.

Morphology. It is a stout herb, with a woody stem, of about 2 feet in height. Nearly whole body of the plant is covered with yellow hairs (hirsute). The stem bears two types of prickles, hooked and straight. Stem is cymosely branched.

Leaf is 7" to 5" long and is deeply lobed, hirsute and bears only straight prickles.

Flowers are white in colour, and are produced in large number. The berry is about 2.5 cms. in diameter which contains about 350 seeds.

Alkaloid and medicinal uses. The berries of *S. viarum* contain the alkaloid *Solasodine*. It is a glycon which is used by pharmaceutical companies for the preparation of many important drugs. It is a nitrogen analogue of diosyenin and is a good source of *sapogenin*. Sapogenin is used as a base for the preparation of *Cortisone* and allied products. The synthetic substitutes are not known for these drugs, hence the importance of this source.

The gluco-alkaloid percentage of the berries is about 6 per cent which in turn produces one-third or 2 per cent of *Solasodine*. It can be extracted from dry as well as fresh berries.

Cortisone, a steroidal hormone prepared from solasodine, is found to be effective in the treatment of acute stages of rheumatoid arthritis, chronic cases of asthma, leukemia, obesity and many skin diseases.

Induced mutations have yielded favourable results regarding the increases in alkaloid content of the berries. (Dhyansagar and Pingle). *Material received from P.N. Purekar.*

DRUGS OBTAINED FROM SEEDS

Carrot

Daucus carota Linn. var ***sativa*** D.C. of family Umbelliferae (Apiaceae); Eng. Carrot; Hindi—*Gajar*.

An annual or biennial herb, grown chiefly in the Punjab, Uttar Pradesh and Madhya Pradesh.

Uses. The seeds are aromatic, stimulant, carminative, useful in the diseases of the kidney and in dropsy, nervine tonic, aphrodisiac, given in uterine pain.

Lesser Cardamom

Elettaria cardamomum Maton., of family Zingiberaceae; Eng. Lesser cardamom; Hindi—*Chotti ilaichi*.

This is a herb; native of India, cultivated in Kerala, Karnataka, Maharashtra, Assam and Tamil Nadu.

Uses. The seeds are aromatic, stimulant, stomachic, carminative, diuretic, cooling, abortifacient, emmenagogue and antidotal; they are useful in asthma, bronchitis, piles, diseases of the bladder, kidney, liver, uterus, rectum and throat. They are also remedy of scabies, headache, earache and toothache.

Blonde Psyllium

Plantago ovata Forsk., of family Plantaginaceae. Blonde psyllium; Hindi—*Isafgol*.

This is an annual herb. It is cultivated in parts of Rajasthan and Maharashtra.

Uses. The seeds are mildly laxative, demulcent, mucilaginous cooling, emollient, astringent and diuretic. They are used in the treatment of dysentery, constipation and disorders of the digestive system. The mucilaginous matter is found in the seed coat, and therefore, the seed coat or husk is used.

Castor Oil Plant

Ricinus communis Linn. of family Euphorbiaceae; Eng. Castor oil plant; Hindi—*Arand*.

This is a small tree, cultivated chiefly in Andhra Pradesh, Maharashtra, Karnataka and Orissa.

Uses. The seeds contain the alkaloid ricinine; the cold drawn oil from the seed is medicinally the most useful part of the plant; it is an effective purgative generally given in acute diarrhoea. As an enema the oil is given in constipation. The oil is locally applied in conjunctivitis. The leaves are applied to the head to relieve headache and as poultice for boils.

Nux-Vomica

Strychnos nux-vomica Linn., of family Loganiaceae; Eng. Nuxvomica; Hindi—*Kuchla*.

A tree, found in Orissa, Bihar, Andhra Pradesh and Tamil Nadu.

Uses. The dried ripe seeds are the source of a drug nux vomica which is used as a tonic stimulant and in the treatment of paralysis and nervous disorders. The powdered seeds are also given in general debility, chronic rheumatism, dyspepsia, intermittent fevers, diarrhoea, hysteria, general constipation, hydrophobia, impotence, cholera and epilepsy.

Fenugreek

Tirgonella foenumgraecum Linn., of family Papilionaceae; Eng. Fenugreek; Hindi—*Methi*.

A herb; native of South Europe, grown mainly in Northern India.

Uses. The seeds are aromatic, diuretic, nutritive, tonic, lactagogue, astringent, emollient, carminative and aphrodisiac. The seeds are given, boiled or roasted in dyspepsia, diarrhoea, dysentery, colic, flatulence, dropsy, rheumatism, chronic cough, liver and spleen enlargement. The seeds contain the alkaloid trigonelline.

DRUGS OBTAINED FROM ALL PARTS OF PLANTS

Absinthe

Artemisia absinthum Linn. of family Compositae (Asteraceae); Eng. Absinthe; Verna; *Vilayati Afsanthin*.

This is an aromatic, bitter, herbaceous hairy perennial herb. It is found in Kashmir from 5,000 to 7,000 feet elevation.

Uses. The whole herb is used medicinally but the leaves are preferred. The fresh plant is more efficacious than the dry plant. It is a tonic, stomachic and anthelmintic. Small doses of its infusion are given for dispersing the yellow bile of jaundice from the skin. It is also given in the disease of the digestive system, nocturnal pollution, anaemia, wasting diseases, etc. The poultice of its leaves is applied for fomentation to gouty or rheumatic joints. It is a useful antiseptic and detergent in skin diseases. An infusion of the leaves is given as an enema for killing worms in the rectum.

Neem Tree

Azadirachta indica A. Juss., (Syn. *Melia azadirachta* Linn)., of family Meliaceae; Eng. Margosa tree, Neem tree; Hindi—*Nim*.

It is native of Myanmar but grown all over India.

Uses. The leaves are carminative, expectorant, anthelmintic, antidotal, diuretic and insecticidal. The fresh juice of the leaves is given for intestinal worms; with honey the juice is prescribed for jaundice and skin diseases. An infusion of the fresh leaves is a bitter vegetable tonic and alterative, especially in chronic malarial fevers, because of its action on the liver. Externally the leaves are applied over skin diseases as a discutient, stimulant and antiseptic; they are specially used for boils, chronic ulcers, eruptions of small pox, syphilitic sores, glandular swellings, wounds, etc. The bark is also a bitter tonic, astringent, alterative, anthelmintic, antispasmodic and stimulant. It is used in the same way as the leaves. The root bark has the same properties as the bark of the stem. The nim gum is a stimulant and demulcent tonic. The fermented sap that exudes from the trunk is a refrigerant, nutrient, stomachic and alterative tonic.

The fruit is recommended for urinary diseases, piles, leprosy, intestinal worms, etc. The seeds are emetic, laxative and anthelmintic. Seed oil is applied as an antiseptic dressing in leprosy and chronic skin diseases.

Spiny Bamboo

Bambusa bambos Voss., of family Gramineae (Poaceae); Eng. Spiny bamboo; Hindi—*Bans*.

This is a tall woody grass, found throughout India, particularly along river valleys and in moist situations.

Uses. The leaf buds are administered for thread-worms. The leaf juice is given with aromatics in vomiting of blood. A decoction of the leaves is used to induce lochia after childbirth.

The young shoots contain hydrocyanic and benzoic acids; they are stomachic and stimulant; they are used in lung diseases. A decoction of the joints of bamboo stem is a useful emmenagogue; it is also used as an abortifacient.

Banslochan, a silicious crystalline secretion found in the culms of the female plants, is a febrifuge, expectorant, tonic, aphrodisiac, demulcent and pectoral. It is given in hectic fever, phthisis, asthma, paralytic complaints, etc.

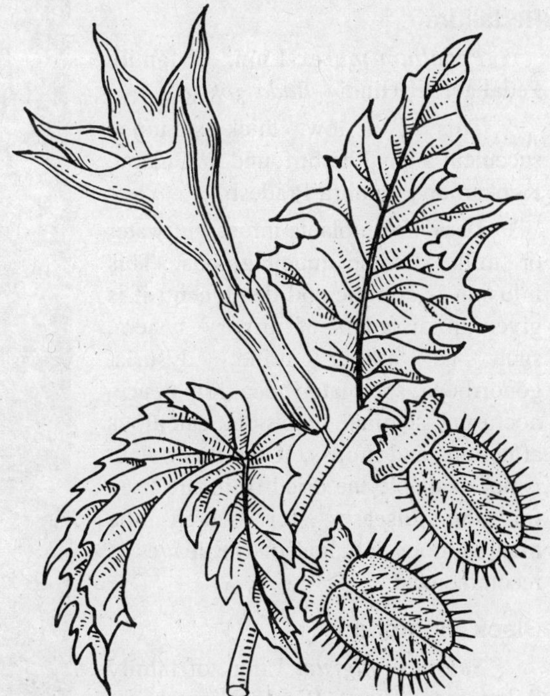

Fig. 23.15. *Datura.* A twig with leaves, flower and fruits. All parts are intoxicant, narcotic, aphrodisiac, toxic, antispasmodic and anodyne.

Apple of Peru

Datura metel Linn., of family Solanaceae, Eng. Apple of Peru; Hindi—*Dhatura.*

This is a herbaceous plant covered with fine, minute hairs. Commonly found throughout India.

Uses. All parts of the plant are strongly intoxicant, narcotic, aphrodisiac, toxic, antispasmodic and anodynous. The young leaves and seeds contain the drugs hyoscine, hyoscyamine and atropine. The dried leaves and twigs of the plants are smoked as an antispasmodic in asthma, whooping cough, bronchitis, etc. The juice of the fruits is a useful dressing for the scalp to check dandruff and falling of the hair. The seeds are astringent, antispasmodic, narcotic, anodynous, intoxicating, aphrodisiac, bitter, carminative and stomachic. The paste of seeds is used for decaying teeth, piles, fistula, tumors and parasitic skin diseases.

Achyranthes

Achyranthes aspera Linn., Hindi—*Chirchita*, *Latzira*, *Apamarg*; family Amarantaceae.

This is a common roadside weed with spike inflorescence. Flowering throughout year. It is of common occurrence throughout India.

Uses. The plants are used medicinally for several diseases such as piles, colic, boils, etc. It is pungent, purgative, diuretic and astringent. Roots are used for pyrrhoea. Also used in cough and fevers.

Pedalium

Pedalium murex Linn., of family Pedaliaceae; Hindi—*Bada gokru.*

This is a low, thick stemmed succulent annual herb; found in Gujarat, Konkan and Andhra Pradesh.

Uses. Fresh plants infused in water or milk become mucilaginous. This infusion is diuretic and demulcent; it is given in disorders of urinary system, such as ardour urine, dysuria, gonorrhoea, spermatorrhoea, impotence, nocturnal seminal emission, calculous affections and dropsy. If the fresh plant is not available the dried fruits are used for the diseases for which the mucilaginous infusion of the leaves is recommended.

Black Nightshade

Solanum nigrum Linn., of family Solanaceae; Eng. Black Nightshade; Hindi—*Makoy*.

This herb is found throughout India.

Uses. The herb is a cardiac tonic, alterative, diuretic, sedative, diaphoretic,

Fig. 23.16. *Solanum nigrum*. (Black Nightshade). A common medicinal plant. It makes the medicine for liver enlargement and jaundice.

cathartic, anodyne, expectorant and hydragogue. It is used as a decoction in dropsy; enlargement of the liver and jaundice. The syrup of herb is given as an expectorant. In fevers it is given as a cooling drink. The leaf juice is given in inflammation of the kidneys and bladder and in gonorrhoea, dropsy, heart diseases, piles and enlargement of spleen. Hot leaves are applied over the swollen and painful testicles. The berries are alterative, and tonic. They are given in fevers, diarrhoea and heart diseases.

Indian Solanum

Solanum xanthocarpum Schrad & Wendl., of family Solanaceae; Eng. Indian Solanum; Hindi—*Kateli*.

A common prickly herb found throughout India.

Uses. The herb is expectorant, bitter, stomachic, astringent, diuretic, alterative and anthelmintic. Its decoction is given in fevers, coughs, asthma, dropsy, flatulence, gonorrhoea, pain in the chest and heart disease.

Chiretta

Swertia chirata Buch-Ham., of family Gentianaceae; Eng. Chiretta; Hindi—*Charayatah*.

This is a shrub commonly found in temperate Himalayas from Kashmir to Bhutan and the Khasia hills.

Uses. The shrub contains two bitter principles ophelic acid and chiratin. It is a bitter tonic, stomachic, febrifuge, appetizer, anthelmintic, alterative, laxative, antidiarrhoeic, and antiperiodic. It makes a good drug for skin diseases, intermittent fevers, intestinal worms, bronchial asthma and regulating the bowels. It is given as an infusion or tincture.

Heart-leaved Moonseed

Tinospora cordifolia (Willd.) Miers of family Menispermaceae; Eng. Heart-leaved moonseed; Hindi—*Gulancha*.

A common climbing shrub (liane). Found throughout tropical India and the Andamans.

Uses. The fresh plant is antiperiodic, alterative, tonic, hepatic, stimulant and diuretic; its watery extract is very effective in fevers. The plant is commonly used in rheumatism, urinary diseases, dyspepsia, general debility, syphilis, skin diseases, piles, bronchitis, impotence, gonorrhoea and jaundice. The fecula prepared from the roots and stem, known as *sat giloe* is a valued drug for intermittent fevers, chronic diarrhoea, chronic dysentery, jaundice and rheumatism.

Fig. 23.17. *Catharanthus roseus* (**Sadabahar**). Commonly grown in gardens. It is propagated by seeds or cuttings. It blooms almost throughout the year. The alkaloids possess hypotensive, sedative and tranquillizing properties.

Lochnera

Lochnera rosea (Linn.) Reichb.,—*Catharanthus roseus* G. Don. Syn. *Vinca rosea* Hindi—*Sadabahar*; family Apocynaceae.

An erect annual or perennial herb; native of Madagascar but now commonly grown in India as an ornamental plant. Leaves opposite, oval, obovate or oblong, glossy; flowers usually 2-3, in cymose axillary clusters; fruit a cylindrical follicle, manyseeded. Flowers white or pink.

Alkaloids. All parts of the plant, particularly the root bark contain alkaloids; these include three alkaloids of the Rauvolfia group; *viz.*, ajmalicine, serpentine and reserpine. The alkaloids of this plant possess hypotensive, sedative and tranquillizing properties. They also cause relaxation of plain muscles and depression of the central nervous system.

Uses. The plant has been used as a folk remedy for diabetes in various parts of South Africa and also in India and Sri Lanka. It is reported to be toxic to cattle. The juice of the leaves is used as an application for wasp stings. An infusion of the leaves is given in the treatment of menorrhagia. The root is considered tonic and stomachic. The alkaloids possess hypotensive, sedative and tranquillizing properties. The alkaloids inhibit the growth of Vibrio cholerae and Micrococcus pyogenes var. aureus. Vindoline and the other alkaloids from leaves are active against M. pyogenes var. aureus and var. albus, Streptococcus haemolyticus, Corynebacterium diphtheriae and a few other bacteria; leaf extracts form a useful anti-bacterial agent for the treatment of staphylococcal and streptococcal infections.

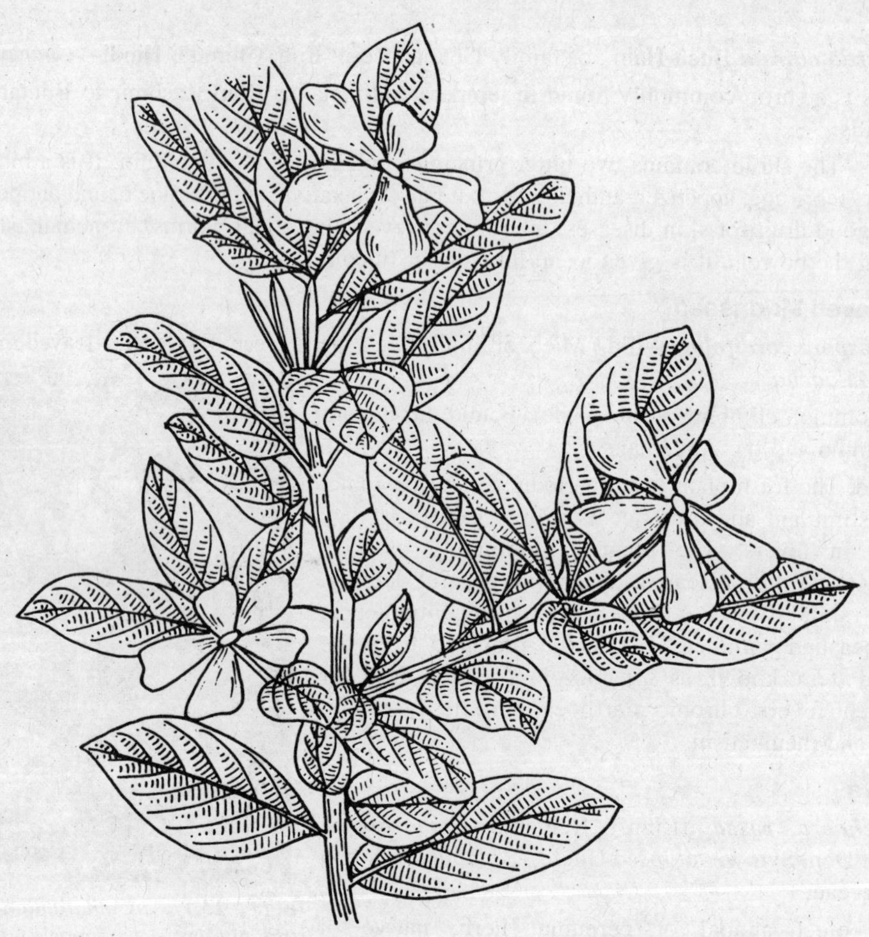

Fig. 23.18. *Catharanthus roseus* (a line diagram). A twig with leaves and flowers. The plant is useful as a medicine in cancer and diabetes.

24

CHAPTER

Non-Alcoholic Beverages and Beverage Plants

The beverages containing caffeine are used all over the world for their stimulating and refreshing qualities. Caffeine is an alkaloid, which has definite medicinal values and acts as a diuretic and nerve stimulant. The most important non-alcoholic beverages are—*tea*, *coffee* and *cocoa*. Caffeine is harmful in large quantities, it is present in these beverages in very small amounts, not exceeding two per cent. The other beverages which do not contain alcohol are commonly known as *soft drinks*. They contain high sugar content and make a good source of energy. The fruit juices are the simplest soft drinks.

INDIAN TEA

Tea

Camellia sinensis (Linn.) O. Kuntze; Eng.—Tea; Hindi—*Cha*; Family—Theaceae.

History. "According to Japanese folklore, the discovery of tea is attributed to an Indian Buddhist monk *Daruma* who lived hundreds of years ago. He left India to preach the message of the Enlightened One—The Lord Buddha—in China. Arriving at Peking, he vowed not to sleep for nine years till he finished his prayers. But, during his meditation he felt tired and his eyes closed. Waking up in disgust, he tore off his eyelids and threw them away. Where the latter fell there sprang up an evergreen bush. Surprised, *Daruma* plucked the leaves, boiled them and drank the liquid. Soon, he felt refreshed and was able to continue his prayers. *Daruma* eventually left China and went to Japan where he died."

In China the people have been drinking tea probably since the 4th century A.D. and the people of Upper Burma, along the Indo-Burmese border and Upper Assam have been used to tea for just as long a period.

Tea was brought to the notice of Europe in 1559 by Ramusio, the noted Venetian writer on voyages and travels, long after it was accepted beverage of the people of Asia. Half a century after Ramusio's notice of tea, the enterprising Dutch began to import it into Europe. England was strangely casual to the beverage during the first half of the 17th century. Tea then was far beyond the reach of the average person. The 2 lbs, 2 oz. tea, brought by the Governors of the East India Company in 1664, as a gift to King Charles II, cost them 85 shillings. Two years later the Company gave him a more ample present of *thea*, consisting of 22¾ lbs., costing 50 shillings per lb. Half a century later the retail price of tea in London was between 20-30 shillings per lb. for *Bohea* (black tea) and 12-30 shillings for the finest *Hyson* (green tea). It took another hundred years before consumption began to spread to the rural areas. According to Prof. Northcote Parkinson, it was presumably between 1784 and 1789 that tea "may literally be said to have descended from the palace to the cottage". About that time it became the breakfast beverage of practically the whole population. It was also drunk at other times which led Raynal (the French historian. 1713-96) to exclaim that tea had done more for English sobriety than any laws, sermons or moral treatises could have achieved.

Tea was sold publicly in London in 1657 and two years later, on 14th November 1659, Mercurius Politicus carried the news that tea was sold "almost in every street" of London. The first order by the East India Company for 100 lb. of the best tea was sent to its agent at Bantam in Java in 1667. In 1669 import of tea into England from Continental Europe was prohibited, thus making absolute the East Indian Company's monopoly in tea.

The British imported China tea from Bantam (Java) when they were dispossessed in 1684. From this period Madras and Surat became centres for the import of "secondhand" Chinese tea. Though, direct British trade with China did not weaken till the 1780 when the merchants of Canton decided to organise the conduct of their country's foreign tea trade.

Meanwhile, the Company's officials in India had realised the possibilities of raising this valuable commodity of trade in India. In 1780, Warren Hastings, the Governor-General, passed on some seeds of *Hyson* for propagation in Bhutan. A tea enthusiast Col. Robert Kyd, who later became Superintendent of the Botanical Gardens in Calcutta, similarly experimented with tea cultivation. In 1788, Sir Joseph Banks tried to interest the East India Company in the possibilities of growing tea in India; but the Company was not agreed to the suggestion as such a venture was against the monopoly of its tea trade with China. However, after five years, Lord Macartney's Embassy was making arrangements to send seeds of the tea locally cultivated to Calcutta. Although imported tea seeds gave the idea and the sustenance for the commercial cultivation of tea in India, the rise of the Tea Industry, as it is today, owes origin to the momentous discovery of the indigenous tea plant in Assam in India.

Many have claimed the discovery of the indigenous Indian tea plant which eventually became the foundation of the world's Tea Trade and Industry during the next century. There were some who believed that Moneram Dewan was the discoverer who led Major Robert Bruce to it; but it was Charles Bruce, his brother, who played a significant role in the development of tea culture in India during the ensuing twenty years.

As we have realised, the East India Company at first was averse to the idea of starting any tea plantation in India; but its officials in the country were eager and the discovery of the indigenous growth of tea in Assam acted as a catalyst in the great movement to develop the industry in the country. If there was difficulty in the early stages, it came largely from the application of Chinese tea labour since the general belief persisted that only such imported methods of tea manufacture were the safe ones to follow. This belief, however, lasted for a while till experience with Assam tea and the response to the indigenous tea in London overcame existing influences and the foundation of a truly Indian Tea Industry was laid.

An index of the steady rise in Britain's consumption of Indian tea can be seen from the figures of her tea imports. In 1836, the import of tea into Britain reached 49 million lbs. and all of it was China tea. Two years later, India sent her first consignment of Assam tea to London to be sold there, on January 10, 1839. In 1859, the last year of the East India Company before the crown took over, Britain imported 69 million lbs. of which only 2 million or 3 per cent came from India. By the close of the century Britain was importing about 250 million lbs. of tea of which 55 per cent came from India, 37 per cent from Ceylon and only 5 per cent from China.

In 1900, Indian production reached 170.5 million lbs., with exports at 164.6 million lbs. and out of which 154 million lbs. went to Great Britain. Within another five years Indian exports reached 214 million lbs., and India became the leading tea supplier to the world and she has retained that position.

Today the total output of this industry is over 779 million lbs.—176 million lbs. in South India and the rest in North-East India. It accounts for about 20% of the country's total exports in value.

Tea is the queen of beverages, and no other beverage except water is consumed by so many. Tea, a treasure of the world, is the most important non-alcoholic beverage; it is a pure, safe and helpful stimulant and one of the chief joys of life. A billion cups of tea all over the world add daily to the zest of life. They bring good cheer with health, stimulation without intoxication, and are refreshing and energising without any exacting demand on the family budget.

The tea plant is considered to be a native of Assam and the adjoining areas of Upper Burma, some regard it to be a native of southern Yunnan and Upper Indo-China. The major areas of production are situated in the northeast region comprising the Brahmaputra and Surma valleys of Assam, and Darjeeling and Jalpaiguri districts of West Bengal. These areas are the Terai, the Dooars and the Brahmaputra valley where tea is grown on the north and south banks of the river Brahmaputra. In addition, considerable quantities of tea are sown in the Cachar district in Assam and in Tripura. In North-West India a fair amount of tea is grown in Kangra, Mandi and a smaller quantity in Dehradun. Down South tea is grown on the Nilgiris, on the Kanan Devan or the High Range, Anamallais and on the slopes of the mountain stretching down to the plains of Kerala.

Brahmaputra valley. Tea is grown mostly on flat land, between 160 to 400 feet above sea level, on both banks of the river Brahmaputra. Soils are alluvial and the mostly sandy loam with some red clay. The average annual rainfall is 80–120 inches and it is well distributed. The cropping season is from April to December, May/June and October/ November.

Teas of orthodox manufacture are black, clean, flinty well-twisted leaf with a good show of golden tip in the best grades, throwing creamy, strong, pungent, full-bodied top quality liquors. Teas of C.T.C. manufacture from this region are brown, rich, grainy, small leaf grades throwing exceptionally strong, coloury, pungent liquors. These teas are in great demand almost all over the world and are particularly suited for blending with flavoury teas to improve strength and colour.

The Terai and Dooars. Tea is grown on flat land and hill slopes bordering the Himalayan foot-hills, from 100–1000 feet above sea level. The average rainfall is 120–200 inches but it is not very evenly distributed. The cropping season is from March to November, with the vintage periods in spring and autumn.

The Terai and Dooars produce coloury, full creamy liquoring teas. The orthodox teas are black, stylish and well-twisted, while the C.T.C. tea are hard, grainy and even. A number of gardens produce very small brown leaf Leg-cut teas which throw bright colour liquors. These teas are in great demand by blenders for mixing with flavoury teas to contribute to colour and body.

Cachar and Tripura. Tea is mostly grown on hillocks peculiar to the region and to some extent on the flat lands surrounding them. Soils are red sand and clay with some areas of peat. The annual rainfall varies from 120–200 inches but distribution is uneven and severe winter droughts are common. The cropping season is from March to November.

Fig. 24.1. *Camellia sinensis* : Tea. The garden looks best, brightest and of course, busiest during the plucking season.

The districts of Cachar and Tripura produce black, well-made teas throwing colloury, sweet liquors very popular with blenders.

Darjeeling Tea

The town of Darjeeling in India is situated in the extreme north of the State of West Bengal, at a height of 7,000 feet, surrounded by the lower hills and valleys of the Himalayas and overlooks the State of Sikkim.

The town is clustered in a bowl among the hills, looking north toward Kanchenjungha, a mere 35 miles away and a very magnificent sight when it is to be seen. Most of Darjeeling's livelihood is derived from the tea gardens which surrounded the town on all sides, at lower levels.

Early in the nineteenth century tea seed was imported from China and planted both on the plains of N.E. India and also in these hills. But it was soon discovered than an indigenous tea plant was to be found in Assam. This proved to be more vigorous, in 'plains' conditions; it gave a large crop and the leaf was more easily harvested, and so the old 'China' bushes were gradually uprooted and replaced. The original bushed plantes in Darjeeling have still remained, they are up to 100 years old and still yielding good crops.

We have continued to use these bushes because they, and they alone, can produce the flavour for which Darjeeling tea is renowned. The crop yield per acre is small and the leaf is small in size and expensive to harvest. And yet, it is this climate and high altitude, the slow growth of leaf, which help to give to Darjeeling teas the flavour for which they are known throughout the world.

In South India, Sri Lanka, Indonesia and Africa the tea bush produces leaf throughout the whole year, but in N.E. India it hibernates in winter. Growth starts again in the spring when the tea bushes of Darjeeling are covered with a mass of small soft bushes. This is the first 'flush' when the tea made is of good quality. When this 'flush' is finished there is a lull in growth until the second 'flush' appears; this is usually small in appearance, the quantity of leaf on the bushes is not great, but this is the vintage period of the whole year.

The monsoon reaches Darjeeling about mid June and from July onwards the quality of tea is of standard character until the rains end in October. Then for a short period until the end of the season in November a small crop of 'autumnal' leaf is harvested which has a flavoury but 'thin' liquor.

The quantity of fine Darjeeling tea produced each year is comparatively small. Of the total crop of 16,000,000 lbs. from this district, the first and second flushes and the autumnal teas are highly prized. Most of these good, flavoury teas are used for blending with teas from other places; the latter add strength and colour; while the lighter liquor of the Darjeeling tea supplies the flavour to the blended mixture.

Public demand in many countries requires a 'strong colour' tea which the 'Assam' variety of bush contains; but for those who appreciate flavour they will find this in full measure in a pure Darjeeling tea.

In countries where tea is drunk without milk or sugar—as for example, in Russia—tea is appreciated for its flavour, and indeed it is true that the delicate flavour of a Darjeeling tea can best be appreciated by drinking the liquor 'straight'.

Good Darjeeling teas are indeed the champagne of teas. There is no other area in the world which can produce the 'muscatel' flavour of these fine teas produced on the mountain sides of this small district.

Darjeeling tea is grown on the mountain slopes of the Himalayas at elevations ranging from 1000 to 6000 feet above sea level. Soils are red loams and sandy loams. Annual rainfall varies between 100 and 200 inches; the winter season being fairly cold and dry. The cropping season is from April to November and vintage teas are produced in April/May and October.

Teas with exquisite aroma come from the Darjeeling district. Stylish, large leaf teas are the main feature with golden tips in the top grades. The liquors are famous for their flavour derived from the high elevation at which the teas are grown.

Kangra and Mandi. The tea plantations are on the mountain slopes of the Kangra Valley in the Himalayas, at elevations from 2000-5000 feet above sea level. The average annual rainfall but it is not evenly distributed; the winter and early summer being dry and droughty. The cropping season is from April to November.

Kangra and Mandi mostly produce green teas which are in demand in many parts of the world.

South India. Tea is grown on hill slopes in the States of Tamil Nadu, Kerala and Karnataka, the main tea districats being the Nilgiris, Nilgiri—Wynaad, High Range and Central Kerala. The plantations are at elevations ranging from 1,000 to 7,000 feet above sea level. Soils are red and yellow loams and are lateritic in origin. Most tea plantations in South India get two monsoons and the annual rainfall ranges from 50–275 inches and it is fairly well distributed throughout the year. The cropping season lasts throughout the year, and vintage teas are produced in December/January.

Fine flavoury teas are produced in the Nilgiris—stylish leaf throwing very bright, brisk, quality liquors with pungency and pronounced flavour; cup character is the dominating feature of these teas. Flavour is derived from the high elevation and prevails throughout the year in varying degrees.

Anamallai produces good, black, heavy, flinty well-made teas with a sprinkling of tips in the top grades during the tipping season and are the pride of this district. Liquors are coloury heavy, full-bodied, brisk, malty with quite useful strength and character, with cup quality during December/January.

Kerala teas are black or blackish, clean, heavy well-made grades, throwing exceptionally coloury, strong liquors with briskness and character and occasional quality. These are very useful for blending with flavoury teas of comparatively lighter liquors, and with others of different liquoring properties.

Karnataka produces black, well-made teas with malty medium liquors.

Nilgiri Tea

The Nilgiris, which literally means the "Blue Mountains" are situated in the state of Tamil Nadu in South India. The place derives its name from the blue haze that covers its wooded slopes. This range, with its half English air and rolling down is undoubtedly one of the most beautiful hill resorts in the world.

Apart from its scenic charms the region constitutes one of the most important tea-growing districts in India. It is the home of famed 'Nilgiri tea'. The Nilgiri district has about 23,000 hectares of land under tea cultivation producing nearly 34 to 35 million kg. of tea every year. The other main districts in the South are Coimbatore and the Anamallai hills in Tamil Nadu producing about 18.50 million kg. of tea and the Idukki high ranges in Kerala producing about 32 million kg. Nilgiri teas are, however, a class by themselves. They have a flavour unique to the high grown teas of South India.

The plantations are spread out over every slope, valley and plateau at elevations ranging from 30 to 2,000 metres above sea level on a soil that is predominantly red and yellow loam and lateritic in origin.

Unlike the gardens in Northern India, where tea growth slides off during a period of "winter dormancy", the gardens of the South enjoy a steady growth throughout the year. The two monsoons are a determining factor in facilitating throughout the year plucking. There are, however, two big flushing periods the first from April to May when about 25% of the annual crop is gathered, and the second from September to December fetching about 32 to 40 per cent of the annual crop. Most of the manufacturing is done in the 'Orthodox' process.

No other tea has the unique Nilgiri flavour. The characteristic is a gift of nature—a result of various natural stimuli on the plant. It almost appears as if the leaves take their special qualities from the breath-taking beauty that surrounds them. This assumption is not far from the truth. The flavour prevails throughout the year in varying degrees. Nilgiri teas are famous for their stylish leaves that produce very bright, brisk and quality liquors having body and strength. Cup character is the dominant feature of these teas.

The importance of South India as a tea-producing region is tremendous, Cochin serving as its nearest auction centre and port. The largest tea-packing factory in India belonging to the Tata-Finlay group is situated in Bangalore. Besides, both the instant tea manufacturing units are in Munar and Cholari in the South. The United Planters Association of South India (UPASI) Tea Research Station in the Anamallai hills in Coimbatore, is yet another factor contributing towards the phenomenal growth in the tea industry in this region.

Nilgiri teas a blender's dream, and a large portion of these teas find their way into overseas markets in Russia, U.K., the Middle East, U.S.A., Canada and Australia to meet the ever growing demand for the Nilgiri flavour. With the emergence of India as a packet tea exporter, packets of fresh high grown Nilgiri teas are available for consumers all over the world.

The Story of Indian Tea

The fascinating story of tea from the seed to your cup is one filled with colourful legend and romance of modern organisation and commerce in India. It is end product of the patient toil of tens of thousands of tea-growers, the ingenuity of scientists, the expert skill of the blenders, business acumen of traders, care and promptness of shippers and the service and courtesy of retailers. It has passed through many phases, thousands of hands, and across the oceans of the world to reach you and to give you satisfaction.

The tea plant is an evergreen of the *Camellia* genus and is known as *Camellia sinensis* and flourishes in warm tropical and subtropical rainy regions. The highest yield will be obtained from

Fig. 24.2. *Camellia sinensis*. Indian tea. The cured leaves are used as a beverage.

tea which is grown in tropical climate. Paradoxically, the finest quality teas are produced from leaves grown in the cooler altitudes of 1,000 to 2,000 metres. The great 'Darjeeling teas' are grown against the background of *Mount Kanchenjungha* of the Himalayas; nowhere else in the world has the 'Darjeeling flavour' been duplicated. From here come the most delicate, the most fragrant teas ever produced.

Our story starts with the tea seeds, round and hard like brown marbles. The seed itself is taken from the tea trees. Many estates have a plot of selected trees which are allowed to grow to their full size of 5 metres to 10 metres. During July to October the trees blossom and tea flowers appear. But only after a year will the seed be ripe. When the shells crack the seeds are removed and planted in the nursery beds. Within a month they germinate into seedlings and green shoots start to grow. In 9 to 12 months the young plants are ready for removal to the tea garden—their permanent home. For another four years the plants must be expertly pruned before they are ready for the first plucking. By this time they have become wide and flat-topped like a waist-high table to make the plucking easier. If tea plants were left unpruned, they would grow as high as 30 feet.

Plucking. Plucking takes place over 7 to 14 days and the deft fingers of the expert plucker (mostly women) can gather as much as 60 lb. of leaf in one day, or enough to make 15 lb. of black tea. Only two small leaves and a bud are gathered from each shoot.

In North India, during March or April the spring or 'First Flush' materialises; the 'Second Flush' comes a month later in May. After the 'Second Flush', the other flushes are not so well marked and the bush gives more or less a continuous supply of green leaf until November.

In India, plucking is carried on throughout the year. Even then, there are two big flushing periods, from April to May when about 25 per cent of the annual crop is gathered, and a second flush in September to December when about 35 to 40 per cent of the annual crop is harvested.

Commercially, tea may be divided into three basic classes:

(1) Black or fermented (2) Green or unfermented and (3) Oblong or semi-fermented.

Black tea is most important commercially and the production of major world supplies—viz., India, Ceylon, Indonesia and recently East Africa—is mostly of this type.

There are four principal operations in the preparation of black tea: (1) withering, (2) rolling; (3) fermenting, and (4) firing.

Withering. About 75% of the weight of a fresh leaf consists of water even on a dry day and nearly half of this has to be removed before the fibre of the leaf and stalk will stand the strain of rolling, without breaking up. The leaf is spread thinly over withering racks arranged one above the other horizontally, and allowed to remain there for 12 to 18 hours to lose its excessive moisture. Sometimes, heated air is forced over these racks if the atmosphere around is humid. The leaf slowly and evenly becomes soft and flaccid like soft leather and is ready for rolling.

Rolling. It imparts the characteristic twist to the leaf, breaks the leaf cells, exposes the juices to the air for fermentation to set in. After half-an-hour's rolling the leaf is removed in aluminium trolleys to a sifter and ball breaker. This machine consists of a long and flat metal sheet with perforations fixed on a frame which makes reciprocating motion. The movement causes the broken leaf and fine particles to fall below and the rest is taken out after sieving. The latter is again rolled for the second time with increased pressure. After the second roll, all the leaf still more or less green and quite flaccid removed to the fermenting room.

Fermentation. During fermentation, the tannin in tea is partly oxidised and the leaf changes colour and turns bright coppery-red. The rolled leaf is spread on tiles or aluminium or even glass sheets and the oxidation which has commenced during rolling continues, the leaf coming into contact with air. The period of fermentation generally extends from 3 to 3 1/2 hours, but this includes rolling time as well. As a general rule, the *shorter* the fermentation the more pungent the liquor: and the *longer* the fermentation the softer the liquor and deeper the colour.

Drying or firing. After the leaf has changed its colour, comes the final operation; namely *drying* or *firing*. The essential function of this process is to arrest further oxidation of the leaf and to remove all moisture excepting a small amount of roughly 3 to 5 per cent.

The driving machine is the largest machine required in a tea estate factory. The automatic tea drier is a large steel oven inside which the leaves, spread on tray, travel slowly from top to bottom while hot dry air is continuously forced into the oven. Careful regulation of temperature is essential as excessive heat will scorch the leaves while lack of it will result in improper drying.

Generally the process is known as first firing, removes only three-fourths of the total moisture of the fermented leaves. After a period of cooling, they are fed into another drier. The latter known as second firing removes the remaining moisture. The tea from this second drier is the finished product. There now remains only sorting or grading and packing before the tea is ready to be sent from the factory.

As the leaves come out of the driers, large and small broken and unbroken leaves are mixed together. Now they are sent through sieves, with graduated mesh, and the different sizes of tea are removed carefully so that final packages are of uniform-sized leaves or particles.

After sorting, the tea is graded and given various names such as *Orange Pekoe*, *Pekoe*, *Fannings* and *Dust*; but these are only indicative of size and not of quality.

The manufacturing process being now complete, the different grades of tea that come from the sorters are bulked, packed in plywood tea chests with linings of aluminium-foil and paper, and are ready for their adventurous journey to all corners of the world.

The process of making tea already described is generally known as the 'orthodox' method and is in vogue in most gardens in India. Recently, however, some import variations have been introduced in the process of manufacturing especially in North India. The variations consist of the use of a machine named C.T.C. or, to give its full name a *Crushing*, *Tearing* and *Curling machine*, all in one. Here, the rollers are either wholly or partly eliminated. The leaf after withering is lightly rolled without any pressure. Then the fine leaf is separated and the coarse leaf is fed into this machine two, three or even four times.

Leg-cut teas. Manufacture is vogue in the Dooars, Terai and in some Cachar gardens of North East India. This process eliminates withering, which is often difficult in these highly humid areas, and also produces a type of tea which is quick brewing and gives an instant liquor of good strength and colour.

Green tea. Most of the teas made in China and Japan are of this type. A small quantity is made in North India also but mainly for sale in Afghanistan, Iran and some for the American market. The essential difference between *green* and *black* varieties of tea is in the fermentation—green tea being completely unfermented. The leaf meant for the manufacture of green tea is generally plucked without stalk.

Oolong tea. It can be described as a cross between *black* and *green*. The fresh leaf is slightly withered before panning; during the process a light ferment is allowed to develop, the leaf is then rolled and fired. Its colour is slightly changed and the resultant teas are of greenish-brown colour. The Oolongs which have long enjoyed a special market in America, are now almost exclusively manufactured in Formosa.

Leaf Grades

These are made up of larger leaves after the broken grades have been sifted out. In brewing, flavour and colour come out of these grades more slowly but, given the necessary brewing time, they produce flavoury liquor. 'Orange Pekoe', 'Pekoe', and 'Pekoe Souchong' are the three generally used grades in this category.

Orange Pekoe (*O.P.*): This consists of long, thin, wiry leaves containing tip or bud leaf. These teas generally come from the finer portions of the shoot that is the bud, first leaf and softer parts of the stalk. The liquors are generally light or pale in colour and, it produced from high grown teas, usually flavoury.

Pekoe (*P*): The leaves are slightly shorter and so not wiry as those of O.P., the buds also are present. The liquors, however, have more colour and this grade is generally quicker brewing than the O.P.

Pekoe Souchong (*P.S.*): A bold and round leaf with pale liquors.

Broken Grades

These are of smaller and broken leaves and are roughly 80% of the crop including the dust varieties, which are also strictly to be classified under broken grades. There are 4 main grades in this category.

Broken Orange Pekoe (*B.O.P.*): They are much smaller than any of the leaf grades and usually contain tips, hence the word orange is added. The liquors have good colour and strength and this is one of the chief attractions.

Broken Pekoe (*B.P*): They are slightly larger than B.O.P. but tips or buds will be absent. The liquors are paler and are less coloury in the cup. These are mainly used as 'fillers'.

Broken Pekoe Souchong (*B.P.S.*): They are a little large or bolder than B.P. and in consequence lighter in the cup, but useful as 'fillers'.

Fannings or Pekoe Fannings: These are much smaller than B.P.S. and are quick brewing and give good coloured liquors.

Dusts

Finally, the smallest particles excluding 'fluff' and 'stalk' are graded as Dusts. These teas are useful in quick brewing, the liquors produced have both strength and colour. They are in a good demand for catering purposes.

Tea Board of India

The Tea Board's major functions at present are: promotion of the tea habit; helping the development of the Industry, more especially of its marginal units; mediation to secure essential goods and services; provision of educational facilities for the dependants of garden workers. For the buyers from abroad it is enabled by the Act to render effective assistance in ensuring the standard of tea shipments from India. But, the Tea Board is more important than its functions would signify. For it embodies the organised knowledge and experience of all tea interests, reflects the attitudes of the industry as a whole and is the only statutorily recognised link between the Industry and the Government.

India continues to be world's largest exporter of tea, and more cups of tea are drunk throughout the world than any other beverage—except water—because it is a good tranquilliser and at the same time acts as a mild stimulant. To relax over a cup of tea is a virtue worth cultivating before the human system, through stress and tension, is allowed to reach a stage when life seems distorted and the mind refuses to adjust itself. Tea's contribution to reduce tension and instil a sense of general comfort has been recognised today by eminent medical experts, though it is by no means a new claim for this little plant whose written origin dates back to some 5000 years. From the distant past an ancient philosopher has left this image of tea to posterity:

"When I drink tea, I am conscious of peace,
The cool breath of heaven rises in my sleeves
And blows my cares away."

Five Golden Steps to a Good Cup of Tea

1. Take fresh water from the cold tap and boil.
2. Warm teapot by rinsing out with hot water.
3. Put into teapot one teaspoonful of tea leaf for each cup.
4. Pour boiling water into the teapot. Cover and wait for three minutes. Give more time for bigger leaf teas.
5. Pour liquid tea from teapot into cup. Add milk or a slice of lemon and sugar to taste. Enjoy the taste and flavour of Indian tea.

INDIAN COFFEE

History. It has been stated that the coffee plant of commerce is native of Abyssinia, where it is called *bun* or *boun*, and coffee must have been used in that country from very early times. There is very fair evidence in Arabic literature that the use of coffee was introduced into Aden by a certain Sheikh Shihabuddin Dhabhani, sometime about the middle of the fifteenth century. From Yemen it spread to Mecca in 1511, then to Cairo to Damascus and Aleppo, and to Constantinople, where the first coffee house was established in 1554. The first European mention of coffee of Aleppo is in 1573. The first mention of a coffee shop in Great Britain occurs in 1652. Coffee is spoken of as being in use in France in 1640, and the first public cafe was opened in Paris in 1669. Shortly after it became general throughout Europe.

Fig. 24.3. *Coffee berries.* There was a record production of coffee in India, one lakh and twenty thousand tonnes valued at Rs. 89 crores were exported in 1977-78.

It was introduced into Ceylon by the Arabs, prior to the invasion of this island by the Portuguese. Its systematic cultivation about 1690 was undertaken by the Dutch but on the cession of their territory its cultivation was continued by the natives of Ceylon (Sri Lanka).

As regards the history of its introduction into India, most writers agree that it was brought to Mysore (now Karnataka) some three centuries ago by a Muslim pilgrim named Baba Budan, who on his return from Mecca, brought seven seeds with him. George Watt writes in his 'Dictionary of the Economic Products of India': "Coffee planting in India at the present day is concentrated in the Madras Presidency, and as a European Industry it may be said to date from Mr. Cannon's plantation at Chikmuglur in Mysore. This was established in 1830, but as a curiosity Major Bevan grew coffee in the Wynaad in 1882. It was cultivated by Mr. Cockburn on the Shevaroys in 1830; Mr. Glasson formed a plantation at Manantoddy in 1840; the plant was taken to the Nilgiris in 1846 and to Darjeeling in 1856."

Location and area. Coffee is grown in India on the sunny slopes of the Western Ghats and it spurs extending eastwards where the ecological conditions are ideal for the growth of firstclass coffee. The F.A.O. has listed India as one of the eight countries in the world possessing optimum conditions for the growing of coffee.

The total area under coffee in India is 3,20,570 acres (or 1,29,730 hectares). About 58% of this coffee area is in the Karnataka State, 20% is in the Tamil Nadu State and 21% in the Kerala State. The remaining 1% is scattered, in small pockets in Andhra Pradesh, Maharashtra, Orissa, Assam and West Bengal States, besides the Andamans.

India produces about 100,000 metric tonnes of coffee every year. About 60,000 metric tonnes are exported to different destinations like U.S.A., Russia, West Germany, France, Italy, Netherlands, Sweden, U.K., Belgium and Canada. Indian Coffee in all grades—is among the best in the world and is used to blend with coffees of other origin to improve the quality.

Two varieties. Two main varieties of coffee are grown in India. They are *Coffea arabica* Linn.; and *Coffea robusta* Linden.

Indian Coffee is processed in two ways, i.e., by wet process (washed coffee) and by dry process (unwashed coffee). The washed coffee is known as Plantation or Parchment coffee. The unwashed coffee is called cherry coffee. Thus according to the process, coffee under arabica and robusta is described as under:

	ARABICA	ROBUSTA
Wet Process (*i.e.*, washed coffee)	Plantation	Parchment
Dry Process (*i.e.*, unwashed coffee)	Cherry	Cherry

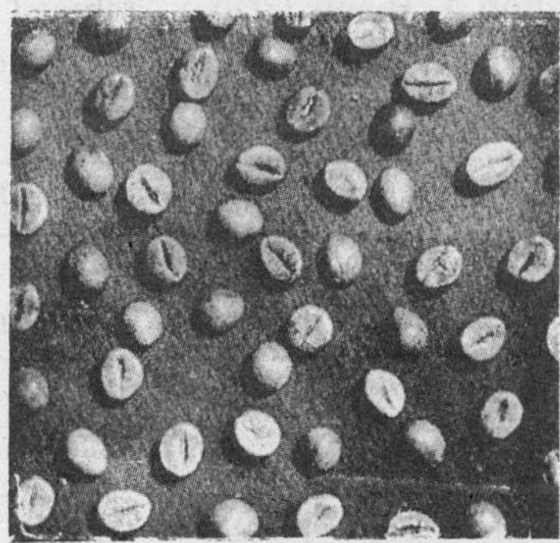

Fig. 24.4. Indian coffee. Coffee beans of *Coffea arabica*.

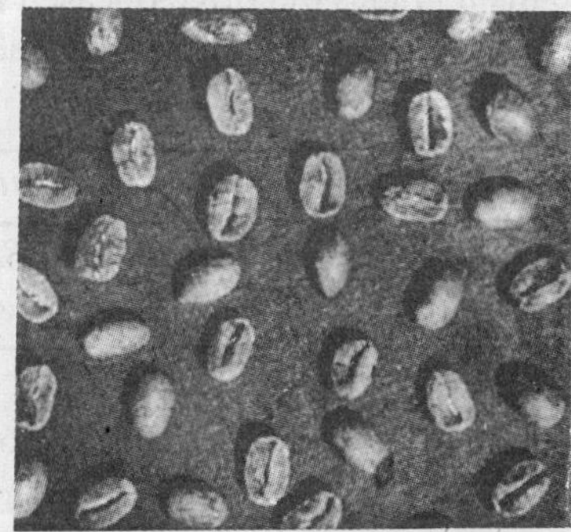

Fig. 24.5. Indian coffee. Coffee beans of *Coffea robusta*.

Plantation coffee. Arabica coffee processed by the wet or washed method is called Plantation coffee. This is an elaborate and time-consuming process in which the outer skin of the fruit and the mucilaginous matter beneath it are removed, by pulping and subsequent washing with fresh water, channelled from the springs in coffee-growing areas. The pulping and washing release the coffee beans from the coffee berries. But each bean is still encased in a soft parchment covering. The parchment-covered coffee is then dried in the sun on barbecues in the estates. Coffee in this form is known as Parchment coffee.

Curing. Parchment coffee is sent to curing works where it is further dried in the sun, if necessary, to a moisture content of not more than 10 percent. The well-dried parchment coffee is fed through the peeling machine to remove the thin brittle parchment cover.

The beans obtained are then graded according to shape and size, and carefully garbled by hand, or using electronic sorters and pneumatic separators to remove blemished and defective beans. The final product of these processes gives the Plantation coffee of commerce.

Cherry coffee. The other method of preparing Arabica coffee for the market is simpler. The berries are harvested, dried in the sun till their outer skin, the parchment cover and pulpy matter in between, dry up to form a thick husk leaving the beans loose inside. The dried cherry is then decorticated by passing through a huller, graded and garbled according to prescribed standards. The final produce of this process is known as Arabica cherry coffee or simply cherry coffee.

Robusta coffee. Robusta coffee is also processed by both wet and dry methods, and the resultant beans are known in trade parlance as Robusta Parchment or washed Robusta and Robusta cherry respectively.

Characteristics. The two types of coffee have certain characteristics:

Variety	*Type*	*Characteristics*
Arabica	Plantation	Long and wide beans, greyish or bluish in colour.
	Cherry	Same as above, but colour golden brown to brown.
Robusta	Plantation	Broad oval beans, smaller in size than (Parchment) Arabica, greyish in colour.
	Cherry	Same as above, but colour golden brown to brown.

Grading. Both Arabica and Robusta can have round beans, called PEABERRY (PB) or flat seeds called FLATS. Classification into A or B or AB is done according to the size of seeds.

The main grades of coffee released for export are Plantation A, B and Triage; Arabica Cherry AB and Triage; Robusta Cherry AB and Triage; Robusta Parchment AB. The main grades are tabulated below:

Arabica Plantation		*Arabica Cherry*		*Robusta Parchment*		*Robusta Cherry*	
Grade	*Sieve Size (mm)*	*Grade*	*Sieve Size (mm)*	*Grade*	*Sieve Size (mm)*	*Grade*	*Sieve Size (mm)*
PB	—	PB	—	PB	—	PB	—
A	6.65 and above	AB	6.00	AB	6.00	AB	6.00
B	6.00						
T	5.50	T	5.50	T	5.50	T	5.50

Coffee board. The production, internal marketing, export marketing, quality control, research, etc., are controlled by the Coffee Board which is a Statutory Body under the Ministry of Commerce of the Government of India. All coffee produced in India is delivered to the Coffee Board, Bangalore, a statutory organization, comprising representatives of coffee growers, curers, traders, workers, consumers and others connected with the industry.

Marketing. One of the major functions of the Coffee Board is the marketing of Indian coffee, through a common pool (as prescribed by the Indian Coffee Act). At the beginning of each season the coffee crop is estimated and after making provision for the internal market, the export allotment is made, so that the coffee trade may operate on the assurance of known quantities of available for export. This quantity is subsequently increased, if warranted by actual crop realisations.

Exports. Coffee for export is sold in open auctions in which exporters who have registered with the Board, bid for the lots. Foreign firms can participate in these export sales by registering their names with the Coffee Board, through their Agents or Branch offices in India.

Export auctions are held once in two to three weeks. The dates of the sales are widely notified in advance. Facilities are provided for drawing the samples of coffee offered for sale and the despatch of the coffee to prospective buyers abroad.

Coffee is normally exported from the West Coast ports of Mangalore and Cochin. The normal packing is in new gunnies of 60 kg. net. Prices are normally quoted in rupees per 50 kg., f.o.b. New Mangalore/Cochin.

There was a record production of coffee, one lakh and twenty thousand tonnes in 1977-78. Thirty-six thousand tonnes of coffee worth Rs. 89 crores were exported in 1977-78.

Quality. Many steps are taken by the Coffee Board to ensure the export of only the best quality of Indian coffee. The measures are aimed at maintaining and improving the quality of coffee right from the stage when the coffee is a tender plant in plantations to the time it is fully cured and exported. Use of improved strains of coffee, scientific control of pests, application of balanced nutrition and extension of coffee research results to the field ensure growth of quality coffees.

Curing of the coffee is done so as to conform to rigid standards set for curing. The cured coffee is again examined by a screening panel before being made available for exports. Production of quality coffee on estates is encouraged by awarding premium points for good quality coffee. A cup testing unit has been established by the Coffee Board to select the best coffee for exports, on the basis of cup test.

Instant Coffee. India can also offer instant coffee of the highest standards in 50 gms., 100 gms. and 25 kg. bulk packings. This instant coffee is manufactured from Green Coffees of other origin to raise the quality of the latter.

Monsooned coffee. A speciality of Indian coffee is the Monsooned coffee, which is in good demand in several countries of Europe, particularly the Scandinavian countries.

Monsooning of coffee is done by coffee curers on the West Coast of India. Monsooned coffee had its origin in the days of sailing ships, when it took about six months for unwashed coffee to be sent to Europe. During this period of voyage, the coffee in the damp holds of the ship lost its original colour and acquired a special flavour which has been liked by consumers in Norway, France and Switzerland. With the opening of the Suez and speedy transport of steamships, consumers in Europe complained that they missed the special flavour of the unwashed Indian coffee ! The curers on the West Coast rose to the occasion and perfected the process known as 'monsooning', by which coffee acquired the special flavour. Indian monsooned coffee is coffee with a golden colour.

With the outbreak of the South-West Monsoon, late in May or early in June, unwashed coffee (Arabica Cherry or Robusta Cherry coffees) is evenly spread, about 4 to 6 inches thick, in airy godowns, open on all sides. It is racked, from time to time, for about four or five days. It is packed loose in gunny bags and stacked in piles with sufficient space between rows, to enable the monsoon air to circulate freely around each bag.

The coffee beans are bulked and repacked, once a week, or they are poured from one leaf to another, prevent their getting mouldy and ensure even monsooning. In about a month and a half, the coffee assumes golden colour, when it is considered to be fully monsooned. At this stage, the coffee beans are carefully hand-garbled to remove those beans that are not up to the required standard. After mild fumigation treatment to prevent attacks by weevils, the monsooned coffee is ready for shipment.

Monsooned coffee is generally sold for export during October to December, every year. The grades offered for sale are *Monsooned Malabar* AA, *Monsooned Basanally* and *Monsooned Robusta* AB.

Why Indian Coffee is one of the Finest? Lovers of good coffee often wonder how Indian coffee is distinctive, delicious and distinguishable. To comprehend fully the principal reasons for Indian coffee being one of the world's finest coffees is to understand pretty well a whole range of coffee facts and operations in the country.

Nature with its generous Indian climate and rich soil, pampers coffee's growth. Careful scientific cultivation of the coffee on hill slopes, programmed picking of the coffee fruits by deft human hands and the gentle painstaking manner in which they are processed, unlock the full flavour latent in every bean. Rigid assessment of quality leads to the export of only the finest beans accounting for the superiority and fine aroma of India's high grown mild coffees.

Many of the different growths of Indian coffee, like Biligiris, Bababudans, other Mysores, Coorgs, Nilgiris, Naidubattams, Shevaroys, Anamallais, Pulneys, Nelliampathis, and Nilagiris Wynaad are familiar to importers of coffee in different countries of the world, as being synonymous, with supreme quality coffee. This is because the Indian Coffee Industry has to its credit three hundred years of continuous service in growing and making available high grown, good quality, mild coffee.

Indian coffee has a high reputation in the world market. This is because of the high quality of Indian coffee and the infinite care taken by the Coffee Board to maintain high standard at each stage of processing Indian coffee from the seed to the cup and special attention given to the screening and selection of only the best coffee for export, so that coffee connoisseurs around the world can always enjoy the superior and fine aroma of India's high-grown mild coffees.

25

CHAPTER

Rubber and its Products

Rubber is obtained from the *latex*, of various plants of tropical or sub-tropical regions. Most of the rubber plants belong to the Moraceae, Euphorbiaceae and Apocynaceae families. Though several species are available as sources of rubber, yet *Hevea brasiliensis* supersedes all. The cultivated *Hevea* trees, the so-called plantation rubber furnish about 98% of the supply. Rubber is the most recent of the major crops of the world. The industry is little more than a century old, and cultivation has been carried on about 80 years or so.

The latex occurs in latex tubes or latex vessels found in the various parts of the plants. Generally the commercial rubber is obtained from the lower part of the stem of tree. Latex is a gummy white liquid full of minute globules. It is mixture of water, hydrocarbons, resins, oils, proteins, acids, salts, sugars and caoutchouc. The physical and chemical properties of rubber have been considered separately in the following paragraph:

PROPERTIES OF RUBBER

Physical properties. The physical properties of rubber vary with temperature. Natural rubber is soft and translucent at 20°C ; when chilled to 0–10°C, it becomes hard and opaque ; at the temperature of liquid air, it is brittle and transparent like glass. At temperatures exceeding 25°, rubber loses elasticity and becomes sticky. Rubber melts to a viscous fluid at about 200°C. Rubber is insoluble in water and is unaffected by alkalies or moderately strong acids. It is soluble in benzene, naphtha, carbon disulphide, ether, chloroform and chlorinated hydrocarbons. When rubber is dissolved in a solvent, it first swells to gel-like consistency and then forms a solution. The elastic property of rubber is closely related to its molecular structure. Rubber is amorphous in the unstretched state, but shows crystalline behaviour when kept at 0° for a long time or when stretched. Rubber is one of the best insulating and dielectric materials available.

Chemical properties. First grade latex plantation rubber contains : rubber hydrocarbon 92–94%, resin 3%, protein 2% and ash 0.2% in Hevea rubber. Rubber hydrocarbon, also known as caoutchouc, is colourless, odourless, transparent and elastic. It consists of long chains of isoprene units connected end to end through 1 : 4 linkages. Rubber forms compounds with halogens, halogen acids and metallic halides, ozone and oxides of nitrogen. Hydrogenation at elevated temperatures and pressures in the presence of a catalyst gives fully saturated hydrorubber. Rubber is oxidised by nitric acid, potassium permanganate and hydrogen peroxide.

Hevea Aubl. (Euphorbiaceae).

A small genus of trees distributed chiefly in the Amazon region of South America. One species *Hevea brasiliensis* ; the source of Para Rubber, has been introduced in India and cultivated as a plantation crop.

Hevea brasiliensis (H.B.K.) Muell. Arg. (Eng. Para rubber ; Verna. *Rabar*). It is a large tree attaining a height of 60–100 ft. or more and a girth of 8–12 ft. The stem is smooth, straight, generally unbranched up to a considerable height, with a much branched leafy canopy. It is considered indigenous to the Amazon valley of Brazil, Venezuela, Peru, Ecuador and Colombia.

The rubber plantation industry in South-East Asia dates back to 1876 ; when 1,900 seedlings raised at Kew, were despatched to Ceylon. Plantations were raised and by 1930 a number of strains had been developed by breeding and selection, some of them superior to the stock originally obtained from Kew.

In India, rubber plantations were first started in Kerala state in 1905. Since then many other areas have been brought under rubber cultivation. At present, Kerala is the most important rubber producing state in India with plantations concentrated particularly in Kottayam and Quilon districts. Kanyakumari district in Tamil Nadu state is also an important rubber-growing area; there are small areas in Coimbatore and Nilgiris in Tamil Nadu state, Coorg in Karnataka state and in Andaman Islands. Rubber is produced in India, in plantations ranging from less than one acre to about 3,000 acres in area. Areas above 50 acres are classified as large estates and those of 50 acres and below, as Holdings.

Fig. 25.1. The rubber plantations in India registered phenomenal growth since independence. Production increased ten-fold in a span of 30 years and area enhanced four-fold while average yield shot up 2½ times. As a result from an importer of 35,000 tonnes of natural rubber in 1963, India gradually attained self-sufficiency and emerged a net exporter since 1973. Natural rubber, has spread now to over 230,000 hectares producing annually over 150,000 tonnes valued at Rs. 1200 million. Over a million people depend on this industry for their livelihood.

Hevea brasiliensis thrives in the tropical belt of the equator, in which the climate is warm, humid and equable and the temperature ranges from 74°F to 95°F. It requires a well distributed rainfall of 70 to 100 inches per annum. In South India, the plantations are located mostly in areas with a prolonged dry season and severe South West monsoon. The temperature ranges from 60°F to 95°F and there is a large variation between winter and summer temperatures.

Hevea brasiliensis thrives best in deep, well drained loamy soils. The soils in the rubber plantations of South India are red lateritic or clayey loams. The soils are generally deep and well drained. They are fairly rich in nitrogen, but poor in mineral constituents.

Rubber is a long duration tree crop which stays on the ground for several years. The economic life of rubber tree is 30–35 years. Newly opened jungle is considered to be the most suitable for rubber cultivation. The area to be planted is cleared of the jungle by cutting down the undergrowth and felling the trees. The cleared area is then lined and marked and terraces prepared. Provision is made for roads and drainage.

Tapping. Para rubber (*Hevea brasiliensis*) is valued for the latex obtained by tapping, *i.e.*, opening up the latex vessels situated in the bark with a sharp incision on the main trunk. When the vessels are cut, the latex flows out, quickly at first, then slowly, and finally coagulating on the cut surface. The tree is rested after each cut for varying periods according to age, climate and condition of the tree. When the tree is tapped again, the coagulated latex of 'plug' from the old cut surface is removed and a thin strip of bark is cut off. During the first tapping, only a small amount of viscous latex exudes. The flow increases with each successive tapping, the cuts being opened always at an angle to the horizontal. Usually, the cut extends half way around the trunk, but quite often, may completely encircle it.

Yield. The average yield of latex per acre in India from unselected ordinary seedling trees is 300 lb. The yield from budgrafted trees of approved clones and from clonal seedlings of approved parentage is twice, or more. An average yield of 700–800 lb. per acre has been obtained with improved planting materials in India.

Latex. Fresh latex is usually milky white in colour. The latex drawn from a tree which has been rested for sometime is yellow. The latex obtained by tapping is essentially a colloidal suspension of rubber particles in an aqueous serum.

Fresh latex is readily susceptible to enzymic and bacterial action leading to acid formation, and preservatives and bactericides are usually added to latex at the time of collection to inhibit acid development and consequent coagulation.

Raw rubber. Rubber of commerce is obtained from latex of *Hevea brasiliensis* by coagulation; rolling and drying. Two forms of raw rubber are known in trade, namely, *Smoked Sheet Rubber* and *Crepe Rubber*.

Grading. The system adopted in our country for grading raw rubber is based on the type description of the Rubber Manufacturers' Association and endorsed by the Rubber Trade Association of New York. These types are recognised as universal standards by producing and consuming countries alike.

Packing. The raw rubber is packed in bales in accordance with R.M.A. packing specifications.

Marketing. The raw rubber is marketed in the form of dry rubber. About two-thirds of total production is marketed by managing agents after grading. Petty merchants and dealers sell the collection to the nearest depot of a purchasing firm or take them to dealers in Kottayam, Trivandrum, Trichur or other market centres.

Uses. The raw rubber as such has few important applications. Its chief uses are in making insulation tapes, shoe soles, adhesives and erasers.

Rubber is used for the production of a wide variety of products utilized in industries and services, and for domestic purposes. It has been estimated that about 50,000 different products are fabricated from rubber directly or indirectly. Tyres and tubes for automobiles and cycles account for about 75% of the total rubber consumption. About 6% is used for footwear ; about 4% for wire and cable insulations. Among miscellaneous manufactured articles mention may be made of rubberised fabrics, motor mountings for absorbing vibrations and shocks, washers and gaskets, transmission and conveyor belting, garden hose, fire hose and hose for gasoline, pneumatic tools, paints, sports goods, household and hospital supplies, such as sheets, hot water bags, ice bags, surgeon's gloves, bathing caps and prophylactics, toys, such as balls, dolls and balloons, erasers and rubber bands and adhesives. Metals are coated with rubber to protect them from wear and corrosion. Sponge rubber from foamed latex finds uses in cushioning, seating and bedding. Elastic fabrics are made from latex threads. It is also employed in the fabrication of battery boxes, fountain pen barrels, tobacco pipe stems, telephones and combs.

Natural rubber latex, in preserved and concentrated forms, finds many commercial applications. Compounded latex is largely replacing rubber solutions for the manufacture of such articles as balloons, gloves, contraceptive appliances, finger stalls and teats. Over 65% of the latex is used in the manufacture of foam rubber and the rest for dipped goods, fabric coatings, impregnation and moulded goods. Uncured latex is used as an adhesive in footwear manufacture. The bulk of latex goods is made from vulcanised latex.

Reclaimed rubber prepared from discarded rubber products and rubber scrap, is used alone or in admixture with raw rubber in the manufacture of heels, soles, mechanical rubber goods, tyres and adhesives for use in fabric dipping, carpet backing, flooring, etc., and hard rubber used for battery containers, tubes and rods.

OTHER RUBBERS

Besides *Hevea*, several plants have been tried from time to time as sources of rubber as shown in the table :

Table 25.1.

Rubber plants percentage	*Rubber hydro-carbon*	*Resin*	*Protein*	*Ash*
Hevea brasiliensis	94	3	2	0.2
Castilla elastica	93	6	0.5	0.1
Landolphia thollonii	93	5	1	0.5
L. owariensis	93	6	0.1	1.0
L. kirkii	91	8	0.1	1.0
Taraxacum kok-saghyz	90	2	2	3.0
Funtumia elastica	88	8	3	0.5
Manihot glaziovii	88	4	6	1.5
Cryptostegia grandiflora	88	9	2	0.3
Ficus elastica	78	20	1	0.5
Parthenium argentatum	72	22	3	2.5

Of these the Russian Dandelion (*Taraxacum kok-saghyz* Rodin) and the Mexican Guayule *Parthenium argentatum*. A. Gray) may specially be mentioned. The Russian Dandelion is grown on C. 2,000,000 acres of land in Russia and rubber of good quality is reported to be obtained from the roots of the plant. The Mexican Guayule is extracted from the whole plant; it is of inferior quality. Other rubber producing plants of local importance are — *Manihot glaziovii* Muell. Arg. (Ceara rubber), *Castilla elastica* Cerv. (Castilla or Panama rubber), *Hancornia speciosa* Gomez (Mangabeira rubber), *Ficus elastica* Roxb. (India or Assam rubber) and *Funtumia elastica* Stapf. (Lagos silk rubber).

Question Bank

N.B. Questions from UGC Question Bank

LONG ANSWER QUESTIONS

1. Give an account of drugs obtained from plants (15 minutes; 15 marks).
2. Assume that you and a small group of your friends are transported to an uninhabited island in the Indian Ocean to live and that you are permitted to take the seeds or propagative organs of 12 species of plants with you. List them in order of preference (mention the botanical name and family) and give reasons for your selection (10 mts.; 10 mks.).
3. By means of outline sketches only, show the morphology of the economically important of any five of the following plants: saffron, opium, poppy, cocoa, black pepper, citrus, coffee, and clove; give the botanical name of each source and comment on the uses (15 mts.; 15 mks.).
4. State the botanical names, families and morphology of the parts from which the following are obtained: (*a*) cotton, (*b*) jute, (*c*) coir, (*d*) hemp. How would you differentiate one from the other? (15 mts.; 15 mks.).
5. Name five important drug plants of India and write detailed account of any two of them. (15 mts.; 15 mks.).
6. List the chief fatty oil-yielding crops in India. Classify the oils on the basis of their drying properties and mention their uses. What do you mean by the term hydrogenation. (15 mts.; 15 mks.).
7. By simple sketches only, show the morphology of the economically important parts of the following: clove, cashew, coriander and black pepper. Mention their latin names and uses. (15 mts.; 15 mks.).
8. What are essential oils? Give a brief account of the botany, cultivation, extraction and uses of important essential oils produced in India. (15 mts.; 15 mks.).
9. Give the method of extraction, properties and economic uses of the following: (*a*) groundnut oil, (*b*) linseed oil and (*c*) castor oil. (10 mts.; 10 mks.).
10. Name any *four* fibre yielding plants, giving their families. Describe the methods of fibre extraction from them and their commercial uses. (15 mts.; 15 mks.).
11. Give the methods of cultivation of any *one* of the following: (*a*) cotton; (*b*) rubber. (10 mts.; 10 mks.).
12. Explain how the latex is processed to get rubber? (10 mts.; 10 mks.).
13. The whole coconut tree is economically useful. Justify the above statement. (20 mts.; 10 mks.).

SHORT ANSWER QUESTIONS

1. Write what do you know of the source and economic importance of the cork. (3 minutes; 3 marks).
2. Write what do you know of the source and economic importance of coconut. (5 mts.; 5 mks.).
3. List the botanical names of six plants from Cruciferae (Brassicaceae) used as vegetables and mention the morphology of the useful part in each. (5 mts.; 5 mks.).
4. Name the species of cottons exploited for commercial purposes. How would you distinguish between the Old World and New World cottons? (5 mts.; 5 mks.).
5. How is opium obtained and how is it used? (5 mts.; 5 mks.).
6. Write note on any *one* of the following:
   (*a*) *Oryza*. (5 mts.; 5 mks.).
   (*b*) *Rauwolfia*. (5 mts.; 5 mks.).
   (*c*) *Cannabis*. (6 mts.; 6 mks.).

7. Give botanical names of *three* important timber-yielding plants of India. (3 mts.; 3 mks.).
8. Name the sources of any *one* of the following products; mention the plant part from which they are obtained and indicate the main areas of cultivation of these plants in India.
   (*a*) Ephedrine. (3 mts.; 3 mks.).
   (*b*) Tobacco. (3 mts.; 3 mks.).
   (*c*) Clove. (3 mts.; 3 mks.).
   (*d*) Saffron. (3 mts.; 3 mks.).
9. Describe the medicinal importance of *Adhatoda*, *Zingiber* and *Penicillium*. (6 mts.; 6 mks.).
10. What is the morphology of the edible part in (*i*) orange, (*ii*) potato, (*iii*) papaya? (3 mts.; 3 mks.).
11. Distinguish between any *one* pair of the following:
    (*a*) Ring porous and diffuse porous. (3 mts.; 3 mks.).
    (*b*) Sap wood and heart wood. (3 mts.; 8 mks.).
    (*c*) Beedi and cigarette. (3 mts.; 3 mks.).
    (*d*) Green tea and black tea. (3 mts.; 3 mks.).
    (*e*) Lint and fuzz. (3 mts.; 3 mks.).
12. Name a plant each yielding a worldwide commodity which owes its origin to (*i*) epidermis, (*ii*) tuberous stem, (*iii*) periderm, and (*iv*) whole leaf.
13. what plant materials (mention Latin names) are used in manufacture of essential oils. (3 mts.; 3 mks.).
14. Write short note on any *one* of the following:
    (*a*) Indigenous raw materials of paper production. (5 mts.; 5 mks.).
    (*b*) Wood seasoning. (5 mts.; 5 mks.).
    (*c*) Molasses. (5 mts.; 5 mks.).
    (*d*) Papain. (5 mts.; 5 mks.).
15. How would you distinguish any *one* of the following pairs:
    (*a*) *Curcuma* from *Zingiber*. (3 mts.; 3 mks.).
    (*b*) Starch of wheat from that of rice and maize (3 mts.; 3 mks.).
    (*c*) Essential oil from fatty oil. (3 mts.; 3 mks.).
    (*d*) Fibres of cotton from those of jute. (3 mts.; 3 mks.).
16. Give the botanical names, morphological nature of plant parts yielding economic products, and uses of any *one* of the following:
    (*a*) Sugarcane. (3 mts.; 3 mks.).
    (*b*) Jute. (3 mts.; 3 mks.).
    (*c*) Potato. (3 mts.; 3 mks.).
    (*d*) Wheat. (3 mts.; 3 mks.).
17. Write botanical names of any *six* plants in which either the stem or the leaf is of economic importance.
18. Explain giving suitable reasons for the following:
    Farmers generally use leguminous plants, with nodulated roots, for rotation of crops. (5 mts.; 5 mks.).
19. Write briefly: Extraction of essential oils from rose and Sandal wood. (3 mts.; 3 mks.).
20. In India cotton textile industry is largely confined to Bombay and other parts of Western India, whereas jute industry is mostly restricted to Calcutta and other parts of West Bengal. Comment. (5 mts.; 5 mks.).
21. Write short note on any one of the following:
    (*a*) Bye-products of sugar industry. (5 mts.; 5 mks.).
    (*b*) Extraction of essential oils. (5 mts.; 5 mks.).
    (*c*) Important Indian spices. (5 mts.; 5 mks.).

22. *Cannabis sativa* hits the front lines in many of the leading magazines of the world. Comment. (5 mts.; 5 mks.).
23. Write a short note on Soybean (5 mts.; 5 mks.).
24. Distinguish between any *one* of the following pairs:
    (*a*) Pine wood and teak wood. (3 mts.; 3 mks.).
    (*b*) Fennel and Coriander. (3 mts.; 3 mks.).
    (*c*) Essential oil and fatty oil. (3 mts.; 3 mks.).
    (*d*) Flue curing and sun curing. (3 mts.; 3 mks.).
25. Name the principal states in India in which the following are cultivated:
    (*i*) Jute, (*ii*) coconut, (*iii*) tea, (*iv*) mango, and (*v*) black pepper. (3 mts.; 3 mks.).
26. How does milling affect the quality of rice, (3 mts.; 3 mks.).
27. Name drugs of plant origin which could be used in the following:
    Malaria, hypertension, rheumatism, heart disorders and dysentery. (3 mts.; 3 mks.).
28. Write what you know of the source and economic importance of Ginger. (3 mts.; 3 mks.).
29. Write briefly: Tobacco and health hazards. (5 mts.; 5 mks.).
30. Write briefly: Harvesting, grading and processing of tea. (5 mts.; 5 mks.).
31. Write briefly: Origin of wheat (5 mts.; 5 mks.).
32. Write briefly: Principal uses of wood. (5 mts.; 5 mks.).
33. Write briefly: Extraction of sugar and by-products of sugar industry. (5 mts.; 5 mks.).
34. What are the principal sources of vegetable oils in India. List the four major uses of oils.
35. Give the botanical name of the plant yielding any *one* of the following product and explain its morphology and economic importance:
    (*a*) Reserpine. (5 mts.; 5 mks.).
    (*b*) Coffee. (5 mts.; 5 mks.).
    (*c*) Rose-wood. (5 mts.; 5 mks.).
36. Name a useful plant product that is derived from (*i*) epidermis, (*ii*) whole leaf, (*iii*) mesocarp, (*iv*) periderm, (*v*) tuberous root and (*vi*) heart wood.
37. Mention some algae as source of food to human beings. (3 mts.; 3 mks.).
38. Mention the importance of Gramineae (Poaceae) as a source of food. (3 mts.; 3 mks.).
39. Write notes on: Economic uses of tannins. (6 mts.; 6 mks.).
40. Write any *three* economic uses of Euphorbiaceae members. (3 mts.; 3 mks.).
41. Outline the processing of coffee by wet method. (3 mts.; 3 mks.).
42. What kind of soil will be preferred for growing rubber. (2 mts.; 2 mks.).
43. How will you start to process the Paddy grains soon after harvesting them (Mention only the first two steps.). (2 mts.; 3 mks.).
44. Mention 5 rubber yielding plants and their families, cultivated in India. (2 mts.; 5 mks.).
45. The natural habit of tea is tree but under cultivation it is a shrub. How is it so? (2 mts.; 2 mks.).
46. Mention the other uses of Paddy, other than the grains being used as a staple food. (2 mts.; 2 mks.).
47. Give the technical name of the fruit and mention the morphology of the useful parts and their actual use in the fruits of the following:
    (*a*) Mango, (*b*) Cashew and (*c*) Jack fruit. (8 mts.; 12 mks.).
48. State what you know about the morphological nature of the following: (i) Apple, (ii) Cauliflower, (iii) Radish.
49. Give the technical name of the fruit and mention the morphology of the edible portion of the following:
    (*i*) Tomato, (*ii*) Rice, (*iii*) Coco, (*iv*) Apple and (*v*) Groundnut. (5 mts.; 10 mks.).

50. Which is the active principle in coffee? (2 mts.; 2 mks.).
51. How is the mesocarp of coconut fruit used in industry? (3 mts.; 2 mks.).
52. Name any two economic uses of cotton other than using it in textile industry. (2 mts.; 3 mks.).

SIMPLE QUESTIONS

1. Arrange the following in the order of their food value: tapioca, groundnut, soybean, potato, rice and Citrus. (2 mts.; 2 mks.).
2. Name five plant products which earn foreign exchange for our country (2 mts.; 2 mks.).
3. Mention the titles and names of authors of four books on economic botany. (2 mts.; 2 mks.).
4. What is the nature of the chief food reserve in (*i*) tapioca, (*ii*) mango, (*iii*) soybean, (*iv*) maize. (2 mts.; 2 mks.).
5. Which of the Indian states are the leading producers for Cane sugar, Sandal wood, Potato, Jute and Teak. (2 mts.; 2 mks.).
6. What are the 'active' chemical principles in the following: Coffee, Ergot, Opium, Rauwolfia and Tobacco? (3 mts.; 3 mks.).
7. What plant materials (mention Latin names) are used in the manufacture of any one of the following:
   (*a*) Sugar. (2 mts.; 2 mks.).
   (*b*) Spice powder for flavouring foods. (2 mts.; 2 mks.).
   (*c*) Paper. (2 mts.; 2 mks.).
   (*d*) Soap. (2 mts.; 2 mks.).
8. Name the Research Institute along with its location where improvement work in any one of the following crop plants is being carried out in India:
   (*a*) Jute. (1 mt.; 1 mk.).
   (*b*) Wheat. (1 mt.; 1 mk.).
   (*c*) Sugarcane. (1 mt.; 1 mk.).
   (*d*) Rice. (1 mt.; 1 mk.).
   (*e*) Potato. (1 mt.; 1 mk.).
9. Give the botanical name from which the following is obtained:
   (*a*) Caffeine. (1 mt.; 1 mk.).
   (*b*) Papain. (1 mt.; 1 mk.).
   (*c*) Morphine. (1 mt.; 1 mk.).
   (*d*) Hashish. (1 mt.; 1 mk.).
10. Which of the Indian states are leading producers of the following:
   (*a*) Tea. (1 mt.; 1 mk.).
   (*b*) Tobacco. (1 mt.; 1 mk.).
   (*c*) Rubber. (1 mt.; 1 mk.).
   (*d*) Banana. (1 mt.; 1 mk.).
   (*e*) Belladonna. (1 mt.; 1 mk.).
   (*f*) Deodar. (1 mt.; 1 mk.).
11. What is the centre of the origin of the following:
   (*a*) Coffee. (1 mt.; 1 mk.).
   (*b*) Cocoa. (1 mt.; 1 mk.).
   (*c*) Clove. (1 mt.; 1 mk.).
   (*d*) Cinchona. (1 mt.; 1 mk.).
   (*e*) Groundnut. (1 mt.; 1 mk.).
   (*f*) Black pepper (1 mt.; 1 mk.).

(*g*) Eucalyptus. (1 mt.; 1 mk.

(*h*) Mango. (1 mt.; 1 mk.).

(*i*) Red pepper. (1 mt.; 1 mk.).

(*j*) Soybean. (1 mt.; 1 mk.).

12. Mention one important source (botanical name) for the following:

(*a*) Atropine. (1 mt.; 1 mk.).

(*b*) Cigar. (1 mt.; 1 mk.).

(*c*) Match sticks. (1 mt.; 1 mk.).

(*d*) Tapioca. (1 mt.; 1 mk.).

13. How is hydrogenation of oil done? (2 mts.; 2 mks.).
14. In what ways does paddy differ from rice? (2 mts.; 2 mks.).
15. Where did maize originate? (2 mts.; 2 mks.).
16. What makes a wood heavy? (2 mts.; 2 mks.).
17. Give the botanical name and the morphology of the product used in

(*a*) Cinchona. (1 mt.; 1 mk.).

(*b*) Groundnut. (1 mt.; 1 mk.).

(*c*) Rubber. (1 mt.; 1 mk.).

(*d*) Coir. (1 mt.; 1 mk.).

(*e*) Sorghum. (1 mt.; 1 mk.).

(*f*) Cardamum. (1 mt.; 1 mk.).

(*g*) Castor. (1 mt.; 1 mk.).

18. Describe how Cichorium intybus is used as an economically important product. (2 mts.; 2 mks.).
19. Name any two fibre-yielding plants. (2 mts.; 2 mks.).
20. Give the botanical name which yields the following: (*a*) Rose wood, (*b*) Satin wood, (*c*) Sandal wood, (*d*) Laurel wood. (3 mts.; 4+ mks.).
21. Name two essential oil-yielding plants. (2 mts.; 2 mks.).
22. Describe the morphology of the useful part in: (*a*) Ganja. (2 mts.; 2 mks.), (*b*) Taro. (2 mts.; 2 mks.), (*c*) Cardamom (2 mts.; 2 mks.), (*d*) Groundnut (2 mts.; 2 mks.), (*e*) Pepper (2 mts.; 1 mk.), (*f*) Coriander (2 mts.; 1 mk.), (g) Cloves. (2 mts.; 1 mk.).
23. Name the plant which yields Quinine. How is it extracted? (5 mts.; 5 mks.).
24. Give the botanical name for the following:

(*a*) Bengal gram, (*b*) Green gram, (*c*) Black gram, (*d*) Red gram. (1 mt.; 4+ mks.).

25. Give the botanical name of the following:

(*a*) The plant produces Aconitum, (*b*) The plant produces cabbage, (*c*) The plant produces Melia oil, (*d*) The plant produces Bengal gram. (2 mts.; 2 mks.).

26. Name a plant for which desuckering is done. (1 mt.; 1 mk.).
27. Mention one cultivated species and one wild species of sugarcane. (1 mt.; 1 mk.).
28. Give the use of bagasse. (1 mt.; 2 mks.).
29. What is meant by retting? (2 mts.; 1 mk.).
30. Which plant produces chicory? (1 mt.; 1 mk.).
31. Give the name of the plant which produces coffee. (1 mt.; 1 mk.).
32. Describe the morphology of the useful parts of cotton. (2 mts.; 2 mks.).
33. Give the names of any two major rice research stations, their location in India. (2 mts.; 2 mks.).
34. Describe the morphology of the useful part in groundnut and cotton. (2 mts.; 2 mks.).
35. Arrange in order of highest protein value: (*a*) Soybean, (*b*) Potato, (*c*) Rice, (*d*) Wheat, (*e*) Sugarcane.
36. Give one example each of plant/plant-products used in the manufacture of following: (*a*) Macroni, (*b*) rum, (*c*) morphine, (*d*) haematoxylin. (2 mts.; 2 mks.).

MULTIPLE CHOICE QUESTIONS

1. Botanical name of jute is:
(*a*) *Cannabis sativa* (*b*) *Linum usitatissimum*
(*c*) *Corchorus capsularis* (*d*) None of the above (1 minute; 1 mark)

2. Teak wood which is obtained from Tectona grandis belongs to the family of:
(*a*) Labiatae (lamiaceae) (*b*) Verbenaceae (*c*) Rubiaceae (*d*) Apocynaceae
(1 mt.; 1 mk.).

3. Digitalis purpurea belongs to the family of:
(*a*) Papilionanceae (*b*) Acanthaceae (*c*) Scrophulariaceae (*d*) Solanaceae
(1 mt.; 1 mk.).

4. Sunhemp is obtained from:
(*a*) *Crotolaria juncea* (*b*) *Cannabis sativa* (*c*) *Musa textiles* (*d*) *Cocos nucifera*
(1 mt.; 1 mk.).

5. Cloves are obtained from Eugenia aromatica and the morphology of the product being:
(*a*) product petioles (*b*) dried pedicels (*c*) dried flower buds (*d*) dried seeds
(1 mts.; 1 mk.).

6. Botanical name of the plant that yields pararubber:
(*a*) *Ficus elastica* (*b*) *Hevea brasiliensis* (*c*) *Manihot glaziovii* (*d*) *Castilla elastica*
(1 mt.; 1 mk.).

7. Reserpine, a drug is extracted from:
(*a*) *Rauwolfia serpentina* (*b*) *Ferula asafoetida*
(*c*) *Atropa belladona* (*d*) *Digitalis purpurea* (1 mt.; 1 mk.).

8. Plant parts useful for extraction of opium from Papaver somniferum are:
(*a*) young seedlings (*b*) old leaves (*c*) unripe fruits (*d*) ripened seeds
(1 mt.; 1 mk.).

9. Botanical name of cauliflower is:
(*a*) *Brassica oleracea* var. *botrytis* (*b*) Brassica oleracea var. *gongylodes*
(*c*) *Brassica oleracea* var. *capitata* (*d*) Brassica oleracea var. *gemmifera*
(1 mt.; 1 mk.).

10. The common name of *Eleusine coracana is:*
(*a*) Ragi (*b*) Barley (*c*) Wheat (*d*) Oats

11. The cotton fibre from cotton is obtained from:
(*a*) roots (*b*) stems (*c*) seeds (*d*) leaves

12. The commercial jute fibres are:
(*a*) Phloem fibre (*b*) Xylem fibres (*c*) Interxylary fibres (1 mt.; 1 mk.).

13. Cotton fibre is derived from:
(*a*) Phloem fibres (*b*) Epidermal hairs on seed
(*c*) Outgrowth on the stem (*d*) Sclerenchymatous cells (1 mt.; 1 mk.).

14. Paddy is suitable for cultivation in:
(*a*) Red soils (*b*) Dry soils (*c*) Black soils (*d*) Irrigated soils
(1 mt.; 1 mk.).

15. The economic product of tobacco plant is:
(*a*) Flowers (*b*) Leaves (*c*) Roots (*d*) Stems
(1 mt:; 1 mk.).

16. Fibre yielding plant is.
(*a*) *Triticum* (*b*) *Pennisetum* (*c*) *Gossypium* (*d*) *Rauwolfia*
(1 mt:; 1 mk.).

17. The quality of tobacco depends mainly on:
(*a*) Curing process (*b*) Variety and curing process
(*c*) Climatic conditions in which plant grows. (*d*) Nutrition of the plant in the field
(1 mt.; 1 mk.).

18. Opium is obtained from:
(*a*) dried leaves (*b*) roots
(*c*) latex from unripe capsules (*d*) seeds that are fried (1 mt.; 1 mk.).

19. Coir is obtained from
(*a*) roots of date palm tree (*b*) leaf bases of areca nut palm
(*c*) mesocarp of coconut (*d*) leaves of Cocos nucifera (1 mt.; 1 mk.).

20. The centre of origin of rice plant is:
(*a*) India (*b*) Indo-Malayan region
(*c*) India and Africa (*d*) Africa (1 mt.; 1 mk.).

21. Rubber is collected from:
(*a*) Crushing the stem of *Euphorbias.*
(*b*) Tapping the stem of *Carica papaya*
(*c*) Tapping the stem of *Hevea brasiliensis*
(*d*) Crushing the fruits and collecting the latex of *Achras sapota* (1 mt.; 1 mk.).

22. Essential oils are those:
(*a*) oils which are essential for human beings.
(*b*) oils which are essential to the plants which produce them
(*c*) oils which are used as lubricants
(*d*) oils which yield perfumes (1 mt.; 1 mk.).

23. Asafoetida is obtained by:
(*a*) exudation from the stem (*b*) extraction from the fruits
(*c*) extraction from the root (*d*) extracted from the leaves. (1 mt.; 1 mk.).

24. Short fibres are known as:
(*a*) lint (*b*) fluff (*c*) fuzz (*d*) flint (1 mt.; 1 mk.).

25. Long fibres are known as:
(*a*) flint (*b*) lint (*c*) fluff (*d*) fuzz (1 mt.; 1 mk.).

26. The bark of *Cinnamomum zeylanicum* is used as spice because of:
(*a*) A sweet flavour (*b*) Aromatic oils secreted
(*c*) Ethereal oils secreted (1 mt.; 1 mk.).

27. Corms and root tubers of Araceae have a pungent acrid taste because they contain:
(*a*) Aromatic oil (*b*) Oxalate crystals (*c*) Alkaloid (*d*) Latex (1 mt.; 1 mk.).

28. The composition of the cotton fibre is:
(*a*) cellulose (*b*) callose (*c*) chitin (*d*) pectin (1 mt.; 1 mk.).

29. Tea can be grown:
(*a*) at sea level (*b*) sea shore (*c*) in elevated areas (*d*) in dry places
(1 mt.; 1 mk.).

30. If tea leaves are kept in hot water for longer period, the liquid becomes bitter because:
(*a*) of the defect in the tea leaves (*b*) the volatile oil in the leaves dissolves out
(*c*) the thein dissolves out (*d*) the tannin dissolves out (1 mt.; 1 mk.).

31. Tea and coffee can be classified as:
(*a*) distilled beverages (*b*) non-alcoholic beverages
(*c*) fermented beverages (*d*) alcoholic beverages (1 mt.; 1 mk.).

32. The products of commercial importance, yielded by coconut palm are:
(*a*) latex and oil (*b*) rubber and wood (*c*) oil and fibre (*d*) fibre and rubber
(*e*) medicines (1 mt., 1 mk.).

33. Potatoes are usually propagated by vegetative means because:
(*a*) they do not produce seeds
(*b*) by this method it is possible to maintain genetic quality
(*c*) by this method incidence of diseases may be reduced
(*d*) potato seeds have long dormancy period (1 mt.; 1 mk.).

34. The banana plant is:
(*a*) tree (*b*) shrub (*c*) herb (1 mt.; 1 mk.).

35. The true stem of banana is:
(*a*) Bulb (*b*) Rhizome (*c*) Corm (1 mt.; 1 mk.).

36. The major composition of banana fruit is:
(*a*) starch (*b*) protein (*c*) glucose (*d*) fat (1 mt.; 1 mk.).

37. Commercial tea is prepared from:
(*a*) stem (*b*) leaves (*c*) flowers (*d*) root
(*e*) bark (1 mt.; 1 mk.).

38. Rubber is obtained from:
(*a*) cell sap (*b*) gum (*c*) resin (*d*) latex (1 mt.; 1 mk.).

39. The commercial coffee is prepared from:
(*a*) leaves (*b*) flowers (*c*) fruits (*d*) root
(*e*) seed (1 mt.; 1 mk.).

40. Most of the rubber plants belong to the family
(*a*) Euphorbiaceae (*b*) Cannaceae (*c*) Rubiaceae (*d*) Rutaceae
(1 mt.; 1 mk.).

41. Latex cells occur in:
(*a*) xylem (*b*) cambium (*c*) bark (*d*) cortex
(1 mt.; 1 mk.).

42. The artificial ripening of banana is done by:
(*a*) Keeping them in a room having temperature below 0°C.
(*b*) Keeping them at room temperature.
(*c*) Keeping them in a room where the temperature is high. (1 mt.; 1 mk.).

43. Coconut oil derived from the copra can be classified as:
(*a*) essential oil (*b*) drying oil (*c*) semi-drying oil (*d*) vegetable fat.
(1 mt.; 1 mk.).

44. What is the source of chewing gum:
(*a*) Sugar from sugar-cane (*b*) Gum arabic from *Acacia* spp.
(*c*) Latex from *Achras sapota* (*d*) Fluid from *Musa* spp. (1 mt.; 1 mk.).

ANSWERS

1. (*c*)	2. (*b*)	3. (*c*)	4. (*a*)	5. (*c*)	6. (*b*)	7. (*a*)	8. (*c*)	9. (*a*)	10. (*a*)
11. (*c*)	12. (*a*)	13. (*b*)	14. (*d*)	15. (*b*)	16. (*c*)	17. (*a*)	18. (*c*)	19. (*c*)	20. (*b*)
21. (*c*)	22. (*d*)	23. (*c*)	24. (*c*)	25. (*b*)	26. (*a*)	27. (*b*)	28. (*a*)	29. (*c*)	30. (*d*)
31. (*b*)	32. (*c*)	33. (*b*)	34. (*c*)	35. (*b*)	36. (*a*)	37. (*b*)	38. (*d*)	39. (*d*)	40. (*a*)
41. (*c*)	42. (*c*)	43. (*d*)	44. (*c*)						